Second Edition

Mechanics of Structures
Variational and Computational Methods

Second Edition

Mechanics of Structures
Variational and Computational Methods

Walter Wunderlich

Professor of Structural Mechanics
Technische Universität München
Munich, Germany

Walter D. Pilkey

Morse Professor of Engineering
School of Engineering and Applied Sciences
University of Virginia
Charlottesville, Virginia, USA

CRC Press
Taylor & Francis Group
Boca Raton London New York

CRC Press is an imprint of the
Taylor & Francis Group, an **informa** business

NW, Suite 300

Group, LLC
Taylor & Francis Group, an Informa business

9

vernment works

(pbk)
(hbk)

Web site at
is.com

e at

of Congress Cataloging-in-Publication Data

31-
res : variational and computational methods / Walter Wunderlich,
d ed.

rs first on the previous edition.
cal references and index.

als. 2. Mechanics, Applied. 3. Engineering mathematics. I. Pilkey,

Preface to the First Edition

The development of the digital computer has led to substantive changes in computational structural mechanics and in the corresponding methods of analysis. The finite element method has been developed and refined to such a level that it is now a standard academic offering. It is, perhaps, time to reflect on the conventional approach for presenting modern structural mechanics, including the fundamentals and even the nomenclature. This book is the authors' response to this need. We believe that the approach presented here will make the well-established teaching patterns of structural mechanics more rational. This book is directed towards the courses that are typically labeled as fundamentals of finite elements: advanced strength of materials; matrix structural analysis; computational solid mechanics; variational methods of mechanics; and rods, plates, and shells. As a minimum, prerequisites for understanding this material are statics and strength of materials. Courses in structural analysis, calculus of variations, and linear algebra are helpful but not required. Students with more complete backgrounds can proceed more rapidly. In the United States, this material is used in senior and introductory graduate courses for students of civil, mechanical, and aerospace engineering.

The contents permit considerable flexibility in presentation and coverage. Although we recommend proceeding through the book in the order provided, a variety of traditional courses can be based on the material and the order of presentation varied to fit the course. Smaller print is employed to identify material that is of an advanced nature, which can be bypassed without losing continuity.

This book deals with the formulation and methods of analysis for modern structural and solid mechanics. Because they are not relevant to the primary theme of this work, only limited consideration is given to numerical procedures for solving equations, either linear or eigenvalue; this subject is satisfactorily covered in numerous textbooks. It is our belief that a reasonable understanding of the material presented in this book will provide sufficient technical literacy to permit the reader to consult, without difficulty, references for details of numerical solution procedures as well as to be able to comprehend contemporary literature of structural mechanics.

The behavior of deformable solids is often modeled with differential equations derived from fundamental (nonvariational) principles. These same differential equations can be obtained using variational principles. Approximate techniques can then be applied to solve the differential equations. As an alternative, approximate solution formulations can be derived directly from the variational principles. In this book, unified and generalized fundamental formulations, based on either the differential equations or the direct use of variational principles, are presented for the computational solution of structural mechanics problems. Although classical approaches, finite difference, and boundary element methods are considered, primary emphasis is given to finite element formulations. The generalized variational principles presented permit many aspects of the finite element method to be covered in a unified, comprehensive fashion. This includes the displacement, force, mixed, and hybrid formulations.

We begin with a brief introduction in Chapter 1 to the equations of elasticity. Then, because variational methods play such an important role here, fundamental and generalized variational principles are developed in Chapter 2, where the common structure of the differential and integral forms of the governing equations is delineated. Following a chapter (Chapter 3) on the derivations and applications of classical energy principles, matrix methods of structural mechanics are treated in Chapters 4 and 5. Finite element (Chapter 6) and weighted residual (Chapter 7) methods are then introduced, along with brief descriptions of the finite difference method (Chapter 8) and the boundary element method (Chapter 9). Formulations for the dynamic and stability behavior of solids are covered in Chapters 10 and 11. Chapter 12 presents the governing equations for beams and bars, in a form suitable for solution using the generalized formulations of the previous chapters. Also included is computational solution methodology for cross-sectional characteristics of beams. The final chapter contains both the governing equations and the solution methodology, analytical and computational, for plates. In several chapters, general solutions for the static, stability, and dynamic response of rods and plates are provided in tabular form. There are brief appendices covering the classical integral theorems and the fundamentals of the calculus of variations.

This book began as an elaboration of lectures given by one of the authors (Wunderlich). We are indebted to the computational solid mechanics students at the University of Virginia for their forbearance in using preliminary versions of this material. Levent Kitis, Yongquan Liu, and Weize Kang have read and contributed substantially to this work.

Many of the figures were carefully crafted by Jim Houston, Huiyan Wu, Allen Hawkins, and Wei Wei Ding. Instrumental in preparing this book has been a monumental effort by Barbara Pilkey.

Walter D. Pilkey
Walter Wunderlich

Preface to the Second Edition

Continued development of computer-based methods in structural mechanics has helped confirm the value of the original concept of this book. The acceptance of the first edition has led to the need for a second edition. Although the table of contents of this edition appears quite similar to that of the first edition, several substantial changes have been incorporated. For example, the presentation of the initial chapters is improved and the coverage broadened. The additions include a discussion of the principal values of geometric properties and stresses. The book should now be easier to study for students with backgrounds only in fundamental statics and strength of materials.

The chapter on stability has been rewritten and supplemented with new detailed example problems. The appendix (III) in the first edition on numerical solution techniques for equilibrium and eigenvalue problems contained material that is widely available in more detail in numerous sources. Hence, it has been discarded. A new appendix (III) contains a summary of the fundamental equations in differential and variational form. Throughout the book, changes have been made and material updated or highlighted, in part in response to suggestions by users. New homework problems have been added to many chapters.

The authors hope that this second edition of the book continues to provide the fundamentals of computational solid mechanics in a unified form. Like the first edition, it is directed to senior and introductory-level graduate students, as well as to engineers in practice and in research working in the field of computational solid mechanics and engineering.

We wish to acknowledge the contributions to this edition by Levent Kitis, Check Kam, Wei Wei Ding, Yongquan Liu, Adam Ziemba, Matt Kindig, Yasmina Abdelilah and Barbara Pilkey.

Walter Wunderlich
Walter D. Pilkey

Authors

Walter Wunderlich, Dr.-Ing., lives in Germany, where he is Professor of Structural Mechanics at the Technische Universität München. He received the degree of Diplom-Ingenieur in Civil Engineering from the Technische Hochschule Hannover in 1958, after which he worked for 3 years in bridge engineering. He returned to the Technische Hochschule Hannover as a scientific assistant and was awarded his Dr.-Ing. in 1966. Thereafter, he moved to the Technische Hochschule Braunschweig, where he became an associate professor in 1970. After spending a year of sabbatical leave at the University of California at Berkeley with R. W. Clough and P. Naghdi, he joined the Ruhr Universität Bochum as a Professor of Structural Engineering. The idea for this book came from his lectures, given at the Ruhr Universität in the early 1970s. In 1988, Professor Wunderlich moved to his position in Munich, from which he retired recently. He has worked in the field of structural mechanics and finite elements analysis since 1961, with an emphasis on the theory of shells and rods and also on nonlinear and mixed finite element methods, and has authored about 150 contributions to scientific journals and books.

Walter Pilkey, Ph.D., grew up mostly in the Richland, Washington area and received his B.A. degree in humanities at Washington State University, his M.S. degree in Engineering Sciences from Purdue University, and his Ph.D. in Mechanics from Pennsylvania State University. While in college, he worked as a smokejumper for the U.S. Forest Service. He was a research scientist at the IIT Research Institute in Chicago from 1962 to 1969, during which time he also spent 18 months as a visiting professor at Kabul University in Afghanistan. He joined the School of Engineering and Applied Sciences at the University of Virginia in 1969, and for a time was chairman of the Applied Mechanics Division. He is the Frederick Tracy Morse Professor of Mechanical Engineering and founder of the Automobile Safety Laboratory and impact biomechanics program. His research interests include the mechanics of solids, structural dynamics, crashworthiness, and injury biomechanics. He has been an author or editor of more than 25 books and has authored more than 250 technical papers. He has also been the editor-in-chief of several journals.

Contents

Section A

Formulations for Linear Problems of Elasticity

1

Basic Equations: Differential Form

The fundamental equations for describing the behavior of a solid can be classified into the three categories:

- The *conditions of equilibrium*
- The *material law*
- The *conditions of geometric fit* (strain-displacement relations).

As will be shown in this chapter, these relationships are often expressed as local differential equations. They can also be written in global or integral form, corresponding to work or variational principles, which are treated in Chapter 2. Both the local and global forms of these relationships are used throughout this text.

To demonstrate that the local basic equations can be grouped into three categories, it is worthwhile to consider a simple structure.

EXAMPLE 1.1 Illustration of the Basic Equations by a Simple Example
Consider a rigid block of weight W attached to two wires of the same initial length (Fig. 1.1). The two wires are equal distances from the center point 0. The conditions of equilibrium ($\Sigma F_{\text{vert}} = 0$, $\Sigma M_0 = 0$) give $P_1 + P_2 = W$, $P_1 = P_2$. Hence, $P_1 = P_2 = W/2$. The equations of equilibrium applied to the undeformed geometry were sufficient to find the forces P_1, P_2. This is not always the case, as an increase in the number of constraints in the problem means that the forces cannot be obtained using equilibrium alone. For a system with a third wire (Fig. 1.2), the number of unknown forces increases, while the number of conditions for equilibrium remains the same. Thus, equilibrium requirements give the two relationships

$$P_1 + P_2 + P_3 = W, \quad P_1 = P_2 \tag{1}$$

which cannot be solved uniquely for the three forces P_1, P_2, and P_3. (Witness that $P_1 = P_2 = P_3 = W/3$, and $P_1 = P_2 = W/4$, $P_3 = W/2$ are possible solutions.) Clearly there are infinitely many *statically* acceptable or *admissible* solutions. Considerations other than equilibrium must be introduced to identify the correct solution.

For the material at hand, observed or documented information of the deformation caused by loads is usually available. The force-deformation effect is described by the *constitutive* or material response *relation*. Often this is referred to as the *material law*. Assume in the present problem that the elongation of each wire is proportional to the force in the wire, i.e.,

$$\Delta_1 = f_1 P_1, \quad \Delta_2 = f_2 P_2, \quad \Delta_3 = f_3 P_3 \tag{2}$$

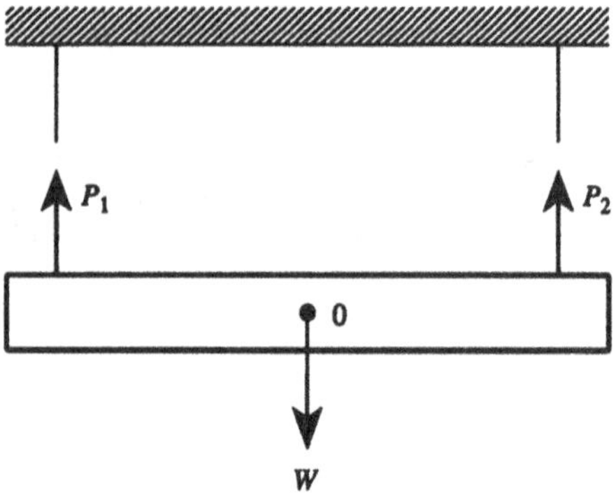

FIGURE 1.1
Statically determinate system.

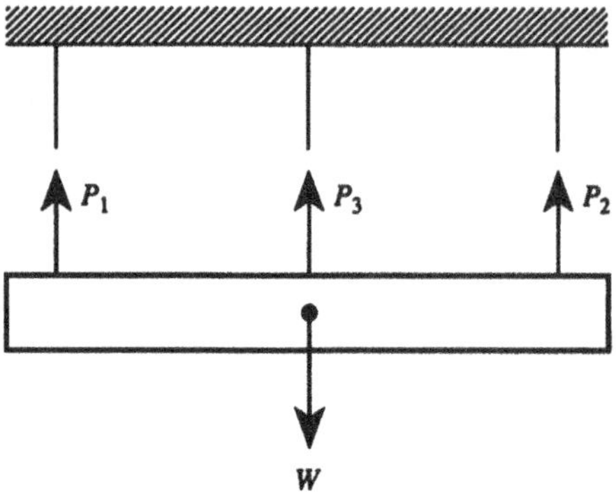

FIGURE 1.2
Statically indeterminate system.

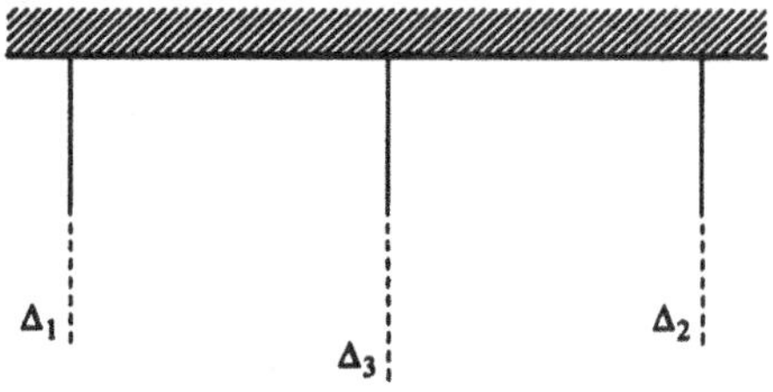

FIGURE 1.3
Displacements of the system of Fig. 1.2.

where Δ_1, Δ_2, and Δ_3 are the elongations of the wires, and f_1, f_2, and f_3 are material constants. Although three more equations have been gained, the problem still remains unsolvable since three additional unknowns Δ_1, Δ_2, and Δ_3 have been introduced. That is, there are now six unknowns P_1, P_2, P_3, Δ_1, Δ_2, and Δ_3 and five relationships of Eqs. (1) and (2). Therefore, a relationship must be sought in addition to the equilibrium and force-deformation equations. Suppose that the admissible solution $P_1 = P_2 = W/4$, $P_3 = W/2$ is correct. Then, on the basis of (2), deformations as shown in Fig. 1.3 would be obtained if each of the wires is made of the same material. However, this solution is incorrect since the pictured deformations are incompatible with the fact that the wires are attached to a rigid block. From this rigid block configuration a relationship between the deformations Δ_1, Δ_2, and Δ_3 can be found. The introduction of this kinematic or deformation condition provides the additional equation necessary to complete the solution. As a simple case, suppose wires 1 and 2 are made of identical material so that the deformation is symmetrical about the center wire. Then the deformation requirement is

$$\Delta_1 = \Delta_3 \tag{3}$$

There are now six unknown and six relationships [Eqs. (1), (2), and (3)]. Equation (3) is called the equation of geometric fit or compatible deformation. It is also referred to as the *kinematic relation*. Equations (1), (2), and (3) give

$$P_1 = P_2 = \frac{Wf_3}{f_1 + 2f_3},$$
$$P_3 = \frac{Wf_1}{f_1 + 2f_3} \tag{4}$$

This solution satisfies both the equations of equilibrium and the equations of compatible deformations; it is said to be *statically and kinematically admissible.*

Trivial as it may appear, the solution of this example, with the three categories of equations, the equations of equilibrium, the material law, and the conditions of geometric fit, contain all of the ingredients of the solution to any general problem in the mechanics of deformable solids. ∎

The complete set of basic equations (equilibrium equations, strain-displacement relations, material law equations, and boundary conditions) should be solved. However, for arbitrary configurations general solutions in closed form do not exist. Nevertheless, for certain types of structures or structural members, such as beams and plates, approximate theories have been constructed on the basis of assumptions reflecting their special properties. Usually these assumptions are made with respect to the distribution of strains or displacements within the structure, and often they are supplemented by assumptions on the relative influence of certain stress components. Thus, for the extension and bending of a bar, it is assumed that transverse cross-sections simply translate and remain planar, respectively, during deformation (Fig. 1.4). In the case of elementary torsion, it is assumed that the shape of the cross section remains unchanged, with no distortional effects.

These approximate theories may contain some inconsistencies, such as in the case of the bending of a bar in which shear stresses are introduced. The strains corresponding to these stresses lead to deformations that violate the basic deformation assumptions. Solutions based on the governing equations can do no better than the deformation model permits.

In order to be considered as satisfactory, the approximate theories must lead to solutions that are within acceptable engineering accuracy for a large class of problems. The respective

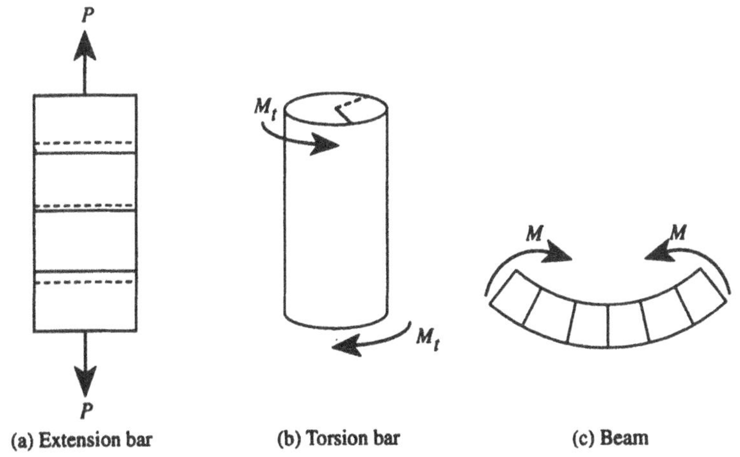

FIGURE 1.4
Some basic structural members.

assumptions usually have been confirmed by experiment, and generally the limitations of these theories are reasonably well known and documented.

Before considering approximate theories for various structures, the general equations of the theory of elasticity which are the basis of all elastic engineering models will be discussed.

1.1 Fundamental Definitions

To describe structural problems it will be useful to employ one of the two right-handed Cartesian[1] coordinate systems of Fig. 1.5. The coordinate system of Fig. 1.5b will normally be used in connection with structural members, e.g., for beams, plates, and shells.

Scalar, vector, and matrix notations are to be used. Einstein's[2] summation convention will often be employed to simplify the notation. This involves omitting summation signs and agreeing to add over *repeated indices*. These indices are "dummy indices" in that they can be arbitrarily renamed.

Thus,

$$a_i b_i = \sum_{i=1}^{3} a_i b_i = a_1 b_1 + a_2 b_2 + a_3 b_3 = a_j b_j = a_k b_k$$

$$\sigma_{ik}\epsilon_{ik} = \sum_{i=1}^{3}\sum_{k=1}^{3} \sigma_{ik}\epsilon_{ik} = \sigma_{11}\epsilon_{11} + \sigma_{12}\epsilon_{12} + \sigma_{13}\epsilon_{13} + \sigma_{21}\epsilon_{21} + \sigma_{22}\epsilon_{22} + \sigma_{23}\epsilon_{23} + \sigma_{31}\epsilon_{31} + \sigma_{32}\epsilon_{32} + \sigma_{33}\epsilon_{33} \tag{1.1}$$

The indices assume the values 1, 2, and 3 for a three-dimensional continuum, whereas they are 1 and 2 for a structural member with two significant directions.

Abbreviations for derivatives will be

Ordinary derivatives:

$$\frac{da_k}{dx} = d_x a_k = a_{k,x} = a'_k \tag{1.2a}$$

Partial derivatives:

$$\frac{\partial a_k}{\partial x_i} = \partial_i a_k = a_{k,i} \quad \text{or} \quad \frac{\partial a_k}{\partial x} = \partial_x a_k = a_{k,x} = a'_k \tag{1.2b}$$

The notation used to define the state of stress in a three-dimensional continuum is indicated in Fig. 1.6. The stress components normal to a coordinate plane are denoted by σ_x, σ_y, and σ_z and the shear stress components by τ_{xy}, τ_{xz}, τ_{yz}, τ_{yx}, τ_{zx}, and τ_{zy}. The single subscript

[1]Rene Descartes (1596–1650) was a French philosopher, physicist, and mathematician. He was a contemporary of Galileo and is probably best known for his philosophical works. His mathematical prowess was first recognized when, as a soldier, he solved a geometrical problem which was posted as a challenge on a placard on a street in Holland. His chief contributions in mathematics were his theories of analytical geometry and vortices.

[2]Albert Einstein (1879–1955), the father of the general theories of relativity and gravitation, was one of the first scientists to make extensive use of the index notation. His reputation as a slow child, pacifist, nuclear bomb politician, Zionist, physicist, and Nobel prize winner is well-known.

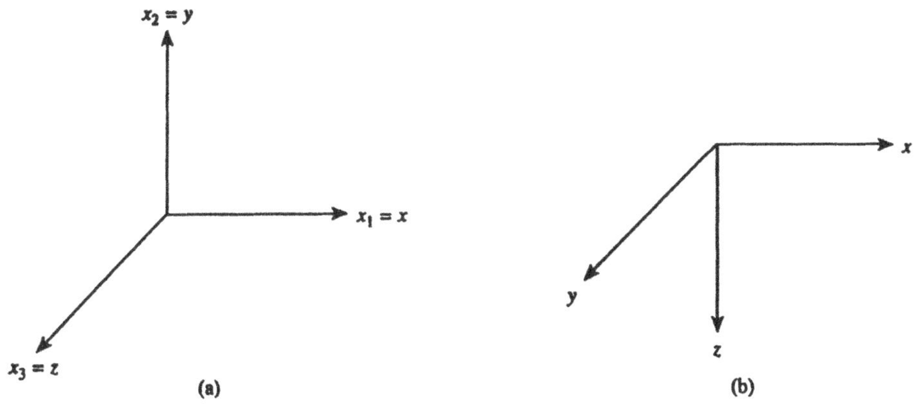

FIGURE 1.5
Right-handed Cartesian coordinate systems.

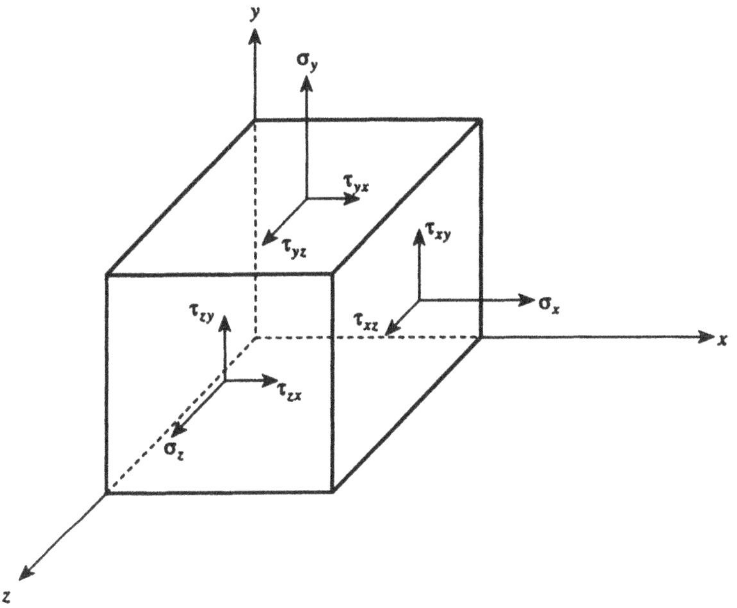

FIGURE 1.6
Definition of the components of the Cartesian stress tensor.

of a normal stress indicates that the stress acts on a plane normal to the axis in the subscript direction. The first letter of the double subscript employed with stresses designates that the plane on which the stress acts is normal to the axis in the subscript direction. The second letter in the stress subscript indicates the coordinate direction in which the stress acts.

There are three normal stress components. The shear stress components are symmetric, so that

$$\tau_{xy} = \tau_{yx}, \quad \tau_{xz} = \tau_{zx}, \quad \tau_{yz} = \tau_{zy}$$

This is a consequence of the equilibrium of moments, which is demonstrated in Section 1.4. Thus the state of stress at a point is completely described by six components. In matrix notation, the stresses are gathered together as

$$\boldsymbol{\sigma} = \begin{bmatrix} \sigma_x \\ \sigma_y \\ \sigma_z \\ \tau_{xy} \\ \tau_{xz} \\ \tau_{yz} \end{bmatrix} = [\sigma_x \quad \sigma_y \quad \sigma_z \quad \tau_{xy} \quad \tau_{xz} \quad \tau_{yz}]^T \tag{1.3a}$$

It is sometimes convenient to write the stresses as

$$\boldsymbol{\sigma} = [\sigma_{xx} \quad \sigma_{yy} \quad \sigma_{zz} \quad \sigma_{xy} \quad \sigma_{xz} \quad \sigma_{yz}]^T \tag{1.3b}$$

or

$$\boldsymbol{\sigma} = [\sigma_{11} \quad \sigma_{22} \quad \sigma_{33} \quad \sigma_{12} \quad \sigma_{13} \quad \sigma_{23}]^T \tag{1.3c}$$

where $x = 1$, $y = 2$, $z = 3$ have been used.

The deformed shape of a solid subjected to loading can be described by the three displacement components

$$\begin{aligned} u_x &= u_x(x, y, z) = u = u_1 \\ u_y &= u_y(x, y, z) = v = u_2 \\ u_z &= u_z(x, y, z) = w = u_3 \end{aligned} \tag{1.4}$$

or in vector notation

$$\mathbf{u} = \begin{bmatrix} u_x \\ u_y \\ u_z \end{bmatrix} = [u_x \quad u_y \quad u_z]^T = [u \quad v \quad w]^T = [u_1 \quad u_2 \quad u_3]^T \tag{1.5}$$

As indicated in Problem 1.35, rotary displacements can be derived from the translations u_x, u_y, and u_z. The normal strains are designated by ϵ_x, ϵ_y, and ϵ_z, while the shear strains are γ_{xy}, γ_{xz}, and γ_{yz}. As in the case of stresses, the strains are symmetric, i.e., $\gamma_{ij} = \gamma_{ji}$. In matrix notation,

$$\boldsymbol{\epsilon} = \begin{bmatrix} \epsilon_x \\ \epsilon_y \\ \epsilon_z \\ \gamma_{xy} \\ \gamma_{xz} \\ \gamma_{yz} \end{bmatrix} = [\epsilon_x \quad \epsilon_y \quad \epsilon_z \quad \gamma_{xy} \quad \gamma_{xz} \quad \gamma_{yz}]^T$$

or

$$\boldsymbol{\epsilon} = [\epsilon_{xx} \quad \epsilon_{yy} \quad \epsilon_{zz} \quad 2\epsilon_{xy} \quad 2\epsilon_{xz} \quad 2\epsilon_{yz}]^T \tag{1.6}$$

or

$$\boldsymbol{\epsilon} = [\epsilon_{11} \quad \epsilon_{22} \quad \epsilon_{33} \quad 2\epsilon_{12} \quad 2\epsilon_{13} \quad 2\epsilon_{23}]^T$$

where the relationship $\gamma_{ij} = 2\epsilon_{ij}, i \neq j$, will be explained in Section 1.2.

Note that the strain components are placed in the matrix in the same order as the corresponding stress quantities. Then a measure of the internal work $\sigma_{ik}\epsilon_{ik}$ is simply equal to the vector product $\boldsymbol{\sigma}^T\boldsymbol{\epsilon}$. Note that the definitions of Eq. (1.6) in which $\gamma_{yz} = 2\epsilon_{yz}$, etc., do not violate the equality $\sigma_{ik}\epsilon_{ik} = \boldsymbol{\sigma}^T\boldsymbol{\epsilon}$. To show this, return to $\sigma_{ik}\epsilon_{ik}$, of Eq. (1.1). Since the stress and strain components are symmetric, e.g., $\epsilon_{13} = \epsilon_{31}$, $\sigma_{ik}\epsilon_{ik}$ of Eq. (1.1) becomes

$\sigma_{11}\epsilon_{11} + \sigma_{22}\epsilon_{22} + \sigma_{33}\epsilon_{33} + 2\sigma_{12}\epsilon_{12} + 2\sigma_{13}\epsilon_{13} + 2\sigma_{23}\epsilon_{23} = \sigma_{xx}\epsilon_{xx} + \sigma_{yy}\epsilon_{yy} + \sigma_{zz}\epsilon_{zz} + \sigma_{xy}\gamma_{xy} + \sigma_{xz}\gamma_{xz} + \sigma_{yz}\gamma_{yz}$. This is the same expression obtained using $\sigma^T \epsilon$.

To produce positive work by positively defined stresses and strains, it is important to adopt consistent sign conventions. Stresses and strains are defined to be positive as indicated in Fig. 1.6, i.e., components on the positive face of an element are positive when they are acting along the positive direction of the coordinates. In addition, stress and strain components are defined to be positive when their components on the negative face are acting in the negative direction of the axis. An element's face with its outward normal along the positive direction of a coordinate axis is defined to be a positive face. A face with its normal in the opposite direction is said to be a negative face.

It is essential that the reader is able to visualize matrix multiplication. It is helpful to be familiar with a matrix multiplication scheme to aid in organizing practical multiplication calculations. To illustrate one such procedure, consider a column vector x and a row matrix y. The product yx can be obtained in the form

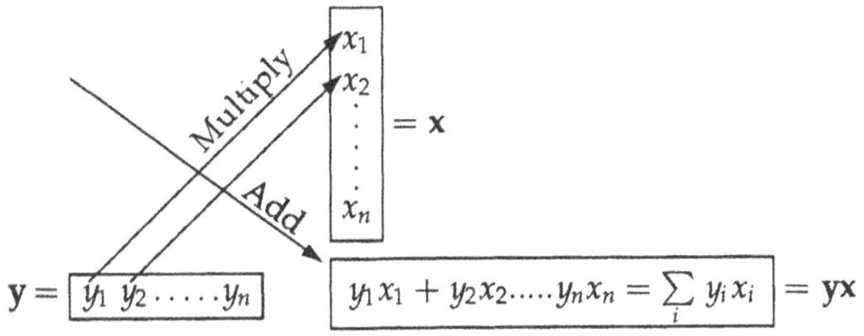

Similarly, $\mathbf{AB} = \mathbf{C}$ can be written as

$$\mathbf{B} = \downarrow \begin{vmatrix} b_{11} & b_{12} \\ b_{21} & b_{22} \end{vmatrix}$$

$$\mathbf{A} = \begin{vmatrix} a_{11} & a_{12} \\ a_{21} & a_{22} \end{vmatrix} \begin{vmatrix} c_{11} & c_{12} \\ c_{21} & c_{22} \end{vmatrix}$$

$$= \mathbf{C}$$

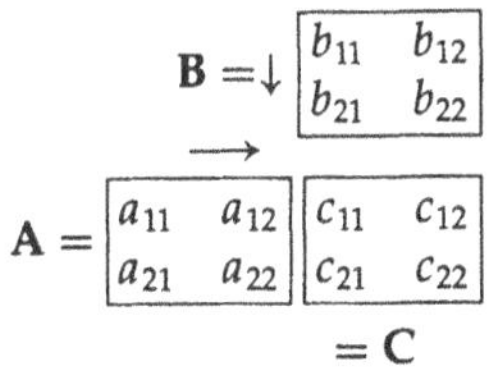

$$c_{11} = a_{11}b_{11} + a_{12}b_{21}$$
$$c_{12} = a_{11}b_{12} + a_{12}b_{22}$$
$$c_{21} = a_{21}b_{11} + a_{22}b_{21}$$
$$c_{22} = a_{21}b_{12} + a_{22}b_{22}$$

This scheme is suitable for organizing the matrix calculations that so frequently occur in structural analysis. For example, to find **ABCD** use either

---

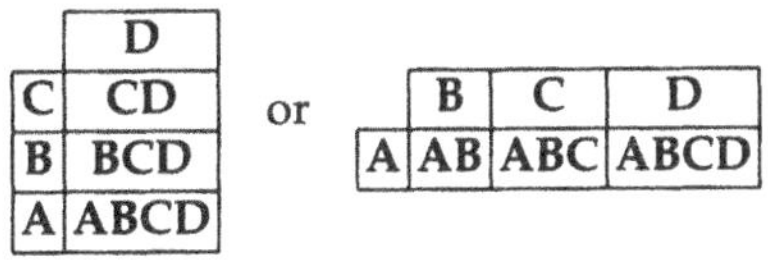

1.2 Deformation Relationships

1.2.1 Kinematical Equations

We begin by formulating the relationships between displacements and strains in a solid. The three components, u_x, u_y, u_z or u, v, w, of the displacement vector at a point in a solid are mutually orthogonal in a Cartesian coordinate system and they are taken to be positive in the direction of the positive coordinate axes.

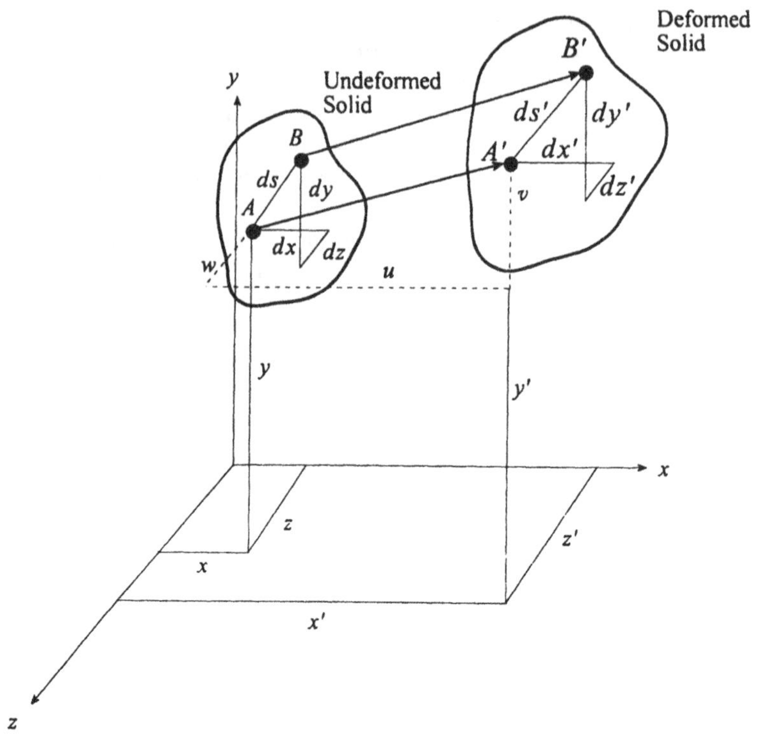

FIGURE 1.7
Deformation of a three-dimensional solid.

Consider two points A and B in a three-dimensional solid, the distance between which is of (infinitesimal) length ds. As a result of deformation, point A moves to A' and point B to B'. Let the initial coordinates of A and B be x, y, z and $x + dx, y + dy, z + dz$ which after deformation become x', y', z', and $x' + dx', y' + dy', z' + dz'$ as shown in Fig. 1.7. The length ds of a line element in the undeformed body connecting A and B is given by

$$ds^2 = dx^2 + dy^2 + dz^2 \tag{1.7}$$

This line element is elongated and rotated during deformation. The new line element connecting A' and B' and the deformed solid is of length ds', with

$$ds'^2 = dx'^2 + dy'^2 + dz'^2 \tag{1.8}$$

The displacement of point A to A' is characterized by the vector $\mathbf{u}$, which has the components

$$u = x' - x, \quad v = y' - y, \quad w = z' - z \tag{1.9}$$

Similarly the displacement of point B to B' is given by $\mathbf{u} + d\mathbf{u}$ where, from the chain rule of differentiation, the components of $d\mathbf{u}$ can be expressed as

$$\begin{aligned} du &= \frac{\partial u}{\partial x} dx + \frac{\partial u}{\partial y} dy + \frac{\partial u}{\partial z} dz \\ dv &= \frac{\partial v}{\partial x} dx + \frac{\partial v}{\partial y} dy + \frac{\partial v}{\partial z} dz \\ dw &= \frac{\partial w}{\partial x} dx + \frac{\partial w}{\partial y} dy + \frac{\partial w}{\partial z} dz \end{aligned} \tag{1.10}$$

Substitution of x', y', z' from Eq. (1.9) into Eq. (1.8) gives

$$ds'^2 = dx^2 + dy^2 + dz^2 + 2\,du\,dx + 2\,dv\,dy + 2\,dw\,dz + du^2 + dv^2 + dw^2 \tag{1.11}$$

or, using Eq. (1.7),

$$ds'^2 - ds^2 = 2(du\,dx + dv\,dy + dw\,dz) + du^2 + dv^2 + dw^2 \tag{1.12}$$

Substitution of the total differentials of Eq. (1.10) into Eq. (1.12) leads to

$$\begin{aligned} ds'^2 - ds^2 = {} & 2\left\{\partial_x u + \frac{1}{2}[(\partial_x u)^2 + (\partial_x v)^2 + (\partial_x w)^2]\right\} dx^2 \\ & + 2\left\{\partial_y v + \frac{1}{2}[(\partial_y u)^2 + (\partial_y v)^2 + (\partial_y w)^2]\right\} dy^2 \\ & + 2\left\{\partial_z w + \frac{1}{2}[(\partial_z u)^2 + (\partial_z v)^2 + (\partial_z w)^2]\right\} dz^2 \\ & + 2(\partial_x v + \partial_y u + \partial_x u\,\partial_y u + \partial_x v\,\partial_y v + \partial_x w\,\partial_y w)\,dx\,dy \\ & + 2(\partial_x w + \partial_z u + \partial_x u\,\partial_z u + \partial_x v\,\partial_z v + \partial_x w\,\partial_z w)\,dx\,dz \\ & + 2(\partial_y w + \partial_z v + \partial_y u\,\partial_z u + \partial_y v\,\partial_z v + \partial_y w\,\partial_z w)\,dy\,dz \end{aligned} \tag{1.13}$$

Note that $ds'^2 - ds^2$ is zero if no relative displacement occurs between the points A and B as they move to A' and B'. This would correspond to a rigid body motion. For $ds'^2 - ds^2$ not equal to zero, the line element ds has changed in length, i.e., the solid is strained. Therefore $ds'^2 - ds^2$ can be chosen as an appropriate measure of deformation of the solid. To define the strain components we write Eq. (1.13) as

$$ds'^2 - ds^2 = 2\epsilon_x\,dx^2 + 2\epsilon_y\,dy^2 + 2\epsilon_z\,dz^2 + 4\epsilon_{xy}\,dx\,dy + 4\epsilon_{xz}\,dx\,dz + 4\epsilon_{yz}\,dy\,dz \tag{1.14}$$

where

$$\epsilon_x = \frac{\partial u}{\partial x} + \frac{1}{2}\left[\left(\frac{\partial u}{\partial x}\right)^2 + \left(\frac{\partial v}{\partial x}\right)^2 + \left(\frac{\partial w}{\partial x}\right)^2\right] \tag{1.15a}$$

$$\epsilon_y = \frac{\partial v}{\partial y} + \frac{1}{2}\left[\left(\frac{\partial u}{\partial y}\right)^2 + \left(\frac{\partial v}{\partial y}\right)^2 + \left(\frac{\partial w}{\partial y}\right)^2\right] \tag{1.15b}$$

$$\epsilon_z = \frac{\partial w}{\partial z} + \frac{1}{2}\left[\left(\frac{\partial u}{\partial z}\right)^2 + \left(\frac{\partial v}{\partial z}\right)^2 + \left(\frac{\partial w}{\partial z}\right)^2\right] \tag{1.15c}$$

$$\epsilon_{xy} = \frac{1}{2}\left(\frac{\partial v}{\partial x} + \frac{\partial u}{\partial y} + \frac{\partial u}{\partial x}\frac{\partial u}{\partial y} + \frac{\partial v}{\partial x}\frac{\partial v}{\partial y} + \frac{\partial w}{\partial x}\frac{\partial w}{\partial y}\right) \tag{1.15d}$$

$$\epsilon_{xz} = \frac{1}{2}\left(\frac{\partial w}{\partial x} + \frac{\partial u}{\partial z} + \frac{\partial u}{\partial x}\frac{\partial u}{\partial z} + \frac{\partial v}{\partial x}\frac{\partial v}{\partial z} + \frac{\partial w}{\partial x}\frac{\partial w}{\partial z}\right) \tag{1.15e}$$

$$\epsilon_{yz} = \frac{1}{2}\left(\frac{\partial w}{\partial y} + \frac{\partial v}{\partial z} + \frac{\partial u}{\partial y}\frac{\partial u}{\partial z} + \frac{\partial v}{\partial y}\frac{\partial v}{\partial z} + \frac{\partial w}{\partial y}\frac{\partial w}{\partial z}\right) \tag{1.15f}$$

which can be abbreviated with the help of the notation of Section 1.1

$$\epsilon_{ik} = \frac{1}{2}(u_{i,k} + u_{k,i} + u_{l,i}u_{l,k}) \tag{1.16}$$

If strain components are given, these strain-displacement relations provide a system of nonlinear partial differential equations in the unknown displacements. The quantity ϵ_{ik} is referred to as the *Green strain tensor* although it is usually considered to have been introduced by Green and Saint-Venant.[3]

A slightly different definition of the shear strains is given by

$$\gamma_{ik} = 2\epsilon_{ik} \quad (i \neq k) \tag{1.17}$$

where γ_{ik} is sometimes called the "engineering" shear strain. These strains have been expressed in terms of the coordinates x, y, z in the undeformed state, i.e., the so-called *Lagrangian*[4] *coordinates*. Alternatively, $ds'^2 - ds^2$ can be written in terms of coordinates of the deformed state, i.e., the *Eulerian*[5] *coordinates* x', y', and z', leading to different expressions for strain, which are referred to as the *Almansi strain tensor*. This strain tensor was proposed by Almansi and Hamel (1877–1954). The Lagrangian description is preferred in structural mechanics because it is much easier to characterize an undeformed body than a deformed body. The Eulerian description is preferred in fluid mechanics. Note that Eqs. (1.15) are nonlinear. Although they are difficult to use, these nonlinear equations are essential to describe certain types of structural behavior which result in large deformations. They are particularly helpful in analyzing stability problems of some structural members.

In many cases of practical importance the deformations in a structure are small enough that the quadratic terms in Eqs. (1.15) make no significant contribution, and neglecting these terms is justified. The resulting linear expressions, which simplify the theories considerably, are

$$\begin{aligned}
\epsilon_x &= \frac{\partial u}{\partial x} = \partial_x u_x = u_{x,x} & \gamma_{xy} &= \frac{\partial v}{\partial x} + \frac{\partial u}{\partial y} = \partial_y u_x + \partial_x u_y = u_{x,y} + u_{y,x} \\
\epsilon_y &= \frac{\partial v}{\partial y} = \partial_y u_y = u_{y,y} & \gamma_{xz} &= \frac{\partial w}{\partial x} + \frac{\partial u}{\partial z} = \partial_z u_x + \partial_x u_z = u_{x,z} + u_{z,x} \\
\epsilon_z &= \frac{\partial w}{\partial z} = \partial_z u_z = u_{z,z} & \gamma_{yz} &= \frac{\partial w}{\partial y} + \frac{\partial v}{\partial z} = \partial_z u_y + \partial_y u_z = u_{y,z} + u_{z,y}
\end{aligned} \tag{1.18}$$

[3]Barré de Saint-Venant (1797–1886) was a French mathematician and engineer who made several important contributions to the theory of elasticity. During his studies, he was ostracized by his fellow students due to his proclaiming opposition to the "usurper" when the students of the Ecole Polytechnique were mobilized in 1814. A student of Navier, he went on to several applied mathematical developments so significant that his name is still associated with them today. Some of his developments experienced immediate application in civil engineering projects, such as in road and bridge building.

[4]Joseph Louis Lagrange (1736–1813) is often considered to be the greatest mathematician of the 18th century. He was born in Turin, Italy, and died in Paris. It was not until he was seventeen that he showed any interest in mathematics. In a letter to Euler which he wrote at the age of nineteen, he enunciated the principles underlying the calculus of variations in connection with the solution of an isoperimetric problem. At the same age, he was made a professor of mathematics at the Royal Artillery School in Turin. On the recommendation of Euler and D'Alembert, he was invited in 1766 to replace Euler at the Berlin Academy. It was there that much of his important work was accomplished, including the preparation of his *Mécanique analytique* and his efforts in differential and integral calculus. He is considered to be one of the fathers of classical mechanics. In 1787, he became a professor at the Ecole Normale and Ecole Polytechnique in Paris, where he settled into a lengthy period of didactic writings on his philosophies of mathematics. Napoleon honored him by giving him a title and making him a senator.

[5]Leonhard Euler (1707–1783) was a Swiss mathematician who entered the University of Basel at the age of 13 and studied with Johann Bernoulli. When Johann Bernoulli's sons Daniel and Nicholas went to Russia in 1725 at the invitation of the empress, they obtained a position for Euler. There he eventually succeeded Daniel Bernoulli in the chair for mathematics at the Russian Academy. In 1735, he lost the use of an eye due, he thought, to the severe climate. At the insistence of Frederick the Great, he moved to Berlin in 1741, and remained there until 1766, when he returned to Russia. There he soon became completely blind. In spite of this and other adversities, he continued his scientific work. His significant contributions to pure and applied mathematics were numerous.

or

$$\epsilon_{ik} = \frac{1}{2}(u_{i,k} + u_{k,i}) \tag{1.19}$$

which is called the *Cauchy*[6] *strain tensor.* In matrix form, these become

$$\begin{bmatrix} \epsilon_x \\ \epsilon_y \\ \epsilon_z \\ \gamma_{xy} \\ \gamma_{xz} \\ \gamma_{yz} \end{bmatrix} = \begin{bmatrix} \partial_x & 0 & 0 \\ 0 & \partial_y & 0 \\ 0 & 0 & \partial_z \\ \partial_y & \partial_x & 0 \\ \partial_z & 0 & \partial_x \\ 0 & \partial_z & \partial_y \end{bmatrix} \begin{bmatrix} u_x \\ u_y \\ u_z \end{bmatrix} \tag{1.20}$$

or

$$\epsilon \quad = \quad \mathbf{D} \quad \mathbf{u} \tag{1.21}$$

with the differential operator matrix

$$\mathbf{D} = \begin{bmatrix} \partial_x & 0 & 0 \\ 0 & \partial_y & 0 \\ 0 & 0 & \partial_z \\ \partial_y & \partial_x & 0 \\ \partial_z & 0 & \partial_x \\ 0 & \partial_z & \partial_y \end{bmatrix} \tag{1.22}$$

Two-Dimensional Problems

For the special case of two-dimensional behavior in the xy plane,

$$\mathbf{u} = [u_x \quad u_y]^T, \qquad \epsilon = [\epsilon_x \quad \epsilon_y \quad \gamma_{xy}]^T \tag{1.23}$$

and Eq. (1.20) becomes

$$\begin{aligned} \begin{bmatrix} \epsilon_x \\ \epsilon_y \\ \gamma_{xy} \end{bmatrix} &= \begin{bmatrix} \partial_x & 0 \\ 0 & \partial_y \\ \partial_y & \partial_x \end{bmatrix} \begin{bmatrix} u_x \\ u_y \end{bmatrix} \\ \epsilon \quad &= \quad \mathbf{D} \quad \mathbf{u} \end{aligned} \tag{1.24}$$

EXAMPLE 1.2 Physical Interpretation of the Strain Components
The strain-displacement relations will be rederived here in a manner that may provide more physical insight into the strain components.

Consider the projection of a deformed differential element onto the xy plane as shown in Fig. 1.8. Points 0, A, and B in the unstrained element move to points $0'$, A', and B' of the

[6]Augustin Louis Cauchy (1789–1857) was a French mathematician and engineer. His father fled Paris during the French revolution, going to a nearby village where many outstanding scientists often met at Laplace's house. There the exceptional mathematical prowess of the young Cauchy was noticed by Lagrange. Beginning in 1815, he held positions at the Ecole Polytechnique, Faculté des Sciences and the Collège de France. In 1830 his refusal to sign an oath of allegiance cost him his positions and he left for the University of Turin. He returned to Paris in 1838. In the theory of elasticity he introduced the concept of stress and derived the conditions of equilibrium, including the surface conditions called Cauchy's formula.

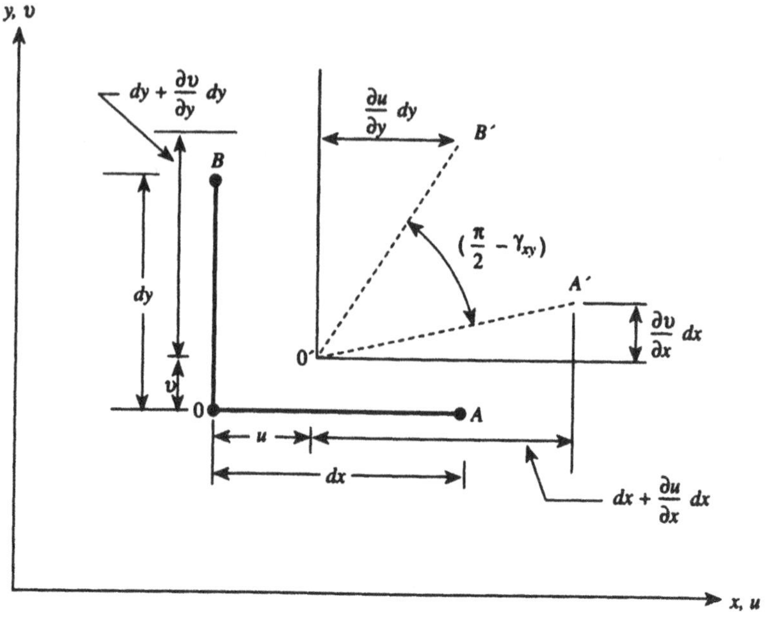

FIGURE 1.8
Displacements of points 0, A, and B. These points move to $0'$, A', and B'.

strained element. Note that two types of changes have occurred during the deformation: (1) the distances between such points as $0A$ have varied, and (2) the angles between such lines as $0A$ and $0B$ are no longer right angles. The x, y displacement components of point $0'$ are u, v. Since y is constant along line $0A$, when point A moves to A' it is displaced $u + (\partial u/\partial x)\,dx$ and $v + (\partial v/\partial x)\,dx$. Thus, along x, point A is displaced $(\partial u/\partial x)\,dx$ more than point 0, so that the x projection of $0'A'$ is $dx + (\partial u/\partial x)\,dx$. The y projection of $0'A'$ is $(\partial v/\partial x)\,dx$. The displacement components of point B moving to B' are $u + (\partial u/\partial y)\,dy$ and $v + (\partial v/\partial y)\,dy$.

The strains associated with the changes in length are called the *normal*, longitudinal, or extensional strains, and those associated with the changes in angle are the *shear* strains. To find the relationship between normal strain and displacement, accept the *engineering* definition of the normal strain ϵ_x which is the unit change in length of a line which was originally oriented in the x direction. Thus,

$$\epsilon_x = \frac{0'A' - 0A}{0A} = \frac{0'A' - dx}{dx} \tag{1}$$

It should be noted that $0'A'$ is chosen to be not aligned with the x direction. From Fig. 1.8.

$$(0'A')^2 = \left(dx + \frac{\partial u}{\partial x}dx\right)^2 + \left(\frac{\partial v}{\partial x}dx\right)^2$$

Then

$$\begin{aligned}\epsilon_x &= \frac{dx\sqrt{(1 + \partial u/\partial x)^2 + (\partial v/\partial x)^2} - dx}{dx} \\ &= \sqrt{1 + 2\partial u/\partial x + (\partial u/\partial x)^2 + (\partial v/\partial x)^2} - 1 \end{aligned} \tag{2}$$

Use of the binomial expansion

$$(1+a)^n = 1 + na + \frac{n(n-1)}{2!}a^2 + \cdots \text{ for } a^2 < 1$$

permits the radical in (2) to be rewritten

$$\epsilon_x = 1 + \frac{1}{2}\left[2\frac{\partial u}{\partial x} + \left(\frac{\partial u}{\partial x}\right)^2 + \left(\frac{\partial v}{\partial x}\right)^2\right] + \cdots - 1$$

$$\approx \frac{\partial u}{\partial x} + \frac{1}{2}\left[\left(\frac{\partial u}{\partial x}\right)^2 + \left(\frac{\partial v}{\partial x}\right)^2\right] \tag{3}$$

where powers of the derivatives higher than the second are neglected. If the displacement in the z direction were also taken into account, (3) would be replaced by

$$\epsilon_x = \frac{\partial u}{\partial x} + \frac{1}{2}\left[\left(\frac{\partial u}{\partial x}\right)^2 + \left(\frac{\partial v}{\partial x}\right)^2 + \left(\frac{\partial w}{\partial x}\right)^2\right] \tag{4}$$

Similarly, the other normal strains are found to be the expressions of Eqs. (1.15b) and (1.15c).

The shear strain at a point is defined as the change due to deformation in the value of the cosine of an angle that in the unstrained state was a right angle. In Fig. 1.8, let γ_{xy} be the shear strain referred to in the xy coordinate system. Consider the cosine of angle $A'0'B'$.

$$\cos A'0'B' = \cos\left(\frac{\pi}{2} - \gamma_{xy}\right) = \sin \gamma_{xy} \tag{5}$$

For small angles, the sine of an angle may be replaced by the angle. Thus, $\cos A'0'B' = \gamma_{xy}$. From analytical geometry, the cosine of the angle between two lines can be replaced by the sum of the products of their direction cosines, i.e.,

$$\gamma_{xy} = \cos A'0'B' = a_{x1}a_{y1} + a_{x2}a_{y2} \tag{6}$$

where a_{x1}, a_{x2} are the direction cosines of $0'A'$, and a_{y1}, a_{y2} are the direction cosines of $0'B'$. Using Fig. 1.8, these direction cosines are found to be

$$\begin{aligned} a_{x1} &= \frac{(1 + \partial u/\partial x)\,dx}{0'A'} \\ a_{x2} &= \frac{(\partial v/\partial x)\,dx}{0'A'} \\ a_{y1} &= \frac{(\partial u/\partial y)\,dy}{0'B'} \\ a_{y2} &= \frac{(1 + \partial v/\partial y)\,dy}{0'B'} \end{aligned} \tag{7}$$

Then

$$\gamma_{xy} = \left[\left(1 + \frac{\partial u}{\partial x}\right)\frac{\partial u}{\partial y} + \frac{\partial v}{\partial x}\left(1 + \frac{\partial v}{\partial y}\right)\right]\frac{dx\,dy}{(0'A')(0'B')} \tag{8}$$

Recall that

$$0'A' = \sqrt{\left(dx + \frac{\partial u}{\partial x}dx\right)^2 + \left(\frac{\partial v}{\partial x}dx\right)^2}$$

$$= dx\sqrt{1 + 2\frac{\partial u}{\partial x} + \left(\frac{\partial u}{\partial x}\right)^2 + \left(\frac{\partial v}{\partial x}\right)^2} \tag{9}$$

If the derivatives of u and v are small with respect to unity, then $0'A' \approx dx$. Also, $0'B' \approx dy$. Therefore,

$$\gamma_{xy} = \frac{\partial v}{\partial x} + \frac{\partial u}{\partial y} + \frac{\partial u}{\partial x}\frac{\partial u}{\partial y} + \frac{\partial v}{\partial x}\frac{\partial v}{\partial y} \tag{10}$$

In three dimensions, this becomes

$$\gamma_{xy} = \frac{\partial v}{\partial x} + \frac{\partial u}{\partial y} + \frac{\partial u}{\partial x}\frac{\partial u}{\partial y} + \frac{\partial v}{\partial x}\frac{\partial v}{\partial y} + \frac{\partial w}{\partial x}\frac{\partial w}{\partial y} \tag{11}$$

The strains γ_{yz} and γ_{zx} follow in a similar fashion. ∎

1.2.2 Compatibility Conditions

Although six strain components are required to describe the state of strain at a point, the above equations contain only three displacement components. Hence, the equations will not possess a unique solution for displacements if the strains are arbitrarily prescribed. In practical terms, this means that the displacements computed could exhibit tears, kinks, cracks, and overlaps, none of which should occur in reality. It would appear that some of the strains should not be independent of some other strains; therefore, the strains should have to satisfy an additional condition. To seek relationships between the strains, differentiate γ_{xy} of Eq. (1.18) with respect to x and y.

$$\frac{\partial^2 \gamma_{xy}}{\partial x\,\partial y} = \frac{\partial^2}{\partial x\,\partial y}\frac{\partial u}{\partial y} + \frac{\partial^2}{\partial x\,\partial y}\frac{\partial v}{\partial x} \tag{1.25}$$

From the calculus, it is known that for single-valued and continuous functions f the condition

$$\frac{\partial^2 f}{\partial x\,\partial y} = \frac{\partial^2 f}{\partial y\,\partial x} \tag{1.26}$$

holds. Apply this condition to displacements, and rewrite the right-hand side of Eq. (1.25), giving

$$\frac{\partial^2 \gamma_{xy}}{\partial x\,\partial y} = \frac{\partial^2}{\partial y^2}\frac{\partial u}{\partial x} + \frac{\partial^2}{\partial x^2}\frac{\partial v}{\partial y}$$

or

$$\frac{\partial^2 \gamma_{xy}}{\partial x\,\partial y} = \frac{\partial^2 \epsilon_x}{\partial y^2} + \frac{\partial^2 \epsilon_y}{\partial x^2} \quad \text{or} \quad R_{zz} = \frac{\partial^2 \epsilon_x}{\partial y^2} + \frac{\partial^2 \epsilon_y}{\partial x^2} - \frac{\partial^2 \gamma_{xy}}{\partial x\,\partial y} = 0 \tag{1.27}$$

This relation between the strains does indeed indicate that to obtain unique, continuous displacements, the strains cannot all be independent of each other. Equation (1.27) is called a strain *compatibility* condition. This relation is also sometimes referred to as the *integrability condition* for Eq. (1.24).

For a three-dimensional problem there will be a set of six *compatibility* equations, which in matrix form, using the compatibility functions R_{ij}, appear as [Leipholz, 1968, p. 75]

$$\begin{bmatrix} R_{xx} \\ R_{yy} \\ R_{zz} \\ \cdots \\ R_{xy} \\ R_{xz} \\ R_{yz} \end{bmatrix} = \begin{bmatrix} 0 & \partial_z^2 & \partial_y^2 & \vdots & 0 & 0 & -\partial_y\partial_z \\ \partial_z^2 & 0 & \partial_x^2 & \vdots & 0 & -\partial_x\partial_z & 0 \\ \partial_y^2 & \partial_x^2 & 0 & \vdots & -\partial_x\partial_y & 0 & 0 \\ \cdots & \cdots & \cdots & \cdot & \cdots & \cdots & \cdots \\ 0 & 0 & -\partial_x\partial_y & \vdots & -\frac{1}{2}\partial_z^2 & \frac{1}{2}\partial_y\partial_z & \frac{1}{2}\partial_z\partial_x \\ 0 & -\partial_x\partial_z & 0 & \vdots & \frac{1}{2}\partial_y\partial_z & -\frac{1}{2}\partial_y^2 & \frac{1}{2}\partial_x\partial_y \\ -\partial_y\partial_z & 0 & 0 & \vdots & \frac{1}{2}\partial_x\partial_z & \frac{1}{2}\partial_x\partial_y & -\frac{1}{2}\partial_x^2 \end{bmatrix} \begin{bmatrix} \epsilon_x \\ \epsilon_y \\ \epsilon_z \\ \cdots \\ \gamma_{xy} \\ \gamma_{xz} \\ \gamma_{yz} \end{bmatrix} = \mathbf{0} \quad (1.28)$$

$$\mathbf{R} \quad = \quad \mathbf{D}_1 \quad \boldsymbol{\epsilon} \quad = \mathbf{0} \quad (1.29)$$

If the strain vector is reordered as $\boldsymbol{\epsilon} = [\epsilon_x\ \epsilon_y\ \epsilon_z\ \gamma_{yz}\ \gamma_{zx}\ \gamma_{xy}]^T$ and the variables R_{ij} are reordered similarly, $\mathbf{D}_1$ will be a symmetric matrix. As we will see, there are methods other than using equations such as Eqs. (1.27) and (1.28) to assure that the displacements are single-valued and continuous. In fact, most matrix methods of approximate structural analysis avoid Eq. (1.28) and deal directly with the displacements, rather than with the strains.

The homogeneous equations of Eq. (1.28) can be shown to be related to each other. It can be verified readily that [Leipholz, 1968, p. 76] $R_{ij,j} = 0$, or using the operator matrix $\mathbf{D}$ of Eq. (1.22),

$$\mathbf{D}^T\mathbf{R} = \mathbf{0} \quad (1.30)$$

and $R_{xy} = R_{yx}$, $R_{xz} = R_{zx}$, $R_{yz} = R_{zy}$ where

$$\mathbf{R} = [R_{xx} \quad R_{yy} \quad R_{zz} \quad R_{xy} \quad R_{xz} \quad R_{yz}]^T$$

Thus, the six equations of Eq. (1.28) are reduced to the three independent equations of Eq. (1.30). Note that Eq. (1.30) has the same form as the homogeneous equations of equilibrium. For this formulation in Cartesian coordinates, these relations merely express that the partial derivatives can be interchanged, e.g.,

$$\partial_x\partial_y\partial_z f(x, y, z) = \partial_z\partial_y\partial_x f(x, y, z) \quad (1.31)$$

which can be considered as a basic statement of compatibility. In curvilinear coordinates embedded in an Euclidean[7] space, the compatibility conditions are related to the so-called Bianchi[8] identities.

1.3 Material Laws

The material law or constitutive equation provides a relationship between the stresses σ and the strains ϵ. In the linear, elastic range of a homogeneous solid, the material law is usually called *Hooke's*[9] *law*. A material is said to be *isotropic* if the material properties at a point are independent of orientation. In the case of an element of isotropic material with

[7]Euclid of Alexandria (c. 300 B.C.) was an early Greek philosopher-mathematician, perhaps the most celebrated mathematician ever. So little is known of his life that he has often been mistaken for another Euclid, a philosopher who studied under Socrates. It is presumed that the mathematician Euclid studied with the students of Plato or perhaps at Plato's academy. He wrote the thirteen books *Elements* which influenced the study of geometry for centuries.
[8]Luigi Bianchi (1856–1928) was Italy's most resourceful 19th century mathematician dealing with differential geometry.
[9]Robert Hooke (1635–1702) was an English physicist who devised numerous mechanics experiments. In 1662, he became curator of experiments of the Royal Society, and in 1664, he became professor of geometry in

uniaxial loading (σ_x) along the x direction, the strain in the x direction is simply σ_x/E, where E is called *Young's*[10] *modulus* or the modulus of elasticity. The lateral strains due to the σ_x are the contractions $\epsilon_y = \epsilon_z = -\nu\sigma_x/E$, where ν is a material constant, which is referred to as *Poisson's*[11] *ratio*. In turn, imposed normal stresses σ_y and σ_z would produce strains in the x direction of magnitudes $\epsilon_x = -\nu\sigma_y/E$, $\epsilon_x = -\nu\sigma_z/E$. As the normal strains are not influenced by shear deformation, the normal strains in a direction can be obtained by superimposing these effects. For ϵ_x we find

$$\epsilon_x = \frac{1}{E}[\sigma_x - \nu(\sigma_y + \sigma_z)]$$

The shear strain corresponding to τ_{xy} is simply $\gamma_{xy} = \tau_{xy}/G$, where G is called the *shear modulus of elasticity*.

The modulus of elasticity and Poisson's ratio are material constants that can be determined experimentally. Typical values of E are 70 GN/m^2 (10.1×10^6 psi) for aluminum and 210 GN/m^2(29×10^6 psi) for steel. Poisson's ratio is usually between 0.25 and 0.35 for metals. As demonstrated in elementary textbooks, the theoretical lower bound for ν is 0, which corresponds to a very compressible material such as cork. Certain concretes can have a ν as low as 0.1. The upper bound of 0.5 applies for very incompressible materials, e.g., water, such as is often assumed to be the case for materials in the plastic range. Rubber, which is quite incompressible, is sometimes incorrectly considered to be compressible because a change in shape may be mistaken for a change in volume.

It can be shown that for an isotropic material there are only two independent material constants. Thus, G, ν, and E are interrelated. More specifically (Problem 1.33), $G = E/2(1+\nu)$.

If all of the equations are placed together in matrix notation, we have

$$\begin{bmatrix} \epsilon_x \\ \epsilon_y \\ \epsilon_z \\ \cdots \\ \gamma_{xy} \\ \gamma_{xz} \\ \gamma_{yz} \end{bmatrix} = \frac{1}{E} \left[\begin{array}{ccc:ccc} 1 & -\nu & -\nu & & & \\ -\nu & 1 & -\nu & & \mathbf{0} & \\ -\nu & -\nu & 1 & & & \\ \cdots & \cdots & \cdots & \cdots & \cdots & \cdots \\ & & & 2(1+\nu) & 0 & 0 \\ & \mathbf{0} & & 0 & 2(1+\nu) & 0 \\ & & & 0 & 0 & 2(1+\nu) \end{array} \right] \begin{bmatrix} \sigma_x \\ \sigma_y \\ \sigma_z \\ \cdots \\ \tau_{xy} \\ \tau_{xz} \\ \tau_{yz} \end{bmatrix} \tag{1.32a}$$

$$\boldsymbol{\epsilon} = \mathbf{E}^{-1} \boldsymbol{\sigma} \tag{1.32b}$$

Gresham College. In 1679, in his brief collection of his studies, *Lectiones Cutleriane*, he enunciated Hooke's law. It is believed that his correspondence with Isaac Newton formed the basis of the theory of gravitation. He was an important geologist and even more important instrumentation inventor.

[10] Thomas Young (1773–1829) was an English physicist who started early as somewhat of a prodigy. He was largely self-educated in modern languages and literature and was well-read in physics and mathematics. In 1819, he proposed the key to decipher Egyptian hieroglyphics. He became a physician, but continued his work in physics, including considerable efforts in the mechanics of solids. He discovered the principle of interference of light and employed what is now known as Young's modulus.

[11] Siméon-Denis Poisson (1781–1840) was a French mathematician who solved many mechanics problems of practical importance. After an unsuccessful attempt to learn surgery, he entered the École Polytechnique, where his work attracted the attention of Lagrange and Laplace. In terms of fundamental ideas, he introduced the concept of tension of a bar being accompanied by lateral contraction. His efforts in mathematical physics were diverse and significant.

Using summation convention notation, this may be written as

$$\epsilon_{ij} = \frac{1+\nu}{E}\sigma_{ij} - \frac{\nu}{E}\delta_{ij}\sigma_{kk} \quad \text{with } \delta_{ij} = \begin{cases} 1 & \text{if } i = j \\ 0 & \text{if } i \neq j \end{cases}$$

and

$$\sigma_{kk} = \sigma_{xx} + \sigma_{yy} + \sigma_{zz} = \sigma_x + \sigma_y + \sigma_z \tag{1.33a}$$

More generally, this is often placed in the form

$$\epsilon_{ij} = F_{ijkl}\,\sigma_{kl} \tag{1.33b}$$

where F_{ijkl} is the index notation form of the 6×6 matrix $\mathbf{E}^{-1}$.

From Eq. (1.32) [or Eq. (1.33)], the stresses can be found by inversion as functions of strains

$$\begin{bmatrix} \sigma_x \\ \sigma_y \\ \sigma_z \\ \cdots \\ \tau_{xy} \\ \tau_{xz} \\ \tau_{yz} \end{bmatrix} = \frac{E}{(1+\nu)(1-2\nu)} \left[\begin{array}{ccc:ccc} 1-\nu & \nu & \nu & & & \\ \nu & 1-\nu & \nu & & \mathbf{0} & \\ \nu & \nu & 1-\nu & & & \\ \cdots & \cdots & \cdots & \cdots & \cdots & \cdots \\ & & & \frac{1-2\nu}{2} & 0 & 0 \\ & \mathbf{0} & & 0 & \frac{1-2\nu}{2} & 0 \\ & & & 0 & 0 & \frac{1-2\nu}{2} \end{array}\right] \begin{bmatrix} \epsilon_x \\ \epsilon_y \\ \epsilon_z \\ \cdots \\ \gamma_{xy} \\ \gamma_{xz} \\ \gamma_{yz} \end{bmatrix} \tag{1.34a}$$

$$\boldsymbol{\sigma} \quad = \quad \mathbf{E} \quad \boldsymbol{\epsilon} \tag{1.34b}$$

In index notation, this can be written as

$$\sigma_{ij} = \frac{E\nu}{(1+\nu)(1-2\nu)}\delta_{ij}\,\epsilon_{kk} + \frac{E}{(1+\nu)}\epsilon_{ij} = \lambda\,\delta_{ij}\,\epsilon_{kk} + 2\mu\,\epsilon_{ij} \tag{1.35a}$$

or

$$\sigma_{ij} = E_{ijkl}\,\epsilon_{kl} \tag{1.35b}$$

where $\lambda = E\nu/[(1+\nu)(1-2\nu)]$ and $\mu = G = E/[2(1+\nu)]$ are the Lamé's[12] constants, and $\epsilon_{kk} = \epsilon_{xx} + \epsilon_{yy} + \epsilon_{zz} = \epsilon_x + \epsilon_y + \epsilon_z$.

It is sometimes convenient, for example, in soil mechanics, to express the material law in terms of the bulk modulus K. To do so, define the *deviatoric* stress and strain components

$$\begin{aligned} \sigma'_{ij} &= \sigma_{ij} - \frac{1}{3}\sigma_{kk}\delta_{ij} \\ \epsilon'_{ij} &= \epsilon_{ij} - \frac{1}{3}\epsilon_{kk}\delta_{ij} \end{aligned} \tag{1.36}$$

[12]Gabriel Lamé (1795–1870) was a French engineer who, after graduation from the Ecole Polytechnique, worked at a Russian railroad institute and helped establish a new engineering school in St. Petersburg. He introduced and applied curvilinear coordinates, and contributed to number theory, applied mechanics, and thermodynamics. In 1852, he co-authored (with Clapeyron) the first book on the theory of elasticity.

Then the constitutive equations are expressed as

$$\sigma'_{ij} = 2G\epsilon'_{ij} \qquad p = -K\,\epsilon_{kk} \tag{1.37}$$

where p is the mean pressure, $p = -\sigma_{kk}/3$.

The matrix **E** is symmetric and, in general, positive definite. Only for incompressible materials for which $\nu = 1/2$ is matrix **E** positive semi-definite.

Sometimes **E** is referred to as the *material stiffness matrix*, while $\mathbf{E}^{-1}$ is called the *material flexibility matrix*.

1.3.1 Two-Dimensional Problems

In two-dimensional planar problems, two approximate cases are of frequent interest: a state of *plane stress* and a state of *plane strain*. Consider a flat element of thickness t lying in the xy plane with the z direction perpendicular to it. For such a solid it is convenient to define *stress resultants* as an alternative expression for stresses

$$n_x = \int_{-t/2}^{t/2} \sigma_x\, dz \qquad n_y = \int_{-t/2}^{t/2} \sigma_y\, dz \qquad n_{xy} = \int_{-t/2}^{t/2} \tau_{xy}\, dz \tag{1.38}$$

The units of these stress resultants are force per length.

In the case of plane stress, it is assumed that $\sigma_z = \tau_{xz} = \tau_{yz} = 0$, and that all remaining variables do not vary through the element thickness. In spite of the fact that these assumptions lead to the violation of some compatibility conditions, they provide a basis for a useful theory that can be shown to be quite accurate for thin plates, e.g., a thin element with loading transverse to the plane of the element.

If the conditions $\sigma_z = \tau_{xz} = \tau_{yz} = 0$ are imposed on Eq. (1.34a), the strains are found to be $\gamma_{yz} = \gamma_{xz} = 0$, $\epsilon_z = -\nu(\epsilon_x + \epsilon_y)/(1-\nu)$. Equation (1.34a) reduces to

$$\underset{\boldsymbol{\sigma}}{\begin{bmatrix} \sigma_x \\ \sigma_y \\ \tau_{xy} \end{bmatrix}} = \underset{\mathbf{E}}{\frac{E}{1-\nu^2}\begin{bmatrix} 1 & \nu & 0 \\ \nu & 1 & 0 \\ 0 & 0 & \frac{1-\nu}{2} \end{bmatrix}} \underset{\boldsymbol{\epsilon}}{\begin{bmatrix} \epsilon_x \\ \epsilon_y \\ \gamma_{xy} \end{bmatrix}} \tag{1.39a}$$

or

$$\underset{\mathbf{s}}{\begin{bmatrix} n_x \\ n_y \\ n_{xy} \end{bmatrix}} = \underset{\mathbf{E}}{D\begin{bmatrix} 1 & \nu & 0 \\ \nu & 1 & 0 \\ 0 & 0 & \frac{1-\nu}{2} \end{bmatrix}} \underset{\boldsymbol{\epsilon}}{\begin{bmatrix} \epsilon_x \\ \epsilon_y \\ \gamma_{xy} \end{bmatrix}} \tag{1.39b}$$

with $D = Et/(1-\nu^2)$ as the extensional rigidity. Note that **E** of $\boldsymbol{\sigma} = \mathbf{E}\boldsymbol{\epsilon}$ differs from the **E** of $\mathbf{s} = \mathbf{E}\boldsymbol{\epsilon}$ by a factor of t. Equation (1.32) becomes

$$\underset{\boldsymbol{\epsilon}}{\begin{bmatrix} \epsilon_x \\ \epsilon_y \\ \gamma_{xy} \end{bmatrix}} = \underset{\mathbf{E}^{-1}}{\frac{1}{E}\begin{bmatrix} 1 & -\nu & 0 \\ -\nu & 1 & 0 \\ 0 & 0 & 2(1+\nu) \end{bmatrix}} \underset{\boldsymbol{\sigma}}{\begin{bmatrix} \sigma_x \\ \sigma_y \\ \tau_{xy} \end{bmatrix}} \tag{1.40a}$$

or

$$\underset{\boldsymbol{\epsilon}}{\begin{bmatrix} \epsilon_x \\ \epsilon_y \\ \gamma_{xy} \end{bmatrix}} = \underset{\mathbf{E}^{-1}}{\frac{1}{Et}\begin{bmatrix} 1 & -\nu & 0 \\ -\nu & 1 & 0 \\ 0 & 0 & 2(1+\nu) \end{bmatrix}} \underset{\mathbf{s}}{\begin{bmatrix} n_x \\ n_y \\ n_{xy} \end{bmatrix}} \tag{1.40b}$$

Although the stress σ_z is zero by definition of plane stress, the corresponding strain ϵ_z (and u_z) does not vanish.

For plane strain, it is assumed that $\epsilon_z = \gamma_{xz} = \gamma_{yz} = 0$. An example where the plane strain assumption is often imposed is the case of a long cylinder under internal or external pressure. It is then assumed that the strains along the long axis of the cylinder are zero. A plane strain section can be relatively thick in the direction of t. Another example of the occurrence of plane strain would be a slice of thickness t taken from a dam between rigid end walls. Use $\epsilon_z = \gamma_{xz} = \gamma_{yz} = 0$ in Eq. (1.32) to obtain $\tau_{yz} = \tau_{xz} = 0$ and $\sigma_z = \nu(\sigma_x + \sigma_y)$. Thus, in the plane strain situation, the stress in the z direction is not zero.

Then

$$\begin{bmatrix} \epsilon_x \\ \epsilon_y \\ \gamma_{xy} \end{bmatrix} = \frac{1+\nu}{E} \begin{bmatrix} 1-\nu & -\nu & 0 \\ -\nu & 1-\nu & 0 \\ 0 & 0 & 2 \end{bmatrix} \begin{bmatrix} \sigma_x \\ \sigma_y \\ \tau_{xy} \end{bmatrix} \tag{1.41a}$$

$$\boldsymbol{\epsilon} \quad = \quad \mathbf{E}^{-1} \quad \boldsymbol{\sigma}$$

or

$$\begin{bmatrix} \epsilon_x \\ \epsilon_y \\ \gamma_{xy} \end{bmatrix} = \frac{1+\nu}{Et} \begin{bmatrix} 1-\nu & -\nu & 0 \\ -\nu & 1-\nu & 0 \\ 0 & 0 & 2 \end{bmatrix} \begin{bmatrix} n_x \\ n_y \\ n_{xy} \end{bmatrix} \tag{1.41b}$$

$$\boldsymbol{\epsilon} \quad = \quad \mathbf{E}^{-1} \quad \mathbf{s}$$

Also

$$\begin{bmatrix} \sigma_x \\ \sigma_y \\ \tau_{xy} \end{bmatrix} = \frac{E}{(1+\nu)(1-2\nu)} \begin{bmatrix} 1-\nu & \nu & 0 \\ \nu & 1-\nu & 0 \\ 0 & 0 & \frac{1-2\nu}{2} \end{bmatrix} \begin{bmatrix} \epsilon_x \\ \epsilon_y \\ \gamma_{xy} \end{bmatrix} \tag{1.42a}$$

$$\boldsymbol{\sigma} \quad = \quad \mathbf{E} \quad \boldsymbol{\epsilon}$$

or

$$\begin{bmatrix} n_x \\ n_y \\ n_{xy} \end{bmatrix} = \frac{Et}{(1+\nu)(1-2\nu)} \begin{bmatrix} 1-\nu & \nu & 0 \\ \nu & 1-\nu & 0 \\ 0 & 0 & \frac{1-2\nu}{2} \end{bmatrix} \begin{bmatrix} \epsilon_x \\ \epsilon_y \\ \gamma_{xy} \end{bmatrix} \tag{1.42b}$$

$$\mathbf{s} \quad = \quad \mathbf{E} \quad \boldsymbol{\epsilon}$$

Replace E by $(1-\nu^2)E$ and then ν by $\nu/(1+\nu)$ to obtain the plane stress formulation from the plane strain equations.

1.3.2 Thermal and Initial Strains

The stress-strain relations of Eqs. (1.32b) and (1.34b) can be generalized to include thermal effects and other types of initial strains that may occur, for example, in lack of fit problems when a structure is assembled from parts. Denote the initial strains by $\boldsymbol{\epsilon}^0$. Then $\boldsymbol{\epsilon}$ should be replaced by $\boldsymbol{\epsilon} - \boldsymbol{\epsilon}^0$ and Eqs. (1.32b) and (1.34b) become

$$\boldsymbol{\epsilon} = \mathbf{E}^{-1}\boldsymbol{\sigma} + \boldsymbol{\epsilon}^0 \tag{1.43}$$

$$\boldsymbol{\sigma} = \mathbf{E}\,\boldsymbol{\epsilon} - \mathbf{E}\,\boldsymbol{\epsilon}^0 \tag{1.44}$$

In the case of thermal strains, consider an element of an elastic solid subjected to a temperature change ΔT. If an element of length dx is not constrained, it expands to a new length of $dx + dx\,\alpha\,\Delta T$, where α is the *coefficient of thermal expansion* which may depend on the temperature. In an isotropic body, the thermal expansion is the same in all directions, so that an unrestrained three-dimensional element experiences a uniform expansion but no angular distortions. Thus, in the unrestrained isotropic body, the temperature change leads to normal thermal strains while not producing shear strains, i.e.,

$$\epsilon_x^0 = \epsilon_y^0 = \epsilon_z^0 = \alpha\,\Delta T, \qquad \gamma_{yz}^0 = \gamma_{xz}^0 = \gamma_{xy}^0 = 0 \tag{1.45}$$

In Eqs. (1.43) and (1.44), these thermal strains lead to the terms

$$\boldsymbol{\epsilon}^0 = \alpha\,\Delta T \begin{bmatrix} 1 \\ 1 \\ 1 \\ 0 \\ 0 \\ 0 \end{bmatrix} \qquad -\mathbf{E}\boldsymbol{\epsilon}^0 = -\frac{E\,\alpha\,\Delta T}{1-2\nu} \begin{bmatrix} 1 \\ 1 \\ 1 \\ 0 \\ 0 \\ 0 \end{bmatrix} \tag{1.46}$$

1.3.3 Anisotropic Material

For the most general case of an elastic anisotropic body, all components in the matrix of the constitutive law are non-zero, but still the symmetry is preserved [Timoshenko and Goodier, 1970; Leipholz, 1968, p. 96]:

$$\begin{bmatrix} \epsilon_x \\ \epsilon_y \\ \epsilon_z \\ \gamma_{xy} \\ \gamma_{xz} \\ \gamma_{yz} \end{bmatrix} = \begin{bmatrix} a_{11} & a_{12} & a_{13} & \bar{a}_{14} & \bar{a}_{15} & \bar{a}_{16} \\ & a_{22} & a_{23} & \bar{a}_{24} & \bar{a}_{25} & \bar{a}_{26} \\ & & a_{33} & \bar{a}_{34} & \bar{a}_{35} & \bar{a}_{36} \\ & & & a_{44} & \bar{a}_{45} & \bar{a}_{46} \\ & \text{Symmetric} & & & a_{55} & \bar{a}_{56} \\ & & & & & a_{66} \end{bmatrix} \begin{bmatrix} \sigma_x \\ \sigma_y \\ \sigma_z \\ \tau_{xy} \\ \tau_{xz} \\ \tau_{yz} \end{bmatrix} \tag{1.47}$$

$$\boldsymbol{\epsilon} = \mathbf{E}^{-1} \boldsymbol{\sigma}$$

$$\begin{bmatrix} \sigma_x \\ \sigma_y \\ \sigma_z \\ \tau_{xy} \\ \tau_{xz} \\ \tau_{yz} \end{bmatrix} = \begin{bmatrix} c_{11} & c_{12} & c_{13} & \bar{c}_{14} & \bar{c}_{15} & \bar{c}_{16} \\ & c_{22} & c_{23} & \bar{c}_{24} & \bar{c}_{25} & \bar{c}_{26} \\ & & c_{33} & \bar{c}_{34} & \bar{c}_{35} & \bar{c}_{36} \\ & & & c_{44} & \bar{c}_{45} & \bar{c}_{46} \\ & \text{Symmetric} & & & c_{55} & \bar{c}_{56} \\ & & & & & c_{66} \end{bmatrix} \begin{bmatrix} \epsilon_x \\ \epsilon_y \\ \epsilon_z \\ \gamma_{xy} \\ \gamma_{xz} \\ \gamma_{yz} \end{bmatrix} \tag{1.48}$$

$$\boldsymbol{\sigma} = \mathbf{E} \boldsymbol{\epsilon}$$

This is, however, an extreme case for which it is very difficult to identify all of the coefficients. There are many materials with a simpler structure, such as those found in rolled sheet metals, wood, and honeycomb fabrications. If the properties of a material differ only in three orthogonal directions it is called *orthotropic*, and nine independent parameters suffice to describe the material. In Eqs. (1.47) and (1.48), the barred quantities are zero for orthotropic (or isotropic) materials.

1.4 Equations of Equilibrium

The description of equilibrium at any point in a body is characterized by (local) differential equations involving stresses and internal (volume or body) forces such as those generated by gravity, acceleration, or magnetic fields. They can be derived for the two-dimensional case by considering the configuration of Fig. 1.9. Over the infinitesimal distance dx, the change

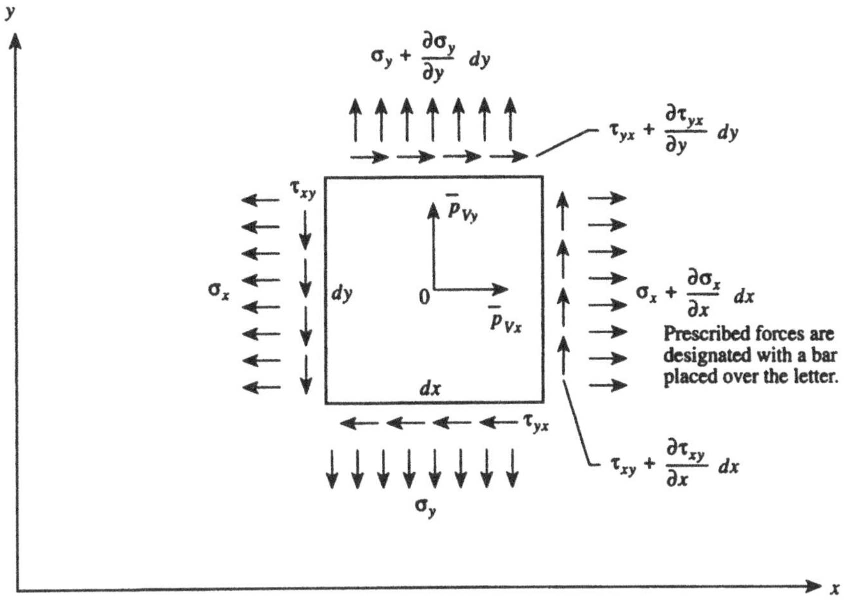

FIGURE 1.9
Planar element used in the derivation of the differential equations of equilibrium.

in the stress σ_x is $\partial_x \sigma_x \, dx$. The changes of the shear components τ_{xy}, τ_{yx} and the stress σ_y in the coordinate directions have a similar form as indicated in Fig. 1.9. The prescribed body forces with units of force/volume are denoted by $\overline{p}_{Vx}$, $\overline{p}_{Vy}$*. For an element of thickness dz, the resultant forces in the x direction must vanish

$$\sum F_x = 0: \quad \left(\sigma_x + \frac{\partial \sigma_x}{\partial x} dx\right) dy\, dz + \left(\tau_{yx} + \frac{\partial \tau_{yx}}{\partial y} dy\right) dx\, dz - \sigma_x\, dy\, dz$$
$$-\tau_{yx}\, dx\, dz + \overline{p}_{Vx}\, dx\, dy\, dz = 0$$

or, for each volume element $dx\, dy\, dz$

$$\frac{\partial \sigma_x}{\partial x} + \frac{\partial \tau_{yx}}{\partial y} + \overline{p}_{Vx} = 0 \tag{1.49a}$$

In the y direction, we find

$$\frac{\partial \sigma_y}{\partial y} + \frac{\partial \tau_{xy}}{\partial x} + \overline{p}_{Vy} = 0 \tag{1.49b}$$

In matrix form,

$$\begin{bmatrix} \partial_x & 0 & \partial_y \\ 0 & \partial_y & \partial_x \end{bmatrix} \begin{bmatrix} \sigma_x \\ \sigma_y \\ \tau_{xy} \end{bmatrix} + \begin{bmatrix} \overline{p}_{Vx} \\ \overline{p}_{Vy} \end{bmatrix} = 0$$
$$\mathbf{D}^T \quad \boldsymbol{\sigma} \quad + \quad \overline{\mathbf{p}}_V = 0 \tag{1.50a}$$

These relations apply for both plane stress and plane strain. For a thin, flat element in which the stresses are replaced by stress resultants, i.e., integrals of the stresses over the element

*Quantities with an overbar are applied (prescribed).

thickness [Eq. (1.38)], these equilibrium conditions can be written as

$$\begin{bmatrix} \partial_x & 0 & \partial_y \\ 0 & \partial_y & \partial_x \end{bmatrix} \begin{bmatrix} n_x \\ n_y \\ n_{xy} \end{bmatrix} + \begin{bmatrix} \overline{p}_{Vx} \\ \overline{p}_{Vy} \end{bmatrix} = 0 \tag{1.50b}$$

$$\mathbf{D}^T \quad \mathbf{s} \quad + \quad \overline{\mathbf{p}}_V \quad = 0$$

where the units of $\overline{p}_{Vx}$ and $\overline{p}_{Vy}$ are force/area.

The equilibrium of moments about point 0 leads to

$$\sum M_0 = 0: \quad (\tau_{xy}\, dy\, dz)\frac{dx}{2} + \left[\left(\tau_{xy} + \frac{\partial \tau_{xy}}{\partial x} dx\right) dy\, dz\right]\frac{dx}{2} - (\tau_{yx}\, dx\, dz)\frac{dy}{2}$$
$$- \left[\left(\tau_{yx} + \frac{\partial \tau_{yx}}{\partial y} dy\right) dx\, dz\right]\frac{dy}{2} = 0$$

or

$$\tau_{xy} + \frac{\partial \tau_{xy}}{\partial x}\frac{dx}{2} - \tau_{yx} - \frac{\partial \tau_{yx}}{\partial y}\frac{dy}{2} = 0$$

The second and fourth terms vanish as dx and dy approach zero. This yields the symmetry property of the shear stresses $\tau_{yx} = \tau_{xy}$ or, in general, $\sigma_{ij} = \sigma_{ji}, i \neq j$, a result which was mentioned earlier.

These expressions generalized for the three-dimensional case are

$$\begin{aligned} \partial_x \sigma_x + \partial_y \tau_{yx} + \partial_z \tau_{zx} + \overline{p}_{Vx} &= 0, & \tau_{xy} &= \tau_{yx} \\ \partial_x \tau_{xy} + \partial_y \sigma_y + \partial_z \tau_{zy} + \overline{p}_{Vy} &= 0, & \tau_{xz} &= \tau_{zx} \\ \partial_x \tau_{xz} + \partial_y \tau_{yz} + \partial_z \sigma_z + \overline{p}_{Vz} &= 0, & \tau_{yz} &= \tau_{zy} \end{aligned} \tag{1.51}$$

Using the summation convention, the differential relations can be written as

$$\sigma_{ji,j} + \overline{p}_{Vi} = 0, \quad i = 1, 2, 3 \quad \text{or} \quad \sigma_{ij,j} + \overline{p}_{Vi} = 0 \tag{1.52}$$

or in matrix form with body forces $\overline{\mathbf{p}}_V = [\overline{p}_{Vx} \quad \overline{p}_{Vy} \quad \overline{p}_{Vz}]^T$

$$\begin{bmatrix} \partial_x & 0 & 0 & \vdots & \partial_y & \partial_z & 0 \\ 0 & \partial_y & 0 & \vdots & \partial_x & 0 & \partial_z \\ 0 & 0 & \partial_z & \vdots & 0 & \partial_x & \partial_y \end{bmatrix} \begin{bmatrix} \sigma_x \\ \sigma_y \\ \sigma_z \\ \cdots \\ \tau_{xy} \\ \tau_{xz} \\ \tau_{yz} \end{bmatrix} + \begin{bmatrix} \overline{p}_{Vx} \\ \overline{p}_{Vy} \\ \overline{p}_{Vz} \end{bmatrix} = \begin{bmatrix} 0 \\ 0 \\ 0 \end{bmatrix} \tag{1.53}$$

$$\mathbf{D}^T \quad \boldsymbol{\sigma} \quad + \quad \overline{\mathbf{p}}_V \quad = \quad \mathbf{0} \tag{1.54}$$

where the matrix of differential operators is the transpose of that of Eq. (1.22).

1.4.1 Stress Functions

It is frequently convenient to express the stress components as derivatives of so-called *stress functions*, which are chosen such that they automatically satisfy the conditions of equilibrium. For example, in the two-dimensional case in the absence of body forces

$(\overline{p}_{Vx} = 0, \overline{p}_{Vy} = 0)$, stresses defined as

$$\sigma_x = \frac{\partial^2 \psi}{\partial y^2} \qquad \sigma_y = \frac{\partial^2 \psi}{\partial x^2} \qquad \tau_{xy} = -\frac{\partial^2 \psi}{\partial x\, \partial y} \tag{1.55}$$

identically satisfy the equilibrium equations of Eq. (1.50a). Here ψ is the appropriate stress function, also called the *Airy*[13] *stress function*. Observe that in this case all stresses σ_x, σ_y, and τ_{xy} in the body are derivable from a single stress function.

A variety of other stress functions can be chosen which are similar to the Airy stress function in that they are selected such that the equilibrium conditions are satisfied. For the general three-dimensional case, the equations of equilibrium of Eq. (1.51) can be satisfied if six stress functions $\boldsymbol{\psi} = [\psi_x\ \psi_y\ \psi_z\ \psi_{xy}\ \psi_{xz}\ \psi_{yz}]^T$ are defined such that

$$\boldsymbol{\sigma} = \mathbf{D}_1\boldsymbol{\psi} - \begin{bmatrix} \int \overline{p}_{Vx}\, dx \\ \int \overline{p}_{Vy}\, dy \\ \int \overline{p}_{Vz}\, dz \\ 0 \\ 0 \\ 0 \end{bmatrix} = \mathbf{D}_1\boldsymbol{\psi} - \boldsymbol{\sigma}^0 \tag{1.56a}$$

where $\mathbf{D}_1$ is defined in Eq. (1.28). The six functions in $\boldsymbol{\psi}$ are a combination of Maxwell's[14] and Morera's stress functions. The first three components of the vector $\boldsymbol{\psi}$ belong to Maxwell's stress function and the final three compose Morera's stress function. These functions, which are not independent, can be expressed as three stress functions $\boldsymbol{\phi}$. The relationship between stresses and stress functions takes the form

$$\boldsymbol{\sigma} = \mathbf{D}\boldsymbol{\phi} - \boldsymbol{\sigma}^0 \tag{1.56b}$$

where $\mathbf{D}$ is the same operator matrix $\mathbf{D}$ as in the kinematical relations of Eq. (1.21).

1.5 Surface Forces and Boundary Conditions

The stress components on the surface, i.e., the boundary, of a body must be in equilibrium with the forces applied to the surface. The equilibrium conditions are obtained by considering the state of stress at a point on the surface. Suppose a small element lies on the surface of a body (Fig. 1.10) with unit normal vector **a** (positive outward) defining its orientation with

[13]George Biddel Airy (1801–1892) was an English mathematician with a primary interest in astronomy. He was a professor at Cambridge University. Although most of his efforts were directed to astronomical work, he showed an interest in the application of mathematics to structural mechanics problems. He proposed the stress function named after him while trying to solve a beam problem of rectangular cross-section. He chose a polynomial form for ψ, with the coefficients selected such that the boundary conditions were satisfied. His solution was incomplete as the ψ he used did not satisfy compatibility requirements.

[14]James Clerk Maxwell (1831–1879) was a Scottish mathematician who was interested in photoelasticity and analytical solid mechanics. He did his work in developing the science of photoelasticity, along with solving numerous problems of the torsion of bars and cylinders and the bending of beams and plates, before he was nineteen. Then he began his studies at Cambridge University. He remained at Cambridge after his 1855 graduation and broadened his interests to include electricity, magnetism, and the kinetic theory of gases. In 1865 he retired to write the celebrated *Treatise on Electricity and Magnetism*. He returned to Cambridge in 1871 where he developed the Cavendish Laboratory. His reciprocal theorem (Chapter 3, Section 3.3) appeared eight years before the more general theorem of Betti.

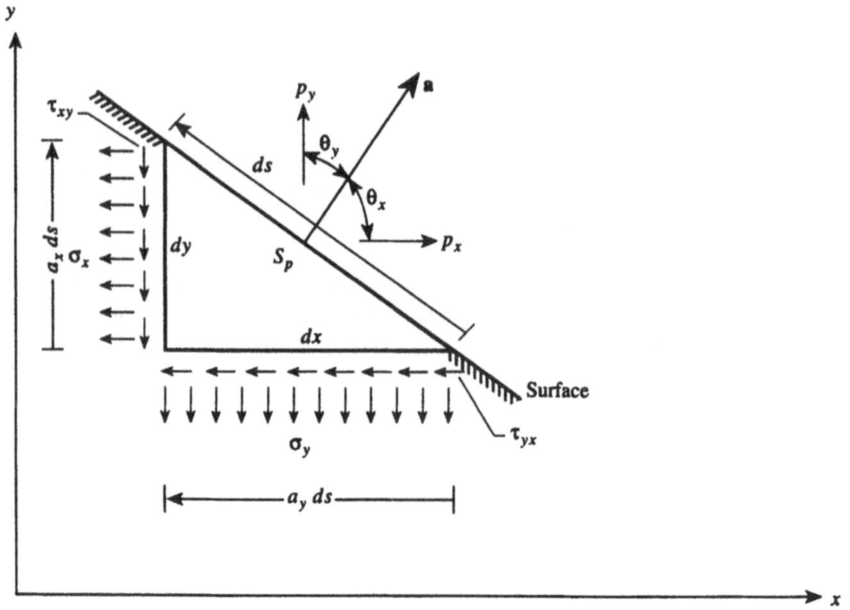

FIGURE 1.10
Stresses and surface forces acting on an element with length ds on the surface. Consider the element to be of unit thickness into the paper so that ds has the units of area. The quantities a_x and a_y are the direction cosines with respect to the x and y axes of the normal to the surface. By definition, $a_x = \cos\theta_x = dy/ds$ and $a_y = \cos\theta_y = -dx/ds$.

respect to the coordinate system.* The direction cosines of the normal are $\mathbf{a} = [a_x\ a_y\ a_z]^T$. The surface stress (force) vector $\mathbf{p} = [p_x\ p_y\ p_z]^T$ resulting from the stresses in the element can be expressed in terms of the stresses and direction cosines by

$$p_x\,ds = \sigma_x\,(a_x\,ds) + \tau_{yx}\,(a_y\,ds)$$

for the x component of the two-dimensional situation of Fig. 1.10 or, in general,

$$p_x\,ds = \sigma_x(a_x\,ds) + \tau_{yx}(a_y\,ds) + \tau_{zx}(a_z\,ds)$$

These relationships hold for each point on the surface. Similar relations apply for the other components of $\mathbf{p}$. In summary,

$$\begin{bmatrix} p_x \\ p_y \\ p_z \end{bmatrix} = \begin{bmatrix} a_x & 0 & 0 & \vdots & a_y & a_z & 0 \\ 0 & a_y & 0 & \vdots & a_x & 0 & a_z \\ 0 & 0 & a_z & \vdots & 0 & a_x & a_y \end{bmatrix} \boldsymbol{\sigma} \tag{1.57a}$$

$$\mathbf{p} = \mathbf{A}^T \boldsymbol{\sigma} \tag{1.57b}$$

or

$$p_j = a_i\,\sigma_{ij} \tag{1.58}$$

* The most common symbol for the unit normal vector is **n**. However, in this book **n** is being used to represent stress resultants for elements in plane stress and strain.

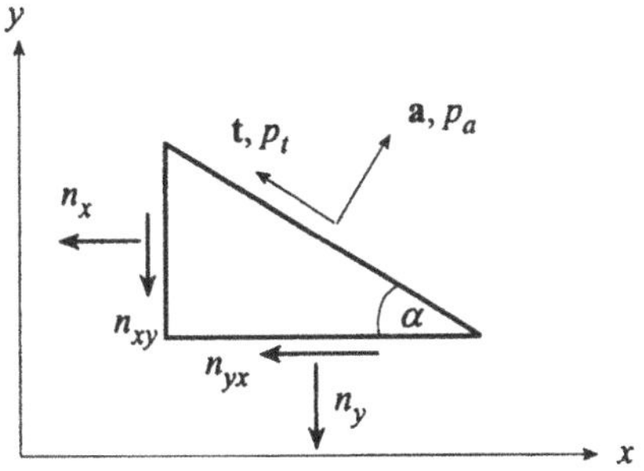

FIGURE 1.11
Surface force conditions for stress resultants n_x, n_y, $n_{xy} = n_{yx}$ for a thin element lying in the xy plane.

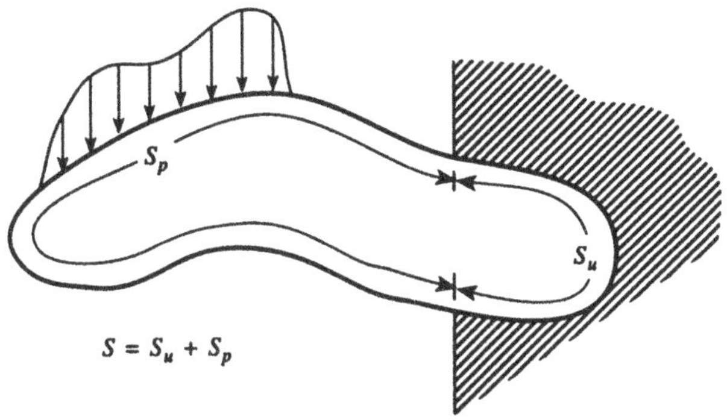

FIGURE 1.12
The surface S of a solid is considered to be made up of two surfaces S_u and S_p. The symbol S_u designates regions with known displacements, whereas S_p denotes everything else, including those portions of the surface where applied forces occur.

Note that $\mathbf{A}^T$ is of the same form as $\mathbf{D}^T$ of Eq. (1.53) in that the derivatives in $\mathbf{D}^T$ correspond to the projection directions in $\mathbf{A}^T$. Equations (1.57), which are surface stress conditions for a point on the boundary, are often referred to as Cauchy's formula.

In particular, for the thin, flat element used for plane stress and strain, the surface condition takes the form (Fig. 1.11)

$$\begin{bmatrix} p_a \\ p_t \end{bmatrix} = \begin{bmatrix} \sin^2\alpha & \cos^2\alpha & 2\sin\alpha\cos\alpha \\ -\sin\alpha\cos\alpha & \sin\alpha\cos\alpha & -\cos^2\alpha + \sin^2\alpha \end{bmatrix} \begin{bmatrix} n_x \\ n_y \\ n_{xy} \end{bmatrix} \tag{1.59}$$

where n_x, n_y, and $n_{xy} = n_{yx}$ are stress resultants for the thin element, and p_a, p_t are the tractions normal and tangential to the boundary.

If surface forces (per unit area) are applied externally they are referred to as prescribed surface *tractions* $\overline{\mathbf{p}}$. Suppose the whole surface of the body is designated by S and that those portions of the surface with prescribed tractions are designated as S_p (Fig. 1.12). Let the remainder of the surface, i.e., $S - S_p$, be denoted as S_u to indicate where prescribed displacements $\overline{\mathbf{u}}$ appear.

Equilibrium requires that the resultant stress $\mathbf{p}$ be equal to the applied surface tractions on S_p:

$$\mathbf{p} = \overline{\mathbf{p}} \quad \text{on} \quad S_p \tag{1.60}$$

These are the *mechanical* or *static (force, stress) boundary conditions*. Continuity requires that on the remaining portion of the surface S_u the displacement $\mathbf{u}$ be equal to the prescribed displacements $\overline{\mathbf{u}}$:

$$\mathbf{u} = \overline{\mathbf{u}} \quad \text{on} \quad S_u \tag{1.61}$$

These are the *(kinematic) displacement boundary conditions*.

We are now able to define more precisely the concepts of admissible stresses and displacements. *Statically admissible* (or consistent) *stresses* (forces) are the stresses that satisfy the conditions of equilibrium [Eq. (1.54)] and the static boundary conditions [Eq. (1.60)]. *Kinematically admissible* (or consistent) *displacements* are displacements that satisfy the kinematic (strain-displacement) conditions [Eq. (1.21)] and the kinematic boundary conditions [Eq. (1.61)].

1.6 Other Forms of the Governing Differential Equations

In the previous sections, the three types of fundamental equations used in the mechanics of solids have been derived. These are the static equations, the kinematic equations, and the constitutive equations. The static equations $\mathbf{D}^T\boldsymbol{\sigma} + \overline{\mathbf{p}}_V = \mathbf{0}$ [Eq. (1.54)] are written in terms of the six static (stress or force) variables, and the kinematic equations $\boldsymbol{\epsilon} = \mathbf{D}\mathbf{u}$ [Eq. (1.21)] are expressed in terms of the kinematic (three displacement and six strain) variables. The constitutive equations $\boldsymbol{\sigma} = \mathbf{E}\boldsymbol{\epsilon}$ [Eq. (1.34b)] provide unique relations between the static and kinematic variables. The general problem of elasticity theory is to calculate the stresses and strains, as well as the displacements, throughout a body. Although in theory solutions for the fifteen unknowns which satisfy these fifteen equations in their present form and the boundary conditions [Eqs. (1.60) and (1.61)] can be found, in practice it is convenient to combine some of the equations to obtain alternative forms of governing equations.

1.6.1 Displacement Formulation

The first formulation involves the development of a system of differential equations that are referred to as the displacement, stiffness, or equilibrium formulations. These equations, which are expressed in terms of displacements, are obtained by forming stress displacement equations and then substituting these into the differential equations of equilibrium. That is, substitute the strain displacement relations $\boldsymbol{\epsilon} = \mathbf{D}\,\mathbf{u}$ [Eq. (1.21)] into $\boldsymbol{\sigma} = \mathbf{E}\,\boldsymbol{\epsilon}$ [Eq. (1.34b)] to obtain the stress displacement relations

$$\boldsymbol{\sigma} = \mathbf{E}\,\boldsymbol{\epsilon} = \mathbf{E}\,\mathbf{D}\,\mathbf{u} \tag{1.62}$$

Substitute this into the differential equations of equilibrium of Eq. (1.54) to obtain

$$\mathbf{D}^T\boldsymbol{\sigma} + \overline{\mathbf{p}}_V = \mathbf{D}^T\mathbf{E}\,\mathbf{D}\,\mathbf{u} + \overline{\mathbf{p}}_V = \mathbf{0} \tag{1.63}$$

These are the governing differential equations of the displacement formulation. Note that Eq. (1.63) applies as well to solids made of anisotropic material if $\mathbf{E}$ is taken from Eq. (1.48). It is of interest that Eq. (1.63), which gives three equations for the three unknown displacements, involves various combinations of first and second derivatives of the displacements and the material characteristics.

The desired solution of Eq. (1.63) would be a set of single-valued, continuous displacements that satisfy all boundary conditions. The existence of such a solution is due to the *uniqueness principle* [Timoshenko and Goodier, 1970, p. 269].

In scalar form, the governing equations of Eq. (1.63) for an isotropic solid with constant E, ν are

$$\begin{aligned}
\nabla^2 u_x + \frac{1}{1-2\nu}\frac{\partial}{\partial x}\left(\frac{\partial u_x}{\partial x} + \frac{\partial u_y}{\partial y} + \frac{\partial u_z}{\partial z}\right) + \frac{\overline{p}_{Vx}}{G} &= 0 \\
\nabla^2 u_y + \frac{1}{1-2\nu}\frac{\partial}{\partial y}\left(\frac{\partial u_x}{\partial x} + \frac{\partial u_y}{\partial y} + \frac{\partial u_z}{\partial z}\right) + \frac{\overline{p}_{Vy}}{G} &= 0 \\
\nabla^2 u_z + \frac{1}{1-2\nu}\frac{\partial}{\partial z}\left(\frac{\partial u_x}{\partial x} + \frac{\partial u_y}{\partial y} + \frac{\partial u_z}{\partial z}\right) + \frac{\overline{p}_{Vz}}{G} &= 0
\end{aligned} \tag{1.64}$$

where $\nabla^2 = \partial_x^2 + \partial_y^2 + \partial_z^2$ is the *Laplacian*[15] or *harmonic* operator. Frequently, these equations are written in the form

$$G\nabla^2 u_i + (\lambda + G)u_{k,ki} + \overline{p}_{Vi} = 0 \tag{1.65}$$

with the Lamé constants

$$\lambda = \frac{E\nu}{(1+\nu)(1-2\nu)} \qquad G = \frac{E}{2(1+\nu)} \tag{1.66}$$

Equation (1.65) is called the *Navier*[16] or *Lamé–Navier* equations of elasticity. In the absence of body forces $\overline{p}_{V_i}$, Eq. (1.65) can be written as the *biharmonic (differential) equation*

$$\nabla^2\nabla^2 u_i = 0 \tag{1.67}$$

which is a frequently occurring equation in mathematical physics and, in particular, in the theory of elasticity.

For this displacement formulation, in which the governing differential equations for the displacements are the conditions of equilibrium, the compatibility requirements are often satisfied trivially, i.e., by inspection or by dealing only with single-valued, continuous displacements. Equation (1.67) is sometimes simplified by choosing displacement

---

[15]Pierre Simon Laplace (1749–1827) was a French astronomer and mathematician. He was born into an upper middle-class family. At sixteen he entered the University of Caën and shortly thereafter he became a professor of mathematics at École Militaire in Paris. Among his numerous remarkable achievements was his work on the idea of a potential. He showed that the potential satisfied what is now known as Laplace's equation. In addition to his efforts in astronomy, he made important contributions in the mathematical theory of probability. He is considered to be one of the most influential scientists of all time. Although he was referred to as "the Newton of France," questions have often been raised as to his sense of "honor." As evidenced by his dedications to the different volumes of his five volume work *Mécanique celeste*, published between 1799 and 1825, Laplace was politically flexible during the French revolution and the times of Napoleon. More "honorable" scientists such as Lavoisier, who collaborated with Laplace in the study of specific heats, met their fate on the guillotine.

[16]Claude-Louis-Marie-Henri Navier (1785–1836) was a French engineer who made wide-ranging contributions to mechanics. Some of his efforts were guided by the somewhat earlier works of Coulomb. Navier worked on torsional and bending stress formulations; however, both were based on erroneous suppositions. He designed bridges in France and Italy and became a professor of calculus and mechanics. His published material enjoyed considerable popularity among French engineers for many years.

functions, e.g., Papkovich[17] Neuber[18] functions, (similar to stress functions) which automatically satisfy the compatibility conditions [Neuber, 1985]. Then Eq. (1.67), the conditions of equilibrium, can be written in a more tractable form.

1.6.2 Force Formulation

Another alternative formulation of the governing equations entails the development of force, flexibility, or compatibility formulations. These equations are found by writing the compatibility equations in terms of stresses. If the constitutive relationship of Eq. (1.32) is substituted in the compatibility relation of Eq. (1.29), then

$$\mathbf{D}_1\epsilon = \mathbf{D}_1\mathbf{E}^{-1}\sigma = 0 \tag{1.68}$$

These six equations involve the six (σ_x, σ_y, σ_z, τ_{xy}, τ_{xz}, τ_{yz}) unknown stresses.

Frequently, the equilibrium conditions are also included in a force (stress) formulation. Differentiate Eq. (1.52) to find $\sigma_{ij,jk} = -\overline{p}_{Vi,k}$. Substitution of this into Eq. (1.68) leads to [Sokolnikoff, 1956].

$$\nabla^2\sigma_{ij} + \frac{1}{1+\nu}\sigma_{kk,ij} - \frac{\nu}{1+\nu}\delta_{ij}\nabla^2\sigma_{kk} = -(\overline{p}_{Vi,j} + \overline{p}_{Vj,i})$$

It can be verified from the compatibility relations of Eq. (1.68) that $\sigma_{ij,ij} = \frac{1-\nu}{1+\nu}\nabla^2\sigma_{kk}$. Also, from the equilibrium conditions $\sigma_{ij,ij} = -\overline{p}_{Vj,j}$. Then, $\nabla^2\sigma_{kk} = -\frac{1+\nu}{1-\nu}\overline{p}_{Vj,j}$ and the compatibility relations in terms of the stress components become

$$\nabla^2\sigma_{ij} + \frac{1}{1+\nu}\sigma_{kk,ij} = \frac{\nu}{1-\nu}\delta_{ij}\overline{p}_{Vk,k} - (\overline{p}_{Vi,j} + \overline{p}_{Vj,i}) \tag{1.69}$$

where

$$\nabla^2\sigma_{ij} = \sigma_{ij,kk}, \quad \delta_{ij} = 1 \quad \text{if} \quad i = j \quad \text{and} \quad \delta_{ij} = 0 \quad \text{if} \quad i \neq j$$

These are *Michell's*[19] *equations*, which together with prescribed static boundary conditions can be used to find the unknown stresses. If the volume (body) forces are constant, we obtain *Beltrami's*[20] *equations*

$$\nabla^2\sigma_{ij} + \frac{1}{1-\nu}\sigma_{kk,ij} = 0 \tag{1.70}$$

These can be rewritten as a biharmonic differential equation

$$\nabla^2\nabla^2\sigma_{ij} = 0 \tag{1.71}$$

---

[17]Petr Fedorovich Papkovich (1887–1946) was a Russian-Soviet shipbuilding engineer, and was a corresponding member of the Soviet Academy of Science. His developments in the vibration and strength analysis of ship structures are considered to be significant contributions to the foundations for the theory of shipbuilding. He published the displacement functions for the solution of the biharmonic equation in 1932.

[18]Heinz Neuber (1906–1989) was a German machanical engineer who received his doctorate under August Föppl in Munich. After lengthy service in the aircraft industry, he became a faculty member of the Technische Hochschule in Dresden in 1946. In 1955, he replaced Ludwig Föppl (the son of August) as a chaired professor at the Technische Hochschule in Munich.

[19]John Henry Michell (1863–1943) was a Cambridge-educated Australian mathematician. He was a professor at the University of Melbourne. His papers were collected into one volume in 1964, along with those of his mechanical engineer brother Anthony-George-Maldon Michell (1870–1959).

[20]Eugenio Beltrami (1835–1900) was a versatile Italian mathematician who served as a professor of mechanics at several Italian universities.

where the relationship $\nabla^2\sigma_{kk} = -\frac{1+\nu}{1-\nu}\nabla^2\overline{p}_{Vj,j} = 0$ has been employed. The solution of these equations (for simply connected domains) provides stresses that satisfy the equilibrium conditions and can lead to strains that can be derived from unique, continuous displacements.

Equation (1.68) may be transformed into equations for unknown stress functions, which are defined in terms of stresses so that the equilibrium conditions are identically satisfied. To accomplish this place $\boldsymbol{\sigma}$ from Eq. (1.56) in Eq. (1.68) and solve

$$\mathbf{D}_1\mathbf{E}^{-1}\mathbf{D}_1\boldsymbol{\psi} - \mathbf{D}_1\mathbf{E}^{-1}\boldsymbol{\sigma}^0 = \mathbf{0} \tag{1.72}$$

which are equations often used in the force formulation. These differential equations represent the conditions of compatibility.

Equation (1.72) is much less complicated for some cases that are frequently of interest. In the two-dimensional stress state, only one stress function ψ is required to replace the three unknown stresses $\sigma_x, \sigma_y, \tau_{xy}$. If the body forces are zero, a very simple expression for ψ can be derived from Eq. (1.72) or by following the process employed in deriving Eq. (1.72). In the latter case, substitution of the material law of Eq. (1.40a) for plane stress into the compatibility equation of Eq. (1.27) gives

$$2(1+\nu)\frac{\partial^2\tau_{xy}}{\partial x\,\partial y} = \frac{\partial^2}{\partial y^2}(\sigma_x - \nu\sigma_y) + \frac{\partial^2}{\partial x^2}(\sigma_y - \nu\sigma_x) \tag{1.73}$$

Replace the stresses here by the Airy stress function relations of Eq. (1.55), giving

$$\frac{\partial^4\psi}{\partial x^4} + 2\frac{\partial^4\psi}{\partial x^2\partial y^2} + \frac{\partial^4\psi}{\partial y^4} = 0 \quad \text{or} \quad \nabla^2\nabla^2\psi = \nabla^4\psi = 0 \tag{1.74}$$

where $\nabla^2 = \partial_x^2 + \partial_y^2$. The exact solution of plane stress problem must satisfy Eq. (1.74), a biharmonic equation, as well as the boundary conditions.

1.6.3 Mixed Formulation

The differential governing equations considered in the two previous sections were written in terms of either displacements or forces (stresses). Equations formulated in terms of both displacement and force variables are referred to as *mixed* equations.

As might be expected, there are a variety of mixed equations that can be derived. Choose as the fundamental unknowns the three components of the displacements $\mathbf{u}$ and the six components of the stress vector $\boldsymbol{\sigma}$. The constitutive relations of Eq. (1.32) or Eq. (1.43) can be rewritten in terms of displacements $\mathbf{u}$ rather than strains $\boldsymbol{\epsilon}$ by using the strain-displacement relations of Eq. (1.21). Thus, $\boldsymbol{\epsilon} = \mathbf{D}\mathbf{u} = \mathbf{E}^{-1}\boldsymbol{\sigma} + \boldsymbol{\epsilon}^0$ or $\mathbf{D}\mathbf{u} - \mathbf{E}^{-1}\boldsymbol{\sigma} = \boldsymbol{\epsilon}^0$. Rewrite the equations of equilibrium of Eq. (1.54) as $\mathbf{D}^T\boldsymbol{\sigma} = -\overline{\mathbf{p}}_V$. Place these relations together in the form

$$\begin{bmatrix} \mathbf{0} & \vdots & \mathbf{D}^T \\ \cdots & \cdot & \cdots \\ \mathbf{D} & \vdots & -\mathbf{E}^{-1} \end{bmatrix} \begin{bmatrix} \mathbf{u} \\ \boldsymbol{\sigma} \end{bmatrix} = \begin{bmatrix} -\overline{\mathbf{p}}_V \\ \boldsymbol{\epsilon}^0 \end{bmatrix} \tag{1.75}$$

The operators in this matrix are partitioned as

$$\begin{bmatrix} \mathbf{0} & \vdots & \text{Equilibrium Equations} \\ \cdots\cdots & \cdot & \cdots\cdots \\ \text{Strain-Displacement Relations} & \vdots & \text{Material Law} \end{bmatrix} \tag{1.76}$$

Equation (1.75) is a mixed governing differential equation. Note that in contrast to the governing equations for the displacement and force formulations, Eq. (1.75) for the mixed method contains no derivatives of the unknowns ($\mathbf{u}$ and $\boldsymbol{\sigma}$) higher than the first and does not involve derivatives of the material parameters.

1.7 Analysis of Stress

1.7.1 Principal Stresses

Figure (1.6) shows the state of stress at a point P in a body by specifying the stresses acting on three coordinate planes passing through point P. Define three stress vectors on these planes as

$$\begin{aligned} \boldsymbol{\sigma}_x &= \sigma_x\,\mathbf{e}_x + \tau_{xy}\,\mathbf{e}_y + \tau_{xz}\,\mathbf{e}_z \\ \boldsymbol{\sigma}_y &= \tau_{yx}\,\mathbf{e}_x + \sigma_y\,\mathbf{e}_y + \tau_{yz}\,\mathbf{e}_z \\ \boldsymbol{\sigma}_z &= \tau_{zx}\,\mathbf{e}_x + \tau_{zy}\,\mathbf{e}_y + \sigma_z\,\mathbf{e}_z \end{aligned} \tag{1.77}$$

where $\mathbf{e}_x$, $\mathbf{e}_y$ and $\mathbf{e}_z$ are the unit vectors along the x, y, and z axes. The vector $\boldsymbol{\sigma}_x$ defines the stress on the face of the cube whose outward normal is $\mathbf{e}_x$, and similarly for $\boldsymbol{\sigma}_y$ and $\boldsymbol{\sigma}_z$. We now calculate the stress vector $\boldsymbol{\sigma}_a$ on an arbitrarily oriented plane passing through the point P. The orientation of this plane is specified by a unit vector $\mathbf{a}$ normal to the plane

$$\mathbf{a} = a_x\mathbf{e}_x + a_y\mathbf{e}_y + a_z\mathbf{e}_z \tag{1.78}$$

where a_x, a_y, and a_z are the direction cosines with respect to the x, y, and z axes of the normal to the plane. To find $\boldsymbol{\sigma}_a$, a tetrahedron element of volume ΔV is isolated from the body (Fig. 1.13), with three triangular sides parallel to the negative x, y, and z coordinate

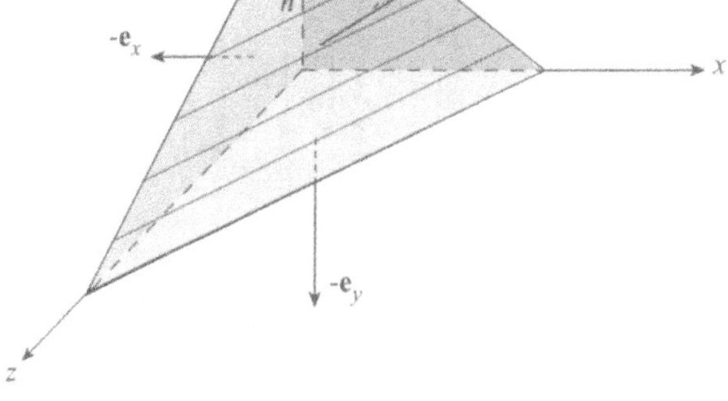

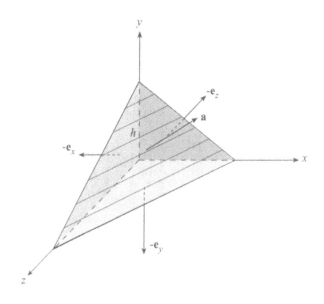

FIGURE 1.13
Tetrahedron of volume ΔV with sides of areas A, $a_x A$, $a_y A$, and $a_z A$.

faces, whose unit normals are $-\mathbf{e}_x$, $-\mathbf{e}_y$, and $-\mathbf{e}_z$. The fourth side of the tetrahedron has the outward normal parallel to $\mathbf{a}$. If the area of this fourth side is A, then the areas of the other three triangular faces are $a_x A$, $a_y A$ and $a_z A$. The force equilibrium equation for the isolated element of volume ΔV is

$$\mathbf{p}_V\, \Delta V + \boldsymbol{\sigma}_a A - \boldsymbol{\sigma}_x\, a_x A - \boldsymbol{\sigma}_y\, a_y A - \boldsymbol{\sigma}_z\, a_z A = 0$$

where $\mathbf{p}_V$ is the body force. The negative signs are necessary because the stress vectors acting on planes with unit normals $-\mathbf{e}_x$, $-\mathbf{e}_y$, and $-\mathbf{e}_z$ are $-\boldsymbol{\sigma}_x$, $-\boldsymbol{\sigma}_y$, and $-\boldsymbol{\sigma}_z$. The volume ΔV is

$$\Delta V = \frac{Ah}{3}$$

where h is the height of the tetrahedron. In the limit as h approaches zero, the equilibrium equation becomes

$$\boldsymbol{\sigma}_a = a_x\, \boldsymbol{\sigma}_x + a_y\, \boldsymbol{\sigma}_y + a_z\, \boldsymbol{\sigma}_z \tag{1.79}$$

Thus, the stress vector on any arbitrarily oriented plane through P is a linear combination of the stress vectors $\boldsymbol{\sigma}_x$, $\boldsymbol{\sigma}_y$, and $\boldsymbol{\sigma}_z$. These three vectors, or equivalently the six stress components in terms of which these vectors are written, define the state of stress at point P, in the sense that the stresses on any cutting plane through P are expressible in terms of them.

The stress vector $\boldsymbol{\sigma}_a$ can be decomposed into normal and shear components. The normal component σ is parallel to the unit vector $\mathbf{a}$ and is given by

$$\sigma = \boldsymbol{\sigma}_a \cdot \mathbf{a} = \sigma_x\, a_x^2 + \sigma_y\, a_y^2 + \sigma_z\, a_z^2 + 2(\tau_{xy}\, a_x\, a_y + \tau_{yz}\, a_y\, a_z + \tau_{zx}\, a_z\, a_x) \tag{1.80}$$

If the orientation of the cutting plane is varied, by allowing the components of $\mathbf{a}$ to vary, the normal stress σ takes on different values. Of particular interest are the minimum and maximum values attained by σ and the corresponding orientations of the cutting plane. Thus, extreme values of σ are sought under the constraint that $\mathbf{a}$ is a unit vector

$$a_x^2 + a_y^2 + a_z^2 = 1 \tag{1.81}$$

This extremum problem can be solved by introducing the Lagrange multiplier λ (Appendix I) and finding the extreme points of the function

$$F(a_x, a_y, a_z) = \sigma - \lambda(a_x^2 + a_y^2 + a_z^2 - 1)$$

The conditions for F to assume its extreme values

$$\frac{\partial F}{\partial a_x} = 0 \qquad \frac{\partial F}{\partial a_y} = 0 \qquad \frac{\partial F}{\partial a_z} = 0$$

are

$$\begin{aligned} (\sigma_x - \lambda)a_x + \tau_{xy}\, a_y + \tau_{xz}\, a_z &= 0 \\ \tau_{xy}\, a_x + (\sigma_y - \lambda)a_y + \tau_{yz}\, a_z &= 0 \\ \tau_{xz}\, a_x + \tau_{yz}\, a_y + (\sigma_z - \lambda)a_z &= 0 \end{aligned}$$

where Eq. (1.80) has been introduced. A fourth condition for F to assume its extreme values, $\partial F/\partial\lambda = 0$, leads to $a_x^2 + a_y^2 + a_z^2 - 1 = 0$ which assures that $\mathbf{a}$ is a unit vector. This set of homogeneous linear equations for a_x, a_y, a_z has a nontrivial solution if and only if the determinant of the coefficients is zero

$$\begin{vmatrix} \sigma_x - \lambda & \tau_{xy} & \tau_{xz} \\ \tau_{xy} & \sigma_y - \lambda & \tau_{yz} \\ \tau_{xz} & \tau_{yz} & \sigma_z - \lambda \end{vmatrix} = 0 \tag{1.82}$$

The problem of finding the extreme values of σ is seen to be algebraically identical to the eigenvalue problem for the 3 × 3 symmetric stress matrix (tensor) **T**

$$\mathbf{T} = \begin{bmatrix} \sigma_x & \tau_{xy} & \tau_{xz} \\ \tau_{xy} & \sigma_y & \tau_{yz} \\ \tau_{xz} & \tau_{yz} & \sigma_z \end{bmatrix} \tag{1.83}$$

The eigenvectors of **T** are the normal vectors defining the planes on which σ assumes its extreme values, and eigenvalues of **T** are the Lagrange multipliers λ.

Since **T** is symmetric, its three eigenvalues $\lambda_1, \lambda_2, \lambda_3$ are real numbers, and the eigenvectors corresponding to distinct eigenvalues are perpendicular. If exactly two of the eigenvalues are identical, then there are two mutually perpendicular eigenvectors corresponding to these two eigenvalues. In this case, any two mutually perpendicular vectors lying in the plane defined by these two eigenvectors are also eigenvectors corresponding to the two identical eigenvalues. If all three eigenvalues are identical, then any three mutually perpendicular vectors are eigenvectors. In the following it will be shown that the eigenvalues are the minimum and maximum values assumed by normal stresses as a function of cutting plane orientation. These eigenvalues of the matrix **T** are called the *principal stresses.* It is always possible to choose the eigenvectors $\mathbf{a}_1, \mathbf{a}_2, \mathbf{a}_3$ such that they form a right-handed triad of unit vectors. The coordinate axes defined by such a triad are termed *principal axes.*

The columns of the matrix **T** are made up of the x, y, z components of the stress vectors $\boldsymbol{\sigma}_x, \boldsymbol{\sigma}_y$, and $\boldsymbol{\sigma}_z$. Therefore, for any unit normal vector **a**,

$$\mathbf{T}\,\mathbf{a} = [\boldsymbol{\sigma}_x \quad \boldsymbol{\sigma}_y \quad \boldsymbol{\sigma}_z]\,\mathbf{a} = a_x\,\boldsymbol{\sigma}_x + a_y\,\boldsymbol{\sigma}_y + a_z\,\boldsymbol{\sigma}_z = \boldsymbol{\sigma}_a \tag{1.84}$$

In this equation, all vectors represent vectors of coordinates with respect to the unit vectors $\mathbf{e}_x, \mathbf{e}_y, \mathbf{e}_z$. Thus, if the x, y, z coordinates of a unit normal vector **a** are multiplied by the matrix **T**, the x, y, z coordinates of the stress vector on the plane defined by **a** are obtained. In particular, the expression for the normal stress σ on the plane defined by **a** may be written in terms of the matrix **T** as

$$\sigma = \mathbf{a} \cdot \boldsymbol{\sigma}_a = \mathbf{a} \cdot \mathbf{T}\,\mathbf{a} \tag{1.85}$$

Suppose that, given the state of stress by the matrix **T**, the three eigenvalues $\lambda_1, \lambda_2, \lambda_3$ and the corresponding unit eigenvectors $\mathbf{a}_1, \mathbf{a}_2, \mathbf{a}_3$ of **T**, have been determined. Then the normal stresses on the planes defined by the eigenvectors are

$$\sigma_k = \mathbf{a}_k \cdot \mathbf{T}\,\mathbf{a}_k = \mathbf{a}_k \cdot \lambda_k\,\mathbf{a}_k = \lambda_k \quad k = 1, 2, 3 \tag{1.86}$$

Thus, the three values of the Lagrange multiplier λ, which are the eigenvalues of **T**, are the extreme values of the normal stress.

If **a** is a unit eigenvector of **T**, with the corresponding eigenvalue λ, then the stress vector $\boldsymbol{\sigma}_a$ is given by

$$\boldsymbol{\sigma}_a = \mathbf{T}\,\mathbf{a} = \lambda\mathbf{a}$$

This equation shows that the stress vector $\boldsymbol{\sigma}_a$ consists only of a component along **a**. In other words, the shear stress on the plane defined by **a** is zero. Hence, principal stress planes are free of shear stress. Conversely, if **a** is to be the normal to a plane that is free of shear stress, then the condition for finding **a** is just the eigenvalue problem for the matrix **T**

$$\mathbf{T}\,\mathbf{a} = \boldsymbol{\sigma}_a = \lambda\mathbf{a}$$

This establishes that a cutting plane at a point of the body is a principal stress plane if and only if it is free of shear stress.

The eigenvalues (principal stresses) of **T** are the three real roots of the cubic equation given by Eq. (1.82)

$$|\mathbf{T} - \lambda \mathbf{I}| = -\lambda^3 + I_1 \lambda^2 - I_2 \lambda + I_3 = 0 \tag{1.87}$$

where

$$\begin{aligned} I_1 &= \sigma_x + \sigma_y + \sigma_z \\ I_2 &= \sigma_x\, \sigma_y - \tau_{xy}^2 + \sigma_y\, \sigma_z - \tau_{yz}^2 + \sigma_x\, \sigma_z - \tau_{xz}^2 \\ I_3 &= \sigma_x\, \sigma_y\, \sigma_z - \sigma_x\, \tau_{yz}^2 - \sigma_y\, \tau_{xz}^2 - \sigma_z\, \tau_{xy}^2 + 2\tau_{xy}\, \tau_{xz}\, \tau_{yz} \end{aligned}$$

Since the principal stresses must be independent of the choice of the coordinate system, the coefficients of this cubic polynomial are uniquely determined. This means that the quantities I_1, I_2, I_3, called *stress invariants*, have the same values regardless of the choice of axes x, y, z in which the state of stress is given. If, in particular, these axes are chosen to be a set of principal axes, then the shear stresses are all zero and the invariants are expressed in terms of the principal stresses $\sigma_1, \sigma_2, \sigma_3$ as

$$\begin{aligned} I_1 &= \sigma_1 + \sigma_2 + \sigma_3 \\ I_2 &= \sigma_1\, \sigma_2 + \sigma_2\, \sigma_3 + \sigma_1\, \sigma_3 \\ I_3 &= \sigma_1\, \sigma_2\, \sigma_3 \end{aligned} \tag{1.88}$$

1.7.2 Extreme Shear Stresses

The extreme values assumed by the shear stresses as cutting plane orientation varies will be determined in this section. The calculations are simplified if the x, y, z axes are chosen to be a set of principal axes at the point in question. We relabel these axes 1, 2, 3, and write the stress vectors on coordinate planes as

$$\begin{aligned} \boldsymbol{\sigma}_1 &= \sigma_1\, \mathbf{e}_1 \\ \boldsymbol{\sigma}_2 &= \sigma_2\, \mathbf{e}_2 \\ \boldsymbol{\sigma}_3 &= \sigma_3\, \mathbf{e}_3 \end{aligned}$$

These are the stress vectors of Eq. (1.77) expressed along the principal axes for which the corresponding shear stresses are zero. As in Eq. (1.79), the stress vector on a plane whose unit normal is the vector **a** is written as

$$\boldsymbol{\sigma}_a = a_1\, \sigma_1\, \mathbf{e}_1 + a_2\, \sigma_2\, \mathbf{e}_2 + a_3\, \sigma_3\, \mathbf{e}_3$$

The stress vector $\boldsymbol{\sigma}_a$ can be decomposed into two orthogonal components $\boldsymbol{\sigma}_a \cdot \mathbf{a}$ and a shear component of magnitude τ. Then, this shear component is given by

$$\tau^2 = \boldsymbol{\sigma}_a \cdot \boldsymbol{\sigma}_a - (\boldsymbol{\sigma}_a \cdot \mathbf{a})^2 = a_1^2\, \sigma_1^2 + a_2^2\, \sigma_2^2 + a_3^2\, \sigma_3^2 - \left(a_1^2\, \sigma_1 + a_2^2\, \sigma_2 + a_3^2\, \sigma_3\right)^2$$

A more convenient, equivalent expression for calculating extreme values is

$$\tau^2 = (\sigma_1 - \sigma_2)^2 a_1^2\, a_2^2 + (\sigma_2 - \sigma_3)^2 a_2^2\, a_3^2 + (\sigma_3 - \sigma_1)^2 a_3^2\, a_1^2 \tag{1.89}$$

For τ to attain an extreme value the function

$$F(a_1, a_2, a_3) = \tau^2 - \lambda\left(a_1^2 + a_2^2 + a_3^2 - 1\right)$$

in which λ is a Lagrange multiplier that must assume its extremum. The conditions

$$\frac{\partial F}{\partial a_1} = 0 \qquad \frac{\partial F}{\partial a_2} = 0 \qquad \frac{\partial F}{\partial a_3} = 0 \qquad \frac{\partial F}{\partial \lambda} = 0$$

give the equations

$$
\begin{aligned}
(\sigma_1-\sigma_2)^2 a_1\, a_2^2+(\sigma_1-\sigma_3)^2 a_1\, a_3^2-\lambda a_1&=0\\
(\sigma_2-\sigma_3)^2 a_2\, a_3^2+(\sigma_2-\sigma_1)^2 a_2\, a_1^2-\lambda a_2&=0\\
(\sigma_3-\sigma_1)^2 a_3\, a_1^2+(\sigma_3-\sigma_2)^2 a_3\, a_2^2-\lambda a_3&=0\\
a_1^2+a_2^2+a_3^2-1&=0
\end{aligned}
$$

These equations have the following three solutions

$$
\begin{aligned}
&a_1=0 \qquad a_2=a_3=\frac{1}{\sqrt{2}} \qquad \lambda=\frac{(\sigma_2-\sigma_3)^2}{2}\\
&a_2=0 \qquad a_3=a_1=\frac{1}{\sqrt{2}} \qquad \lambda=\frac{(\sigma_3-\sigma_1)^2}{2}\\
&a_3=0 \qquad a_1=a_2=\frac{1}{\sqrt{2}} \qquad \lambda=\frac{(\sigma_1-\sigma_2)^2}{2}
\end{aligned}
\tag{1.90}
$$

The extreme values of the shear stresses are found by substituting the solutions of Eq. (1.90) into Eq. (1.89). For example, for the third case with $a_1 = a_2 = \frac{1}{\sqrt{2}}$ and $a_3 = 0$, Eq. (1.89) gives $\tau = \frac{1}{2}(\sigma_1 - \sigma_2)$. Thus, in general the extreme values of the shear stresses

$$
\tau=\frac{|\sigma_2-\sigma_3|}{2} \qquad \tau=\frac{|\sigma_3-\sigma_1|}{2} \qquad \tau=\frac{|\sigma_1-\sigma_2|}{2} \tag{1.91}
$$

It is seen that the planes on which shear stresses reach their extreme values make a 45° angle with the principal directions. These planes are, in general, not free of normal stress.

EXAMPLE 1.3 Plane Stress Problems

For the plane stress problem of Section 1.3.1, $\sigma_z = \tau_{xz} = \tau_{yz} = 0$. The stress matrix of Eq. (1.83) becomes

$$
\mathbf{T}=\begin{bmatrix}\sigma_x & \tau_{xy} & 0\\ \tau_{xy} & \sigma_y & 0\\ 0 & 0 & 0\end{bmatrix} \tag{1}
$$

The principal stresses are obtained from Eq. (1.87) as

$$
\sigma_{\max,\min}=\frac{\sigma_x+\sigma_y}{2}\pm\sqrt{\left(\frac{\sigma_x-\sigma_y}{2}\right)^2+\tau_{xy}^2} \tag{2}
$$

These are expressions for the principal stresses in the x, y plane. The out-of-plane normal stress $\sigma_z = 0$ is the third principal stress. The magnitudes and signs of $\sigma_{\max}$ and $\sigma_{\min}$ can vary. As a consequence there are several possibilities for choosing which principal stress corresponds to σ_1, σ_2, or σ_3, which usually are ordered from the algebraically largest (σ_1) to the algebraically smallest (σ_3). The possibilities for the plane stress problem are

$$
\begin{aligned}
&\sigma_1=\sigma_{\max}, \qquad \sigma_2=\sigma_{\min}, \qquad \sigma_3=0\\
&\sigma_1=\sigma_{\max}, \qquad \sigma_2=0, \qquad \sigma_3=\sigma_{\min}\\
&\sigma_1=0, \qquad \sigma_2=\sigma_{\max}, \qquad \sigma_3=\sigma_{\min}
\end{aligned}
\tag{3}
$$

The extreme values of the shear stress are given by Eq. (1.91), with the maximum shear stress being the largest of these values. ∎

EXAMPLE 1.4 Calculation of Principal Stresses
Suppose that the stresses at a point of the body have been computed so that the matrix **T** is

$$\mathbf{T} = \begin{bmatrix} 151 & -30 & -27 \\ -30 & 190 & 10 \\ -27 & 10 & 79 \end{bmatrix} \tag{1}$$

Find the principal stresses and directions.

The characteristic equation for this matrix is

$$|\mathbf{T}-\lambda\mathbf{I}| = -\lambda^3 + 420\lambda^2 - 53\,900\lambda + 2\,058\,000 = 0 \tag{2}$$

In factored form this equation becomes

$$-(\lambda - 70)(\lambda - 140)(\lambda - 210) = 0 \tag{3}$$

so that the eigenvalues are

$$\lambda_1 = 70 \quad \lambda_2 = 140 \quad \lambda_3 = 210 \tag{4}$$

which are equal to the principal stresses σ_1, σ_2, and σ_3. The extreme values of the shear stresses are found using Eq. (1.91).

The principal directions will be specified by determining a right-handed orthogonal triad of unit vectors. A vector **b** is an eigenvector corresponding to the first eigenvalue λ_1 if and only if

$$\mathbf{T}\,\mathbf{b} = \lambda_1\mathbf{b} \tag{5}$$

or

$$\begin{aligned} 151b_x - 30b_y - 27b_z &= 70b_x \\ -30b_x + 190b_y + 10b_z &= 70b_y \\ -27b_x + 10b_y + 79b_z &= 70b_z \end{aligned} \tag{6}$$

or

$$\begin{aligned} 81b_x - 30b_y - 27b_z &= 0 \\ -30b_x + 120b_y + 10b_z &= 0 \\ -27b_x + 10b_y + 9b_z &= 0 \end{aligned} \tag{7}$$

The first and third equations are dependent. The first and second equations force the value of b_y to be zero, which may be verified by multiplying the first equation by 10, the second by 27 and adding the resulting equations. This leaves one condition to be satisfied, namely

$$-30b_x + 10b_z = 0 \quad \text{or} \quad b_z = 3b_x \tag{8}$$

Thus, any vector

$$\mathbf{b} = b_x\,\mathbf{e}_x + b_y\,\mathbf{e}_y + b_z\,\mathbf{e}_z = b\mathbf{e}_x + 3b\mathbf{e}_z \tag{9}$$

with b a nonzero real number is an eigenvector corresponding to the eigenvalue λ_1. Usually, b is set equal to one.

Similar considerations show that any vector

$$\mathbf{c} = 3c\mathbf{e}_x + 2c\mathbf{e}_y - c\mathbf{e}_z \tag{10}$$

where c any nonzero real number is an eigenvector for λ_2, and that any vector

$$\mathbf{d} = 3d\mathbf{e}_x + 5d\mathbf{e}_y - d\mathbf{e}_z \tag{11}$$

where d any nonzero real number is an eigenvector for λ_3. The numbers c and d are usually taken as being equal to one.

Since the eigenvectors serve only to define principal directions, the full collection of eigenvectors defined above, by the presence of scalars b, c, d is not needed. We choose the first two eigenvectors by requiring that they have unit magnitude and define the third as the cross product of the first two. This choice produces a right-handed orthogonal triad of unit vectors

$$\begin{aligned}
\mathbf{b}_1 &= \frac{b}{\sqrt{b^2 + (3b)^2}}(\mathbf{e}_x + 3\mathbf{e}_z) = \frac{\mathbf{e}_x + 3\mathbf{e}_z}{\sqrt{10}} \\
\mathbf{b}_2 &= \frac{3\mathbf{e}_x + 2\mathbf{e}_y - \mathbf{e}_z}{\sqrt{14}} \\
\mathbf{b}_3 &= \mathbf{b}_1 \times \mathbf{b}_2 = \frac{-3\mathbf{e}_x + 5\mathbf{e}_y + \mathbf{e}_z}{\sqrt{35}}
\end{aligned} \tag{12}$$

The matrix $\mathbf{T}$ in this case has three distinct eigenvalues. The vector subspace spanned by the set of eigenvectors corresponding to any one eigenvalue is one dimensional. Therefore, the normalization of the eigenvectors means that the only possible choices for the first eigenvector are $\mathbf{b}_1$ or $-\mathbf{b}_1$, and the only possible choices for the second eigenvector are $\mathbf{b}_2$ or $-\mathbf{b}_2$. Once the first two eigenvectors have been chosen, the third is uniquely determined as the right-handed cross product of the first two, because a right-handed orthogonal triad is stipulated. In physical terms, if the principal planes are visualized as the faces of a cube, the orientation of this cube in three-dimensional space is uniquely determined. This uniqueness is not obtained when $\mathbf{T}$ has repeated eigenvalues. ■

1.7.3 Strength Theories

The stress-strain curve characteristics normally are determined by a tensile test of a bar. At a certain level of stress, such as the yield stress σ_{ys} or the ultimate stress σ_u, the bar undergoes a transition to inelastic behavior. Often the bar is considered to have "failed", hence the following theories are sometimes referred to as *failure theories* or *strength theories*. For complex states of stress, failure theories have been developed that provide relationships between stresses in complicated (two or three dimensional) situations and the behavior of a material in simple tension. Although a brief discussion of some failure theories are given in this section, it may be helpful to consult an appropriate reference for a more detailed discussion. The literature contains a variety of other theories, as well as specialized criteria for phenomena such as anisotropic materials.

Maximum Stress Theory

The maximum stress, or *Rankine*[21], theory is based on the maximum stress being chosen as the criterion of failure. Yield (or some other measure of "failure") occurs when loading on

[21] William John Macquorn Rankine (1820–1872) was a Scottish engineer at Glasgow University. Although he is best known for his work in thermodynamics (Rankine cycle and Rankine absolute temperature scale), he made numerous contributions to the theory of elasticity and to a variety of other fields. He developed the stress transformation equations when he was 32. He was also interested in the theory of structural analysis and such members as restraining walls and arches.

the structure is such that the maximum principal stress at a point reaches the stress at yield (or other "failure" level) in a tensile test for the material. Choose the principal directions such that $\sigma_1 > \sigma_2 > \sigma_3$. Then, yield occurs when

$$\sigma_1 = \sigma_{ys} \tag{1.92}$$

A similar expression is obtained if a yield stress for a compressive test is available. Also a maximum strain theory is readily derived.

Maximum Shear Theory

According to maximum shear theory, failure occurs in a body in a complex state of stress when the maximum shear stress at a point (Eq. 1.91), e.g., $|\sigma_1 - \sigma_2|/2$, reaches the value of the shear yield stress of the material in a tensile test, $\sigma_{ys}/2$. Thus yield (failure) for the complex state of stress occurs if

$$\max(|\sigma_1 - \sigma_2|, |\sigma_2 - \sigma_3|, |\sigma_3 - \sigma_1|) = \sigma_{ys} \tag{1.93}$$

The term *stress intensity* is sometimes used to indicate the highest of these absolute values. The failure theory of Eq. (1.93) is also called the *Tresca theory*. Frequently, the theory of Eq. (1.93) is expressed as

$$\sigma_{\max} - \sigma_{\min} = \sigma_{ys} \tag{1.94}$$

where $\sigma_{\max}$ and $\sigma_{\min}$ are the maximum and minimum principal stresses, respectively. For the case $\sigma_1 > \sigma_2 > \sigma_3$, Eq. (1.94) would be $\sigma_1 - \sigma_3 = \sigma_{ys}$.

von Mises[22] Criterion

Another failure criterion is based on the *maximum distortion energy* at a location in the structure reaching the maximum distortion energy at yield in a tensile test. This leads to the following expression as a criterion of failure by yielding:

$$\sqrt{\frac{(\sigma_1 - \sigma_2)^2 + (\sigma_2 - \sigma_3)^2 + (\sigma_1 - \sigma_3)^2}{2}} = \sigma_{ys} \tag{1.95}$$

For another failure mode, such as fatigue, ultimate stress, or fracture stress, simply replace σ_{ys} by the appropriate tensile stress level. The quantity on the left-hand side of Eq. (1.95) is sometimes referred to as the *equivalent stress* and is often available as output of structural analysis software. The underlying theory for Eq. (1.95) can also be based on the maximum shear stress on an octachedral plane (a plane which intersects the principal axes at equal angles) and hence this criterion is sometimes referred to as the *octachedral shear stress theory*. Other names for this failure criterion are the *von Mises theory* or the *Maxwell-Huber-Hencky-von Mises theory*. If $\sigma_3 = 0$, Eq. (1.95) reduces to

$$\sqrt{\sigma_1^2 - \sigma_1 \sigma_2 + \sigma_2^2} = \sigma_{ys} \tag{1.96}$$

[22]Richard von Mises (1883–1953), an aerodynamicist and mathematician, was born and educated in Austria. He did research in flight mechanics and was the founding editor of the journal *Zeitschrift für angewandte Mathematik und Mechanik*. Early in his career he proposed two fundamental axioms in probability theory: the axioms of convergence and randomness. Later he proposed the famous birthday problem which asks for a probability of at least 50%, how many people must be in a room so that some have the same birthday.

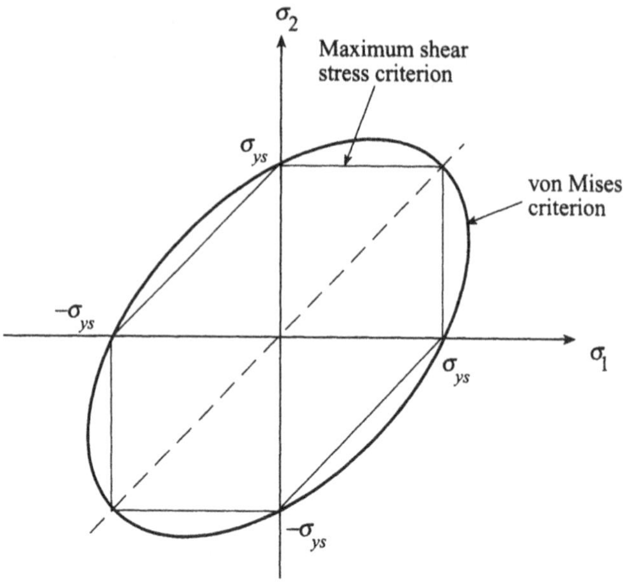

FIGURE 1.14
The von Mises ellipse and the maximum shear stress hexagon.

Figure 1.14 shows the plot of Eq. (1.96) and the corresponding expression for the maximum shear stress theory. Note that both theories intersect the axes at the same points. They also have in common the two points of intersection with the bisectors of the first and third quadrants. It is clear from the figure that the von Mises criterion is less conservative than the maximum shear stress criterion. On one hand the maximum shear stress criterion is linear, or more accurately a set of bilinear curves, and is easy to manipulate, on the other hand the von Mises criterion involves a single expression in contrast to the three expressions of Eq. (1.93).

1.8 Engineering Beam Theory

The technical or engineering theories for structural members are distinguished from the theory of elasticity in that their governing equations are highly tractable. This is primarily the result of imposing simplifying geometric assumptions on the theory, e.g., for the bending of beams it is assumed that planar cross-sections remain planar throughout the bending process. In this chapter, it is shown that the engineering beam theory equations can be developed and expressed in a fashion similar to the elasticity equations just considered.

1.8.1 Kinematical Relationships

The kinematical relationships are the strain-displacement equations. For a beam, the bending strain is taken to be the curvature $\kappa = 1/\rho$, where ρ is the radius of curvature of the beam axis through the centroids of the cross-section. From analytical geometry, the definition of

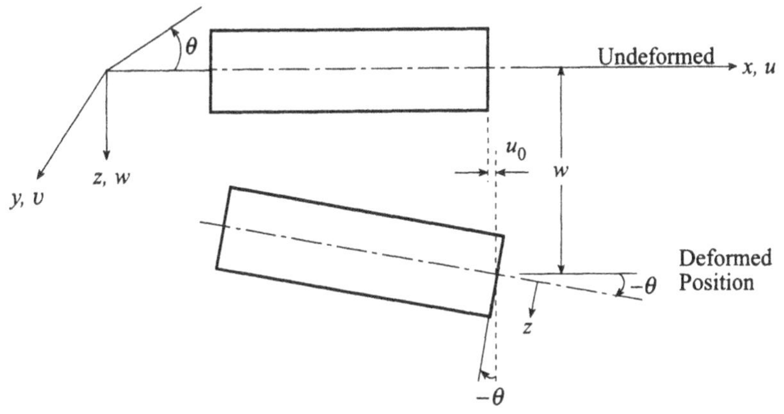

FIGURE 1.15
Element of a beam in bending and extension.

the curvature of the deflection curve is

$$\kappa = \frac{d\theta/dx}{(1+\theta^2)^{3/2}}$$

or for θ small relative to unity

$$\kappa = \frac{d\theta}{dx} \tag{1.97}$$

where θ is the slope of the deflection curve.

We assume that regardless of the type of applied loading, the beam deforms as though it were undergoing pure bending (a constant moment along the beam). This contention, which is referred to as Bernoulli's[23] hypothesis, implies that cross-sections of the beam remain planar and perpendicular to the longitudinal axis under bending. That is, for a beam with arbitrary supports, changes in cross-section, or applied loading, it is assumed that a "flat" cross-section remains "flat" as it deforms, which is how the beam would deform if it were uniform and were subjected only to a constant moment along the beam. See an elementary strength of materials text for an in-depth discussion of this beam theory. It can be seen (Fig. 1.15) that for this deformation the axial displacement of a point on a cross-sectional plane is

$$u(x, z) = u_0(x) + z\theta(x) \tag{1.98}$$

where u_0 is the axial displacement of the point at which the x axis intersects the cross-sectional plane and $\theta = \tan\alpha$ is the slope of the beam axis. For small α, it follows that $\theta \approx \alpha$.

[23]Daniel Bernoulli (1700–1782) was the son of Johann Bernoulli. He studied medicine and philosophy. Mathematics instruction came from his father and his uncle Jakob. He is best known for his book *Hydrodynamica,* and for his efforts in the mechanics of flexible bodies. He suggested by letter to Euler that the calculus of variations be used to derive the equations governing the elastic curve. As a consequence, the governing equations for engineering beam theory are often referred to as Euler-Bernoulli equations.

Displacements in the positive coordinate directions are taken to be positive. Rotations (slopes) are positive if their corresponding vectors, according to the right-hand rule, lie in a positive coordinate direction. The shear strain γ_{xz} takes the form

$$\gamma_{xz} = \frac{\partial u}{\partial z} + \frac{\partial w}{\partial x} = \theta + \frac{\partial w}{\partial x} = \gamma \tag{1.99}$$

In order for the cross-sections to remain planar, it is necessary that this shear strain be zero. This corresponds to neglecting the shear deformation effects. Then

$$\theta = -\frac{dw}{dx} \quad \text{or} \quad \kappa = -\frac{d^2w}{dx^2} \tag{1.100}$$

This is the strain-displacement relation for bending. The component w is the displacement of the beam axis, i.e., of the centerline of the beam. This displacement is referred to as the *deflection* of the beam.

The axial displacement of the centerline of the beam is u_0. The corresponding strain is $\epsilon_{0x} = du_0/dx$. This strain-displacement relation can be combined with that for bending in the matrix form

$$\begin{array}{ccccc} \begin{bmatrix} \epsilon_{0x} \\ \kappa \end{bmatrix} & = & \begin{bmatrix} d_x & 0 \\ 0 & -d_x^2 \end{bmatrix} & \begin{bmatrix} u_0 \\ w \end{bmatrix} \\ \boldsymbol{\epsilon} & = & \mathbf{D}_u & \mathbf{u} \end{array} \tag{1.101}$$

where the subscript u has been added so that it is clear that the differential operator $\mathbf{D}$ belongs to the strain-displacement relations.

If shear deformation effects over the cross-section are retained, then $\gamma \neq 0$ and the strain-displacement relations can be written as

$$\begin{array}{ccccc} \begin{bmatrix} \epsilon_{0x} \\ \gamma \\ \kappa \end{bmatrix} & = & \begin{bmatrix} d_x & 0 & 0 \\ 0 & d_x & 1 \\ 0 & 0 & d_x \end{bmatrix} & \begin{bmatrix} u_0 \\ w \\ \theta \end{bmatrix} \\ \boldsymbol{\epsilon} & = & \mathbf{D}_u & \mathbf{u} \end{array} \tag{1.102}$$

Engineering beam theory equations will be derived here from the theory of elasticity without imposing the above deformation assumptions. For strength of material theories of beams, plates, and shells, the bending due to transverse applied loads is usually considered to be governed primarily by the extension and contraction of longitudinal "fibers." Strains not associated with this extension and contraction may generally be neglected. Thus, transverse normal and shear strains are set equal to zero. This is equivalent to assuming that the beam material reacts rigidly when subjected to transverse normal and shear strains. For our beam, then, the strains ϵ_z and ϵ_{xz} are considered to vanish. Thus,

$$\epsilon_z = \frac{\partial w}{\partial z} = 0, \qquad \gamma_{xz} = \frac{\partial u}{\partial z} + \frac{\partial w}{\partial x} = 0 \tag{1.103}$$

As mentioned previously, $\gamma_{xz} = 0$ corresponds to ignoring shear deformation effects. Integration of Eq. (1.103) gives

$$w = w_0(x), \qquad u = u_0(x) - z\frac{dw}{dx} \tag{1.104}$$

where u_0 and w_0 are the horizontal and vertical displacement components along the x axis of the beam. As noted in the previous derivation, these components apply for a beam with a cross-section that remains planar during bending and simply rotates about the y axis.

1.8.2 Material Laws

The constitutive relations for the material of the beam should reflect the assumption that the extension and contraction of longitudinal fibers are the dominant deformations. This is equivalent to assuming that the material is rigid in the z direction. In terms of fibers being deformed in the longitudinal (x) direction, this rigidity means that there will be no contribution to the longitudinal strain ϵ_x by stresses in the z direction. Thus, ϵ_x of Eq. (1.32) reduces to $\epsilon_x = (\sigma_x - \nu\sigma_y)/E$. Since the loading is in the xz plane, it is reasonable to assume that $\sigma_y = 0$. Thus, $\epsilon_x = \sigma_x/E$ or $\sigma_x = E\epsilon_x$. Recall from Eq. (1.98) that for the beam $u = u_0 + z\theta$, so that

$$\epsilon_x = \frac{du}{dx} = \frac{du_0}{dx} + z\frac{d\theta}{dx} = \frac{du_0}{dx} + z\kappa \tag{1.105}$$

The stress distribution over the cross-section gives rise to the stress resultants or the net internal forces:

The axial force N

$$N = \int_A \sigma_x \, dA = \int_A E\epsilon_x \, dA = \int_A E\left(\frac{du_0}{dx} + z\kappa\right) dA = EA\frac{du_0}{dx} = EA\epsilon_{0x} \tag{1.106a}$$

and the bending moment M

$$M = \int_A \sigma_x z \, dA = \int_A E\epsilon_x z \, dA = \int_A E\left(\frac{du_0}{dx} + z\kappa\right) z \, dA = \kappa E \int_A z^2 \, dA = \kappa EI \tag{1.106b}$$

where I is the moment of inertia about the y axis. The integral involving z, i.e., $\int z \, dA$, is zero if z is measured from a centroidal axis of the beam. In matrix notation, Eqs. (1.106a) and (1.106b) appear as

$$\begin{bmatrix} N \\ M \end{bmatrix} = \begin{bmatrix} EA & 0 \\ 0 & EI \end{bmatrix} \begin{bmatrix} \epsilon_{0x} \\ \kappa \end{bmatrix} \tag{1.107}$$

$$\mathbf{s} \quad = \quad \mathbf{E} \quad \boldsymbol{\epsilon}$$

or

$$\epsilon = \begin{bmatrix} 1/EA & 0 \\ 0 & 1/EI \end{bmatrix} \mathbf{s} \tag{1.108}$$

$$\epsilon = \quad \mathbf{E}^{-1} \quad \mathbf{s}$$

where $\mathbf{s}$ is chosen to be equal to $[N \; M]^T$.

If shear deformation effects are to be taken into account, the constitutive equation relating the shear strain and the net internal shear force is needed. Hooke's law for shear is $\tau_{xz} = \tau = G\gamma$. Choose $V = \tau_{\text{average}} A$ as a stress-force relationship where V is the shear force and A is the cross-sectional area. If τ is the shear stress at the centroid of the cross-section, then select $\tau_{\text{average}} = k_s\tau$, where k_s is a dimensionless *shear form* or *shear stiffness factor* that depends on the cross-sectional shape. Later we will discuss techniques for finding k_s. The reciprocal of k_s is the *shear correction factor* that is often tabulated in handbooks of formulas for stress analysis. The desired material law relationship becomes

$$V = k_s G A \gamma \tag{1.109}$$

Thus, the complete material law is

$$\begin{bmatrix} N \\ V \\ M \end{bmatrix} = \begin{bmatrix} EA & 0 & 0 \\ 0 & k_s GA & 0 \\ 0 & 0 & EI \end{bmatrix} \begin{bmatrix} \epsilon_{0x} \\ \gamma \\ \kappa \end{bmatrix} \tag{1.110}$$

$$\mathbf{s} = \mathbf{E} \quad \boldsymbol{\epsilon}$$

or

$$\boldsymbol{\epsilon} = \begin{bmatrix} \frac{1}{EA} & 0 & 0 \\ 0 & \frac{1}{k_s GA} & 0 \\ 0 & 0 & \frac{1}{EI} \end{bmatrix} \mathbf{s} \tag{1.111}$$

$$\boldsymbol{\epsilon} = \mathbf{E}^{-1} \quad \mathbf{s}$$

1.8.3 Equations of Equilibrium

Figure 1.16 shows a beam element isolating the internal forces. The sign convention for the theory of elasticity stresses applies as well for the forces and moments on a beam cross-section. Thus, the forces and moments shown in Fig. 1.16 are positive. Applied loads are positive if their corresponding vectors lie in positive coordinate directions.

For the purpose of applying the conditions of equilibrium, the applied distributed load $\overline{p}_z$ (force/length) is replaced by its resultant $\overline{p}_z\, dx$. The summation of forces in the vertical direction gives

$$-V + \overline{p}_z\, dx + V + \frac{\partial V}{\partial x} dx = 0 \quad \text{or} \quad \frac{\partial V}{\partial x} + \overline{p}_z = 0 \tag{1.112a}$$

Remember that a variable with an overbar, e.g., $\overline{p}_z$, is an applied quantity. In a similar fashion, it is found that

$$\frac{\partial N}{\partial x} + \overline{p}_x = 0 \tag{1.112b}$$

is the equilibrium relation for the axial (x) direction. To establish the moment-shear equation, sum moments about the left face of the element

$$-M + M + \frac{\partial M}{\partial x} dx - dx\left(V + \frac{\partial V}{\partial x} dx\right) - \frac{dx}{2} \overline{p}_z\, dx = 0$$

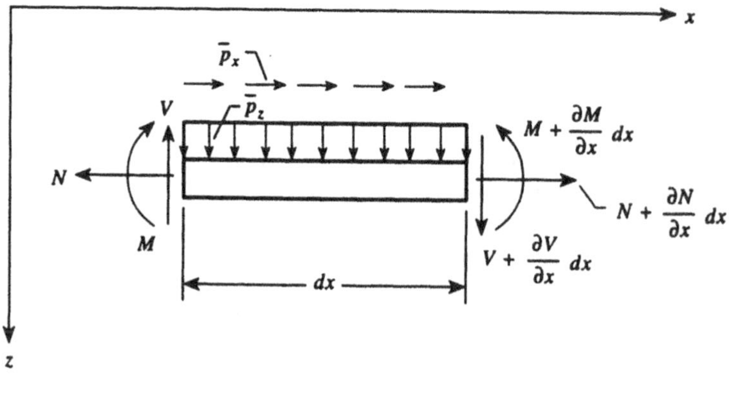

FIGURE 1.16
Beam element with internal forces and applied loading in positive directions, all applied at the centroid of the cross-section.

or

$$\frac{\partial M}{\partial x} - V - \frac{\partial V}{\partial x}dx - \frac{1}{2}\overline{p}_z\,dx = 0$$

The terms $\frac{\partial V}{\partial x}dx$ and $\frac{1}{2}\overline{p}_z\,dx$ approach zero in the limit leaving

$$\frac{dM}{dx} - V = 0 \tag{1.113}$$

The shear force V can be eliminated from Eqs. (1.112a) and (1.113) giving

$$\frac{d^2M}{dx^2} + \overline{p}_z = 0 \tag{1.114}$$

These equilibrium relations can be placed in the matrix form

$$\begin{bmatrix} d_x & 0 \\ 0 & d_x^2 \end{bmatrix} \begin{bmatrix} N \\ M \end{bmatrix} + \begin{bmatrix} \overline{p}_x \\ \overline{p}_z \end{bmatrix} = \mathbf{0} \tag{1.115}$$

$$\mathbf{D}_s^T \quad \mathbf{s} \quad + \quad \overline{\mathbf{p}} \quad = \mathbf{0}$$

if shear deformation effects are not considered. If shear deformation effects are included, then

$$\begin{bmatrix} d_x & 0 & 0 \\ 0 & d_x & 0 \\ 0 & -1 & d_x \end{bmatrix} \begin{bmatrix} N \\ V \\ M \end{bmatrix} + \begin{bmatrix} \overline{p}_x \\ \overline{p}_z \\ 0 \end{bmatrix} = \mathbf{0} \tag{1.116}$$

$$\mathbf{D}_s^T \quad \mathbf{s} \quad + \quad \overline{\mathbf{p}} \quad = \mathbf{0}$$

An important characteristic of $\mathbf{D}_s$ is its functional relationship to $\mathbf{D}_u$. The operators $\mathbf{D}_u$ and $\mathbf{D}_s^T$ are formally adjoint in the sense that they satisfy the relationship

$$\int_V (\mathbf{D}_u\mathbf{u})^T\mathbf{s}\,dV = \int_V \mathbf{u}^T{}_u\mathbf{D}^T\mathbf{s}\,dV = \oint_S \mathbf{u}^T\mathbf{A}^T\mathbf{s}\,dS - \int_V \mathbf{u}^T\mathbf{D}_s^T\mathbf{s}\,dV \tag{1.117}$$

where the arbitrary vectors **u** and **s** are smooth in the region considered, and $\mathbf{A}^T$ is the matrix of direction cosines in Eq. (1.57) for the tractions on an oblique surface. This adjoint relationship, which holds for both the case of general elasticity and for specialized structures, is a member of a group of formulas referred to collectively as the divergence theorem. An example is Appendix II, Eq. (II.7), a statement of the divergence theorem for vectors. Equation (1.117) with **s** replaced by $\boldsymbol{\sigma}$ holds for the elasticity equations. In this case $\mathbf{D}^T$ of the equilibrium equation is denoted by $\mathbf{D}_\sigma^T$ and **D** of the kinematic equations by $\mathbf{D}_u$. The subscript index (u) to the left of the operator matrix **D**, i.e., u of ${}_u\mathbf{D}^T$, indicates that the operator is applied to the preceding quantity, i.e., to $\mathbf{u}^T$. Thus, $\mathbf{u}^T{}_u\mathbf{D}^T = (\mathbf{D}_u\mathbf{u})^{T*}$. This notation has the advantage of revealing clearly the structure of the resulting operator. For example, the symmetry properties of the equations are apparent in the operator form.

The adjoint relationship of Eq. (1.117) can be demonstrated for particular cases using integration by parts. Consider the two-dimensional elasticity equations for a rectangle of

---

*The transpose of a product of matrices is equal to the product of the transpose matrices in reversed order, i.e., if

$$\mathbf{H} = \mathbf{AB}\cdots\mathbf{EF}$$

then

$$\mathbf{H}^T = \mathbf{F}^T\mathbf{E}^T\cdots\mathbf{B}^T\mathbf{A}^T$$

area A, bounded by the lines $x = 0$, $x = a$, $y = 0$, and $y = b$. Begin with the final integral on the right-hand side of Eq. (1.117). Introduce $\mathbf{D}^T$ of Eq. (1.50a) and set $\mathbf{D}_\sigma^T = \mathbf{D}^T$

$$
\begin{aligned}
\int_A \mathbf{u}^T \mathbf{D}_\sigma^T \boldsymbol{\sigma}\, dA &= \int_A [u_x \;\; u_y] \begin{bmatrix} \partial_x & 0 & \partial_y \\ 0 & \partial_y & \partial_x \end{bmatrix} \begin{bmatrix} \sigma_x \\ \sigma_y \\ \tau_{xy} \end{bmatrix} dA \\
&= \int_0^b \int_0^a u_x\, \sigma_{x,x}\, dx\, dy + \int_0^a \int_0^b u_x \tau_{xy,y}\, dy\, dx + \int_0^a \int_0^b u_y\, \sigma_{y,y}\, dy\, dx \\
&\quad + \int_0^b \int_0^a u_y\, \tau_{xy,x}\, dx\, dy \\
&= \int_0^b \left(u_x\, \sigma_x \Big|_0^a - \int_0^a u_{x,x}\, \sigma_x\, dx \right) dy + \int_0^a \left(u_x\, \tau_{xy} \Big|_0^b - \int_0^b u_{x,y} \tau_{xy}\, dy \right) dx \\
&\quad + \int_0^a \left(u_y \sigma_y \Big|_0^b - \int_0^b u_{y,y}\, \sigma_y\, dy \right) dx + \int_0^b \left(u_y\, \tau_{xy} \Big|_0^a - \int_0^a u_{y,x}\, \tau_{xy}\, dx \right) dy \\
&= \int_0^b [u_x \;\; u_y] \begin{bmatrix} 1 & 0 & 0 \\ 0 & 0 & 1 \end{bmatrix} \begin{bmatrix} \sigma_x \\ \sigma_y \\ \tau_{xy} \end{bmatrix} \Bigg|_0^a dy + \int_0^a [u_x \;\; u_y] \begin{bmatrix} 0 & 0 & 1 \\ 0 & 1 & 0 \end{bmatrix} \begin{bmatrix} \sigma_x \\ \sigma_y \\ \tau_{xy} \end{bmatrix} \Bigg|_0^b dx \\
&\quad - \int_0^b \int_0^a u_{x,x}\, \sigma_x\, dx\, dy - \int_0^b \int_0^a u_{y,x}\, \tau_{xy}\, dx\, dy \\
&\quad - \int_0^a \int_0^b u_{x,y}\, \tau_{xy}\, dy\, dx - \int_0^a \int_0^b u_{y,y}\, \sigma_y\, dy\, dx \\
&= \oint_S [u_x \;\; u_y] \begin{bmatrix} a_x & 0 & a_y \\ 0 & a_y & a_x \end{bmatrix} \begin{bmatrix} \sigma_x \\ \sigma_y \\ \tau_{xy} \end{bmatrix} ds - \int_A [u_x \;\; u_y] \begin{bmatrix} {}_x\partial & 0 & {}_y\partial \\ 0 & {}_y\partial & {}_x\partial \end{bmatrix} \begin{bmatrix} \sigma_x \\ \sigma_y \\ \tau_{xy} \end{bmatrix} dA \\
&= \oint_S \mathbf{u}^T \mathbf{A}^T \boldsymbol{\sigma}\, ds - \int_A \mathbf{u}^T{}_u \mathbf{D}^T \boldsymbol{\sigma}\, dA
\end{aligned} \tag{1.118}
$$

where a_x and a_y represent the direction cosines of the unit normal to s and ${}_x\partial$ and ${}_y\partial$ are partial derivatives of the preceding quantities (u_x and u_y as appropriate). Thus, $\mathbf{D}_u$ and $\mathbf{D}_\sigma^T$ are adjoints in the sense of Eq. (1.117) for the case of two-dimensional elasticity.

The situation is similar for the case of beams with shear deformation. For a beam element from $x = a$ to $x = b$,

$$
\begin{aligned}
\int_a^b \mathbf{u}^T \mathbf{D}_s^T \mathbf{s}\, dx &= \int_a^b [u_0 \;\; w \;\; \theta] \begin{bmatrix} d_x & 0 & 0 \\ 0 & d_x & 0 \\ 0 & -1 & d_x \end{bmatrix} \begin{bmatrix} N \\ V \\ M \end{bmatrix} dx \\
&= \int_a^b [u_0 N' + w V' + \theta (M' - V)]\, dx \\
&= (u_0 N + w V + \theta M)|_a^b - \int_a^b (u_0' N + w' V + \theta' M + \theta V)\, dx \\
&= [u_0 \;\; w \;\; \theta] \begin{bmatrix} 1 & 0 & 0 \\ 0 & 1 & 0 \\ 0 & 0 & 1 \end{bmatrix} \begin{bmatrix} N \\ V \\ M \end{bmatrix} \Bigg|_b + [u_0 \;\; w \;\; \theta] \begin{bmatrix} -1 & 0 & 0 \\ 0 & -1 & 0 \\ 0 & 0 & -1 \end{bmatrix} \begin{bmatrix} N \\ V \\ M \end{bmatrix} \Bigg|_a \\
&\quad - \int_a^b [u_0 \;\; w \;\; \theta] \begin{bmatrix} {}_x d & 0 & 0 \\ 0 & {}_x d & 0 \\ 0 & 1 & {}_x d \end{bmatrix} \begin{bmatrix} N \\ V \\ M \end{bmatrix} dx \\
&= \mathbf{u}^T \mathbf{A}^T \mathbf{s}|_a^b - \int_a^b \mathbf{u}^T{}_u \mathbf{D}^T \mathbf{s}\, dx
\end{aligned}
$$

where ${}_x d$ is a derivative, with respect to x, that operates on the preceding variables u_0, w, and θ. This derivation shows that, for one-dimensional problems, the adjoint relationship takes the form

$$\int_a^b \mathbf{u}^T{}_u\mathbf{D}^T\mathbf{s}\,dx = \mathbf{u}^T\mathbf{A}^T\mathbf{s}|_a^b - \int_a^b \mathbf{u}^T\mathbf{D}_s^T\mathbf{s}\,dx \tag{1.119}$$

The equilibrium equations of Eqs. (1.112) and (1.113) for the beam element of Fig. 1.16 can be derived from the theory of elasticity. For our beam, which lies in and is loaded in the xz plane, we can safely consider the y direction stresses to be zero, i.e., $\sigma_y = \tau_{xy} = \tau_{yz} = 0$. From the theory of elasticity, the equilibrium equations of Eq. (1.51) for non-zero stresses σ_x, σ_z, and τ_{xz} are

$$\frac{\partial \sigma_x}{\partial x} + \frac{\partial \tau_{xz}}{\partial z} = 0 \tag{1.120a}$$

$$\frac{\partial \tau_{xz}}{\partial x} + \frac{\partial \sigma_z}{\partial z} = 0 \tag{1.120b}$$

where the body forces are taken to be zero. We wish to integrate these relations over the cross-sectional area. Suppose the cross-section is rectangular of width b. The boundary conditions for the upper and lower surfaces of the beam element are $\sigma_z = \overline{p}_z(x)/b$, $\tau_{xz} = 0$ at the upper surface and $\sigma_z = 0$, $\tau_{xz} = 0$ at the lower surface. Suppose the stress σ_z does not vary through the width (y direction) for a prescribed z value. Then, using these boundary conditions, integration over the cross-sectional area A of the above differential equilibrium equations gives

$$\frac{dN}{dx} = 0 \tag{1.121a}$$

$$\frac{dV}{dx} = -\overline{p}_z(x) \tag{1.121b}$$

where N, the axial force, and V, the shear force, are stress resultants defined as

$$N = \int_A \sigma_x\,dA, \qquad V = \int_A \tau_{xz}\,dA$$

Multiply Eq. (1.120a) by z, integrate over A, apply integration by parts, and use the prescribed surface conditions to find

$$\frac{dM}{dx} = V \tag{1.121c}$$

where $M = \int_A z\,\sigma_x\,dA$. Thus, the equilibrium conditions of Eqs. (1.112) and (1.113) have been derived directly from the elasticity equations.

1.8.4 Boundary Conditions

The boundary conditions for a beam are referred to the ends of the beam where either conditions on forces (S_p) or displacements (S_u) can be imposed. For example, if an end of a beam, for which axial effects are ignored, is fixed, then $w = \theta = 0$ on S_u. If the end of the beam is free, then $M = V = 0$ on S_p. And for a simply supported end, $w = M = 0$ on $S_u + S_p = S$. In summary, with

$$\mathbf{s} = \begin{bmatrix} V \\ M \end{bmatrix} \qquad \mathbf{u} = \begin{bmatrix} w \\ \theta \end{bmatrix}$$

the boundary conditions for a beam can be written as

$$\begin{matrix} V = \overline{V} \\ M = \overline{M} \end{matrix} \quad \text{or} \quad \mathbf{s} = \overline{\mathbf{s}} \quad \text{on} \quad S_p \tag{1.122}$$

for static or force conditions and

$$\begin{matrix} w = \overline{w} \\ \theta = \overline{\theta} \end{matrix} \quad \text{or} \quad \mathbf{u} = \overline{\mathbf{u}} \quad \text{on} \quad S_u \tag{1.123}$$

for displacement or kinematic conditions. If axial deformation and force are to be included, then supplement Eq. (1.122) with $N = \overline{N}$ and Eq. (1.123) with $u = \overline{u}$.

1.8.5 Displacement Form of the Governing Differential Equations

It should be apparent that the governing equations for a beam can be written in a form quite similar to the general governing equations for an elastic solid. For the elastic solid, we have

$$\boldsymbol{\epsilon} = \mathbf{D}\mathbf{u}, \quad \boldsymbol{\sigma} = \mathbf{E}\boldsymbol{\epsilon}, \quad \mathbf{D}^T\boldsymbol{\sigma} + \overline{\mathbf{p}}_V = \mathbf{0} \tag{1.124a}$$

and for the beam,

$$\boldsymbol{\epsilon} = \mathbf{D}_u\mathbf{u}, \quad \mathbf{s} = \mathbf{E}\boldsymbol{\epsilon}, \quad \mathbf{D}_s^T\mathbf{s} + \overline{\mathbf{p}} = \mathbf{0} \tag{1.124b}$$

In the case of the general elastic solid, we derived alternative forms to Eq. (1.124) for governing differential equations. The most common forms for the beam equations are in terms of displacements w, or w and θ only, or in terms of all of the *state variables* (a mixed form): displacement w; slope θ; moment M; or shear V.

A familiar form of the beam equations is the Euler-Bernoulli beam in which the shear deformation has been neglected. To derive this form, consider only the equations related to bending, i.e., ignore the axial extension relationships. From the kinematical relations of Eqs. (1.97) and (1.100), and the material law of Eq. (1.106b),

$$M = EI\kappa = EI\frac{d\theta}{dx} = -EI\frac{d^2w}{dx^2} \tag{1.125}$$

Substitution of this relation into the equilibrium conditions, Eqs. (1.112a) and (1.113) results in

$$V = \frac{dM}{dx} = -\frac{d}{dx}EI\frac{d^2w}{dx^2} \tag{1.126a}$$

and

$$-\frac{dV}{dx} = \frac{d^2}{dx^2}EI\frac{d^2w}{dx^2} = \overline{p}_z \tag{1.126b}$$

The full set of governing differential equations is then

$$\frac{d^2}{dx^2}EI\frac{d^2w}{dx^2} = \overline{p}_z \tag{1.127a}$$

$$V = -\frac{d}{dx}EI\frac{d^2w}{dx^2} \tag{1.127b}$$

$$M = -EI\frac{d^2w}{dx^2} \tag{1.127c}$$

$$\theta = -\frac{dw}{dx} \tag{1.127d}$$

Note that these displacement relations were derived following the procedure used in Section 1.6.1 to find the displacement formulation governing differential equations for an elastic solid. That is, the strain displacement relations were substituted into the constitutive equations. The resulting stress (force)-displacement functions were then placed in the equilibrium equations.

A more general form of the displacement formulation equations is obtained if shear deformation effects are retained. Again, substitute the strain-displacement relations of Eq. (1.102), $\gamma = d_x w + \theta$ and $\kappa = d_x\theta$, in the constitutive relations of Eq. (1.110), $V = k_s G A\gamma$ and $M = EI\kappa$. If the resulting force-displacement equations are placed in the conditions of equilibrium of Eq. (1.116),

$$-\frac{dV}{dx} = \overline{p}_z \quad \text{or} \quad -\frac{d}{dx}\left[k_s G A\left(\frac{dw}{dx} + \theta\right)\right] = \overline{p}_z \tag{1.128a}$$

$$\frac{dM}{dx} = V \quad \text{or} \quad \frac{d}{dx}\left(EI\frac{d\theta}{dx}\right) = k_s G A\left(\frac{dw}{dx} + \theta\right) \tag{1.128b}$$

or in matrix form

$$\begin{bmatrix} d_x k_s G A d_x & d_x k_s G A \\ k_s G A d_x & -d_x E I d_x + k_s G A \end{bmatrix}\begin{bmatrix} w \\ \theta \end{bmatrix} + \begin{bmatrix} \overline{p}_z \\ 0 \end{bmatrix} = 0 \tag{1.129}$$

In matrix notation these beam equations appear as

$$\mathbf{D}_s^T \mathbf{E} \mathbf{D}_u \mathbf{u} + \overline{\mathbf{p}} = 0 \tag{1.130}$$

which corresponds to Eq. (1.63) for the general elastic solid.

1.8.6 Mixed Form of the Governing Differential Equations

The other most frequently used form of the governing equations is a mixed form involving both forces and displacements [Wunderlich, 1977]. As in the case of the general elastic solid, these relations for a beam are found in a straight forward fashion by eliminating ϵ between the strain-displacements of Eq. (1.102) and the constitutive relationships of Eq. (1.111). That is, upon equating the strains ϵ and ignoring the axial relations,

$$\frac{dw}{dx} = -\theta + \frac{V}{k_s G A} \tag{1.131a}$$

$$\frac{d\theta}{dx} = \frac{M}{EI} \tag{1.131b}$$

or in matrix form

$$\begin{bmatrix} d_x & 1 \\ 0 & d_x \end{bmatrix}\begin{bmatrix} w \\ \theta \end{bmatrix} = \begin{bmatrix} 1/k_s G A & 0 \\ 0 & 1/EI \end{bmatrix}\begin{bmatrix} V \\ M \end{bmatrix} \tag{1.132}$$

These are then placed together with the equilibrium conditions of Eq. (1.116) in the form

$$\frac{dw}{dx} = -\theta + \frac{V}{k_s G A} \tag{1.133a}$$

$$\frac{d\theta}{dx} = \frac{M}{EI} \tag{1.133b}$$

$$\frac{dV}{dx} = -\overline{p}_z \tag{1.133c}$$

$$\frac{dM}{dx} = V \tag{1.133d}$$

or in matrix form

$$\frac{d\mathbf{z}}{dx} = \mathbf{A}\mathbf{z} + \overline{\mathbf{P}} \tag{1.134}$$

where

$$\mathbf{z} = \begin{bmatrix} w \\ \theta \\ V \\ M \end{bmatrix} \quad \mathbf{A} = \begin{bmatrix} 0 & -1 & \frac{1}{k_s GA} & 0 \\ 0 & 0 & 0 & \frac{1}{EI} \\ 0 & 0 & 0 & 0 \\ 0 & 0 & 1 & 0 \end{bmatrix} \quad \overline{\mathbf{P}} = \begin{bmatrix} 0 \\ 0 \\ -\overline{p}_z \\ 0 \end{bmatrix} \tag{1.135}$$

Equations (1.133) or (1.134) can be solved using the methodology developed in Chapters 4 and 5. It is of interest to note that as in the case of the mixed method equations for the general elastic solids, the mixed method governing equations for the beam of Eq. (1.133) do not involve derivatives of geometrical or material parameters and all derivatives are of the first order. This is not the case for the displacement governing equations of Eqs. (1.127) and (1.128). In many instances, these characteristics are highly advantageous when solving the equations. For example, many numerical integration schemes operate with first order derivatives only, and equations with higher order derivatives must first be reduced to this form.

If the axial terms are included, Eq. (1.134) would be defined using

$$\mathbf{z} = \begin{bmatrix} u_0 \\ w \\ \theta \\ N \\ V \\ M \end{bmatrix} \quad \mathbf{A} = \begin{bmatrix} 0 & 0 & 0 & \frac{1}{EA} & 0 & 0 \\ 0 & 0 & -1 & 0 & \frac{1}{k_s GA} & 0 \\ 0 & 0 & 0 & 0 & 0 & \frac{1}{EI} \\ 0 & 0 & 0 & 0 & 0 & 0 \\ 0 & 0 & 0 & 0 & 0 & 0 \\ 0 & 0 & 0 & 0 & 1 & 0 \end{bmatrix} \quad \overline{\mathbf{P}} = \begin{bmatrix} 0 \\ 0 \\ 0 \\ -\overline{p}_x \\ -\overline{p}_z \\ 0 \end{bmatrix} \tag{1.136}$$

1.8.7 Stress Formulas

The solution of the governing equations of this section provides the state variables w, θ, V, and M along the beam. Given the moment M and shear force V, the major normal and shear stresses in the beam can be computed. Recall from Eq. (1.106a) $du_0/dx = N/(EA)$ and from Eqs. (1.97) and (1.106b) $d\theta/dx = M/(EI)$. Substitution of these into $\sigma_x = E\epsilon_x = E(du_0/dx + z\,d\theta/dx)$ gives

$$\sigma_x = \frac{N}{A} + \frac{M}{I}z \tag{1.137}$$

When the bending moment is zero, $\sigma_x = N/A$, which indicates that the stress due to the axial force N is equal to the average normal stress on the cross-section. That is, this stress is uniformly distributed over the cross-section. The other term, $\sigma_x = M\,z/I$, is referred to as the *flexure formula*. This bending stress is a linearly distributed stress equal to zero at $z = 0$, the location of the centroid. It assumes its maximum value when z reaches its maximum value at an outer edge of the cross section.

To find the shear stress τ_{xz}, substitute $\sigma_x = M\,z/I$ in the first term of Eq. (1.120a) and use Eq. (1.121c). Then $V\,z/I + \partial\tau_{xz}/\partial z = 0$ or $-\partial\tau_{xz}/\partial z = V\,z/I$. Assume the cross-section is rectangular of height t and suppose the shear stresses are distributed uniformly across the width. Integrate $-\partial\tau_{xz}/\partial z = V\,z/I$ with respect to z from the z position where τ_{xz} is to be evaluated, to the level of $z = t/2$, where $t/2$ defines the top (or bottom) surface of the beam. Then $-\tau_{xz}|_z^{t/2} = \frac{V}{I}\int_z^{t/2} z\,dz$. Since there are no loads in the x direction on the upper or lower surfaces, $\tau_{xz} = 0$ at $z = t/2$, and we find for the shear stress τ_{xz} at z

$$\tau_{xz} = \frac{V}{I}\int_z^{t/2} z\,dz = \frac{V}{2I}\left(\frac{t^2}{4} - z^2\right) \tag{1.138}$$

This indicates that the shear stress varies parabolically with z. This stress is zero at the top and bottom of the cross-section ($z = \pm t/2$) and has its maximum value at the centroidal axis ($z = 0$).

1.9 Torsion, An Example of Field Theory Equations

Torsional stresses occurring on the cross-section of a bar present an interesting example of structural behavior. They are of particular concern here because governing equations are similar to those occurring in a variety of mechanics problems, which are often referred to as *field theory* problems. Included in this class are several fluid mechanics problems.

1.9.1 Kinematical Relationships

Coulomb,[23] while studying the resistance of thin wires to torsion, formulated a solution that was based on the assumption that a cross-section of a bar remains plane after deformation and the cross-sections merely rotate. Navier erroneously applied this assumtion to the general case of torsion of bars with noncircular cross-sections. Proper formulation of the twisting bar, however, must account for deformation of cross-sections in the axial direction.

Saint-Venant provided the correct solution for a bar in torsion. He used the *semi-inverse method* in which assumptions are made at the outset as to the deformations. These assumed deformations, which include an unknown function, are then shown to satisfy the equilibrium conditions and the surface boundary conditions. That is, the unknown function is chosen to satisfy these relations.

It then follows from the uniqueness principle for the solution of the elasticity equations that the initial displacement assumptions were correct and the solution is exact. Saint-Venant was guided by the solution for a shaft with a circular cross-section for which the assumed deformation is formed of "rigid" rotations of the plane cross-sections. In addition, deformation of the cross-section in the axial direction was incorporated. This is referred to as the *warping* of the cross-section. However, it is assumed that there are no restraints placed on the warping along the bar.

Consider a section of a shaft having a constant cross-section and a constant torque along the bar axis. Turn first to the case where the cross-sections remain plane and simply rotate as rigid surfaces so that the displacement in the x direction is zero. Suppose displacements are small relative to the cross-sectional dimensions. Also, assume the profile of the cross-section, i.e., the shape of the section, is not distorted due to twisting. As indicated in Fig. 1.17a, a cross-section rotates about an *axis of twist* through an angle ϕ. This is the angle of twist of the bar. One face of a shaft segment of length dx will rotate $d\phi$ relative to the other face.

[23]Charles Augustin Coulomb (1736–1806) probably furthered the mechanics of elastic bodies more than any other 18th century scientist. He was educated as a military engineer in France and served for thirty years in the military. His theories of friction, strength of materials, and torsion are still used. Of his accomplishments in applied mechanics, only his friction theory was accepted prior to the 19th century. As a result of his work in electricity and magnetism, the unit of quantity of electricity, the coulomb, was named for him. Near the end of his life, he devoted his efforts to hospital reform and the improvement of education in his native land, France.

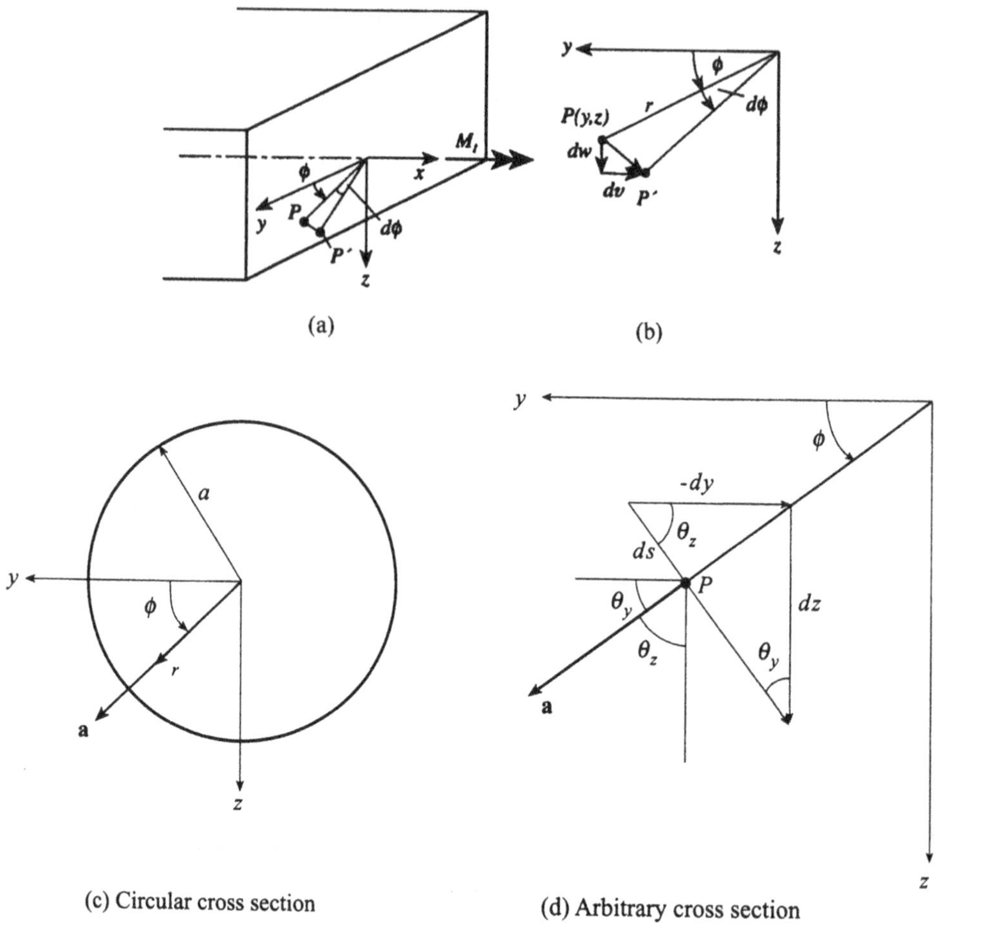

FIGURE 1.17
The cross-sectional rotation of a bar due to torque M_t. Point P moves to point P'.

The point $P(y, z)$ is rotated to $P' = P(y+dv, z+dw)$. Since $d\phi$ is small, $\cos d\phi \approx 1$ and $\sin d\phi \approx d\phi$. It follows from Fig. 1.17b that

$$
\begin{aligned}
dv &= r\cos(\phi + d\phi) - r\cos\phi = r(\cos\phi\ \cos d\phi - \sin\phi\ \sin d\phi) - r\cos\phi \\
&\approx r\cos\phi - d\phi\, r\sin\phi - r\cos\phi = -d\phi\, r\sin\phi \\
&= -d\phi\, z = -\phi' dx\, z \\
dw &= r\sin(\phi + d\phi) - r\sin\phi = r(\cos\phi\ \sin d\phi + \sin\phi\ \cos d\phi) - r\sin\phi \\
&\approx d\phi\, r\cos\phi + r\sin\phi - r\sin\phi = d\phi\, r\cos\phi \\
&= d\phi\, y = \phi' dx\, y
\end{aligned}
$$

or

$$
\begin{aligned}
v &= -\phi' x\, z \\
w &= \phi' x\, y
\end{aligned}
\tag{1.139}
$$

The term $\phi' = d\phi(x)/dx$ is the angle of twist per unit length. Here, it is assumed to be constant along the bar axis.

EXAMPLE 1.5 Torsion of a Shaft of Circular Cross-Section

In the case of a shaft of circular cross-section, where radial symmetry is involved and deformations of a cross section are the same when viewed from either end of the shaft segment, plane cross-sections normal to the x axis will in fact remain normal to the x axis and the radii will remain straight and unextended. The deformations (Eq. 1.139)

$$v = -\phi' xz, \quad w = \phi' xy, \quad u = 0 \tag{1}$$

lead to an exact solution for a twisted bar of circular cross-section.

From Eqs. (1.18), the strain components are

$$\begin{gathered} \epsilon_x = \frac{\partial u}{\partial x} = 0, \quad \epsilon_y = \frac{\partial v}{\partial y} = 0, \quad \epsilon_z = \frac{\partial w}{\partial z} = 0 \\ \gamma_{xy} = \frac{\partial v}{\partial x} + \frac{\partial u}{\partial y} = -\phi' z, \quad \gamma_{xz} = \frac{\partial w}{\partial x} + \frac{\partial u}{\partial z} = \phi' y, \quad \gamma_{yz} = \frac{\partial w}{\partial y} + \frac{\partial v}{\partial z} = 0 \end{gathered} \tag{2}$$

The conditions of compatibility are satisfied as these strains are obtained from an admissible displacement field. These strains reduce the constitutive relations of Eq. (1.34a) to

$$\sigma_x = \sigma_y = \sigma_z = \tau_{yz} = 0, \quad \tau_{xy} = G\gamma_{xy} = -G\phi' z, \quad \tau_{xz} = G\gamma_{xz} = G\phi' y, \tag{3}$$

The stress components of (3) satisfy the equations of equilibrium of Eq. (1.51), provided the body forces are equal to zero.

The surface conditions of Eq. (1.57) must be satisfied on the circumferential surfaces of the bar. The direction cosines of the unit normal to the circumferential surfaces are $(a_x = 0, a_y, a_z)$. For this torsion problem the surface tractions on this surface are zero, i.e., $p_x = p_y = p_z = 0$. Also, the stress components $\sigma_x, \sigma_y, \sigma_z, \tau_{yz}$ are zero. It follows that the second and third of Eqs. (1.57a) are satisfied and the first equation becomes

$$a_z \tau_{xz} + a_y \tau_{xy} = 0 \tag{4}$$

As seen in Fig. 1.17c

$$a_y = \cos\phi = \frac{y}{a}, \qquad a_z = \sin\phi = \frac{z}{a}, \tag{5}$$

Substitution of (3) and (5) into (4) shows that the surface conditions on the circumferential surfaces are satisfied.

On the ends of the bar, the resultant shear forces can be shown to be zero although the resultant moment is not zero. This moment is expressed as (Fig. 1.18)

$$M_t = \int\!\!\int (\tau_{xz}\, y - \tau_{xy}\, z)\, dy\, dz \tag{6}$$

Substitute (3) in this expression to find

$$M_t = G\phi' \int_A (y^2 + z^2)\, dA = G\phi' \int_A r^2\, dA \tag{7}$$

This integral is the polar moment of inertia J of the circular cross section. Thus

$$\frac{d\phi}{dx} = \frac{M_t}{GJ} \tag{8}$$

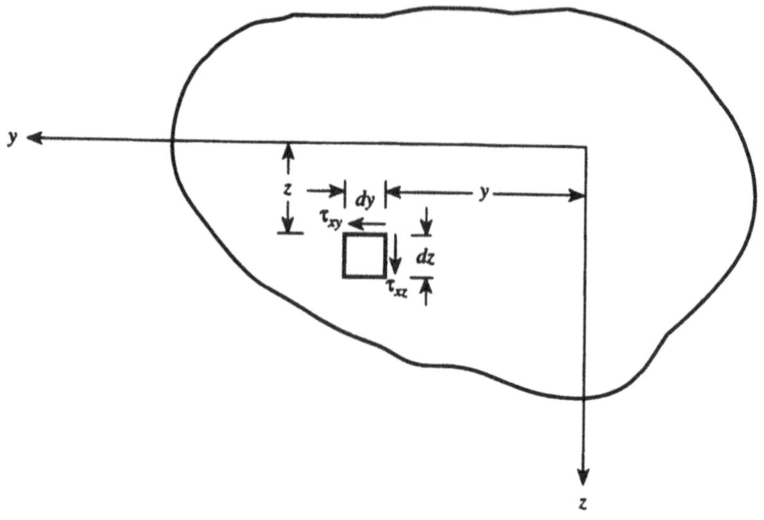

FIGURE 1.18
Shear stresses and notation used to find the resultant moment on a cross-section.

which can be considered as a relationship between the angle of twist and the applied torque. These relations are referred to as the Coulomb solution for the torsion of a bar of circular cross section. The shear stresses of (3) are solutions to the theory of elasticity problem in that the kinematic, constitutive, and equilibrium conditions are satisfied. The shear stresses τ_{xy} and τ_{xz} of (3) probably do not accurately reflect the actual shear stress on the ends of the bar where the torsional loads are applied. In reality, the shear stress components of (3) are usually accurate at a distance of several bar diameters from the ends. This concept that there is a redistribution of the stress values at a distance from the ends is referred to as *Saint-Venant's principle*[24].

The stress distribution represented by (3) is independent of the axial coordinate x, so that the stress distribution remains the same for any cross section along the shaft. From (3) and (8)

$$\tau_{xy} = -\frac{M_t z}{J} \qquad \tau_{xz} = \frac{M_t y}{J} \tag{9}$$

The resultant shear stress lies in the plane of the cross-section and is perpendicular to the radius r from the origin. The magnitude of this resultant stress τ is

$$\tau = \sqrt{\tau_{xy}^2 + \tau_{xz}^2} = \frac{M_t}{J}\sqrt{z^2 + y^2} = \frac{M_t r}{J} \tag{10}$$

which is a relationship between the torque M_t and the shear stress for cylindrical bars of hollow or solid circular cross-sections. ∎

In the case of bars of circular cross-sections, $u(y, z)$ was zero. For cross-sections of arbitrary shape, there is experimental evidence that the axial deformation of each cross-section along

[24]In 1855, Barre de Saint-Venant enunciated a useful principle that now bears his name. In essence, this principle can be stated as the redistribution of loading, resulting from a set of forces acting on a small region of the surface of an elastic body being replaced by a statically equivalent set of forces, causing significant changes in the stress distribution only in the neighborhood of the loading, while stresses remain essentially the same in those portions of the body located at large distances from the applied loading. By "large distances" are meant distances great in comparison with the dimensions of the surface on which the loading is applied. "Statically equivalent" sets of forces mean that the two distributions of loadings have the same resultant force and moment.

the bar is nearly the same. That is,

$$u = f(y, z) \tag{1.140}$$

This assumption is attributed to Saint Venant and hence this formulation leads to what is known as *Saint-Venant's theory of torsion*. It is convenient to introduce ϕ' in the expression for u. Thus, letting $f(y, z) = -\phi'\,\omega(y, z)$

$$u = -\phi'\omega(y, z) \tag{1.141}$$

where ω is referred to as the *warping* function and the minus sign has been inserted for later convenience.

In summary, the displacements for bars of arbitrary cross-sectional shape are

$$u = -\phi'\,\omega(y, z), \quad v = -\phi' xz, \quad w = \phi' xy$$

The strain-displacement relations of Eqs. (1.18) become

$$\begin{aligned}
\epsilon_x &= \frac{\partial u}{\partial x} = -\phi''\omega(y, z) = 0 \\
\epsilon_y &= \frac{\partial v}{\partial y} = 0 \\
\epsilon_z &= \frac{\partial w}{\partial z} = 0 \\
\gamma_{xy} &= \frac{\partial v}{\partial x} + \frac{\partial u}{\partial y} = -\phi'\left(z + \frac{\partial \omega}{\partial y}\right) \\
\gamma_{xz} &= \frac{\partial w}{\partial x} + \frac{\partial u}{\partial z} = -\phi'\left(-y + \frac{\partial \omega}{\partial z}\right) \\
\gamma_{yz} &= \frac{\partial w}{\partial y} + \frac{\partial v}{\partial z} = \phi' x - \phi' x = 0
\end{aligned} \tag{1.142}$$

These kinematic relations will be used to develop governing equations for a bar that has a constant cross-section with constant torque. However, once the relationship between the applied torque and induced stresses is established, this theory can be applied to bars with fewer restrictions, e.g., with nonconstant cross-sections and torques (Wempner, 1973; Pilkey, 2002).

1.9.2 Material Laws

The constitutive relations of Eqs. (1.34a) reduce for the strains of Eq. (1.142) to

$$\begin{aligned}
\sigma_x = \sigma_y = \sigma_z = \tau_{yz} = 0 \\
\tau_{xy} = G\gamma_{xy}, \qquad \tau_{xz} = G\gamma_{xz}
\end{aligned} \tag{1.143}$$

1.9.3 Equations of Equilibrium

Since $\sigma_x = \sigma_y = \sigma_z = \tau_{yz} = 0$, the equations of equilibrium [Eqs. (1.51)] are given by

$$\frac{\partial \tau_{xy}}{\partial y} + \frac{\partial \tau_{xz}}{\partial z} = 0 \tag{1.144a}$$

$$\frac{\partial \tau_{yx}}{\partial x} = \frac{\partial \tau_{zx}}{\partial x} = 0 \tag{1.144b}$$

where the applied loadings have been set equal to zero. Equations (1.144b) imply that the shear stresses τ_{yx} and τ_{zx} are independent of x.

1.9.4 Surface Forces and Boundary Conditions

On the circumferential surfaces of the bar, Eq. (1.57) must be satisfied. There are no prescribed surface tractions on this surface, so that $p_x = p_y = p_z = 0$, and $a_x = 0$. Since $\sigma_x = \sigma_y = \sigma_z = \tau_{yz} = 0$, the second and third Eqs. (1.57a) are identically satisfied and the first equation reduces to

$$a_z\, \tau_{xz} + a_y\, \tau_{xy} = 0 \tag{1.145}$$

where (Fig. 1.17d) $a_y = dz/ds$ and $a_z = -dy/ds$, so that

$$-\frac{dy}{ds}\tau_{xz} + \frac{dz}{ds}\tau_{xy} = 0$$

Next consider the conditions on the ends of the bar. The normals to the cross-sections on the ends are parallel to the x axis. Hence $a_x = \pm 1$, $a_y = a_z = 0$ and the boundary conditions as represented by Eqs. (1.57) are

$$p_y = \pm\tau_{xy}, \qquad p_z = \pm\tau_{xz} \tag{1.146}$$

Here the + sign applies to the end of the bar with an external normal in the direction of the positive x axis. Equation (1.146) indicates that the (shear) forces have the same distribution as the shear stresses on the ends. It can be shown that the resultant of these stresses at the end of a bar is a torque and the resultant forces vanish. With the notation of Fig. 1.18, the resultant forces on an end of area A are

$$\iint_A \tau_{xy}\, dy\, dz, \qquad \iint_A \tau_{xz}\, dy\, dz \tag{1.147}$$

in the y and z directions, respectively. Since these quantities can be shown to be zero (Problem 1.61), the resultant force acting on the end of the bar is zero. However, the resultant moment is not zero; from moment equilibrium requirements on a cross-section (Fig. 1.18)

$$M_t = \iint_A (\tau_{xz} y - \tau_{xy} z)\, dy\, dz \tag{1.148}$$

Suppose the twisting moment M_t and the angle of twist per unit length ϕ' are related by Eq. (8) of Example 1.4, which was shown to apply if the cross-section of the bar is circular. That is

$$M_t = GJ\phi' \tag{1.149}$$

It follows from Eqs. (1.148) and (1.149) that J can be defined as

$$J = \frac{M_t}{G\phi'} = \frac{1}{G\phi'} \iint_A (\tau_{xz} y - \tau_{xy} z)\, dy\, dz \tag{1.150}$$

The quantity J is known as the *torsional constant* and GJ is the *torsional stiffness* of the bar. For bars of circular cross section, J of Eq. (1.150) reduces to the polar moment of inertia.

1.9.5 Displacement Form of the Governing Differential Equations

As outlined in Section 1.6.1, the displacement form of the governing equations is obtained by first forming stress-displacement relations by combining Eqs. (1.142) and (1.143), then placing the result

$$\tau_{xy} = -G\phi'\left(z + \frac{\partial\omega}{\partial y}\right), \qquad \tau_{xz} = -G\phi'\left(-y + \frac{\partial\omega}{\partial z}\right)$$

in the differential equation of equilibrium [Eq. (1.144a)] to obtain

$$\frac{\partial^2 \omega}{\partial y^2} + \frac{\partial^2 \omega}{\partial z^2} = 0 \quad \text{or} \quad \nabla^2 \omega = 0 \tag{1.151}$$

with $\nabla^2 = \partial_y^2 + \partial_z^2$. This is often written in terms of the displacement u [Eq. (1.141)]

$$\frac{\partial^2 u}{\partial y^2} + \frac{\partial^2 u}{\partial z^2} = 0 \quad \text{or} \quad \nabla^2 u = 0 \tag{1.152}$$

This type of differential relationship is called *Laplace's equation*. A solution to Laplace's equation is called a *harmonic function*. Thus, the warping function is a harmonic function. If the expression for stresses obtained from Eqs. (1.142) and (1.143) are substituted in Eq. (1.145), it is apparent that the boundary condition [Eq. (1.145)] in terms of the warping function ω can be written as

$$\frac{\partial \omega}{\partial y} a_y + \frac{\partial \omega}{\partial z} a_z + z a_y - y a_z = 0 \tag{1.153}$$

The expression for the torsional constant J in terms of the warping function ω is [Eq. (1.150)]

$$\begin{aligned} J = \frac{M_t}{G\phi'} &= \frac{1}{G\phi'} \iint_A (\tau_{xz} y - \tau_{xy} z)\, dy\, dz \\ &= \iint_A \left[y\left(y - \frac{\partial \omega}{\partial z} \right) + z\left(z + \frac{\partial \omega}{\partial y} \right) \right] dy\, dz \\ &= \iint_A \left(z\frac{\partial \omega}{\partial y} - y\frac{\partial \omega}{\partial z} + y^2 + z^2 \right) dy\, dz \end{aligned} \tag{1.154}$$

where Eqs. (1.142) and (1.143) have been introduced.

EXAMPLE 1.6 Constant Warping Function

Suppose the warping function ω is constant, that is $\omega(y, z) = c$ where c is a constant.

In this case, $\nabla^2 \omega = 0$ is satisfied by $\omega = c$. From Eqs. (1.142) and (1.143),

$$\tau_{xy} = -G\phi'\left(z + \frac{\partial \omega}{\partial y} \right), \qquad \tau_{xz} = -G\phi'\left(-y + \frac{\partial \omega}{\partial z} \right) \tag{1}$$

Substitute $\omega = c$ into (1) and use the surface condition of Eq. (1.145) to find

$$-\frac{dy}{ds} y - \frac{dz}{ds} z = 0 \tag{2}$$

Then

$$\frac{d}{ds}\left(\frac{y^2 + z^2}{2} \right) = 0 \tag{3}$$

or

$$y^2 + z^2 = r^2 \tag{4}$$

is constant on the boundary. Since (4) is the equation of a circle of radius r, it follows that $\omega = c$ represents the warping function for the torsion of a bar of circular cross-section.

If the displacement $u = 0$ at the bar end ($x = 0$), then $c = 0$ and the solution corresponds to the Coulomb solution of Example 1.6 for the torsion of a shaft of circular cross-section.

A uniqueness study of ω (e.g., see Little (1973)) shows that the warping function may be determined only up to a constant. The shear stresses obtained from ω involve derivatives of ω and, hence, are found uniquely. However, because of the constant, care must be taken in the direct use of ω. ∎

1.9.6 Force Form of the Governing Differential Equations

To obtain a force form of the governing equations, a stress function is introduced such that the conditions of equilibrium are satisfied. Then, the compatibility equations are expressed in terms of the stress function. For the torsion problem, Prandtl[25] introduced a stress function $\psi(y, z)$ defined as

$$\tau_{xy} = \frac{\partial \psi}{\partial z}, \qquad \tau_{xz} = -\frac{\partial \psi}{\partial y} \tag{1.155}$$

This is usually called the *Prandtl stress function.* Note that the condition of equilibrium of Eq. (1.144a) is identically satisfied. The solution to the torsion problem requires the calculation of the stress function ψ.

Based on the strains of Eq. (1.142), the meaningful compatibility equations of Eq. (1.28) are

$$-\frac{1}{2}\partial_y^2 \gamma_{xz} + \frac{1}{2}\partial_y \partial_z \gamma_{xy} = 0, \qquad \frac{1}{2}\partial_z \partial_y \gamma_{zx} - \frac{1}{2}\partial_z^2 \gamma_{xy} = 0 \tag{1.156}$$

Since

$$\gamma_{xy} = \partial_z \psi / G, \qquad \gamma_{xz} = -\partial_y \psi / G$$

Eq. (1.156) can be written

$$\frac{1}{2}\partial_y (\partial_y^2 + \partial_z^2)\psi = 0, \qquad -\frac{1}{2}\partial_z (\partial_y^2 + \partial_z^2)\psi = 0 \tag{1.157}$$

In order for both of the equations in Eq. (1.157) to hold, $(\partial_y^2 + \partial_z^2)\psi$ must be a constant. To find this constant, introduce stress-displacement relations obtained by substituting Eq. (1.142) into Eq. (1.143). Then differentiate τ_{xy} with respect to z and τ_{xz} with respect to y. Thus

$$\begin{aligned} \frac{\partial \tau_{xy}}{\partial z} &= \frac{\partial}{\partial z}\left(\frac{\partial \psi}{\partial z}\right) = \partial_z^2 \psi = -G\phi'\left(1 + \frac{\partial^2 \omega}{\partial z\, \partial y}\right) \\ \frac{\partial \tau_{xz}}{\partial y} &= \frac{\partial}{\partial y}\left(-\frac{\partial \psi}{\partial y}\right) = -\partial_y^2 \psi = -G\phi'\left(-1 + \frac{\partial^2 \omega}{\partial y\, \partial z}\right) \end{aligned} \tag{1.158}$$

so that

$$\frac{\partial^2 \psi}{\partial z^2} + \frac{\partial^2 \psi}{\partial y^2} = -2G\phi' \quad \text{or} \quad \nabla^2 \psi = -2G\phi' \tag{1.159}$$

[25] Ludwig Prandtl (1875–1953) was a German engineer who is best known for his pioneering work in aerodynamics. In 1900, he received his doctorate under August Föppl in Munich. He established engineering mechanics at the University of Göttingen, where he guided many students who have since played leading roles in mechanics. He initiated the development of several areas of mechanics with fundamental studies. He is the founder of the boundary layer theory of fluid mechanics. He cooperated with Theodore von Karman in significant research on airfoil theory, drag, and turbulent flows. One of his first papers points out that by using a soap film, information on the distribution of torsional stresses on a cross-section can be obtained. This is the *membrane analogy*.

This same equation can be obtained more directly using the requirement for the warping to be single-valued and continuous

$$\frac{\partial^2 \omega}{\partial z\, \partial y} = \frac{\partial^2 \omega}{\partial y\, \partial z} \tag{1.160}$$

Insertion of Eq. (1.158) into Eq. (1.160) leads to Eq. (1.159). A partial differential equation of the form of Eq. (1.159) is called *Poisson's equation.*

Upon introduction of Eq. (1.155), the boundary condition of Eq. (1.145) becomes

$$\frac{\partial \psi}{\partial y}\frac{dy}{ds} + \frac{\partial \psi}{\partial z}\frac{dz}{ds} = 0 \tag{1.161}$$

By the chain rule of differentiation, $d\psi/ds = 0$ or ψ is constant along the boundary of the cross-section. Since the stresses are defined in terms of derivatives of ψ rather than ψ itself, the magnitude of the constant ψ is arbitrary. Therefore, without loss of generality, it is common to assume that $\psi = 0$ along the boundary.

The resultant moment condition of Eq. (1.148) now appears as

$$M_t = \iint_A \left(-\frac{\partial \psi}{\partial y} y - \frac{\partial \psi}{\partial z} z \right) dy\, dz \tag{1.162}$$

where Eq. (1.155) has been employed. Integration by parts and invoking $\psi = 0$ on the boundaries yields

$$M_t = 2 \iint_A \psi\, dy\, dz \tag{1.163}$$

and hence the torsional constant is

$$J = \frac{M_t}{G\phi'} = \frac{2}{G\phi'} \iint_A \psi\, dy\, dz \tag{1.164}$$

EXAMPLE 1.7 Bar of Elliptical Cross-Section

The equation for the ellipse of Fig. 1.19 is

$$\frac{y^2}{a^2} + \frac{z^2}{b^2} = 1 \tag{1}$$

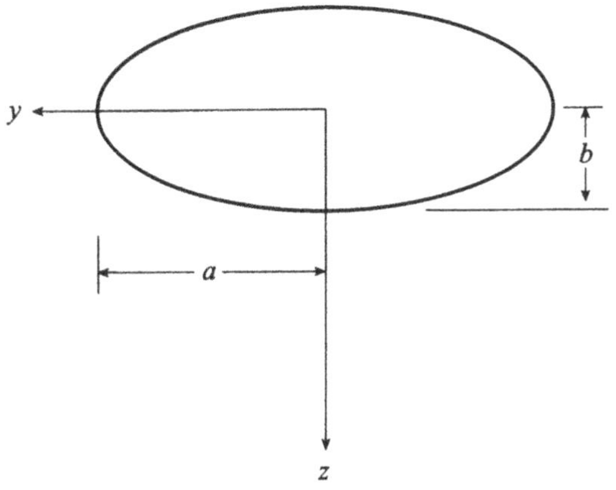

FIGURE 1.19
Elliptical cross-section of bar.

where a and b are the semi-axes. The stress function ψ must be constant along the boundary of the cross-section. We choose to let $\psi = 0$ on the boundary. A stress function that would satisfy this condition would be

$$\psi = C\left(\frac{y^2}{a^2} + \frac{z^2}{b^2} - 1\right) \tag{2}$$

where C is a constant to be determined. Substitution of this stress function into Poisson's equation of Eq. (1.159) gives

$$C = -\frac{a^2 b^2}{a^2 + b^2} G\phi' \tag{3}$$

From (2) and (3), the stress function is now

$$\psi = -\frac{a^2 b^2}{a^2 + b^2} G\phi' \left(\frac{y^2}{a^2} + \frac{z^2}{b^2} - 1\right) \tag{4}$$

The stress components for this elliptical cross-section are given by Eq. (1.155) with (4)

$$\tau_{xy} = \frac{\partial \psi}{\partial z} = -\frac{2a^2 z}{a^2 + b^2} G\phi' \qquad \tau_{xz} = -\frac{\partial \psi}{\partial y} = \frac{2b^2 y}{a^2 + b^2} G\phi' \tag{5}$$

The maximum shear stress occurs at the boundary closest to the centroid, i.e., at $z = b$.

$$\tau_{\max} = \tau_{xy}|_{z=b} = -\frac{2a^2 b}{a^2 + b^2} G\phi' \tag{6}$$

From Eq. (1.163) the torque becomes

$$\begin{aligned} M_t = 2\iint_A \psi\, dy\, dz &= -\frac{a^2 b^2}{a^2 + b^2} G\phi' \left[\frac{1}{a^2}\iint_A y^2\, dy\, dz + \frac{1}{b^2}\iint_A z^2\, dy\, dz \right. \\ &\quad \left. - \iint_A dy\, dz\right] \\ &= -\frac{a^2 b^2}{a^2 + b^2} G\phi' \left[\frac{I_{yy}}{a^2} + \frac{I_{zz}}{b^2} - A\right] \end{aligned} \tag{7}$$

where

$$I_{yy} = \iint_A y^2\, dy\, dz, \qquad I_{zz} = \iint_A z^2\, dy\, dz, \tag{8}$$

and $A = \iint_A dy\, dz$. For an ellipse, $I_{yy} = \pi a^3 b/4$, $I_{zz} = \pi a b^3/4$ and $A = \pi a b$. If these values are placed in (7), we find

$$M_t = \frac{\pi a^3 b^3}{(a^2 + b^2)} G\phi' \tag{9}$$

Then the stresses of (5) become

$$\tau_{xy} = -\frac{2}{ab^3\pi} M_t z \qquad \tau_{xz} = \frac{2}{a^3 b\pi} M_t y \tag{10}$$

From $\phi' = M_t/GJ$ and (9),

$$J = \frac{\pi a^3 b^3}{a^2 + b^2} \tag{11}$$

These results are the same obtained using the displacement formulation with a warping function (Problem 1.63). ∎

The Membrane Analogy

It was noted by Prandtl that the governing equations for the Saint-Venant torsion problem closely resemble the equilibrium equations for a flat membrane lying in the yz plane subjected to a lateral pressure. If N is the uniform tension per unit length in the membrane, p is the lateral pressure, and u is the lateral displacement of the membrane, the differential equation of equilibrium is

$$\frac{\partial^2 u}{\partial y^2} + \frac{\partial^2 u}{\partial z^2} = -\frac{p}{N} \tag{1.165}$$

Comparison of this with Eq. (1.159) shows that the relations are the same if u replaces ψ and p/N replaces $2G\phi'$. Since the volume between a deformed membrane and the yz plane is $\int\int_A u\, dy\, dz$, it follows from Eq. (1.163) that the twisting moment is equal to twice the volume of the membrane. It can be shown that the contour lines showing constant u correspond to lines of constrant stress function and that the slope of the membrane is equal to the value of the shear stress. The membrane analogy opens the door to a wide range of experimental studies of torsional problems. The membrane is stretched over a hole with a boundary the same as the boundary of a cross-section of the bar subject to torsion.

1.9.7 Mixed Form of the Governing Differential Equations

The equations as originally derived in Sections 1.9.1, 1.9.2, and 1.9.3 are almost in mixed form already. Normally, a mixed form is established by first writing the constitutive relations in terms of displacements rather than strains. Thus, if the equations of Eq. (1.142) are substituted into the relations of Eq. (1.143),

$$\tau_{xy} = -G\phi'\left(z + \frac{\partial \omega}{\partial y}\right), \qquad \tau_{xz} = -G\phi'\left(-y + \frac{\partial \omega}{\partial z}\right) \tag{1.166}$$

The combination of this and the differential equilibrium conditions forms the desired mixed relations. To gather these relations together, as was done in Eq. (1.75), consider the state vector of displacement and stress variables

$$\begin{bmatrix} \omega \\ \tau_{xz} \\ \tau_{xy} \end{bmatrix} \tag{1.167}$$

Then Eqs. (1.166) and (1.144a) take the form

$$\begin{bmatrix} 0 & \partial_z & \partial_y \\ \partial_z & 1 & 0 \\ \partial_y & 0 & 1 \end{bmatrix} \begin{bmatrix} \omega G\phi' \\ \tau_{xz} \\ \tau_{xy} \end{bmatrix} + G\phi' \begin{bmatrix} 0 \\ -y \\ z \end{bmatrix} = \begin{bmatrix} 0 \\ 0 \\ 0 \end{bmatrix} \tag{1.168}$$

These exhibit the usual characteristics of mixed formulations of having unknowns with no higher than first order derivatives and of not involving derivatives of the material parameters.

Equation (1.168) along with the boundary conditions of Eq. (1.145) and the resultant moment condition of Eq. (1.148) form a complete set of equations for the mixed formulation of the Saint-Venant torsion problem. Equations (1.168) provide the stresses τ_{xz}, τ_{xy}, and the warping function as a function of ϕ'. By setting M_t of Eq. (1.148) equal to the applied torque $\overline{M}_t$, ϕ' can be computed.

References

Leipholz, H., 1968, *Einfuehrung in die Elastizitaetstheorie*, G. Braun-Verlag, Karlsruhe, Germany.
Little, R.W., 1973, *Elasticity*, Prentice Hall, New Jersey.
Neuber, H., 1985, *Kerbspannungslehre*, 3rd Ed., Springer-Verlag, Berlin, Germany.
Pilkey, W.D., 2002, *Analysis and Design of Elastic Beams Computational Methods*, Wiley, NY.
Pilkey, W.D. and Pilkey, O.H., 1986, *The Mechanics of Solids*, Krieger Publishers, Melbourne, FL.
Sokolnikoff, I.S., 1956, *Mathematical Theory of Elasticity*, McGraw Hill, NY.
Timoshenko, S.P. and Goodier, J.N., 1970, *Theory of Elasticity*, 3rd Ed., McGraw Hill, NY.
Wempner, G.H., 1973, *Mechanics of Solids*, McGraw Hill, NY.
Wunderlich, W., 1977, Incremental Formulation for Geometrically Nonlinear Problems, in *Formulations and Computational Algorithms in Finite Element Analysis*, Bathe, K.J., Oden, J.T., Wunderlich, W. (Eds.), MIT Press, Cambridge, MA.

Problems

Theory of Elasticity

1.1 For the two-dimensional element described in polar coordinates, show that the strain-displacement relations are

$$\epsilon_r = \frac{\partial u}{\partial r} \qquad \epsilon_\phi = \frac{u}{r} + \frac{1}{r}\frac{\partial v}{\partial \phi} \qquad \gamma_{r\phi} = \frac{1}{r}\frac{\partial u}{\partial \phi} + \frac{\partial v}{\partial r} - \frac{v}{r}$$

where the polar coordinate displacements are u and v.

1.2 Show for the strains and displacements of Problem 1.1 that a compatibility requirement would be

$$\frac{\partial^2 \epsilon_\phi}{\partial r^2} + \frac{1}{r^2}\frac{\partial^2 \epsilon_r}{\partial \phi^2} + \frac{2}{r}\frac{\partial \epsilon_\phi}{\partial r} - \frac{1}{r}\frac{\partial \epsilon_r}{\partial r} = \frac{1}{r}\frac{\partial^2 \gamma_{r\phi}}{\partial r \partial \phi} + \frac{1}{r^2}\frac{\partial \gamma_{r\phi}}{\partial \phi}$$

1.3 Suppose the displacements

$$\begin{aligned} u &= A_1x^2 + B_1y^2 + C_1z^2 \quad A_i, B_i, C_i, i = 1, 2, 3 \text{ are constants} \\ v &= A_2x^2 + B_2y^2 + C_2z^2 \\ w &= A_3x^2 + B_3y^2 + C_3z^2 \end{aligned}$$

occur in an elastic solid.

(a) Find the strains ϵ_{ij}.

(b) Do these strains constitute a compatible state of strain?

Answer:

(a)

$$\begin{aligned} \epsilon_x &= \frac{\partial u}{\partial x} = 2A_1x \qquad & \gamma_{xy} &= 2\epsilon_{xy} = \frac{1}{2}(\partial_x v + \partial_y u) = 2A_2x + 2B_1y \\ \epsilon_y &= 2B_2y & \gamma_{xz} &= 2A_3x + 2C_1z \\ \epsilon_z &= 2C_3z & \gamma_{yz} &= 2B_3y + 2C_2z \end{aligned}$$

(b) Yes

1.4 Do the following strains constitute a possible (compatible) state of strain?

$$\epsilon_x = A\,y^2 + Bxy \qquad \epsilon_y = A(x^2 + y^2) + By$$
$$\gamma_{xy} = Bxy \qquad \epsilon_z = \gamma_{xz} = \gamma_{yz} = 0$$

Answer: Yes if $A = B/4$.

1.5 Measured strains for a deformed solid are

$$\epsilon_x = A(x^2 + z^2) \quad \epsilon_y = 0 \qquad \epsilon_z = Az^2$$
$$\gamma_{yz} = 0 \qquad \gamma_{zx} = 2Axz \quad \gamma_{xy} = 0$$

Find the corresponding displacements.

Hint: Integrate $\epsilon_x = \partial_x u$ and $\epsilon_z = \partial_z w$. Then $\gamma_{zx} = 2Axz$ leads to $\frac{\partial f_1(y,z)}{\partial z} = \frac{-\partial f_2(x,y)}{\partial x} = g(y)$, where $f_1(y, z)$ and $f_2(x, y)$ are functions of integration. Apply $\gamma_{yz} = 0$ and $\gamma_{xy} = 0$ to show that $g(y) = C_1$. It follows that $f_1 = C_1 z + C_2$, $f_2 = -C_1 x + C_3$, where C_1, C_2, and C_3 are constants.

Answer: $u = A(\frac{x^3}{3} + xz^2) + C_1 z + C_2$, $v = C_3$, $w = \frac{Az^3}{3} - C_1 x + C_3$.

1.6 The components of a displacement field are given by $u = (x^2 + 10)10^{-4}$, $v = (yz)10^{-3}$, $w = (z^2 - 2xy)10^{-3}$. Compute the distance after deformation between the two points (3, 5, 7) and (2, 4, 6) in the undeformed body. Also, find expressions for the strains ϵ_{ij} and calculate the strains at (3, −2, 5). Do the strains satisfy compatibility?

Answer: $ds' = 1.738$, $\epsilon_x = 6 \cdot 10^{-4}$, The strains do not satisfy compatibility.

1.7 The strain components for the unrestrained thermal expansion of a solid are

$$\epsilon_x^0 = \epsilon_y^0 = \epsilon_z^0 = \alpha\,\Delta T, \qquad \gamma_{xy}^0 = \gamma_{yz}^0 = \gamma_{xz}^0 = 0$$

Does this state of strain lead to unique, continuous displacements regardless of how ΔT varies with x, y, and z? What restrictions are placed on ΔT as a function of x, y, and z?

Answer: ΔT can be at most a linear function of x, y, and z.

1.8 Derive constitutive relations for a planar (x, y) solid in plane stress and in plane strain if the material is orthotropic.

1.9 Show that in general the plane stress assumptions for an isotropic solid of elastic material can lead to a violation of some compatibility conditions.

1.10 Show that the equilibrium equations in polar coordinates for a two-dimensional element are

$$\frac{\partial \sigma_r}{\partial r} + \frac{1}{r}\frac{\partial \tau_{r\phi}}{\partial \phi} + \frac{\sigma_r - \sigma_\phi}{r} + \overline{p}_{Vr} = 0 \qquad \frac{1}{r}\frac{\partial \sigma_\phi}{\partial \phi} + \frac{\partial \tau_{r\phi}}{\partial r} + 2\frac{\tau_{r\phi}}{r} + \overline{p}_{V\phi} = 0$$

Figure P1.10 shows notation for this problem, but does not provide a complete free-body diagram for the application of the conditions of equilibrium.

1.11 The equilibrium equations contain more unknowns than equations so that alone they cannot be solved for the stresses. However, it is possible to test whether a stress field satisfies the equilibrium equations. Suppose the body forces are zero. Are the equations of equilibrium satisfied by the following stress field?

$$\sigma_x = x^2 + y^2 + z^2 \qquad \tau_{xy} = \tau_{yx} = -xy + z^3$$
$$\sigma_y = x^2 + y^2 + 2z^2 \qquad \tau_{xz} = \tau_{zx} = y^2 - xz$$
$$\sigma_z = -2x^2 + y^2 + z^2 \qquad \tau_{yz} = \tau_{zy} = x^2 - yz$$

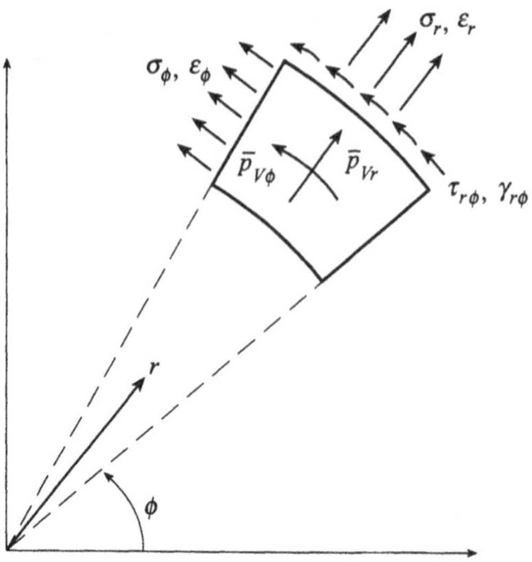

FIGURE P1.10

Hint: The three equations of equilibrium lead to

$$2x - x - x = 0 \quad -y + 2y - y = 0 \quad -z - z + 2z = 0$$

Answer: yes

1.12 Is $\sigma_x = C_1 + C_2 y$, $\sigma_y = C_3 + C_4 x$, and $\tau_{xy} = C_5$ a possible two-dimensional stress field from the standpoint of equilibrium? Is $\sigma_x = C_1 + C_2 x$, $\sigma_y = C_3 + C_4 y$, and $\tau_{xy} = C_5$ a possible stress field?

Hint: Use the homogeneous form of the equilibrium equations.

Answer: Yes. No unless $C_2 = C_4 = 0$.

1.13 Is the solid subject to the stresses

$$\begin{array}{lll} \sigma_x = xy^2 + z & \sigma_y = \sin x + 3z & \sigma_z = 10z \\ \tau_{xy} = 3xz + 4y & \tau_{yz} = -3z^2/2 & \tau_{xz} = -z(y^2 + 4) \end{array}$$

in equilibrium? If not, what movement will the body experience?

Answer: No. The solid will accelerate in the z direction.

1.14 Given the stress field

$$\sigma_x = 10x^3 + \frac{1}{2}y^2 \qquad \sigma_y = 15x^3 + 50 \qquad \sigma_z = 15y^2 + 15z^3$$

$$\tau_{xy} = 50 + 40y^2 \qquad \tau_{yz} = 0 \qquad \tau_{xz} = \frac{1}{2}xz^3 + 15x^2 y$$

Find the body forces needed for equilibrium.

Answer:

$$\overline{p}_{Vx} = -(\partial_x \sigma_x + \partial_y \tau_{xy} + \partial_z \tau_{xz}) = -\left(30x^2 + 80y + \frac{3}{2}xz^2\right)$$

$$\overline{p}_{Vy} = 0, \qquad \overline{p}_{Vz} = -\left(\frac{1}{2}z^3 + 30xy + 45z^2\right)$$

1.15 Show that Hooke's law for isotropic materials can be written as (Eq. 1.35)

$$\sigma_{ij} = \lambda\,\delta_{ij}\,\epsilon_{kk} + 2\mu\,\epsilon_{ij}$$

Show that the compatibility relations can be expressed in terms of stresses as (Eq. 1.69)

$$\Delta^2\sigma_{ij} + \frac{1}{1+\nu}\sigma_{kk,ij} = \frac{\nu}{1+\nu}\delta_{ij}\overline{p}_{Vk,k} - (\overline{p}_{Vi,j} + \overline{p}_{Vj,i})$$

1.16 Frequently in the literature the material relationship for an isotropic material is written as

$$\epsilon_{ij} = \frac{1}{2\mu}\sigma_{ij} - \frac{\lambda}{2\mu(2\mu+3\lambda)}\delta_{ij}\,\sigma_{kk}$$

Is this expression correct?

1.17 Show that, for linearly elastic isotropic materials, the following properties of the material constants apply $E > 0, G > 0, \lambda > 0, 0 < \nu < 1/2$.

Hint: The conditions $E > 0$ and $\nu > 0$ follow from Hooke's law for uniaxial tension in the absence of body forces $\epsilon_x = \sigma_0/E,\ \epsilon_y = \epsilon_z = -\nu\sigma_0/E$ and the normal stress in the x direction given a constant positive value σ_0. Under this loading condition, the solid must undergo extension (positive) in the axial direction x and contraction (negative) in the transverse direction y and z. In order for this to be true, $E > 0$ and $\nu > 0$.

The condition $0 < \nu < 1/2$ can be established by imposing hydrostatic compression so that $\sigma_x = \sigma_y = \sigma_z = -p,\ p > 0$, while all shear stresses are zero. Then, the change in volume per unit volume (the dilitation) $e = \epsilon_x + \epsilon_y + \epsilon_z$ will be $e = -3p/(3\lambda + 2G) = -3p(1-2\nu)/E$. The condition $0 < \nu < 1/2$ follows because the change in hydrostatic compression is negative. Also, it should be recognized that a negative value implies the unlikely situation where an elongational strain in a tension bar corresponds to an expansion in a perpendicular direction.

1.18 Show that a material with Poisson's ratio equal to one half is incompressible.

1.19 Show that for an isotropic material the stress-strain relationship $\sigma_{ij} = E_{ijk\ell}\epsilon_{k\ell}$ is defined by

$$E_{ijk\ell} = G(\delta_{ik}\,\delta_{j\ell} + \delta_{i\ell}\,\delta_{jk}) + \left(K - \frac{2}{3}G\right)\delta_{ij}\,\delta_{k\ell}, \qquad K = \frac{E}{3(1-2\upsilon)}$$

or

$$E_{ijk\ell} = \lambda\delta_{ij}\,\delta_{\ell m} + \mu\delta_{i\ell}\,\delta_{jm} + \mu\delta_{im}\,\delta_{j\ell}$$

1.20 Show that in the absence of body forces, the equilibrium equations of Problem 1.10 are identically satisfied by an Airy stress function $\psi(r,\phi)$ defined as

$$\sigma_r = \frac{1}{r}\frac{\partial\psi}{\partial r} + \frac{1}{r^2}\frac{\partial^2\psi}{\partial\phi^2} \qquad \sigma_\phi = \frac{\partial^2\psi}{\partial r^2} \qquad \tau_{r\phi} = \frac{1}{r^2}\frac{\partial\psi}{\partial\phi} - \frac{1}{r}\frac{\partial^2\psi}{\partial r\,\partial\phi}$$

1.21 Verify that the Airy stress functions of Eq. (1.55) satisfy the equations of equilibrium.

1.22 Show that $\psi = Ax^2 + Ay^2 - 2x^3 + 6xy^2 + B$ is a permissible Airy stress function for a two-dimensional elastic solid with no body forces.

1.23 Suppose the triangular element of Fig. P1.23 is in a state of plane stress with no body forces. If the Airy stress function is given by $\psi(x,y) = C(x^4/12 + x^2y^2/4 - y^4)$, where C is a constant, find the stress distributions in the element. Also find the normal stress distribution along the edges.

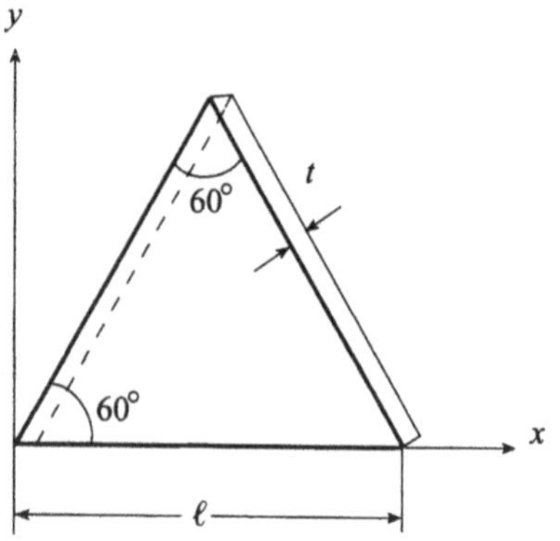

FIGURE P1.23
Triangular element with in-plane stress.

1.24 Develop compatibility differential equations corresponding to three-dimensional stress functions defined as

$$\sigma_x = \frac{\partial^2 \psi_1}{\partial z\, \partial y}$$

$$\sigma_y = \frac{\partial^2 \psi_2}{\partial x\, \partial z}$$

$$\sigma_z = \frac{\partial^2 \psi_3}{\partial x\, \partial y}$$

$$\tau_{xy} = -\frac{1}{2}\frac{\partial}{\partial z}\left(\frac{\partial \psi_1}{\partial x} + \frac{\partial \psi_2}{\partial y} - \frac{\partial \psi_3}{\partial z}\right)$$

$$\tau_{yz} = -\frac{1}{2}\frac{\partial}{\partial x}\left(-\frac{\partial \psi_1}{\partial x} + \frac{\partial \psi_2}{\partial y} + \frac{\partial \psi_3}{\partial z}\right)$$

$$\tau_{zx} = -\frac{1}{2}\frac{\partial}{\partial y}\left(\frac{\partial \psi_1}{\partial x} - \frac{\partial \psi_2}{\partial y} + \frac{\partial \psi_3}{\partial z}\right)$$

1.25 Use index notation to verify that Eq. (1.63) represents the displacement formulation equations for a general, isotropic elastic solid.

1.26 Show that the governing differential equations for the displacement formulation of the plane stress problem excluding body forces are

$$\frac{E}{(1-\nu^2)}\left(\frac{\partial^2 u}{\partial x^2} + \frac{1-\nu}{2}\frac{\partial^2 u}{\partial y^2}\right) + \frac{E}{2(1-\nu)}\frac{\partial^2 v}{\partial x\, \partial y} = 0$$

$$\frac{E}{(1-\nu^2)}\left(\frac{1-\nu}{2}\frac{\partial^2 v}{\partial x^2} + \frac{\partial^2 v}{\partial y^2}\right) + \frac{E}{2(1-\nu)}\frac{\partial^2 u}{\partial x\, \partial y} = 0$$

1.27 Derive the governing differential equations for the displacement formulation of the plane strain problem.

1.28 Derive the governing differential equations for the mixed formulation of the plane stress problem.

1.29 Develop as many forms as you can of the governing differential equations for the two-dimensional polar coordinate element of Fig. P1.10 that is considered in Problems 1.1, 1.2, 1.10, and 1.20. The constitutive relations will be the same as in the rectangular coordinate case.

1.30 A solid bar with a rectangular cross-section is subjected to a uniform axial tension of σ_x. No other stresses are present. Find the strain in the axial and lateral directions

Answer:

$$\epsilon_x = [\sigma_x - \nu(\sigma_y + \sigma_z)]/E = \sigma_x/E$$
$$\epsilon_y = -\nu\sigma_x/E, \epsilon_z = -\nu\sigma_x/E$$

1.31 Suppose a flat plate lies in the xy plane. The applied forces shown in Fig. P1.31 are uniformly distributed along the edges. Also, there is a temperature increase of 100°K. If this is an aluminum plate with $E = 70\,\text{GPa}$, $\nu = 0.33$, $\alpha = 23 \times 10^6\,\text{K}^{-1}$, calculate the strains and the changes in dimensions if

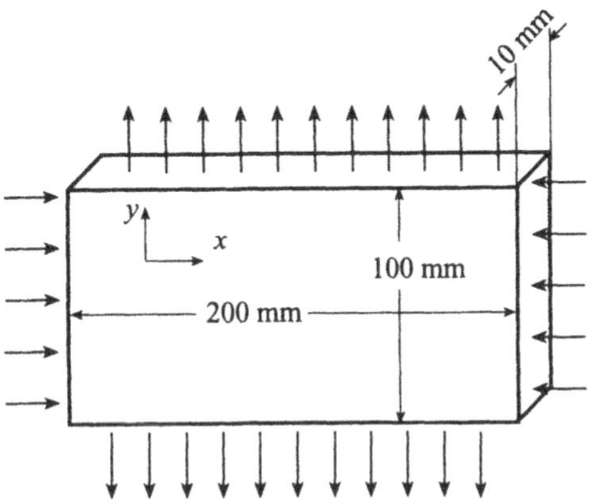

FIGURE P1.31

(a) the plate is unrestrained in the z direction, and
(b) the plate is placed firmly between lubricated dies which prevent expansion in the z direction. Compare the results of (a) and (b).

Hint: Use plane stress for (a) and plane strain for (b). Also, $\tau_{xy} = 0$.

Answer: For case (a), $\epsilon_x = [\sigma_x - \nu(\sigma_y + \sigma_z)]/E + \alpha\ \Delta T = 1.92 \times 10^{-3}$, $\epsilon_y = 3.06 \times 10^{-3}$, $\epsilon_z = 2.11 \times 10^{-3}$, $\gamma_{xy} = \gamma_{yz} = \gamma_{xz} = 0$, $\Delta x = 200\epsilon_x$, $\Delta y = 100\epsilon_y$, $\Delta z = 10\epsilon_z$.

1.32 For an isotropic elastic body with a temperature change ΔT, the stress-strain relationship takes the form [from Eq. (1.35a)]

$$\sigma_{ij} = \lambda\delta_{ij}\epsilon_{kk} + 2\mu\epsilon_{ij} - (3\lambda + 2\mu)\Delta T\delta_{ij}$$

Show that these relations for plane stress ($\sigma_{zz} = \tau_{xz} = \tau_{yz} = 0$) reduce to

$$\sigma_{xx} = \frac{E}{(1-\nu^2)}(\epsilon_{xx} + \nu\epsilon_{yy}) - \frac{E}{1+\nu}\alpha\ \Delta T \quad \tau_{xy} = 2\mu\epsilon_{xy}$$
$$\sigma_{yy} = \frac{E}{(1-\nu^2)}(\nu\epsilon_{xx} + \epsilon_{yy}) - \frac{E}{1+\nu}\alpha\ \Delta T$$

1.33 Show that for an isotropic material there are only two independent material constants.

Hint: Show that $G = E/(2(1+\nu))$. See a basic mechanics of solids text.

$$\sigma_{ij} = \lambda\delta_{ik}\epsilon_{kk} + 2\mu\epsilon_{ij}, \quad \mu = G \tag{1}$$

For a tensile test in the x direction $\sigma_y = \sigma_z = 0$ and (1) gives

$$(\lambda + 2\mu)\epsilon_y + \lambda\epsilon_z = -\lambda\epsilon_x, \qquad \lambda\epsilon_y + (\lambda + 2\mu)\epsilon_z = -\lambda\epsilon_x$$

or

$$\epsilon_y = \epsilon_z = -\frac{\lambda}{2(\lambda + \mu)}\epsilon_x$$

Since $\epsilon_y = \epsilon_z = -\nu\epsilon_x$, $\nu = (\lambda + \mu)\lambda/2$, or $\lambda = 2\nu\mu/(1 - 2\nu)$
From (1) and the fact that $\sigma_x = E\epsilon_x$

$$E = \frac{2\nu\mu}{1 - \nu}(1 - 2\nu) + 2\mu \quad \text{or} \quad G = \mu = \frac{E}{2(1 + \nu)}$$

1.34 As indicated in Eq. (1.37), the two constants G and K can be utilized as the two independent constants for an isotropic material. Find K as a function of E and ν.

Answer: $K = \lambda + \frac{2}{3}G = \frac{E}{3(1-2\nu)}$

1.35 Show that a rotation

$$\omega_z = \frac{1}{2}\left(\frac{\partial v}{\partial x} - \frac{\partial u}{\partial y}\right)$$

would move lines OA' and OB' of Fig. 1.8 such that the angles they make with the x and y directions are equal. These angles are of magnitude

$$\epsilon_{xy} = \frac{1}{2}\left(\frac{\partial v}{\partial x} + \frac{\partial u}{\partial y}\right)$$

Furthermore, show that the total derivative du can be expressed as

$$du = \epsilon_x\, dx + \epsilon_{xy}\, dy - \omega_z\, dy$$

For given strain and rotation components, this relationship can be integrated to give the displacement u.

Hint: For the final request, begin with

$$du = \frac{\partial u}{\partial x}dx + \frac{\partial u}{\partial y}dy$$

1.36 Prove that in the absence of body forces $\nabla^2\epsilon_{kk} = 0$ and further that $\nabla^2\nabla^2 u_i = 0$ [Eq. (1.67)].

Hint: From Eq. (1.69) show that $\nabla^2\sigma_{ii} = 0$. This leads to $\nabla^2\epsilon_{kk} = 0$.

1.37 Given the expression $u_{1,i}\, u_1 + u_{2,i}\, u_2 + u_{1,1}\, u_{1,i1} + u_{1,2}\, u_{2,i1} + u_{2,1}\, u_{1,i2} + u_{2,2}\, u_{2,i2}$, $i = 1$

(a) Write this in a more compact form using index notation and the summation convention.

(b) Express this in matrix notation, i.e., with $\mathbf{u}^T = [u_1 \;\; u_2]$, find $\mathbf{D}$ of $\mathbf{u}^T\mathbf{D}\mathbf{u}$.

Answer:

(a) $u_{j,i}\, u_j + u_{j,k}\, u_{k,ij} \quad i = 1,\; j, k = 1$ or 2

(b) $\mathbf{D} = \begin{bmatrix} {}_i\partial + {}_1\partial\partial_{i1} & {}_2\partial\partial_{i1} \\ {}_1\partial\partial_{i2} & {}_i\partial + {}_2\partial\partial_{i2} \end{bmatrix}$

1.38 For the following descriptions of various types of deformation, determine the strains ϵ_{ij}

(a) Simple dilitation $x' = ax,\ y' = y,\ z' = z$

(b) Pure deformation $x' = a_1 x,\ y' = a_2 y,\ z' = a_3 z$

(c) Simple shear $x' = x + by,\ y' = y,\ z' = z$

Hint: Use Eqs. (1.9) to (1.15).

1.39 Suppose a thin, round, flat element lies in the xy plane. An external compressive surface stress (force) p_a occurs on the outer radius. The outward normal **a** is along a radius at a counterclockwise angle α' from the x axis. Find the surface condition for this element, comparable to Eq. (1.59).

Answer:

$$\begin{bmatrix} -p_a \\ 0 \end{bmatrix} = \begin{bmatrix} \sin^2(90-\alpha') & \cos^2(90-\alpha') & 2\sin(90-\alpha')\cos(90-\alpha') \\ -\sin(90-\alpha')\cos(90-\alpha') & \sin(90-\alpha')\cos(90-\alpha') & -\cos^2(90-\alpha')+\sin^2(90-\alpha') \end{bmatrix} \times \begin{bmatrix} n_x \\ n_y \\ n_{xy} \end{bmatrix}$$

Principal Stresses

1.40 For a plane inclined equally toward three principal axes, show that the normal stress on the plane in terms of the principal stresses is $(\sigma_1 + \sigma_2 + \sigma_3)/3$.

1.41 At a point in a solid the stress matrix **T** has been determined as

$$\mathbf{T} = \begin{bmatrix} 21 & -6 & -6 \\ -6 & 37 & 18 \\ -6 & 18 & 37 \end{bmatrix}$$

Calculate the principal stresses and the extreme values of the shear stresses. Specify a set of principal directions by a right-handed orthogonal triad of unit vectors.

1.42 From sketches of the planes on which shear stresses attain their extreme values, verify that these planes make 45° angles with the principal directions. For example, when the normal to the plane is

$$a_1 = 0 \qquad a_2 = a_3 = \frac{1}{\sqrt{2}}$$

the plane is parallel to the a_1 axis and intersects the $a_2 a_3$ axis at 45°.

1.43 At a point O of a solid, the stress tensor referred to a right-handed coordinate system with origin O and axes x, y, z is

$$\mathbf{T} = \begin{bmatrix} 320 & 15 & 0 \\ 15 & 280 & 0 \\ 0 & 0 & 360 \end{bmatrix}$$

Determine the normal and shear stresses at point O on a surface whose outward normal is the bisector of the angle between the lines Ox and Oz. Determine the principal stresses, and principal axes as a right-handed coordinate system with origin

O and axes $\overline{x}, \overline{y}, \overline{z}$. Calculate the normal and shear stresses at point O on a surface whose outward normal is the bisector of the angle between the lines $O\overline{x}$ and $O\overline{z}$.

1.44 Determine the extreme values of the shear stresses for the stress tensor

$$\mathbf{T} = \begin{bmatrix} 151 & -30 & -27 \\ -30 & 190 & 10 \\ -27 & 10 & 79 \end{bmatrix}$$

1.45 At a given point in a body all the normal stresses have the same value

$$\sigma_x = \sigma_y = \sigma_z = -p \quad p > 0$$

Show that any three mutually orthogonal directions are principal directions for this state of stress (hydrostatic compression). Hydrostatic (fluid) pressure p is often shown as $p = -(\sigma_1 + \sigma_2 + \sigma_3)/3$. Show that hydrostatic pressure is an invariant.

1.46 An octahedral plane is one which intersects the principal axes at equal angles, that is, the normal to this plane is

$$\mathbf{a} = \frac{1}{\sqrt{3}}(\mathbf{e}_1 + \mathbf{e}_2 + \mathbf{e}_3)$$

where $\mathbf{e}_1$, $\mathbf{e}_2$, $\mathbf{e}_3$ are unit vectors parallel to the principal axes. Show that the shear stress on this plane, the *octahedral shear stress*, is given by

$$\tau_{oct} = \frac{1}{3}\sqrt{(\sigma_1 - \sigma_2)^2 + (\sigma_2 - \sigma_3)^2 + (\sigma_3 - \sigma_1)^2}$$

1.47 The *principal directions* for strain are those directions for which the shear strain becomes an extremum. If the stress and strain tensors are related by Hooke's law, prove that the principal axes of stress coincide with the principal axes of strain.

1.48 A circle of radius 100 mm is inscribed on the surface of a thin sheet, which lies in the xy plane and is stress-free. The plate is then subjected to the uniform stress field $\sigma_x = 200$ MPa, $\sigma_y = 25$ MPa, $\sigma_z = 0$, $\tau_{xy} = 125$ MPa, $\tau_{yz} = \tau_{xz} = 0$. The material properties are $E = 210$ GPa and $\nu = 0.3$. In the stress state, the circle deforms into an ellipse. Determine the lengths and orientations of the minor and major axes of this ellipse.

1.49 A displacement field has been determined for an isotropic, linearly elastic solid as

$$u = K(x - y) \quad v = K(x + y) \quad w = Ky$$

where K is a positive constant. Determine the principal stress directions.

1.50 For the pressure vessel shown in Fig. P1.50 the circumferential normal stress (hoop stress) is $\sigma_y = pr/t$, the longitudinal normal stress (axial stress) is $\sigma_x = pr/(2t)$, and assume the radial direction stress varies linearly such that $\sigma_z = -p(1 - z/t)$. Determine the location, orientation, and magnitude of the maximum shear stress.

1.51 A closed thin-walled cylinder is subject to an internal pressure of 1000 psi. The diameter of the cylinder is 40 in. and the wall thickness is 1/2 in. What maximum shear stress will occur? See Problem 1.50 for the formula for the cylinder stress.

Hint: The radial stress is much smaller than the axial and hoop stresses, and can be neglected.

Answer: $\tau_{\max} = 20{,}000$ psi

1.52 The thin rectangular element of Fig. P1.52 is in the state of plane stress. If the stress components have magnitudes in the proportions $\sigma_x/\sigma_y/\tau_{xy} = 1/\frac{1}{2}/\frac{1}{4}$ identify the planes of maximum shear stress.

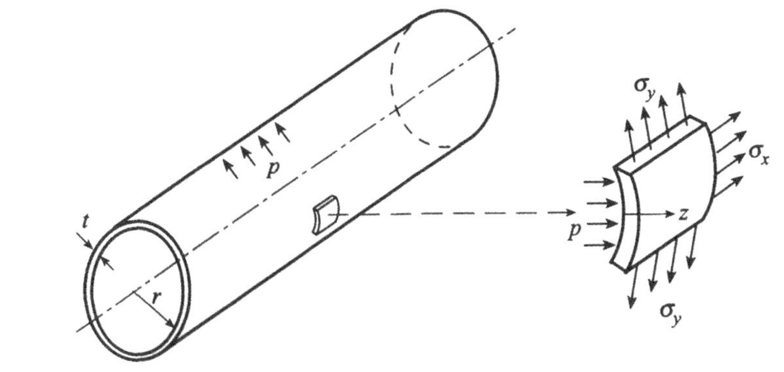

FIGURE P1.50
Cylindrical pressure vessel.

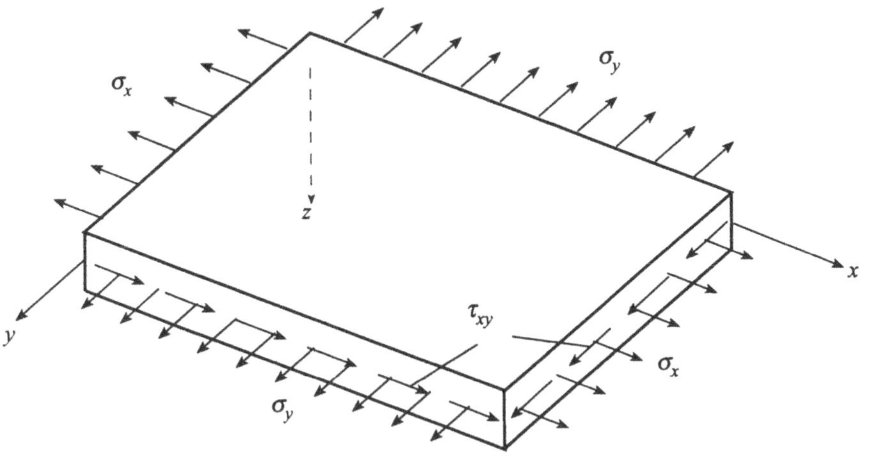

FIGURE P1.52
A thin rectangular element in plane stress.

Strength Theories

1.53 Find the torque necessary to initiate yielding in a 10 mm diameter steel shaft. The yield strength in tension is 180 MN/m^2, and the shear yield stress is one-half of this value. Use (a) maximum shear stress theory and (b) the von Mises theory.

Answer: (a) 17.67 Nm (b) 20.39 Nm

1.54 A thin-walled cylinder of radius 1 m is subject to an internal pressure of 700 kN/m^2. The yield stress of the material in tension is 250 MN/m^2. Find the required wall thickness according to (a) maximum stress theory, (b) maximum shear stress theory, and (c) von Mises theory. See Problem 1.50 for the fundamental stress formulas for a cylinder.

Answer: (a) 2.8 mm, (b) 2.8 mm, (c) 2.43 mm

1.55 A material is found to have a yield stress in a tensile test of $\sigma_{ys} = 40{,}000$ psi. If a sheet of this material is in a state of plane stress and is subject to the stresses $\sigma_x = 150{,}000$ psi, $\sigma_y = -30{,}000$ psi, and τ_{xy}. Find the value of τ_{xy} that corresponds to yield according to (a) the Tresca theory and (b) the von Mises theory.

Engineering Beam Theory

1.56 According to the engineering beam theory, the stresses in a beam of rectangular cross-section are

$$\sigma_x = \frac{M}{I}z, \quad \sigma_y = 0, \quad \sigma_z = 0, \quad \tau_{xz} = \frac{V}{2I}\left(\frac{h^2}{4} - z^2\right)$$

where z is measured from the neutral axis (a centroidal axis) and h is the height of the cross-section. Do these stresses satisfy equilibrium and boundary conditions for a cantilevered beam with

(a) a uniform load along the beam?

(b) a concentrated force at the free end?

1.57 Suppose that the stress function for a uniformly loaded ($\overline{p}_z$) beam of rectangular cross-section of depth h is given by

$$\psi = \overline{p}_z\left\{x^2\left[\left(\frac{z}{h}\right)^3 - \frac{3}{4}\frac{z}{h} - \frac{1}{4}\right] - \frac{h^2}{5}\left[\left(\frac{z}{h}\right)^5 - \frac{1}{2}\left(\frac{z}{h}\right)^3\right]\right\}$$

Note that the plane of interest is now xz and not xy.

(a) Find the stresses in the beam.

(b) Compute the resultant moment and shear on the boundaries at $x = 0$, $x = L$. Can you determine what the boundary conditions are?

1.58 Use the stress formulas of Problem 1.56 to determine the distribution of stresses in a cantilevered beam with a force $\overline{P}$ applied at the free end in the xz plane at an angle of α with the beam axis (x axis).

1.59 Verify that the displacement formulation governing equations for a shear beam are given by Eqs. (1.130).

Torsion

1.60 Show that the stress functions defined by Eq. (1.155) satisfy the equations of equilibrium.

1.61 Show that for a bar undergoing torsion, the resultant forces

$$\int\int_A \tau_{xy}\, dy\, dz, \quad \int\int_A \tau_{xz}\, dy\, dz$$

are zero.

Hint: To prove the first integral is zero, use

$$\begin{aligned}\left(\frac{\partial\omega}{\partial y} - z\right) &= \left(\frac{\partial\omega}{\partial y} - z\right) + y\left(\frac{\partial^2\omega}{\partial z^2} + \frac{\partial^2\omega}{\partial y^2}\right)\\ &= \frac{\partial}{\partial y}\left[y\left(\frac{\partial\omega}{\partial y} - z\right)\right] + \frac{\partial}{\partial z}\left[y\left(\frac{\partial\omega}{\partial z} + y\right)\right]\end{aligned}$$

Then use Green's theorem and Eq. (1.153).

1.62 Utilize the force formulation to solve the torsion problem of a solid shaft of circular cross-section. Use a stress function $\psi = c(r^2 - a^2)$, where a is the radius of the outer surface of the shaft. Verify that your solution agrees with that given in an elementary treatment of strength of materials.

1.63 Use the displacement form of the governing equations to compute the torsional constant and stresses for a uniform bar of elliptical cross-section (Fig. 1.19). Assume the warping function is of the form $\omega = cyz$, where c is a constant.

Hint: From Eq. (1.145),

$$a_z \tau_{xz} + a_y \tau_{xy} = -\frac{dy}{ds}\tau_{xz} + \frac{dz}{ds}\tau_{xy} = 0 \tag{1}$$

Introduce τ_{xz} and τ_{xy} of Eq. (1.166). Calculate the direction cosines from the equation of an ellipse (Fig. 1.19) and find

$$\left(-y + \frac{\partial \omega}{\partial z}\right)a^2 z + \left(z + \frac{\partial \omega}{\partial y}\right) b^2 y = 0$$

Substitute $\omega = cyz$, which satisfies Laplace's equation, into this expression and solve for c.

Alternatively, substitute $\omega = cyz$ into (1) to find $\frac{dy}{ds} y(-1+c) - \frac{dz}{ds} z(1+c) = 0$. Integrate, giving $y^2 \frac{(1-c)}{(1+c)} + z^2 = \text{constant}$. Comparison with $y^2 \frac{b^2}{a^2} + z^2 = b^2$ for the ellipse of Fig. 1.19 leads to the value of c.

Answer:

$$c = \frac{a^2 - b^2}{a^2 + b^2}, \qquad J = \frac{a^3 b^3 \pi}{a^2 + b^2}$$

$$\tau_{xy} = -\frac{2}{ab^3\pi} M_t z, \qquad \tau_{xz} = \frac{2}{a^3 b\pi} M_t y$$

1.64 Compare the torsion of a shaft of elliptical cross-section with that of a shaft of circular cross-section with a radius equal to the minor axis b of the ellipse. Show that for the same angle of twist the maximum shear stress will be greater in the elliptical cross-section.

1.65 A bar subject to torsion has a cross-section in the shape of an equilateral triangle as shown in Fig. P1.65. Compute the maximum shear stress. Also find the angle of twist per unit length, the torsional constant J, and the torsional rigidity GJ. Begin with the stress function

$$\psi = \frac{G\phi'}{2h}\left(y - \sqrt{3}z - \frac{2h}{3}\right)\left(y + \sqrt{3}z - \frac{2h}{3}\right)\left(y + \frac{h}{3}\right)$$

Answer:

$$\tau_{\max} = \frac{15\sqrt{3}M_t}{2h^3}$$

$$J = \frac{h^4}{15\sqrt{3}\, G\theta} \qquad M_x = \frac{h^4}{15\sqrt{3}}$$

1.66 Find the warping function for a prismatic bar with a rectangular-shaped cross-section (Fig. P1.66). Begin with a warping function of the form

$$\omega = yz + c \sin kz \, \sinh ky$$

where c and k are constants to be identified such that the boundary conditions are satisfied.

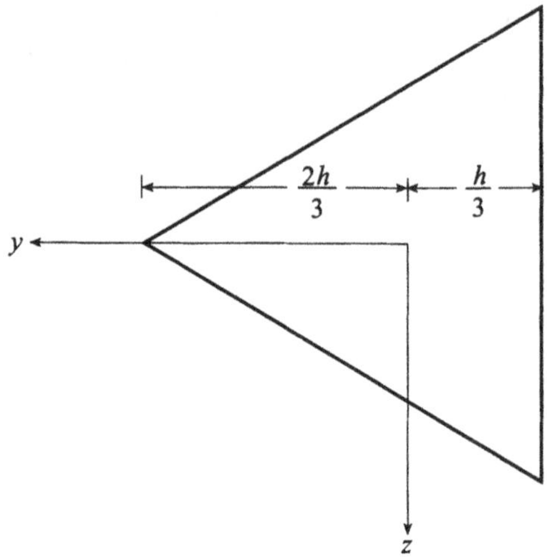

FIGURE P1.65
Equilateral triangle.

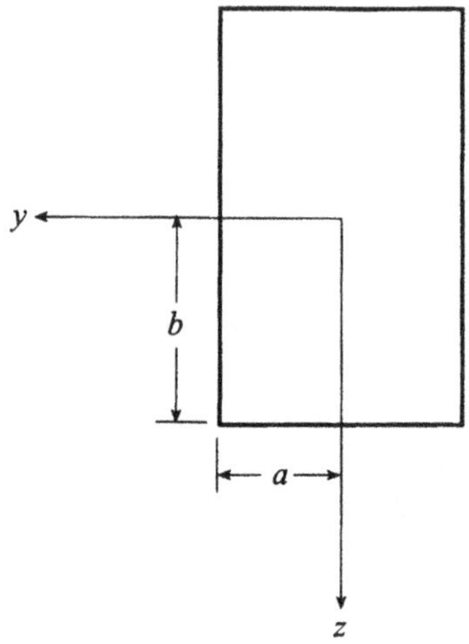

FIGURE P1.66
Bar of rectangular cross-section.

Hint: The boundary conditions lead to

$$\omega = yz + \sum_{n=1,3,5\ldots}^{\infty} c_n \sin\left(\frac{n\pi}{2b} z\right) \sinh\left(\frac{n\pi}{2b} y\right)$$

$$c_n = (-1)^{(n+1)/2} \frac{32b^2}{n^3\pi^3} \frac{1}{\cosh(n\pi a/2b)}$$

1.67 Show that for a bar of rectangular cross-section, Fig. P1.66

$$M_t = \frac{16}{3} G\phi' b^3 a - \left(\frac{4}{\pi}\right)^5 G\phi' b^4 \sum_{n=1,3,5\ldots}^{\infty} \frac{1}{n^5} \tanh \frac{n\pi a}{2b}$$

$$\tau_{xy} = 2G\phi' z + \frac{16G\phi' b}{\pi^2} \sum_{n=1,3,5\ldots}^{\infty} (-1)^{(n+1)/2} \frac{1}{n^2} \frac{\cosh(n\pi y/2b)}{\cosh(n\pi a/2b)} \sin \frac{n\pi z}{2b}$$

$$\tau_{xz} = \frac{16G\phi' b}{\pi^2} \sum_{n=1,3,5\ldots}^{\infty} (-1)^{(n+1)/2} \frac{1}{n^2} \frac{\sinh(n\pi y/2b)}{\cosh(n\pi a/2b)} \cos \frac{n\pi z}{2b}$$

1.68 Show that the maximum torsional stress for a bar of rectangular cross-section occurs at the midpoints of the longer sides of the rectangle.

Hint: See Problem 1.67.

1.69 If the stress function for the torsion of a bar of rectangular cross-section (Fig. P1.66) is

$$\psi = c_1 \cos\frac{\pi y}{2a}\cos\frac{\pi z}{2b} + c_2 \cos\frac{3\pi y}{2a}\cos\frac{3\pi z}{2b}$$

(a) Find the stresses τ_{xy} and τ_{xz}.

(b) Are the conditions of equilibrium satisfied?

(c) Find τ_{max}, M_t, and ϕ'.

Hint: Use Eqs. (1.154), (1.144), (1.148), and (1.149)

1.70 Find relationships between the Prandtl stress function ψ and the warping function ω.

1.71 Find the torsional characteristics of the thin strip of Fig. P1.71, including, $d\phi/dx$, J, τ_{max}. The results apply to many thin-walled open cross sections.

Hint: Use a Prandtl stress function that is a function of y only. Poisson's equation becomes

$$\frac{d^2\psi}{dy^2} = -2G\frac{d\phi}{dx}$$

for which a stress funtion can be constructed

Answer:

$$M_t = 2\int\int \psi \, dy \, dz = \frac{1}{3}bt^3G\frac{d\phi}{dx}$$

$$J = \frac{1}{3}bt^3$$

$$\tau_{max} = -\left.\frac{\partial\psi}{\partial y}\right|_{y=\pm t/2} = \pm tG\frac{d\phi}{dx} = \pm\frac{tM_t}{J}$$

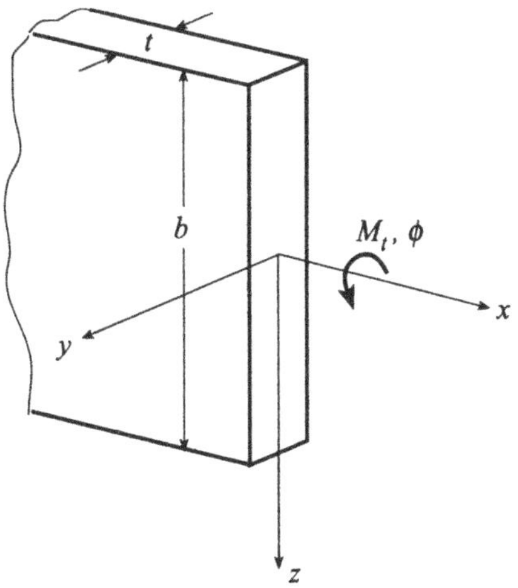

FIGURE P1.71
Thin strip under torsion.

1.72 If $G\phi' = 1$, the force formulation for the torsion problem leads to the governing equation

$$\frac{\partial^2 \psi}{\partial y^2} + \frac{\partial^2 \psi}{\partial z^2} = -2$$

with the requirement that $\psi = 0$ on the boundary. Show that this problem formulation involving Poisson's equation can be reformulated in terms of Laplace's equation.

Hint: Let $\psi = \psi^* - \frac{1}{2}(y^2 + z^2)$

Answer: With $\psi = \psi^* - \frac{1}{2}(y^2 + z^2)$, the governing formula becomes

$$\frac{\partial^2 \psi^*}{\partial y^2} + \frac{\partial^2 \psi^*}{\partial z^2} = 0$$

with $\psi^* = \frac{1}{2}(y^2 + z^2)$ on the boundary.

2

Principles of Virtual Work: Integral Form of the Basic Equations

The *variational* or *energy* methods of structural mechanics have been important tools for the development of basic equations for more than a century. In particular, they have been useful, and sometimes essential, for the derivation of governing differential equations of motion. It has long been recognized that the fundamental relationships of the previous chapter, e.g., the equations of equilibrium or the strain-displacement equations, have equivalent energy representations. Now, the variationally based integral forms of the basic equations are emerging as important foundations for computational techniques of structural mechanics. This chapter considers the classical variational principles of the theory of elasticity and then treats the generalized variational principles which are particularly helpful in achieving a better understanding of the interrelationships between the various methods of structural mechanics. The study of variational methods should begin with a brief look at the basics of the calculus of variations, a branch of mathematics dealing with the extremal values of integrals. Numerous sources covering the calculus of variations are available; a short summary is provided in Appendix I.

2.1 Fundamental Definitions of Work and Energy

Before proceeding to the variational principles of the mechanics of solids, it is essential to understand the concepts of work and energy. It is useful to define work and energy in what may appear to be somewhat abstract terms. The subsequent development of the variational principles will clarify why such definitions are given.

2.1.1 Work and Energy

Work is defined as the product of a force and the displacement of its point of application in the direction of the force. More specifically, define the differential work dW done, while the force F moves through a differential displacement ds as the product of ds and F_s, the component of F in the direction of ds

$$dW = F_s \, ds \tag{2.1}$$

Energy is defined as a quantity representing the ability or capacity to perform work. We say a structural system "possesses" energy, whereas the forces in the system may "perform" work. A linear elastic spring (or rod) that is stretched has acquired potential energy. Work is done as the energy is stored or diminished due to stretching or relaxing the stretch.

As a solid deforms, the internal forces perform work in moving through displacements until reaching a final configuration. If the strained elastic solid were permitted to return slowly to its unstrained state, the solid would be capable of returning the work performed by the external forces. This capacity of the internal forces to do work in a strained solid is due to the *strain energy* or the internal energy stored in the body. Thus, in an elastic body (with no initial strains) the strain energy (U_i) is equal to but opposite in sign to the work done by the internal forces, i.e., $U_i = -W_i$. This kind of energy is also called the *potential energy* of the internal forces or the *internal energy* as it describes the behavior of the structural system due to the material properties of its members. Another kind of energy is related to the position of the body with respect to the gravity effect. A rigid solid situated at some height h above some reference plane has the potential energy (with respect to a given plane) of h times the weight of the solid.

Under certain conditions, the quantity $d\Pi$ is the exact differential of some functional, say Π. In a two-dimensional space, if Π is the energy stored,

$$d\Pi = -dW = -(F_y\,dy + F_z\,dz) \tag{2.2}$$

where F_y and F_z are the forces in the y and z directions. The necessary and sufficient condition for $d\Pi$ to be an exact differential is (see an elementary calculus textbook)

$$\frac{\partial F_y}{\partial z} = \frac{\partial F_z}{\partial y}$$

Suppose that the force moves along a closed contour, corresponding to an area S. According to Green's theorem [Appendix II, Eq. (II.3)]

$$\oint (R\,dy + T\,dz) = \int_S \left(\frac{\partial T}{\partial y} - \frac{\partial R}{\partial z}\right) dy\,dz$$

Let R be F_y and T be F_z, so that the energy stored would be

$$\Pi = -W = -\oint (F_y\,dy + F_z\,dz) = -\int_S \left(\frac{\partial F_z}{\partial y} - \frac{\partial F_y}{\partial z}\right) dy\,dz \tag{2.3}$$

If the necessary and sufficient condition is satisfied, the energy as the forces move along a closed contour is zero. Equivalently, the integral

$$\int_A^B (F_y\,dy + F_z\,dz)$$

is independent of the path on which the forces move. Furthermore, the energy stored in moving around a closed path will be zero ($\Pi_A - \Pi_A = 0$).

In a system where the force F is moved from point A to B, the potential energy is the energy stored or possessed as a result of the position of the force. More precisely, the potential energy is the capacity of a conservative force system to perform work by virtue of its position with respect to a reference level. The function Π is the potential energy.

If no net work is done in moving on a closed path, the system is said to be *conservative*. Forces associated with pure elastic deformations are conservative, whereas those associated with friction and with plastic or damped deformations are termed *nonconservative*.

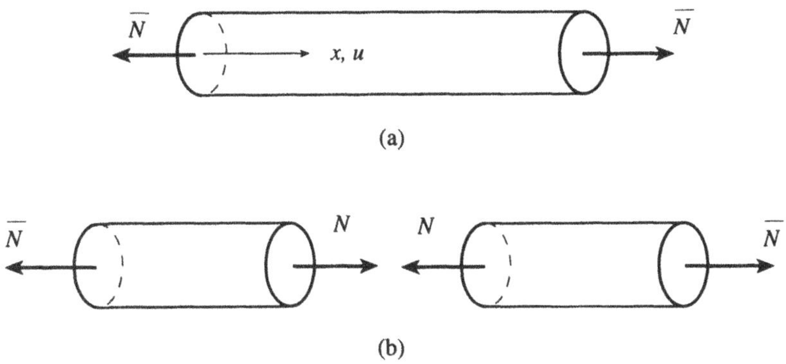

FIGURE 2.1
Element of an extension bar.

2.1.2 Work and Potential Energy of the Internal Forces

Consider an elastic extension bar, a one-dimensional system, of length L and cross-sectional area A. Suppose that it is loaded with an axial force $\overline{N}$ which induces the axial stress σ_x. Applied quantities are identified with overbars. What is the work done during the deformation of an element of length dx of uniform cross-section (Fig. 2.1a)?

The internal force $N = \int_A \sigma_x \, dA$ is by equilibrium equal to the applied force $\overline{N}$ but opposite in sign. Refer to the fictitious cut in the bar of Fig. 2.1b. Then, the internal work done in a differential element dx is $-\sigma_x \, d\epsilon_x A \, dx = -\sigma_x \, d\epsilon_x \, dV$ with the strain $d\epsilon_x = d(du/dx)$ and the volume $dV = A \, dx$ of the element. The negative sign is due to the expression being formed on the ficticious cut of Fig. 2.1b where the stresses and stress resultant N are defined as internal variables. Thus, the total work of the bar as the strains (defined as being positive in the direction of the x-axis) increase from zero to their final values has a negative sign and the value

$$W_i = -\int_0^L \int_0^{\epsilon_x} \sigma_x \, d\epsilon_x \, A \, dx \tag{2.4}$$

Upon substitution of Hooke's law, $\sigma_x = E\epsilon_x$, which is the relationship between stress σ_x and strain ϵ_x

$$W_i = -\int_0^L A \int_0^{\epsilon_x} E\epsilon_x \, d\epsilon_x \, dx = -\frac{1}{2}\int_0^L EA\,\epsilon_x^2 \, dx = -\frac{1}{2}\int_0^L N\epsilon_x \, dx \tag{2.5}$$

The expression for the work of the internal forces is similar in the case of a three-dimensional body with six independent components of the stress tensor σ_{ij} and the strain tensor ϵ_{ij}. For an infinitesimal rectangular parallelepiped of volume $(dx\,dy\,dz)$, the work done during the strain increments $d\epsilon_x, d\epsilon_y, \ldots, d\gamma_{yz}$ by the stresses would be

$$(\sigma_x \, d\epsilon_x + \sigma_y \, d\epsilon_y + \cdots + \tau_{yz} \, d\gamma_{yz}) \, dx \, dy \, dz$$

where

$$d\epsilon_x = d\left(\frac{\partial u}{\partial x}\right), \qquad d\epsilon_y = d\left(\frac{\partial v}{\partial y}\right), \ldots, d\gamma_{yz} = d\left(\frac{\partial w}{\partial y} + \frac{\partial v}{\partial z}\right)$$

The work for the whole body of volume V then amounts to

$$W_i = -\int_V \int_{(0,0,\ldots 0)}^{(\epsilon_x,\epsilon_y,\ldots,\gamma_{yz})} (\sigma_x \, d\epsilon_x + \sigma_y \, d\epsilon_y + \cdots + \tau_{yz} \, d\gamma_{yz}) \, dV$$

$$= -\int_V \int_0^{\epsilon_{ij}} \sigma_{ij} \, d\epsilon_{ij} \, dV = -\int_V \int_0^{\epsilon} \boldsymbol{\sigma}^T d\boldsymbol{\epsilon} \, dV \tag{2.6}$$

with [Chapter 1, Eqs. (1.3) and (1.6)]

$$\begin{aligned} \boldsymbol{\sigma} &= [\sigma_x \quad \sigma_y \quad \sigma_z \quad \tau_{xy} \quad \tau_{xz} \quad \tau_{yz}]^T = [\sigma_{11} \quad \sigma_{22} \quad \sigma_{33} \quad \sigma_{12} \quad \sigma_{13} \quad \sigma_{23}]^T \\ \boldsymbol{\epsilon} &= [\epsilon_{xx} \quad \epsilon_{yy} \quad \epsilon_{zz} \quad 2\epsilon_{xy} \quad 2\epsilon_{xz} \quad 2\epsilon_{yz}]^T = [\epsilon_{11} \quad \epsilon_{22} \quad \epsilon_{33} \quad \gamma_{12} \quad \gamma_{13} \quad \gamma_{23}]^T \end{aligned} \tag{2.7}$$

The generalized Hooke's law takes the form [Chapter 1, Eqs. (1.34) and (1.35)] $\sigma_{ij} = E_{ijkl}\,\epsilon_{kl}$ or $\boldsymbol{\sigma} = \mathbf{E}\boldsymbol{\epsilon}$, so that for elastic bodies

$$W_i = -\frac{1}{2}\int_V E_{ijkl}\,\epsilon_{ij}\,\epsilon_{kl} \, dV = -\frac{1}{2}\int_V \boldsymbol{\epsilon}^T \mathbf{E}\,\boldsymbol{\epsilon} \, dV$$

$$= -\frac{1}{2}\int_V \sigma_{ij}\,\epsilon_{ij} \, dV = -\frac{1}{2}\int_V \boldsymbol{\sigma}^T \boldsymbol{\epsilon} \, dV \tag{2.8}$$

where $\boldsymbol{\sigma}^T = (\mathbf{E}\,\boldsymbol{\epsilon})^T = \boldsymbol{\epsilon}^T \mathbf{E}^T$ and, because $\mathbf{E}$ is symmetric, $\mathbf{E} = \mathbf{E}^T$.

For a three-dimensional linear elastic body for which the strain energy is equal to but opposite in sign to the work of the internal forces,

$$U_i = -W_i = \int_V U_0(\epsilon) \, dV = \frac{1}{2}\int_V E_{ijkl}\,\epsilon_{ij}\,\epsilon_{kl} \, dV = \frac{1}{2}\int_V \boldsymbol{\epsilon}^T \mathbf{E}\,\boldsymbol{\epsilon} \, dV \tag{2.9}$$

in which $U_0(\epsilon) = \frac{1}{2}\boldsymbol{\epsilon}^T \mathbf{E}\,\boldsymbol{\epsilon}$ is the specific potential of a unit volume or the *strain energy density*. With a symmetric matrix $\mathbf{E}$, U_0 has the quadratic form

$$U_0(\epsilon) = \frac{1}{2}\boldsymbol{\epsilon}^T \mathbf{E}\,\boldsymbol{\epsilon}$$

$$= G\left[\epsilon_{11}^2 + \epsilon_{22}^2 + \epsilon_{33}^2 + \frac{\nu}{1-2\nu}(\epsilon_{11} + \epsilon_{22} + \epsilon_{33})^2 + \frac{1}{2}\left(\gamma_{12}^2 + \gamma_{13}^2 + \gamma_{23}^2\right)\right] \tag{2.10}$$

Here, it has been assumed that the material is isotropic, and $\mathbf{E}$ is taken from Eq. (1.34). A quadratic function which is never negative for arbitrary values of the variables is called a *positive definite* (quadratic) function. Such a function is zero only when each variable is zero. As is shown in introductory mechanics of solids textbooks for isotropic materials, the shear modulus of elasticity G must be positive, and Poisson's ratio ν can vary between zero and one half, i.e., $G > 0$, $0 \le \nu \le 0.5$. Thus, the strain energy density of Eq. (2.10) is a positive definite function in the strain components.

It is useful to visualize again the strain energy in terms of a stress-strain curve of an element, such as an extension bar. From Eq. (2.4)

$$\frac{U_0}{A} = \int_0^{\epsilon_x} \sigma_x \, d\epsilon_x \tag{2.11}$$

is the area under the curves of Fig. 2.2. The area above the curve would be $\int \epsilon_x \, d\sigma_x$. Using integration by parts, it follows that

$$\int \epsilon_x \, d\sigma_x = \sigma_x\,\epsilon_x - \int \sigma_x \, d\epsilon_x = \sigma_x\,\epsilon_x - \frac{U_0}{A} \tag{2.12}$$

The area above the curve is referred to as the *complement* of the area U_0/A. Set

$$\int_0^{\sigma_x} \epsilon_x \, d\sigma_x = \frac{U_0^*}{A} \tag{2.13}$$

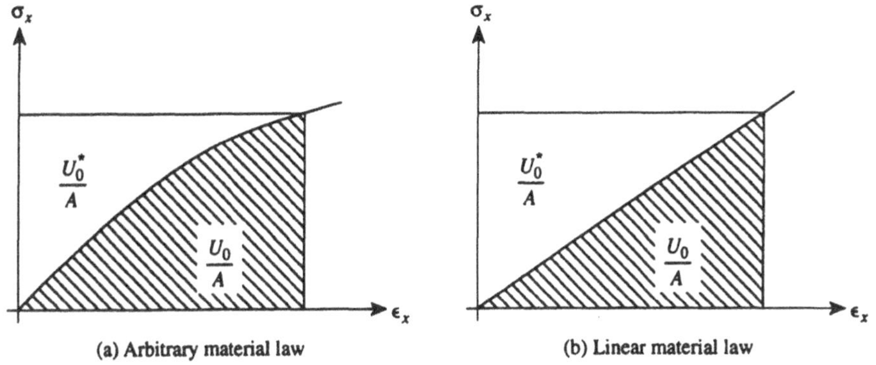

FIGURE 2.2
Strain energy and complementary strain energy densities for an axially stressed bar.

where U_0^* is called the *complementary strain energy density.* For the linear material law of Fig. 2.2b, $U_0 = U_0^*$, whereas for the nonlinear material law of Fig. 2.2a, $U_0 \neq U_0^*$.

For the general solid, this gives a useful alternative representation for the strain energy; the complementary strain energy density $U_0^*(\sigma)$ is

$$U_0^*(\sigma) = \sigma_{ij}\,\epsilon_{ij} - U_0(\epsilon) \tag{2.14}$$

This relationship, which is sometimes referred to as Friedrichs'[1] transformation, implies that if the strain energy density exists, then the existence of the complementary strain energy density is assured. In the case of a linear material law [Chapter 1, Eqs. (1.32)], $\boldsymbol{\epsilon} = \mathbf{E}^{-1}\boldsymbol{\sigma}$ or $\epsilon_{ij} = F_{ijkl}\,\sigma_{kl}$, $U_0^*(\sigma)$ can be expressed as a function of the stress components alone.

$$\begin{aligned} U_0^*(\sigma) &= \boldsymbol{\sigma}^T\boldsymbol{\epsilon} - \frac{1}{2}\boldsymbol{\sigma}^T\boldsymbol{\epsilon} = \frac{1}{2}\boldsymbol{\sigma}^T\mathbf{E}^{-1}\boldsymbol{\sigma} \\ &= \frac{1}{2E}\left[(\sigma_{11}+\sigma_{22}+\sigma_{33})^2 + 2(1+\nu)\left(\sigma_{12}^2+\sigma_{13}^2+\sigma_{23}^2-\sigma_{11}\sigma_{22}-\sigma_{11}\sigma_{33}-\sigma_{22}\sigma_{33}\right)\right] \end{aligned} \tag{2.15}$$

With a linear material law, there is no difference between U_0, a function of strain components, and U_0^*, a function of stress components. This contention follows from

$$U_0 = \frac{1}{2}\boldsymbol{\epsilon}^T\mathbf{E}\,\boldsymbol{\epsilon} = \frac{1}{2}\boldsymbol{\epsilon}^T\boldsymbol{\sigma} = \frac{1}{2}\boldsymbol{\sigma}^T\boldsymbol{\epsilon} = \frac{1}{2}\boldsymbol{\sigma}^T\mathbf{E}^{-1}\boldsymbol{\sigma} = U_0^*$$

In general, for most plastic materials, U_0 is not equal to U_0^*.

The complementary strain energy, that is, the complementary potential of the internal forces, U_i^*, for the whole body can be written as

$$U_i^* = \int_V U_0^*(\sigma)\,dV = \frac{1}{2}\int_V F_{ijkl}\,\sigma_{ij}\,\sigma_{kl}\,dV = \frac{1}{2}\int_V \boldsymbol{\sigma}^T\mathbf{E}^{-1}\boldsymbol{\sigma}\,dV \tag{2.16}$$

[1]Kurt Otto Friedrichs was born in 1901 in Kiel, Germany. He worked with Richard Courant, John von Neumann, Hans Lewy, and J.J. Stoker and is credited with many developments in mathematics and mechanics. He received his doctoral degree from the University of Göttingen in 1925. He was at Göttingen in 1925–27, Aachen in 1927–29, and Braunschweig in 1930–37. He departed Germany, along with other scientists in the late 30s. In 1937 he joined the faculty of New York University, where he eventually became the director of the Courant Institute.

The complementary work of the internal forces is given by

$$W_i^* = -\int_V \int^{\sigma_{ij}} \epsilon_{ij}\, d\sigma_{ij}\, dV = -\int_V \int^{\sigma} \epsilon^T d\sigma\, dV \tag{2.17}$$

If during loading and unloading U_0 and U_0^* are independent of the path of deformation (but depend only on the initial and final states), the differentials dU_0 and dU_0^* will be exact differentials, and U_0, U_0^* are then potential functions. Accordingly, from the chain rule of differentiation

$$dU_0 = \frac{\partial U_0}{\partial \epsilon_x} d\epsilon_x + \frac{\partial U_0}{\partial \epsilon_y} d\epsilon_y + \cdots + \frac{\partial U_0}{\partial \gamma_{yz}} d\gamma_{yz} \tag{2.18a}$$

or

$$dU_0 = \frac{\partial U_0}{\partial \epsilon_{kl}} d\epsilon_{kl} = \left(\frac{\partial U_0}{\partial \epsilon}\right)^T d\epsilon \tag{2.18b}$$

Similarly,

$$dU_0^* = \frac{\partial U_0^*}{\partial \sigma_{kl}} d\sigma_{kl} = \left(\frac{\partial U_0^*}{\partial \sigma}\right)^T d\sigma \tag{2.19}$$

Since (Eq. 2.6), $U_0 = \int^{\epsilon_{ij}} \sigma_{ij}\, d\epsilon_{ij} = \int^{\epsilon} \sigma^T d\epsilon$, it follows that dU_0 can be expressed as

$$dU_0 = \sigma_x\, d\epsilon_x + \sigma_y\, d\epsilon_y + \cdots + \tau_{yz}\, d\gamma_{yz} \tag{2.20}$$

A comparison of Eqs. (2.18a) and (2.20) indicates that

$$\frac{\partial U_0}{\partial \epsilon_x} = \sigma_x, \quad \frac{\partial U_0}{\partial \epsilon_y} = \sigma_y, \ldots, \frac{\partial U_0}{\partial \gamma_{yz}} = \tau_{yz} \tag{2.21}$$

or

$$\left(\frac{\partial U_0}{\partial \epsilon}\right)^T = \sigma^T \tag{2.22}$$

Similarly,

$$\left(\frac{\partial U_0^*}{\partial \sigma}\right)^T = \epsilon^T \tag{2.23}$$

Thus, for example, in the case of Eqs. (2.21) or (2.22), the partial derivative of the strain energy density with respect to the strain component is equal to the corresponding stress component. In verifying that U_0 is a potential function and that Eq. (2.21) is valid, it is necessary to utilize a form of the strain energy density U_0 that is expressed in terms of strains only, e.g., Eq. (2.10) where $U_0 = U_0(\epsilon)$, rather than one in terms of both strains and stresses.

It can be reasoned that Eqs. (2.21) or (2.22) establish a unique relationship between stress and strain, i.e., the stress-strain relationship (the material law) is uniquely determined by the strain energy density and vice versa. Moreover, Eq. (2.21) can be used to show that the material law matrix **E** is symmetric (and so is $\mathbf{E}^{-1}$). By differentiation of Eq. (2.21),

$$\frac{\partial^2 U_0}{\partial \epsilon_y\, \partial \epsilon_x} = \frac{\partial \sigma_x}{\partial \epsilon_y} = E_{xy} \qquad \frac{\partial^2 U_0}{\partial \epsilon_x\, \partial \epsilon_y} = \frac{\partial \sigma_y}{\partial \epsilon_x} = E_{yx}$$

Since

$$\frac{\partial^2 U_0}{\partial \epsilon_y\, \partial \epsilon_z} = \frac{\partial^2 U_0}{\partial \epsilon_x\, \partial \epsilon_y}$$

it follows that $E_{xy} = E_{yx}$. Thus, **E** is symmetric.

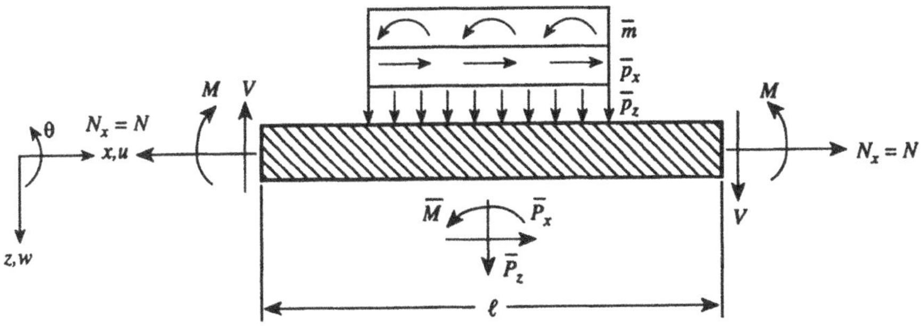

FIGURE 2.3
Positive state variables (u, w, θ, M, V, N) and positive applied loadings (variables with superscript bars) for a beam.

EXAMPLE 2.1 Beam Theory
Consider the use of Eq. (2.9) to calculate the strain energy in a beam with an internal axial force and bending moment.

For the engineering beam theory of Chapter 1, it is assumed that the deformations are small, the material obeys Hooke's law, and the length of the beam is much greater than the cross-sectional dimensions, e.g., width and depth. All of the state variables of the beam are referred to the beam axis (the centroid), so that we are dealing with a one-dimensional problem. The positive directions of applied loadings and state variables are indicated in Fig. 2.3. For the Euler-Bernoulli beam theory, in which shear deformation effects are not taken into account, the only non-zero strain is ϵ_x. Thus,

$$U_i = \frac{1}{2}\int_V \epsilon^T \mathbf{E}\, \epsilon\, dV = \frac{1}{2}\int_V \epsilon_x E\, \epsilon_x\, dV \tag{1}$$

As indicated in Chapter 1, the use of Bernoulli's hypothesis on the cross-section leads to [Chapter 1, Eq. (1.98)] $u = u_0(x) + z\theta$ and $\theta = -dw/dx$,

$$\epsilon_x = \frac{du}{dx} = \frac{du_0}{dx} - z\frac{d^2w}{dx^2} \tag{2}$$

With $dV = dA\, dx$ and since by definition $\int_A z\, dA = 0$ if the quantities are referred to the centroid, for a beam of length L, the strain energy of (1) reduces to

$$\begin{aligned} U_i &= \frac{1}{2}\int_V \epsilon_x E\, \epsilon_x\, dV = \frac{1}{2}\int_0^L \int_A E\left(\frac{du_0}{dx} - z\frac{d^2w}{dx^2}\right)^2 dA\, dx \\ &= \frac{1}{2}\int_0^L \left[EA\left(\frac{du_0}{dx}\right)^2 + E\left(\frac{d^2w}{dx^2}\right)^2 \int_A z^2\, dA\right] dx \\ &= \int_0^L \left[\frac{EA}{2}\left(\frac{du_0}{dx}\right)^2 + \frac{EI}{2}\left(\frac{d^2w}{dx^2}\right)^2\right] dx \end{aligned} \tag{3}$$

This final expression can also be written in terms of internal forces, leading to the expression for the complementary potential energy. From Chapter 1, Eq. (1.125), $EI\,(d^2w/dx^2) = -M$. For the axial extension of the centerline of the beam (Chapter 1, Eq. 1.106a), $\epsilon_x|_{\text{axial}} = \epsilon_{0x} = du_0/dx$ or $N = EA\epsilon_{0x} = EA(du_0/dx)$. Thus, (3) becomes

$$U_i^* = \int_0^L \frac{1}{2}\left(\frac{N^2}{EA} + \frac{M^2}{EI}\right) dx \tag{4}$$

The first term in (4) corresponds to the complementary strain energy for the stretching displacement u_0, and the second term corresponds to that for the lateral displacement w.

The complementary strain energy expression of (4) can also be obtained by simply substituting the expression for the normal stress [(Chapter 1, Eq. (1.137)]

$$\sigma_x = \frac{N}{A} + \frac{M}{I} z \tag{5}$$

into

$$U_i^* = \frac{1}{2} \int_V \frac{\sigma_x^2}{E} dV \tag{6}$$

Thus,

$$U_i^* = \frac{1}{2} \int_0^L \int_A \frac{1}{E} \left(\frac{N^2}{A^2} + \frac{2NMz}{AI} + \frac{M^2 z^2}{I^2} \right) dA\, dx = \int_0^L \frac{1}{2} \left(\frac{N^2}{EA} + \frac{M^2}{EI} \right) dx \tag{7}$$

where $\int_A z\, dA = 0$ and $\int_A z^2\, dA = I$. ▮

EXAMPLE 2.2 Beam Theory Including Shear Deformation

In the previous example, it was assumed that shear deformation effects can be ignored. If this is not the case, $\gamma_{xz} = \gamma \neq 0$, so that from Eq. (2.7)

$$U_i = \frac{1}{2} \int_V (\sigma_x \epsilon_x + \tau_{xz} \gamma_{xz})\, dV \tag{1}$$

Use

$$\epsilon_x = \frac{\sigma_x}{E}, \quad \sigma_x = \frac{N}{A} + \frac{M}{I} z, \quad \tau_{xz} = \tau = \frac{V}{A} = k_s G \gamma, \quad \gamma_{xz} = \gamma = \frac{V}{k_s G A} \tag{2}$$

The expressions for τ and γ_{xz} are taken from Chapter 1, Eq. (1.109). Substitute (2) into (1) to obtain

$$U_i^* = \int_0^L \frac{1}{2} \left(\frac{N^2}{EA} + \frac{M^2}{EI} + \frac{V^2}{k_s G A} \right) dx \tag{3}$$

▮

2.1.3 Work and Potential Energy of the Applied Loading

In order to obtain the total work, it is necessary to take into account the work done by the body forces $\overline{\mathbf{p}}_V$, the external forces $\overline{\mathbf{p}}$ applied to the surface area S_p, and the displacements $\overline{\mathbf{u}}$ applied on the surface area $S_u = S - S_p$. These are referred to as the *applied loadings*.

Consider again the simple extension bar with axial force $\overline{N}$ and axial extension. As the axial force increases from zero to its final value $\overline{N}_F$, the displacement u in the axial direction increases from zero to u_F. "External" work W_e is performed until the final configuration is reached. From the definition of work, we have

$$W_e = \int_0^{u_F} \overline{N}\, du \tag{2.24}$$

If the displacement is proportional to the load, i.e., $u = k\overline{N}$, where k is a constant of proportionality, then

$$W_e = \int_0^{u_F} \overline{N}\, du = \int_0^{u_F} \frac{u}{k}\, du = \frac{1}{2} \frac{u_F^2}{k} = \frac{1}{2} \overline{N}_F u_F \tag{2.25}$$

In general, the external work done by the prescribed forces $\overline{p}_i, \overline{p}_{V_i}$ can be expressed as

$$W_e = \int_{S_p} \int_0^{u_i} \overline{p}_i \, du_i \, dS + \int_V \int_0^{u_i} \overline{p}_{V_i} \, du_i \, dV \tag{2.26}$$

In the instance where the solid is linear in its response (zero to u_i) to a load applied from a zero value to its final value $\overline{p}_i$ or $\overline{p}_{Vi}$,

$$W_e = \frac{1}{2}\int_{S_p} \overline{p}_i \, u_i \, dS + \frac{1}{2}\int_V \overline{p}_{Vi} \, u_i \, dV = \frac{1}{2}\int_{S_p} \mathbf{u}^T \overline{\mathbf{p}} \, dS + \frac{1}{2}\int_V \mathbf{u}^T \overline{\mathbf{p}}_V \, dV \tag{2.27}$$

To establish the quantity *complementary external work* performed by prescribed displacements $\overline{u}_i$, a similar pattern can be followed. For the bar with prescribed elongation $\overline{u}$,

$$W_e^* = \int_0^{N_F} \overline{u} \, dN \tag{2.28}$$

where N_F should be interpreted as the reactive force developed in the bar. In general,

$$W_e^* = \int_{S_u} \int_0^{p_i} \overline{u}_i \, dp_i \, dS \tag{2.29}$$

If the relationship between displacements and forces is linear, then

$$W_e^* = \frac{1}{2}\int_{S_u} \overline{u}_i \, p_i \, dS = \frac{1}{2}\int_{S_u} \mathbf{p}^T \overline{\mathbf{u}} \, dS \tag{2.30}$$

which is referred to as the complementary work of the prescribed displacements. The change in the work from the initial to the final configuration is given by Eq. (2.26), where the integration from zero to u_i is to be made from the initial state to the final state. Similar to the case of internal energy with Eq. (2.22), the conditions for the existence of potential functions for applied loads are

$$\left(\frac{\partial \mathcal{U}_{0e(1)}}{\partial \mathbf{u}}\right)^T = -\overline{\mathbf{p}}^T \qquad \left(\frac{\partial \mathcal{U}_{0e(2)}}{\partial \mathbf{u}}\right)^T = -\overline{\mathbf{p}}_V^T \tag{2.31}$$

where the $\mathcal{U}_{0e(k)}$, $k = 1, 2$ are density functions with the subscript k distinguishing between those portions of the potential related to forces $\overline{\mathbf{p}}$ and those resulting from $\overline{\mathbf{p}}_V$. In order to understand better the relations of Eq. (2.31), consider the related energy expression

$$\begin{aligned} \mathcal{U}_e &= -\int_{S_p} \int_0^{u_i} \overline{p}_i \, du_i \, dS - \int_V \int_0^{u_i} \overline{p}_{Vi} \, du_i \, dV \\ &= \int_{S_p} \mathcal{U}_{0e(1)} \, dS + \int_V \mathcal{U}_{0e(2)} \, dV \end{aligned} \tag{2.32a}$$

with the energy density of the prescribed forces

$$\mathcal{U}_{0e(1)} = -\int_0^{u_i} \overline{p}_i \, du_i \quad \text{and} \quad \mathcal{U}_{0e(2)} = -\int_0^{u_i} \overline{p}_{Vi} \, du_i$$

Equation (2.32) satisfies Eq. (2.31), i.e.,

$$\begin{aligned} \frac{\partial \mathcal{U}_{0e(1)}}{\partial u_i} &= \frac{\partial}{\partial u_i}\left(-\int_0^{u_i} \overline{p}_i \, du_i\right) = -\overline{p}_i \\ \frac{\partial \mathcal{U}_{0e(2)}}{\partial u_i} &= \frac{\partial}{\partial u_i}\left(-\int_0^{u_i} \overline{p}_{Vi} \, du_i\right) = -\overline{p}_{Vi} \end{aligned}$$

If $\overline{p}_i$ and $\overline{p}_{Vi}$ are constant during deformation, Eq. (2.32a) gives

$$\mathcal{U}_e = -\int_{S_p} \overline{p}_i\, u_i\, dS - \int_V \overline{p}_{Vi}\, u_i\, dV \tag{2.32b}$$

By comparison with Eq. (2.27), it follows that for linear response and conservative loads

$$\mathcal{U}_e = -2W_e \tag{2.33}$$

The complementary potential functions are obtained in a similar fashion. For prescribed displacements,

$$\left[\frac{\partial \mathcal{U}^*_{0e}}{\partial \mathbf{p}}\right]^T = -\overline{\mathbf{u}}^T \tag{2.34}$$

and $\mathcal{U}^*_e = -\int_{S_u} \overline{u}_i\, p_i\, dS = -\int_{S_u} \mathbf{p}^T \overline{\mathbf{u}}\, dS$, where $\mathcal{U}^*_{0e}$ is the complementary energy density of the prescribed displacements.

2.1.4 Virtual Work

It is useful to define the work done by the loads on a body during a small, admissible change in the displacements. An admissible or possible change is a displacement which varies continuously as a function of the coordinates and does not violate displacement boundary conditions. Although the actual displacements may be large, the *change* in the displacements must be small. Traditionally, these infinitesimal, possible changes in displacements have been named *virtual displacements*. Virtual displacements are designated by δu_i, which indicates that they correspond to a variation of a function as defined in Appendix I.

The use of virtual displacements is usually traced to Johann Bernoulli, although the concept of work as the basis for investigating mechanical problems dates back to Leonardo da Vinci,[2] and some believe it can be traced to Aristotle.[3] A brief historical survey is given in Lanczos (1970). A more detailed discussion of the historical developments of variational formulations is provided in Oravas and McLean (1966).

The definition of virtual work follows directly from the definitions of work considered earlier in this chapter. Thus, from Eq. (2.6), *internal virtual work* is given by

$$\delta W_i = -\int_V \sigma_{ij}\, \delta\epsilon_{ij}\, dV \tag{2.35}$$

and *external virtual work* would be (Eq. 2.26)

$$\delta W_e = \int_{S_p} \overline{p}_i\, \delta u_i\, dS + \int_V \overline{p}_{V_i}\, \delta u_i\, dV \tag{2.36}$$

[2]Leonardo da Vinci (1452–1519) was an Italian artist and scientist. An illegitimate son who was raised by his father, he served an apprenticeship in art and studied anatomy, astronomy, botany, mathematics, engineering, and music. His accomplishments are legendary. He is famous not only for his paintings, but also for his engineering, including the design of a glider, a parachute, an elevator, a canal, an irrigation system, and a cathedral. He studied stresses in columns, arches, and walls. His dissection of cadavers led to significant improvements in the understanding of anatomy, including the heart and valves.

[3]Aristotle (384–322 B.C.), the Greek philosopher, contributed to the fields of physics, astronomy, meteorology, mathematics, biology, and psychology. At 17 he joined Plato's academy in Athens, where he remained until Plato's death in 347. In 342 he began to tutor the future Alexander the Great. On the death of Alexander in 323, Aristotle was charged with impiety. This led to voluntary exile. He died a few months later. At the Academy, Aristotle is credited with initiating a fundamental change in philosophy when he steered away from Plato's dialectic, Socratic examination in mathematics in favor of an axiomatic approach in which theorems are based on basic principles (hypotheses, definitions, and axioms). This proof-theory approach of Greek mathematics is still followed today.

This represents the actual external loads moving through the corresponding virtual displacements.

Also useful is the concept of *complementary virtual work* in which the variation applies to stress rather than displacement. The *internal complementary virtual work* is defined as

$$\delta W_i^* = -\int_V \epsilon_{ij}\, \delta\sigma_{ij}\, dV \tag{2.37}$$

while the *external complementary virtual work* would be

$$\delta W_e^* = \int_{S_u} \bar{u}_i\, \delta p_i\, dS \tag{2.38}$$

which corresponds to the complementary work of Eqs. (2.17) and (2.29), respectively.

Surface integrals can be transformed into volume integrals and vice versa using the Gauss (divergence) integral theorem described in Appendix II, Section II.3. Typically, the transformation of a surface integral of the sort

$$\int_S p_i\, \delta u_i\, dS \tag{2.39}$$

into a volume integral is desired. Use Chapter 1, Eq. (1.58), $p_i = \sigma_{ij}\, a_j$, and the Gauss integral theorem of Appendix II, Eq. (II.8) to obtain

$$\int_S p_i\, \delta u_i\, dS = \int_V (\sigma_{ij}\, \delta u_i)_{,j}\, dV = \int_V \sigma_{ij,j}\, \delta u_i\, dV + \int_V \sigma_{ij}\, \delta u_{i,j}\, dV \tag{2.40}$$

If the variation in stress is considered, Gauss' theorem takes the form

$$\begin{aligned}\int_S u_i\, \delta p_i\, dS &= \int_S u_i\, \delta\sigma_{ij}\, a_i\, dS = \int_V (u_i\, \delta\sigma_{ij})_{,j}\, dV \\ &= \int_V u_i\, \delta\sigma_{ij,j}\, dV + \int_V u_{i,j}\, \delta\sigma_{ij}\, dV\end{aligned} \tag{2.41}$$

2.2 Classical Variational Principles of Elasticity

In this section, the classical variational principles which are essential for pursuing variational formulations of structural mechanics are presented. At first glance, such principles tend to appear very comprehensive due, in part, to the general nomenclature and terminology that are used to express them. As will be seen, this generality is necessary to develop a proper foundation. In order to understand the basic concepts better, several simple examples are given.

Before considering the variational principles, it is worthwhile to mention the principle of conservation of energy, which is a familiar energy theorem. For solids of interest to us, this principle is the first law of thermodynamics for adiabatic processes. As an example, consider an elastic solid with static loading for which there is no loss of energy through the conversion of mechanical work into heat or through friction or other dissipative forces. Assume a potential exists for the internal forces. For this conservative system, the principle of conservation of energy is as follows: work done by the applied forces is equal to the strain energy stored in the solid. It is important to understand that this principle deals with changes in energy because the work being done corresponds to a change in energy.

2.2.1 Principle of Virtual Work

The principle of virtual work for a solid can be derived from the equations of equilibrium and vice versa. They are, in a sense, equivalent because the principle of virtual work is a global (integral) form of the conditions of equilibrium and the static boundary conditions. To look at this principle, consider a solid under prescribed body forces for which the conditions of equilibrium hold at all points throughout the body, i.e., [Chapter 1, Eq. (1.53)]

$$\sigma_{ij,j} + \overline{p}_{Vi} = 0 \quad \text{in} \quad V$$

or

$$\mathbf{D}^T \boldsymbol{\sigma} + \overline{\mathbf{p}}_V = \mathbf{0} \quad \text{in} \quad V \tag{2.42}$$

and for which the force (mechanical or static) boundary conditions are given on the surface, i.e., [Chapter 1, Eq. (1.60)]

$$p_i = \overline{p}_i \quad \text{or} \quad p_i - \overline{p}_i = 0 \quad \text{on} \quad S_p$$

or

$$\mathbf{p} = \overline{\mathbf{p}} \quad \text{or} \quad \mathbf{p} - \overline{\mathbf{p}} = \mathbf{0} \quad \text{on} \quad S_p \tag{2.43}$$

where $\overline{\mathbf{p}}$ are tractions applied to the surface area S_p. Recall that overbars signify applied quantities. Equations (2.42) and (2.43) constitute the definition of a statically admissible stress field.

The local conditions of equilibrium [Eq. (2.42)] and the static boundary conditions [Eq. (2.43)] are now to be expressed in global (integral) form. To do so, multiply Eqs. (2.42) and (2.43) by the virtual displacements δu_i, and integrate the first relation over V and the second over S_p. Take the sum of these two integrals, each of which is equal to zero, to form

$$-\int_V (\sigma_{ij,j} + \overline{p}_{Vi})\, \delta u_i \, dV + \int_{S_p} (p_i - \overline{p}_i)\, \delta u_i \, dS = 0$$

or

$$-\int_V \delta \mathbf{u}^T (\mathbf{D}^T \boldsymbol{\sigma} + \overline{\mathbf{p}}_V)\, dV + \int_{S_p} \delta \mathbf{u}^T (\mathbf{p} - \overline{\mathbf{p}})\, dS = 0 \tag{2.44}$$

It is necessary to introduce the negative sign in order to obtain consistent relations later. Note that according to the fundamental lemma of the calculus of variations (Appendix I), the integral relations of Eq. (2.44) are equivalent to the local conditions of Eqs. (2.42) and (2.43).

Now change the form of Eq. (2.44) such that the integrals can be identified as being virtual work. This is accomplished using Gauss' integral theorem of Appendix II, Eq. (II.8) [which was employed to derive Eq. (2.40)]. Since $S = S_p + S_u$, where S_u is the surface area on which displacements $\overline{u}$ are applied, it is possible to write the integral $\int_{S_p} p_i \, \delta u_i \, dS$ of Eq. (2.44) as

$$\int_{S_p} p_i \, \delta u_i \, dS = \int_S p_i \, \delta u_i \, dS - \int_{S_u} p_i \, \delta u_i \, dS \tag{2.45}$$

Substitute

$$\int_V \sigma_{ij,j}\, \delta u_i\, dV = \int_S p_i\, \delta u_i\, dS - \int_V \sigma_{ij}\, \delta u_{i,j}\, dV$$

of Eq. (2.40) into Eq. (2.44). Also, introduce Eq. (2.45) to find

$$\int_V \sigma_{ij}\, \delta u_{i,j}\, dV - \int_V \overline{p}_{Vi}\, \delta u_i\, dV - \int_{S_p} \overline{p}_i\, \delta u_i\, dS - \int_{S_u} p_i\, \delta u_i\, dS = 0 \tag{2.46}$$

To introduce strain into Eq. (2.46), we recall that in Appendix I, the virtual displacement δu_i is defined by introducing a family of neighboring functions,

$$\widehat{u}_i = u_i + \delta u_i$$

The corresponding strain-displacement relationship of Chapter 1, Eq. (1.21) is $\widehat{\epsilon} = \mathbf{D}\widehat{\mathbf{u}}$. Then

$$\begin{aligned} \widehat{\epsilon}_{ij} &= \frac{1}{2}(\widehat{u}_{i,j} + \widehat{u}_{j,i}) = \frac{1}{2}(u_{i,j} + u_{j,i}) + \frac{1}{2}\left[(\delta u_i)_{,j} + (\delta u_j)_{,i}\right] \\ &= \frac{1}{2}(u_{i,j} + u_{j,i}) + \frac{1}{2}\delta(u_{i,j} + u_{j,i}) = \epsilon_{ij} + \delta\epsilon_{ij} \end{aligned} \tag{2.47}$$

This uses Eq. (I.6) in Appendix I, $(\delta u_i)_{,j} = \delta(u_{i,j})$. It follows that if the displacements $\mathbf{u}$ and $\widehat{\mathbf{u}}$ obey their respective strain-displacement relationships, then

$$\delta\epsilon_{ij} = \frac{1}{2}\delta(u_{i,j} + u_{j,i}) \quad \text{in} \quad V \tag{2.48}$$

By using the summation convention and Eq. (2.48) it can be verified that (Problem 2.13) $\sigma_{ij}\, \delta u_{i,j} = \sigma_{ij}\, \delta\epsilon_{ij}$ and Eq. (2.46) can be written as

$$\int_V \sigma_{ij}\, \delta\epsilon_{ij}\, dV - \int_V \overline{p}_{Vi}\, \delta u_i\, dV - \int_{S_p} \overline{p}_i\, \delta u_i\, dS - \int_{S_u} p_i\, \delta u_i\, dS = 0$$

or

$$\int_V \delta\epsilon^T \boldsymbol{\sigma}\, dV - \int_V \delta\mathbf{u}^T \overline{\mathbf{p}}_V\, dV - \int_{S_p} \delta\mathbf{u}^T \overline{\mathbf{p}}\, dS - \int_{S_u} \delta\mathbf{u}^T \mathbf{p}\, dS = 0 \tag{2.49}$$

If u_i and $\widehat{u}_i$ both satisfy the geometric boundary conditions ($u_i = \widehat{u}_i$ on S_u),

$$\delta u_i = 0 \quad \text{on} \quad S_u \tag{2.50}$$

Virtual displacements that satisfy both Eqs. (2.48) and (2.50) are said to be kinematically admissible. Applying Eq. (2.50) to Eq. (2.49) causes the integral over S_u [in Eq. (2.49)] to be equal to zero. Thus,

$$\int_V \sigma_{ij}\, \delta\epsilon_{ij}\, dV - \int_V \overline{p}_{Vi}\, \delta u_i\, dV - \int_{S_p} \overline{p}_i\, \delta u_i\, dS = 0$$

or (2.51)

$$\int_V \delta\epsilon^T \boldsymbol{\sigma}\, dV - \int_V \delta\mathbf{u}^T \overline{\mathbf{p}}_V\, dV - \int_{S_p} \delta\mathbf{u}^T \overline{\mathbf{p}}\, dS = 0$$

The integrals in Eq. (2.51) can be identified as work expressions. The first integral corresponds to the internal virtual work of Eq. (2.35), and the final two integrals are [Eq. (2.36)] expressions for the work of the actual external forces moving through the corresponding virtual displacements δu_i. Hence,

$$-\delta W_i - \delta W_e = 0 \tag{2.52}$$

This can also be expressed as

$$\delta(W_i + W_e) = 0 \quad \text{or} \quad \delta W = 0 \tag{2.53}$$

with

$$W = W_i + W_e$$

Equations (2.51), (2.52), or (2.53) are expressions for the *principle of virtual work*, if the additional requirement of kinematically admissible virtual displacements is satisfied.

In summary, the equations representing the principle of virtual work appear as

$$\int_V \sigma_{ij}\,\delta\epsilon_{ij}\,dV - \int_V \overline{p}_{Vi}\,\delta u_i\,dV - \int_{S_p} \overline{p}_i\,\delta u_i\,dS = 0$$

or

$$\underbrace{\int_V \delta\boldsymbol{\epsilon}^T \boldsymbol{\sigma}\,dV}_{-\delta W_i} \underbrace{- \int_V \delta\mathbf{u}^T \overline{\mathbf{p}}_V\,dV - \int_{S_p} \delta\mathbf{u}^T \overline{\mathbf{p}}\,dS}_{-\delta W_e} = 0 \tag{2.54}$$

with kinematically admissible δu_i, i.e.,

$$\delta\epsilon_{ij} = \frac{1}{2}\delta(u_{i,j} + u_{j,i}) \quad \text{or} \quad \epsilon_{ij} = \frac{1}{2}(u_{i,j} + u_{j,i}) \quad \text{in} \quad V$$
$$\delta u_i = 0 \quad \text{or} \quad u_i = \overline{u}_i \quad \text{on} \quad S_u$$

or

$$\delta\boldsymbol{\epsilon} = \mathbf{D}\,\delta\mathbf{u} \quad \text{or} \quad \boldsymbol{\epsilon} = \mathbf{D}\mathbf{u} \quad \text{in} \quad V$$
$$\delta\mathbf{u} = \mathbf{0} \quad \text{or} \quad \mathbf{u} = \overline{\mathbf{u}} \quad \text{on} \quad S_u$$

This principle can be stated as the following: *A deformable system is in equilibrium if the sum of the external virtual work and the internal virtual work is zero for virtual displacements δu_i that satisfy the kinematic equations and kinematic boundary conditions, i.e., for δu_i that are kinematically admissible.*

This principle is independent of the material properties of the solid. Moreover, it has not been necessary to assume that a potential function exists for the internal or external virtual work.

The fundamental unknowns for the principle of virtual work are displacements. Although stresses or forces often appear in the equations representing the principle, these variables should be considered as being expressed as functions of the displacements. Also, the variations are always taken on the displacements in the principle of virtual work. In fact, this principle is also known as the *principle of virtual displacements.*

It has been shown in this section that for any structural system in equilibrium, Eq. (2.54) holds. This derivation can be reversed by beginning with Eq. (2.54) and then proving that the equilibrium conditions are satisfied. This implies that the principle of virtual work [Eq. (2.54)] is a necessary and sufficient condition for a structure to be in equilibrium.

Briefly, the term $\int_V \delta\boldsymbol{\epsilon}^T \boldsymbol{\sigma}\, dV$ in Eq. (2.54) can, with the aid of Eq. (2.48) [a condition underlying Eq. (2.54)], be transformed to

$$\int_V \sigma_{ij}\, \delta\epsilon_{ij}\, dV = \int_V \sigma_{ij}\, \delta u_{i,j}\, dV = \int_V (\sigma_{ij}\, \delta u_i)_{,j}\, dV - \int_V \sigma_{ij,j}\, \delta u_i\, dV \tag{2.55}$$

With the help of the divergence theorem [Eq. (2.40)] the first integral on the right-hand side can be transformed to a surface integral. Then

$$\int_V \sigma_{ij}\, \delta\epsilon_{ij}\, dV = \int_{S_u} p_i\, \delta u_i\, dS + \int_{S_p} p_i\, \delta u_i\, dS - \int_V \sigma_{ij,j}\, \delta u_i\, dV \tag{2.56}$$

Invoking the remaining kinematic boundary condition in Eq. (2.54), i.e., $\delta u_i = 0$ on S_u, the integral over S_u in Eq. (2.56) is zero. Substitution of Eq. (2.56) into the first relation of Eq. (2.54) gives Eq. (2.44), which can only be zero if the system is in equilibrium [Eq. (2.42)] and if the static boundary conditions are satisfied. That is, according to the calculus of variations, the principle of virtual work leads to the equilibrium conditions as Euler's equations with the static (force) boundary conditions as the natural boundary conditions. Thus, the principle of virtual work is basically a global formulation of the equilibrium conditions and the static boundary conditions.

The principle of virtual work as expressed by Eq. (2.54) formally contains both stresses and displacements, although only the displacements are the fundamental unknowns. This equation can be written in a pure displacement form by replacing the stresses by displacement gradients with the assistance of the stress-strain relations (a material law), which, of course, makes the principle material dependent. Note that no use of the stress-strain relations has been made in deriving the principle of virtual work of Eq. (2.54). Thus, this form is valid for systems with nonlinear materials, as well as for linear materials. For the case of linearly elastic material [Chapter 1, Eq. (1.34)], $\boldsymbol{\sigma} = \mathbf{E}\boldsymbol{\epsilon}$. Using the kinematic relation $\boldsymbol{\epsilon} = \mathbf{D}\mathbf{u}$, we find

$$\boldsymbol{\sigma} = \mathbf{E}\boldsymbol{\epsilon} = \mathbf{E}(\mathbf{D}\mathbf{u}) = \mathbf{E}\mathbf{D}_u\mathbf{u} \tag{2.57a}$$

Since $\boldsymbol{\epsilon}^T = (\mathbf{D}\mathbf{u})^T$, we have

$$\delta\boldsymbol{\epsilon}^T = \delta(\mathbf{D}\mathbf{u})^T = \delta\mathbf{u}^T{}_u\mathbf{D}^T \tag{2.57b}$$

In Eq. (2.57a), the subscript u has been added to $\mathbf{D}$ to indicate clearly that the differential operator matrix $\mathbf{D}$ operates on $\mathbf{u}$. In Eq. (2.57b), however, as explained in Chapter 1, Section 1.8.3, the subscript index to the left of the operator matrix, ${}_u\mathbf{D}^T$, signifies the application of the operator to the preceding quantity which, in this case, is $\delta\mathbf{u}^T$. Substitution of Eqs. (2.57a) and (2.57b) into the second relationship of Eq. (2.54) gives

$$\int_V \delta\mathbf{u}^T \left[({}_u\mathbf{D}^T\mathbf{E}\mathbf{D}_u)\mathbf{u} - \overline{\mathbf{p}}_V\right] dV - \int_{S_p} \delta\mathbf{u}^T\overline{\mathbf{p}}\, dS = 0 \tag{2.58a}$$

Define

$$\mathbf{k}^D = {}_u\mathbf{D}^T\mathbf{E}\mathbf{D}_u \tag{2.58b}$$

where, due to the definition of ${}_u\mathbf{D}$ and $\mathbf{D}_u$, the operator matrix $\mathbf{k}^D$ is recognized as a symmetric operator matrix. Then the displacement form for the principle of virtual work relation is

$$-\delta W = \int_V \delta\mathbf{u}^T(\mathbf{k}^D\mathbf{u} - \overline{\mathbf{p}}_V)\, dV - \int_{S_p} \delta\mathbf{u}^T\overline{\mathbf{p}}\, dS = 0 \tag{2.58c}$$

The expanded operator matrix $\mathbf{k}^D$ can be formed using the expressions for **D** and **E** from Chapter 1, Eqs. (1.22) and (1.34).

$$\mathbf{D}_u = \begin{bmatrix} \partial_x & 0 & 0 \\ 0 & \partial_y & 0 \\ 0 & 0 & \partial_z \\ \cdots & \cdots & \cdots \\ \partial_y & \partial_x & 0 \\ \partial_z & 0 & \partial_x \\ 0 & \partial_z & \partial_y \end{bmatrix}$$

$$\mathbf{E} = \frac{2G}{1-2\nu} \begin{bmatrix} 1-\nu & \nu & \nu & \vdots & 0 & 0 & 0 \\ \nu & 1-\nu & \nu & \vdots & 0 & 0 & 0 \\ \nu & \nu & 1-\nu & \vdots & 0 & 0 & 0 \\ \cdots & \cdots & \cdots & \cdots & \cdots & \cdots & \cdots \\ 0 & 0 & 0 & \vdots & (1-2\nu)/2 & 0 & 0 \\ 0 & 0 & 0 & \vdots & 0 & (1-2\nu)/2 & 0 \\ 0 & 0 & 0 & \vdots & 0 & 0 & (1-2\nu)/2 \end{bmatrix}$$

$$G = \frac{E}{2(1+\nu)}$$

It is then found that

$$-\delta W = \int_V \delta[u_x \;\; u_y \;\; u_z]$$

$$\times \left\{ \underbrace{\begin{bmatrix} {}_x\partial \frac{2G(1-\nu)}{1-2\nu} \partial_x & \vdots & {}_x\partial \frac{2G\nu}{1-2\nu} \partial_y & \vdots & {}_x\partial \frac{2G\nu}{1-2\nu} \partial_z \\ + {}_y\partial G \partial_y + {}_z\partial G \partial_z & \vdots & + \partial_y G \partial_x & \vdots & + {}_z\partial G \partial_x \\ \cdots & \cdots & \cdots & \cdots & \cdots \\ {}_y\partial \frac{2G\nu}{1-2\nu} \partial_x & \vdots & {}_y\partial \frac{2G(1-\nu)}{1-2\nu} \partial_y & \vdots & {}_y\partial \frac{2G\nu}{1-2\nu} \partial_z \\ + {}_x\partial G \partial_y & \vdots & + {}_x\partial G \partial_x + {}_z\partial G \partial_z & \vdots & + {}_z\partial G \partial_y \\ \cdots & \cdots & \cdots & \cdots & \cdots \\ {}_z\partial \frac{2G\nu}{1-2\nu} \partial_x & \vdots & {}_z\partial \frac{2G\nu}{1-2\nu} \partial_y & \vdots & {}_z\partial \frac{2G(1-\nu)}{1-2\nu} \partial_z \\ + {}_x\partial G \partial_z & \vdots & + {}_y\partial G \partial_z & \vdots & + {}_x\partial G \partial_x + {}_y\partial G \partial_y \end{bmatrix}}_{\mathbf{k}^D} \begin{bmatrix} u_x \\ u_y \\ u_z \end{bmatrix} - \begin{bmatrix} \overline{p}_{Vx} \\ \overline{p}_{Vy} \\ \overline{p}_{Vz} \end{bmatrix} \right\} dV$$

$$- \int_{S_p} \delta[u_x \;\; u_y \;\; u_z] \begin{bmatrix} \overline{p}_x \\ \overline{p}_y \\ \overline{p}_z \end{bmatrix} dS = 0 \tag{2.59}$$

where $\partial_x = \frac{\partial}{\partial x}$, $\partial_y = \frac{\partial}{\partial y}$, $\partial_z = \frac{\partial}{\partial z}$ and ${}_x\partial$, ${}_y\partial$, ${}_z\partial$ are partial derivatives acting on the preceding variable, i.e., u_x, u_y, or u_z. In Eq. (2.59), the only unknowns are the displacements. As will be seen later, this form of the principle of virtual work embodies the displacement method of structural analysis. The introduction of the operator matrix $\mathbf{k}^D$ is very useful for a unified description of several different problem types. It reveals the structure of a corresponding

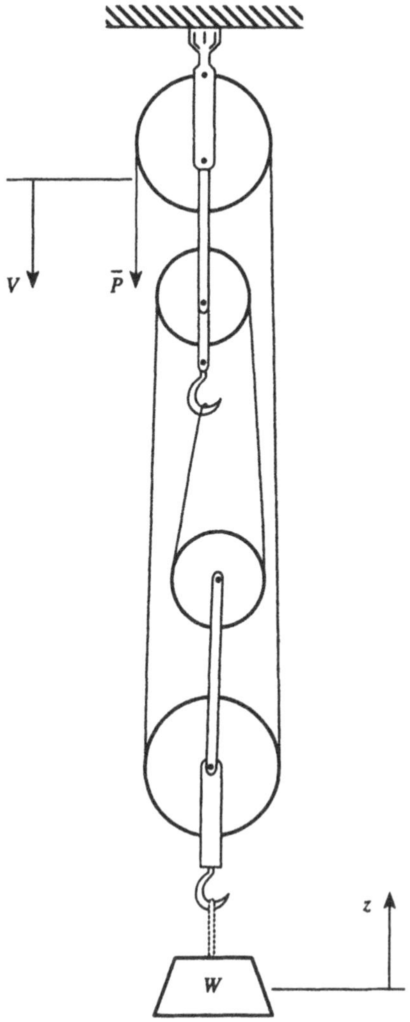

FIGURE 2.4
Pulley system.

nonoperator matrix (the stiffness matrix) which is the central ingredient of the displacement method. As indicated in Chapter 4, the stiffness matrix follows from $\mathbf{k}^D$ upon introduction of unit displacement functions or appropriate trial functions.

EXAMPLE 2.3 Pulley

Determine the force $\overline{P}$ required to raise a weight W if the pulley of Fig. 2.4 consists of four frictionless rollers.

From the principle of virtual work or virtual displacements,

$$-\delta W_i - \delta W_e = W\,\delta z - \overline{P}\,\delta V = 0 \quad \text{or} \quad \overline{P}\,\delta V = W\,\delta z \tag{1}$$

The mechanics of the pulley system dictates that geometrically admissible virtual displacements δV and δz are related according to $\delta V = 4\,\delta z$. Substitution of this relationship into (1) gives $\overline{P} = W/4$ as the force required for equilibrium. Any greater force will raise the weight. ∎

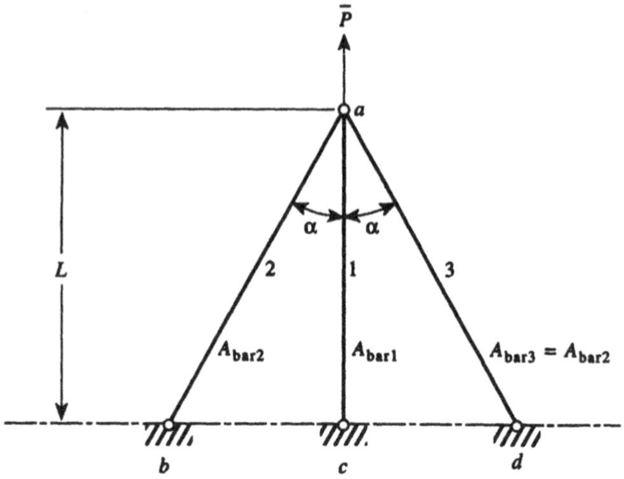

FIGURE 2.5
Statically indeterminate structure.

EXAMPLE 2.4 Truss

Find the displacement of point a and the forces in the members of the statically indeterminate structure of Fig. 2.5.

For the principle of virtual work, apply a kinematically admissible virtual displacement and use Eq. (2.54) to find a condition relating the unknowns. Choose a virtual displacement of δV of point a in the vertical direction. The external virtual work is $\delta W_e = \overline{P}\,\delta V$. Then the first equation in Eq. (2.54) becomes

$$\sum_i (\sigma\ \delta\epsilon\ L\ A)^i - \overline{P}\,\delta V = 0 \tag{1}$$

where $i = 1, 2, 3$ identifies the number of each truss element as shown in Fig. 2.5. Suppose that the geometry of the undeformed truss is valid for the deformed configuration. The elongation of bar 1 is δV, and the kinematically admissible displacement of bar 2 or 3 is $\delta V \cos\alpha$. Then the virtual strain in bar 1 is $\delta V/L$, and in bar 2 or 3, it is

$$\delta\epsilon = \frac{\delta V \cos\alpha}{L/\cos\alpha} = \delta V \frac{\cos^2\alpha}{L} \tag{2}$$

The stresses can be expressed in terms of the unknown displacement V. Since the strains are $\epsilon_{\text{bar1}} = \frac{V}{L}$, $\epsilon_{\text{bar2}} = (V\cos\alpha)\frac{\cos\alpha}{L} = \epsilon_{\text{bar3}}$, Hooke's law gives the stresses

$$\sigma_{\text{bar1}} = \frac{EV}{L}, \quad \sigma_{\text{bar2}} = \sigma_{\text{bar3}} = EV\frac{\cos^2\alpha}{L} \tag{3}$$

where E is the modulus of elasticity for each of these bars. Substitution of the virtual strains (2) and the stresses (3) into the principle of virtual work (1) yields

$$\left(\frac{EV}{L}\right)\left(\frac{\delta V}{L}\right) L\,A_{\text{bar1}} + 2\left(\frac{EV\cos^2\alpha}{L}\right)\left(\frac{\delta V\cos^2\alpha}{L}\right)\left(\frac{L}{\cos a}\right) A_{\text{bar2}} - \overline{P}\,\delta V = 0 \tag{4}$$

Factoring out δV, one sees that for an arbitrary virtual displacement δV

$$\frac{EAV}{L} + 2\left(\frac{EAV\cos^3\alpha}{L}\right) - \overline{P} = 0 \tag{5}$$

for $A_{\text{bar1}} = A_{\text{bar2}} = A$.

Thus, the displacement of point a is

$$V = \frac{\overline{P}L}{EA(1 + 2\cos^3\alpha)} \tag{6}$$

The internal forces N in each bar are found from the stresses (3) in terms of V (which is given by (6) as a function of the known applied load $\overline{P}$) by using $N = \sigma A$.

An alternative solution to this problem is obtained by using the principle in the form $-\delta W_i - \delta W_e = 0$, in which δW_i is written in terms of the internal forces. For the kinematically admissible virtual displacement, the internal virtual work is $-\delta W_i = N_{\text{bar1}}\delta V + N_{\text{bar2}}\delta V\cos\alpha + N_{\text{bar3}}\delta V\cos\alpha$. From $-\delta W_i = \delta W_e$,

$$\overline{P} = N_{\text{bar1}} + N_{\text{bar2}}\cos\alpha + N_{\text{bar3}}\cos\alpha \tag{7}$$

This, as is to be anticipated from the principle of virtual work, is the equilibrium relationship that would be obtained by summing forces in the vertical direction. As noted previously in this section, in order to evaluate the internal forces, one turns to Hooke's law. The internal forces in terms of the unknown displacement V are

$$N_{\text{bar1}} = \frac{EAV}{L}, \quad N_{\text{bar2}} = N_{\text{bar3}} = V\cos\alpha\left(\frac{EA}{L/\cos\alpha}\right) = V\frac{EA}{L}\cos^2\alpha \tag{8}$$

Substitution of the relations of (8) into (7) gives (6) again. ∎

EXAMPLE 2.5 In-Plane Deformation of a Flat Element

The principle of virtual work as represented by Eq. (2.54) can be specialized for two-dimensional problems such as the in-plane deformation of flat element (plate) lying in the xy plane. For this case, the virtual work relation [Eq. (2.54)]

$$-\delta W = \int_V \delta\boldsymbol{\epsilon}^T\boldsymbol{\sigma}\,dV - \int_V \delta\mathbf{u}^T\overline{\mathbf{p}}_V\,dV - \int_{S_p}\delta\mathbf{u}^T\overline{\mathbf{p}}\,dS = 0 \tag{1}$$

becomes

$$-\delta W = \int_A \delta\boldsymbol{\epsilon}^T\mathbf{s}\,dA - \int_A \delta\mathbf{u}^T\overline{\mathbf{p}}_V\,dA - \int_{S_p}\delta\mathbf{u}^T\overline{\mathbf{p}}\,ds = 0 \tag{2}$$

or

$$-\delta W = \int_A \delta\mathbf{u}^T({}_u\mathbf{D}^T\mathbf{s} - \overline{\mathbf{p}}_V)\,dA - \int_{S_p}\delta\mathbf{u}^T\overline{\mathbf{p}}\,ds = 0$$

where, from Chapter 1,

$$n_x = \int_{-\frac{t}{2}}^{\frac{t}{2}}\sigma_x\,dz, \quad n_y = \int_{-\frac{t}{2}}^{\frac{t}{2}}\sigma_y\,dz, \quad n_{xy} = \int_{-\frac{t}{2}}^{\frac{t}{2}}\tau_{xy}\,dz$$

$$\delta\boldsymbol{\epsilon}^T = \delta(\mathbf{D}\mathbf{u})^T = \delta\mathbf{u}^T{}_u\mathbf{D}^T, \quad \mathbf{u} = \begin{bmatrix} u_x \\ u_y \end{bmatrix} = \begin{bmatrix} u \\ v \end{bmatrix},$$

$$\mathbf{D} = \begin{bmatrix} \partial_x & 0 \\ 0 & \partial_y \\ \partial_y & \partial_x \end{bmatrix}, \quad \mathbf{s} = \begin{bmatrix} n_x \\ n_y \\ n_{xy} \end{bmatrix}$$

$$\overline{\mathbf{p}}_V = \begin{bmatrix} \overline{p}_{V_x} \\ \overline{p}_{V_y} \end{bmatrix} \quad \overline{\mathbf{p}} = \begin{bmatrix} \overline{p}_x \\ \overline{p}_y \end{bmatrix}$$

Also, A is the surface area of the flat element in the xy plane, s is a coordinate along the perimeter, S_p is the total length of the perimeter, $\overline{\mathbf{p}}_V$ is the weight, and $\overline{\mathbf{p}}$ contains the edge loads. Introduce the material law $\mathbf{s} = \mathbf{E}\boldsymbol{\epsilon} = \mathbf{E}\mathbf{D}_u\mathbf{u}$

$$-\delta W = \int_A \delta\mathbf{u}^T(\mathbf{k}^D\mathbf{u} - \overline{\mathbf{p}}_V)\,dA - \int_{S_p} \delta\mathbf{u}^T\overline{\mathbf{p}}\,ds = 0 \tag{3}$$

where, for plane stress [Eq. (1.39)],

$$\mathbf{E} = D\begin{bmatrix} 1 & \nu & 0 \\ \nu & 1 & 0 \\ 0 & 0 & \frac{1-\nu}{2} \end{bmatrix} \qquad D = \frac{Et}{1-\nu^2} \tag{4}$$

so that

$$\mathbf{k}^D = {}_u\mathbf{D}^T\mathbf{E}\mathbf{D}_u = \begin{bmatrix} {}_x\partial D\,\partial_x & \vdots & {}_x\partial\,\nu D\,\partial_y \\ +{}_y\partial\frac{D(1-\nu)}{2}\partial_y & \vdots & +{}_y\partial\frac{D(1-\nu)}{2}\partial_x \\ \cdots\cdots\cdots & \vdots & \cdots\cdots\cdots \\ {}_y\partial\,\nu D\,\partial_x & \vdots & {}_y\partial D\,\partial_y \\ +{}_x\partial\frac{D(1-\nu)}{2}\partial_y & \vdots & +{}_x\partial\frac{D(1-\nu)}{2}\partial_x \end{bmatrix} \tag{5}$$

■

EXAMPLE 2.6 Differential Equations of Equilibrium for a Beam

In the same manner that the equations of equilibrium of an elastic solid were shown to be equivalent to the principle of virtual work, the principle can be used to establish governing differential equilibrium equations for a structural member such as a beam. According to the principle,

$$-\delta W = \int_V \delta\boldsymbol{\epsilon}^T\boldsymbol{\sigma}\,dV - \int_V \delta\mathbf{u}^T\overline{\mathbf{p}}_V\,dV - \int_{S_p} \delta\mathbf{u}^T\overline{\mathbf{p}}\,dS = 0 \tag{1}$$

For the engineering beam theory of Chapter 1, Section 1.8, the first integral reduces to

$$\int_V \delta\boldsymbol{\epsilon}^T\boldsymbol{\sigma}\,dV = \int_V (\sigma_x\,\delta\epsilon_x + \tau_{xz}\,\delta\gamma_{xz})\,dV \tag{2}$$

The strain-displacement relations, which in accordance with the principle of virtual work must be satisfied by the displacements, are

$$\epsilon_x = \frac{\partial u}{\partial x} = u_0' + z\theta' \tag{3}$$

$$\gamma_{xz} = \gamma = \frac{\partial u}{\partial z} + \frac{\partial w}{\partial x} = \theta + w' \tag{4}$$

where from Chapter 1, Eq. (1.98), $u = u_0(x) + z\,\theta(x)$. Then (2) becomes

$$\begin{aligned} &\int_V (\sigma_x\,\delta\epsilon_x + \tau_{xz}\,\delta\gamma_{xz})\,dV \\ &\quad= \int_V \sigma_x\,\delta u_0'\,dV + \int_V \sigma_x z\,\delta\theta'\,dV + \int_V \tau_{xz}(\delta\theta + \delta w')\,dV \\ &\quad= \int_x [N\,\delta u_0' + M\,\delta\theta' + V(\delta\theta + \delta w')]\,dx \end{aligned} \tag{5}$$

where the internal axial force N, bending moment M, and shear force V have been defined as

$$N = \int_A \sigma_x \, dA \qquad M = \int_A \sigma_x \, z \, dA \qquad V = \int_A \tau_{xz} \, dA \tag{6}$$

Thus, (2) reduces to

$$\int_V \delta\boldsymbol{\epsilon}^T \boldsymbol{\sigma} \, dV = \int_x \delta\boldsymbol{\epsilon}^T \mathbf{s} \, dx \tag{7}$$

where, as in Chapter 1, Section 1.8,

$$\begin{array}{c} \begin{bmatrix} \epsilon_{0x} \\ \gamma \\ \kappa \end{bmatrix} \\ \boldsymbol{\epsilon} \end{array} \begin{array}{c} = \\ = \end{array} \begin{array}{c} \begin{bmatrix} d_x & 0 & 0 \\ 0 & d_x & 1 \\ 0 & 0 & d_x \end{bmatrix} \\ \mathbf{D}_u \end{array} \begin{array}{c} \begin{bmatrix} u_0 \\ w \\ \theta \end{bmatrix} \\ \mathbf{u} \end{array} \qquad \mathbf{s} = \begin{bmatrix} N \\ V \\ M \end{bmatrix} \tag{8}$$

In a similar fashion, the integral over the body force $\overline{\mathbf{p}}_V$ in (1) can be rewritten as

$$\int_V \delta\mathbf{u}^T \overline{\mathbf{p}}_V \, dV = \int_x \delta\mathbf{u}^T \overline{\mathbf{p}} \, dx \tag{9}$$

where

$$\overline{\mathbf{p}} = \begin{bmatrix} \overline{p}_x \\ \overline{p}_z \\ \overline{m} \end{bmatrix} \tag{10}$$

Here, $\overline{p}_x$ and $\overline{p}_z$ are force per unit length, and $\overline{m}$ is the moment intensity, i.e., the moment per unit length.

The final integral of (1) represents the virtual work performed by surface loads. If these surface forces are applied at the ends $(0, L)$ of the beam, they would be concentrated forces, and the integral would reduce to

$$\int_{S_p} \delta\mathbf{u}^T \overline{\mathbf{p}} \, dS = \left[\delta\mathbf{u}^T \, \overline{\mathbf{s}}\right]_0^L \tag{11}$$

where

$$\overline{\mathbf{s}} = \begin{bmatrix} \overline{N} \\ \overline{V} \\ \overline{M} \end{bmatrix} \tag{12}$$

are the concentrated forces at the ends. This expression is readily adjusted to account for in-span concentrated loads.

In summary, the principle of virtual work expression for beams would appear as

$$\begin{aligned} -\delta W &= \int_x \delta\boldsymbol{\epsilon}^T \mathbf{s} \, dx - \int_x \delta\mathbf{u}^T \overline{\mathbf{p}} \, dx - \left[\delta\mathbf{u}^T \, \overline{\mathbf{s}}\right]_0^L \\ &= \int_x [\delta u_{0,x} \, N + (\delta\theta + \delta w_{,x})V + \delta\theta_{,x} \, M] \, dx \\ &\quad - \int_x (\delta u_0 \, \overline{p}_x + \delta w \, \overline{p}_z + \delta\theta \, \overline{m}) \, dx - \left(\delta u_0 \, \overline{N} + \delta w \, \overline{V} + \delta\theta \, \overline{M}\right)_0^L \\ &= 0 \end{aligned} \tag{13}$$

In order to extract the corresponding Euler's equations, the first integral of (13) should be rewritten in a form such that the variations are taken on the displacements and not on the displacements *and* their derivatives, as is the case with (13). This will permit (13) to be expressed such that each term contains the factor $\delta \mathbf{u}^T$. The desired transformation is accomplished by integration by parts, which, in serving to switch the derivatives from one variable to another, is the one-dimensional equivalent of the multi-dimensional divergence theorem (Appendix II). This changes the initial integral to

$$\left[\delta u_0\, N + \delta w\, V + \delta\theta\, M\right]_0^L - \int_x (\delta u_0\, N' + \delta w\, V' - \delta\theta\, V + \delta\theta\, M')\, dx \tag{14}$$

The complete expression for the virtual work is now

$$\begin{aligned}-\delta W = &- \int_x [(N' + \overline{p}_x)\delta u_0 + (V' + \overline{p}_z)\delta w + (-V + M' + \overline{m})\delta\theta]\, dx \\ &+ \left[(N - \overline{N})\delta u_0 + (V - \overline{V})\delta w + (M - \overline{M})\delta\theta\right]_0^L\end{aligned} \tag{15}$$

Since the variations are arbitrary, $-\delta W = 0$ gives the equilibrium equations

$$N' = -\overline{p}_x, \qquad V' = -\overline{p}_z, \qquad V - M' = \overline{m} \quad 0 \le x \le L \tag{16}$$

and the boundary terms

$$N = \overline{N}, \qquad V = \overline{V}, \qquad M = \overline{M} \quad \text{at } x = 0, L \tag{17}$$

Thus, the principle of virtual work has led to the differential equations of equilibrium and the static boundary conditions for a beam.

In matrix notation, (15) could be expressed as

$$-\delta W = -\int_x \delta \mathbf{u}^T (\mathbf{D}_s^T \mathbf{s} + \overline{\mathbf{p}}) dx + \left[\delta \mathbf{u}^T (\mathbf{s} - \overline{\mathbf{s}})\right]_0^L \tag{18}$$

where from Chapter 1, Eq. (1.116)

$$\mathbf{D}_s^T = \begin{bmatrix} d_x & 0 & 0 \\ 0 & d_x & 0 \\ 0 & -1 & d_x \end{bmatrix} \tag{19}$$

The equivalent local differential equilibrium equations would be $\mathbf{D}_s^T \mathbf{s} + \overline{\mathbf{p}} = \mathbf{0}$ with the force boundary conditions $\mathbf{s} = \overline{\mathbf{s}}$. Of course, these are the same relations obtained in Chapter 1, Section 1.8.

These manipulations are typical of those employed in using the principle of virtual work to establish local governing differential equations. The divergence theorem is used to transform the integrals, so that the fundamental unknowns (the displacements) on which the variations are taken can be factored out, thus permitting Euler's equations to be extracted. This, of course, is the same procedure used to show the equivalence of the global form of the equilibrium equations and the principle of virtual work. ∎

EXAMPLE 2.7 The Principle of Virtual Work in Terms of Displacements for Beams

In the principle of virtual work, displacements are treated as the unknowns. If $\boldsymbol{\epsilon} = \mathbf{D}_u \mathbf{u}$ ($\delta\boldsymbol{\epsilon}^T = \delta(\mathbf{D}_u \mathbf{u})^T = \delta \mathbf{u}^T{}_u \mathbf{D}^T$) and $\mathbf{s} = \mathbf{E}\boldsymbol{\epsilon} = \mathbf{E}\mathbf{D}_u \mathbf{u}$ are substituted in Eq. (13) of Example 2.6, the displacement form of the principle of virtual work, specialized for beams, takes the form

$$
\begin{aligned}
-\delta W &= \int_x \delta \mathbf{u}^T{}_u\mathbf{D}^T\mathbf{E}\mathbf{D}_u\mathbf{u}\,dx - \int_x \delta\mathbf{u}^T\overline{\mathbf{p}}\,dx - \left[\delta\mathbf{u}^T\,\overline{\mathbf{s}}\right]_0^L \\
&= \int_x \delta\mathbf{u}^T\left({}_u\mathbf{D}^T\mathbf{E}\mathbf{D}_u\mathbf{u} - \overline{\mathbf{p}}\right)dx - \left[\delta\mathbf{u}^T\,\overline{\mathbf{s}}\right]_0^L = 0
\end{aligned}
\tag{1}
$$

For engineering beam theory [Chapter 1, Eq. (1.110)],

$$
\mathbf{E} = \begin{bmatrix} EA & 0 & 0 \\ 0 & k_s GA & 0 \\ 0 & 0 & EI \end{bmatrix} \tag{2}
$$

Then, with $\mathbf{D}_u$ from Eq. (8) of Example 2.6,

$$
\mathbf{k}^D = {}_u\mathbf{D}^T\mathbf{E}\mathbf{D}_u = \begin{bmatrix} {}_xd\,EA\,d_x & \vdots & 0 & \vdots & 0 \\ \cdots & & \cdots & & \cdots \\ 0 & \vdots & {}_xd\,k_sGA\,d_x & \vdots & {}_xd\,k_sGA \\ \cdots & & \cdots & & \cdots \\ 0 & \vdots & k_sGA\,d_x & \vdots & {}_xd\,EI\,d_x + k_sGA \end{bmatrix} \tag{3}
$$

where $d_x = d/dx$ and ${}_xd$ is the ordinary derivative on the preceding variable. If shear deformation effects are not included, matrix $\mathbf{E}$ reduces to [Chapter 1, Eq. (1.107)]

$$
\mathbf{E} = \begin{bmatrix} EA & 0 \\ 0 & EI \end{bmatrix} \tag{4}
$$

and $\mathbf{D}_u$ is given by [Chapter 1, Eq. (1.101)]

$$
\mathbf{D}_u = \begin{bmatrix} d_x & 0 \\ 0 & -d_x^2 \end{bmatrix} \tag{5}
$$

Then $\mathbf{k}^D$ becomes

$$
\mathbf{k}^D = {}_u\mathbf{D}^T\mathbf{E}\mathbf{D}_u = \begin{bmatrix} {}_xd\,EA\,d_x & 0 \\ 0 & {}_xd^2\,EI\,d_x^2 \end{bmatrix} \tag{6}
$$

It is important to remember that the operator ${}_u\mathbf{D}$ acts on $\delta\mathbf{u}^T$ in (1). Hence, $\delta\mathbf{u}^T$ has not been factored out of the integrand of the integral of (1) and the quantity $({}_u\mathbf{D}^T\mathbf{E}\mathbf{D}_u\mathbf{u} - \overline{\mathbf{p}})$ does not represent Euler's equations for the beam. The operator $\mathbf{k}^D$ is most useful in establishing a discrete model on which a computational solution can be based. In order to reduce (1) to a form from which the local equations can be extracted, it is necessary to first apply integration by parts (or the divergence theorem) to transfer the derivatives away from $\delta\mathbf{u}^T$ so that $\delta\mathbf{u}^T$ can be correctly factored out. This immediately provides the local beam equations

$$
\begin{aligned}
\frac{d}{dx}\left(EA\frac{du_0}{dx}\right) - \overline{p}_x &= 0 \\
\frac{d^2}{dx^2}\left(EI\frac{d^2w}{dx^2}\right) - \overline{p}_z &= 0
\end{aligned}
\tag{7}
$$

along with appropriate boundary conditions. ∎

EXAMPLE 2.8 Torsion, An Example of Field Theory Equations

The governing equations for the Saint Venant torsion problem were considered in Chapter 1, Section 1.9, as an example of the theory of elasticity. The principle of virtual work will now be used to provide equilibrium relations and static boundary conditions by assuming kinematically admissible displacements at the outset.

As in Chapter 1, Section 1.9, an element of shaft from $x = a$ to $x = b$ with a constant torque M_t along the bar axis will be examined. Suppose this torque is caused by an applied torque $\overline{M}_t$ at the $x = a$ end. As shown in Chapter 1, Eq. (1.142), the kinematic conditions are

$$\epsilon_x = 0, \quad \epsilon_y = 0, \quad \epsilon_z = 0, \quad \gamma_{yz} = 0$$
$$\gamma_{xy} = -\phi'\left(z + \frac{\partial\omega}{\partial y}\right), \quad \gamma_{xz} = -\phi'\left(-y + \frac{\partial\omega}{\partial z}\right) \tag{1}$$

From Eq. (2.54),

$$-\delta W = -\delta(W_i + W_e) = \int_V (\tau_{xy}\,\delta\gamma_{xy} + \tau_{xz}\,\delta\gamma_{xz})\,dV - \overline{M}_t\,\delta\phi|_a^b = 0 \tag{2}$$

Substitution of the kinematic relations (1) into (2) yields

$$\int_x\int_A \left\{-\tau_{xy}\left[\left(z + \frac{\partial\omega}{\partial y}\right)\delta\phi' + \phi'\,\delta\frac{\partial\omega}{\partial y}\right] + \tau_{xz}\left[\left(y - \frac{\partial\omega}{\partial z}\right)\delta\phi' - \phi'\,\delta\frac{\partial\omega}{\partial z}\right]\right\} dA\,dx - \overline{M}_t\,\delta\phi|_a^b$$
$$= \int_x\left[\int_A (\tau_{xz}\,y - \tau_{xy}\,z)\,dA - \int_A\left(\tau_{xz}\frac{\partial\omega}{\partial z} + \tau_{xy}\frac{\partial\omega}{\partial y}\right)dA\right]\delta\phi'\,dx$$
$$- \int_x\int_A \phi'\left(\tau_{xy}\,\delta\frac{\partial\omega}{\partial y} + \tau_{xz}\,\delta\frac{\partial\omega}{\partial z}\right)dA\,dx - \overline{M}_t\,\delta\phi|_a^b = 0 \tag{3}$$

where x is the coordinate along the axis of the bar, and A is the cross-sectional area. We will apply Green's theorem to the second and third surface integrals.

Green's theorem of Appendix II, Eq. (II.6), applied to a typical term in (3) appears as

$$\int_A \tau_{xy}\left(\frac{\partial\omega}{\partial y}\right)dA = \oint \tau_{xy}\,\omega\frac{dz}{ds}ds - \int_A \frac{\partial\tau_{xy}}{\partial y}\omega\,dA \tag{4}$$

Then, with Chapter 1, Eq. (1.148),

$$\left[(M_t - \overline{M}_t)\delta\phi\right]_a^b + \int_A\left(\frac{\partial\tau_{xz}}{\partial z} + \frac{\partial\tau_{xy}}{\partial y}\right)\delta(\omega\phi')\,dA\,dx$$
$$-L\oint_{S_p}\left(-\tau_{xz}\frac{dy}{ds} + \tau_{xy}\frac{dz}{ds}\right)\delta(\omega\phi')\,ds = 0 \tag{5}$$

where L is the length of the bar segment. Since the variations are arbitrary, it can be concluded that

$$\tau_{xz,z} + \tau_{xy,y} = 0 \quad \text{on} \quad A \quad \text{(equation of equilibrium)} \tag{6a}$$
$$-\tau_{xz}\frac{dy}{ds} + \tau_{xy}\frac{dz}{ds} = 0 \quad \text{or} \quad -\tau_{xz}a_z + \tau_{xy}a_y = 0 \quad \text{on} \quad S_p \tag{6b}$$
$$\text{(static boundary condition)}$$
$$M_t = \int_A (\tau_{xz}\,y - \tau_{xy}\,z)\,dA = \overline{M}_t \quad \text{on} \quad a, b \tag{6c}$$

The equation of equilibrium of (6a) reduces immediately with the assistance of Chapter 1, Eqs. (1.142) and (1.143), to the displacement form of the governing differential equation:

$$\frac{\partial^2 \omega}{\partial y^2} + \frac{\partial^2 \omega}{\partial z^2} = 0 \quad \text{or} \quad \nabla^2 \omega = 0 \tag{7}$$

where $\nabla^2 = \partial_y^2 + \partial_z^2$. This is the Laplace equation. As indicated in Chapter 1, Section 1.9.4, (6c) can be reduced to the governing equation along the bar by introducing the torsional constant of Chapter 1, Eq. (1.150). Thus

$$GJ\,\phi' = M_t \tag{8}$$

■

2.2.2 Principle of Stationary Potential Energy

The principle of virtual work can be specialized for systems for which a potential exists for both the internal and external forces. In the case of internal forces, it follows from Eqs. (2.18b) and (2.22) that

$$\delta U_0 = \frac{\partial U_0(\epsilon)}{\partial \epsilon_{ij}}\,\delta\epsilon_{ij} = \sigma_{ij}\,\delta\epsilon_{ij} \tag{2.60}$$

Then the left-hand integral of the first equation in Eq. (2.54) can be written as

$$\int_V \sigma_{ij}\,\delta\epsilon_{ij}\,dV = \int_V \delta U_0\,dV = \delta \int_V U_0\,dV = \delta U_i \tag{2.61}$$

Now the first relation in Eq. (2.54) appears as

$$\int_V \delta U_0\,dV - \int_V \overline{p}_{Vi}\,\delta u_i\,dV - \int_{S_p} \overline{p}_i\,\delta u_i\,dS = 0 \tag{2.62}$$

Suppose that in the final two integrals of Eq. (2.62) the surface tractions and body forces do not alter their magnitudes and directions during deformation, i.e., they are derivable from a potential, so that we can write

$$\delta\left[\int_V U_0\,dV - \int_V \overline{p}_{Vi}u_i\,dV - \int_{S_p} \overline{p}_i u_i\,dS\right] = 0 \tag{2.63}$$

or $\delta\Pi = 0$ where

$$\Pi = \int_V U_0\,dV - \int_V \overline{p}_{Vi}\,u_i\,dV - \int_{S_p} \overline{p}_i\,u_i\,dS = U_i + U_e \tag{2.64}$$

with U_e defined by Eq. (2.32b). It is known from the calculus of variations that a zero first variation $\delta\Pi$ is equivalent to a stationary value Π. In other words, Π assumes an extremal value. We have thus arrived at the *principle of a stationary value of the total potential energy* or simply the *principle of stationary potential energy.*

Of all kinematically admissible deformations, the actual deformations (those which correspond to stresses which satisfy equilibrium) are the ones for which the total potential energy assumes a stationary value, i.e., an extremal value.

The principle of stationary potential energy and the principle of virtual work appear to be the same in form. The difference between them is that for the principle of stationary potential energy, the internal and external forces must be derivable from a potential. For the principle of virtual work, there is no such limitation.

The principle of stationary potential energy will be quite useful. For example, one way of determining the relative merits of approximate solutions is to compare their total potential energies. The approximate solution for which the potential energy has an extreme value is, in some sense, the best approximation.

The principle can also be written as

$$\delta\Pi = 0 \quad \text{or} \quad \Pi \rightarrow \text{extremum} \tag{2.65}$$

where

$$\left.\begin{aligned} \boldsymbol{\epsilon} &= \mathbf{Du} \text{ in } V \\ \mathbf{u} &= \overline{\mathbf{u}} \text{ on } S_u \end{aligned}\right\} \text{are assumed to be satisfied.}$$

Information on the type of extremum in Eq. (2.65) can be found by studying the second variation of Π. From Π of Eq. (2.64), we can form $\Pi(u_i + \delta u_i) - \Pi(u_i) = \Delta\Pi$, the difference between the potential energies corresponding to the displacement state u_i and the neighboring state $u_i + \delta u_i$, as

$$\begin{aligned} \Delta\Pi &= \Pi(u_i + \delta u_i) - \Pi(u_i) \\ &= \int_V [U_0(\epsilon_{ij} + \delta\epsilon_{ij}) - U_0(\epsilon_{ij})]\, dV - \int_V \overline{p}_{Vi}\, \delta u_i\, dV - \int_{S_p} \overline{p}_i\, \delta u_i\, dS \end{aligned}$$

Expand $U_0(\epsilon_{ij} + \delta\epsilon_{ij})$ in a Taylor series, giving

$$U_0(\epsilon_{ij} + \delta\epsilon_{ij}) = U_0(\epsilon_{ij}) + \frac{\partial U_0}{\partial \epsilon_{ij}} \partial\epsilon_{ij} + \frac{1}{2} \frac{\partial^2 U_0}{\partial \epsilon_{kl} \partial \epsilon_{ij}} \delta\epsilon_{kl}\, \delta\epsilon_{ij} + \cdots$$

Using Eq. (2.60), we get

$$\begin{aligned} U_0(\epsilon_{ij} + \delta\epsilon_{ij}) - U_0(\epsilon_{ij}) &= \delta U_0 + \frac{1}{2}\left(\frac{\partial}{\partial \epsilon_{kl}} \sigma_{ij}\right) \delta\epsilon_{kl}\, \delta\epsilon_{ij} + \cdots \\ &= \delta U_0 + \frac{1}{2} \delta\sigma_{ij}\, \delta\epsilon_{ij} + \cdots \\ &= \delta U_0 + \frac{1}{2} E_{ijkl}\, \delta\epsilon_{kl}\, \delta\epsilon_{ij} + \cdots \\ &= \delta U_0 + \frac{1}{2} \delta\boldsymbol{\epsilon}^T \mathbf{E}\, \delta\boldsymbol{\epsilon} + \cdots \end{aligned}$$

Substitution of this expression and the definition of $\delta\Pi$ in $\Delta\Pi$ leads to

$$\Delta\Pi = \delta\Pi + \delta^2\Pi + \cdots \text{terms of higher order} \tag{2.66}$$

with $\delta^2\Pi$, the *second variation* of Π, defined as

$$\delta^2\Pi(u_i) = \frac{1}{2} \int_V \delta\boldsymbol{\epsilon}^T \mathbf{E}\, \delta\boldsymbol{\epsilon}\, dV \tag{2.67}$$

According to the principle of stationary potential energy, $\delta\Pi = 0$ for a system in equilibrium. Since $\delta\boldsymbol{\epsilon}^T \mathbf{E}\, \delta\boldsymbol{\epsilon}$ is a positive definite quadratic function,

$$\Delta\Pi = \delta^2\Pi(u_i) = \frac{1}{2} \int_V \delta\boldsymbol{\epsilon}^T \mathbf{E}\, \delta\boldsymbol{\epsilon}\, dV \geq 0 \tag{2.68}$$

Since for a linearly elastic solid the change in potential energy away from the state of equilibrium is always positive, i.e., $\Delta\Pi \geq 0$ or $\Pi(u_i + \delta u_i) \geq \Pi(u_i)$, we conclude that for linear problems the extremum of Π is an absolute minimum. Thus, this principle, in the case of a solid for which Hooke's law is applicable, is often referred to as the *principle of minimum potential energy*. It is also known as *Dirichlet's*[4] *principle.*

2.2.3 Principle of Complementary Virtual Work

For kinematically admissible displacements, the principle of virtual work and its associated theorems hold for a system in equilibrium. A "dual" to the principle of virtual work is the *principle of complementary virtual work.* For stresses (forces) that satisfy the conditions of equilibrium and the static boundary conditions—so-called statically admissible stresses—the principle of complementary virtual work and its corollary theorems hold for a kinematically compatible system. The kinematic (strain-displacement) conditions plus the displacement boundary conditions and this complementary principle are equivalent in the sense that the principle of complementary virtual work is a global (integral) form of the kinematic equation and the kinematic boundary conditions. In Section 2.2.1, variations in the displacements for fixed external forces were prescribed; now the stresses and forces for fixed displacements will be varied. Thus, *the stresses and forces are the fundamental unknowns for the principle of complementary virtual work.*

To establish the principle of complementary virtual work, proceed in a fashion similar to that used for the principle of virtual work. However, rather than dealing with a solid in equilibrium, begin with a body for which the local kinematic conditions [Chapter 1, Eqs. (1.19) or (1.21) and (1.61)] are satisfied, i.e.,

$$\epsilon_{ij} = \frac{1}{2}(u_{i,j} + u_{j,i}) \quad \text{or} \quad \boldsymbol{\epsilon} = \mathbf{Du} \quad \text{in} \quad V \tag{2.69a}$$

$$u_i = \overline{u}_i \quad \text{or} \quad \mathbf{u} = \overline{\mathbf{u}} \quad \text{on} \quad S_u \tag{2.69b}$$

where $\overline{u}_i$ or $\overline{\mathbf{u}}$ are prescribed displacements on surface area S_u.

Multiply Eq. (2.69a) by a virtual stress field $\delta\sigma_{ij}$, and integrate over the volume. Multiply Eq. (2.69b) by the virtual force δp_i, and integrate over the surface. The sum of the two expressions gives

$$\int_V (\epsilon_{ij} - u_{i,j})\, \delta\sigma_{ij}\, dV - \int_{S_u} (\overline{u}_i - u_i)\, \delta p_i\, dS = 0 \tag{2.70}$$

or

$$\int_V \delta\boldsymbol{\sigma}^T (\boldsymbol{\epsilon} - \mathbf{Du})\, dV - \int_{S_u} \delta\mathbf{p}^T (\overline{\mathbf{u}} - \mathbf{u})\, dS = 0$$

According to the fundamental lemma of the calculus of variations, Eq. (2.70) is equivalent to the kinematic conditions of Eq. (2.69).

To alter Eq. (2.70) so that the integrals can be identified as work expressions, use Gauss' integral theorem of Eq. (2.41) along with the condition of statically admissible stresses. As a "dual" to $\widehat{u}_i$, introduce the concept of statically admissible stresses $\widehat{\sigma}_{ij} = \sigma_{ij} + \delta\sigma_{ij}$ which, like the actual stresses σ_{ij}, satisfy the equilibrium conditions [Chapter 1, Eqs. (1.52) or (1.54)]

[4]Lejeune Dirichlet (1805–1859) studied mathematics at Jesuit college in Bonn, Germany, where one of his teachers was Ohm. He illuminated the works of his friend and father-in-law Jacobi and of Gauss whom he replaced in Göttingen. His own original efforts in mathematics, especially his celebrated Fourier theorem, were very significant. During his studies of mathematical physics, he treated a boundary value problem, now referred to as *Dirichlet's problem*, in which a solution is sought to Laplace's equation having prescribed values on a given surface.

in V and the static (mechanical) boundary conditions [Chapter 1, Eq. (1.60)] on S_p. Then, for the first variation of the stresses, it follows that

$$\delta\sigma_{ij,j} = 0 \quad \text{in} \quad V \tag{2.71}$$

$$\delta p_i = 0 \quad \text{on} \quad S_p \tag{2.72}$$

In Eq. (2.71), no $\overline{\mathbf{p}}_V$ term appears, since $\overline{\mathbf{p}}_V$ is prescribed and its variation would be zero. Equation (2.72) follows from [Chapter 1, Eq. (1.58)] $\sigma_{ij}\, a_i = p_j$ or $\delta\sigma_{ij}\, a_i = \delta p_j$. If $\delta\sigma_{ij} = 0$ on S_p, then $\delta p_j = 0$ on S_p.

The conditions of Eqs. (2.71) and (2.72) require that two of the integrals in the Gauss integral expression of Eq. (2.41) be zero, leaving

$$\int_{S_u} u_i\, \delta p_i\, dS = \int_V u_{i,j}\, \delta\sigma_{ij}\, dV \tag{2.73}$$

Substitution of Eq. (2.73) into Eq. (2.70) gives

$$\int_V \epsilon_{ij}\, \delta\sigma_{ij}\, dV - \int_{S_u} \overline{u}_i\, \delta p_i\, dS = 0 \tag{2.74}$$

or

$$\int_V \delta\boldsymbol{\sigma}^T \boldsymbol{\epsilon}\, dV - \int_{S_u} \delta\mathbf{p}^T \overline{\mathbf{u}}\, dS = 0$$

The integrals in Eq. (2.74) can be interpreted as virtual work expressions. The first integral corresponds to the definition of complementary work of internal forces [Eq. (2.37)]

$$-\delta W_i^* = \int_V \epsilon_{ij}\, \delta\sigma_{ij}\, dV = \int_V \delta\boldsymbol{\sigma}^T \boldsymbol{\epsilon}\, dV \tag{2.75}$$

The external complementary virtual work [Eq. (2.38)], in terms of prescribed displacements, is

$$\delta W_e^* = \int_{S_u} \overline{u}_i\, \delta p_i\, dS = \int_{S_u} \delta\mathbf{p}^T \overline{\mathbf{u}}\, dS \tag{2.76}$$

Thus, Eq. (2.74) can be written as

$$-\delta W_i^* - \delta W_e^* = -\delta W^* \quad \text{or} \quad -\delta(W_i^* + W_e^*) = 0 \tag{2.77}$$

with $W^* = W_i^* + W_e^*$. Equation (2.74) or (2.77) and the requirement of statical admissibility constitute the *principle of complementary virtual work*. This is also known as the *principle of virtual stresses* and as the *principle of virtual forces*.

In summary,

$$\int_V \epsilon_{ij}\, \delta\sigma_{ij}\, dV - \int_{S_u} \overline{u}_i\, \delta p_i\, dS = 0$$

$$\text{or} \int_V \delta\boldsymbol{\sigma}^T \boldsymbol{\epsilon}\, dV - \int_{S_u} \delta\mathbf{p}^T \overline{\mathbf{u}}\, dS = 0$$

$$-\delta W_i^* \qquad -\delta W_e^* = 0 \tag{2.78}$$

with statically admissible $\delta\sigma_{ij}$, i.e.,

$$\delta\sigma_{ij,j} = \mathbf{0} \quad \text{or} \quad \mathbf{D}^T \boldsymbol{\sigma} = \mathbf{0} \quad \text{in} \quad V$$

$$\delta p_i = 0 \quad \text{or} \quad \mathbf{p} = \overline{\mathbf{p}} \quad \text{on} \quad S_p$$

This principle can be stated as the following: *a deformable system satisfies all kinematical requirements if the sum of the external complementary virtual work and the internal complementary virtual work is zero for all statically admissible virtual stresses* $\delta\sigma_{ij}$.

The fundamental unknowns for the principle of complementary virtual work are stresses (forces). The variations are always taken on the stresses or forces.

As indicated previously, Eq. (2.78) holds if the displacement field of a structural system is kinematically admissible. Conversely, it can be shown that the conditions of Eq. (2.78) lead to kinematically admissible displacements. To do so, transform the first integral in Eq. (2.74) as follows:

$$\begin{aligned}\int_V \epsilon_{ij}\,\delta\sigma_{ij}\,dV &= \int_V (\epsilon_{ij} - u_{i,j})\delta\sigma_{ij}\,dV + \int_V u_{i,j}\,\delta\sigma_{ij}\,dV \\ &= \int_V (\epsilon_{ij} - u_{i,j})\delta\sigma_{ij}\,dV + \int_V (u_i\,\delta\sigma_{ij})_{,j}\,dV - \int_V u_i\,\delta\sigma_{ij,j}\,dV \qquad (2.79)\end{aligned}$$

Apply Gauss' theorem to the second integral on the right-hand side giving

$$\int_V (u_i\,\delta\sigma_{ij})_{,j}\,dV = \int_S u_i\,\delta p_i\,dS = \int_{S_p} u_i\,\delta p_i\,dS + \int_{S_u} u_i\,\delta p_i\,dS \qquad (2.80)$$

With the aid of Eqs. (2.79) and (2.80) and the conditions of Eqs. (2.71) and (2.72), Eq. (2.70) can be recovered from Eq. (2.74). Thus, the Euler equations of the variational principle (2.78) are the strain-displacement relations [Chapter 1, Eqs. (1.19) or (1.21)] and the displacement (geometrical) boundary conditions [Chapter 1, Eq. (1.61)] are the natural boundary conditions.

Thus, the fulfillment of the principle of complementary virtual work is an alternative statement of the conditions for kinematic admissibility of the displacement field.

Finally, a material law can be introduced to express the strains in Eq. (2.78) in terms of the stresses, and the surface tractions $\mathbf{p}$ can be written as functions of the stresses. The stresses or forces and not the displacements are the fundamental unknowns for the principle of complementary virtual work. Thus, setting $\boldsymbol{\epsilon} = \mathbf{E}^{-1}\boldsymbol{\sigma}$ [Chapter 1, Eq. (1.32)] and $\mathbf{p} = \mathbf{A}^T\boldsymbol{\sigma}$ [Eq. (1.57)], the second relation of Eq. (2.78) becomes

$$\int_V \delta\boldsymbol{\sigma}^T\mathbf{E}^{-1}\boldsymbol{\sigma}\,dV - \int_{S_u} \delta\boldsymbol{\sigma}^T\mathbf{A}\overline{\mathbf{u}}\,dS = 0 \qquad (2.81)$$

This expression can be used as the basis for the force method of structural analysis.

EXAMPLE 2.9 Torsion, An Example of Field Theory Equations

Derive the force form of the governing equations for Saint Venant torsion. See Chapter 1, Section 1.9, and Example 2.8 for related material, including notation and definitions. Consider a shaft of length L with the $x = a$ end fixed and with an angle of twist of magnitude $\phi' L$ imposed at the $x = b$ end.

The principle of complementary virtual work [Eq. (2.78)] takes the form

$$\begin{aligned}&\int_V \epsilon_{ij}\,\delta\sigma_{ij}\,dV - \int_{S_u} \overline{u}_i\,\delta p_i\,dS \\ &= \int_V (\gamma_{xy}\,\delta\tau_{xy} + \gamma_{xz}\,\delta\tau_{xz})\,dV - \int_{S_u} (-\phi' Lz\,\delta\tau_{xy} + \phi' Ly\,\delta\tau_{xz})\,dS = 0 \qquad (1)\end{aligned}$$

where, from Chapter 1, Eq. (1.139), $\overline{u}_2 = \overline{v} = -\phi' Lz$ and $\overline{u}_3 = \overline{w} = \phi' Ly$. Also, from Chapter 1, Eq. (1.146), $p_y = \tau_{xy}$ and $p_z = \tau_{xz}$ on the $x = b$ end. Introduce the Prandtl stress function ψ of Chapter 1, Eq. (1.155). This implies that the conditions of equilibrium are satisfied as required by the principle of complementary virtual work. Equation (1), with the help of the material law [Chapter 1, Eq. (1.143)], now appears as

$$\frac{L}{G}\int_A\left(\frac{\partial\psi}{\partial z}\frac{\partial}{\partial z}\delta\psi+\frac{\partial\psi}{\partial y}\frac{\partial}{\partial y}\delta\psi\right)dA-L\phi'\int_A\left(z\frac{\partial}{\partial z}\delta\psi+y\frac{\partial}{\partial y}\delta\psi\right)dA=0 \tag{2}$$

Integrate by parts the second integral in a fashion similar to that shown in Eq. (4) of Example 2.8. Use of $\psi=0$ (Chapter 1, Section 1.9.6), along the boundary of the cross-section will reduce (2) to

$$\int_A\left(\frac{\partial\psi}{\partial z}\frac{\partial\,\delta\psi}{\partial z}+\frac{\partial\psi}{\partial y}\frac{\partial\,\delta\psi}{\partial y}-2G\phi'\,\delta\psi\right)dA=0 \tag{3}$$

Integrate the first two terms by parts, and use the condition $\delta\psi=0$ on the boundary. Thus,

$$\int_A(\nabla^2\psi+2G\phi')\delta\psi\,dA=0 \tag{4}$$

This provides the force form of the governing equations as given by Chapter 1, Eq. (1.159). ∎

2.2.4 Principle of Stationary Complementary Energy

An energy principle will be provided which is "dual" to the principle of stationary potential energy. Beginning with the principle of complementary virtual work and assuming that the appropriate potential functions exists, the integral expressions in Eq. (2.78) can be written as

$$\delta\left[\int_V U_0^*(\sigma)\,dV-\int_{S_u}\mathbf{p}^T\overline{\mathbf{u}}\,dS\right]=0 \tag{2.82}$$

or

$$\delta\Pi^*=0$$

with

$$\Pi^*=U_i^*+U_e^*=\int_V U_0^*(\sigma)\,dV-\int_{S_u}\mathbf{p}^T\overline{\mathbf{u}}\,dS \tag{2.83}$$

where U_i^* is the complementary strain energy [Eq. (2.16)] with density U_0^*, and U_e^* is the potential of the prescribed displacements $\overline{\mathbf{u}}$ [Eq. (2.34)].

This is the principle of stationary complementary energy which states that *for all statically admissible states of stress, the actual state of stress (the one corresponding to kinematically compatible displacements) leads to an extremal value for the total complementary energy.* With Π^* defined by Eq. (2.83), this principle can be rewritten as

$$\delta\Pi^*=0\quad\text{or}\quad\Pi^*\rightarrow\text{extremum} \tag{2.84}$$

where

$$\left.\begin{aligned}\mathbf{D}^T\boldsymbol{\sigma}+\overline{\mathbf{p}}_V&=\mathbf{0}\text{ in }V\\ \mathbf{p}=\overline{\mathbf{p}}&\quad\text{on}\quad S_p\end{aligned}\right\}\quad\begin{aligned}&\text{are assumed to}\\&\text{be satisfied.}\end{aligned}$$

Similar to the second variation for the principle of potential energy, the second variation $\delta^2\Pi^*$ can be used to characterize the extremum.

$$\delta^2\Pi^*(\sigma_{ij})=\frac{1}{2}\int_V\delta\boldsymbol{\sigma}^T\mathbf{E}^{-1}\delta\boldsymbol{\sigma}\,dV\geq 0 \tag{2.85}$$

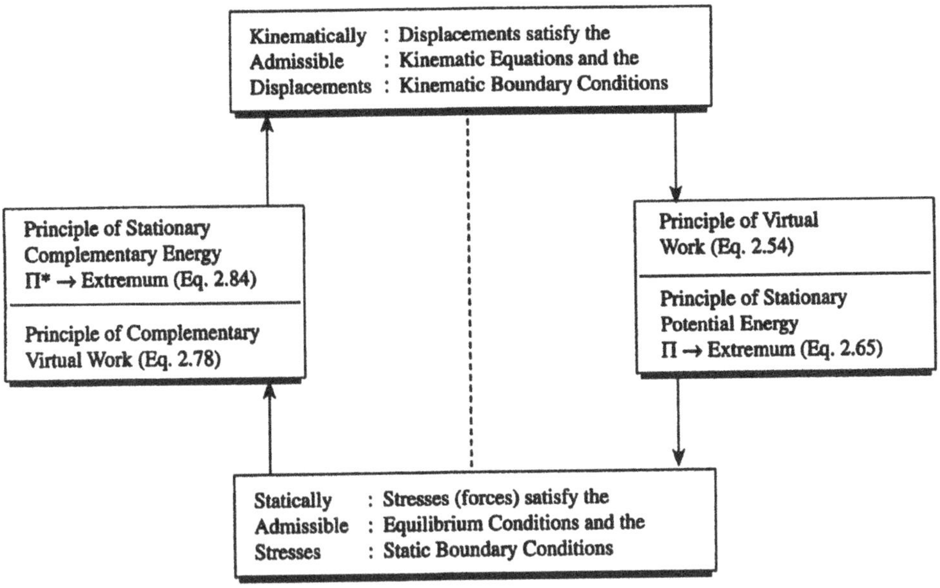

FIGURE 2.6
Duality of variational principles. Beginning with kinematically admissible displacements, the principle of virtual work provides statically admissible solutions. On the other hand, for statically admissible stresses the principle of complementary virtual work leads to kinematically admissible solutions.

Thus, for a linearly elastic body, the stationary value of Π^* is an absolute minimum. If no boundary displacements are prescribed, i.e., if $\bar{\mathbf{u}} = \mathbf{0}$ on S_u, the principle embodied by Eq. (2.84) is often referred to as the *Principle of Menabrea*[5]*/Castigliano.*[6]

The duality between the principle of Eq. (2.54) and the complementary principle of Eq. (2.78) is evident; for each principle, the underlying required initial conditions are the final results of the other principle. This duality is illustrated in Fig. 2.6.

EXAMPLE 2.10 Beam Theory

Show how the principle of stationary complementary energy as expressed by Eq. (2.84) can be applied to the one-dimensional problem of a beam. This principle should lead to the local kinematic relations for a beam.

From Eq. (2.16), the total potential complementary energy for an elastic solid can be written as

$$\Pi^* = \frac{1}{2}\int_V \boldsymbol{\sigma}^T \mathbf{E}^{-1} \boldsymbol{\sigma}\, dV - \int_{S_u} \mathbf{p}^T \bar{\mathbf{u}}\, dS \tag{1}$$

[5]Count Luigi Federico Menabrea (1809–1896) was an Italian statesman (prime minister), soldier (general), diplomat (ambassador to Paris and London), and professor of strength of materials. He is considered by many to have made the most significant contributions to the underlying theory of the energy methods for structural mechanics. Much more recognition was given to the efforts of Castigliano.

[6]Carlo Alberto Castigliano (1847–1884) was an Italian mathematician and engineer. Although he was from a poor family, he was able to receive formal education at the Industrial Engineering Institute of Asti and the Royal University of Torino. In 1873, he completed the famous dissertation "Stress in Elastic Systems." This laid a rigorous foundation for the least energy principle and presented the first of his two theorems. In 1881, his highly acclaimed book on the theory of elasticity was published. His death at 37 terminated the career of a promising engineer.

Then

$$\delta\Pi^* = \frac{1}{2}\int_V (\delta\boldsymbol{\sigma}^T\mathbf{E}^{-1}\boldsymbol{\sigma} + \boldsymbol{\sigma}^T\mathbf{E}^{-1}\delta\boldsymbol{\sigma})\,dV - \int_{S_u} \delta\mathbf{p}^T\overline{\mathbf{u}}\,dS$$
$$= \int_V \delta\boldsymbol{\sigma}^T\boldsymbol{\epsilon}\,dV - \int_{S_u} \delta\mathbf{p}^T\overline{\mathbf{u}}\,dS = 0 \tag{2}$$

This expression is the same as $-\delta W_i^* - \delta W_e^* = 0$ of Eq. (2.78). For the beam, using the notation of Example 2.6,

$$\delta\Pi^* = \int_x \delta[N\ V\ M]\begin{bmatrix}\epsilon_{0x}\\ \gamma\\ \kappa\end{bmatrix}dx - \sum_{S_u}\delta[N\ V\ M]\begin{bmatrix}\overline{u}_0\\ \overline{w}\\ \overline{\theta}\end{bmatrix} \tag{3}$$

where $\boldsymbol{\sigma}$ and $\mathbf{p}$ of (2) are replaced by the stress resultants $\mathbf{s}$, with $\mathbf{s} = [N\ V\ M]^T$. In order to derive the proper relations, it is necessary to add the identity

$$\int_V \delta\mathbf{s}^T\mathbf{D}_u\,\mathbf{u}\,dV - \int_V \delta\mathbf{s}^T\mathbf{D}_u\,\mathbf{u}\,dV = 0, \quad \mathbf{D}_u = \begin{bmatrix}\partial_x & 0 & 0\\ 0 & \partial_x & 1\\ 0 & 0 & \partial_x\end{bmatrix} \tag{4}$$

to the right-hand side of (3). This leads to

$$\delta\Pi^* = \int_x [\delta N(\epsilon_{0x} - \underline{u_0'}) + \delta V(\gamma - \underline{w' - \theta}) + \delta M(\kappa - \underline{\theta'})]\,dx$$
$$+ \int_x \underline{[\delta N u_0' + \delta V(w' + \theta) + \delta M\,\theta']\,dx} - \sum_{S_u}(\delta N\,\overline{u}_0 + \delta V\,\overline{w} + \delta M\,\overline{\theta})_0^L = 0 \tag{5}$$

where the terms due to (4) are indicated by an underline. Integration by parts of the second integral gives

$$\int_x [\delta N u_0' + \delta V(w' + \theta) + \delta M\theta']\,dx$$
$$= \sum_{S_u}(\delta N\,u_0 + \delta V\,w + \delta M\,\theta)_0^L - \int_x [\delta N' u_0 + \delta V' w + (\delta M' - \delta V)\theta]\,dx \tag{6}$$

Since for equilibrium (Chapter 1, Section 1.8.3) $\delta N' = 0$, $\delta V' = 0$, $\delta(M' - V) = 0$ along x, (6) becomes

$$\int_x [\delta N\,u_0' + \delta V(w' + \theta) + \delta M\,\theta']\,dx = \sum_{S_u}(\delta N\,u_0 + \delta V\,w + \delta M\,\theta)_0^L \tag{7}$$

Substitution of (7) into (5) gives $\delta\Pi^*$ in the form

$$\delta\Pi^* = \int_x [\delta N(\epsilon_{0x} - u_0') + \delta V(\gamma - w' - \theta) + \delta M(\kappa - \theta')]\,dx$$
$$- \sum_{S_u}[\delta N(\overline{u}_0 - u_0) + \delta V(\overline{w} - w) + \delta M(\overline{\theta} - \theta)]_0^L \tag{8}$$

or

$$\delta\Pi^* = \int_V \delta\mathbf{s}^T(\boldsymbol{\epsilon} - \mathbf{D}_u\mathbf{u}) - \int_{S_u} \delta\mathbf{p}^T(\overline{\mathbf{u}} - \mathbf{u})\,dS = 0 \tag{9}$$

Thus, for a beam the principle of stationary complementary energy leads to the kinematical relations and the geometrical boundary conditions, i.e.,

$$\begin{aligned} &\epsilon_{0x} = u_0', \quad \gamma = w' + \theta, \quad \kappa = \theta' \quad \text{along } x \\ &u_0 = \overline{u}_0, \quad w = \overline{w}, \quad \theta = \overline{\theta} \quad \text{at} \quad x = 0, L \end{aligned} \tag{10}$$

■

2.3 Generalized Variational Principles

In Section 2.2, the most important classical variational principles were developed. Now, it will be shown how the integral expressions of the classical principles can be modified to achieve more comprehensive forms which are referred to as *generalized variational principles*. Whereas the classical variational principles can be considered as single field principles involving either displacements [Eq. (2.58a)] or forces [Eq. (2.81)] as unknowns, the generalized principles may involve two fields such as displacements and forces simultaneously as unknowns. These derived principles, like the classical principles, are useful for formulations and numerical solutions of structural mechanics problems and, in some cases, offer advantages over the classical principles (Wunderlich, 1970 and 1973).

In establishing generalized variational principles, it is instructive to start with the set of fundamental equations of the theory of elasticity and to write them as shown in Table 2.1, where a distinction is made between equations expressed in terms of stress and displacement variables. As indicated, the two sets of equations are related through the material law. Typically, with a generalized principle, the equations of equilibrium, the kinematical relations, and the collective boundary conditions will all be fulfilled simultaneously. Displacements and forces will be varied independently of each other.

TABLE 2.1

Local Form of the Fundamental Equations for an Elastic Continuum

Force or Stress Variables	Displacement Variables
Equilibrium equations (Chapter 1, Eq. 1.54):	Kinematical (strain-displacement) equations (Chapter 1, Eq. 1.21):
$\mathbf{D}^T\boldsymbol{\sigma} + \overline{\mathbf{p}}_V = \mathbf{0}$ in V	$\boldsymbol{\epsilon} = \mathbf{Du}$ in V
Force (static or mechanical) boundary conditions (Chapter 1, Eqs. 1.57 and 1.60):	Kinematic (displacement) boundary conditions (Chapter 1, Eq. 1.61):
$\mathbf{A}^T\boldsymbol{\sigma} = \mathbf{p} = \overline{\mathbf{p}}$ on S_p	$\mathbf{u} = \overline{\mathbf{u}}$ on S_u
Material Law (linear) (Chapter 1, Eq. 1.34): $\boldsymbol{\sigma} = \mathbf{E}\boldsymbol{\epsilon}$ or $\boldsymbol{\epsilon} = \mathbf{E}^{-1}\boldsymbol{\sigma}$	

In Section 2.2, it was shown that the equilibrium equations and the force (stress) boundary conditions are equivalent to [Eq. (2.44)]

$$\boxed{\int_{S_p} \delta \mathbf{u}^T(\mathbf{p}-\overline{\mathbf{p}})\,dS = \int_V \delta \mathbf{u}^T(\mathbf{D}^T\boldsymbol{\sigma}+\overline{\mathbf{p}}_V)\,dV} \qquad \text{(A)**}$$

The δu_i are the variations of the displacement field. Similarly, the strain-displacement equations and the kinematic boundary conditions follow from [Eq. (2.70)]

$$\boxed{\int_V \delta\boldsymbol{\sigma}^T(\mathbf{D}\mathbf{u}-\boldsymbol{\epsilon})\,dV = \int_{S_u} \delta \mathbf{p}^T(\mathbf{u}-\overline{\mathbf{u}})\,dS} \qquad \text{(B)**}$$

where $\delta\sigma_{ij}$ and δp_i are the variations of the stresses and forces. Use of Gauss' integral theorem, along with the kinematic admissibility conditions $\delta\boldsymbol{\epsilon} = \mathbf{D}\delta\mathbf{u}$ in V and $\delta\mathbf{u} = \mathbf{0}$ on S_u, converts (A) into the principle of virtual work [Eq. (2.54)]

$$\boxed{\int_V \delta\boldsymbol{\epsilon}^T\boldsymbol{\sigma}\,dV - \int_V \delta\mathbf{u}^T\overline{\mathbf{p}}_V\,dV - \int_{S_p} \delta\mathbf{u}^T\overline{\mathbf{p}}\,dS = 0} \qquad \text{(C)**}$$

In a similar fashion, Gauss' integral theorem and the static admissibility conditions of $\delta\sigma_{ij,j} = 0$ in V and $\delta p_i = 0$ on S_p convert (B) into the principle of complementary virtual work [Eq. (2.78)]

$$\boxed{-\int_V \delta\boldsymbol{\sigma}^T\boldsymbol{\epsilon}\,dV + \int_{S_u} \delta\mathbf{p}^T\overline{\mathbf{u}}\,dS = 0} \qquad \text{(D)**}$$

It is possible to relax the underlying assumptions that were required for each of these principles. For example, in establishing (C), it was assumed that $\delta\mathbf{u} = 0$ on S_u. If this assumption is abandoned, then the term $\int_{S_u} \delta\mathbf{u}^T\mathbf{p}\,dS$ must be included in (C) giving

$$\int_V \delta(\mathbf{D}\mathbf{u})^T\boldsymbol{\sigma}\,dV - \int_V \delta\mathbf{u}^T\overline{\mathbf{p}}_V\,dS - \int_{S_p} \delta\mathbf{u}^T\overline{\mathbf{p}}\,dS - \int_{S_u} \underline{\delta\mathbf{u}^T\mathbf{p}\,dS} = 0 \qquad \text{(C)}$$

where the new term is underlined, and the kinematic condition $\boldsymbol{\epsilon} = \mathbf{D}\mathbf{u}$ has been inserted. Equation (C) follows directly from (A) if the underlined integral is not set equal to zero when applying Gauss' theorem. Refer to Section 2.2.1, where the underlined integral is set equal to zero in establishing Eq. (2.51).

Similarly, if it is *not* assumed that $\delta\sigma_{ij,j} = 0$ in V and $\delta p_i = 0$ on S_p, then expression (D) must be appended as [see Eq. (2.73) versus Eq. (2.41)]

$$-\int_V \delta\boldsymbol{\sigma}^T\boldsymbol{\epsilon}\,dV + \int_{S_u} \delta\mathbf{p}^T\overline{\mathbf{u}}\,dS - \int_V \underline{\delta(\mathbf{D}^T\boldsymbol{\sigma})\mathbf{u}\,dV} + \int_{S_p} \underline{\delta\mathbf{p}^T\mathbf{u}\,dS} = 0 \qquad \text{(D)}$$

These modified equations are summarized in Table 2.2. Recall that cases C and D can be returned to the forms of cases A and B, respectively, by applying Gauss' integral theorem.

**These four relationships are so important to the fundamental theme of this work that they are given the special labels A, B, C, and D, rather than equations numbers.

TABLE 2.2

Global Forms of the Fundamental Equations and the Classical Variational Principles

A

The force (static) boundary conditions and the conditions of equilibrium (Eq. 2.44)

$$\int_{S_p} \delta\mathbf{u}^T(\mathbf{p} - \overline{\mathbf{p}})\, dS = \int_V \delta\mathbf{u}^T(\mathbf{D}^T\boldsymbol{\sigma} + \overline{\mathbf{p}}_V)\, dV$$

B

The kinematical equations and displacement (kinematical) boundary conditions (Eq. 2.70)

$$\int_V \delta\boldsymbol{\sigma}^T(\mathbf{D}\mathbf{u} - \boldsymbol{\epsilon})\, dV = \int_{S_u} \delta\mathbf{p}^T(\mathbf{u} - \overline{\mathbf{u}})\, dS$$

A → Gauss' Integral (Divergence) Theorem → C

B → Gauss' Integral (Divergence) Theorem → D

C

The principle of virtual work (Eq. 2.54)

$$\int_V \delta(\mathbf{D}\mathbf{u})^T\boldsymbol{\sigma}\, dV - \int_V \delta\mathbf{u}^T\overline{\mathbf{p}}_V\, dV - \int_{S_p} \delta\mathbf{u}^T\overline{\mathbf{p}}\, dS - \int_{S_u} \underline{\delta\mathbf{u}^T\mathbf{p}\, dS} = 0$$

D

The principle of complementary virtual work (Eq. 2.78)

$$-\int_V \delta\boldsymbol{\sigma}^T\boldsymbol{\epsilon}\, dV + \int_{S_u} \delta\mathbf{p}^T\overline{\mathbf{u}}\, dS - \int_V \underline{\delta(\mathbf{D}^T\boldsymbol{\sigma})\mathbf{u}\, dV} + \int_{S_p} \underline{\delta\mathbf{p}^T\mathbf{u}\, dS} = 0$$

-- C and D are usually written without the underlined terms and then require admissible displacements and stresses, respectively.

Observe that a complete description in variational form of the solid is provided by summing, for example, Eqs. (A) and (B)

$$-\int_V \delta\mathbf{u}^T(\mathbf{D}^T\boldsymbol{\sigma} + \overline{\mathbf{p}}_V)\, dV + \int_V \delta\boldsymbol{\sigma}^T(\mathbf{D}\mathbf{u} - \boldsymbol{\epsilon})\, dV \qquad (A + B = AB)$$

$$+\int_{S_p} \delta\mathbf{u}^T(\mathbf{p} - \overline{\mathbf{p}})\, dS - \int_{S_u} \delta\mathbf{p}^T(\mathbf{u} - \overline{\mathbf{u}})\, dS = 0 \qquad (2.86)$$

That is, this expression is equivalent to

$$\mathbf{D}^T\boldsymbol{\sigma} + \overline{\mathbf{p}}_V = \mathbf{0} \quad \text{in} \quad V \qquad (2.87a)$$

$$\boldsymbol{\epsilon} = \mathbf{D}\mathbf{u} \quad \text{in} \quad V \qquad (2.87b)$$

with boundary conditions for forces:

$$\mathbf{A}^T\boldsymbol{\sigma} = \mathbf{p} = \overline{\mathbf{p}} \quad \text{on} \quad S_p \qquad (2.87c)$$

for displacements:

$$\mathbf{u} = \overline{\mathbf{u}} \quad \text{on} \quad S_u \qquad (2.87d)$$

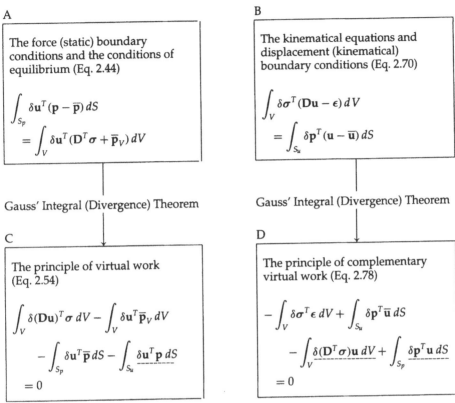

Essentially, this combination, i.e., A + B which we will denote as AB, along with the material law is a global form of all the basic equations of Table 2.1. The combined form AB can be thought of as being derived from a general variational form (functional) by employing variations of the displacements and stresses simultaneously. Note that for the displacements and stresses, derivatives of at least the first order must exist.

Combinations, such as AB, of the variational principles of Table 2.2 are referred to as *generalized variational principles*. They can be considered to be extensions of classical variational principles to which additional terms corresponding to governing equations not yet considered have been added with the aid of Lagrange multipliers (Appendix I). Thus, the functional corresponding to A + B can be considered to have been constructed from the functional corresponding to B which incorporates the kinematical equations and displacement boundary conditions to which the constraints for the force boundary conditions and the equilibrium equations are appended with the displacements **u** (or $\delta\mathbf{u}$) as Lagrange multipliers.

Suppose the expression AB is represented by $\delta\Pi_{AB}$, where Π_{AB} is the corresponding functional. Then Eq. (2.86) implies that the relationships that render Π_{AB} stationary are the complete governing equations of elasticity. Note that $\delta\Pi_{AB} = 0$ leads to only a stationary value of the functional. It cannot be proven that the stationary value is a minimum. Normally this same situation occurs with other generalized variational principles.

Table 2.3 shows three other possible combinations: $C + D = CD$, $C + B = CB$, and $A + D = AD$ [Wunderlich, 1973].

The four combinations may be summarized as follows:

$A + B = AB$	Global form of the fundamental equations
$C + D = CD$	Combination of the principle of virtual work and the principle of complementary virtual work
$C + B = CB$	Extended principle of virtual work
$A + D = AD$	Extended principle of complementary virtual work

2.3.1 Matrix Form

The generalized variational principles are displayed with index notation in Table 2.3. In order to utilize these principles for numerical solutions, it is useful to write them in matrix notation [Wunderlich, 1972]. Previously, the combination $A + B = AB$ was treated in this way; now the case of $C + B = CB$ will be discussed. From C and B of Table 2.2,

$$-\int_V \delta\boldsymbol{\sigma}^T\boldsymbol{\epsilon}\,dV + \underbrace{\int_V \left[\delta\boldsymbol{\sigma}^T\mathbf{D}\mathbf{u} + \delta(\mathbf{D}\mathbf{u})^T\boldsymbol{\sigma}\right]dV}_{=\int_V \delta(\sigma_{ij}u_{i,j})dV \text{ in index notation}} - \int_V \delta\mathbf{u}^T\overline{\mathbf{p}}_V\,dV$$

$$-\int_{S_p} \delta\mathbf{u}^T\overline{\mathbf{p}}\,dS - \underbrace{\int_{S_u} \delta\left[\mathbf{p}^T(\mathbf{u}-\overline{\mathbf{u}})\right]dS}_{\substack{\text{This term vanishes if the}\\ \text{displacement boundary}\\ \text{conditions are satisfied.}}} = 0 \tag{2.88}$$

In order to convert this relationship to a more useful matrix form, introduce the material law $\boldsymbol{\epsilon} = \mathbf{E}^{-1}\boldsymbol{\sigma}$, and make use of the notation

$$\delta\boldsymbol{\sigma}^T\mathbf{D}\mathbf{u} = \delta\boldsymbol{\sigma}^T\mathbf{D}_u\mathbf{u}, \qquad \delta(\mathbf{D}\mathbf{u})^T\boldsymbol{\sigma} = \delta(\mathbf{u}^T\mathbf{D}^T)\boldsymbol{\sigma} = \delta\mathbf{u}^T{}_u\mathbf{D}^T\boldsymbol{\sigma}$$
$$\mathbf{p} = \mathbf{A}^T\boldsymbol{\sigma}, \quad \mathbf{p}^T = \boldsymbol{\sigma}^T\mathbf{A} \tag{2.89}$$

where $\mathbf{D}_u$ has been used specifically to indicate that **D** operates on **u**, and ${}_u\mathbf{D}^T$ signifies the application of the operator $\mathbf{D}^T$ to the preceding variable $\mathbf{u}^T$. Then Eq. (2.88) can be rewritten

TABLE 2.3

Generalized Variational Principles

A: Force Boundary Conditions and Equilibrium

$$\int_{S_p}(p_i - \overline{p}_i)\delta u_i\, dS = \int_V(\sigma_{ij,j} + \overline{p}_{Vi})\delta u_i\, dV$$

B: Kinematical Equations and Displacement Boundary Conditions

$$\int_V(u_{i,j} - \epsilon_{ij})\delta\sigma_{ij}\, dV = \int_{S_u}(u_i - \overline{u}_i)\delta p_i\, dS$$

$A + B = AB$ (Fundamental Equations)

$$-\int_V(\sigma_{ij,j} + \overline{p}_{Vi})\delta u_i\, dV + \int_V(u_{i,j} - \epsilon_{ij})\delta\sigma_{ij}\, dV + \int_{S_p}(p_i - \overline{p}_i)\delta u_i\, dS - \int_{S_u}(u_i - \overline{u}_i)\delta p_i\, dS = 0$$

Gauss' Theorem (A → C; B → D; AB → CD)

C: Virtual Work

$$\int_V \sigma_{ij}\,\delta u_{i,j}\, dV - \int_V \overline{p}_{Vi}\,\delta u_i\, dV - \int_{S_p} \overline{p}_i\,\delta u_i\, dS - \underline{\int_{S_u} p_i\,\delta u_i\, dS} = 0$$

D: Complementary Virtual Work

$$-\int_V \epsilon_{ij}\,\delta\sigma_{ij}\, dV + \int_{S_u} \overline{u}_i\,\delta p_i\, dS - \underline{\int_V u_i\,\delta\sigma_{ij,j}\, dV + \int_{S_p} u_i\,\delta p_i\, dS} = 0$$

$C + D = CD$

$$\int_V \sigma_{ij}\,\delta u_{i,j}\, dV - \int_V \epsilon_{ij}\,\delta\sigma_{ij}\, dV - \int_V p_{Vi}\,\delta u_i\, dV - \int_V u_i\,\delta\sigma_{ij,j}\, dV - \int_{S_p} \overline{p}_i\delta u_i\, dS + \int_{S_p} u_i\,\delta p_i\, dS - \int_{S_u} p_i\,\delta u_i\, dS + \int_{S_u} \overline{u}_i\,\delta p_i\, dS = 0$$

$C + B = CB$ (Extended Principle of Virtual Work)

C: $\int_V \sigma_{ij}\,\delta u_{i,j}\, dV - \int_V \overline{p}_{Vi}\,\delta u_i\, dV - \int_{S_p} \overline{p}_i\,\delta u_i\, dS - \int_{S_u} p_i\delta u_i\, dS$

\+

B: $-\int_V \epsilon_{ij}\,\delta\sigma_{ij}\, dV + \int_V u_{i,j}\,\delta\sigma_{ij}\, dV \qquad - \int_{S_u}(u_i - \overline{u}_i)\delta p_i\, dS = 0$

CB: $-\delta\int_V U_0^*(\sigma)\, dV + \int_V \delta(u_{i,j}\sigma_{ij})\, dV - \int_V \delta(\overline{p}_{Vi}u_i)\, dV - \int_{S_p} \delta(\overline{p}_i u_i)\, dS - \int_{S_u} \delta[(u_i - \overline{u}_i)p_i]\, dS = 0$

$A + D = AD$ (Extended Principle of Complementary Virtual Work)

A: $-\int_V \overline{p}_{Vi}\,\delta u_i\, dV - \int_V \sigma_{ij,j}\,\delta u_i\, dV + \int_{S_p}(p_i - \overline{p}_i)\delta u_i\, dS$

\+

D: $-\int_V \epsilon_{ij}\,\delta\sigma_{ij}\, dV + \int_{S_u} \overline{u}_i\,\delta p_i\, dS \qquad - \int_V u_i\,\delta\sigma_{ij,j}\, dV + \int_{S_p} u_i\,\delta p_i\, dS = 0$

AD: $-\delta\int_V U_0^*(\sigma)\, dV + \int_{S_u} \overline{u}_i\,\delta p_i\, dS - \int_V \delta[(\sigma_{ij,j} + \overline{p}_{Vi})u_i]\, dV + \int_{S_p} \delta[(p_i - \overline{p}_i)u_i]\, dS = 0$

_ C and D normally appear without the underlined terms. Then C and D require admissible displacements and stresses, respectively.

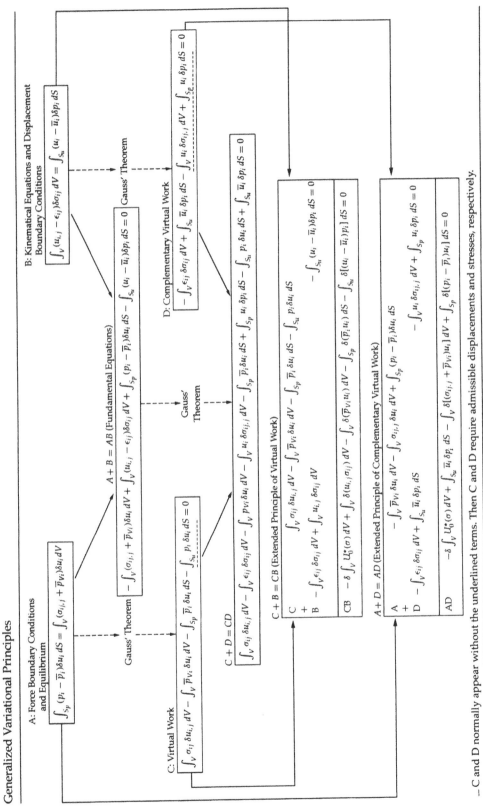

as

$$\int_V [\delta\mathbf{u}^T \quad \delta\boldsymbol{\sigma}^T]\left(\begin{bmatrix}\mathbf{0} & {}_u\mathbf{D}^T \\ \mathbf{D}_u & -\mathbf{E}^{-1}\end{bmatrix}\begin{bmatrix}\mathbf{u} \\ \boldsymbol{\sigma}\end{bmatrix} - \begin{bmatrix}\overline{\mathbf{p}}_V \\ \mathbf{0}\end{bmatrix}\right)dV - \int_{S_p} \delta\mathbf{u}^T\overline{\mathbf{p}}\,dS$$
$$- \int_{S_u} [\delta\mathbf{u}^T \quad \delta\boldsymbol{\sigma}^T]\left(\begin{bmatrix}\mathbf{0} & \mathbf{A}^T \\ \mathbf{A} & \mathbf{0}\end{bmatrix}\begin{bmatrix}\mathbf{u} \\ \boldsymbol{\sigma}\end{bmatrix} - \begin{bmatrix}\mathbf{0} \\ \mathbf{A}\overline{\mathbf{u}}\end{bmatrix}\right) dS = 0 \tag{2.90}$$

If a state vector $\mathbf{z}$ is defined as

$$\mathbf{z} = \begin{bmatrix}\mathbf{u} \\ \boldsymbol{\sigma}\end{bmatrix} \tag{2.91}$$

then expression *CB* of Eq. (2.90) appears as

$$\int_V \delta\mathbf{z}^T\left(\begin{bmatrix}\mathbf{0} & {}_u\mathbf{D}^T \\ \mathbf{D}_u & -\mathbf{E}^{-1}\end{bmatrix}\mathbf{z} - \begin{bmatrix}\overline{\mathbf{p}}_V \\ \mathbf{0}\end{bmatrix}\right) dV - \int_{S_p} \delta\mathbf{u}^T\overline{\mathbf{p}}\,dS$$
$$- \int_{S_u} \delta\mathbf{z}^T\left(\begin{bmatrix}\mathbf{0} & \mathbf{A}^T \\ \mathbf{A} & \mathbf{0}\end{bmatrix}\mathbf{z} - \begin{bmatrix}\mathbf{0} \\ \mathbf{A}\overline{\mathbf{u}}\end{bmatrix}\right) dS = 0 \tag{2.92}$$

The other generalized principles can be converted to matrix notation in a similar fashion. Definitions similar to those in Eq. (2.89) and identities of the form

$$\delta u_i\, \sigma_{ij,j} = \delta\mathbf{u}^T(\mathbf{D}^T\boldsymbol{\sigma}) = \delta\mathbf{u}^T\mathbf{D}_\sigma^T\boldsymbol{\sigma} = (\boldsymbol{\sigma}^T\mathbf{D})\delta\mathbf{u} = \boldsymbol{\sigma}^T{}_\sigma\mathbf{D}\,\delta\mathbf{u} \tag{2.93a}$$

$$\delta\sigma_{ij}\frac{1}{2}(u_{i,j} + u_{j,i}) = \delta\boldsymbol{\sigma}^T(\mathbf{D}\mathbf{u}) = \delta\boldsymbol{\sigma}^T\mathbf{D}_u\,\mathbf{u} = (\mathbf{D}\mathbf{u})^T\delta\boldsymbol{\sigma} = \mathbf{u}^T{}_u\mathbf{D}^T\delta\boldsymbol{\sigma} \tag{2.93b}$$

$$\delta u_i\, p_i = \delta\mathbf{u}^T\mathbf{p} = \delta\mathbf{u}^T\mathbf{A}^T\boldsymbol{\sigma} = \mathbf{p}^T\delta\mathbf{u} = \boldsymbol{\sigma}^T\mathbf{A}\,\delta\mathbf{u} \tag{2.93c}$$

must be used. The resulting generalized principles are shown in Table 2.4. In these expressions, the material law of Chapter 1, Eq. (1.43), which incorporates the effects of initial strains, has been employed, i.e.,

$$\boldsymbol{\epsilon} = \mathbf{E}^T\boldsymbol{\sigma} + \boldsymbol{\epsilon}^0 \tag{2.94}$$

It is of interest to observe in the variational forms of Table 2.4 that the matrices appearing in functions *CB* and *AD* are symmetric, whereas this is not the case, in general, for *AB* and *CD*. Whether or not this symmetry exists in particular instances can play an important role in the development of numerical solution techniques.

2.3.2 Related Forms

A number of useful expressions can be derived from the generalized principles. For example, if it is assumed that the applied loads can be expressed in terms of a potential, then form *CB* can be written as

$$\delta\left[-\int_V U_0^*(\sigma)\,dV + \int_V \sigma_{ij}\,u_{i,j}\,dV - \int_V u_i\,\overline{p}_{Vi}\,dV\right.$$
$$\left. - \int_{S_p} u_i\,\overline{p}_i\,dS - \int_{S_u} p_i(u_i - \overline{u}_i)\,dS\right] = 0 \tag{2.95}$$

which can be obtained from case *CB* of Table 2.3 and can be considered as the variation of a functional which will be denoted by Π_R. Then Eq. (2.95) is equivalent to setting $\delta\Pi_R$ equal to zero. The functional

$$\Pi_R = -\int_V U_0^*(\sigma)\,dV + \int_V \sigma_{ij}\,u_{i,j}\,dV - \int_V u_i\,\overline{p}_{Vi}\,dV$$
$$- \int_{S_p} u_i\,\overline{p}_i\,dS - \int_{S_u} p_i(u_i - \overline{u}_i)\,dS \tag{2.96}$$

TABLE 2.4

The Generalized Variational Principles in Matrix Form

Form AB:

$$\int_V \delta\mathbf{z}^T \left(\begin{bmatrix} 0 & -\mathbf{D}_\sigma^T \\ \mathbf{D}_u & -\mathbf{E}^{-1} \end{bmatrix} \mathbf{z} - \begin{bmatrix} \overline{\mathbf{p}}_V \\ \epsilon^0 \end{bmatrix} \right) dV + \int_{S_p} \delta\mathbf{z}^T \left(\begin{bmatrix} 0 & \mathbf{A}^T \\ 0 & 0 \end{bmatrix} \mathbf{z} + \begin{bmatrix} -\overline{\mathbf{p}} \\ 0 \end{bmatrix} \right) dS$$
$$+ \int_{S_u} \delta\mathbf{z}^T \left(\begin{bmatrix} 0 & 0 \\ -\mathbf{A} & 0 \end{bmatrix} \mathbf{z} + \begin{bmatrix} 0 \\ \mathbf{A}\overline{\mathbf{u}} \end{bmatrix} \right) dS = 0$$

Form CD:

$$\int_V \delta\mathbf{z}^T \left(\begin{bmatrix} 0 & {}_u\mathbf{D}^T \\ -{}_\sigma\mathbf{D} & -\mathbf{E}^{-1} \end{bmatrix} \mathbf{z} - \begin{bmatrix} \overline{\mathbf{p}}_V \\ \epsilon^0 \end{bmatrix} \right) dV + \int_{S_p} \delta\mathbf{z}^T \left(\begin{bmatrix} 0 & 0 \\ \mathbf{A} & 0 \end{bmatrix} \mathbf{z} + \begin{bmatrix} -\overline{\mathbf{p}} \\ 0 \end{bmatrix} \right) dS$$
$$+ \int_{S_u} \delta\mathbf{z}^T \left(\begin{bmatrix} 0 & -\mathbf{A}^T \\ 0 & 0 \end{bmatrix} \mathbf{z} + \begin{bmatrix} 0 \\ \mathbf{A}\overline{\mathbf{u}} \end{bmatrix} \right) dS = 0$$

Form CB:

$$\int_V \delta\mathbf{z}^T \left(\begin{bmatrix} 0 & {}_u\mathbf{D}^T \\ \mathbf{D}_u & -\mathbf{E}^{-1} \end{bmatrix} \mathbf{z} - \begin{bmatrix} \overline{\mathbf{p}}_V \\ \epsilon^0 \end{bmatrix} \right) dV + \int_{S_p} \delta\mathbf{z}^T \begin{bmatrix} -\overline{\mathbf{p}} \\ 0 \end{bmatrix} dS$$
$$+ \int_{S_u} \delta\mathbf{z}^T \left(\begin{bmatrix} 0 & -\mathbf{A}^T \\ -\mathbf{A} & 0 \end{bmatrix} \mathbf{z} + \begin{bmatrix} 0 \\ \mathbf{A}\overline{\mathbf{u}} \end{bmatrix} \right) dS = 0$$

Form AD:

$$\int_V \delta\mathbf{z}^T \left(\begin{bmatrix} 0 & -\mathbf{D}_\sigma^T \\ -{}_\sigma\mathbf{D} & -\mathbf{E}^{-1} \end{bmatrix} \mathbf{z} - \begin{bmatrix} \overline{\mathbf{p}}_V \\ \epsilon^0 \end{bmatrix} \right) dV + \int_{S_p} \delta\mathbf{z}^T \left(\begin{bmatrix} 0 & \mathbf{A}^T \\ \mathbf{A} & 0 \end{bmatrix} \mathbf{z} - \begin{bmatrix} \overline{\mathbf{p}} \\ 0 \end{bmatrix} \right) dS$$
$$+ \int_{S_u} \delta\mathbf{z}^T \begin{bmatrix} 0 \\ \mathbf{A}\overline{\mathbf{u}} \end{bmatrix} dS = 0$$

is often referred to as the *Hellinger*[7]*-Reissner*[8] *functional* [Hellinger, 1914 and Reissner, 1950]. It commonly forms the basis of the so-called *mixed methods* of analysis.

[7]Ernst Hellinger (1883–1950) was a German pioneer in operator theory. He attended the same high school (gymnasium) in Breslau as Richard Courant and Max Born and then became Hilbert's student in Göttingen, as did Courant and Born. Born is sometimes also given credit in the development of the fundamentals of the mixed variational theorems. Hellinger was a long-term champion of Hilbert's philosophies. After some time in Marburg, Hellinger in 1914 received a professorship at what is now the Johann Wolfgang Goethe Universität Frankfurt am Main. After a brief 1938 stay in Dachau he departed from Germany. In 1939 a temporary position, involving no university financial support, was created at Northwestern University. His position was continued for a second year with private support, and, finally, he was funded as a lecturer in mathematics by Northwestern. He retired from Northwestern in 1949 at the mandatory retirement age of 65 and, because of financial difficulties, took a temporary position at the Illinois Institute of Technology.

[8]Max Erich Reissner (1913–1996) was born in Aachen, Germany, and educated primarily in Germany, receiving his Dipl. Ing. (1935) and Dr. Ing. (1936) from the Technische Hochschule, Berlin and his Ph.D. (1938) from MIT. He began to publish technical papers as a student in 1934. Because of the political developments in Germany he departed for the United States in 1936. His professional career was spent at MIT in the Department of Mathematics and as a Professor of Applied Mechanics and Mathematics at the University of California at San Diego. He and his father as well as his son have worked in similar areas of solid mechanics. Reissner contributions to the literature on thin-walled beams, plates, and shells, the theory of elasticity, and structural mechanics were immense. Some of the ideas he proposed have become familiar theories, e.g., Reissner's plate theory and his variational theorem. His work with thin-walled structures has been influential in design. For example his shear lag theory for box beams has been important in the design of the wing-body intersection of jet aircraft.

Further transformation of Eq. (2.95) leads to a generalized form of the principle of stationary potential energy. Add

$$\delta\left[\int_V \epsilon_{ij}\,\sigma_{ij}\,dV - \int_V \epsilon_{ij}\,\sigma_{ij}\,dV\right] \tag{2.97}$$

to Eq. (2.95) and use Eq. (2.14) to obtain

$$-\int_V U_0^*(\sigma)\,dV + \int_V \sigma_{ij}\,\epsilon_{ij}\,dV = \int_V U_0(\epsilon)\,dV \tag{2.98}$$

Then Eq. (2.95) appears as

$$\delta\Bigg[\underbrace{\int_V U_0(\epsilon)\,dV - \int_V \mathbf{u}^T\overline{\mathbf{p}}_V\,dV - \int_{S_p}\mathbf{u}^T\overline{\mathbf{p}}\,dS}_{\text{Principle of virtual work C}}$$
$$\underbrace{-\int_V \underline{\boldsymbol{\sigma}^T(\boldsymbol{\epsilon} - \mathbf{D}\mathbf{u})\,dV} - \int_{S_u}\underline{\mathbf{p}^T(\mathbf{u} - \overline{\mathbf{u}})\,dS}}_{\text{Kinematics B}}\Bigg] = 0 \tag{2.99}$$

The underlined terms can be dropped for a kinematically admissible $\mathbf{u}$. The potential function implied by Eq. (2.99) is

$$\Pi = \int_V U_0(\epsilon)\,dV - \int_V \overline{p}_{Vi}\,u_i\,dV - \int_{S_p}\overline{p}_i\,u_i\,dS$$
$$+ \int_V \underline{(u_{i,j} - \epsilon_{ij})\sigma_{ij}\,dV} - \int_{S_u}\underline{(u_i - \overline{u}_i)p_i\,dS} \tag{2.100}$$

which, by comparison with Eq. (2.64), is an expanded form of the principle of stationary potential energy.

In a similar fashion, $AD = A + D$ leads to an expanded form of the functional of the complementary energy:

$$\Pi^* = -\int_V U_0^*(\sigma)\,dV + \int_{S_u}\overline{u}_i\,p_i\,dS - \int_V \underline{(\sigma_{ij,j} + \overline{p}_{Vi})u_i\,dV} + \int_{S_p}\underline{(p_i - \overline{p}_i)u_i\,dS} \tag{2.101}$$

Another important form of the generalized principles leads to the so-called hybrid functionals which are the basis of *hybrid methods of analysis*. Suppose, for example, that the displacements $\mathbf{u}$ that satisfy the kinematical relations in the body V are introduced, i.e., $\boldsymbol{\epsilon} = \mathbf{D}\mathbf{u}$. Then Eq. (2.99) reduces to

$$\delta\Bigg[\underbrace{\int_V U_0(\epsilon)\,dV - \int_V \mathbf{u}^T\overline{\mathbf{p}}_V\,dV - \int_{S_p}\mathbf{u}^T\overline{\mathbf{p}}\,dS}_{\text{Principle of virtual work C}} \underbrace{- \int_{S_u}\mathbf{p}^T(\mathbf{u} - \overline{\mathbf{u}})\,dS = 0}_{\substack{\text{Additional term for the} \\ \text{boundary } \delta u \neq 0 \text{ on } S_u}}\Bigg] \tag{2.102}$$

The corresponding functional is the hybrid functional

$$\Pi_H = \int_V U_0(\epsilon)\,dV - \int_V \mathbf{u}^T\overline{\mathbf{p}}_V\,dV - \int_{S_p}\mathbf{u}^T\overline{\mathbf{p}}\,dS - \int_{S_u}\mathbf{p}^T(\mathbf{u} - \overline{\mathbf{u}})\,dS \tag{2.103}$$

The corresponding complementary hybrid functional is obtained by assuming that the stresses $\boldsymbol{\sigma}$ satisfy equilibrium in V. Then from Eq. (2.101),

$$\Pi_H^* = \underbrace{-\int_V U_0^*(\sigma)\, dV + \int_{S_u} \mathbf{p}^T \overline{\mathbf{u}}\, dS}_{\text{Complementary virtual work D}} + \underbrace{\int_{S_p} \mathbf{u}^T(\mathbf{p} - \overline{\mathbf{p}})\, dS}_{\text{Additional term for the boundary } \delta p \neq 0 \text{ on } S_p} \tag{2.104}$$

All of the principles considered thus far involve the variation of $\delta\mathbf{u}$ or $\delta\boldsymbol{\sigma}$ or both. An interesting form, requiring the variation of $\delta\mathbf{u}$, $\delta\boldsymbol{\sigma}$, or $\delta\mathbf{e}$, is obtained if the material law is introduced explicitly as an additional condition. The four global generalized forms can be extended by including the term

$$-\int \delta\epsilon_{ij}(\sigma_{ij} - E_{ijkl}\,\epsilon_{kl})\, dV \tag{2.105}$$

In the literature, the resulting functional is sometimes referred to as the *Hu*[9]*-Washizu*[10] *functional* [Washizu, 1982]. As an example of how this modifies the generalized principles, consider the form $CB = C + B$. This first two terms of form CB in Table 2.3 would be altered. Introduce (Eq. 2.14) $-U_0^* = U_0 - \sigma_{ij}\,\epsilon_{ij}$, with the strains ϵ_{ij} being treated as independent unknowns. Then

$$\begin{aligned} -\int_V U_0^*(\sigma)\, dV + \int_V (u_{i,j}\,\sigma_{ij})\, dV &= \int_V U_0(\epsilon)\, dV - \int_V \sigma_{ij}\,\epsilon_{ij}\, dV + \int_V \sigma_{ij}\, u_{i,j}\, dV \\ &= \int_V U_0(\epsilon)\, dV + \int_V \sigma_{ij}(u_{i,j} - \epsilon_{ij})\, dV \end{aligned} \tag{2.106}$$

or rewrite the first term of form CB of Table 2.4 as

$$\int [\delta\mathbf{u}^T \quad \delta\boldsymbol{\sigma}^T \quad \delta\boldsymbol{\epsilon}^T] \begin{bmatrix} \mathbf{0} & {}_u\mathbf{D}^T & \mathbf{0} \\ \mathbf{D}_u & \mathbf{0} & -\mathbf{I} \\ \mathbf{0} & -\mathbf{I} & \mathbf{E} \end{bmatrix} \begin{bmatrix} \mathbf{u} \\ \boldsymbol{\sigma} \\ \boldsymbol{\epsilon} \end{bmatrix} dV \tag{2.107}$$

where $\mathbf{I}$ is a unit diagonal matrix.

The generalized principles AB, BC, CD and AD can be further modified by replacing the variation in the stresses by a variation in the strains with the help of the material law, i.e., by making $\delta\boldsymbol{\sigma} = \mathbf{E}\,\delta\mathbf{e}$ [Wunderlich, 1972]. In this case, the first two terms of form CB in Table 2.3 become

$$\int_V (U_0(\epsilon) - \epsilon_{ij}\, E_{ijkl}\,\epsilon_{kl} + u_{i,j}\, E_{ijkl}\,\epsilon_{kl})\, dV = \int_V (-U_0(\epsilon) + u_{i,j}\, E_{ijkl}\,\epsilon_{kl})\, dV \tag{2.108}$$

or the first term of form CB is altered to

$$\int_V [\delta\mathbf{u}^T \quad \delta\mathbf{e}^T] \begin{bmatrix} \mathbf{0} & {}_u\mathbf{D}^T\mathbf{E} \\ \mathbf{E}\mathbf{D}_u & -\mathbf{E} \end{bmatrix} \begin{bmatrix} \mathbf{u} \\ \boldsymbol{\epsilon} \end{bmatrix} dV \tag{2.109}$$

---

[9] Hai-Chang Hu was born in Zhejiang Province, China, in 1928 and graduated from the Department of Civil Engineering, Zhejiang University, in 1950. He proposed the Hu-Washizu principle in 1954. Much of his career was spent at positions in the Chinese Academy of Sciences, the Chinese Academy of Space Technology, and Beijing University.

[10] Kyuichoro Washizu (1924–1981) graduated in 1942 from Tokyo University where he subsequently became a professor of Aerospace Engineering. He contributed to such areas as variational methods and plasticity. Research with T. Pian at MIT led to a 1955 report that proposed the Hu-Washizu principle. At the time of his death he was a professor of the Faculty of Engineering Science, Osaka University.

This variational principle is useful in problems in which the strains are of immediate concern, such as in numerical comparisons with experimental results in which strains are measured directly. Also, this form is useful in the numerical analysis of large structures in which continuity requirements at member intersections are given in terms of strains rather than stresses.

EXAMPLE 2.11 Beam Theory

Use the Hellinger-Reissner functional to derive the governing equations for a beam. Include the effects of shear deformation.

We will employ Π_R of Eq. (2.96). There are only two stresses or strains of significance in engineering beam theory (see Chapter 1, Section 1.8). These are σ_x, τ_{xz} and ϵ_x, γ_{xz}. From Eq. (2.15)

$$U_0^*(\sigma) = \frac{\sigma_x^2}{2E} + \frac{2(1+v)}{2E}\tau_{xz}^2 = \frac{\sigma_x^2}{2E} + \frac{\tau_{xz}^2}{2G} \tag{1}$$

The integrand of the second term of Π_R [Eq. (2.96)] is

$$\begin{aligned} \sigma_{ij}\, u_{i,j} &= \sigma_{11}u_{1,1} + \sigma_{13}u_{1,3} + \sigma_{31}u_{3,1} \\ &= \sigma_x \frac{\partial u}{\partial x} + \tau_{xz}\left(\frac{\partial u}{\partial z} + \frac{\partial w}{\partial x}\right) = \sigma_x\epsilon_x + \tau_{xz}\gamma_{xz} = \sigma_{ij}\,\epsilon_{ij} \end{aligned} \tag{2}$$

Suppose the only applied loading is the distributed force $\overline{p}_z$. Then Π_R becomes

$$\Pi_R = \int_V \left[-\frac{\sigma_x^2}{2E} - \frac{\tau_{xz}^2}{2G} + \sigma_x\frac{\partial u}{\partial x} + \tau_{xz}\left(\frac{\partial u}{\partial z} + \frac{\partial w}{\partial x}\right)\right] dV - \int_0^L \overline{p}_z w\, dx \tag{3}$$

To proceed, introduce the kinematic relations for engineering beam theory into (3). From Chapter 1, Eq. (1.98), if the extension of the centerline is ignored, $\epsilon_x = \partial_x u + z\,\partial_x\theta$. From Eq. (1.99), $\gamma_{xz} = \partial_x u + \partial_x w = \theta + \partial_x w$. The stresses and net resultant forces are related by $\sigma_x = Mz/I$, $\tau_{xz} = \tau_{\text{average}} = \frac{V}{A}$ (see Chapter 1, Section 1.8). Then

$$\begin{aligned} \Pi_R &= \int_0^L \int_A \left[-\frac{1}{2E}\left(\frac{Mz}{I}\right)^2 - \frac{1}{2G}\left(\frac{V}{A}\right)^2 + \frac{Mz^2\partial_x\theta}{I} + \frac{V(\theta+\partial_x w)}{A}\right] dA\, dx - \int_0^L \overline{p}_z\, w\, dx \\ &= \int_0^L \left[-\frac{M^2}{2EI} - \frac{V^2}{2Gk_s A} + M\partial_x\theta + V(\theta + \partial_x w)\right] dx - \int_0^L \overline{p}_z\, w\, dx \end{aligned} \tag{4}$$

where the area A has been modified by the shear factor k_s. Now, set $\delta\Pi_R = 0$, where variations are taken independently with respect to the forces (M, V) and displacements (w, θ). Thus,

$$\begin{aligned} \delta\Pi_R = \int_0^L \bigg[&-\frac{M}{EI}\delta M - \frac{V}{Gk_s A}\delta V + \partial_x\theta\,\delta M + M\delta\partial_x\theta + (\theta + \partial_x w)\delta V \\ &+ V(\delta\theta + \delta\partial_x w) - \overline{p}_z\,\delta w \bigg] dx = 0 \end{aligned} \tag{5}$$

Integrate by parts the integrals containing $\delta\,\partial_x\theta$ and $\delta\,\partial_x w$. Then (5) becomes

$$\begin{aligned} \delta\Pi_R = [M\delta\theta + V\,\delta w]_0^L + \int_0^L \bigg[&\left(-\frac{M}{EI} + \partial_x\theta\right)\delta M + \left(-\frac{V}{Gk_s A} + \theta + \partial_x w\right)\delta V \\ &+ (-\partial_x M + V)\delta\theta + (-\partial_x V - \overline{p}_z)\delta w \bigg] dx = 0 \end{aligned} \tag{6}$$

In matrix form, with $\mathbf{z} = [w\ \theta\ V\ M]^T$ and upon integration by parts applied to some of the terms, this appears as

$$\int_0^L \delta\mathbf{z}^T \left\{ \begin{bmatrix} 0 & 0 & {}_x\partial & 0 \\ 0 & 0 & 1 & {}_x\partial \\ \partial_x & 1 & -\frac{1}{k_s GA} & 0 \\ 0 & \partial_x & 0 & -\frac{1}{EI} \end{bmatrix} \underbrace{\begin{bmatrix} w \\ \theta \\ V \\ M \end{bmatrix}}_{\mathbf{z}} - \begin{bmatrix} \overline{p}_z \\ 0 \\ 0 \\ 0 \end{bmatrix} \right\} dx + [\text{Boundary terms}] = 0 \tag{7}$$

Euler's equations of (6) are

Equilibrium conditions

$$\partial_x M = V \qquad \partial_x V = -\overline{p}_z \tag{8}$$

Kinematical equations including the material law

$$M = EI\partial_x\theta = EI\kappa \qquad V = k_s GA(\theta + \partial_x w) = k_s GA\,\gamma \tag{9}$$

Boundary conditions at $x = 0$ and $x = L$

$$M = 0 \quad \text{or} \quad \delta\theta = 0 \qquad V = 0 \quad \text{or} \quad \delta w = 0 \tag{10}$$

In summary, by utilizing the kinematic conditions initially, the remaining governing beam equations have been derived by using a variational theorem in which both displacements and forces are varied independently. ∎

EXAMPLE 2.12 Torsion, A Field Theory Example

The global form of the governing equations for the torsion problem will be derived using a generalized variational principle formed as a combination of the principles of virtual work and complementary virtual work [Zeller, 1979]. To do so, consider a bar of uniform cross-section and length L subjected to a torque $\overline{M}_t$. The cross-sectional shape is arbitrary. Suppose the left end ($x = a = 0$) is fixed and the right end ($x = b = L$), where the torque is applied, is free.

Begin with principle C, the principle of virtual work, as given by Eq. (2) of Example 2.8. This includes the term $\overline{M}_t\,\delta\phi$ for the applied torque at the boundary $x = L$, as well as a term $M_t\,\delta\phi$ that "extends" the principle for the displacement boundary condition ($\delta\phi = 0$) at $x = 0$. This condition is usually satisfied for kinematically admissible displacements. Then C becomes

$$\int_V (\tau_{xy}\,\delta\gamma_{xy} + \tau_{xz}\,\delta\gamma_{xz})\,dV - [\overline{M}_t\,\delta\phi]_L + \underline{[M_t\,\delta\phi]_0} = 0 \qquad \text{(C) or (1)}$$

The underline is used to indicate the term that is "extending" the principle.

Integration by parts will connect C to A. We choose to repeat some of the formulation presented in Example 2.8. First substitute the strain-displacement relations of Chapter 1, Eq. (1.142) into (1), giving

$$\int_V \left\{ \tau_{xy}\left[-\delta\phi'\left(\frac{\partial\omega}{\partial y} + z\right) - \phi'\,\delta\frac{\partial\omega}{\partial y}\right] + \tau_{xz}\left[-\delta\phi'\left(\frac{\partial\omega}{\partial z} - y\right) - \phi'\,\delta\frac{\partial\omega}{\partial z}\right]\right\} dV$$
$$-[\overline{M}_t\,\delta\phi]_L + [M_t\,\delta\phi]_0 = 0 \tag{2}$$

Now apply Green's theorem in the form of [Appendix II, Eq. (II.6)], as in Example 2.8.

$$-\int_x \phi' \int_A \left(\tau_{xy}\,\delta\frac{\partial\omega}{\partial y} + \tau_{xz}\,\delta\frac{\partial\omega}{\partial z}\right) dA\,dx$$
$$= -\int_x \left[\oint \delta\omega\,\phi'(\tau_{xz}a_z + \tau_{xy}\,a_y)\,ds - \int_A \delta\omega\,\phi'\left(\frac{\partial\tau_{xy}}{\partial y} + \frac{\partial\tau_{xz}}{\partial z}\right) dA\right] dx \tag{3}$$

$$-\int_x \delta\phi' \int_A \left(\tau_{xy}\frac{\partial\omega}{\partial y} + \tau_{xz}\frac{\partial\omega}{\partial z}\right) dA\,dx$$
$$= -\int_x \left[\oint \omega\,\delta\phi'(\tau_{xz}\,a_z + \tau_{xy}\,a_y)\,ds - \int_A \omega\,\delta\phi'\left(\frac{\partial\tau_{xy}}{\partial y} + \frac{\partial\tau_{xz}}{\partial z}\right) dA\right] dx \tag{4}$$

Furthermore [Chapter 1, Eq. (1.148)],

$$-\int_x \delta\phi' \int_A (\tau_{xy}\,z - \tau_{xz}\,y)\,dA\,dx = M_t\,\delta\phi|_0^L = M_t\,\delta\phi|_L - M_t\,\delta\phi|_0 \tag{5}$$

Substitution of (3), (4), and (5) into (2) gives

$$-\int_x \left[\oint \delta(\omega\phi')(\tau_{xz}\,a_z + \tau_{xy}\,a_y)\,ds + \int_A \delta(\omega\phi')\left(\frac{\partial\tau_{xy}}{\partial y} + \frac{\partial\tau_{xz}}{\partial z}\right) dA\right] dx$$
$$+ [(\overline{M}_t - \overline{M}_t)\,\delta\phi]_L = 0 \qquad \text{(A) or (6)}$$

Note that (6) contains the equilibrium conditions and the statical boundary conditions of Eq. (6) of Example 2.8. Thus (6) is identical to principle A.

Consider now this same torsion bar starting from the principle of complementary virtual work. For torsion, Eq. (2.78) can be expressed as

$$\int_V (\gamma_{xy}\,\delta\tau_{xy} + \gamma_{xz}\,\delta\tau_{xz})\,dV - \overline{\phi}\,\delta M_t = 0 \tag{7}$$

Extend this relation by including the force boundary condition at $x = L(M_t = \overline{M}_t$ or $\delta M_t = 0)$, the surface force condition $\tau_{xz}\,a_z + \tau_{xy}\,a_y = 0$ or $\delta\tau_{xz}\,a_z + \delta\tau_{xy}\,a_y = 0$ on S_p, and the conditions of equilibrium $\partial_y\tau_{xy} + \partial_z\tau_{xz} = 0$ or $\delta\partial_y\tau_{xy} + \delta\partial_z\tau_{xz} = 0$ in V. As explained in Chapter 1, Section 1.9.3, these are the static admissibility requirements.

$$\int_V (\gamma_{xy}\,\delta\tau_{xy} + \gamma_{xz}\,\delta\tau_{xz})\,dV + [\overline{\phi}\,\delta M_t]_0 - \underline{[\phi\,\delta M_t]_L}$$
$$\underline{+\int_x \left[\oint \phi'\omega(\delta\tau_{xz}\,a_z + \delta\tau_{xy}\,a_y)\,ds + \int_A \phi'\omega\left(\delta\frac{\partial\tau_{xy}}{\partial y} + \delta\frac{\partial\tau_{xz}}{\partial z}\right) dA\right] dx} = 0 \qquad \text{(D) or (8)}$$

The extended terms are underlined. This can be rewritten by noting that

$$[\phi\,\delta M_t]_L = \int_x \phi' \int_A (-\delta\tau_{xy}\,z + \delta\tau_{xz}\,y)\,dA\,dx + [\phi\,\delta M_t]_0 \tag{9}$$

Then

$$\int_x \left[\int_A (\gamma_{xy}\,\delta\tau_{xy} + \gamma_{xz}\,\delta\tau_{xz})\,dA - \phi' \int_A (-\delta\tau_{xy}\,z + \delta\tau_{xz}\,y)\,dA + \oint \phi'\omega(\delta\tau_{xz}\,a_z + \delta\tau_{xy}\,a_y)\,ds\right.$$
$$\left.+ \int_A \phi'\omega\left(\delta\frac{\partial\tau_{xy}}{\partial y} + \delta\frac{\partial\tau_{xz}}{\partial z}\right) dA\right] dx + [(\overline{\phi} - \phi)\,\delta M_t]_0 = 0 \tag{10}$$

Integration by parts gives

$$\int_V \left\{ \left[\gamma_{xy} + \phi'\left(\frac{\partial \omega}{\partial y} + z\right)\right]\delta\tau_{xy} + \left[\gamma_{xz} + \phi'\left(\frac{\partial \omega}{\partial z} - y\right)\right]\delta\tau_{xz} \right\} dV + [(\overline{\phi} - \phi)\delta M_t]_0 = 0 \qquad \text{(B) or (11)}$$

which is identical to form *B* and from which the strain-displacement relations and displacement boundary conditions follow.

Now that *A, B, C,* and *D* have been defined for the torsion problem, combinations of these as outlined in Table 2.3 can be formed as generalized variational principles. For example, $AB = A + B$ can provide all the fundamental equations. Form *AB* corresponds to Chapter 1, Eq. (1.168). To show this, multiply *A* by $-G$, use the constitutive relations of Chapter 1, Eq. (1.143), and form $A + B = AB$, giving

$$\begin{aligned} -\int_V &\left\{ \delta(G\phi'\omega)(\partial_y \tau_{xy} + \partial_z \tau_{xz}) + \left[\tau_{xy} + G\phi'\left(\frac{\partial \omega}{\partial y} + z\right)\right]\delta\tau_{xy} \right. \\ &\left. + \left[\tau_{xz} + G\phi'\left(\frac{\partial \omega}{\partial z} - y\right)\right]\delta\tau_{xz} \right\} dV + \int_x \oint \delta(G\phi'\omega)(\tau_{xz}\, a_z + \tau_{xy}\, a_y)\, ds\, dx \\ &- [(M_t - \overline{M}_t)\,\delta\phi]_L G + [(\overline{\phi} - \phi)\,\delta M_t]_0 G = 0 \end{aligned} \qquad (12)$$

In matrix notation,

$$\begin{aligned} &\int_V \delta[G\phi'\omega \;\; \tau_{xz} \;\; \tau_{xy}] \begin{bmatrix} 0 & \partial_z & \partial_y \\ \partial_z & 1 & 0 \\ \partial_y & 0 & 1 \end{bmatrix} \begin{bmatrix} \omega G\phi' \\ \tau_{xz} \\ \tau_{xy} \end{bmatrix} dV + \int_V \delta[G\phi'\omega \;\; \tau_{xz} \;\; \tau_{xy}] \begin{bmatrix} 0 \\ -y \\ z \end{bmatrix} G\phi'\, dV \\ &+ \int_x \oint \delta[G\phi'\omega \quad \tau_{xz} \quad \tau_{xy}] \begin{bmatrix} 0 & a_z & a_y \\ 0 & 0 & 0 \\ 0 & 0 & 0 \end{bmatrix} \begin{bmatrix} \omega G\phi' \\ \tau_{xz} \\ \tau_{xy} \end{bmatrix} ds\, dx \\ &- [(M_t - \overline{M}_t)\delta\phi]_L G + [(\overline{\phi} - \phi)\delta M_t]_0 G = 0 \end{aligned} \qquad \text{(AB) or (13)}$$

Note the lack of symmetry in one of the matrix expressions. Euler's equations for the volume terms are the mixed governing differential equations of Chapter 1, Eq. (1.168).

Other combinations of *A, B, C,* and *D* lead to different generalized variational principles. Computational considerations can make it advantageous to employ a generalized principle that results in symmetric equations. One such combination is $AD = A + D$. To derive this principle, multiply *D* by -1 and form $A + D = AD$, giving

$$\begin{aligned} -\delta &\int_x \oint \omega\phi'(\tau_{xz}\, a_z + \tau_{xy}\, a_y)\, ds\, dx + \delta \int_x \int_A \omega\phi'\left(\frac{\partial \tau_{xy}}{\partial y} + \frac{\partial \tau_{xz}}{\partial z}\right) dA\, dx \\ &- \int_x \int_A (\gamma_{xy}\, \delta\tau_{xy} + \gamma_{xz}\, \delta\tau_{xz})\, dA\, dx + \int_x \phi' \int_A (-\delta\tau_{xy}\, z + \delta\tau_{xz}\, y)\, dA\, dx \\ &+ [(M - \overline{M}_t)\delta\phi]_L - [(\overline{\phi} - \phi)\delta M_t]_0 = 0 \end{aligned} \qquad \text{(AD) or (14)}$$

that is,

$$
\begin{aligned}
&-\int_x \delta\phi' \oint \omega(\tau_{xz}\,a_z + \tau_{xy}\,a_y)\,ds\,dx + \int_x \delta\phi' \int_A \omega\left(\frac{\partial \tau_{xy}}{\partial y} + \frac{\partial \tau_{xz}}{\partial z}\right) dA\,dx \\
&+\int_x \phi'\left[\delta \int_A \omega\left(\frac{\partial \tau_{xy}}{\partial y} + \frac{\partial \tau_{xz}}{\partial z}\right) dA - \delta \oint \omega(\tau_{xz}\,a_z + \tau_{xy}\,a_y)\,ds\right. \\
&\left.-\int_A \left(\tau_{xy}\frac{1}{G}\delta\tau_{xy} + \tau_{xz}\frac{1}{G}\delta\tau_{xz}\right) dA + \int_A (-\delta\tau_{xy}\,z + \delta\tau_{xz}\,y)\,dA\right] dx \\
&+[(M - \overline{M}_t)\delta\phi]_L - [(\overline{\phi} - \phi)\delta M_t]_0 = 0
\end{aligned}
\tag{15}
$$

From Eq. (4), and Chapter 1, Eqs. (1.142) and (1.143),

$$
\begin{aligned}
&-\int_x \oint \delta\phi'\omega(\tau_{xz}\,a_z + \tau_{xy}\,a_y)\,ds\,dx \\
&\qquad = -\int_x \delta\phi' \int_A \left(\tau_{xy}\frac{\partial \omega}{\partial y} + \tau_{xz}\frac{\partial \omega}{\partial z}\right) dA\,dx \\
&-\int_x \delta\phi' \int_A \omega\left(\frac{\partial \tau_{xy}}{\partial y} + \frac{\partial \tau_{xz}}{\partial z}\right) dA\,dx \\
&\qquad = -\int_x \delta\phi' \int_A \left[\tau_{xy}\left(\frac{\partial \omega}{\partial y} + z\right) + \tau_{xz}\left(\frac{\partial \omega}{\partial z} - y\right)\right] dA\,dx \\
&\qquad\quad + \int_x \delta\phi' \int_A (\tau_{xy}\,z - \tau_{xz}\,y)\,dA\,dx - \int_x \delta\phi' \int_A \omega\left(\frac{\partial \tau_{xy}}{\partial y} + \frac{\partial \tau_{xz}}{\partial z}\right) dA\,dx \\
&\qquad = \int_x \delta\phi'\,G \int_A \left[\left(\frac{\partial \omega}{\partial y} + z\right)^2 + \left(\frac{\partial \omega}{\partial z} - y\right)^2\right] dA\,dx - \int_x \delta\phi' M_t\,dx \\
&\qquad\quad + \int_x \delta\phi' \int_A \omega\left(\frac{\partial \tau_{xy}}{\partial y} + \frac{\partial \tau_{xz}}{\partial z}\right) dA\,dx \\
&\qquad = \int_x \delta\phi'\,G\,\widehat{J}\,\phi' dx - \delta\phi\, M_t|_0^L - \int_x \delta\phi' \int_A \omega\left(\frac{\partial \tau_{xy}}{\partial y} + \frac{\partial \tau_{xz}}{\partial z}\right) dA\,dx
\end{aligned}
\tag{16}
$$

with $\widehat{J} = \int_A \left[(\partial\omega/\partial y + z)^2 + (\partial\omega/\partial z - y)^2\right] dA$.

Substitution of (16) into (15) gives

$$
\int_x \delta\phi'\,GJ\,\phi' dx - [\overline{M}_t\,\delta\phi]_L - [\overline{\phi}\,\delta M_t]_0 + [\delta(M_t\phi)]_0 \qquad \text{Applicable to the direction of the bar axis.}
$$

$$
\left.
\begin{aligned}
&+\int_x \phi'\left[\delta \int_A \omega\left(\frac{\partial \tau_{xy}}{\partial y} + \frac{\partial \tau_{xz}}{\partial z}\right) dA - \delta \oint \omega(\tau_{xz}\,a_z + \tau_{xy}\,a_y)\,ds\right. \\
&\left.-\int_A \left(\tau_{xy}\frac{1}{G}\delta\tau_{xy} + \tau_{xz}\frac{1}{G}\delta\tau_{xz}\right) dA + \int_A (-\delta\tau_{xy}\,z + \delta\tau_{xz}\,y)\,dA\right] dx = 0
\end{aligned}
\right\} \text{Applicable to the bar's cross-section.}
\tag{17}
$$

The axial and cross-sectional parts must be equal to zero separately. For the cross-sectional terms, using

$$
\tau_{xy}^* = \frac{\tau_{xy}}{(G\phi')} \quad \text{and} \quad \tau_{xz}^* = \frac{\tau_{xz}}{(G\phi')}
$$

one obtains

$$\delta \int_A \omega \left(\frac{\partial \tau_{xy}^*}{\partial y} + \frac{\partial \tau_{xz}^*}{\partial z} \right) dA - \delta \oint \omega (\tau_{xz}^* \, a_z + \tau_{xy}^* \, a_y) \, ds$$
$$- \int_A (\tau_{xy}^* \, \delta\tau_{xy}^* + \tau_{xz}^* \, \delta\tau_{xz}^*) \, dA + \int_A (-\delta\tau_{xy}^* \, z + \delta\tau_{xz}^* \, y) \, dA = 0 \tag{18}$$

The symmetry of the relations can be observed in the matrix form of (18)

$$\int_A \delta [\omega \quad \tau_{xz}^* \quad \tau_{xy}^*] \begin{bmatrix} 0 & \partial_z & \partial_y \\ {}_z\partial & -1 & 0 \\ {}_y\partial & 0 & -1 \end{bmatrix} \begin{bmatrix} \omega \\ \tau_{xz}^* \\ \tau_{xy}^* \end{bmatrix} dA$$
$$- \oint \delta [\omega \quad \tau_{xz}^* \quad \tau_{xy}^*] \begin{bmatrix} 0 & a_z & a_y \\ a_z & 0 & 0 \\ a_y & 0 & 0 \end{bmatrix} \begin{bmatrix} \omega \\ \tau_{xz}^* \\ \tau_{xy}^* \end{bmatrix} ds$$
$$- \int_A \delta [\omega \quad \tau_{xz}^* \quad \tau_{xy}^*] \begin{bmatrix} 0 \\ -y \\ z \end{bmatrix} dA = 0 \tag{19}$$

Note that the relationships *AB* and *AD* are equivalent in the sense that one can be transformed into the other. To observe this, rewrite *AB* by multiplying *B* by −1 to obtain

$$\int_x \int_A \delta(\phi' \omega) \left(\frac{\partial \tau_{xy}}{\partial y} + \frac{\partial \tau_{xz}}{\partial z} \right) dA \, dx - \int_x \int_A (\gamma_{xy} \, \delta\tau_{xy} + \gamma_{xz} \, \delta\tau_{xz}) \, dA \, dx$$
$$- \int_x \oint \delta(\phi' \omega)(\tau_{xy} \, a_y + \tau_{xz} \, a_z) \, ds \, dx + [(M - \overline{M}_t)\delta\phi]_L - [(\overline{\phi} - \phi)\delta M_t]_0$$
$$- \int_x \phi' \int_A \left[\left(\frac{\partial \omega}{\partial y} + z \right) \delta\tau_{xy} + \left(\frac{\partial \omega}{\partial z} - y \right) \delta\tau_{xz} \right] dA \, dx = 0 \tag{20}$$

The final term of (20) can be written as

$$\int_x \phi' \int_A \left[\frac{\partial \omega}{\partial y} \delta\tau_{xy} + \frac{\partial \omega}{\partial z} \delta\tau_{xz} + z \, \delta\tau_{xy} - y \, \delta\tau_{xz} \right] dA \, dx$$
$$= \int_x \phi' \left\{ \int_A \left[\frac{\partial}{\partial y} (\omega \, \delta\tau_{xy}) + \frac{\partial}{\partial z} (\omega \, \delta\tau_{xz}) \right] dA - \int_A \omega \left(\frac{\partial \delta\tau_{xy}}{\partial y} + \frac{\partial \delta\tau_{xz}}{\partial z} \right) dA \right.$$
$$\left. + \int_A (z \, \delta\tau_{xy} - y \, \delta\tau_{xz}) \, dA \right\} dx$$
$$= \int_x \phi' \left\{ \oint \omega(\delta\tau_{xy} \, a_y + \delta\tau_{xz} \, a_z) ds - \int_A \omega \left(\frac{\partial \delta\tau_{xy}}{\partial y} + \frac{\partial \delta\tau_{xz}}{\partial z} \right) dA \right.$$
$$\left. + \int_A (z \, \delta\tau_{xy} - y \, \delta\tau_{xz}) \, dA \right\} dx \tag{21}$$

Substitution of (21) into (20) results in (14). This indicates the equivalence of *AB* and *AD*. Since (20) is essentially the same as (12), the two matrix expressions of (13) and (19) are equivalent. ∎

2.4 Engineering Beam Theory

The classical and generalized variational principles apply for beam theory if the notation of Section 2.3 is changed from that of three-dimensional elasticity to the notation of beam theory. Often, for example, terms such as $\boldsymbol{\sigma}$, the stress vector for elasticity, are replaced by $\mathbf{s}$, the vector containing the shear force and bending moment. It is instructive, however, to know how the variational principles for beams can be derived by following the procedures used in deriving the three-dimensional elasticity versions of the principles.

The beam theory relations of Chapter 1, Section 1.8, will be utilized. The conversion of the local governing differential equations for beams to their equivalent global (integral) forms will be discussed first.

2.4.1 Equations of Equilibrium and Force Boundary Conditions (A)

The conditions of equilibrium for a beam [Chapter 1, Eq. (1.116)]

$$\mathbf{D}_s^T\mathbf{s} + \overline{\mathbf{p}} = \mathbf{0} \tag{2.110a}$$

and the force boundary conditions [Chapter 1, Eq. (1.122)]

$$\mathbf{s} = \overline{\mathbf{s}} \quad \text{on} \quad S_p \tag{2.110b}$$

can be placed in the equivalent global form

$$\int_0^L \delta\mathbf{u}^T\left(\mathbf{D}_s^T\mathbf{s} + \overline{\mathbf{p}}\right)dx = \left[\delta\mathbf{u}^T(\mathbf{s} - \overline{\mathbf{s}})\right]_{\substack{0\\ \text{on } S_p}}^L \qquad (2.111) \text{ or (A)}$$

where, from Chapter 1, Section 1.8, if axial deformation terms are ignored,

$$\mathbf{s} = \begin{bmatrix} V \\ M \end{bmatrix} \qquad \mathbf{u} = \begin{bmatrix} w \\ \theta \end{bmatrix} \qquad \mathbf{D}_s = \begin{bmatrix} \partial_x & 0 \\ -1 & \partial_x \end{bmatrix} \qquad \overline{\mathbf{p}} = \begin{bmatrix} \overline{p}_z \\ 0 \end{bmatrix}$$

and L is the length of the beam. In component form, this appears as

$$\int_0^L [(V' + \overline{p}_z)\delta w + (M' - V)\delta\theta]\,dx = \left[(V - \overline{V})\,\delta w + (M - \overline{M})\,\delta\theta\right]_{\substack{0\\ \text{on } S_p}}^L \qquad (2.112) \text{ or (A)}$$

2.4.2 Strain-Displacement Relations and Displacement Boundary Conditions (B)

The local (differential) form of the strain-displacement relations [Chapter 1, Eq. (1.102)]

$$\boldsymbol{\epsilon} = \mathbf{D}_u\mathbf{u} \tag{2.113a}$$

and the displacement boundary conditions [Chapter 1, Eq. (1.123)]

$$\mathbf{u} = \overline{\mathbf{u}} \quad \text{on} \quad S_u \tag{2.113b}$$

are equivalent to the global (integral) form

$$\int_0^L \delta\mathbf{s}^T(\mathbf{D}_u\,\mathbf{u} - \boldsymbol{\epsilon})\,dx = \left[\delta\mathbf{s}^T(\mathbf{u} - \overline{\mathbf{u}})\right]_{\substack{0\\ \text{on } S_u}}^L \qquad (2.114) \text{ or (B)}$$

where

$$\epsilon = \begin{bmatrix} \gamma \\ \kappa \end{bmatrix} \qquad \mathbf{D}_u = \begin{bmatrix} \partial_x & 1 \\ 0 & \partial_x \end{bmatrix} \tag{2.115}$$

If the material laws of Chapter 1, Eqs. (1.106b) and (1.109), are introduced, the component form becomes

$$\int_0^L \left\{ \left[(w' + \theta) - \frac{V}{k_s GA} \right] \delta V + \left[\theta' - \frac{M}{EI} \right] \delta M \right\} dx = \underset{\text{on } S_u}{\left[(w - \overline{w})\, \delta V + (\theta - \overline{\theta})\, \delta M \right]_0^L}$$

(2.116) or (B)

Proper application of integration by parts transforms A and B into the principles of virtual work and complementary virtual work, respectively.

2.4.3 Principle of Virtual Work (C)

In order to transform A into the principle of virtual work, begin by rewriting $\delta \mathbf{u}^T \mathbf{s}$ as

$$\underset{\text{on } S_p}{\left[\delta \mathbf{u}^T \mathbf{s}\right]_0^L} = \underset{\text{on } S_p}{\left[V\,\delta w + M\,\delta\theta\right]_0^L} = \underset{\text{on } S}{\left[V\,\delta w + M\,\delta\theta\right]_0^L} - \underset{\text{on } S_u}{\left[V\,\delta w + M\,\delta\theta\right]_0^L}$$

$$= \int_0^L [V\,\delta w + M\,\delta\theta]\, dx - \underset{\text{on } S_u}{\left[V\,\delta w + M\,\delta\theta\right]_0^L} \tag{2.117}$$

where $S = S_p + S_u$. Insert this expression in the right-hand side of (A), giving

$$\int_0^L [(V' + \overline{p}_z)\,\delta w + (M' - V)\,\delta\theta]\, dx$$

$$= \int_0^L [V'\,\delta w + V\,\delta w' + M'\,\delta\theta + M\,\delta\theta']\, dx - \underset{\text{on } S_u}{[V\,\delta w + M\,\delta\theta]_0^L} - \underset{\text{on } S_p}{\left[\overline{V}\,\delta w + \overline{M}\,\delta\theta\right]_0^L}$$

or

$$\int_0^L [-V\,\delta(\theta + w') - M\,\delta\theta' + \overline{p}_z\,\delta w]\, dx = -\underset{\text{on } S_p}{\left[\overline{M}\,\delta\theta + \overline{V}\,\delta w\right]_0^L} - \underset{\text{on } S_u}{\underline{\left[V\,\delta w + M\,\delta\theta\right]_0^L}}$$

(2.118) or (C)

The same result can be derived by applying integration by parts to Eq. (2.111). Since the classical principle of virtual work utilizes kinematically admissible displacements, the underlined terms would be dropped because they contain displacement boundary conditions which must be satisfied.

2.4.4 Principle of Complementary Virtual Work (D)

The global form (B) of the strain-displacement relations and the displacement boundary conditions can be reformed into an integral relationship representing the principle of complementary virtual work. Apply integration by parts to some of the terms of Eq. (2.116), giving

$$\int_0^L w'\,\delta V\, dx = -\int_0^L w\,\delta V'\, dx + \left[w\,\delta V\right]_0^L$$

$$\int_0^L \theta'\,\delta M dx = -\int_0^L \theta\,\delta M'\, dx + \left[\theta\,\delta M\right]_0^L$$

Then Eq. (2.116) becomes

$$\int_0^L \left[-w\,\delta V' + \theta\,\delta V - \frac{V}{k_s GA}\,\delta V - \theta\,\delta M' - \frac{M}{EI}\,\delta M\right] dx$$
$$= -\underbrace{\left[w\,\delta V\right]_0^L}_{\text{on } S} - \underbrace{\left[\theta\,\delta M\right]_0^L}_{\text{on } S} + \underbrace{\left[(w-\overline{w})\,\delta V + (\theta - \overline{\theta})\,\delta M\right]_0^L}_{\text{on } S_u}$$

or

$$-\int_0^L \left(\frac{V}{k_s GA}\,\delta V + \frac{M}{EI}\,\delta M\right) dx - \int_0^L \underline{(\delta V'\,w - \delta V\,\theta + \delta M'\,\theta)\,dx}$$
$$= -\underbrace{\underline{\left[w\,\delta V + \theta\,\delta M\right]_0^L}}_{\text{on } S_p} - \underbrace{\left[\overline{w}\,\delta V + \overline{\theta}\,\delta M\right]_0^L}_{\text{on } S_u} \qquad \text{(2.119) or (D)}$$

The classical theorem would require the underlined term to be dropped because they represent statically admissible boundary conditions and equilibrium.

2.4.5 Generalized Principles

The classical variational principles for a beam are summarized in Table 2.5. These fundamental forms can be combined to form the generalized principles as follows:

$$A + B = AB \qquad C + D = CD$$
$$A + D = AD \qquad C + B = CB$$

For example, consider the $C + B$ combination,

$$\int_0^L \left[V\,(\delta\theta + \delta w') + M\,\delta\theta' + \theta'\,\delta M + \delta V(w' + \theta) - \frac{V\,\delta V}{k_s GA} - \frac{M\,\delta M}{EI}\right] dx$$
$$-\int_0^L \overline{p}_z\,\delta w\,dx - \underbrace{\left[\overline{V}\,\delta w + \overline{M}\,\delta\theta\right]_0^L}_{\text{on } S_p}$$
$$-\delta\underbrace{\left[V\,(w-\overline{w}) + M\,(\theta - \overline{\theta})\right]_0^L}_{\text{on } S_u} = 0 \qquad \text{(2.120) or (CB)}$$

In matrix notation this symmetric principle is the same as Eq. (7) of Example 2.11. Other generalized forms can be derived in a similar fashion.

2.5 Structure of the Differential and Integral Forms of the Governing Equations

The general form of the basic equations given in Chapter 1 and in this chapter is shown in Table 2.6. The equations are arranged such that the dual character of the principle of virtual work C and the principle of complementary virtual work D is evident. The local equations corresponding to principle C are the equilibrium conditions A and the force boundary conditions A_B, with the kinematic equations B and the displacement boundary conditions B_B as side conditions to be satisfied a priori. The reverse holds for principle D in a fully dual

TABLE 2.5

The Classical Variational Principles for a Beam

A

The conditions of equilibrium and the force boundary conditions

$$\int_0^L [(V' + \overline{p}_z)\,\delta w + (M' - V)\,\delta\theta]\,dx = \underbrace{[(V - \overline{V})\,\delta w + (M - \overline{M})\,\delta\theta]_0^L}_{\text{on } S_p}$$

B

The strain-displacement relations and displacement boundary conditions

$$\int_0^L \left\{ \left[(w' + \theta) - \frac{V}{k_s G A}\right]\delta V + \left[\theta' - \frac{M}{EI}\right]\delta M \right\} dx = \underbrace{[(w - w')\,\delta V + (\theta - \overline{\theta})\,\delta M]_0^L}_{\text{on } S_u}$$

A → C: Gauss' Integral Theorem (Integration by Parts)

B → D: Gauss' Integral Theorem (Integration by Parts)

C

The principle of virtual work

$$\int_0^L [-V\,\delta(\theta + w') - M\,\delta\theta' + \overline{p}_z\,\delta w]\,dx + \underbrace{[\overline{V}\,\delta w + \overline{M}\,\delta\theta]_0^L}_{\text{on } S_p} + \underbrace{\underline{[V\,\delta w + M\,\delta\theta]_0^L}}_{\text{on } S_u} = 0$$

D

The principle of complementary virtual work

$$\int_0^L \left[\frac{V}{k_s G A}\delta V + \frac{M}{EI}\delta M\right] dx \;\underbrace{-[\overline{w}\,\delta V + \overline{\theta}\,\delta M]_0^L}_{\text{on } S_u} \;\underline{+ \int_0^L [\delta V' w - \delta V\,\theta + \delta M'\,\theta]\,dx} \;\underbrace{\underline{-[w\,\delta V + \theta\,\delta M]_0^L}}_{\text{on } S_p} = 0$$

__C and D are usually written without the underlined terms and then require admissible displacements and forces, respectively.

manner. That is, the kinematic equations B and displacement boundary conditions B_B are the local equations corresponding to principle D, while the equilibrium conditions A and the force boundary conditions A_B are the side conditions.

Principle C provides the basis of the displacement method as all equations are ultimately written with displacements as unknowns, whereas principle D forms the basis of the force method as the principle and the corresponding equations are written with stresses or stress-resultants as unknowns.

Note also in Table 2.6 the stress functions and strain compatibility conditions. The similarities in form of the static admissibility conditions (A and A_B) with the strain compatibility relations as well as the kinematic admissibility conditions (B and B_B) with the stress function relationship are particularly noteworthy. In order to emphasize the similarities in form of the equilibrium conditions with the compatibility relationships, the compatibility terms of Chapter 1, Eq. (1.30), have been supplemented with the vector of incompatibilities (misfits) η, as displayed in Table 2.6.

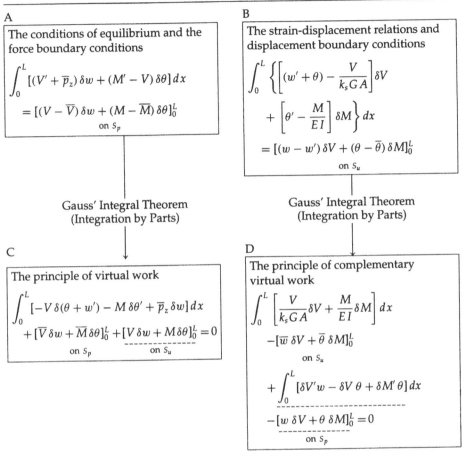

TABLE 2.6

Interrelationships Between the Basic Equations

$\boldsymbol{\sigma} = \mathbf{D}_1\boldsymbol{\psi}$ (Chapter 1, Eq. 1.56a)	$\mathbf{R} = \mathbf{D}_1\boldsymbol{\epsilon}$ (Chapter 1, Eq. 1.29)
Stress Functions	Compatibility
$\phi = \overline{\phi}$ on $\mathbf{S_P}$; $\boldsymbol{\sigma} = \mathbf{D}\boldsymbol{\phi}$ in $\mathbf{V}$ (Chapter 1, Eq. 1.56b)	$\mathbf{D}^T\mathbf{R} + \overline{\boldsymbol{\eta}}_V = \mathbf{0}$ in V; $\boldsymbol{\eta} = \overline{\boldsymbol{\eta}}$ on S_u
$\mathbf{s} = \mathbf{E}\boldsymbol{\epsilon}$ (Chapter 1, Eq. 1.34)	$\boldsymbol{\epsilon} = \mathbf{E}^{-1}\mathbf{s}$ (Chapter 1, Eq. 1.32)
A_B Static Boundary Conditions: $\mathbf{p} = \overline{\mathbf{p}}$ on S_p (Chapter 1, Eq. 1.60)	B Strain Displacement Relationship: $\boldsymbol{\epsilon} = \mathbf{D}\mathbf{u}$ in V (Chapter 1, Eq. 1.20)
A Equilibrium: $-(\mathbf{D}^T\mathbf{s} + \overline{\mathbf{p}}_V) = \mathbf{0}$ in V (Chapter 1, Eq. 1.53)	B_B Displacement Boundary Conditions: $\mathbf{u} = \overline{\mathbf{u}}$ on S_u (Chapter 1, Eq. 1.61)
C Virtual Work (Eq. 2.54)	D Complementary Virtual Work (Eq. 2.78)
$\int_V \delta\boldsymbol{\epsilon}^T\mathbf{s}\,dV - \int_V \delta\mathbf{u}^T\overline{\mathbf{p}}_V\,dV - \int_{S_p} \delta\mathbf{u}^T\overline{\mathbf{p}}\,dS = 0$	$-\int_V \delta\mathbf{s}^T\boldsymbol{\epsilon}\,dV + \int_{S_u} \delta\mathbf{p}^T\overline{\mathbf{u}}\,dS = 0$

References

Hellinger, E., 1914, Die allgemeinen Ansätze der Mechanik der Kontinua, *Enzyklopädie der Mathematischen Wissenschaften* 4, Vol. 30, pp. 601–694.

Lanczos, C., 1970, *The Variational Principles of Mechanics*, University of Toronto Press, Canada.

Oravas, G. and McLean, L., 1966, Historical development of energetical principles in elastomechanics, Part 1 - From Heraclitos to Maxwell, *Appl. Mech. Rev.*, Vol. 19, pp. 647–658.

Oravas, G. and McLean, L., 1966, Historical development of energetical principles in elastomechanics, Part 2 - From Cotterill to Prange, *Appl. Mech. Rev.*, Vol. 19, pp. 919–933.

Pilkey, W.D. and Pilkey, O.H., 1986, *The Mechanics of Solids*, Krieger Publishers, Melbourne, FL.

Reissner, E., 1950, On a variational theorem in elasticity, *J. Math. Phys.*, Vol. 29, pp. 90–95.

Washizu, K., 1982, *Variational Methods in Elasticity and Plasticity*, 3rd ed., Pergamon Press, Oxford.

Wunderlich, W., 1970, Ein verallgemeinertes Variationsverfahren zur vollen order teilweisen Diskretisierung mehrdimensionaler Elastizitaetsprobleme (Generalized variational method for the full or partial discretization of problems in elasticity), *Ing. Arch.*, Vol. 39, pp. 230–247.

Wunderlich, W., 1972, Incremental formulation of the generalized variational approach in structural mechanics, *Proc. Int. Conf. on Variational Methods in Engineering, Southampton*, Chapter 7, pp. 111–125.

Wunderlich, W., 1973, Grundlagen und Anwendung eines verallgemeinerten Variationsverfahrens (Foundations and applications of generalized variational methods), in *Finite Elemente in der Statik*, Wilhelm Ernst & Sohn, Berlin, pp. 126–144.

Zeller, Ch., 1979, Eine Finite-Element-Methode zur Berechnung der Verwölbungen und Profilverformungen von Stäben mit beliebiger Querschnittsform, Techn.-Wiss. Mitteilungen No. 79-7, Institut für Konstruktiven Ingenieurbau, Ruhr-Universität.

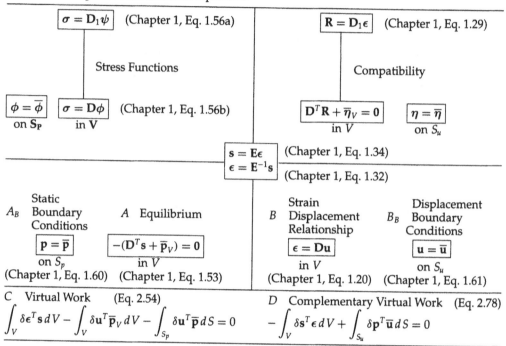

Problems

Work and Energy

2.1 Verify that $\mathcal{U}_0$ of Eq. (2.10) satisfies $(\partial \mathcal{U}_0/\partial \epsilon)^T = \sigma^T$ (Eq. 2.22). Also check Eq. (2.23) using $\mathcal{U}_0^*$ of Eq. (2.15).

2.2 For a nonlinear elastic material with an isotropic coefficient of linear thermal expansion, show that the strain energy per unit volume can be written as

$$\mathcal{U}_0 = \int_0^{\epsilon_{ij}} \sigma_{ij}\, d\epsilon_{ij} - \int_0^{\Delta T} (\sigma_x + \sigma_y + \sigma_z)\, \alpha\, d\Delta T$$

and the complementary energy density as

$$\mathcal{U}_0^* = \int_0^{\sigma_{ij}} \epsilon_{ij}\, d\sigma_{ij} - \int_0^{\sigma_{ii}} \left(\int_0^{\Delta T} \alpha\, d\Delta T \right) d\sigma_{ii} + (\sigma_x + \sigma_y + \sigma_z) \int_0^{\Delta T} \alpha\, d\Delta T$$

If the material is linearly elastic, show that these reduce to

$$\mathcal{U}_0 = \frac{1}{2}\sigma_{ij}\,\epsilon_{ij} - \frac{\alpha\,\Delta T\,\sigma_{ii}}{2} \quad \text{and} \quad \mathcal{U}_0^* = \frac{1}{2}\sigma_{ij}\,\epsilon_{ij} + \frac{\alpha\,\Delta T\,\sigma_{ii}}{2}$$

2.3 Show that for a linearly elastic solid in plane stress

$$\begin{aligned}\mathcal{U}_0 &= \frac{E}{2(1+\nu)}\left[\frac{1}{1-\nu}(\epsilon_x + \epsilon_y)^2 - 2\epsilon_x\epsilon_y + \frac{1}{2}\epsilon_{xy}^2\right] \\ &\quad - \frac{\alpha\, E\,\Delta T}{1-\nu}(\epsilon_x + \epsilon_y) + \frac{E(\alpha\,\Delta T)^2}{1-\nu} \\ \mathcal{U}_0^* &= \frac{1}{2E}\left[(\sigma_x + \sigma_y)^2 - 2(1+\nu)\left(\sigma_x\sigma_y - \sigma_{xy}^2\right)\right] + \alpha\,\Delta T(\sigma_x + \sigma_y)\end{aligned}$$

2.4 Write the complementary energy density $\mathcal{U}_0^*$ in terms of the Airy stress function for a solid in plane stress.

Answer:

$$\begin{aligned}\mathcal{U}_0^* = \frac{1}{2E}\Bigg\{&\left(\frac{\partial^2\psi}{\partial x^2} + \frac{\partial^2\psi}{\partial y^2}\right)^2 - 2(1+\nu)\left[\left(\frac{\partial^2\psi}{\partial x^2}\right)\left(\frac{\partial^2\psi}{\partial y^2}\right) - \left(\frac{\partial^2\psi}{\partial x\,\partial y}\right)^2\right] \\ &+ \alpha\,\Delta T\left(\frac{\partial^2\psi}{\partial x^2} + \frac{\partial^2\psi}{\partial y^2}\right)\Bigg\}\end{aligned}$$

2.5 Show that for a material following a generalized Hooke's law, if a potential function exists for the internal work then the number of material constants is reduced from 36 to 21.

Hint: Use Eq. (2.22)

2.6 Show that the complementary strain energy for a rod of circular cross section subjected to torsion is $\mathcal{U}_i^* = \int_0^L M_t^2/2GJ\, dx$, where M_t is the net torque on a cross-section, L is the length of the rod, G is the shear modulus, and J is the polar moment of inertia.

2.7 Show that the complementary strain energy density for a two-dimensional linearly elastic isotropic body is

$$U_0^* = \frac{1}{2E}\left(\sigma_x^2 + \sigma_y^2\right) - \frac{\nu}{E}\sigma_x\sigma_y + \frac{1}{2G}\tau_{xy}^2$$

2.8 Show that the complementary strain energy density for a three-dimensional linearly elastic isotropic body can be written as

$$\begin{aligned} U_0^* &= \frac{1}{2E}\left(\sigma_x^2 + \sigma_y^2 + \sigma_z^2\right) - \frac{2\nu}{E}(\sigma_x\sigma_y + \sigma_y\sigma_z + \sigma_x\sigma_z) + \frac{1+\nu}{E}\left(\tau_{xy}^2 + \tau_{yz}^2 + \tau_{xz}^2\right) \\ &= \frac{1}{2}(\sigma_x\epsilon_x + \sigma_y\epsilon_y + \sigma_z\epsilon_z + \tau_{xy}\gamma_{xy} + \tau_{yz}\gamma_{yz} + \tau_{xz}\gamma_{xz}) \end{aligned}$$

2.9 Determine an integral expression for the total potential energy of a beam on an elastic foundation (modulus k_w) with a compressive axial force (N) and a transverse loading intensity $\overline{p}_z$.

2.10 In Chapter 13, it is shown that the stresses in a thin elastic plate are

$$\sigma_x = \frac{E}{1-\nu^2}(\epsilon_x + \nu\epsilon_y), \quad \sigma_y = \frac{E}{1-\nu^2}(\epsilon_y + \nu\epsilon_x), \quad \tau_{xy} = \frac{E}{2\,(1+\nu)}\gamma_{xy}$$

with

$$\epsilon_x = -z\frac{\partial^2 w}{\partial x^2}, \quad \epsilon_y = -z\frac{\partial^2 w}{\partial y^2}, \quad \gamma_{xy} = -2z\frac{\partial^2 w}{\partial x\,\partial y}$$

where w is the transverse displacement and z is the transverse coordinate measured from the middle plane of the plate. Assume that the contribution to the strain energy of other stress components is negligible. If t is the plate thickness and A is the area of the plate, show that the strain energy developed in the plate is

$$U_i = \frac{Et^3}{24(1-\nu^2)}\iint_A \left\{\left(\frac{\partial^2 w}{\partial x^2} + \frac{\partial^2 \omega}{\partial y^2}\right)^2 - 2(1-\nu)\left[\frac{\partial^2 w}{\partial x^2}\frac{\partial^2 w}{\partial y^2} - \left(\frac{\partial^2 w}{\partial x\,\partial y}\right)^2\right]\right\} dx\,dy$$

Principle of Virtual Work and Related Theorems

2.11 Find the displacement V of point a and the elongation of each bar of the truss of Fig. P2.11.

Answer: $V = 2\overline{P}L/(3EA)$

2.12 Show that Eq. (2.58a) is equivalent to the displacement form of the governing differential equations of motion of Chapter 1, Eq. (1.63). To do so it is necessary to utilize the divergence theorem.

2.13 Verify the identity $\sigma_{ij}\,\delta\epsilon_{ij} = \sigma_{ij}\,\delta u_{i,j}$, which is used in deriving Eq. (2.49).

Hint: Since i and j are dummy indices, they can be interchanged in an expression such as $\sigma_{ij}\,\delta u_{i,j}$. Hence, $\sigma_{ij}\,\delta u_{i,j} = \frac{1}{2}(\sigma_{ij}\,\delta u_{i,j} + \sigma_{ji}\,\delta u_{j,i})$ and since $\sigma_{ij} = \sigma_{ji}$, $\sigma_{ij}\,\delta u_{i,j} = \sigma_{ij}\frac{1}{2}(\delta u_{i,j} + \delta u_{j,i}) = \sigma_{ij}\,\delta\epsilon_{ij}$. Another form of proof would be verification obtained by simply expanding with the summation convention both sides of the identity in question.

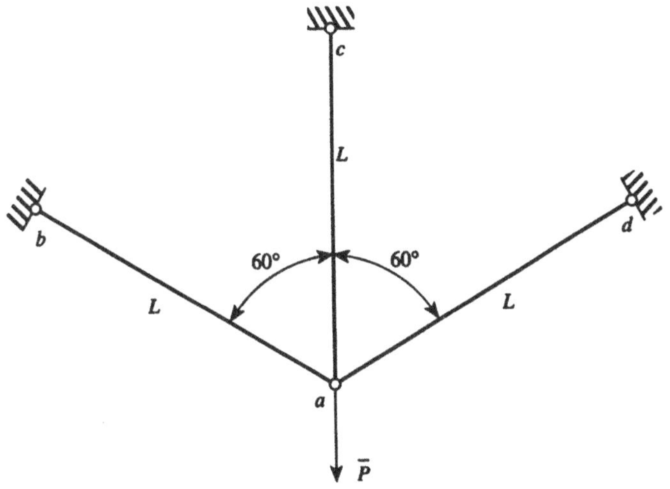

FIGURE P2.11
Statically indeterminate truss.

2.14 Consider the quadratic function in the n variables $x_1, x_2, \ldots, x_n$:

$$\mathbf{x}^T\mathbf{A}\mathbf{x} = \sum_{i,k=1}^{n} a_{ik}x_i x_k$$

where

$$\mathbf{A} = \begin{bmatrix} a_{11} & \cdots & a_{1n} \\ \vdots & & \vdots \\ a_{n1} & \cdots & a_{nn} \end{bmatrix} \qquad \mathbf{x} = \begin{bmatrix} x_1 \\ \vdots \\ x_n \end{bmatrix}$$

Assume that $\mathbf{A}$ is symmetic. Show that this quadratic function is positive definitive if and only if

$$D_1 > 0, \quad D_2 > 0, \quad \ldots, \quad D_n > 0$$

where $D_1, D_2, \ldots, D_n$ are the principal minors of $\mathbf{A}$

$$D_1 = a_{11}, \quad D_2 = \begin{vmatrix} a_{11} & a_{12} \\ a_{21} & a_{22} \end{vmatrix}, \quad D_n = \begin{vmatrix} a_{11} & \cdots & a_{1n} \\ \vdots & & \vdots \\ a_{n1} & \cdots & a_{nn} \end{vmatrix}$$

This theorem is useful in studying inequality relations for material constants using the positive definitive strain energy density written in terms of the strain components.

Hint: See F. R. Gantmacher, *The Theory of Matrices*, Vol. I, Chelsea, 1959, p. 306.

2.15 Use the principle of virtual work to find the forces in the bars of the system shown in Fig. P2.15 if the material is linearly elastic. Find the horizontal displacement of point 0.

Answer: $N_a = -303$ kN, $N_b = 175$ kN, $N_c = 289$ kN, Horizontal disp. $= \frac{303}{AE}$ to the right

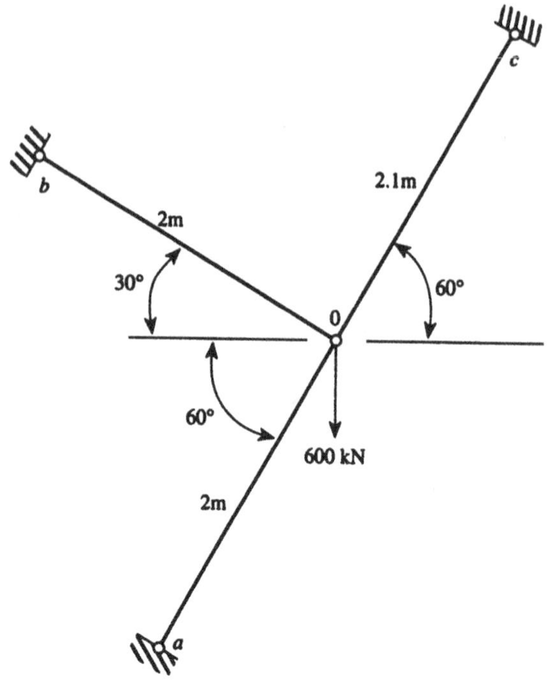

FIGURE P2.15
A bar structure. Each bar has the same modulus of elasticity and cross-sectional area.

2.16 Resolve Problem 2.15 if the material is nonlinearly elastic with $\sigma = 20 \times 10^7 \epsilon^{0.6}$.

2.17 Solve Problem 2.15 using the principle of stationary potential energy.

2.18 Use the principle of stationary potential energy to find the equation of equilibrium for a flexible string of length L with tension N and transverse loading intensity $\overline{p}_z(x)$. Let $w(x)$ be the transverse displacement.

2.19 An elastic string of length L, modulus E, and cross-sectional area A is fixed at its ends and subjected to a transverse force $\overline{p}_z$. The strain is given by $\epsilon_x = u_{,x} + \frac{1}{2}w_{,x}^2$, where u is the axial component of displacement and w is the transverse displacement. Use the principle of stationary potential energy to show that the governing equations are

$$\frac{d}{dx}\left[EA\left(u_{,x} + \frac{1}{2}w_{,x}^2\right)\right] = 0 \quad \text{and} \quad u|_{x=0} = u|_{x=L} = 0$$

$$\frac{d}{dx}\left[EA\left(u_{,x} + \frac{1}{2}w_{,x}^2\right)w_{,x}\right] + \overline{p}_z = 0 \qquad w|_{x=0} = w|_{x=L} = 0$$

Hint:

$$\Pi = \frac{1}{2}\int \sigma_x \epsilon_x \, dV - \int_L \overline{p}_z w \, dx = \int_L \left[\frac{1}{2}EA\left(u_{,x} + \frac{1}{2}w_{,x}^2\right)^2 - \overline{p}_z w\right] dx$$

$\delta\Pi = 0$ leads to the governing equations.

2.20 The principle of virtual work expression for a flat member of thickness t with in-plane loading is

$$\delta W = \int_A \left(\delta\epsilon_{xx}\,\sigma_{xx} + \delta\epsilon_{yy}\,\sigma_{yy} + \delta\gamma_{xy}\,\sigma_{xy}\right) t\,dA - \int_{S_p} \delta u_x\,\overline{p}_x\,dS - \int_{S_p} \delta u_y\,\overline{p}_y\,dS = 0$$

(a) Derive the equation of equilibrium in V.

(b) Derive the statical boundary condition on S_p.

(c) Formulate the virtual work expression along with the derived relations in matrix notation.

2.21 For a rod subject to extension (Chapter 1), the kinematic relation is $\epsilon_{0x} = \partial_x u_0$, the force-displacement relation is $N = EA\epsilon_{0x}$, and the condition of equilibrium is $\partial_x N + \overline{p}_x = 0$.

(a) Derive $\mathbf{A}$ of the first-order governing equations $\partial_x \mathbf{z} = \mathbf{A}\mathbf{z} + \overline{\mathbf{P}}$.

(b) Find $\mathbf{D}_u$ of $\epsilon = \mathbf{D}_u\mathbf{u}$.

(c) Derive $\mathbf{k}^D$ of the principle of virtual work.

(d) Derive $\mathbf{A}$ from $\mathbf{k}^D$. Note that, since $\mathbf{k}^D$ embodies the principle of virtual work, it leads to the equilibrium condition only. This must be supplemented with the kinematic and material relations in order to obtain $\mathbf{A}$.

Generalized Principles

2.22 Use form AD of Section 2.3 to derive the governing differential equations for a bar under extension.

2.23 Use the Hellinger-Reissner functional to derive the governing differential equations for a bar under extension.

2.24 Obtain an expression in matrix form for the hybrid functional of Eq. (2.104) for a 2D plane stress problem.

3

Related Variational and Energy Principles

The two fundamental variational theorems of solid mechanics are the principles of virtual work and complementary virtual work. Important corollaries to these theorems, the principles of stationary potential and complementary energies, respectively, were derived in Chapter 2 based on the existence of appropriate potentials for the forces. Further useful corollaries will be presented in this chapter. These additional variational and energy principles are usually considered to be the classical energy techniques for practical problem solving in solid mechanics.

3.1 The Principle of Virtual Work Related Theorems

The principle of virtual work contends that a solid is in equilibrium if the sum of the external and internal virtual work is zero for kinematically admissible virtual displacements. This principle is expressed by the equations

$$\underbrace{\int_V \sigma_{ij}\,\delta\epsilon_{ij}\,dV}_{-\delta W_i} \underbrace{- \int_V \overline{p}_{V_i}\delta u_i\,dV - \int_{S_p} \overline{p}_i\,\delta u_i\,dS}_{-\delta W_e} = 0$$

or

$$\int_V \delta\boldsymbol{\epsilon}^T \boldsymbol{\sigma}\,dV - \int_V \delta\mathbf{u}^T \overline{\mathbf{p}}_V\,dV - \int_{S_p} \delta\mathbf{u}^T \overline{\mathbf{p}}\,dS = 0 \tag{3.1}$$

with kinematically admissible δu_i, i.e., δu_i that satisfy the kinematical equations

$$\boldsymbol{\epsilon} = \mathbf{Du} \quad \text{in } V \qquad \mathbf{u} = \overline{\mathbf{u}} \quad \text{on } S_u$$

If a potential exists for both the internal and external forces, the principle of virtual work can be specialized to the principle of stationary potential energy. This principle states that among the kinematically admissible deformations, the actual deformations which correspond to forces in equilibrium are the ones for which the total potential energy Π is stationary. This is expressed by the equation

$$\delta\Pi = 0 \quad \text{or} \quad \Pi = U_i + U_e \longrightarrow \text{Stationary} \tag{3.2}$$

with

$$\epsilon = \mathbf{Du} \text{ in } V \quad \text{and} \quad \mathbf{u} = \overline{\mathbf{u}} \text{ on } S_u$$

Some useful theorems related to these principles will be derived now.

3.1.1 Castigliano's Theorem, Part I

A frequently used classical theorem related to virtual work is *Castigliano's first theorem* or *Castigliano's theorem, part I*. This is a global version of the locally applicable relation [Chapter 2, Eq. (2.21)] $\partial U_0/\partial \epsilon_{kj} = \sigma_{kj}$. The principle of stationary potential energy will be used here to derive Castigliano's theorem. To begin the process, consider a general, three-dimensional solid subjected to a system of external forces (or moments) $\overline{P}_1, \overline{P}_2, \ldots \overline{P}_k, \ldots \overline{P}_n$, with corresponding displacements $V_1, V_2, \ldots, V_k, \ldots V_n$. The term V_k is the displacement of $\overline{P}_k$ in the direction of $\overline{P}_k$. The total potential energy is given by [Chapter 2, Eqs. (2.64)]

$$\Pi = U_i + U_e = U_i - \sum_{k=1}^{n} \overline{P}_k V_k \tag{3.3}$$

The strain energy U_i can be expressed in terms of the displacements V_k so that the potential energy is a function of the V_k which are often referred to as *generalized coordinates* or *generalized displacements*. If the solid is in equilibrium, $\delta\Pi$ must vanish [Eq. (3.2)]. Thus,

$$\delta\Pi = \frac{\partial U_i}{\partial V_1}\delta V_1 + \frac{\partial U_i}{\partial V_2}\delta V_2 + \cdots + \frac{\partial U_i}{\partial V_n}\delta V_n - \overline{P}_1\delta V_1 - \overline{P}_2\delta V_2 \cdots - \overline{P}_n\delta V_n = 0$$

or

$$\left(\frac{\partial U_i}{\partial V_1} - \overline{P}_1\right)\delta V_1 + \left(\frac{\partial U_i}{\partial V_2} - \overline{P}_2\right)\delta V_2 + \cdots + \left(\frac{\partial U_i}{\partial V_n} - \overline{P}_n\right)\delta V_n = 0$$

The variations δV_k may be considered to be arbitrary, so that the factors in parentheses must each be zero for the complete expression to vanish. Then $\partial U_i/\partial V_1 = \overline{P}_1, \ldots, \partial U_i/\partial V_n = \overline{P}_n$, or

$$\frac{\partial U_i}{\partial V_k} = \overline{P}_k \tag{3.4}$$

This is Castigliano's theorem, part I. This theorem asserts that:

> If the strain energy of a body is expressed in terms of displacement components in the direction of the prescribed forces, then the first partial derivative of the strain energy, with respect to a displacement, is equal to the corresponding force.

The subscript k is arbitrary in that it can denote any force from $\overline{P}_1$ to $\overline{P}_n$. If all of the n derivatives are taken, then a system of equations

$$\frac{\partial U_i}{\partial \mathbf{V}} = \overline{\mathbf{P}} \qquad \begin{aligned} \mathbf{V} &= [V_1\ V_2 \ldots V_n]^T \\ \overline{\mathbf{P}} &= [\overline{P}_1\ \overline{P}_2 \ldots \overline{P}_n]^T \end{aligned} \tag{3.5}$$

is obtained. To use these relations, it is necessary to express U_i in terms of the displacements $\mathbf{V}$.

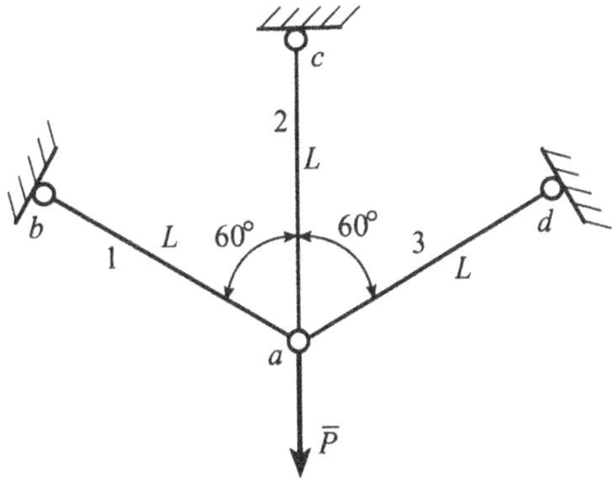

FIGURE 3.1
Statically indeterminate bar system.

If the structural system is subjected to a specified temperature distribution, the temperatures would have to be kept constant if Castigliano's first theorem is to apply. Then Eq. (3.5) could be written $(\partial U_i/\partial \mathbf{V})_{T=\text{constant}} = \overline{\mathbf{P}}$.

This theorem applies also to structural systems with nonlinear stress-strain laws and also to structures with large displacements. In the latter case it is necessary to calculate the strain energy using the expressions for large strains (Chapter 1, Eq. 1.16).

It should be emphasized that by "forces" are meant both forces and moments. For example, if the rotation corresponding to the moment $\overline{M}_k$ is θ_k, then Castigliano's theorem, part I would state that

$$\frac{\partial U_i}{\partial \theta_k} = \overline{M}_k \tag{3.6}$$

EXAMPLE 3.1 Force-Displacement Relationships for a Structure

Determine the force $\overline{P}$ that will cause the joint at a of the bar system (truss) of Fig. 3.1 to displace vertically a certain distance V.

Castigliano's theorem, part I is represented by Eq. (3.4), where $\overline{P}_k$ is the force in the direction of the displacement V_k. Here, for this simple structure, we can drop the subscript k. In order to employ this formula, which leads to a relationship between force and displacement, the strain energy U_i must first be expressed in terms of the displacement V. The theorem can be used for statically determinate or indeterminate systems; ours is statically indeterminate.

From Chapter 2, Eq. (2.5), the strain energy for an extension bar of length L with constant cross-sectional area A and axial force N can be expressed as

$$\int_0^L \frac{A}{2} E\epsilon_x^2\, dx = \frac{EA}{2}\frac{v^2}{L} \tag{1}$$

where it is recognized that the axial strain along the bar is constant $\epsilon_x = \partial u_0/\partial x = v/L$, where v is the total elongation.

The total strain energy stored in the three rods is the sum of the energy in each rod. This still must be converted to a function of V, the vertical displacement of point a. Since the displacements are small, it will be assumed that the bars remain at the same angles with respect to each other throughout the deformation process. Then kinematically admissible V of any of the bars can be expressed by $v = V\cos\alpha$, where α is the angle measured from the vertical (0, 60°, or −60°), so that

$$U_i = \frac{EA}{2L}\left[V^2 + \left(\frac{1}{2}V\right)^2 + \left(\frac{1}{2}V\right)^2\right] = \frac{3EAV^2}{4L} \tag{2}$$

Then the force $\overline{P}$ necessary to generate vertical displacement V is

$$\overline{P} = \frac{\partial U_i}{\partial V} = \frac{3}{2}\frac{EAV}{L} \tag{3}$$

This force-displacement relationship can be "post-processed" to obtain information about the member forces, stresses, and deformations. Equation (3) gives $V = 2\overline{P}L/3EA$. Use of $v = V\cos\alpha$ gives the elongation of any bar. Also, the axial force N in each bar is found from

$$N = \frac{EAv}{L} = \frac{EAV}{L}\cos\alpha \tag{4}$$

■

EXAMPLE 3.2 Stiffness Matrix for a Spring

Suppose a linear elastic spring is placed between points a and b in a structure. The strain energy for this spring will be

$$U_i = \frac{N\,\Delta u}{2} \tag{1}$$

where N is the force in the spring, and Δu is the deformation experienced by the spring, i.e., $\Delta u = u_b - u_a$, where u_a and u_b are the extensions at a and b. From the definition of an elastic spring, the force in the spring is given by

$$N = k\,\Delta u = k(u_b - u_a) \tag{2}$$

where k is the spring constant or spring rate. Substituting (2) into (1), the strain energy becomes

$$U_i = \frac{N\,\Delta u}{2} = \frac{k}{2}(u_b - u_a)^2 \tag{3}$$

From Castigliano's theorem, part I, the tensile forces N_a and N_b at the ends a and b are

$$\begin{aligned} N_a &= \frac{\partial U_i}{\partial u_a} = ku_a - ku_b \\ N_b &= \frac{\partial U_i}{\partial u_b} = -ku_a + ku_b \end{aligned} \tag{4}$$

These are the expected results, since, from the conditions of equilibrium, $N = N_b = -N_a$.

In (4), Castigliano's theorem, part I has been applied to a case where the forces, N_a and N_b, are not applied loads. These forces should be interpreted as the forces that would have to be applied to generate the displacements u_a and u_b. In matrix notation, (4) can be expressed as

$$\begin{bmatrix} N_a \\ N_b \end{bmatrix} = \begin{bmatrix} k & -k \\ -k & k \end{bmatrix} \begin{bmatrix} u_a \\ u_b \end{bmatrix} = \begin{bmatrix} k_{aa} & k_{ab} \\ k_{ba} & k_{bb} \end{bmatrix} \begin{bmatrix} u_a \\ u_b \end{bmatrix} \tag{5}$$
$$\mathbf{p} \quad = \quad \mathbf{k} \quad \mathbf{v}$$

The constants k_{ij} are called the *stiffnesses* of the structure (spring), while the associated matrix $\mathbf{k}$ is referred to as the *stiffness matrix*. Several characteristics of the stiffnesses are of interest. Note that if u_a is zero and u_b is unity, $N_a = k_{ab}$ and $N_b = k_{bb}$. We conclude that k_{ij}

is the force at i due to a unit displacement at j, with the displacement at i equal to zero. Also, observe that $k_{ij} = k_{ji}$, i.e., the stiffness matrix is symmetric. ∎

EXAMPLE 3.3 Strain Energy in Terms of the Stiffness Matrix

Suppose a linear solid is acted upon by a system of gradually applied forces $\overline{P}_1, \overline{P}_2, \ldots \overline{P}_k, \ldots \overline{P}_n$, with corresponding displacements $V_1, V_2, \ldots V_k, \ldots V_n$. The strain energy is equal to the external work.

$$U_i = \frac{1}{2}\sum_{k=1}^{n} \overline{P}_k V_k \tag{1}$$

In matrix notation,

$$U_i = \frac{1}{2}\mathbf{V}^T\overline{\mathbf{P}} = \frac{1}{2}\overline{\mathbf{P}}^T\mathbf{V} \tag{2}$$

where $\mathbf{V} = [V_1\ V_2, \ldots, V_n]^T, \overline{\mathbf{P}} = [\overline{P}_1\ \overline{P}_2, \ldots, \overline{P}_n]^T$. The stiffness coefficients K_{ij} for this solid are defined in the same fashion as in Example 3.2. Upper case letter **K** is used to indicate that the stiffness matrix is for the whole system, whereas a lower case **k** denotes the stiffness matrix for a single element. The force $\overline{P}_k$ at coordinate k would be

$$\overline{\mathbf{P}}_k = \sum_{j=1}^{n} K_{kj} V_j \quad \text{or} \quad \overline{\mathbf{P}} = \mathbf{KV} \tag{3}$$

The strain energy of (2) would then appear as

$$U_i = \frac{1}{2}\mathbf{V}^T\mathbf{KV} \tag{4}$$

Transpose both sides of (4) to obtain

$$U_i = \frac{1}{2}\mathbf{V}^T\mathbf{K}^T\mathbf{V} \tag{5}$$

Equating (4) and (5) gives

$$\mathbf{K} = \mathbf{K}^T \tag{6}$$

Thus, the system matrix **K**, is symmetric, i.e., $K_{kj} = K_{jk}$.

The partial derivative of (4), with respect to an arbitrary displacement V_k, gives

$$\frac{\partial U_i}{\partial V_k} = \sum_{j=1}^{n} K_{kj} V_j \tag{7}$$

which, according to (3), is the same as Castigliano's theorem, part I. The partial derivative of (7), with respect to a displacement V_h, gives

$$\frac{\partial^2 U_i}{\partial V_h\, \partial V_k} = K_{kh} \tag{8}$$

which is a stiffness coefficient. The stiffness coefficient is equal to the second derivative of the strain energy with respect to the displacements at k and h.

Equation (8) applies only if U_i is expressed in terms of independent displacements V_k. If some of the displacements are dependent on other displacements, the partial derivatives of (8) cannot be taken because this implies that V_j other than V_k and V_h are to be held constant. Then the stiffness matrix **K** does not exist. Dependent displacements are treated in some detail in Chapter 4. ∎

EXAMPLE 3.4 Properties of Stiffness Matrices

The definitions of Examples 3.2 and 3.3 permit some important properties of stiffness matrices to be identified. In summation form, the strain energy of Eq. (4) of Example 3.3 would appear as

$$U_i = \frac{1}{2} \sum_{k=1}^{n} \sum_{j=1}^{n} K_{kj} V_k V_j \tag{1}$$

This double summation is said to be a *quadratic form* in variables V_k, provided that coefficients K_{kj} are symmetric—which they are for a stiffness matrix.

A quadratic form such as (1) is defined as being *positive definite* if U_i is positive for arbitrary non-zero values of V_k and V_j and is zero if all V_k and V_j are zero. The strain energy satisfies this definition, and, hence, it is positive definite, and the associated stiffness matrix **K** is referred to as a *positive definite matrix*. In Example 3.3, the positive definite **K** considered is defined only if the displacements V_k are independent variables. It will be shown in Chapter 4 that this corresponds to a system for which rigid body motion has been eliminated. A stiffness matrix from which the rigid body motion has not been removed is positive semi-definite.

Another property of **K** is that the elements on the principal diagonal, K_{ii}, are always positive because, by definition, K_{ii} is the force at i corresponding to a displacement i. The force is in the same direction as the displacement and hence K_{ii} is positive. This property of K_{ii} is not valid if the structure becomes unstable. ∎

EXAMPLE 3.5 Stiffness Matrix for a Truss

Find a relationship between the displacements and the applied loads of the truss of Fig. 3.2a. All of the bars are elastic and have the same cross-sectional areas.

The desired load-displacement relation can be obtained by calculating the stiffness coefficients using Eq. (8) of Example 3.3. The elongation v of the jth bar is (Fig. 3.2b)

$$v = (u_{xb} - u_{xa}) \cos\theta + (u_{zb} - u_{xa}) \sin\theta \tag{1}$$

The strain energy of the jth bar with extension v is

$$U_i = \frac{EA}{2} \frac{V^2}{L} \tag{2}$$

Invoke the displacement boundary conditions and the conditions of compatibility as is required by the principle of the virtual work method.

$$\begin{aligned}
&u_{xd|5} = u_{zd|5} = u_{xd|3} = u_{zd|3} = 0, \qquad u_{xa|1} = u_{za|1} = u_{xa|4} = u_{za|4} = 0\\
&u_{xb|5} = u_{xb|1} = u_{xb|2} = V_{xb} = V_1\\
&u_{zb|5} = u_{zb|1} = u_{zb|2} = V_{zb} = V_2\\
&u_{xc|4} = u_{xc|3} = u_{xc|2} = V_{xc} = V_3\\
&u_{zc|4} = u_{zc|3} = u_{zc|2} = V_{zc} = V_4
\end{aligned} \tag{3}$$

where $u_{xd|5} = u_{xd}$ of bar 5, etc.

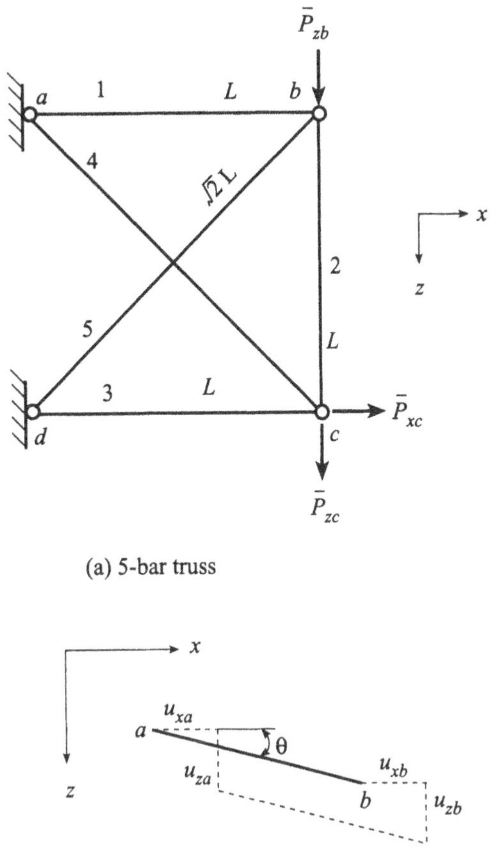

FIGURE 3.2
Truss.

(b) Elongation of the jth bar

Use (1), (2), and (3), to form the total strain energy

$$U_i = \frac{EA}{L}\left[\frac{1}{2}V_1^2 + \frac{1}{2}(V_2 - V_4)^2 + \frac{1}{2}V_3^2 + \frac{1}{2\sqrt{2}}\left(\frac{V_1 + V_2}{\sqrt{2}}\right)^2 + \frac{1}{2\sqrt{2}}\left(\frac{V_3 - V_4}{\sqrt{2}}\right)^2\right] \tag{4}$$

From Eq. (8) of Example 3.3

$$\frac{\partial^2 U_i}{\partial V_h\, \partial V_k} = K_{kh} \tag{5}$$

Thus, for example,

$$\begin{aligned} K_{11} &= \frac{\partial^2 U_i}{\partial V_1\, \partial V_1} = \frac{EA}{L}\left(1 + \frac{\sqrt{2}}{4}\right) \\ K_{12} &= \frac{\partial^2 U_i}{\partial V_2\, \partial V_1} = \frac{EA}{L}\frac{\sqrt{2}}{4} \end{aligned} \tag{6}$$

The remaining stiffness coefficients are obtained in a similar fashion. The relationship between the displacements at b and c and the applied forces at b and c becomes

$$\frac{EA}{L}\begin{bmatrix} 1+\frac{\sqrt{2}}{4} & \frac{\sqrt{2}}{4} & 0 & 0 \\ \frac{\sqrt{2}}{4} & 1+\frac{\sqrt{2}}{4} & 0 & -1 \\ 0 & 0 & 1+\frac{\sqrt{2}}{4} & \frac{-\sqrt{2}}{4} \\ 0 & -1 & \frac{-\sqrt{2}}{4} & 1+\frac{\sqrt{2}}{4} \end{bmatrix}\begin{bmatrix} V_{xb} \\ V_{zb} \\ V_{xc} \\ V_{zc} \end{bmatrix} = \begin{bmatrix} \overline{P}_{xb} \\ \overline{P}_{zb} \\ \overline{P}_{xc} \\ \overline{P}_{zc} \end{bmatrix} \tag{7}$$

$$\mathbf{K} \qquad \mathbf{V} = \overline{\mathbf{P}}$$

Of course, the same result can be obtained directly from the principle of virtual work, or the principle of stationary potential energy. For example, consider the use of the principle of stationary potential energy. The potential energy is given by

$$\Pi = U_i + U_e = U_i - \sum_{j=1}^{4} \overline{P}_j V_j \tag{8}$$

with $\overline{P}_{xb}, \overline{P}_{zb}, \overline{P}_{xc}, \overline{P}_{zc} = \overline{P}_1, \overline{P}_2, \overline{P}_3, \overline{P}_4$. Then use of

$$\delta\Pi = 0 = \frac{\partial \Pi}{\partial V_1}\delta V_1 + \frac{\partial \Pi}{\partial V_2}\delta V_2 + \frac{\partial \Pi}{\partial V_3}\delta V_3 + \frac{\partial \Pi}{\partial V_4}\delta V_4 \tag{9}$$

along with the strain energy of (4), leads to the stiffness relations of (7) again.

The relationship of (7) i.e., $\mathbf{KV} = \overline{\mathbf{P}}$, can be solved for the displacements $\mathbf{V}$. ∎

3.1.2 The Unit Displacement Method

The *unit displacement* or *dummy displacement method* serves almost the same purpose as Castigliano's theorem, part I. This method, which is described in Chapter 3 of the first edition of this book, determines the force $\overline{P}_j$ at a given point necessary to maintain equilibrium in a structure under a known state of stress $\boldsymbol{\sigma}$.

3.2 The Principle of Complementary Virtual Work Related Theorems

The principle of complementary virtual work and its corollaries are very useful for hand calculations of structural analysis, but they are somewhat difficult to systematize for computer solutions of large-scale systems. This is in stark contrast to the principle of virtual work theorems which, because of the need for imposing geometric compatibility, are not well suited for hand calculations, yet—as will be seen in Chapter 5—are readily systematized for the solution of large-scale systems.

According to the principle of complementary virtual work, a solid satisfies the kinematical conditions if the sum of the external complementary virtual work and the internal complementary virtual work is zero for statically admissible virtual stresses. The principle is embodied in the equations

$$
\begin{aligned}
-\delta W_i^* \;\; - \delta W_e^* \qquad &= 0 \\
\int_V \epsilon_{ij}\, \delta\sigma_{ij}\, dV - \int_{S_u} \bar{u}_i\, \delta p_i\, dS &= 0 \\
\int_V \delta\boldsymbol{\sigma}^T \boldsymbol{\epsilon}\, dV - \int_{S_u} \delta\mathbf{p}^T \bar{\mathbf{u}}\, dS &= 0
\end{aligned}
\tag{3.7}
$$

with statically admissible $\delta\sigma_{ij}$, i.e., $\delta\sigma_{ij}$ that satisfy the equilibrium conditions and static boundary conditions

$$
\begin{aligned}
\mathbf{D}^T\boldsymbol{\sigma} + \bar{\mathbf{p}}_V = \mathbf{0} \quad &\text{in } V \\
\mathbf{p} = \bar{\mathbf{p}} \quad &\text{on } S_p
\end{aligned}
$$

If the potential functions exist, the principle of complementary virtual work is specialized as the principle of stationary complementary energy. This principle contends that among the statically admissible states of stress, the actual state of stress which corresponds to kinematically admissible deformations is the one for which the total complementary energy Π^* is stationary. This is expressed as

$$
\delta\Pi^* = 0 \quad \text{or} \quad \Pi^* = U_i^* + U_e^* \longrightarrow \text{Stationary} \tag{3.8}
$$

where

$$
\begin{aligned}
\mathbf{D}^T\boldsymbol{\sigma} + \bar{\mathbf{p}}_V = \mathbf{0} \quad &\text{in } V \\
\mathbf{p} = \bar{\mathbf{p}} \quad &\text{on } S_p
\end{aligned}
$$

Other useful theorems can be derived from these principles.

3.2.1 Castigliano's Theorem, Part II

The complementary energy theorem in a form similar to Eq. (3.5) is *Castigliano's theorem, part II* or *Castigliano's second theorem*. A general three-dimensional solid acted on by a system of forces $P_1, P_2, \ldots, P_k, \ldots P_n$ with corresponding displacements $V_1, V_2, \ldots, V_k, \ldots V_n$ at their points of application will be used to illustrate this. Here, the forces can be regarded as reactions generated by prescribed displacements V_k, which more precisely should be written $\overline{V}_k$. That is, these are the displacements that would have to be applied to create the forces $P_1, P_2, \ldots P_n$. If the V_k's are considered as being independent of the P_k's, the total complementary potential energy can be expressed as

$$
\Pi^* = U_i^* + U_e^* = U_i^* - \sum_{k=1}^{n} P_k V_k \tag{3.9}
$$

The complementary energy is a function of the P_k, which are sometimes referred to as the *generalized forces* of the system. Recall that for the complementary virtual work theorems, the variations are taken on the forces. According to the principle of stationary complementary energy, $\delta\Pi^* = 0$. Therefore,

$$
\delta\Pi^* = 0 = \frac{\partial U_i^*}{\partial P_1}\delta P_1 + \frac{\partial U_i^*}{\partial P_2}\delta P_2 + \cdots - V_1\,\delta P_1 - V_2\,\delta P_2 \cdots - V_n\,\delta P_n
$$

or

$$
\sum_{k=1}^{n}\left(\frac{\partial U_i^*}{\partial P_k}\delta P_k - V_k\,\delta P_k\right) = \sum_{k=1}^{n}\left[\left(\frac{\partial U_i^*}{\partial P_k} - V_k\right)\delta P_k\right] = 0
$$

The variations δP_k are arbitrary, except that they must satisfy the equilibrium conditions, so that each term in the parentheses must vanish. Then

$$\frac{\partial U_i^*}{\partial P_k} = V_k \tag{3.10}$$

In vector notation,

$$\frac{\partial U_i^*}{\partial \mathbf{P}} = \mathbf{V} \tag{3.11}$$

This is Castigliano's theorem, part II. This theorem asserts that:

> If the complementary energy of a body is expressed in terms of the forces, then the first partial derivative of the complementary energy, with respect to any one of the forces, is equal to the corresponding displacement at the point where the force is located.

The subscript k is arbitrary in that it can represent any force from 1 to n. To utilize this theorem, it is necessary to express U_i^* in terms of the forces $\mathbf{P}$.

The theorem implies that any temperature distribution to which the structure is subjected must remain constant. This can be indicated by writing Eq. (3.11) as $(\partial U_i^*/\partial \mathbf{P})_{\Delta T=\text{constant}} = \mathbf{V}$.

This theorem applies for linear and nonlinear elastic bodies. If large displacements are to be taken into account, it is necessary to appropriately redefine U_i^*. Sometimes the theorem is referred to as *Engesser's*[1] *first theorem* and if the structure is linearly elastic, so that $U_i = U_i^*$ and

$$\frac{\partial U_i}{\partial \mathbf{P}} = \mathbf{V} \tag{3.12}$$

then the theorem is almost always called Castigliano's second theorem.

"Forces" as used here include both forces and moments. Thus, the theorem may also take the form

$$\frac{\partial U_i^*}{\partial M_k} = \theta_k \tag{3.13}$$

In summary, Castigliano's theorem, part II can be used to compute displacements and slopes of surface displacements at the locations of concentrated forces or moments.

EXAMPLE 3.6 Deflection of a Cantilevered Beam

Find the deflection and slope at the free end of the linearly elastic cantilevered beam shown in Fig. 3.3a.

For a linearly elastic structure, $U_i^* = U_i$. Then Castigliano's theorem, part II in the form $\partial U_i/\partial P = w_0$ and $\partial U_i/\partial M_0 = \theta_0$ applies, where w_0 and θ_0 are the deflection and slope at the free end. As shown in Chapter 2, Example 2.1, the complementary strain energy in a

[1] Friedrich Engesser (1848–1931) was a German engineer who made many significant contributions to the analysis of statically indeterminate systems. He began his career as a railroad design engineer specializing in bridge design. He worked extensively in developing theories for the buckling of members, in particular, lateral instability. He presented the general form of Castigliano's theorem, part II, in 1889 in "Über Statisch Unbestimmte Träger bei beliebigem Formänderungs-Gesetz," *Z. Arch. Ing. Ver. Hannover*, 35, 733–744, 1889.

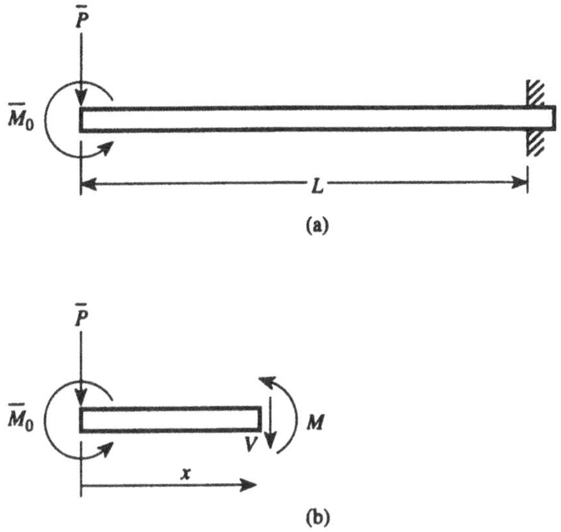

FIGURE 3.3
Cantilevered beam.

beam is given by

$$U_i^* = \int_0^L \frac{M^2}{2EI}\,dx \tag{1}$$

Since Castigliano's theorem, part II is related to the principle of complementary virtual work, the forces must be in equilibrium.

For our beam, the conditions of equilibrium give (Fig. 3.3b) $M = -\overline{M}_0 - \overline{P}x$. In utilizing $\partial U_i^*/\partial \overline{P}$ or $\partial U_i^*/\partial \overline{M}_0$ with U_i^* defined by (1), it is generally simpler to differentiate the integrand first and then to carry out the integration rather than the opposite. Thus,

$$w_0 = \frac{\partial U_i^*}{\partial \overline{P}} = \frac{1}{EI}\int_0^L M\,\frac{\partial M}{\partial \overline{P}}\,dx \tag{2}$$

With $M = -\overline{M}_0 - \overline{P}x$ and $\partial M/\partial \overline{P} = -x$,

$$w_0 = \frac{1}{EI}\int_0^L (\overline{M}_0 + \overline{P}x)x\,dx = \frac{\overline{M}_0 L^2}{2EI} + \frac{\overline{P}L^3}{3EI} \tag{3}$$

For the end slope, $\partial M/\partial \overline{M}_0 = -1$ and

$$\theta_0 = \frac{\partial U_i^*}{\partial \overline{M}_0} = \frac{1}{EI}\int_0^L M\frac{\partial M}{\partial \overline{M}_0}\,dx = \frac{1}{EI}\int_0^L (\overline{M}_0 + \overline{P}x)\,dx$$

$$= \frac{\overline{M}_0 L}{EI} + \frac{\overline{P}L^2}{2EI} \tag{4}$$

Positive (negative) values of w_0 and θ_0 indicate that these variables are in the same (opposite) direction as the corresponding force $\overline{P}$ and moment $\overline{M}_0$. ∎

EXAMPLE 3.7 Truss Analysis

Find the vertical displacement of point b of the simple truss of Fig. 3.4.

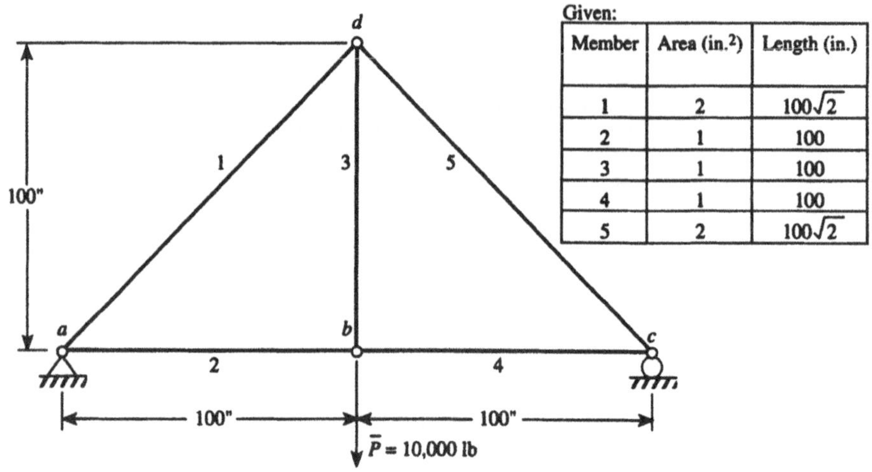

Given:

Member	Area (in.²)	Length (in.)
1	2	$100\sqrt{2}$
2	1	100
3	1	100
4	1	100
5	2	$100\sqrt{2}$

FIGURE 3.4
Truss.

For this structure, Castigliano's theorem, part II states that the displacement due to $\overline{P}$ follows from the differentiation of the total complementary strain energy with respect to the force $\overline{P}$. Here the total complementary strain energy, which is equal to the total strain energy, i.e., $U_i^* = U_i$, for linear elastic structures, is the sum of the strain energies of each member. This is a statically determinate structure in which all members are assumed to be pin-connected bars and are subjected only to axial forces with no bending. The complementary strain energy in bar j is that due to its axial force N and has the value (Chapter 2, Example 2.1) $N^2L/(2EA)$. The total complementary strain energy is then

$$U_i^* = \sum_{\text{All bars}} N^2L/(2EA) \tag{1}$$

As in the case of the beam of Example 3.6, the simplest procedure is to take the derivative of the complementary strain energy before integration (summation), rather than after the total complementary strain energy is computed. Then

$$V_b = \frac{\partial U_i^*}{\partial \overline{P}} = \sum_{\text{All bars}} \frac{L}{EA} N \frac{\partial N}{\partial \overline{P}}$$

is the vertical displacement at b.

In almost all cases, the most rapid computation technique is to take advantage of the fact that $\partial N/\partial \overline{P}$ in each bar is equal to the axial load produced in that bar by a unit load at b in the direction of $\overline{P}$. This follows from the interpretation of the partial derivative $\partial N/\partial \overline{P}$ as the rate at which the internal axial force N in the jth member changes as $\overline{P}$ changes. Numerically, this rate of change is equal to the internal axial force produced by $\overline{P}$ set equal to unity with all actual loads on the structure removed. This calculation involves only the conditions of equilibrium, as is to be expected for this complementary energy-related theorem. This approach is sometimes referred to as the unit load method which is discussed in a subsequent section.

The required calculations can be done in tabular form. The N, $\partial N/\partial \overline{P}$ entries are determined using the conditions of equilibrium for the truss. Assume the bars are made of steel ($E = 3 \times 10^7$ psi).

Member	Length L (in.)	Area A (in.^{2})	Internal Axial Force N in Each Member Due to the Actual Loads (here $\overline{P}$)	$\partial N/\partial \overline{P}$ Internal Axial Force in Each Member Due to a Unit Load at b in the Direction of $\overline{P}$. All Actual Applied Loads are Removed	$\frac{N \frac{\partial N}{\partial \overline{P}} L}{EA}$
1	$100\sqrt{2}$	2	$-5000\sqrt{2}$	$-\sqrt{2}/2$	$(10^{-2})5\sqrt{2}/6$
2	100	1	5000	1/2	$(10^{-2})5/6$
3	100	1	10,000	1	$(10^{-1})1/3$
4	100	1	5000	1/2	$(10^{-2})5/6$
5	$100\sqrt{2}$	2	$-5000\sqrt{2}$	$-\sqrt{2}/2$	$(10^{-2})5\sqrt{2}/6$

$$V_b = \sum_{5 \text{ bars}} \frac{N \frac{\partial N}{\partial \overline{P}} L}{EA} = 0.00736 \text{ in.}$$

Although the use of the unit load is recommended in most cases, for this simple truss the N column could just as well have been expressed in terms of $\overline{P}$ (rather than using $\overline{P} = 10{,}000$) and then differentiated to find $\partial N/\partial \overline{P}$ directly. These entries would then have appeared as

Member	N	$\partial N/\partial \overline{P}$
1	$-\overline{P}\sqrt{2}/2$	$-\sqrt{2}/2$
2	$\overline{P}/2$	1/2
3	$\overline{P}$	1
4	$\overline{P}/2$	1/2
5	$-\overline{P}\sqrt{2}/2$	$-\sqrt{2}/2$

∎

EXAMPLE 3.8 Frame Analysis

Compute the reactions of the statically indeterminate frame of Fig. 3.5a. Consider only the effects of bending in each of the members which form the frame.

Use Castigliano's theorem, part II in the form

$$
\begin{aligned}
u_d &= \frac{\partial U^*}{\partial R_{Vd}} = 0 = \frac{1}{EI}\int M \frac{\partial M}{\partial R_{Vd}}\,dx \\
w_d &= \frac{\partial U^*}{\partial R_{Hd}} = 0 = \frac{1}{EI}\int M \frac{\partial M}{\partial R_{Hd}}\,dx \qquad (1) \\
\theta_d &= \frac{\partial U^*}{\partial M_d} = 0 = \frac{1}{EI}\int M \frac{\partial M}{\partial M_d}\,dx
\end{aligned}
$$

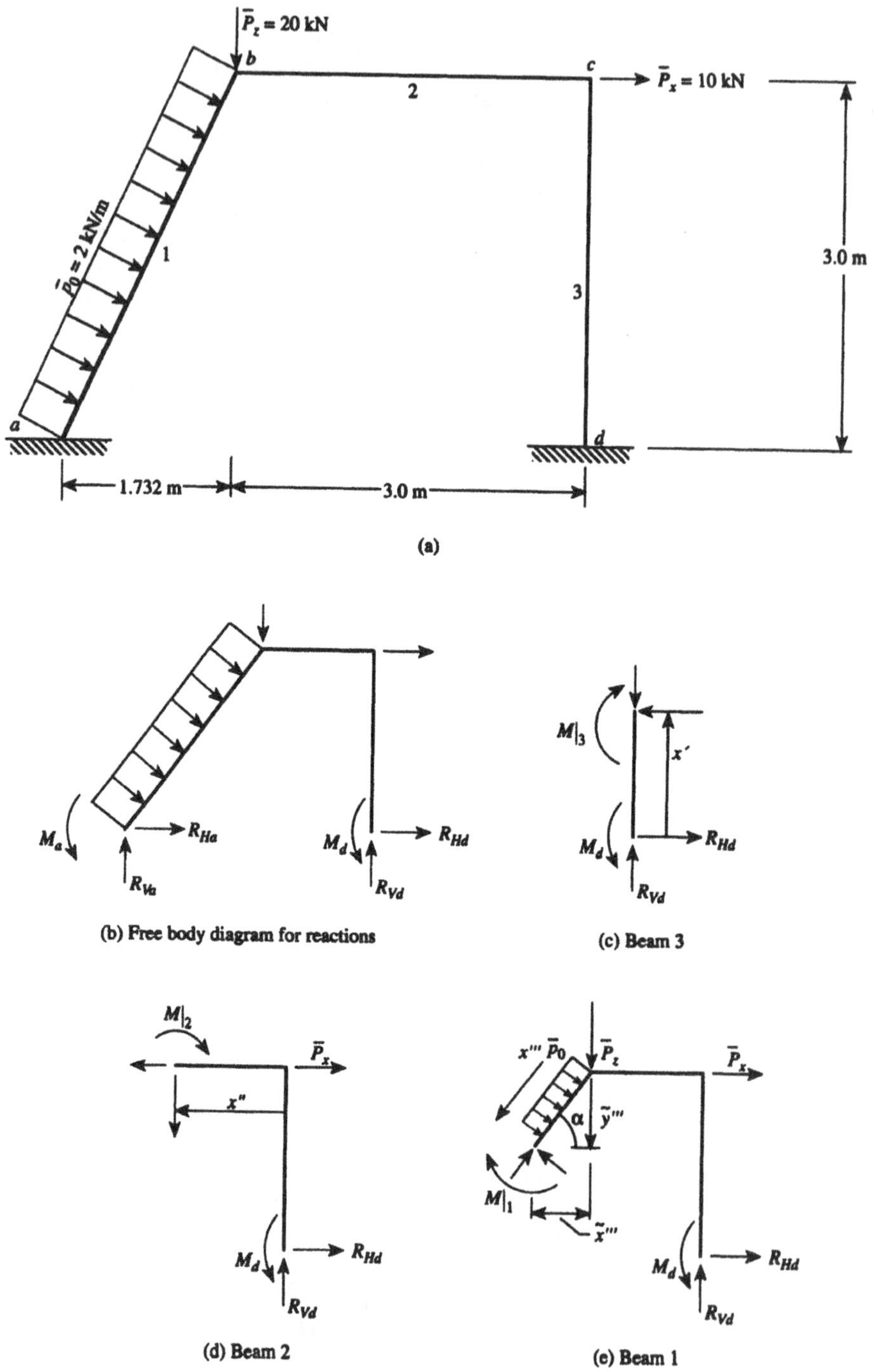

FIGURE 3.5
Frame. Notation: $M|_i = M_{\text{beam } i}$.

These three relationships, plus three conditions of equilibrium, can be solved for the six reactions M_a, R_{Va}, R_{Ha}, M_d, R_{Vd}, and R_{Hd} of Fig. 3.5b.

Begin by computing the moment in each member using the free-body diagrams of Fig. 3.5c, d, and e. From a summation of moments about the cut ends of these figures,

$$\begin{aligned} x' R_{Hd} + M_d &= M|_3 \\ 3R_{Hd} + x'' R_{Vd} + M_d &= M|_2 \\ (3 - \tilde{y}''') R_{Hd} + (3 + \tilde{x}''') R_{Vd} + M_d & \\ -\overline{P}_z \tilde{x}''' - \overline{P}_x \tilde{y}''' - \overline{p}_0 \frac{(x''')^2}{2} &= M|_1 \end{aligned} \tag{2}$$

where $\tilde{y}''' = x''' \sin\alpha$, $\tilde{x}''' = x''' \cos\alpha$, $\alpha = 60°$, $0 \le x' \le 3$, $0 \le x'' \le 3$, $0 \le x''' \le 3.464$, and $M|_i = M_{\text{beam}\,i}$, $i = 1, 2, 3$.

To apply the first relationship of (1), form the strain energy for each of the three members, sum the results to find U, and set $\partial U / \partial R_{Vd} = 0$. Better still, utilize

$$\frac{1}{EI} \int M \frac{\partial M}{\partial R_{Vd}}\bigg|_{\text{beam 1}} dx + \frac{1}{EI} \int M \frac{\partial M}{\partial R_{Vd}}\bigg|_{\text{beam 2}} dx + \frac{1}{EI} \int M \frac{\partial M}{\partial R_{Vd}}\bigg|_{\text{beam 3}} dx = 0 \tag{3}$$

with

$$\frac{\partial M}{\partial R_{Vd}}\bigg|_3 = 0, \qquad \frac{\partial M}{\partial R_{Vd}}\bigg|_2 = x'', \qquad \frac{\partial M}{\partial R_{Vd}}\bigg|_1 = 3 + \tilde{x}''' \tag{4}$$

This leads to

$$32.088\ R_{Hd} + 61.638\ R_{Vd} + 17.89\ M_d - 524.694 = 0 \tag{5}$$

For the second equation of (1), follow the above procedure using

$$\frac{\partial M}{\partial R_{Hd}}\bigg|_3 = x', \qquad \frac{\partial M}{\partial R_{Hd}}\bigg|_2 = 3, \qquad \frac{\partial M}{\partial R_{Hd}}\bigg|_1 = 3 - \tilde{y}''' \tag{6}$$

to find

$$46.392\ R_{Hd} + 32.0884\ R_{Vd} + 18.696\ M_d - 122.35 = 0 \tag{7}$$

For the third equation of (1), use

$$\frac{\partial M}{\partial M_d}\bigg|_1 = \frac{\partial M}{\partial M_d}\bigg|_2 = \frac{\partial M}{\partial M_d}\bigg|_3 = 1 \tag{8}$$

This gives

$$18.696\ R_{Hd} + 17.892\ R_{Vd} + 9.464\ M_d - 125.810 = 0 \tag{9}$$

Solve (5), (7), and (9) to find

$$R_{Hd} = -10.205\ \text{kN}, \qquad R_{Vd} = 9.118\ \text{kN}, \qquad M_d = 16.217\ \text{kN} \tag{10}$$

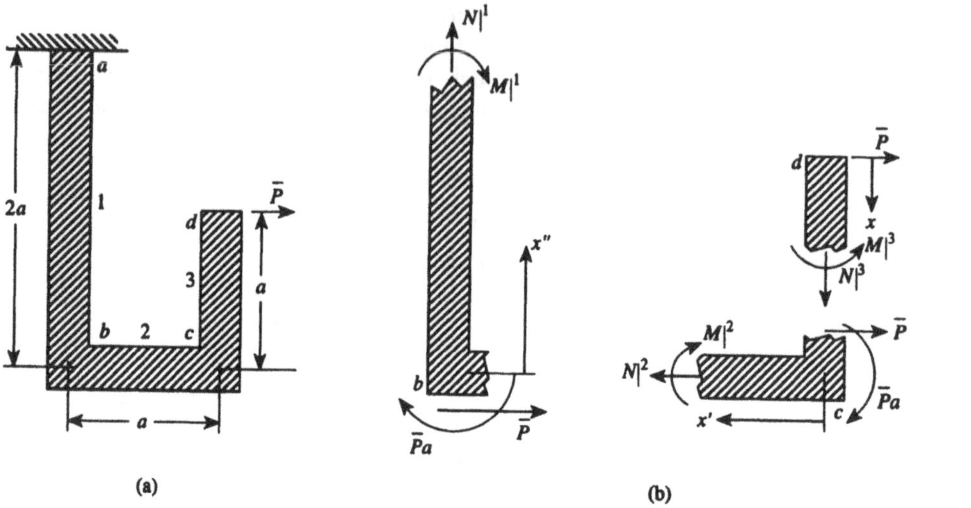

FIGURE 3.6
Frame.

The remaining reactions are found by applying the conditions of equilibrium to the free-body diagram of Fig. 3.5b. This gives

$$R_{Ha} = 5.795 \text{ kN}, \qquad R_{Va} = 14.346 \text{ kN}, \qquad M_a = 17.275 \text{ kN} \cdot \text{m} \tag{11}$$

∎

EXAMPLE 3.9 Combined Effects on Bar Systems

Determine the horizontal displacement of point d on the frame of Fig. 3.6a.

This structure, along with the structures of Fig. 3.7, is loaded such that more than either simple extension or simple bending is generated. Apart from shear deformation, which can always be included in structures built of beam members, a portion of the frame of Fig. 3.6a, i.e., element 2, is subject to both extension and bending. The displacement at point b due to the extension of element 2 is considered to be insignificant relative to the displacement caused by bending and is sometimes ignored, but it will be considered here. Some sections of the structures in Fig. 3.7 are also subjected to extension (or compression) and bending, e.g., element 1 in Fig. 3.7b and all of the bar in Fig. 3.7c, while others are subjected to both torsion and bending, e.g., element 1 in Fig. 3.7a and all of the bar of Fig. 3.7d. Member 1 in Fig. 3.7e is deformed by bending, torsion, and compression.

First the internal bending moment and axial force in each bar of the frame of Fig. 3.6a will be determined. The equilibrium equations yield (Fig. 3.6b) $M|^3 = \overline{P}x$, $N|^3 = 0$,

$$M|^2 = -\overline{P}a, \quad N|^2 = \overline{P}, \quad M|^1 = -\overline{P}a + \overline{P}x'', \quad N|^1 = 0$$

where $M|^3 = M_{\text{bar}\,3}$, $N|^2 = N_{\text{bar}\,2}$, etc.

To find the displacement of point d in the direction of $\overline{P}$, take the derivative of the total complementary strain energy with respect to $\overline{P}$: $V_{d\,\text{horizontal}} = \partial U_i^*/\partial \overline{P}$. Since this leads to integrals of the form $\int \frac{M}{EI}\frac{\partial M}{\partial \overline{P}}dx + \int \frac{N}{EA}\frac{\partial N}{\partial \overline{P}}dx$ compute

$$\begin{aligned} &\frac{\partial M|^3}{\partial \overline{P}} = x, \quad \frac{\partial N|^3}{\partial \overline{P}} = 0, \quad \frac{\partial M|^2}{\partial \overline{P}} = -a, \\ &\frac{\partial N|^2}{\partial \overline{P}} = 1, \quad \frac{\partial M|^1}{\partial \overline{P}} = -a + x'', \quad \frac{\partial N|^1}{\partial \overline{P}} = 0 \end{aligned} \tag{1}$$

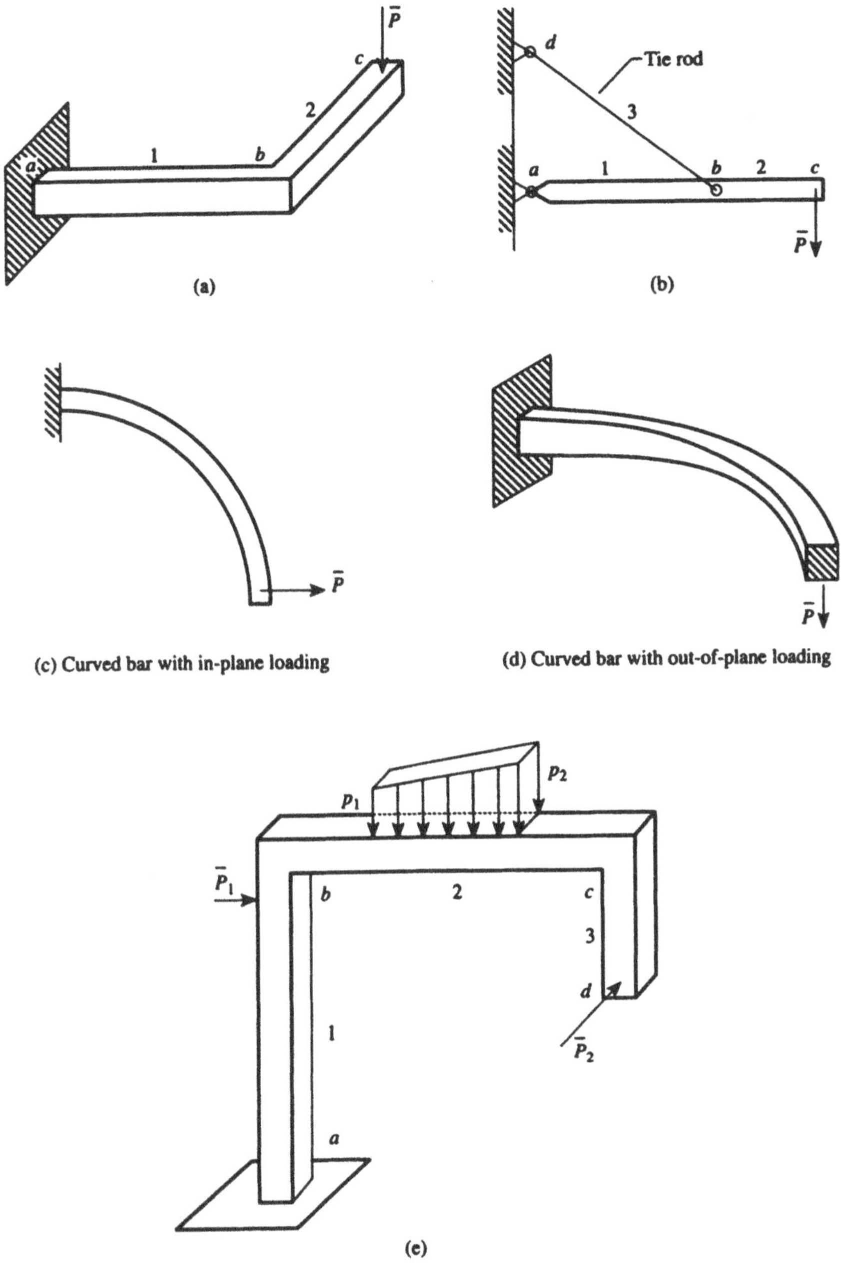

FIGURE 3.7
Several structural systems whose members are subjected to combined loadings.

Then

$$V_{d\,\text{horizontal}} = \int_0^a \frac{M|^3}{EI}\frac{\partial M|^3}{\partial \overline{P}}\,dx + \int_0^a \frac{M|^2}{EI}\frac{\partial M|^2}{\partial \overline{P}}\,dx'$$
$$+ \int_0^a \frac{N|^2}{EA}\frac{\partial N|^2}{\partial \overline{P}}\,dx + \int_0^{2a} \frac{M|^1}{EI}\frac{\partial M|^1}{\partial \overline{P}}\,dx'' = \frac{2\overline{P}a^3}{EI} + \frac{\overline{P}a}{EA} \qquad (2)$$

The final term is due to the extension of segment bc.

Shear deformation effects, which are frequently small, are easily included in a solution. In the above solution, the expressions for the moments and axial forces derived previously are still valid. The internal shear forces can be written as $V_{\text{bar}\,3} = \overline{P}$, $V_{\text{bar}\,2} = 0$, and $V_{\text{bar}\,1} = -\overline{P}$. Shear deformation contributes terms of the form (see Chapter 2, Example 2.2)

$$\int \frac{V^2}{2Gk_s A}\,dx$$

to the total strain energy, where k_s is the shear shape factor. Then

$$V_{d\,\substack{\text{horizontal}\\ \text{(due to shear)}}} = \int \frac{V}{Gk_s A}\frac{\partial V}{\partial \overline{P}}\,dx = \int_0^a \frac{\overline{P}}{Gk_s A}\,dx + \int_0^{2a} \frac{\overline{P}}{Gk_s A}\,dx'' = \frac{3\overline{P}a}{Gk_s A} \tag{3}$$

The total horizontal displacement of d is then

$$V_{d\,\text{horizontal}} = \frac{2\overline{P}a^3}{EI} + \frac{\overline{P}a}{EA} + \frac{3\overline{P}a}{Gk_s A} \tag{4}$$

■

EXAMPLE 3.10 In-Plane Deformation of a Curved Bar

Determine the vertical displacement of point a of the curved bar of Fig. 3.8a. Segment 2 is a circular arc.

This type of problem is similar to the ones treated in Example 3.9 in that several kinds of internal forces (stress resultants) occur simultaneously. An axial force acts on element 1; a bending moment, shear force, and an axial force act on segment 2; and a bending moment and shear force act on member 3. For this problem, only the bending moment, which is usually the dominant force, will be considered.

Castigliano's theorem, part II in the form

$$V_{a\,\text{vertical}} = \frac{\partial U_i^*}{\partial \overline{P}} = \int_a^d \frac{M}{EI}\frac{\partial M}{\partial \overline{P}}\,dx \tag{1}$$

provides the vertical movement of point a. Use of formula (1), which has been derived for straight bars, is justifiable for curved bars if the curved bars cross-sectional dimensions are very small in comparison with the radius of curvature of the middle line. For the various bar segments, the distribution of M and $\partial M/\partial \overline{P}$ are found to be (Fig. 3.8b)

$$\begin{aligned}
M|^1 &= 0, & \frac{\partial M|^1}{\partial \overline{P}} &= 0\\
M|^2 &= -\overline{P}R(1-\cos\alpha), & \frac{\partial M|^2}{\partial \overline{P}} &= -R(1-\cos\alpha)\\
M|^3 &= -\overline{P}x' - \overline{P}R, & \frac{\partial M|^3}{\partial \overline{P}} &= -x' - R
\end{aligned} \tag{2}$$

where $M|^1 = M_{\text{bar}\,1}$, $M|^2 = M_{\text{bar}\,2}$, $M|^3 = M_{\text{bar}\,3}$

Then

$$\begin{aligned}
V_{a\,\text{vertical}} &= \int_0^{a_1} \frac{M|^1}{EI}\frac{\partial M|^1}{\partial \overline{P}}\,dx + \int_0^{\pi/2} \frac{M|^2}{EI}\frac{\partial M|^2}{\partial \overline{P}}R\,d\alpha + \int_0^{a_2} \frac{M|^3}{EI}\frac{\partial M|^3}{\partial \overline{P}}\,dx'\\
&= \frac{\overline{P}R^3(3\pi-8)}{4EI} + \frac{\overline{P}a_2^3}{3EI} + \frac{\overline{P}a_2^2 R}{EI} + \frac{\overline{P}R^2 a_2}{EI}
\end{aligned} \tag{3}$$

■

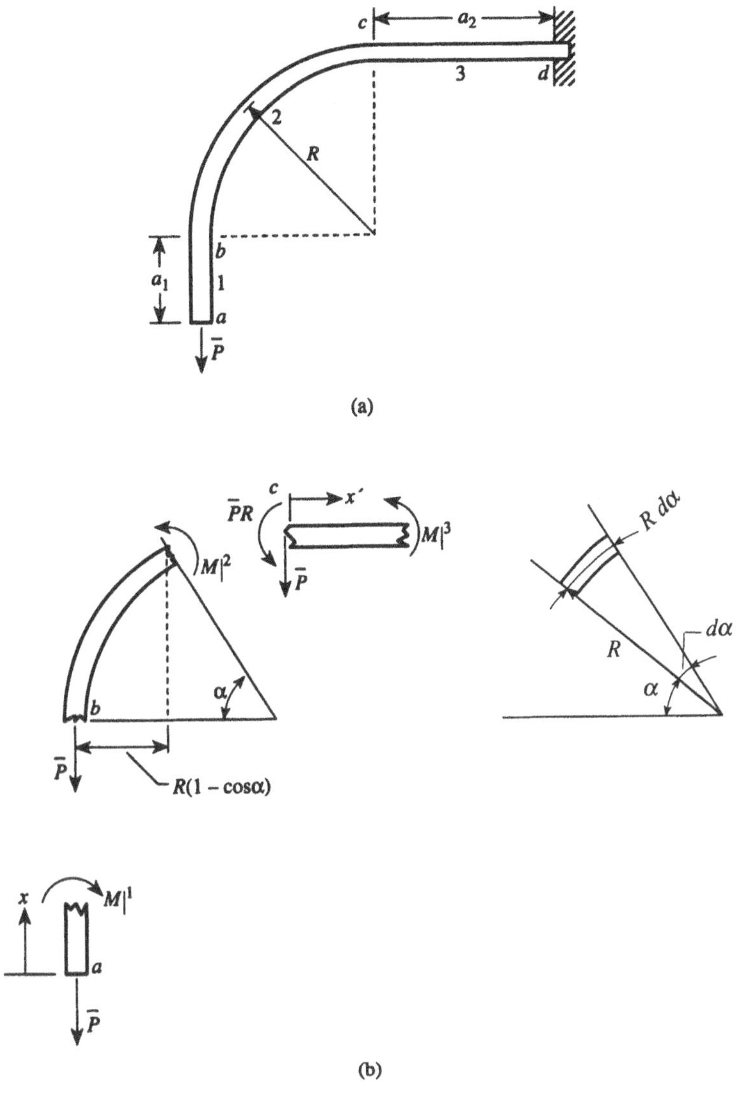

FIGURE 3.8
Curved bar with in-plane loading.

EXAMPLE 3.11 Flexibility and Stiffness of a Beam
In Example 3.2, it was shown that a stiffness coefficient k_{ij} can be considered to be the force developed at i due to a unit displacement at j with other end displacements set to zero. This definition lends itself to implementation using Castigliano's second theorem. Equation (3.12) provides a displacement in terms of the applied loading. This type of relationship is referred to as a *flexibility* equation. In contrast to flexibility, *stiffness* is an expression for a force in terms of the displacements. Castigliano's theorem, part II, leads to flexibility relationships, which can be converted to stiffness equations.

To illustrate the use of Castigliano's theorem, part II in deriving a stiffness matrix, consider a beam. It would appear that there are two independent displacements*—one deflection

*Chapter 4 contains a discussion on the subject of the independence of these two displacements.

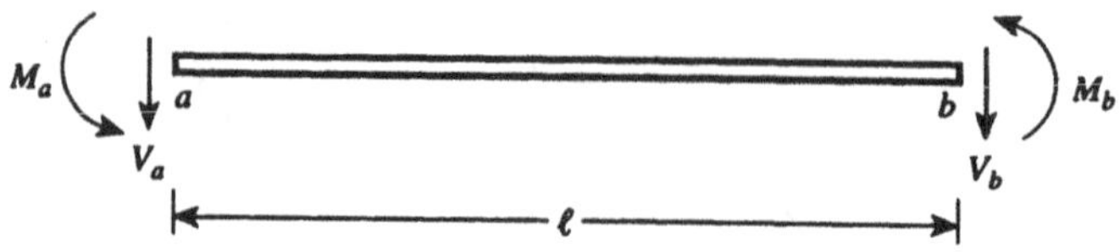

FIGURE 3.9
A beam element.

and one slope—at each end of a beam. Such displacements are called *degrees of freedom* (DOF). It is to be expected then that the vector $\mathbf{v}$ of displacements for the stiffness matrix will contain these four DOF.

For the beam element of Fig. 3.9, the four independent displacements are $w_a, \theta_a, w_b, \theta_b$. Castigliano's theorem, part II can be employed to find force-displacement relations when each displacement is in turn given the value of unity, while the other displacements are set to zero. From the theorem, the four end displacements are given by $w_a = \partial U_i^*/\partial V_a, \theta_a = \partial U_i^*/\partial M_a, w_b = \partial U_i^*/\partial V_b, \theta_b = \partial U_i^*/\partial M_b$, with $U_i^* = \int_0^\ell M^2/(2EI)\,dx$. Consider first the forces and moments at the ends of the beam defined by $w_a = 1, \theta_a = \theta_b = w_b = 0$. The moment of the beam of Fig. 3.9 in terms of M_a, V_a at a section at coordinate x from the left end would be $M = -M_a - V_a x$. Then $\partial M/\partial M_a = -1, \partial M/\partial V_a = -x$. Use Castigliano's theorem, part II for $w_a = 1, \theta_a = 0$, giving

$$
\begin{aligned}
w_a = 1 &= \frac{1}{EI}\int_0^\ell M\frac{\partial M}{\partial V_a}\,dx = \frac{1}{EI}\int_0^\ell (-M_a - V_a x)(-x)\,dx = \frac{M_a\ell^2}{2EI} + \frac{V_a\ell^3}{3EI} \\
\theta_a = 0 &= \frac{1}{EI}\int_0^\ell M\frac{\partial M}{\partial M_a}\,dx = \frac{M_a\ell}{EI} + \frac{V_a\ell^2}{2EI}
\end{aligned}
\tag{1}
$$

These two relations can be solved for M_a and V_a, giving

$$
M_a = -\frac{6EI}{\ell^2} \qquad V_a = \frac{12EI}{\ell^3} \tag{2}
$$

By definition of the stiffness coefficients, $M_a = k_{21}, V_a = k_{11}$. The equilibrium conditions applied to the element of Fig. 3.9 give

$$
M_b = -V_a\ell - M_a, \qquad V_b = -V_a \tag{3}
$$

Thus,

$$
\begin{aligned}
M_b &= -\frac{12EI\ell}{\ell^3} + \frac{6EI}{\ell^2} = -\frac{6EI}{\ell^2} = k_{41} \\
V_b &= -\frac{12EI}{\ell^3} = k_{31}
\end{aligned}
\tag{4}
$$

The remaining stiffness coefficients are computed in a similar manner, i.e., by using Castigliano's second theorem applied for $w_a = 0$ and $\theta_a = 1$, $w_b = 1$ and $\theta_b = 0$, $w_b = 0$ and $\theta_b = 1$, with the displacements at the opposite end set equal to zero. The resulting force-displacement relations in stiffness matrix form are

$$\begin{bmatrix} V_a \\ M_a \\ \cdots \\ V_b \\ M_b \end{bmatrix} = \begin{bmatrix} 12EI/\ell^3 & -6EI/\ell^2 & \vdots & -12EI/\ell^3 & -6EI/\ell^2 \\ -6EI/\ell^2 & 4EI/\ell & \vdots & 6EI/\ell^2 & 2EI/\ell \\ \cdots & \cdots & \cdot & \cdots & \cdots \\ -12EI/\ell^3 & 6EI/\ell^2 & \vdots & 12EI/\ell^3 & 6EI/\ell^2 \\ -6EI/\ell^2 & 2EI/\ell & \vdots & 6EI/\ell^2 & 4EI/\ell \end{bmatrix} \begin{bmatrix} w_a \\ \theta_a \\ \cdots \\ w_b \\ \theta_b \end{bmatrix} \tag{5}$$

$$\mathbf{p} = \mathbf{k} \qquad \mathbf{v}$$

■

3.2.2 The Unit Load Method

This method is used to determine the displacement $\overline{V}_k$ at a given point in a given direction of a structure for which the state of stress, and through the material law also the state of strain ϵ, is known. To proceed, apply a virtual force δP_k in the direction of $\overline{V}_k$. Then the external complementary virtual work is $\delta W_e^* = \overline{V}_k\, \delta P_k$. The virtual force corresponds to a system of virtual stresses $\delta\sigma_k$, and the complementary virtual work δW_i^* becomes $\int_V \delta\sigma_k^T \epsilon\, dV$. From the principle of complementary virtual work,

$$\delta P_k\, \overline{V}_k = \int_V \delta\sigma_k^T \epsilon\, dV = \int_V \epsilon_{ij}(\delta\sigma_{ij})_k\, dV \tag{3.14}$$

Since δP_k is arbitrary, for simplicity it can be set equal to unity. Thus,

$$1 \cdot \overline{V}_k = \int_V \delta\sigma_k^T \epsilon\, dV = \int_V \epsilon_{ij}(\delta\sigma_{ij})_k\, dV \tag{3.15}$$

This is a statement of the *unit load* or the *dummy load* method. The stresses $\delta\sigma_k$ are due to a unit force applied in the direction of $\overline{V}_k$. They can be chosen to be the same as for a similar but statically determinate system as the work done by the unknowns of the corresponding indeterminate system through the stresses of the real system is zero. If initial strains, e.g., thermal strains, are present, then ϵ should include these. Equation (3.15) remains valid for nonlinear structures.

Physically, $\overline{V}_k$ can be interpreted as the *flexibility* of the solid. The method is helpful in calculating flexibility properties of structures for use in matrix methods. In Example 3.14, the stiffness characteristics for a beam will be obtained from beam flexibilities found using the unit load method.

For a truss or beam element, use can be made of Eq. (3.15), or an equivalent theorem in more indigenous notation can be derived. In the case of a beam, the external complementary virtual work done by the virtual (unit) load at location k moving through the real deflection w_k is

$$w_k\, \delta P_k = w_k(1)$$

Let δM designate the internal bending moment generated by the virtual (unit) load. Real forces cause a rotation of an element of a beam of $d\theta = M\,dx/(EI)$. Hence, the internal complementary virtual work done on an element of a beam by the moment δM is $\delta M\, M\,dx/(EI)$, and the total internal complementary virtual work for a beam of length L is

$$\int_0^L \frac{M}{EI}\, \delta M\, dx \tag{3.16}$$

Thus, by setting $\delta P_k = 1$, Eq. (3.15) takes the form

$$w_k = \int_0^L \frac{M}{EI}\,\delta M\,dx \tag{3.17}$$

Similar relations apply for the effects of axial extension and shear deformation. In using Eq. (3.17), remember that the moment M is due to the actual loads on the structure, whereas the moment δM is due to the virtual (unit) load. If the slope at location k is desired, δM in Eq. (3.17) should be due to a virtual (unit) moment at k.

EXAMPLE 3.12 Statically Determinate Beam

Use the unit load method to find the vertical deflection at the free end of a cantilevered beam carrying a distributed load p_0 (Fig. 3.10a). Show how this solution should be extended to encompass the beam of Fig. 3.10c with a sudden jump in cross-section. Also, find the slope at the free end.

Apply a virtual (unit) force at the point where the deflection is sought (Fig. 3.10b). From the equilibrium condition, the moment δM generated by this force is $\delta M = -x$. Similarly, the summation of moments for a segment with the actual loading p_0 (Fig. 3.10a) gives $M = -p_0x^2/2$. We find

$$w_0 = \int_0^L \frac{\delta M\,M}{EI}\,dx = \int_0^L \frac{x(p_0x^2/2)}{EI}\,dx = \left.\frac{p_0x^4}{8EI}\right|_0^L = \frac{p_0L^4}{8EI} \tag{1}$$

as the deflection at the free end.

In utilizing the unit load method, both the real and the virtual internal forces were chosen to satisfy equilibrium.

Extend this solution to a beam of piecewise constant varying cross-section by computing the internal complementary virtual work for each segment of the beam separately and

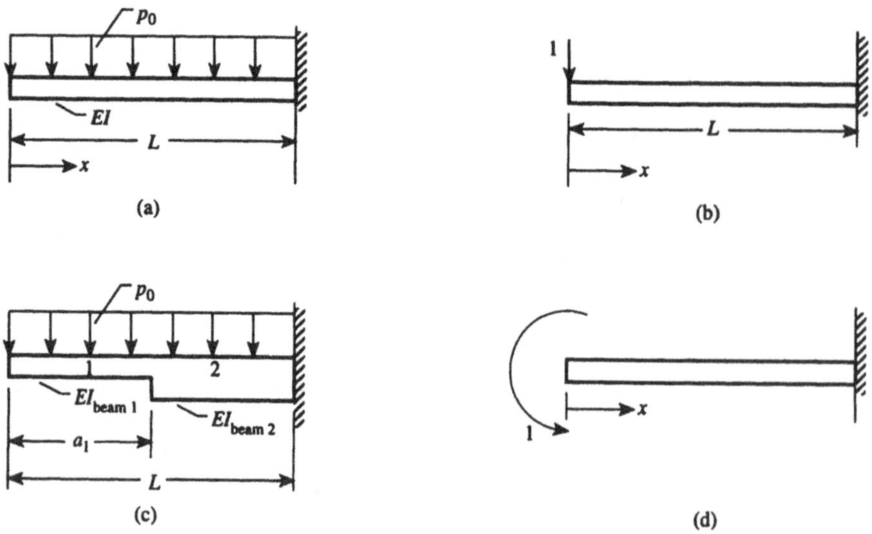

FIGURE 3.10
Cantilevered beam for unit load method examples.

summing the result. Thus, for the beam of Fig. 3.10c,

$$w_0 = \int_0^{a_1} \frac{\delta M\, M}{EI_{\text{beam 1}}}\, dx + \int_{a_1}^{L} \frac{\delta M\, M}{EI_{\text{beam 2}}}\, dx \tag{2}$$

where moments δM and M remain unchanged.

The slope θ of a beam at any location is found by using

$$\theta(1) = \int \frac{M}{EI}\, \delta M\, dx \tag{3}$$

where δM is now the bending moment due to a virtual (unit) moment at the position where the slope is sought. To find the slope at the free end of the cantilevered beam of Fig. 3.10a, place a unit moment at $x = 0$ (Fig. 3.10d), and compute the corresponding internal moment δM to be $\delta M = -1$, so that

$$\theta_{x=0} = \theta_0 = \int_0^L \frac{\delta M\, M}{EI}\, dx = \int_0^L \frac{p_0 x^2}{2EI}\, dx = \frac{p_0 L^3}{6EI} \tag{4}$$

∎

EXAMPLE 3.13 Statically Indeterminate Beam with Linearly Varying Loading

Compute the reactions of the statically indeterminate beam of Fig. 3.11.

This beam, with its three reactions, is said to be statically indeterminate to the first degree. Apply the unit load method to take advantage of the fact that the deflection of the beam at the right-hand support is zero. The bending moment at any section in terms of the coordinate $L - x$ from the right end is $M = R_L(L - x) - p_0(L - x)^3/(6L)$. The moment due

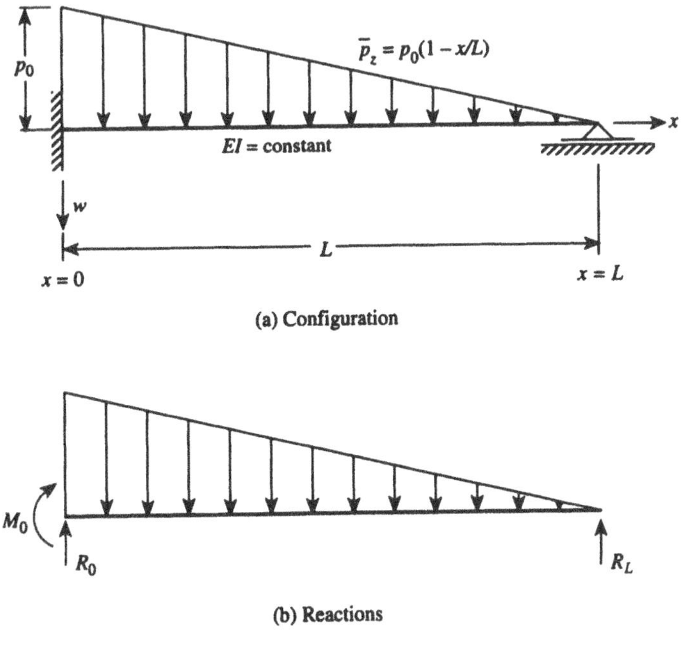

FIGURE 3.11
Beam with ramp loading.

to a unit load at the right support is $\delta M = L - x$. The unit load method formula of Eq. (3.17) appears as

$$w_R = 0 = \int_L^0 \frac{\delta M\, M}{EI}\, d(L-x) = -\int_L^0 \left[R_L (L-x)^2 - p_0 \frac{(L-x)^4}{6L} \right] dx \tag{1}$$

This gives $R_L = p_0 L/10$. The left-hand reactions are found from the conditions of equilibrium for the beam configuration of Fig. 3.11.

$$R_0 = \frac{2}{5} p_0 L, \qquad M_0 = -\frac{1}{15} p_0 L^2$$

∎

EXAMPLE 3.14 Stiffness Matrix for a Beam

The unit load method provides a straightforward technique for establishing a stiffness matrix. As with the other complementary virtual work theorems, flexibility relationships are obtained initially. These relationships are reorganized in stiffness matrix form. Return to the beam element of Fig. 3.9. To find the deflection at a in terms of M_a and V_a, the unit load method uses the moment at x due to a unit (downwards) force at a. This moment is of magnitude $-x$. The moment at x due to M_a and V_a is $M = -M_a - V_a x$. Thus,

$$w_a = \frac{1}{EI} \int_0^\ell -x(-M_a - V_a x)\, dx = \frac{M_a \ell^2}{2EI} + \frac{V_a \ell^3}{3EI} \tag{1}$$

The calculation of the slope at a involves the moment at x due to a unit counterclockwise moment at a. This moment is equal to -1. Then

$$\theta_a = \frac{1}{EI} \int_0^\ell (-1)(-M_a - V_a x)\, dx = \frac{M_a \ell}{EI} + \frac{V_a \ell^2}{2EI} \tag{2}$$

From (1) and (2),

$$\begin{bmatrix} w_a \\ \theta_a \end{bmatrix} = \begin{bmatrix} \ell^3/3EI & \ell^2/2EI \\ \ell^2/2EI & \ell/EI \end{bmatrix} \begin{bmatrix} V_a \\ M_a \end{bmatrix} \tag{3}$$

Relationships such as this which relate forces to displacements are the *flexibility* equations. If (3) is solved for V_a and M_a we obtain the stiffnesses $\mathbf{k}_{aa}$ as

$$\begin{bmatrix} V_a \\ M_a \end{bmatrix} = \begin{bmatrix} 12EI/\ell^3 & -6EI/\ell^2 \\ -6EI/\ell^2 & 4EI/\ell \end{bmatrix} \begin{bmatrix} w_a \\ \theta_a \end{bmatrix} = \mathbf{k}_{aa} \begin{bmatrix} w_a \\ \theta_a \end{bmatrix} \tag{4}$$

From the conditions of equilibrium and (4),

$$V_b = -V_a = -\frac{12EI}{\ell^3} w_a + \frac{6EI}{\ell^2} \theta_a$$

$$M_b = -V_a \ell - M_a = -\frac{6EI}{\ell^2} w_a + \frac{2EI}{\ell} \theta_a$$

or

$$\begin{bmatrix} V_b \\ M_b \end{bmatrix} = \begin{bmatrix} -12EI/\ell^3 & 6EI/\ell^2 \\ -6EI/\ell^2 & 2EI/\ell \end{bmatrix} \begin{bmatrix} w_a \\ \theta_a \end{bmatrix} = \mathbf{k}_{ba} \begin{bmatrix} w_a \\ \theta_a \end{bmatrix} \tag{5}$$

Similarly, use the unit load method to find w_b and θ_b in terms of V_a and M_a. This leads to

$$\begin{bmatrix} V_a \\ M_a \end{bmatrix} = \begin{bmatrix} -12EI/\ell^3 & -6EI/\ell^2 \\ 6EI/\ell^2 & 2EI/\ell \end{bmatrix} \begin{bmatrix} w_b \\ \theta_b \end{bmatrix} = \mathbf{k}_{ab} \begin{bmatrix} w_b \\ \theta_b \end{bmatrix} \tag{6}$$

and

$$\begin{bmatrix} V_b \\ M_b \end{bmatrix} = \begin{bmatrix} 12EI/\ell^3 & 6EI/\ell^2 \\ 6EI/\ell^2 & 4EI/\ell \end{bmatrix} \begin{bmatrix} w_b \\ \theta_b \end{bmatrix} = \mathbf{k}_{bb} \begin{bmatrix} w_b \\ \theta_b \end{bmatrix} \tag{7}$$

Equations (4), (5), (6), and (7) form the stiffness matrix of Eq. (5) of Example 3.11 or

$$\mathbf{p} = \begin{bmatrix} \mathbf{k}_{aa} & \mathbf{k}_{ab} \\ \mathbf{k}_{ba} & \mathbf{k}_{bb} \end{bmatrix} \mathbf{v} \tag{8}$$

∎

EXAMPLE 3.15 Truss Analysis

Find the horizontal displacement of point a of the truss in Fig. 3.12a. All members have equal lengths, cross-sectional areas, and moduli of elasticity.

The unit load method proceeds with the application of a virtual (unit) force at a in the direction of the desired displacement. Equation (3.15), appropriately modified to account for axial extension, leads to an expression for the displacement V_a, giving

$$V_a = \sum_{\text{All bars}} \left(\frac{L}{EA}\right) N\,\delta N \tag{1}$$

where A, E, and L are the respective area, modulus of elasticity, and length of the jth member; N is the axial force in the jth member due to the applied loading (Fig. 3.12b); δN is the axial force in the jth member due to the virtual (unit) force at a applied to the structure without the actual loadings (Fig. 3.12c); and the summation is taken over all members. The N and δN can be determined from equilibrium alone since the truss is statically determinate.

The necessary calculations are readily performed in tabular form (Fig. 3.13). Refer to Fig. 3.12b for the member numbering scheme. Thus, $V_a = -2PL/(9\sqrt{3}EA)$. The negative sign indicates that V_a is in the direction opposite to the applied unit force. ∎

Equivalence of the Unit Load Method and Castigliano's Theorem, Part II

In Example 3.7, it was shown that the application of Castigliano's theorem, part II to a truss can be equivalent to the use of the unit load method. This equivalence, of course, applies to other structures as well, as is illustrated in the following table.

	Unit Load Method	Castigliano's Theorem, Part II
Extension	$u = \int \delta N(N/EA)\,dx$	$u = \int (\partial N/\partial \overline{P})(N/EA)\,dx$
Bending	$w = \int \delta M(M/EI)\,dx$	$w = \int (\partial M/\partial \overline{P})(M/EI)\,dx$
Torsion	$\phi = \int \delta M_t(M_t/GJ)\,dx$	$\phi = \int (\partial M_t/\partial \overline{P})(M_t/GJ)\,dx$

In this table, δN is the axial force in a member due to a virtual (unit) force applied at the point and in the direction of the desired displacement. The other variables are defined

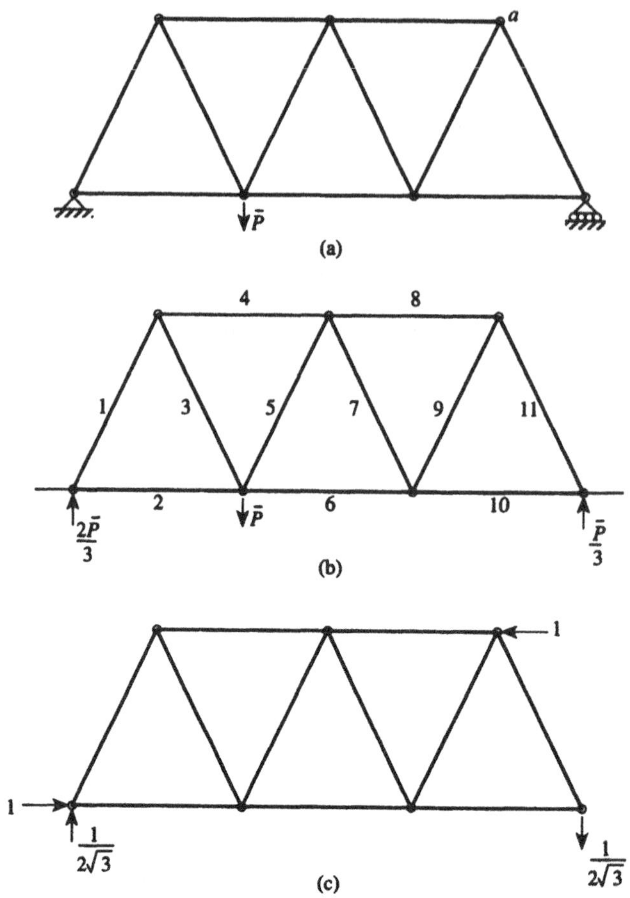

FIGURE 3.12
Truss.

similarly: δM is the internal moment due to a virtual (unit) applied moment, and δM_t is the net internal torque due to a virtual (unit) applied torque.

This equivalence of the unit load term and the rate of change of force, with respect to the applied load, e.g., $\delta N = \partial N/\partial \overline{P}$, permits problems in which no applied loads occur at the position where the response is sought to be treated by the addition of a dummy or "fictitious" load at this point. Then, after the application of this load, the fictitious load is set to zero. This is then an alternative, equivalent procedure for the use of a unit load at the location where the response is desired. The fictitious load approach to the unit load method is discussed in Chapter 3 of the first edition of this book.

3.2.3 Theorem of Least Work

The *theorem of least work* can be considered as a special case of Castligliano's theorem, part II. It is sometimes also referred to as *Engesser's second theorem* or *Engesser's theorem of compatibility* and it is usually employed to study statically indeterminate structures. The quantities (state variables) in a structure that cannot be determined by using the conditions of equilibrium alone are referred to as *redundants*. The number of redundant stress resultants in a solid is equal to the number of independent forces and/or moments less the number

Member (J)	Axial force N in each member due to the actual loads (here $\overline{P}$)	Axial force in each member due to a unit load at a in the horizontal direction. All actual loads are removed	$\sum\left(\frac{L}{EA}\right)N\,\delta N$
1	$-\frac{4}{3\sqrt{3}}\overline{P}$	$-\frac{1}{3}$	$\frac{4}{9\sqrt{3}}\frac{\overline{P}L}{EA}$
2	$\frac{2}{3\sqrt{3}}\overline{P}$	$-\frac{5}{6}$	$-\frac{5}{9\sqrt{3}}\frac{\overline{P}L}{EA}$
3	$\frac{4}{3\sqrt{3}}\overline{P}$	$\frac{1}{3}$	$\frac{4}{9\sqrt{3}}\frac{\overline{P}L}{EA}$
4	$-\frac{4}{3\sqrt{3}}\overline{P}$	$-\frac{1}{3}$	$\frac{4}{9\sqrt{3}}\frac{\overline{P}L}{EA}$
5	$\frac{2}{3\sqrt{3}}\overline{P}$	$-\frac{1}{3}$	$-\frac{2}{9\sqrt{3}}\frac{\overline{P}L}{EA}$
6	$\frac{1}{\sqrt{3}}\overline{P}$	$-\frac{1}{2}$	$-\frac{1}{2\sqrt{3}}\frac{\overline{P}L}{EA}$
7	$-\frac{2}{3\sqrt{3}}\overline{P}$	$\frac{1}{3}$	$-\frac{2}{9\sqrt{3}}\frac{\overline{P}L}{EA}$
8	$-\frac{2}{3\sqrt{3}}\overline{P}$	$-\frac{2}{3}$	$\frac{4}{9\sqrt{3}}\frac{\overline{P}L}{EA}$
9	$\frac{2}{3\sqrt{3}}\overline{P}$	$-\frac{1}{3}$	$-\frac{2}{9\sqrt{3}}\frac{\overline{P}L}{EA}$
10	$\frac{1}{3\sqrt{3}}\overline{P}$	$-\frac{1}{6}$	$-\frac{1}{18\sqrt{3}}\frac{\overline{P}L}{EA}$
11	$-\frac{2}{3\sqrt{3}}\overline{P}$	$\frac{1}{3}$	$-\frac{2}{9\sqrt{3}}\frac{\overline{P}L}{EA}$
			$\sum_{11\text{ bars}}\left(\frac{L}{EA}\right)N\,\delta N = -\frac{2}{9\sqrt{3}}\frac{\overline{P}L}{EA}$

FIGURE 3.13
Tabular solution for Example 3.15.

of available conditions of equilibrium. The theorem of least work is obtained by setting the right-hand side of Eq. (3.11) equal to zero, i.e.,

$$\frac{\partial U_i^*}{\partial \mathbf{P}} = \mathbf{0} \tag{3.18}$$

where the forces **P** are the redundant forces. Typically, the redundants are chosen to be reactions, although they can also be internal forces. In the notation of the force method of Chapter 5, Eq. (3.18) would appear as $\partial U_i^*/\partial \mathbf{X} = \mathbf{0}$.

For linearly elastic solids, Eq. (3.18) is sometimes referred to as *Castigliano's theorem of least work* or *Castligliano's theorem of compatibility*. The first edition of this book provides applications of the theorem of least work.

3.3 Reciprocal Theorems

Consider a linear elastic body subjected to two different sets of forces. Suppose the body is first subjected to surface forces $\overline{\mathbf{p}}_1$ and then, at the same locations as $\overline{\mathbf{p}}_1$ and in the same directions, is subjected to applied surface forces $\overline{\mathbf{p}}_2$. For the sake of brevity, body forces will be ignored. From Chapter 2, Eq. (2.27), the external work done by $\overline{\mathbf{p}}_1$ is

$$W_{e11} = \frac{1}{2} \int_{S_p} \mathbf{u}_1^T \, \overline{\mathbf{p}}_1 \, dS \tag{3.19}$$

where $\mathbf{u}_1$ are the displacements which result from the application of $\overline{\mathbf{p}}_1$. The notation W_{e11} is used to indicate the work of the force set 1 moving through the displacement set 1. Now let additional forces $\overline{\mathbf{p}}_2$ be applied, causing displacements $\mathbf{u}_2$, while force set $\overline{\mathbf{p}}_1$ is held constant. The work of $\overline{\mathbf{p}}_2$ moving through $\mathbf{u}_2$ is

$$W_{e22} = \frac{1}{2} \int_{S_p} \mathbf{u}_2^T \, \overline{\mathbf{p}}_2 \, dS \tag{3.20}$$

An additional increment of work, say W_{e12}, will be done by the first forces $\overline{\mathbf{p}}_1$ moving through the displacements $\mathbf{u}_2$ of the second set of forces. Since the first forces $\overline{\mathbf{p}}_1$ remain constant during these further displacements $\mathbf{u}_2$, we have

$$W_{e12} = \int_{S_p} \mathbf{u}_2^T \, \overline{\mathbf{p}}_1 \, dS \tag{3.21}$$

Thus, the total external work performed by the two sets of forces is

$$W_e = W_{e11} + W_{e22} + W_{e12} \tag{3.22}$$

Now, remove the loads and apply them again, but in reverse order. For this second case, the total work done will be

$$W_e = W_{e22} + W_{e11} + W_{e21} \tag{3.23}$$

where W_{e22} is again given by Eq. (3.20), W_{e11} by Eq. (3.19), and W_{e21} is now

$$W_{e21} = \int_{S_p} \mathbf{u}_1^T \, \overline{\mathbf{p}}_2 \, dS \tag{3.24}$$

That is, W_{e21} is the work performed by force set 2 moving through the incremental displacements associated with force set 1.

The principle of superposition is applicable for this linearly elastic body. Since in this case the total work of the applied forces must be independent of the order of application of the loading, it follows by equating W_e of Eqs. (3.22) and (3.23) that

$$W_{e12} = W_{e21} \tag{3.25}$$

This is the *reciprocal theorem of Betti*[2]. It can be stated as the following:

> If a linearly elastic body is subjected to two force systems, the work done by the first system of forces in moving through the displacements produced by the second system of forces is equal to the work performed by the second system of forces in moving through the displacements due to the first system of forces.

A theory of elasticity form of Betti's theorem is also derived in a straightforward manner. Because of the already complex index notation of the theory of elasticity, we choose to abandon the 1 and 2 subscripts used above and to replace them with no index for the first system and with a superscript * for the second system. Begin with the equilibrium condition

$$\sigma_{ij,j} + \overline{p}_{Vi} = 0 \quad \text{in} \quad V \tag{3.26}$$

Follow the reasoning of Chapter 2, Section 2.2.1, and form the global condition

$$\int_V (\sigma_{ij,j} + \overline{p}_{Vi}) u_i^* \, dV = 0 \tag{3.27}$$

This process can be considered as having weighted the equilibrium equations for one set of forces with displacements u_i^* due to a second set of forces and having orthogonalized the product. From Gauss' integral theorem (Appendix II),

$$\int_S p_j u_j^* \, dS = \int_S \sigma_{ij} \, a_i \, u_j^* \, dS = \int_S (\sigma_{ij} \, u_j^*) a_i \, dS = \int_V (\sigma_{ij} \, u_j^*)_{,i} \, dV \tag{3.28}$$

Since (Chapter 2, Problem 2.13)

$$\sigma_{ij} \, u_{i,j} = \sigma_{ij} \, \epsilon_{ij} \tag{3.29}$$

$$\int_V \sigma_{ij} \, \epsilon_{ij}^* \, dV = \int_V \sigma_{ij} \, u_{i,j}^* \, dV = \int_V (\sigma_{ij} \, u_i^*)_{,j} \, dV - \int_V \sigma_{ij,j} \, u_i^* \, dV \tag{3.30}$$

From Eqs. (3.28) and (3.30) ,

$$\int_V \sigma_{ij} \, \epsilon_{ij}^* \, dV + \int_V \sigma_{ij,j} \, u_i^* \, dV = \int_S p_j \, u_j^* \, dS \tag{3.31}$$

Since, from Eq. (3.26), $\sigma_{ij,j} = -\overline{p}_{Vi}$ Eq. (3.31) becomes

$$-\int_V \sigma_{ij} \, \epsilon_{ij}^* \, dV + \int_V \overline{p}_{Vi} \, u_i^* \, dV = -\int_S p_j \, u_j^* \, dS \tag{3.32}$$

Use[§]

$$\sigma_{ij}\epsilon_{ij}^* = \sigma_{ij}^*\epsilon_{ij} \tag{3.33}$$

---

[2]Enrico Betti (1823–1892) was an Italian mathematician who for many years was a professor of mathematical physics at the University of Pisa. He made major contributions to algebra, topology, and elasticity. Betti's reciprocal theorem appeared in 1872.

[§] To show that $\sigma_{ij} \, \epsilon_{ij}^* = \sigma_{ij}^* \, \epsilon_{ij}$, note that $\boldsymbol{\sigma}^T \, \boldsymbol{\epsilon}^*$ is a scalar quantity, so that

$$\boldsymbol{\sigma}^T \, \boldsymbol{\epsilon}^* = (\boldsymbol{\sigma}^T \, \boldsymbol{\epsilon}^*)^T = \boldsymbol{\epsilon}^{*T} \boldsymbol{\sigma} = \boldsymbol{\epsilon}^{*T} \mathbf{E} \boldsymbol{\epsilon}$$

For a symmetric **E**,

$$\boldsymbol{\epsilon}^{*T} \mathbf{E} \boldsymbol{\epsilon} = \boldsymbol{\epsilon}^{*T} \mathbf{E}^T \boldsymbol{\epsilon} = (\mathbf{E} \boldsymbol{\epsilon}^*)^T \, \boldsymbol{\epsilon} = \boldsymbol{\sigma}^{*T} \boldsymbol{\epsilon}$$

to rewrite this expression as

$$-\int_V \sigma_{ij}^* \epsilon_{ij}\, dV + \int_V \overline{p}_{Vi}\, u_i^*\, dV = -\int_S p_j\, u_j^*\, dS \tag{3.34}$$

It follows from Gauss' integral theorem in the form of Eq. (3.32) (switch the superscript * to the adjoining variables) that Eq. (3.34) can be written as

$$-\int_V \overline{p}_{Vi}^*\, u_i\, dV + \int_V \overline{p}_{Vi}\, u_i^*\, dV = -\int_S p_i\, u_i^*\, dS + \int_S p_i^*\, u_i\, dS \tag{3.35}$$

or

$$\int_V \overline{p}_{Vi}\, u_i^*\, dV + \int_S p_i\, u_i^*\, dS = \int_V \overline{p}_{Vi}^*\, u_i\, dV + \int_S p_i^*\, u_i\, dS \tag{3.36}$$

Typically, this expression is referred to as Betti's theorem for linearly elastic problems. If desired, the surface integrals can be written as $\int_S = \int_{S_p} + \int_{S_u}$ so that a distinction between p_i and $\overline{p}_i$ or u_i and $\overline{u}_i$ can be made.

We proceed to consider a special case of Betti's theorem. Equation (3.25) appears as

$$\int_{S_p} \mathbf{u}_2^T\, \overline{\mathbf{p}}_1\, dS = \int_{S_p} \mathbf{u}_1^T\, \overline{\mathbf{p}}_2\, dS \tag{3.37}$$

Suppose that for our linearly elastic body, each force system contains only a single non-zero force. Designate these forces as $\overline{P}_1$ and $\overline{P}_2$. First, $\overline{P}_1$ is applied at point 1 and causes a displacement $\overline{P}_1 f_{21}$ at point 2 (in the direction in which $\overline{P}_2$ is to be applied), where f_{ij} is the displacement at i due to a unit force at j. Next, $\overline{P}_2$ is applied at point 2, causing a displacement $\overline{P}_2 f_{12}$ at point 1 (in the direction in which $\overline{P}_1$ was applied). According to Eq. (3.37), $\overline{P}_1(\overline{P}_2 f_{12}) = \overline{P}_2(\overline{P}_1 f_{21})$ or $f_{12} = f_{21}$. In general,

$$f_{ij} = f_{ji} \tag{3.38}$$

This represents *Maxwell's reciprocal theorem* which can be expressed as the following:

> For a linearly elastic body subjected to two unit (or equal in magnitude) forces, the displacement at the location of (and in the direction of) the first force caused by the second force is equal to the displacement at the location of (and in the direction of) the second force which is due to the first force.

By definition, f_{ij} are the *influence* or *flexibility coefficients*. They form the elements of what later will be called the flexibility matrix. Maxwell's theorem shows that flexibility matrices must be symmetric for linear structures.

The forces and displacements in Maxwell's theorem may be the usual forces and displacements, as well as moments and corresponding rotations, or combinations of forces and moments and respective displacements and rotations. For example, suppose that $\overline{P}_1$ is a force and $\overline{P}_2$ is a moment (of the same magnitude, say unity, but with different units) applied to a beam. Then f_{21} is the rotation at point 2 due to a unit transverse force at 1. And f_{12} is the deflection at point 1 due to a unit moment at point 2.

EXAMPLE 3.16 Betti's Reciprocal Theorem

If the deflection is known along a cantilevered beam with a force $\overline{P}$ at the free end (Fig. 3.14a), find the free end deflection of the beams of Figs. 3.14b, c, and d.

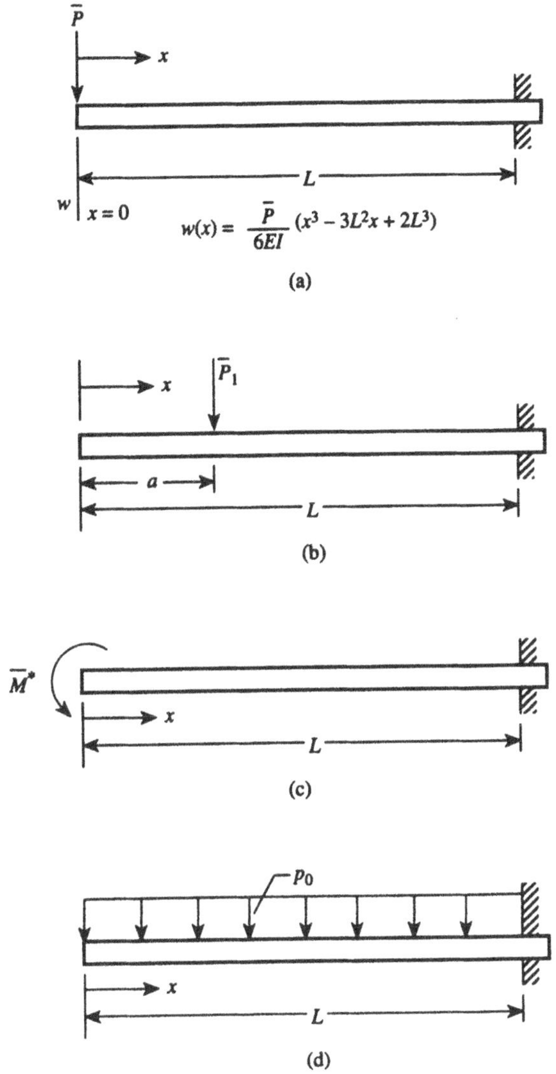

FIGURE 3.14
Cantilevered beam.

This is a direct application of Betti's reciprocal theorem. Thus, when applied to the beam of Fig. 3.14b,

$$\overline{P} \cdot w|_{x=0,\text{ due to }\overline{P}_1} = \overline{P}_1 \cdot w|_{x=a,\text{ due to }\overline{P}} \tag{1}$$

As noted in Fig. 3.14a, $w|_{x=a}$ due to $\overline{P}$ is

$$w|_{x=a} = \frac{\overline{P}}{6EI}(a^3 - 3L^2a + 2L^3).$$

From (1),

$$w|_{x=0,\text{ due to }\overline{P}_1} = \frac{\overline{P}_1}{\overline{P}} w\Big|_{x=a,\text{ due to }\overline{P}} = \frac{\overline{P}_1}{6EI}(a^3 - 3L^2a + 2L^3) \tag{2}$$

This is the requested result.

We can treat the beam of Fig. 3.14c in a similar fashion. According to the reciprocal theorem,

$$\overline{P} \cdot w|_{x=0,\text{ due to }\overline{M}^*} = \overline{M}^* \cdot \theta|_{x=0,\text{ due to }\overline{P}}$$

where $\theta = -dw/dx = -\overline{P}(x^2 - L^2)/(2EI)$.
Thus,

$$w|_{x=0,\text{ due to }M^*} = \frac{\overline{M}^*}{\overline{P}} \theta \bigg|_{x=0,\text{ due to }\overline{P}} = \frac{\overline{M}^* L^2}{2EI} \tag{3}$$

In the case of the beam of Fig. 3.14d,

$$\overline{P} w|_{x=0,\text{ due to }p_0} = \int_0^L p_0 \frac{\overline{P}}{6EI}(x^3 - 3L^2 x + 2L^2)\, dx = p_0 \frac{\overline{P} L^4}{8EI} \tag{4}$$

or

$$w|_{x=0,\text{ due to }p_0} = \frac{p_0 L^4}{8EI} \tag{5}$$

■

An influence line is a response, such as a reaction force, produced by a unit applied loading, such as a force, as it traverses the member. Then, for any other value of the applied loading, say $\overline{P}$, the response at a point in the system is obtained by multiplying the value of the influence line at that point by $\overline{P}$. As demonstrated by example in Chapter 3 of the first edition of this book, reciprocal theorems are useful in finding influence lines.

Problems

Principle of Virtual Work and Related Methods

3.1 Use the principle of virtual work to find the reactions in the beam of Fig. P3.1.

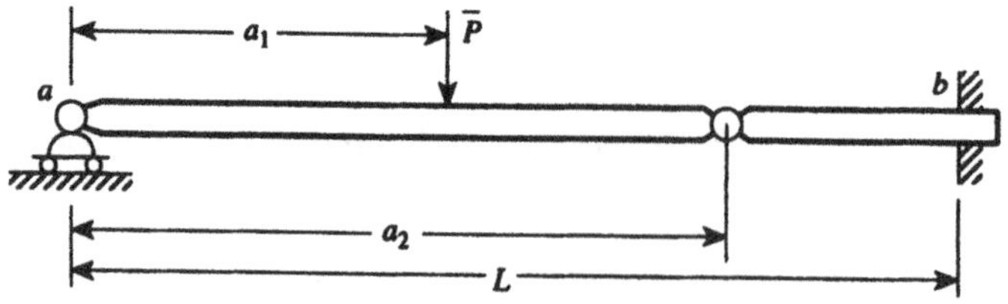

FIGURE P3.1
Beam with hinge.

Answer:

$$R_a = \frac{(a_2 - a_1)}{a_2}\overline{P}, \quad R_b = \overline{P} - R_a, \quad M_b = a_1 \frac{L - a_2}{a_2}\overline{P}$$

3.2 Use the principle of virtual work to find the forces in the bars of the system shown in Fig. P3.2 if the material is linearly elastic. Find the horizontal displacement of point a.

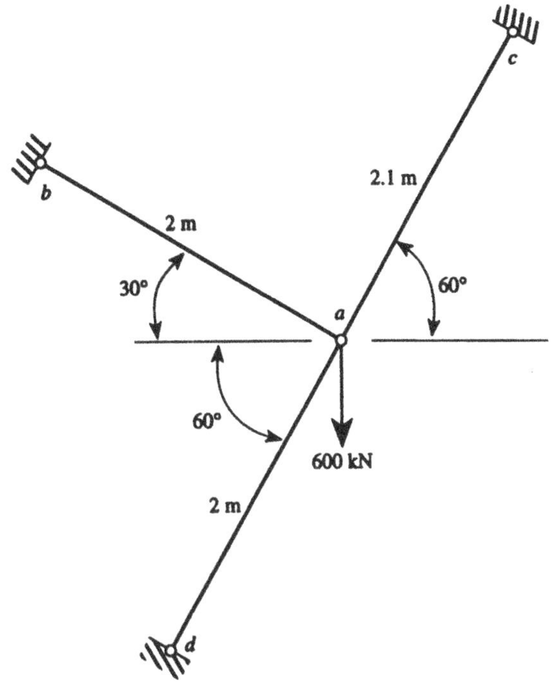

FIGURE P3.2
A bar structure. Each bar has the same modulus of elasticity and cross-sectional area.

3.3 Solve Problem 3.1 using the principle of stationary potential energy.

3.4 Solve Problem 3.1 using Castigliano's theorem, part I.

3.5 Determine the vertical displacement V at point a of the truss shown. The cross-sectional area of each member is A.

Answer: $V = \overline{P}/(KE)$, where $K = (2A/L)\cos^2\alpha + A/(L\cos\alpha)$

Hint: The strain in bars ab and ad is $V\cos\alpha/L$ and the strain in ac is $V/(L\cos\alpha)$.

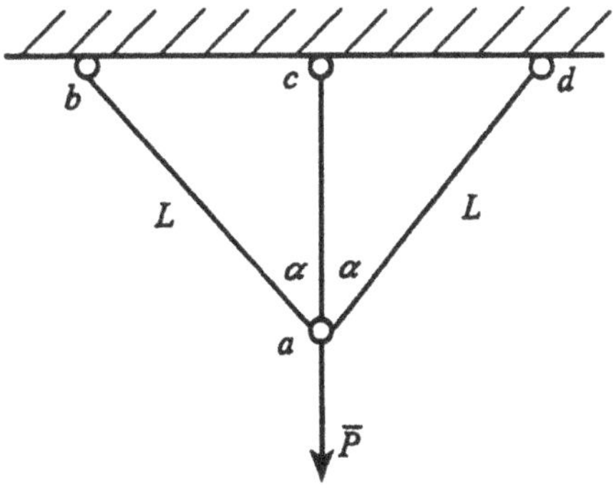

FIGURE P3.5

Castigliano's Theorem, Part II

3.6 For a cantilevered beam of length L with a load $\overline{P}$ applied at the free end, find the deflection at the point where the load is applied.

Answer: $w_{x=L} = \overline{P}L^3/(3EI)$

3.7 Compute the vertical displacement of point c of the beam, tie rod system of Fig. P3.7. Both members are made of steel. Neglect the effects of shear deformation. Also, determine the midpoint deflection and find the slope at 2.5 ft from the left end.

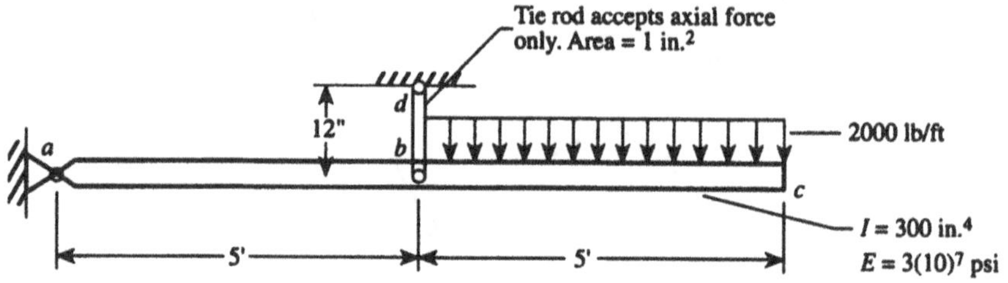

FIGURE P3.7
Beam with tie rod.

Answer: 0.08 in., 0.006 in., 0.0000167 radian clockwise

3.8 Compute the deflection of the beam of Fig. 3.3 at $L/4$ distance from the left end. Let $\overline{P} = 1000$ lb, $\overline{M}_0 = 300$ lb-in., $L = 110$ in., $E = 11(10^6)$ psi, and $I = 480$ in.^{4}

Answer: 0.053 in.

3.9 Find the vertical and horizontal movement of point a of the pin-connected truss of Fig. P3.9. Assume no members will buckle. Tension members are 1 in.^{2} in area while compression members are 1.5 in.^{2} Also, $E = 30(10^6)$ psi.

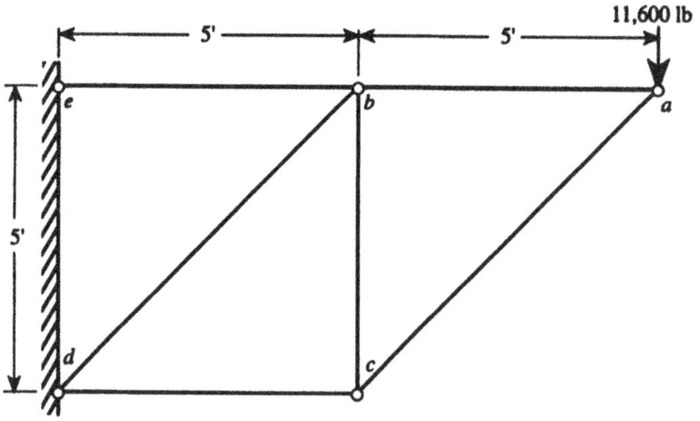

FIGURE P3.9
Truss.

Answer: 0.24 in., 0.07 in.

3.10 Determine the vertical displacement of point d of the frame shown in Fig. 3.6a.

Hint: Apply a unit load in the direction of the desired displacement.

Answer: $V_{d \text{ vertical}} = \overline{P}a^3/(2EI)$

3.11 Find the horizontal, out-of-plane displacement of point d of the frame of Fig. P3.11.

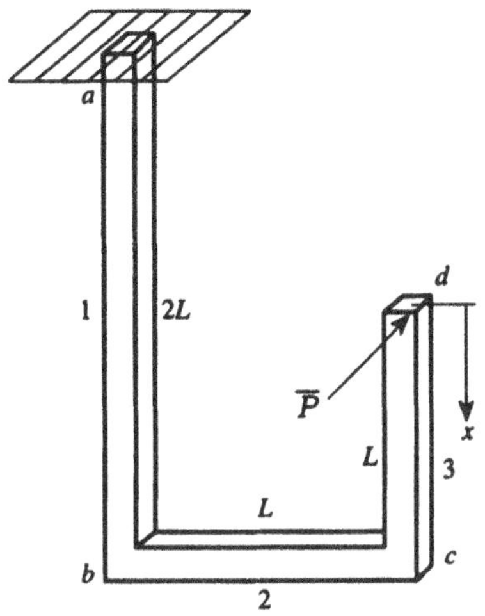

FIGURE P3.11
Out-of-plane displacement of frame.

Hint: Use

$$V_{d\ \text{horizontal}} = \int \frac{M}{EI}\frac{\partial M}{\partial \overline{P}}\,dx + \int \frac{M_t}{GJ}\frac{\partial M_t}{\partial \overline{P}}\,dx + \int \frac{V}{k_s G A}\frac{\partial V}{\partial \overline{P}}\,dx$$

Answer: $V_{d\ \text{horizontal}} = \frac{4}{3}\frac{\overline{P}L^3}{EI} + 3\frac{\overline{P}L^3}{GJ} + \frac{4\overline{P}L}{k_s AG}$

3.12 Determine the horizontal displacement of point a in the abc plane of the bar of Fig. P3.12. Consider bending and torsion of the members. Let $G = 0.4E$. The cross-sections are constant and circular. Also, compute the rotation of point c in the abc plane.

Answer: $3500/(EI) + 2400/(EJ)$, $2100/(EI) + 2400/(EJ)$

3.13 Find the relative displacement between points a and b of the truss of Fig. P3.13a. Also, indicate how the rotation of member 1 can be determined.

Hint: Apply a pair of opposing forces Q at a and b (Fig. P3.13b). Then Castigliano's second theorem gives $V_a + V_b = \partial U_i^* / \partial Q|_{Q=0}$. To compute $\partial N/\partial Q$ in U_i^*, remove $\overline{P}_e$, $\overline{P}_d$, and $\overline{P}_b$ and calculate N in each bar due to $Q = 1$ at a and b.
The rotation of bar 1 is found by applying a couple M^* as shown in Fig. P3.13c. The rotation is then $\theta = \partial U_i^* / \partial M^*|_{M^*=0}$.

3.14 Consider the curved bar of Fig. P3.14 with $\overline{P}$ acting perpendicular to the plane of the bar. Find the horizontal, out-of-plane deflection and angle of twist of end a of the bar.

Hint: If bending and twisting effects are taken into account,

$$V_{a\ \text{horizontal}} = \int_0^{\pi/2} \frac{M}{EI}\frac{\partial M}{\partial \overline{P}} R\,d\alpha + \int_0^{\pi/2} \frac{M_t}{GJ}\frac{\partial M_t}{\partial \overline{P}} R\,d\alpha$$

with $M = \overline{P}R\sin\alpha$, $M_t = \overline{P}R(1-\cos\alpha)$.
Add fictitious torque M_t^* at a, then

$$\phi_a = \left.\frac{\partial U}{\partial M_t^*}\right|_{M_t^*=0} = \left[\int_0^{\pi/2} \frac{M}{EI}\frac{\partial M}{\partial M_t^*} R\,d\alpha + \int_0^{\pi/2} \frac{M_t}{GJ}\frac{\partial M_t}{\partial M_t^*} R\,d\alpha\right]_{M_t^*=0}$$

with $M = \overline{P}R\sin\alpha - M_t^*\sin\alpha$, $M_t = \overline{P}R(1-\cos\alpha) + M_t^*\cos\alpha$.

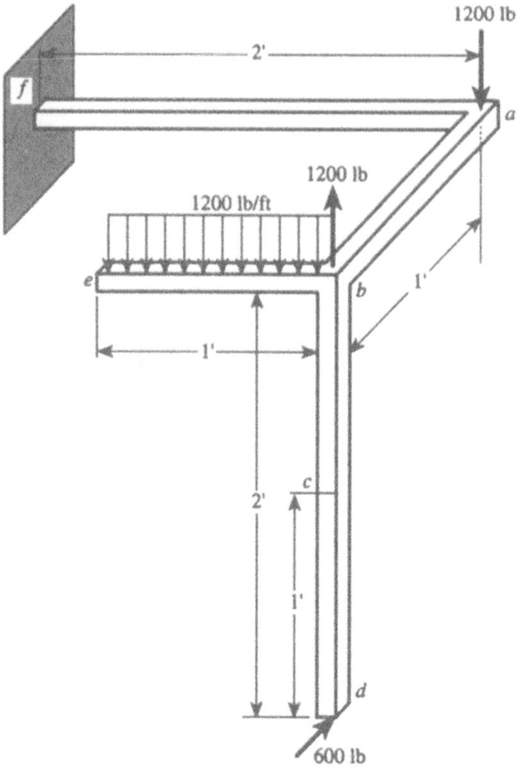

FIGURE P3.12
Bar system.

Answer:

$$V_{a\ \text{horizontal}} = \frac{\pi \overline{P} R^3}{4EI} + \frac{\overline{P} R^3}{JG}\frac{(3\pi - 8)}{4}, \quad \phi_a = -\frac{\pi \overline{P} R^2}{4EI} + \frac{\overline{P} R^2}{JG}\left(1 - \frac{\pi}{4}\right)$$

3.15 A concentrated load $\overline{P}$ is applied at the free end of a cantilevered beam. The cross-section of the beam is a rectangle of height h. The stress-strain relation is $\sigma = k\sqrt{|\epsilon|}$ $sgn\,\epsilon$, where k is a constant and $sgn\,\epsilon$ (read "signum ϵ") is defined as $+1$ if $\epsilon > 0$, -1 if $\epsilon < 0$, and 0 if $\epsilon = 0$. Find the deflection of the free end of the beam under the applied load.

Answer: $25\overline{P}^2 L^4/(2b^2k^2h^5)$

Unit Load Method

3.16 Find the displacement of point a of the truss shown.

Answer: $V_x = -0.096\frac{\overline{P}L}{AE}, \quad V_z = 0.728\frac{\overline{P}L}{AE}$

3.17 Assume a beam of length L is pinned on the left end and fixed on the right end. Compute the reactions at the fixed end if the beam is loaded with forces $\overline{P}$ at $x = L/3$ and $x = 2L/3$, where x is measured from the left end.

Answer: $R = 4\overline{P}/3, \quad M = \overline{P}L/3$

3.18 Find the deflection of the free end of the beam of Fig. 3.10c if $L = 0.5$ m, $p_0 = 1$ kN/m, $E = 75.84$ GN/m^2, $I_{\text{beam1}} = 10\,000$ cm^4, $I_{\text{beam2}} = 33\,750$ cm^4, and $a_1 = 0.2$ m.

Answer: 4.0857 mm

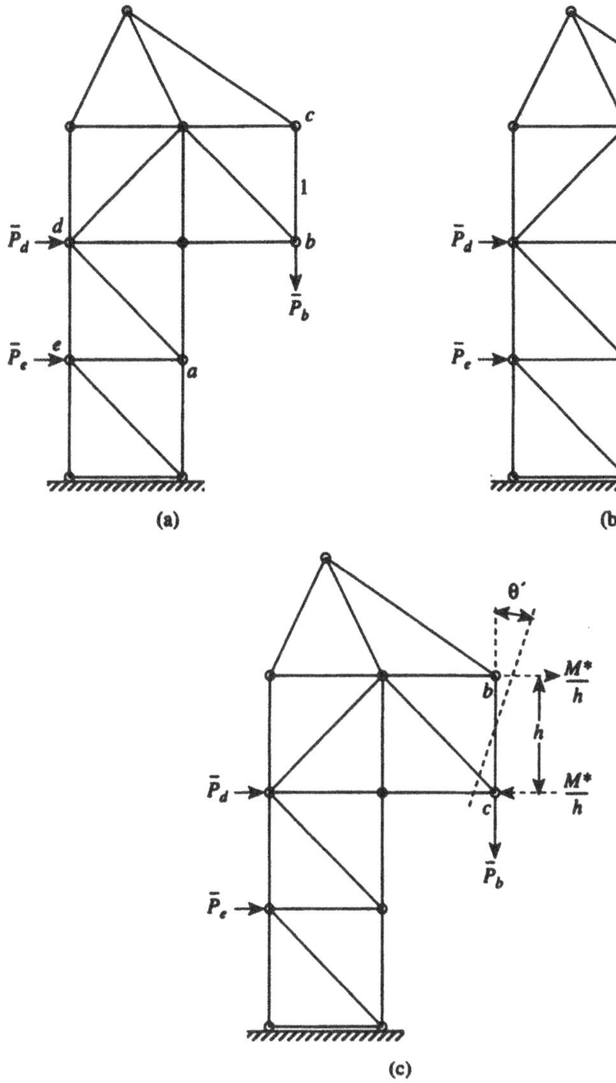

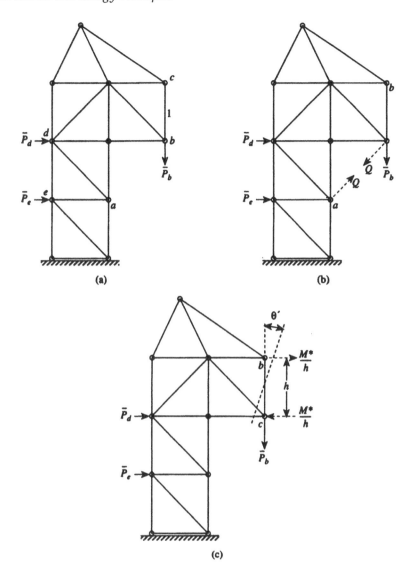

FIGURE P3.13
Truss.

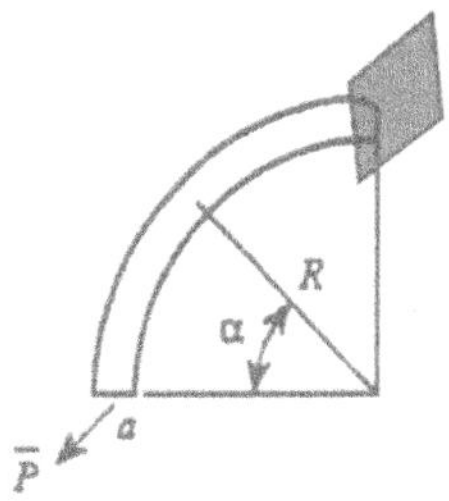

FIGURE P3.14
Curved bar with out-of-plane loading.

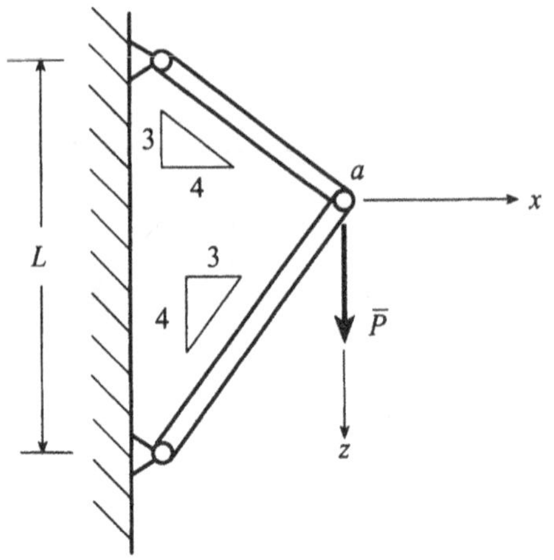

FIGURE P3.16

3.19 Consider a beam of length 100 in. which is free on the left end and fixed on the right end. A linearly varying load begins at 0 lb/in. at the free end and increases to 500 lb/in. at the fixed end. Find the deflection at the free end if $E = 30(10^6)$ psi and $I = 150$ in.^{4}

Answer: 0.37 in.

3.20 Solve again the beam of Problem 3.19 if an extension spring with a constant of 1000 lb/in. is connected to the free end.

Answer: 0.34 in.

3.21 Compute the vertical displacement of point d of the frame of Fig. P3.21.

Answer: $V_d = -p_0 L^4/(6EI)$

3.22 Determine the displacements of the end of the curved bar of Fig. P3.22a. Consider axial, shear deformation, and bending effects.

Hint: Make use of the geometry of Fig. P3.22b. Apply unit horizontal and vertical forces and a moment at point a.

Answer:

$$V_{\text{horizontal}} = \frac{\overline{P}R}{2}\left(-\frac{1}{EA} + \frac{1}{k_s GA} + \frac{R^2}{EI}\right)$$

$$V_{\text{vertical}} = \frac{\pi \overline{P}R}{4}\left(\frac{1}{EA} + \frac{1}{k_s GA} + \frac{R^2}{EI}\right), \quad \text{Rotation} = \frac{\overline{P}R^2}{EI}$$

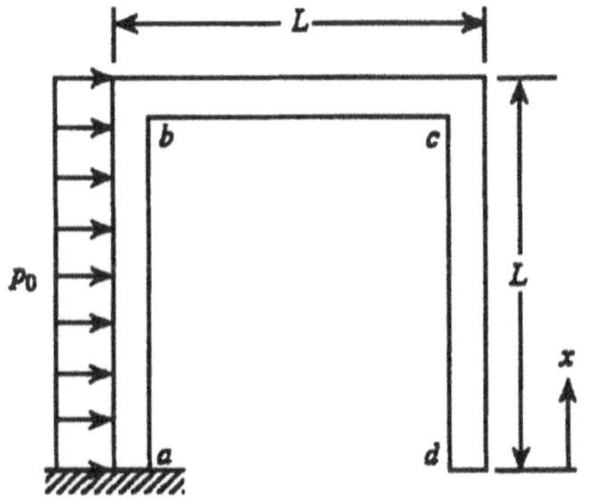

FIGURE P3.21
Frame.

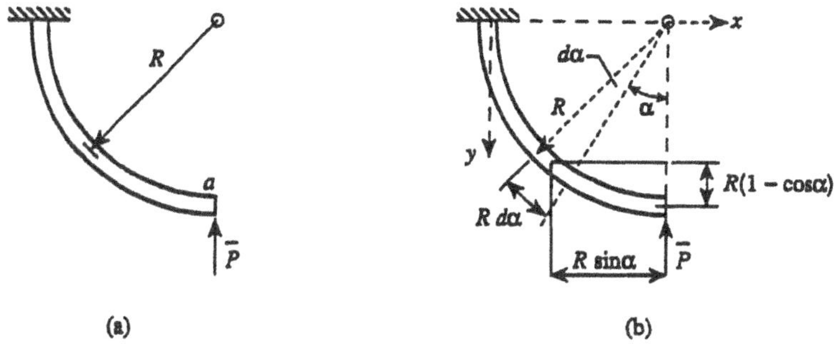

FIGURE P3.22
In-plane motion of circular bar.

3.23 For the tapered cantilever beam (Fig. P3.23) the moment of inertia varies as $I = (c_1 x + c_2)^{-1}$ where c_1 and c_2 are constants. Derive a formula for the deflection of the free end.

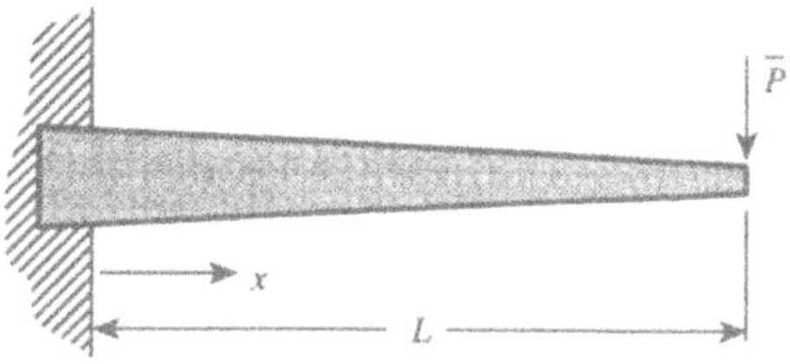

FIGURE P3.23

Miscellaneous Problems

3.24 Show how to find the reactions in the continuous beam of Fig. P3.24a.

Hint: Choose $R_a, R_b, \ldots, R_n$, the reactions at supports at $a, b, \ldots, n$, as redundant forces and use the relations

$$\frac{\partial U_i^*}{\partial R_a} = 0, \quad \frac{\partial U_i^*}{\partial R_b} = 0, \ldots, \quad \frac{\partial U_i^*}{\partial R_n} = 0$$

3.25 Indicate how to find the forces in the springs of the beam of Fig. P3.24b.

Hint: Instead of being equal to zero as in Problem 3.24, the derivative of the complementary strain energy with respect to a reaction R_s is now equal to $-R_s/k_s$, i.e., $\partial U_i^*/\partial R_s = -R_s/k_s$, $s = a, b, \ldots, n$. The negative sign indicates that the displacement in the spring is opposite in sense to the reaction on the beam.

3.26 Compute the displacement and reaction at the left end of the beam of Fig. P3.26. How should this problem be approached if the spring does not fit properly?

Answer: $R = \dfrac{p_0 L^4/8}{EI/k + L^3/3}$, $\quad w_{x=0} = R/k$

If the spring is V_s units too long (replace V_s by $-V_s$ if it is too short) before p_0 is applied, then force the beam in place and use

$$\frac{\partial U_i^*}{\partial R} = V_s - \frac{R}{k}$$

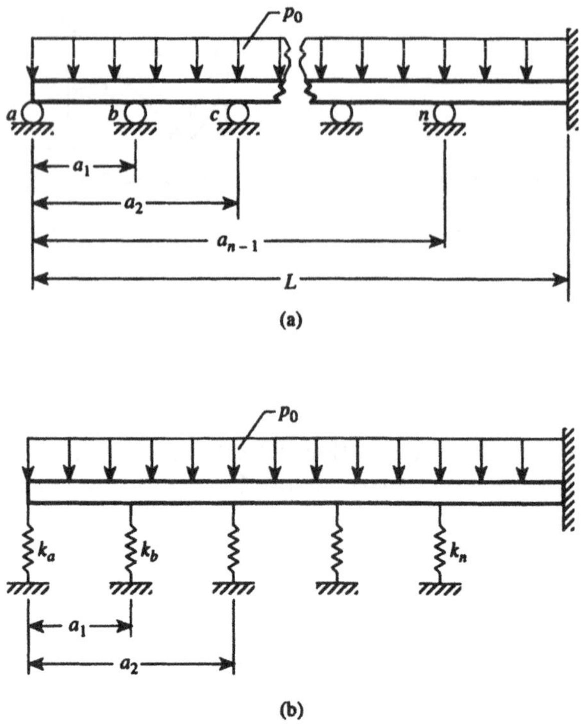

FIGURE P3.24
Continuous and spring-supported beams.

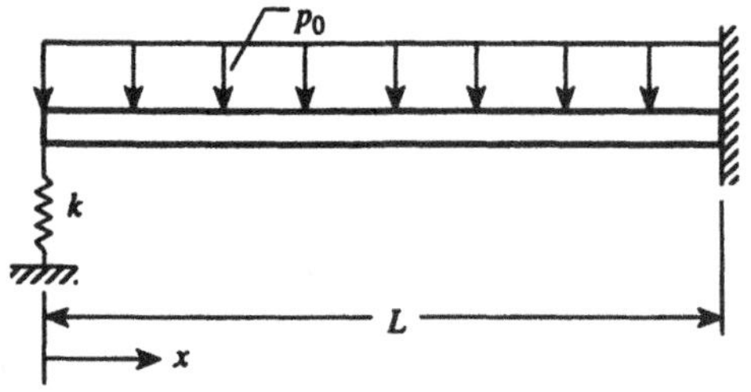

FIGURE P3.26
Spring-supported beam.

3.27 Find the deflection at $x = a_1/2$ of the beam of Fig. P3.27 if $L = 1$ m, $p_0 = 5$ kN/m, $a_1 = 0.5$ m, $I = 450\ \mu\text{m}^4$, and $E = 207$ GN/m^2. Find the reactions if the in-span support is removed and the distributed load varies linearly from zero at the left end to p_0 at the right end.

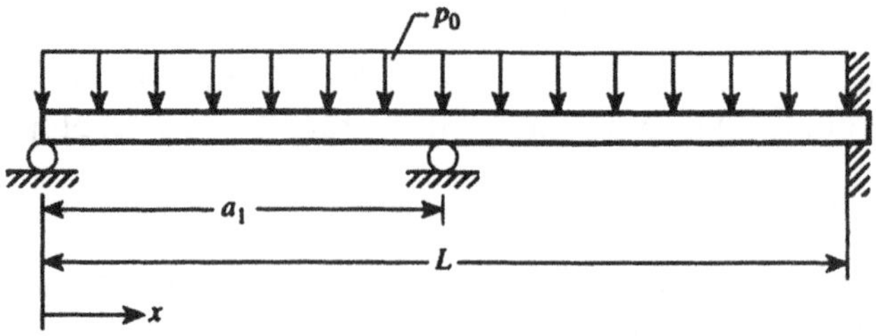

FIGURE P3.27
Statically indeterminate beam.

Answer: 0.212 mm, $R_{\text{right end}} = -1.16$ kN, $M_{\text{right end}} = -89.28$ kN·m, $R_{\text{left end}} = 982.14$ kN

3.28 Determine the force in the middle spring of the beam of Fig. P3.24b with five springs. Let $k_a = 50$ kN/m, $k_b = 60$ kN/m, $k_c = 55$ kN/m, $k_d = 100$ kN/m, $k_e = k_n = 10$ kN/m, $L = 2.5$ m, $p_0 = 5$ kN/m, $I = 9000$ cm^4, $E = 75.84$ GN/m^2, $a_1 = 0.5$ m, $a_2 = 1$ m, $a_3 = 1.5$ m, and $a_4 = 2$ m.

Answer: 1.307 kN

3.29 Find the redundant force R between two simply supported beams crossing each other and loaded by force $\overline{P}$ (Fig. P3.29).

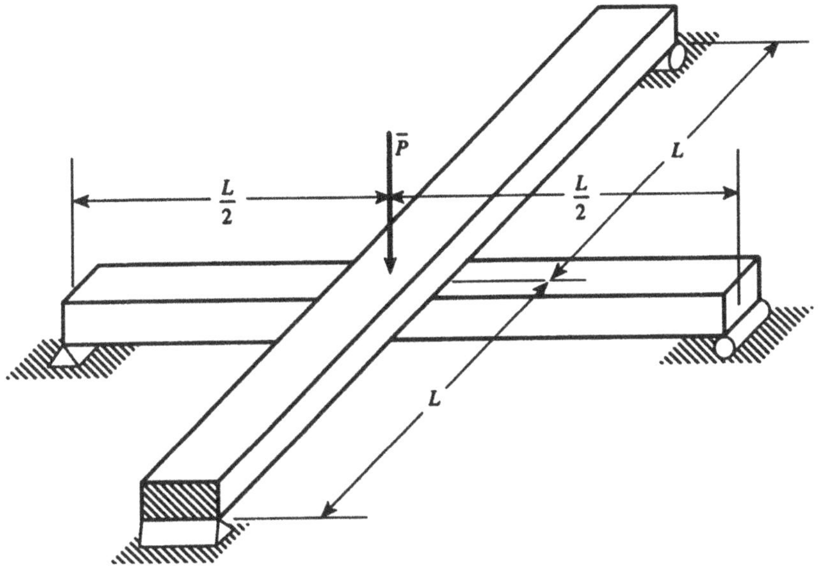

FIGURE P3.29
Crossed beams.

Answer: $R = 8\overline{P}/9$

3.30 A thin circular ring is subjected to equal and opposite loads $\overline{P}$ as shown in Fig. P3.30a. Determine the moments in the ring. Consider only bending in the bar.

Hint: Choose M^* of Fig. P3.30b as the redundant force. Use $\partial U_i^*/\partial M^* = 0$ to find M^*. Because of symmetry,

$$U_i^* = 2\int_0^{\pi/2} \frac{M^2 R}{2EI}\, d\alpha, \quad M = M^* - \frac{\overline{P}R(1-\cos\alpha)}{2}$$

Answer: $M = 0.182\overline{P}R - \frac{\overline{P}R}{2}(1-\cos\alpha)$, $M_{\max} = -0.318\overline{P}R$ under the load.

3.31 Find the vertical displacement in the ring of Fig. P3.30a at the position of loading.

Hint: Since the internal moments in the ring are known from the previous problem, a direct application of Castigliano's theorem, part II will yield the displacements $2V_P = \partial U_i^*/\partial \overline{P}$.

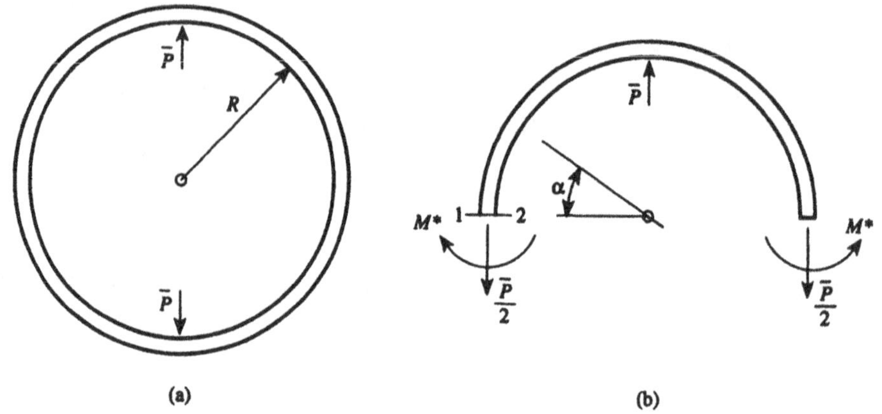

FIGURE P3.30
A thin circular ring.

Answer: $2V_P = 0.149\overline{P}R^3/(EI)$

3.32 Find the horizontal displacement of the ring of Fig. P3.30a at a point 90° from $\overline{P}$.

Hint: Employ a unit load at the point at which the displacement is sought.

Answer: $V = \frac{1}{2}\left(\frac{2}{\pi} - \frac{1}{2}\right)\frac{\overline{P}R^3}{EI}$

3.33 Determine the axial forces in the members of the pin-connected, statically indeterminate structure of Fig. P3.33. The cross-sectional area of each bar is 4 in.^{2}

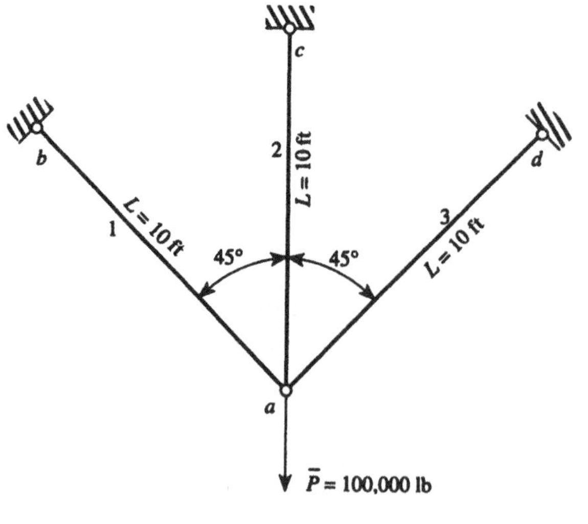

FIGURE P3.33

Hint: Treat $N|^2$ as a redundant force. From the condition of equilibrium express $N|^1$ and $N|^3$ in terms of the unknown $N|^2$. Use $\partial U_i^*/\partial N|^2 = 0$.

Answer: $N|^2 = \overline{P}/2 = 50{,}000$ lb, $N|^1 = N|^3 = (\overline{P} - N|^2)/\sqrt{2} = 35{,}360$ lb.

3.34 Find the vertical displacement at point b of the truss of Fig. P3.34. The members are made of the same material with the same areas.

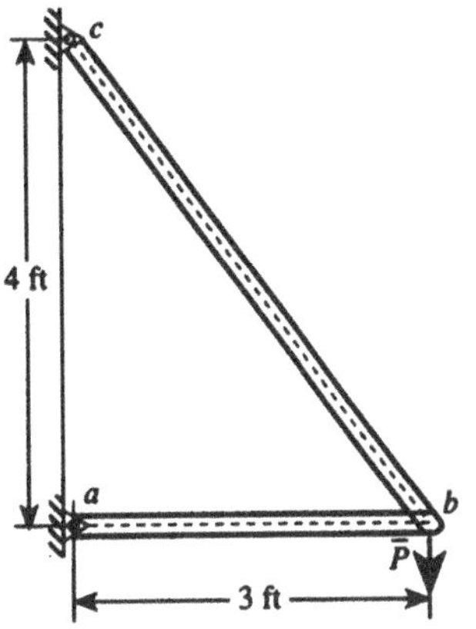

FIGURE P3.34
A simple truss.

Answer: $V = 9.5\overline{P}/(EA)$

3.35 Derive the deformation formula $V = 2\pi N\overline{P}R^3/(GJ)$ for the stretching of a closely coiled helical spring (Fig. P3.35). J is the polar moment of inertia of the cross-section of the wire.

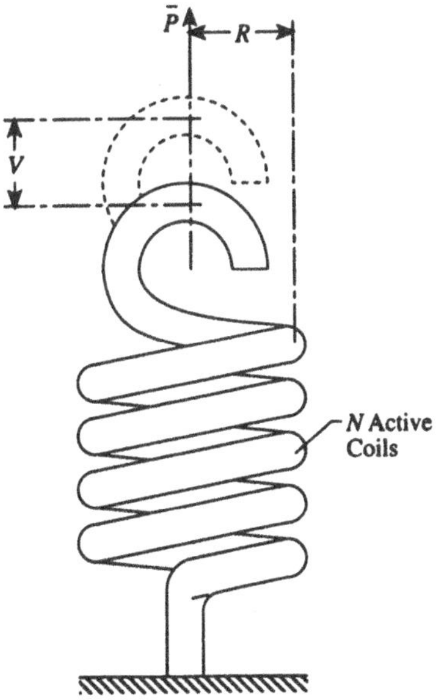

FIGURE P3.35
Helical spring.

Hint: Assume the displacement V of the spring is due only to the twisting of the wire. If there are N active coils, the total length of the wire being twisted is $(2\pi R)N$. The torque in the wire is $\overline{P}R$. $U_i^* = M_t^2 L/(2GJ)$, where M_t is the torque.

3.36 A cantilevered beam of length L and of rectangular cross section is loaded at the free end with a concentrated force $\overline{P}$. Determine the maximum deflection produced by

(a) bending

(b) the additional deflection caused by shear deformation.

Answer: $w_{max} = w_{bending} + w_{shear} = \frac{\overline{P}L^3}{3EI} + \frac{\overline{P}L}{k_s GA}$

3.37 Solve Problem 3.11 using Castigliano's theorem, part I.

3.38 Determine the effect of shear deformation on the maximum deflection of a simply supported beam of length L with a concentrated force $\overline{P}$ at $x = L/2$.

Answer: $w_{max} = w_{bending} + w_{shear} = \frac{\overline{P}L^3}{48EI} + \frac{\overline{P}L}{4k_s GA}$

Reciprocal Theorems

3.39 Suppose a cantilevered beam of length L is loaded with a linearly varying distributed force beginning with magnitude 0 at the free end and increasing to magnitude p_0 at the fixed end. Use Betti's reciprocal theorem to find the deflection at the free end (a) and at the midspan (b).

Answer: $w_a = p_0 L^4/(30EI), \quad w_b = 0.01276 p_0 L^4/EI$

3.40 Consider a uniformly loaded (magnitude p_0) beam (length L) with one end fixed and one end simply-supported. Use Betti's reciprocal theorem to calculate the deflection at $L/2$.

Answer: $p_0 L^4/(192EI)$

3.41 For a cantilevered beam with a concentrated force $\overline{P}$ at the free end, use Maxwell's theorem to calculate the deflection at $x = 2L/3$ from the fixed end, where L is the beam length.

Answer: $w = 14\overline{P}L^3/(81EI)$

3.42 Calculate the vertical displacement of point a of the structure of Fig. P3.42. Let $E = 200 \text{ GN/m}^2$, $I = 6000 \text{ cm}^4$. Use Maxwell's reciprocal theorem.

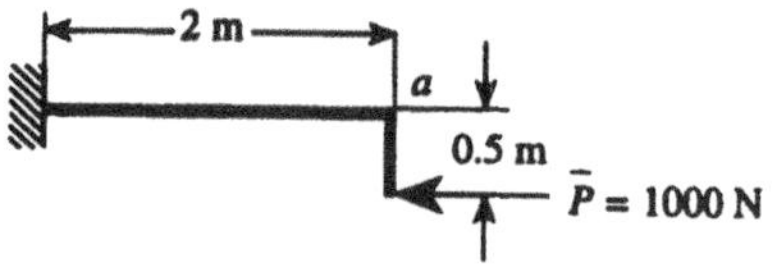

FIGURE P3.42

Answer: $w_a = 0.0833$ mm.

3.43 Use Maxwell's theorem to find the midspan deflection of the beam described in Problem 3.40.

Section B

Solution Methods

4

Structural Analysis Methods I: Beam Elements

The sources for many contemporary computational methods of solid mechanics can be traced to structural analysis techniques. In this and the following chapter, we will outline structural analysis methodology. We begin with the study of structural members, with primary emphasis given to the plane beam element. In the next chapter, the mixed, displacement, and force methods of joining the elements into structural systems will be outlined.

4.1 Sign Convention

The sign convention of Chapter 1, which is frequently employed for structural members, where the distribution of the *internal* bending moment and shear force are of concern, is illustrated in Fig. 4.1a and labeled Sign Convention 1. It is convenient to introduce another sign convention which is better suited for use in the stiffness methods of analysis of network structures where values of the bending moment and shear force at the ends of the elements are to be calculated. This new sign convention, which will be referred to as Sign Convention 2, is shown in Fig. 4.1b. For this second convention, on both ends of the beam, the forces and moments along the positive coordinate directions are considered to be positive.

Designate the forces on the ends of the beam element by **s** for Sign Convention 1 and by **p** for Sign Convention 2. More specifically, define for **Sign Convention 1**

$$\mathbf{s}_a = \begin{bmatrix} V_a \\ M_a \end{bmatrix} \qquad \mathbf{s}_b = \begin{bmatrix} V_b \\ M_b \end{bmatrix} \qquad \mathbf{s} = \begin{bmatrix} \mathbf{s}_a \\ \mathbf{s}_b \end{bmatrix} \tag{4.1}$$

and for **Sign Convention 2**

$$\mathbf{p}_a = \begin{bmatrix} V_a \\ M_a \end{bmatrix} \quad \mathbf{p}_b = \begin{bmatrix} V_b \\ M_b \end{bmatrix} \quad \mathbf{p} = \begin{bmatrix} \mathbf{p}_a \\ \mathbf{p}_b \end{bmatrix} \tag{4.2}$$

The sign conventions are related by $V_a|_{\text{Sign Convention 1}} = -V_a|_{\text{Sign Convention 2}}$ and $M_a|_{\text{Sign Convention 1}} = -M_a|_{\text{Sign Convention 2}}$. In matrix notation,

$$\begin{bmatrix} \mathbf{p}_a \\ \cdots \\ \mathbf{p}_b \end{bmatrix} = \begin{bmatrix} -1 & 0 & \vdots & 0 & 0 \\ 0 & -1 & \vdots & 0 & 0 \\ \cdots & \cdots & & \cdots & \cdots \\ 0 & 0 & \vdots & 1 & 0 \\ 0 & 0 & \vdots & 0 & 1 \end{bmatrix} \begin{bmatrix} \mathbf{s}_a \\ \cdots \\ \mathbf{s}_b \end{bmatrix} \tag{4.3a}$$

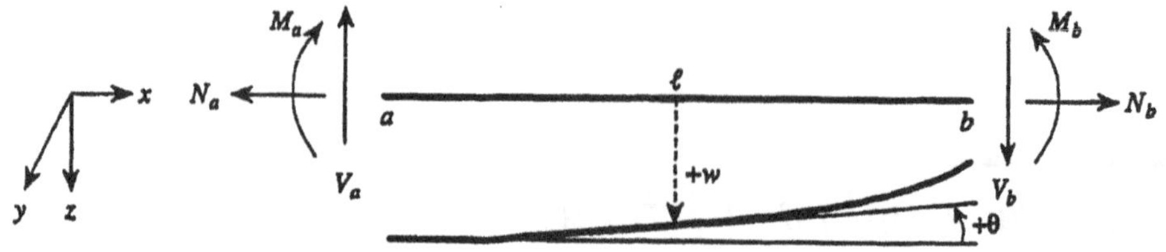

(a) Sign Convention 1. Positive forces, moments, slopes, and displacements are shown. This sign convention is used for structural members and has been employed in most of the work in the previous chapters. The underlying logic for this convention is explained in detail in Chapter 1.

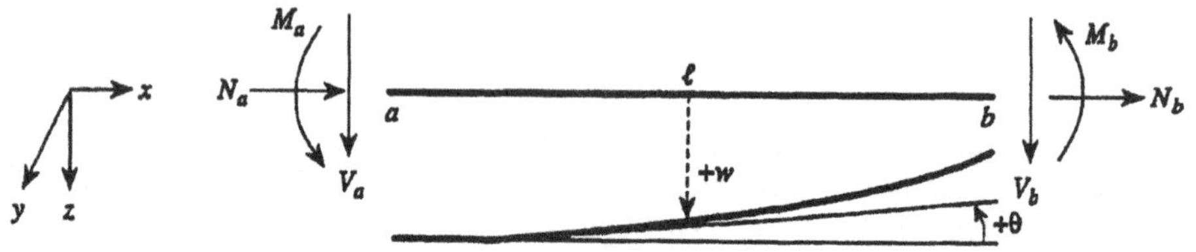

(b) Sign Convention 2. Forces and moments on both ends of the beam element are positive if they (their vectors) lie in the positive coordinate directions. Positive forces, moments, slopes, and displacements are shown. Note that positive deflection and slope are the same as for Sign Convention 1. Sign Convention 2 is convenient to use in the study of network structural systems.

FIGURE 4.1
Sign conventions for a beam element.

and

$$\begin{bmatrix} \mathbf{s}_a \\ \cdots \\ \mathbf{s}_b \end{bmatrix} = \begin{bmatrix} -\mathbf{I} & \vdots & \mathbf{0} \\ \cdots & & \cdots \\ \mathbf{0} & \vdots & \mathbf{I} \end{bmatrix} \begin{bmatrix} \mathbf{p}_a \\ \cdots \\ \mathbf{p}_b \end{bmatrix} \tag{4.3b}$$

where **I** is the unit diagonal matrix. For the axial force $N_a|_{\text{Sign Convention 1}} = -N_a|_{\text{Sign Convention 2}}$.

Deflections and slopes remain the same according to both sign conventions, and, hence, no special displacement transformation is required.

4.2 Fundamental Relations for a Beam Element

There are several possibilities for developing relations between the forces and displacements on both ends of a beam segment. We first consider the pure bending of a beam element with no loads applied between the ends. Begin with the element of Fig. 4.1a in which the net forces (and moments) are shown on ends a and b using Sign Convention 1.

4.2.1 The Equations of Equilibrium

From Fig. 4.1a, the conditions of equilibrium appear as

$$\sum M_{x=b} = 0 : M_b = V_a \ell + M_a$$

$$\sum F_z = 0 : V_b = V_a$$

If the axial effects are taken into account, $\Sigma F_x = 0 : N_b = N_a$.

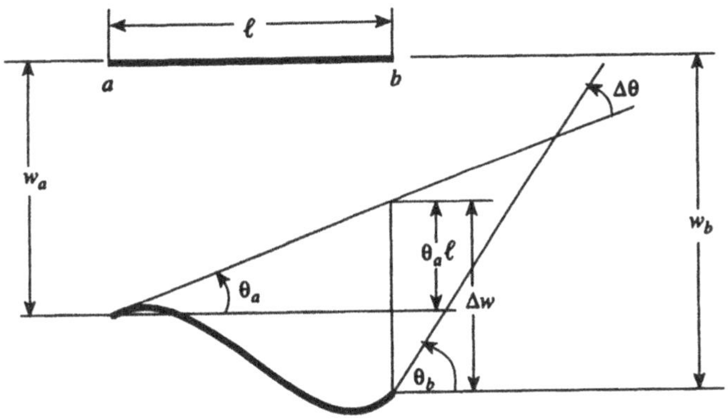

FIGURE 4.2
Geometry of deformation for a beam element.

4.2.2 The Geometry of Deformation

It follows from the geometry of the deformed beam shown in Fig. 4.2 that the deflection w and slope θ at b in terms of these variables at a are

$$\begin{array}{llcl} w_b = w_a & -\ell\theta_a & \vdots & +\Delta w \\ \theta_b = & \theta_a & \vdots & +\Delta\theta \\ & \text{Rigid body} & \vdots & \text{Deformation} \\ & \text{displacements} & \vdots & \end{array} \tag{4.4}$$

where the deflection Δw and slope $\Delta\theta$ are due to the deformation.

4.2.3 The Material Law

The deformation quantities Δw and $\Delta\theta$ can be determined using the material law relationships. Suppose the Δw and $\Delta\theta$ terms are separated into the effects of shear force and bending moment, giving

$$\begin{aligned} \Delta w &= \Delta w_V + \Delta w_M \\ \Delta\theta &= \Delta\theta_V + \Delta\theta_M \end{aligned} \tag{4.5}$$

Various strength of material methods, such as the energy methods of Chapter 3, can be employed to find these variables in terms of the bending moment and shear force at some point along the beam. The resulting expressions are *force-deformation relations*. First the rigid body motion should be eliminated. To do so, fix the left end of the beam of Fig. 4.2, which has already undergone rigid body displacements. This gives the configuration of Fig. 4.3a.

Rotate the beam of Fig. 4.3a clockwise an angle θ_a about end a. Then the beam can be viewed as a deformed horizontal beam with the left end clamped (Fig. 4.3b). If w_a and θ_a are now considered to be zero, the quantities Δw and $\Delta\theta$ are the deflection and rotation, respectively, of the right end. Both Δw and $\Delta\theta$ can be computed using strength of materials techniques.

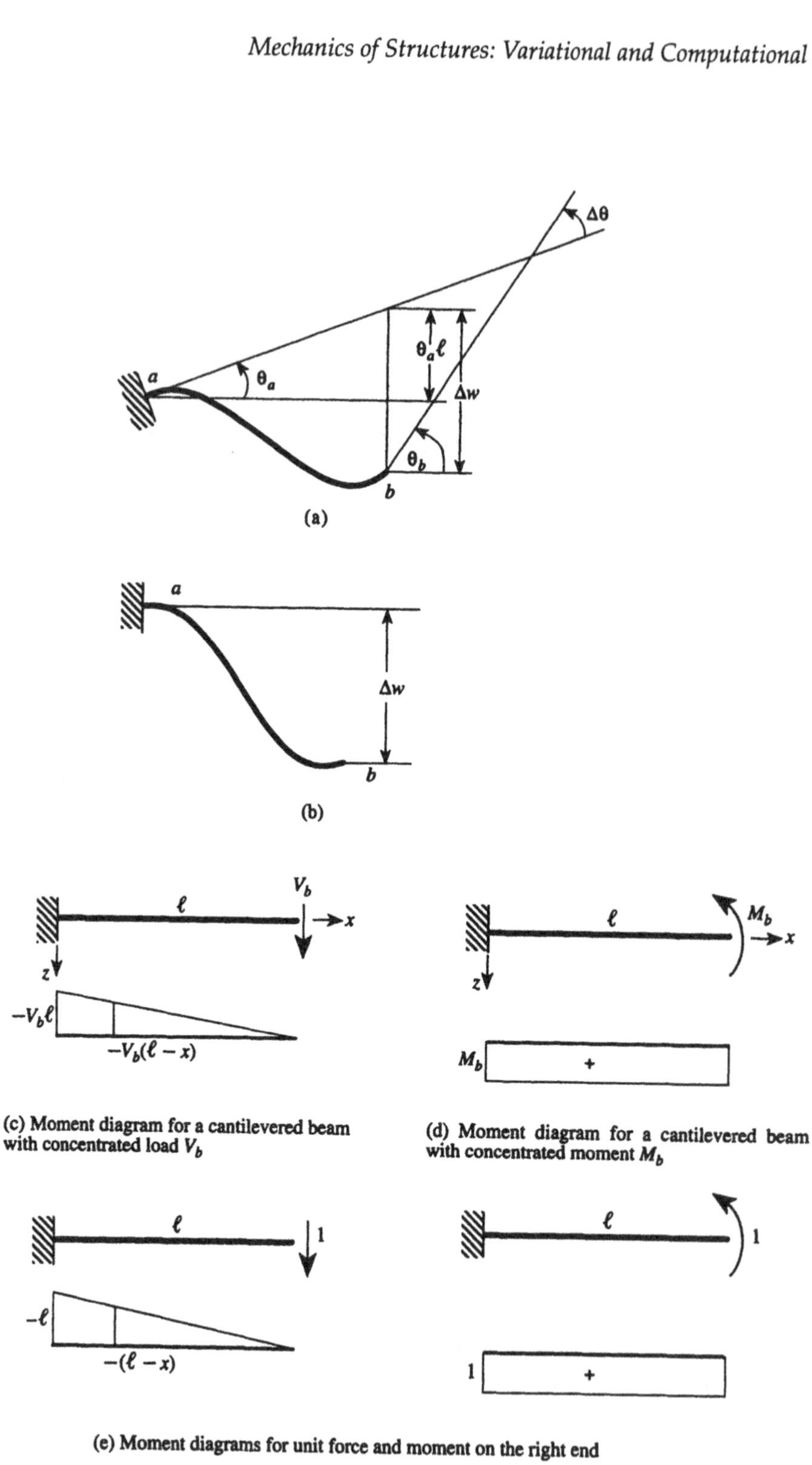

FIGURE 4.3
Diagrams illustrating the determination of Δw and $\Delta\theta$.

Unit Load Method

A simple technique for finding Δw and $\Delta\theta$ is the unit load method of Chapter 3. To compute the displacement (Δw_V) and slope ($\Delta\theta_V$) due to V_b at the right end, apply a unit force downward (for obtaining Δw_V) and a unit moment (for obtaining $\Delta\theta_V$), and use the moment diagrams of Figs. 4.3c and e. Assume EI is constant. It follows from Chapter 3, Eq. (3.17) that

$$\Delta w_V = \int_0^\ell \frac{V_b(\ell - x)}{EI}(\ell - x)\,dx = \frac{\ell^3 V_b}{3EI}$$

$$\Delta\theta_V = -\int_0^\ell \frac{V_b(\ell - x)}{EI} 1\,dx = -\frac{\ell^2 V_b}{2EI}$$

For the displacement (Δw_M) and slope ($\Delta\theta_M$) due to M_b at the right end, use the moment diagrams of Figs. 4.3d and e. From Eq. (3.17)

$$\Delta w_M = \int_0^\ell -\frac{M_b}{EI}(\ell - x)\,dx = -\frac{\ell^2 M_b}{2EI}$$

$$\Delta\theta_M = \int_0^\ell \frac{M_b}{EI} 1\,dx = \frac{\ell M_b}{EI}$$

Then

$$\Delta w = \Delta w_V + \Delta w_M = \frac{\ell^3 V_b}{3EI} - \frac{\ell^2 M_b}{2EI}$$

$$\Delta\theta = \Delta\theta_V + \Delta\theta_M = -\frac{\ell^2 V_b}{2EI} + \frac{\ell M_b}{EI}$$

Substitute the equilibrium conditions, $V_b = V_a$, $M_b = M_a + V_a\ell$, into these relations to obtain expressions for Δw and $\Delta\theta$ as functions of V_a and M_a.

$$\begin{aligned} \Delta w &= -\frac{\ell^3}{6EI} V_a - \frac{\ell^2}{2EI} M_a \\ \Delta\theta &= \frac{\ell^2}{2EI} V_a + \frac{\ell}{EI} M_a \end{aligned} \tag{4.6}$$

4.2.4 Summary

In matrix notation, the fundamental relations using Sign Convention 1 appear as

Equilibrium:

$$\begin{bmatrix} V_b \\ M_b \end{bmatrix} = \begin{bmatrix} 1 & 0 \\ \ell & 1 \end{bmatrix} \begin{bmatrix} V_a \\ M_a \end{bmatrix}$$
$$\mathbf{s}_b = \mathbf{U}_{ss} \quad \mathbf{s}_a \tag{4.7a}$$

Geometry:

$$\begin{bmatrix} w_b \\ \theta_b \end{bmatrix} = \begin{bmatrix} 1 & -\ell \\ 0 & 1 \end{bmatrix} \begin{bmatrix} w_a \\ \theta_a \end{bmatrix} + \begin{bmatrix} \Delta w \\ \Delta\theta \end{bmatrix}$$
$$\mathbf{v}_b = \mathbf{U}_{vv} \quad \mathbf{v}_a + \Delta\mathbf{v} \tag{4.7b}$$

Material Law:

$$\begin{bmatrix} \Delta w \\ \Delta\theta \end{bmatrix} = \begin{bmatrix} -\ell^3/6EI & -\ell^2/2EI \\ \ell^2/2EI & \ell/EI \end{bmatrix} \begin{bmatrix} V_a \\ M_a \end{bmatrix}$$
$$\Delta \mathbf{v} = \mathbf{U}_{vs} \qquad \mathbf{s}_a \tag{4.7c}$$

4.3 Element Matrices, Definitions

4.3.1 Transfer Matrix

Combine the matrices of Eqs. (4.7a), (4.7b), and (4.7c) such that all the variables (forces and displacements) at a are on one side and all variables at b are on the other. Then

$$\begin{aligned} \mathbf{v}_b &= \mathbf{U}_{vv}\mathbf{v}_a + \mathbf{U}_{vs}\mathbf{s}_a \\ \mathbf{s}_b &= \mathbf{U}_{ss}\mathbf{s}_a \end{aligned} \tag{4.8a}$$

or

$$\begin{bmatrix} \mathbf{v}_b \\ \cdots \\ \mathbf{s}_b \end{bmatrix} = \begin{bmatrix} \mathbf{U}_{vv} & \vdots & \mathbf{U}_{vs} \\ \cdots\cdots & & \cdots\cdots \\ \mathbf{0} & \vdots & \mathbf{U}_{ss} \end{bmatrix} \begin{bmatrix} \mathbf{v}_a \\ \cdots \\ \mathbf{s}_a \end{bmatrix} \tag{4.8b}$$

or

$$\begin{bmatrix} w \\ \theta \\ V \\ M \end{bmatrix}_b = \begin{bmatrix} 1 & -\ell & \vdots & -\ell^3/6EI & -\ell^2/2EI \\ 0 & 1 & \vdots & \ell^2/2EI & \ell/EI \\ \cdots & \cdots & & \cdots\cdots & \cdots\cdots \\ 0 & 0 & \vdots & 1 & 0 \\ 0 & 0 & \vdots & \ell & 1 \end{bmatrix} \begin{bmatrix} w \\ \theta \\ V \\ M \end{bmatrix}_a \tag{4.8c}$$

$$\mathbf{z}_b = \mathbf{U}^i \qquad \mathbf{z}_a \tag{4.8d}$$

The matrix $\mathbf{U}^i$, which is sometimes denoted by $\mathbf{U}^i(\ell) = \mathbf{U}^i(x_b - x_a)$, is referred to as a *transfer matrix* since it "transfers" the variables w, θ, V, and M from $x = x_a$ to $x = x_b$. The vector $\mathbf{z}$ of displacements and forces is called the *state vector* because these variables fully describe the response or "state" of the beam.

Note from Eqs. (4.7a), (4.7b), and (4.7c) that the partitions of Eqs. (4.8b) and (4.8c) can be identified with the basic equations for a beam.

$$\mathbf{U}^i = \begin{bmatrix} \text{Geometry} & \vdots & \text{Material} \\ \text{(Rigid body} & \vdots & \text{law} \\ \text{displacements)} & \vdots & \\ \cdots\cdots\cdots\cdots & & \cdots\cdots \\ \mathbf{0} & \vdots & \\ \text{(Influence of} & \vdots & \text{Equilibrium} \\ \text{springs, foundations,} & \vdots & \\ \text{etc.)} & \vdots & \end{bmatrix} = \begin{bmatrix} & \vdots & \\ \mathbf{U}_{vv} & \vdots & \mathbf{U}_{vs} \\ & \vdots & \\ \cdots\cdots & & \cdots \\ & \vdots & \\ \mathbf{0} & \vdots & \mathbf{U}_{ss} \\ (\mathbf{U}_{sv}) & \vdots & \\ & \vdots & \end{bmatrix} \tag{4.9}$$

4.3.2 Stiffness Matrix

A stiffness matrix for the beam element relates all of the displacements at a and b, i.e., w_a, θ_a, w_b, and θ_b, to all the forces, i.e., V_a, M_a, V_b, and M_b. The stiffness matrix $\mathbf{k}^i$ for element i is defined as

$$\mathbf{p}^i = \mathbf{k}^i \mathbf{v}^i \tag{4.10}$$

where

$$\mathbf{p}^i = \begin{bmatrix} \mathbf{p}_a \\ \mathbf{p}_b \end{bmatrix} = \begin{bmatrix} V_a \\ M_a \\ V_b \\ M_b \end{bmatrix} \qquad \mathbf{v}^i = \begin{bmatrix} \mathbf{v}_a \\ \mathbf{v}_b \end{bmatrix} = \begin{bmatrix} w_a \\ \theta_a \\ w_b \\ \theta_b \end{bmatrix}$$

$$\mathbf{k}^i = \begin{bmatrix} \mathbf{k}_{aa} & \mathbf{k}_{ab} \\ \mathbf{k}_{ba} & \mathbf{k}_{bb} \end{bmatrix} = \begin{bmatrix} k_{11} & k_{12} & k_{13} & k_{14} \\ k_{21} & k_{22} & k_{23} & k_{24} \\ k_{31} & k_{32} & k_{33} & k_{34} \\ k_{41} & k_{42} & k_{43} & k_{44} \end{bmatrix}$$

The stiffness matrix is defined using $\mathbf{p}$ of Sign Convention 2, Eq. (4.2). The stiffness matrix is an important building block for the analysis of structural systems. The fundamental relations of Eqs. (4.7a), (4.7b), and (4.7c) were placed together in a special manner to form the transfer matrix. The same fundamental relations can be reorganized to form the stiffness matrix. This is to be expected since both the transfer and stiffness matrices are relationships between the same eight variables $w_a, \theta_a, w_b, \theta_b, V_a, M_a, V_b$, and M_b. Of course, there are numerous other methods for finding the stiffness matrix, some of which were treated in Chapter 3, while others will be considered in this section.

Consider first the arrangement of the fundamental relations of Eqs. (4.7a), (4.7b), and (4.7c) into a stiffness matrix. From Eq. (4.8a), written in terms of Sign Convention 2 [replace $\mathbf{s}$ by $\mathbf{p}$ and use Eq. (4.3)],

$$\mathbf{p}_b = \mathbf{U}_{pp}\mathbf{p}_a, \qquad \mathbf{v}_b = \mathbf{U}_{vv}\mathbf{v}_a + \mathbf{U}_{vp}\mathbf{p}_a$$

where $\mathbf{U}_{pp} = -\mathbf{U}_{ss}$ and $\mathbf{U}_{vp} = -\mathbf{U}_{vs}$. It follows that

$$\mathbf{p}_a = \mathbf{U}_{vp}^{-1}\mathbf{v}_b - \mathbf{U}_{vp}^{-1}\mathbf{U}_{vv}\mathbf{v}_a$$
$$\mathbf{p}_b = \mathbf{U}_{pp}\mathbf{p}_a = \mathbf{U}_{pp}\mathbf{U}_{vp}^{-1}\mathbf{v}_b - \mathbf{U}_{pp}\mathbf{U}_{vp}^{-1}\mathbf{U}_{vv}\mathbf{v}_a$$

or

$$\begin{bmatrix} \mathbf{p}_a \\ \cdots \\ \mathbf{p}_b \end{bmatrix} = \underbrace{\begin{bmatrix} -\mathbf{U}_{vp}^{-1}\mathbf{U}_{vv} & \vdots & \mathbf{U}_{vp}^{-1} \\ \cdots\cdots & & \cdots\cdots \\ -\mathbf{U}_{pp}\mathbf{U}_{vp}^{-1}\mathbf{U}_{vv} & \vdots & \mathbf{U}_{pp}\mathbf{U}_{vp}^{-1} \end{bmatrix}}_{\mathbf{k}^i} \begin{bmatrix} \mathbf{v}_a \\ \cdots \\ \mathbf{v}_b \end{bmatrix} \tag{4.11}$$

where $\mathbf{U}_{vp}^{-1}$ is given by (Sign Convention 2)

$$\mathbf{U}_{vp}^{-1} = \begin{bmatrix} -12EI/\ell^3 & -6EI/\ell^2 \\ 6EI/\ell^2 & 2EI/\ell \end{bmatrix}$$

Note that all displacements are now located on the right-hand side while the forces are on the left-hand side. Equation (4.11) can be written for a beam element as

$$\begin{bmatrix} V_a \\ M_a \\ \\ V_b \\ M_b \end{bmatrix} = \begin{bmatrix} \frac{12EI}{\ell^3} & -\frac{6EI}{\ell^2} & \vdots & -\frac{12EI}{\ell^3} & -\frac{6EI}{\ell^2} \\ -\frac{6EI}{\ell^2} & \frac{4EI}{\ell} & \vdots & \frac{6EI}{\ell^2} & \frac{2EI}{\ell} \\ \cdots\cdots & \cdots & & \cdots\cdots & \cdots\cdots \\ -\frac{12EI}{\ell^3} & \frac{6EI}{\ell^2} & \vdots & \frac{12EI}{\ell^3} & \frac{6EI}{\ell^2} \\ -\frac{6EI}{\ell^2} & \frac{2EI}{\ell} & \vdots & \frac{6EI}{\ell^2} & \frac{4EI}{\ell} \end{bmatrix} \begin{bmatrix} w_a \\ \theta_a \\ \\ w_b \\ \theta_b \end{bmatrix} = \begin{bmatrix} \mathbf{k}_{aa} & \mathbf{k}_{ab} \\ \mathbf{k}_{ba} & \mathbf{k}_{bb} \end{bmatrix} \mathbf{v}^i$$

$$\mathbf{p}^i \quad = \quad \mathbf{k}^i \quad \mathbf{v}^i \tag{4.12}$$

This is, of course, the same as the beam stiffness matrix derived in Chapter 3, Example 3.11.

It follows from the form of this relationship that a stiffness element k_{ij}, e.g., $k_{11} = 12EI/\ell^3$, can be considered to be the force developed at coordinate i due to a unit displacement at coordinate j, with all other displacements equal to zero. These "coordinates" are usually referred to as *degrees of freedom* (DOF). More precisely, the DOF are the independent displacement components necessary to fully describe the spatial position of a structure. The number of DOF depends on the modeling of the structure for analysis. In static analyses, we analyze each element before the overall structure is treated, and thus, reduce the behavior of the element to selected DOF at each end of the element. Some of these end DOF can be ignored if it is known that the response of the structure does not depend heavily on these DOF. This situation occurs with rigid frames, for example, where displacements due to uniform axial strain are usually significantly smaller than the displacements resulting from bending.

It is possible to rewrite the stiffness matrix of Eq. (4.12) giving a form that is often more convenient. A redefinition of the vectors $\mathbf{p}^i$ and $\mathbf{v}^i$ leads to the modified form

$$\begin{bmatrix} V_a \\ M_a/\ell \\ V_b \\ M_b/\ell \end{bmatrix} = \frac{EI}{\ell^3} \begin{bmatrix} 12 & -6 & -12 & -6 \\ -6 & 4 & 6 & 2 \\ -12 & 6 & 12 & 6 \\ -6 & 2 & 6 & 4 \end{bmatrix} \begin{bmatrix} w_a \\ \ell\theta_a \\ w_b \\ \ell\theta_b \end{bmatrix} \tag{4.13}$$

$$\mathbf{p}^i \quad = \quad \mathbf{k}^i \quad \mathbf{v}^i$$

The complete description of an element should include a vector representing applied loads on the element. They can be calculated as the reactions of a beam element with fixed ends. Another possibility, which is presented in the following paragraph, is to obtain the loading vector by a transformation of the transfer matrix. The element stiffness matrix including its element loading vector can be expressed as

$$\begin{bmatrix} V_a \\ M_a \\ V_b \\ M_b \end{bmatrix} = \begin{bmatrix} k_{11} & k_{12} & k_{13} & k_{14} \\ k_{21} & k_{22} & k_{23} & k_{24} \\ k_{31} & k_{32} & k_{33} & k_{34} \\ k_{41} & k_{42} & k_{43} & k_{44} \end{bmatrix} \begin{bmatrix} w_a \\ \theta_a \\ w_b \\ \theta_b \end{bmatrix} + \begin{bmatrix} \overline{V}_a^0 \\ \overline{M}_a^0 \\ \overline{V}_b^0 \\ \overline{M}_b^0 \end{bmatrix} \tag{4.14}$$

$$\mathbf{p}^i \quad = \quad \mathbf{k}^i \quad \mathbf{v}^i \quad - \quad \overline{\mathbf{P}}$$

As mentioned in the previous paragraph, it is necessary to include with the stiffness matrix a vector to account for applied loading. Normally, this vector would account for only the loading applied between the ends, since end loadings are inserted using the vector $\mathbf{p}^i$. To derive a stiffness matrix with a loading vector appended, begin by writing a transfer matrix in the notation of Sign Convention 2, giving

$$\begin{bmatrix} \mathbf{v}_b \\ \mathbf{p}_b \end{bmatrix} = \begin{bmatrix} \mathbf{U}_{vv} & \mathbf{U}_{vp} \\ \mathbf{U}_{pv} & \mathbf{U}_{pp} \end{bmatrix} \begin{bmatrix} \mathbf{v}_a \\ \mathbf{p}_a \end{bmatrix} + \begin{bmatrix} \mathbf{F}_v^0 \\ \mathbf{F}_p^0 \end{bmatrix} \tag{4.15}$$

where

$$\mathbf{F}_v^0 = \begin{bmatrix} w_b^0 \\ \theta_b^0 \end{bmatrix} \qquad \mathbf{F}_p^0 = \begin{bmatrix} V_b^0 \\ M_b^0 \end{bmatrix}$$

Expressions for the entries of the vectors $\mathbf{F}_v^0$ and $\mathbf{F}_p^0$ will be given later. Follow the procedure utilized for forming the stiffness matrix of Eq. (4.11) to rearrange Eq. (4.15) into stiffness matrix form. This gives

$$\begin{bmatrix} \mathbf{p}_a \\ \\ \mathbf{p}_b \end{bmatrix} = \begin{bmatrix} -\mathbf{U}_{vp}^{-1}\mathbf{U}_{vv} & \vdots & \mathbf{U}_{vp}^{-1} \\ \cdots\cdots\cdots & & \cdots \\ \mathbf{U}_{pv} - \mathbf{U}_{pp}\mathbf{U}_{vp}^{-1}\mathbf{U}_{vv} & \vdots & \mathbf{U}_{pp}\mathbf{U}_{vp}^{-1} \end{bmatrix} \begin{bmatrix} \mathbf{v}_a \\ \cdots \\ \mathbf{v}_b \end{bmatrix} + \begin{bmatrix} -\mathbf{U}_{vp}^{-1}\mathbf{F}_v^0 \\ \\ \mathbf{F}_p^0 - \mathbf{U}_{pp}\mathbf{U}_{vp}^{-1}\mathbf{F}_v^0 \end{bmatrix} \tag{4.16}$$

The contents of this matrix are easily identified with the nomenclature of the matrix of Eq. (4.14).

4.3.3 Flexibility Matrix

A *flexibility matrix* relates the forces $\mathbf{p}$ at a and b of an element to the displacements $\mathbf{v}$ at a and b. From this definition, it would appear that the flexibility matrix is simply the inverse of the stiffness matrix, which relates the displacement $\mathbf{v}$ at a and b to the forces $\mathbf{p}$ at a and b. Certainly, this would be the case if the stiffness matrix were nonsingular. However, it will be shown in this section that the stiffness matrix for an unconstrained beam element is singular, and, hence, the corresponding flexibility matrix does not exist.

The stiffness matrices defined in Section 4.3.2 relate all of the degrees of freedom (displacements) to all of the end forces, without regard as to how the element is supported or constrained. Thus, the beams are treated as being free or unconstrained bodies, and rigid body displacements can occur. Although rigid body displacements may not affect the deformation of the beam, they cause the rows or columns of the stiffness matrix to be linearly dependent.

We wish to observe the problems arising from the unconstrained beam element. From Eqs. (4.7a), (4.7b), and (4.7c), written in the nomenclature of Sign Convention 2 (replace $\mathbf{s}$ by $\mathbf{p}$), the fundamental relations for a beam element can be written as

$$\Delta\mathbf{v} = \mathbf{U}_{vp}\mathbf{p}_a \tag{4.17a}$$

$$\mathbf{v}_b - \mathbf{U}_{vv}\mathbf{v}_a = \Delta\mathbf{v} \tag{4.17b}$$

$$\mathbf{p}_b = \mathbf{U}_{pp}\mathbf{p}_a \tag{4.17c}$$

From Eq. (4.17a) for given forces $\mathbf{p}_a$, $\Delta\mathbf{v}$ can be determined uniquely, as can $\mathbf{p}_b$ from Eq. (4.17c). It follows from Eq. (4.17b) that a particular combination of displacements $\mathbf{v}_b$ and $\mathbf{v}_a$ can be determined uniquely, but not $\mathbf{v}_b$ and $\mathbf{v}_a$ themselves. Thus, $\mathbf{v}_b$ and $\mathbf{v}_a$ are not single-valued, as one depends on the other. In other terms, the deformation of the beam as characterized by $\Delta\mathbf{v}$ is not influenced by certain combinations of displacements.

These are so-called rigid body displacements of the bar. That is, the beam can undergo rigid body motion without introducing elastic forces.

As a result of these dependent variables, the stiffness matrix is singular. This is evident from

$$\begin{bmatrix} \mathbf{p}_a \\ \mathbf{p}_b \end{bmatrix} = \begin{bmatrix} \mathbf{k}_{aa} & \mathbf{k}_{ab} \\ \mathbf{k}_{ba} & \mathbf{k}_{bb} \end{bmatrix} \begin{bmatrix} \mathbf{v}_a \\ \mathbf{v}_b \end{bmatrix} \tag{4.18}$$
$$\mathbf{p}^i = \mathbf{k}^i \quad \mathbf{v}^i$$

which is a system of linear equations. Because $\mathbf{v}_a$ and $\mathbf{v}_b$ are dependent, the solution to Eq. (4.18) is not unique. From Cramer's[1] rule, a set of simultaneous linear equations ($\mathbf{k}^i\mathbf{v}^i = \mathbf{p}^i$) has a unique solution if, and only if, the determinant of the coefficient matrix $\mathbf{k}^i$ is not zero, i.e., det $\mathbf{k}^i \neq 0$. Thus, if $\mathbf{k}^i\mathbf{v}^i = \mathbf{p}^i$ does not have a unique solution, the determinant of $\mathbf{k}^i$ must be zero. It follows that matrix $\mathbf{k}^i$ is singular, and its rows (columns) are linearly dependent.

The conclusion of the previous paragraph can also be reached by scruitiny of Eq. (4.12). Note from the stiffness matrix of Eq. (4.12) that the sum of rows 1 and 3 is $[0 \;\; 0 \;\; 0 \;\; 0]$. Thus, the determinant of $\mathbf{k}^i$ is zero and $\mathbf{k}^i$ is singular. Also, rows 1 and 3 are linearly dependent. The singular property of the 4×4 stiffness matrix is also evident by noting that the sum of columns 1 and 3 is zero. Again the determinant of $\mathbf{k}^i$ will be zero. Furthermore, the sum of columns 2 and 4 is equal to ℓ times column 1. This too leads to a singular matrix.

It is possible to ascertain analytically the number of rigid body displacements contained in an elemental stiffness matrix. This is accomplished by transforming the stiffness matrix into diagonal form. The number of rigid body displacements is then equal to the number of zero terms in the diagonal. The transformation involves establishing the eigenvalues and eigenvectors of the stiffness matrix, with the number of zero eigenvalues being equal to the number of rigid body motions. It is also possible to scrutinize the strain energy, since the lack of deformation of a rigid body motion should correspond to zero strain energy. Thus, if the strain energy contributed by a certain eigenvector of the stiffness matrix is zero, then the associated eigenvalue is zero which corresponds to rigid body motion.

Elimination of the rigid body displacements will produce a nonsingular stiffness matrix and permit the flexibility matrix to be formed by inversion. This elimination of the rigid body displacements, e.g., through the consideration of supports or constraints, reduces the number of degrees of freedom of the element. Thus, a flexibility matrix $\mathbf{f}$ relates a reduced set of forces $\mathbf{p}_R$ at a and b to a reduced set of displacements $\mathbf{v}_R$ at a and b through

$$\mathbf{v}_R = \mathbf{f}\,\mathbf{p}_R \tag{4.19}$$

The flexibility matrix is defined only for restrained systems. Otherwise rigid body motion would occur and the magnitude of the displacements could be unlimited. This follows from the definition of a flexibility coefficient f_{ij} being the displacement at i due to a unit load at j. Because of the need for a restraining condition, the flexibility matrix for a beam element is not unique. Rather, it depends on which DOF are chosen to be unrestrained.

To illustrate these principles, consider a beam element simply supported at both ends. The displacement boundary conditions are $w_a = 0$, $w_b = 0$. With these restraints taken into

[1]Gabriel Cramer (1704–1752) was an early 18th century Swiss mathematician and professor who was probably not the originator of the popular rule named after him. Cramer's rule is a simple scheme for solving a set of linear equations. Maclaurin, whose name is attached to a series (Maclaurin's series) which he did not discover, appears to have originated Cramer's rule. Cramer is also known for Cramer's paradox which he also did not originate. Cramer's paradox deals with the intersection of cubic curves.

consideration, the second and fourth rows of the stiffness matrix of Eq. (4.12) become

$$\begin{bmatrix} M_a \\ M_b \end{bmatrix} = \underbrace{\begin{bmatrix} k_{22} & k_{24} \\ k_{42} & k_{44} \end{bmatrix}}_{\mathbf{k}_R} \begin{bmatrix} \theta_a \\ \theta_b \end{bmatrix} = \underbrace{\begin{bmatrix} \frac{4EI}{\ell} & \frac{2EI}{\ell} \\ \frac{2EI}{\ell} & \frac{4EI}{\ell} \end{bmatrix}}_{\mathbf{k}_R} \begin{bmatrix} \theta_a \\ \theta_b \end{bmatrix} \tag{4.20}$$

Define the *reduced force* and *displacement vectors*

$$\mathbf{p}_R = \begin{bmatrix} M_a \\ M_b \end{bmatrix} \quad \text{and} \quad \mathbf{v}_R = \begin{bmatrix} \theta_a \\ \theta_b \end{bmatrix} \tag{4.21}$$

Then Eq. (4.20) becomes $\mathbf{p}_R = \mathbf{k}_R\mathbf{v}_R$, where $\mathbf{k}_R$ is the *reduced stiffness matrix*. We wish to find $\mathbf{f}$ of $\mathbf{v}_R = \mathbf{f}\,\mathbf{p}_R$. Since $\mathbf{k}_R$ is nonsingular, we can obtain the flexibility matrix by inversion. Thus,

$$\mathbf{f} = \mathbf{k}_R^{-1} = \begin{bmatrix} k_{22} & k_{24} \\ k_{42} & k_{44} \end{bmatrix}^{-1} = \frac{\begin{bmatrix} k_{44} & -k_{24} \\ -k_{42} & k_{22} \end{bmatrix}}{k_{22}k_{44} - k_{24}k_{42}} = \frac{\ell}{EI}\begin{bmatrix} 1/3 & -1/6 \\ -1/6 & 1/3 \end{bmatrix} \tag{4.22}$$

The relationship between the stiffness and flexibility matrices can be illustrated further by the following two cases which are examples of constrained beam elements. The flexibility matrix depends upon which variables are chosen as independent variables.

Case 1

Suppose the shear force V_b and the moment M_b are chosen to be independent variables, and it is desired to eliminate or to suppress the displacements associated with rigid body motion. From Eq. (4.17b),

$$\Delta\mathbf{v} = \begin{bmatrix} \Delta w \\ \Delta\theta \end{bmatrix} = \mathbf{v}_b - \mathbf{U}_{vv}\mathbf{v}_a = [-\mathbf{U}_{vv}\ \mathbf{I}]\begin{bmatrix} \mathbf{v}_a \\ \mathbf{v}_b \end{bmatrix} = \mathbf{g}\mathbf{v} \tag{4.23a}$$

with

$$\mathbf{g} = \begin{bmatrix} -1 & \ell & 1 & 0 \\ 0 & -1 & 0 & 1 \end{bmatrix} \tag{4.23b}$$

Recall from the derivation of Section 4.2.2 that the rigid body displacements are $w_a - \ell\theta_a$ and θ_a. Physically, this can be considered to correspond to a beam with the right end free and the left end cantilevered (Fig. 4.4), where the cantilevered end can translate and rotate. Remove the rigid body displacement $(w_a - \ell\theta_a, \theta_a)$ from w_b and θ_b, thereby defining new displacement variables, say w_R and θ_R, which, according to Eq. (4.4), are identical to Δw and $\Delta\theta$, respectively.

$$\begin{aligned} w_R &= w_b - (w_a + \ell\theta_a) = \Delta w \\ \theta_R &= \theta_b - \theta_a = \Delta\theta \end{aligned} \tag{4.24}$$

Define

$$\mathbf{v}_R = \begin{bmatrix} w_R \\ \theta_R \end{bmatrix} \tag{4.25a}$$

Set $V_R = V_b$, $M_R = M_b$ or

$$\mathbf{p}_R = \mathbf{p}_b = \begin{bmatrix} V_b \\ M_b \end{bmatrix} \tag{4.25b}$$

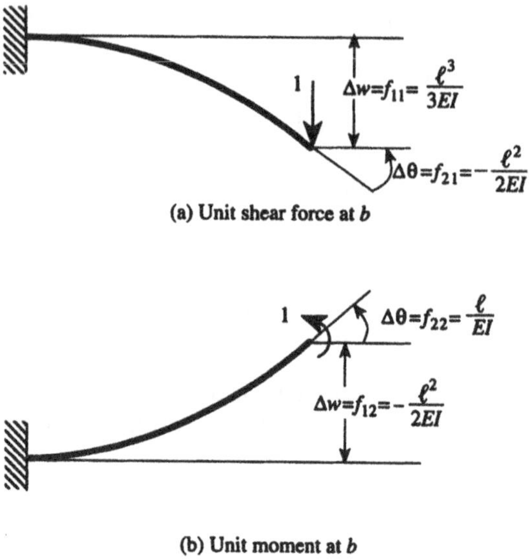

FIGURE 4.4
Calculation of flexibility coefficients.

Since V_b and M_b are to be the independent variables, the variables V_a and M_a (or $\mathbf{p}_a$) will be eliminated. From Eq. (4.17c),

$$\mathbf{p} = \begin{bmatrix} \mathbf{p}_a \\ \mathbf{p}_b \end{bmatrix} = \begin{bmatrix} \mathbf{U}_{pp}^{-1} \\ \mathbf{I} \end{bmatrix} \mathbf{p}_b = \begin{bmatrix} -1 & 0 \\ \ell & -1 \\ 1 & 0 \\ 0 & 1 \end{bmatrix} \mathbf{p}_b = \mathbf{g}^T \mathbf{p}_R \tag{4.26}$$

In summary,

$$\begin{aligned} \mathbf{v}_R &= \mathbf{g}\,\mathbf{v} \\ \mathbf{p} &= \mathbf{g}^T \mathbf{p}_R \end{aligned} \tag{4.27}$$

For a properly established **g**, these general relations are valid when other variables are selected as the independent variables.

Virtual Work

Since rigid body movement does not contribute to the internal virtual work, the virtual work expressed in terms of the complete force and displacement vectors, **p** and **v**, has the same value as when it is expressed in terms of the vectors $\mathbf{p}_R$ and $\mathbf{v}_R$ from which the rigid body displacements have been removed. Thus

$$\mathbf{p}_R^T\,\delta\mathbf{v}_R = \mathbf{p}_R^T\,\delta(\mathbf{g}\,\mathbf{v}) = \mathbf{p}_R^T\,\mathbf{g}\,\delta\mathbf{v} = \mathbf{p}^T\,\delta\mathbf{v} \tag{4.28}$$

Case 2

As a second possibility, suppose M_a and M_b are selected to be the independent variables. Physically, this corresponds to a beam segment with end moments (Fig. 4.5). Again, the rigid-body displacements are to be suppressed. For this case, set

$$\mathbf{p}_R = \begin{bmatrix} M_a \\ M_b \end{bmatrix} \tag{4.29}$$

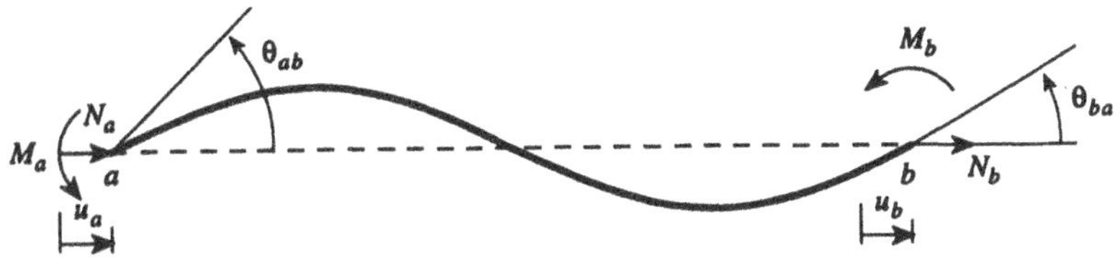

(a) Unrestrained beam segment with end moments

or

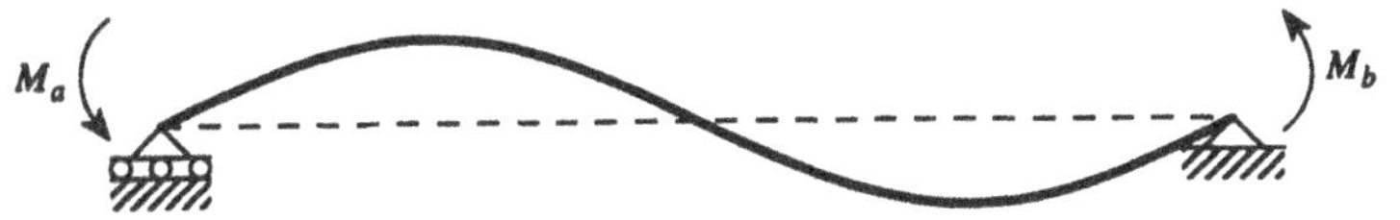

(b) The segment of (a) restrained by end supports

FIGURE 4.5
Element for an alternative (reduced) form of the stiffness matrix.

To find **g**, apply the conditions of equilibrium to the beam element of Fig. 4.1b, giving

$$\begin{aligned} \sum M_{x=b} &= 0 \rightarrow V_a = -(M_a + M_b)/\ell \\ \sum M_{x=a} &= 0 \rightarrow V_b = (M_a + M_b)/\ell \end{aligned} \tag{4.30}$$

Thus,

$$\begin{bmatrix} V_a \\ V_b \\ M_a \\ M_b \end{bmatrix} = \begin{bmatrix} -1/\ell & -1/\ell \\ 1/\ell & 1/\ell \\ 1 & 0 \\ 0 & 1 \end{bmatrix} \begin{bmatrix} M_a \\ M_b \end{bmatrix} \tag{4.31}$$
$$\mathbf{p} \quad = \quad \mathbf{g}^T \quad \mathbf{p}_R$$

where the definition of **p** has been altered from that used previously. The corresponding independent displacements $\mathbf{v}_R$ are (from $\mathbf{v}_R = \mathbf{g}\mathbf{v}$)

$$\begin{aligned} \mathbf{v}_R &= \begin{bmatrix} -1/\ell & 1/\ell & 1 & 0 \\ -1/\ell & 1/\ell & 0 & 1 \end{bmatrix} \begin{bmatrix} w_a \\ w_b \\ \theta_a \\ \theta_b \end{bmatrix} \\ &= \begin{bmatrix} (w_b - w_a)/\ell + \theta_a \\ (w_b - w_a)/\ell + \theta_b \end{bmatrix} = \begin{bmatrix} \theta_{ab} \\ \theta_{ba} \end{bmatrix} \end{aligned} \tag{4.32}$$

where it was necessary to rearrange the elements of **v** to correspond to those of **p**. It is observed that the newly defined displacement variables are tangent angles of the element.

Flexibility and Stiffness Matrices for a Cantilevered Beam Element

As mentioned, the first case of the section, in which the rigid body displacements of the left end were removed, corresponds to a beam element fixed on the left end. Since M_b and V_b are independent non-zero variables, the right end can be treated as being free. The displacements Δw and $\Delta\theta$ (w_R and θ_R) are to be related to the forces M_b and V_b. The stiffness coefficients are thus relating the variables at end b.

The flexibility coefficient, f_{ij}, is the displacement at DOF i due to a unit force applied at DOF j. The coefficients f_{ii} and $f_{ji}, i \neq j$ are referred to as direct and cross-flexibility coefficients, respectively. In the definition of a flexibility coefficient, force and displacement are used in a generalized sense. That is, for example, a unit load applied at j may cause a rotation at i, and a unit moment at j may cause a deflection at i. For the case at hand, where $\mathbf{v}_R = \mathbf{f}\,\mathbf{p}_R$, with $\mathbf{p}_R = [V_b \;\; M_b]^T$ and $\mathbf{v}_R = [w_R \;\; \theta_R]^T = [\Delta w \;\; \Delta\theta]^T$, or

$$\begin{bmatrix} \Delta w \\ \Delta\theta \end{bmatrix} = \frac{1}{EI}\begin{bmatrix} \ell^3/3 & -\ell^2/2 \\ -\ell^2/2 & \ell \end{bmatrix}\begin{bmatrix} V_b \\ M_b \end{bmatrix} \tag{4.33}$$

where the flexibility coefficients are given in Fig. 4.4. It should be clear from the f_{ij} shown in Fig. 4.4 that for this case both i and j of f_{ij} refer to the same location of the beam, i.e., the right end. The stiffness coefficients can be found using $\mathbf{k}_R = \mathbf{f}^{-1}$ or

$$\mathbf{k}_R = \left(\frac{1}{EI}\right)^{-1}\begin{bmatrix} \ell^3/3 & -\ell^2/2 \\ -\ell^2/2 & \ell \end{bmatrix}^{-1} = EI\begin{bmatrix} 12/\ell^3 & 6/\ell^2 \\ 6/\ell^2 & 4/\ell \end{bmatrix} \tag{4.34}$$

If the axial effects are included, the reduced stiffness matrix for a cantilevered beam element would be

$$\begin{bmatrix} N_b \\ V_b \\ M_b \end{bmatrix} = \begin{bmatrix} EA/\ell & 0 & 0 \\ 0 & 12EI/\ell^3 & 6EI/\ell^2 \\ 0 & 6EI/\ell^2 & 4EI/\ell \end{bmatrix}\begin{bmatrix} \Delta u \\ \Delta w \\ \Delta\theta \end{bmatrix} \tag{4.35}$$

where (Fig. 4.5a) $\Delta u = u_b - u_a$, the extension of the bar element.

Flexibility and Stiffness Matrices in Terms of End Moments

Case 2 of this section corresponds to the element of Fig. 4.5. For this configuration, the flexibility matrix and corresponding stiffness matrix are found to be

$$\mathbf{v}_R = \mathbf{f}\,\mathbf{p}_R = \begin{bmatrix} -\Delta u \\ \theta_{ab} \\ \theta_{ba} \end{bmatrix} = \begin{bmatrix} \frac{\ell}{EA} & 0 & 0 \\ 0 & \frac{\ell}{3EI} & -\frac{\ell}{6EI} \\ 0 & -\frac{\ell}{6EI} & \frac{\ell}{3EI} \end{bmatrix}\begin{bmatrix} N_a \\ M_a \\ M_b \end{bmatrix} \tag{4.36a}$$

$$\mathbf{p}_R = \mathbf{k}_R\mathbf{v}_R = \begin{bmatrix} N_a \\ M_a \\ M_b \end{bmatrix} = \begin{bmatrix} \frac{EA}{\ell} & 0 & 0 \\ 0 & \frac{4EI}{\ell} & \frac{2EI}{\ell} \\ 0 & \frac{2EI}{\ell} & \frac{4EI}{\ell} \end{bmatrix}\begin{bmatrix} -\Delta u \\ \theta_{ab} \\ \theta_{ba} \end{bmatrix} \tag{4.36b}$$

with $\Delta u = u_b - u_a$.

If the bar is subject to bending and torsion, the reduced stiffness matrix would appear in nondimensional form as

$$\begin{matrix} \begin{bmatrix} M_a \\ M_b \\ M_{tb} \end{bmatrix} & = & \frac{EI}{\ell}\begin{bmatrix} 4 & 2 & 0 \\ 2 & 4 & 0 \\ 0 & 0 & J^* \end{bmatrix} & \begin{bmatrix} \theta_{ab} \\ \theta_{ba} \\ \Delta\phi \end{bmatrix} \\ \mathbf{p}_R & = & \mathbf{k}_R & \mathbf{v}_R \end{matrix} \tag{4.37}$$

TABLE 4.1

Transfer Matrix, Stiffness Matrix, Reduced Stiffness Matrices, and Flexibility Matrices for Beams (Sign Convention 2)

Transfer Matrix

$$\begin{bmatrix} w_b \\ \theta_b \ell \\ V_b \\ M_b/\ell \end{bmatrix} = \underbrace{\begin{bmatrix} 1 & -1 & \frac{\ell^3}{6EI} & \frac{\ell^3}{2EI} \\ 0 & 1 & -\frac{\ell^3}{2EI} & -\frac{\ell^3}{EI} \\ 0 & & -1 & 0 \\ 0 & & -1 & -1 \end{bmatrix}}_{\mathbf{U}^i} \begin{bmatrix} w_a \\ \theta_a \ell \\ V_a \\ M_a/\ell \end{bmatrix}$$

Stiffness Martix

$$\begin{bmatrix} V_a \\ M_a/\ell \\ V_b \\ M_b/\ell \end{bmatrix} = \frac{EI}{\ell^3} \underbrace{\begin{bmatrix} 12 & -6 & -12 & -6 \\ -6 & 4 & 6 & 2 \\ -12 & 6 & 12 & 6 \\ -6 & 2 & 6 & 4 \end{bmatrix}}_{\mathbf{k}^i} \begin{bmatrix} w_a \\ \ell\theta_a \\ w_b \\ \ell\theta_b \end{bmatrix}$$

Flexibility Matrices | Reduced Stiffness Matrices

In Terms of End Moments and Tangents

$$\begin{bmatrix} \theta_{ab} \\ \theta_{ba} \end{bmatrix} = \underbrace{\frac{\ell}{6EI} \begin{bmatrix} 2 & -1 \\ -1 & 2 \end{bmatrix}}_{\mathbf{f}^i} \begin{bmatrix} M_a \\ M_b \end{bmatrix} \qquad \mathbf{g} = \begin{bmatrix} -1/\ell & 1/\ell & \vdots & 1 & 0 \\ -1/\ell & 1/\ell & \vdots & 0 & 1 \end{bmatrix} \qquad \begin{bmatrix} M_a \\ M_b \end{bmatrix} = \underbrace{\frac{2EI}{\ell} \begin{bmatrix} 2 & 1 \\ 1 & 2 \end{bmatrix}}_{\mathbf{k}^i_R} \begin{bmatrix} \theta_{ab} \\ \theta_{ba} \end{bmatrix}$$

In Terms of Variables at One End

$$\begin{bmatrix} \Delta w \\ \Delta\theta\,\ell \end{bmatrix} = \underbrace{\frac{\ell^3}{6EI} \begin{bmatrix} 2 & -3 \\ -3 & 6 \end{bmatrix}}_{\mathbf{f}^i} \begin{bmatrix} V_b \\ M_b/\ell \end{bmatrix} \qquad \mathbf{g} = \begin{bmatrix} -1 & \ell & \vdots & 1 & 0 \\ 0 & -1 & \vdots & 0 & 1 \end{bmatrix} \qquad \begin{bmatrix} V_b \\ M_b/\ell \end{bmatrix} = \underbrace{\frac{2EI}{\ell^3} \begin{bmatrix} 6 & 3 \\ 3 & 2 \end{bmatrix}}_{\mathbf{k}^i_R} \begin{bmatrix} \Delta w \\ \Delta\theta\,\ell \end{bmatrix}$$

with

$$J^* = \frac{GJ}{EI} = \frac{J}{2(1+\nu)I} \qquad \Delta\phi = \phi_b - \phi_a$$

$$\mathbf{g} = \begin{bmatrix} -1 & 1 & 0 & \vdots & 1 & 0 & 0 \\ -1 & 1 & 0 & \vdots & 0 & 1 & 0 \\ 0 & 0 & -1 & \vdots & 0 & 0 & 1 \end{bmatrix}$$

A summary of some of the matrices discussed in this chapter is provided in Table 4.1.

Flexibility Matrices with Applied Loads

The flexibility matrices discussed here can be generalized to include the effects of applied loading:

Case 1 (Fig. 4.4):

$$\begin{bmatrix} \Delta w \\ \Delta\theta \end{bmatrix} = \frac{1}{EI} \begin{bmatrix} \ell^3/3 & -\ell^2/2 \\ -\ell^2/2 & \ell \end{bmatrix} \begin{bmatrix} V_b \\ M_b \end{bmatrix} + \begin{bmatrix} \Delta w^0 \\ \Delta\theta^0 \end{bmatrix} \tag{4.38a}$$

Case 2 (Fig. 4.5):

$$\begin{bmatrix} \theta_{ab} \\ \theta_{ba} \end{bmatrix} = \frac{1}{EI} \begin{bmatrix} \ell/3 & -\ell/6 \\ -\ell/6 & \ell/3 \end{bmatrix} \begin{bmatrix} M_a \\ M_b \end{bmatrix} + \begin{bmatrix} \theta^0_{ab} \\ \theta^0_{ba} \end{bmatrix} \tag{4.38b}$$

where the superscript 0 indicates the terms due to prescribed loadings.

Note that the flexibility matrix was introduced for a problem of a beam with a fixed end. It should be evident by now that a *flexibility matrix can be defined only for restrained systems.* Otherwise, rigid body motion will result from the applied load, and the response will be unbounded. In general, the element flexibility matrix can be obtained by first establishing a statically determinate system of supports, extracting from the complete element stiffness matrix the rows and columns corresponding to the supports, and, finally, inverting the remaining matrix.

Relationships Between Reduced and Complete Stiffness Matrices

It is readily shown that by using **g**, the complete stiffness matrix **k** can be formed from the reduced stiffness matrix $\mathbf{k}_R$, which is defined by

$$\mathbf{p}_R = \mathbf{k}_R \mathbf{v}_R$$

Thus

$$\begin{aligned} \mathbf{p}_R &= \mathbf{k}_R \mathbf{v}_R = \mathbf{k}_R\, \mathbf{g}\, \mathbf{v} \\ \mathbf{p} &= \mathbf{g}^T \mathbf{p}_R = \mathbf{g}^T \mathbf{k}_R\, \mathbf{g}\, \mathbf{v} = \mathbf{k}\, \mathbf{v} \end{aligned}$$

or

$$\mathbf{k} = \mathbf{g}^T \mathbf{k}_R\, \mathbf{g} \tag{4.39}$$

4.4 Stiffness Matrices

4.4.1 Determination of Stiffness Matrices

The conversion of the transfer matrix for a beam element into the corresponding stiffness matrix was described in Section 4.3.2, while the unit load method and Castigliano's second theorem for obtaining the same stiffness matrix were presented in Chapter 3. Of particular interest here is the use of trial functions to derive a beam stiffness matrix, since this approach can be employed as well with other structural elements.

Before considering the trial function approach in the following subsection, we will outline the direct evaluation of a stiffness matrix using the differential equations for a beam. This is accomplished by applying a unit displacement (deflection or rotation) at each of the two degrees of freedom on each of the two ends of a beam element.

To find the first column of the stiffness matrix, that is, to compute k_{i1}, $i = 1, 2, 3, 4$, the configuration of Fig. 4.6 can be used. For these prescribed displacements, i.e., $w_a = 1$, $\theta_a = 0$, $w_b = 0$, and $\theta_b = 0$, which according to the definition of a stiffness matrix correspond to the values of the degrees of freedom required to find the forces k_{i1}, the stiffness coefficients

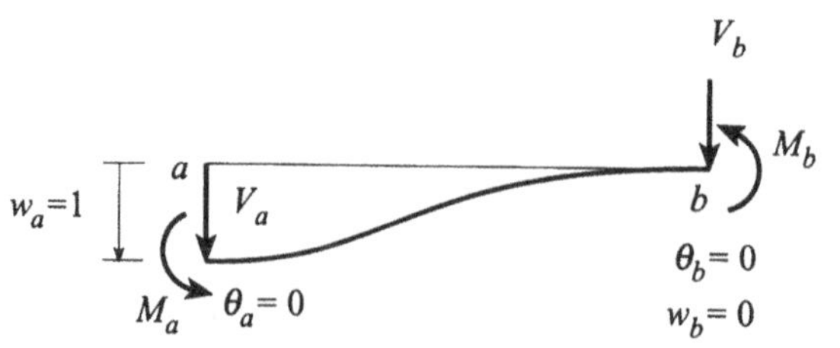

FIGURE 4.6
Beam element configuration for computing the first column of stiffness coefficients. Sign Convention 2.

k_{i1} become

$$k_{11} = V_a, \quad k_{21} = M_a, \quad k_{31} = V_b, \quad k_{41} = M_b$$

The beam of Fig. 4.6 can be used to find V_a, M_a, V_b, and M_b, based on the specified conditions $w_a = 1, \theta_a = 0, w_b = 0$, and $\theta_b = 0$. Integrate $d^2w/dx^2 = -M/EI$ with $M = -M_a - V_a x$ (Sign Convention 2), giving

$$\frac{dw}{dx} = \frac{1}{EI}\left(M_a x + V_a \frac{x^2}{2}\right) + C_1$$
$$w = \frac{1}{EI}\left(M_a \frac{x^2}{2} + V_a \frac{x^3}{6}\right) + C_1 x + C_2$$

Apply the boundary conditions $w_a = 1, \theta_a = 0, w_b = 0$, and $\theta_b = 0$ to fix the arbitrary constants C_1 and C_2 and then solve for M_a and V_a. From $\theta_a = 0$ and $w_b = 0$, $C_1 = 0$ and $C_2 = -M_a\ell^2/2EI - V_a\ell^3/6EI$. Apply $\theta_b = 0$ to the first equation and $w_a = 1$ to the second giving

$$V_a = 12EI/\ell^3 = k_{11}, \quad M_a = -6EI/\ell^2 = k_{21}$$

The conditions of equilibrium applied to the beam of Fig. 4.6 yield M_b and V_b. We find

$$k_{31} = V_b = -V_a = -12EI/\ell^3, \qquad k_{41} = M_b = -V_a\ell - M_a = -6EI/\ell^2$$

It is evident from this derivation of one column of stiffness coefficients that $k_{i1}, i = 1, 2, 3, 4$ are a set of equilibrated forces on the element. In a sense, V_a and M_a are the force and the moment required to generate the unit displacement $w_a = 1$ with slope $\theta_a = 0$, whereas V_b and M_b are the reactive forces for this configuration. Each column has a similar interpretation, with the coefficients satisfying equilibrium. In the case of the first column, note that $\Sigma F_z = 0(V_a + V_b = k_{11} + k_{31} = (12 - 12)EA/\ell^3 = 0)$ and the moments about any point must be zero (e.g., $\Sigma M|_{x=b} = 0, M_a + V_a\ell + M_b = k_{21} + k_{11}\ell + k_{41} = (-6 + 12 - 6)EI/\ell^2 = 0$).

The beam configurations for computing the second, third, and fourth columns of the stiffness matrix are shown in Figs. 4.7a, b, and c, respectively.

4.4.2 Stiffness Matrices Based on Polynomial Trial Functions

For each of the above methods, and those of Chapter 3, for finding the stiffness matrix for a beam segment, the exact solution of the engineering theory of beams has been arranged in a stiffness matrix format. For other structural elements, it is not always possible to find an exact solution to place in stiffness matrix form. In such cases, a method involving an assumed or trial series solution leading to an approximate solution can be employed. Typically, this approach is used to evaluate stiffness matrices within the finite element method.

Even though the exact stiffness matrix is readily derived for a beam, it is useful to employ the beam element to illustrate the general procedure for using trial-functions. As will be seen, for an Euler-Bernoulli beam element, a judiciously chosen series will result in the exact rather than an approximate stiffness matrix. We continue to employ Sign Convention 2.

Derivation of Interpolation Functions

For a beam element extending from $x = a$ to $x = b$, assume the deflection can be approximated by a polynomial

$$w = C_1 + C_2 x + C_3 x^2 + C_4 x^3 + \cdots = \widehat{w}_1 + \widehat{w}_2 x + \widehat{w}_3 x^2 + \widehat{w}_4 x^3 + \cdots \tag{4.40}$$

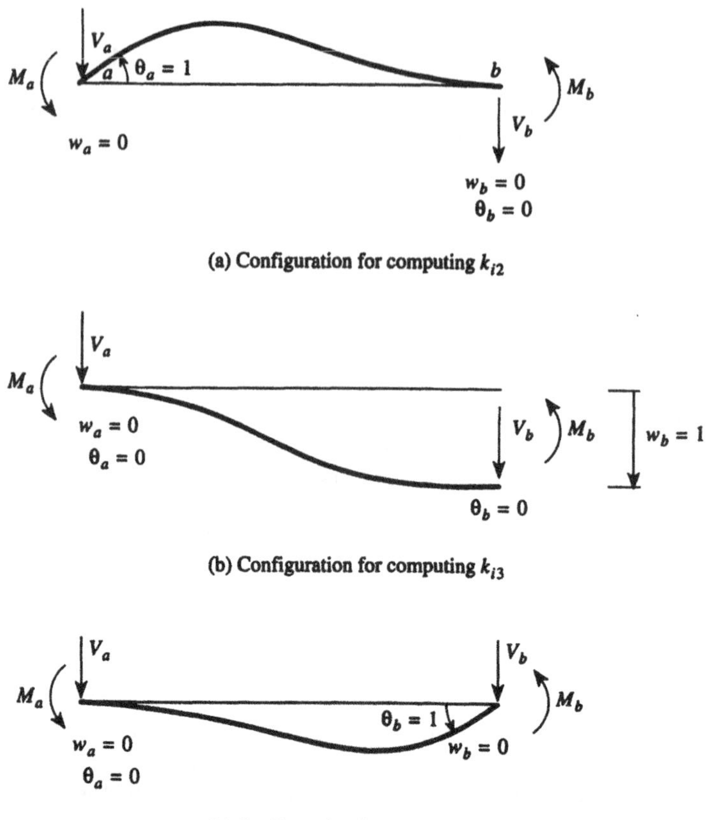

(c) Configuration for computing k_{i4}

FIGURE 4.7
Beam elements for computing the second, third, and fourth columns of stiffness coefficients.

where $C_i = \widehat{w}_i$, $i = 1, 2, \ldots$ are the unknown constants of the assumed series that are chosen so as to obtain a good approximation. Such trial functions are sometimes referred to as *basis functions*. Often the variable representation by a trial series will be supplemented with a superscript tilde, e.g., $\tilde{w}$, to indicate that it is being approximated. For this beam element, we choose to retain only the first four terms. Write the polynomial in the form

$$w = [1 \quad x \quad x^2 \quad x^3] \begin{bmatrix} \widehat{w}_1 \\ \widehat{w}_2 \\ \widehat{w}_3 \\ \widehat{w}_4 \end{bmatrix} = \mathbf{N}_u \widehat{\mathbf{w}} = \widehat{\mathbf{w}}^T \mathbf{N}_u^T \tag{4.41}$$

where

$$\mathbf{N}_u = [1 \quad x \quad x^2 \quad x^3]$$

and

$$\widehat{\mathbf{w}} = [\widehat{w}_1 \quad \widehat{w}_2 \quad \widehat{w}_3 \quad \widehat{w}_4]^T$$

It is convenient to rewrite the assumed series in terms of physically meaningful parameters (unknown displacements) at the ends of the beam element, rather than in terms of the unknown constants $\widehat{w}_i$ as in Eq. (4.41). Thus, the vector of unknowns $\widehat{\mathbf{w}}$ can be transformed into the unknown nodal displacement vector

$$\mathbf{v} = \begin{bmatrix} w_a \\ \theta_a = -w'_a \\ w_b \\ \theta_b = -w'_b \end{bmatrix} \tag{4.42}$$

The assumed series is then referred to as an *interpolation* or *shape function*. The derivative of w, which is needed to form Eq. (4.42), is given by

$$w' = [0 \quad 1 \quad 2x \quad 3x^2] \begin{bmatrix} \widehat{w}_1 \\ \widehat{w}_2 \\ \widehat{w}_3 \\ \widehat{w}_4 \end{bmatrix} = \mathbf{N}'_u \widehat{\mathbf{w}} = \widehat{\mathbf{w}}^T (\mathbf{N}'_u)^T \tag{4.43}$$

Now, evaluate w and $\theta = -w'$ at $x = a$ and $x = b$, giving

$$\begin{bmatrix} w_a \\ \theta_a \\ w_b \\ \theta_b \end{bmatrix} = \begin{bmatrix} w(0) \\ -w'(0) \\ w(\ell) \\ -w'(\ell) \end{bmatrix} = \begin{bmatrix} 1 & 0 & 0 & 0 \\ 0 & -1 & 0 & 0 \\ 1 & \ell & \ell^2 & \ell^3 \\ 0 & -1 & -2\ell & -3\ell^2 \end{bmatrix} \begin{bmatrix} \widehat{w}_1 \\ \widehat{w}_2 \\ \widehat{w}_3 \\ \widehat{w}_4 \end{bmatrix} \tag{4.44}$$

$$\mathbf{v} \quad = \quad \widehat{\mathbf{N}}_u \quad \widehat{\mathbf{w}}$$

The constants $\widehat{\mathbf{w}}$ are found in terms of the (unknown) displacements at a and b by using the inverse of $\widehat{\mathbf{N}}_u$,

$$\widehat{\mathbf{w}} = \widehat{\mathbf{N}}_u^{-1}\mathbf{v} = \mathbf{G}\mathbf{v} \tag{4.45}$$

where

$$\mathbf{G} = \widehat{\mathbf{N}}_u^{-1} = \begin{bmatrix} 1 & 0 & 0 & 0 \\ 0 & -1 & 0 & 0 \\ -3/\ell^2 & 2/\ell & 3/\ell^2 & 1/\ell \\ 2/\ell^3 & -1/\ell^2 & -2/\ell^3 & -1/\ell^2 \end{bmatrix}$$

The relationship between w and $\mathbf{v}$, i.e., between the deflection w and the values of displacements w and θ at the ends of the element, is

$$w = \mathbf{N}_u \widehat{\mathbf{w}} = \mathbf{N}_u \mathbf{G}\mathbf{v} = \mathbf{N}\mathbf{v} \tag{4.46a}$$

or

$$w(x) = \left(1 - 3\frac{x^2}{\ell^2} + 2\frac{x^3}{\ell^3}\right) w_a + \left(-x + 2\frac{x^2}{\ell} - \frac{x^3}{\ell^2}\right)\theta_a + \left(3\frac{x^2}{\ell^2} - 2\frac{x^3}{\ell^3}\right) w_b + \left(\frac{x^2}{\ell} - \frac{x^3}{\ell^2}\right)\theta_b \tag{4.46b}$$

where $\mathbf{N} = \mathbf{N}_u\mathbf{G}$. This expression is the desired form of the assumed series, where the matrix $\mathbf{N}$, which is often called a *shape function matrix*, characterizes the "interpolation" or "shape" between the nodes. The components of this expression are often referred to as *shape, basis,* or *interpolation functions*.

Interpolation Functions Based on a Normalized Coordinate

Some mathematical handbooks tabulate various interpolation functions. The Hermitian[2] interpolation polynomials of Fig. 4.8 can be employed when derivatives of the displacements at the nodes are involved, as with beams. The Hermitian polynomials can be derived

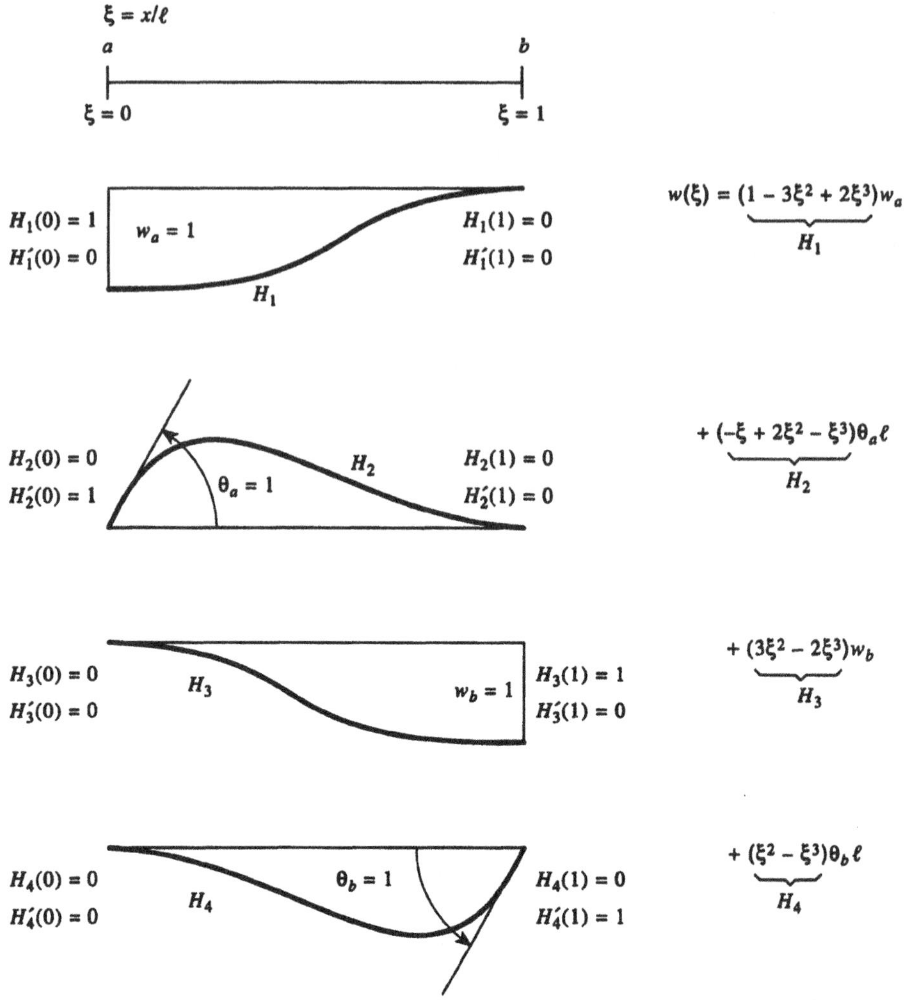

FIGURE 4.8
Third-degree Hermitian interpolation polynomials.

from Eq. (4.46) using the normalized coordinate $\xi = x/\ell$. From Eq. (4.46)

$$w(\xi) = [1 \quad \xi \quad \xi^2 \quad \xi^3]\begin{bmatrix} 1 & 0 & 0 & 0 \\ 0 & -1 & 0 & 0 \\ -3 & 2 & 3 & 1 \\ 2 & -1 & -2 & -1 \end{bmatrix}\begin{bmatrix} w_a \\ \ell\theta_a \\ w_b \\ \ell\theta_b \end{bmatrix}$$

$$= \underbrace{[1 \quad \xi \quad \xi^2 \quad \xi^3]}_{\mathbf{N}_u(\xi)}\underbrace{\begin{bmatrix} 1 & 0 & 0 & 0 \\ 0 & -\ell & 0 & 0 \\ -3 & 2\ell & 3 & \ell \\ 2 & -\ell & -2 & -\ell \end{bmatrix}}_{\mathbf{G}}\underbrace{\begin{bmatrix} w_a \\ \theta_a \\ w_b \\ \theta_b \end{bmatrix}}_{\mathbf{v}^i}$$

$$= \mathbf{N}_u(\xi)\mathbf{G}\,\mathbf{v}^i = \mathbf{N}\,\mathbf{v}^i = \mathbf{N}(\xi)\mathbf{v}^i \tag{4.47a}$$

or

$$\begin{aligned} w(\xi) &= (1-3\xi^2+2\xi^3)w_a + (-\xi+2\xi^2-\xi^3)\theta_a\ell + (3\xi^2-2\xi^3)w_b + (\xi^2-\xi^3)\theta_b\ell \\ &= H_1(\xi)w_a + H_2(\xi)\,\theta_a\ell + H_3(\xi)\,w_b + H_4(\xi)\,\theta_b\ell \end{aligned} \tag{4.47b}$$

The terms in brackets are Hermitian[2] polynomials. Note that due to scaling, **G** of Eq. (4.45) differs from **G** of Eq. (4.47).

The goal in this subsection is to derive the element stiffness matrix using an assumed series. Although it will be illustrated below that for beams this can be accomplished directly, the more useful and more important technique is to employ the principle of virtual work.

Direct Evaluation of the Stiffness Matrix

Thus far, the deflection has been expressed in terms of the end displacements using $w = \mathbf{N}_u\,\mathbf{G}\mathbf{v}$. Since we seek a stiffness relationship of the form $\mathbf{p} = \mathbf{k}\mathbf{v}$, we still need to relate the end forces $\mathbf{p}$ to the deflection w. A force-displacement relationship is provided by the material law [Chapter 1, Eq. (1.106)] $M = EI\kappa = -EIw''$, which relates the internal moments to the second derivative of the deflection. Use $w = \mathbf{N}_u\,\mathbf{G}\mathbf{v}$, so that $M = -EI\mathbf{N}_u''\,\mathbf{G}\mathbf{v}$. With $\mathbf{N}_u = [1\ \ \xi\ \ \xi^2\ \ \xi^3]$ and $\mathbf{N}_u' = (d\mathbf{N}_u/d\xi)(d\xi/dx)$, $\mathbf{N}_u''$ becomes

$$\mathbf{N}_u'' = [0\quad 0\quad 2\quad 6\xi]/\ell^2 \tag{4.48}$$

Define

$$\mathbf{N}_u'' = \mathbf{B}_u \tag{4.49}$$

Then

$$M = -EI\mathbf{B}_u\,\mathbf{G}\mathbf{v} \tag{4.50}$$

It is a simple matter to use the conditions of equilibrium to find the shear force V in terms of the displacements and thereby to complete the derivation of the stiffness matrix. The moment M of Eq. (4.50) is an internal moment which adheres to Sign Convention 1 as shown in Fig. 4.9. Let a superscript 1 indicate Sign Convention 1. Then the internal bending moments at a and b are $M_{x=a} = M_a^1$ and $M_{x=b} = M_b^1$, respectively. It can be observed in Fig. 4.9 that the moments defined according to the two sign conventions are related by $M_a = -M_a^1$ and $M_b = M_b^1$. Furthermore, from the element equilibrium condition of the summation of moments being zero first about a and then about b, it is found that $V_b = (M_a + M_b)/\ell = (-M_a^1 + M_b^1)/\ell$ and $V_a = (-M_a - M_b)/\ell = (M_a^1 - M_b^1)/\ell$. Thus, the set of equations relating the end moments and shear forces in Sign Convention 2 to the internal end moments is

$$\begin{bmatrix} V_a \\ M_a \\ V_b \\ M_b \end{bmatrix} = \begin{bmatrix} \mathbf{p}_a \\ \\ \mathbf{p}_b \end{bmatrix} = \frac{1}{\ell}\begin{bmatrix} 1 & -1 \\ -\ell & 0 \\ -1 & 1 \\ 0 & \ell \end{bmatrix} \begin{bmatrix} M_a^1 \\ M_b^1 \end{bmatrix} \tag{4.51}$$

Finally, evaluate M_a^1 and M_b^1 in Eq. (4.51) by setting $M_a^1 = M_{\xi=0}$ and $M_b^1 = M_{\xi=1}$ in Eq. (4.50). Substitution of these values for M_a^1 and M_b^1 into the expression of Eq. (4.51) gives the desired stiffness relation $\mathbf{p} = \mathbf{k}\mathbf{v}$.

---

[2]Charles Hermite (1822–1901) was a great French algebraist, probably the leading French mathematician of the second half of the 19th century. From 1869, he was a professor of mathematics at the Sorbonne University. His work and that of his students exercised a profound influence on contemporary mathematics. Henri Poincaré, his student, said, "Talk with Hermite: he never evokes a concrete image; yet you soon perceive that the most abstract entities are for him like living creatures."

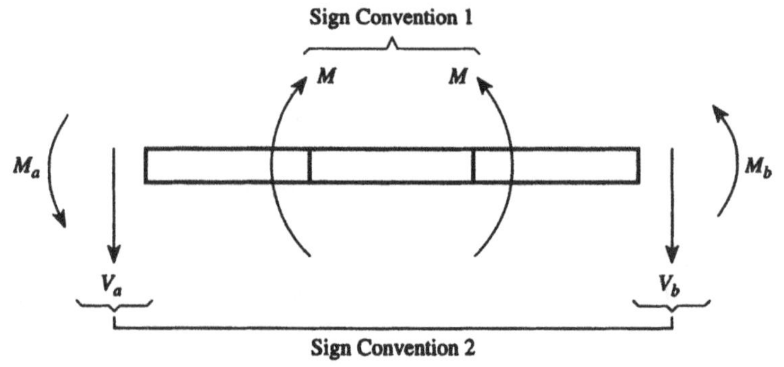

FIGURE 4.9
A beam element with positive internal bending moment M (Sign Convention 1) and nodal moments and shear forces (Sign Convention 2).

Evaluation of the Stiffness Matrix Using the Principle of Virtual Work

The purpose of this section on the use of a trial series is to show how the principle of virtual work can be employed to evaluate an element stiffness matrix. This is a very general procedure which can be used to derive stiffness matrices for any element. Recall from Chapter 2 that for kinematically admissible virtual displacements, the principle of virtual work relation $-\delta W_i - \delta W_e = 0$ assures that the system is in equilibrium.

For a beam element with no consideration of either axial or shear deformation effects, the element contribution to the principle can be written in terms of the operator $\mathbf{k}^D$ as (Chapter 2, Example 2.7)

$$\int_a^b \delta w\, \mathbf{k}^D w\, dx - \int_a^b \delta w\, \overline{p}_z\, dx \tag{4.52}$$

with $\mathbf{k}^D = k^D = {}_x d^2\; EI\; d_x^2$ and $\overline{p}_z$ is the applied loading intensity along the beam. The bar over p_z indicates that this is a prescribed or applied quantity. It is possible to include the term $-[\overline{M}\,\delta\theta + \overline{V}\,\delta w]_a^b$ in Eq. (4.52), where $\overline{M}$ and $\overline{V}$ are concentrated loads on the ends a, b of the element. However, it is assumed here that applied loads on the boundary are included in the term $\int_a^b \delta w\ \overline{p}_z\, dx$. Also, a derivative with a subscript on the left side, e.g., ${}_x d$, means that the derivative is taken on the preceeding variable. Thus, $\delta w\ {}_x d$ is the same as $\frac{d}{dx}\delta w$. The trial series satisfies the displacement boundary conditions. Then Eq. (4.52) is the element contribution to the satisfaction of the conditions of equilibrium for the whole system. Since the assumed displacement series is approximate, an "approximate" fulfillment of equilibrium should be expected.

The variational quantities δw and $\delta w''$ expressed in terms of the trial series are needed to proceed with Eq. (4.52). Note that in Eq. (4.46a), $\mathbf{v}$ contains unknown discrete values of displacements, $\mathbf{G}$ contains constant factors for the polynomials, and only $\mathbf{N}_u$ is a function of the axial coordinate x. Then

$$\delta w = \delta(\mathbf{N}_u\, \mathbf{G}\, \mathbf{v}) = \mathbf{N}_u\, \mathbf{G}\, \delta\mathbf{v} = \delta\mathbf{v}^T\, \mathbf{G}^T\, \mathbf{N}_u^T = \delta\mathbf{v}^T\, \mathbf{N}^T \tag{4.53}$$

where $\mathbf{N} = \mathbf{N}_u\ \mathbf{G}$ and $\delta\mathbf{v}^T$ contains the virtual end displacements. Also, set $\mathbf{N}_u'' = \mathbf{B}_u = [0\ \ 0\ \ 2\ \ 6x]$ and form

$$\begin{aligned} w'' &= \mathbf{N}_u''\, \mathbf{G}\, \mathbf{v} = \mathbf{B}_u\, \mathbf{G}\, \mathbf{v} = \mathbf{N}''\mathbf{v} = \mathbf{B}\, \mathbf{v} \\ \delta w'' &= \mathbf{B}\, \delta\mathbf{v} = \delta\mathbf{v}^T\, \mathbf{B}^T \end{aligned} \tag{4.54}$$

Introduction of these expressions into Eq. (4.52) gives

$$\int_a^b \overbrace{\delta\mathbf{v}^T\mathbf{B}^T}^{\delta w''}\, EI\, \overbrace{\mathbf{B}\mathbf{v}}^{w''}\, dx - \int_a^b \overbrace{\delta\mathbf{v}^T\mathbf{N}^T}^{\delta w}\, \overline{p}_z\, dx$$

$$= \delta\mathbf{v}^T \left[\overbrace{\underbrace{\int_a^b \mathbf{B}^T(x)\, EI\, \mathbf{B}(x)\, dx}_{\mathbf{k}}\ \underbrace{\mathbf{v}}_{\mathbf{v}}}^{\mathbf{p}} - \underbrace{\int_a^b \mathbf{N}^T(x)\, \overline{p}(x)\, dx}_{\overline{\mathbf{p}}^0} \right] \tag{4.55}$$

Since $\mathbf{B}(x) = \mathbf{N}''(x) = \mathbf{B}_u(x)\,\mathbf{G},\ \mathbf{N}(x) = \mathbf{N}_u(x)\,\mathbf{G},\ \mathbf{B}^T = \mathbf{G}^T\mathbf{B}_u^T,\ \mathbf{N}^T = \mathbf{G}^T\mathbf{N}_u^T$, Eq. (4.56) can be expressed as

$$\delta\mathbf{v}^T \left[\overbrace{\underbrace{\mathbf{G}^T \int_a^b \mathbf{B}_u^T(x)\, EI\, \mathbf{B}_u(x)\, dx\, \mathbf{G}}_{\mathbf{k}}\ \underbrace{\mathbf{v}}_{\mathbf{v}}}^{\mathbf{p}} - \underbrace{\mathbf{G}^T \int_a^b \mathbf{N}_u^T(x)\overline{p}_z(x)\, dx}_{\overline{\mathbf{p}}^0} \right] \tag{4.56}$$

As indicated

$$\mathbf{k} = \int_a^b \mathbf{B}^T(x)\, EI\, \mathbf{B}(x)\, dx \tag{4.57}$$

$$\overline{\mathbf{p}}^0 = \mathbf{G}^T \int_a^b \mathbf{N}_u^T\, \overline{p}_z\, dx \tag{4.58}$$

In summary, Eq. (4.52) can be expressed as

$$\delta\mathbf{v}^T(\mathbf{p} - \overline{\mathbf{p}}^0) \tag{4.59}$$

where

$$\delta\mathbf{v}^T \left\{ \overbrace{\underbrace{\int_a^b \mathbf{B}^T\, EI\, \mathbf{B}\, dx}_{\mathbf{k}}\ \mathbf{v}}^{\mathbf{p}} - \overline{\mathbf{p}}^0 \right\} \quad \mathbf{v} - \overline{\mathbf{p}}^0 \tag{4.60}$$

This reflects that the principle of virtual work expresses the conditions of equilibrium ($\mathbf{p} = \overline{\mathbf{p}}^0$) between the forces $\mathbf{p}$ representing the element properties and the load vector $\overline{\mathbf{p}}^0$ containing the influence of the applied element loading at the ends of the element. Since, by definition of a stiffness matrix, $\mathbf{p} = \mathbf{k}\mathbf{v}$, this virtual work relationship becomes

$$\delta\mathbf{v}^T(\mathbf{p} - \overline{\mathbf{p}}^0) = \delta\mathbf{v}^T(\mathbf{k}\mathbf{v} - \overline{\mathbf{p}}^0) \tag{4.61}$$

Thus, we observe that the stiffness matrix $\mathbf{k}$ can be obtained by introducing the trial function into the operator $\mathbf{k}^D$.

If this beam element is part of a structural system, then Eq. (4.61) gives the contribution of the element to the equilibrium of the whole system, expressed as the virtual work of the element which forms a portion of the virtual work of the whole system. In this derivation, the element indices have been neglected. The element contribution of Eq. (4.61) more properly should appear as

$$\delta \mathbf{v}^{iT}(\mathbf{p}^i - \overline{\mathbf{p}}^{i0}) = \delta \mathbf{v}^{iT}(\mathbf{k}^i \mathbf{v}^i - \overline{\mathbf{p}}^{i0}) \tag{4.62}$$

We wish to evaluate the stiffness matrix $\mathbf{k}^i$ of Eq. (4.57). Begin with $\mathbf{N}_u = [1 \;\; x \;\; x^2 \;\; x^3]$ of Eq. (4.41). Then

$$\mathbf{B}_u(x) = \mathbf{N}_u''(x) = \frac{d^2\mathbf{N}_u(x)}{dx^2} = [0 \;\; 0 \;\; 2 \;\; 6x] \tag{4.63}$$

For constant EI, the integration in the expression for $\mathbf{k}$ of Eq. (4.56) leads to

$$\int_0^\ell \mathbf{B}_u^T \, EI \, \mathbf{B}_u \, dx = EI \int_0^\ell \begin{bmatrix} 0 & 0 & 0 & 0 \\ 0 & 0 & 0 & 0 \\ 0 & 0 & 4 & 12x \\ 0 & 0 & 12x & 36x^2 \end{bmatrix} dx \tag{4.64}$$

$$= EI \begin{bmatrix} 0 & 0 & 0 & 0 \\ 0 & 0 & 0 & 0 \\ 0 & 0 & 4\ell & 6\ell^2 \\ 0 & 0 & 6\ell^2 & 12\ell^3 \end{bmatrix}$$

Finally, use $\mathbf{G}$ of Eq. (4.45) in Eq. (4.57) to obtain

$$\mathbf{k}^i = \frac{EI}{\ell^3} \begin{bmatrix} 12 & -6\ell & -12 & -6\ell \\ -6\ell & 4\ell^2 & 6\ell & 2\ell^2 \\ -12 & 6\ell & 12 & 6\ell \\ -6\ell & 2\ell^2 & 6\ell & 4\ell^2 \end{bmatrix} \tag{4.65}$$

Stiffness Matrix Based on Normalized Coordinate

To calculate the stiffness matrix for an Euler-Bernoulli beam element using the normalized coordinate $\xi = x/\ell$, begin by defining the assumed displacement (Eq. 4.47a)

$$w(\xi) = \mathbf{N}_u(\xi)\mathbf{G}\mathbf{v}^e \tag{4.66}$$

with $\mathbf{G}$ given in Eq. (4.47a), and

$$\mathbf{N}_u = [1 \;\; \xi \;\; \xi^2 \;\; \xi^3] \tag{4.67a}$$

$$\mathbf{v}^i = [w_a \;\; \theta_a \;\; w_b \;\; \theta_b]^{iT} \tag{4.67b}$$

The corresponding force vector is

$$\mathbf{p}^i = [V_a \;\; M_a \;\; V_b \;\; M_b]^{iT} \tag{4.67c}$$

We have chosen to use the same $\mathbf{v}^i$ and $\mathbf{p}^i$ that were employed in evaluating $\mathbf{k}^i$ of Eq. (4.57). The stiffness matrix $\mathbf{k}^i$ as a function of x is calculated using

$$\mathbf{k}^i = \int_a^b \mathbf{B}^T(x) \, EI \, \mathbf{B}(x) \, dx = \int_0^\ell \mathbf{B}^T(x) \, EI \, \mathbf{B}(x) \, dx$$

$$= \mathbf{G}^T \int_0^\ell \mathbf{B}_u^T(x) \, EI \, \mathbf{B}_u(x) \, dx \, \mathbf{G} \tag{4.68}$$

with $\mathbf{G}$ from Eq. (4.45). In terms of the normalized coordinate, $dx = \ell \, d\xi$ and

$$\mathbf{B}_u(\xi) = \mathbf{N}_u''(\xi) = \frac{d^2}{dx^2}\mathbf{N}_u(\xi) = \frac{d^2\mathbf{N}_u(\xi)}{\ell^2 d\xi^2} = \frac{1}{\ell^2}[0 \;\; 0 \;\; 2 \;\; 6\xi] \tag{4.69}$$

Then

$$\mathbf{k}^i = \mathbf{G}^T \int_0^\ell \mathbf{B}_u^T(x)\, EI\, \mathbf{B}_u(x)\, dx\, \mathbf{G} = \mathbf{G}^T \int_0^1 \mathbf{B}_u^T(\xi)\, EI\, \mathbf{B}_u(\xi)\ell\, d\xi\, \mathbf{G}$$

$$\begin{matrix} \mathbf{G} \text{ from Eq. (4.45)} & \mathbf{G} \text{ from Eq. (4.47a)} \\ \mathbf{B}_u(x) \text{ from Eq. (4.54)} & \mathbf{B}_u(\xi) \text{ from Eq. (4.69)} \end{matrix} \tag{4.70}$$

$$= \mathbf{G}^T EI \int_0^1 \frac{1}{\ell^4} \begin{bmatrix} 0 & 0 & 0 & 0 \\ 0 & 0 & 0 & 0 \\ 0 & 0 & 4 & 12\xi \\ 0 & 0 & 12\xi & 36\xi^2 \end{bmatrix} \ell\, d\xi\, \mathbf{G} = \mathbf{G}^T \begin{bmatrix} 0 & 0 & 0 & 0 \\ 0 & 0 & 0 & 0 \\ 0 & 0 & 4 & 6 \\ 0 & 0 & 6 & 12 \end{bmatrix} \frac{EI}{\ell^3} \mathbf{G}$$

Pre-and post multiplication by $\mathbf{G}$ of Eq. (4.47a) leads to the stiffness matrix

$$\mathbf{k}^i = \mathbf{G}^T \int_0^1 \mathbf{B}_u^T(\xi)\, EI\, \mathbf{B}_u(\xi)\ell\, d\xi\, \mathbf{G} = \begin{bmatrix} 12 & -6\ell & -12 & -6\ell \\ -6\ell & 4\ell^2 & 6\ell & 2\ell^2 \\ -12 & 6\ell & 12 & 6\ell \\ -6\ell & 2\ell^2 & 6\ell & 4\ell^2 \end{bmatrix} \frac{EI}{\ell^3} \tag{4.71}$$

which, of course, is the same as $\mathbf{k}^i$ of Eq. (4.65).

Note that for the case of a simple beam element for which the polynomials of Eq. (4.41) or (4.47) are employed as trial series, the resulting stiffness matrix of Eq. (4.71) is the correct stiffness matrix [of Eq. (4.12)], rather than an approximate stiffness matrix. If fewer terms in the polynomial are retained, a different stiffness matrix results.

Loading Vector

To evaluate the loading vector $\overline{\mathbf{p}}^{i0}$ use Eq. (4.58), which corresponds to $\overline{\mathbf{p}}^{i0} = [\overline{V}_a\ \overline{M}_a\ \overline{V}_b\ \overline{M}_b]^T$. We choose to utilize the normalized coordinate at $\xi = x/\ell$.

$$\overline{\mathbf{p}}^{i0} = \int_a^b \mathbf{N}^T(x)\, \overline{p}_z(x)\, dx = \mathbf{G}^T \int_0^\ell \mathbf{N}_u^T(x)\, \overline{p}_z(x)\, dx = \ell \mathbf{G}^T \int_0^1 \mathbf{N}_u^T(\xi)\, \overline{p}_z(\xi)\, d\xi$$

$$\begin{matrix} \mathbf{G} \text{ from Eq. (4.45)} & \mathbf{N}_u \text{ and } \mathbf{G} \\ \mathbf{N}_u \text{ from Eq. (4.54)} & \text{from Eq. (4.47a)} \end{matrix}$$

$$= \ell \int_0^1 \mathbf{N}^T(\xi)\, \overline{p}_z(\xi)\, d\xi = \ell \int_0^1 \begin{bmatrix} 1 - 3\xi^2 + 2\xi^3 \\ (-\xi + 2\xi^2 - \xi^3)\ell \\ 3\xi^2 - 2\xi^3 \\ (\xi^2 - \xi^3)\ell \end{bmatrix} \overline{p}_z(\xi)\, d\xi \tag{4.72}$$

If the applied distributed load $\overline{p}_z$ is constant of magnitude p_0, then the integral of Eq. (4.72) results in

$$\overline{\mathbf{p}}^{i0} = p_0\ell \begin{bmatrix} 1/2 \\ -\ell/12 \\ 1/2 \\ \ell/12 \end{bmatrix} \tag{4.73}$$

If $\overline{p}_z$ varies linearly from $\xi = 0$ to $\xi = 1$, where its magnitude is p_0, then $\overline{p}_z = p_0\xi$ and Eq. (4.72) provides

$$\overline{\mathbf{p}}^{i0} = \frac{p_0\ell}{60} \begin{bmatrix} 9 \\ -2\ell \\ 21 \\ 3\ell \end{bmatrix} \tag{4.74}$$

Table 4.2 lists the vector $\overline{\mathbf{p}}^{i0}$ for a variety of loading conditions.

TABLE 4.2

Loading Vector $\overline{\mathbf{p}}^{i0}$ of a Beam Stiffness Matrix for Applied Distributed Loading (Sign Convention 2)

Definitions	Uniform Force	Uniform Moment	Linearly Varying Force	Linearly Varying Moment
	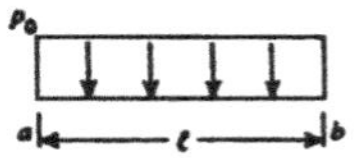	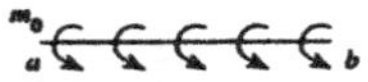	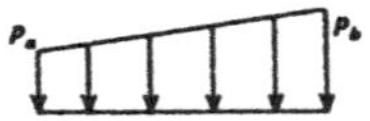	
Force and Displacement Vectors $\mathbf{v}_i = \begin{bmatrix} u_a \\ w_a \\ \theta_a \\ u_b \\ w_b \\ \theta_k \end{bmatrix}$ $\mathbf{p}^i = \begin{bmatrix} N_a \\ V_a \\ M_a \\ N_b \\ V_b \\ M_b \end{bmatrix}$ $\overline{\mathbf{p}}^{i0} =$	$\begin{bmatrix} 0 \\ p_0\ell/2 \\ -p_0\ell^2/12 \\ 0 \\ p_0\ell/2 \\ p_0\ell^2/12 \end{bmatrix}$	$\begin{matrix} 0 \\ m_0 \\ 0 \\ 0 \\ -m_0 \\ 0 \end{matrix}$	$\begin{matrix} 0 \\ \ell(7p_a+3p_b)/20 \\ -\ell^2(p_a/20+p_b/30) \\ 0 \\ \ell(3p_a+7p_b)/20 \\ \ell^2(p_a/30+p_b/20) \end{matrix}$	$\begin{matrix} 0 \\ (m_a+m_b)/2 \\ \ell(m_a-m_b)/12 \\ 0 \\ -(m_a+m_b)/2 \\ -\ell(m_a-m_b)/12 \end{matrix}$
Normalised Force and Displacement Vectors $\mathbf{v}^i = \begin{bmatrix} u_a \\ w_a \\ \ell\theta_a \\ u_b \\ w_b \\ \ell\theta_b \end{bmatrix}$ $\mathbf{p}^i = \begin{bmatrix} N_a \\ V_a \\ M_a/\ell \\ N_b \\ V_b \\ M_b/\ell \end{bmatrix}$ $\overline{\mathbf{p}}^{i0} =$	$\ell p_0 \begin{bmatrix} 0 \\ 1/2 \\ -1/12 \\ 0 \\ 1/2 \\ 1/12 \end{bmatrix}$	$m_0 \begin{bmatrix} 0 \\ 1 \\ 0 \\ 0 \\ -1 \\ 0 \end{bmatrix}$	$\ell \begin{bmatrix} 0 & 0 \\ 7/20 & 3/20 \\ -1/20 & -1/30 \\ 0 & 0 \\ 3/20 & 7/20 \\ 1/30 & 1/20 \end{bmatrix} \begin{bmatrix} p_a \\ p_b \end{bmatrix}$	$\begin{bmatrix} 0 & 0 \\ 1/2 & 1/2 \\ 1/12 & -1/12 \\ 0 & 0 \\ -1/2 & -1/2 \\ -1/12 & 1/12 \end{bmatrix} \begin{bmatrix} m_a \\ m_b \end{bmatrix}$
	Linear Thermal Loading along z Direction 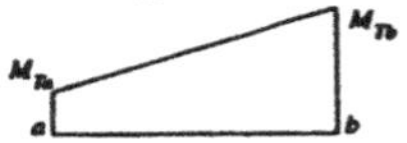	Uniform Axial Force 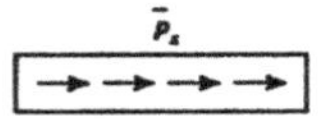	Linearly Varying Axial Force 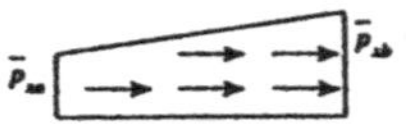	Uniform Temperature Change ΔT $\alpha =$ Thermal Coefficient
$\frac{\Delta M_T}{\Delta\ell} = \frac{M_{T_b} - M_{T_a}}{\ell}$ $M_T = \int_A \Delta T\, z\, dA$	$\begin{matrix} 0 \\ -\frac{\Delta M_T}{\Delta\ell} \\ -M_{T_a} \\ 0 \\ \frac{\Delta M_T}{\Delta\ell} \\ M_{T_b} \end{matrix}$	$\begin{matrix} \bar{p}_x\ell/2 \\ 0 \\ 0 \\ \bar{p}_x\ell/2 \\ 0 \\ 0 \end{matrix}$	$\begin{matrix} \ell(2\bar{p}_{xa}+\bar{p}_{xb})/6 \\ 0 \\ 0 \\ \ell(\bar{p}_{xa}+2\bar{p}_{xb})/6 \\ 0 \\ 0 \end{matrix}$	$\begin{matrix} -(\ell/2)\alpha EA\,\Delta T \\ 0 \\ 0 \\ (\ell/2)\alpha EA\,\Delta T \\ 0 \\ 0 \end{matrix}$
	$\frac{1}{\ell} \begin{bmatrix} 0 \\ M_{T_a} - M_{T_b} \\ -M_{T_a} \\ 0 \\ M_{T_b} - M_{T_a} \\ M_{T_b} \end{bmatrix}$	$\frac{\bar{p}_x\ell}{2} \begin{bmatrix} 1 \\ 0 \\ 0 \\ 1 \\ 0 \\ 0 \end{bmatrix}$	$\frac{\ell}{6} \begin{bmatrix} 2 & 1 \\ 0 & 0 \\ 0 & 0 \\ 1 & 2 \\ 0 & 0 \\ 0 & 0 \end{bmatrix} \begin{bmatrix} \bar{p}_{xa} \\ \bar{p}_{xb} \end{bmatrix}$	$\frac{\ell}{2}\alpha EA\,\Delta T \begin{bmatrix} -1 \\ 0 \\ 0 \\ 1 \\ 0 \\ 0 \end{bmatrix}$

Note: This table applies to the ith element of a simple Euler-Bernoulli beam. For more complex beam elements, see Table 4.4.

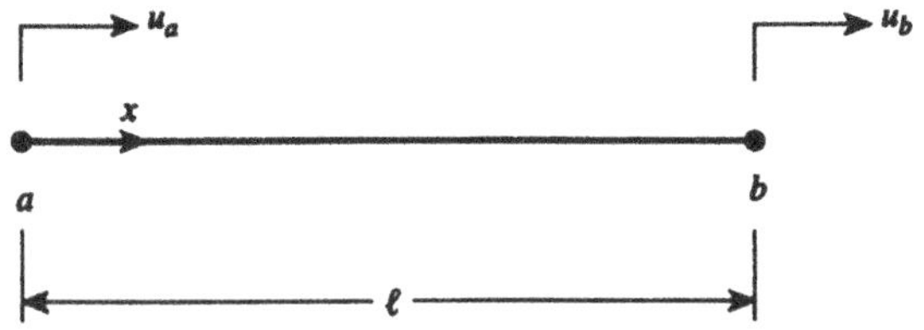

FIGURE 4.10
Bar undergoing axial extension.

EXAMPLE 4.1 Axial Deformation
Derive the stiffness matrix for a bar of cross-sectional area A undergoing axial deformation.

The axial end displacements are given by u_a and u_b as shown in Fig. 4.10. Choose the polynomial shape function

$$u = \mathbf{N}_u\,\hat{\mathbf{u}} = [1 \quad x]\begin{bmatrix}\hat{u}_1\\ \hat{u}_2\end{bmatrix} \tag{1}$$

In terms of the displacements at ends a and b,

$$u_a = \hat{u}_1, \qquad u_b = \hat{u}_1 + \ell\hat{u}_2 \tag{2}$$

or in matrix notation,

$$\begin{bmatrix}u_a\\ u_b\end{bmatrix} = \mathbf{v} = \hat{\mathbf{N}}_u\,\hat{\mathbf{u}} = \begin{bmatrix}1 & 0\\ 1 & \ell\end{bmatrix}\begin{bmatrix}\hat{u}_1\\ \hat{u}_2\end{bmatrix} \tag{3}$$

Then

$$\hat{\mathbf{u}} = \hat{\mathbf{N}}_u^{-1}\mathbf{v} = \mathbf{G}\mathbf{v}, \quad \mathbf{G} = \hat{\mathbf{N}}_u^{-1} = \begin{bmatrix}1 & 0\\ -1/\ell & 1/\ell\end{bmatrix} \tag{4}$$

and $u = \mathbf{N}_u\,\mathbf{G}\mathbf{v} = \mathbf{N}\mathbf{v} = [1 - x/\ell \quad x/\ell]\mathbf{v}$.

The internal virtual work term for axial motion takes the form

$$-\delta W_i = \int_0^\ell \delta u\,\mathbf{k}^D\,u\,dx \tag{5}$$

To express this in discrete form,

$$\delta u = \delta(\mathbf{N}_u\,\mathbf{G}\mathbf{v}) = \mathbf{N}_u\,\mathbf{G}\,\delta\mathbf{v} = \delta\mathbf{v}^T\mathbf{G}^T\,\mathbf{N}_u^T \tag{6}$$

From Chapter 2, Example 2.7,

$$\mathbf{k}^D = {}_x d\,EA\,d_x \tag{7}$$

The element stiffness matrix $\mathbf{k}^i$ is given by

$$\mathbf{k}^i = \mathbf{G}^T\int_0^\ell \mathbf{N}_u^T\,\mathbf{k}^D\,\mathbf{N}_u\,dx\,\mathbf{G} \tag{8}$$

The term $d_x\mathbf{N}_u$ would be equal to $[0 \quad 1]$ and

$$\mathbf{N}_u^T\,{}_x d = \begin{bmatrix}0\\ 1\end{bmatrix}$$

With the integral

$$\int_0^\ell \mathbf{N}_u^T\,\mathbf{k}^D\,\mathbf{N}_u\,dx = EA\int_0^\ell \begin{bmatrix}0\\ 1\end{bmatrix}[0 \quad 1]\,dx = EA\begin{bmatrix}0 & 0\\ 0 & \ell\end{bmatrix} \tag{9}$$

and with $\mathbf{G}$ taken from (4), the stiffness matrix of (8) becomes

$$\mathbf{k}^i = \frac{EA}{\ell}\begin{bmatrix} 1 & -1 \\ -1 & 1 \end{bmatrix} \tag{10}$$

■

EXAMPLE 4.2 Loading Vector Due to the Weight of the Bar

Find the loading vector for the axial deformation of a vertical bar loaded by its own weight.

The weight per unit length along the bar is γA, where γ is the weight density of the material. Assume the bar is in a vertical position and x is positive upwards. The external virtual work would be

$$\begin{aligned} \delta W_e &= \int_0^\ell \delta u(-\gamma A)\, dx = \delta\hat{\mathbf{u}}^T \int_0^\ell \mathbf{N}_u^T(-\gamma A)\, dx \\ &= \delta\mathbf{v}^T\mathbf{G}^T \int_0^\ell \mathbf{N}_u^T(-\gamma A)\, dx = \delta\mathbf{v}^T\, \overline{\mathbf{p}}^{i0} \end{aligned} \tag{1}$$

where $\hat{\mathbf{u}}$, $\mathbf{G}$, $\mathbf{N}_u$, and $\mathbf{v}$ are given in Example 4.1. The same expression, i.e.,

$$\delta W_e = \int_0^\ell \delta u\, \overline{p}_x\, dx \tag{2}$$

with $\overline{p}_x = -\gamma A$, follows from Chapter 2, Example 2.6. Also, it is analogous to $\int_a^b \delta w\, \overline{p}_z\, dx$ of Eq. (4.52), which is intended for transverse motion. Upon integration, (1) leads to

$$\overline{\mathbf{p}}^{i0} = -\frac{\gamma A\ell}{2}\begin{bmatrix} 1 \\ 1 \end{bmatrix} \tag{3}$$

■

EXAMPLE 4.3 Loading Vector Due to Thermal Loading

Find the loading vector $\overline{\mathbf{p}}^{i0}$ for axial displacement due to a temperature change along the bar of $\Delta T_x = \Delta T = T_a + (T_\ell - T_a)\frac{x}{\ell}$ where T_ℓ and T_a are reference temperature changes and α is the thermal expansion coefficient.

The imposed strain due to ΔT would be $\epsilon^0 = \alpha\, \Delta T$. Equation (1.43) can be inserted in Eq. (2.35) to obtain an expression for virtual work. This "external" virtual work takes the form

$$\delta W_e = -\int_0^\ell \delta\epsilon\, EA\, \epsilon^0\, dx \tag{1}$$

From Example 4.1, Eqs. (1) and (4)

$$\delta u = \mathbf{N}_u\, \delta\hat{\mathbf{u}} = \mathbf{N}_u\mathbf{G}\, \delta\mathbf{v} \quad \text{and} \quad \delta\epsilon = \mathbf{N}_u'\, \mathbf{G}\, \delta\mathbf{v} = [0 \quad 1]\mathbf{G}\, \delta\mathbf{v} = \delta\mathbf{v}^T\, \mathbf{G}^T\begin{bmatrix} 0 \\ 1 \end{bmatrix} \tag{2}$$

Then (1) becomes

$$\delta W_e = \delta\mathbf{v}^T\mathbf{G}^T \int_0^\ell \begin{bmatrix} 0 \\ 1 \end{bmatrix} (EA\, \alpha\, \Delta T)\, dx \tag{3}$$

Also, $\delta W_e = \delta\mathbf{v}^T\overline{\mathbf{p}}^{i0}$. Finally,

$$\begin{aligned} \overline{\mathbf{p}}^{i0} &= -\mathbf{G}^T \int_0^\ell \begin{bmatrix} 0 \\ 1 \end{bmatrix} EA\, \alpha\left[T_a + (T_\ell - T_a)\frac{x}{\ell}\right] dx \\ &= EA\, \alpha\left[T_a + \frac{1}{2}(T_\ell - T_a)\right]\begin{bmatrix} -1 \\ 1 \end{bmatrix} \end{aligned} \tag{4}$$

■

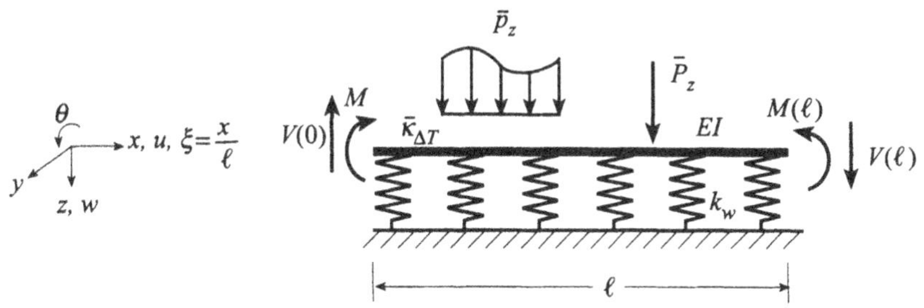

FIGURE 4.11
Beam element on elastic foundation. Sign Convention 1.

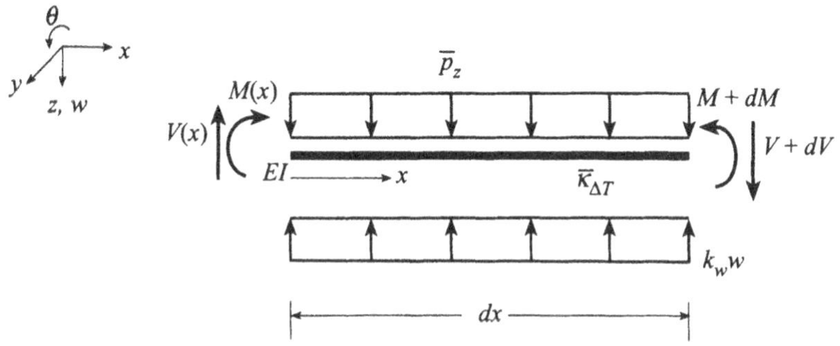

FIGURE 4.12
Differential element. Sign Convention 1.

EXAMPLE 4.4 Beam Element on Elastic Foundation

A beam on elastic foundation provides an example of the sort of approximation involved in the use of polynomial trial functions to develop element stiffness matrices. Whereas the simple Euler-Bernoulli beam of this chapter can be solved "exactly" using a cubic polynomial as the (assumed) trial function, the same polynomial leads to approximate element matrices for a beam on an elastic foundation.

Begin with notation for the beam element of Fig. 4.11.

k_w is the modulus of the elastic (Winkler) foundation (force/length^2)

$\overline{\kappa}_{\Delta T}$ is the prescribed curvature from the temperature difference ΔT, e.g., $\overline{\kappa}_{\Delta T} = \frac{\alpha \Delta T}{h}$ where α is the coefficient of linear thermal expansion and h is the height of the beam cross section

Fundamental Relations for a Differential Element

The elastic foundation imposes a force of magnitude $k_w w$ on the beam element. This is introduced as a distributed reaction (force/length), opposite in sign to $\overline{p}_z$ (Fig. 4.12). The fundamental differential equations for the response of this beam element are derived using the same procedure employed in Chapter 2 for the simple beam element. The resulting equations are the same as given by Eq. (1.133) with the addition of the effects of temperature and the elastic foundation.

Conditions of Equilibrium

$$\frac{dV}{dx} + \overline{p}_z - k_w w = 0$$
$$\frac{dM}{dx} - V = 0 \tag{1}$$

Material Law

$$M = EI(\kappa - \overline{\kappa}_{\Delta T})$$
$$V = GA_s \gamma \tag{2}$$

Kinematics

$$\kappa = \frac{d\theta}{dx}$$
$$\gamma = \theta + \frac{dw}{dx} \tag{3a}$$

If shear deformation is not taken into account

$$\gamma = 0 \quad \text{or} \quad \theta = -\frac{dw}{dx} \qquad \kappa = -\frac{d^2w}{dx^2} \tag{3b}$$

Boundary Conditions

$$\begin{array}{lll} \text{Force} & V = \overline{V} & M = \overline{M} \\ \text{Displacement} & w = \overline{w} & \theta = \overline{\theta} \end{array} \tag{4}$$

where the variables with superscript bars are specified quantities.

Equations (1), (2), and (3b) lead to the differential equation

$$\frac{d^2}{dx^2}\left(EI\frac{d^2w}{dx^2}\right) + k_w w - \overline{p}_z + \frac{d^2}{dx^2}(EI)\,\overline{\kappa}_{\Delta T} = 0 \tag{5}$$

This is a rather simple fourth order differential equation, for which an exact transcendental solution is readily established if EI and k_w are constant (Hetenyi, 1946). This solution takes the form

$$w(x) = C_1\, e^{\lambda x} \cos \lambda x + C_2\, e^{\lambda x} \sin \lambda x + C_3\, e^{-\lambda x} \cos \lambda x + C_4\, e^{-\lambda x} \sin \lambda x \tag{6}$$

where $\lambda = \sqrt[4]{k_w/4EI}$. However, in this example the principle of virtual work is utilized to form an approximate solution for this beam element. The procedure is the same as that developed in Section 4.4.2.

Approximate Element Matrices Obtained with the Help of the Principle of Virtual Work

The stiffness matrix is obtained from the principle of virtual work and, in particular, from the internal virtual work expressions, which for a beam element on an elastic foundation appears as

$$\delta W_i = \int_x [\delta\kappa\; M + \delta\gamma\; V - \delta w\; \overline{p}_z - \delta w(-k_w w)]\, dx \tag{7}$$

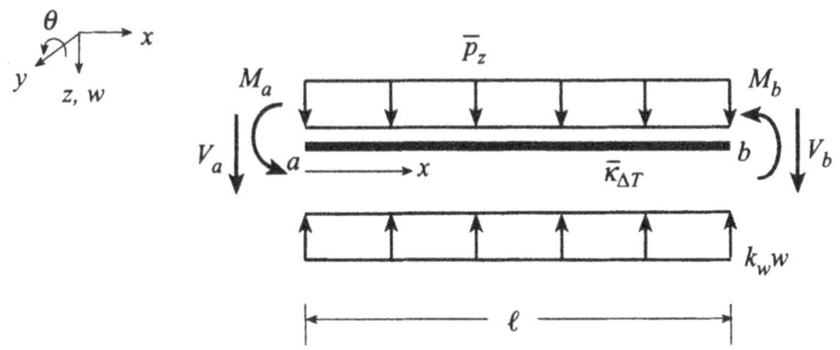

FIGURE 4.13
Beam element with Sign Convention 2.

This expression comes from Example 2.6 with the addition of the beam on elastic foundation term in which $k_w w$ is the reactive pressure due to the foundation, which is assumed to react linearly elastic.

Introduction of the material law and the kinematics will permit the virtual work to be expressed in terms of displacements. As indicated by (3a), if shear deformation is neglected $\kappa = -d^2w/dx^2$.
Then

$$\delta W_i = \int_x [\delta w'' EI\ (w'' + \overline{\kappa}_{\Delta T}) - \delta w\ \overline{p}_z + \delta w\ k_w w]\, dx \tag{8}$$

where $w' = dw/dx$ and the contribution of a temperature differential has been included.

Choice of Approximate Trial Displacements

Choose as an approximate deflection a four term polynomial for this element with four nodal degrees of freedom. The element, with Sign Convention 2, is shown in Fig. 4.13. The proposed trial function for the displacement $w(x)$ of this beam on elastic foundation element is the same one utilized in this chapter for a simple beam element. That is, assume (Eq. 4.40)

$$w(x) = C_1 + C_2x + C_3x^2 + C_4x^3 = \mathbf{N}_u\widehat{\mathbf{w}} \tag{9}$$

with

$$\mathbf{N}_u(x) = [1 \quad x \quad x^2 \quad x^3]$$
$$\widehat{\mathbf{w}}^T = [C_1 \quad C_2 \quad C_3 \quad C_4]$$

where the generalized degrees of freedom are the constants C_1, C_2, C_3, C_4. This leads to the deflection expressed in terms of the unknown displacements $\mathbf{v}^i$ as

$$w(x) = \mathbf{N}_u(x) \quad \widehat{\mathbf{w}} = \mathbf{N}_u(x)\mathbf{G}\mathbf{v} = \mathbf{N}(x)\mathbf{v} \tag{10}$$

where $\mathbf{N}(x)$, which is defined in Eq. (4.46), contains the shape functions for this polynomial approximation of $w(x)$.

Substitution of the Assumed Displacement in the Fundamental Beam Equations

The beam element state variables w, θ, M, and V can be expressed in terms of the shape functions $\mathbf{N} = \mathbf{N}_u\mathbf{G}$. Thus, introduce $w(x)$ of (10) into the expressions of (1), (2), and (3)

$$\begin{aligned}
w(x) &= \mathbf{N}_u(x)\mathbf{G}\mathbf{v}^i \\
\theta(x) &= -w'(x) &&= -\mathbf{N}_u'(x)\mathbf{G}\mathbf{v}^i \\
\kappa(x) &= -w''(x) &&= -\mathbf{N}_u''(x)\mathbf{G}\mathbf{v}^i \\
M(x) &= EI(\kappa - \overline{\kappa}_{\Delta T}) &&= -EI\mathbf{N}_u''(x)\mathbf{G}\mathbf{v}^i - EI\,\overline{\kappa}_{\Delta T} \\
V(x) &= M'(x) &&= -EI\mathbf{N}_u'''(x)\mathbf{G}\mathbf{v}^i
\end{aligned} \tag{11}$$

where

$$\mathbf{N}_u''(x) = [0 \quad 0 \quad 2 \quad 6x]$$

$$\mathbf{N}_u'''(x) = [0 \quad 0 \quad 0 \quad 6]$$

Since the matrices $\mathbf{G}$ and $\mathbf{v}^i$ contain only discrete values (and are not functions of x), the variables $w(x), \theta(x), M(x)$, and $V(x)$ along the element can be expressed as functions of the assumed displacement and its derivatives.

Substitution of the Assumed Displacement into the Principle of Virtual Work

Substitute $w = \mathbf{N}_u(x)\mathbf{G}\mathbf{v}^i$ into the virtual work of (8) for a single element i.

$$\begin{aligned}
\delta W^i = {} & \underbrace{\delta\mathbf{v}^{iT}\,\mathbf{G}^T \int_a^b \mathbf{N}_u''^T(x)\; EI\; \mathbf{N}_u''(x)\,dx\;\mathbf{G}\,\mathbf{v}^i}_{\mathbf{k}_B} + \underbrace{\delta\mathbf{v}^{iT}\,\mathbf{G}^T \int_a^b \mathbf{N}_u^T(x)\; k_w\; \mathbf{N}_u(x)\,dx\;\mathbf{G}\,\mathbf{v}^i}_{\mathbf{k}_w} \\
& + \underbrace{\delta\mathbf{v}^{iT}\,\mathbf{G}^T \int_a^b \mathbf{N}_u''^T(x)\; EI\;\overline{\kappa}_{\Delta T}\,dx}_{\overline{\mathbf{p}}^0_{\Delta T}} - \underbrace{\delta\mathbf{v}^{iT}\,\mathbf{G}^T \int_a^b \mathbf{N}_u^T(x)\;\overline{p}_z(x)\,dx}_{\overline{\mathbf{p}}^0_{p_z}}
\end{aligned} \tag{12}$$

where $\mathbf{k}_B$ is the stiffness matrix for bending of the element. This was developed in Section 4.4.2.

- $\mathbf{k}_w$ is the element stiffness matrix to account for the elastic foundation.
- $\overline{\mathbf{p}}^0_{\Delta T}$ is the loading vector for thermal effects.
- $\overline{\mathbf{p}}^0_{p_z}$ is the loading vector for the line loading along the element.

A more common notation is to use $\mathbf{N}_u\mathbf{G} = \mathbf{N}$. Also define $\mathbf{B} = \mathbf{N}''$ and $\mathbf{B}_u = \mathbf{N}_u''$. Then

$$\begin{aligned}
\delta W^i = {} & \underbrace{\delta\mathbf{v}^{iT} \int_a^b \mathbf{B}^T EI\,\mathbf{B}\,dx\;\mathbf{v}^i}_{\mathbf{k}_B} + \underbrace{\delta\mathbf{v}^{iT} \int_a^b \mathbf{N}^T k_w\,\mathbf{N}\,dx\;\mathbf{v}^i}_{\mathbf{k}_w} \\
& + \underbrace{\delta\mathbf{v}^{iT} \int_a^b \mathbf{B}_u^T EI\,\overline{\kappa}_{\Delta T}\,dx}_{\overline{\mathbf{p}}^0_{\Delta T}} - \underbrace{\delta\mathbf{v}^{iT} \int_a^b \mathbf{B}^T\,\overline{p}_z\,dx}_{\overline{\mathbf{p}}^0_{p_z}}
\end{aligned} \tag{13}$$

Calculation of the Element Stiffness Matrix $\mathbf{k}_w$ for the Elastic Foundation

Most of the matrices in (12) were developed previously in this chapter. An exception is $\mathbf{k}_w$ for the elastic foundation, which will be treated here. From (12)

$$\mathbf{k}_w = \mathbf{G}^T \int_o^\ell (\mathbf{N}_u^T\,k_w\,\mathbf{N}_u\,dx)\mathbf{G} \tag{14}$$

where $\mathbf{N}_u$ and $\mathbf{G}$ are taken from Eqs. (4.41) and (4.45), respectively. Assume k_ω, the coefficient for the elastic foundation, is constant. Then

$$\int_0^\ell \mathbf{N}_u^T k_w \mathbf{N}_u \, dx = k_w \int_0^\ell \begin{bmatrix} 1 & x & x^2 & x^3 \\ x & x^2 & x^3 & x^4 \\ x^2 & x^3 & x^4 & x^5 \\ x^3 & x^4 & x^5 & x^6 \end{bmatrix} dx = \begin{bmatrix} \ell & \ell^2/2 & \ell^3/3 & \ell^4/4 \\ \ell^2/2 & \ell^3/3 & \ell^4/4 & \ell^5/5 \\ \ell^3/3 & \ell^4/4 & \ell^5/5 & \ell^6/6 \\ \ell^4/4 & \ell^5/5 & \ell^6/6 & \ell^7/7 \end{bmatrix}$$

and finally,

$$\mathbf{k}_w = \mathbf{G}^T \int_0^\ell (\mathbf{N}_u^T k_w \mathbf{N}_u \, dx)\mathbf{G}$$

$$= \frac{k_w \ell}{420} \begin{bmatrix} 156 & -22\ell & 54 & 13\ell \\ -22\ell & 4\ell^2 & -13\ell & -3\ell^2 \\ 54 & -13\ell & 156 & 22\ell \\ 13\ell & -3\ell^2 & 22\ell & 4\ell^2 \end{bmatrix} \tag{15}$$

■

4.4.3 Properties of Stiffness Matrices

Some properties of stiffness matrices were derived in Chapter 3. It was shown that the stiffness matrix should be symmetric, i.e.,

$$k_{ij} = k_{ji} \quad \text{and} \quad \mathbf{k}_{ab} = \mathbf{k}_{ba}^T \tag{4.75}$$

Note that $\mathbf{k}^i$ of Eq. (4.42) is indeed symmetric.

It was also shown that the diagonal elements of a stiffness matrix are positive. Furthermore, it was illustrated in Section 4.3 that a stiffness matrix is singular and, hence, cannot be inverted. Also, after elimination of rigid body motion, a stiffness matrix is positive definite.

4.4.4 Response of Simple Beams

The stiffness matrix as developed in this chapter can be used to solve problems concerning uniform beams. In the next chapter, structural systems formed of beam elements will be treated.

It is shown in Section 4.3 that, in general $\mathbf{k}^i$ is a singular matrix. However, the application of the boundary conditions has the effect of rendering $\mathbf{k}^i$ nonsingular and providing the desired solution for the response of the beam.

EXAMPLE 4.5 Beam with Linearly Varying Loading

Find the response of the beam of Fig. 4.14. This is the same beam treated in Chapter 3, Example 3.13 and Fig. 3.11.

The stiffness matrix for an Euler-Bernoulli beam is provided by Eq. (4.12), and the loading vector $\overline{\mathbf{p}}^{i0}$ can be taken from Table 4.2 with $p_a = p_0$ and $p_b = 0$, giving

$$\overline{\mathbf{p}}^{i0} = \begin{bmatrix} 7\ell p_a/20 \\ -\ell^2 p_a/20 \\ 3\ell p_a/20 \\ \ell^2 p_a/30 \end{bmatrix} \tag{1}$$

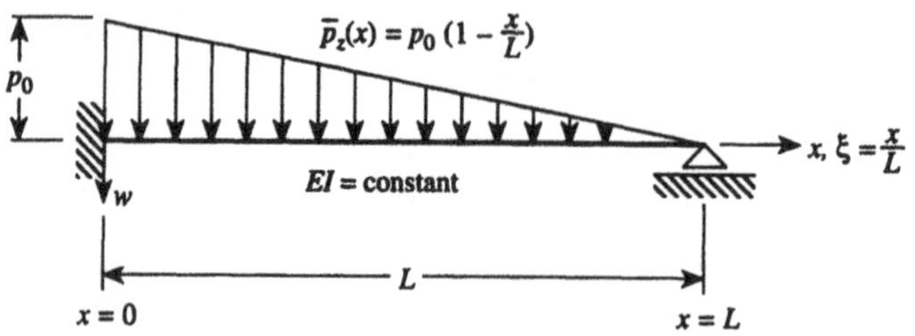

FIGURE 4.14
Beam with ramp loading. Examples 4.5, 4.9, and 4.10.

The boundary conditions are $w_0 = \theta_0 = w_L = M_L = 0$. Insert the displacement conditions ($w_0 = \theta_0 = w_L = 0$) in the right-hand side of

$$\begin{bmatrix} V_0 \\ M_0 \\ V_L \\ M_L \end{bmatrix} = \mathbf{k}^i \begin{bmatrix} w_0 \\ \theta_0 \\ w_L \\ \theta_L \end{bmatrix} - \overline{\mathbf{p}}^{i0}$$
$$\mathbf{p}^i = \mathbf{k}^i \quad \mathbf{v}^i \quad - \overline{\mathbf{p}}^{i0} \tag{2}$$

Solve for θ_L from the final row of this relationship with $M_L = 0$. Substitute this value of θ_L back into (2) to compute the left end reactions (Sign Convention 2)

$$V_0 = -(2/5)p_0L, \qquad M_0 = (1/15)p_0L^2 \tag{3}$$

This example is provided to illustrate that stiffness equations in the form developed in this chapter can be used for the solution of simple problems. However, the most important use of the stiffness relations is in the form of $\mathbf{k}^i\mathbf{v}^i - \overline{\mathbf{p}}^{i0}$, wherein the distributed loadings on an element are applied as end loads and inserted in $\overline{\mathbf{p}}^i$. Then this stiffness relationship is combined (assembled) with those of the other elements of the structure to form a set of global stiffness equations. See Chapter 5. ∎

4.4.5 Conversion of a Stiffness Matrix to a Transfer Matrix

The transformation of a transfer matrix into a stiffness matrix is a useful technique for deriving stiffness matrices. Of course, the process can be reversed to convert a stiffness matrix into a transfer matrix. To see this, begin with Eq. (4.12) which was established using Sign Convention 2.

$$\begin{aligned} \mathbf{p}_a &= \mathbf{k}_{aa}\mathbf{v}_a + \mathbf{k}_{ab}\mathbf{v}_b \\ \mathbf{p}_b &= \mathbf{k}_{ba}\mathbf{v}_a + \mathbf{k}_{bb}\mathbf{v}_b \end{aligned} \tag{4.76}$$

We wish to reorganize these equations so that they are in transfer matrix form. The first relationship is readily changed to

$$\mathbf{v}_b = \mathbf{k}_{ab}^{-1}\mathbf{p}_a - \mathbf{k}_{ab}^{-1}\mathbf{k}_{aa}\mathbf{v}_a \tag{4.77}$$

Substitute $\mathbf{v}_b$ of Eq. (4.77) into the second of Eqs. (4.76) and find

$$\mathbf{p}_b = \left(\mathbf{k}_{ba} - \mathbf{k}_{bb}\mathbf{k}_{ab}^{-1}\mathbf{k}_{aa}\right)\mathbf{v}_a + \mathbf{k}_{bb}\mathbf{k}_{ab}^{-1}\mathbf{p}_a \tag{4.78}$$

The desired transfer matrix in Sign Convention 2 notation is obtained from Eqs. (4.77) and (4.78) as

$$\begin{bmatrix} \mathbf{v}_b \\ \cdots \\ \mathbf{p}_b \end{bmatrix} = \begin{bmatrix} -\mathbf{k}_{ab}^{-1}\mathbf{k}_{aa} & \vdots & \mathbf{k}_{ab}^{-1} \\ \cdots\cdots & & \cdots\cdots \\ \mathbf{k}_{ba} - \mathbf{k}_{bb}\mathbf{k}_{ab}^{-1}\mathbf{k}_{aa} & \vdots & \mathbf{k}_{bb}\mathbf{k}_{ab}^{-1} \end{bmatrix} \begin{bmatrix} \mathbf{v}_a \\ \cdots \\ \mathbf{p}_a \end{bmatrix} = \begin{bmatrix} \mathbf{U}_{vv} & \mathbf{U}_{vp} \\ \mathbf{U}_{pv} & \mathbf{U}_{pp} \end{bmatrix} \begin{bmatrix} \mathbf{v}_a \\ \mathbf{p}_a \end{bmatrix} \tag{4.79}$$

4.4.6 Inclusion of Axial and Torsional Motion

Combine the stiffness matrices of Eq. (4.12) and Example 4.1 to give

$$\begin{bmatrix} N_a \\ V_a \\ M_a \\ \cdots \\ N_b \\ V_b \\ M_b \end{bmatrix} = \begin{bmatrix} \frac{EA}{\ell} & & & \vdots & -\frac{EA}{\ell} & & \\ & \frac{12EI}{\ell^3} & -\frac{6EI}{\ell^2} & \vdots & & -\frac{12EI}{\ell^3} & -\frac{6EI}{\ell^2} \\ & -\frac{6EI}{\ell^2} & \frac{4EI}{\ell} & \vdots & & \frac{6EI}{\ell^2} & \frac{2EI}{\ell} \\ \cdots & \cdots & \cdots & & \cdots & \cdots & \cdots \\ -\frac{EA}{\ell} & & & \vdots & \frac{EA}{\ell} & & \\ & -\frac{12EI}{\ell^3} & \frac{6EI}{\ell^2} & \vdots & & \frac{12EI}{\ell^3} & \frac{6EI}{\ell^2} \\ & -\frac{6EI}{\ell^2} & \frac{2EI}{\ell} & \vdots & & \frac{6EI}{\ell^2} & \frac{4EI}{\ell} \end{bmatrix} \begin{bmatrix} u_a \\ w_a \\ \theta_a \\ \cdots \\ u_b \\ w_b \\ \theta_b \end{bmatrix} \tag{4.80}$$

$$\mathbf{p}^i \quad = \quad \mathbf{k}^i \quad \mathbf{v}^i$$

as the stiffness matrix for a beam undergoing bending and extension. The sign convention of Fig. 4.1b (Sign Convention 2) was used in constructing this stiffness matrix.

In the case of a bar undergoing bending and torsion, the stiffness matrix in a nondimensional displacement form is (Sign Convention 2)

$$\begin{bmatrix} V_a\ell \\ M_a \\ M_{ta} \\ V_b\ell \\ M_b \\ M_{tb} \end{bmatrix} = \frac{EI}{\ell} \begin{bmatrix} 12 & -6 & 0 & -12 & -6 & 0 \\ -6 & 4 & 0 & 6 & 2 & 0 \\ 0 & 0 & J^* & 0 & 0 & -J^* \\ -12 & 6 & 0 & 12 & 6 & 0 \\ -6 & 2 & 0 & 6 & 4 & 0 \\ 0 & 0 & -J^* & 0 & 0 & J^* \end{bmatrix} \begin{bmatrix} w_a/\ell \\ \theta_a \\ \phi_a \\ w_b/\ell \\ \theta_b \\ \phi_b \end{bmatrix} \tag{4.81}$$

$$\mathbf{p}^i \quad = \quad \mathbf{k}^i \quad \mathbf{v}^i$$

where

$$J^* = \frac{GJ}{EI} = \frac{J}{2(1+\nu)I}$$

Note that this form differs from that of the stiffness matrix of Eq. (4.13) due to different scaling.

4.5 Transfer Matrices

4.5.1 Determination of Transfer Matrices

There are many approaches for deriving transfer matrices such as the one leading to Eq. (4.8c). Some of the techniques are described in the references for this chapter. One of these techniques is the direct integration of the governing equations, either using the higher order or the first order system of governing equations.

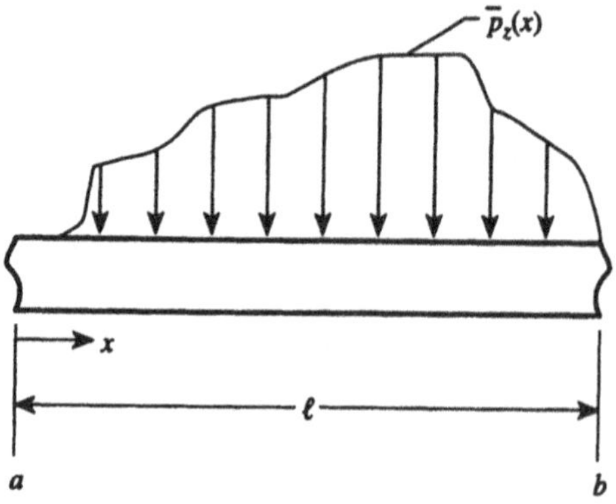

FIGURE 4.15
A general distributed loading $\overline{p}_z$.

Higher Order Form (Displacement Form) of the Governing Equations

From Chapter 1, Eq. (1.127), the higher order form of the governing differential equations for an Euler-Bernoulli beam is

$$\begin{aligned} \frac{dw}{dx} &= -\theta \\ \frac{d^2w}{dx^2} &= -\frac{M}{EI} \\ \frac{d^3w}{dx^3} &= -\frac{V}{EI} \\ \frac{d^4w}{dx^4} &= \frac{\overline{p}_z}{EI} \end{aligned} \tag{4.82}$$

Four integrations of the final relationship ($d^4w/dx^4 = \overline{p}_z/EI$) lead to

$$\begin{aligned} V &= -EI\frac{d^3w}{dx^3} = -C_1 - \int_0^x \overline{p}_z(\tau)d\tau \\ M &= -EI\frac{d^2w}{dx^2} = -C_2 - C_1x - \int_0^x \int \overline{p}_z(\tau)d\tau \\ \theta &= -\frac{dw}{dx} = -\frac{C_3}{EI} - \frac{C_2x}{EI} - \frac{C_1}{EI}\frac{x^2}{2} - \int_0^x \iint \frac{\overline{p}_z(\tau)}{EI}d\tau \\ w &= \frac{C_4}{EI} + \frac{C_3}{EI}x + \frac{C_2}{EI}\frac{x^2}{2} + \frac{C_1}{EI}\frac{x^3}{3!} + \int_0^x \iiint \frac{\overline{p}_z(\tau)}{EI}d\tau \end{aligned} \tag{4.83}$$

where $\overline{p}_z(x)$ is the transverse distributed loading as shown in Fig. 4.15 and, by definition,

$$\int_0^x \int \overline{p}_z(\tau)d\tau = \int_0^x \left[\int_0^\xi \overline{p}_z(\tau)d\tau\right]d\xi$$

and so on. Suppose $x = 0$ corresponds to the left end, i.e., the "a" end, of the beam element. To rewrite Eq. (4.83) in transfer matrix form, it is necessary to reorganize the constants of integration C_1, C_2, C_3, and C_4 in terms of the state variables at a: w_a, θ_a, V_a, and M_a. Assume there is no loading at $x = 0$ so that the integrals of $\overline{p}_z$ vanish at $x = 0$. From Eq. (4.83) for $x = 0$,

$$\begin{aligned} w_a &= w_{x=0} = \frac{C_4}{EI}, \qquad \theta_a = \theta_{x=0} = -\frac{C_3}{EI}, \\ M_a &= M_{x=0} = -C_2, \qquad V_a = V_{x=0} = -C_1 \end{aligned} \tag{4.84}$$

Use Eq. (4.84) to replace the constants C_1, C_2, C_3, and C_4 in Eq. (4.83) by the state variables and set $x = \ell$, giving

$$\begin{aligned}
w_b &= w_a - \theta_a \ell - V_a \frac{\ell^3}{3!EI} - M_a \frac{\ell^2}{2EI} + \int_0^\ell \int\int\int \frac{\overline{p}_z(\tau)}{EI} d\tau \\
\theta_b &= \theta_a + V_a \frac{\ell^2}{2EI} + M_a \frac{\ell}{EI} - \int_0^\ell \int\int \frac{\overline{p}_z(\tau)}{EI} d\tau \\
V_b &= V_a - \int_0^\ell \overline{p}_z(\tau) d\tau \\
M_b &= V_a \ell + M_a - \int_0^\ell \int \overline{p}_z(\tau) d\tau
\end{aligned} \tag{4.85}$$

In matrix notation this appears as $\mathbf{z}_b = \mathbf{U}^i \mathbf{z}_a + \overline{\mathbf{z}}^i$, where $\mathbf{z} = [w \;\; \theta \;\; V \;\; M]^T$, giving

$$\mathbf{U}^i = \mathbf{U}^i(\ell) = \begin{bmatrix} 1 & -\ell & -\frac{\ell^3}{6EI} & -\frac{\ell^2}{2EI} \\ 0 & 1 & \frac{\ell^2}{2EI} & \frac{\ell}{EI} \\ 0 & 0 & 1 & 0 \\ 0 & 0 & \ell & 1 \end{bmatrix} \tag{4.86}$$

$$\overline{\mathbf{z}}^i = \begin{bmatrix} \int_0^\ell \int\int\int \frac{\overline{p}_z(\tau)}{EI} d\tau \\ -\int_0^\ell \int\int \frac{\overline{p}_z(\tau)}{EI} d\tau \\ -\int_0^\ell \overline{p}_z(\tau) d\tau \\ -\int_0^\ell \int \overline{p}_z(\tau) d\tau \end{bmatrix} = \begin{bmatrix} w_b^0 \\ \theta_b^0 \\ V_b^0 \\ M_b^0 \end{bmatrix} \tag{4.87}$$

First Order Form of the Governing Equations

A typical method of developing transfer matrices, which applies to both simple and difficult problems, is that of integration of first order equations in the state variables [Wunderlich, 1966 and 1967]. In Chapter 1, Eq. (1.133), first order governing equations for a beam were derived. Retain the sign convention of Chapter 1, i.e., use Sign Convention 1. If shear deformation is ignored, then Chapter 1, Eq. (1.135) becomes

$$\begin{aligned}
\frac{d}{dx}\begin{bmatrix} w \\ \theta \\ V \\ M \end{bmatrix} &= \begin{bmatrix} 0 & -1 & 0 & 0 \\ 0 & 0 & 0 & 1/EI \\ 0 & 0 & 0 & 0 \\ 0 & 0 & 1 & 0 \end{bmatrix} \begin{bmatrix} w \\ \theta \\ V \\ M \end{bmatrix} + \begin{bmatrix} 0 \\ 0 \\ -\overline{p}_z \\ 0 \end{bmatrix} \\
\mathbf{z}' &= \qquad \mathbf{A} \qquad\qquad \mathbf{z} \;\; + \;\; \overline{\mathbf{P}}
\end{aligned} \tag{4.88}$$

Integration of these relations ($\mathbf{z}' = \mathbf{A}\,\mathbf{z} + \overline{\mathbf{P}}$) gives*

$$\mathbf{z} = e^{\mathbf{A}x}\mathbf{z}_a + e^{\mathbf{A}x} \int_{x=x_a}^{x} e^{-\mathbf{A}\tau} \overline{\mathbf{P}}(\tau) d\tau \tag{4.89}$$

*The solution to the matrix relation in Eq. (4.88) is analogous to the familiar solution of the first order scalar differential equation

$$z' = Az + \overline{p} : z = z_0 e^{Ax} + e^{Ax} \int_0^x e^{-A\tau} \overline{p}(\tau) d\tau$$

In more general terms for an element of length ℓ, extending from $x = x_a$ to $x = x_b$,

$$\mathbf{z}_b = \mathbf{U}^i \left[\mathbf{z}_a + \int_{x_a}^{x_b} (\mathbf{U}^i)^{-1} \overline{\mathbf{P}}\, d\tau \right] = \mathbf{U}^i \mathbf{z}_a + \overline{\mathbf{z}}^i \tag{4.90a}$$

where

$$\overline{\mathbf{z}}^i = \overline{\mathbf{z}}^i_b = \mathbf{U}^i \int_0^{\ell} (\mathbf{U}^i(\tau))^{-1} \overline{\mathbf{P}}(\tau) d\tau \tag{4.90b}$$

and, as indicated above for a constant coefficient matrix **A**,

$$\mathbf{U}^i = \mathbf{U}^i(\ell) = e^{\mathbf{A}(x_b - x_a)} \tag{4.90c}$$

with $x_b - x_a = \ell$. The exponential relation of Eq. (4.90c) can be expanded in the series

$$\mathbf{U}^i = e^{\mathbf{A}\ell} = \mathbf{I} + \frac{\mathbf{A}\ell}{1!} + \frac{\mathbf{A}^2 \ell^2}{2!} + \cdots = \sum_{s=0}^{\infty} \frac{\mathbf{A}^s \ell^s}{s!} \tag{4.91}$$

where **I** is the identity matrix, a diagonal matrix with diagonal values of unity. Such an expansion lends itself well for numerical calculations for more complicated members than simple beams, as it is often possible to control the error. From Eq. (4.90c), the exponential solution leads to a loading term of the form

$$\overline{\mathbf{z}}^i = e^{\mathbf{A}(x_b - x_a)} \int_{x_a}^{x_b} e^{-\mathbf{A}(\tau - x_a)} \overline{\mathbf{P}}\, d\tau \tag{4.92}$$

It follows from $(\mathbf{U}^i(x))^{-1} = e^{-\mathbf{A}x}$ that for constant **A**

$$(\mathbf{U}^i(x))^{-1} = \mathbf{U}^i(-x) \tag{4.93}$$

a result that can be useful when finding the loading vector $\overline{\mathbf{z}}^i$.

EXAMPLE 4.6 Transfer Matrix for an Euler-Bernoulli Beam

For the Euler-Bernoulli beam, with the governing equations of Eq. (4.88), the transfer matrix is obtained from Eq. (4.91) using

$$\mathbf{A} = \begin{bmatrix} 0 & -1 & 0 & 0 \\ 0 & 0 & 0 & 1/EI \\ 0 & 0 & 0 & 0 \\ 0 & 0 & 1 & 0 \end{bmatrix}$$

$$\mathbf{A}^2 = \mathbf{A}\mathbf{A} = \begin{bmatrix} 0 & 0 & 0 & -1/EI \\ 0 & 0 & 1/EI & 0 \\ 0 & 0 & 0 & 0 \\ 0 & 0 & 0 & 0 \end{bmatrix} \tag{1}$$

$$\mathbf{A}^3 = \mathbf{A}\mathbf{A}^2 = \begin{bmatrix} 0 & 0 & -1/EI & 0 \\ 0 & 0 & 0 & 0 \\ 0 & 0 & 0 & 0 \\ 0 & 0 & 0 & 0 \end{bmatrix}$$

Continued multiplication of **A** shows that $\mathbf{A}^4 = 0$ and that $\mathbf{A}^s = 0$ for any s greater than 3. The termination of the expansion is to be expected since the analytical solution for this beam is simply a polynomial of limited order (3). The transfer matrix of Eq. (4.91) is now fully defined. ∎

A multitude of methods for computing transfer matrices are treated in the references. The solution $e^{\mathbf{A}\ell}$ can be represented as a matrix polynomial using the Cayley[3]-Hamilton[4] theorem (Pestel and Leckie, 1963) i.e., the minimal polynomial, which requires knowledge of the eigenvalues of **A**. For a large matrix **A**, these can be difficult to obtain. Padé[5] approximations can be useful in reducing the number of terms needed in an expansion of $e^{\mathbf{A}\ell}$.

For a nonconstant **A**, methods such as the Picard[6] iteration (Pestel and Leckie, 1963) are available. Also important are the numerical integration techniques, such as Runge[7]-Kutta (1867–1944) (Problem 4.23) that can be employed to solve differential equations. State space control methods also often involve the solution of a system of first order differential equations. Hence the relevant control theory literature is a fruitful source of information on the calculation of transfer matrices.

Two General Analytical Techniques

Two procedures are to be presented which are suitable for finding the transfer matrices for general forms of the governing equations of motion (Pilkey, 1994). These techniques, based on the Cayley-Hamilton theorem mentioned earlier and on the Laplace transform, apply to any set of governing equations for which an analytical solution can be obtained. A beam will be used to demonstrate the method.

The first order differential equations for the static response of a beam with axial load N, displacement foundation $k_w = k$, and rotary foundation k^*, are derived from the relations of Example 4.4 as

$$\begin{aligned} \frac{dw}{dx} &= -\theta + \frac{V}{k_s G A} & \frac{d\theta}{dx} &= \frac{M}{EI} \\ \frac{dV}{dx} &= kw - \overline{p}_z & \frac{dM}{dx} &= V + (k^* - N)\theta \end{aligned} \tag{4.94}$$

Thus, Eq. (4.88), when expanded to include a compressive axial force N (Chapter 11), shear deformation effects, and elastic foundations, becomes

$$\mathbf{z}' = \mathbf{A}\,\mathbf{z} + \overline{\mathbf{P}} \tag{4.95}$$

---

[3]Arthur Cayley (1821–1895) was a prolific (almost one thousand papers) English mathematician and Cambridge University professor. He was elected a fellow of Trinity College, Cambridge, the year (1842) he graduated from Cambridge. He also practiced law for several years.

[4]Sir William Rowan Hamilton (1805–1865) was a brilliant Irish mathematician who discovered the theory of quaternions and made major contributions to optics and dynamics. He was elected to the chair of Astronomy at Trinity College, Dublin, while he was still an undergraduate. Although appointed astronomer royal at Dunsink Observatory, he was not considered to be a successful practical astronomer.

[5]Henri Eugene Padé (1863–1953) was a French mathematician who received some of his higher education in Leipzig and Göttingen, Germany. His PhD thesis was written with Charles Hermite as the advisor and Picard on his committee and is the best known of his writings. It dealt with approximations to functions.

[6]Charles Émile Picard (1856–1941) was a Frenchman who turned to mathematics at the end of his secondary school tenure. After an interview with Pasteur, he attended the École Normale Supérieure, where he was permitted to devote himself exclusively to research. He had a brilliant career in mathematical analysis and algebraic geometry, including two theorems that bear his name. He married the daughter of his mentor Charles Hermite. He was a mountain climber whose family life was marred by tragedy. His daughter and two sons died in World War I and his grandsons were wounded and taken prisoner in World War II. The occupation of France clouded his final two years.

[7]Carl David Tolmé Runge (1856–1927) was a German professor of mathematics in Hannover and Göttingen. He performed research on spectroscopy, the theory of functions, and the numerical solution of differential equations.

with

$$\mathbf{A} = \begin{bmatrix} 0 & -1 & 1/k_sGA & 0 \\ 0 & 0 & 0 & 1/EI \\ k & 0 & 0 & 0 \\ 0 & k^* - N & 1 & 0 \end{bmatrix} \qquad \overline{\mathbf{P}} = \begin{bmatrix} 0 \\ 0 \\ -\overline{p}_z \\ 0 \end{bmatrix}$$

As can be observed in Eq. (4.90b), the loading elements $\overline{\mathbf{z}}^i$ can be computed if the transfer matrix $\mathbf{U}^i$ is available. Since, for a complete solution, it is necessary only to find $\mathbf{U}^i$, we can set $\overline{p}_z$ equal to zero in Eqs. (4.94) and (4.95).

Thus, we seek to solve the homogeneous differential equations

$$\frac{d\mathbf{z}}{dx} = \mathbf{A}\,\mathbf{z} \tag{4.96}$$

The solution can be in the form [Eq. (4.90c)]

$$\mathbf{z} = \mathbf{U}^i\mathbf{z}_0 = e^{\mathbf{A}\ell}\mathbf{z}_0 \quad \text{or} \quad \mathbf{z}_b = \mathbf{U}^i\mathbf{z}_a \tag{4.97}$$

A function of a square matrix $\mathbf{A}$ of order n is equal to a polynomial in $\mathbf{A}$ of order $n-1$. Hence, for the 4×4 matrix $\mathbf{A}$ for a beam,

$$e^{\mathbf{A}\ell} = c_0\mathbf{I} + c_1\mathbf{A}\ell + c_2(\mathbf{A}\ell)^2 + c_3(\mathbf{A}\ell)^3 \tag{4.98}$$

The Cayley-Hamilton Theorem ("a matrix satisfies its own characteristic equation") permits $\mathbf{A}$ of Eq. (4.98) to be replaced by its characteristic values λ_i. Thus,

$$e^{\lambda_i\ell} = c_0 + c_1\lambda_i\ell + c_2(\lambda_i\ell)^2 + c_3(\lambda_i\ell)^3, \quad i = 1, 2, 3, 4 \tag{4.99}$$

represents four equations that can be solved for the functions $c_0, c_1, c_2,$ and c_3. Substitution of these values into Eq. (4.98) leads to the desired transfer matrix.

To find the characteristic values λ_i, substitute $e^{\lambda_i\ell}$ into the homogeneous first order governing equations, yielding the characteristic equation

$$|\mathbf{I}\lambda_i - \mathbf{A}| = 0 \tag{4.100}$$

With $\mathbf{A}$ given by Eq. (4.95)

$$|\mathbf{I}\lambda_i - \mathbf{A}| = \begin{bmatrix} \lambda_i & 1 & -(1/k_sGA) & 0 \\ 0 & \lambda_i & 0 & -(1/EI) \\ -k & 0 & \lambda_i & 0 \\ 0 & -(k^* - N) & -1 & \lambda_i \end{bmatrix} = 0 \tag{4.101}$$

The roots $\lambda_1, \lambda_2, \lambda_3, \lambda_4 = (\pm n_1, \pm i n_2)$, of this determinant are readily found to be

$$n_{1,2} = (((\zeta + \eta)^2/4 - \lambda)^{1/2} \pm (\zeta - \eta)/2)^{1/2} \tag{4.102}$$

where

$$\lambda = k/EI, \quad \eta = k/(k_sGA), \quad \zeta = (N - k^*)/EI$$

Substitute these characteristic values into Eq. (4.99), giving

$$\begin{aligned} e^{n_1\ell} &= c_0 + c_1n_1\ell + c_2(n_1\ell)^2 + c_3(n_1\ell)^3 \\ e^{-n_1\ell} &= c_0 - c_1n_1\ell + c_2(n_1\ell)^2 - c_3(n_1\ell)^3 \\ e^{in_2\ell} &= c_0 + ic_1n_2\ell - c_2(n_2\ell)^2 - ic_3(n_2\ell)^3 \\ e^{-in_2\ell} &= c_0 - ic_1n_2\ell - c_2(n_2\ell)^2 + ic_3(n_2\ell)^3 \end{aligned} \tag{4.103}$$

Solve these four equations for c_0, c_1, c_2, and c_3.

$$\begin{aligned} c_0 &= \left(n_2^2 \cosh n_1\ell + n_1^2 \cos n_2\ell\right)/\left(n_1^2 + n_2^2\right) \\ c_1 &= \left[\left(n_2^2/n_1\right) \sinh n_1\ell + \left(n_1^2/n_2\right) \sin n_2\ell\right] / \left[\ell\left(n_1^2 + n_2^2\right)\right] \\ c_2 &= \left[\left(\cosh n_1\ell - \cos n_2\ell\right)\right] / \left[\ell^2\left(n_1^2 + n_2^2\right)\right] \\ c_3 &= \left[(1/n_1) \sinh n_1\ell - (1/n_2) \sin n_2\ell\right] / \left[\ell^3\left(n_1^2 + n_2^2\right)\right] \end{aligned} \tag{4.104}$$

Finally, the transfer matrix is given by Eq. (4.98) as

$$\mathbf{U}^i = c_0\mathbf{I} + c_1(\mathbf{A}\ell) + c_2(\mathbf{A}\ell)^2 + c_3(\mathbf{A}\ell)^3 \tag{4.105}$$

or, after some algebraic manipulations

$$\mathbf{U}^i = \begin{bmatrix} c_0 + \ell^2 c_2 \eta & -\ell c_1 - \ell^3 c_3(\eta - \zeta) & \frac{c_1\ell + c_3\ell^3\eta}{k_s G A} - \frac{c_3\ell^3}{EI} & -\ell^2 c_2/EI \\ \lambda c_3 \ell^3 & c_0 - \ell^2 c_2 \zeta & \frac{\ell^2 c_2}{EI} & \frac{\ell c_1 - c_3\ell^3\zeta}{EI} \\ \lambda EI(\ell c_1 + \eta\ell^3 c_3) & -\lambda EI c_2 \ell^2 & c_0 + c_2\ell^2\eta & -\ell^3 c_3 \lambda \\ \lambda EI c_2 \ell^2 & EI[-c_1\ell\zeta + c_3\ell^3(\zeta^2 - \lambda)] & \ell c_1 + c_3\ell^3(\eta - \zeta) & c_0 - c_2\ell^2\zeta \end{bmatrix} \tag{4.106}$$

It is convenient, from the standpoint of installing this transfer matrix as a subroutine in a computer program, to redefine c_0, c_1, c_2, and c_3 in terms of new funtions e_0, e_1, e_2, e_3, and e_4, where

$$\begin{aligned} e_0 &= \ell(\eta - \zeta)c_1 + \ell^3[(\zeta^2 - \lambda) + \eta(\eta - \zeta)]c_3 \\ e_1 &= c_0 + \ell^2(\eta - \zeta)c_2 \\ e_2 &= \ell c_1 + \ell^3(\eta - \zeta)c_3 \\ e_3 &= \ell^2 c_2 \\ e_4 &= \ell^3 c_3 \end{aligned}$$

or

$$\begin{aligned} e_0 &= \left(n_1^3 \sinh n_1\ell - n_2^3 \sin n_2\ell\right)/\left(n_1^2 + n_2^2\right) \\ e_1 &= \left(n_1^2 \cosh n_1\ell + n_2^2 \cos n_2\ell\right)/\left(n_1^2 + n_2^2\right) \\ e_2 &= (n_1 \sinh n_1\ell + n_2 \sin n_2\ell)/\left(n_1^2 + n_2^2\right) \\ e_3 &= c_2\ell^2 \\ e_4 &= c_3\ell^3 \end{aligned} \tag{4.107}$$

The above transfer matrix then becomes

$$\mathbf{U}^i = \begin{bmatrix} e_1 + \zeta e_3 & -e_2 & -e_4/EI + (e_2 + \zeta e_4)/k_s G A & -e_3/EI \\ \lambda e_4 & e_1 - \eta e_3 & e_3/EI & (e_2 - \eta e_4)/EI \\ \lambda EI(e_2 + \zeta e_4) & -\lambda EI e_3 & e_1 + \zeta e_3 & -\lambda e_4 \\ \lambda EI e_3 & EI(e_0 - \eta e_2) & e_2 & e_1 - \eta e_3 \end{bmatrix} \tag{4.108}$$

This transfer matrix is presented in Table 4.3, where the functions e_1, e_2, e_3, and e_4 are given in a form suitable for use in a computer program. This table yields the same transfer matrix as in Eq. (4.86) if k, k^*, $1/k_s G A$, and N are set equal to zero.

TABLE 4.3

Transfer Matrix for a Beam Element, Part a:

Definitions: $\lambda = k/EI$, $\eta = k/(k_s GA)$, $\zeta = (N - k^*)/EI$ Sign Convention 1

$$\mathbf{U}^i = \begin{bmatrix} e_1 + \zeta e_3 & -e_2 & -e_4/EI + (e_2 + \zeta e_4)/k_s GA & -e_3/EI \\ \lambda e_4 & e_1 - \eta e_3 & e_3/EI & (e_2 - \eta e_4)/EI \\ \lambda EI(e_2 + \zeta e_4) & -\lambda EI e_3 & e_1 + \zeta e_3 & -\lambda e_4 \\ \lambda EI e_3 & EI(e_0 - \eta e_2) & e_2 & e_1 - \eta e_3 \end{bmatrix}$$

$$\mathbf{z}_b = \mathbf{U}^i \mathbf{z}_a + \bar{\mathbf{z}}^i \qquad \bar{\mathbf{z}} = \begin{bmatrix} w \\ \theta \\ V \\ M \end{bmatrix} \qquad \bar{\mathbf{z}}^i = \begin{bmatrix} w_b^0 \\ \theta_b^0 \\ V_b^0 \\ M_b^0 \end{bmatrix}$$

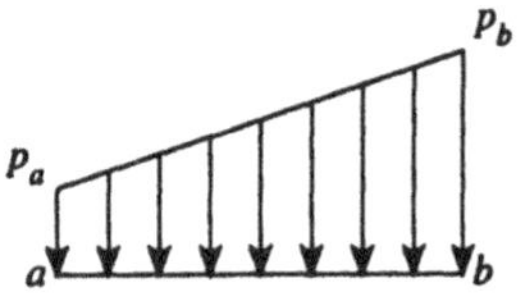

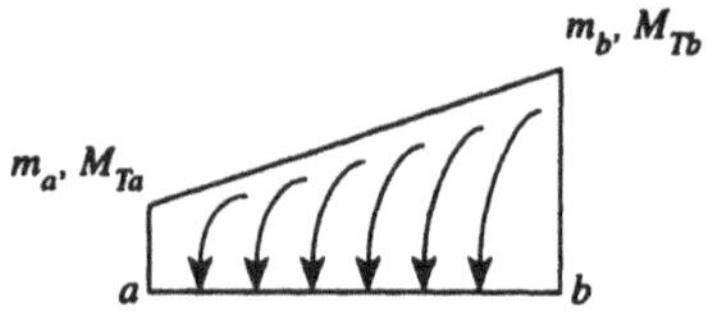

$$w_b^0 = (p_a e_5 + \tfrac{\Delta p}{\Delta \ell} e_6 + m_a e_4 + \tfrac{\Delta m}{\Delta \ell} e_5)/EI - p_a(e_3 + \zeta e_5)/k_s GA - \tfrac{\Delta p}{\Delta \ell}(e_4 + \zeta e_6)/k_s GA$$
$$- (M_{Ta} e_3 + \tfrac{\Delta M_T}{\Delta \ell} e_4)/EI;$$

$$\theta_b^0 = -[p_a e_4 + \tfrac{\Delta p}{\Delta \ell} e_5 + m_a(e_3 - \eta e_5) + \tfrac{\Delta m}{\Delta \ell}(e_4 - \eta e_6)]/EI + [M_{Ta}(e_2 - \eta e_4)$$
$$+ \tfrac{\Delta M_T}{\Delta \ell}(e_3 - \eta e_5)]/EI;$$

$$V_b^0 = -p_a(e_2 + \zeta e_4) - \tfrac{\Delta p}{\Delta \ell}(e_3 + \zeta e_5) + \lambda(m_a e_5 + \tfrac{\Delta m}{\Delta \ell} e_6) - M_{Ta}\lambda e_4 - \tfrac{\Delta M_T}{\Delta \ell}\lambda e_5;$$

$$M_b^0 = -p_a e_3 - \tfrac{\Delta p}{\Delta \ell} e_4 - m_a(e_2 - \eta e_4) - \tfrac{\Delta m}{\Delta \ell}(e_3 - \eta e_5) + M_{Ta}(e_1 - 1 - \eta e_3)$$
$$+ \tfrac{\Delta M_T}{\Delta \ell}(e_2 - \ell - \eta e_4)$$

$$\tfrac{\Delta M_T}{\Delta \ell} = \tfrac{M_{Tb} - M_{Ta}}{b-a} \quad \tfrac{\Delta p}{\Delta \ell} = \tfrac{p_b - p_a}{b-a} \quad \tfrac{\Delta m}{\Delta \ell} = \tfrac{m_b - m_a}{b-a}$$

p_a is the magnitude of the distributed applied force at $x = a$ (force/length).

m_a is the magnitude of the distributed applied moment at $x = a$ (force-length/length).

$\frac{\Delta p}{\Delta \ell}$ is the gradient of the distributed applied force, linearly varying in the x direction (force/length^2).

$\frac{\Delta m}{\Delta \ell}$ is the gradient of the distributed applied moment, linearly varying in the x direction (force-length/length^2).

k_s is the shear form factor.

k is the Winkler (elastic) foundation modulus (force/length^2).

k^* is the rotary foundation modulus (force-length/length).

N is a compressive axial force.

E is the modulus of elasticity of the material.

I is the moment of inertia taken about the neutral axis.

G is the shear modulus of elasticity.

A is the cross-sectional area.

M_{Ta} is the magnitude of the distributed thermal moment at $x = a$.

$\Delta M_T/\Delta \ell$ is the gradient of the thermal moment, linearly varying in the x-direction. The moment is defined as follows:

$$M_T = \int_A E\,\alpha\,\Delta T\,z\,dA$$

where ΔT is the temperature change and α is the coefficient of thermal expansion.

TABLE 4.3

Transfer Matrix for a Beam Elements, Part a: (*Continued*)

Note: The column 5 definitions of Part b along with case 6 or 7 of Part c of this table apply for any magnitude of λ, ζ, η. However, usually the transfer matrix elements will then be complex quantities, and the computer calculations may be quite difficult.

To use this general transfer matrix, follow the steps:

1. Calculate the three parameters λ, ζ, and η. If shear deformation is not to be considered, set $1/k_s GA = 0$
2. Compare the magnitude of these parameters and look up the appropriate e_i functions in Parts b and c of this table.
3. Substitute these e_i expressions in the general transfer matrix above.

A second general, viable technique for deriving transfer matrices is based on the use of the Laplace transform. The homogeneous form of Eq. (4.94) can be combined into a single fourth-order equation

$$\frac{d^4w}{dx^4} + (\zeta - \eta)\frac{d^2w}{dx^2} + (\lambda - \zeta\eta)w = 0 \tag{4.109}$$

where

$$\zeta = (N - k^*)/EI, \quad \eta = k/(k_s GA), \quad \lambda = k/EI$$

The Laplace transform of Eq. (4.109) is

$$\begin{aligned} &w(s)\,[s^4 + (\zeta - \eta)s^2 + (\lambda - \zeta\eta)] \\ &= s^3w(0) + s^2w'(0) + sw''(0) + w'''(0) + (\zeta - \eta)w'(0) + (\zeta - \eta)sw(0) \end{aligned} \tag{4.110}$$

where s is the transform variable. The inverse transform gives

$$\begin{aligned} w(x) = {} & [e_1(x) + (\zeta - \eta)e_3(x)]w(0) + [e_2(x) + (\zeta - \eta)e_4(x)]w'(0) \\ & + e_3(x)\,w''(0) + e_4(x)w'''(0) \end{aligned} \tag{4.111}$$

where

$$e_i(x) = L^{-1}\left[\frac{s^{4-i}}{s^4 + (\zeta - \eta)s^2 + \lambda - \zeta\eta}\right] \tag{4.112}$$

The quantity L^{-1} indicates the inverse Laplace transform. It follows from Eq. (4.112) that several useful identities hold:

$$\begin{aligned} e_i(x) &= \frac{d}{dx}e_{i+1}(x) \qquad i = -2, -1, 0, 1, 2, 3 \\ e_{i+1}(x) &= \int_0^x e_i(u)\,du \qquad i = 4, 5, 6 \end{aligned} \tag{4.113}$$

Take the first three derivatives of $w(x)$ and arrange the results as

$$\begin{bmatrix} w(x) \\ w'(x) \\ w''(x) \\ w'''(x) \end{bmatrix} = \begin{bmatrix} e_1 + (\zeta - \eta)e_3 & e_2 + (\zeta - \eta)e_4 & e_3 & e_4 \\ e_0 + (\zeta - \eta)e_2 & e_1 + (\zeta - \eta)e_3 & e_2 & e_3 \\ e_{-1} + (\zeta - \eta)e_1 & e_0 + (\zeta - \eta)e_2 & e_1 & e_2 \\ e_{-2} + (\zeta - \eta)e_0 & e_{-1} + (\zeta - \eta)e_1 & e_0 & e_1 \end{bmatrix} \begin{bmatrix} w(0) \\ w'(0) \\ w''(0) \\ w'''(0) \end{bmatrix} \tag{4.114}$$

$$\mathbf{w}(x) = \mathbf{Q}(x)\,\mathbf{w}(0)$$

TABLE 4.3

Transfer Matrix for a Beam Element, Part b:

		$\lambda = 0, \lambda - \zeta\eta = 0$		$\lambda > 0, \lambda - \zeta\eta > 0$		
	1.* $\lambda < 0$	2. $\zeta = \eta = 0$	3. $\eta = 0, \zeta \neq 0$	4. $\lambda - \zeta\eta = \frac{1}{4}(\zeta - \eta)^2$	5. $\lambda - \zeta\eta < \frac{1}{4}(\zeta - \eta)^2$, $\zeta - \eta \neq 0$	6. $\lambda - \zeta\eta > \frac{1}{4}(\zeta - \eta)^2$
e_0	$\frac{1}{g}(a^3C - b^3D)$	0	$-\zeta B$	$-\frac{\zeta-\eta}{4}(3C + A\ell)$	$-\frac{1}{g}(b^3D - a^3C)$	$-(\lambda - \zeta\eta)e_4 - (\zeta - \eta)e_2$
e_1	$\frac{1}{g}(a^2A + b^2B)$	1	A	$\frac{1}{2}(2A - B\ell)$	$\frac{p}{g}(b^2B - a^2A)$	$AB - \frac{b^2-a^2}{2ab}CD$
e_2	$\frac{1}{g}(aC + bD)$	ℓ	B	$\frac{1}{2}(C + A\ell)$	$\frac{p}{g}(bD - aC)$	$\frac{1}{2ab}(aAD + bBC)$
e_3	$\frac{1}{g}(A - B)$	$\frac{\ell^2}{2}$	$\frac{1}{\zeta}(1 - A)$	$\frac{C\ell}{2}$	$\frac{1}{g}(A - B)$	$\frac{1}{2ab}CD$
e_4	$\frac{1}{g}\left(\frac{C}{a} - \frac{D}{b}\right)$	$\frac{\ell^3}{6}$	$\frac{1}{\zeta}(\ell - B)$	$\frac{1}{(\zeta-\eta)}(C - A\ell)$	$\frac{1}{g}\left(\frac{C}{a} - \frac{D}{b}\right)$	$\frac{1}{2(a^2+b^2)}\left(\frac{AD}{b} - \frac{BC}{a}\right)$
e_5	$\frac{1}{g}\left(\frac{A}{a^2} + \frac{B}{b^2}\right) - \frac{1}{a^2b^2}$	$\frac{\ell^4}{24}$	$\frac{1}{\zeta}\left(\frac{\ell^2}{2} - e_3\right)$	$\frac{2}{(\zeta-\eta)^2}(-2A - B\ell + 2)$	$\frac{p}{g}\left(\frac{B}{b^2} - \frac{A}{a^2}\right) + \frac{1}{a^2b^2}$	$\frac{1-e_1}{\lambda-\zeta\eta} - \frac{\zeta-\eta}{\lambda-\zeta\eta}e_3$
e_6	$\frac{1}{g}\left(\frac{C}{a^3} + \frac{D}{b^3}\right) - \frac{\ell}{a^2b^2}$	$\frac{\ell^5}{120}$	$\frac{1}{\zeta}\left(\frac{\ell^3}{6} - e_4\right)$	$\frac{2}{(\zeta-\eta)^2}(-3C + A\ell + 2\ell)$	$\frac{p}{g}\left(\frac{D}{b^3} - \frac{C}{a^3}\right) + \frac{\ell}{a^2b^2}$	$\frac{\ell-e_2}{\lambda-\zeta\eta} - \frac{\zeta-\eta}{\lambda-\zeta\eta}e_4$

*This case may appear to be of little practical value, but it is retained for later use with dynamics problems.

TABLE 4.3

Transfer Matrix for a Beam Element, Part c:

$\lambda < 0$	$\lambda = 0$, $\lambda - \zeta\eta = 0$	$\lambda > 0, \lambda - \zeta\eta > 0$: $\lambda - \zeta\eta = \frac{1}{4}(\zeta - \eta)^2$	$\lambda > 0, \lambda - \zeta\eta > 0$: $\lambda - \zeta\eta < \frac{1}{4}(\zeta - \eta)^2, \zeta - \eta \neq 0$	$\lambda > 0, \lambda - \zeta\eta > 0$: $\lambda - \zeta\eta > \frac{1}{4}(\zeta - \eta)^2$
1. $A = \cosh a\ell,\ B = \cos b\ell$ $C = \sinh a\ell,\ D = \sin b\ell$ $g = a^2 + b^2$ $a^2 = \sqrt{\beta^4 + \frac{1}{4}(\zeta + \eta)^2} - \frac{1}{2}(\zeta - \eta)$ $b^2 = \sqrt{\beta^4 + \frac{1}{4}(\zeta + \eta)^2} + \frac{1}{2}(\zeta - \eta)$ $\beta^4 = -\lambda$	2. $\zeta > 0: \alpha^2 = \zeta$ $A = \cos a\ell$ $B = (\sin a\ell)/\alpha$	4. $\zeta - \eta > 0: \beta^2 = \frac{1}{2}(\zeta - \eta)$ $A = \cos \beta\ell,\ B = \beta \sin \beta\ell$ $C = (\sin \beta\ell)/\beta$	6. $\zeta - \eta > 0: g = b^2 - a^2,\ p = 1$ $A = \cos a\ell,\ B = \cos b\ell$ $C = \sin a\ell,\ D = \sin b\ell$ $a^2 = \frac{1}{2}(\zeta - \eta) - \sqrt{\frac{1}{4}(\zeta + \eta)^2 - \lambda}$ $b^2 = \frac{1}{2}(\zeta - \eta) + \sqrt{\frac{1}{4}(\zeta + \eta)^2 - \lambda}$	8. $A = \cosh a\ell,\ B = \cos b\ell$ $C = \sinh a\ell,\ D = \sin b\ell$ $a^2 = \frac{1}{2}\sqrt{\lambda - \zeta\eta} - \frac{1}{4}(\zeta - \eta)$ $b^2 = \frac{1}{2}\sqrt{\lambda - \zeta\eta} + \frac{1}{4}(\zeta - \eta)$
	3. $\zeta < 0: \alpha^2 = -\zeta$ $A = \cosh \alpha\ell$ $B = (\sinh \alpha\ell)/\alpha$	5. $\zeta - \eta < 0: \beta^2 = -\frac{1}{2}(\zeta - \eta)$ $A = \cosh \beta\ell,\ B = -\beta \sinh \beta\ell$ $C = (\sinh \beta\ell)/\beta$	7. $\zeta - \eta < 0: g = a^2 - b^2,\ p = -1$ $A = \cosh a\ell,\ B = \cosh b\ell$ $C = \sinh a\ell,\ D = \sinh b\ell$ $a^2 = -\frac{1}{2}(\zeta - \eta) + \sqrt{\frac{1}{4}(\zeta + \eta)^2 - \lambda}$ $b^2 = -\frac{1}{2}(\zeta - \eta) - \sqrt{\frac{1}{4}(\zeta + \eta)^2 - \lambda}$	

From Eq. (4.94), by finding the derivatives d^2w/dx^2 and d^3w/dx^3 from $dw/dx = -\theta + V/k_sGA$, it is possible to form $\mathbf{w}(x) = \mathbf{R}\,\mathbf{z}(x)$, which relates the deflection $w(x)$ and its derivatives to the state variables $\mathbf{z}(x)$, where

$$\mathbf{R} = \begin{bmatrix} 1 & 0 & 0 & 0 \\ 0 & -1 & 1/k_sGA & 0 \\ \eta & 0 & 0 & -1/EI \\ 0 & \zeta - \eta & -1/EI + \eta/k_sGA & 0 \end{bmatrix} \tag{4.115}$$

An expression for the transfer matrix is obtained by substituting Eq. (4.114) into $\mathbf{z}(x) = \mathbf{R}^{-1}\mathbf{w}(x)$. Thus,

$$\mathbf{z}(x) = \mathbf{R}^{-1}\mathbf{Q}(x)\mathbf{w}(0) = \mathbf{R}^{-1}\mathbf{Q}(x)\mathbf{R}\,\mathbf{z}(0)$$

or

$$\mathbf{z}_b = \mathbf{R}^{-1}\mathbf{Q}(\ell)\mathbf{R}\;\mathbf{z}_a = \mathbf{U}^i\mathbf{z}_a \tag{4.116}$$

This relationship, i.e., $\mathbf{U}^i = \mathbf{R}^{-1}\mathbf{Q}(\ell)\mathbf{R}$, forms the basis for a very useful method for finding the transfer matrix.

EXAMPLE 4.7 Transfer Matrix for a Beam Element with Axial Force

For a beam element with no foundations and no shear deformation effects,

$$\lambda = 0, \quad \eta = 0, \quad \zeta = N/EI, \quad 1/k_sGA = 0 \tag{1}$$

To develop $\mathbf{U}^i$ using Eq. (4.110), it is necessary to set up $\mathbf{R}$ and $\mathbf{R}^{-1}$, and to calculate the elements of $\mathbf{Q}$. From Eq. (4.115),

$$\mathbf{R} = \begin{bmatrix} 1 & 0 & 0 & 0 \\ 0 & -1 & 0 & 0 \\ 0 & 0 & 0 & -1/EI \\ 0 & \zeta & -1/EI & 0 \end{bmatrix} \qquad \mathbf{R}^{-1} = \begin{bmatrix} 1 & 0 & 0 & 0 \\ 0 & -1 & 0 & 0 \\ 0 & -\zeta EI & 0 & -EI \\ 0 & 0 & -EI & 0 \end{bmatrix} \tag{2}$$

From Eqs. (4.112) and (4.113), with $a^2 = \zeta$,

$$\begin{aligned}
e_1 &= L^{-1}\left(\frac{s^3}{s^4 + \zeta s^2}\right) = L^{-1}\left(\frac{s}{s^2+\zeta}\right) = \cos a\ell \\
e_0 &= -a\sin a\ell \;\; e_{-1} = -a^2\cos a\ell \;\; e_{-2} = a^3 \sin a\ell \\
e_2 &= L^{-1}\left(\frac{1}{s^2+\zeta}\right) = \frac{1}{a}\sin a\ell \\
e_3 &= L^{-1}\left(\frac{1}{s^2+\zeta}\right)\frac{1}{s} = \int_0^\ell \frac{1}{a}\sin ax\,dx = \frac{1}{a^2}(1-\cos a\ell) \\
e_4 &= \int_0^\ell \frac{1}{a^2}(1-\cos ax)\,dx = \frac{\ell}{a^2} - \frac{1}{a^3}\sin a\ell = \frac{1}{a^3}(a\ell - \sin a\ell)
\end{aligned} \tag{3}$$

For the case of a beam element with no axial force, i.e., $\zeta = 0$, the functions e_i, $i = -2, -1, 0, 1, \ldots, 4$, can be obtained by going to the limit in (3) as $a \to 0$. These e_i relations are substituted in Eq. (4.114) to form $\mathbf{Q}(\ell)$. To simplify $\mathbf{Q}$, note that $e_1 + \zeta e_3 = 1$, $e_2 + \zeta e_4 = \ell$, $e_0 + \zeta e_2 = 0$, $e_{-1} + \zeta e_1 = 0$, and, $e_{-2} + \zeta e_0 = 0$.

Now we can proceed to find the transfer matrix. First establish $\mathbf{R}^{-1}\mathbf{Q}(\ell)$ as

$$\mathbf{R}^{-1}\mathbf{Q}(\ell) = \begin{bmatrix} 1 & \ell & e_3 & e_4 \\ 0 & -1 & -e_2 & -e_3 \\ 0 & -\zeta EI & 0 & -EI \\ 0 & 0 & -EIe_1 & -EIe_2 \end{bmatrix} \tag{4}$$

And from Eq. (4.116), the transfer matrix is given by

$$\mathbf{R}^{-1}\mathbf{Q}(\ell)\mathbf{R} = \begin{bmatrix} 1 & -\ell+\zeta e_4 & -e_3/EI & -e_4/EI \\ 0 & 1-\zeta e_4 & e_2/EI & e_3/EI \\ 0 & -\zeta EIe_2 & e_1 & e_2 \\ 0 & 0 & 0 & 1 \end{bmatrix} \tag{5}$$

Some of the terms in (5) can be simplified using

$$\begin{aligned} -\ell+\zeta e_4 &= -\ell+\ell-\frac{\sin a\ell}{a} = -\frac{\sin a\ell}{a} = -e_2 \\ 1-\zeta e_3 &= \cos a\ell = e_1, \quad \zeta EI = N \end{aligned} \tag{6}$$

Then

$$\mathbf{U}^i = \mathbf{R}^{-1}\mathbf{Q}(\ell)\mathbf{R} = \begin{bmatrix} 1 & -e_2 & -e_4/EI & -e_3/EI \\ 0 & e_1 & e_3/EI & e_2/EI \\ 0 & 0 & 1 & 0 \\ 0 & -Ne_2 & e_2 & e_1 \end{bmatrix} \tag{7}$$

It is useful to check to see if this transfer matrix reduces to that for the Euler-Bernoulli beam with no axial force. If N is zero, then $\zeta = a^2 = 0$. Also, in the limit as $a \to 0$, $e_1 = 1$, $e_2 = \ell$, $e_3 = \ell^2/2$, and $e_4 = \ell^3/3!$ Then (7) reduces to Eq. (4.86) as desired. ∎

The procedure, as detailed in Example 4.7, readily provides the general transfer matrix for a beam element that is given in Table 4.3.

A General Stiffness Matrix for Beams

A very general stiffness matrix, including the effect of elastic foundations and shear deformation, can be obtained by inserting the general transfer matrix components of Table 4.3 in Eq. (4.16). This leads to the stiffness matrix of Table 4.4.

4.5.2 The Effect of Applied Loading

The effect on the response of a prescribed loading $\overline{\mathbf{P}}$ can be determined using Eq. (4.90b), i.e., $\overline{\mathbf{z}}^i = \mathbf{U}^i \int_0^\ell (\mathbf{U}^i)^{-1}\overline{\mathbf{P}}\, d\tau$. It is apparent that this effect can be calculated if the transfer matrix for the element is known either analytically or numerically.

EXAMPLE 4.8 Calculation of Loading Functions

To illustrate the use of Eq. (4.90b) to compute loading functions, suppose an Euler-Bernoulli beam segment of constant cross-section and length ℓ is loaded with a linearly increasing force described by $\overline{p}_z = p_0 x/\ell$.

As indicated in Eq. (4.93), for a beam with constant $\mathbf{A}$, $(\mathbf{U}^i(x))^{-1} = \mathbf{U}^i(-x)$. Then

$$\int_0^\ell (\mathbf{U}^i)^{-1}\overline{\mathbf{P}}\, dx = \int_0^\ell \mathbf{U}^i(-x)\overline{\mathbf{P}}\, dx = \int_0^\ell \begin{bmatrix} 1 & x & \frac{x^3}{6EI} & -\frac{x^2}{2EI} \\ 0 & 1 & \frac{x^2}{2EI} & -\frac{x}{EI} \\ 0 & 0 & 1 & 0 \\ 0 & 0 & -x & 1 \end{bmatrix} \begin{bmatrix} 0 \\ 0 \\ -p_0 x/\ell \\ 0 \end{bmatrix} dx$$

$$= \int_0^\ell \frac{1}{\ell} \begin{bmatrix} -\frac{p_0 x^4}{6EI} \\ -\frac{p_0 x^3}{2EI} \\ -p_0 x \\ p_0 x^2 \end{bmatrix} dx \tag{1}$$

TABLE 4.4

Stiffness Matrix for a Beam Element

$$\begin{bmatrix} V_a \\ M_a \\ V_b \\ M_b \end{bmatrix} = \begin{bmatrix} k_{11} & k_{12} & k_{13} & k_{14} \\ k_{21} & k_{22} & k_{23} & k_{24} \\ k_{31} & k_{32} & k_{33} & k_{34} \\ k_{41} & k_{42} & k_{43} & k_{44} \end{bmatrix} \begin{bmatrix} w_a \\ \theta_a \\ w_b \\ \theta_b \end{bmatrix} - \begin{bmatrix} \overline{V}_a^0 \\ \overline{M}_a^0 \\ \overline{V}_b^0 \\ \overline{M}_b^0 \end{bmatrix}$$

$$\mathbf{p}^i = \mathbf{k}^i \quad \mathbf{v}^i - \overline{\mathbf{P}}^{i0}$$

Definitions:

Sign Convention 2	$k_{11} = [(e_2 - \eta e_4)(e_1 + \zeta e_3) + \lambda e_3\, e_4]EI/\nabla$
$\lambda = k/EI$	$k_{12} = [e_3(e_1 - \eta e_3) - e_2(e_2 - \eta e_4)]EI/\nabla$
$\eta = k/k_s GA$	$k_{13} = -(e_2 - \eta e_4)EI/\nabla$
$\zeta = (N - k^*)/EI$	$k_{14} = -e_3 EI/\nabla$
$\xi = EI/k_s GA$	$k_{21} = k_{12}$
See Table 4.3 for the definitions of	$k_{22} = \{-(e_1 - \eta e_3)[e_4 - \xi(e_2 + \zeta e_4)] + e_2\, e_3\}EI/\nabla$
$e_i, i = 1, 2, 3, 4$ and	$k_{23} = e_3 EI/\nabla = -k_{14}$
$w_b^0, \theta_b^0, V_b^0, M_b^0$	$k_{24} = [e_4 - \xi(e_2 + \zeta e_4)]EI/\nabla$
$\nabla = e_3^2 - (e_2 - \eta e_4)[e_4 - \xi(e_2 + \zeta e_4)]$	$k_{31} = k_{13}$
EI is the bending stiffness, $k_s GA$ is the shear stiffness	$k_{32} = k_{23}$
	$k_{33} = [(e_1 + \zeta e_3)(e_2 - \eta e_4) + \lambda e_3\, e_4]EI/\nabla = k_{11}$
Set $1/k_s GA = 0$ if shear deformation is not to be considered.	$k_{34} = \{(e_1 + \zeta e_3)e_3 + \lambda e_4[e_4 - \xi(e_2 + \zeta e_4)]\}EI/\nabla$
	$k_{41} = k_{14}$
	$k_{42} = k_{24}$
$\overline{V}_a^0 = k_{13}\, w_b^0 + k_{14}\, \theta_b^0$	$k_{43} = k_{34}$
$\overline{M}_a^0 = k_{23}\, w_b^0 + k_{24}\, \theta_b^0$	$k_{44} = \{e_2\, e_3 - (e_1 - \eta e_3)[e_4 - \xi(e_2 + \zeta e_4)]\}EI/\nabla$
$\overline{V}_b^0 = -V_b^0 + k_{33}\, w_b^0 + k_{34}\, \theta_b^0$	$= k_{22}$
$\overline{M}_b^0 = -M_b^0 + k_{43}\, w_b^0 + k_{44}\, \theta_b^0$	

Finally, Eq. (4.90b) becomes

$$\overline{\mathbf{z}}^i = \mathbf{U}^i(\ell) \int_0^\ell (\mathbf{U}^i(x))^{-1} \overline{\mathbf{P}}\, dx = p_0 \begin{bmatrix} \dfrac{\ell^4}{120EI} & -\dfrac{\ell^3}{24EI} & -\dfrac{\ell}{2} & -\dfrac{\ell^2}{6} \end{bmatrix}^T \tag{2}$$

which is the result at $x = \ell$. If $x < \ell$,

$$\overline{\mathbf{z}}^i = \mathbf{U}^i((x)) \int_0^x (\mathbf{U}^i(\tau))^{-1} \overline{\mathbf{P}}(\tau) d\tau$$

$$= \frac{p_0}{\ell} \begin{bmatrix} \dfrac{x^5}{120EI} & -\dfrac{x^4}{24EI} & -\dfrac{x^2}{2} & -\dfrac{x^3}{6} \end{bmatrix}^T \tag{3}$$

With the assistance of $\mathbf{U}^i = e^{\mathbf{A}x}$, the loading vector $\overline{\mathbf{z}}^i$ can be expressed in the series form

$$\overline{\mathbf{z}}^i = \sum_{j=0}^{\infty} \frac{\mathbf{A}^j x^{(j+1)}}{(j + k + 1)!} k!\, \overline{\mathbf{P}} \tag{4}$$

where $k = 0$ for a uniform load, $k = 1$ for a linearly varying load, etc. For example, if $k = 1$, then

$$\overline{\mathbf{z}}^i = \left(\frac{\mathbf{I}x}{2} + \frac{\mathbf{A}^1 x^2}{3!} + \frac{\mathbf{A}^2 x^3}{4!} + \frac{\mathbf{A}^3 x^4}{5!} \right) \overline{\mathbf{P}} \tag{5}$$

where use has been made of $\mathbf{A}^j = 0$ for $j \geq 4$ and $\mathbf{A}^0 = \mathbf{I}$, the unit diagonal matrix. Equation (5) at $x = \ell$ gives (2). ∎

As indicated in the following chapter, the inclusion of applied loadings in the transfer matrix form of the response can be accomplished in several other ways. A simple approach which is particularly useful if the transfer matrix elements are known analytically will be discussed briefly here.

It is convenient to introduce a general notation for a transfer matrix:

$$\begin{bmatrix} w \\ \theta \\ V \\ M \end{bmatrix}_b = \begin{bmatrix} U_{ww} & U_{w\theta} & U_{wV} & U_{wM} \\ U_{\theta w} & U_{\theta\theta} & U_{\theta V} & U_{\theta M} \\ U_{Vw} & U_{V\theta} & U_{VV} & U_{VM} \\ U_{Mw} & U_{M\theta} & U_{MV} & U_{MM} \end{bmatrix} \begin{bmatrix} w \\ \theta \\ V \\ M \end{bmatrix}_a + \begin{bmatrix} w_b^0 \\ \theta_b^0 \\ V_b^0 \\ M_b^0 \end{bmatrix} \tag{4.117}$$

$$\mathbf{z}_b \quad = \quad \mathbf{U}^i(\ell) \qquad \mathbf{z}_a \quad + \quad \overline{\mathbf{z}}^i$$

where $\ell = x_b - x_a$ and U_{ij}, i and $j = w, \theta, V, M$, represents a transfer matrix element. For example, by comparison of Eqs. (4.86) and (4.117), $U_{w\theta}$ for a simple beam segment is $-\ell$. The loading functions w_b^0, θ_b^0, V_b^0, and M_b^0 are defined in Eq. (4.87) for a simple uniform beam segment or by Eq. (4.90b) for a more general element.

The form of a transfer matrix is such that the effect of various types of loading can be identified readily. For example, it is apparent from Eq. (4.117) that a shear force V at $x = a$ contributes the magnitude $VU_{wV}(\ell)$ to the deflection w at $x = b$. An applied concentrated load would have the effect of a shear force. Thus, the effect on the deflection at $x = b$ of a downward concentrated force $\overline{P}$ at $x = a$ would be expressed as $-\overline{P}U_{wV}(\ell)$ for Sign Convention 1. This same reasoning may be applied to the other responses, θ, V, and M, giving the loading function vector for a concentrated load at $x = a$ as

$$\begin{bmatrix} w_b^0 \\ \theta_b^0 \\ V_b^0 \\ M_b^0 \end{bmatrix} = \begin{bmatrix} -\overline{P}U_{wV}(\ell) \\ -\overline{P}U_{\theta V}(\ell) \\ -\overline{P}U_{VV}(\ell) \\ -\overline{P}U_{MV}(\ell) \end{bmatrix} \tag{4.118}$$

This procedure is readily extended [Pilkey and Chang, 1978] to distributed loads with the loading functions given by the Duhamel[8] or convolution integral

$$j_b^0 = -\int_0^\ell \overline{p}_z(x)\, U_{jV}(\ell - x)\, dx = -\int_0^\ell \overline{p}_z(\ell - x)\, U_{jV}(x)\, dx \tag{4.119}$$

with $j = w, \theta, V, M$. Two forms of the convolution integral are provided in Eq. (4.119) since one or the other can be more convenient for a particular problem. Since many types of loading can be expressed in terms of the loading intensity $\overline{p}_z(x)$, including both distributed and concentrated loadings, e.g., applied forces, moments, and thermal loading, Eqs. (4.90b) or (4.119) are very versatile for computing the loading terms in the transfer matrix.

EXAMPLE 4.9 Loading Functions for a Linearly Varying Load

Find the loading function component w_b^0 for the linearly varying distributed applied load of Fig. 4.14.

[8]Jean Marie Constant Duhamel (1797–1872) was a French professor of higher algebra at the Sorbonne University. Hermite was his successor in this position. Using Poisson's theory of elasticity, Duhamel investigated mathematically the influence of temperature on stress.

As indicated in Fig. 4.14, the distributed load is represented analytically as

$$\overline{p}_z(x) = \frac{p_0}{L}(L - x) \tag{1}$$

Then, from Eq. (4.119),

$$\begin{aligned} w_b^0 &= -\int_0^L \overline{p}_z(x)\, U_{wV}(L - x)\, dx \\ &= -\int_0^L \frac{p_0}{L}(L - x)\left[-\frac{(L - x)^3}{3!EI}\right] dx = \frac{p_0 L^4}{30EI} \end{aligned} \tag{2}$$

where ℓ in Eq. (4.119) has been set equal to L. ▮

4.5.3 Response of Simple Beams

The transfer matrix developed in Sections 4.5.1 and 4.5.2 can be used to find the displacement and other state variables along uniform beams. For prescribed loading, Eq. (4.90a) can be employed to provide these variables. The boundary conditions are utilized to evaluate the state variables at the left end, i.e., to find $\mathbf{z}_a$.

EXAMPLE 4.10 Beam with Linearly Varying Loading

Return to the beam of Fig. 4.14, which was treated in Example 4.5 using a stiffness matrix. The boundary conditions for this fixed-hinged beam are

$$w_{x=0} = 0, \quad \theta_{x=0} = 0, \quad w_{x=L} = 0, \quad M_{x=L} = 0 \tag{1}$$

These conditions can be inserted in $\mathbf{z}_b = \mathbf{U}^i \mathbf{z}_a + \overline{\mathbf{z}}^i$ to evaluate $\mathbf{z}_a$. Let x_a correspond to the location of the left end of the beam so that the elements of $\mathbf{z}_a$ are the state variables at the origin, i.e., $x = x_a = 0$, of the beam. Also, set $x_b = L$ so that the elements of $\mathbf{z}_b$ are the state variables at the right end of the beam. Then, using the transfer matrix of Eq. (4.86),

$$\underset{\mathbf{z}_b}{\begin{bmatrix} w = 0 \\ \theta \\ V \\ M = 0 \end{bmatrix}_{x=L}} = \underset{\mathbf{U}^i}{\begin{bmatrix} 1 & -L & -\frac{L^3}{6EI} & -\frac{L^2}{2EI} \\ 0 & 1 & \frac{L^2}{2EI} & \frac{L}{EI} \\ 0 & 0 & 1 & 0 \\ 0 & 0 & L & 1 \end{bmatrix}} \underset{\mathbf{z}_a}{\begin{bmatrix} w = 0 \\ \theta = 0 \\ V \\ M \end{bmatrix}_{x=0}} + \underset{\overline{\mathbf{z}}^i}{\begin{bmatrix} p_0L^4/30EI \\ -p_0L^3/8EI \\ -p_0L/2 \\ -p_0L^2/3 \end{bmatrix}}$$

where the loading function column is calculated using Eq. (4.113). In fact, w_b^0 for this loading is given by Eq. (2) of Example 4.9. The unknown initial parameters $V_{x=0}$, $M_{x=0}$ of the above expression are evaluated by solving the equations $w_{x=L} = 0$, $M_{x=L} = 0$. Thus, with $V_{x=0} = V_0$, $M_{x=0} = M_0$,

$$\begin{aligned} w_{x=L} &= 0 = -V_0L^3/6EI - M_0L^2/2EI + p_0L^4/30EI \\ M_{x=L} &= 0 = V_0L + M_0 - p_0L^2/3 \end{aligned} \tag{2}$$

Cramer's rule or another solution technique gives

$$V_0 = (2/5)p_0L, \qquad M_0 = (-1/15)p_0L^2 \tag{3}$$

The initial shear force and bending moment in Sign Convention 1 have been utilized here. As expected, the signs of V_0 and M_0 of (3) differ from those in Example 4.5, where Sign Convention 2 applies.

Since $w_0 = \theta_0 = 0$ and V_0, M_0 are given by (3), all of the elements of $\mathbf{z}_a$ are known. Then the deflection, slope, shear, and moment at $x = L$ are given by $\mathbf{z}_b = \mathbf{U}^i\ \mathbf{z}_a + \overline{\mathbf{z}}^i$, where $\mathbf{U}^i$ is taken from Eq. (4.86). The response $\mathbf{z}$ at x less than L is evaluated by assigning x_b in $\mathbf{U}^i \mathbf{z}_a + \overline{\mathbf{z}}^i$ to be any location along the beam at which the state variables are sought. Of course, ℓ is chosen to correspond to this location also, i.e., $\ell = x_b - x_a = x_b$. ∎

4.5.4 Inclusion of Axial Extension and Torsion

The transfer matrix can be expanded to include the effects of axial extension. If the axial displacement is u, the governing differential equations for extension of a bar of cross-sectional area A are (Chapter 1)

$$\frac{du}{dx} = \frac{N}{EA}, \qquad \frac{dN}{dx} = -\overline{p}_x \tag{4.120}$$

In transfer matrix form, the solution to these equations is

$$\begin{aligned} u &= u_0 + N_0 \frac{x}{EA} - \int \int_0^x \frac{\overline{p}_x(\tau)}{EA} d\tau \\ N &= N_0 - \int_0^x \overline{p}_x(\tau)\, d\tau \end{aligned} \tag{4.121}$$

or, in terms of x_a, x_b, and $\ell = x_b - x_a$,

$$\begin{array}{ccccccc} \begin{bmatrix} u \\ N \end{bmatrix}_b & = & \begin{bmatrix} 1 & \ell/EA \\ 0 & 1 \end{bmatrix} & \begin{bmatrix} u \\ N \end{bmatrix}_a & + & \begin{bmatrix} u_b^0 \\ N_b^0 \end{bmatrix} \\ \mathbf{z}_b & = & \mathbf{U}^i & \mathbf{z}_a & + & \overline{\mathbf{z}}^i \end{array} \tag{4.122}$$

If the effects of extension and bending are combined, then, for the sign convention of Fig. 4.1a (Sign Convention 1),

$$\begin{array}{ccccc} \begin{bmatrix} u \\ w \\ \theta \\ N \\ V \\ M \end{bmatrix}_b & = & \begin{bmatrix} 1 & & & \ell/EA & & \\ & 1 & -\ell & & -\ell^3/6EI & -\ell^2/2EI \\ & & 1 & & \ell^2/2EI & \ell/EI \\ & & & 1 & & \\ & & & & 1 & \\ & & & & \ell & 1 \end{bmatrix} & \begin{bmatrix} u \\ w \\ \theta \\ N \\ V \\ M \end{bmatrix}_a \\ \mathbf{z}_b & = & \mathbf{U}^i & \mathbf{z}_a \end{array} \tag{4.123}$$

Table 4.5 provides loading vectors $\overline{\mathbf{z}}^i$ for several common applied forces.

The governing first order equations for the torsion of a shaft are (Chapter 1)

$$\frac{d\phi}{dx} = \frac{M_t}{GJ}, \qquad \frac{dM_t}{dx} = -\overline{m}_x \tag{4.124}$$

where ϕ is the angle of twist, M_t is the twisting moment, G is the shear modulus of elasticity, J is the torsional constant, and $\overline{m}_x$(force-length/length) is the magnitude of the applied distributed torque. Because of the similarity of Eq. (4.124) with the governing equations [Eq. (4.120)] for the extension of a bar, the transfer matrices of Eqs. (4.122) and (4.123) apply to the torsion of a bar if u, N, EA, and $\overline{p}_x$ are replaced by ϕ, M_t, GJ, and $\overline{m}_x$, respectively.

TABLE 4.5

Loading Vactor $\overline{\mathbf{z}}^i$ for the Transfer Matrix of a Beam.* (Sign Convention 1)

$\overline{\mathbf{z}}^i$	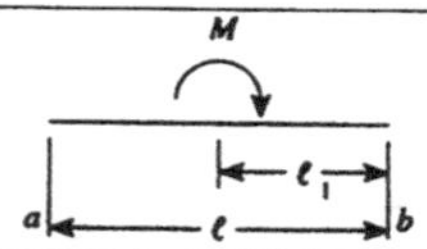	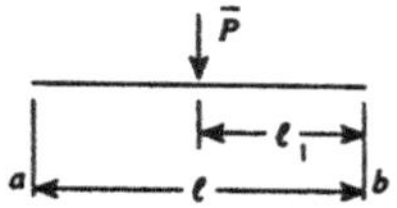	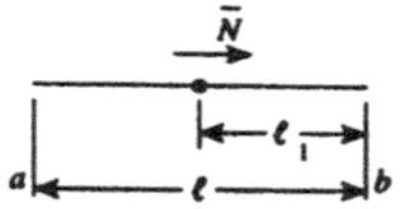	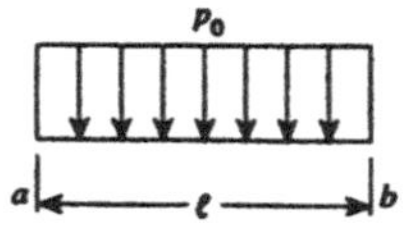	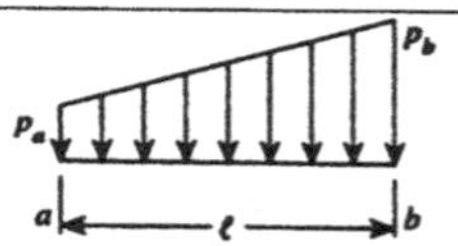
u_b^0	0	0	$\frac{\overline{N}\ell_1}{EI}$	0	0
w_b^0	$-\frac{\overline{M}\ell_1^2}{2EI}$	$\frac{\overline{P}\ell_1^3}{6EI}$	0	$\frac{p_0\ell^4}{24EI}$	$\frac{(4p_a+p_b)\ell^4}{120EI}$
θ_b^0	$\frac{\overline{M}\ell_1}{EI}$	$-\frac{\overline{P}\ell_1^2}{2EI}$	0	$-\frac{p_0\ell^3}{6EI}$	$-\frac{(3p_a+p_b)\ell^3}{24EI}$
N_b^0	0	0	$-\overline{N}$	0	0
V_b^0	0	$-\overline{P}$	0	$-p_0\ell$	$-\frac{(p_a+p_b)\ell}{2}$
M_b^0	$\overline{M}$	$-\overline{P}\ell_1$	0	$\frac{-p_0\ell^2}{2}$	$-\frac{(2p_a+p_b)\ell^2}{6}$

$\overline{\mathbf{z}}^i$	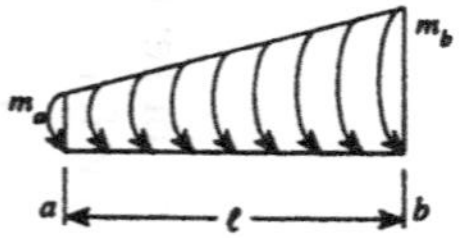	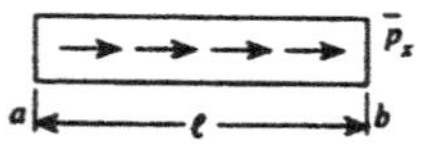	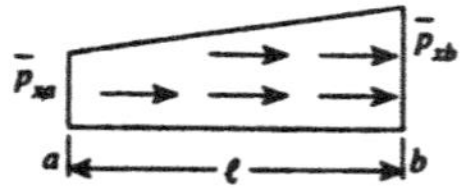	Uniform Temperature Change ΔT $\alpha =$ Thermal Coefficient	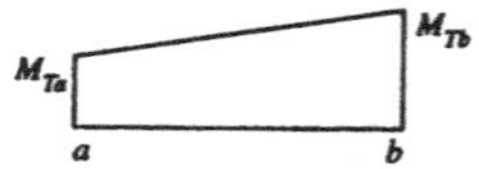
u_b^0	0	$-\frac{\overline{p}_x\ell^2}{2EA}$	$-\frac{\ell^2}{6EA}(2\overline{p}_{xa}+\overline{p}_{xb})$	$\alpha\ \ell\ \Delta T$	0
w_b^0	$\frac{\ell^3(3m_a+m_b)}{24\ EI}$	0	0	0	$-\frac{\ell^2}{2EI}\left(M_{Ta}+\frac{\ell}{3}\frac{\Delta M_T}{\Delta\ell}\right)$
θ_b^0	$-\frac{\ell^2(2m_a+m_b)}{6\ EI}$	0	0	0	$\frac{\ell}{EI}\left(M_{Ta}+\frac{\ell}{2}\frac{\Delta M_T}{\Delta\ell}\right)$
N_b^0	0	$-\overline{p}_x\ell$	$-\frac{\ell(\overline{p}_{xa}+\overline{p}_{xb})}{2}$	0	0
V_b^0	0	0	0	0	0
M_b^0	$-\frac{(m_a+m_b)\ell}{2}$	0	0	0	0

* This table applies to the ith element of Euler-Bernoulli beams with uncoupled extension. For more complex beam elements, see Table 4.3.
Definitions: $M_T = \int_A E\alpha\ \Delta T\ z\,dA$

$\Delta M_T/\Delta\ell = \frac{M_{T_b}-M_{T_a}}{b-a}$

References

Hetenyi, M., 1946, *Beams on Elastic Foundation*, University of Michigan Press, Ann Arbor, MI.
Pestel, E. and Leckie, F., 1963, *Matrix Methods in Elastomechanics*, McGraw-Hill, NY.
Pilkey, W.D., 1994, *Formulas for Stress, Strain, and Structural Matrices*, Wiley, NY.
Pilkey, W.D., 2002, *Analysis and Design of Elastic Beams, Computational Methods*, Wiley, NY.
Pilkey, W.D. and Chang, P.Y., 1978, *Modern Formulas for Statics and Dynamics*, McGraw-Hill, NY.
Pilkey, W.D. and Pilkey, O.H., 1986, *Mechanics of Solids*, Krieger Publishers, Melbourne, FL.
Wunderlich, W., 1967, On the analysis of shells of revolution by transfer matrices, *Ing.-Arch.*, No. 36, pp. 262–279.
Wunderlich, W., 1966, Calculation of transfer matrices applied to the bending theory of shells of Revolution, Proc. Int. Symposium The Use of Electronic Digital Computers in Structural Engineering, Newcastle upon Tyne.
Zurmühl, R. and Falk, S., 1986, *Matrizen and ihre Anwendungen, Numerische Methoden, Teil 2*, 5th ed., Springer-Verlag, Berlin.

Problems

Stiffness matrices

4.1 Several stiffness coefficients for a beam were found in Section 4.4.1 using a direct evaluation procedure. Complete the direct evaluation of the stiffness coefficients for a beam using the configurations of Fig. 4.7.

4.2 Find the stiffness matrix for an extension bar element (Fig. P4.2) using the approximate series approach. Begin with the series $\tilde{u}(\xi) = \hat{u}_1 + \hat{u}_2\xi$, where $\xi = x/\ell$.

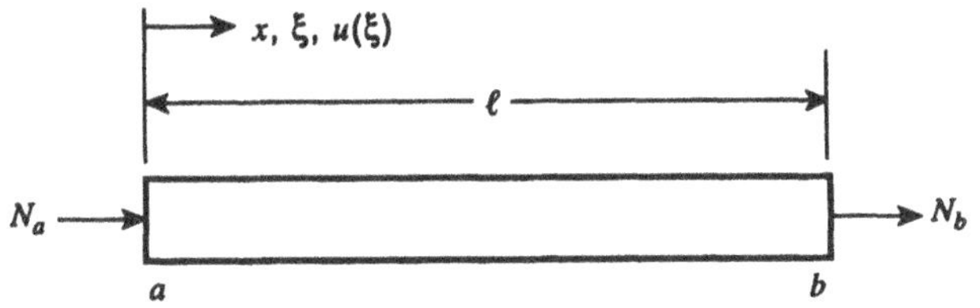

FIGURE P4.2

Answer: See the stiffness matrix for extension in Example 4.1.

4.3 Repeat the previous problem, except add a new DOF at the point which lies halfway between a and b, and use a quadratic series

$$\tilde{u}(\xi) = \hat{u}_1 + \hat{u}_2\xi + \hat{u}_3\xi^2$$

Answer:

$$\mathbf{k}^i = \frac{EA}{3\ell}\begin{bmatrix} 7 & -8 & 1 \\ -8 & 16 & -8 \\ 1 & -8 & 7 \end{bmatrix}$$

4.4 Derive the 2×2 stiffness matrix of the stepped truss element (extension bar) shown in Fig. P4.4.

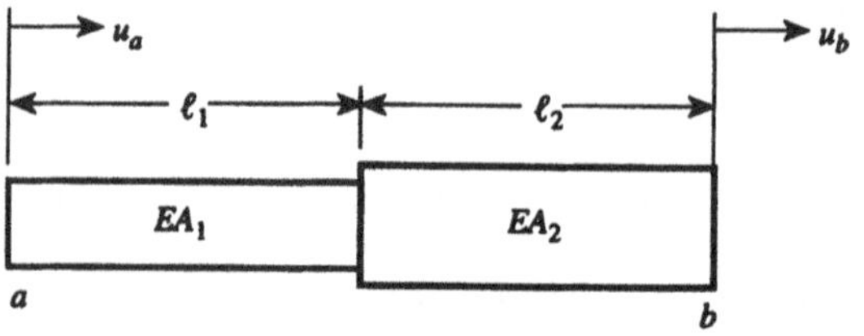

FIGURE P4.4

Answer:

$$\mathbf{k}^i = \begin{bmatrix} \alpha & -\alpha \\ -\alpha & \alpha \end{bmatrix} \quad \text{where} \quad \alpha = \frac{1}{\ell_1/EA_1 + \ell_2/EA_2}$$

4.5 Show that the stiffness and flexibility matrices associated with the DOF of the spring chain shown in Fig. P4.5 are given by

$$\mathbf{k} = \begin{bmatrix} k_1+k_2 & -k_2 & 0 & \vdots & \vdots & \vdots & 0 \\ -k_2 & k_2+k_3 & -k_3 & \vdots & \vdots & \vdots & \vdots \\ 0 & -k_3 & k_3+k_4 & \vdots & \vdots & \vdots & \vdots \\ 0 & 0 & -k_4 & \vdots & \vdots & \vdots & \vdots \\ 0 & \vdots & 0 & \vdots & \vdots & \vdots & \vdots \\ \vdots & \vdots & \vdots & \vdots & \vdots & \vdots & \vdots \\ \vdots & \vdots & \vdots & \vdots & \vdots & \vdots & \vdots \\ \vdots & \vdots & \vdots & \vdots & -k_{n-1} & k_n+k_{n-1} & -k_n \\ 0 & 0 & 0 & \vdots & \vdots & -k_n & k_n \end{bmatrix}$$

$$\mathbf{f} = \begin{bmatrix} \alpha_1 & & & & \\ \alpha_1 & \alpha_2 & & & \\ \cdots & \cdots & \text{Symmetric} & & \\ \cdots & \cdots & & & \\ \alpha_1 & \alpha_2 & \cdots\cdots\cdots\cdots & \cdots & \alpha_n \end{bmatrix} \qquad \alpha_i = \sum_{j=1}^{i} \frac{1}{k_j}$$

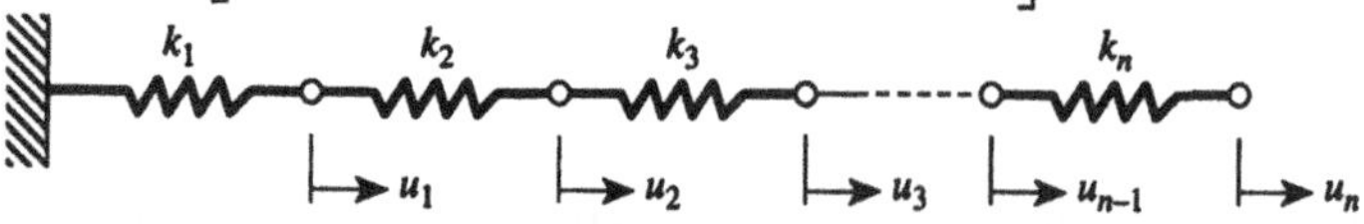

FIGURE P4.5

4.6 Suppose the moment of inertia of a beam element varies as $I(x) = I_0(1 + x^2/\ell^2)$. Use the approximate series method to find the stiffness matrix.

Answer: If the third order Hermitian interpolation polynomials of Fig. 4.8 are used,

$$\mathbf{k}^i = \frac{EI_0}{\ell^3} \begin{bmatrix} 16.8 & -7.4 & -16.8 & -9.4 \\ & 4.533 & 7.4 & 2.867 \\ & & 16.8 & 9.4 \\ & \text{Symmetric} & & 6.533 \end{bmatrix}$$

4.7 Derive a stiffness matrix for a beam element if the moment of interia is given by $I(x) = I_0(1 + 2x/\ell)$. Use an approximate series along with the principle of virtual work in this derivation.

4.8 A stiffness matrix which provides a relationship between forces $\mathbf{p}^i$ and displacements $\mathbf{v}^i$ on both ends of a member, is usually considered to have the following properties

- It is singular and, hence, cannot be inverted.
- It is symmetric.
- All diagonal elements are positive.
- After elimination of rigid body motion, it is not singular.

Suppose the stiffness matrix for a beam is

$$\mathbf{k}^i = \frac{EI}{\ell}\begin{bmatrix} 12 & -6\ell & 12\ell & -6\ell \\ -6\ell & 4\ell^2 & 6\ell & 2\ell^2 \\ 12 & -6\ell & 12\ell & -6\ell \\ -6\ell & 4\ell & 6\ell & 2\ell \end{bmatrix}$$

and for a bar is

$$\mathbf{k}^i = \frac{EA}{\ell}\begin{bmatrix} -2 & -1 \\ -1 & 2 \end{bmatrix}$$

Are these matrices proper stiffness matrices? If not, what is wrong with them?

Answer: The first matrix is not symmetic. Some elements do not have correct units. It is singular. The second matrix does not have positive diagonal units. It is not singular.

4.9 Can

$$\begin{bmatrix} 1 & -2 \\ -1 & 1 \end{bmatrix} EA/\ell$$

be a valid stiffness matrix for a bar in extension?

4.10 Consider the stiffness matrix

$$\begin{bmatrix} 12 & -6 & -12 & -6\ell \\ 6 & -4\ell^2 & 6\ell & 2\ell^2 \\ -11.99 & 6\ell & 12 & 6\ell \\ -6\ell & 2\ell^2 & 6\ell & 4\ell^2 \end{bmatrix} EI/\ell^3$$

Which elements may not be correct? Why? What is the meaning of the k_{22} element?

4.11 For a simple beam, i.e., beam theory with no consideration taken for shear deformation, show that the stiffness matrix loading vector of Eq. (4.16) is the same as obtained using [Eq. (4.66)] $\mathbf{G}^T \int_a^b \mathbf{N}_u^T \overline{p}_z\, dx$.

4.12 Derive an element stiffness matrix for a beam with axial compressive force N, shear deformation, and elastic foundations k, k^*. To do so, convert the transfer matrix of Eq. (4.102) into a stiffness matrix $\mathbf{k}^i$.

4.13 Convert the 6×6 transfer matrix of a beam with axial extension [Eq. (4.123)] into a stiffness matrix $\mathbf{k}^i$.

Hint: Use Eq. (4.16).

4.14 Derive the element stiffness matrix $\mathbf{k}^i$ for a rod of circular cross-section subject to torsional loading.

4.15 Convert the stiffness matrix $\mathbf{k}^i$ for an Euler-Bernoulli beam into a transfer matrix $\mathbf{U}^i$.

Hint: Use Eq. (4.79).

4.16 Derive the element stiffness matrix for a bar in extension on an elastic foundation (modulus k_x).

Hint: Your answer can be checked by comparison with $\mathbf{k}^i$ of Problem 4.26, which was obtained by converting the transfer matrix into a stiffness matrix.

Answer:

$$\mathbf{k}^i = \begin{bmatrix} EA\,\beta \cosh \beta\ell / (\sinh \beta\ell) & -EA\,\beta/(\sinh \beta\ell) \\ -EA\,\beta/(\sinh \beta\ell) & EA\,\beta \cosh \beta\ell/(\sinh \beta\ell) \end{bmatrix} \qquad \beta^2 = k_x/EA$$

4.17 Use the transfer matrix for an Euler-Bernoulli beam element [Eq. (4.8c)] to derive the shape function $\mathbf{N}$ of $w(x) = \mathbf{N}\mathbf{v}$.

Hint: Condense V_a and M_a out of the standard transfer matrix.

Answer: Your $\mathbf{N}$ should be the same as $\mathbf{N}$ of Eq. (4.47a).

4.18 For a rod subject to extension, the kinematic relation is $\epsilon_{0x} = d_x u_0$, the material law is $N = EA\epsilon_{0x}$, and the condition of equilibrium is $d_x N + \overline{p}_x = 0$. See Chapter 2, Problem 2.21.

(a) Derive $\mathbf{A}$ of the first-order governing equations $d_x \mathbf{z} = \mathbf{A}\mathbf{z} + \overline{\mathbf{P}}$.

(b) Use $\mathbf{A}$ to compute the transfer matrix $\mathbf{U}^i$.

(c) Convert the transfer matrix $\mathbf{U}^i$ into the element stiffness matrix $\mathbf{k}^i$.

Hint: Use Eq. (4.16).

(d) Derive $\mathbf{k}^D$ of the principle of virtual work.

(e) Derive the stiffness matrix $\mathbf{k}^i$ using the differential stiffness operator $\mathbf{k}^D$.

(f) Convert the stiffness matrix $\mathbf{k}^i$ into the transfer matrix $\mathbf{U}^i$.

Hint: Use Eq. (4.79).

Transfer Matrices

4.19 Calculate the e_i functions of Eqs. (4.112) and (4.113). Use can be made of the roots of the denominator. The four roots of

$$s^4 + (\zeta - \eta)\, s^2 + \lambda - \zeta\eta = 0$$

are

$$s_{1,2,3,4} = \pm\sqrt{-(\zeta - \eta)/2 \pm \sqrt{(\zeta - \eta)^2/4 - (\lambda - \eta\zeta)}}$$

Find $e_4(\ell)$ for

(a) $\zeta = \eta = \lambda = 0$

(b) $\eta = \lambda = 0,\ \zeta < 0$

(c) $\eta = \lambda = 0,\ \zeta > 0$

(d) $\eta = \zeta = 0,\ \lambda < 0$

(e) $\eta = \zeta = 0,\ \lambda > 0$

Answer:

(a) $e_4(\ell) = L^{-1}\left(\frac{1}{s^4}\right) = \frac{\ell^3}{6}$

(b) $e_4(\ell) = L^{-1}\left[\frac{1}{s^2(s^2-a^2)}\right] = \frac{1}{a^3}\,(\sin a\ell - a\ell)$

(c) $e_4(\ell) = L^{-1}\left[\frac{1}{s^2(s^2+a^2)}\right] = \frac{1}{a^3}(a\ell - \sin a\ell)$

(d) $e_4(\ell) = L^{-1}\frac{1}{s^4-a^4} = \frac{1}{2a^3}\,(\sinh a\ell - \sin a\ell)$

(e) $e_4(\ell) = L^{-1}\frac{1}{s^4+a^4} = \frac{1}{4a^4}\,(\cosh a\ell\ \sin a\ell - \sinh a\ell\ \cos a\ell)$

4.20 Find the initial parameters for a transfer matrix solution of a beam that is simply supported on both ends.

Answer:

$$w_0 = M_0 = 0, \quad \theta_0 = w_b^0/L + LM_b^0/(6EI), \quad V_0 = -M_b^0/L$$

4.21 Suppose a beam has a variable moment of inertia $I = I_0(1+x/\ell)$. Place the governing differential equations in first order form. Integrate these equations to find an expression for the state of variables at position x. That is, find the transfer matrix for this beam.

Hint: **A** is not constant.

4.22 Find the loading functions w_b^0, θ_b^0, V_b^0, and M_b^0 for a transfer matrix of a beam element with a parabolically distributed applied loading.

4.23 The Runge-Kutta method is a numerical integration technique for the solution of initial value problems. For the system of differential equations

$$z_1' = f_1(x, z_1, z_2)$$
$$z_2' = f_2(x, z_1, z_2)$$

with the initial values $z_1(x_0) = z_{1,0}$ and $z_2(x_0) = z_{2,0}$, one Runge-Kutta approximation for moving one step (0 to h) is

$$z_{1,1} = z_{1,0} + \frac{1}{6}(k_1 + 2k_2 + 2k_3 + k_4)$$
$$z_{2,1} = z_{2,0} + \frac{1}{6}(\ell_1 + 2\ell_2 + 2\ell_3 + \ell_4)$$

with

$$k_1 = hf_1(x_0, z_{1,0}, z_{2,0}) \qquad \ell_1 = hf_2(x_0, z_{1,0}, z_{2,0})$$

$$k_2 = hf_1\left(x_0 + \frac{h}{2}, z_{1,0} + \frac{k_1}{2}, z_{2,0} + \frac{\ell_1}{2}\right) \quad \ell_2 = hf_2\left(x_0 + \frac{h}{2}, z_{1,0} + \frac{k_1}{2}, z_{2,0} + \frac{\ell_1}{2}\right)$$

$$k_3 = hf_1\left(x_0 + \frac{h}{2}, z_{1,0} + \frac{k_2}{2}, z_{2,0} + \frac{\ell_2}{2}\right) \quad \ell_3 = hf_2\left(x_0 + \frac{h}{2}, z_{1,0} + \frac{k_2}{2}, z_{2,0} + \frac{\ell_2}{2}\right)$$

$$k_4 = hf_1(x_0 + h, z_{1,0} + k_3, z_{2,0} + \ell_3) \qquad \ell_4 = hf_2(x_0 + h, z_{1,0} + k_3, z_{2,0} + \ell_3)$$

As an example, integrate the beam equations

$$\frac{d}{dx}\begin{bmatrix} w \\ \theta \\ V \\ M \end{bmatrix} = \begin{bmatrix} 0 & -1 & 0 & 0 \\ 0 & 0 & 0 & \frac{1}{EI} \\ 0 & 0 & 0 & 0 \\ 0 & 0 & 1 & 0 \end{bmatrix} \begin{bmatrix} w \\ \theta \\ V \\ M \end{bmatrix} + \begin{bmatrix} 0 \\ 0 \\ -\overline{p}_z(x) \\ 0 \end{bmatrix}$$

Hint: For this system of four differential equations, use the first step of length ℓ

$$
\begin{aligned}
k_1 &= \ell f_1(x_0, w_0, \theta_0, V_0, M_0)\\
k_2 &= \ell f_1\left(x_0 + \frac{\ell}{2}, w_0 + \frac{k_1}{2}, \theta_0 + \frac{m_1}{2}, V_0 + \frac{n_1}{2}, M_0 + \frac{r_1}{2}\right)\\
k_3 &= \ell f_1\left(x_0 + \frac{\ell}{2}, w_0 + \frac{k_2}{2}, \theta_0 + \frac{m_2}{2}, V_0 + \frac{n_2}{2}, M_0 + \frac{r_2}{2}\right)\\
k_4 &= \ell f_1(x_0 + \ell, w_0 + k_3, \theta_0 + m_3, V_0 + n_3, M_0 + r_3)\\
m_1 &= \ell f_2(x_0, w_0, \theta_0, V_0, M_0)\\
&\vdots\\
&\text{etc.}
\end{aligned}
$$

with

$$
f_1 = -\theta, \qquad f_2 = \frac{1}{EI}M, \qquad f_3 = -\overline{p}_z(x) = p_0, \qquad f_4 = V
$$

Initial conditions:

$$
w(0) = w_0, \qquad \theta(0) = \theta_0, \qquad V(0) = V_0, \qquad M(0) = M_0
$$

$$
\begin{aligned}
k_1 &= \ell\,(-\theta_0) & m_1 &= \frac{\ell}{EI}M_0\\
k_2 &= \left(-\theta_0 - \frac{1}{2}\frac{\ell}{EI}M_0\right) & m_2 &= \tfrac{\ell}{EI}\left(M_0 + \tfrac{1}{2}\ell p_0\right)\\
k_3 &= \ell\left(-\theta_0 - \frac{1}{2}\frac{\ell}{EI}\left(M_0 + \frac{1}{2}\ell V_0\right)\right) & m_3 &= \frac{\ell}{EI}\left(M_0 + \frac{1}{2}\ell\left(V_0 + \frac{1}{2}\ell p_0\right)\right)\\
k_4 &= \ell\left(-\theta_0 - \frac{\ell}{EI}\left(M_0 + \frac{1}{2}\ell\left(V_0 + \frac{1}{2}\ell p_0\right)\right)\right) & m_4 &= \tfrac{\ell}{EI}\left(M_0 + \ell\left(V_0 + \tfrac{1}{2}\ell p_0\right)\right)\\
n_1 &= \ell p_0 & r_1 &= \ell V_0\\
n_2 &= \ell p_0 & r_2 &= \ell\left(V_0 + \tfrac{1}{2}\ell p_0\right)\\
n_3 &= \ell p_0 & r_3 &= \ell\left(V_0 + \tfrac{1}{2}\ell p_0\right)\\
n_4 &= \ell p_0 & r_4 &= \ell\left(V_0 + \ell p_0\right)
\end{aligned}
$$

Answer:

$$
\begin{aligned}
w_\ell &= w_0 + \frac{1}{6}(k_1 + 2k_2 + 2k_3 + k_4)\\
&= w_0 - \ell\theta_0 - \frac{\ell^3}{6EI}V_0 - \frac{\ell^2}{2EI}M_0 - \frac{\ell^4}{24EI}p_0\\
\theta_\ell &= \theta_0 + \frac{1}{6}(m_1 + 2m_2 + 2m_3 + m_4)\\
&= \theta_0 + \frac{\ell^2}{2EI}V_0 + \frac{\ell}{EI}M_0 + \frac{\ell^3}{6EI}p_0\\
V_\ell &= V_0 + \frac{1}{6}(n_1 + 2n_2 + 2n_3 + n_4)\\
&= V_0 + \ell p_0\\
M_\ell &= M_0 + \frac{1}{6}(r_1 + 2r_2 + 2r_3 + r_4)\\
&= \ell V_0 + M_0 + \frac{1}{2}\ell^2 p_0
\end{aligned}
$$

In transfer matrix form

$$\begin{bmatrix} w \\ \theta \\ V \\ M \end{bmatrix} = \begin{bmatrix} 1 & -\ell & -\frac{\ell^3}{6EI} & -\frac{\ell^2}{2EI} \\ 0 & 1 & \frac{\ell^2}{2EI} & \frac{\ell}{EI} \\ 0 & 0 & 1 & 0 \\ 0 & 0 & \ell & 1 \end{bmatrix} \begin{bmatrix} w \\ \theta \\ V \\ M \end{bmatrix}_0 + p_0 \begin{bmatrix} -\frac{\ell^4}{24EI} \\ \frac{\ell^3}{6EI} \\ \ell \\ \frac{\ell^2}{2} \end{bmatrix}$$

4.24 Find the deflection, slope, moment, and shear along a uniform beam fixed on both ends.

Answer: $\mathbf{z}_b = \mathbf{U}^i\mathbf{z}_a + \overline{\mathbf{z}}^i$, with $\mathbf{U}^i$ taken from Eq. (4.8c) and $w_0 = \theta_0 = 0$, $V_0 = -EI(\frac{12}{L^3}w_b^0 + \frac{6}{L^2}\theta_b^0)$, and $M_0 = EI(\frac{2}{L}\theta_b^0 + \frac{6}{L^2}w_b^0)$

4.25 Use the Laplace transform to derive the transfer matrix for a bar in extension.

4.26 The governing differential equations for a bar in extension on an elastic foundation (modulus k_x) are $du/dx = N/EA$, $dN/dx = k_x u$ or $du/dx = N/EA$, $d^2u/dx^2 = k_x u/EA$. Do not include the effects of applied loading. Derive the element transfer matrix, using

(a) An exponential series expansion

(b) The Cayley-Hamilton theorem

(c) The Laplace transform

Use this transfer matrix to derive the element stiffness matrix.

Answer:

$$\mathbf{U}^i = \begin{bmatrix} \cosh\beta\ell & \sinh\beta\ell/(EA\beta) \\ EA\beta\sinh\beta\ell & \cosh\beta\ell \end{bmatrix} \qquad \beta^2 = k_x/EA$$

$$\mathbf{k}^i = \frac{EA\beta}{\sinh\beta\ell}\begin{bmatrix} \cosh\beta\ell & -1 \\ -1 & \cosh\beta\ell \end{bmatrix}$$

Flexibility Matrices

4.27 Show that the flexibility matrix for a beam element that is hinged at the left end ($x = a$) and guided at the right end ($x = b$) is given by

$$\begin{bmatrix} w_b \\ \theta_a \end{bmatrix} = \frac{\ell}{6EI}\begin{bmatrix} 2\ell^2 & -3\ell \\ -3\ell & 6 \end{bmatrix}\begin{bmatrix} V_b \\ M_a \end{bmatrix}$$

4.28 Derive the complete (not reduced) element stiffness matrix of a beam beginning with any of the flexibility matrices discussed in this chapter.

4.29 For an extension bar on an elastic foundation (modulus k_x) with the left end ($x = a$) fixed, find the nonsingular stiffness matrix and then find the flexibility matrix for an element of length ℓ.

Hint: Remove a dependent DOF from the stiffness matrix of Problem 4.16.

Answer:

Reduced Stiffness Matrix	Flexibility Matrix
$N_b = EA\beta\frac{\cosh\beta l}{\sinh\beta l}\Delta u$	$\Delta u = \frac{\sinh\beta l}{EA\beta\cosh\beta l}N_b$
$\Delta u = u_b - u_a = u_b$	$\beta^2 = k_x/EA$

5

Structural Analysis Methods II: Structural Systems

A structure can be considered to be a system composed of structural elements connected at joints or *nodes*. This structural system is analyzed by assembling information gained from an analysis of each element. In the previous chapter, the beam element was studied as a representative structural element. In the present chapter, a beam system formed of an assembly of many beam elements will be treated, and the appropriate *global* analysis procedures investigated. The question arises as to how the element characteristics should be assembled to permit an effective and efficient computational solution of the whole structure. The correct solution must fulfill the following conditions:

1. Equilibrium. The applied forces and the resulting internal forces must be in equilibrium at each node.
2. Force-displacement relationship. The internal forces and deformations in each element must satisfy the appropriate stress-strain relationship (material law). These relations will be linear in this chapter.
3. Compatibility or kinematics. The ends of the elements must fit together at the nodes, and each element must be continuous.

We begin the study of global analysis procedures of structural systems by considering a *mixed method*, the *transfer matrix method*. This is a progressive matrix multiplication scheme that applies to a system whose geometry is line-like, such as several beam elements placed end-to-end to form a long beam or curved bar. The transfer matrix method is characterized by a sequence of matrix multiplications along the line system, a procedure which leads to the same size final matrix regardless of the number of elements in the system. It is referred to as a mixed method, since both displacement and force variables, i.e., all of the state variables, are retained throughout the computations.

Network structures, e.g., a framework, are normally treated using *force* or *displacement methods*. These techniques will be considered following the discussion of the transfer matrix method. Rather than the matrix multiplication of the transfer matrix method, the system response using the force and displacement methods is found by assembling the element response characteristics through addition. Both the force and displacement methods lead to final matrices which depend in size on the number of elements composing the system. Frequently, a combination of methods is useful. For example, it is sometimes convenient to compute the transfer matrix for some of the more complex portions of a system. Using the transformation of Chapter 4, these transfer matrices are then converted to stiffness matrices which are placed in a displacement method analysis. Also, the displacement method can be used to compute displacement and forces at the nodes, followed by the transfer matrix method to calculate the state variables between the nodes.

The force method is also referred to as the *flexibility* or *influence coefficient method* and occasionally as the *compatibility method*. This method is based on the principle of complementary virtual work, which provides global compatibility conditions that lead to a system flexibility matrix relating redundant forces (forces in excess of those that can be determined using the equations of equilibrium) to applied loadings.

The dominant method in use today in the practice of structural analysis is the displacement method which is also called the *stiffness method* or, somewhat less often, the *equilibrium method*. The displacements (at the joints or nodes of the elements into which the structure is idealized) essential for describing the deformed state of the structure are selected as the unknowns. The principle of virtual work, which is the basis for this method, provides global equilibrium conditions and corresponding global stiffness equations. These equations are solved for the nodal displacements in terms of applied forces. This method is normally considered to be much easier to automate than the force method for the solution of large structural systems [Argyris, 1954; Przemieniecki, 1968].

A number of considerations determine the most suitable method for a particular problem. These considerations will be discussed as the various methods are presented.

5.1 Transfer Matrix Method

A transfer matrix $\mathbf{U}^i$ relating the state variables $\mathbf{z}$ at point a to the state variables at point b of a structure can be written as

$$\mathbf{z}_b = \mathbf{U}^i \mathbf{z}_a + \overline{\mathbf{z}}^i \tag{5.1}$$

in which $\mathbf{U}^i$ for a beam element is given by Chapter 4, Eq. (4.8c) for Sign Convention 1. The bar over $\overline{\mathbf{z}}^i$ indicates that these are applied loads.

Note from Chapter 4, Eq. (4.9) that all of the basic relations, i.e., conditions of equilibrium, geometry, and material law, are included in the transfer matrix. All three will be satisfied simultaneously as the element matrices are placed together to represent a whole system.

From Chapter 4, Eqs. (4.86) and (4.87) (Sign Convention 1), the transfer matrix for a simple beam element appears as

$$\begin{bmatrix} w \\ \theta \\ V \\ M \end{bmatrix}_b = \begin{bmatrix} 1 & -\ell & -\ell^3/6EI & -\ell^2/2EI \\ 0 & 1 & \ell^2/2EI & \ell/EI \\ 0 & 0 & 1 & 0 \\ 0 & 0 & \ell & 1 \end{bmatrix} \begin{bmatrix} w \\ \theta \\ V \\ M \end{bmatrix}_a + \begin{bmatrix} w_b^0 \\ \theta_b^0 \\ V_b^0 \\ M_b^0 \end{bmatrix}^i \tag{5.2}$$

$$\mathbf{z}_b = \qquad \mathbf{U}^i \qquad \mathbf{z}_a \quad + \quad \overline{\mathbf{z}}^i$$

It is often useful to incorporate the loading terms in the transfer matrix. In so doing, an *extended state vector* $\mathbf{z}$ and an *extended transfer matrix* $\mathbf{U}^i$ are defined.

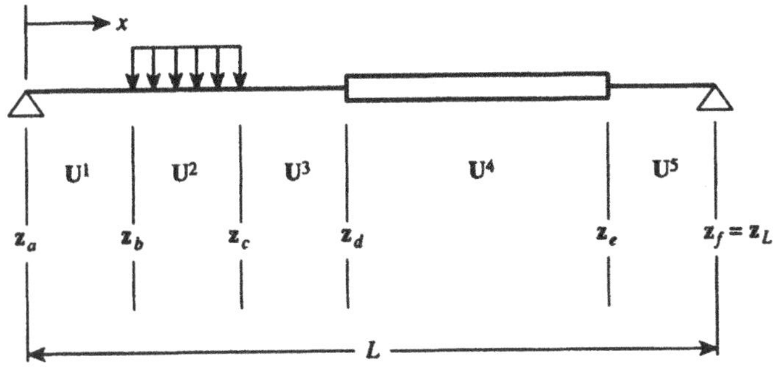

FIGURE 5.1
A beam and its corresponding transfer matrices and state vectors.

$$
\begin{bmatrix} w \\ \theta \\ V \\ M \\ \cdots \\ 1 \end{bmatrix}_b = \begin{bmatrix} 1 & -\ell & -\ell^3/6EI & -\ell^2/2EI & \vdots & w_b^0 \\ 0 & 1 & \ell^2/2EI & \ell/EI & \vdots & \theta_b^0 \\ 0 & 0 & 1 & 0 & \vdots & V_b^0 \\ 0 & 0 & \ell & 1 & \vdots & M_b^0 \\ \cdots & \cdots & \cdots & \cdots & \vdots & \cdots \\ 0 & 0 & 0 & 0 & \vdots & 1 \end{bmatrix} \begin{bmatrix} w \\ \theta \\ V \\ M \\ \cdots \\ 1 \end{bmatrix}_a \tag{5.3}
$$

$$
\mathbf{z}_b = \mathbf{U}^i \mathbf{z}_a
$$

Note that these relations, i.e., Eqs. (5.2) and (5.3), are equivalent.

Consider now the system of Fig. 5.1 formed of several beam elements joined end to end. By definition of the transfer matrices,

$$\mathbf{z}_b = \mathbf{U}^1\mathbf{z}_a \tag{5.4a}$$
$$\mathbf{z}_c = \mathbf{U}^2\mathbf{z}_b \tag{5.4b}$$
$$\mathbf{z}_d = \mathbf{U}^3\mathbf{z}_c \tag{5.4c}$$
$$\mathbf{z}_e = \mathbf{U}^4\mathbf{z}_d \tag{5.4d}$$
$$\mathbf{z}_f = \mathbf{U}^5\mathbf{z}_e \tag{5.4e}$$

where $\mathbf{z}_a$ is the state vector at the origin, i.e., at the left end. As illustrated in Fig. 5.1, $\mathbf{U}^1$ is the transfer matrix for the first element at the left end of the beam, i.e., use Eq. (5.3) with ℓ equal to the length of the first element.

Vector $\mathbf{z}_b$ is the state variable at the right end of the first element. Matrix $\mathbf{U}^2$ represents the second element and so on. Note in Eqs. (5.4a to e) that each of the state vectors $\mathbf{z}_b$, $\mathbf{z}_c$, $\mathbf{z}_d$, $\mathbf{z}_e$, and $\mathbf{z}_f$ can be expressed in terms of the initial state vector $\mathbf{z}_a$ by progressively replacing $\mathbf{z}_b$ in Eq. (5.4b) by $\mathbf{z}_b$ of Eq. (5.4a), $\mathbf{z}_c$ of Eq. (5.4c) by $\mathbf{z}_c$ of Eq. (5.4b), etc. Thus, from Eq. (5.4b), $\mathbf{z}_c = \mathbf{U}^2\mathbf{z}_b$. Equation (5.4a) permits this to be expressed as $\mathbf{z}_c = \mathbf{U}^2\mathbf{U}^1\mathbf{z}_a$. In summary, the state vector at locations, i.e., stations or nodes, b, c, d, e, and f, can be written as

$$\mathbf{z}_b = \mathbf{U}^1\mathbf{z}_a \tag{5.5a}$$
$$\mathbf{z}_c = \mathbf{U}^2\mathbf{z}_b = \mathbf{U}^2\mathbf{U}^1\mathbf{z}_a \tag{5.5b}$$
$$\mathbf{z}_d = \mathbf{U}^3\mathbf{z}_c = \mathbf{U}^3\mathbf{U}^2\mathbf{U}^1\mathbf{z}_a \tag{5.5c}$$
$$\mathbf{z}_e = \mathbf{U}^4\mathbf{z}_d = \mathbf{U}^4\mathbf{U}^3\mathbf{U}^2\mathbf{U}^1\mathbf{z}_a \tag{5.5d}$$
$$\mathbf{z}_f = \mathbf{U}^5\mathbf{z}_e = \mathbf{U}^5\mathbf{U}^4\mathbf{U}^3\mathbf{U}^2\mathbf{U}^1\mathbf{z}_a \tag{5.5e}$$

We conclude that the state vector $\mathbf{z}$ at any point along the beam is found by progressive multiplication of the transfer matrices for all elements from left to right up to that point. That is, the state variables at any point j are given by

$$\mathbf{z}_j = \mathbf{U}^j \mathbf{U}^{j-1} \dots \mathbf{U}^2 \mathbf{U}^1 \mathbf{z}_a \tag{5.6}$$

If there are M elements along the beam, the state variables at the right end of the beam become

$$\mathbf{z}_{x=L} = \mathbf{z}_L = \mathbf{U}^M \mathbf{U}^{M-1} \dots \mathbf{U}^2 \mathbf{U}^1 \mathbf{z}_a = \mathbf{U}\mathbf{z}_a \tag{5.7}$$

where $\mathbf{U}$ is the *global* or *overall transfer matrix* extending from the left to the right end of the beam.

The beam response problem is a boundary value problem. Of the four initial state variables w_a, θ_a, V_a, and M_a of $\mathbf{z}_a$, two are known by observation of the left-hand ($x = a$) boundary. Thus, the only unknowns in Eqs. (5.6) or (5.7) are half of the four state variables w_a, θ_a, V_a, and M_a of $\mathbf{z}_a$ which are determined by using the two boundary conditions that occur at the right hand end ($x = L$) of the beam. This situation is typical of more general problems in that conditions for half of the state variables of $\mathbf{z}$ occur at each boundary. As a consequence of the solution being expressed in terms of the initial state variables, the transfer matrix procedure is sometimes considered to be an initial value solution of a boundary value problem. Two "sweeps" along the beam are required for a complete transfer matrix solution. First, the overall or global transfer matrix $\mathbf{U}$ of Eq. (5.7) is formed, usually by a computer program that calls up stored transfer matrices and performs the matrix multiplications of Eq. (5.7). The four boundary conditions are applied to Eq. (5.7), with the left-hand boundary conditions leading to the direct identification of half of the four unknown state variables w_a, θ_a, V_a, and M_a. The right-hand boundary conditions provide two equations for the remaining two unknown state variables. Then, with $\mathbf{z}_a$ known, a second "sweep" along the member using Eq. (5.6) is made to compute and to print out the state variables w, θ, V, and M along the beam.

For the beam of Fig. 5.1, Eq. (5.6) is written to provide the state variables $\mathbf{z}_b, \mathbf{z}_c, \mathbf{z}_d, \mathbf{z}_e$, and $\mathbf{z}_f$. Between stations, the state variables are computed by appropriately adjusting the x coordinate (ℓ) in the transfer matrix for that element. This transfer matrix technique is simple and schematic and involves only small element matrices and system matrices of equally small dimensions. Both force and displacement state variables are computed simultaneously. The transfer matrix method does, however, have the disadvantage of being numerically sensitive, especially when the boundaries are far enough apart to have little influence on each other or if there are large variations in some parameters. The transfer matrix method is best suited to structural systems possessing a chain-like topology.

5.1.1 Loading and In-Span Conditions

The incorporation in the transfer matrix of the effects of loading and such in-span occurrences as springs and supports requires special attention. Formulas for the calculation of the loading vector $\overline{\mathbf{z}}^i$ for distributed applied loading were derived in Chapter 4; the introduction of concentrated applied loadings will be treated here.

Point Occurrences

Consider a concentrated transverse force $\overline{P}$ at point (node) j. A short segment spanning j is shown in Fig. 5.2. Since no abrupt change in displacement occurs at j, the deflection w and slope θ will be continuous across j. That is, $w_+ = w_-, \theta_+ = \theta_-$. Summation of moments about j for this infinitesimally short segment shows that the bending moment

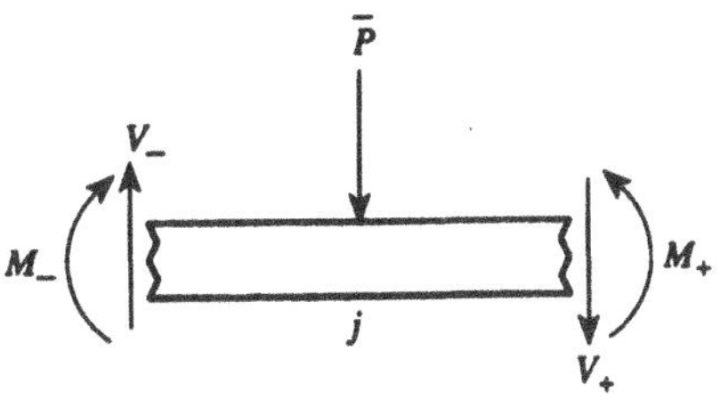

FIGURE 5.2
A short beam segment with concentrated force $\overline{P}$ at location j. Net internal forces and moments just to each side of j are shown. Sign Convention 1 is used.

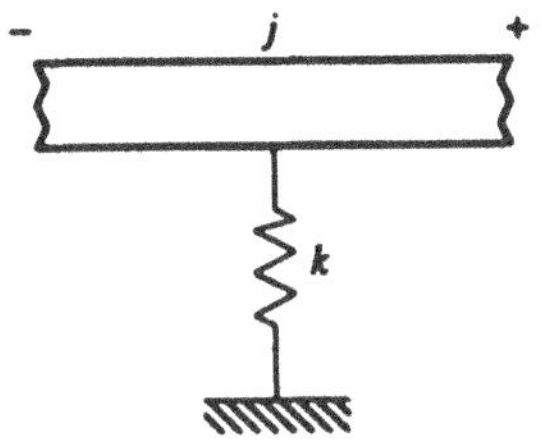

FIGURE 5.3
A beam supported by a spring with stiffness k.

is also continuous here, i.e., $M_+ = M_-$. Now, sum the vertical forces: $V_- - \overline{P} - V_+ = 0$ or $V_+ = V_- - \overline{P}$. Thus, it is seen that the shear force changes by a magnitude $\overline{P}$ in moving across the load. In summary, these relations are

$$\begin{bmatrix} w \\ \theta \\ V \\ M \end{bmatrix}_j^+ = \begin{bmatrix} w \\ \theta \\ V \\ M \end{bmatrix}_j^- + \begin{bmatrix} 0 \\ 0 \\ -\overline{P} \\ 0 \end{bmatrix}_j \qquad (5.8)$$
$$\mathbf{z}_j^+ = \mathbf{z}_j^- + \overline{\mathbf{z}}_j$$

or in the equivalent transfer matrix form

$$\begin{bmatrix} w \\ \theta \\ V \\ M \\ \cdots \\ 1 \end{bmatrix}_j^+ = \begin{bmatrix} 1 & 0 & 0 & 0 & \vdots & 0 \\ 0 & 1 & 0 & 0 & \vdots & 0 \\ 0 & 0 & 1 & 0 & \vdots & -\overline{P} \\ 0 & 0 & 0 & 1 & \vdots & 0 \\ \cdots & \cdots & \cdots & \cdots & \cdot & \cdots \\ 0 & 0 & 0 & 0 & \vdots & 1 \end{bmatrix} \begin{bmatrix} w \\ \theta \\ V \\ M \\ \cdots \\ 1 \end{bmatrix}_j^- \qquad (5.9)$$
$$\mathbf{z}_j^+ = \mathbf{U}_j \qquad \mathbf{z}_j^-$$

This transfer matrix is often referred to as a *point matrix* to distinguish it from the transfer matrix for an element of finite length, which is called a *field matrix*.

EXAMPLE 5.1 Point Matrix for Extension Spring
Many point matrices can be derived in a fashion similar to that accounting for a concentrated force. Consider a beam supported by an extension spring at j (Fig. 5.3). The force in the spring is proportional to the beam deflection at j, i.e., the force is kw_j. The point matrices of Eqs. (5.8) and (5.9) apply with $V_+ = V_- + kw$; the sign indicates that the force

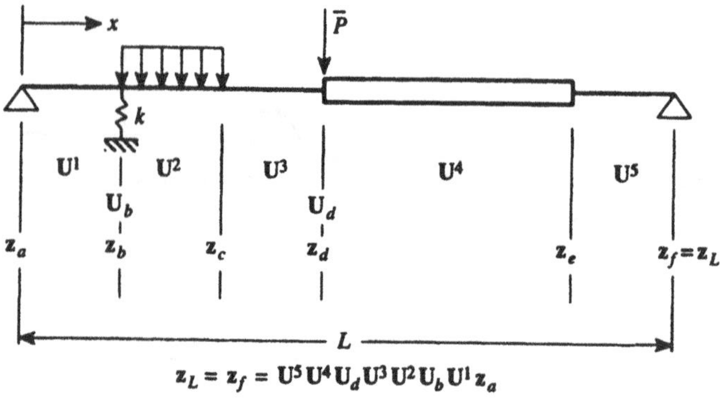

FIGURE 5.4
Incorporation of point occurrences into a transfer matrix solution.

due to the spring is upwards, while $\overline{P}$ of Fig. 5.2 was downwards. Thus,

$$\begin{bmatrix} w \\ \theta \\ V \\ M \\ \cdots \\ 1 \end{bmatrix}_j^+ = \begin{bmatrix} 1 & 0 & 0 & 0 & \vdots & 0 \\ 0 & 1 & 0 & 0 & \vdots & 0 \\ k & 0 & 1 & 0 & \vdots & 0 \\ 0 & 0 & 0 & 1 & \vdots & 0 \\ \cdots & \cdots & \cdots & \cdots & \cdot & \cdots \\ 0 & 0 & 0 & 0 & \vdots & 1 \end{bmatrix} \begin{bmatrix} w \\ \theta \\ V \\ M \\ \cdots \\ 1 \end{bmatrix}_j^- \tag{1}$$

■

No distinction is made between the use of point and field matrices in the progressive matrix multiplications of a transfer matrix solution. Thus, for example, for the beam of Fig. 5.4, the state vector $\mathbf{z}_f$ appears as

$$\mathbf{z}_{x=L} = \mathbf{z}_f = \mathbf{U}^5\, \mathbf{U}^4\, \mathbf{U}_d\, \mathbf{U}^3\, \mathbf{U}^2 \mathbf{U}_b\, \mathbf{U}^1\, \mathbf{z}_a \tag{5.10}$$

5.1.2 Transfer Matrix Catalogue

Catalogues of transfer matrices for various structural elements with arbitrary loading are available in many sources, e.g., Pestel and Leckie (1963), Pilkey (1994), Pilkey and Chang (1978). One of the most useful transfer matrices, which was derived in Chapter 4 and is displayed in Table 4.3, is that for the beam element. If extension, as well as bending, is to be included, the transfer matrix is expanded as in Chapter 4, Eq. (4.123). Then, the state vector is

$$\mathbf{z} = [u \quad w \quad \theta \quad N \quad V \quad M]^T \tag{5.11}$$

where u is the axial displacement, and N is the axial force. Torsion is included in a similar manner. Table 5.1 provides a variety of point matrices, including one necessary to transfer the state variables across a corner.

5.1.3 Incorporation of Boundary Conditions

The transfer matrices developed thus far can be used to solve boundary value problems for beams. Once the overall or global transfer matrix is formed as in Eq. (5.7), the boundary conditions can be applied to find the unknown initial parameters of $\mathbf{z}_a$. As mentioned

TABLE 5.1

Transfer Matrices for Bar with Extension and Bending (Sign Convention 1)

Transfer Matrix

$$\begin{bmatrix} u \\ w \\ \theta \\ N \\ V \\ M \end{bmatrix}_b = \begin{bmatrix} 1 & & & \ell/EA & & \\ & 1 & -\ell & & -\ell^3/6EI & -\ell^2/2EI \\ & & 1 & & \ell^2/2EI & \ell/EI \\ & & & 1 & & \\ & & & & 1 & \\ & & & & \ell & 1 \end{bmatrix} \begin{bmatrix} u \\ w \\ \theta \\ N \\ V \\ M \end{bmatrix}_a$$

$$\mathbf{z}_b \quad = \quad \mathbf{U}^i \quad \mathbf{z}_a$$

Point Matrix for Joint

$$\begin{bmatrix} u \\ w \\ \theta \\ N \\ V \\ M \end{bmatrix}_k^+ = \begin{bmatrix} \cos\alpha & \sin\alpha & & & & \\ -\sin\alpha & \cos\alpha & & & & \\ & & 1 & & & \\ & & & -\cos\alpha & -\sin\alpha & \\ & & & \sin\alpha & -\cos\alpha & \\ & & & & & -1 \end{bmatrix} \begin{bmatrix} u \\ w \\ \theta \\ N \\ V \\ M \end{bmatrix}_k^-$$

$$\mathbf{z}_k^+ \quad = \quad \mathbf{U}_k \quad \mathbf{z}_k^-$$

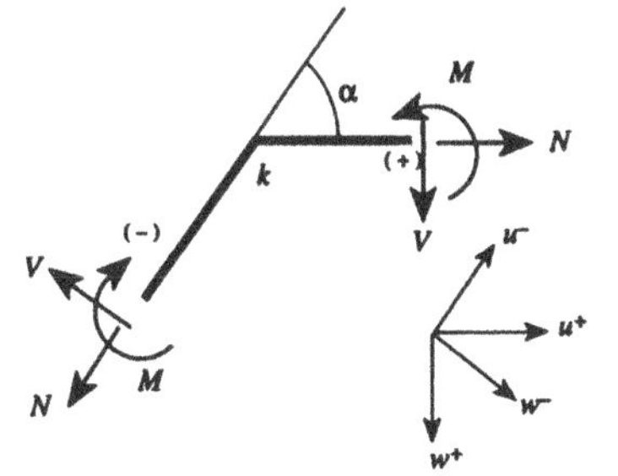

Point Matrix for Spring Supports

$$\begin{bmatrix} u \\ w \\ \theta \\ N \\ V \\ M \end{bmatrix}_k^+ = \begin{bmatrix} 1 & & & & & \\ & 1 & & & & \\ & & 1 & & & \\ k_u & & & 1 & & \\ & k_w & & & 1 & \\ & & k_\theta & & & 1 \end{bmatrix} \begin{bmatrix} u \\ w \\ \theta \\ N \\ V \\ M \end{bmatrix}_k^-$$

$$\mathbf{z}_k^+ \quad = \quad \mathbf{U}_k \quad \mathbf{z}_k^-$$

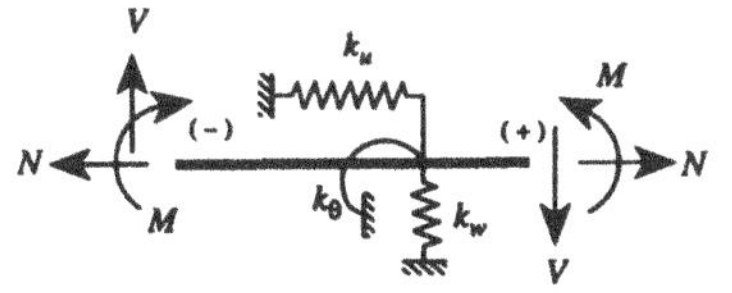

Point Matrix Extended for Applied Concentrated Forces

$$\begin{bmatrix} u \\ w \\ \theta \\ N \\ V \\ M \\ 1 \end{bmatrix}_k^+ = \begin{bmatrix} 1 & & & & & & \\ & 1 & & & & & \\ & & 1 & & & & \\ & & & 1 & & & -\overline{N} \\ & & & & 1 & & -\overline{P} \\ & & & & & 1 & -\overline{M} \\ & & & & & & 1 \end{bmatrix} \begin{bmatrix} u \\ w \\ \theta \\ N \\ V \\ M \\ 1 \end{bmatrix}_k^-$$

$$\mathbf{z}_k^+ \quad = \quad \mathbf{U}_k \quad \mathbf{z}_k^-$$

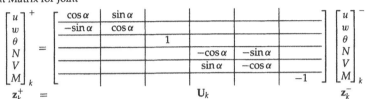

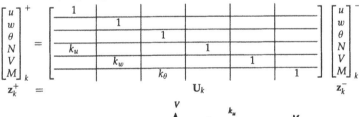

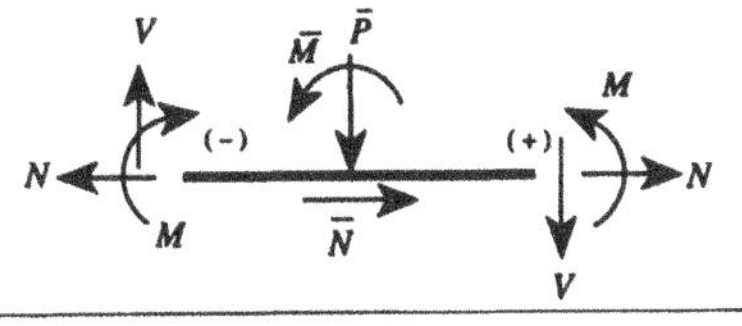

earlier, this transfer matrix procedure requires two passes of progressive multiplications of the transfer matrices. In the first pass, Eq. (5.7), is formed, and the initial parameters $\mathbf{z}_a$ are evaluated using the boundary conditions. Usually, the values of two of the four unknown initial parameters w_a, θ_a, V_a, M_a are known by simple observation of the left end of the beam. In the second pass, using the just determined $\mathbf{z}_a$ and Eq. (5.6), the state variables w, θ, V, and M are printed out along the beam. The solution procedure is illustrated using a particular beam in the next example problem.

Formulas for the initial parameters for the common boundary conditions are listed in tables in references such as Pilkey (1994).

5.1.4 Summary of the Transfer Matrix Solution Procedure

The notation employed for the transfer matrix method is summarized in Table 5.2. A solution should begin with the modeling of the beam system in terms of elements that connect locations (stations) of point occurrences (for example, applied concentrated forces) or abrupt changes (for example, a jump in cross-sectional area). For each element, compute the required section properties such as the moment of inertia I. Then calculate the elements of the field matrices for each segment, as well as the point matrices for the concentrated occurrences. Now compute the global transfer matrix by multiplying, in sequence, all transfer matrices from the left end to the right end of the beam. That is, calculate $\mathbf{U}$ of

$$\mathbf{z}_{x=L} = \mathbf{z}_L = \mathbf{U}^M\,\mathbf{U}^{M-1}\cdots\mathbf{U}_k\cdots\mathbf{U}^2\mathbf{U}^1\mathbf{z}_a = \mathbf{U}\mathbf{z}_{x=0} = \mathbf{U}\mathbf{z}_0 \tag{5.12a}$$

Use this expression to evaluate the unknown initial variables of

$$\mathbf{z}_a = \begin{bmatrix} w \\ \theta \\ V \\ M \\ 1 \end{bmatrix}_a \tag{5.12b}$$

by applying the boundary conditions to Eq. (5.12a). This can be accomplished by eliminating the unnecessary rows and columns of Eq. (5.12a) and solving the remaining equations. Finally, the deflection, slope, shear force, and internal moment are computed at all points of interest using

$$\mathbf{z}_j = \mathbf{U}^j\mathbf{U}^{j-1}\cdots\mathbf{U}_k\cdots\mathbf{U}^2\mathbf{U}^1\mathbf{z}_a \tag{5.12c}$$

TABLE 5.2

Notation for the Transfer Matrix Method

Symbol	Definition	Remarks
$\mathbf{U}^i$	Transfer matrix for the ith element (field)	$\mathbf{z}_k = \mathbf{U}^i\mathbf{z}_j$
$\mathbf{U}$	Global or overall transfer matrix that spans several elements	$\mathbf{U} = \mathbf{U}^M\mathbf{U}^{M-1}\ldots\mathbf{U}^2\mathbf{U}^1$
$\mathbf{U}_k$	Point matrix to account for concentrated occurrence, e.g., a point force or discrete spring, at location k	
$\mathbf{z}_k$	State vector at location k, contains all displacement and force state variables	$\mathbf{z}_k = \begin{bmatrix} \mathbf{v}_k \\ \mathbf{s}_k \end{bmatrix} = \begin{bmatrix} \text{Displacements} \\ \text{Forces} \end{bmatrix}$
$\bar{\mathbf{z}}^i$	Applied loading function vector for the ith element	
or		
$\bar{\mathbf{z}}^i_k$	Vector of applied loading functions for the ith field, evaluated at point k	

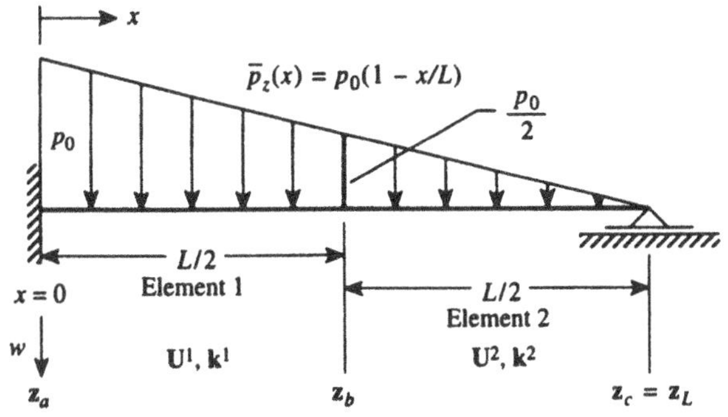

FIGURE 5.5
Beam with ramp loading (Examples 5.2 and 5.6).

The state variables can be computed between nodes by adjusting the x coordinate (ℓ) in the transfer matrix for that element.

Procedures to improve the computational efficiency and the numerical stability are treated in Section 5.1.5.

EXAMPLE 5.2 Beam with Linearly Varying Loading

The fixed-simply supported uniform beam of Fig. 5.5 has been treated frequently in this book. Suppose it is modeled with two elements and the response is to be found using the transfer matrix method.

The transfer matrices for each element can be taken from Chapter 4, Table 4.3 as

$$\mathbf{U}^1 = \begin{bmatrix} 1 & -\ell & -\ell^3/6EI & -\ell^2/2EI & \vdots & (4p_0 + p_0/2)\ell^4/120EI \\ 0 & 1 & \ell^2/2EI & \ell/EI & \vdots & -(3p_0 + p_0/2)\ell^3/24EI \\ 0 & 0 & 1 & 0 & \vdots & -(p_0 + p_0/2)\ell/2 \\ 0 & 0 & \ell & 1 & \vdots & -(2p_0 + p_0/2)\ell^2/6 \\ \cdots & \cdots & \cdots & \cdots & \cdot & \cdots \\ 0 & 0 & 0 & 0 & \vdots & 1 \end{bmatrix}_{\ell=\frac{L}{2}} \tag{1}$$

$$\mathbf{U}^2 = \begin{bmatrix} 1 & -\ell & -\ell^3/6EI & -\ell^2/2EI & \vdots & p_0\ell^4/60EI \\ 0 & 1 & \ell^2/2EI & \ell/EI & \vdots & -p_0\ell^3/16EI \\ 0 & 0 & 1 & 0 & \vdots & -p_0\ell/4 \\ 0 & 0 & \ell & 1 & \vdots & -p_0\ell^2/6 \\ \cdots & \cdots & \cdots & \cdots & \cdot & \cdots \\ 0 & 0 & 0 & 0 & \vdots & 1 \end{bmatrix}_{\ell=\frac{L}{2}} \tag{2}$$

The overall transfer matrix $\mathbf{U}$ would be

$$\mathbf{z}_{x=L} = \mathbf{z}_c = \mathbf{U}^2\mathbf{U}^1\mathbf{z}_a = \mathbf{U}\mathbf{z}_a \tag{3}$$

The initial parameter vector $\mathbf{z}_a$ is determined by applying boundary conditions to (3). As is always the case, two of the four initial parameters are evaluated by observation. For the beam of Fig. 5.5, we simply note that $w_a = 0$ and $\theta_a = 0$ since the left end is fixed. The conditions at the simply supported right end are $w_{x=L} = 0$ and $M_{x=L} = 0$. These are applied to the first and fourth rows of (3). Equation (3), that is, $\mathbf{z}_{x=L} = \mathbf{U}\mathbf{z}_a$, can be written as

$$\begin{bmatrix} w=0 \\ \theta \\ V \\ M=0 \\ 1 \end{bmatrix}_{x=L} = \begin{bmatrix} & \vdots & \vdots & U_{wV} & U_{wM} & w^0 \\ \cdots & \vdots & \vdots & \cdots & \cdots & \cdots \\ \cdots & \vdots & \vdots & \cdots & \cdots & \cdots \\ & \vdots & \vdots & U_{MV} & U_{MM} & M^0 \\ & \vdots & \vdots & 0 & 0 & 1 \end{bmatrix}_{x=L} \begin{bmatrix} w=0 \\ \theta=0 \\ V \\ M \\ 1 \end{bmatrix}_{x=0} \tag{4}$$

Cancel columns 1 and 2 because $w_0 = \theta_0 = 0$

Ignore rows 2 and 3 because $\theta_{x=L}$ and $V_{x=L}$ are unknown

where U_{kj} and the loading vector components are the elements of $\mathbf{U}$ of (3). The equations $w_{x=L} = 0$ and $M_{x=L} = 0$ are used to compute M_a and V_a, the remaining unknown initial parameters. Thus from (4)

$$\begin{bmatrix} w \\ M \end{bmatrix}_{x=L} = \begin{bmatrix} 0 \\ 0 \end{bmatrix} = \begin{bmatrix} U_{wV} & U_{wM} \\ U_{MV} & U_{MM} \end{bmatrix}_{x=L} \begin{bmatrix} V_a \\ M_a \end{bmatrix} + \begin{bmatrix} w^0 \\ M^0 \end{bmatrix}_{x=L} \tag{5}$$

or

$$\begin{aligned} 0 &= V_a U_{wV} + M_a U_{wM} + w_L^0 \\ 0 &= V_a U_{MV} + M_a U_{MM} + M_L^0 \end{aligned} \tag{6}$$

Equations (5) and (6) are solved for M_a, V_a, giving

$$\begin{aligned} V_a &= (M^0 U_{wM} - w^0 U_{MM})|_{x=L}/\nabla \\ M_a &= (w^0 U_{MV} - M^0 U_{wV})|_{x=L}/\nabla \end{aligned} \tag{7}$$

where U_{ij} and the loading components are the coefficients of the overall transfer matrix $\mathbf{U}$ and

$$\nabla = (U_{wV} U_{MM} - U_{wM} U_{MV})|_{x=L} \tag{8}$$

For the beam of Fig. 5.5, (3) gives

$$\mathbf{U} = \begin{bmatrix} 1 & -2\ell & -\frac{4}{3}\frac{\ell^3}{EI} & \frac{-2\ell^2}{EI} & \vdots & \frac{8}{15}\frac{p_0\ell^4}{EI} \\ 0 & 1 & \frac{2\ell^2}{EI} & \frac{2\ell}{EI} & \vdots & \frac{-p_0\ell^3}{EI} \\ 0 & 0 & 1 & 0 & \vdots & -p_0\ell \\ 0 & 0 & 2\ell & 1 & \vdots & \frac{-4p_0\ell^2}{3} \\ \cdots & \cdots & \cdots & \cdots & \vdots & \cdots \\ 0 & 0 & 0 & 0 & \vdots & 1 \end{bmatrix} \tag{9}$$

Introduce the boundary conditions to (9) or use (7) with $w_L^0 = \frac{8}{15}\frac{p_0\ell^4}{EI}$ and $M_L^0 = \frac{-4p_0\ell^2}{3}$ to find the initial parameters V_a and M_a. Thus,

$$\begin{bmatrix} -\frac{4}{3}\frac{\ell^3}{EI} & \frac{-2\ell^2}{EI} \\ 2\ell & 1 \end{bmatrix} \begin{bmatrix} V_a \\ M_a \end{bmatrix} = \begin{bmatrix} -\frac{8}{15}\frac{p_0\ell^4}{EI} \\ \frac{4p_0\ell^2}{3} \end{bmatrix} \tag{10}$$

and, with $\ell = L/2$,

$$V_a = \frac{4p_0\ell}{5} = \frac{2p_0L}{5} \qquad M_a = -\frac{4p_0\ell^2}{15} = -\frac{p_0L^2}{15} \tag{11}$$

Since $w_a = 0$, $\theta_a = 0$ and M_a, V_a are given by (11), $\mathbf{z}_a$ is now known. The variables w, θ, V, and M are calculated using Eq. (5.6) and can be printed out at designated locations. For example, the displacements and forces at nodes b and c are given by

$$\begin{aligned} \mathbf{z}_b &= \mathbf{z}_{x=L/2} = \mathbf{U}^1\mathbf{z}_a \\ \mathbf{z}_c &= \mathbf{z}_{x=L} = \mathbf{U}^2\mathbf{U}^1\mathbf{z}_a = \mathbf{U}\mathbf{z}_a \end{aligned} \tag{12}$$

Between nodes, results are computed by adjusting the coordinate in the transfer matrix for that element. If, for example, the state variables are sought at the midpoint of the second element, then

$$\mathbf{z}_{x=L/2+(L/2)/2} = \mathbf{U}^2((L/2)/2)\mathbf{U}^1(L/2)\mathbf{z}_a$$

$$= \mathbf{U}^2\left(\frac{\ell}{2}\right)\mathbf{U}^1(\ell)\begin{bmatrix} 0 \\ 0 \\ \frac{4p_0\ell}{5} \\ -\frac{4p_0\ell^2}{15} \\ 1 \end{bmatrix} = \begin{bmatrix} \frac{7p_0\ell^4}{240EI} \\ \frac{17p_0\ell^3}{384EI} \\ \frac{-3p_0\ell}{40} \\ \frac{p_0\ell^2}{10} \\ 1 \end{bmatrix} = \begin{bmatrix} \frac{7p_0L^4}{3840EI} \\ \frac{17p_0L^3}{3072EI} \\ \frac{-3p_0L}{80} \\ \frac{p_0L^2}{40} \\ 1 \end{bmatrix} \tag{13}$$

■

5.1.5 Some Computational Considerations

In Chapter 4 it was demonstrated that a transfer matrix can be transformed into a stiffness matrix. Much of the present chapter deals with the displacement method of structural analysis in which element stiffness matrices are assembled into a global stiffness matrix and the resulting system of equations is solved for the displacements. This displacement method of analysis is probably the most reasonable approach for solving structural mechanics problems for which element transfer matrices are available for the elements, that is, rather than preparing a computer program to implement the progressive matrix multiplication which characterizes the transfer matrix method, it is usually better to transform the transfer matrices into stiffness matrices and utilize the displacement method. After the displacements at the nodes are computed and the forces at the nodes are determined using the stiffness matrices, the transfer matrices can be used to print out the displacements and forces along the member.

Numerical Difficulties

The use of the pure transfer matrix method, with its progressive multiplications of element transfer matrices as it proceeds along the member, tends to encounter numerical difficulties. It should not be surprising that serious numerical difficulties would arise for chain-like structures of such a nature that an occurrence, e.g., an applied loading, at one location has little effect on the response at a distant location. There are actually several causes of numerical difficulties and several seemingly effective corrective measures that can be taken. Detailed discussions of the sources of the numerical problems and techniques for overcoming them are given in Horner and Pilkey (1978), Marguerre and Uhrig (1964), and Pestel and Leckie (1963).

Among the characteristics of the transfer matrix method that can cause numerical difficulties is the build-up of roundoff and truncation errors by the progressive multiplication of the form

$$\mathbf{z}_j = \mathbf{U}^j \cdots \mathbf{U}^2\mathbf{U}^1\mathbf{z}_a = \mathbf{U}\mathbf{z}_a \tag{5.13}$$

Rather than converging to $\mathbf{z}_j$, this can converge to $\tilde{\mathbf{z}}$, the eigenvector of the first eigenvalue of $\mathbf{U}$. This is the result of the vectors in $\mathbf{U}$ becoming linearly dependent and the determinant of the system of equations approaching the value zero. Even when $\mathbf{z}_a$ is known exactly, the solution can converge to $\tilde{\mathbf{z}}$.

Numerical difficulties also occur when the transfer matrix manipulations involve differences of large numbers, which can lead to inaccuracies in the computations. This can occur, for example, if a very stiff spring is included in the model. Also, this difficulty can be expected if the effect of occurrences on one boundary is small on the other boundary, i.e., the solution appears to die out rapidly. In such cases, the calculation of the initial conditions ($\mathbf{z}_a$) can involve differences between large numbers, a hazardous numerical operation at best.

Conclusions

There are two principal reasons for implementing a transfer matrix solution in combination with a displacement method. First, in many cases, the switch to the combination approach yields a procedure that is more efficient computationally than the pure progressive transfer matrix multiplication. Second, the transfer matrix method tends to be numerically unstable for many practical problems, whereas the displacement method may eliminate such difficulties. There are many circumstances under which it is desirable to use a combination of mixed (e.g., transfer matrix) and displacement methods. For example, the displacement method can be used to compute the responses at the nodes, followed by the use of the transfer matrix method to print out the responses along the element between the nodes.

5.2 General Structural Systems

Structural systems of arbitrary geometry are usually analyzed with force or displacement methods, whereas the transfer matrix method is appropriate only for structural systems with a line-like configuration. However, transfer matrices are also useful for the development of stiffness matrices or for computing displacements and forces along a member if the nodal responses have been calculated with another method. In contrast to the transfer matrix method, wherein the system matrix resulting from progressive multiplication of element matrices remains small regardless of the system complexity, the force and displacement methods develop system matrices whose size depends on the complexity of the system model. Before considering details of the force and displacement methods, we will define what may appear to be cumbersome notation. It is essential, however, to have generic notation so that arbitrary configurations can be handled. Table 5.3 gives a summary of the most frequently occurring notation in the matrix analysis of general structural systems. The structural analysis methods to be developed here also provide the foundation for the finite element method which can be applied to very general systems.

TABLE 5.3
Notation for Classical Matrix Techniques of Structural Analysis

Indices: Superscript for element index, subscript for node index, e.g., $\mathbf{v}_k^i$ is the displacement **v** of the ith element at the kth node.
Local coordinate system: Whenever it is essential to make the distinction, variables referred to a local coordinate system will be designated by a tilde $\tilde{}$, e.g., $\tilde{\mathbf{k}}^i$ is the local stiffness matrix of the ith element.
Prescribed (applied) variables will be indicated with a line over the letter, e.g., $\overline{\mathbf{V}}_k$ is the prescribed displacement at node k.

Symbol	Definition	Remarks
$\mathbf{V}$	Vector of nodal displacements of the complete system	$\mathbf{V}$ and $\mathbf{P}$ are considered only in the global coordinate system
$\mathbf{P}$	Vector of nodal forces of the complete system	
$\overline{\mathbf{V}}$	Vector of applied displacements of the complete system	
$\overline{\mathbf{P}}$	Vector of applied forces of the complete system	
$\mathbf{v}^i$	Vector of the end displacements of the ith element, e.g., $\tilde{\mathbf{v}}^3$ is for element 3 in the local coordinates	
$\mathbf{p}^i$	Column matrix of end forces of the ith element	
$\overline{\mathbf{v}}^i, \overline{\mathbf{p}}^i$	Element vectors containing the effects of applied loading	
$\mathbf{k}^i$	Stiffness matrix of the ith element e.g., $\tilde{\mathbf{k}}^3$ is the stiffness matrix of element 3 in local coordinates	
$\mathbf{K}$	Stiffness matrix of the complete system	
$\mathbf{f}^i$	Flexibility matrix of the ith element	
$\mathbf{F}$	Flexibility matrix of the complete system	
$\mathbf{a}$	Kinematic transformation matrix	$\mathbf{v} = \mathbf{a}\,\mathbf{V}$
$\mathbf{b}$	Static transformation matrix	$\mathbf{p} = \mathbf{b}\,\mathbf{P}$
$\mathbf{T}$	Transformation matrix, $\mathbf{T}^i$ corresponds to the ith element	e.g., $\tilde{\mathbf{v}}^i = \mathbf{T}^i\mathbf{v}^i$
$\mathbf{T}_{kk}^i$	Transformation submatrix, element i, node k	$\mathbf{T}^i = \left[\begin{array}{c\|c} \mathbf{T}_{aa} & 0 \\ \hline 0 & \mathbf{T}_{bb} \end{array}\right]^i$. For nodes a, b of element i

5.2.1 Basic Definitions of Elements, Nodes, Forces, Displacements, and Coordinate Systems

To achieve computational tractability, network structures are usually modeled by a finite number of elements connected at nodes. A solution procedure is established to compute forces and displacements at the nodes. Since only these discrete nodal variables will appear in the governing equations, the structure is said to be discretized spatially. As indicated in Fig. 5.6, the element may be one, two or three-dimensional, as required by the structure. These structural models are also referred to as *finite element* models, and the associated structural analysis methodology forms the basis of the finite element method. In the finite element method the element characteristics may be obtained by numerical approximation.

System State Variables at the Nodes

The elements (Fig. 5.6) of a system may possess a finite number of common nodal points. It is convenient to describe the location of the nodes and elements in a single global coordinate system (X, Y, Z). Define at each node forces and displacements as shown in Fig. 5.7. These system or global forces and displacements serve as the unknowns in the respective

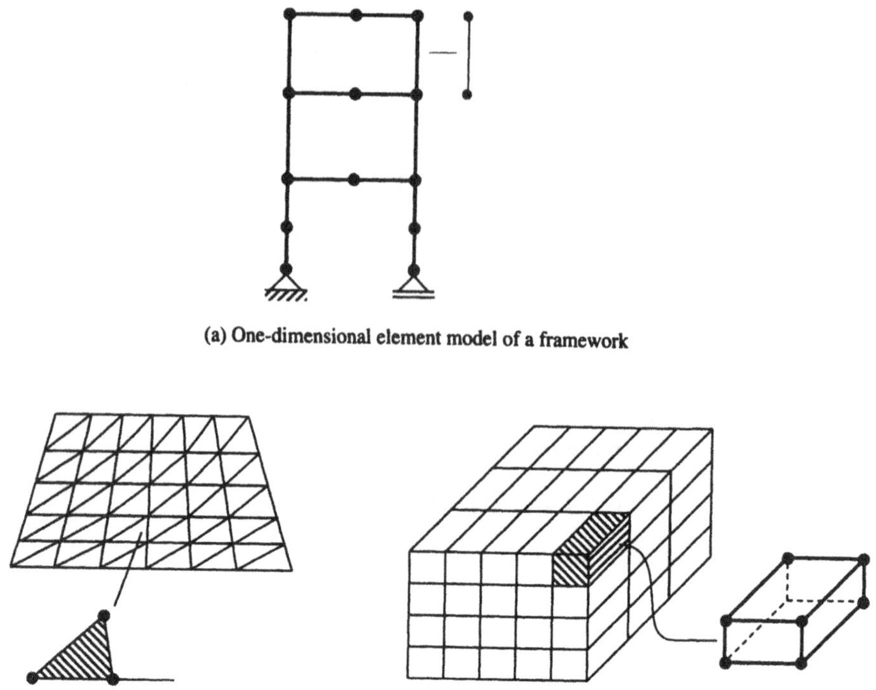

(a) One-dimensional element model of a framework

(b) Two-dimensional elements

(c) Three-dimensional elements

FIGURE 5.6
Structural discretization in terms of elements.

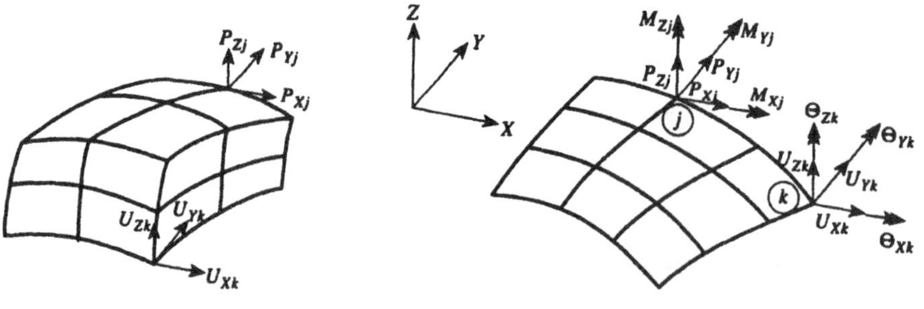

(a) Solid element

(b) Shell element

FIGURE 5.7
Global coordinates X, Y, Z and the nodal displacements and forces for the whole system.

structural analysis method. The internal forces and displacements between the nodes of an element, e.g., a bar, are evaluated after the nodal responses are computed.

As explained in Chapter 4, the independent coordinates (displacements) essential for completely describing the motion of a node are called the nodal *degrees of freedom* (DOF). The number of DOF can vary from one element type to another. For the general solid elements of Fig. 5.7a, each node can have three DOF: three translations (shown at node k) U_{Xk}, U_{Yk}, and U_{Zk}. There are also three corresponding forces (shown at node j) P_{Xj}, P_{Yj}, and P_{Zj}. For the shell elements of Fig. 5.7b, each node can have six DOF: three translations and three rotations. Corresponding to these DOF will be three forces and three moments.

Thus, six displacements and six (generalized) forces occur at each node

$$\begin{array}{ll} U_X, U_Y, U_Z & \text{3 translations} \\ \Theta_X, \Theta_Y, \Theta_Z & \text{3 rotations} \\ P_X, P_Y, P_Z & \text{3 forces} \\ M_X, M_Y, M_Z & \text{3 moments} \end{array} \tag{5.14}$$

As in the case of the coordinates X, Y, Z, these system forces and displacements are designated by capital letters.

In vector form, the forces and displacements at each node are written as

$$\mathbf{P}_j = \begin{bmatrix} P_X \\ P_Y \\ P_Z \\ M_X \\ M_Y \\ M_Z \end{bmatrix}_j \qquad \mathbf{V}_j = \begin{bmatrix} U_X \\ U_Y \\ U_Z \\ \Theta_X \\ \Theta_Y \\ \Theta_Z \end{bmatrix}_j \tag{5.15}$$

where the subscript j indicates the jth node. The nodal forces $\mathbf{P}$ and nodal displacements $\mathbf{V}$ for the whole structure are expressed as

$$\mathbf{P} = \begin{bmatrix} \mathbf{P}_1 \\ \mathbf{P}_2 \\ \vdots \\ \mathbf{P}_j \\ \vdots \\ \mathbf{P}_N \end{bmatrix} \qquad \mathbf{V} = \begin{bmatrix} \mathbf{V}_1 \\ \mathbf{V}_2 \\ \vdots \\ \mathbf{V}_j \\ \vdots \\ \mathbf{V}_N \end{bmatrix} \tag{5.16}$$

where N is the number of nodes.

State Variables for an Element

A global reference frame X, Y, Z was established in the previous section in order to define system forces and displacements. It is also necessary to represent the displacements and forces on an element in the directions of the global X, Y, Z system coordinates. We shall continue the practice here of using generalized nomenclature of letting the term *forces* include moments and *displacements* include rotations. Suppose an arbitrary element i, with ends a and b, belongs to a structural model. Element forces $\mathbf{p}^i$ and corresponding displacements $\mathbf{v}^i$ at the ends a and b are shown in Fig. 5.8, where

$$\mathbf{p}^i = \begin{bmatrix} \mathbf{p}_a^i \\ \mathbf{p}_b^i \end{bmatrix} = \begin{bmatrix} F_{Xa}^i \\ F_{Ya}^i \\ F_{Za}^i \\ F_{Xb}^i \\ F_{Yb}^i \\ F_{Zb}^i \end{bmatrix} \qquad \mathbf{v}^i = \begin{bmatrix} \mathbf{v}_a^i \\ \mathbf{v}_b^i \end{bmatrix} = \begin{bmatrix} u_{Xa}^i \\ u_{Ya}^i \\ u_{Za}^i \\ u_{Xb}^i \\ u_{Yb}^i \\ u_{Zb}^i \end{bmatrix} \tag{5.17}$$

As derived in Chapter 4, the fundamental relationships of solid mechanics—equilibrium conditions, material law, and kinematical conditions (compatibility)—provide the functional connection between the end cross-sectional forces and the end displacements for an element. These relationships can be pictured as

$$\begin{aligned} \mathbf{p}^i &\rightarrow \textit{Equilibrium Conditions} \rightarrow \boldsymbol{\sigma}^i \rightarrow \textit{Material Law} \rightarrow \\ &\rightarrow \boldsymbol{\epsilon}^i \rightarrow \textit{Strain-Displacement (Kinematic) Relations} \rightarrow \mathbf{v}^i \end{aligned} \tag{5.18}$$

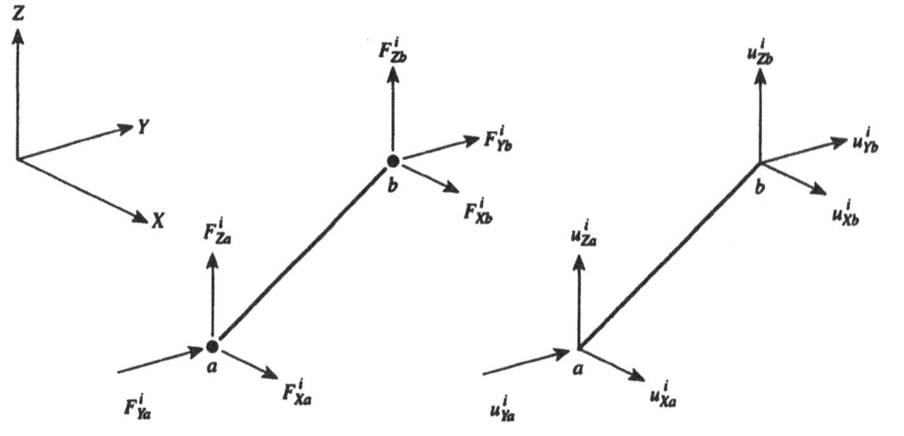

FIGURE 5.8
Element i with its forces and displacements (aligned along the global coordinate directions) at the ends a and b. For simplicity, only three forces and three displacements are indicated at each end. Although only pure forces and displacements are shown, moments and rotations are implied.

They are included in the basic relationships for an element

$$\begin{bmatrix} \mathbf{p}_a \\ \mathbf{p}_b \end{bmatrix}^i = \begin{bmatrix} \mathbf{k}_{aa} & \mathbf{k}_{ab} \\ \mathbf{k}_{ba} & \mathbf{k}_{bb} \end{bmatrix}^i \begin{bmatrix} \mathbf{v}_a \\ \mathbf{v}_b \end{bmatrix}^i \qquad (5.19)$$
$$\mathbf{p}^i \quad = \quad \mathbf{k}^i \quad \mathbf{v}^i$$

where $\mathbf{k}^i$ is a measure of the stiffness of an element and, hence, was referred to in Chapter 4 as the element stiffness matrix. Recall that the stiffness coefficients indicate the magnitude of the forces corresponding to unit displacements at the nodes.

Note that the element stiffness matrix coefficients are assigned double subscripts, e.g., $\mathbf{k}_{jk}$. It follows from Eq. (5.19) that the first subscript (j) designates the node or location for which the equation is established, while the second subscript (k) identifies the DOF "causing" (or corresponding to) the forces $\mathbf{p}_j$. In general, the $\mathbf{k}_{jk}$ of Eq. (5.19) are submatrices, and not scalar coefficients.

It is often helpful to begin the representation of forces and displacements in a coordinate system whose orientation may differ from that of the global coordinates. Define a new reference frame x, y, z, along with corresponding forces and displacements, aligned in a convenient, natural direction along the element. The coordinates x, y, z are referred to as a local reference frame. Consider the bar of Fig. 5.9, where only pure forces and displacements are shown. This chapter deals primarily with such line elements, although often moments and rotations will occur at the nodes. Figure 5.9 shows a global reference frame X, Z, as well as a local reference frame x, z. The stiffness matrix is usually readily and very naturally established in the local coordinate system. This would appear as

$$\begin{bmatrix} \tilde{\mathbf{p}}_a \\ \tilde{\mathbf{p}}_b \end{bmatrix}^i = \begin{bmatrix} \tilde{\mathbf{k}}_{aa} & \tilde{\mathbf{k}}_{ab} \\ \tilde{\mathbf{k}}_{ba} & \tilde{\mathbf{k}}_{bb} \end{bmatrix}^i \begin{bmatrix} \tilde{\mathbf{v}}_a \\ \tilde{\mathbf{v}}_b \end{bmatrix}^i \qquad (5.20)$$
$$\tilde{\mathbf{p}}^i \quad = \quad \tilde{\mathbf{k}}^i \quad \tilde{\mathbf{v}}^i$$

whereas for the global reference frame the stiffness matrix of Eq. (5.19) is a applicable. Note that local coordinate quantities are indicated with a tilde ($\tilde{\ }$).

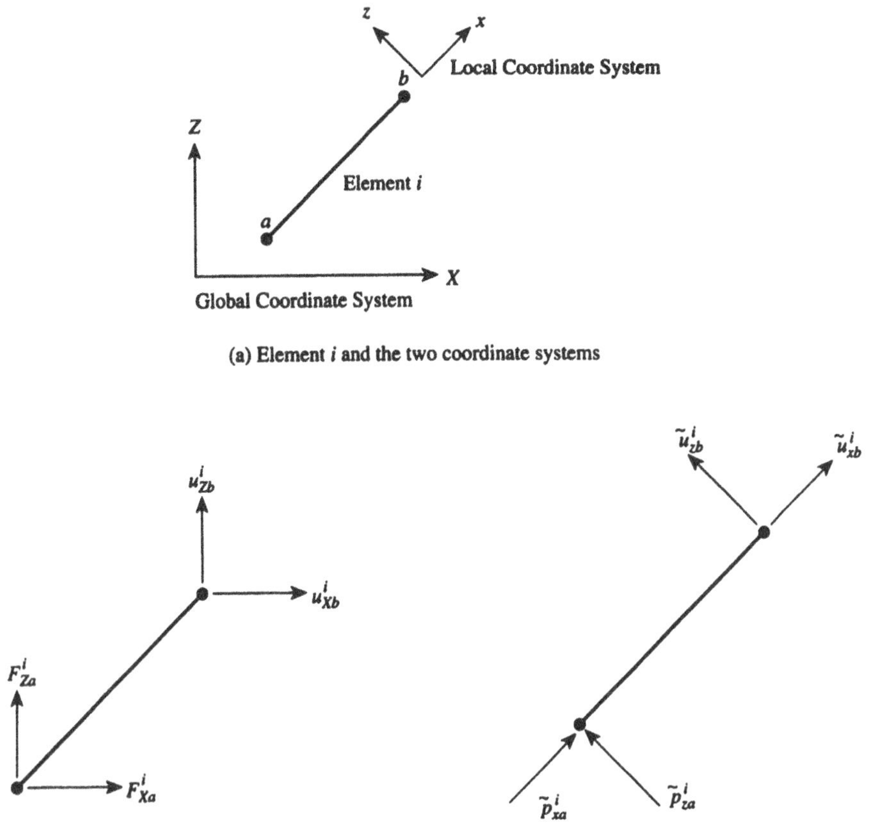

(a) Element *i* and the two coordinate systems

(b) Forces and displacements referred to the global coordinates

(c) Forces and displacements referred to the local coordinates. Moments and rotations are not shown.

FIGURE 5.9
Forces and displacements at the two ends of a line element.

5.2.2 Coordinate Transformations

In order to combine element stiffness matrices to perform an analysis of an entire structure, it is helpful to be able to have all forces and displacements for all elements referred to a common reference frame, the global coordinates. To accomplish this, those nodal forces and displacements expressed in the local coordinates must first be transformed to the global coordinates.

A typical coordinate transformation is illustrated in Fig. 5.10.

Global to Local Coordinates: Local to Global Coordinates:

$$\begin{bmatrix} x \\ z \end{bmatrix} = \begin{bmatrix} \cos\alpha & -\sin\alpha \\ \sin\alpha & \cos\alpha \end{bmatrix} \begin{bmatrix} X \\ Z \end{bmatrix} \qquad \begin{bmatrix} X \\ Z \end{bmatrix} = \begin{bmatrix} \cos\alpha & \sin\alpha \\ -\sin\alpha & \cos\alpha \end{bmatrix} \begin{bmatrix} x \\ z \end{bmatrix} \tag{5.21}$$

Forces and displacements are transformed in the same manner. Moments and rotations may have a different form of transformation.

The transformation for forces and displacements appears as

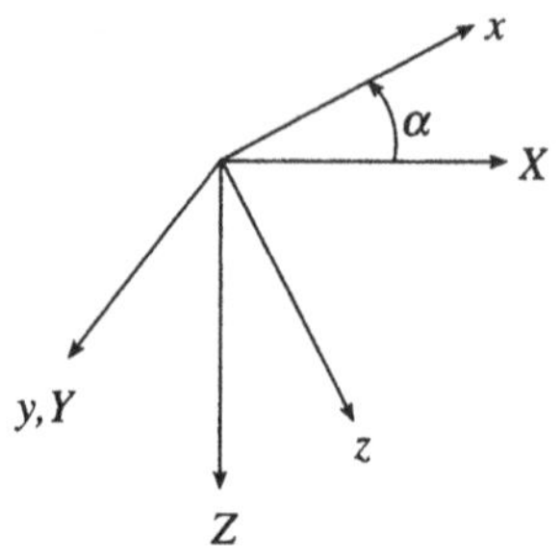

FIGURE 5.10
Right-handed global (X, Y, Z) and local (x, y, z) coordinate systems. The angle α is positive as shown (vector corresponding to α is positive along positive Y axis).

Global to Local Forces and Displacements:

$$\begin{bmatrix} \tilde{\mathbf{p}}_a \\ \tilde{\mathbf{p}}_b \end{bmatrix}^i = \begin{bmatrix} \mathbf{T}_{aa} & \mathbf{0} \\ \mathbf{0} & \mathbf{T}_{bb} \end{bmatrix}^i \begin{bmatrix} \mathbf{p}_a \\ \mathbf{p}_b \end{bmatrix}^i \quad \text{and} \quad \tilde{\mathbf{v}}^i = \mathbf{T}^i\mathbf{v}^i \tag{5.22}$$
$$\tilde{\mathbf{p}}^i \quad = \quad \mathbf{T}^i \quad \mathbf{p}^i$$

with, in more general notation, the transformation matrix $\mathbf{T}^i$

$$\mathbf{T}^i = \begin{bmatrix} \mathbf{T}_{aa} & \mathbf{T}_{ab} \\ \mathbf{T}_{ba} & \mathbf{T}_{bb} \end{bmatrix}^i \tag{5.23}$$

$$\mathbf{T}_{ba} = \mathbf{T}_{ab} = \mathbf{0} \quad \text{and} \quad \mathbf{T}_{aa} = \mathbf{T}_{bb} = \begin{bmatrix} \cos\alpha & -\sin\alpha \\ \sin\alpha & \cos\alpha \end{bmatrix} \tag{5.24}$$

where α is the angle between the X (global coordinate) and the x (local) coordinate.

Observe that for $\mathbf{T}^i_{jj}$, $j = a$ or b, defined by Eq. (5.24)

$$\mathbf{T}^{iT}_{jj}\mathbf{T}^i_{jj} = \mathbf{I}, \qquad \mathbf{T}^{iT}\mathbf{T}^i = \mathbf{I} \tag{5.25}$$

where superscript T designates the transpose and $\mathbf{I}$ is the unit diagonal matrix. The coordinate transformation for orthogonal coordinate axes always possesses this property. Since $(\mathbf{T}^i)^{-1}\mathbf{T}^i = \mathbf{I}$, it follows that

$$(\mathbf{T}^i)^{-1} = \mathbf{T}^{iT} \tag{5.26}$$

Hence, the transformation from local to global coordinates can be expressed as

Local to Global Forces and Displacements:

$$\begin{bmatrix} \mathbf{p}_a \\ \mathbf{p}_b \end{bmatrix}^i = \begin{bmatrix} \mathbf{T}^T_{aa} & \mathbf{0} \\ \mathbf{0} & \mathbf{T}^T_{bb} \end{bmatrix}^i \begin{bmatrix} \tilde{\mathbf{p}}_a \\ \tilde{\mathbf{p}}_b \end{bmatrix}^i \quad \text{and} \quad \mathbf{v}^i = \mathbf{T}^{iT}\tilde{\mathbf{v}}^i \tag{5.27}$$
$$\mathbf{p}^i \quad = \quad \mathbf{T}^{iT} \quad \tilde{\mathbf{p}}^i$$

Stiffness matrices are readily transformed from one coordinate system to another. This is accomplished by noting that

$$\mathbf{p}^i = \mathbf{T}^{iT}\tilde{\mathbf{p}}^i = \mathbf{T}^{iT}\tilde{\mathbf{k}}^i\tilde{\mathbf{v}}^i = \mathbf{T}^{iT}\tilde{\mathbf{k}}^i\mathbf{T}^i\mathbf{v}^i$$

Since $\mathbf{p}^i = \mathbf{k}^i\mathbf{v}^i$, it follows that

$$\mathbf{k}^i = \mathbf{T}^{iT}\tilde{\mathbf{k}}^i\mathbf{T}^i \tag{5.28}$$

A triple matrix product of this form is called a *congruent transformation*. The product $\mathbf{k}^i$ will be a symmetric matrix if the $\tilde{\mathbf{k}}^i$ is symmetric. As indicated in the previous two chapters, stiffness matrices are symmetric.

5.3 Displacement Method

The most commonly used method today for the analysis of large structural systems is the displacement method. Although the displacement method can be employed without reference to its roots as a variational method (in such a case it is sometimes called the *direct stiffness method*), it is perhaps best understood when it is considered to be a variationally based approach. The basis of the displacement method is the principle of virtual work. Since, as shown in Chapter 2, the principle of virtual work is equivalent to the global form of the equations of equilibrium, it is understandable that the displacement method is also referred to as the *equilibrium method*.

5.3.1 Nodal Displacement Equations Based on the Principle of Virtual Work

The principle of virtual work relations, which are designated as equations C in Chapter 2, are expressed in terms of displacements for an elastic solid by [Chapter 2, Eq. (2.58c)]

$$\int_V \delta\mathbf{u}^T \mathbf{k}^D \mathbf{u}\, dV - \int_V \delta\mathbf{u}^T \overline{\mathbf{p}}_V\, dV - \int_{S_p} \delta\mathbf{u}^T \overline{\mathbf{p}}\, dS = 0 \tag{5.29}$$

where

$$\mathbf{k}^D = {}_u\mathbf{D}^T \mathbf{E} \mathbf{D}_u \tag{5.30}$$

A different form of the principle of virtual work is useful in establishing the fundamentals of the displacement method. The structural system is to be modeled in terms of elements for which the responses and applied loading are represented by forces and displacements at the nodes. This modeling amounts to spatial discretization of the structure. From Chapter 4, Eq. (4.62), the virtual work for element i is

$$-(\delta W_i + \delta W_e)^i = \delta\mathbf{v}^{iT}(\mathbf{k}^i\mathbf{v}^i - \overline{\mathbf{p}}^{i0}) \tag{5.31}$$

The element nodal displacements are $\mathbf{v}^i$ and the nodal forces representing the effects of applied loading are $\overline{\mathbf{p}}^{i0}$. If the structural system is modeled as M elements, the principle of virtual work for all elements becomes

$$-(\delta W_i + \delta W_e) = -\sum_{i=1}^{M}(\delta W_i + \delta W_e)^i = \sum_{i=1}^{M} \delta\mathbf{v}^{iT}(\mathbf{k}^i\mathbf{v}^i - \overline{\mathbf{p}}^{i0}) = 0 \tag{5.32}$$

It should be understood that Eq. (5.32) represents the summation of internal and external virtual work done by the element forces at the nodes, i.e., this equation contains the total virtual work of all elements of the system.

A useful form of the principle of virtual work is obtained by expressing the element displacements $\mathbf{v}^i$ in terms of the unknown global nodal displacements $\mathbf{V}$. This can be accomplished by enforcing the nodal compatibility conditions. The nodal displacements of the various elements joined at a particular node must match the values of the system displacements of the node. This requirement should not be surprising, since the principle of virtual work (Chapter 2) corresponds to the equilibrium equations provided that the displacements are kinematically admissible. In order to implement these compatibility conditions, the local (element) end displacements must be transformed to the directions of the global coordinate system. Assume $\widetilde{\mathbf{v}}^i$ has been transformed to $\mathbf{v}^i$. The compatibility conditions for node k of Fig. 5.11, where four elements meet, are

$$\mathbf{v}_k^1 = \mathbf{v}_k^2 = \mathbf{v}_k^3 = \mathbf{v}_k^4 = \mathbf{V}_k \tag{5.33}$$

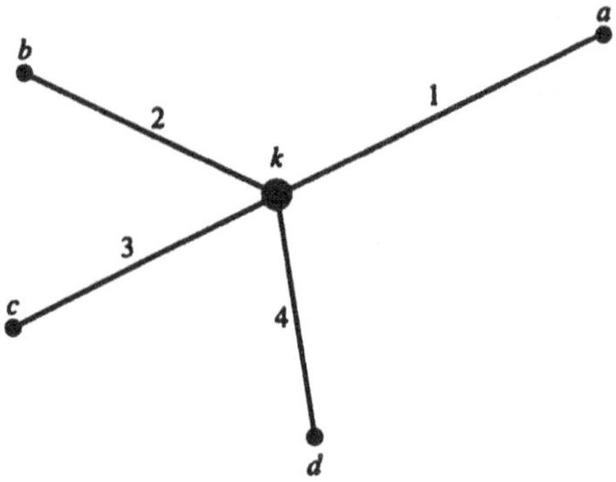

FIGURE 5.11
Node k with four elements.

If nodes a, b, c, and d as well as node k, are considered, the compatibility conditions would appear as

$$\begin{bmatrix} \mathbf{v}_a^1 \\ \mathbf{v}_k^1 \\ \cdots \\ \mathbf{v}_b^2 \\ \mathbf{v}_k^2 \\ \cdots \\ \mathbf{v}_c^3 \\ \mathbf{v}_k^3 \\ \cdots \\ \mathbf{v}_d^4 \\ \mathbf{v}_k^4 \end{bmatrix} = \underset{\mathbf{v}}{\begin{bmatrix} \mathbf{v}^1 \\ \cdots \\ \mathbf{v}^2 \\ \cdots \\ \mathbf{v}^3 \\ \cdots \\ \mathbf{v}^4 \end{bmatrix}} = \underset{\mathbf{a}}{\left[\begin{array}{c:c:c:c:c} \mathbf{I} & & & & \\ & & & & \mathbf{I} \\ \hdashline & \mathbf{I} & & & \\ & & & & \mathbf{I} \\ \hdashline & & \mathbf{I} & & \\ & & & & \mathbf{I} \\ \hdashline & & & \mathbf{I} & \\ & & & & \mathbf{I} \end{array}\right]} \underset{\mathbf{V}}{\begin{bmatrix} \mathbf{V}_a \\ \mathbf{V}_b \\ \mathbf{V}_c \\ \mathbf{V}_d \\ \mathbf{V}_k \end{bmatrix}} = \begin{bmatrix} \mathbf{a}^1 \\ \cdots \\ \mathbf{a}^2 \\ \cdots \\ \mathbf{a}^3 \\ \cdots \\ \mathbf{a}^4 \end{bmatrix} \mathbf{V} \tag{5.34}$$

where $\mathbf{I}$ is a unit diagonal matrix and, for example,

$$\mathbf{a}^2 = \begin{bmatrix} 0 & \mathbf{I} & 0 & 0 & 0 \\ 0 & 0 & 0 & 0 & \mathbf{I} \end{bmatrix}$$

is a submatrix of $\mathbf{a}$.

In general, for a system with M elements, in which $\mathbf{v}^i$ contains all element nodal displacements of element i, $\mathbf{a}^i$ is formed for all nodes of element i, and $\mathbf{V}$ includes all system displacements.

$$\begin{bmatrix} \mathbf{v}^1 \\ \mathbf{v}^2 \\ \vdots \\ \mathbf{v}^M \end{bmatrix} = \begin{bmatrix} \mathbf{a}^1 \\ \mathbf{a}^2 \\ \vdots \\ \mathbf{a}^M \end{bmatrix} \mathbf{V} \qquad \mathbf{v} = \mathbf{a}\ \mathbf{V} \tag{5.35}$$

The matrix $\mathbf{a}$, which is given many labels such as *global kinematic, connectivity, locator*, or *incidence matrix*, contains information designating which element is connected to which

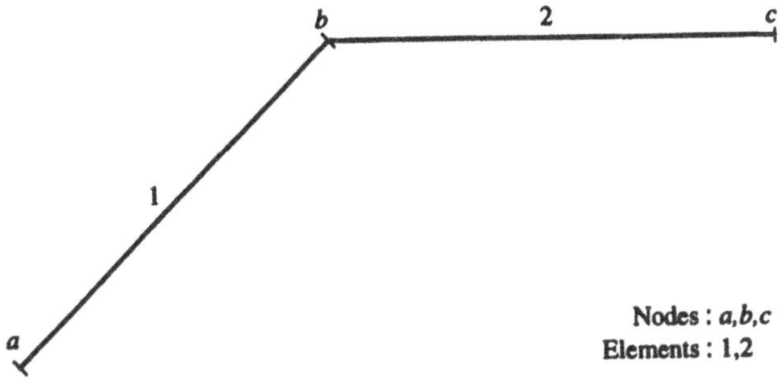

FIGURE 5.12
A three-node, two-element framework.

node. This matrix would be populated with null or unit values. Such a matrix is termed a *Boolean*[1] *matrix*.

To illustrate the compatibility conditions between the element and system displacements for a structure, consider the three-node, two-element system of Fig. 5.12. The compatibility conditions at nodes a, b, and c are

$$\begin{aligned} \mathbf{v}_a^1 &= \mathbf{V}_a \\ \mathbf{v}_b^1 = \mathbf{v}_b^2 &= \mathbf{V}_b \\ \mathbf{v}_c^2 &= \mathbf{V}_c \end{aligned} \tag{5.36}$$

or, in the notation of Eq. (5.35),

$$\begin{bmatrix} \mathbf{v}^1 \\ \\ \mathbf{v}^2 \end{bmatrix} = \begin{bmatrix} \mathbf{v}_a^1 \\ \mathbf{v}_b^1 \\ \mathbf{v}_b^2 \\ \mathbf{v}_c^2 \end{bmatrix} = \begin{bmatrix} \mathbf{I} & & \\ & \mathbf{I} & \\ & \mathbf{I} & \\ & & \mathbf{I} \end{bmatrix} \begin{bmatrix} \mathbf{V}_a \\ \mathbf{V}_b \\ \mathbf{V}_c \end{bmatrix} \tag{5.37}$$
$$\mathbf{v} \quad = \quad \mathbf{a} \quad \mathbf{V}$$

where $\mathbf{I}$ is the unit matrix. Note that these compatibility conditions serve to transform a displacement vector $\mathbf{v}$ containing displacements at the same node from different elements, i.e., $\mathbf{v}_b^1$ and $\mathbf{v}_b^2$, into the nodal displacement vector $\mathbf{V}$.

If Eq. (5.32) were written in matrix rather than in summation form, it would appear as

$$\sum_{i=1}^{M} \delta\mathbf{v}^{iT}(\mathbf{k}^i\mathbf{v}^i - \overline{\mathbf{p}}^{io}) = \delta\mathbf{v}^T(\mathbf{k}\,\mathbf{v} - \overline{\mathbf{p}}) = 0 \tag{5.38}$$

where $\mathbf{v}$ as defined in Eq. (5.35) is an unassembled displacement vector,

$$\overline{\mathbf{p}} = [\overline{\mathbf{p}}^{10}\ \overline{\mathbf{p}}^{20}\ \cdots\ \overline{\mathbf{p}}^{M0}]^T \tag{5.39}$$

[1]George Boole (1815–1864) was a self-taught British mathematician. Among his works was a brief, influential book written in 1847, *The Mathematical Analysis of Logic*, wherein he asserted that logic should be grounded in mathematics, rather than be associated with metaphysics. He founded symbolic logic with the book *An Investigation of the Laws of Thought*, which appeared in 1854. At the time he was a professor of mathematics in Queen's College, Cork.

is an unassembled vector of elements loads, and

$$\mathbf{k} = \begin{bmatrix} \mathbf{k}^1 & & & \\ & \mathbf{k}^2 & & \\ & & \ddots & \\ & & & \mathbf{k}^M \end{bmatrix} = \text{diagonal } [\mathbf{k}^i] \tag{5.40}$$

is an unassembled global stiffness matrix.

If $\mathbf{v} = \mathbf{a}\mathbf{V}$ of Eq. (5.35) is inserted in Eq. (5.38), the principle of virtual work expression would then be written in terms of the system nodal displacements. Thus,

$$\delta\mathbf{v}^T(\mathbf{k}\mathbf{v} - \overline{\mathbf{p}}) = \delta\mathbf{V}^T\mathbf{a}^T(\mathbf{k}\,\mathbf{a}\mathbf{V} - \overline{\mathbf{p}}) = \delta\mathbf{V}^T(\mathbf{a}^T\mathbf{k}\,\mathbf{a}\mathbf{V} - \mathbf{a}^T\overline{\mathbf{p}}) = 0 \tag{5.41}$$

Define

$$\mathbf{K} = \mathbf{a}^T\mathbf{k}\,\mathbf{a} \tag{5.42a}$$

as the assembled system stiffness matrix, and

$$\overline{\mathbf{P}} = \mathbf{a}^T\overline{\mathbf{p}} \tag{5.42b}$$

as the assembled applied load vector, so that Eq. (5.41), the principle of virtual work, becomes

$$\delta\mathbf{V}^T(\mathbf{K}\mathbf{V} - \overline{\mathbf{P}}) = 0 \tag{5.43}$$

or for arbitrary nodal displacements

$$\mathbf{K}\mathbf{V} = \overline{\mathbf{P}} \tag{5.44}$$

which is a set of algebraic equations for the unknown nodal displacements. The load vector $\overline{\mathbf{P}}$ also includes loads applied directly at the nodes.

In terms of the system nodal forces $\mathbf{P}$, the principle of virtual work can be expressed as $-(\delta W_i + \delta W_e) = \delta\mathbf{V}^T(\mathbf{P} - \overline{\mathbf{P}}) = 0$, so that

$$-(\delta W_i + \delta W_e) = \delta\mathbf{V}^T(\mathbf{P} - \overline{\mathbf{P}}) = \delta\mathbf{V}^T(\mathbf{K}\mathbf{V} - \overline{\mathbf{P}}) = 0 \tag{5.45}$$

Since the principle of virtual work is equivalent to the conditions of equilibrium, it follows that $\mathbf{K}\mathbf{V} = \overline{\mathbf{P}}$ is an expression of the global statement of equilibrium, $\mathbf{P} = \overline{\mathbf{P}}$. The matrices are "assembled" in the sense that the duplications occurring in $\mathbf{v}$ [e.g., in Eq. (5.37) where $\mathbf{v}_b^1 = \mathbf{v}_b^2$] are removed by invoking the compatibility conditions. Thus, the vector $\mathbf{v}$ of unassembled unknown displacements is replaced in the governing equations of equilibrium by the vector $\mathbf{V}$ of assembled unknown displacements.

The relations $\mathbf{K}\mathbf{V} = \overline{\mathbf{P}}$ are equilibrium equations that form the core of the displacement method and which can be solved for the system nodal displacements. These displacements can then be used in computing forces, stresses, and other displacements.

The topological information for the assembly of the stiffness matrix is contained in the connectivity matrix $\mathbf{a}$. The congruent transformation $\mathbf{a}^T\mathbf{k}\,\mathbf{a} = \mathbf{K}$ is rarely used in practice to form $\mathbf{K}$. The matrix $\mathbf{a}$ of this "assembly by multiplication" may contain many zero coefficients. In practice, this assembly of $\mathbf{K}$ as a congruent transformation is avoided in favor of an "assembly by addition" procedure. Thus, the assembled matrix relationships $\mathbf{K} = \mathbf{a}^T\mathbf{k}\,\mathbf{a}$ and $\overline{\mathbf{P}} = \mathbf{a}^T\overline{\mathbf{p}}$ are more of conceptual than practical value.

The assembly of $\mathbf{K}$ by an addition or superposition technique is illustrated using the two-element, three-node system of Fig. 5.12. For element number 1, which spans between global nodes a and b, the stiffness matrix can be written in terms of submatrices as [Eq. (5.19)]

$$\mathbf{k}^1 = \begin{bmatrix} \mathbf{k}_{aa} & \mathbf{k}_{ab} \\ \mathbf{k}_{ba} & \mathbf{k}_{bb} \end{bmatrix}^1 \tag{5.46a}$$

For the second element, which begins at node b and ends at node c,

$$\mathbf{k}^2 = \begin{bmatrix} \mathbf{k}_{bb} & \mathbf{k}_{bc} \\ \mathbf{k}_{cb} & \mathbf{k}_{cc} \end{bmatrix}^2 \tag{5.46b}$$

The global stiffness matrix $\mathbf{K}$ formed using $\mathbf{K} = \mathbf{a}^T\mathbf{k}\,\mathbf{a}$ is obtained in terms of the submatrices $\mathbf{k}^i_{jk}$ as

$$\underset{\mathbf{a}^T}{\begin{bmatrix} \mathbf{I} & & & \\ & \mathbf{I} & \mathbf{I} & \\ & & & \mathbf{I} \end{bmatrix}} \underset{\mathbf{k}}{\begin{bmatrix} \mathbf{k}^1 & 0 \\ 0 & \mathbf{k}^2 \end{bmatrix}} \underset{\mathbf{a}}{\begin{bmatrix} \mathbf{I} & & \\ & \mathbf{I} & \\ & \mathbf{I} & \\ & & \mathbf{I} \end{bmatrix}} \underset{=}{=} \underset{\mathbf{K}}{\begin{bmatrix} \mathbf{k}^1_{aa} & \mathbf{k}^1_{ab} & \\ \mathbf{k}^1_{ba} & \mathbf{k}^1_{bb} + \mathbf{k}^2_{bb} & \mathbf{k}^2_{bc} \\ & \mathbf{k}^2_{cb} & \mathbf{k}^2_{cc} \end{bmatrix}} \tag{5.47}$$

The process of forming an assembled matrix can be observed in Eq. (5.47). The number of columns in the global connectivity matrix $\mathbf{a}$ is equal to the number of system DOF. Postmultiplication of $\mathbf{k}$ by $\mathbf{a}$ places the coefficients of $\mathbf{k}$ in the proper columns of the assembled system stiffness matrix, whereas premultiplication by $\mathbf{a}^T$ locates the coefficients of $\mathbf{k}$ in the proper rows of the global stiffness matrix. The same process performed separately on each element stiffness matrix expands it into its proper location in the global stiffness matrix.

In contrast to the unassembled stiffness equations which appear as

$$\underset{\mathbf{p}}{\begin{bmatrix} \mathbf{p}^1 \\ \\ \mathbf{p}^2 \end{bmatrix}} \underset{=}{=} \begin{bmatrix} \mathbf{p}^1_a \\ \mathbf{p}^1_b \\ \mathbf{p}^2_b \\ \mathbf{p}^2_c \end{bmatrix} = \underset{\mathbf{k}}{\begin{bmatrix} \boxed{\mathbf{k}^1} & \\ & \boxed{\mathbf{k}^2} \end{bmatrix}} \underset{\mathbf{v}}{\begin{bmatrix} \mathbf{v}^1 \\ \\ \mathbf{v}^2 \end{bmatrix}} = [\mathbf{k}] \begin{bmatrix} \mathbf{v}^1_a \\ \mathbf{v}^1_b \\ \mathbf{v}^2_b \\ \mathbf{v}^2_c \end{bmatrix} \tag{5.48}$$

the assembled global stiffness equations are of the form [Eq. (5.47)]

$$\begin{bmatrix} \mathbf{p}^1_a \\ \mathbf{p}^1_b + \mathbf{p}^2_b \\ \mathbf{p}^2_c \end{bmatrix} = \underset{\mathbf{K}}{\begin{bmatrix} \boxed{\mathbf{k}^1} & \\ & \boxed{\mathbf{k}^2} \end{bmatrix}} \underset{\mathbf{V}}{\begin{bmatrix} \mathbf{V}_a \\ \mathbf{V}_b \\ \mathbf{V}_c \end{bmatrix}} = \begin{bmatrix} \mathbf{k}^1_{aa} & \mathbf{k}^1_{ab} & 0 \\ \mathbf{k}^1_{ba} & \mathbf{k}^1_{bb} + \mathbf{k}^2_{bb} & \mathbf{k}^2_{bc} \\ 0 & \mathbf{k}^2_{cb} & \mathbf{k}^2_{cc} \end{bmatrix} \begin{bmatrix} \mathbf{V}_a \\ \mathbf{V}_b \\ \mathbf{V}_c \end{bmatrix} \tag{5.49}$$

with

$$\begin{bmatrix} \mathbf{V}_a \\ \mathbf{V}_b \\ \mathbf{V}_c \end{bmatrix} = \begin{bmatrix} \mathbf{v}^1_a \\ \mathbf{v}^1_b = \mathbf{v}^2_b \\ \mathbf{v}_c \end{bmatrix} \tag{5.50}$$

Inspection of the assembled stiffness matrix of Eq. (5.49) provides insight as to why a stiffness matrix should be assembled using addition rather than the multiplication implied by $\mathbf{K} = \mathbf{a}^T\mathbf{k}\,\mathbf{a}$. In Eq. (5.49), it is apparent that all coefficients of $\mathbf{K}$ either are taken directly from $\mathbf{k}^1$ or $\mathbf{k}^2$, or, as in the case of the overlapping boxes, are the sum of $\mathbf{k}^1$ and $\mathbf{k}^2$ coefficients. The summation process shown in Eq. (5.49) can be programmed without much difficulty.

Although this summation assembly process is considered in depth in Section 5.3.4, the summation procedure is evident from Eq. (5.49). The assembled stiffness matrix $\mathbf{K}$ is formed by summation of those element stiffness matrix coefficients with identical subscripts.

Thus, in Eq. (5.49), $\mathbf{K}_{ij} = \mathbf{k}^1_{ij} + \mathbf{k}^2_{ij}$. Before carrying out the summation, it is necessary to fit each element stiffness matrix into the global nodal numbering system. That is, the subscripts of $\mathbf{k}^1$ are the node identifiers a and b, and the subscripts for $\mathbf{k}^2$ are b and c. The *incidence table* of Section 5.3.4 can be helpful in associating each element with the nodal numbering of the global system. We observe that the incidence table contains the same transformation information as **a**, except there are no zero coefficients.

5.3.2 Direct Derivation of the Displacement Equilibrium Equations

It is important to understand that the principle of virtual work leads essentially to equations of equilibrium. As in the case of Chapter 2, the displacements must satisfy kinematic or compatibility requirements. To emphasize that the displacement relations of Eq. (5.44) are based on equilibrium, we also can derive these equations directly from the conditions of equilibrium. Consider again the four-element, one-node system of Fig. 5.11. At this node, the forces must satisfy equilibrium. This is accomplished by summing all forces contributed by the elements joined at the node. Before this addition can take place, all forces have to be referred to the same reference frame by applying the local to global coordinate transformation of Eq. (5.27). For the node k of Fig. 5.11, the condition of equilibrium will be

$$\mathbf{p}^1_k + \mathbf{p}^2_k + \mathbf{p}^3_k + \mathbf{p}^4_k = \overline{\mathbf{P}}_k \tag{5.51}$$

For the three-node, two-element system of Fig. 5.12, the nodal equilibrium relations can be expressed as

$$\begin{aligned} \mathbf{p}^1_a &= \overline{\mathbf{P}}_a \\ \mathbf{p}^1_b + \mathbf{p}^2_b &= \overline{\mathbf{P}}_b \\ \mathbf{p}^2_c &= \overline{\mathbf{P}}_c \end{aligned} \quad \text{or} \quad \underset{\mathbf{b}^*}{\begin{bmatrix} \mathbf{I} & & & \\ & \mathbf{I} & \mathbf{I} & \\ & & & \mathbf{I} \end{bmatrix}} \underset{\mathbf{p}}{\begin{bmatrix} \mathbf{p}^1_a \\ \mathbf{p}^1_b \\ \mathbf{p}^2_b \\ \mathbf{p}^2_c \end{bmatrix}} = \underset{\overline{\mathbf{P}}}{\begin{bmatrix} \overline{\mathbf{P}}_a \\ \overline{\mathbf{P}}_b \\ \overline{\mathbf{P}}_c \end{bmatrix}} \tag{5.52}$$

where $\mathbf{b}^*$ is the *global statics* or *equilibrium matrix* that defines the conditions of equilibrium between $\mathbf{p}$, the unassembled element forces, and $\overline{\mathbf{P}}$. It is possible to establish a reciprocal relation of the form

$$\mathbf{p} = \mathbf{b}\overline{\mathbf{P}} \tag{5.53}$$

However, some care must be exercised, since, as can be observed in Eq. (5.52), $\mathbf{b}^*$ is not necessarily a square matrix and, hence, cannot be obtained as the inverse of $\mathbf{b}$. It is the case, however, that

$$\mathbf{b}^*\,\mathbf{b} = \mathbf{I} \tag{5.54}$$

whereas $\mathbf{b}\,\mathbf{b}^* \neq \mathbf{I}$. In general,

$$\underset{\mathbf{p}}{\begin{bmatrix} \mathbf{p}^1 \\ \mathbf{p}^2 \\ \vdots \\ \mathbf{p}^M \end{bmatrix}} = \underset{\mathbf{b}}{\begin{bmatrix} \mathbf{b}^1 \\ \mathbf{b}^2 \\ \vdots \\ \mathbf{b}^M \end{bmatrix}} \underset{\overline{\mathbf{P}}}{\overline{\mathbf{P}}} \tag{5.55}$$

By comparison of Eqs. (5.37) and (5.52), we observe that

$$\mathbf{b}^* = \mathbf{a}^T \tag{5.56}$$

It can also be shown that $\mathbf{a}^* = \mathbf{b}^T$, where $\mathbf{a}^*$ is defined by $\mathbf{V} = \mathbf{a}^*\mathbf{v}$. Although $\mathbf{a}^*\mathbf{a} = \mathbf{I}$, the product $\mathbf{a}\mathbf{a}^*$ is not the identity matrix.

This relationship between the equilibrium matrix $\mathbf{b}^*$ and the kinematic matrix $\mathbf{a}^T$ for discrete systems is akin to the relationship between the kinematic operator $\mathbf{D}$ and the equilibrium operator $\mathbf{D}^T$ for continuous systems (Chapter 1). With the help of these new definitions, the equilibrium equations in the displacement form of Eq. (5.44) are readily obtained. From the nodal equilibrium relations $\mathbf{p} = \mathbf{b}\overline{\mathbf{P}}$ and the identities of Eqs. (5.52) and (5.56), we find

$$\mathbf{b}^*\mathbf{p} = \overline{\mathbf{P}} = \mathbf{a}^T\mathbf{p} \tag{5.57}$$

The unassembled set of stiffness equations would be $\mathbf{p} = \mathbf{k}\,\mathbf{v}$. Introduce this into Eq. (5.57), giving

$$\mathbf{a}^T\mathbf{k}\,\mathbf{v} = \overline{\mathbf{P}} \tag{5.58}$$

From the nodal connectivity equation (Eq. 5.37) it follows that

$$\mathbf{a}^T\mathbf{k}\,\mathbf{a}\mathbf{V} = \overline{\mathbf{P}} \quad \text{or} \quad \mathbf{K}\mathbf{V} = \overline{\mathbf{P}} \tag{5.59}$$

where $\mathbf{K}$ is again given by $\mathbf{a}^T\mathbf{k}\;\mathbf{a}$. Thus, without direct reference to the principle of virtual work, it is evident that the equations of equilibrium provide the basis for the displacement relations of Eq. (5.44).

With the derivation of the congruent transformation representation of $\mathbf{K}$, it can be observed that the system matrix $\mathbf{K}$ contains all the basic equations:

$$\mathbf{K} = \mathbf{a}^T\mathbf{k}\;\mathbf{a} \tag{5.60}$$

$\mathbf{a}^T$: Conditions of Equilibrium
$\mathbf{k}$: Material Law
$\mathbf{a}$: Compatibility

5.3.3 Transformation of Coordinates

In the previous development of the transformation $\mathbf{K} = \mathbf{a}^T\mathbf{k}\;\mathbf{a}$, it was assumed that all of the element forces, displacements, and stiffness matrices were available in the form referred to the global coordinates. The transformation $\mathbf{a}^T\mathbf{k}\;\mathbf{a}$ can be generalized to include a transformation from local to global coordinates for the element stiffness matrix. From Eq. (5.28), the stiffness matrix $\tilde{\mathbf{k}}^i$ in local coordinates is transformed to global coordinates using

$$\mathbf{k}^i = \mathbf{T}^{iT}\tilde{\mathbf{k}}^i\mathbf{T}^i \tag{5.61}$$

Then $\mathbf{K} = \mathbf{a}^T\mathbf{k}\;\mathbf{a}$ becomes

$$\mathbf{K} = \mathbf{a}^T \text{ diag } [\mathbf{T}^{iT}\tilde{\mathbf{k}}^i\mathbf{T}^i]\mathbf{a} = \tilde{\mathbf{a}}^T\tilde{\mathbf{k}}\,\tilde{\mathbf{a}} \tag{5.62}$$

with

$$\tilde{\mathbf{a}} = \begin{bmatrix} \mathbf{T}^i & & & \\ & \mathbf{T}^i & & \\ & \mathbf{T}^i & & \\ & & \mathbf{T}^i & \end{bmatrix} \tag{5.63}$$

where $\mathbf{T}^i$ has now replaced $\mathbf{I}$ in Eq. (5.37).

5.3.4 Assembly of a System Stiffness Matrix by a Summation Process

As indicated previously, the congruent transformation $\mathbf{K} = \mathbf{a}^T \mathbf{k}\, \mathbf{a}$ is not normally utilized explicitly to assemble the system stiffness matrix. Rather, this matrix should be recognized as being a judiciously formed superposition of element stiffness matrices and, as such, should be calculated by a summation process. This can be accomplished with the aid of an *incidence table* which replaces the connectivity or incidence matrix $\mathbf{a}$. In contrast to $\mathbf{a}$ which, in describing the topology of a system, is burdened by the frequent occurrence of zero coefficients, the incidence table contains no zero coefficients. It simply identifies the end nodes of each element with the corresponding global nodes. For the two-bar element, three-node system of Fig. 5.12, the system nodes are numbered a, b, and c. An incidence table would then appear as

Element No.	Global Node Numbers Corresponding to Element End Numbers: Element Begins at System Node No.	Element Ends at System Node No.
1	a	b
2	b	c

(5.64)

Since for each element (bar member) beginning and end system nodes are indicated, an incidence table provides a sense of direction, a feature that can be important in interpreting the responses that are calculated. Sometimes the beginning node of a bar element is said to be the + or +1 node, and the end node is said to be the − or −1 node.

The incidence table is used to associate each end of each element with a particular system node. Thus, this table tags each element end with a system number to indicate where it should be placed in $\mathbf{K}$. After all of the element ends are assigned to their correct system nodes, all of the element stiffness coefficients, which are now associated with a particular set of system node numbers, are summed to provide the corresponding system stiffness coefficient.

It may be helpful to describe this stiffness matrix assembly process in terms of particular stiffness coefficients. Suppose the stiffness matrix for the ith (1 or 2 in the case of Fig. 5.12) element in terms of submatrices is of the form

$$\mathbf{k}^i = \begin{bmatrix} \mathbf{k}_{jj} & \mathbf{k}_{jk} \\ \mathbf{k}_{kj} & \mathbf{k}_{kk} \end{bmatrix}^i \tag{5.65}$$

where j and k are system node numbers (a, b, or c in the case of Fig. 5.12) provided by the incidence table for element i. The contribution of the ith element to the system stiffness matrix is

Row Index of System Matrix	Column Index of System Matrix: j	k
j	$\mathbf{k}^i_{jj}$	$\mathbf{k}^i_{jk}$
k	$\mathbf{k}^i_{kj}$	$\mathbf{k}^i_{kk}$

(5.66)

All element coefficients (or submatrices) thus identified as belonging to the same location of the system matrix are then summed. In practial terms, this means that stiffness matrix coefficients (or submatrices) of like subscripts are summed to form the corresponding (same subscripts) coefficient of the global stiffness matrix.

This process of forming $\mathbf{K}$ is equivalent to a loop summation calculation over all elements i using

$$\mathbf{K}_{jk} \leftarrow \mathbf{K}_{jk} + \mathbf{k}^i_{jk} \tag{5.67}$$

where j and k are taken from the incidence table for each element i.

For the system of Fig. 5.12, the incidence table of Eq. (5.64) shows that the indices of the stiffness matrix $\mathbf{k}^1$ for bar 1 are a and b, while for bar 2 the stiffness matrix indices are b and c. Summation of element stiffness matrix coefficients or submatrices of like subscripts gives

$$\mathbf{K} = \begin{bmatrix} \mathbf{k}_{aa}^1 & \mathbf{k}_{ab}^1 & \\ \mathbf{k}_{ba}^1 & \mathbf{k}_{bb}^1 + \mathbf{k}_{bb}^2 & \mathbf{k}_{bc}^2 \\ & \mathbf{k}_{cb}^2 & \mathbf{k}_{cc}^2 \end{bmatrix} = \begin{bmatrix} \mathbf{K}_{aa} & \mathbf{K}_{ab} & \mathbf{K}_{ac} \\ \mathbf{K}_{ba} & \mathbf{K}_{bb} & \mathbf{K}_{bc} \\ \mathbf{K}_{ca} & \mathbf{K}_{cb} & \mathbf{K}_{cc} \end{bmatrix} \tag{5.68}$$

As can be imagined, numerous schemes have been devised for the efficient handling of data during the assembly process. In some cases, all element stiffness matrices are generated and stored in a background storage unit. They are called up as needed to create $\mathbf{K}$ in active storage. In other cases, the information from each element is processed into the system matrix as each element stiffness matrix is created, thus eliminating the use of background storage.

5.3.5 Incorporation of Boundary Conditions, Reactions

The global stiffness matrix, like the element stiffness matrix, is singular. This singularity can be illustrated using the structure of Fig. 5.12. The global stiffness can be expressed as [Eq. (5.68)]

$$\begin{matrix} \begin{bmatrix} \mathbf{K}_{aa} & \mathbf{K}_{ab} & \mathbf{K}_{ac} \\ \mathbf{K}_{ba} & \mathbf{K}_{bb} & \mathbf{K}_{bc} \\ \mathbf{K}_{ca} & \mathbf{K}_{cb} & \mathbf{K}_{cc} \end{bmatrix} & \begin{bmatrix} \mathbf{V}_a \\ \mathbf{V}_b \\ \mathbf{V}_c \end{bmatrix} & = & \begin{bmatrix} \overline{\mathbf{P}}_a \\ \overline{\mathbf{P}}_b \\ \overline{\mathbf{P}}_c \end{bmatrix} \\ \mathbf{K} & \mathbf{V} & = & \overline{\mathbf{P}} \end{matrix} \tag{5.69}$$

If the structure is not constrained, the displacements can occur in two forms: elastic deformation of the structure due to the applied loading and movement as a rigid body. Due to this rigid body motion, two of the nodal displacements, say $\mathbf{V}_a$ and $\mathbf{V}_c$, can have arbitrary values, while the third nodal displacement ($\mathbf{V}_b$) depends on $\mathbf{V}_a$ and $\mathbf{V}_c$. The stiffness relations of Eq. (5.69) form a set of simultaneous linear equations. From Cramer's rule, if the equations do not have a unique solution, the determinant of the coefficient matrix $\mathbf{K}$ must be zero, $\mathbf{K}$ is singular. A unique solution exists only after the structure is constrained, i.e., displacement boundary conditions are imposed, so that rigid body motion is prevented.

The introduction of boundary (support) conditions is readily portrayed by appropriately partitioning the global equilibrium equations $\mathbf{KV} = \overline{\mathbf{P}}$, although, in practice, matrix rearrangement operations are avoided. Blind partitioning of the global equations would normally lead to an increase in the bandwidth, which, as explained in Section 5.3.7, is undesirable. To illustrate the partitioning approach, suppose the displacement vector $\mathbf{V}$ contains both prescribed displacements, which will be denoted by $\overline{\mathbf{V}}$, and unknown nodal displacements which will be designated by $\mathbf{V}_y$. Typically, the prescribed displacements are joint displacements that are zero. Often these displacements are referred to as constrained DOF. The remaining unconstrained displacements are called the active DOF. Also, the force vector can contain applied forces $\overline{\mathbf{P}}$ and unknown nodal forces $\mathbf{P}_v$, i.e., the support reactions corresponding to prescribed displacements $\overline{\mathbf{V}}$. Then the global equations can be rearranged to achieve the partitioned form

$$\begin{bmatrix} \mathbf{K}_{11} & \mathbf{K}_{12} \\ \mathbf{K}_{21} & \mathbf{K}_{22} \end{bmatrix} \begin{bmatrix} \overline{\mathbf{V}} \\ \mathbf{V}_y \end{bmatrix} = \begin{bmatrix} \mathbf{P}_v \\ \overline{\mathbf{P}} \end{bmatrix} \tag{5.70}$$

It follows that the unknown displacements are given by

$$\mathbf{V}_y = \mathbf{K}_{22}^{-1}(\overline{\mathbf{P}} - \mathbf{K}_{21}\overline{\mathbf{V}}) \tag{5.71}$$

and the unknown forces (reactions) are

$$\mathbf{P}_v = \mathbf{K}_{11}\overline{\mathbf{V}} + \mathbf{K}_{12}\mathbf{V}_y \tag{5.72}$$

However, this form of the solution is only formal, since rearrangement and inversion operations are usually circumvented. That is, rather than expressing the unknown displacements in the form of Eq. (5.71), the system of equations

$$\mathbf{K}_{22}\mathbf{V}_y = \overline{\mathbf{P}} - \mathbf{K}_{21}\overline{\mathbf{V}} \tag{5.73}$$

is solved for $\mathbf{V}_y$.

A straightforward approach to the introduction of boundary conditions is to ignore those columns in the system matrix that correspond to zero (prescribed) displacements and those rows for the corresponding unknown reactions. Then solve the remaining equations for the unknown nodal displacements and compute the reactions $\mathbf{P}_v$ through Eq. (5.72). This is the technique employed in most of the example solutions of the displacement method in this chapter.

Schemes have been developed whereby the boundary conditions are applied to the element stiffness matrices prior to assembly of the global matrices (see, for example, Bathe (1996) and Cook, et al. (1989)).

5.3.6 Internal Forces, Stress Resultants, and Stresses

The internal force, stress resultant, or stress distribution in an element has to be computed using an additional procedure, since only nodal displacements $\mathbf{V}$ are calculated directly. For beams, the nodal forces can be obtained using

$$\mathbf{p}^i = \mathbf{k}^i\mathbf{v}^i \tag{5.74}$$

where the displacement vector $\mathbf{v}^i$ follows from $\mathbf{V}$ through the compatibility relationships $\mathbf{v} = \mathbf{a}\mathbf{V}$ of Eq. (5.37). The nodal forces of Eq. (5.74) are component forces along the global coordinate axes which can be transformed into local components by relations such as Eq. (5.22), after which desired responses such as stresses can be calculated.

If the distribution of response variables along an element is needed, it is often convenient to use the transfer matrix method for performing the calculations. Since both end displacements $\mathbf{v}^i$ and forces $\mathbf{p}^i$ are obtained by postprocessing the results of a global displacement analysis, i.e., the state vector is known at the ends of an element, it is a simple task to use transfer matrices to compute the distribution of these variables along a member.

5.3.7 Some Characteristics of Stiffness Matrices

Several properties of stiffness matrices were developed in Chapter 3. As indicated in Chapter 3, both element and global stiffness matrices are symmetric and positive definite. The symmetry property is important in practice, since only terms on and to one side of the main diagonal need to be generated and retained in a computer program. The positive definite property applies only to a stiffness matrix to which constraints have been applied, e.g., the rigid body motion has been removed. Singular matrices can be positive semi-definite, but not positive definite.

Observe the structure of the stiffness equations. For example, consider a particular global stiffness matrix $\mathbf{K}$, and note that each row of the matrix corresponds to a force at a node.

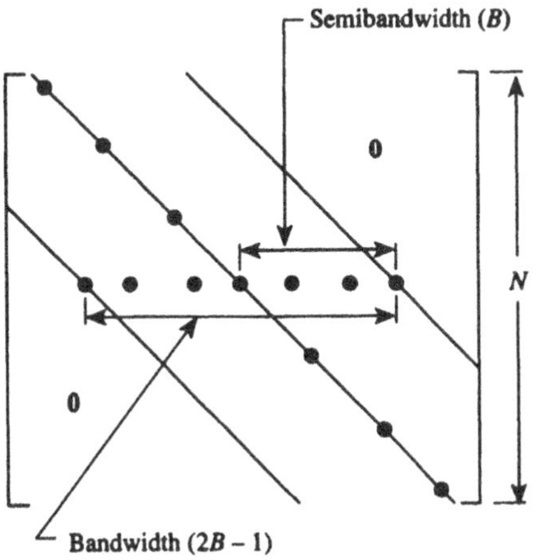

FIGURE 5.13
A banded stiffness matrix. Coefficients within the band are usually non-zero.

This row contains non-zero terms, in addition to those on the main diagonal, for the DOF at the node and for the DOF of the other nodes of the elements connected to the primary node. All other coefficients in this row of the stiffness matrix are zero. It can be concluded, then, that each row of the stiffness matrix contains non-zero terms only for the DOF belonging to the elements meeting at the node containing the force for which this row is written. For a structure of many nodes and elements, the stiffness matrix may appear to contain mostly zero terms with relatively few non-zero terms. In such cases, the matrix is said to be *sparse* or *weakly populated.*

Some solution procedures for systems of linear equations can take advantage of the sparseness of a matrix, particularly if the non-zero terms are clustered close to the diagonal, i.e., if the matrix is *banded*. Also, storage of the stiffness matrix in a computer is simplified and efficient for such an arrangement. The non-zero terms can be placed close to the diagonal by judiciously choosing the numbering system for the DOF. The DOF should be numbered such that a columnar distance from the main diagonal to the most remote non-zero term in a particular row is minimized. This is referred to as minimizing the *bandwidth* (Fig. 5.13). The two numberings of the joints (which, in terms of minimizing the bandwidth, correspond closely to the DOF) of the frame of Fig. 5.14 illustrate the influence of the numbering scheme for a structure. Usually, a small bandwidth results if nodes across the shorter dimension of a structure are numbered consecutively. This certainly holds for the numbering schemes of Fig. 5.14.

Non-zero elements of the stiffness matrix of Fig. 5.13 are contained in the NB coefficients of the semi-band. Since, in practice, N is much greater then B, it can be important to avoid storing N^2 coefficients by retaining only the NB essential coefficients. In terms of equation-solving efficiency, it has been shown [Cook, et al., 1989] that the savings achieved by utilizing NB coefficients rather than all coefficients in the upper triangle is proportional to $(N/B)^2$.

A format for the storage of a banded matrix is shown in Fig. 5.15. Rows of the upper semi-band are simply shifted to the left. The first row remains in place, the second row is shifted 1 space, the third row 2 spaces, and the kth row $k-1$ spaces. This places the diagonal coefficients of the matrix of Fig. 5.13 in the first column of the array of Fig. 5.15.

There is considerable literature (see, for example, Cook, et al., (1989)) on economical methods of retaining only the essential information of a stiffness matrix. Although the band format is an efficient method, other techniques have been designed which are even more efficient (e.g., see Everstine (1979)). Also, schemes have been devised (e.g., Everstine (1979) and Gibbs, et al., (1976)) for the automatic renumbering of nodes, so that a criterion such

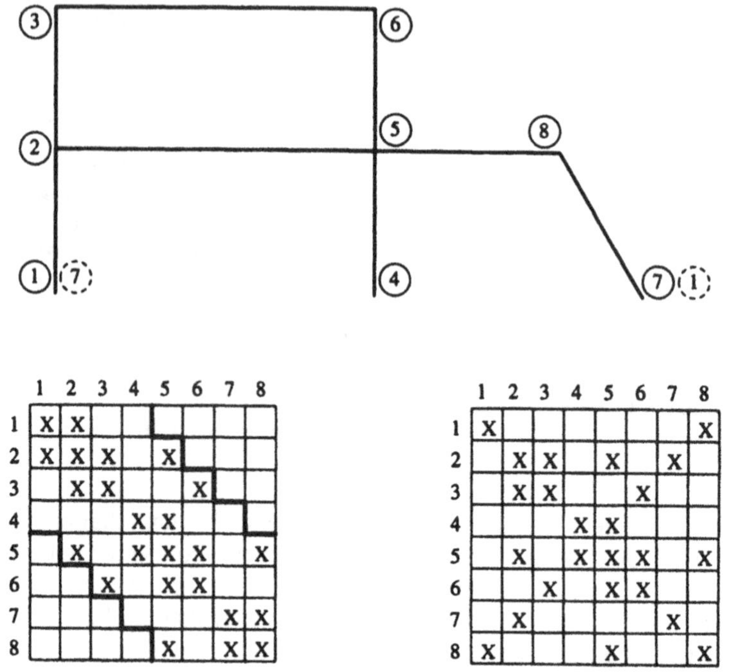

(a) A favorable numbering leads to a band-like structure of the stiffness matrix

(b) Replace ⑦ and ① by ⟨1⟩ and ⟨7⟩ and note that the band structure of the stiffness matrix is lost

FIGURE 5.14
Example of the effect of numbering DOF to achieve a banded stiffness matrix.

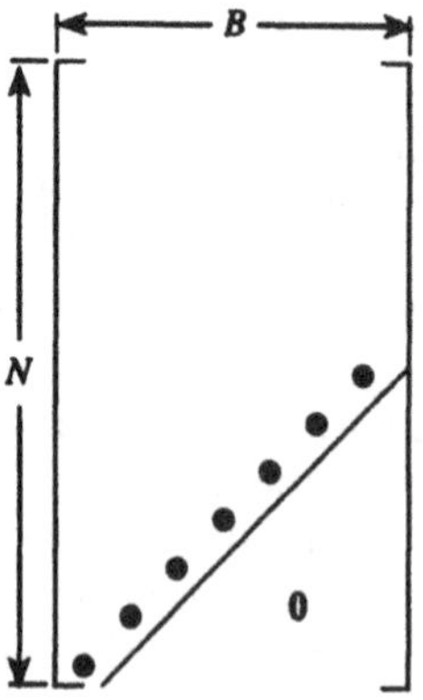

FIGURE 5.15
Form for computer storage of the banded matrix.

as bandwidth is reduced. These procedures are particularly useful when several problems involving modifications to a basic structure or mesh system are being solved.

5.3.8 Static Condensation—Substructuring

In practice, structures can be very large and complicated. These structures may require models of many thousands of elements, nodes, and DOF. If the elements of the model are assembled in the manner described previously, a large global stiffness matrix may result, requiring a major effort to achieve a solution. Often substructuring can be employed to avoid excessive computational cost. This technique is especially suitable for structures in which parts of the system (substructures) with the same number and types of elements are repeated. Then the stiffness matrix of a repeated part needs to be evaluated only once. Substructuring also helps one to retain mechanical insight into the behavior of complicated systems despite a very large number of unknowns.

The first step in substructuring is to divide the large structure into several smaller parts referred to as "substructures." Then discretize each substructure into elements, obtain the element stiffness matrices and loading vectors, and assemble the element matrices into a set of global stiffness equations for each substructure. At the boundaries where the substructures connect, there will be common nodes called *exterior nodes*. The substructures can be viewed as being formed of elements with many interior nodes along with some exterior nodes. The global stiffness matrix and the loading vector of the complete structure is assembled from the stiffness matrices and loading vectors of the substructures, with the displacements of the exterior nodes as unknowns. The degrees of freedom of the interior nodes are *condensed out* of the stiffness equations for each substructure. Let $\widehat{\mathbf{K}}^i$ and $\widehat{\overline{\mathbf{P}}}^i$ be the stiffness matrix and applied loading vector for the ith substructure, $\widehat{\mathbf{V}}_1^i$ and $\widehat{\mathbf{V}}_2^i$ be the displacements of the exterior nodes and interior nodes, respectively, and $\widehat{\overline{\mathbf{P}}}_1^i$ and $\widehat{\overline{\mathbf{P}}}_2^i$ be the loading vectors associated with $\widehat{\mathbf{V}}_1^i$ and $\widehat{\mathbf{V}}_2^i$. The stiffness equations for this substructure can be expressed as

$$\begin{bmatrix} \widehat{\mathbf{K}}_{11}^i & \widehat{\mathbf{K}}_{22}^i \\ \widehat{\mathbf{K}}_{21}^i & \widehat{\mathbf{K}}_{22}^i \end{bmatrix} \begin{bmatrix} \widehat{\mathbf{V}}_1^i \\ \widehat{\mathbf{V}}_2^i \end{bmatrix} = \begin{bmatrix} \widehat{\overline{\mathbf{P}}}_1^i \\ \widehat{\overline{\mathbf{P}}}_2^i \end{bmatrix} \tag{5.75}$$

$$\widehat{\mathbf{K}}^i \quad \widehat{\mathbf{V}}^i = \widehat{\overline{\mathbf{P}}}^i$$

In expanded form,

$$\begin{aligned} \widehat{\mathbf{K}}_{11}^i \widehat{\mathbf{V}}_1^i + \widehat{\mathbf{K}}_{12}^i \widehat{\mathbf{V}}_2^i &= \widehat{\overline{\mathbf{P}}}_1^i \\ \widehat{\mathbf{K}}_{21}^i \widehat{\mathbf{V}}_1^i + \widehat{\mathbf{K}}_{22}^i \widehat{\mathbf{V}}_2^i &= \widehat{\overline{\mathbf{P}}}_2^i \end{aligned} \tag{5.76}$$

Solve the second equation of Eq. (5.76) for $\widehat{\mathbf{V}}_2^i$, giving

$$\widehat{\mathbf{V}}_2^i = \left(\widehat{\mathbf{K}}_{22}^i\right)^{-1} \widehat{\overline{\mathbf{P}}}_2^i - \left(\widehat{\mathbf{K}}_{22}^i\right)^{-1} \widehat{\mathbf{K}}_{21}^i \widehat{\mathbf{V}}_1^i \tag{5.77}$$

Substitute the expression of $\widehat{\mathbf{V}}_2^i$ into the first equation of Eq. (5.76), giving

$$\mathbf{K}^i \mathbf{V}^i = \overline{\mathbf{P}}^i \tag{5.78}$$

with $\mathbf{K}^i = \widehat{\mathbf{K}}_{11}^i - \widehat{\mathbf{K}}_{12}^i (\widehat{\mathbf{K}}_{22}^i)^{-1} \widehat{\mathbf{K}}_{21}^i$, $\mathbf{V}^i = \widehat{\mathbf{V}}_1^i$, and $\overline{\mathbf{P}}^i = \widehat{\overline{\mathbf{P}}}_1^i - \widehat{\mathbf{K}}_{12}^i (\widehat{\mathbf{K}}_{22}^i)^{-1} \widehat{\overline{\mathbf{P}}}_2^i$. Equation (5.78) contains only the displacements of the exterior nodes as the displacements of the interior nodes have been condensed out. This process is called *static condensation*.

The stiffness equations in the form of Eq. (5.78) for the substructures can be assembled into the global stiffness equations for the whole structure, with the displacement vector formed of the displacements of the exterior nodes. After these displacements are obtained, they can be substituted into the substructure stiffness equations, which can then be used to compute the displacements of the interior nodes.

The concept of substructuring can also be used to manipulate the stiffness equations for individual elements which contain interior nodes, that is, elements that are not common with the nodes of other elements. The displacements of the interior nodes can be condensed out.

The effectiveness of substructuring analysis can be improved by defining higher levels of substructures. At each level, the stiffness equations for the substructure are formed and condensed, and then assembled to form the stiffness equations of the next level substructure. Multi level substructuring can greatly reduce the dimensions of the global stiffness equations for the complete structure.

5.3.9 Summary of the Solution Procedure

The computation implementation of the displacement method proceeds as follows:

1. Idealize the structure by subdividing it conceptually into elements. Define global coordinates, the number and location of the nodes, and the nodal variables **V**, **P**. Read in the loading cases and the number of elements (members). Construct an incidence table.
2. For each element i, input properties, e.g., A, E, I, and J for a beam element. Compute the element stiffness matrix $\widetilde{\mathbf{k}}^i$ in local coordinates. Establish the global to local coordinate transformation $\mathbf{T}^i$, and calculate the element stiffness matrix in global coordinates using $\mathbf{k}^i = \mathbf{T}^{iT}\widetilde{\mathbf{k}}^i\mathbf{T}^i$.
3. In accordance with the incidence table, assemble the element matrices $\mathbf{k}^i$ into a system stiffness matrix **K**, summing all submatrices of like subscript. Also assemble the loading vector $\overline{\mathbf{P}}$.
4. In essence, the system equations $\mathbf{KV} = \overline{\mathbf{P}}$ have been established. Incorporate the boundary conditions by removing each column that corresponds to a displacement that is specified to be zero. Also, temporarily discard the rows corresponding to the unknown support reactions. In practice, the corresponding rows and columns are not established at all, as their absence can be accounted for in the incidence table. The result will be a square, non singular stiffness matrix. What remains of the applied load vector $\overline{\mathbf{P}}$ should contain the values of applied forces including zeros corresponding to DOF where no loads are applied.
5. Solve the set of equations $\mathbf{KV} = \overline{\mathbf{P}}$ for the system nodal displacements **V** for each case of applied loading, $\overline{\mathbf{P}}$.
6. Considerable information can now be obtained as a postprocessing computation. The support reactions can be calculated using the equations discarded in Step 4. For each member, the end displacements are calculated using the compatibility conditions ($\mathbf{v} = \mathbf{aV}$), and the end forces are found from $\mathbf{p} = \mathbf{kv}$. Member displacements and forces can be transformed from global to local coordinates. If desired, in-span displacements and forces can be computed using, for example, the transfer matrix method. For beams, cross-sectional stresses can now be computed using the formulas e.g., $\sigma = Mz/I$, that relate stresses to forces.
7. It is essential that the results of these computations be carefully scrutinized. At the outset, the plausibility of the responses should be studied. Further controls are provided by checking the conditions of equilibrium for the whole system, for particular nodes, and for particular parts. In the case of elements, the compatibility conditions should be verified.

5.3.10 Trusses

An important example of network structures is a truss system which is formed of pin-connected members. It is assumed that the bars only extend or compress and transmit only axial forces; no bending moments or shear forces are generated. For an ideal truss with ideal, frictionless pins or hinges, this is not an assumption as each bar, in fact, according to equilibrium requirements, can only transmit axial forces. All forces and motion will be (assumed to be) in the xz plane, i.e., this is a planar problem with no out-of-plane motion.

Element Coordinate Transformations and Stiffness Matrices

Local and global coordinate systems, forces, and displacements are shown in Fig. 5.16. Because of the fundamental truss assumption of bars being pin-connected, with the

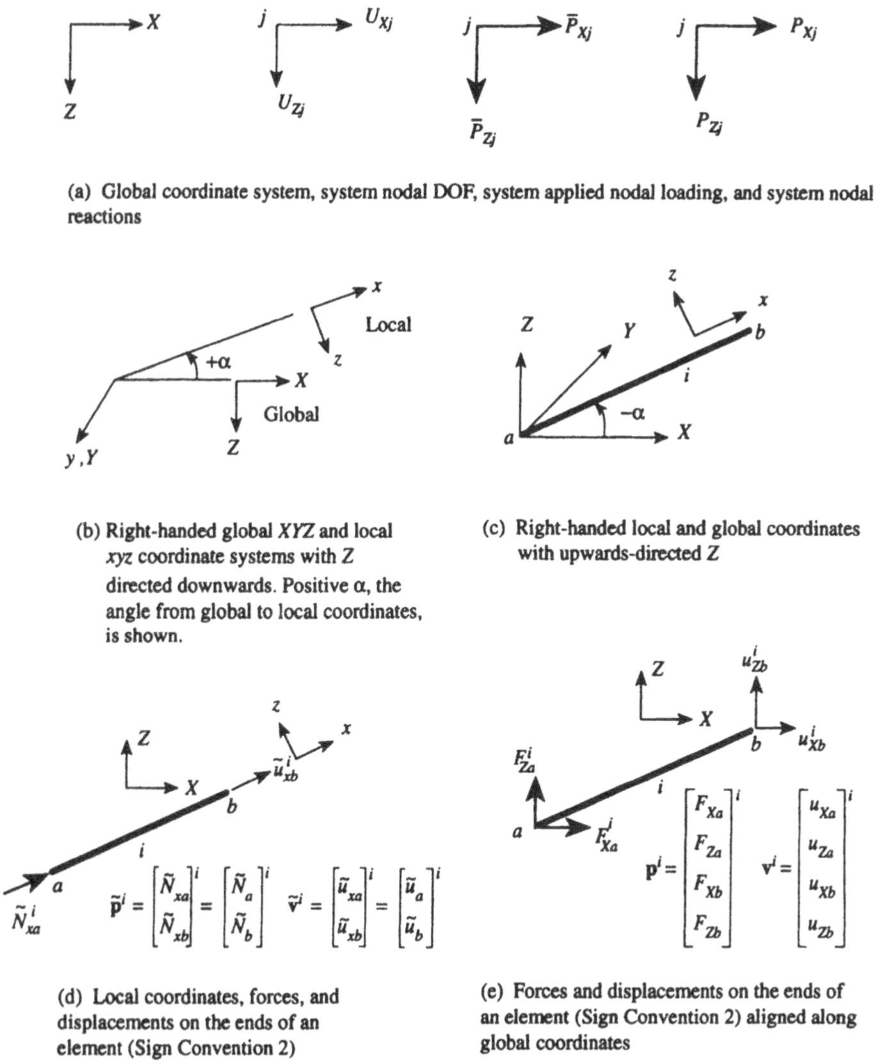

FIGURE 5.16
Local and global orientation of coordinates, displacements and forces for trusses. For simplicity, only forces are displayed at one end of the bar, while only displacements are shown at the other end.

resulting alignment of the forces along the bar axis, the element force and displacement vectors in the local xz coordinates will be

$$\tilde{\mathbf{p}}^i = \begin{bmatrix} \tilde{N}_{xa} \\ \tilde{N}_{xb} \end{bmatrix}^i = \begin{bmatrix} \tilde{N}_a \\ \tilde{N}_b \end{bmatrix}^i \qquad \tilde{\mathbf{v}}^i = \begin{bmatrix} \tilde{u}_{xa} \\ \tilde{u}_{xb} \end{bmatrix}^i = \begin{bmatrix} \tilde{u}_a \\ \tilde{u}_b \end{bmatrix}^i \tag{5.79}$$

where coordinate x is along the bar axis. However, these same forces and displacements in the global XZ coordinate systems will appear as

$$\mathbf{p}^i = \begin{bmatrix} F_{Xa} \\ F_{Za} \\ F_{Xb} \\ F_{Zb} \end{bmatrix}^i \qquad \mathbf{v}^i = \begin{bmatrix} u_{Xa} \\ u_{Za} \\ u_{Xb} \\ u_{Zb} \end{bmatrix}^i \tag{5.80}$$

The expanded vectors are necessary, since there will be more components in the global system than in the local system, where the forces and displacements are aligned with the x axis. In Section 5.2.1 element end forces in both the local and global coordinates are designated by lower case letters and capital letters are used for the system (nodal) forces. Unfortunately, tradition dictates that capital letters should be used for the element forces of truss members, e.g., Eqs. (5.79) or (5.80). The same situation occurs with beam elements, where traditionally the bending moment and shear force are represented by upper case letters.

The uniaxial characteristics of the forces and displacements referred to the local coordinate system simplify the coordinate transformations. For example, the displacements (or forces) would transform as

$$\tilde{u}^i_{xa} = u^i_{Xa}\cos\alpha - u^i_{Za}\sin\alpha$$

or, more generally,

$$\begin{array}{ccccc} \begin{bmatrix} \tilde{u}_{xa} \\ \tilde{u}_{xb} \end{bmatrix}^i & = & \left[\begin{array}{cc|cc} \cos\alpha & -\sin\alpha & 0 & 0 \\ 0 & 0 & \cos\alpha & -\sin\alpha \end{array}\right] & \begin{bmatrix} u_{Xa} \\ u_{Za} \\ u_{Xb} \\ u_{Zb} \end{bmatrix}^i \\ \tilde{\mathbf{v}}^i & = & \mathbf{T}^i & \mathbf{v}^i \end{array} \tag{5.81}$$

with

$$\mathbf{T}^i = \begin{bmatrix} \mathbf{T}_{aa} & \mathbf{0} \\ \mathbf{0} & \mathbf{T}_{bb} \end{bmatrix}^i \tag{5.82}$$

and

$$\mathbf{T}_{aa} = \mathbf{T}_{bb} = [\cos\alpha \quad -\sin\alpha]$$

Similarly, $\mathbf{T}^{iT}$ can be employed to transform variables from local to global coordinates. Thus, for example,

$$\begin{array}{ccccc} \begin{bmatrix} F_{Xa} \\ F_{Za} \\ \cdots\cdots \\ F_{Xb} \\ F_{Zb} \end{bmatrix}^i & = & \begin{bmatrix} \cos\alpha & 0 \\ -\sin\alpha & 0 \\ \cdots\cdots & \cdots\cdots \\ 0 & \cos\alpha \\ 0 & -\sin\alpha \end{bmatrix} & \begin{bmatrix} \tilde{N}_{xa} \\ \tilde{N}_{xb} \end{bmatrix}^i \\ \mathbf{p}^i & = & \mathbf{T}^{iT} & \tilde{\mathbf{p}}^i \end{array} \tag{5.83}$$

The stiffness matrix for a bar element with the uniaxial stiffness property $EA/\ell = k$ is given by (Chapter 4, Example 4.1)

$$\underbrace{\begin{bmatrix} \widetilde{N}_{xa} \\ \widetilde{N}_{xb} \end{bmatrix}^i}_{\widetilde{\mathbf{p}}^i} = \frac{EA}{\ell}\begin{bmatrix} 1 & -1 \\ -1 & 1 \end{bmatrix}\begin{bmatrix} \widetilde{u}_{xa} \\ \widetilde{u}_{xb} \end{bmatrix}^i = \underbrace{\begin{bmatrix} k & -k \\ -k & k \end{bmatrix}}_{\widetilde{\mathbf{k}}^i}\underbrace{\begin{bmatrix} \widetilde{u}_{xa} \\ \widetilde{u}_{xb} \end{bmatrix}^i}_{\widetilde{\mathbf{v}}^i} \tag{5.84}$$

which is written in terms of the local coordinates of a truss member. Equation (5.28), $\mathbf{k}^i = \mathbf{T}^{iT}\widetilde{\mathbf{k}}^i\mathbf{T}^i$, permits the stiffness matrix to be expressed in terms of the global coordinates. Thus,

$$\mathbf{k}^i = \mathbf{T}^{iT}\,\widetilde{\mathbf{k}}^i\,\mathbf{T}^i = \frac{EA}{\ell}\begin{bmatrix} \mathbf{T}_{aa}^T\mathbf{T}_{aa} & -\mathbf{T}_{aa}^T\mathbf{T}_{aa} \\ -\mathbf{T}_{aa}^T\mathbf{T}_{aa} & \mathbf{T}_{aa}^T\mathbf{T}_{aa} \end{bmatrix}^i \tag{5.85}$$

with

$$\mathbf{T}_{aa}^T\mathbf{T}_{aa} = \begin{bmatrix} \cos^2\alpha & -\cos\alpha\,\sin\alpha \\ -\cos\alpha\,\sin\alpha & \sin^2\alpha \end{bmatrix}^i \tag{5.86}$$

The element stiffness matrix represented in global coordinates is of size 4×4. It is convenient to employ notation for the element stiffness matrix which indicates that Eq. (5.85) can be partitioned into 2×2 matrices:

$$\mathbf{k}^i = \begin{bmatrix} \mathbf{k}_{jj}^i & \mathbf{k}_{jk}^i \\ \mathbf{k}_{kj}^i & \mathbf{k}_{kk}^i \end{bmatrix} \tag{5.87}$$

where $\mathbf{k}_{jj}^i$, $\mathbf{k}_{jk}^i$, $\mathbf{k}_{kj}^i$, and $\mathbf{k}_{kk}^i$ are 2×2 matrices and j, k are the initial and final nodes of element i.

With these relationships for the element, we can proceed to use the displacement method to solve plane truss problems.

EXAMPLE 5.3 Three-Bar Truss

The solution for the displacements and forces in the three-bar truss of Fig. 5.17 will illustrate the use of the displacement method.

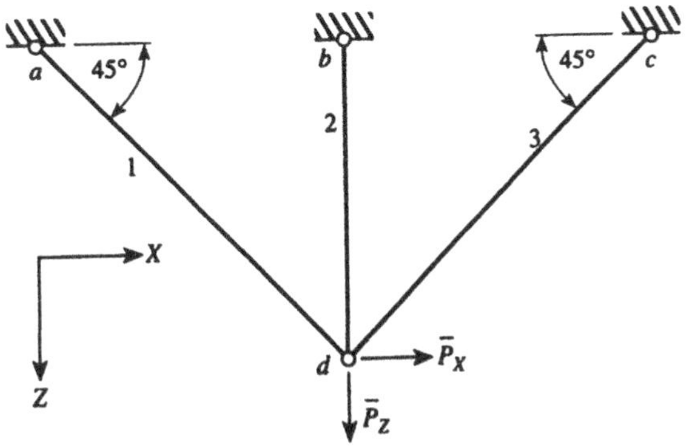

FIGURE 5.17
Three-bar truss.

The nodes and elements (bars) are numbered as indicated in Fig. 5.17. The global coordinate system is also shown in this figure. The ends of the bars must be identified in terms of the node numbers. This can be accomplished by constructing an incidence table.

Bar No.	Bar Begins at System Node No.	Bar Ends at System Node No.
1	d	a
2	b	d
3	d	c

It is important to observe that the incidence table does not contain unique entries. For example, bar 2 could be considered as beginning at node d and ending at node b, instead of beginning at b and ending at d as indicated in the above table. Of course, the orientation of the angle α will vary according to the choice of beginning and end nodes for a bar. Regardless of the selection scheme for nodal numbering of the element ends, the final solution will, of course, be the same.

The incidence table permits the element stiffness matrices to be identified in terms of the global node numbers. Use the element stiffness matrix of Eq. (5.85), which includes the transformation from local to global coordinates. Then

Bar 1:
$\alpha = 135°$(Fig. 5.18a)

$$\mathbf{k}^1 = \begin{bmatrix} \mathbf{k}^1_{dd} & \vdots & \mathbf{k}^1_{da} \\ \cdots & & \cdots \\ \mathbf{k}^1_{ad} & \vdots & \mathbf{k}^1_{aa} \end{bmatrix} \tag{1a}$$

Matrices $\mathbf{k}^1_{kj}$ are of size 2 × 2.

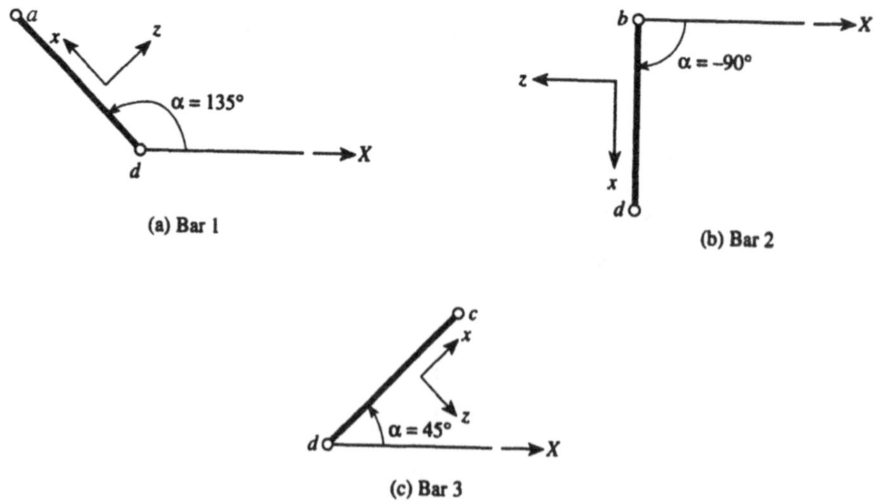

FIGURE 5.18
Element stiffness matrix calculations.

Bar 2:
$\alpha = -90°$(Fig. 5.18b)

$$\mathbf{k}^2 = \begin{bmatrix} \mathbf{k}_{bb}^2 & \vdots & \mathbf{k}_{bd}^2 \\ \cdots & & \cdots \\ \mathbf{k}_{db}^2 & \vdots & \mathbf{k}_{dd}^2 \end{bmatrix} \tag{1b}$$

Another equivalent representation of $\mathbf{k}^2$ is found by designating d as the beginning node and b as the end node. Then x would be up and z would be to the left, giving $\alpha = +90°$.

Bar 3:
$\alpha = 45°$(Fig. 5.18c)

$$\mathbf{k}^3 = \begin{bmatrix} \mathbf{k}_{dd}^3 & \vdots & \mathbf{k}_{dc}^3 \\ \cdots & & \cdots \\ \mathbf{k}_{cd}^3 & \vdots & \mathbf{k}_{cc}^3 \end{bmatrix} \tag{1c}$$

Since the subscripts in the element stiffness matrices have been identified with the global nodal numbering scheme, the global stiffness matrix is readily assembled by summation of submatrices (or stiffness coefficients, if appropriate) using

$$\mathbf{K}_{jk} = \sum_i \mathbf{k}_{jk}^i \tag{2}$$

where the summation is taken over all the elements. Thus, the global stiffness matrix would appear as

$$\underset{\mathbf{K}}{\begin{bmatrix} \mathbf{K}_{aa} & \mathbf{K}_{ab} & \mathbf{K}_{ac} & \mathbf{K}_{ad} \\ \mathbf{K}_{ba} & \mathbf{K}_{bb} & \mathbf{K}_{bc} & \mathbf{K}_{bd} \\ \mathbf{K}_{ca} & \mathbf{K}_{cb} & \mathbf{K}_{cc} & \mathbf{K}_{cd} \\ \mathbf{K}_{da} & \mathbf{K}_{db} & \mathbf{K}_{dc} & \mathbf{K}_{dd} \end{bmatrix}} \underset{\mathbf{V}}{\begin{bmatrix} \mathbf{V}_a \\ \mathbf{V}_b \\ \mathbf{V}_c \\ \mathbf{V}_d \end{bmatrix}} = \underset{\overline{\mathbf{P}}}{\begin{bmatrix} \overline{\mathbf{P}}_a \\ \overline{\mathbf{P}}_b \\ \overline{\mathbf{P}}_c \\ \overline{\mathbf{P}}_d \end{bmatrix}} \tag{3}$$

with

$$\mathbf{K}_{jk} = \sum_{i=1}^{3} \mathbf{k}_{jk}^i = \mathbf{k}_{jk}^1 + \mathbf{k}_{jk}^2 + \mathbf{k}_{jk}^3 \tag{4}$$

and

$$\mathbf{V}_j = \begin{bmatrix} U_{xj} \\ U_{zj} \end{bmatrix} \qquad \overline{\mathbf{P}}_j = \begin{bmatrix} \overline{P}_{xj} \\ \overline{P}_{zj} \end{bmatrix}$$

Now that the global stiffness equations $\mathbf{KV} = \overline{\mathbf{P}}$ have been established, the boundary conditions can be introduced. From the constraints shown in Fig. 5.17, it is apparent that the boundary conditions are

$$\begin{aligned} U_{Xa} &= U_{Za} = 0 \quad \text{or} \quad \mathbf{V}_a = \mathbf{0} \\ U_{Xb} &= U_{Zb} = 0 \quad \text{or} \quad \mathbf{V}_b = \mathbf{0} \\ U_{Xc} &= U_{Zc} = 0 \quad \text{or} \quad \mathbf{V}_c = \mathbf{0} \end{aligned} \tag{5}$$

The forces at nodes a, b, and c are the unknown reactions. Thus, the equilibrium equations of (3) become

$$\begin{bmatrix} \mathbf{K}_{aa} & \mathbf{K}_{ab} & \mathbf{K}_{ac} & \vdots & \mathbf{K}_{ad} \\ \mathbf{K}_{ba} & \mathbf{K}_{bb} & \mathbf{K}_{bc} & \vdots & \mathbf{K}_{bd} \\ \mathbf{K}_{ca} & \mathbf{K}_{cb} & \mathbf{K}_{cc} & \vdots & \mathbf{K}_{cd} \\ \cdots & \cdots & \cdots & \cdot & \cdots \\ \mathbf{K}_{da} & \mathbf{K}_{db} & \mathbf{K}_{dc} & \vdots & \mathbf{K}_{dd} \end{bmatrix} \begin{bmatrix} \mathbf{0} \\ \mathbf{0} \\ \mathbf{0} \\ \cdots \\ \mathbf{V}_d \end{bmatrix} = \begin{bmatrix} \mathbf{P}_a \\ \mathbf{P}_b \\ \mathbf{P}_c \\ \cdots \\ \overline{\mathbf{P}}_d \end{bmatrix} \begin{bmatrix} ? \\ ? \\ ? \\ \cdots \\ \overline{\mathbf{P}}_d \end{bmatrix} \tag{6}$$

If the rows corresponding to the unknown reactions are ignored for the moment, then (6) reduces to

$$\mathbf{K}_{dd}\mathbf{V}_d = \overline{\mathbf{P}}_d \tag{7}$$

where

$$\mathbf{K}_{dd} = \sum_{i=1}^{3} \mathbf{k}_{dd}^{i} = \mathbf{k}_{dd}^{1} + \mathbf{k}_{dd}^{2} + \mathbf{k}_{dd}^{3} \tag{8}$$

$$\overline{\mathbf{P}}_d = \begin{bmatrix} \overline{P}_{Xd} \\ \overline{P}_{Zd} \end{bmatrix} = \begin{bmatrix} \overline{P}_X \\ \overline{P}_Z \end{bmatrix} \tag{9}$$

To complete the calculations, it is apparent that it is necessary to find only the element stiffness matrices $\mathbf{k}_{dd}^{i}$. Suppose the nodes are located at the XZ coordinates: Node a : (0, 0), Node b : (1, 0), Node c : (2, 0), Node d : (1, 1). Then, if EA is assigned a unit magnitude, $\mathbf{k}_{dd}^{i}$ are given by [Eqs. (5.87), (5.86), and (5.85)]

Bar 1:

$$\ell = \sqrt{2}, \qquad EA/\ell = 1/\sqrt{2}$$
$$\alpha = 135^\circ, \qquad \sin\alpha = \sqrt{2}/2, \qquad \cos\alpha = -\sqrt{2}/2$$

$$k_{dd}^{1} = \begin{bmatrix} 1 & \vdots & 1 \\ \cdots & \cdot & \cdots \\ 1 & \vdots & 1 \end{bmatrix} \sqrt{2}/4 \tag{10a}$$

Bar 2:

$$\ell = 1, \qquad EA/\ell = 1$$
$$\alpha = -90^\circ, \qquad \sin\alpha = -1, \qquad \cos\alpha = 0$$

$$k_{dd}^{2} = \begin{bmatrix} 0 & \vdots & 0 \\ \cdots & \cdot & \cdots \\ 0 & \vdots & 1 \end{bmatrix} \tag{10b}$$

Bar 3:

$$\ell = \sqrt{2}, \qquad EA/\ell = 1/\sqrt{2}$$
$$\alpha = 45^\circ, \qquad \sin\alpha = \cos\alpha = \sqrt{2}/2$$

$$k_{dd}^{3} = \begin{bmatrix} 1 & \vdots & -1 \\ \cdots & \cdot & \cdots \\ -1 & \vdots & 1 \end{bmatrix} \sqrt{2}/4 \tag{10c}$$

Then, if the applied loads take the values $\overline{\mathbf{P}}_X = \overline{\mathbf{P}}_Z = 1$, the global equations (7) become

$$\begin{bmatrix} \frac{1}{2}\sqrt{2} & \vdots & 0 \\ \cdots & \cdot & \cdots \\ 0 & \vdots & 1+\frac{1}{2}\sqrt{2} \end{bmatrix} \begin{bmatrix} U_{Xd} \\ \cdots \\ U_{Zd} \end{bmatrix} = \begin{bmatrix} 1 \\ \cdots \\ 1 \end{bmatrix} \tag{11}$$

from which the displacements of node d are found to be

$$\mathbf{V}_d = \begin{bmatrix} U_{Xd} \\ \cdots \\ U_{Zd} \end{bmatrix} = \begin{bmatrix} 1.414 \\ \cdots \\ 0.586 \end{bmatrix} \tag{12}$$

Now that the nodal displacements are known, other displacements, forces, and stresses can be computed. For example, the reactions at nodes a, b, and c are found by placing the values of the displacements $\mathbf{V}_d$ from (12) in the first three equations of (6). The displacements at the ends of the bars are found by recognizing, i.e., using $\mathbf{v} = \mathbf{aV}$, that the global displacements of the nodes (upper case U) are the same as the corresponding element end displacements referred to the global coordinates (lower case u). Thus, for example,

$$U_{Xa} = u_{Xa}|_{\text{bar 1}} = 0 \quad \text{and} \quad U_{Xd} = u_{Xd}|_{\text{bar 1}} = u_{Xd}|_{\text{bar 2}} = u_{Xd}|_{\text{bar 3}} = 1.414$$

Equation (5.81) provides the displacements referred to the local coordinates x. Then the forces at the ends of each bar can be computed using Eq. (5.84). For a two-force member such as the bar of a truss, the end forces in local coordinates are the same as the internal forces. For the truss of Fig. 5.17, we obtain

$$N|_{\text{bar 1}} = 1.0, \qquad N|_{\text{bar 2}} = 0.586, \qquad N|_{\text{bar 3}} = -0.4142 \tag{13}$$

Alternatively, the bar forces or, equivalently, the stresses can be calculated using the material law. If the displacements of the joints of the truss have been computed, then the elongation of an element, for example element 2, would be $(\widetilde{u}_{xb} - \widetilde{u}_{xd})|_{bar\,2}$ and the stress σ in bar 2 becomes

$$\sigma = E\epsilon = \frac{E}{\ell}(\widetilde{u}_{xb} - \widetilde{u}_{xd})|_{\text{bar 2}} \tag{14}$$

The force in a bar is obtained by multiplying the stress by the cross-sectional area. ▮

EXAMPLE 5.4 Stiffness Matrix for a Five-Bar Truss

In Chapter 3, Example 3.5, the global equilibrium equations were established for the five-bar truss of Fig. 5.19 using the principle of virtual work. The same relationships are easily established by assembling the element stiffness matrices.

The most fundamental step in implementing a displacement solution is identifying where an element stiffness matrix fits in the global stiffness matrix. The goal is to assign appropriate global nodal numbers as subscripts for the element stiffness matrices. As in Example 5.3,

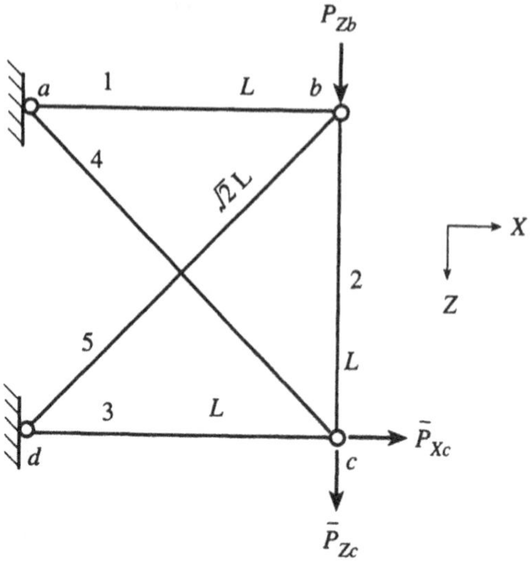

FIGURE 5.19
Five-bar truss (Chapter 3, Example 3.5 and Example 5.4).

one technique for accomplishing this is to use an incidence table. For the truss of Fig. 5.19,

Bar No.	Bar Begins at System Node No.	Bar Ends at System Node No.
1	*a*	*b*
2	*b*	*c*
3	*c*	*d*
4	*a*	*c*
5	*d*	*b*

The element stiffness matrices for each bar can now be assigned subscripts corresponding to the global node numbers in the incidence table. Equation (5.85) gives the element stiffness matrix referred to global coordinates. For each bar, in addition to the assignment of subscripts corresponding to the global node numbers, values for ℓ, A, E, and the angle α (see Fig. 5.10) are needed to utilize Eq. (5.85). We find

Bar 1 : $\alpha = 0°$ Bar 2 : $\alpha = 90°$ Bar 3 : $\alpha = 180°$

$$\mathbf{k}^1 = \begin{bmatrix} \mathbf{k}^1_{aa} & \mathbf{k}^1_{ab} \\ \mathbf{k}^1_{ba} & \mathbf{k}^1_{bb} \end{bmatrix} \quad (1a) \qquad \mathbf{k}^2 = \begin{bmatrix} \mathbf{k}^2_{bb} & \mathbf{k}^2_{bc} \\ \mathbf{k}^2_{cb} & \mathbf{k}^2_{cc} \end{bmatrix} \quad (1b) \qquad \mathbf{k}^3 = \begin{bmatrix} \mathbf{k}^3_{cc} & \mathbf{k}^3_{cd} \\ \mathbf{k}^3_{dc} & \mathbf{k}^3_{dd} \end{bmatrix} \quad (1c)$$

Bar 4 : $\alpha = 45°$ Bar 5 : $\alpha = -45°$ (1)

$$\mathbf{k}^4 = \begin{bmatrix} \mathbf{k}^4_{aa} & \mathbf{k}^4_{ac} \\ \mathbf{k}^4_{ca} & \mathbf{k}^4_{cc} \end{bmatrix} \quad (1d) \qquad \mathbf{k}^5 = \begin{bmatrix} \mathbf{k}^5_{dd} & \mathbf{k}^5_{db} \\ \mathbf{k}^5_{bd} & \mathbf{k}^5_{bb} \end{bmatrix} \quad (1e)$$

The global stiffness matrix is assembled by summation of the submatrices with

$$\mathbf{K}_{jk} = \sum_i \mathbf{k}^i_{jk} \tag{2}$$

where the summation is taken over all bars. For example,

$$\mathbf{K}_{aa} = \mathbf{k}^1_{aa} + \mathbf{k}^4_{aa}, \qquad \mathbf{K}_{bb} = \mathbf{k}^1_{bb} + \mathbf{k}^2_{bb} + \mathbf{k}^5_{bb} \tag{3}$$

The boundary conditions are $\mathbf{V}_a = \mathbf{V}_d = 0$. These constraints reduce the stiffness (equilibrium) relations to

$$\begin{bmatrix} \mathbf{K}_{bb} & \mathbf{K}_{bc} \\ \mathbf{K}_{cb} & \mathbf{K}_{cc} \end{bmatrix} \begin{bmatrix} \mathbf{V}_b \\ \mathbf{V}_c \end{bmatrix} = \begin{bmatrix} \overline{\mathbf{P}}_b \\ \overline{\mathbf{P}}_c \end{bmatrix} \tag{4}$$

where

$$\overline{\mathbf{P}}_b = \begin{bmatrix} \overline{P}_{Xb} \\ \overline{P}_{Zb} \end{bmatrix} \qquad \overline{\mathbf{P}}_c = \begin{bmatrix} \overline{P}_{Xc} \\ \overline{P}_{Zc} \end{bmatrix}$$

Substitution of the appropriate values of α in the element stiffness equations gives

$$\mathbf{K}_{bb} = \frac{EA}{L} \begin{bmatrix} 1+\sqrt{2}/4 & \sqrt{2}/4 \\ \sqrt{2}/4 & 1+\sqrt{2}/4 \end{bmatrix}$$

$$\mathbf{K}_{bc} = \frac{EA}{L} \begin{bmatrix} 0 & 0 \\ 0 & -1 \end{bmatrix} = \mathbf{K}_{cb}$$

$$\mathbf{K}_{cc} = \frac{EA}{L} \begin{bmatrix} 1+\sqrt{2}/4 & -\sqrt{2}/4 \\ -\sqrt{2}/4 & 1+\sqrt{2}/4 \end{bmatrix} \tag{5}$$

These are the same results obtained in Chapter 3, Example 3.5. ∎

5.3.11 Frames

A *frame* or rigid frame is composed of beam elements in which both transverse (bending) and axial (extension and torsion) effects occur. This differs from the truss of Section 5.3.10, where each bar element could only extend or compress.

Element Coordinate Transformations

Two-dimensional frames loaded in their planes are to be considered in this section. The frame is assumed to lie in the XZ plane. Local and global force and displacement components in the local xz and global XZ coordinates systems are shown in Fig. 5.20.

The transformation relations to change from local to global components can be established using the diagrams of Fig. 5.20. In global coordinates, the forces and displacements at the a end of the ith element of a plane frame are denoted as (Fig. 5.20c)

$$\mathbf{p}_a^i = \begin{bmatrix} F_X \\ F_Z \\ M \end{bmatrix}_a^i \qquad \mathbf{v}_a^i = \begin{bmatrix} u_X \\ u_Z \\ \theta \end{bmatrix}_a^i \tag{5.88}$$

where $M_a^i = M_{Ya}^i$, and $\theta_a^i = \theta_{Ya}^i$. Although according to Eq. (5.17) lower case letters should be employed to designate element forces referred to global coordinates, it is customary to use upper case letters for element forces of a framework. The corresponding forces referred to the local coordinates are designated in the usual fashion, e.g., $\widetilde{N}_{xa}$. The transformations from global to local coordinates for the forces at the end of a beam element are readily established. From Fig. 5.20d, ignoring the superscript i,

$$\widetilde{N}_{xa} = \widetilde{N}_a = F_{Xa}\cos\alpha - F_{Za}\sin\alpha \tag{5.89a}$$

and

$$\widetilde{V}_{za} = \widetilde{V}_a = F_{Xa}\sin\alpha + F_{Za}\cos\alpha \tag{5.89b}$$

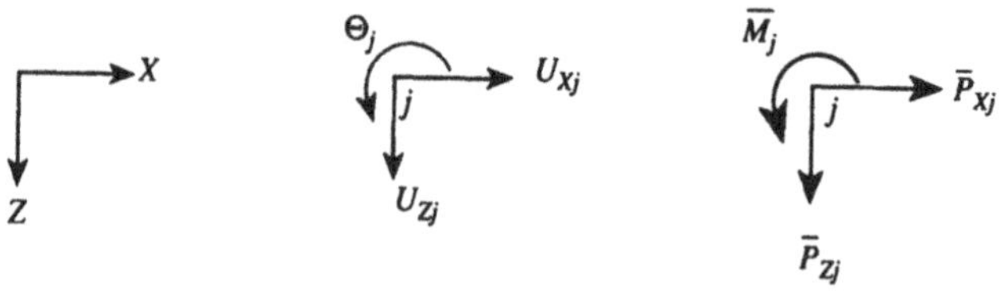

(a) Global coordinate system, system nodal DOF, and system applied nodal loading

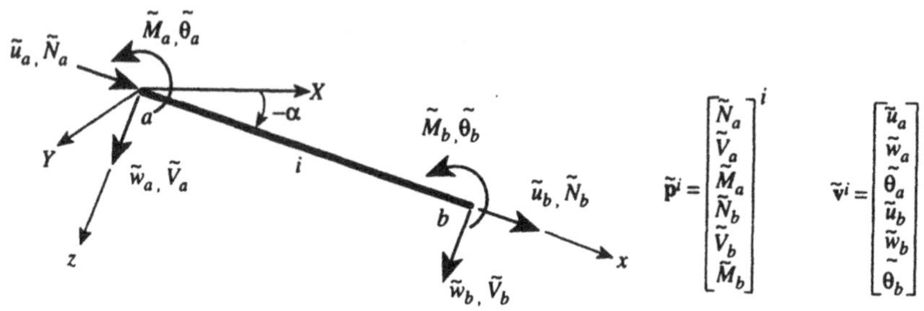

(b) Local coordinates, forces, and displacements on the ends of an element (Sign Convention 2). For right-handed global (*XYZ*) and local (*xyz*) coordinate systems, the vector corresponding to a positive α is along the *y* axis

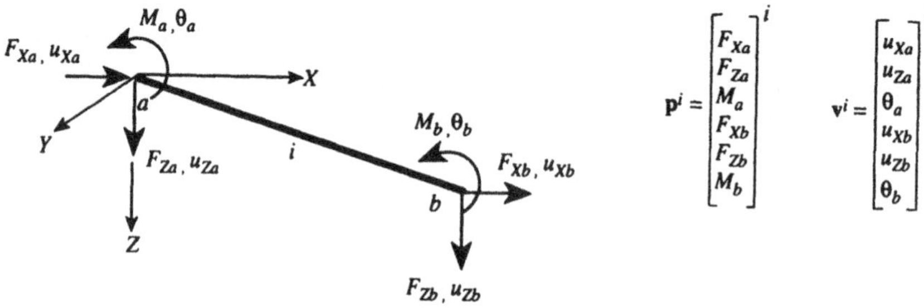

(c) Forces and displacements on the ends of an element (Sign Convention 2) aligned along global coordinates

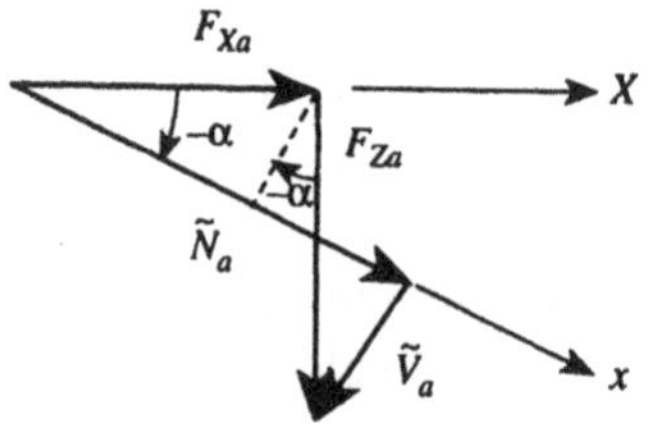

(d) Components of forces

FIGURE 5.20
Forces and displacements represented in the local and global coordinate systems.

Also,

$$\widetilde{M}_{Ya} = \widetilde{M}_a = M_a \tag{5.89c}$$

In matrix notation, for end a of the ith element,

$$\widetilde{\mathbf{p}}_a^i = \mathbf{T}_{aa}^i \mathbf{p}_a^i \tag{5.90}$$

where

$$\mathbf{T}_{aa}^i = \begin{bmatrix} \cos\alpha & -\sin\alpha & 0 \\ \sin\alpha & \cos\alpha & 0 \\ 0 & 0 & 1 \end{bmatrix}^i \tag{5.91}$$

The same transformation applies also to displacements. The transformation of Eq. (5.23) is appropriate when both ends (a and b) of element i are to be treated together. Then:

$$\begin{array}{ll} \text{Global to Local} & \text{Local to Global} \\ \widetilde{\mathbf{v}}^i = \mathbf{T}^i \mathbf{v}^i & \mathbf{v}^i = \mathbf{T}^{iT} \widetilde{\mathbf{v}}^i \\ \widetilde{\mathbf{p}}^i = \mathbf{T}^i \mathbf{p}^i & \mathbf{p}^i = \mathbf{T}^{iT} \widetilde{\mathbf{p}}^i \end{array} \tag{5.92}$$

where

$$\mathbf{T}^i = \begin{bmatrix} \mathbf{T}_{aa}^i & 0 \\ 0 & \mathbf{T}_{bb}^i \end{bmatrix}$$

with $\mathbf{T}_{aa}^i$ given by Eq. (5.91) and $\mathbf{T}_{bb}^i = \mathbf{T}_{aa}^i$.

The stiffness matrix transformation from local to global coordinates is given by [Eq. (5.28)]

$$\mathbf{k}^i = \mathbf{T}^{iT} \widetilde{\mathbf{k}}^i \mathbf{T}^i = \begin{bmatrix} \mathbf{k}_{jj}^i & \mathbf{k}_{jk}^i \\ \mathbf{k}_{kj}^i & \mathbf{k}_{kk}^i \end{bmatrix} \tag{5.93}$$

where $\mathbf{k}_{jj}^i$, $\mathbf{k}_{jk}^i$, $\mathbf{k}_{kj}^i$, and $\mathbf{k}_{kk}^i$ are 3×3 matrices, and $\widetilde{\mathbf{k}}^i$ is the element stiffness matrix referred to local coordinates. From Chapter 4, Eq. (4.80),

$$\widetilde{\mathbf{k}}^i = \begin{bmatrix} \frac{EA}{\ell} & 0 & 0 & -\frac{EA}{\ell} & 0 & 0 \\ 0 & \frac{12EI}{\ell^3} & -\frac{6EI}{\ell^2} & 0 & -\frac{12EI}{\ell^3} & -\frac{6EI}{\ell^2} \\ 0 & -\frac{6EI}{\ell^2} & \frac{4EI}{\ell} & 0 & \frac{6EI}{\ell^2} & \frac{2EI}{\ell} \\ -\frac{EA}{\ell} & 0 & 0 & \frac{EA}{\ell} & 0 & 0 \\ 0 & -\frac{12EI}{\ell^3} & \frac{6EI}{\ell^2} & 0 & \frac{12EI}{\ell^3} & \frac{6EI}{\ell^2} \\ 0 & -\frac{6EI}{\ell^2} & \frac{2EI}{\ell} & 0 & \frac{6EI}{\ell^2} & \frac{4EI}{\ell} \end{bmatrix} \tag{5.94}$$

Now that the coordinate transformations have been established for an element, the displacement method analysis of a frame proceeds in the same fashion as for a truss.

EXAMPLE 5.5 Displacement Method for a Frame

The three-element frame of Fig. 5.21a is modeled and labeled as shown in Fig. 5.21b. Suppose it is subjected to the concentrated loadings shown in Fig. 5.22. Find the response if both legs are treated as being fixed. The element properties are $E = 200\ \text{GN/m}^2$ and

$$\text{Elements 1 and 2: } I = 2356\ \text{cm}^4, \quad A = 32\ \text{cm}^2$$
$$\text{Elements 3: } I = 5245\ \text{cm}^4, \quad A = 66\ \text{cm}^2$$

The displacement method analysis of a frame is basically the same as that used for a truss; only the stiffness matrices differ. Fundamental to the method is the identification

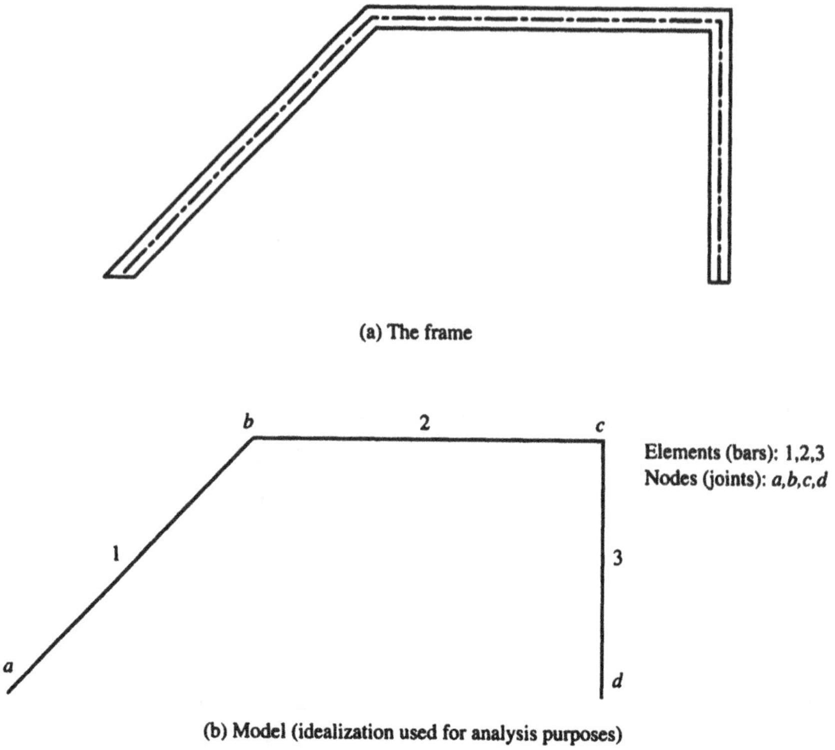

FIGURE 5.21
Plane frame used as an example of a network structural system.

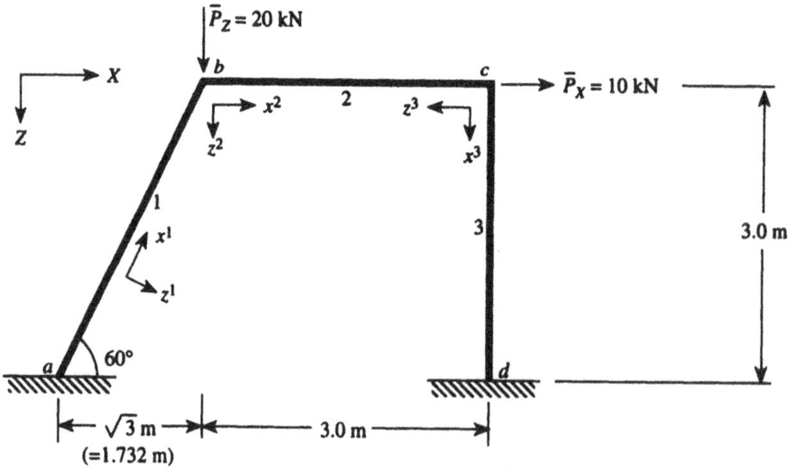

FIGURE 5.22
A plane frame.

of the element stiffness matrix with the global node numbers. This permits assembly of the global stiffness matrix by summation. An incidence table assists in correlating elements with the global nodes. For this three-element frame, the incidence table appears as the following:

Beam No.	Beam Begins at System Node No.	Beam Ends at System Node No.
1	*a*	*b*
2	*b*	*c*
3	*c*	*d*

The entries of the incidence table permit the element stiffness matrix for any beam element to be assigned subscripts corresponding to the global node numbers.

The next step is to compute the element stiffness matrix referred to global coordinates.

Element 1:

Use $\ell = 2\sqrt{3} = 3.464$ m, $EI = 4.712$ MN $\cdot$ m^2, and $EA = 640$ MN.
From Eq. (5.94),

$$\tilde{\mathbf{k}}^1 = \begin{bmatrix} 184.76 & 0 & 0 & -184.76 & 0 & 0 \\ 0 & 1.3604 & -2.3561 & 0 & -1.3604 & -2.3561 \\ 0 & -2.3561 & 5.4411 & 0 & 2.3561 & 2.7206 \\ -184.76 & 0 & 0 & 184.76 & 0 & 0 \\ 0 & -1.3604 & 2.3561 & 0 & 1.3604 & 2.3561 \\ 0 & -2.3561 & 2.7206 & 0 & 2.3561 & 5.4411 \end{bmatrix} \tag{1}$$

The next step is to transform the stiffness matrix from a local to the global coordinate system. From the incidence table, bar 1 begins at a and ends at b. For a global coordinate system XZ placed at a, α is a positive 60° (Figs. 5.10 and 5.20b). With $\cos\alpha = 0.5$ and $\sin\alpha = 0.866$,

$$\mathbf{T}^1 = \left[\begin{array}{ccc:ccc} 0.5 & -0.866 & 0 & & & \\ 0.866 & 0.5 & 0 & & 0 & \\ 0 & 0 & 1 & & & \\ \hdashline & & & 0.5 & -0.866 & 0 \\ & 0 & & 0.866 & 0.5 & 0 \\ & & & 0 & 0 & 1 \end{array}\right] \tag{2}$$

The stiffness matrix referred to global coordinates is then formed as [Eq. (5.93)]

$$\mathbf{k}^1 = \mathbf{T}^{1T}\tilde{\mathbf{k}}^1\mathbf{T}^1 = \left[\begin{array}{c:c} \mathbf{k}^1_{aa} & \mathbf{k}^1_{ab} \\ \hdashline \mathbf{k}^1_{ba} & \mathbf{k}^1_{bb} \end{array}\right]$$

$$= \left[\begin{array}{ccc:ccc} 47.210 & -79.412 & -2.0404 & -47.210 & 79.412 & -2.0404 \\ -79.412 & 138.90 & -1.1781 & 79.412 & -138.90 & -1.1781 \\ -2.0404 & -1.1781 & 5.4411 & 2.0404 & 1.1781 & 2.7206 \\ \hdashline -47.210 & 79.412 & 2.0404 & 47.210 & -79.412 & 2.0404 \\ 79.412 & -138.90 & 1.1781 & -79.412 & 138.90 & 1.1781 \\ -2.0404 & -1.1781 & 2.7206 & 2.0404 & 1.1781 & 5.4411 \end{array}\right] \tag{3}$$

where the subscripts of the submatrices have been taken from the incidence table.

Element 2:

Here the local and global coordinates coincide so that $\mathbf{k}^2 = \widetilde{\mathbf{k}}^2$. Substitute $\ell = 3.0$ m, $EI = 4.712$ MN · m², and $EA = 640$ MN in Eq. (5.94) giving

$$\mathbf{k}^2 = \begin{bmatrix} \mathbf{k}^2_{bb} & \mathbf{k}^2_{bc} \\ \mathbf{k}^2_{cb} & \mathbf{k}^2_{cc} \end{bmatrix}$$

$$= \left[\begin{array}{ccc:ccc} 213.33 & 0 & 0 & -213.33 & 0 & 0 \\ 0 & 2.0942 & -3.1413 & 0 & -2.0942 & -3.1413 \\ 0 & -3.1413 & 6.2827 & 0 & 3.1413 & 3.1413 \\ \hdashline -213.33 & 0 & 0 & 213.33 & 0 & 0 \\ 0 & -2.0942 & 3.1413 & 0 & 2.0942 & 3.1413 \\ 0 & -3.1413 & 3.1413 & 0 & 3.1413 & 6.2827 \end{array}\right] \tag{4}$$

Element 3:

Use $\ell = 3.0$ m, $EI = 10.49$ MN · m, and $EA = 1320$ MN in Eq. (5.94). Also, for a global coordinate system XZ placed at point c, $\alpha = -90°$. This leads to

$$\mathbf{k}^3 = \mathbf{T}^{3T}\widetilde{\mathbf{k}}^3\mathbf{T}^3 = \begin{bmatrix} \mathbf{k}^3_{cc} & \mathbf{k}^3_{cd} \\ \mathbf{k}^3_{dc} & \mathbf{k}^3_{dd} \end{bmatrix}$$

$$= \left[\begin{array}{ccc:ccc} 4.6622 & 0 & 6.9933 & -4.6622 & 0 & 6.9933 \\ 0 & 440 & 0 & 0 & -440 & 0 \\ 6.9933 & 0 & 13.987 & -6.9933 & 0 & 6.9933 \\ \hdashline -4.6622 & 0 & -6.9933 & 4.6622 & 0 & -6.9933 \\ 0 & -440 & 0 & 0 & 440 & 0 \\ 6.9933 & 0 & 6.9933 & -6.9933 & 0 & 13.987 \end{array}\right] \tag{5}$$

The global stiffness matrix is assembled by superimposing the element stiffness matrices using

$$\mathbf{K}_{jk} = \sum_{i=1}^{M} \mathbf{k}^i_{jk} \tag{6}$$

where the summation is taken over all beam elements (M). By observation of $\mathbf{k}^i_{jk}$ of (3), (4), and (5), we see that

$$\mathbf{K}_{jk} = \mathbf{k}^i_{jk} \qquad i = 1, 2, \text{ or } 3 \tag{7}$$

with the exception of $\mathbf{K}_{bb}$ and $\mathbf{K}_{cc}$ which are given by

$$\begin{aligned} \mathbf{K}_{bb} &= \mathbf{k}^1_{bb} + \mathbf{k}^2_{bb} \\ \mathbf{K}_{cc} &= \mathbf{k}^2_{cc} + \mathbf{k}^3_{cc} \end{aligned} \tag{8}$$

Thus, assembly leads to a global stiffness matrix that appears as

$$\mathbf{K} = \begin{bmatrix} \boxed{\mathbf{k}^1} & & \\ & \boxed{\mathbf{k}^2} & \\ & & \boxed{\mathbf{k}^3} \end{bmatrix} \tag{9}$$

The full displacement vector and corresponding load vector can be split into parts with given displacement boundary conditions at nodes a and d to which the reactions $\mathbf{P}_a$ and $\mathbf{P}_d$ are conjugates, and into another part with the unknown nodal displacements to which the applied loads $\overline{\mathbf{P}}$ are conjugates.

$$\begin{bmatrix} 0 \\ 0 \\ 0 \\ \cdots \\ U_{Xb} \\ U_{Zb} \\ \Theta_b \\ \cdots \\ U_{Xc} \\ U_{Zc} \\ \Theta_c \\ \cdots \\ 0 \\ 0 \\ 0 \end{bmatrix} \begin{matrix} \left.\vphantom{\begin{matrix}0\\0\\0\end{matrix}}\right\} \overline{\mathbf{V}}_a \\ \\ \left.\vphantom{\begin{matrix}0\\0\\0\\0\\0\\0\\0\end{matrix}}\right\} \mathbf{V} \\ \\ \left.\vphantom{\begin{matrix}0\\0\\0\end{matrix}}\right\} \overline{\mathbf{V}}_d \end{matrix} \iff \begin{bmatrix} P_{Xa} \\ P_{Za} \\ M_a \\ \cdots \\ 0 \\ \overline{P}_Z \\ 0 \\ \cdots \\ \overline{P}_X \\ 0 \\ 0 \\ \cdots \\ P_{Xd} \\ P_{Zd} \\ M_d \end{bmatrix} \begin{matrix} \left.\vphantom{\begin{matrix}0\\0\\0\end{matrix}}\right\} \mathbf{P}_a \\ \\ \left.\vphantom{\begin{matrix}0\\0\\0\\0\\0\\0\\0\end{matrix}}\right\} \overline{\mathbf{P}} \\ \\ \left.\vphantom{\begin{matrix}0\\0\\0\end{matrix}}\right\} \mathbf{P}_d \end{matrix} \tag{10}$$

If the columns corresponding to the zero displacements at the supports are deleted and the rows corresponding to the reactions are temporarily ignored, the equilibrium equations become

$$\begin{bmatrix} 260.54 & -79.412 & 2.0404 & -213.33 & 0 & 0 \\ -79.412 & 140.99 & -1.9632 & 0 & -2.0942 & -3.1413 \\ 2.0404 & -1.9632 & 11.724 & 0 & 3.1413 & 3.1413 \\ -213.33 & 0 & 0 & 217.99 & 0 & 6.9933 \\ 0 & -2.0942 & 3.1413 & 0 & 442.09 & 3.1413 \\ 0 & -3.1413 & 3.1413 & 6.9933 & 3.1413 & 20.270 \end{bmatrix} \begin{bmatrix} U_{Xb} \\ U_{Zb} \\ \Theta_b \\ U_{Xc} \\ U_{Zc} \\ \Theta_c \end{bmatrix} = \begin{bmatrix} 0 \\ 0.02 \\ 0 \\ 0.01 \\ 0 \\ 0 \end{bmatrix} \tag{11}$$

The displacements found by solving these relations are

$$\begin{bmatrix} U_{Xb} \\ U_{Zb} \\ \Theta_b \\ U_{Xc} \\ U_{Zc} \\ \Theta_c \end{bmatrix} = \begin{bmatrix} 3.660 \text{ mm} \\ 2.1828 \text{ mm} \\ -29.127\ \mu\text{rad} \\ 3.6575 \text{ mm} \\ 0.0171 \text{ mm} \\ -921.73\ \mu\text{rad} \end{bmatrix} \tag{12}$$

Now that the global displacements are known, we can calculate a considerable amount of useful information. The reactions are found by returning to the rows of the full system matrix (9) which corresponds to $\mathbf{P}_a$ and $\mathbf{P}_d$ of (10). This gives the reactions (in global coordinates)

$$\begin{aligned} P_{Xa} &= 0.597 \text{ kN} & P_{Xd} &= -10.606 \text{ kN} \\ P_{Za} &= -12.485 \text{ kN} & P_{Zd} &= -7.524 \text{ kN} \\ M_a &= 9.961 \text{ kN}\cdot\text{m} & M_d &= 19.132 \text{ kN}\cdot\text{m} \end{aligned} \tag{13}$$

To find the forces on the end of an element, first calculate the forces in global coordinates and then transform them into local coordinates. Since the nodal displacements are equal to the end displacements referred to global coordinates, we can obtain the forces at the ends of element 1 using

$$\mathbf{p}^1 = \mathbf{k}^1\mathbf{v}^1 = \mathbf{k}^1 \begin{bmatrix} u_{Xa} \\ u_{Za} \\ \theta_a \\ u_{Xb} \\ u_{Zb} \\ \theta_b \end{bmatrix} = \mathbf{k}^1 \begin{bmatrix} U_{Xa} \\ U_{Za} \\ \Theta_a \\ U_{Xb} \\ U_{Zb} \\ \Theta_b \end{bmatrix}$$

$$= [\text{Eq. } 3] \begin{bmatrix} 0 \\ 0 \\ 0 \\ 3.6603 \times 10^{-3} \\ 2.1828 \times 10^{-3} \\ -2.9127 \times 10^{-3} \end{bmatrix} = \begin{bmatrix} 0.597\text{ kN} \\ -12.485\text{ kN} \\ 9.961\text{ kN}\cdot\text{m} \\ -0.597\text{ kN} \\ 12.485\text{ kN} \\ 9.882\text{ kN}\cdot\text{m} \end{bmatrix} \tag{14}$$

The forces of element 1 in local coordinates are

$$\tilde{\mathbf{p}}^1 = \mathbf{T}^1\mathbf{p}^1 = \begin{bmatrix} \tilde{N}_a \\ \tilde{V}_a \\ \tilde{M}_a \\ \tilde{N}_b \\ \tilde{V}_b \\ \tilde{M}_b \end{bmatrix} = \begin{bmatrix} 11.111\text{ kN} \\ -5.725\text{ kN} \\ 9.961\text{ kN}\cdot\text{m} \\ -11.111\text{ kN} \\ 5.725\text{ kN} \\ 9.882\text{ kN}\cdot\text{m} \end{bmatrix} \tag{15}$$

Forces and displacements in other members of the frame are found in a similar fashion. The responses are sketched in Fig. 5.23. ∎

5.3.12 Structures with Distributed Loads

Thus far, in formulating the displacement method, structures have been modeled such that all loads are applied at the nodes. Special consideration must be given if loads are applied between the nodes. An alternative to adding new nodes, a practice which increases the size of the system of equations to be solved, is to include these effects in the loading vector of the stiffness equations. This is a particularly appropriate technique for distributed loads applied between the nodes. For a single beam element, this approach was discussed in Chapter 4, Section 4.4.2. The incorporation of this loading vector into the assembly procedure for global stiffness matrices will be considered here.

In Chapter 4, element stiffness matrices were defined using $\mathbf{p}^i = \mathbf{k}^i\mathbf{v}^i$, where $\mathbf{p}^i$ is a vector containing element end (nodal) forces and moments. Also, a vector $\overline{\mathbf{p}}^{i0} = \int_a^b \mathbf{N}^T \overline{p}_z\, dx$ containing the effect on the nodes of applied distributed loading $\overline{p}_z$ was derived. For those elements the forces at the element i boundaries are

$$\underbrace{\mathbf{p}^i}_{\substack{\text{Vector}\\ \text{containing}\\ \text{element end}\\ \text{forces and moments}}} + \underbrace{\overline{\mathbf{p}}^{i0}}_{\substack{\text{Vector, with}\\ \text{entries at the}\\ \text{element ends (nodes),}\\ \text{due to applied}\\ \text{distributed}\\ \text{loading}}} = \mathbf{k}^i\mathbf{v}^i \quad \text{or} \quad \mathbf{p}^i = \mathbf{k}^i\mathbf{v}^i - \overline{\mathbf{p}}^{i0} \tag{5.95}$$

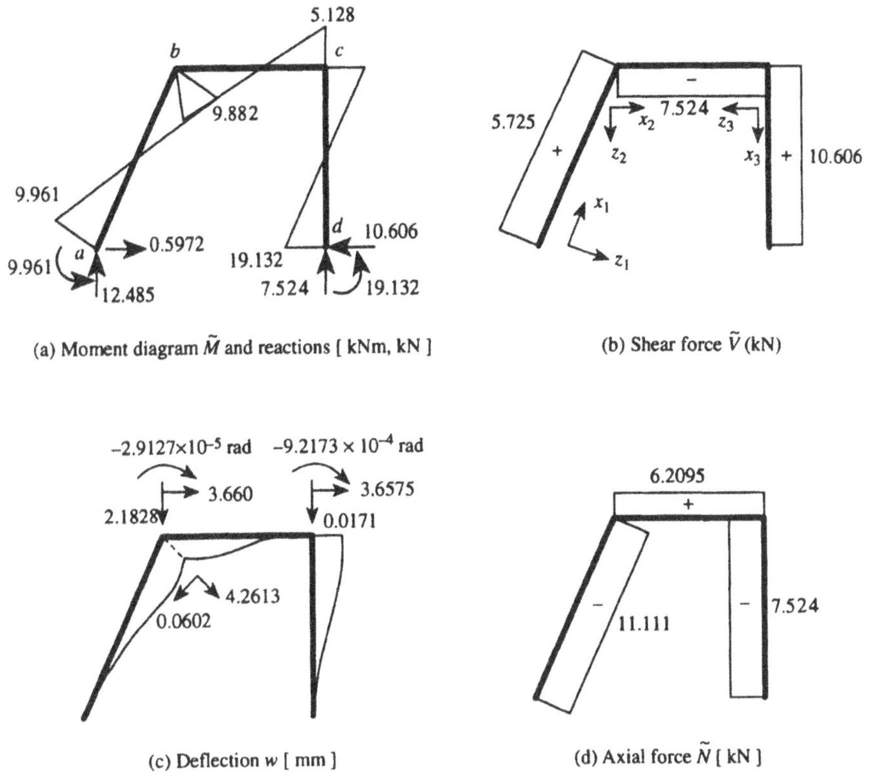

FIGURE 5.23
Results for Example 5.5.

for each element i. Direct nodal loads are not introduced at the element level. Rather, they are inserted in the assembled global stiffness equations.

Tables 4.2 and 4.4 provide general expressions for the components of the loading vector $\overline{\mathbf{p}}^{i0}$. For example, for a beam element with a uniformly distributed load of magnitude p_0, this vector appears as (Table 4.2)

$$\overline{\mathbf{p}}^{i0} = \frac{p_0\ell}{2}\begin{bmatrix} 1 \\ -\ell/6 \\ 1 \\ \ell/6 \end{bmatrix} \tag{5.96}$$

If the displacements and forces are referred to a local coordinate system, then Eq. (5.95) would appear as

$$\widetilde{\mathbf{p}}^i = \widetilde{\mathbf{k}}^i\widetilde{\mathbf{v}}^i - \widetilde{\overline{\mathbf{p}}}^{i0} \tag{5.97}$$

As expected, the coordinate transformation matrix $\mathbf{T}^i$ can be used to move from local to global coordinates and vice versa. Thus,

$$\mathbf{p}^i = \mathbf{T}^{iT}\widetilde{\mathbf{p}}^i, \qquad \mathbf{k}^i = \mathbf{T}^{iT}\widetilde{\mathbf{k}}^i\mathbf{T}^i, \qquad \overline{\mathbf{p}}^{i0} = \mathbf{T}^{iT}\,\widetilde{\overline{\mathbf{p}}}^{i0} \tag{5.98}$$

The assembly of the global stiffness equations $\mathbf{KV} = \overline{\mathbf{P}}$ must now be adjusted to include the loading. By definition vector $\overline{\mathbf{P}}^*$ contains prescribed nodal loading. Inclusion of an element loading vector $\overline{\mathbf{p}}^{i0}$ to account for effects distributed between the nodes will involve the special assembly of a new global loading vector. The global stiffness matrix is still assembled in terms of stiffness coefficients, or submatrices as appropriate, using

$$\mathbf{K}_{jk} = \sum_{i=1}^{M} \mathbf{k}_{jk}^{i} \tag{5.99a}$$

where M is the number of elements in the structural model, while the global vector at node j is formed as

$$\overline{\mathbf{P}}_j^0 = \sum_{i=1}^{M} \overline{\mathbf{p}}_j^{i0} \tag{5.99b}$$

For the whole system, the $\overline{\mathbf{P}}_j^0$ due to applied distributed loads forms a global nodal vector $\overline{\mathbf{P}}^0$. This can be incorporated directly into $\overline{\mathbf{P}}$ of the global stiffness equations $\mathbf{KV} = \mathbf{P}$. It is advisable to distinguish between the direct nodal loads $\overline{\mathbf{P}}^*$ and the influence of the elements on the nodes due to distributed loading, represented by $\overline{\mathbf{P}}^0$. Then we may write the system equilibrium equations as

$$\mathbf{KV} = \underbrace{\overline{\mathbf{P}}^*}_{\substack{\text{Nodal vector}\\ \text{containing}\\ \text{direct nodal}\\ \text{loads}}} + \underbrace{\overline{\mathbf{P}}^0}_{\substack{\text{Nodal vector}\\ \text{due to applied}\\ \text{distributed}\\ \text{loading}}} = \overline{\mathbf{P}} \tag{5.100}$$

Upon solution of the system equations for $\mathbf{V}$, the usual procedure is followed to compute the local displacements and forces. Use the incidence conditions ($\mathbf{v} = \mathbf{aV}$) to relate the member end displacements to the global node displacements. After transformation of $\mathbf{v}$ to local coordinates using $\widetilde{\mathbf{v}}^i = \mathbf{T}^i\mathbf{v}^i$, Eq. (5.97) can be employed to calculate member forces.

EXAMPLE 5.6 Beam with Linearly Varying Loading

Return to the fixed-simply supported beam of Fig. 5.5. This beam has been treated extensively in Chapters 3 and 4 as well as in the present chapter.

The solution procedure for this structure with loading distributed between the nodes follows the outline of Section 5.3.9. For this straight beam, the local and global coordinate systems coincide; consequently, no transformation of variables from local to global coordinates is required. Hence, the stiffness matrix of Chapter 4, Eq. (4.13) is assembled directly into the global stiffness matrix $\mathbf{K}$.

The element loading vectors $\overline{\mathbf{p}}^{i0}$ can be taken from Chapter 4, Table 4.2. For the first element, the distributed load begins with a magnitude of p_0 and ends with $p_0/2$. The second element begins with a magnitude of $p_0/2$ and ends with a zero. With this information, the loading vectors are provided directly by Chapter 4, Table 4.2. If desired, the loading vectors can be calculated using Chapter 4, Eq. (4.58). We choose to treat it here for the quite general case of a distributed load $\overline{p}_z(\xi)$ varying linearly as in Fig. 5.24. Then $\overline{p}_z$ can be described as

$$\overline{p}_z(\xi) = p_a + (p_b - p_a)\xi \tag{1}$$

Rewrite this in the form

$$\overline{p}_z(\xi) = \mathbf{N}_p\mathbf{G}_p\overline{\mathbf{p}}_p \tag{2}$$

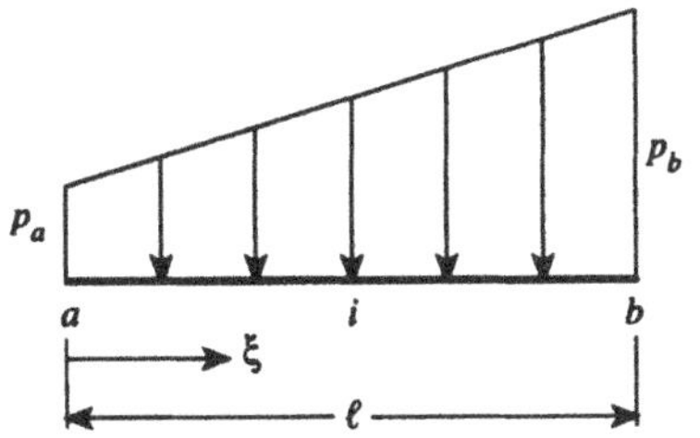

FIGURE 5.24
Linearly varying load.

where

$$\mathbf{N}_p = [1 \quad \xi] \qquad \mathbf{G}_p = \begin{bmatrix} 1 & 0 \\ -1 & 1 \end{bmatrix} \qquad \overline{\mathbf{p}}_p = \begin{bmatrix} p_a \\ p_b \end{bmatrix} \tag{3}$$

We wish to calculate [Eq. (4.58)]

$$\overline{\mathbf{p}}^{i0} = \mathbf{G}^T \int_a^b \mathbf{N}_u^T \overline{p}_z \, dx = \mathbf{G}^T \int_0^1 \mathbf{N}_u^T \mathbf{N}_p \ell \mathbf{G}_p \overline{\mathbf{p}}_p \, d\xi \tag{4}$$

where **G** and $\mathbf{N}_u$ are taken from Chapter 4, Eq. (4.47a). We find

$$\int_0^1 \mathbf{N}_u^T \mathbf{N}_p \ell \, d\xi = \int_0^1 \begin{bmatrix} 1 \\ \xi \\ \xi^2 \\ \xi^3 \end{bmatrix} [1 \quad \xi] \, d\xi \; \ell = \begin{bmatrix} 1 & \vdots & 1/2 \\ 1/2 & \vdots & 1/3 \\ 1/3 & \vdots & 1/4 \\ 1/4 & \vdots & 1/5 \end{bmatrix} \ell \tag{5}$$

Carry out the multiplications indicated in (4) and find

$$\overline{\mathbf{p}}^{i0} = \ell \begin{bmatrix} 7/20 & 3/20 \\ -1/20 & -1/30 \\ 3/20 & 7/20 \\ 1/30 & 1/20 \end{bmatrix} \begin{bmatrix} p_a \\ p_b \end{bmatrix} \tag{6}$$

For the two-element beam of Fig. 5.5:

Element 1, with $p_a = p_0, \; p_b = p_0/2$ Element 2, with $p_a = p_0/2, \; p_b = 0$

$$(\overline{\mathbf{p}}^1)^0 = \frac{p_0 \ell}{120} \begin{bmatrix} 51 \\ -8 \\ 39 \\ 7 \end{bmatrix} = \begin{bmatrix} \overline{\mathbf{P}}_a^1 \\ \overline{\mathbf{P}}_b^1 \end{bmatrix} \qquad (\overline{\mathbf{p}}^2)^0 = \frac{p_0 \ell}{120} \begin{bmatrix} 21 \\ -3 \\ 9 \\ 2 \end{bmatrix} = \begin{bmatrix} \overline{\mathbf{P}}_b^2 \\ \overline{\mathbf{P}}_c^2 \end{bmatrix} \tag{7}$$

The global matrix will appear as

$$\mathbf{KV} - \overline{\mathbf{P}}^0 = \overline{\mathbf{P}}^* \tag{8}$$

where

$$\mathbf{K} = \begin{bmatrix} \mathbf{k}^1_{aa} & \vdots & \mathbf{k}^1_{ab} & \vdots & 0 \\ \cdots & & \cdots & & \cdots \\ \mathbf{k}^1_{ba} & \vdots & \mathbf{k}^1_{bb} + \mathbf{k}^2_{bb} & \vdots & \mathbf{k}^2_{bc} \\ \cdots & & \cdots & & \cdots \\ 0 & \vdots & \mathbf{k}^2_{cb} & \vdots & \mathbf{k}^2_{cc} \end{bmatrix} \qquad \mathbf{V} = \begin{bmatrix} w_a \\ \theta_a \ell \\ w_b \\ \theta_b \ell \\ w_c \\ \theta_c \ell \end{bmatrix}$$

$$= \frac{EI}{\ell^3} \begin{bmatrix} 12 & -6 & -12 & -6 & 0 & 0 \\ -6 & 4 & 6 & 2 & 0 & 0 \\ -12 & 6 & 24 & 0 & -12 & -6 \\ -6 & 2 & 0 & 8 & 6 & 2 \\ 0 & 0 & -12 & 6 & 12 & 6 \\ 0 & 0 & -6 & 2 & 6 & 4 \end{bmatrix} \tag{9}$$

The global loading vector $\overline{\mathbf{P}}^0$ is assembled in a fashion similar to the global stiffness matrix. Use Eq. (5.99b)

$$\overline{\mathbf{P}}^0_j = \sum_{i=1}^{2} \overline{\mathbf{P}}^i_j$$

Thus,

$$\overline{\mathbf{P}}^0 = \begin{bmatrix} \overline{\mathbf{P}}^0_a \\ \overline{\mathbf{P}}^0_b \\ \overline{\mathbf{P}}^0_c \end{bmatrix} = \begin{bmatrix} \overline{\mathbf{P}}^1_a \\ \overline{\mathbf{P}}^1_b + \overline{\mathbf{P}}^2_b \\ \overline{\mathbf{P}}^2_c \end{bmatrix}^0 = \frac{p_0 \ell}{120} \begin{bmatrix} 51 \\ -8 \\ 39 + 21 \\ 7 - 3 \\ 9 \\ 2 \end{bmatrix} = \frac{p_0 \ell}{120} \begin{bmatrix} 51 \\ -8 \\ 60 \\ 4 \\ 9 \\ 2 \end{bmatrix} \tag{10}$$

The boundary conditions $w_a = \theta_a = w_c = 0$ imposed on $\mathbf{KV} = \overline{\mathbf{P}}^0$ give the nonsingular system

$$\frac{EI}{\ell^3} \begin{bmatrix} 24 & 0 & -6 \\ 0 & 8 & 2 \\ -6 & 2 & 4 \end{bmatrix} \begin{bmatrix} w_b \\ \theta_b \ell \\ \theta_c \ell \end{bmatrix} = \frac{p_0 \ell}{120} \begin{bmatrix} 60 \\ 4 \\ 2 \end{bmatrix} \tag{11}$$

Here the rows corresponding to the unknown reactions (resulting from the prescribed displacements w_a, θ_a, and w_c) are ignored. Thus, (11) is equivalent to $\mathbf{K}_{22}\mathbf{V}_y = \overline{\mathbf{P}}$ of Eq. (5.73). The solution to these relations is

$$\begin{bmatrix} w_b \\ \theta_b \\ \theta_c \end{bmatrix} = \frac{p_0 \ell^3}{120 EI} \begin{bmatrix} 4.5\ell \\ -1.5 \\ 8.0 \end{bmatrix} \tag{12}$$

The internal forces are computed using

$$\mathbf{k}^i \mathbf{v}^i - \overline{\mathbf{p}}^{i0} = \mathbf{p}^i \tag{13}$$

The displacements at the ends of the elements are equal to the global node displacements. Thus, (13) gives

$$\mathbf{p}^1 = \begin{bmatrix} V_a \\ M_a \\ V_b \\ M_b \end{bmatrix} = \mathbf{k}^1\mathbf{v}^1 - (\overline{\mathbf{p}}^1)^0$$

$$= \frac{p_0\ell^3}{120EI}\frac{EI}{\ell^3}\begin{bmatrix} -12 & -6\ell \\ 6\ell & 2l^2 \\ 12 & 6\ell \\ 6\ell & 4\ell^2 \end{bmatrix}\begin{bmatrix} 4.5\ell \\ -1.5 \end{bmatrix} - \frac{p_0\ell}{120}\begin{bmatrix} 51 \\ -8\ell \\ 39 \\ 7\ell \end{bmatrix} = \frac{p_0\ell}{120}\begin{bmatrix} -96 \\ 32\ell \\ 6 \\ 14\ell \end{bmatrix} \tag{14}$$

$$\mathbf{p}^2 = \begin{bmatrix} V_b \\ M_b \\ V_c \\ M_c \end{bmatrix} = \mathbf{k}^2\mathbf{v}^2 - (\overline{\mathbf{p}}^2)^0$$

$$= \frac{p_0\ell^3}{120EI}\frac{EI}{\ell^3}\begin{bmatrix} 12 & -6\ell & -6\ell \\ -6\ell & 4\ell^2 & 2\ell^2 \\ -12 & 6\ell & 6\ell \\ -6\ell & 2\ell^2 & 4\ell^2 \end{bmatrix}\begin{bmatrix} 4.5\ell \\ -1.5 \\ 8.0 \end{bmatrix} - \frac{p_0\ell}{120}\begin{bmatrix} 21 \\ -3\ell \\ 9 \\ 2\ell \end{bmatrix} = \frac{p_0\ell}{120}\begin{bmatrix} -6 \\ -14\ell \\ -24 \\ 0 \end{bmatrix} \tag{15}$$

While using these remember that $\ell = L/2$.

Although the deflection along the beam can be obtained using stiffness relations, it is perhaps simpler to use transfer matrices. Thus,

$$\mathbf{z}_j = \mathbf{U}^1\mathbf{z}_a \quad \text{for element 1} \tag{16}$$

$$\mathbf{z}_j = \mathbf{U}^2\mathbf{z}_b \quad \text{for element 2} \tag{17}$$

The x coordinate (ℓ) in $\mathbf{U}^1$ and $\mathbf{U}^2$ can be adjusted, e.g., set equal to x, such that $\mathbf{z}_j$ represents the state variables at any x coordinate between the nodes. The state vectors at a and b have already been calculated above, so that these readily lead to the state variables at any coordinate, e.g., with $\xi = x/L$,

$$w(\xi) = \begin{cases} \frac{p_0\ell^4}{120EI}(16\xi_1^2 - 16\xi_1^3 + 5\xi_1^4 - 0.5\xi_1^5) & \xi_1 = 2\xi \quad 0 \le x \le L/2 \\ \frac{p_0\ell^4}{120EI}(4.5 + 1.5\xi_2 - 7\xi_2^2 - \xi_2^3 + 2.5\xi_2^4 - 0.5\xi_2^5) & \xi_2 = 2\xi - 1 \quad L/2 \le x \le L \end{cases}$$

$$= \frac{p_0L^4}{120EI}\xi^2(4 - 8\xi + 5\xi^2 - \xi^3)$$

$$= -\frac{p_0L^4}{120EI}(\xi - 2)^2(\xi - 1)\xi^2 \quad \text{for any } \xi \tag{18}$$

These are exact results. These deflections, along with the bending moment and shear force, are plotted in Fig. 5.25.

It is rather straightforward to increase the number of elements, a step one may wish to take if a less complete trial solution is employed. Here, of course, we are dealing with exact trial solutions and the resulting exact stiffness matrices and loading vector, so that the exact final solution is obtained regardless of the number of elements utilized. Still, it is instructive to consider briefly the solution when more elements are introduced. Suppose this beam is divided into four elements of equal length. The loading function $\overline{\mathbf{p}}^i$ of (7) for each element is fully evaluated if p_a and p_b for the element are identified.

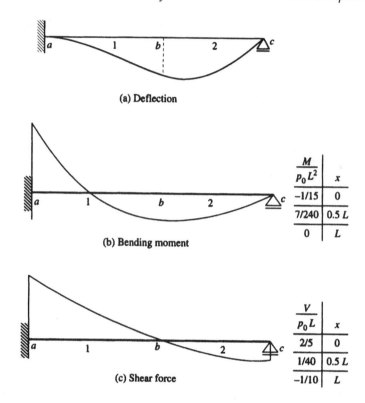

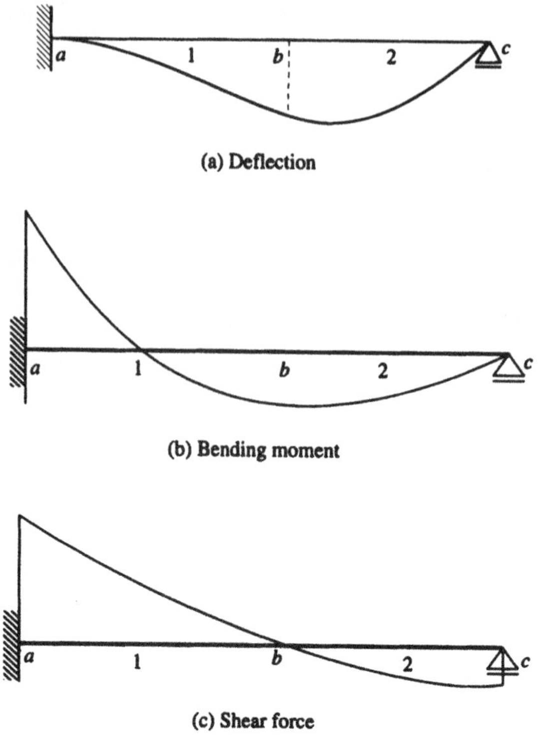

$\frac{M}{p_0 L^2}$	x
−1/15	0
7/240	0.5 L
0	L

$\frac{V}{p_0 L}$	x
2/5	0
1/40	0.5 L
−1/10	L

FIGURE 5.25
Response of the beam of Fig. 5.5 with linearly varying loading.

Element 1:

$$\overline{p}_a^1 = p_0, \qquad \overline{p}_b^1 = \frac{3}{4} p_0$$

$$\overline{\mathbf{P}}^{10} = \ell \begin{bmatrix} 7/20 & 3/20 \\ -1/20 & -1/30 \\ 3/20 & 7/20 \\ 1/30 & 1/20 \end{bmatrix} \begin{bmatrix} p_a^{-1} \\ p_b^{-1} \end{bmatrix} = p_0 \ell \begin{bmatrix} 37/80 \\ -6/80 \\ 33/80 \\ 17/240 \end{bmatrix} \tag{19}$$

Element 2:

$$\overline{p}_a^2 = \frac{3}{4} p_0, \qquad \overline{p}_b^2 = \frac{1}{2} p_0$$

$$\overline{\mathbf{P}}^{20} = p_0 \ell \begin{bmatrix} \frac{27}{80} & -\frac{13}{240} & \frac{23}{80} & \frac{1}{20} \end{bmatrix}^T \tag{20}$$

Element 3:

$$\overline{p}_a^3 = \frac{1}{2} p_0, \qquad \overline{p}_b^3 = \frac{1}{4} p_0$$

$$\overline{\mathbf{P}}^{30} = p_0 \ell \begin{bmatrix} \frac{17}{80} & -\frac{1}{30} & \frac{13}{80} & \frac{7}{240} \end{bmatrix}^T \tag{21}$$

Element 4:

$$\overline{p}_a^4 = \frac{1}{4}p_0, \qquad \overline{p}_b^4 = 0$$

$$\overline{\mathbf{P}}^{40} = p_0\ell\begin{bmatrix}\frac{7}{80} & -\frac{1}{80} & \frac{3}{80} & \frac{1}{120}\end{bmatrix}^T \tag{22}$$

The complete set of equations $\mathbf{KV} - \overline{\mathbf{P}}^0 = 0$ would appear as

$$\underbrace{\frac{EI}{\ell^3}\begin{bmatrix} 12 & -6 & -12 & -6 & & & & & & \\ -6 & 4 & 6 & 2 & & & & & & \\ -12 & 6 & 12+12 & 6-6 & -12 & -6 & & & & \\ -6 & 2 & 6-6 & 4+4 & 6 & 2 & & & & \\ & & -12 & 6 & 12+12 & 6-6 & -12 & -6 & & \\ & & -6 & 2 & 6-6 & 4+4 & 6 & 2 & & \\ & & & & -12 & 6 & 12+12 & 6-6 & -12 & -6 \\ & & & & -6 & 2 & 6-6 & 4+0 & 6 & 2 \\ & & & & & & -12 & 6 & 12 & 6 \\ & & & & & & -6 & 2 & 6 & 4 \end{bmatrix}}_{\mathbf{K}} \underbrace{\begin{bmatrix} w_a \\ \theta_a\ell \\ w_b \\ \theta_b\ell \\ w_c \\ \theta_c\ell \\ w_d \\ \theta_d\ell \\ w_e \\ \theta_e\ell \end{bmatrix}}_{\mathbf{V}}$$

$$\underbrace{-\begin{bmatrix} 37/80 \\ -6/80 \\ 33/80+27/80 \\ 17/240-13/240 \\ 23/80+17/80 \\ 1/20-1/30 \\ 13/80+7/80 \\ 7/240-1/80 \\ 3/80 \\ 1/120 \end{bmatrix}\frac{p_0\ell}{4}}_{-\overline{\mathbf{P}}^0} = 0 \tag{23}$$

For this beam, the displacement boundary conditions are on the left end, $w_a = \theta_a = 0$, and on the right end, $w_e = 0$.

The reduced simultaneous equations of (23) are readily found to have the solution with $\ell = L/4$.

$$\mathbf{V} = [w_b \ \ \theta_b\ell \ \ w_c \ \ \theta_c\ell \ \ w_d \ \ \theta_d\ell \ \ \theta_e\ell]^T$$

$$= \frac{p_0\ell^4}{EI}[0.3063 \ \ -0.4229 \ \ 0.6000 \ \ -0.1000 \ \ 0.4688 \ \ 0.3437 \ \ 0.5333]^T \tag{24}$$

■

5.3.13 Hinges and Other Indeterminate Nodal Conditions

The conditions, e.g., hinges or supports, illustrated in Fig. 5.26, which constrain one response variable (usually the value is zero) while generating a discontinuity (a reaction) in the complementary variable, occur frequently in structural models.

In-Span Support Condition	Fixed State Variable (often equal to zero)	Discontinuous State Variable (of unknown magnitude at the support condition)
1. Rigid Support 	w	V
2. Moment Release (Hinge) 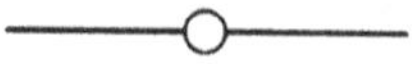	M	θ
3. Shear Release 	V	w
4. Angle Guide 	θ	M

FIGURE 5.26
In-span indeterminate conditions.

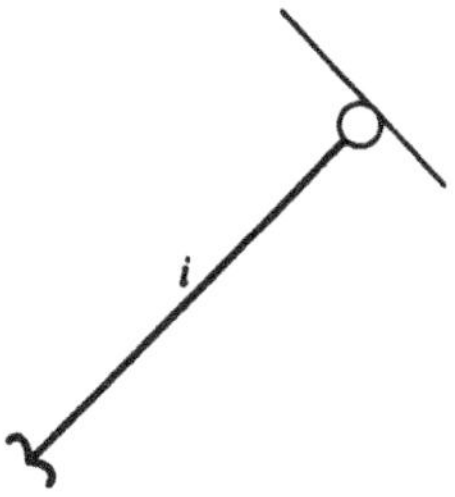

FIGURE 5.27
A "half" hinge.

This type of occurrence is usually more readily taken into account in using the displacement method than it is with the transfer matrix method. Often the effect of such conditions is to constrain a global DOF, e.g., when a rigid support at a node completely restrains the displacement in the direction of a component of **V**, and then this displacement is simply a prescribed global displacement. In such cases, set one displacement in **V** equal to zero.

On the other hand, it is possible that such indeterminate conditions affect only one of several elements meeting at a certain node. The hinge shown in Fig. 5.27 is such a case. To set a global DOF equal to zero would be improper modeling. However, the effect can be accounted for on the element level. To see this, suppose the forces $\tilde{\mathbf{p}}_2$ of element i are constrained to be zero at a node. Rearrange the element stiffness matrix as

$$\underbrace{\begin{bmatrix} \tilde{\mathbf{p}}_1 \\ \cdots\cdots \\ \tilde{\mathbf{p}}_2 \end{bmatrix}}_{\tilde{\mathbf{p}}^i} = \begin{bmatrix} \tilde{\mathbf{p}}_1 \\ \cdots\cdots \\ 0 \end{bmatrix} = \underbrace{\begin{bmatrix} \tilde{\mathbf{k}}_{11} & \vdots & \tilde{\mathbf{k}}_{12} \\ \cdots & \cdots & \cdots \\ \tilde{\mathbf{k}}_{21} & \vdots & \tilde{\mathbf{k}}_{22} \end{bmatrix}}_{\tilde{\mathbf{k}}^i} \underbrace{\begin{bmatrix} \tilde{\mathbf{v}}_1 \\ \cdots\cdots \\ \tilde{\mathbf{v}}_2 \end{bmatrix}}_{\tilde{\mathbf{v}}^i} \tag{5.101}$$

Such a reordering of the stiffness matrix is for conceptual purposes only. If $\tilde{\mathbf{p}}_2$ is a single force, then $\tilde{\mathbf{v}}_2$ is a single displacement, and $\tilde{\mathbf{k}}_{22}$, for example, is a scalar. Thus, in the case of a hinge at b, $M_b = 0 = \tilde{\mathbf{p}}_2$ and $\tilde{\mathbf{v}}_2 = \theta_b$ would be the unknown variable. From Eq. (5.101)

$$\tilde{\mathbf{v}}_2 = -\tilde{\mathbf{k}}_{22}^{-1}\tilde{\mathbf{k}}_{21}\,\tilde{\mathbf{v}}_1 \tag{5.102}$$

so that forces $\tilde{\mathbf{p}}_1$ are related to the displacements $\tilde{\mathbf{v}}_1$, by

$$\tilde{\mathbf{p}}_1 = \left(\tilde{\mathbf{k}}_{11} - \tilde{\mathbf{k}}_{12}\tilde{\mathbf{k}}_{22}^{-1}\tilde{\mathbf{k}}_{21}\right)\tilde{\mathbf{v}}_1 \tag{5.103}$$

This serves as a set of springs for which forces have known dependencies (spring rates) on displacements. The displacement dependent forces of Eq. (5.103) are readily incorporated into the analysis. That is, the effect of $\tilde{\mathbf{p}}_1$ can be assembled into the global stiffness matrix $\mathbf{K}$ using

$$\begin{bmatrix} \tilde{\mathbf{p}}_1 \\ \cdots \\ 0 \end{bmatrix} = \begin{bmatrix} \left(\tilde{\mathbf{k}}_{11} - \tilde{\mathbf{k}}_{12}\tilde{\mathbf{k}}_{22}^{-1}\tilde{\mathbf{k}}_{21}\right) & \vdots & 0 \\ \cdots\cdots & & \cdots \\ 0 & \vdots & 0 \end{bmatrix} \begin{bmatrix} \tilde{\mathbf{v}}_1 \\ \cdots \\ 0 \end{bmatrix} \tag{5.104}$$

for the ith element. After the global displacements have been computed, the variables, e.g., reactions, $\tilde{\mathbf{v}}_2$ can be determined using Eq. (5.102).

There are a number of other effects that can be accounted for in a similar fashion. For example, it may be desirable to model a joint as having a certain degree of flexibility. As formulated above, an appropriate dependency between forces and displacements at the joint can be incorporated into the displacement method analysis.

5.3.14 Rigid Elements

Rigid elements can be the source of numerical instabilities in an analysis. For example, for a beam element that is rigid against axial deformation, EA approaches infinity and the determinant $(\mathbf{K}) \to 0$. Thus, a rigid element, which implies that some of its displacements will be the same, i.e., some displacements are dependent, can lead to a singular stiffness matrix. This singularity is different from the singularity property exhibited by all element stiffness matrices (Chapter 4, Section 4.3). As is to be illustrated in Example 5.7, this problem can be corrected by taking into account the dependency of certain variables. For example, if a bar element, which extends from $x = a$ to $x = b$, is rigid in the axial direction, set $u_b = u_a$. Then certain rows and columns of the stiffness matrix should be reorganized.

EXAMPLE 5.7 Frame with Rigid Members and Distributed Loading

Return to the frame of Figs. 5.21 and 5.22, and suppose that a distributed load is placed between nodes a and b as shown in Fig. 5.28. Furthermore, assume that members 2 and 3 cannot extend and compress, i.e., they are rigid with respect to axial deformation. Other than these two characteristics, the material and geometric properties are the same as employed in Example 5.5. The response due to two loading cases is sought.

Loading Case 1: $p_0 = 2.0$ kN/m (Fig. 5.28)
Loading Case 2: $\overline{P}_Z = 20$ kN, $\overline{P}_X = 10$ kN, the same as in Example 5.5

The element stiffness matrix of Eq. (5.94) applies and can be inserted in $\tilde{\mathbf{p}}^i = \tilde{\mathbf{k}}^i\tilde{\mathbf{v}}_i - \tilde{\tilde{\mathbf{p}}}^{i0}$, where $\tilde{\tilde{\mathbf{p}}}^{i0}$ accounts for the distributed loading applied to element i.

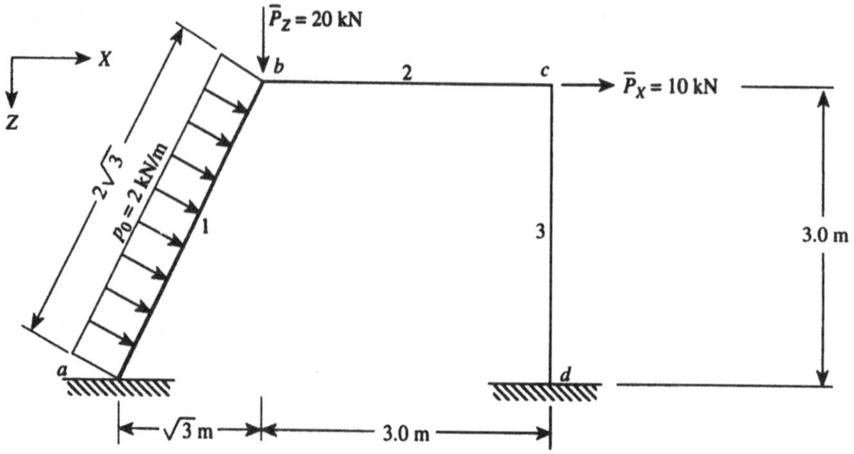

FIGURE 5.28
Frame of Example 5.7.

Element 1:

Use $\tilde{\mathbf{k}}^1$ and $\mathbf{k}^1$ as given by Eqs. (1) and (3) of Example 5.5. For the first loading case, take $\tilde{\tilde{\mathbf{p}}}^{i0}$ from Chapter 4, Table 4.2.

$$\tilde{\tilde{\mathbf{p}}}^{i0} = -\frac{p_0 \ell}{2}\begin{bmatrix} 0 \\ -1 \\ \ell/6 \\ 0 \\ -1 \\ -\ell/6 \end{bmatrix} = -\begin{bmatrix} 0 \\ -2\sqrt{3} \\ 2.0 \\ 0 \\ -2\sqrt{3} \\ -2.0 \end{bmatrix} \tag{1}$$

Using Eq.(2) of Example 5.5, the loading is transformed to global coordinates as

$$\mathbf{T}^{1T}\,\tilde{\tilde{\mathbf{p}}}^{i0} = -\begin{bmatrix} -3.0 \\ -\sqrt{3} \\ 2.0 \\ -3.0 \\ -\sqrt{3} \\ -2.0 \end{bmatrix} = \overline{\mathbf{p}}^{10} \tag{2}$$

Element 2:

Since there is no distributed loading on element 2, $\tilde{\tilde{\mathbf{p}}}^{20} = 0$. The local and global coordinate systems coincide so that $\mathbf{k}^2 = \tilde{\mathbf{k}}^2$. Use $\ell = 3.0$ m, $EI = 4712$ kN $\cdot$ m^2, and $EA = \infty$, giving

$$\mathbf{k}^2 = \begin{bmatrix} \infty & 0 & 0 & -\infty & 0 & 0 \\ 0 & 2094.2 & -3141.3 & 0 & -2094.2 & -3141.3 \\ 0 & -3143.3 & 6282.7 & 0 & 3141.3 & 3141.3 \\ -\infty & 0 & 0 & \infty & 0 & 0 \\ 0 & -2094.2 & 3143.3 & 0 & 2094.2 & 3141.3 \\ 0 & -3141.3 & 3141.3 & 0 & 3141.3 & 6282.7 \end{bmatrix} \tag{3}$$

Element 3:

The distributed loading vector $\tilde{\bar{\mathbf{p}}}^{30} = \mathbf{0}$, as there is no distributed load on this element. Use $\ell = 3.0$ m, $EI = 10\,490$ kN $\cdot$ m^2, and $EA = \infty$, giving

$$\tilde{\mathbf{k}}^3 = \begin{bmatrix} \infty & 0 & 0 & -\infty & 0 & 0 \\ 0 & 4662.2 & -6993.3 & 0 & -4662.2 & -6993.3 \\ 0 & -6993.3 & 13987 & 0 & 6993.3 & 6993.3 \\ -\infty & 0 & 0 & \infty & 0 & 0 \\ 0 & -4662.2 & 6993.3 & 0 & 4662.2 & 6993.3 \\ 0 & -6993.3 & 6993.3 & 0 & 4662.2 & 6993.3 \end{bmatrix} \tag{4}$$

Use the same transformation employed in Example 5.5 to convert this to global coordinates,

$$\mathbf{k}^3 = \begin{bmatrix} 4662.2 & 0 & 6993.3 & -4662.2 & 0 & 6993.3 \\ 0 & \infty & 0 & 0 & -\infty & 0 \\ 6993.3 & 0 & 13987 & -6993.3 & 0 & 6993.3 \\ -4662.2 & 0 & -6993.3 & 4662.2 & 0 & -6993.3 \\ 0 & -\infty & 0 & 0 & \infty & 0 \\ 6993.3 & 0 & 6993.3 & -6993.3 & 0 & 13987 \end{bmatrix} \tag{5}$$

Formation of the Global Equations

The global stiffness matrix $\mathbf{K}$ is assembled as it was in Example 5.5. The global equilibrium equations are $\mathbf{KV} = \overline{\mathbf{P}}^* + \overline{\mathbf{P}}^0 = \overline{\mathbf{P}}$. For the first loading case, with $p_0 = 2.0$ on beam 1 only and no concentrated applied forces, $\overline{\mathbf{P}}^* = \mathbf{0}$ and $\overline{\mathbf{P}}^0$ is taken simply from $\overline{\mathbf{p}}^{10}$. Thus, $\overline{\mathbf{P}}_j^0 = \overline{\mathbf{p}}_j^{10}$ with $j = a, b$. For $j = c$ or d, $\overline{\mathbf{P}}_j^0 = \mathbf{0}$. For the second loading case, with no loading distributed between nodes, $\overline{\mathbf{P}}^0 = \mathbf{0}$, and $\overline{\mathbf{P}}^*$ is formed from the applied forces concentrated at the nodes. Since the beams are fixed at the bases, all displacements are equal to zero at nodes a and d. Delete the columns corresponding to these displacements, and ignore the rows corresponding to the reactions at a and d. Then

$$\underbrace{\begin{bmatrix} 47\,210+\infty & -79\,412 & 2040.4 & -\infty & 0 & 0 \\ -79\,412 & 140\,994.2 & -1963.2 & 0 & -2094.2 & -3141.3 \\ 2040.4 & -1963.2 & 11\,723.8 & 0 & 3141.3 & 3141.3 \\ -\infty & 0 & 0 & 4662.2+\infty & 0 & -6993.3 \\ 0 & -2094.2 & 3141.3 & 0 & 2094.2+\infty & 3141.3 \\ 0 & -3141.3 & 3141.3 & 6993.3 & 3141.3 & 20\,269.7 \end{bmatrix}}_{\mathbf{K}} \underbrace{\begin{bmatrix} U_{Xb} \\ U_{Zb} \\ \Theta_b \\ U_{Xc} \\ U_{Zc} \\ \Theta_c \end{bmatrix}}_{\mathbf{V}}$$

$$= \begin{array}{c} \text{Load Case 1} \quad \text{Load Case 2} \\ \begin{bmatrix} 3.0 & \vdots & 0 \\ \sqrt{3} & \vdots & 20 \\ 2.0 & \vdots & 0 \\ 0 & \vdots & 10 \\ 0 & \vdots & 0 \\ 0 & \vdots & 0 \end{bmatrix} \\ \overline{\mathbf{P}}^0 \ \vdots \ \overline{\mathbf{P}}^* \end{array} \tag{6}$$

This stiffness matrix representing a frame with some of its members rigid will be singular, i.e., determinant ($\mathbf{K}$) $= 0$. However, this problem is readily removed by taking advantage of geometrical conditions resulting from the rigid members. Since bar 2 is rigid in the axial direction, U_{Xb} must be equal to U_{Xc}. Also, because bar 3 is rigid, $U_{Zc} = U_{Zd} = 0$. Use these conditions in (6) to help sort out the dependent rows and columns. Set $U_{Zc} = 0$ in (6), delete the corresponding column, and ignore the corresponding row. Set U_{Xc} equal to U_{Xb} in each equation of (6). Add the first and fourth equations, and we find

$$\begin{bmatrix} 51872.2 & -79412 & 2040.4 & 6993.3 \\ -79412 & 140994.2 & -1963.2 & -3141.3 \\ 2040.4 & -1963.2 & 11723.8 & 3141.3 \\ 6993.3 & -3141.2 & 3141.2 & 20269.7 \end{bmatrix} \begin{bmatrix} U_{Xb} \\ U_{Zb} \\ \Theta_b \\ \Theta_c \end{bmatrix} = \begin{matrix} \text{Load} & \text{Load} \\ \text{Case 1} & \text{Case 2} \end{matrix} \begin{bmatrix} 3.0 & \vdots & 10 \\ \sqrt{3} & \vdots & 20 \\ 2.0 & \vdots & 0 \\ 0 & \vdots & 0 \end{bmatrix} \tag{7}$$

The solutions to this set of equations are

$$\begin{bmatrix} U_{Xb} \\ U_{Zb} \\ \Theta_b \\ U_{Xc} \\ U_{Zc} \\ \Theta_c \end{bmatrix} = \begin{bmatrix} \text{Load Case 1} & & \text{Load Case 2} \\ 0.6802 \text{ mm} & \vdots & 3.6497 \text{ mm} \\ 0.3933 \text{ mm} & & 2.1767 \text{ mm} \\ 1.7167 \times 10^{-4} \text{ rad} & \vdots & -2.5057 \times 10^{-5} \text{ rad} \\ 0.6802 \text{ mm} & \vdots & 3.6497 \text{ mm} \\ 0.0 \text{ mm} & & 0.0 \text{ mm} \\ -2.0006 \times 10^{-4} \text{ rad} & \vdots & -9.1674 \times 10^{-4} \text{ rad} \end{bmatrix} \tag{8}$$

Calculation of Elementary Forces

Consider the forces on bar 1. The force vector $\mathbf{p}^1$ involves $\mathbf{k}^1\mathbf{v}^1$, which we will compute using the displacements of (8) for each of the loading cases. Because end a is fixed, $U_{Xa} = U_{Za} = \Theta_a = 0$. Also, due to compatibility, the element displacements referred to the global coordinates at end b are equal to the global displacements at b of (8). Thus,

$$\mathbf{k}^1\mathbf{v}^1 = \begin{bmatrix} \vdots & \vdots & \vdots & -47\,210 & 79\,412 & -2040.4 \\ \vdots & \vdots & \vdots & 79\,412 & -138\,900 & -1178.1 \\ \vdots & \vdots & \vdots & 2040.4 & 1178.1 & 2720.6 \\ \vdots & \vdots & \vdots & 47\,210 & -79\,412 & 2040.4 \\ \vdots & \vdots & \vdots & -79\,412 & 138\,900 & 1178.1 \\ \vdots & \vdots & \vdots & 2040.4 & 1178.1 & 5441.1 \end{bmatrix} \begin{bmatrix} 0 \\ 0 \\ 0 \\ U_{Xb} \\ U_{Zb} \\ \Theta_b \end{bmatrix}$$

$$= \begin{matrix} \text{Load Case 1} & \text{Load Case 2} \end{matrix} \begin{bmatrix} -1.2277 \text{ kN} & \vdots & 0.6043 \text{ kN} \\ -0.8192 \text{ kN} & \vdots & -12.4831 \text{ kN} \\ 2.3183 \text{ kN}\cdot\text{m} & \vdots & 9.9429 \text{ kN}\cdot\text{m} \\ 1.2277 \text{ kN} & \vdots & -0.6043 \text{ kN} \\ 0.8192 \text{ kN} & \vdots & 12.4831 \text{ kN} \\ 2.7853 \text{ kN}\cdot\text{m} & \vdots & 9.8748 \text{ kN}\cdot\text{m} \end{bmatrix} \tag{9}$$

In order to compute $\mathbf{p}^1$ for the case with the distributed load located on bar 1, i.e., for load case 1, it is necessary to use $\mathbf{p}^1 = \mathbf{k}^1\mathbf{v}^1 - \overline{\mathbf{p}}^{10}$ with $\overline{\mathbf{p}}^{10}$ given by (2). For load case 2, $\mathbf{p}^1$ is simply equal to $\mathbf{k}^1\mathbf{v}^1$. Thus, $\mathbf{p}^1$ becomes

$$\mathbf{p}^1 = \begin{array}{c} \text{Load Case 1} \qquad \text{Load Case 2} \\ \begin{bmatrix} -1.2277 - 2.9998 & \vdots & 0.6043 \\ -0.8192 - 1.7320 & \vdots & -12.4831 \\ 2.3183 - 1.9999 & \vdots & 9.9429 \\ 1.2277 - 2.9998 & \vdots & -0.6043 \\ 0.8192 - 1.7320 & \vdots & 12.4831 \\ 2.7853 - 1.9999 & \vdots & 9.8748 \end{bmatrix} \end{array}$$

$$= \begin{array}{c} \text{Load Case 1} \qquad \text{Load Case 2} \\ \begin{bmatrix} -4.2275\text{ kN} & \vdots & 0.6043\text{ kN} \\ -2.5512\text{ kN} & \vdots & -12.4831\text{ kN} \\ 4.3182\text{ kN}\cdot\text{m} & \vdots & 9.9429\text{ kN}\cdot\text{m} \\ -1.7721\text{ kN} & \vdots & -0.6043\text{ kN} \\ -0.9128\text{ kN} & \vdots & 12.4831\text{ kN} \\ 0.7854\text{ kN}\cdot\text{m} & \vdots & 9.8748\text{ kN}\cdot\text{m} \end{bmatrix} \end{array} \tag{10}$$

In terms of the local coordinates, the forces are found from $\widetilde{\mathbf{p}}^1 = \mathbf{T}^1\mathbf{p}^1$ to be

$$\begin{bmatrix} \widetilde{N}_a^1 \\ \widetilde{V}_a^1 \\ \widetilde{M}_a^1 \\ \widetilde{N}_b^1 \\ \widetilde{V}_b^1 \\ \widetilde{M}_b^1 \end{bmatrix} = \begin{array}{c} \text{Load Case 1} \qquad \text{Load Case 2} \\ \begin{bmatrix} 0.09559\text{ kN} & \vdots & 11.1125\text{ kN} \\ -4.9366\text{ kN} & \vdots & -5.7182\text{ kN} \\ 4.3182\text{ kN}\cdot\text{m} & \vdots & 9.9429\text{ kN}\cdot\text{m} \\ -0.09557\text{ kN} & \vdots & -11.1125\text{ kN} \\ -1.9910\text{ kN} & \vdots & 5.7182\text{ kN} \\ 0.7854\text{ kN}\cdot\text{m} & \vdots & 9.8748\text{ kN}\cdot\text{m} \end{bmatrix} \end{array} \tag{11}$$

The local forces on bar 1 for load case 1 are pictured in Fig. 5.29, along with other responses on all of the bars. ■

5.3.15 Symmetry

Many structures such as bridges, buildings, and ships exhibit some form of symmetry. For symmetrical structures, an analysis of the whole structure can often be avoided by considering only a portion of the structure. This will simplify the problem and decrease the cost of the computation.

There are three steps involved in taking advantage of symmetry in the analysis of a structure: (1) Recognition of the type of symmetry, (2) Use of superposition to reorganize

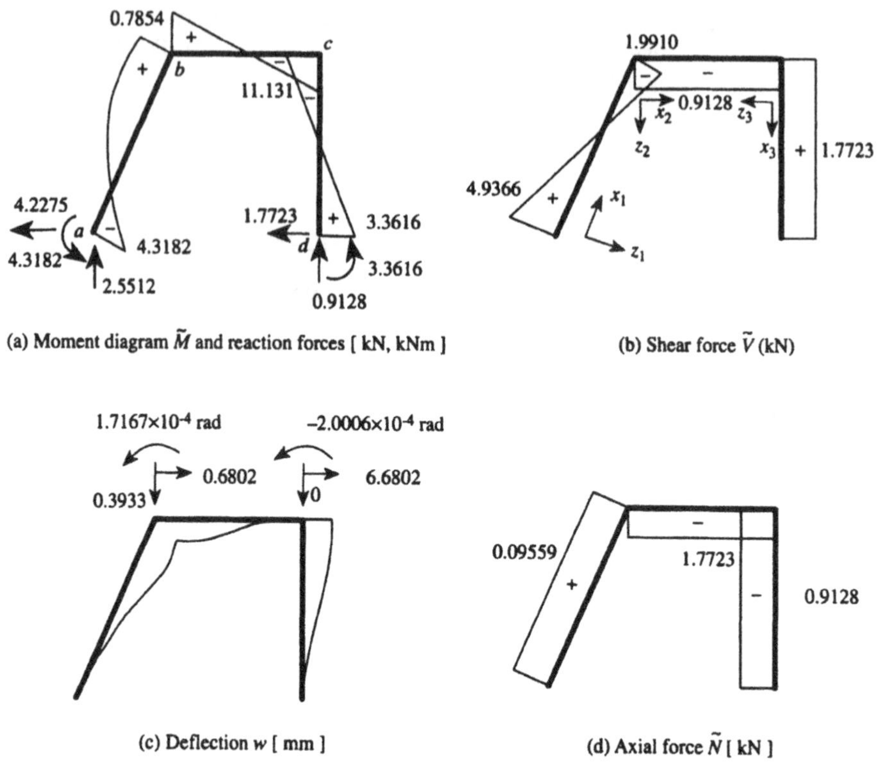

FIGURE 5.29
Results for a frame with distributed loading. Example 5.7.

the loading in order to be able to exploit the symmetry, and (3) Designation of appropriate boundary conditions on the portion of the structure to be analyzed.

Structural symmetry means that a structure is symmetric around one or more planes or axes of geometry, material properties and boundary conditions. Apart from the applied loading, the structure in Fig. 5.30a is symmetric about the Z axis. It can be seen that by rotating the right hand side of the structure about the Z axis by 180 degrees, the left- and right-hand sides of the structure can be made to coincide. In general, if one part of the structure coincides with another part after being rotated around one or more axes or planes, the structure is symmetric.

In analyzing symmetrical structures, the concept of symmetric and antisymmetric loads is important. If, when one part of a symmetric structure is rotated about the axis or plane of symmetry to coincide with another part, the loads on the two parts coincide, the system of loads is said to be *symmetric*. After the rotation, if the loads on the two parts have the same magnitudes but reversed directions, the load system is called *antisymmetric*. A load system which is neither symmetric nor antisymmetric on a symmetric structure can be transformed into a superposition of symmetric and antisymmetric load systems. Figures 5.30b and c illustrate how the unsymmetric loading on the symmetric structure of Fig. 5.30a can be converted into the superposition of symmetric and antisymmetric load systems. The load system on Fig. 5.30c is antisymmetric because by rotating the right part of the structure by 180 degrees around the Z axis, the loads on the right part will have the same magnitudes but reversed directions as those on the left part of the structure. Analyze the symmetric structure with these symmetric and antisymmetric load systems and use the principle of superposition to obtain the response of the whole structure.

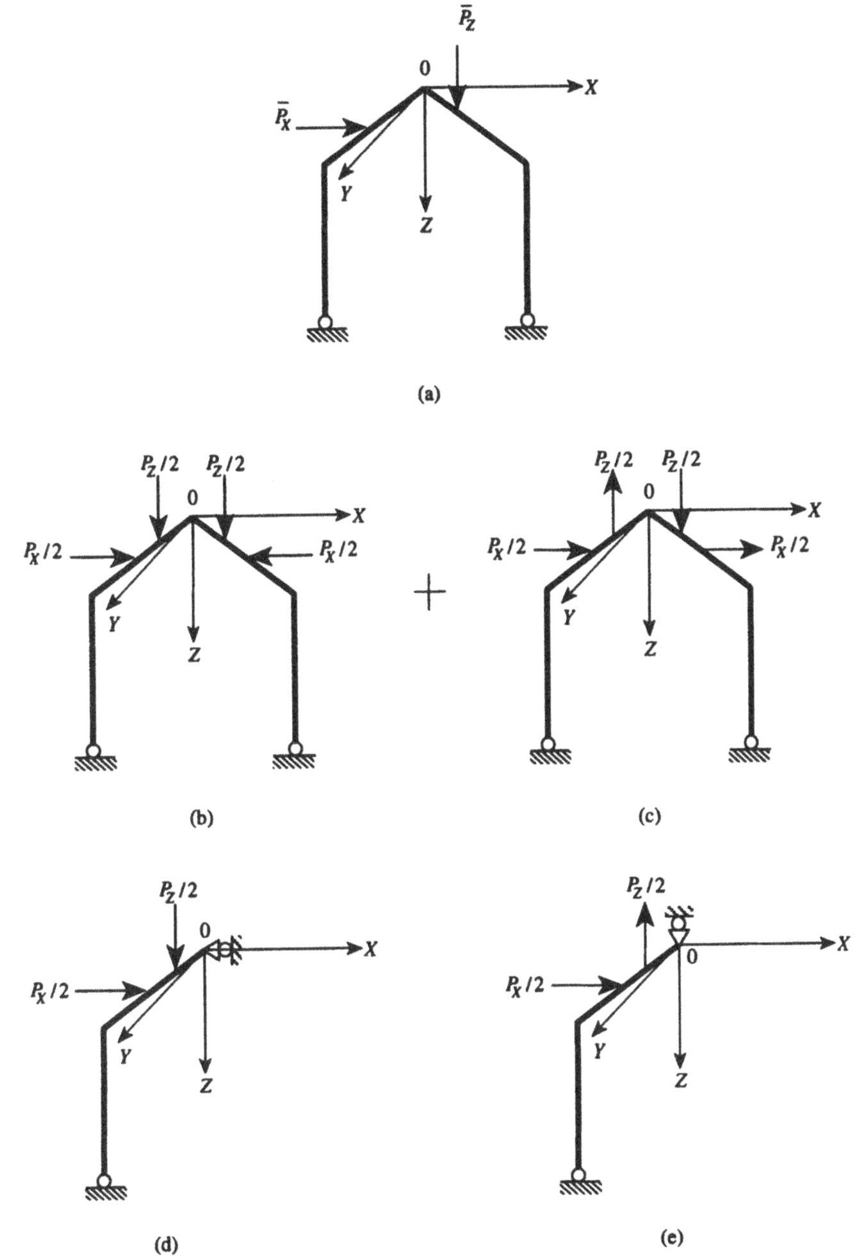

FIGURE 5.30
Loads decomposed into symmetric and antisymmetric load systems.

The analysis of a symmetric structure is efficient because only one portion of the structure needs to be analyzed. New boundary conditions are needed at the connecting points of the symmetric parts of the structure. For symmetric and antisymmetric load systems, the boundary conditions of the portions of the symmetric structure to be analyzed can be different. Generally, the boundary conditions are obtained from physical observations. In terms of displacements at the new boundaries, for the symmetric loading systems, there should be no translation normal to the plane of symmetry and no rotations about the axis of symmetry and other axes orthogonal to the axis of symmetry. For the antisymmetric loading systems, there should be no translation in the plane of symmetry and no rotation about an axis normal to the plane of symmetry at the new boundary. Figures 5.30d and e show the boundary conditions for the symmetric and the antisymmetric load systems of Figs. 5.30b and c. Since the structure lies in one plane and the plane of symmetry is the *YOZ* plane, the translation in the *X* direction and the rotations about the *Y* and *Z* axes are constrained for the symmetric load system. Rotation about the *X* axis is not shown because there is no deformation out of the *XOZ* plane. For the antisymmetric loading, translation in the *Z* direction and rotation about the *Y* axis are zero. These boundary conditions can be identified by predicting the displacement patterns of the structures in Figs. 5.30b and c.

After the symmetry of the structure is recognized, the loads processed and boundary conditions imposed, the structure can be analyzed separately for symmetric and antisymmetric load systems and then the principle of superposition can be used to obtain the response of the whole structure.

5.3.16 Reanalysis

The objective of structural reanalysis is to compute the responses of a modified structure efficiently by utilizing the response of the original (unmodified) structure. Typically, these modifications are a result of proposed changes in design. Reanalysis methodology applies to structures with either a modest number of localized changes of arbitrary magnitude or a widely distributed change of limited magnitude. The former case, which is treated in this section, is normally handled with an exact reanalysis and the latter case with an approximate reanalysis [Pilkey and Wang, 1988]. Both cases are formulated for solution as problems of much lower order than the original problem. Thus, reanalysis technology avoids the cost of a complete analysis of a structure that has been modified. This economy is achieved by exploiting the linearity properties of the structure, and by expressing the responses of the structure as functions of the modifications. The availability of reanalysis technology is often important in the effective implementation of iterative structural optimization.

Begin with a linear structural system described by the equation

$$\mathbf{K}\mathbf{V} = \overline{\mathbf{P}} \tag{5.105}$$

Suppose the system is modified so that

$$\mathbf{K}' = \mathbf{K} + \Delta\mathbf{K} \tag{5.106}$$

The governing equation for the modified system is

$$\mathbf{K}'\mathbf{V}' = \overline{\mathbf{P}} \tag{5.107}$$

where $\mathbf{V}'$ is the solution of the modified system. Substitute Eq. (5.106) into Eq. (5.107) and rearrange

$$\mathbf{K}\mathbf{V}' = \overline{\mathbf{P}} - \Delta\mathbf{K}\,\mathbf{V}' \tag{5.108}$$

Premultiply Eq. (5.108) by $\mathbf{K}^{-1}$,

$$\mathbf{V}' = \mathbf{K}^{-1}\overline{\mathbf{P}} - \mathbf{K}^{-1}\Delta\mathbf{K}\,\mathbf{V}' \tag{5.109}$$

For local modifications, $\Delta\mathbf{K}$ will be a sparse matrix, that is, $\Delta\mathbf{K}$ contains a small number of non-zero columns and rows. Let $\Delta\widehat{\mathbf{K}}$ be a submatrix of $\Delta\mathbf{K}$ which contains only the non-zero columns of $\Delta\mathbf{K}$. Furthermore, let the non-zero columns of $\Delta\mathbf{K}$ be the $i, j, k, \ldots, \ell$

columns, and define $\widehat{\mathbf{V}}' = [\mathbf{V}_i \; \mathbf{V}_j \ldots \mathbf{V}_\ell]^T$ with ℓ elements and $\widehat{\mathbf{V}}'_S$ as the column vector formed by deleting $\widehat{\mathbf{V}}'$ from $\mathbf{V}'$.

Similar definitions apply to $\widehat{\mathbf{V}}$ and $\widehat{\mathbf{V}}_S$ as subsets of $\mathbf{V}$. Equation (5.109) can be simplified to

$$\mathbf{V}' = \mathbf{V} - \mathbf{K}^{-1}\Delta\widehat{\mathbf{K}}\,\widehat{\mathbf{V}}' \tag{5.110}$$

Let $\mathbf{Y} = \mathbf{K}^{-1}\Delta\widehat{\mathbf{K}}$ so that Eq. (5.110) can be written as

$$\mathbf{V}' = \mathbf{V} - \mathbf{Y}\,\widehat{\mathbf{V}}' \tag{5.111}$$

Rearrange Eq. (5.111) into the following partitioned set of equations

$$\begin{bmatrix} \widehat{\mathbf{V}}' \\ \widehat{\mathbf{V}}'_S \end{bmatrix} = \begin{bmatrix} \widehat{\mathbf{V}} \\ \widehat{\mathbf{V}}_S \end{bmatrix} - \begin{bmatrix} \widehat{\mathbf{Y}} \\ \widehat{\mathbf{Y}}_S \end{bmatrix} \widehat{\mathbf{V}}' \tag{5.112}$$

where $\mathbf{Y}$ has been partitioned into $\widehat{\mathbf{Y}}$ and $\widehat{\mathbf{Y}}_S$ to correspond with the rearrangement of $\mathbf{V}$ into $\widehat{\mathbf{V}}$ and $\widehat{\mathbf{V}}_S$. From the first ℓ equations of Eq. (5.112)

$$\widehat{\mathbf{V}}' = \widehat{\mathbf{V}} - \widehat{\mathbf{Y}}\,\widehat{\mathbf{V}}' \tag{5.113}$$

or

$$\widehat{\mathbf{V}}' = (\mathbf{I} + \widehat{\mathbf{Y}})^{-1}\widehat{\mathbf{V}} \tag{5.114}$$

and from the lower portion of Eq. (5.112)

$$\widehat{\mathbf{V}}'_S = \widehat{\mathbf{V}}_S - \widehat{\mathbf{Y}}_S\widehat{\mathbf{V}}' \tag{5.115}$$

Equations (5.114) and (5.115) constitute the reanalysis solution to the modified structure problem. In Eq. (5.114), the response $\widehat{\mathbf{V}}'$ of the modified system is expressed in terms of the available response $\widehat{\mathbf{V}}$ of the original structure. The computational efficiency in analyzing the modified structure stems from the fact that Eq. (5.114) is a system of ℓ equations, rather than a much greater number of equations that occur if the modified system of Eq. (5.107) were to be solved.

5.4 Force Method

The force method, although it is used consistently for hand calculations of small problems, does not enjoy much popularity as an approach for solving large-scale problems. This method was the subject of intense investigation during the early evolution of computer automated structural analysis. If the currently available general purpose computer programs are used as a measure of popularity, the displacement method completely overshadows the force method. For reasons that will be delineated in this section, the force method is not as easily automated for large-scale problems as the displacement method and the menu of elements available for a force method analysis is quite limited. Variations on the force method are still under development. See, for example, Gallagher (1993). The basis of the force method is the principle of complementary virtual work. Recall from Chapter 2 that the principle of complementary virtual work is equivalent to the global form of the kinematic admissibility conditions. Hence, the force method is sometimes referred to as the *compatibility* method, as well as the *flexibility* or *influence coefficient* method. The derivation of the force method equations follows closely the derivation (Section 5.3.1) of the displacement method equations, since the two methods, as will be demonstrated later, are "dual" approaches.

5.4.1 Nodal Force Equations Based on the Principle of Complementary Virtual Work

In Chapter 2 the principle of complementary virtual work relations for an elastic solid, which are designated as Eq. (D), take the form [Chapter 2, Eq. (2.78)]

$$-\int_V \delta\boldsymbol{\sigma}^T \mathbf{E}^{-1}\boldsymbol{\sigma}\, dV + \int_{S_u} \delta\mathbf{p}^T\, \overline{\mathbf{u}}\, dS = 0 \tag{5.116}$$

For a system of beam (bar) elements, this becomes

$$\underbrace{\underbrace{-\int_x \delta\mathbf{p}^T\, \mathbf{E}^{-1}\mathbf{p}\, dx}_{\substack{\text{Element}\\ \text{Contributions}}}}_{\text{Internal Work}} + \underbrace{\underbrace{\int_x \delta\mathbf{p}^T\, \overline{\mathbf{u}}\, dx}_{\substack{\text{Applied}\\ \text{displacements}\\ \text{along the}\\ \text{elements}}} - \underbrace{[\delta\mathbf{p}^T\, \overline{\mathbf{u}}]_k}_{\substack{\text{Applied}\\ \text{displacements}\\ \text{at the}\\ \text{nodes}}}}_{\text{External work}} = 0 \tag{5.117}$$

As with the displacement method, for the force method the structure is modeled by M elements. Forces $\mathbf{p}^i$ and displacements $\mathbf{v}^i$ at the ends of the elements compose the responses. The principle of complementary virtual work, $-\delta W_i^* - \delta W_e^* = 0$, for this discretized representation would take the form

$$\sum_{i=1}^{M} \delta\mathbf{p}^{iT}(\mathbf{v}^i - \overline{\mathbf{v}}^i) = \sum_{i=1}^{M} \delta\mathbf{p}_R^{iT}\left(\mathbf{f}^i\mathbf{p}_R^i - \overline{\mathbf{v}}^i\right) = 0 \tag{5.118}$$

which is analogous to Eq. (5.32) for the displacement method. The notation of Chapter 4 is utilized here, e.g., the element flexibility matrix $\mathbf{f}^i$ is defined by $\mathbf{v}_R^i = \mathbf{f}^i\mathbf{p}_R^i$ in which the reduced force and displacement vectors are introduced. The terms in parentheses in Eq. (5.118) give the compatibility contributions for the individual elements. As expected by analogy with the displacement method, the internal complementary virtual work provides a relationship for the element flexibility matrix. As indicated in Chapter 4, the flexibility matrix is formed only for restrained systems. The applied displacement vector $\overline{\mathbf{v}}^i$ is the result of either influences, like thermal loading, distributed along the member or imposed displacements at the nodes. If the summations of Eq. (5.118) were to be expressed as system matrices, the following equation would be expected:

$$\delta\mathbf{P}^T(\mathbf{V} - \overline{\mathbf{V}}) = 0 \tag{5.119}$$

The summation of the nodal compatibility is expressed in terms of the unknown global nodal forces $\mathbf{P}$. The requirement that has to be satisfied is that the nodal forces of the various elements joined at each node must be in equilibrium with the applied nodal forces. Imposition of the equilibrium conditions is an expected requirement, since the principle of complementary virtual work corresponds to the kinematic conditions provided that the forces are in equilibrium. Of course, the local (element) nodal forces may need transformation to the global coordinate system to ease the establishment of equilibrium conditions.

The equations of equilibrium for the nodes were expressed in Eq. (5.52) in the form

$$[\mathbf{b}^{*1}\,\mathbf{b}^{*2}\ldots\mathbf{b}^{*M}]\begin{bmatrix}\mathbf{p}^1\\ \mathbf{p}^2\\ \vdots\\ \mathbf{p}^M\end{bmatrix}=\begin{bmatrix}\overline{\mathbf{P}}_1\\ \overline{\mathbf{P}}_2\\ \vdots\\ \overline{\mathbf{P}}_N\end{bmatrix} \tag{5.120}$$

$$\mathbf{b}^* \qquad \mathbf{p} = \overline{\mathbf{P}}$$

where N is the number of nodes. The matrix $\mathbf{b}^*$ contains information detailing which element is connected to which node.

A comparison of Eqs. (5.38) and (5.52) indicates that $\mathbf{b}^* = \mathbf{a}^T$ (Eq. 5.56). This relationship can be verified in general. Recall that from the principle of virtual work

$$\delta\mathbf{v}^T\mathbf{p} = \sum_{i=1}^{M}\delta\mathbf{v}^{iT}\mathbf{p}^i = \delta\mathbf{V}^T\mathbf{P} = \delta\mathbf{V}^T\overline{\mathbf{P}} \tag{5.121}$$

where $\mathbf{v}$ and $\mathbf{p}$ are the unassembled vectors of element displacements and forces. Substitute Eq. (5.120) into the right-hand side of this expression and Eq. (5.35) into the left-hand side, giving

$$\delta\mathbf{V}^T\mathbf{b}^*\mathbf{p} = \delta(\mathbf{aV})^T\mathbf{p} = \delta\mathbf{V}^T\mathbf{a}^T\mathbf{p} \tag{5.122}$$

so that $\mathbf{b}^* = \mathbf{a}^T$. This equality is an interesting relationship between the global equilibrium and kinematic matrices.

The reciprocal relationship $\mathbf{p} = \mathbf{b}\overline{\mathbf{P}}$ [Eq. (5.55)] is needed to superimpose the element relationships. Recall that since $\mathbf{b}^*$ is usually not a square matrix, $\mathbf{b}$ cannot be obtained by inversion of $\mathbf{b}^*$. However, $\mathbf{b}^*$ is a square matrix for statically determinate structures, such as a tree-like structure containing no meshes, i.e., closed branches.

For statically indeterminate systems, it is convenient to distinguish between a statically determinate set of forces and the remaining *redundant* forces. Thus, the equilibrium relations $\mathbf{p} = \mathbf{b}\overline{\mathbf{P}}$ will be split into two parts. Define

$$\mathbf{b} = \mathbf{b}_0 + \mathbf{b}_1\mathbf{X} \tag{5.123}$$

where $\mathbf{b}_0$ is obtained from a statically determinate system, $\mathbf{X}$ is formed of dimensionless forces corresponding to those forces that are selected as redundants and comprises the unknowns for the force method, and $\mathbf{b}_1$ is the equilibrium state derived for unit conditions associated with the redundants of $\mathbf{X}$. The equilibrium conditions then appear as

$$\mathbf{p} = (\mathbf{b}_0 + \mathbf{b}_1\mathbf{X})\overline{\mathbf{P}} = \mathbf{p}_0 + \mathbf{p}_x \tag{5.124}$$

with

$$\mathbf{p}_0 = \mathbf{b}_0\overline{\mathbf{P}} \quad \text{and} \quad \mathbf{p}_x = \mathbf{b}_1\mathbf{X}\overline{\mathbf{P}} = \mathbf{b}_1\mathbf{P}_x$$

The introduction of $\mathbf{P}_x = \mathbf{X}\overline{\mathbf{P}}$ permits Eq. (5.124) to be expressed as

$$\mathbf{p} = \mathbf{b}_0\overline{\mathbf{P}} + \mathbf{b}_1\mathbf{P}_x \tag{5.125}$$

where $\mathbf{P}_x$ now represents the unknown redundant forces of the force method.

Equation (5.123) provides us with the procedure for establishing $\mathbf{b}_0$ and $\mathbf{b}_1$. If $\mathbf{P}_x$ is set equal to zero, $\mathbf{b}_0$ can be calculated column by column as the elements of $\overline{\mathbf{P}}$ are set equal to zero except for the element corresponding to the column being calculated. This entry should be set equal to one. Then this column of $\mathbf{b}_0$ is equal to $\mathbf{p}$. Thus, to set up $\mathbf{b}_0$, a unit value is employed one at a time for each of the applied loads of $\overline{\mathbf{P}}$, while the redundant forces $\mathbf{P}_x$ are set equal to zero. The forces $\mathbf{p}$ (which are now equal to a column of $\mathbf{b}_0$) can then be computed using equilibrium alone, since the structure under this loading is statically determinate.

Matrix $\mathbf{b}_1$ can be computed in a similar manner. This time, set the applied loads $\overline{\mathbf{P}}$ equal to zero, and give the redundants $\mathbf{P}_x$ unit values one by one. Those redundants not assuming a unit value are set to zero. This leads to the equilibrium states $\mathbf{b}_1 = \mathbf{p}$, column by column.

In theory, this procedure for determining $\mathbf{b}_0$ and $\mathbf{b}_1$ applies to any problem. In practice, however, it is often difficult to establish $\mathbf{b}_0$ and $\mathbf{b}_1$ because it is not easy to determine the equilibrium states automatically.

Insertion of the equilibrium conditions $\mathbf{p} = \mathbf{b}\overline{\mathbf{P}}$ in the form of Eq. (5.125) into the principle of complementary virtual work expression will provide equations in terms of the system nodal forces. Thus,

$$\sum_{i=1}^{M} \delta \mathbf{p}^{iT}(\mathbf{v}^i - \overline{\mathbf{v}}^i) = \delta \mathbf{p}^T(\mathbf{v} - \overline{\mathbf{v}}) = \delta \mathbf{p}^T(\mathbf{f}\,\mathbf{p} - \overline{\mathbf{v}})$$

$$= \delta \mathbf{P}_x^T \mathbf{b}_1^T[\mathbf{f}(\mathbf{b}_0\overline{\mathbf{P}} + \mathbf{b}_1\mathbf{P}_x) - \overline{\mathbf{v}}] = 0 \tag{5.126}$$

where

$$\delta\mathbf{p} = \delta\mathbf{p}_0 + \delta\mathbf{p}_x = \delta\mathbf{p}_x = \mathbf{b}_1\,\delta\mathbf{P}_x$$

$$\mathbf{f} = \begin{bmatrix} \mathbf{f}^1 & & & \\ & \mathbf{f}^2 & & \\ & & \ddots & \\ & & & \mathbf{f}^M \end{bmatrix} = \text{diagonal } [\mathbf{f}^i] \text{ is an unassembled global flexibility matrix}$$

$\mathbf{v}, \overline{\mathbf{v}}$ are unassembled displacement vectors, e.g., $\mathbf{v} = [\mathbf{v}^1\ \mathbf{v}^2 \ldots \mathbf{v}^M]^T$.

Define

$$\mathbf{F} = \mathbf{b}_1^T \mathbf{f}\,\mathbf{b}_1 \tag{5.127}$$

as the assembled system flexibility matrix and

$$\overline{\mathbf{V}} = -\mathbf{b}_1^T(\mathbf{f}\,\mathbf{b}_0\,\overline{\mathbf{P}} - \overline{\mathbf{v}}) \tag{5.128}$$

as the assembled applied displacement vector, so that Eq. (5.126) becomes

$$\delta\mathbf{P}_x^T(\mathbf{F}\,\mathbf{P}_x - \overline{\mathbf{V}}) = 0$$

or

$$\mathbf{F}\,\mathbf{P}_x = \overline{\mathbf{V}} \tag{5.129}$$

which is a set of algebraic equations for the unknown nodal forces. These equations represent the global statement of compatibility $\mathbf{V} = \overline{\mathbf{V}}$ for all nodes of the system.

If there are no non-zero prescribed displacements $\overline{\mathbf{v}}$, i.e., $\overline{\mathbf{v}} = 0$, then from Eqs. (5.128) and (5.129) with $\mathbf{P}_x = \mathbf{X}\,\overline{\mathbf{P}}$, the nodal force equations become

$$\mathbf{F}\mathbf{X} = \overline{\mathbf{V}}' \tag{5.130}$$

with

$$\mathbf{F} = \mathbf{b}_1^T \, \mathbf{f} \, \mathbf{b}_1 \tag{5.131a}$$

$$\overline{\mathbf{V}}' = -\mathbf{b}_1^T \, \mathbf{f} \, \mathbf{b}_0 \tag{5.131b}$$

Once **X** is evaluated using Eq. (5.130), several responses of interest can be computed. For example,

$$\mathbf{p} = \mathbf{b} \, \overline{\mathbf{P}} \quad \text{with} \quad \mathbf{b} = \mathbf{b}_0 + \mathbf{b}_1 \mathbf{X} \tag{5.132a}$$

$$\mathbf{v} = \mathbf{f} \, \mathbf{p} = \mathbf{f} \, \mathbf{b} \overline{\mathbf{P}} \tag{5.132b}$$

$$\mathbf{V} = \mathbf{b}^T \mathbf{v} = \mathbf{b}_0^T \mathbf{v} = \mathbf{b}_0^T \mathbf{f} \, \mathbf{b} \overline{\mathbf{P}} \tag{5.132c}$$

See Example 5.8 for a demonstration of the validity of

$$\mathbf{b}^T \mathbf{v} = \mathbf{b}_0^T \mathbf{v} \tag{5.133}$$

EXAMPLE 5.8 The Force Method: A Continuous Beam

Use the force method to analyze the beam resting on two supports of Fig. 5.31a. For this continuous beam with two applied moments, find the internal bending moments and the corresponding slopes at the supports.

As shown in Fig. 5.31b, the beam is considered to be formed of three elements (1, 2, and 3) with the ends designated by a, b, c, and d. For this beam the global and local coordinate systems will coincide; consequently, no coordinate transformation will be introduced. The variables that are sought are the deformations **v** and forces **p**.

$$\mathbf{v} = \begin{bmatrix} \theta_a^1 \\ \theta_b^1 \\ \theta_b^2 \\ \theta_c^2 \\ \theta_c^3 \\ \theta_d^3 \end{bmatrix} \qquad \mathbf{p} = \begin{bmatrix} M_a^1 \\ M_b^1 \\ M_b^2 \\ M_c^2 \\ M_c^3 \\ M_d^3 \end{bmatrix} \tag{1}$$

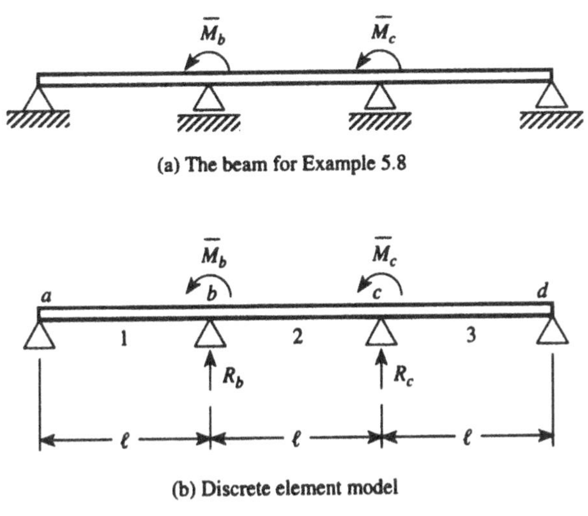

FIGURE 5.31
A continuous beam.

These are the unassambled displacement and force vectors. The applied loadings are

$$\overline{\mathbf{P}} = \begin{bmatrix} \overline{M}_b \\ \overline{M}_c \end{bmatrix} \tag{2}$$

We begin the analysis by developing the fundamental equations. Use Sign Convention 2. For a single bar element, the flexibility matrix is given by Chapter 4, Eq. (4.36a) as

$$\begin{aligned} \begin{bmatrix} \theta_a \\ \theta_b \end{bmatrix}^i &= \frac{\ell}{12EI} \begin{bmatrix} 4 & -2 \\ -2 & 4 \end{bmatrix}^i \begin{bmatrix} M_a \\ M_b \end{bmatrix}^i \\ \mathbf{v}_R^i &= \qquad \mathbf{f}^i \qquad \mathbf{p}_R^i \end{aligned} \tag{3}$$

Place the element matrices into a single unassembled global flexibility matrix $\mathbf{f}$ such that $\mathbf{v} = \mathbf{f}\mathbf{p}$ where

$$\mathbf{f} = \frac{\ell}{12EI} \left[\begin{array}{cc:cc:cc} 4 & -2 & & & & \\ -2 & 4 & \multicolumn{2}{c:}{0} & \multicolumn{2}{c}{0} \\ \hdashline & & 4 & -2 & & \\ \multicolumn{2}{c:}{0} & -2 & 4 & \multicolumn{2}{c}{0} \\ \hdashline & & & & 4 & -2 \\ \multicolumn{2}{c:}{0} & \multicolumn{2}{c:}{0} & -2 & 4 \end{array}\right] \tag{4}$$

As the next step, we will establish $\mathbf{b}_0$ and $\mathbf{b}_1$. Although numerous choices are possible, we select the reactions at b and c as the redundants. Note that the beam would be statically determinate if the supports at b and c were removed. Matrices $\mathbf{b}_0$ and $\mathbf{b}_1$ are found from the relationship [Eq. (5.125)] $\mathbf{p} = \mathbf{b}_0\overline{\mathbf{P}} + \mathbf{b}_1\mathbf{P}_x$ with

$$\mathbf{P}_x = \begin{bmatrix} R_b \\ R_c \end{bmatrix} \tag{5}$$

To find $\mathbf{b}_0$, set $\mathbf{P}_x$ equal to zero. This means that $\mathbf{p} = \mathbf{b}_0\,\overline{\mathbf{P}}$ or

$$\begin{bmatrix} M_a^1 \\ M_b^1 \\ M_b^2 \\ M_c^2 \\ M_c^3 \\ M_d^3 \end{bmatrix} = \mathbf{b}_0 \begin{bmatrix} \overline{M}_b \\ \overline{M}_c \end{bmatrix} \tag{6}$$

Let $\overline{M}_b = 1, \overline{M}_c = 0$ to form the first column of $\mathbf{b}_0$, and let $\overline{M}_b = 0, \overline{M}_c = 1$ to form the second column. Use the configurations (Sign Convention 2) of Fig. 5.32 and summation of moments to calculate the forces $\mathbf{p}$ which correspond to the first column of $\mathbf{b}_0$. The second

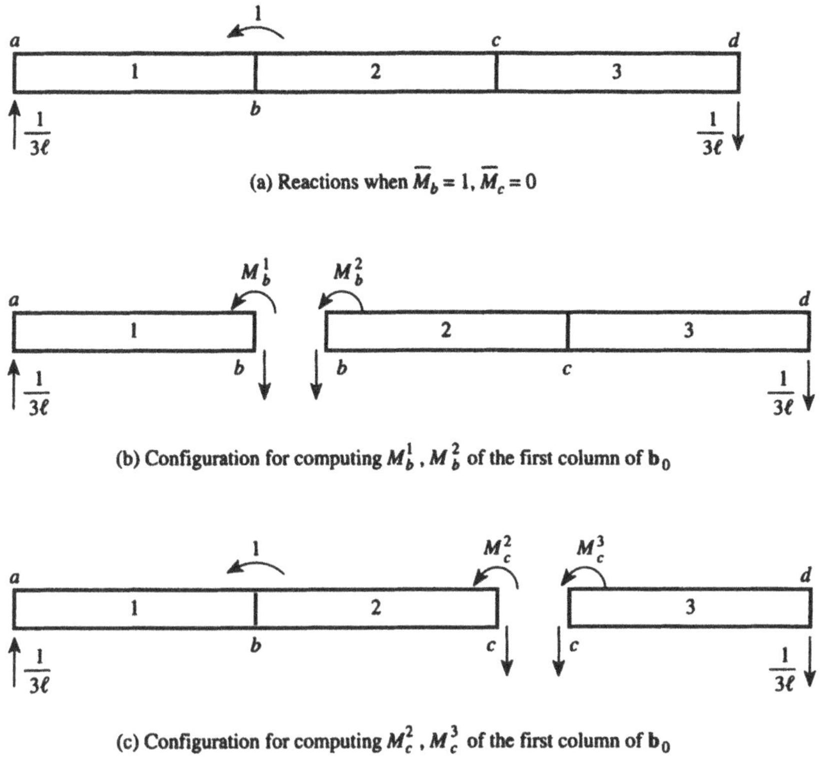
(a) Reactions when $\overline{M}_b = 1, \overline{M}_c = 0$
(b) Configuration for computing M_b^1, M_b^2 of the first column of $\mathbf{b}_0$
(c) Configuration for computing M_c^2, M_c^3 of the first column of $\mathbf{b}_0$

FIGURE 5.32
Equilibrium states $\mathbf{b}_0$ for the loading.

column of $\mathbf{b}_0$ is found in a similar fashion. These equilibrium states for the loads result in

$$\mathbf{b}_0 = \begin{bmatrix} 0 & 0 \\ 1/3 & 1/3 \\ 2/3 & -1/3 \\ -1/3 & 2/3 \\ 1/3 & 1/3 \\ 0 & 0 \end{bmatrix} \tag{7}$$

where the first and second columns correspond to nodes $b(\overline{M}_b)$ and $c(\overline{M}_c)$, respectively.

Matrix $\mathbf{b}_1$ is found by setting $\overline{\mathbf{P}}$ equal to zero in $\mathbf{p} = \mathbf{b}_0\overline{\mathbf{P}} + \mathbf{b}_1\mathbf{P}_x$ and using $R_b = 1$, $R_c = 0$, followed by $R_b = 0$, $R_c = 1$. Apply a summation of moments for obtaining equilibrium of the configurations of Fig. 5.33. The resulting moments ($\mathbf{p}$) at the end of the bar form the first column of $\mathbf{b}_1$. The second column of $\mathbf{b}_1$ is determined similarly. These calculations lead to

$$\mathbf{b}_1 = \begin{bmatrix} 0 & 0 \\ -2\ell/3 & -\ell/3 \\ 2\ell/3 & \ell/3 \\ -\ell/3 & -2\ell/3 \\ \ell/3 & 2\ell/3 \\ 0 & 0 \end{bmatrix} \tag{8}$$

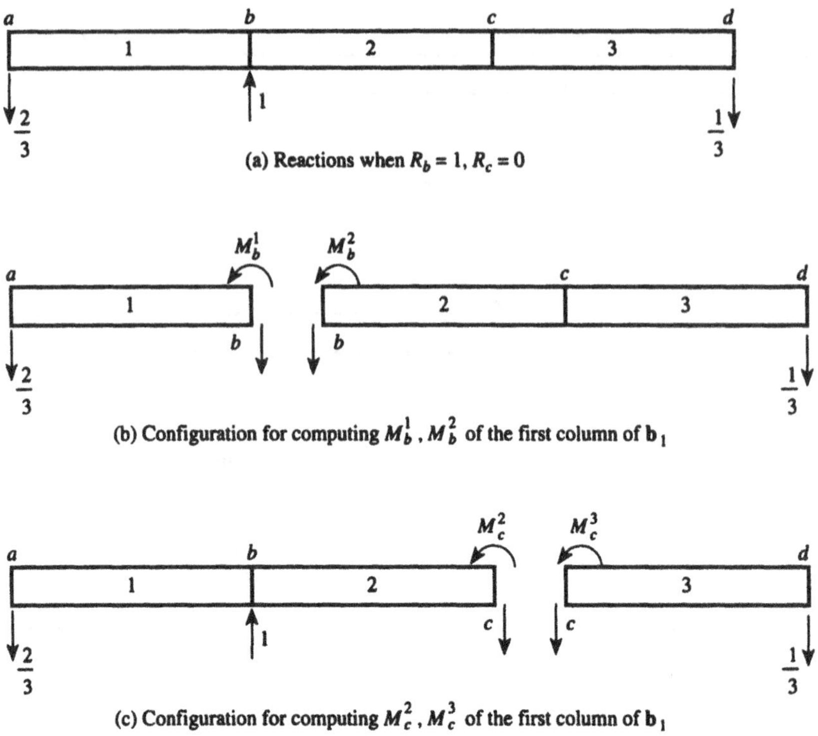

a b c d
1 2 3
2/3
1
1/3
(a) Reactions when $R_b = 1, R_c = 0$
M_b^1 M_b^2
a c d
1 2 3
b b
2/3
1/3
(b) Configuration for computing M_b^1, M_b^2 of the first column of $\mathbf{b}_1$
M_c^2 M_c^3
a b d
1 2 3
c c
1
2/3
1/3
(c) Configuration for computing M_c^2, M_c^3 of the first column of $\mathbf{b}_1$

FIGURE 5.33
Equilibrium states $\mathbf{b}_1$ for the redundants.

Now that $\mathbf{f}$, $\mathbf{b}_0$, and $\mathbf{b}_1$ have been established, $\mathbf{P}_x$ and $\mathbf{X}$ can be computed using Eqs. (5.129) and (5.130), respectively. Vector $\mathbf{P}_x$ contains the redundant forces as noted in (5), and matrix $\mathbf{X}$ is a 2×2 matrix.

Before introducing specific calculations, it may be of interest to observe how the nodal force equations [Eq. (5.129) or (5.130)] can be derived directly without reference to the principle of complementary virtual work. Equation (5.129) is based on the compatibility conditions, which require that all end displacements of the various elements joined at a particular node must be equal to the value of the displacement of the node. That is, $\mathbf{v} = \mathbf{aV}$. Since $\mathbf{a}^* = \mathbf{b}^T$, where $\mathbf{a}^*$ is defined by $\mathbf{V} = \mathbf{a}^*\mathbf{v}$, the compatibility can be expressed as

$$\mathbf{V} = \mathbf{b}^T\mathbf{v} = \left[\mathbf{b}_0^T + (\mathbf{b}_1\mathbf{X})^T\right]\mathbf{v} = \mathbf{b}_0^T\mathbf{v} + \mathbf{X}^T\mathbf{b}_1^T\mathbf{v} \tag{9}$$

The compatibility conditions that we must impose across the in-span supports, the locations of the redundant reactions, are the continuity of the slopes, i.e., $\theta_b^1 = \theta_b^2$ and $\theta_c^2 = \theta_c^3$. From (1) and (8), these conditions are

$$\mathbf{0} = \mathbf{b}_1^T\mathbf{v} \tag{10}$$

Then, with $\mathbf{v} = \mathbf{f}\,\mathbf{p}$ and $\mathbf{p} = (\mathbf{b}_0 + \mathbf{b}_1\mathbf{X})\overline{\mathbf{P}}$, (10) can be written as

$$\mathbf{0} = \mathbf{b}_1^T\mathbf{v} = \mathbf{b}_1^T\mathbf{f}\,\mathbf{p} = \mathbf{b}_1^T\mathbf{f}(\mathbf{b}_0 + \mathbf{b}_1\mathbf{X})\overline{\mathbf{P}} \tag{11}$$

or

$$\mathbf{b}_1^T \mathbf{f}\, \mathbf{b}_1 \mathbf{X} = -\mathbf{b}_1^T \mathbf{f}\, \mathbf{b}_0$$
$$\mathbf{F X} = \overline{\mathbf{V}}' \tag{12}$$

This is the same result given by Eqs. (5.130) and (5.131).

This completes the direct derivation of the fundamental equations for the force method. We can now return to the example problem to complete the calculations for the response of the beam of Fig. 5.31. The essential ingredients of the force method formulation are the matrices $\mathbf{b}_0$, $\mathbf{b}_1$, and $\mathbf{f}$. Unfortunately, the computation of $\mathbf{b}_0$ and $\mathbf{b}_1$, which contain equilibrium information, is difficult to accomplish systematically. This is a shortcoming that retards the use of the force method in general purpose structural analysis computer programs. This contrasts with the displacement method which is readily systematized due to the ease in formulating the equivalent of the kinematic matrix $\mathbf{a}$ (using assembly by superposition of element matrices).

For our beam, the calculations for $\mathbf{P}_x$ or $\mathbf{X}$ involve $\mathbf{F}$ and $\overline{\mathbf{V}}'$. We find

$$\mathbf{F} = \mathbf{b}_1^T \mathbf{f}\, \mathbf{b}_1 = \frac{\ell}{12EI} \begin{bmatrix} 48\ell^2/9 & 42\ell^2/9 \\ 42\ell^2/9 & 48\ell^2/9 \end{bmatrix} \tag{13}$$

$$\overline{\mathbf{V}}' = -\mathbf{b}_1^T \mathbf{f}\, \mathbf{b}_0 = -\frac{\ell}{12EI} \begin{bmatrix} 24\ell/9 & -30\ell/9 \\ 30\ell/9 & -24\ell/9 \end{bmatrix} \tag{14}$$

With Eq. (5.129) ($\mathbf{F P}_x = \overline{\mathbf{V}}$), the redundants are evaluated from

$$\underset{\mathbf{F}}{\ell \begin{bmatrix} 48 & 42 \\ 42 & 48 \end{bmatrix}} \underset{\mathbf{P}_x}{\begin{bmatrix} R_b \\ R_c \end{bmatrix}} = -\begin{bmatrix} 24 & -30 \\ 30 & -24 \end{bmatrix} \underset{\overline{\mathbf{P}}}{\begin{bmatrix} \overline{M}_b \\ \overline{M}_c \end{bmatrix}} \tag{15}$$

Alternatively, if $\mathbf{X}$ is to be utilized, it can be obtained from [Eq. (5.130)]

$$\underset{\mathbf{F}}{\ell \begin{bmatrix} 48 & 42 \\ 42 & 48 \end{bmatrix}} \mathbf{X} = \underset{\overline{\mathbf{V}}'}{-\begin{bmatrix} 24 & -30 \\ 30 & -24 \end{bmatrix}} \tag{16}$$

This gives

$$\mathbf{X} = \begin{bmatrix} 0.2 & 0.8 \\ -0.8 & -0.2 \end{bmatrix} \tag{17}$$

The displacements and forces of interest can now be computed. If the redundants are determined using (15), the forces can be found from $\mathbf{p} = \mathbf{b}_0 \overline{\mathbf{P}} + \mathbf{b}_1 \mathbf{P}_x$. If $\mathbf{X}$ of (17) is to be utilized, first obtain $\mathbf{b} = \mathbf{b}_0 + \mathbf{b}_1 \mathbf{X}$ and then use this to find the forces $\mathbf{p} = \mathbf{b}\overline{\mathbf{P}}$. Finally, displacements are available using

$$\mathbf{v} = \mathbf{f}\, \mathbf{p} = \mathbf{f}\, \mathbf{b}\, \overline{\mathbf{P}} \quad \text{and} \quad \mathbf{V} = \mathbf{b}^T \mathbf{v} = \mathbf{b}_0^T \mathbf{v} = \mathbf{b}_0^T \mathbf{f}\, \mathbf{b}\, \overline{\mathbf{P}} \tag{18}$$

∎

5.5 The Duality of the Force and Displacement Methods

The similarity of the formulations of the displacement and force methods of Sections 5.3 and 5.4 illustrates the dual nature of the two techniques. This is the same duality that exists with the principle of virtual work and the principle of complementary virtual work. The displacement method and the principle of virtual work require kinematically admissible displacements, i.e., **a** must be formed, and provide equilibrium equations. On the other hand, the force method and the principle of complementary virtual work begin with equilibrium conditions, i.e., **b** must be formed, and lead to kinematic equations. These relationships are illustrated in Fig. 5.34.

The dominant method in use today is the displacement method because it can be implemented in a systematic fashion with the global stiffness matrix being assembled by a summation process. This is strikingly different from the force method for which the formation of **b** is difficult to systematize.

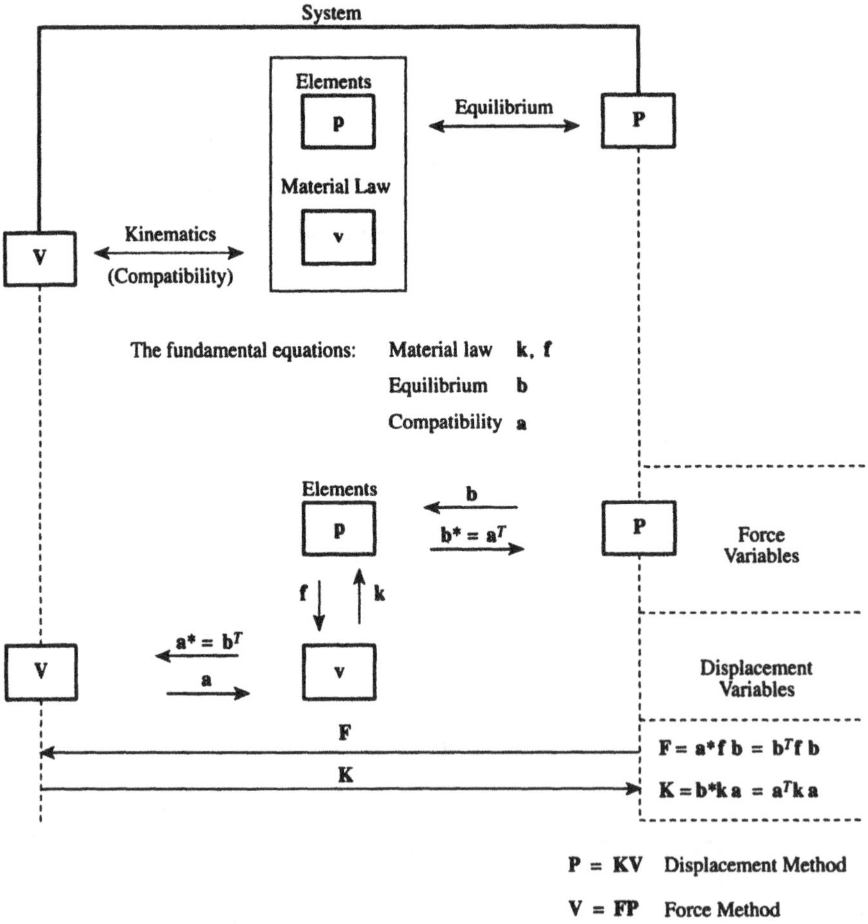

FIGURE 5.34
Element and system matrices.

TABLE 5.4

Comparison of the Transfer Matrix, Displacement, and Force Methods

Method	Transfer Matrix Method	Displacement Method	Force Method
Unknowns	Displacement and force variables	Displacement variables	Force variables
To characterize the ith element	Transfer matrix $\mathbf{U}^i$	Stiffness matrix $\mathbf{k}^i$	Flexibility matrix $\mathbf{f}^i$
To characterize the system	$\mathbf{U} = \prod^{i} \mathbf{U}^i$	$\mathbf{K} = \sum_i k^i = \mathbf{a}^T\mathbf{k}\mathbf{a}$	$\mathbf{F} = \mathbf{b}^T\mathbf{f}\,\mathbf{b}$
Conditions fulfilled at the outset	—	Compatibility (of the geometrically determinate system)	Equilibrium (of the statically determinate system)
Resulting relations satisfy	Equilibrium and compatibility	Equilibrium	Compatibility

In contrast to the force and displacement methods, the transfer matrix approach does not involve the assembly of a system matrix whose size increases with the DOF of the system. Rather, the system matrix for the transfer matrix method is characterized by progressive element matrix multiplications instead of by superposition. As a result, the system matrix for the transfer matrix method is the same size as the element matrix. Of course, the transfer matrix is suitable primarily for the solution of line-like systems. The three methods are compared in Table 5.4.

5.5.1 Indeterminacy

In the development of the force method, a distinction was made between determinate and indeterminate systems. Static indeterminacy is usually defined in terms of the number of equations in addition to the equations of equilibrium that are necessary for an analysis. This extra number of equations is employed as a measure of the degree of static indeterminacy, which is equal to the number of *redundant forces.* It should be noticed, however, that it was not necessary to distinguish between statically determinate and indeterminate structures in setting up a displacement method solution. However, to understand better the relationship between the displacement and force methods, it is helpful to define the concept of *kinematic indeterminacy.*

Kinematic indeterminacy may appear to be a significantly different concept than static indeterminacy, although it can be considered to be analogous. The number of DOF (displacements) necessary to provide the response of a structure is a measure of kinematic indeterminacy. More specifically, the degree of kinematic indeterminacy or redundancy is equal to the number of nodal displacements that would have to be constrained, in addition to those displacements constrained by the boundary conditions, in order to impose a value of zero for each DOF. A system so constrained is said to be kinematically determinate. In the displacement method, the degree of kinematic indeterminacy is equal to the number of equations needed for an analysis, i.e., it is the number of columns or rows in a stiffness matrix in which the boundary constraints have been taken into account.

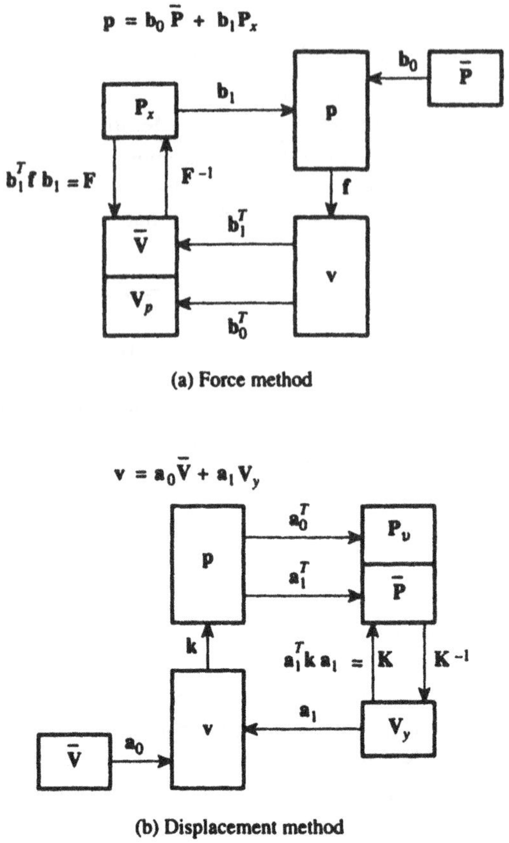

FIGURE 5.35
Force and displacement methods for applied forces $\overline{\mathbf{P}}$ and displacement $\overline{\mathbf{V}}$.

For geometrically determinate structures, **a** follows from the compatibility conditions alone. For geometrically indeterminate systems, define

$$\mathbf{a} = \mathbf{a}_0 + \mathbf{a}_1\mathbf{Y} \tag{5.134}$$

where $\mathbf{a}_0$ is obtained from a geometrically determinate primary system, **Y** contains indeterminate displacements and $\mathbf{a}_1$ is derived from geometrically indeterminate displacements corresponding to the components of **Y**. Alternatively, define

$$\mathbf{v} = \mathbf{v}_0 + \mathbf{v}_y = \mathbf{a}_0\overline{\mathbf{V}} + \mathbf{a}_1\mathbf{V}_y \tag{5.135}$$

Here, $\overline{\mathbf{V}}$ represents the prescribed displacements.

The force method, as outlined in Example 5.8, is summarized in Table 5.5. The equivalent form of the displacement formulation is provided in the same table.

5.5.2 General Loading

The formulations of Table 5.5 can be extended to include the simultaneous application of prescribed forces $\overline{\mathbf{P}}$ and displacements $\overline{\mathbf{V}}$. This generalization is outlined in Fig. 5.35 and Table 5.6.

TABLE 5.5

Outline of the Force and Displacement Methods

Unknowns	Force Method	Displacement Method
Global coordinates	Forces **X**	Displacements **Y**
	$\mathbf{v}_i = (\mathbf{T}^i)^{-1}\tilde{\mathbf{v}}^i \to \mathbf{v}^i = \mathbf{f}^i\mathbf{p}^i$	$\mathbf{p}_i = (\mathbf{T}^i)^{-1}\tilde{\mathbf{p}}^i \to \mathbf{p}^i = \mathbf{f}^i\mathbf{v}^i$
For a network structure	(I) $\mathbf{v} = \begin{bmatrix}\mathbf{v}^1\\ \mathbf{v}^2\\ \vdots\end{bmatrix} = \begin{bmatrix}\mathbf{f}^1 & & \\ & \mathbf{f}^2 & \\ & & \ddots\end{bmatrix}\begin{bmatrix}\mathbf{p}^1\\ \mathbf{p}^2\\ \vdots\end{bmatrix} = \mathbf{f}\,\mathbf{p}$	(I) $\mathbf{p} = \begin{bmatrix}\mathbf{p}^1\\ \mathbf{p}^2\\ \vdots\end{bmatrix} = \begin{bmatrix}\mathbf{k}^1 & & \\ & \mathbf{k}^2 & \\ & & \ddots\end{bmatrix}\begin{bmatrix}\mathbf{v}^1\\ \mathbf{v}^2\\ \vdots\end{bmatrix} = \mathbf{k}\,\mathbf{v}$
Primary system, e.g.	Statistically determinate	Kinematically determinate
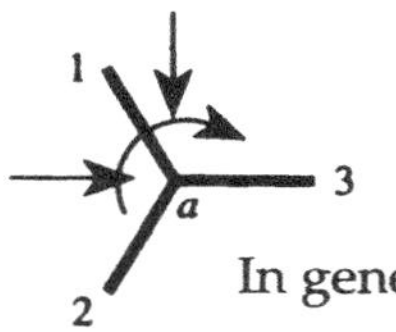	Equilibrium: $\mathbf{p}_a^1 + \mathbf{p}_a^2 + \mathbf{p}_a^3 = \overline{\mathbf{P}}_a$	Compatibility: $\mathbf{v}_a^1 = \mathbf{v}_a^2 = \mathbf{v}_a^3 = \overline{\mathbf{V}}_a$
In general	(II) $\mathbf{p} = (\mathbf{b}_0 + \mathbf{b}_1\,\mathbf{X})\overline{\mathbf{P}} = \mathbf{b}\,\overline{\mathbf{P}}$	(II) $\mathbf{v} = (\mathbf{a}_0 + \mathbf{a}_1\,\mathbf{Y})\overline{\mathbf{V}} = \mathbf{a}\,\overline{\mathbf{V}}$
Conditions to be satisfied	Compatibility: $\mathbf{V} = \mathbf{a}^*\mathbf{v}\ (\mathbf{a}^*\mathbf{a} = \mathbf{I})$	Equilibrium: $\mathbf{P} = \mathbf{b}^*\mathbf{p}\ (\mathbf{b}^*\mathbf{b} = \mathbf{I})$
	(III) $\mathbf{V} = \mathbf{b}^T\mathbf{v} = [\mathbf{b}_0^T + (\mathbf{b}_1\mathbf{X})^T]\mathbf{v}$ $\mathbf{0} = \mathbf{b}_1^T\mathbf{v}$ $\mathbf{0} = \mathbf{b}_1^T\mathbf{f}\,\mathbf{p} = \mathbf{b}_1^T\mathbf{f}(\mathbf{b}_0 + \mathbf{b}_1\mathbf{X})\overline{\mathbf{P}}$ $\mathbf{b}_1^T\mathbf{f}\,\mathbf{b}_0 + \mathbf{b}_1^T\mathbf{f}\,\mathbf{b}_1\,\mathbf{X} = \mathbf{0} \to \mathbf{X} \to \mathbf{b}$	(III) $\mathbf{P} = \mathbf{a}^T\mathbf{p} = [\mathbf{a}_0^T + (\mathbf{a}_1\mathbf{Y})^T]\mathbf{p}$ $\mathbf{0} = \mathbf{a}_1^T\mathbf{p}$ $\mathbf{0} = \mathbf{a}_1^T\mathbf{k}\,\mathbf{v} = \mathbf{a}_1^T\mathbf{k}(\mathbf{a}_0 + \mathbf{a}_1\mathbf{Y})\overline{\mathbf{V}}$ $\mathbf{a}_1^T\mathbf{k}\,\mathbf{a}_0 + \mathbf{a}_1^T\mathbf{k}\,\mathbf{a}_1\,\mathbf{Y} = \mathbf{0} \to \mathbf{Y} \to \mathbf{a}$
Results	(II) $\mathbf{p} = \mathbf{b}\,\overline{\mathbf{P}}$	(III) $\mathbf{v} = \mathbf{a}\overline{\mathbf{V}}$
	(I) $\mathbf{v} = \mathbf{f}\,\mathbf{p} = \mathbf{f}\,\mathbf{b}\,\overline{\mathbf{P}}$	(I) $\mathbf{p} = \mathbf{k}\,\mathbf{v} = \mathbf{k}\,\mathbf{a}\,\overline{\mathbf{V}}$
	(III) $\mathbf{V} = \mathbf{b}^T\mathbf{v} = [\mathbf{b}_0^T + (\mathbf{b}_1\mathbf{X})^T]\mathbf{f}\,\mathbf{b}\,\overline{\mathbf{P}} = \mathbf{b}_0^T\mathbf{f}\,\mathbf{b}\,\overline{\mathbf{P}}$	(II) $\mathbf{P} = \mathbf{a}^T\mathbf{p} = [\mathbf{a}_0^T + (\mathbf{a}_1\mathbf{Y})^T]\mathbf{k}\,\mathbf{a}\,\overline{\mathbf{V}} = \mathbf{a}_0^T\mathbf{k}\,\mathbf{a}\,\overline{\mathbf{V}}$

TABLE 5.6

Force and Displacement Methods for Applied Forces $\overline{\mathbf{P}}$ and Displacements $\overline{\mathbf{V}}$

FORCE METHOD		DISPLACEMENT METHOD
$\mathbf{P}_x$	Unknowns	$\mathbf{V}_y$
Compatibility Conditions: $\mathbf{b}_1^T(\mathbf{v}-\overline{\mathbf{v}})=0$ or $\mathbf{b}_1^T\mathbf{v}=\overline{\mathbf{V}}$		Equilibrium Conditions: $\mathbf{a}_1^T(\mathbf{p}-\overline{\mathbf{p}})=0$ or $\mathbf{a}_1^T\mathbf{p}=\overline{\mathbf{P}}$
$\mathbf{b}_1^T\mathbf{f}(\mathbf{b}_0\overline{\mathbf{P}}+\mathbf{b}_1\mathbf{P}_x)=\overline{\mathbf{V}}$	This leads to	$\mathbf{a}_1^T\mathbf{k}(\mathbf{a}_0\overline{\mathbf{V}}+\mathbf{a}_1\mathbf{V}_y)=\overline{\mathbf{P}}$
$\mathbf{P}_x=\mathbf{F}^{-1}[\overline{\mathbf{V}}-\mathbf{b}_1^T\mathbf{f}(\mathbf{b}_0\overline{\mathbf{P}})]$	(1)	$\mathbf{V}_y=\mathbf{K}^{-1}[\overline{\mathbf{P}}-\mathbf{a}_1^T\mathbf{k}(\mathbf{a}_0\overline{\mathbf{V}})]$
$\mathbf{p}=[\mathbf{b}_0+\mathbf{b}_1\mathbf{F}^{-1}(-\mathbf{b}_1^T\mathbf{f}\,\mathbf{b}_0)]\overline{\mathbf{P}}+\mathbf{b}_1\mathbf{F}^{-1}\overline{\mathbf{V}}$	from which	$\mathbf{v}=[\mathbf{a}_0+\mathbf{a}_1\mathbf{K}^{-1}(-\mathbf{a}_1^T\mathbf{k}\mathbf{a}_0)]\overline{\mathbf{V}}+\mathbf{a}_1\mathbf{K}^{-1}\overline{\mathbf{P}}$
$\mathbf{v}=\mathbf{f}\,\mathbf{p}$	Since then	$\mathbf{p}=\mathbf{k}\,\mathbf{v}$
$\mathbf{V}_p=\mathbf{b}_0^T\mathbf{v}=\mathbf{b}_0^T\mathbf{f}\,\mathbf{b}\,\overline{\mathbf{P}}+(\mathbf{b}_0^T\mathbf{f}\,\mathbf{b}_1)\mathbf{F}^{-1}\overline{\mathbf{V}}$	(2)	$\mathbf{P}_v=\mathbf{a}_0^T\mathbf{p}=\mathbf{a}_0^T\mathbf{k}\,\mathbf{a}\,\overline{\mathbf{V}}+(\mathbf{a}_0^T\mathbf{k}\,\mathbf{a}_1)\mathbf{K}^{-1}\overline{\mathbf{P}}$
	The unknowns can be obtained from (1) and (2):	
$\begin{bmatrix}\mathbf{V}_p\\ \cdots\\ \overline{\mathbf{V}}\end{bmatrix}=\begin{bmatrix}\mathbf{b}_0^T\\ \cdots\\ \mathbf{b}_1^T\end{bmatrix}\mathbf{v}=\begin{bmatrix}\mathbf{b}_0^T\mathbf{f}\,\mathbf{b}_0 & \vdots & \mathbf{b}_0\mathbf{f}\,\mathbf{b}_1\\ \cdots\cdots & \cdots & \cdots\cdots\\ \mathbf{b}_1^T\mathbf{f}\,\mathbf{b}_0 & \vdots & \mathbf{b}_1^T\,\mathbf{f}\,\mathbf{b}_1\end{bmatrix}\begin{bmatrix}\overline{\mathbf{P}}\\ \cdots\\ \mathbf{P}_x\end{bmatrix}$		$\begin{bmatrix}\mathbf{P}_v\\ \cdots\\ \overline{\mathbf{P}}\end{bmatrix}=\begin{bmatrix}\mathbf{a}_0^T\\ \cdots\\ \mathbf{a}_1^T\end{bmatrix}\mathbf{p}=\begin{bmatrix}\mathbf{a}_0^T\mathbf{k}\,\mathbf{a}_0 & \vdots & \mathbf{a}_0\mathbf{k}\,\mathbf{a}_1\\ \cdots\cdots & \cdots & \cdots\cdots\\ \mathbf{a}_1^T\mathbf{k}\,\mathbf{a}_0 & \vdots & \mathbf{a}_1^T\mathbf{k}\,\mathbf{a}_1\end{bmatrix}\begin{bmatrix}\overline{\mathbf{V}}\\ \cdots\\ \mathbf{V}_y\end{bmatrix}$

References

Argyris, J., 1954, Energy theorems and structural analysis part I: general theory, *Aircraft Engineering*, Vol. 26.

Bathe, K.J., 1996, *Finite Element Procedures*, Prentice-Hall, New Jersey.

Cook, R.D., Malkus, D.S. and Plesha, M.E., 1989, *Concepts and Applications of Finite Element Analysis*, 3rd ed., John Wiley & Sons, NY.

Everstine, G.C., 1979, A comparison of three resequencing algorithms for the reduction of matrix profile and wavefront, *IJNME*, Vol. 14, pp. 837–858.

Gallagher, R.H., 1993, Finite element structural analysis and complementary energy, *J. Finite Element Anal. Des.*, Vol. 13, Nos. 2, 3, pp. 115–126.

Gibbs, N.E., Poole, W.G., Jr. and Stockmeyer, P.K., 1976, An algorithm for reducing the bandwidth and profile of a sparse matrix, *SIAM J. Numerical Anal.*, Vol. 13, pp. 236–250.

Horner, G.C. and Pilkey, W.D., 1978, The Riccati transfer matrix method, *J. Mech. Des.*, Vol. 1, pp. 297–302.

Marguerre, V.K. and Uhrig, R., 1964, Berechnung vielgliedriger Gelenk Ketten I. Ubertragungsverfahren und seine Grenzen, Z. *Angew. Math. Mech.*, Vol. 44. pp. 1–21.

Pestel, E. and Leckie, F., 1963, *Matrix Methods in Elastomechanics*, McGraw-Hill, NY.

Pilkey, W.D., 1994, *Formulas for Stress, Strain, and Structural Matrices*, John Wiley & Sons, NY.

Pilkey, W.D. and Chang, P.Y., 1978, *Modern Formulas for Statics and Dynamics*, McGraw-Hill, NY.

Pilkey, W.D. and Pilkey, O.H., 1986, *Mechanics of Solids*, Krieger Publishing Co., Malabar, FL.

Pilkey, W.D. and Wang, B.P., 1988, Structural reanalysis revisited, *Proceedings, 7th SAE International Conference on Vehicle Structural Mechanics*, pp. 297–305.

Przemieniecki, J.S., 1968, *Theory of Matrix Structural Analysis*, McGraw Hill, NY.

Problems

Transfer Matrices

5.1 Find the deflection, slope, shear force, and bending moment along the beam of Fig. P5.1.

Partial Answer:

x (ft)	w (in.)	θ (rad)	V (lb)	M (in. lb)
0	0	-0.2×10^{-2}	83,333	0
10	0.18	-0.583×10^{-3}	-0.1667×10^{5}	4×10^{6}
30	0	0.1417×10^{-2}	-0.1667×10^{5}	0

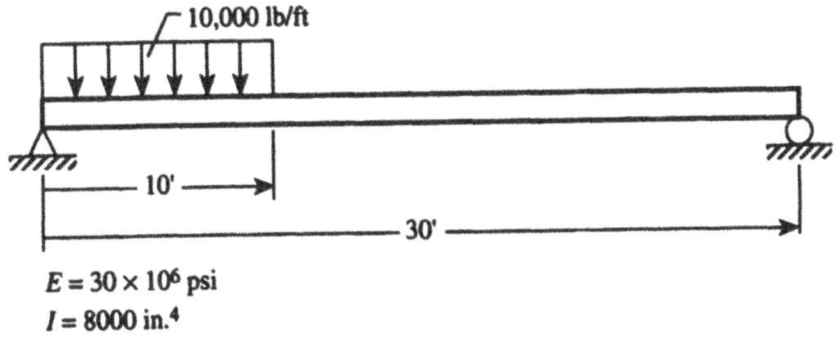

FIGURE P5.1

5.2 Calculate the distribution of displacements and forces at the cantilevered and free ends of the beam of Fig. P5.2.

Partial Answer:

x (ft)	w (in.)	θ	V (lb)	M (in. −lb)
0	0	0	5000	−1,512,000
18	0.112	-0.8748×10^{-3}	3000	−432,000
30	0.250	-0.1004×10^{-2}	0	0

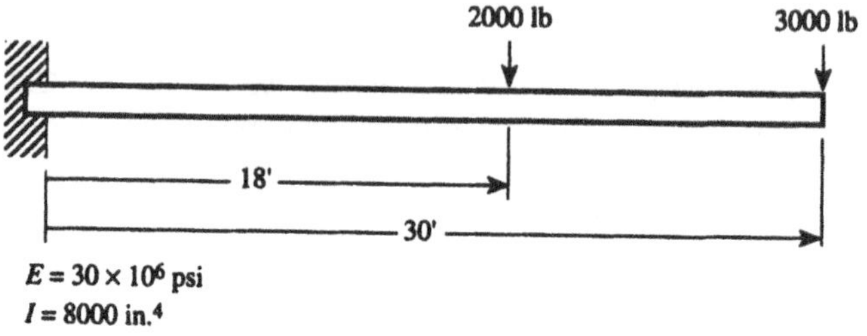

FIGURE P5.2

5.3 Determine the bending moment and deflection at midspan of the beam of Fig. P5.3.

Answer: $M = 104.2$ Nm, $w = 0.00755$ mm

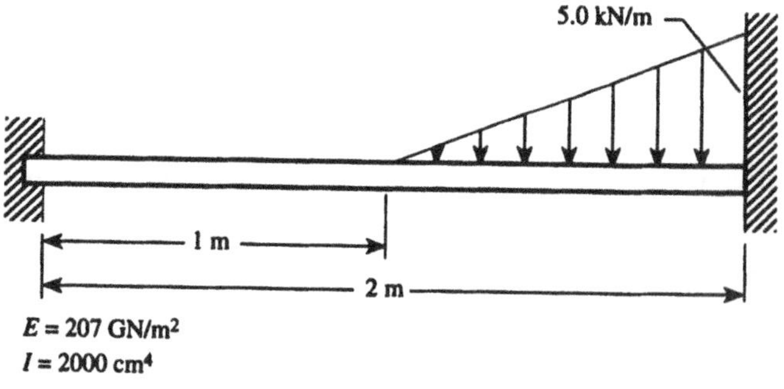

FIGURE P5.3

5.4 Find the displacements and net internal forces along the beam of Fig. P5.4, if $p_0 = 200$ N/m, $\overline{P} = 500$ N, $\overline{M}_b = 300$ Nm, and $\overline{M}_c = 100$ Nm.

Partial Answer:

x (m)	w (mm)	θ (rad)	V (N)	M (Nm)
0	0	0	837.5	−375.0
1	0.0134	0.248×10^{-5}	137.5	62.5
2	0	0.258×10^{-4}	−62.5	100.0

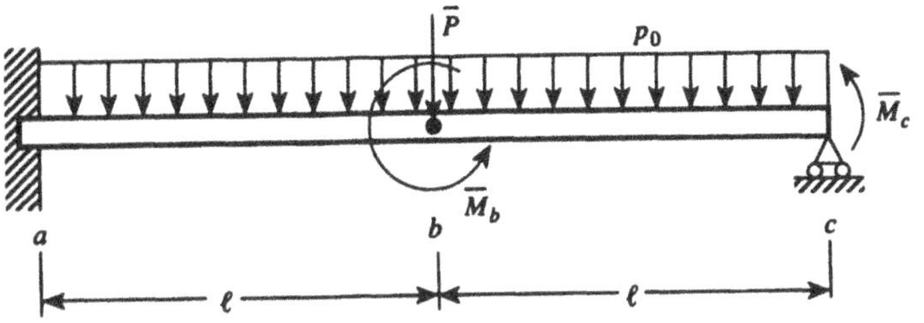

FIGURE P5.4
Beam with multiple loadings.

5.5 Find the state variables (deflection, slope, shear force, and bending moments) along the beam of Fig. P5.5. Let $L = 2$ m, $I = 2000$ cm^4, $E = 207$ GN/m^2, $p_0 = 1$ kN/m, $\overline{P} = 3$ kN, and $\overline{M} = 2$ kNm.

Partial Answer:

x (m)	w (mm)	θ (rad)	V (N)	M (Nm)
0	0	0	800	−183.3
1	0	0.1208×10^{-4}	4000	−1883.0
2	0.0644	0	3000	1617.0

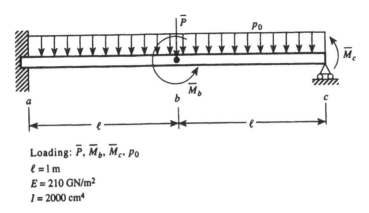

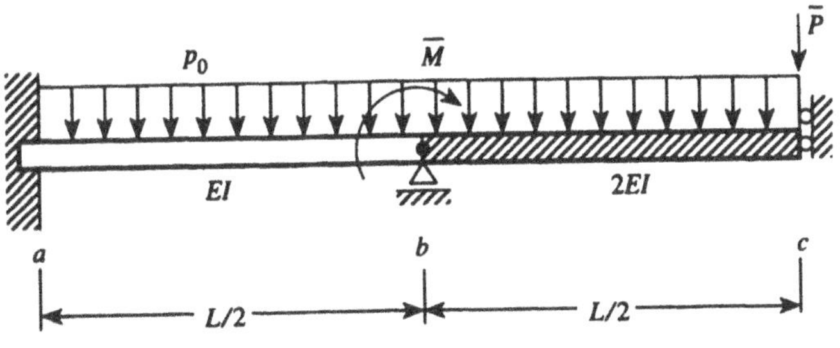

FIGURE P5.5
Beam with guided end and in-span support.

5.6 Find the responses along the beam of Fig. P5.6.

Partial Answer:

x (m)	w (mm)	θ (rad)	V (N)	M (Nm)
0	1.1470	0.2523×10^{-3}	0	0
5	0	0.1608×10^{-3}	5067	−37 500
11	−0.2305	-0.4823×10^{-4}	67.34	−13 100
14	0	-0.1053×10^{-3}	20 680	−12 890
24	0	0	−29 320	−56 050

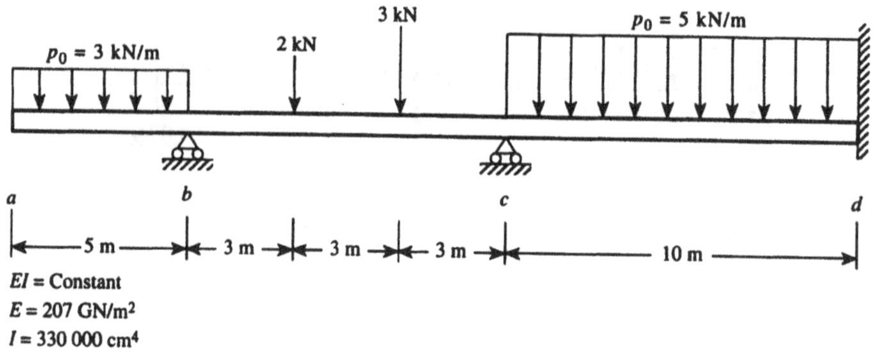

FIGURE P5.6
Beam with two in-span supports.

5.7 Compute the state variables along the beam of Fig. P5.7. Let $E = 207\ \text{GN/m}^2$, $I_{\text{section 1}} = 330\,000\ \text{cm}^4$, $I_{\text{section 2}} = 660\,000\ \text{cm}^4$.

Partial Answer:

x (m)	w (mm)	θ (rad)	V (N)	M (Nm)
0	0	0	15 700	−27 330
10	0	0.1704×10^{-4}	6724	−20 340
13	−0.00625	-0.5490×10^{-5}	1724	−172.4
19	0	0	−3276	−4827

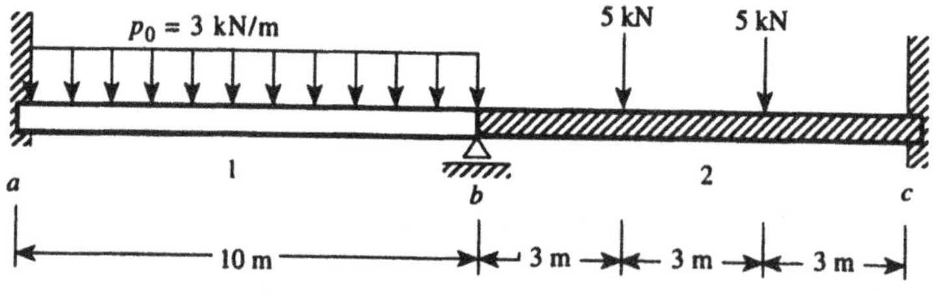

FIGURE P5.7
Beam of variable cross-section. Let $E = 207\ \text{GN/m}^2$, $I_{\text{section 1}} = 330\,000\ \text{cm}^4$, and $I_{\text{section 2}} = 660\,000\ \text{cm}^4$.

5.8 Calculate the deflection and stresses along the beam of Fig. P5.8. This is a 16 m WF beam with a uniformly distributed loading of $\overline{p}_0 = 10\,000\ \text{N/m}$.

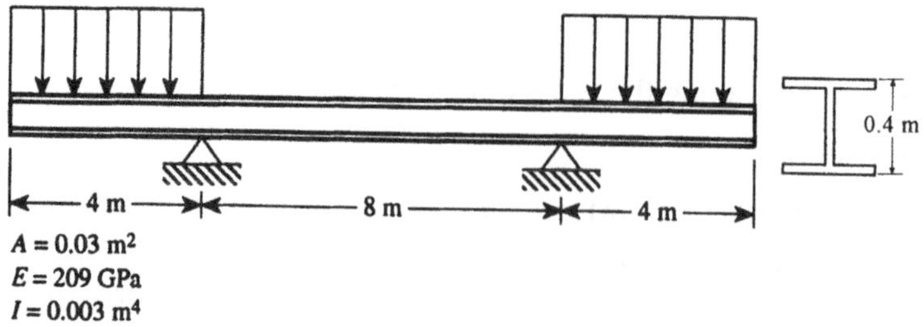

FIGURE P5.8
Overhanging beam.

Partial Answer:
$\sigma_{max} = 5.34$ MPa, $w_{max} = 2.55$ mm at both ends of the beam.

5.9 Use the transfer matrix method to find the displacements and forces along the plane frame of Fig. 5.22.

5.10 Compute, with transfer matrices, the displacement and force responses of the plane frame of Fig. 5.28.

Displacement method

5.11 Find the displacements, reaction forces, and spring forces in the one-dimensional system of springs connected to rigid bars shown in Fig. P5.11. For numerical results, let the spring constants be given by $k^1 = 2$, $k^2 = 1$, $k^3 = 1$, and $k^4 = 1$.

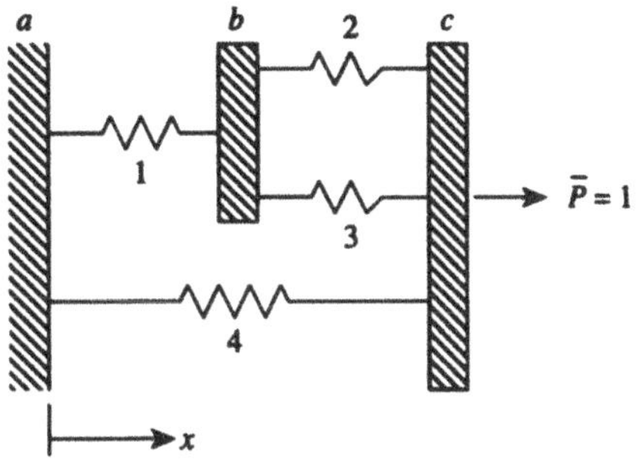

FIGURE P5.11
A spring system.

Hint:
For a direct stiffness, nonvariational formulation, use the following procedure. Take the element stiffness matrix for a spring from Chapter 3, i.e.,

$$\mathbf{p}^i = \begin{bmatrix} k^i & -k^i \\ -k^i & k^i \end{bmatrix} \mathbf{v}^i$$

Equilibrium at the nodes:
Node a: $P_a = p_a^1 + p_a^4 =$ reaction force
Node b: $\overline{P}_b = 0 = p_b^1 + p_b^2 + p_b^3 = 0$
Node c: $\overline{P}_c = 1 = p_c^2 + p_c^3 + p_c^4$

Compatibility at the nodes:
Node a: $u_a^1 = u_a^4 = \overline{U}_a = 0$
Node b: $u_b^1 = u_b^2 = u_b^3 = U_b$
Node c: $u_c^2 = u_c^3 = u_c^4 = U_c$

Global Stiffness Matrix:

$$\begin{bmatrix} K_{11} & K_{12} & K_{13} \\ K_{21} & K_{22} & K_{23} \\ K_{31} & K_{32} & K_{33} \end{bmatrix} \begin{bmatrix} \overline{U}_a \\ U_b \\ U_c \end{bmatrix} = \begin{bmatrix} P_a \\ \overline{P}_b \\ \overline{P}_c \end{bmatrix} = \begin{bmatrix} ? \\ 0 \\ 1 \end{bmatrix}$$

where

$$\mathbf{K} = \begin{bmatrix} k_{11}^1 + k_{11}^4 & k_{12}^1 & k_{13}^4 \\ k_{21}^1 & k_{22}^1 + k_{22}^2 + k_{22}^3 & k_{23}^2 + k_{23}^3 \\ k_{31}^4 & k_{32}^2 + k_{32}^3 & k_{33}^2 + k_{33}^3 + k_{33}^4 \end{bmatrix} = \begin{bmatrix} 3 & -2 & -1 \\ -2 & 4 & -2 \\ -1 & -2 & 3 \end{bmatrix}$$

Introduce boundary conditions:

$$\begin{bmatrix} 4 & -2 \\ -2 & 3 \end{bmatrix} \begin{bmatrix} U_b \\ U_c \end{bmatrix} = \begin{bmatrix} 0 \\ 1 \end{bmatrix}$$

Answer:

$U_b = 1/4,\ U_c = 1/2,\ p_a = -1$

Spring Forces: Calculate p_b^i for each spring. $p_a^i = -p_b^i$, $p_b^1 = 0.5$, $p_c^2 = 0.25$, $p_c^3 = 0.25$, and $p_c^4 = 0.5$.

5.12 Suppose the applied load of the spring system of the previous problem (and Fig. P5.11) is replaced by a prescribed displacement $\overline{U}_c = \mathbf{1}$. Find the response variables again.

Hint:
The global equilibrium equations are

$$\begin{bmatrix} K_{11} & K_{12} & K_{13} \\ K_{21} & K_{22} & K_{23} \\ K_{31} & K_{32} & K_{33} \end{bmatrix} \begin{bmatrix} 0 \\ U_b \\ 1 \end{bmatrix} = \begin{bmatrix} P_a \\ \overline{P}_b \\ P_c \end{bmatrix} = \begin{bmatrix} ? \\ 0 \\ ? \end{bmatrix}$$

Partial Answer:

$U_b = 1/2,\ P_c = 2$

5.13 If $\mathbf{b}^*\mathbf{p} = \overline{\mathbf{P}}$ and $\mathbf{p} = \mathbf{b}\overline{\mathbf{P}}$, show that $\mathbf{b}^* = \mathbf{b}^{-1}$, $\mathbf{b}^*\mathbf{b} = \mathbf{I}$, and $\mathbf{b}\mathbf{b}^* \neq \mathbf{I}$.

Trusses

5.14 Find the vertical and horizontal displacements of joint a of the bar system of Chapter 3, Fig. 3.1. Also, compute the elongations of the individual bars, as well as the axial forces in the bars.

Answer: See Example 3.1.

5.15 Calculate the nodal displacements of the truss of Chapter 3, Fig. 3.4. Also, find the axial forces and displacements for each member.

Answer: See Example 3.7.

5.16 Determine the movement of joint a of the truss of Fig. P5.16. Assume all members behave in a linear elastic manner.

5.17 Calculate the nodal displacements and the bar forces of the truss of Fig. P5.16. Assume linear properties.

5.18 Compute the displacements of joint a of the truss of Chapter 3, Fig. 3.12a. Also find the forces in the bars.

Answer: See Fig. 3.13.

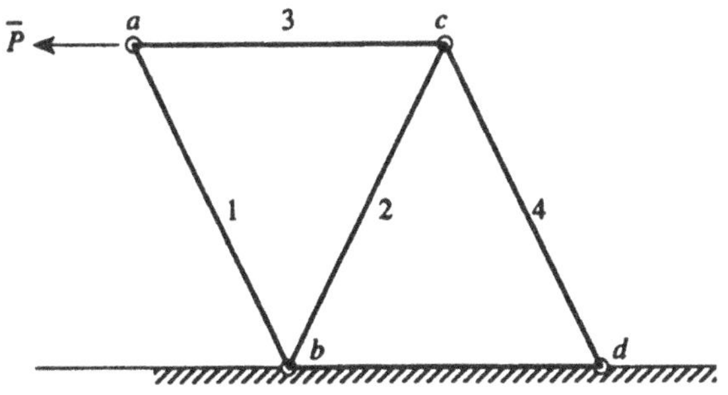

FIGURE P5.16

5.19 Determine the movement of joint a of the truss of Chapter 3, Fig. P3.2.

5.20 Find the displacements at the joints of the truss of Chapter 3, Fig. P3.9.

5.21 Compute the displacements of joints a and b of the truss of Chapter 3, Fig. P3.13a. Express the answer in terms of symbols.

5.22 Find the displacement of joint b of the truss of Chapter 3, Fig. P3.34.

5.23 The structure of Fig. P5.23 is formed of two steel rods, each 4.572 m long. All ends are hinged. Compute the axial stress in the bars and the vertical displacement at the load $\overline{P}$. Let $\overline{P} = 22.246$ kN, $A = 3.2258$ cm^2, and $E = 207$ GN/m^2.

Answer:
$\sigma = 68.96$ MN/m^2, $U_{Zb} = 3.05$ mm

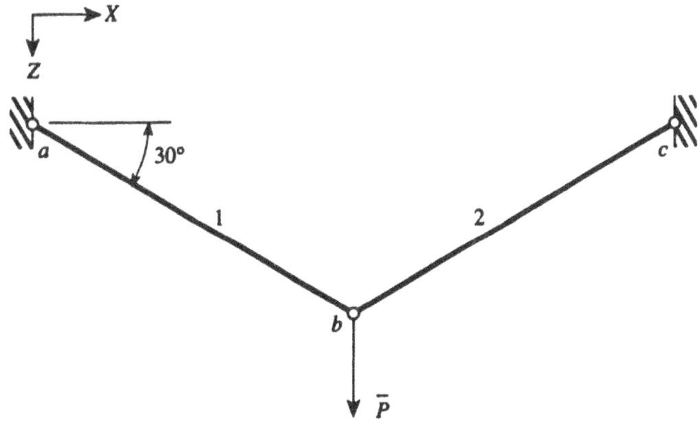

FIGURE P5.23
Hinged rods.

5.24 Find the vertical displacement at the location of the applied force, as well as the forces in the bars of the truss of Fig. P5.24.

Answer: $u_Z|_{\text{at } \overline{P}} = 0.088$ in., $p^1 = +5000$ lb, $p^2 = -5000$ lb

5.25 A truss is loaded as shown in Fig. P5.25. Also, element 3 is heated. Find the displacements of the joints and the forces and stresses in the bars. Let $\overline{P} = 1000$ lb, $(\Delta T)_{\text{element}\,3} = +100$°F, $E = 1.0 \times 10^7$ lb/in.^{2}, $\alpha = 1.0 \times 10^{-6}$/°F, $\ell = 20$ in.; for bars 1, 3, 5, and 6, $A = 1.0$ in.^{2}; for bars 2 and 4, $A = 0.7071068$ in.^{2}

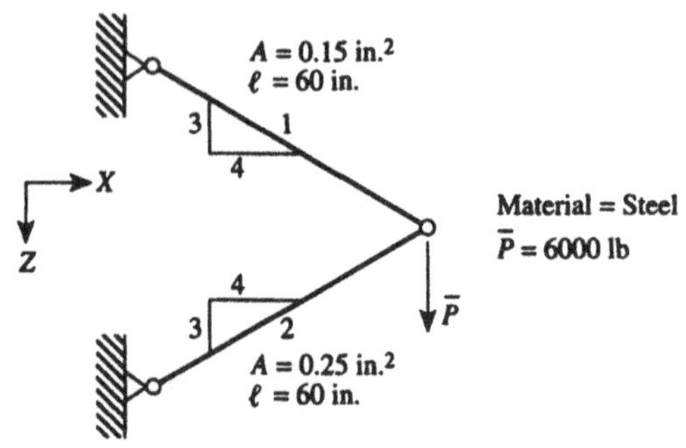

FIGURE P5.24
Plane truss

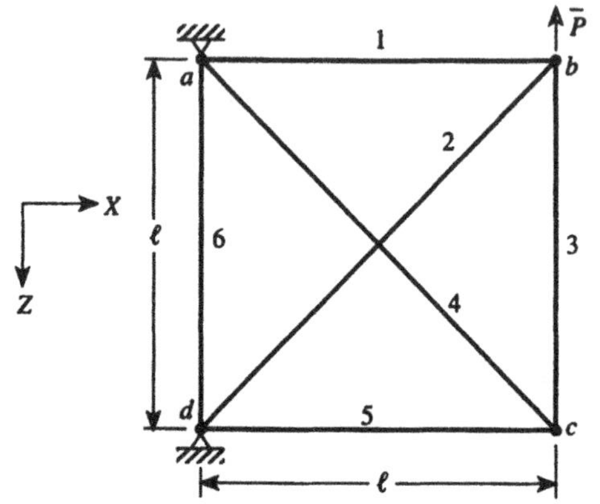

FIGURE P5.25
A pin-jointed truss.

Answer:
Displacements:
$u_{Xb} = -1.273 \times 10^{-3}$ in. $u_{Xc} = 7.273 \times 10^{-4}$ in.
$u_{Zb} = -6.636 \times 10^{-3}$ in. $u_{Zc} = -3.636 \times 10^{-3}$ in.

Bar Forces:	Bar Stresses:
$N^1 = -636.36$ lb (compression)	$\sigma^1 = -636.36$ lb/in.^{2}
$N^2 = +899.95$ lb (tension)	$\sigma^2 = +1272.73$ lb/in.^{2}
$N^3 = +363.64$ lb	$\sigma^3 = +363.64$ lb/in.^{2}
$N^4 = -514.26$ lb	$\sigma^4 = -727.27$ lb/in.^{2}
$N^5 = +363.64$ lb	$\sigma^5 = +363.64$ lb/in.^{2}
$N^6 = 0$	$\sigma^6 = 0$

Beams

5.26 Compute the deflection of the free end of the stepped beam of Chapter 3, Fig. 3.10c. Express your answer in terms of general symbols such as $EI_{\text{beam 1}}$.

5.27 Calculate the reactions of the beam of Chapter 3, Fig. P3.27.

5.28 Find the reactions at the ends and at the hinge of the beam of Chapter 3, Fig. P3.1.

5.29 Compute the displacement and reaction at the left end of the beam of Chapter 3, Fig. P3.26. If the spring is v_s units too long before the load is applied, what will be the magnitude of the reactions?

5.30 Suppose a beam element undergoes a change in temperature ΔT. Show that the loading vector $\overline{\mathbf{p}}^{i0}$ corresponding to a stiffness matrix for a beam element of bisymmetric cross-section of depth h would be

$$\overline{\mathbf{p}}^{i0} = [0 - EI\alpha \quad \Delta T/h \quad 0 \quad EI\alpha \quad \Delta T/h]^T$$

where α is the coefficient of thermal expansion.

5.31 Set up a displacement method computer routine to find the displacements of the beam of Fig. P5.4. Find the displacements at b and c for:

Case 1: Concentrated loads only with $\overline{P} = 500$ N, $\overline{M}_b = 0$, $\overline{M}_c = 100$ Nm
Case 2: Thermal loading only with $\Delta T = 20°\text{C}$, $\alpha = 12 \times 10^{-6}/°\text{C}$

Answer:
Case 1:

$$[w_b \quad \theta_b \quad \theta_c]^T = [0.0117 \text{ mm} \quad -0.670 \times 10^{-5} \text{ rad} \quad 0.268 \times 10^{-4} \text{ rad}]^T$$

Case 2:

$$[w_b \quad \theta_b \quad \theta_c]^T = [-0.183 \text{ mm} \quad 0.183 \times 10^{-3} \text{ rad} \quad -0.733 \times 10^{-3} \text{ rad}]^T$$

5.32 Use the displacement method to find the displacements at the supports of the beam of Fig. P5.5. Use the numerical geometrical and physical values given in Problem 5.5.

Answer: See Problem 5.5.

5.33 Find the slopes of the beam of Fig. P5.6 at nodes b and c.

Answer:

$$\begin{bmatrix} \theta_b \\ \theta_c \end{bmatrix} = 1/(0.11EI) \begin{bmatrix} 44.0 \\ -30.4 \end{bmatrix}$$

5.34 Calculate the bending moments at nodes a, b, and c of the beam of Fig. P5.7.

Answer: (for $EI = 1$)

$$\begin{bmatrix} M_a \\ M_b \\ M_c \end{bmatrix} = \begin{bmatrix} 244.0 \\ -186.9 \\ 72.6 \end{bmatrix}$$

5.35 Find the maximum deflection and bending stress in the overhanging beam of Fig. P5.8.

Answer: See Problem 5.8.

5.36 Derive a 4×4 stiffness matrix for the beam of Fig. P5.36.

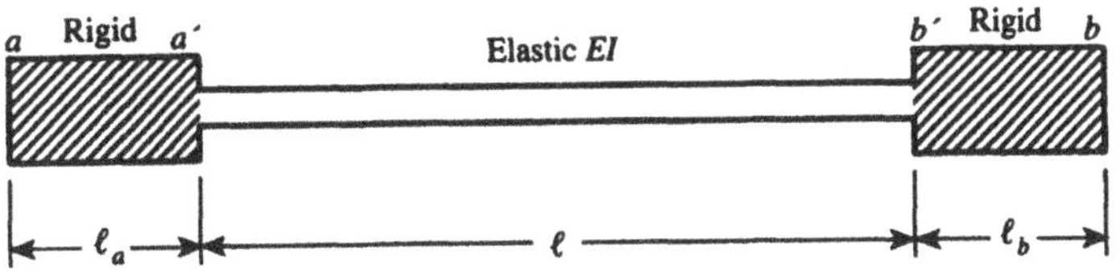

FIGURE P5.36
A beam with rigid end segments.

Hint:

$$\begin{bmatrix} w_a' \\ \theta_a' \end{bmatrix} = \begin{bmatrix} 1 & -\ell_a \\ 0 & 1 \end{bmatrix} \begin{bmatrix} w_a \\ \theta_a \end{bmatrix} \qquad \begin{bmatrix} w_b' \\ \theta_b' \end{bmatrix} = \begin{bmatrix} 1 & \ell_b \\ 0 & 1 \end{bmatrix} \begin{bmatrix} w_b \\ \theta_b \end{bmatrix}$$

Answer:

$$\mathbf{k} = EI \begin{bmatrix} \frac{12}{\ell^2} & -\frac{12}{\ell^2}\ell_a - \frac{6}{\ell} & -\frac{12}{\ell^2} & \frac{12}{\ell^2}\ell_b - \frac{6}{\ell} \\ -\frac{12}{\ell^2}\ell_a - \frac{6}{\ell} & k_{22} & \frac{12}{\ell^2}\ell_a + \frac{6}{\ell} & k_{24} \\ -\frac{12}{\ell^2} & \frac{12}{\ell^2}\ell_a + \frac{6}{\ell} & \frac{12}{\ell^2} & \frac{12}{\ell^2}\ell_b + \frac{6}{\ell} \\ -\frac{12}{\ell^2}\ell_b - \frac{6}{\ell} & k_{42} & \frac{12}{\ell^2}\ell_b + \frac{6}{\ell} & k_{44} \end{bmatrix}$$

$$k_{22} = 12\frac{\ell_a^2}{\ell^2} + 12\frac{\ell_a}{\ell} + 4 \qquad k_{42} = \frac{12}{\ell^2}\ell_a\ell_b + \frac{6}{\ell}\ell_b + \frac{6}{\ell}\ell_a + 2$$

$$k_{24} = \frac{12}{\ell^2}\ell_a\ell_b + \frac{6}{\ell}\ell_a + \frac{6}{\ell}\ell_b + 2 \qquad k_{44} = \frac{12}{\ell^2}\ell_b^2 + 12\frac{\ell_b}{\ell} + 4$$

5.37 Calculate the deflection at $x = 10$ ft of the beam of Fig. P5.1.

Answer: See Problem 5.1.

5.38 Find the displacement under the loads for the beam of Fig. P5.2.

Answer: See Problem 5.2.

5.39 Compute the bending moment and deflection at midspan of the beam of Fig. P5.3.

Frames

5.40 Determine the horizontal and vertical displacements of point d of the frame of Chapter 3, Fig. 3.6a. Assume numerical values as needed.

5.41 Compute the displacements of the free end of the cantilevered angle of Chapter 3, Fig. 3.7a. Express the answer in terms of symbols, e.g., lengths, $\overline{\mathbf{P}}$.

5.42 Find the force in the tie rod of the structural system of Chapter 3, Fig. 3.7b. Assign numerical values as needed.

5.43 Calculate the deflection at the free end of the beam of Chapter 3, Fig. 3.7b. Also find the reactions at the cantilevered end of the beam. Express your answer in terms of symbols, e.g., $E, I, \overline{P}$.

5.44 Find the displacements at the nodes of the frame of Chapter 3, Fig. 3.7e. Assume all members are uniform with the same E and I. Assign numerical values as needed for $\overline{P}_i$, E, I, and locations of applied loads.

5.45 A simple frame is loaded with a moment as shown in Fig. P5.45. List the element stiffness matrices in local and global coordinates. Find the displacements of the nodes. Also, find the global nodal forces, as well as the local element forces. Let $\overline{M} = 1000.0$ in. lb, $E = 1.0 \times 10^7$ lb/in.^{2}, $\nu = 0.3$, $\ell = 1$ in., $A = 1$ in.^{2} and $I = 1.0 \times 10^{-4}$ in.^{4}.

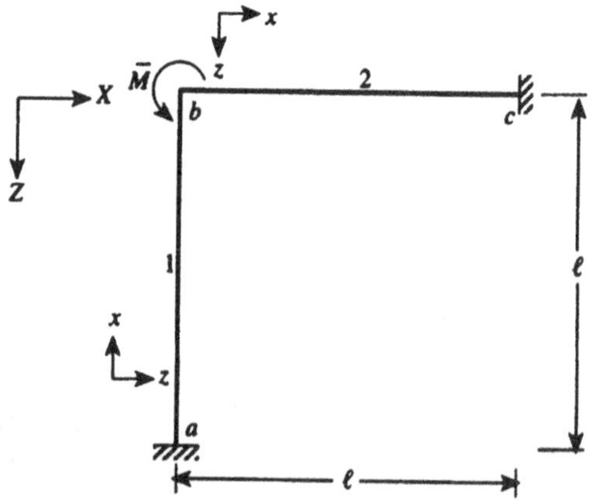

FIGURE P5.45
A plane frame.

Answer:
$U_{Xb} = -7.491909 \times 10^{-5}$ in., $U_{Zb} = 7.491909 \times 10^{-5}$ in., $\Theta_b = 1.254046 \times 10^{-1}$

Global Nodal Forces

Element 1 Element 2

$$\begin{bmatrix} P_{Xa} \\ P_{Za} \\ M_a \\ P_{Xb} \\ P_{Zb} \\ M_b \end{bmatrix} = \begin{bmatrix} -749.78 \\ -749.78 \\ 249.78 \\ 749.78 \\ 749.78 \\ 500.00 \end{bmatrix} \quad \begin{bmatrix} P_{Xb} \\ P_{Zb} \\ M_b \\ P_{Xc} \\ P_{Zc} \\ M_c \end{bmatrix} = \begin{bmatrix} -749.78 \\ -749.78 \\ 500.00 \\ 749.78 \\ 749.78 \\ 249.78 \end{bmatrix}$$

Local Element Forces

Element 1 Element 2

$$\begin{bmatrix} \widetilde{N}_a \\ \widetilde{V}_a \\ \widetilde{M}_a \\ \widetilde{N}_b \\ \widetilde{V}_b \\ \widetilde{M}_b \end{bmatrix} = \begin{bmatrix} 749.78 \\ -749.78 \\ 249.78 \\ -749.78 \\ 749.78 \\ 500.00 \end{bmatrix} \quad \begin{bmatrix} \widetilde{N}_b \\ \widetilde{V}_b \\ \widetilde{M}_b \\ \widetilde{N}_c \\ \widetilde{V}_c \\ \widetilde{M}_c \end{bmatrix} = \begin{bmatrix} -749.78 \\ -749.78 \\ 500.00 \\ 749.78 \\ 749.78 \\ 249.78 \end{bmatrix}$$

5.46 Find the displacements at the nodes of the frame of Fig. P5.46.

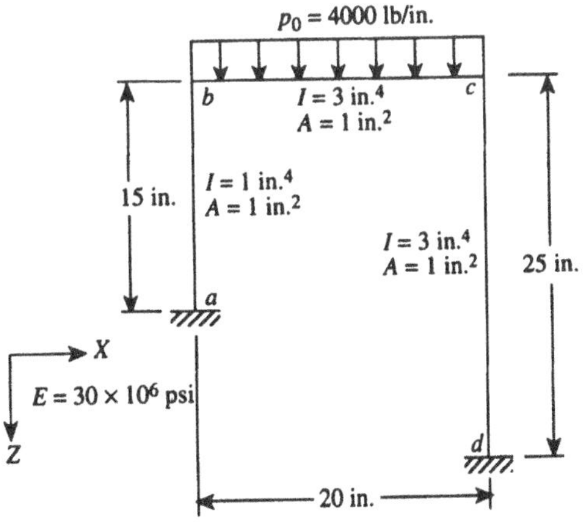

FIGURE P5.46
Frame.

Answer:

$$\mathbf{V} = [U_{Xb} \quad U_{Zb} \quad \theta_{Yb} \quad U_{Xc} \quad U_{Zc} \quad \theta_{Yc}]^T$$
$$= [0.01146 \quad 0.01924 \quad -0.00820 \quad 0.00790 \quad 0.03459 \quad 0.00554]^T$$

5.47 Find the displacement response of the simple framework of Fig. P5.47.

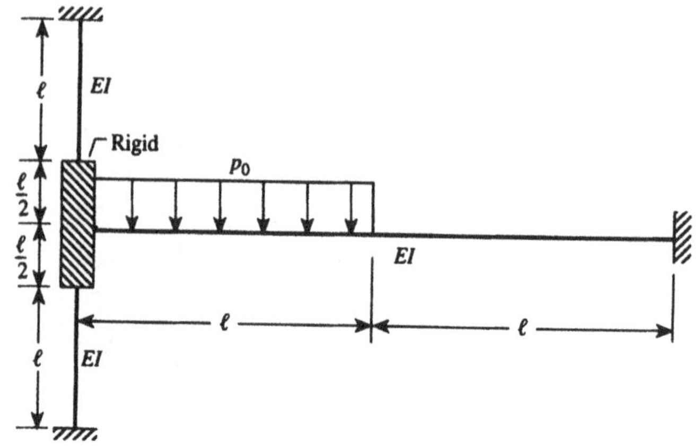

FIGURE P5.47
A plane frame.

5.48 Figure P5.48 shows a crane boom with a uniform beam supported by a tie bar. Consider a model with two DOF (two translations) at a and three DOF (two translations and one rotation) at b and c. Find the two element stiffness matrices and assemble the 8×8 unconstrained global stiffness matrix.

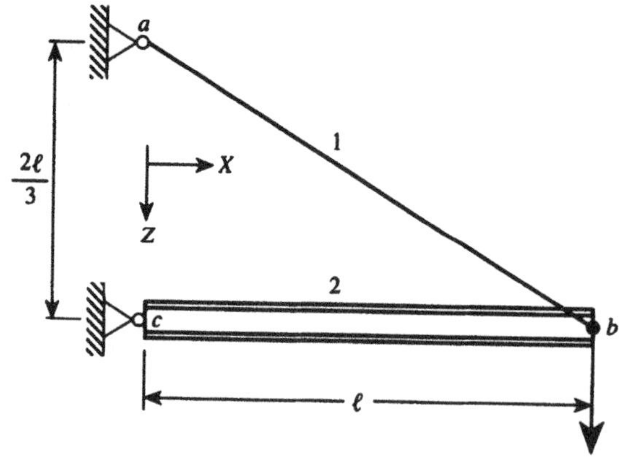

FIGURE P5.48
A crane boom.

5.49 Form the unassembled global stiffness matrix **k** for the structure of Fig. P5.48. Also, find the global kinematics matrix **a** and, finally, assemble the global matrix using $\mathbf{K} = \mathbf{a}^T\mathbf{k}\mathbf{a}$.

5.50 Compute the displacements of node d of the frame of Chapter 3, Fig. P3.11.

5.51 Calculate the displacements of points a and c of the frame of Chapter 3, Fig. P3.12.

5.52 Find the displacements of the free end of the frame of Chapter 3, Fig. P3.21.

5.53 Find the pressure between the two simply supported beams of Chapter 3, Fig. P3.29.

5.54 For the frame of Fig. P5.54, determine the distribution of deflection, moment, shear force, and axial force for the two loading cases given.

Answer: Partial results are displayed in Fig. P5.54.

$\bar{P}_z = 200$ kN

X, U_X

Θ

Z, U_Z

$\bar{P}_X = 50$ kN

x_2, u_2

z_3, w_3

z_2, w_2

x_3, u_3

5.0 m

$\bar{p}_z^1 = 20$ kN/m

x_1, u_1

$60°$

z_1, w_1

$\leftarrow$ 2.89 m $\rightarrow$ $\leftarrow$ 5.0 m $\rightarrow$

$E = 210\,000$ MN/m^2

Load Case 1: $\bar{p}_z^1 = 20$ kN/m, $\bar{P}_z = \bar{P}_X = 0$

Load Case 2: $\bar{p}_z^1 = 0$, $\bar{P}_Z = 200$ kN, $\bar{P}_X = 50$ kN

Bar 1 : $I = 0.00025$ m^4 $A = 0.015$ m^2

Bar 2 : $I = 0.00025$ m^4 $A = 0.015$ m^2

Bar 3 : $I = 0.00015$ m^4 $A = 0.012$ m^2

Results for load case 2 :

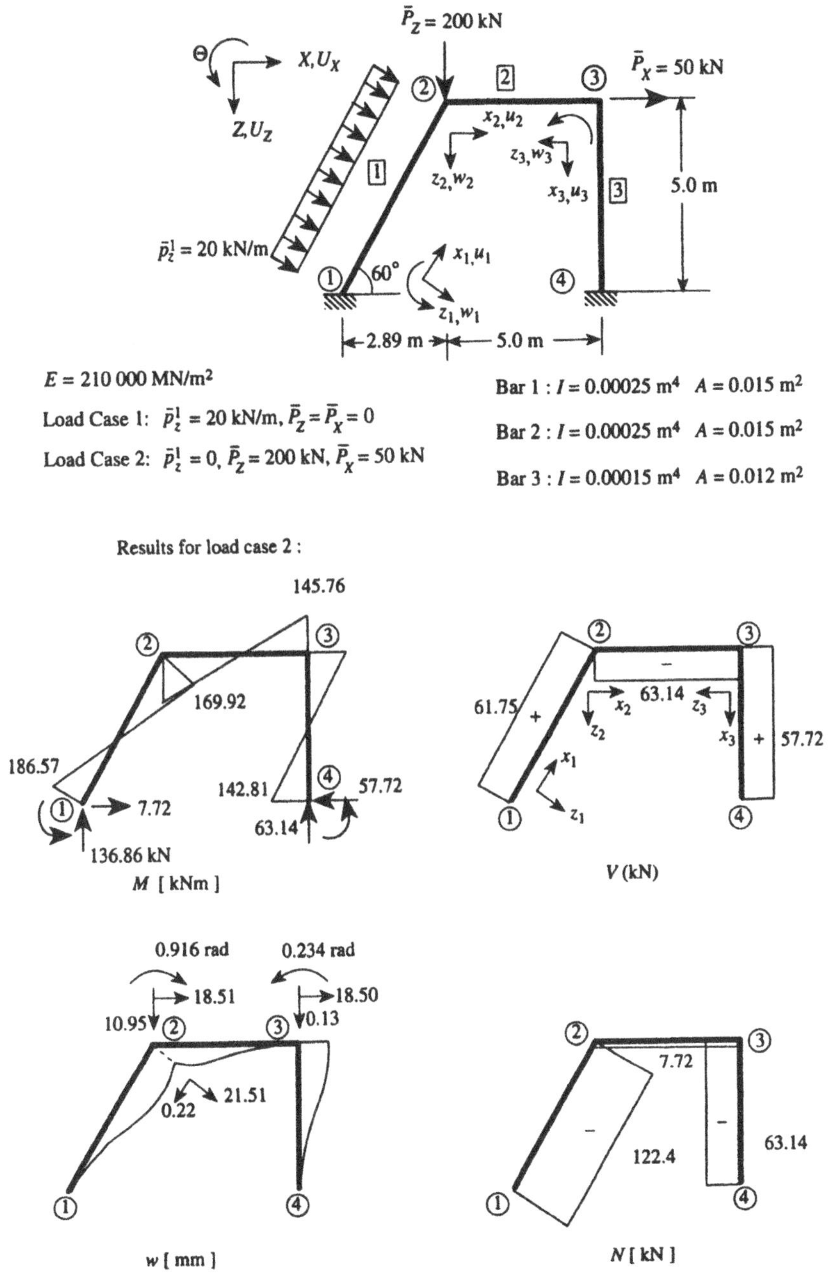

FIGURE P5.54

Force Method

5.55 Find the displacement of joint d of the space frame of Fig. P5.55. Let $\overline{M} = 1.13$ N·m and for each bar $E = 6.90$ GN/m^2, $I_y = I_z = 3.538$ cm^4, $I_x = 5.869$ cm^4, $v = 0.33$, $\ell = 0.254$ m.

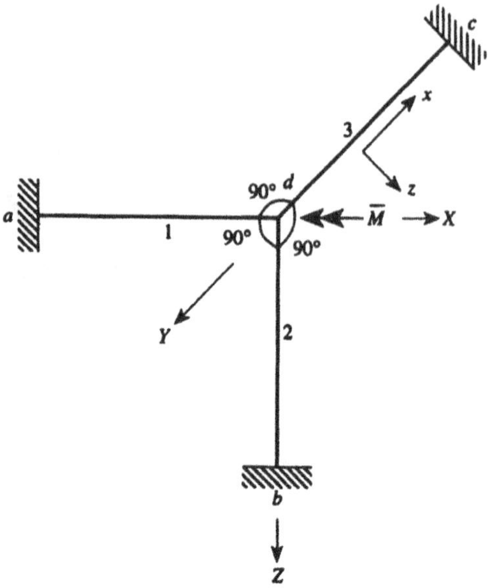

FIGURE P5.55
Frame response.

5.56 Perform a flexibility method analysis of the beam shown in Fig. P5.56.

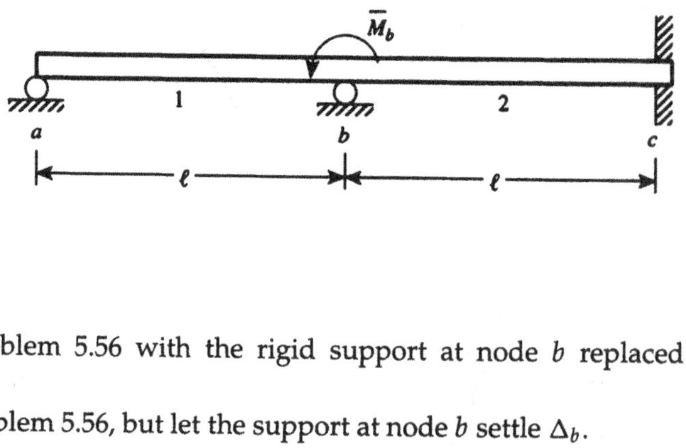

FIGURE P5.56
A continuous beam.

5.57 Repeat Problem 5.56 with the rigid support at node b replaced by a spring of constant k.

5.58 Repeat Problem 5.56, but let the support at node b settle Δ_b.

5.59 Use the force method to solve the beam problem of Example 5.8 with the internal moments at b and c chosen as the two redundants.

Hint: Use $\mathbf{v}_R^i = \mathbf{f}^i \mathbf{p}_R^i$, where $\mathbf{f}^i$ is from Chapter 4. Set up unassembled matrices $\mathbf{f}$ and $\mathbf{p}^T = [M_a^1 = 0 \quad M_b^1 \quad M_b^2 \quad M_c^2 \quad M_c^3 \quad M_d^3 = 0]^T$. Choose M_b^1 and M_c^2 as redundants. In $\mathbf{p} = \mathbf{b}_0\overline{\mathbf{P}} + \mathbf{b}_1\mathbf{P}_x$, $\overline{\mathbf{P}} = [\overline{M}_b \quad \overline{M}_c]^T$ and $\mathbf{P}_x = [M_b^1 \quad M_c^2]^T$. Setting $\mathbf{P}_x = [M_b^1 \quad M_c^2]^T = [0 \quad 0]^T$ is equivalent to adding hinges at b and c. Then

$$\mathbf{b}_0 = \begin{bmatrix} 0 & 0 & 1 & 0 & 0 & 0 \\ 0 & 0 & 0 & 0 & 1 & 0 \end{bmatrix}^T$$

$$\mathbf{b}_1 = \begin{bmatrix} 0 & 1 & -1 & 0 & 0 & 0 \\ 0 & 0 & 0 & 1 & -1 & 0 \end{bmatrix}^T$$

5.60 Solve the continuous beam problem of Example 5.8 using the reactions at a and d as redundants.

6

The Finite Element Method

The finite element method is the dominant computational tool of contemporary structural and solid mechanics. It permits difficult problems of complicated geometry to be solved with relative ease. A number of finite element-based general purpose computer programs for analysis and design are available to the engineer. These programs can be used to solve an array of difficult structural engineering problems. An introduction to finite element technology from the viewpoint of the structural mechanics philosophy of this book is presented in this chapter.

The finite element method has its origins in the matrix methods of structural analysis (Chapter 5). Some four decades ago it was recognized to be an emerging viable computational method for solid mechanics with applications to many problems of structural analysis. Since then, a mathematical basis for finite elements has been developed and applications have been expanded to include fluid mechanics, heat transfer, biomechanics, geomechanics, and acoustics. This ever-widening applicability of the finite element method is due in part to its common formulation based on the variational principles. Also, one-, two-, and three-dimensional problems are handled in like fashion with a uniform notation.

Technical histories of the finite element method are provided in Clough (1990), Felippa (1994), and Gupta and Meek (1996). Biographies of some of the pioneers in the development of finite element technology are available in Robinson (1985). The method can be considered as a natural evolution of the standard methods of structural mechanics for frames modeled as discrete elements, or as an approximate solution technique for continuum mechanics problems utilizing a regionally discretized model with assumed strain patterns for the regions. The concept of regional discretization can be traced to the much earlier work of Courant[1]

[1] Richard Courant (1888–1972) was a German-born mathematician who studied with Hilbert at Göttingen. Beginning with his doctoral thesis, he made several significant contributions to the calculus of variations. He authored several books, including *Methods of Mathematical Physics* with the first volume published in Germany and the second when Courant was working in the United States. Courant suffered through difficult economic times in Germany. His Prussian army service in World War I lasted more than 4 years. He was wounded early in the war, but remained in the army until 1918. After a brief stint as a politician, he took a position at Münster and then returned in 1920 to the University of Göttingen. With the support of the Prussian government and the Rockefeller Foundation, he established the Mathematics Institute at Göttingen. The interference of the National Socialist government led to a breakup of the mathematics "club" in Göttingen, with Courant moving to Cambridge University in 1933 and to New York University in 1934. In New York he collaborated with K.O. Friedrichs and J.J. Stoker, and established the Institute of Mathematical Sciences, which now bears his name.

(1943) and of Prager[2] and Synge[3] (1947). However, computers were not available then to implement the necessary extensive numerical operations.

According to Clough (1990), who was a summer employee of the Structural Dynamics Unit of Boeing Airplane Company in Seattle, WA, "by 1952 aircraft structural analysis had advanced to the point where a complex structure idealized as an assemblage of simple truss, beam, or shear panel elements could be analyzed by either the force or displacement method formulated as a series of matrix operations and using an automatic digital computer to carry out the calculations."

The fundamental breakthrough in the U.S. development occurred in the 1952–1953 winter when M.J. Turner, the head of Boeing's Structural Dynamics Unit, conceived of a novel model of panels for a wing. As described by Clough, "the essential idea in the proposed Turner procedure was that the deformations of any plane stress element be approximated by assuming a combination of simple strain fields acting within the element. The idea is applicable to both rectangular and triangular elements, but the use of triangular elements was given greater emphasis because an assemblage of triangular elements could serve to approximate plates of any shape. In modeling a triangular plate, the deformations were approximated by three constant strain fields: uniform normal strains in the x and y directions combined with a uniform x-y shear strain. Based on these strain patterns, the force-displacement relationships for the corner nodal points could be calculated using Castigliano's theorem, or the equivalent principle of virtual displacements." It was not until September 1956 that this 1953 work appeared in print (Turner, et al, 1956). The name *finite element method* was first used in Clough (1960).

During this development period, there were significant advances in the matrix analysis of structures. In particular, Argyris (1954 and 1955) published a series of articles organizing the matrix formulation of structural mechanics. The dual nature of the transformations of the force and displacement methods was identified, as were the relationships between the displacement method and the principle of virtual work and between the force method and the principle of complementary virtual work.

The finite element method serves as the analysis component of the design process (Fig. 6.1). This, of course, is the same roll played by the structural analyses of Chapter 5. As indicated in Fig. 6.1, the design of a system may involve repeated analyses and the introduction of design criteria. An important initial step is the idealization of the structure, resulting in a model that can be analyzed.

The structure is idealized into a model composed of a number of elements of finite size (Fig. 6.2). The model provides the name of the analysis technique. The connections between these so-called "finite elements" are point locations, i.e., nodes. The locations of the nodes are identified in the global coordinate system X, Y, Z. The distribution of displacements and forces are represented by values of these variables at the nodes, i.e., the finite element method involves a discretization process. The particular names given to the finite element method depend on the variables selected as unknowns at the nodes. As in Chapter 5, if the

---

[2]William Prager (1908–1980) was born in Karlsruhe, Germany, and received his doctorate in engineering sciences in Darmstadt in 1926. Between 1929 and 1932 he worked with Prandtl in Göttingen. He moved to the Technical Institute of Karlsruhe, only to be fired by Hitler. The next eight years were spent in Turkey. In 1941 he accepted a position at Brown University in the United States. The center of gravity of applied mechanics seemed to move to Brown University. It is said that Prager followed Einstein's philosophy of scientific work being "as simple as possible–but no simpler." He is credited with being influential in the introduction of more mathematics and research into US engineering schools.

[3]John Lighton Synge, born in 1897, was from an Anglo-Irish family that can be dated to the fifteenth century. The name Synge has been traced to Henry VIII saying to a choirboy "Synge, Millington, synge." He served on faculties of universities in Dublin, Toronto, and the United States. He was a visiting professor at Brown University in 1941. He is most famous for his geometrical insight into the theory of relativity.

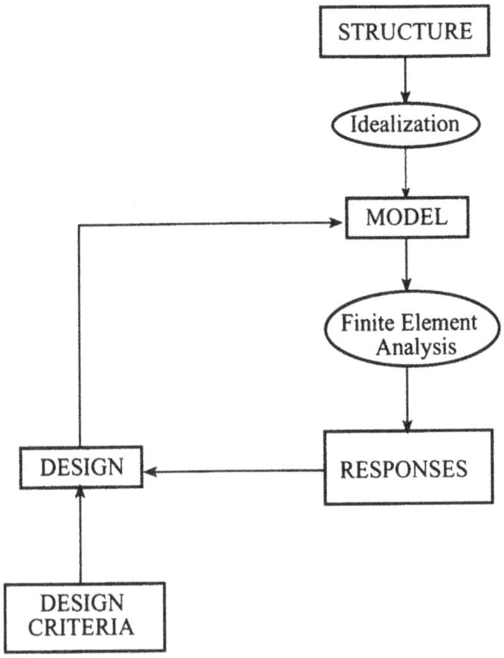

FIGURE 6.1
Design.

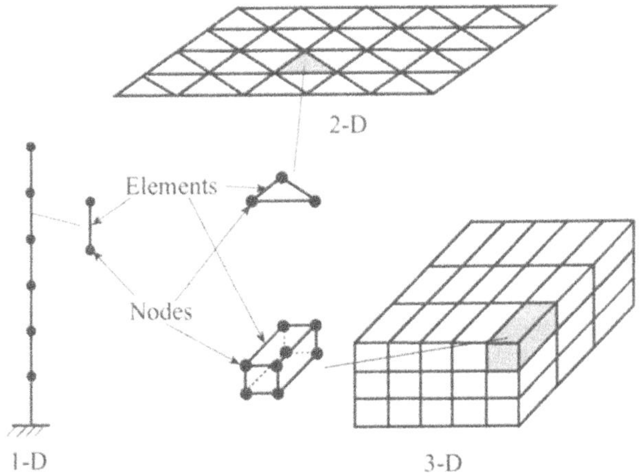

FIGURE 6.2
One-, two-, and three-dimensional systems.

nodal displacements are the unknowns, it is called the displacement method, and if nodal forces are the unknowns, it is the force method. For the mixed method, both forces and displacements are chosen as unknowns.

The finite element method is one of several variational methods (Stein and Wunderlich, 1973) that result in algebraic equations relating undetermined coefficients. Typically, mechanically significant parameters are used as the unknown coefficients. The response variables throughout the elements are related to the nodal unknowns with trial functions formed primarily of polynomials. Interpolation theory provides this relationship. The most essential characteristic of the method of finite elements is that the trial functions need not span the entire system; they apply only for an element. In the case of the displacement method, the displacements $\mathbf{u}^i$ in the ith element are related to the displacements $\mathbf{v}^i$ at the nodes by

$$\mathbf{u}^i = \mathbf{N}^i \mathbf{v}^i$$

where $\mathbf{N}^i$ are the trial or shape functions. Thus, the finite element method is a technique for solving boundary value problems in which the domain is subdivided into small elements over which the solution is approximated by interpolation.

In the finite element method, information characterizing the whole system, e.g., applied loading, is assembled in a form that is independent of the boundary conditions and other properties. This leads to a versatile method that can be used to solve efficiently classes of problems and that is well suited for solution using digital computers (Bathe, et al., 1977; Wunderlich, et al., 1981).

6.1 The Finite Element Method Based on the Displacement Method

The *displacement method* was shown in the previous chapter to be a method for the calculation of the forces and displacements at the nodes of a structure. In Chapter 5 this method was used to analyze structures formed of bar elements. In this chapter, we describe the finite element method, primarily the displacement formulation, beginning with structures modeled with elements of the sort shown in Fig. 6.2. The fundamental step is to utilize an approximate stiffness matrix to characterize an element. The structure is assumed to be modeled with a fine grid of many small elements in order to achieve an acceptably accurate solution in spite of the approximation involved in the stiffness matrix.

The most essential ingredient in a displacement analysis is the element stiffness matrix $\mathbf{k}^i$ which is a relationship between the forces and displacements at the nodes of the element. The equilibrium conditions for the forces at common nodes for a structure modeled with multiple elements leads to system (global) equations assembled by superposition of the element stiffness matrices for connected elements.

In Chapter 4, exact stiffness matrices $\mathbf{k}^i$ and loading vectors were derived for the extension, tension, and bending of simple bar elements. For most structural components it is not possible to find exact expressions for $\mathbf{k}^i$. This is usually accomplished by assuming that the displacements in the element can be replaced by a polynomial. In the case of bars, the exact displacements are already polynomials so the assumed polynomials can lead to exact stiffness matrices (Section 4.4.2). For most other elements the assumed polynomials generate approximate stiffness matrices. The solution procedure, when the approximate stiffness matrices are used, is referred to as the *finite element method*. Thus, the finite element method is a technique for solving boundary value problems in which the domain is subdivided into small elements over which the solution is approximated by polynomials.

As mentioned in Chapter 4, the assumed solution (polynomial) for a response is referred to as a *trial function*. The unknown quantities in the trial functions are calculated using equations established at the nodes. For the displacement method these are nodal equilibrium conditions. In general, the smaller the domain for which the trial function applies, the better the approximation.

In this chapter, we will consider the finite element method primarily as a displacement method, beginning with a summary of the displacement method as developed in Chapter 5.

In fact, the displacement method is only one of many solution methods in which "finite elements" can be utilized. In terms of commercially available general purpose analysis software, the displacement method is the most popular approach.

6.2 Summary of the Displacement Method

The displacement method as developed in Chapter 5 for bar structures is appropriate for the analysis of any structure modeled using any type of element such as those displayed in Fig. 6.2. A brief summary of the displacement method was given in Chapter 5, Section 5.3.9. We choose to expand on this summary here. An outline of the method follows:

1. From the system and loading, develop a model that is discretized into elements.
2. Determine for each element i the element stiffness matrix $\tilde{\mathbf{k}}^i$ and the loading vector $\tilde{\overline{\mathbf{p}}}^{i0}$ for distributed loads.
3. Transform the element properties from local to global coordinates

$$\mathbf{p}^i = \mathbf{T}^{iT}\tilde{\mathbf{p}}^i, \qquad \mathbf{k}^i = \mathbf{T}^{iT}\tilde{\mathbf{k}}^i\mathbf{T}^i, \qquad \overline{\mathbf{p}}^{i0} = \mathbf{T}^{iT}\tilde{\overline{\mathbf{p}}}^{i0} \tag{6.1}$$

   where $\mathbf{T}^i$ is the coordinate transformation matrix. In most of this chapter it is assumed that all local element forces, displacements, and stiffness matrices have been transformed to the global coordinate system.
4. Develop the system equations:
   - Establish the nodal displacements $\mathbf{V}$.
   - Develop the system equations, without consideration of the boundary conditions, by assembling the element stiffness matrices $\mathbf{k}^i$, leading to the global stiffness matrix $\mathbf{K}$. These equations represent the equations of equilibrium in the global coordinate system. If a consistent global node numbering system has been established, then the assembly is accomplished by summing element stiffness coefficients with identical subscripts. Thus, for example, for elements 1 and 2,

$$\mathbf{K}_{jk} = (\mathbf{k}^1_{jk} + \mathbf{k}^2_{jk}) \tag{6.2}$$

   - Develop the loading vector $\overline{\mathbf{P}}$ from the applied nodal forces $\overline{\mathbf{P}}^*$ and an assembled distributed load vector $\overline{\mathbf{P}}^0$.
   - Introduce the boundary conditions by eliminating appropriate columns and rows. Alternatively, introduce the boundary conditions during the assembly of the system matrices.
5. Solve the system of algebraic equations,

$$\mathbf{KV} = \overline{\mathbf{P}}^* + \overline{\mathbf{P}}^0 = \overline{\mathbf{P}} \tag{6.3}$$

   yielding the unknown displacements $\mathbf{V}$ of the nodes. These are in the global coordinate system.
6. Postprocess these displacements by computing the forces and displacements for the elements.

— Find the element displacements in the local coordinate system:

$$\tilde{\mathbf{v}}^i = \mathbf{T}^i \mathbf{v}^i \tag{6.4}$$

— Calculate the nodal element forces in the local coordinate system. Use either

$$\tilde{\mathbf{p}}^i = \tilde{\mathbf{k}}^i \tilde{\mathbf{v}}^i - \tilde{\overline{\mathbf{p}}}^{i0} \tag{6.5}$$

with $\tilde{\mathbf{v}}^i$ from Eq. (6.4) or

$$\mathbf{p}^i = \mathbf{k}^i \mathbf{v}^i - \overline{\mathbf{p}}^{i0} \quad \text{and} \quad \tilde{\mathbf{p}}^i = \mathbf{T}^i \mathbf{p}^i \tag{6.6}$$

— Compute the variables, such as cross-sectional forces and corresponding stresses, along the element.

— Display graphically the responses.

7. Controls for the calculations
   — Scrutinize the results for physical plausibility.
   — Check the overall system equilibrium.
   — Check the equilibrium at particular nodes.
   — Check the equilibrium and compatibility of particular elements.

6.3 A Simple Finite Element Calculation

We will consider an analysis of a beam lying on an elastic foundation as an initial illustration of the finite element method. We choose a simple problem that follows directly from the displacement method for frameworks of Chapter 5, yet contains the essential ingredients of the finite element method. For the simple Euler-Bernoulli beam of Chapter 4, use of a polynomial as the (assumed) trial function can lead to "exact" element matrices and solution. However, the same polynomial leads to approximate element matrices for a beam on an elastic foundation. The approximate element matrices for the beam on elastic foundation are developed in Example 4.4.

Summation of the virtual work for all of the elements leads to, upon assembly of the element matrices, the system equations for the displacement method. These equations are the nodal equilibrium conditions for the system. Let δW_k represent the work of the concentrated forces and moments applied to a system node. Then, introducing Eq. (13) of Example 4.4,

$$\begin{aligned} -\delta W &= \sum_{\textit{Elements}} -\delta W^i - \sum_{\textit{Nodes}} \delta W_k \\ &= \sum_{\textit{Elements}} \delta \mathbf{v}^{iT} \left[(\mathbf{k}_B^i + \mathbf{k}_w^i)\mathbf{v}^i - \overline{\mathbf{p}}^{i0} \right] - \delta \mathbf{V}^T \overline{\mathbf{P}}^* \\ &= \delta \mathbf{V}^T (\mathbf{K}\mathbf{V} - \overline{\mathbf{P}}^0 - \overline{\mathbf{P}}^*) = 0 \end{aligned} \tag{6.7}$$

or

$$\mathbf{K}\mathbf{V} = \overline{\mathbf{P}}^* + \overline{\mathbf{P}}^0 \tag{6.8}$$

where

$\mathbf{k}^i = \mathbf{k}^i_B + \mathbf{k}^i_w$ is the stiffness matrix for element i. $\mathbf{k}^i_B$ and $\mathbf{k}^i_w$ are given in Example 4.4.
$\mathbf{v}^i$ is the vector of displacements on the ends of the element i.
$\overline{\mathbf{p}}^{i0}$ is the nodal force vector due to distributed loading along the element.
$\overline{\mathbf{P}}^*$ is the assembled vector of concentrated forces and moments applied to the system nodes.
$\overline{\mathbf{P}}^0$ is the assembled vector representing the applied distributed loads $\overline{\mathbf{p}}^{i0}$.
$\mathbf{K}$ is the assembled system stiffness matrix.
$\mathbf{V}$ is the assembled vector of system nodal displacements.

That is,

$$\mathbf{V} = \begin{bmatrix} \mathbf{V}_1 & \mathbf{V}_2 & \cdots & \mathbf{V}_N \end{bmatrix}^T \tag{6.9a}$$

where N is the number of nodes. Also, for this beam

$$\mathbf{V}_k = \begin{bmatrix} U_{Zk} & \Theta_{Yk} \end{bmatrix}^T \tag{6.9b}$$

EXAMPLE 6.1 Numerical Example of a Beam on Elastic Foundation
Consider the beam on an elastic foundation, loaded with concentrated forces, as shown in Fig. 6.3. Let $E = 21\ \text{GN/m}^2$, $I = 1.08\ \text{m}^4$, $k_w = 320\ \text{MN/m}^2$.

With the approximate stiffness matrix for the beam element on an elastic foundation given by Eq. (17) of Example 4.4, the problem is solved using Eq. (6.8), following the displacement method. Here we choose to study the effect of refining the mesh (increasing the number of elements) on the accuracy of the solution. Recall that with the displacement method the displacement boundary conditions are applied to the expression $\mathbf{KV} = \overline{\mathbf{P}}^* + \overline{\mathbf{P}}^0$. However, for a beam element resting on an elastic foundation, the boundary conditions are on the forces (M and V) at the ends of the beam. That is, no displacement boundary conditions occur and Eq. (6.8) is solved without reduction. Since, the elastic foundation prevents rigid body motion, the system stiffness matrix K is not singular. At each node, there are two unknown displacements (degrees of freedom). These are the deflection w and slope θ.

The vector $\mathbf{V}$ is established for a straight beam that assures that the deflection w and slope $\theta = -w'$ are continuous (the same as in the case of a simple beam) at a node for each element connected at the node. However, nothing in the interpolation based finite element method of analysis assures that the moment M and shear force V are continuous at a node.

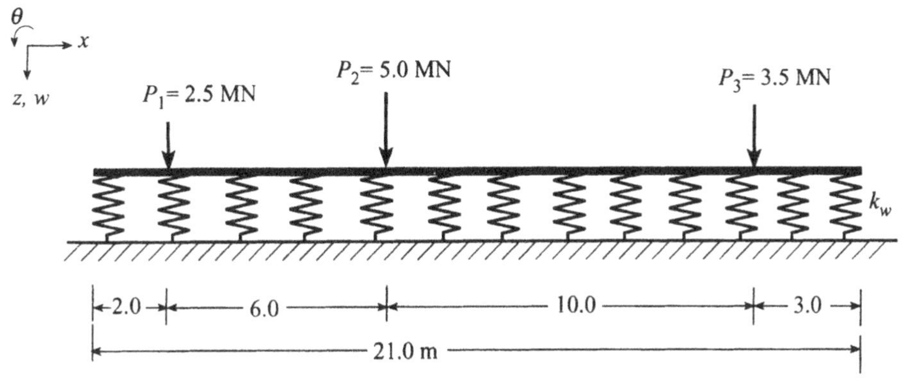

FIGURE 6.3
Beam on elastic foundation.

In fact, in general this method will lead to M and V that are not continuous across a node, even though for the actual beam M and V may be continuous.

It is assumed here that the deflection w for an element can be satisfactorily approximated by the cubic polynomial of Eq. (9) of Example 4.4

$$w(x) = C_1 + C_2x + C_3x^2 + C_4x^3 \tag{1}$$

whereas the exact deflection w (the complementary solution of Eq. (6) of Example 4.4) is

$$w(x) = C_1e^{\lambda x}\cos\lambda x + C_2e^{\lambda x}\sin\lambda x + C_3e^{-\lambda x}\cos\lambda x + C_4e^{-\lambda x}\sin\lambda x \tag{2}$$

where $\lambda = \sqrt[4]{k_w/4EI}$. Both of these representations of $w(x)$ are continuous. If the relationship $\theta(x) = -w'(x)$ is used to calculate $\theta(x)$, then (1) leads to

$$\theta(x) = -C_2 - 2C_3x - 3C_4x^2 \tag{3}$$

whereas the exact solution is derived from (2). The bending moment M calculated using $M(x) = -EIw''$ and (1) gives

$$M(x) = -2EIC_3 - 6EIC_4x \tag{4}$$

which is linear and clearly strays from the exact solution that appears similar to (2). The case of $V = -EI\frac{d^3w}{dx^3}$ and (1) provides a constant V

$$V(x) = -6EIC_4 \tag{5}$$

which obviously differs from the exact solution which is similar to (2). It should be clear from this discussion that there is cause for concern if forces (or stresses) are computed using the derivatives of the assumed approximate displacement.

For the numerical example of the beam on an elastic foundation we will study the solution for the three mesh refinements shown in Fig. 6.4. Begin with a comparison at the point of application of the force P_3. The moment M at the load P_3 can be calculated either using (Eq. 6.5) $\widetilde{\mathbf{p}}^i = \widetilde{\mathbf{k}}^i\widetilde{\mathbf{v}}^i - \overline{\widetilde{\mathbf{p}}}^{i0}$ or from Eqs. (2) and (3b) of Example 4.4 $M = -EIw''$. The results are shown in Fig. 6.5. It seems to be apparent that the solution improves as the number of elements increases. However, note the difference in the moment values for the two different calculation techniques.

Convergence by mesh refinement for the displacement method can only be assured for the total work (or energy) and for the nodal displacements. Remember that *the system equations fulfill the equilibrium conditions only at the nodes.* It is important to recognize that the conditions of equilibrium are satisfied only in an average sense.

It is of interest to study the distributions of the bending moment and shear force for the three mesh refinements.

Bending Moment Distributions

Figure 6.6 shows the bending moment distributions for mesh refinements A, B, and C, as well as for the exact solution taken from Hetenyi (1946). These moments for cases A, B, and C are determined from the nodal forces $\widetilde{\mathbf{p}}^i = \widetilde{\mathbf{k}}^i\widetilde{\mathbf{v}}^i - \overline{\widetilde{\mathbf{p}}}^{i0}$. Since this expression provides the moments only at the nodes, we choose to connect the nodes with straight lines, making the moments piecewise linear. More realistic moment distributions would be obtained by using some of the components of $\widetilde{\mathbf{p}}^i = \widetilde{\mathbf{k}}^i\widetilde{\mathbf{v}}^i - \overline{\widetilde{\mathbf{p}}}^{i0}$ as initial conditions and integrating the governing differential equations of Eqs. (1), (2), and (3) of Example 4.4.

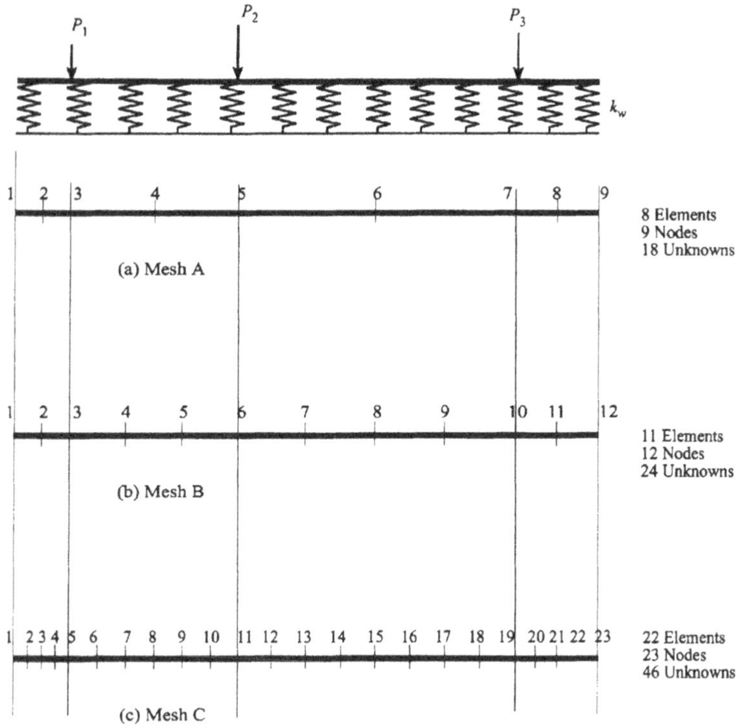

FIGURE 6.4
Three mesh refinements.

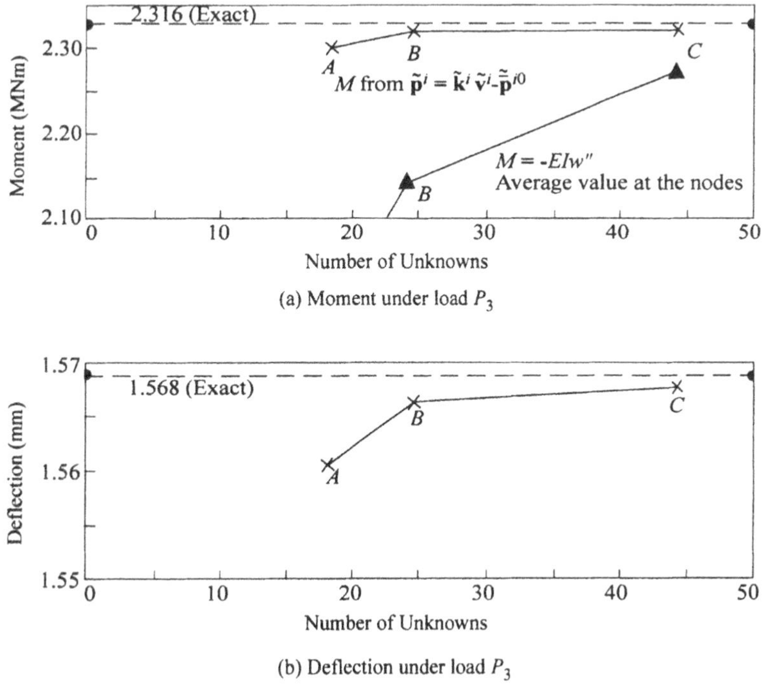

FIGURE 6.5
Responses for mesh refinements *A*, *B*, and *C* for beam on elastic foundation.

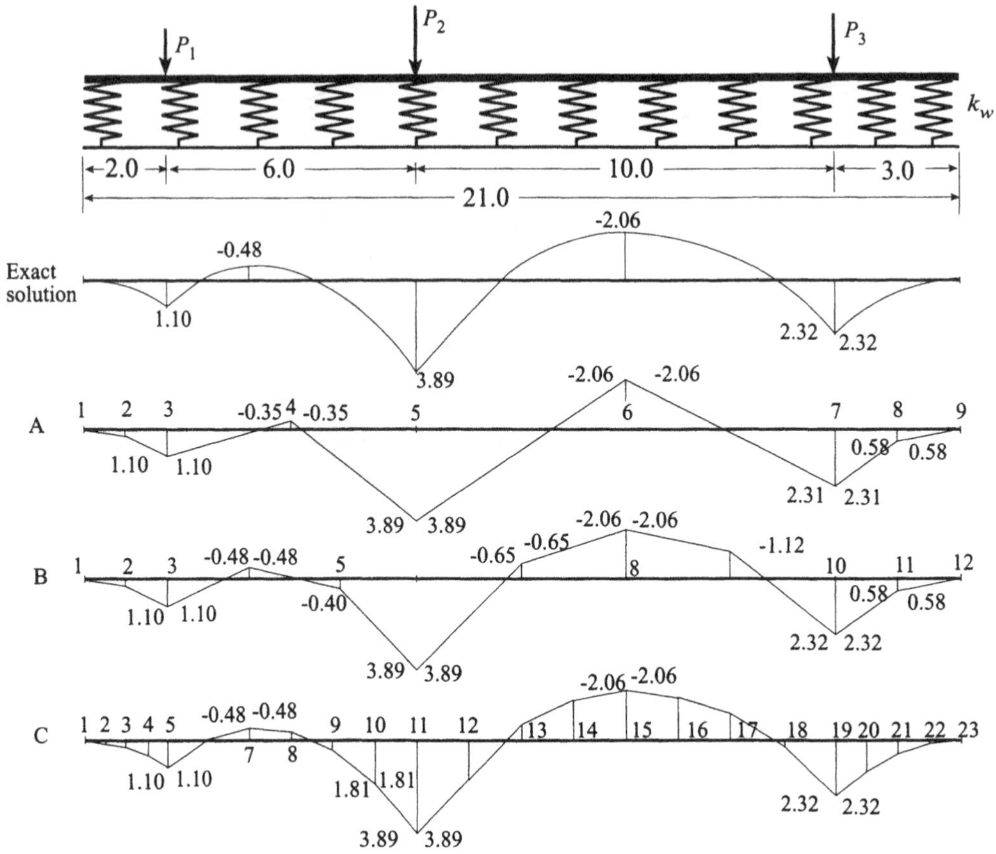

FIGURE 6.6
The exact moment distribution followed by the moment distributions for mesh refinements A, B, and C, using nodal forces $\widetilde{\mathbf{p}}^i = \widetilde{\mathbf{k}}^i\widetilde{\mathbf{v}}^i - \overline{\widetilde{\mathbf{p}}}^{i0}$. For A, B, and C, the nodal values are arbitrarily connected with straight lines.

Consider the case where the moment distributions are determined from the material law $M = -EIw''$. Since the second derivative of the assumed $w(x)$ leads to a linear expression, and only the continuity of w and θ are enforced at the nodes, a piecewise linear distribution of the moment with possible jumps in value at the nodes can occur. See Fig. 6.7. Relative to the case of nodal forces (Fig. 6.6), the approximation here, involving derivatives of the deflection, is not as accurate. Remember that the equilibrium conditions for the element are satisfied only in an average sense.

Shear Force Distributions

Figure 6.8 displays the shear force distributions for the exact solution as well as for the three mesh refinements determined from the nodal forces $\widetilde{\mathbf{p}}^i = \widetilde{\mathbf{k}}^i\widetilde{\mathbf{v}}^i - \overline{\widetilde{\mathbf{p}}}^{i0}$. The nodal values of V are arbitrarily linearly connected.

Shear forces distributions determined from the material law $V = -EIw'''$are shown in Fig. 6.9. The third derivative of the deflection assumption (cubic polynomial) is constant. The use of this method to find the shear force leads to a piecewise constant distribution of shear force between the nodes, whereas the exact solution is linear. Jumps can occur at the element boundaries and may be used as an indication of error level. The use of derivatives of the shape function results in a poorer approximation than that of Fig. 6.8. ∎

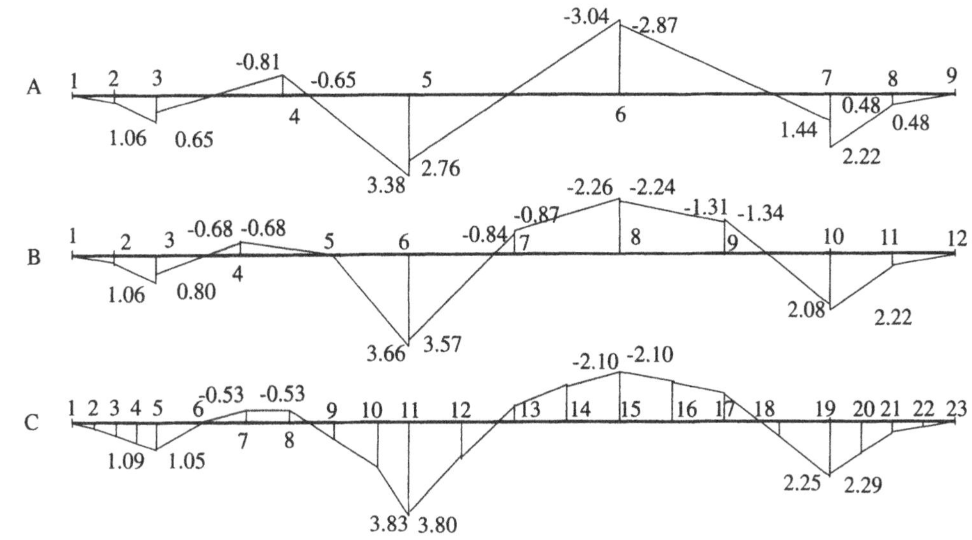

FIGURE 6.7
Moment distributions, for mesh refinements A, B, and C, using the material law $M = -EIw''$.

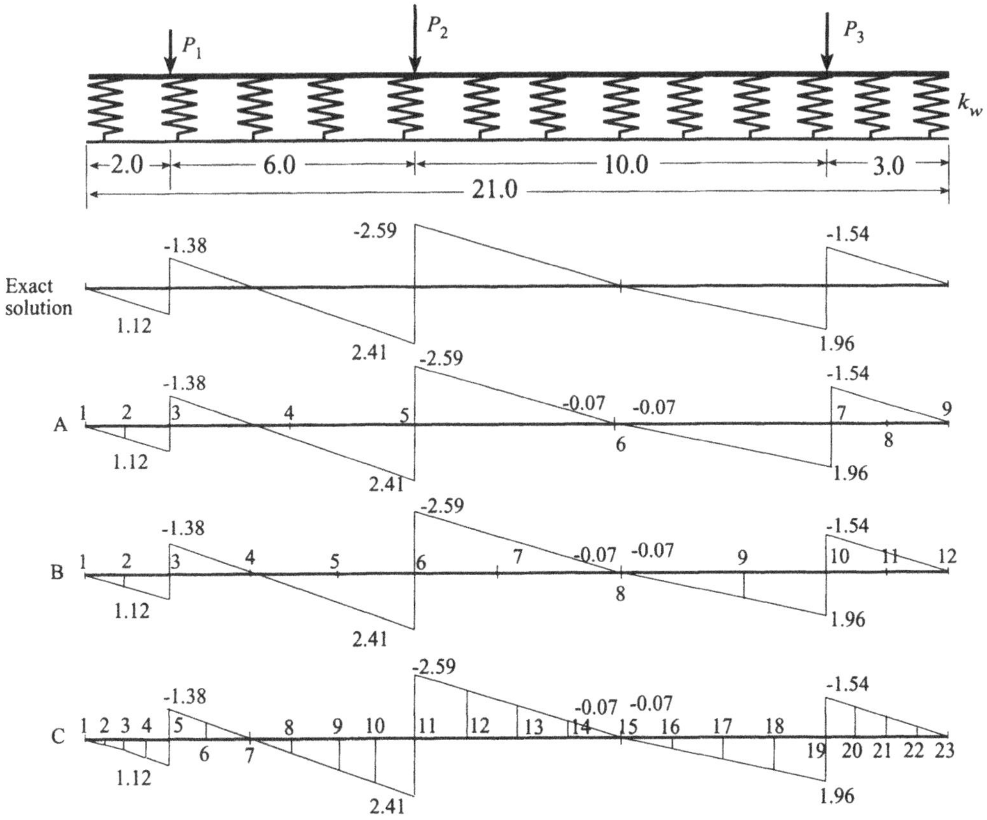

FIGURE 6.8
The exact distribution of shear force and the shear force distributions for mesh refinements A, B, and C, using nodal forces $\widetilde{\mathbf{p}}^i = \widetilde{\mathbf{k}}^i \widetilde{\mathbf{v}}^i - \widetilde{\overline{\mathbf{p}}}^{i0}$. The nodal values are arbitrarily connected with straight lines.

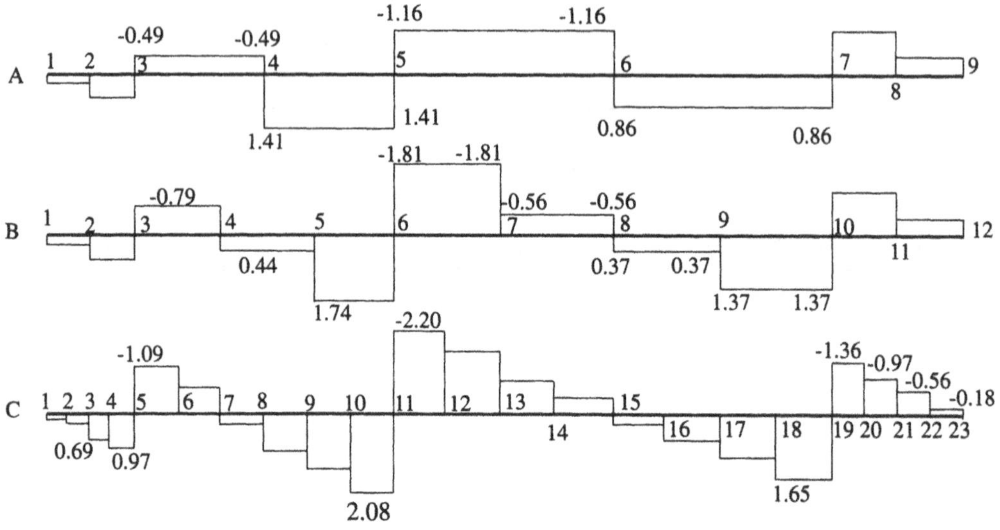

FIGURE 6.9
Shear force distributions for mesh refinements A, B, and C, using the material law $V = -EIw'''$.

6.4 Finite Element Method for Plane Problems

A two-dimensional structure provides the opportunity to describe the finite element method for problems of reasonably general geometry. We begin by establishing a 4-node rectangular element (Melosh, 1963), which will be used here as a fundamental building block for a multi-degree-of-freedom model of a structure.

6.4.1 Rectangular Element

Consider the rectangular element with four nodes and eight degrees of freedom shown in Fig. 6.10. To describe the displacement field in this two-dimensional element, assume dispacements are in the x, y plane, i.e., this is an in-plane problem, and represent the displacement $\mathbf{u}$ with the two rectilinear components u_x, u_y, i.e.,

$$\mathbf{u} = \begin{bmatrix} u_x(\xi, \eta) \\ u_y(\xi, \eta) \end{bmatrix} \tag{6.10}$$

Introduce a trial displacement solution (the trial function) $\mathbf{u}$ with eight unknowns, i.e., with eight degrees-of-freedom (DOF) (coordinates). Assume that a polynomial can adequately represent each of the displacements u_x and u_y. For the displacement in the x direction, choose the bilinear polynomial

$$u_x(\xi, \eta) = \mathbf{N}_{ux}\widehat{\mathbf{u}}_x = \widehat{u}_1 + \widehat{u}_2\xi + \widehat{u}_3\xi\eta + \widehat{u}_4\eta = \begin{bmatrix} 1 & \xi & \xi\eta & \eta \end{bmatrix} \begin{bmatrix} \widehat{u}_1 \\ \widehat{u}_2 \\ \widehat{u}_3 \\ \widehat{u}_4 \end{bmatrix} \tag{6.11}$$

where the origin of the coordinates ξ, η is at node 1. This polynomial contains the four unknowns, $\widehat{u}_1, \widehat{u}_2, \widehat{u}_3, \widehat{u}_4$. Similarly, the displacement in the y direction is chosen to have the

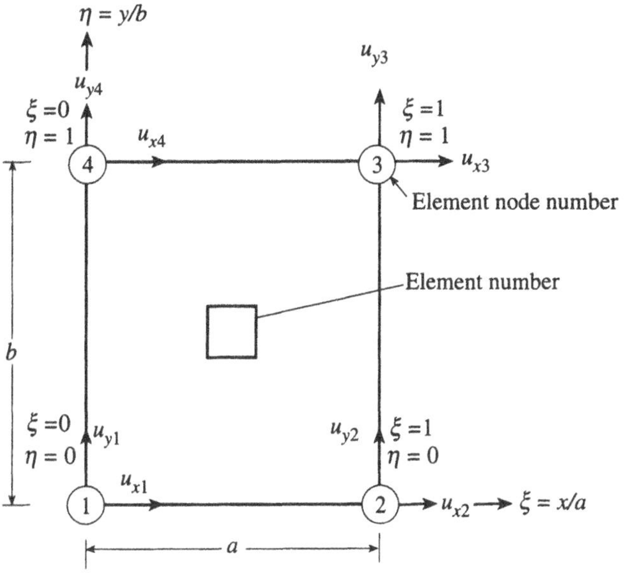

FIGURE 6.10
Notation for a rectangular plane element. Local coordinates are shown. The node are numbered counterclockwise.

form of the polynomial

$$u_y(\xi, \eta) = \mathbf{N}_{uy}\widehat{\mathbf{u}}_y = \widehat{u}_5 + \widehat{u}_6\xi + \widehat{u}_7\xi\eta + \widehat{u}_8\eta \tag{6.12}$$

Thus, the complete assumed displacements are

$$\mathbf{u} = \mathbf{N}_u\widehat{\mathbf{u}}$$

$$\begin{bmatrix} u_x \\ u_y \end{bmatrix} = \begin{bmatrix} 1 & \xi & \xi\eta & \eta & \vdots & & \mathbf{0} & & \\ & & \mathbf{0} & & \vdots & 1 & \xi & \xi\eta & \eta \end{bmatrix} \begin{bmatrix} \widehat{u}_1 \\ \vdots \\ \widehat{u}_8 \end{bmatrix} \tag{6.13}$$

This assumed distribution of displacements is called a bilinear approximation.

The unknown constants $\widehat{\mathbf{u}}$, i.e., $\widehat{u}_1, \widehat{u}_2, \cdots, \widehat{u}_8$, which have no direct mechanical significance are referred to as *generalized displacements*. With the finite element method, the generalized displacements are often replaced by mechanically meaningful unknowns. For this in-plane element, it is convenient to choose as unknowns the eight displacements $u_{x1}, u_{x2}, u_{x3}, u_{x4}$ and $u_{y1}, u_{y2}, u_{y3}, u_{y4}$ at the nodes of the rectangular element. To effect this replacement, begin at node 1 where $\xi = \eta = 0$. Consider only u_x, which at this node takes the value

$$u_{x1} = \begin{bmatrix} 1 & 0 & 0 & 0 \end{bmatrix} \begin{bmatrix} \widehat{u}_1 \\ \widehat{u}_2 \\ \widehat{u}_3 \\ \widehat{u}_4 \end{bmatrix} \tag{6.14}$$

where u_{x1} is the displacement of node 1 in the x direction. For node 2, use $\xi = 1,\ \eta = 0$. Let $\mathbf{v}_x$ be the vector of nodal displacements u_{x1}, u_{x2}, u_{x3}, and u_{x4}. If the relations for the four corners, e.g., Eq. (6.14), are assembled,

$$\mathbf{v}_x = \begin{bmatrix} u_{x1} \\ u_{x2} \\ u_{x3} \\ u_{x4} \end{bmatrix} = \begin{bmatrix} 1 & 0 & 0 & 0 \\ 1 & 1 & 0 & 0 \\ 1 & 1 & 1 & 1 \\ 1 & 0 & 0 & 1 \end{bmatrix} \begin{bmatrix} \widehat{u}_1 \\ \widehat{u}_2 \\ \widehat{u}_3 \\ \widehat{u}_4 \end{bmatrix} = \widehat{\mathbf{N}}_{ux}\widehat{\mathbf{u}}_x \tag{6.15}$$

The 4×4 matrix $\widehat{\mathbf{N}}_{ux}$ is readily inverted.

$$\widehat{\mathbf{N}}_{ux}^{-1} = \begin{bmatrix} 1 & 0 & 0 & 0 \\ -1 & 1 & 0 & 0 \\ 1 & -1 & 1 & -1 \\ -1 & 0 & 0 & 1 \end{bmatrix} = \mathbf{G}_x \tag{6.16}$$

Then

$$\widehat{\mathbf{u}}_x = \widehat{\mathbf{N}}_{ux}^{-1}\mathbf{v}_x = \mathbf{G}_x\mathbf{v}_x$$

and, from Eq. (6.11),

$$u_x(\xi, \eta) = \mathbf{N}_{ux}\mathbf{G}_x\mathbf{v}_x = \mathbf{N}_x\mathbf{v}_x \tag{6.17}$$

with

$$\underset{\mathbf{N}_{ux}}{[1 \quad \xi \quad \xi\eta \quad \eta]} \underset{\mathbf{G}_x(=\widehat{\mathbf{N}}_{ux}^{-1})}{\begin{bmatrix} 1 & 0 & 0 & 0 \\ -1 & 1 & 0 & 0 \\ 1 & -1 & 1 & -1 \\ -1 & 0 & 0 & 1 \end{bmatrix}} = \underset{\mathbf{N}_x}{[(1-\xi)(1-\eta) \quad \xi(1-\eta) \quad \eta\xi \quad \eta(1-\xi)]} \tag{6.18}$$

Thus, in Eq. (6.17) the assumed displacement u_x has been expressed in terms of the unknown nodal displacements.

The same manipulations for u_y lead to similar relationships. For the y coordinate, $\mathbf{N}_{uy}\mathbf{G}_y = \mathbf{N}_y$ can be found, where $\mathbf{N}_y(\xi, \eta) = \mathbf{N}_x(\xi, \eta)$. If the displacements u_x and u_y are placed together,

$$\mathbf{u} = \mathbf{u}^i = \begin{bmatrix} u_x \\ u_y \end{bmatrix} = \begin{bmatrix} \mathbf{N}_x & \mathbf{0} \\ \mathbf{0} & \mathbf{N}_y \end{bmatrix} \begin{bmatrix} \mathbf{v}_x \\ \mathbf{v}_y \end{bmatrix} = \mathbf{N}\mathbf{v} = \mathbf{N}\mathbf{v}^i = \mathbf{N}^i\mathbf{v}^i \tag{6.19a}$$

where superscript i has been included to indicate that this is the ith element, or

$$\underset{\mathbf{u}}{\begin{bmatrix} u_x \\ u_y \end{bmatrix}} = \underset{\mathbf{N}}{\begin{bmatrix} (1-\xi)(1-\eta) \quad \xi(1-\eta) \quad \xi\eta \quad \eta(1-\xi) & \mathbf{0} \\ \mathbf{0} & (1-\xi)(1-\eta) \quad \xi(1-\eta) \quad \xi\eta \quad \eta(1-\xi) \end{bmatrix}} \underset{\mathbf{v}^i}{\begin{bmatrix} u_{x1} \\ u_{x2} \\ u_{x3} \\ u_{x4} \\ \cdots \\ u_{y1} \\ u_{y2} \\ u_{y3} \\ u_{y4} \end{bmatrix}} \tag{6.19b}$$

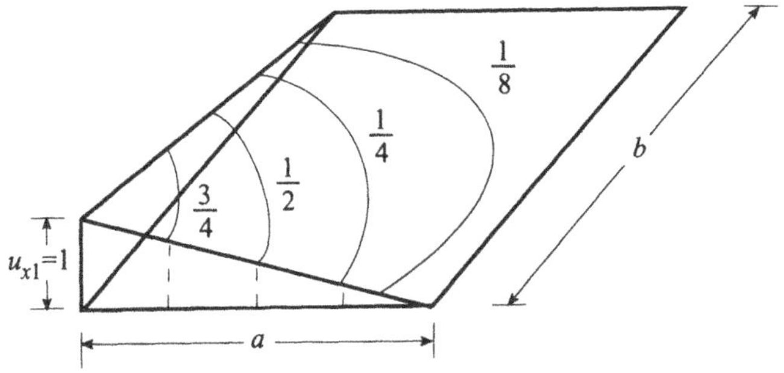

FIGURE 6.11
Plot of u_x for $u_{x1} = 1$ with all other nodal displacements equal to zero.

Often, this is expressed as

$$\mathbf{u} = \begin{bmatrix} N_1 & N_2 & N_3 & N_4 & & & \mathbf{0} & \\ & & \mathbf{0} & & N_1 & N_2 & N_3 & N_4 \end{bmatrix} \mathbf{v}^i \tag{6.19c}$$

The displacements in each element of the model are to be represented by these rather simple polynomials. These assumed displacements contain eight DOF with two DOF per node.

This trial solution matrix **N** has "interpolation" polynomials as components. As can be seen in Eq. (6.19b), the components of **N** represent the displacement across the element due to a unit nodal displacement. Thus, a plot of in-plane displacement **u**, drawn perpendicular to the plane of the element, would appear for the nodal displacement u_{x1} as shown in Fig. 6.11. The curved lines connect points of the element for which the u_{x1} component of **N** has the same value. It is important to note that this procedure of establishing **N** can be avoided if the displacements due to unit nodal displacements are known beforehand. Thus, for a variety of elements the inverse required to form $\widehat{\mathbf{N}}_{ux}^{-1}$ can be avoided if standard interpolation polynomials such as Hermitian or Lagrangian interpolation polynomials are introduced.

Principle of Virtual Work

The stiffness matrix for an element will be derived by substituting the assumed displacement field of Eq. (6.19) into the principle of virtual work (form C of Chapter 2, Section 2.3). In matrix notation, the principle of virtual work can be expressed as

$$-\delta W = \overbrace{\underbrace{\int_V \delta\boldsymbol{\epsilon}^T \boldsymbol{\sigma}\, dV}_{\text{I}}}^{-\delta W_i} \overbrace{- \underbrace{\int_V \delta\mathbf{u}^T \overline{\mathbf{p}}_V\, dV}_{\text{II}} - \underbrace{\int_{S_p} \delta\mathbf{u}^T \overline{\mathbf{p}}\, dS}_{\text{III}}}^{-\delta W_e} = 0 \tag{6.20}$$

Equation (6.20) is valid for the entire structural system. In order to set up a computational method in which the system is subdivided into elements, Eq. (6.20) has to be evaluated for each element. The total virtual work will be obtained as the sum of the internal and external virtual work, each a scalar quantity, for all the elements.

Suppose there is only surface loading so that integral II need not be considered. The surface integral III applies for the element borders, as well as the free surfaces to which loads are applied. Inner or interface boundaries lie between two adjoining elements. Care must be taken in carrying out the surface integral to see that the integration along the element interfaces, which seem to cover the same surface twice, leads to the proper single contribution to the virtual work. No question arises for problems in which the loading is applied only on "outer" boundaries.

In terms of M elements, the virtual work relationship can be expressed as

$$-\delta W = \sum_{i=1}^{M} \left[\underbrace{\int_V \delta\boldsymbol{\epsilon}^T \boldsymbol{\sigma}\, dV}_{\text{I}} - \underbrace{\int_{S_p} \delta\mathbf{u}^T \overline{\mathbf{p}}\, dS}_{\text{III}} \right]^i = 0 \tag{6.21}$$

The variables in this expression are to be written in terms of displacements.

Recall from Chapter 2 that the principle of virtual work requires that the displacements, in this case the assumed displacement field, must be kinematically admissible. Thus, the displacements must satisfy $\boldsymbol{\epsilon} = \mathbf{D}\mathbf{u}$ in V and the displacement boundary conditions $\mathbf{u} = \overline{\mathbf{u}}$ on S_u.

Kinematics

The strains are obtained from the displacement function $\mathbf{u}$ using the kinematic relation

$$\boldsymbol{\epsilon} = \mathbf{D}\mathbf{u} \tag{6.22}$$

where, for the ith element,

$$\mathbf{u} = \begin{bmatrix} u_x \\ u_y \end{bmatrix} = \mathbf{N}\mathbf{v}^i \tag{6.23}$$

For this in-plane case [Chapter 1, Eq. (1.24)], the operator matrix is

$$\mathbf{D} = \mathbf{D}_u = \begin{bmatrix} \partial_x & 0 \\ 0 & \partial_y \\ \partial_y & \partial_x \end{bmatrix}$$

where, since $x = a\xi,\ y = b\eta$,

$$\begin{aligned} \partial_x &= \frac{\partial}{\partial x} = \frac{1}{a}\frac{\partial}{\partial \xi} = \frac{1}{a}\partial_\xi \\ \partial_y &= \frac{\partial}{\partial y} = \frac{1}{b}\frac{\partial}{\partial \eta} = \frac{1}{b}\partial_\eta \end{aligned} \tag{6.24}$$

The strains in terms of the trial functions and nodal unknowns appear as

$$\begin{bmatrix} \epsilon_x \\ \epsilon_y \\ \gamma_{xy} \end{bmatrix} = \begin{bmatrix} \partial_x & 0 \\ 0 & \partial_y \\ \partial_y & \partial_x \end{bmatrix} \begin{bmatrix} u_x \\ \\ u_y \end{bmatrix} = \mathbf{D}\mathbf{N}\mathbf{v}^i = \mathbf{B}\mathbf{v}^i \tag{6.25}$$

where **B** is obtained by applying the derivatives in **D** to the trial functions **N**

$$\underbrace{\begin{bmatrix} \frac{1}{a}\partial_\xi & 0 \\ 0 & \frac{1}{b}\partial_\eta \\ \frac{1}{b}\partial_\eta & \frac{1}{a}\partial_\xi \end{bmatrix}}_{\mathbf{D}} \underbrace{\begin{bmatrix} (1-\xi)(1-\eta) & \xi(1-\eta) & \xi\eta & \eta(1-\xi) & 0 & 0 & 0 & 0 \\ 0 & 0 & 0 & 0 & (1-\xi)(1-\eta) & \xi(1-\eta) & \xi\eta & \eta(1-\xi) \end{bmatrix}}_{\mathbf{N}}$$

$$= \underbrace{\begin{bmatrix} -\frac{1-\eta}{a} & \frac{1-\eta}{a} & \frac{\eta}{a} & -\frac{\eta}{a} & 0 & 0 & 0 & 0 \\ 0 & 0 & 0 & 0 & -\frac{1-\xi}{b} & -\frac{\xi}{b} & \frac{\xi}{b} & \frac{1-\xi}{b} \\ -\frac{1-\xi}{b} & -\frac{\xi}{b} & \frac{\xi}{b} & \frac{1-\xi}{b} & -\frac{1-\eta}{a} & \frac{1-\eta}{a} & \frac{\eta}{a} & -\frac{\eta}{a} \end{bmatrix}}_{\mathbf{B}} \tag{6.26}$$

Material Law

Equation (6.25) is the strain-displacement relationship needed in integral I of Eq. (6.21). Also, in order to express the stress in terms of displacements in integral I, a stress-strain relationship, i.e., the material law, is necessary. Assume a plane stress condition for this case of an in-plane problem. Then [Chapter 1, Eq. (1.39a)] the material law is

$$\underbrace{\begin{bmatrix} \sigma_x \\ \sigma_y \\ \tau_{xy} \end{bmatrix}}_{\boldsymbol{\sigma}} = \frac{E}{1-\nu^2} \underbrace{\begin{bmatrix} 1 & \nu & 0 \\ \nu & 1 & 0 \\ 0 & 0 & \frac{1-\nu}{2} \end{bmatrix}}_{\mathbf{E}} \underbrace{\begin{bmatrix} \epsilon_x \\ \epsilon_y \\ \gamma_{xy} \end{bmatrix}}_{\boldsymbol{\epsilon}} \tag{6.27}$$

Formation of the Element Stiffness Matrix

The material law, the kinematics, and the trial solution can be introduced into Eq. (6.21) to express the principle of virtual work in terms of the nodal displacements. For a single element the volume integral I becomes

$$\begin{aligned} \int_V \delta\epsilon^T \sigma\, dV &= \int_V \delta\epsilon^T \mathbf{E}\,\epsilon\, dV = \int_V \delta(\mathbf{D}_u\,\mathbf{u})^T \mathbf{E}\,\mathbf{D}_u\,\mathbf{u}\, dV \\ &= \int_V \delta\mathbf{u}^T\,{}_u\mathbf{D}^T \mathbf{E}\,\mathbf{D}_u\,\mathbf{u}\, dV = \int_V \delta\mathbf{u}^T \mathbf{k}^D \mathbf{u}\, dV \end{aligned} \tag{6.28}$$

where $\epsilon = \mathbf{D}_u\,\mathbf{u}$ and the operator notation $\mathbf{k}^D$ of Chapter 2 have been introduced. The subscript index to the left of the operator matrix **D** designates that the operator is applied to the preceding quantity which, in this case, is $\delta\mathbf{u}^T$. The element stiffness matrix can be obtained from this expression.

Consider next the details of the development of the element stiffness matrix $\mathbf{k}^i$. For a single element, it follows from Eqs. (6.28) that the contribution of the volume integral I of Eq. (6.21) can be expressed as

$$\int_V \delta\epsilon^T \sigma\, dV = \int_V \delta\mathbf{u}^T \mathbf{k}^D \mathbf{u}\, dV \tag{6.29}$$

Next, the integral containing $\mathbf{k}^D$ needs to be expressed in discrete form. For this two-dimensional problem, let t be the constant thickness of the element and $dV = t\,dx\,dy = t\,a\,b\,d\xi\,d\eta$. The contribution to the volume integral for a single element is obtained by inserting $\mathbf{u} = \mathbf{N}\mathbf{v}^i$ (alternatively expressed as $\mathbf{u}^i = \mathbf{N}^i\mathbf{v}^i$) in Eq. (6.29)

$$\int_V \delta\mathbf{u}^T\mathbf{k}^D\mathbf{u}\,dV = \int_V \delta\mathbf{u}^T{}_u\mathbf{D}^T\mathbf{E}\,\mathbf{D}_u\,\mathbf{u}\,dV = \delta\mathbf{v}^{iT}\underbrace{\int_V \mathbf{N}^T{}_u\mathbf{D}^T\mathbf{E}\,\mathbf{D}_u\,\mathbf{N}\,dV}_{\mathbf{k}^i}\mathbf{v}^i$$

$$= \delta\mathbf{v}^{iT}\underbrace{\int_V (\mathbf{D}_u\mathbf{N})^T\mathbf{E}\,\mathbf{D}_u\mathbf{N}\,dV}_{\mathbf{k}^i}\mathbf{v}^i = \delta\mathbf{v}^{iT}\underbrace{abt\int_0^1\int_0^1 \mathbf{B}^T\mathbf{E}\,\mathbf{B}\,d\xi\,d\eta}_{\text{Stiffness matrix } \mathbf{k}^i}\mathbf{v}^i \tag{6.30}$$

where the element stiffness matrix $\mathbf{k}^i$ is identified in Section 4.4.2 and is equal to

$$\mathbf{k}^i = \int_V \mathbf{B}^T\mathbf{E}\,\mathbf{B}\,dV = abt\int_0^1\int_0^1 \mathbf{B}^T\mathbf{E}\,\mathbf{B}\,d\xi\,d\eta \tag{6.31}$$

Details of the evaluation of the stiffness matrix, for this plane stress element, according to Eq. (6.31) follow. Begin with the integrand $\mathbf{B}^T\mathbf{E}\,\mathbf{B}$.

$$\frac{E}{1-\nu^2}\underbrace{\begin{bmatrix} 1 & \nu & 0 \\ \nu & 1 & 0 \\ 0 & 0 & \frac{1-\nu}{2} \end{bmatrix}}_{\mathbf{E}}\underbrace{\begin{bmatrix} -\frac{1-\eta}{a} & \frac{1-\eta}{a} & \frac{\eta}{a} & -\frac{\eta}{a} & 0 & 0 & 0 & 0 \\ 0 & 0 & 0 & 0 & -\frac{1-\xi}{b} & -\frac{\xi}{b} & \frac{\xi}{b} & \frac{1-\xi}{b} \\ -\frac{1-\xi}{b} & -\frac{\xi}{b} & \frac{\xi}{b} & \frac{1-\xi}{b} & -\frac{1-\eta}{a} & \frac{1-\eta}{a} & \frac{\eta}{a} & -\frac{\eta}{a} \end{bmatrix}}_{\mathbf{B}}$$

$$= \underbrace{\begin{bmatrix} -\frac{1-\eta}{a} & \frac{1-\eta}{a} & \frac{\eta}{a} & -\frac{\eta}{a} & -\frac{\nu(1-\xi)}{b} & -\frac{\nu\xi}{b} & \frac{\nu\xi}{b} & \frac{\nu(1-\xi)}{b} \\ -\frac{\nu(1-\eta)}{a} & \frac{\nu(1-\eta)}{a} & \frac{\nu\eta}{a} & -\frac{\nu\eta}{a} & -\frac{1-\xi}{b} & -\frac{\xi}{b} & \frac{\xi}{b} & \frac{1-\xi}{b} \\ -\frac{1-\nu}{2}\frac{1-\xi}{b} & -\frac{1-\nu}{2}\frac{\xi}{b} & \frac{1-\nu}{2}\frac{\xi}{b} & \frac{1-\nu}{2}\frac{1-\xi}{b} & -\frac{1-\nu}{2}\frac{1-\eta}{a} & \frac{1-\nu}{2}\frac{1-\eta}{a} & \frac{1-\nu}{2}\frac{\eta}{a} & -\frac{1-\nu}{2}\frac{\eta}{a} \end{bmatrix}}_{\mathbf{E}\,\mathbf{B}}\cdot\frac{E}{1-\nu^2}$$

$$\mathbf{B}^T\mathbf{E}\,\mathbf{B} = \begin{bmatrix} \frac{(1-\eta)^2}{a^2} + \frac{1-\nu}{2}\frac{(1-\xi)^2}{b^2} & \vdots & -\frac{(1-\eta)^2}{a^2} + \frac{1-\nu}{2}\frac{\xi-\xi^2}{b^2} & \vdots & -\frac{(\eta-\eta^2)}{a^2} - \frac{1-\nu}{2}\frac{\xi-\xi^2}{b^2} & \vdots & \frac{(\eta-\eta^2)}{a^2} - \frac{1-\nu}{2}\left(\frac{1-\xi}{b}\right)^2 & \vdots & \text{The remaining entries are calculated in the same fashion} \\ \cdots\cdots & & \cdots\cdots & & \cdots\cdots & & \cdots\cdots & & \\ -\frac{(1-\eta)^2}{a^2} + \frac{1-\nu}{2}\frac{\xi-\xi^2}{b^2} & \vdots & \frac{(1-\eta)^2}{a^2} + \frac{1-\nu}{2}\left(\frac{\xi}{b}\right)^2 & \vdots & \frac{(\eta-\eta^2)}{a^2} - \frac{1-\nu}{2}\frac{\xi^2}{b^2} & \vdots & -\frac{(\eta-\eta^2)}{a^2} - \frac{1-\nu}{2}\frac{\xi-\xi^2}{b^2} & \vdots & \\ \cdots\cdots & & \cdots\cdots & & \cdots\cdots & & \cdots\cdots & & \\ & \vdots & & \vdots & \frac{\eta^2}{a^2} + \frac{1-\nu}{2}\frac{\xi^2}{b^2} & \vdots & -\frac{\eta^2}{a^2} + \frac{1-\nu}{2}\frac{\xi-\xi^2}{b^2} & \vdots & \\ \cdots\cdots & & \cdots\cdots & & \cdots\cdots & & \cdots\cdots & & \\ & & & & & \vdots & \frac{\eta^2}{a^2} + \frac{1-\nu}{2}\left(\frac{1-\xi}{b}\right)^2 & \vdots & \\ & & \text{Symmetric} & & & & \cdots\cdots & & \end{bmatrix}\frac{E}{1-\nu^2} \tag{6.32}$$

Next, the integration indicated in Eq. (6.31) over $d\xi$ and $d\eta$ must be carried out for all entries of Eq. (6.32). For example, the term in the first row, second column of $\mathbf{k}^i$ becomes

$$k_{12}^i = \frac{Eabt}{(1-\nu^2)}\int_0^1\int_0^1\left(-\frac{1-2\eta+\eta^2}{a^2} + \frac{1-\nu}{2}\frac{\xi-\xi^2}{b^2}\right)d\xi\,d\eta$$

where the appropriate entry of $\mathbf{B}^T\mathbf{E}\,\mathbf{B}$ of Eq. (6.32) has been employed. Integration first over $d\xi$ yields

$$k_{12}^i = \frac{Eabt}{(1-\nu^2)}\int_0^1\left(-\frac{1-2\eta+\eta^2}{a^2}+\frac{1-\nu}{2}\frac{\frac{1}{2}-\frac{1}{3}}{b^2}\right)d\eta$$
$$= \frac{Eabt}{(1-\nu^2)}\int_0^1\left(-\frac{1-2\eta+\eta^2}{a^2}+\frac{1-\nu}{12b^2}\right)d\eta$$

Now integrate over $d\eta$, giving

$$k_{12}^i = \frac{Eabt}{(1-\nu^2)}\left(-\frac{1-1+\frac{1}{3}}{a^2}+\frac{1-\nu}{12b^2}\right) = \frac{Eabt}{(1-\nu^2)}\left(-\frac{1}{3a^2}+\frac{1-\nu}{12b^2}\right)$$

Finally, we find

$$k_{12}^i = \frac{Et}{24(1-\nu^2)}\left[-8\frac{b}{a}+2(1-\nu)\frac{a}{b}\right] = \frac{Et}{24(1-\nu^2)}[-8\alpha+2\beta(1-\nu)] \tag{6.33}$$

where

$$\alpha = b/a, \qquad \beta = a/b$$

The other stiffness coefficients are obtained in the same fashion.

For this case of rectangular elements with eight DOF, closed form integration can be performed. In general, if more nodes are introduced to the rectangular elements or for elements for other shapes with many DOF, numerical integration is usually employed.

Now, for the element of Fig. 6.10 put together all 64 coefficients using the symmetry of the stiffness matrix.

$$\mathbf{k}^i\mathbf{v}^i = \frac{Et}{24(1-\nu^2)}$$
$$\times\begin{bmatrix} A_{\alpha\beta} & C_{\alpha\beta} & -A_{\alpha\beta}/2 & B_{\alpha\beta} & \nu_2 & -\nu_3 & -\nu_2 & \nu_3 \\ & A_{\alpha\beta} & B_{\alpha\beta} & -A_{\alpha\beta}/2 & \nu_3 & -\nu_2 & -\nu_3 & \nu_2 \\ & & A_{\alpha\beta} & C_{\alpha\beta} & -\nu_2 & \nu_3 & \nu_2 & -\nu_3 \\ & & & A_{\alpha\beta} & -\nu_3 & \nu_2 & \nu_3 & -\nu_2 \\ & & & & A_{\beta\alpha} & B_{\beta\alpha} & -A_{\beta\alpha}/2 & C_{\beta\alpha} \\ & \text{Symmetric} & & & & A_{\beta\alpha} & C_{\beta\alpha} & -A_{\beta\alpha}/2 \\ & & & & & & A_{\beta\alpha} & B_{\beta\alpha} \\ & & & & & & & A_{\beta\alpha} \end{bmatrix}\begin{bmatrix} u_{x1} \\ u_{x2} \\ u_{x3} \\ u_{x4} \\ u_{y1} \\ u_{y2} \\ u_{y3} \\ u_{y4} \end{bmatrix} \tag{6.34}$$

where $\alpha = b/a$, $\beta = a/b$, $\nu_1 = 1-\nu$, $\nu_2 = 3(1+\nu)$, $\nu_3 = 3(1-3\nu)$, $A_{\alpha\beta} = 8\alpha + 4\beta\nu_1$, $B_{\alpha\beta} = 4\alpha - 4\beta\nu_1$, $C_{\alpha\beta} = -8\alpha + 2\beta\nu_1$, and $A_{\beta\alpha}$, $B_{\beta\alpha}$, $C_{\beta\alpha}$ are obtained by interchanging α and β in $A_{\alpha\beta}$, $B_{\alpha\beta}$, $C_{\alpha\beta}$, respectively.

We have now completed the formation of the stiffness matrix for a single element.

The element stiffness matrix of Eq. (6.34) is arranged such that the first four displacements correspond to x direction displacements, while the second four are in the y direction. For computationally assembling the global stiffness matrix, it is more convenient to place together the displacements for each node. Such a rearrangement will be employed for many of the subsequent numerical calculations in this chapter.

Formation of the Element Loading Vector

The surface integral III of Eq. (6.21) remains to be evaluated. The integration is carried out element by element. In terms of the assumed displacements of Eq. (6.19), integral III can be written as

$$\int_{S_p} \delta \mathbf{u}^T \overline{\mathbf{p}} \, dS = \delta \mathbf{v}^{iT} \int_{S_p} \mathbf{N}^T \overline{\mathbf{p}} \, dS \tag{6.35}$$

where $\overline{\mathbf{p}}$ is the applied surface loading. In terms of the nodal displacements $\mathbf{v}^i$ and the equivalent nodal forces $\overline{\mathbf{p}}^{i0}$ (the element loading vector due to $\overline{\mathbf{p}}$), the virtual work due to the applied loading is $\delta \mathbf{v}^{iT} \overline{\mathbf{p}}^{i0}$. This must be equal to the expression of Eq. (6.35), that is

$$\delta \mathbf{v}^{iT} \int_{S_p} \mathbf{N}^T \overline{\mathbf{p}} \, dS = \delta \mathbf{v}^{iT} \, \overline{\mathbf{p}}^{i0} \tag{6.36}$$

It would also be appropriate to include a loading vector due to the interelement forces. However, this will not be considered here. Since $\mathbf{u}$ has x and y components, the applied loads $\overline{\mathbf{p}}$ can be separated into those acting in the x and those acting in the y direction. Thus,

$$\overline{\mathbf{p}}(\xi, \eta) = \begin{bmatrix} \overline{p}_\xi(\xi, \eta) \\ \overline{p}_\eta(\xi, \eta) \end{bmatrix} \tag{6.37}$$

For loading applied on the outer rim boundaries, the integration has to be performed around the circumference (S_p) and not over the broad surface. The loading vector $\overline{\mathbf{p}}^{i0}$ of an element is composed of the sum of the integrals for each of the loaded boundaries. For our elements of Fig. 6.10, we can write

$$\begin{aligned} \overline{\mathbf{p}}^{i0} &= \int_{S_p} \mathbf{N}^T \overline{\mathbf{p}}(\xi, \eta) \, dS \\ &= a \int_0^1 \mathbf{N}^T(\xi, \eta = 0) \, \overline{\mathbf{p}}(\xi, \eta = 0) \, d\xi + a \int_0^1 \mathbf{N}^T(\xi, \eta = 1) \, \overline{\mathbf{p}}(\xi, \eta = 1) \, d\xi \\ &\quad + b \int_0^1 \mathbf{N}^T(\xi = 0, \eta) \, \overline{\mathbf{p}}(\xi = 0, \eta) \, d\eta + b \int_0^1 \mathbf{N}^T(\xi = 1, \eta) \, \overline{\mathbf{p}}(\xi = 1, \eta) \, d\eta \end{aligned} \tag{6.38}$$

To illustrate the application of Eq. (6.38), consider the element of Fig. 6.10 with a linearly distributed load as shown in Fig. 6.12. For this element, all surface loads are zero except

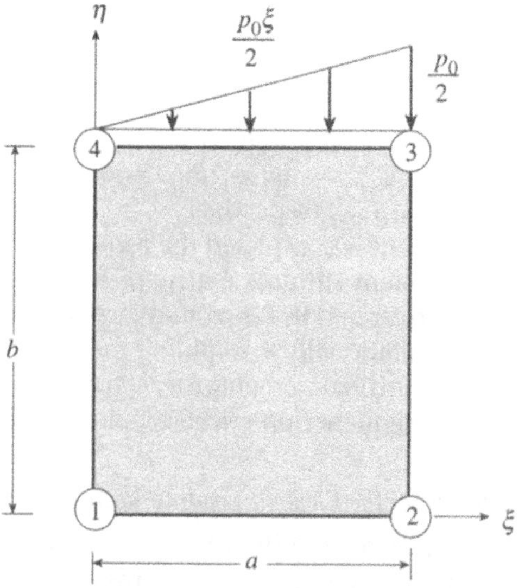

FIGURE 6.12
The element of Fig. 6.10 with linearly varying loading along one edge.

$\overline{\mathbf{p}}_\eta(\xi, \eta = 1)$, which is applied on the boundary in the y direction between nodes 3 and 4. Then

$$\overline{\mathbf{p}}_\eta(\xi, \eta = 1) = -\frac{p_0}{2}\xi \qquad \overline{\mathbf{p}}(\xi, \eta = 1) = \begin{bmatrix} 0 \\ -\frac{p_0}{2}\xi \end{bmatrix} \tag{6.39}$$

and Eq. (6.38) reduces to

$$\overline{\mathbf{p}}^{i0} = a \int_0^1 \mathbf{N}^T(\xi, \eta = 1)\, \overline{\mathbf{p}}(\xi, \eta = 1)\, d\xi \tag{6.40}$$

where the superscript i refers to the ith element.

The integral of Eq. (6.40) can be computed as

$$\underbrace{\begin{bmatrix} 0 & 0 \\ 0 & 0 \\ \xi & 0 \\ 1-\xi & 0 \\ \cdots & \cdots \\ 0 & 0 \\ 0 & 0 \\ 0 & \xi \\ 0 & 1-\xi \end{bmatrix}}_{\mathbf{N}^T(\xi,\, \eta = 1)} \begin{bmatrix} 0 \\ \cdots \\ -\frac{p_0}{2}\xi \end{bmatrix} = \underbrace{\begin{bmatrix} 0 \\ 0 \\ 0 \\ 0 \\ \cdots \\ 0 \\ 0 \\ \xi^2 \\ \xi - \xi^2 \end{bmatrix}}_{\overline{\mathbf{p}}(\xi,\, \eta = 1)} \left(-\frac{p_0}{2}\right) \tag{6.41}$$

Upon carrying out the integration of $a \int_0^1 \mathbf{N}^T \overline{\mathbf{p}}\, d\xi$, we find

$$(\overline{\mathbf{p}}^{i0})^T = -\frac{a p_0}{2} \begin{matrix} u_{x1} & u_{x2} & u_{x3} & u_{x4} & u_{y1} & u_{y2} & u_{y3} & u_{y4} \\ [0 & 0 & 0 & 0 & 0 & 0 & 1/3 & 1/6] \end{matrix} \tag{6.42}$$

Similarly, for an element with the linearly varying distributed load beginning at node 4 with magnitude $p_0/2$ and terminating at node 3 with magnitude p_0, the loading can be expressed as $\overline{p}_\eta(\xi, \eta = 1) = -p_0(1 + \xi)/2$. Then

$$(\overline{\mathbf{p}}^{i0})^T = -\frac{a p_0}{2} \begin{bmatrix} 0 & 0 & 0 & 0 & 0 & 0 & 5/6 & 2/3 \end{bmatrix} \tag{6.43}$$

A more general expression for the loading vector can be derived for the distributed loading of Fig. 6.13, which acts in the y direction between nodes 4 and 3. Then

$$\overline{\mathbf{p}}(\xi, \eta) = \begin{bmatrix} \overline{p}_\xi \\ \overline{p}_\eta \end{bmatrix} = \begin{bmatrix} 0 \\ \overline{p}_\eta(\xi, \eta = 1) \end{bmatrix} \tag{6.44}$$

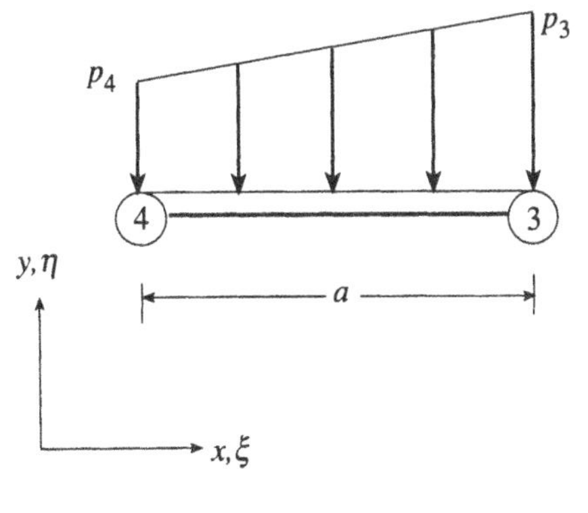

FIGURE 6.13
Applied distributed loading along the element boundary between nodes 4 and 3 of the element of Fig. 6.12.

For the linearly varying load of Fig. 6.13, the distribution of $\overline{p}_\eta$ is a known function of ξ, i.e.,

$$\overline{p}_\eta(\xi, \eta = 1) = p_4 + (p_3 - p_4)\xi = \begin{bmatrix} \xi & \vdots & 1 - \xi \end{bmatrix} \begin{bmatrix} p_3 \\ p_4 \end{bmatrix} = \mathbf{N}_p \begin{bmatrix} p_3 \\ p_4 \end{bmatrix} \tag{6.45}$$

Then the loading vector integral of Eq. (6.38) becomes

$$\overline{\mathbf{p}}^{i0} = \int_0^1 \begin{bmatrix} 0 & 0 \\ 0 & 0 \\ \xi & 0 \\ 1-\xi & 0 \\ 0 & 0 \\ 0 & 0 \\ 0 & \xi \\ 0 & 1-\xi \end{bmatrix} \begin{bmatrix} 0 & 0 \\ \xi & 1-\xi \end{bmatrix} d\xi \cdot a \begin{bmatrix} p_3 \\ p_4 \end{bmatrix} = \begin{bmatrix} 0 & 0 \\ 0 & 0 \\ 0 & 0 \\ 0 & 0 \\ 0 & 0 \\ 0 & 0 \\ 1/3 & 1/6 \\ 1/6 & 1/3 \end{bmatrix} a \begin{bmatrix} p_3 \\ p_4 \end{bmatrix} \tag{6.46}$$

Assembly of the System Stiffness Matrix and Loading Vector

The procedure for assembling the system stiffness matrix and system loading vector will be demonstrated with a specific numerical example.

EXAMPLE 6.2 Planar structure

To illustrate the fundamentals of applying the displacement finite element method to a multidimensional structure, consider the problem of determining the in-plane stresses and displacements in a flat structure of constant thickness t lying in the xy plane as shown in Fig. 6.14a. A linearly varying line load is applied to the top edge. Suppose this two-dimensional planar structure is in a state of plane stress.

Begin the solution by taking advantage of the vertical axis of symmetry and choosing the model of Fig. 6.14b to replace the actual structure of Fig. 6.14a. Note that supports have been added to the model. Support is necessary in order to assure that, at least from a rigid body motion viewpoint, the model is a reasonable idealization of the structure of Fig. 6.14a. Without support, a singular stiffness matrix would be expected, which mathematically corresponds to rigid body motion of the structure.

The selection of the finite element model, along with appropriate constraints, is the most important step in modern structural analyses. The model must be established by the engineer even when readily available general purpose finite element programs are used. The supports (constraints) of Fig. 6.14b were chosen to achieve a symmetric deformation pattern like that of the original system of Fig. 6.14a. Of course, if we do not take advantage of the symmetry, the larger model of the complete original system could be used. However, the desire for computational economy usually prevails, and symmetry is utilized whenever appropriate.

The model of Fig. 6.14b is discretized into four rectangular elements. The use of only four elements can lead to rather inaccurate results. Each of these elements can be illustrated as in Fig. 6.10, which shows the local, natural coordinates ξ, η for this particular element. The nodes, which in this case are the four corners of the element, are numbered counterclockwise. The elements are attached to each other only at the nodes. For this elementary example, the local coordinates used as a reference for displacements and forces for the elements are the same as the global coordinates used for the system displacements and loading. Hence it is not necessary to perform a coordinate transformation for the element matrices and loading vectors. Furthermore, x and y will designate the global coordinate system, rather than X and Y of Chapter 5.

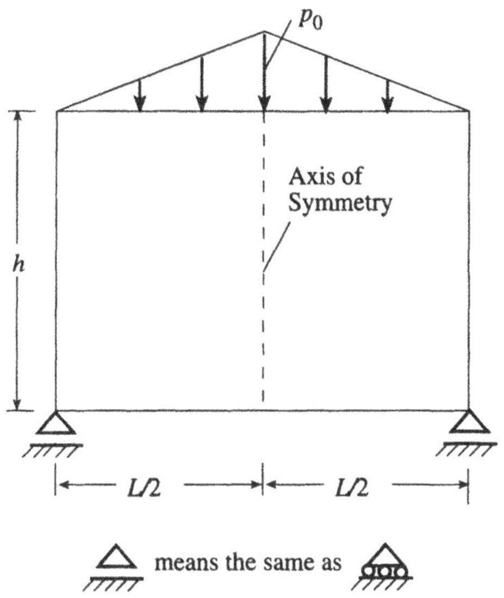

(a) The original system

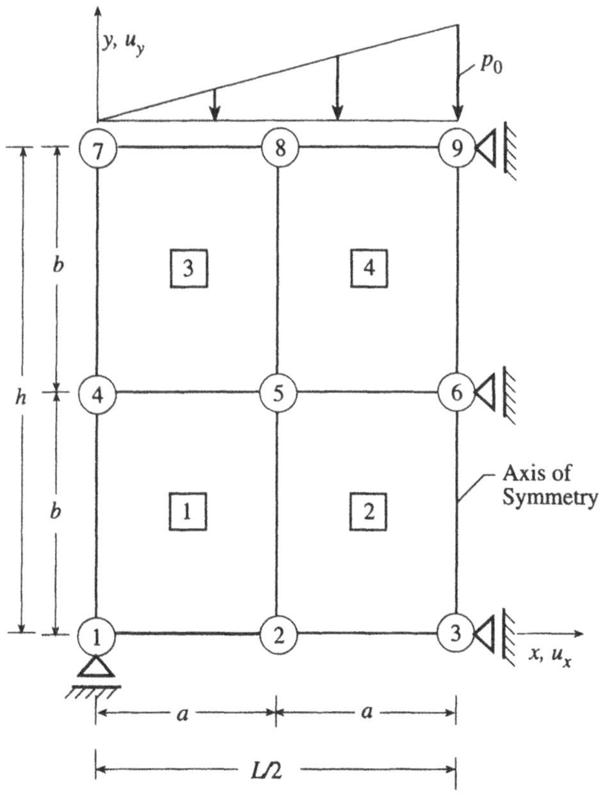

(b) Finite element model chosen to replace the actual structure of (a). This corresponds to the left half of the structure of (a). Rectangular elements are shown. The global node numbers are enclosed with a circle and the element numbers are boxed

FIGURE 6.14
In-plane structure of Example 6.2.

In Chapter 5 care was taken to distinguish between the "ends" of an element and the "nodes" of a system. In the study of multidimensional finite elements, it is traditional and useful to relax this distinction and to refer to the corners of elements as "nodes".

It is convenient at this stage to treat a specific example and to assign numerical values to some of the variables of the problem. Choose

$$a = b = 1.0 \text{ m}, \qquad t = 0.2 \text{ m}, \qquad E = 30 \text{ GN/m}^2, \qquad \nu = 0.0 \tag{1}$$

These lead to

$$\alpha = b/a = \beta = a/b = 1.0$$

Element Stiffness Matrix

The stiffness matrix of Eq. (6.34), which is identical for all four elements, is

$$\mathbf{k}^i\mathbf{v}^i = \begin{bmatrix} 12 & 3 & -6 & -3 & -6 & -3 & 0 & 3 \\ 3 & 12 & 3 & 0 & -3 & -6 & -3 & -6 \\ -6 & 3 & 12 & -3 & 0 & -3 & -6 & 3 \\ -3 & 0 & -3 & 12 & 3 & -6 & 3 & -6 \\ -6 & -3 & 0 & 3 & 12 & 3 & -6 & -3 \\ -3 & -6 & -3 & -6 & 3 & 12 & 3 & 0 \\ 0 & -3 & -6 & 3 & -6 & 3 & 12 & -3 \\ 3 & -6 & 3 & -6 & -3 & 0 & -3 & 12 \end{bmatrix} \begin{bmatrix} u_{x1} \\ u_{y1} \\ u_{x2} \\ u_{y2} \\ u_{x3} \\ u_{y3} \\ u_{x4} \\ u_{y4} \end{bmatrix}$$

$$= \begin{bmatrix} k_{11}^i & k_{12}^i & k_{13}^i & \cdots & k_{17}^i & k_{18}^i \\ k_{21}^i & k_{22}^i & \cdots & & & k_{28}^i \\ k_{31}^i & k_{32}^i & & & & k_{38}^i \\ k_{41}^i & \cdots & & & & k_{48}^i \\ k_{51}^i & \cdots & & & & k_{58}^i \\ k_{61}^i & \cdots & & & & k_{68}^i \\ k_{71}^i & \cdots & & & & k_{78}^i \\ k_{81}^i & k_{82}^i & & & k_{87}^i & k_{88}^i \end{bmatrix} \begin{bmatrix} u_{x1} \\ u_{y1} \\ u_{x2} \\ u_{y2} \\ u_{x3} \\ u_{y3} \\ u_{x4} \\ u_{y4} \end{bmatrix} \tag{2}$$

where the displacements at each node have been placed together.

Assembly of the System Stiffness Matrix

The variational principle of Eq. (6.21) establishes a relationship involving a sum over all element stiffness matrices and loading vectors. This is, of course, the same assembly procedure developed in Chapter 5 for the displacement method applied to the solution of bar and beam systems. The element matrices must be summed such that the nodes common to more than one element are properly taken into account. This is accomplished by identifying each element in terms of the global node numbering system.

The assembly of the global stiffness matrix of this solid with two-dimensional elements involves, as is to be expected, more bookkeeping than that of the examples of Chapter 5, where the elements are simpler. Begin by numbering the global DOF at each node. Table 6.1 gives the assigned numbers for the global DOF for each of the nine global node numbers of Fig. 6.14b. Since each node has two DOF, there are eighteen global DOF.

An incidence table, Table 6.2, is used to relate the numbering system for each element to the global topology, i.e., the global node numbers. Note that the nodes of each element are numbered in the same direction, counterclockwise. The entries of Table 6.2 are obtained by comparing the element of Fig. 6.10 with each of the elements of Fig. 6.14b. For example, the

TABLE 6.1

Numbering of the Global DOF

Global Node Number (Fig. 6.14b)	Assigned Global DOF Number	
	u_x	u_y
1	1	2
2	3	4
3	5	6
4	7	8
5	9	10
6	11	12
7	13	14
8	15	16
9	17	18

TABLE 6.2

Element Topology

Element Number	Node number 1	2	3	4	
	1	2	3	4	Element Node Numbers Fig. 6.10
1	1	2	5	4	Global Node Numbers Fig. 6.14b
2	2	3	6	5	
3	4	5	8	7	
4	5	6	9	8	

relationship between the element nodal numbers and the global node numbers for element 1 is

Element Node Number		Global Node Number
1	⟶	1
2	⟶	2
3	⟶	5
4	⟶	4

The next step of the assembly is to fit the element stiffness matrices into the global matrix so that the element matrices can be superimposed (added) to form the global stiffness matrix. To accomplish this, a relationship between each element node number and a global DOF number has to be established. This can be achieved by the combination of Tables 6.1 and 6.2, i.e., the incidence table of Table 6.2 is expanded to include the global DOF numbers. Each global node number of Table 6.2 is associated with two corresponding DOF as provided in Table 6.1. The result is Table 6.3, which relates the element node numbers and the numbers of the global DOF. From this table, the entries in the element stiffness matrices of (2) can be located in the global stiffness matrix.

Figure 6.15a shows the distribution of the entries of the element stiffness matrix of element 1 in the global stiffness matrix. For example for element 1, it follows from Table 6.3 that element node number 2 corresponds to global node number 2 and global DOF numbers 3 and 4. Hence, k^1_{i3}, k^1_{i4}, k^1_{3i}, and k^1_{4i}, $i = 1, 2, \ldots, 8$, of (2) correspond to node 2 in the element node numbering system, which is node 2 in the global numbering system with global DOF 3 and 4. Thus, these entries (k^1_{i3}, k^1_{i4}, k^1_{3i}, and k^1_{4i}) are placed in the 3rd and 4th rows and

TABLE 6.3
Global DOF Numbers for Each Element

Element Number	Row	Element Node Numbers: 1	2	3	4
1	Global Node Numbers of an Element	1	2	5	4
	Global DOF Numbers	1, 2	3, 4	9, 10	7, 8
2	Global Node Numbers of an Element	2	3	6	5
	Global DOF Numbers	3, 4	5, 6	11, 12	9, 10
3	Global Node Numbers of an Element	4	5	8	7
	Global DOF Numbers	7, 8	9, 10	15, 16	13, 14
4	Global Node Numbers of an Element	5	6	9	8
	Global DOF Numbers	9, 10	11, 12	17, 18	15, 16

columns of the global stiffness matrix. Similarly, in Table 6.3 element node number 3 of element 1 corresponds to global node number 5 and global DOF numbers 9 and 10. The entries of the element stiffness matrix of (2) corresponding to element node 3 are k^1_{5i}, k^1_{6i}, k^1_{i5} and k^1_{i6}, $i = 1, 2, \ldots, 8$. These should be placed in the 9th and 10th rows and columns of the global stiffness matrix. Figure 6.15a displays the location of k^1_{i3}, k^1_{i4}, k^1_{3i}, k^1_{4i}, k^1_{5i}, k^1_{6i}, k^1_{i5}, and k^1_{i6}, $i = 1, 2, \ldots, 8$, in the global stiffness matrix, as well as the rest of the stiffness coefficients for element 1. Figure 6.15b shows the layout of the stiffness matrix of element 2 within the global stiffness matrix.

The global stiffness matrix is formed by adding entries from the various element stiffness matrices that occur in the same place in the global matrix. This can be represented as

$$K_{jk} = \sum_{i=1}^{M} k^i_{jk} \tag{3}$$

where M is the number of elements. If this relationship is to be employed, the subscripts of the entries in Figs. 6.15a and b should be replaced by the appropriate global DOF numbers. Figures 6.15a and b would then appear as in Figs. 6.15c and d, respectively. The global stiffness matrix, formed by superposition of the element stiffness matrices, takes the form shown in Fig. 6.16a.

After the global stiffness matrix is formed, the boundary conditions, which are shown in Fig. 6.14b, should be imposed. For example, nodal displacements such as the horizontal displacement at global node 6 are set equal to zero. Comparison of Fig. 6.14b with Table 6.1 shows that DOF 2, 5, 11, and 17 are constrained. That is, the corresponding displacements are set equal to zero. These zero displacements eliminate the columns in the global stiffness matrix for DOF 2, 5, 11, and 17 and also the equivalent rows, as they represent the unknown reactions at the constrained DOF. The final global stiffness matrix is shown in Fig. 6.16b. It is seen that after the boundary conditions are imposed, the number of DOF in the global stiffness matrix is reduced from eighteen to fourteen and the global stiffness matrix is a square, banded, and symmetric matrix.

Formation of the System Loading Vector

The upper edge of the structure is subjected to a vertical, linearly varying load as shown in Fig. 6.14. This loading is applied to elements 3 and 4 (Fig. 6.17). Equations (6.42) and (6.43) provide the loading vector for each element. These vectors must be assembled into a global vector corresponding to the system stiffness matrix. The loading vectors are reordered such

Element Node No.			1		2				4		3	
Global Node No.			1		2		3		4		5	
Global DOF No.			1	2	3	4	5	6	7	8	9	10
1	1	1	k^1_{11}	k^1_{12}	k^1_{13}	k^1_{14}			k^1_{17}	k^1_{18}	k^1_{15}	k^1_{16}
		2	k^1_{21}	k^1_{22}	k^1_{23}	k^1_{24}			k^1_{27}	k^1_{28}	k^1_{25}	k^1_{26}
2	2	3	k^1_{31}	k^1_{32}	k^1_{33}	k^1_{34}			k^1_{37}	k^1_{38}	k^1_{35}	k^1_{36}
		4	k^1_{41}	k^1_{42}	k^1_{43}	k^1_{44}			k^1_{47}	k^1_{48}	k^1_{45}	k^1_{46}
	3	5										
		6										
4	4	7	k^1_{71}	k^1_{72}	k^1_{73}	k^1_{74}			k^1_{77}	k^1_{78}	k^1_{75}	k^1_{76}
		8	k^1_{81}	k^1_{82}	k^1_{83}	k^1_{84}			k^1_{87}	k^1_{88}	k^1_{85}	k^1_{86}
3	5	9	k^1_{51}	k^1_{52}	k^1_{53}	k^1_{54}			k^1_{57}	k^1_{58}	k^1_{55}	k^1_{56}
		10	k^1_{61}	k^1_{62}	k^1_{63}	k^1_{64}			k^1_{67}	k^1_{68}	k^1_{65}	k^1_{66}

(a) Element 1 with subscripts corresponding to Eq. (2)

Element Node No.			1		2				4		3	
Global Node No.			2		3		4		5		6	
Global DOF No.			3	4	5	6	7	8	9	10	11	12
1	2	3	k^2_{11}	k^2_{12}	k^2_{13}	k^2_{14}			k^2_{17}	k^2_{18}	k^2_{15}	k^2_{16}
		4	k^2_{21}	k^2_{22}	k^2_{23}	k^2_{24}			k^2_{27}	k^2_{28}	k^2_{25}	k^2_{26}
2	3	5	k^2_{31}	k^2_{32}	k^2_{33}	k^2_{34}			k^2_{37}	k^2_{38}	k^2_{35}	k^2_{36}
		6	k^2_{41}	k^2_{42}	k^2_{43}	k^2_{44}			k^2_{47}	k^2_{48}	k^2_{45}	k^2_{46}
	4	7										
		8										
4	5	9	k^2_{71}	k^2_{72}	k^2_{73}	k^2_{74}			k^2_{77}	k^2_{78}	k^2_{75}	k^2_{76}
		10	k^2_{81}	k^2_{82}	k^2_{83}	k^2_{84}			k^2_{87}	k^2_{88}	k^2_{85}	k^2_{86}
3	6	11	k^2_{51}	k^2_{52}	k^2_{53}	k^2_{54}			k^2_{57}	k^2_{58}	k^2_{55}	k^2_{56}
		12	k^2_{61}	k^2_{62}	k^2_{63}	k^2_{64}			k^2_{67}	k^2_{68}	k^2_{65}	k^2_{66}

(b) Element 2 with subscripts corresponding to Eq. (2)

FIGURE 6.15
Placement of elements 1 and 2 into the global stiffness matrix.

Global DOF No.	1	2	3	4	5	6	7	8	9	10
1	k_{11}^1	k_{12}^1	k_{13}^1	k_{14}^1			k_{17}^1	k_{18}^1	k_{19}^1	$k_{1,10}^1$
2	k_{21}^1	k_{22}^1	k_{23}^1	k_{24}^1			k_{27}^1	k_{28}^1	k_{29}^1	$k_{2,10}^1$
3	k_{31}^1	k_{32}^1	k_{33}^1	k_{34}^1			k_{37}^1	k_{38}^1	k_{39}^1	$k_{3,10}^1$
4	k_{41}^1	k_{42}^1	k_{43}^1	k_{44}^1			k_{47}^1	k_{48}^1	k_{49}^1	$k_{4,10}^1$
5										
6										
7	k_{71}^1	k_{72}^1	k_{73}^1	k_{74}^1			k_{77}^1	k_{78}^1	k_{79}^1	$k_{7,10}^1$
8	k_{81}^1	k_{82}^1	k_{83}^1	k_{84}^1			k_{87}^1	k_{88}^1	k_{89}^1	$k_{8,10}^1$
9	k_{91}^1	k_{92}^1	k_{93}^1	k_{94}^1			k_{97}^1	k_{98}^1	k_{99}^1	$k_{9,10}^1$
10	$k_{10,1}^1$	$k_{10,2}^1$	$k_{10,3}^1$	$k_{10,4}^1$			$k_{10,7}^1$	$k_{10,8}^1$	$k_{10,9}^1$	$k_{10,10}^1$

(c) Element 1 with subscripts corresponding to global DOF numbers

Global DOF No.	3	4	5	6	7	8	9	10	11	12
3	k_{33}^2	k_{34}^2	k_{35}^2	k_{36}^2			k_{39}^2	$k_{3,10}^2$	$k_{3,11}^2$	$k_{3,12}^2$
4	k_{43}^2	k_{44}^2	k_{45}^2	k_{46}^2			k_{49}^2	$k_{4,10}^2$	$k_{4,11}^2$	$k_{4,12}^2$
5	k_{53}^2	k_{54}^2	k_{55}^2	k_{56}^2			k_{59}^2	$k_{5,10}^2$	$k_{5,11}^2$	$k_{5,12}^2$
6	k_{63}^2	k_{64}^2	k_{65}^2	k_{66}^2			k_{69}^2	$k_{6,10}^2$	$k_{6,11}^2$	$k_{6,12}^2$
7										
8										
9	k_{93}^2	k_{94}^2	k_{95}^2	k_{96}^2			k_{99}^2	$k_{9,10}^2$	$k_{9,11}^2$	$k_{9,12}^2$
10	$k_{10,3}^2$	$k_{10,4}^2$	$k_{10,5}^2$	$k_{10,6}^2$			$k_{10,9}^2$	$k_{10,10}^2$	$k_{10,11}^2$	$k_{10,12}^2$
11	$k_{11,3}^2$	$k_{11,4}^2$	$k_{11,5}^2$	$k_{11,6}^2$			$k_{11,9}^2$	$k_{11,10}^2$	$k_{11,11}^2$	$k_{11,12}^2$
12	$k_{12,3}^2$	$k_{12,4}^2$	$k_{12,5}^2$	$k_{12,6}^2$			$k_{12,9}^2$	$k_{12,10}^2$	$k_{12,11}^2$	$k_{12,12}^2$

(d) Element 2 with subscripts corresponding to global DOF numbers

FIGURE 6.15
(continued).

that the x and y loading components of a node are placed together (this corresponds to the order of the displacement components in the element stiffness matrix of 2). From Eqs. (6.42) (or 6.46) and (6.43), the loading vector for the element of Fig. 6.10 would appear as

$$\overline{\mathbf{p}}^{i0} = \Big[\overbrace{p_x \quad p_y}^{1} \quad \overbrace{p_x \quad p_y}^{2} \quad \overbrace{p_x \quad p_y}^{3} \quad \overbrace{p_x \quad p_y}^{4}\Big]^T = [p_{x1} \ \ p_{y1} \ \ p_{x2} \ \ p_{y2} \ \ p_{x3} \ \ p_{y3} \ \ p_{x4} \ \ p_{y4}]^T \quad (4)$$

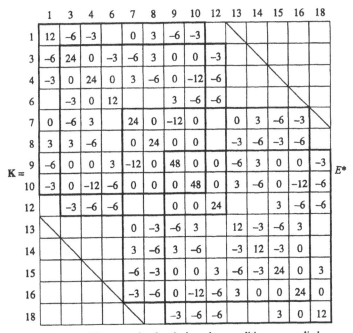

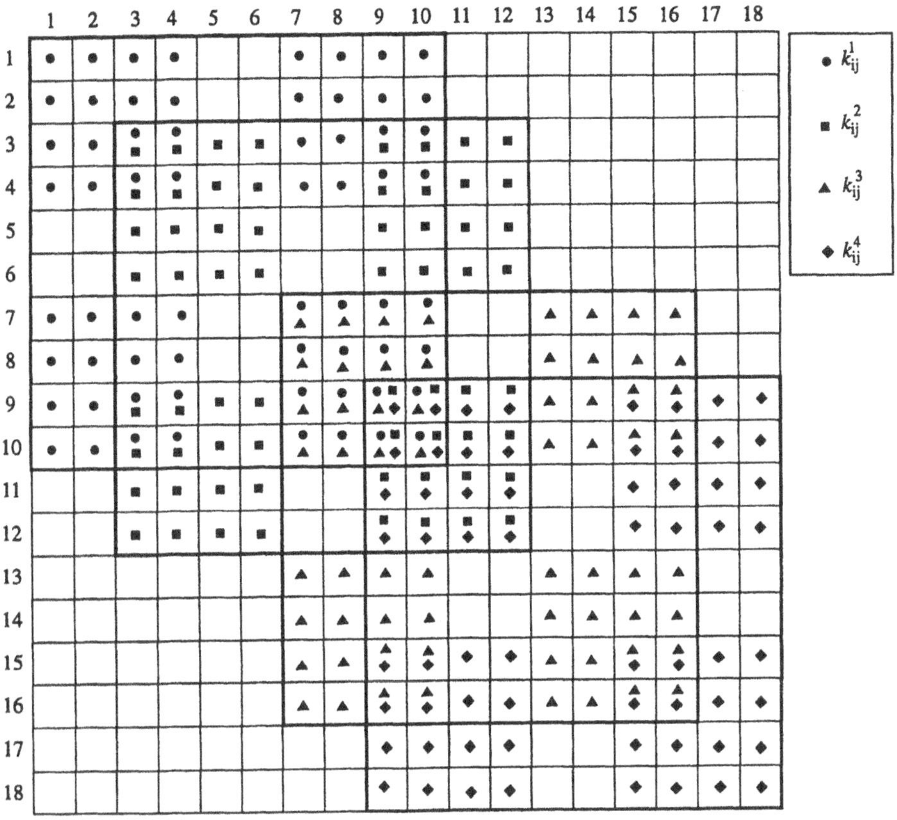

(a) Layout for global stiffness matrix before the boundary conditions are applied

$\mathbf{K} =$ E^*

	1	3	4	6	7	8	9	10	12	13	14	15	16	18
1	12	–6	–3		0	3	–6	–3						
3	–6	24	0	–3	–6	3	0	0	–3					
4	–3	0	24	0	3	–6	0	–12	–6					
6		–3	0	12			3	–6	–6					
7	0	–6	3		24	0	–12	0		0	3	–6	–3	
8	3	3	–6		0	24	0	0		–3	–6	–3	–6	
9	–6	0	0	3	–12	0	48	0	0	–6	3	0	0	–3
10	–3	0	–12	–6	0	0	0	48	0	3	–6	0	–12	–6
12		–3	–6	–6			0	0	24			3	–6	–6
13					0	–3	–6	3		12	–3	–6	3	
14					3	–6	3	–6		–3	12	–3	0	
15					–6	–3	0	0	3	–6	–3	24	0	3
16					–3	–6	0	–12	–6	3	0	0	24	0
18							–3	–6	–6			3	0	12

(b) Global stiffness matrix after the boundary conditions are applied

FIGURE 6.16
Global stiffness matrix **K**.

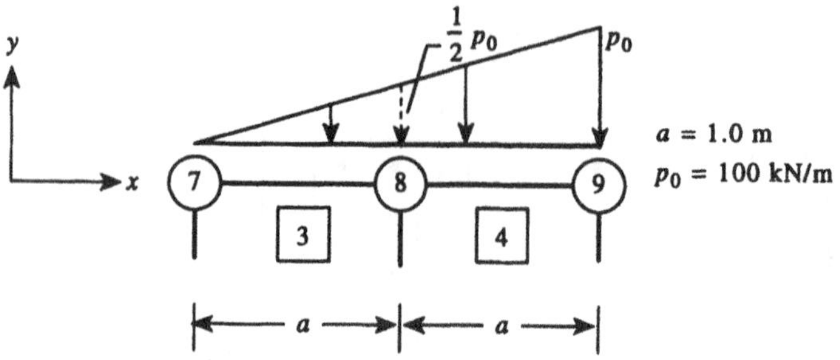

FIGURE 6.17
Loading on the model of Fig. 6.14b.

For elements 3 and 4 of Fig. 6.14, the loading vectors of Eqs. (6.42) and (6.43) become

$$\begin{aligned}\overline{\mathbf{p}}^{30} &= -\begin{bmatrix}0 & 0 & 0 & 0 & 0 & \frac{1}{3} & 0 & \frac{1}{6}\end{bmatrix}^T \frac{a \cdot p_0}{2} \\ \overline{\mathbf{p}}^{40} &= -\begin{bmatrix}0 & 0 & 0 & 0 & 0 & \frac{5}{6} & 0 & \frac{2}{3}\end{bmatrix}^T \frac{a \cdot p_0}{2}\end{aligned} \tag{5}$$

The global loading vector for the whole system is developed from the element loading vector similarly to the development of the system stiffness matrix. Table 6.3 provides the equation numbers of the loading components in the global loading vector $\overline{\mathbf{P}}^0 = \overline{\mathbf{P}}$. For element 3, where only p_{y3} and p_{y4} are non-zero in (5), we find from Table 6.3 that

Loading Vector Component	
On the element level	**On the global level**
p_{y3}	$p_{16} = 1/3(ap_0/2)$
p_{y4}	$p_{14} = 1/6(ap_0/2)$

Similarly, for element 4,

Loading Vector Component	
On the element level	**On the global level**
p_{y3}	$p_{18} = 5/6(ap_0/2)$
p_{y4}	$p_{16} = 2/3(ap_0/2)$

Assemble the element loading vectors in the same manner as the stiffness matrices to form the global loading vector

$$\begin{array}{ccccccccccccccccccc} & 1 & 2 & 3 & 4 & 5 & 6 & 7 & 8 & 9 & 10 & 11 & 12 & 13 & 14 & 15 & 16 & 17 & 18 \\ \overline{\mathbf{P}}^0 = \overline{\mathbf{P}} = -[0 & 0 & 0 & 0 & 0 & 0 & 0 & 0 & 0 & 0 & 0 & 0 & 0 & 1/6 & 0 & 1 & 0 & 5/6]^T a p_0/2 \end{array}$$

After the boundary conditions are imposed, the global loading vector, with $p_0 = 100\,\text{kN/m}$, becomes

$$\begin{array}{cccccccccccccc} & 1 & 2 & 3 & 4 & 5 & 6 & 7 & 8 & 9 & 10 & 11 & 12 & 13 & 14 \\ \overline{\mathbf{P}} = -[0 & 0 & 0 & 0 & 0 & 0 & 0 & 0 & 0 & 0 & 1/6 & 0 & 1 & 5/6]^T a \cdot p_0/2 \\ = -[0 & 0 & 0 & 0 & 0 & 0 & 0 & 0 & 0 & 0 & 50/6 & 0 & 50 & 250/6]^T\,\text{kN} \end{array} \tag{6}$$

Computation of the System Nodal Displacements

The global nodal displacements are obtained from the expression

$$\mathbf{KV} = \overline{\mathbf{P}} \tag{7}$$

From a mechanical standpoint, the individual equations represent equilibrium conditions for a node. The solution of (7) provides for the complete structure an approximation to the conditions of equilibrium and the static boundary conditions. Since the formulation utilizes the principle of virtual work, the approximate solution is optimal relative to the chosen (kinematically admissible) trial displacement functions.

The stiffness matrix **K** is symmetric and positive definite. These properties permit special solution algorithms, e.g., Cholesky decomposition, to be employed. Moreover, normally the system of equations will be banded, a property that depends on the arrangement of the global node numbers and DOF. For this example, a standard linear equation solver was employed for numerical results. The nodal displacements were found to be

Node Number	u_x (m)	u_y (m)
1	-0.3976×10^{-4}	0.0
2	-0.2692×10^{-4}	-0.7152×10^{-4}
3	0.0	-0.8754×10^{-4}
4	0.4568×10^{-5}	-0.3287×10^{-4}
5	0.1913×10^{-5}	-0.7037×10^{-4}
6	0.0	-0.9028×10^{-4}
7	0.3062×10^{-4}	-0.4259×10^{-4}
8	0.2309×10^{-4}	-0.7757×10^{-4}
9	0.0	-0.9951×10^{-4}

(8)

The deformed and undeformed systems are sketched in Fig. 6.18.

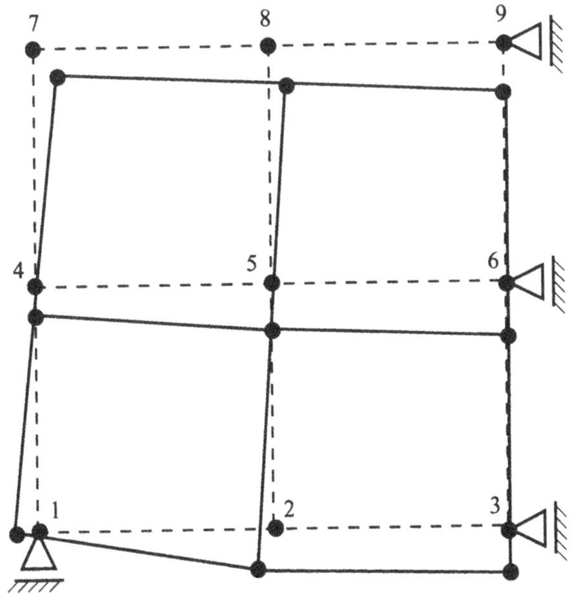

FIGURE 6.18
Deformation pattern of the model of Fig. 6.14b.

Computation of Stresses

If the state of deformation is known, the stress distribution follows from the material law and kinematic relations. From Eq. (6.25) for element i, the stress anywhere, i.e., for any value of ξ and η, is given by

$$\boldsymbol{\sigma} = \mathbf{E}\boldsymbol{\epsilon} = \mathbf{E}\,\mathbf{B}(\xi, \eta)\,\mathbf{v}^i \tag{9}$$

where the vector $\mathbf{v}^i$ contains the nodal displacements, including boundary conditions, of element i. Care must be taken here because, as explained previously, the entries in the vector $\mathbf{v}^i$ have been reordered for part of this example problem. Thus, the entries in $\mathbf{B}$ should be rearranged accordingly.

As an example, compute the stresses at the center of the element where $\xi = \eta = 0.5$. The relationship for the strains in terms of locally numbered displacements is

$$\boldsymbol{\epsilon} = \mathbf{B}(\xi = 0.5, \eta = 0.5)\,\mathbf{v}^i$$

$$\begin{bmatrix} \epsilon_x \\ \epsilon_y \\ \gamma_{xy} \end{bmatrix} = \begin{bmatrix} -0.5 & 0 & 0.5 & 0 & 0.5 & 0 & -0.5 & 0 \\ 0 & -0.5 & 0 & -0.5 & 0 & 0.5 & 0 & 0.5 \\ -0.5 & -0.5 & -0.5 & 0.5 & 0.5 & 0.5 & 0.5 & -0.5 \end{bmatrix} \begin{bmatrix} u_{x1} \\ u_{y1} \\ u_{x2} \\ u_{y2} \\ u_{x3} \\ u_{y3} \\ u_{x4} \\ u_{y4} \end{bmatrix} \tag{10}$$

The displacements of (8) for a particular element can be inserted into (10) to find the strains in the center of that element. From (9) we can compute the stresses in the middle of an element.

Equation (6.27), with $E = 30\ \text{GN/m}^2$ and $\nu = 0$, becomes

$$\underset{\boldsymbol{\sigma}}{\begin{bmatrix} \sigma_x \\ \sigma_y \\ \tau_{xy} \end{bmatrix}} = 3.0(10^{10}) \underset{\mathbf{E}}{\begin{bmatrix} 1 & 0 & 0 \\ 0 & 1 & 0 \\ 0 & 0 & 0.5 \end{bmatrix}} \underset{\boldsymbol{\epsilon}}{\begin{bmatrix} \epsilon_x \\ \epsilon_y \\ \gamma_{xy} \end{bmatrix}} \tag{11}$$

The principal stresses for the center of each element, which are illustrated in Fig. 6.19, are computed to be

Element	σ-Max (kPa)	σ-Min (kPa)	Angle
1	0.2522×10^3	-0.5753×10^3	−20.28
2	0.3820×10^3	-0.3106×10^2	−7.47
3	-0.7267×10^3	-0.3992×10^3	−37.54
4	-0.1427×10^3	-0.4786×10^3	−56.27

(12)

The stress can be expected to experience jumps in value at the boundaries of the elements. The magnitudes of these discontinuities provide one indication of the errors involved in this approximate solution.

Another indication of the accuracy of the solution can be obtained by comparing the results here with the response of this structure as found by other methods. For example, if this structure is treated as the simply supported beam of Fig. 6.20, simple statics gives a bending moment at the center of the beam to be 133 kNm. A theory of elasticity solution provides a center deflection along the bottom of the structure of 1.86×10^{-4} m.

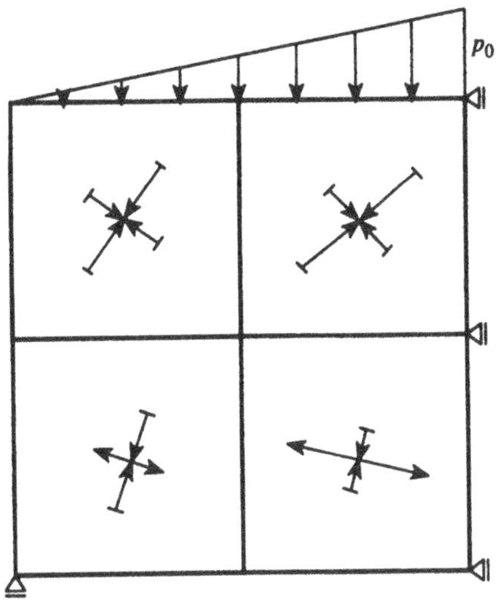

FIGURE 6.19
Principal stresses at the centroid of the elements of the model of Fig. 6.14b.

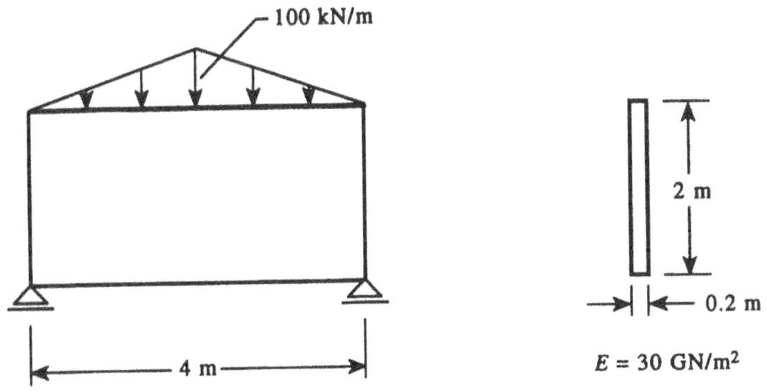

FIGURE 6.20
A beam model of the structure of Fig. 6.14a.

A comparison can be made between these midspan values and those obtained with the finite element solution. For the structure of Fig. 6.14a, the appropriate displacement is u_y of global node 3, which from (8) is -0.88×10^{-4} m. This is significantly smaller than the exact value; consequently, it can be observed that the behavior of the approximate structural model is too stiff. For a comparison of the moment, first compute the σ_x stresses along the vertical line between nodes 3 and 9 using (50): $\sigma_{x9} = -692.53$ kPa, $\sigma_{x6} = -57.40$ kPa, and $\sigma_{x3} = 807.33$ kPa. The resultant moment about node 6 will be calculated. For the sake of this calculation, σ_{x6} will be set to zero since 57.40 is small in comparison to 692.53 and 807.33. The total force in the stress triangle between nodes 6 and 9 is $692.53 \times 1.0 \times 0.2/2 = 69.25$ kN, where the area on which the stress acts is 1.0×0.2. The total force between nodes 6 and 3 would be $807.33 \times 1.0 \times 0.2/2 = 80.73$ kN. These forces are assumed to act through the centroid of the triangles as shown in Fig. 6.21. The moment at node 6 of the beam then would be

$$M = 69.25 \times 2/3 + 80.73 \times 2/3 = 99.99 \text{ kNm} \tag{13}$$

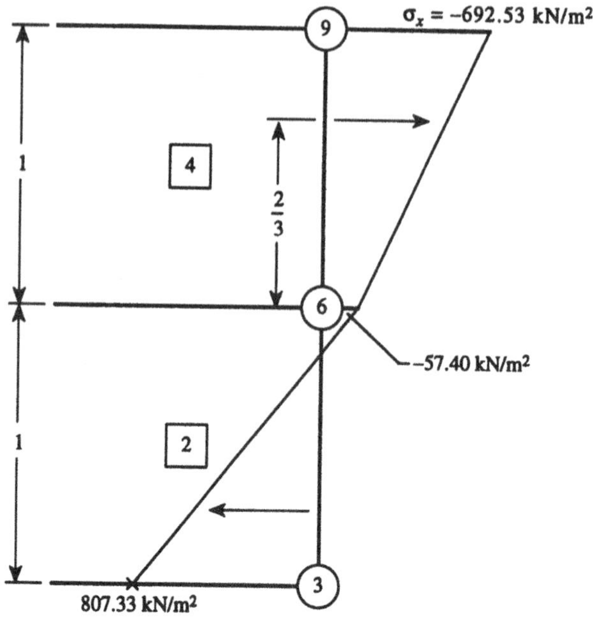

FIGURE 6.21
Stresses at nodes 3, 6, and 9.

Similar to the displacement calculation, this is much smaller than the correct value of 133 kNm.

The finite element solution using four elements does not appear to be very accurate in comparison with the values obtained by other methods. This is hardly a surprising conclusion, considering the choice of simple trial functions and the modeling with only four elements. One would suspect that greater accuracy would be achieved by using a higher order polynomial. For a given element, this requires more nodes if an interpolation polynomial is to be employed. The same problem (Fig. 6.14a) was solved using the elements of Fig. 6.22a. Here, higher order polynomials were chosen in the trial displacements, and nodes between the corners were added to the elements. The resulting displacement pattern remains the same as in Fig. 6.8. The midspan u_y displacement along the bottom edge is computed to be 1.4273×10^{-4} m, and the moment is found to be 140.69 kNm. Both of these values are more accurate than corresponding values obtained using elements with lower order polynomials.

A variety of cases with an increasing number of elements were computed. The results are:

	Elements with nodes at the corners only			Elements with nodes at and between the corners		
No. of Elements	No. of Unknowns	u_y Displacement (10^{-4} m)	Moment (kNm)	No. of Unknowns	u_y Displacement (10^{-4} m)	Moment (kNm)
4	14	0.875	99.99	36	1.4273	140.69
9	27	1.0794	116.68	72	1.5998	138.46
16	44	1.2082	123.48	120	1.6987	136.10
36	90	1.3758	128.95	252	1.8430	134.64

(14)

These results are plotted in Figs. 6.22b and c. As might be expected, the accuracy of the solution increases with an increase in the number of degrees of freedom. ∎

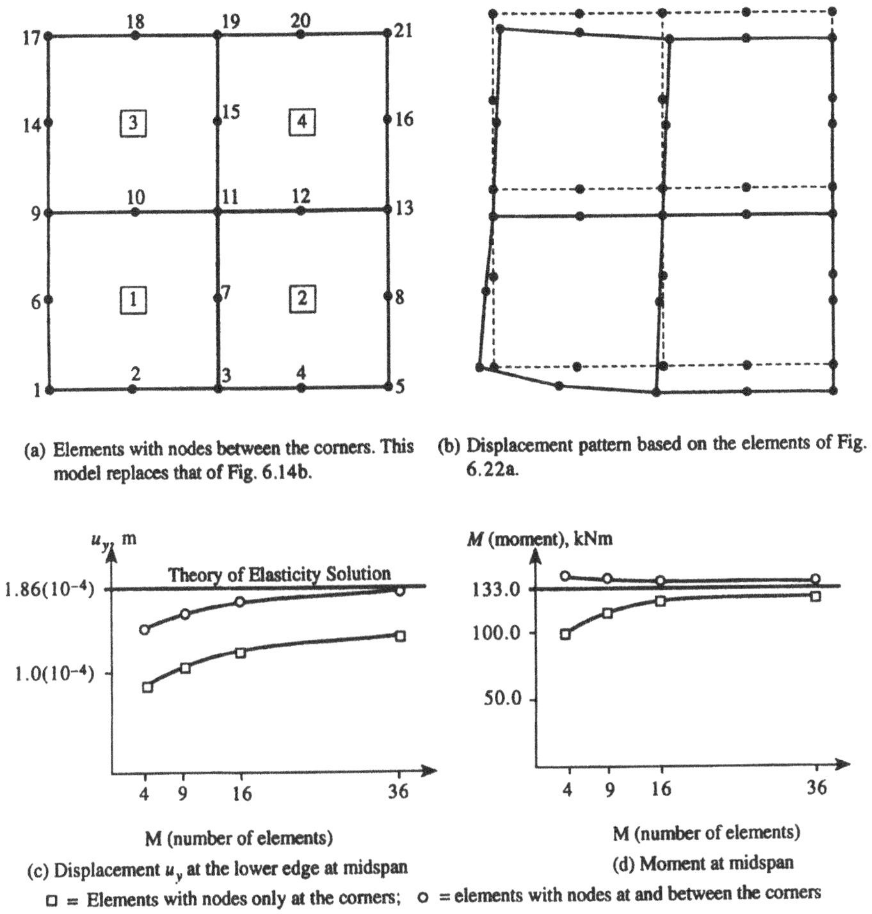

FIGURE 6.22
Displacement pattern and accuracy of the finite element solution of the structure of Fig. 6.14a as the number of elements varies.

6.5 Trial Functions and Formulation of Some Elements

6.5.1 Trial Functions

Fundamental to the successful implementation of the finite element method is the establishment of the element stiffness matrix and loading vector. This entails the evaluation of integrals, a task which, as discussed in the following section, must often be accomplished numerically. Also essential to establishing a useful element is the selection of adequate trial functions for each element [Wunderlich and Redanz, 1995]. For the displacement finite element method, the trial functions are approximate patterns of displacements, rotations, or other fundamental variables often expressed in terms of the same variables at the nodes, i.e., $\mathbf{u}^i = \mathbf{N}^i\mathbf{v}^i$ or if the superscript indicating the ith element is dropped

$$\mathbf{u} = \mathbf{N}\mathbf{v} \tag{6.47}$$

where $\mathbf{v}$ is a vector of values of $\mathbf{u}$, or derivatives of $\mathbf{u}$, at nodal points, and $\mathbf{N}$ is appropriately constructed to permit $\mathbf{u}$ to take the desired values at the nodes. Of course, to insure kinematical continuity, the trial solution of Eq. (6.47) is chosen such that at any node shared by more than one element, a particular variable of **v** is the same physical variable regardless from which element the node is approached.

The functions **N** which make **u** equal to certain variables at prescribed points are called *interpolation functions*. By definition, for two-dimensional problems, [e.g., see Eq. (6.19c)]

$$\begin{aligned} &\text{At node } i: \quad N_i(\xi_i, \eta_i) = 1 \\ &\text{At node } j: \quad N_i(\xi_j, \eta_j) = 0 \end{aligned} \tag{6.48}$$

Equation (6.48) can serve as the basic definition of interpolation functions. It is because of these characteristics that we refer to the functions **N** as "shape functions." The expression of Chapter 4, Eq. (4.47) for the deflection w of a beam element that begins at $\xi = 0$ and ends at $\xi = 1$ can be written as

$$w = \mathbf{N}\mathbf{v} = w_a N_1 + \theta_a N_2 + w_b N_3 + \theta_b N_4 \tag{6.49}$$

Since, as explained in Section 4.4.2, the N_i of this expression are Hermitian polynomials, the polynomials N_1 and N_3 satisfy conditions similar to Eq. (6.48), as well as their derivatives being zero at points a and b. The polynomials N_2 and N_4 are zero at points a and b. For certain two-dimensional problems, the derivative conditions would appear as

$$\begin{aligned} &\text{At node } i: \quad \frac{d}{dx} N_i(\xi_i, \eta_i) = 1 \\ &\text{At node } j: \quad \frac{d}{dx} N_i(\xi_j, \eta_j) = 0 \quad i, j = 2, 4 \end{aligned} \tag{6.50}$$

It should be observed that the quantities N_i represent the contribution of a nodal unit displacement to the total deflection.

6.5.2 Convergence

Presumably, for successful finite element solutions, the interpolation functions should lead to an analysis that monotonically converges to the exact solution as the size of the elements tends to zero, i.e., the accuracy of the solution increases as the finite element mesh is continuously refined. Convergence to the correct solution is critical to the proper use of a finite element analysis and is the topic of numerous papers and books, e.g., see Bathe (1998). To achieve monotonic convergence, the element must be *complete* and *compatible* (or *conforming*) [Bathe, 1996; Zienkiewicz, 1977]. The requirement for completeness means that the displacement functions must be able to represent the rigid body displacements and constant strain states. Compatibility assures that no gaps occur within the elements and between the elements when the system of elements is assembled and loaded.

Consider these characteristics in more detail. For completeness,

1. The trial functions should be able to represent displacements that the element undergoes as a rigid body without developing stress. For example, consider a cantilevered beam with a concentrated force acting at the midpoint. Since stresses will not be generated beyond the location of load application, the trial functions for the elements at the free end must be able to permit the elements to translate and rotate stress free.
2. The displacement functions of an element must be such that the strain in each element approaches a constant value in the limit as the element approaches a very small size.

Then a complex variation of strain within the structure can be approximated. For this constant strain representation, the displacement function must contain those terms that can eventually result in constant strain states. If the structure is actually in a constant strain state, the functions must be able to represent this constant strain.

To satisfy the condition of compatibility, the trial functions should be chosen such that (1) they are continuous within the element, and (2) at the element interfaces at least the first r derivatives are continuous, where $r+1$ is the highest derivative appearing in the functional of the principle of virtual work, i.e., the highest derivative appearing in the $\mathbf{D}_u$ matrix in the principle of virtual work expression $\int_V \delta(\mathbf{D}_u\,\mathbf{u})^T \mathbf{E}\,\mathbf{D}_u\,\mathbf{u}\,dV = \int_V \delta\mathbf{u}^T{}_u\mathbf{D}^T\mathbf{E}\,\mathbf{D}_u\,\mathbf{u}\,dV$. For linear elastic elements where $r = 0$ ($\mathbf{D}_u$ contains first order derivatives for linear elastic solids [see Chapter 1], hence $r+1 = 1$ and $r = 0$), the compatibility condition requires that the trial function be continuous both inside the element and on the interelement boundaries. For bending elements of the sort needed for beams and plates where $r = 1$ ($\mathbf{D}_u$ contains second order derivatives), compatibility means that the slope of the trial function must be continuous inside the element and on its boundaries. This requirement of r continuity will ensure that no contribution is made from the element interface to the total functional of the variational principle [Zienkiewicz, 1977]. This condition is satisfied by a complete polynomial of degree $r+1$. This polynomial is defined in Section 6.5.5.

Trial functions are said to exhibit C^r *continuity* [Bathe, 1996] if their derivatives of order r are continuous. The completeness and compatibility requirement for an element can be stated in terms of the continuity conditions. If the trial function has C^{r+1} continuity inside the element, the element is complete. If the trial function has C^r continuity at the interelement boundaries, the element is compatible. The elements which satisfy both of these continuity conditions are called C^r *elements*. The requirement for C^r continuity at the boundaries was explained above. The C^{r+1} continuity requirement means that a derivative of order $r+1$ in the element is continuous, so that it can approach a constant value as the element size approaches zero. A complete polynomial which has the C^{r+1} continuity satisfies the complete conditions of an element. The constant and linear terms of the polynomial ensure that rigid body motion is permitted, while all constant and linear single terms below the order $r+1$ ensure that the solution and its derivatives in each element can approach a constant value as the elements are refined further [Bathe, 1996].

Elements that do not satisfy the compatibility requirement are called *incompatible* or *nonconforming* [Bazeley, et al., 1965]. If the incompatibility disappears with increasing mesh refinement, the elements can still be acceptable as they may lead to convergence to the correct solution.

EXAMPLE 6.3 Completeness and Compatibility

Investigate the completeness and compatibility of the displacement functions used in Section 6.4.

First consider the completeness. From Eqs. (6.11) and (6.12) the displacement functions show that a rigid body displacement in the x direction can be achieved if $\hat{u}_1 \neq 0$, and that in the y direction can occur if $\hat{u}_5 \neq 0$.

A rigid body rotation can be achieved if (Fig 6.10) $u_{x4}(\xi = 0, \eta = 1) = -u_{y2}(\xi = 1, \eta = 0)$, i.e., $\hat{u}_4 = -\hat{u}_6$, where $\hat{u}_4$ and $\hat{u}_6$ are non-zero. If either $\hat{u}_1$ or $\hat{u}_3$ is not equal to zero, the movement of the element is a translation plus rotation. If they are zero, the movement is a rotation.

For the element in Fig. 6.10, with $\xi = x/a$ and $\eta = y/b$, the displacements of Eqs. (6.11) and (6.12) lead to the strains

$$\begin{aligned}
\epsilon_x &= \frac{\partial u_x}{\partial x} = \frac{\partial u_x}{\partial \xi}\frac{\partial \xi}{\partial x} + \frac{\partial u_x}{\partial \eta}\frac{\partial \eta}{\partial x} = \frac{1}{a}(\widehat{u}_3\eta + \widehat{u}_2) \\
\epsilon_y &= \frac{\partial u_y}{\partial y} = \frac{\partial u_y}{\partial \xi}\frac{\partial \xi}{\partial y} + \frac{\partial u_y}{\partial \eta}\frac{\partial \eta}{\partial y} = \frac{1}{b}(\widehat{u}_7\xi + \widehat{u}_8) \\
\gamma_{xy} &= \frac{\partial u_x}{\partial y} + \frac{\partial u_y}{\partial x} = \frac{\partial u_x}{\partial \xi}\frac{\partial \xi}{\partial y} + \frac{\partial u_x}{\partial \eta}\frac{\partial \eta}{\partial y} + \frac{\partial u_y}{\partial \xi}\frac{\partial \xi}{\partial x} + \frac{\partial u_y}{\partial \eta}\frac{\partial \eta}{\partial x} \\
&= \frac{1}{b}(\widehat{u}_3\xi + \widehat{u}_4) + \frac{1}{a}(\widehat{u}_7\eta + \widehat{u}_6)
\end{aligned} \tag{1}$$

It is apparent that if $\widehat{u}_2$ is non-zero, a constant ϵ_x can be obtained. Similar reasoning holds for ϵ_y and γ_{xy}. Thus, it can be seen that the displacement functions are complete.

Now, consider the compatibility. From Eqs. (6.11) and (6.12) the displacements u_x and u_y are continuous throughout the element. From Eq. (6.30), the highest derivative appearing in the principle of virtual work is 1, so that the displacement function must be continuous on the interelement boundary in order for the element to be compatible. Note that the trial functions for both u_x and u_y have continuous first derivatives, with respect to ξ and η throughout the element.

Consider elements 1 and 2 in Fig. 6.14. These two elements have a common boundary connecting node 2 and node 5. Assume that nodes 2 and 5 have displacements u_2 and u_5 in the x direction. Using the shape functions given in Eq. (6.19), the displacement at the common boundary for element 1, where $u_2 = u_{x2},\ u_5 = u_{x3}$, and $\xi = 1$, is

$$u_x^1 = (1-\eta)u_2 + \eta u_5 \tag{2}$$

and that for element 2, where $u_2 = u_{x1},\ u_5 = u_{x4}$, and $\xi = 0$, is

$$u_x^2 = (1-\eta)u_2 + \eta u_5 \tag{3}$$

It is seen that u_x^1 and u_x^2 are the same, so that the displacements are continuous at this boundary. We conclude that the element is compatible. ∎

6.5.3 Test of Convergence and Accuracy

The Patch Test

A useful test of acceptability of particular elements in practice is the patch test. A patch test is used to check the completeness of a group of elements. It extends the philosophy of the constant strain requirement from the individual element to a group of elements. In this test, a small field (a patch) of elements, with at least one being completely surrounded by elements, is assembled, and a set of displacements or forces are applied at the boundary of the structure such that the constant strain state should occur inside the structure. For certain elements, e.g., quadrilateral and solid elements, some standard finite element meshes and boundary conditions for patch tests have been proposed [MacNeal and Harder, 1985]. Typical proposed meshes are shown in Fig. 6.23, and the boundary conditions along with the theoretical solutions of the problems are shown in Table 6.4. The strains corresponding to these boundary conditions are constant throughout the elements. The elements in Fig. 6.23 are distorted intentionally, as this is essential to the success of the test. The rectangular exterior shapes of the elements ease the task of applying boundary conditions that should lead to the constant strains in the elements.

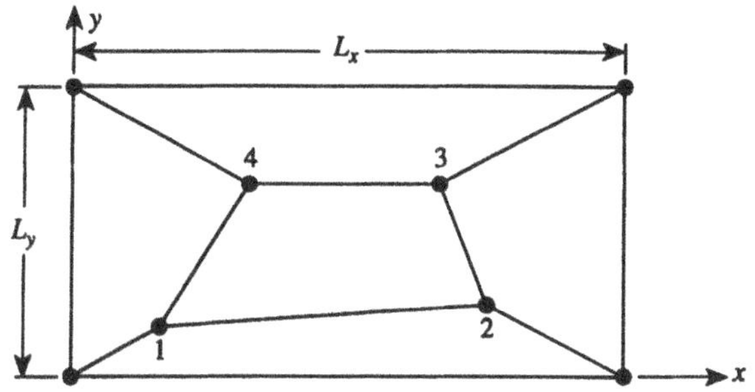

(a) Patch test for the in-plane deformation of flat structures. Let $L_y = 0.24$, $L_x = 0.24$, $t = 0.001$, $E = 1.0 \times 10^6$, and $\nu = 0.25$. Use the boundary conditions of Table 6.4. The inner nodes should be located at

	x	y
1	0.04	0.02
2	0.18	0.03
3	0.16	0.08
4	0.08	0.08

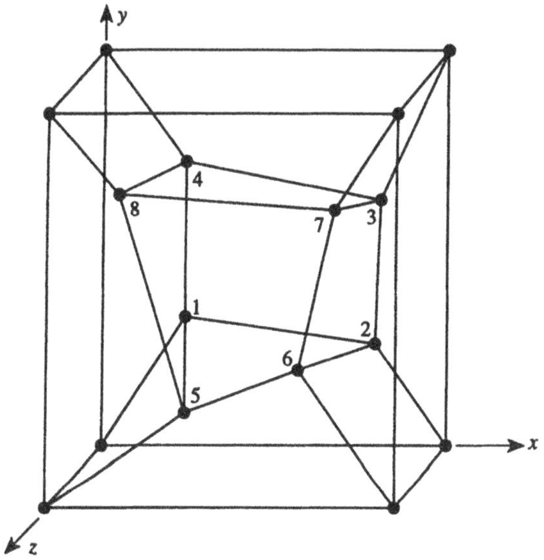

(b) Patch test for solids. Use a unit cube with $E = 1.0 \times 10^6$, $\nu = 0.25$, and the boundary conditions of Table 6.4. Locate interior nodes at

	x	y	z
1	0.249	0.342	0.192
2	0.826	0.288	0.288
3	0.850	0.649	0.263
4	0.273	0.750	0.230
5	0.320	0.186	0.643
6	0.677	0.305	0.683
7	0.788	0.693	0.644
8	0.165	0.745	0.702

FIGURE 6.23
Patch test meshes [MacNeal and Harder, 1985].

TABLE 6.4

Boundary Conditions and Theoretical Solutions for the Patch Tests in Fig. 6.23

a. *Quadrilateral Thin, Flat Element* (In-plane deformation)
Boundary Conditions:

$$\overline{u} = 10^{-3}(x + y/2)$$
$$\overline{v} = 10^{-3}(y + x/2)$$

Theoretical Solutions:*

$$\epsilon_x = \epsilon_y = \gamma_{xy} = 10^{-3}$$
$$\sigma_x = \sigma_y = 1333;\ \tau_{xy} = 400$$

b. *Solid Element*
Boundary Conditions:

$$\overline{u} = 10^{-3}(2x + y + z)/2$$
$$\overline{v} = 10^{-3}(x + 2y + z)/2$$
$$\overline{w} = 10^{-3}(x + y + 2z)/2$$

Theoretical Solutions:*

$$\epsilon_x = \epsilon_y = \epsilon_z = \gamma_{xy} = \gamma_{xz} = 10^{-3}$$
$$\sigma_x = \sigma_y = \sigma_z = 2000;\ \tau_{xy} = \tau_{xz} = \tau_{yz} = 400$$

*The strains are constant throughout the elements.

For the usual patch test, the calculated displacements, strains, and stresses in the interior of the patch should be consistent with the constant strain state. If so, the element is considered to be viable; if not, the element formulation is suspect, and it can be anticipated that the results obtained using it may not converge correctly. The success of the patch test may depend on the geometry of the element, i.e., on the topology of the element layout, and on the boundary conditions. Thus, a given element should be tested using more than one geometry, mesh layout, and strain state. A patch test is considered to be useful in the study of nonconforming elements. If the element cannot pass the patch test, it can still be acceptable.

Accuracy Test [MacNeal and Harder, 1985]

In addition to the patch test, which is designed to check the convergence of the element, other benchmark tests to verify the accuracy of the element may be important. The design of a comprehensive set of tests should take into account the parameters which affect the accuracy, e.g., element geometry, loading, problem geometry, and material properties.

Normally, in an accuracy test, each element has a standard shape: for a two-dimensional element, a square, and for a three-dimensional element, a cube. Of course, the standard shapes cannot always be used in a structural analysis. In some meshes, they have to be distorted. In a test problem, distorted, i.e., nonstandard shaped, elements should also be tested. The distorted shaped elements should be checked with several kinds of loading.

For the loading, the benchmark test problem should account for all load cases that can cause all possible deformations of the structure. For example, for a beam, the load should include a shear force and bending moment in all coordinate directions, along with an axial force and a twisting moment.

In the case of geometry, structures of different shapes, e.g., straight, curved, or twisted, should be tested. An element may give excellent results for one structural geometry, but may behave poorly for another geometry.

Poisson's ratio may have a strong effect on element accuracy as its value approaches 0.5. For some materials, such an effect should be considered.

The basic guideline for the test problem design is to use the element to be tested in different kinds of structures under different loading conditions. Table 6.5 lists several kinds

TABLE 6.5

Elements and Some Structures that can be Formed Using the Elements

	Structure				
Element Type	**Straight Beam**	**Curved Beam**	**Twisted Beam**	**Rectangular Plate**	**Thick-Walled Cylinder**
Beam	X	X	X		
Plate (Membrane) (In-plane loading)	X	X			X*
Plate (Out-of-plane loading)	X	X		X	
Solid	X	X	X	X	X

*Use plane strain option.

of elements and some structures that can be formed with these elements. For example, a solid element can be used in a straight beam, curved beam, or twisted beam, rectangular plate, etc. The loadings in the test problem should be simple yet capable of generating all possible deformations. In MacNeal and Harder (1985), a set of standard benchmark tests for some widely used elements are given. With these tests, the advantages and weaknesses of the elements can be detected. Results of actual tests tend to show considerable variations in accuracy for what may appear to be reasonable elements. In finite element development, these are the kinds of benchmark test problems that are helpful in locating and correcting the element errors and weaknesses.

The following two examples from this reference show the test design for quadrilateral and solid elements.

EXAMPLE 6.4 Accuracy Test for a Four-Node Quadrilateral Element

Consider a four-node quadrilateral element shown in Fig. 6.24a. The following cases should be tested.

Element shape:

1. Regular (the element shape has not been intentionally distorted)—square or rectangular
2. Irregular—distorted

Structures which employ this element are

1. Straight beam
2. Curved beam
3. Thick-walled cylinder

Loading:

1. For cantilevered straight and curved beams—tip loading including extensional and shearing forces
2. For thick-walled cylinder—pressure at the inner radius

The test results should be compared to the theoretical solutions of the test structures with the load cases. The theoretical solutions can be obtained from Pilkey (1994), as well as from other structural mechanics books. Typical problems occur in finding, for example, that the element geometry significantly influences the accuracy for a straight beam under bending load. Also, errors are often large for a thick-walled cylinder when Poisson's ratio approaches 1/2. ▮

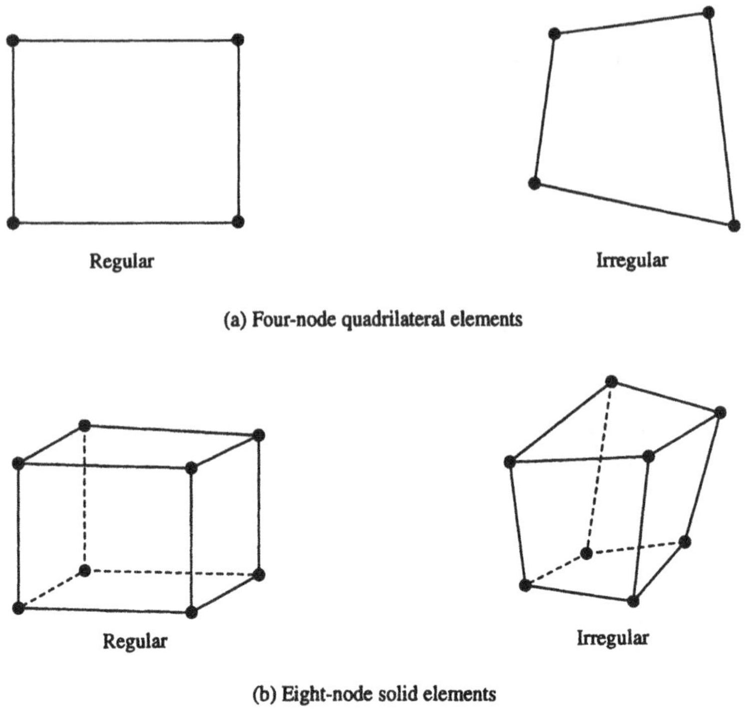

FIGURE 6.24
Quadrilateral and solid elements.

EXAMPLE 6.5 Accuracy Test for an Eight-Node Solid Element
Consider an eight-node solid element as shown in Fig. 6.24b. This type of element is widely used in structural analysis. Two kinds of shapes should be tested.

1. Regular—cube
2. Irregular—distorted

Some structures which employ this element are:

1. Straight beam
2. Curved beam
3. Twisted beam
4. Rectangular plate
5. Thick-walled cylinder

The structures are subject to the following loading:

1. For cantilevered straight, curved, and twisted beams—tip loading, including extensional force and in-plane and out-of-plane shearing forces
2. For rectangular plate—in-plane and out-of-plane pressure or central load
3. For thick-walled cylinder—pressure at inner radius

The results of these tests should be compared to theoretical results. Usually these solid elements work fairly well for straight beams under tensile loads, but significant errors can result under bending loads when the element shape is irregular. ∎

6.5.4 h-Convergence and p-Convergence

The most common method for achieving an improved approximate solution is by successively using finer meshes of elements. This approach is usually called *h-convergence*. Alternatively, the introduction of higher order trial functions can lead to a more accurate solution for a constant set of mesh divisions (*p-convergence*). Although highly problem dependent, often the use of higher order trial functions results in an improved solution at less computational expense than the use of mesh refinement [Szabo and Babuska, 1991]. A combination of h- and p-convergence is often considered as the most favorable approach [Oden, 1990].

Much of the work in this chapter utilizes interpolation functions, wherein the unknown parameters $\widehat{\mathbf{u}}$ of $\mathbf{u} = \mathbf{N}_u\widehat{\mathbf{u}}$ (Eq. 6.13) are replaced by nodal values $\mathbf{v}$. A drawback to the use of interpolation functions is that the introduction of higher order trial functions involves a completely new element matrix. On the other hand, if a trial function expressed in terms of the generalized displacements $\widehat{\mathbf{u}}$ is employed, the effect of successive higher order trial functions can be additive. Such trial functions, which are called *hierarchical*, are illustrated below.

For the two-dimensional plane problem presented earlier in this chapter, the shape function is obtained from (Eq. 6.11)

$$u_x = \widehat{u}_1 + \widehat{u}_2\xi + \widehat{u}_3\xi\eta + \widehat{u}_4\eta \tag{6.51}$$

where $\widehat{u}_i, i = 1, 2, \ldots, 4$ are the unknown constants (generalized displacements). In this case, the number of unknown constants is equal to the number of nodes; consequently, it is convenient to express the shape function in terms of the nodal displacements. However, more terms can be included in the shape function. Instead of using Eq. (6.51), let the shape function u_x be expressed as

$$\begin{aligned} u_x &= \widehat{u}_1 + \widehat{u}_2\xi + \widehat{u}_3\eta + \widehat{u}_4\xi\eta - \widehat{u}_5\xi^2 - \widehat{u}_6\eta^2 \\ &= [1 \quad \xi \quad \eta \quad \xi\eta \quad -\xi^2 \quad -\eta^2] \begin{bmatrix} \widehat{\mathbf{u}}_1 \\ \widehat{\mathbf{u}}_2 \\ \vdots \\ \widehat{u}_6 \end{bmatrix} \\ &= [\mathbf{N}_{u1} \quad \mathbf{N}_{u2}] \begin{bmatrix} \widehat{\mathbf{u}}_1 \\ \widehat{\mathbf{u}}_2 \end{bmatrix} \end{aligned} \tag{6.52}$$

where

$$\mathbf{N}_{u1} = [1 \quad \xi \quad \eta \quad \xi\eta], \qquad \mathbf{N}_{u2} = [-\xi^2 - \eta^2]$$
$$\widehat{\mathbf{u}}_1 = [\widehat{u}_1 \quad \widehat{u}_2 \quad \widehat{u}_3 \quad \widehat{u}_4]^T, \qquad \widehat{\mathbf{u}}_2 = [\widehat{u}_5 \quad \widehat{u}_6]^T$$

Substitute the nodal coordinates into Eq. (6.52), giving

$$\begin{aligned} \begin{bmatrix} u_{x1} \\ u_{x2} \\ u_{x3} \\ u_{x4} \end{bmatrix} &= \begin{bmatrix} 1 & 0 & 0 & 0 & 0 & 0 \\ 1 & 1 & 0 & 0 & -1 & 0 \\ 1 & 1 & 1 & 1 & -1 & -1 \\ 1 & 0 & 1 & 0 & 0 & -1 \end{bmatrix} \begin{bmatrix} \widehat{u}_1 \\ \widehat{u}_2 \\ \vdots \\ \widehat{u}_6 \end{bmatrix} \\ \mathbf{v}_x &= [\widehat{\mathbf{N}}_{u1} \quad \widehat{\mathbf{N}}_{u2}] \begin{bmatrix} \widehat{\mathbf{u}}_1 \\ \widehat{\mathbf{u}}_2 \end{bmatrix} \end{aligned} \tag{6.53}$$

in which

$$\widehat{\mathbf{N}}_{u1} = \begin{bmatrix} 1 & 0 & 0 & 0 \\ 1 & 1 & 0 & 0 \\ 1 & 1 & 1 & 1 \\ 1 & 0 & 1 & 0 \end{bmatrix} \qquad \widehat{\mathbf{N}}_{u2} = \begin{bmatrix} 0 & 0 \\ -1 & 0 \\ -1 & -1 \\ 0 & -1 \end{bmatrix}$$

The 4×4 matrix $\widehat{\mathbf{N}}_{u1}$ can be inverted readily, giving

$$\widehat{\mathbf{N}}_{u1}^{-1} = \begin{bmatrix} 1 & 0 & 0 & 0 \\ -1 & 1 & 0 & 0 \\ -1 & 0 & 0 & 1 \\ 1 & -1 & 1 & -1 \end{bmatrix}$$

Then, from Eq. (6.53) $\widehat{\mathbf{u}}_1$ is given by

$$\widehat{\mathbf{u}}_1 = \widehat{\mathbf{N}}_{u1}^{-1}\, \mathbf{v}_x - \widehat{\mathbf{N}}_{u1}^{-1}\, \widehat{\mathbf{N}}_{u2}\, \widehat{\mathbf{u}}_2 \tag{6.54}$$

Substitute Eq. (6.54) into Eq. (6.52), yielding

$$\begin{aligned} u_x &= \mathbf{N}_{u1}\big(\widehat{\mathbf{N}}_{u1}^{-1}\mathbf{v}_x - \widehat{\mathbf{N}}_{u1}^{-1}\widehat{\mathbf{N}}_{u2}\widehat{\mathbf{u}}_2\big) + \mathbf{N}_{u2}\widehat{\mathbf{u}}_2 \\ &= \mathbf{N}_{u1}\widehat{\mathbf{N}}_{u1}^{-1}\mathbf{v}_x + \big(-\mathbf{N}_{u1}\widehat{\mathbf{N}}_{u1}^{-1}\widehat{\mathbf{N}}_{u2} + \mathbf{N}_{u2}\big)\widehat{\mathbf{u}}_2 \\ &= \mathbf{N}_1\mathbf{v}_x + \mathbf{N}_2\widehat{\mathbf{u}}_2 \end{aligned} \tag{6.55}$$

where

$$\mathbf{N}_1 = [(1-\xi)(1-\eta) \quad \xi(1-\eta) \quad \xi\eta \quad (1-\xi)\eta]$$

and

$$\mathbf{N}_2 = [(1-\xi)\xi \quad \eta(1-\eta)]$$

Since the terms $\mathbf{N}_2\widehat{\mathbf{u}}_2$ do not represent nodal displacements, they are called nodeless variables. Sometimes they are referred to as "extra shapes."

From Eq. (6.55), using the principle of virtual work, the element stiffness equations can be formed as

$$\begin{bmatrix} \mathbf{k}_{11} & \mathbf{k}_{12} \\ \mathbf{k}_{21} & \mathbf{k}_{22} \end{bmatrix} \begin{bmatrix} \mathbf{v}_x \\ \widehat{\mathbf{u}}_2 \end{bmatrix} = \begin{bmatrix} \mathbf{p} \\ \mathbf{r} \end{bmatrix} \tag{6.56}$$

where $\mathbf{r}$ contains the load parameters associated with $\mathbf{N}_2$. The variables $\widehat{\mathbf{u}}_2$ can be condensed out before the element stiffness matrix is assembled into the global stiffness matrix. In the condensation process, the second equation of Eq. (6.59) is used to express $\widehat{\mathbf{u}}_2$ in terms of $\mathbf{k}_{21}$, $\mathbf{k}_{22}$, $\mathbf{v}_x$, and $\mathbf{r}$. Then this $\widehat{\mathbf{u}}_2$ is substituted into the first equation of Eq. (6.56), resulting in a stiffness matrix $\mathbf{k}\mathbf{v}_x = \mathbf{p}'$. In Eq. (6.56), $\mathbf{k}_{11}$ is the same as $\mathbf{k}^i$ in Eq. (6.31). That is, the inclusion of extra shapes only expands the stiffness matrix and the original stiffness matrix remains intact. As this higher order approximation proceeds, the element matrices computed at the previous step of the approximation are used and, hence, need not be re-established. The introduction of $\widehat{\mathbf{u}}_2$ serves the purpose of introducing more DOF, and thereby, the accuracy of an element can be improved. Using the extra shapes can permit, for example, a parabolic deformation along the element edge, and a more realistic deformation shape may be achieved. But because $\widehat{u}_i\,(i = 5, 6)$ do not represent the nodal displacements, it may have different values for the adjacent elements. This raises the question of compatibility. A gap or overlap may develop at the interelement boundaries, but incompatible elements are still acceptable if the incompatibilities disappear and a constant strain state is approached as the mesh is refined. Some modifications on the extra shapes can also help reduce the incompatibility problem.

6.5.5 Polynomial Shape Functions

Polynomials are commonly employed to approximate unknown functions. In the finite element method, the interpolation functions N_i are almost always polynomials. This class of functions is considered to be desirable because of the ease of manipulation. Normally, the higher the degree of the polynomial the better the approximation, and a refined mesh usually improves convergence. Moreover, polynomials are easy to differentiate. Thus, if polynomials approximate the displacements of the structure, the strains can be evaluated with ease.

It is useful to sketch the derivation of interpolation functions, even though similar derivations are given in Chapter 4 and earlier in this chapter. Begin with a two-dimensional trial function for the displacement $u(\xi, \eta)$ in the form

$$u = \widehat{u}_1 + \widehat{u}_2\xi + \widehat{u}_3\eta + \widehat{u}_4\xi^2 + \widehat{u}_5\xi\eta + \cdots \quad = \mathbf{N}_u \ \widehat{\mathbf{u}} \tag{6.57}$$

with $\mathbf{N}_u = [1 \ \ \xi \ \ \eta \ \ \xi^2 \ \ \xi\eta \cdots]$ and the generalized displacements or parameters are

$$\widehat{\mathbf{u}} = \begin{bmatrix} \widehat{u}_1 \\ \widehat{u}_2 \\ \vdots \end{bmatrix}$$

The number n of generalized parameters $\widehat{u}_i$ corresponds to the number of nodal DOF.

The generalized parameters $\widehat{u}_i$ are replaced by the mechanically meaningful nodal variables $\mathbf{v}$, using

$$\mathbf{v} = \widehat{\mathbf{N}}_u \ \widehat{\mathbf{u}} \tag{6.58}$$

where the $n \times n$ matrix $\widehat{\mathbf{N}}_u$ contains the discrete values resulting from evaluating Eq. (6.57), or its derivatives, at the nodes. Then

$$\widehat{\mathbf{u}} = \widehat{\mathbf{N}}_u^{-1} \ \mathbf{v} = \mathbf{G} \ \mathbf{v}$$

and

$$u = \mathbf{N}_u \ \widehat{\mathbf{N}}_u^{-1} \ \mathbf{v} = \mathbf{N} \ \mathbf{v} \tag{6.59}$$

where, in two dimensions,

$$\mathbf{N}(\xi, \eta) = \mathbf{N}_u \ \widehat{\mathbf{N}}_u^{-1} = \mathbf{N}_u \ \mathbf{G} \tag{6.60}$$

is the desired interpolation function.

EXAMPLE 6.6 Interpolation Functions

Suppose the polynomial

$$u = \widehat{u}_1 + \widehat{u}_2\xi = [1 \quad \xi] \begin{bmatrix} \widehat{u}_1 \\ \widehat{u}_2 \end{bmatrix} = \mathbf{N}_u \ \widehat{\mathbf{u}} \tag{1}$$

is to be employed for a two-node element with end displacements as DOF (see Fig. 6.25a).

To replace the constants $\widehat{u}_1$ and $\widehat{u}_2$ by the coordinates v_a and v_b, use

$$\begin{aligned} u(\xi = 0) &= v_a = \widehat{u}_1 + \widehat{u}_2 \cdot 0 \\ u(\xi = 1) &= v_b = \widehat{u}_1 + \widehat{u}_2 \cdot 1 \end{aligned} \tag{2}$$

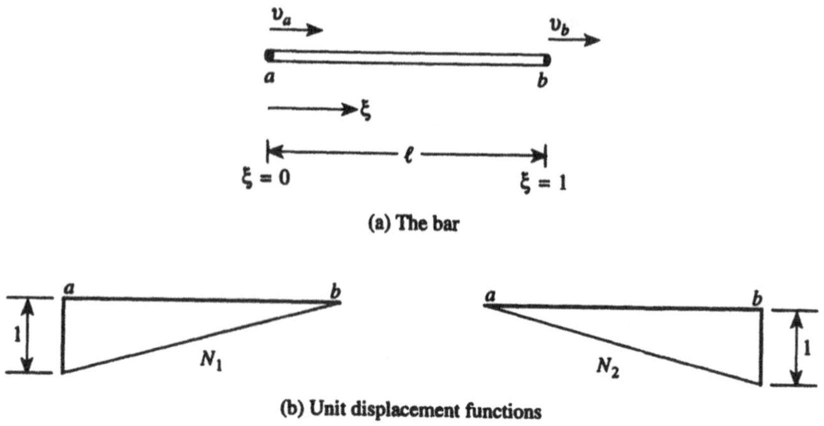

FIGURE 6.25
A two-node system.

or

$$\begin{bmatrix} v_a \\ v_b \end{bmatrix} = \begin{bmatrix} 1 & 0 \\ 1 & 1 \end{bmatrix} \begin{bmatrix} \widehat{u}_1 \\ \widehat{u}_2 \end{bmatrix} = \widehat{\mathbf{N}}_u \, \widehat{\mathbf{u}} \tag{3}$$

From (1),

$$u = [1 \quad \xi] \underbrace{\begin{bmatrix} 1 & 0 \\ -1 & 1 \end{bmatrix}}_{\widehat{\mathbf{N}}_u^{-1}} \begin{bmatrix} v_a \\ v_b \end{bmatrix} = [(1-\xi) \quad \xi] \begin{bmatrix} v_a \\ v_b \end{bmatrix} = [N_1 \quad N_2]\mathbf{v} = \mathbf{N}(\xi) \ \mathbf{v} \tag{4}$$

The unit displacement functions N_1 and N_2 are sketched in Fig. 6.25b. ∎

Complete Polynomials

The *order* (or *degree*) m of a polynomial is the highest power to which the variables are raised. For example, if the "final" term of a two-dimensional polynomial series is $\xi^5\eta^4$, the order is $5 + 4 = 9$. A polynomial is said to be *complete* to a given order if it contains all terms of that order and below. The row matrix $\mathbf{N}_u$ of Eq. (6.57) contains the terms for a polynomial that is complete to a certain degree, say m. The number of terms n present in a complete two-dimensional polynomial of degree m is

$$n = \frac{1}{2}(m+1)(m+2) \tag{6.61}$$

Sometimes it is convenient to use the *Pascal*[4] *triangle* to assist in assuring that some terms are not overlooked. The diagonal lines identify complete polynomials.

[4]Blaise Pascal (1623–1662) was a French mathematician, with an interest in physics and epistemology. His scientific interest was aroused at about the age of 13 when he read Euclid's *Elements*. At 16 he made his first recognized contribution now known as Pascal's "mystic-hexagram." By 1645 he completed the design and development of a calculation machine, which was originally intended to assist his father, an accountant, with addition and subtraction. Pascal oversaw the manufacture and marketing of his machine. When he was in his early 30s, for two years he abandoned his scientific pursuits in favor of religious activities.

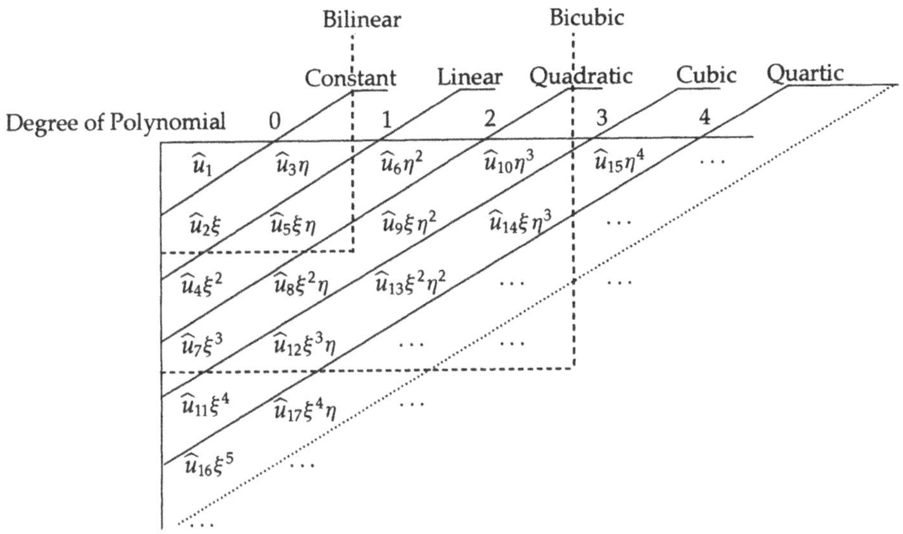

Note, for example, from Eq. (6.61) or the Pascal triangle, that ten terms are needed to form a complete two-dimensional cubic polynomial.

There are several procedures available for using interpolation polynomials to represent the response throughout a two- or three-dimensional element and to place a network of nodes on the element. Three schemes—Lagrangian interpolation, natural coordinates, and Hermitian interpolation—are considered in this section. Each is introduced as a unidirectional polynomial and then extended to the multidimensional case.

6.5.6 Lagrangian Interpolation

In order to study the problem of creating an interpolation trial function for two- and three-dimensional elements, we begin by considering a one-dimensional Lagrange polynomial which provides an interpolation function for a line divided into segments. Products of these polynomials lead to interpolation functions, along with a network of nodes, for multidimensional elements.

One-Dimensional Case in Cartesian Coordinates

Consider a line (Fig. 6.26) divided into segments of equal length by m points (nodes), with the nodes defined by the coordinates $\xi_1, \xi_2, \ldots, \xi_m$. We wish to establish a function $u(\xi)$ that takes on the exact displacements $v_1, v_2, \ldots, v_m$ at these m points and also provides approximate displacement values at points intermediate to these nodal points. In the case of Lagrangian interpolation, a polynomial $u(\xi)$ of order $m-1$ is defined to replace the true displacements such that $u(\xi_i)$ is equal to v_i at the nodal points ξ_i, i.e.,

$$u(\xi_i) = v_i, \quad i = 1, 2, \ldots, m$$

Express $u(\xi)$ as

$$u(\xi) = \sum_i^m N_i(\xi)\, u(\xi_i) = \sum_i^m N_i\; v_i \quad i = 1, 2, \ldots, m \tag{6.62}$$

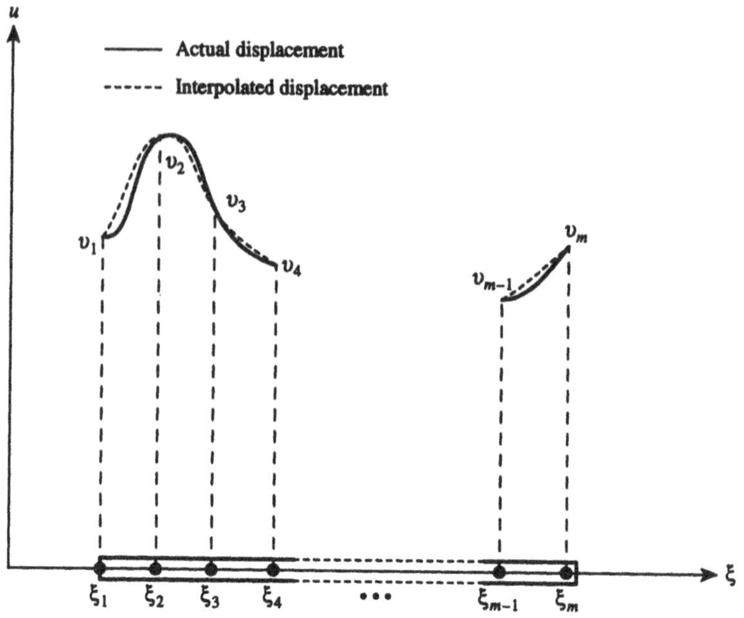

FIGURE 6.26
One-dimensional case with m nodes for the Lagrangian interpolation.

where $N_i(\xi)$ is a polynomial of order $m-1$, and v_i are values of u at the nodal points. For Lagrangian interpolation, N_i must have the properties

$$\begin{aligned} N_i(\xi_i) &= 1 \\ N_i(\xi_k) &= 0 \quad (i \neq k) \end{aligned} \tag{6.63}$$

Note that these two characteristics of N_i are the same as those of the shape functions defined in Eq. (6.48).

From the properties of Eq. (6.63), $N_i(\xi)$ can be of the form (i term omitted in numerator and denominator)

$$N_i(\xi) = \frac{(\xi-\xi_1)(\xi-\xi_2)\cdots(\xi-\xi_{i-1})(\xi-\xi_{i+1})\cdots(\xi-\xi_m)}{(\xi_i-\xi_1)(\xi_i-\xi_2)\cdots(\xi_i-\xi_{i-1})(\xi_i-\xi_{i+1})\cdots(\xi_i-\xi_m)} = \frac{\prod\limits_{\substack{j=1\\ j\neq i}}^{m}(\xi-\xi_j)}{\prod\limits_{\substack{j=1\\ j\neq i}}^{m}(\xi_i-\xi_j)} \tag{6.64}$$

which are called *Lagrangian polynomials*. As an example, for a 3 node, single DOF per node, line element, use Eq. (6.62) with $m=3$ and

$$N_1 = \frac{(\xi-\xi_2)(\xi-\xi_3)}{(\xi_1-\xi_2)(\xi_1-\xi_3)} \qquad N_2 = \frac{(\xi-\xi_1)(\xi-\xi_3)}{(\xi_2-\xi_1)(\xi_2-\xi_3)} \qquad N_3 = \frac{(\xi-\xi_1)(\xi-\xi_2)}{(\xi_3-\xi_1)(\xi_3-\xi_2)}$$

Two-Dimensional Case in Cartesian Coordinates

In the two-dimensional case, we seek functions that are uniquely defined at the DOF on the edge and interior nodes. A simple product of the one-dimensional shape functions for

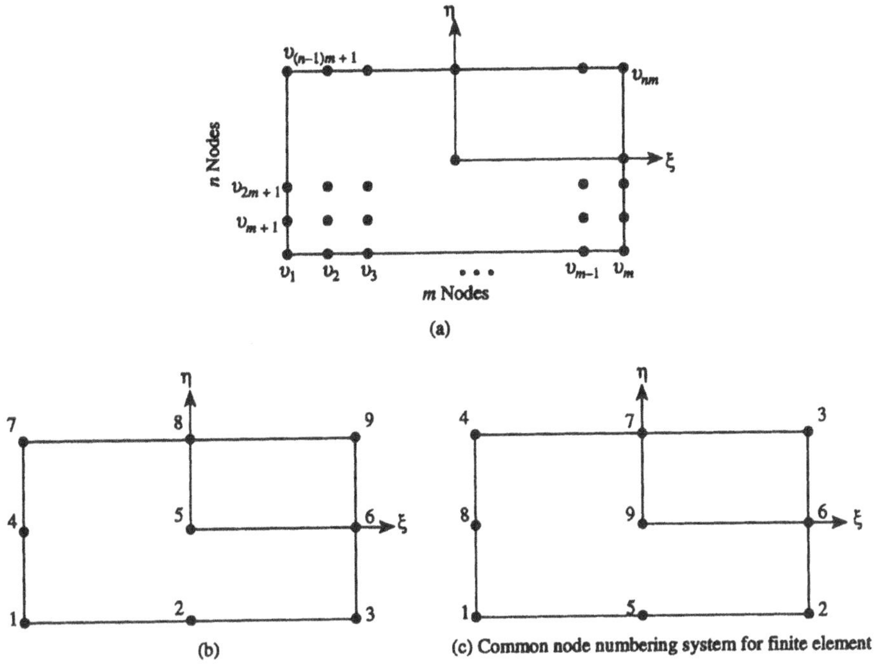

FIGURE 6.27
Rectangle for two-dimensional Lagrangian interpolation.

the ξ and η directions will suffice for such cases. Consider the two-dimensional element shown in Fig. 6.27a, in which there are m equally spaced nodes in each row and n equally spaced nodes in each column. The shape function $u(\xi, \eta)$ can be written as

$$u(\xi, \eta) = \sum N_{ij}\ v_k = \sum_{i=1}^{n} \sum_{j=1}^{m} N_j(\xi) N_i(\eta) v_k \tag{6.65a}$$

where $k = (i-1)m + j$, and $N_j(\xi)$ and $N_i(\eta)$ are defined in Eq. (6.64). In matrix form,

$$\begin{aligned} u(\xi, \eta) &= [N_1(\xi) \quad N_2(\xi) \cdots N_m(\xi)] \begin{bmatrix} v_1 & v_{m+1} & & v_{(n-1)m+1} \\ v_2 & v_{m+2} & & v_{(n-1)m+2} \\ \vdots & \vdots & & \\ v_m & v_{2m} & \cdots & v_{nm} \end{bmatrix} \begin{bmatrix} N_1(\eta) \\ N_2(\eta) \\ \vdots \\ N_n(\eta) \end{bmatrix} \\ &= \mathbf{N}_\xi^T \mathbf{R} \mathbf{N}_\eta \end{aligned} \tag{6.65b}$$

Take the element in Fig. 6.27b as an example.

$$\begin{aligned} u(\xi, \eta) = {} & N_{1\xi} N_{1\eta} v_1 + N_{2\xi} N_{1\eta} v_2 + N_{3\xi} N_{1\eta} v_3 + N_{1\xi} N_{2\eta} v_4 + N_{2\xi} N_{2\eta} v_5 + N_{3\xi} N_{2\eta} v_6 \\ & + N_{1\xi} N_{3\eta} v_7 + N_{2\xi} N_{3\eta} v_8 + N_{3\xi} N_{3\eta} v_9 = \mathbf{N}\,\mathbf{v} \end{aligned} \tag{6.65c}$$

where $N_{i\xi} = N_i(\xi)$ and $N_{i\eta} = N_i(\eta)$ of Eq. (6.64) with $m = 3$. Also, v_k, $k = 1, 2, \ldots, 9$ are nodal displacements. Frequently, the finite element method uses a node numbering scheme of Fig. 6.27c. For this case, Eq. (6.65c) becomes

$$u(\xi,\eta) = N_{1\xi}N_{1\eta}v_1 + N_{2\xi}N_{1\eta}v_5 + N_{3\xi}N_{1\eta}v_2 + N_{1\xi}N_{2\eta}v_8 + N_{2\xi}N_{2\eta}v_9 + N_{3\xi}N_{2\eta}v_6 + N_{1\xi}N_{3\eta}v_4 + N_{2\xi}N_{3\eta}v_7 + N_{3\xi}N_{3\eta}v_3 = \mathbf{N}\,\mathbf{v} \tag{6.65d}$$

In matrix notation, Eq. (6.65c) appears as

$$u(\xi,\eta) = [N_{1\xi} \quad N_{2\xi} \quad N_{2\xi}]\begin{bmatrix} v_1 & v_4 & v_7 \\ v_2 & v_5 & v_8 \\ v_3 & v_6 & v_9 \end{bmatrix}\begin{bmatrix} N_{1\eta} \\ N_{2\eta} \\ N_{3\eta} \end{bmatrix} = \mathbf{N}_\xi^T \mathbf{R}\,\mathbf{N}_\eta \tag{6.65e}$$

and Eq. (6.65d) becomes

$$u(\xi,\eta) = [N_{1\xi} \quad N_{2\xi} \quad N_{3\xi}]\begin{bmatrix} v_1 & v_8 & v_4 \\ v_5 & v_9 & v_7 \\ v_2 & v_6 & v_3 \end{bmatrix}\begin{bmatrix} N_{1\eta} \\ N_{2\eta} \\ N_{3\eta} \end{bmatrix} = \mathbf{N}_\xi^T \mathbf{R}\,\mathbf{N}_\eta. \tag{6.65f}$$

This is referred to a *biquadratic interpolation*. The two-dimensional Lagrangian interpolation function contains a complete order m polynomial plus some individual terms of higher order. Since the convergence of the finite element process can be shown to be related to the highest order complete polynomial, it can be useful to eliminate some terms, e.g., those corresponding to the inner nodes, so that the interpolation functions depend only on corner and boundary points.

EXAMPLE 6.7 Shape Functions of a Bilinear Element

For the bilinear element in Fig. 6.10, construct the shape functions corresponding to u_x, the displacement in the x direction, directly from the Lagrangian interpolation polynomials.

The bilinear polynomial of Eq. (6.11)

$$u_x = \hat{u}_1 + \hat{u}_2\xi + \hat{u}_3\xi\eta + \hat{u}_4\eta \tag{1}$$

This polynomial is, according to the Pascal triangle, not complete in the sense of not including all of the terms of order 2 and below. On the other hand, the element is complete because (1) satisfies all the completeness conditions given in Section 6.5.2. Alternatively, a shape function similar in form to that of Eq. (6.65f) can be assembled for the displacement u_x.

$$u_x = N_{1\xi}N_{1\eta}u_{x1} + N_{2\xi}N_{1\eta}u_{x2} + N_{2\xi}N_{2\eta}u_{x3} + N_{1\xi}N_{2\eta}u_{x4} \tag{2}$$

or

$$u_x = [N_{1\xi} \quad N_{2\xi}]\begin{bmatrix} u_{x1} & u_{x4} \\ u_{x2} & u_{x3} \end{bmatrix}\begin{bmatrix} N_{1\eta} \\ N_{2\eta} \end{bmatrix} \tag{3}$$

where $N_{i\xi}$ and $N_{i\eta}$, $i = 1, 2$ are the Lagrangian polynomials of Eq. (6.64) with $m = 2$. Thus, with $\xi_1 = 0$ and $\xi_2 = 1$ (and $\eta_1 = 0$, $\eta_2 = 1$),

$$\begin{aligned} N_{1\xi} &= \frac{\xi - 1}{0 - 1} = -(\xi - 1) = (1 - \xi), & N_{2\xi} &= \frac{\xi - 0}{1 - 0} = \xi \\ N_{1\eta} &= \frac{\eta - 1}{0 - 1} = -(\eta - 1) = (1 - \eta), & N_{2\eta} &= \frac{\eta - 0}{1 - 0} = \eta \end{aligned} \tag{4}$$

and

$$\begin{aligned} N_{1\xi} N_{1\eta} &= (1-\xi)(1-\eta) \\ N_{2\xi} N_{1\eta} &= \xi(1-\eta) \\ N_{2\xi} N_{2\eta} &= \xi\eta \\ N_{1\xi} N_{2\eta} &= (1-\xi)\eta \end{aligned} \tag{5}$$

Then u_x can be written as

$$u_x = [(1-\xi)(1-\eta) \quad \xi(1-\eta) \quad \xi\eta \quad (1-\xi)\eta] \begin{bmatrix} u_{x1} \\ u_{x2} \\ u_{x3} \\ u_{x4} \end{bmatrix} = \mathbf{N}_x \mathbf{v}_x \tag{6}$$

Note that $\mathbf{N}_x$ here is the same as that in Eq. (6.18). ∎

Lagrangian Elements

The *Lagrange* family of elements is characterized by the nodal unknowns being values of the dependent variables. This is in contrast to the *Hermitian element*, e.g., beam element, which includes also the derivatives (slopes) of the displacement (deflection) among the unknown nodal parameters. With the exception of some lower order members of the family, the Lagrange elements have internal nodes. This is sometimes considered to be disadvantageous. Elements with all nodes on the boundary are often referred to as *serendipity elements*.

One-Dimensional Case in Natural Coordinates

An interesting means of expressing coordinates is by using so-called *natural* (or *homogeneous*) *coordinates*. These coordinates are a mapping of physical coordinates into nondimensional coordinates that assume the values one or zero at the nodes. Natural coordinates are commonly used to create an interpolation function and nodal pattern for two-dimensional elements. We begin the study of natural coordinates by defining them for a single dimension.

For the one-dimensional case, consider a line with points 1, 2, and i defined by coordinates ξ_1, ξ_2, and ξ, respectively, as shown in Fig. 6.28. The natural coordinates L_2 and L_1, which serve as weighting functions, are defined by the ratios

$$L_2 = \frac{\xi - \xi_1}{\xi_2 - \xi_1} \qquad L_1 = \frac{(\xi_2 - \xi_1) - (\xi - \xi_1)}{\xi_2 - \xi_1} = \frac{\xi_2 - \xi}{\xi_2 - \xi_1} \tag{6.66}$$

where $\xi - \xi_1$ is the distance to point i. Note that the natural coordinates are not independent, since they have been nondimensionalized, so that

$$L_1 + L_2 = 1 \tag{6.67}$$

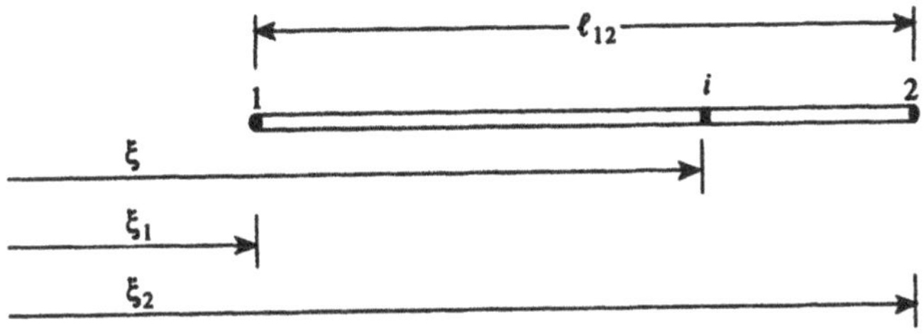

FIGURE 6.28
One-dimensional natural coordinates L_1 and L_2.

From Eqs. (6.66), it follows that the natural coordinates are defined such that

$$L_1 = 1 \quad \text{if } i \text{ falls on point 1,} \qquad L_1 = 0 \quad \text{if } i \text{ falls on point 2}$$
$$L_2 = 1 \quad \text{if } i \text{ falls on point 2,} \qquad L_2 = 0 \quad \text{if } i \text{ falls on point 1}$$

It can be seen that L_1 and L_2 satisfy the properties of Eq. (6.63) and, hence, can be considered to be forms of Lagrangian interpolation functions. The ξ coordinate of any point along the line can be expressed as

$$\xi = L_1\xi_1 + L_2\xi_2 \tag{6.68}$$

a result which is verified by use of Eq. (6.66). Equations (6.67) and (6.68) placed together appear as

$$\begin{bmatrix} 1 \\ \xi \end{bmatrix} = \begin{bmatrix} 1 & 1 \\ \xi_1 & \xi_2 \end{bmatrix} \begin{bmatrix} L_1 \\ L_2 \end{bmatrix}$$

These relationships, i.e., Eqs. (6.67) and (6.68), can be considered to be the definition of the natural coordinates, since they lead to

$$\begin{bmatrix} L_1 \\ L_2 \end{bmatrix} = \frac{1}{(\xi_2 - \xi_1)} \begin{bmatrix} \xi_2 & -1 \\ -\xi_1 & 1 \end{bmatrix} \begin{bmatrix} 1 \\ \xi \end{bmatrix} \tag{6.69}$$

which is the same as Eq. (6.66)

A useful property of these natural coordinates is

$$\int_{\xi_1}^{\xi_2} \begin{bmatrix} L_1 \\ L_2 \end{bmatrix} d\xi = \frac{\ell_{12}}{2} \begin{bmatrix} 1 \\ 1 \end{bmatrix} \tag{6.70}$$

where ℓ_{12} is the length from ξ_1 to ξ_2.

The natural coordinates are useful in representing shape functions for a line divided into segments. It is convenient to utilize a different node numbering scheme. Designate the left end point as 0 and the right end point as m, i.e., these points are numbered $0, 1, 2, \ldots, m$. Using natural coordinates, each point is identified by its location relative to the two end points of the line. Define numbers p and q, where p and q are the number of points to the right and left, respectively, of the point under consideration. Observe Figs. 6.29a, b, and c wherein the three nodes in Fig. 6.29c are denoted as 20, 11, and 02. Thus, the coordinate ξ_2 of Eq. (6.66) is now ξ_{02} and ξ_1 is ξ_{20}. The nodal degrees of freedom will now be labeled with p, q, i.e., v_{pq} or in the case of Fig. 6.29c, v_{20}, v_{11} and v_{02}. The corresponding shape functions are N_{pq}, i.e., N_{20}, N_{11}, and N_{02}.

Shape functions in terms of natural coordinates are given by

$$N_{pq}(L_1, L_2) = N_p(L_1)\, N_q(L_2) \tag{6.71a}$$

where, from the Lagrangian interpolation formula, $N_p(L_1)$ and $N_q(L_2)$ are defined by

$$N_i(L_k) = \begin{cases} \prod_{j=1}^{i} \frac{mL_k - j + 1}{j}, & \text{for } i \geq 1 \\ 1 & \text{for } i = 0 \end{cases} \qquad k = 1, 2, \cdots \tag{6.71b}$$

where m is the number of segments.

EXAMPLE 6.8 Three-Node Element

For the element of Fig. 6.29c, the trial function for the displacement can be expressed as

$$u = N_{20}\, v_{20} + N_{11}\, v_{11} + N_{02}\, v_{02}$$

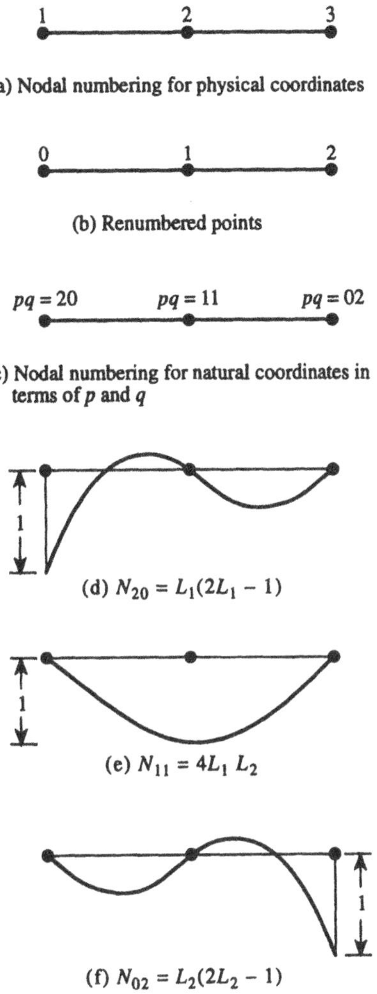

FIGURE 6.29
Shape functions for three-node element.

where, from Eq. (6.71b) with $m = 2$,

$$N_2(L_1) = L_1(2L_1 - 1) \qquad N_2(L_2) = L_2(2L_2 - 1)$$
$$N_1(L_1) = 2L_1 \qquad N_1(L_2) = 2L_2 \qquad N_0(L_1) = N_0(L_2) = 1$$

and from Eq. (6.71a),

$$\begin{aligned} N_{20} &= N_2(L_1)N_0(L_2) = L_1(2L_1 - 1) \\ N_{11} &= N_1(L_1)N_1(L_2) = 4L_1L_2 \\ N_{02} &= N_0(L_1)N_2(L_2) = L_2(2L_2 - 1) \end{aligned}$$

which are plotted in Figs. 6.29d, e, and f. Then the trial displacement function becomes

$$u(L_1,\ L_2) = L_1(2L_1 - 1)v_{20} + 4L_1L_2v_{11} + L_2(2L_2 - 1)v_{02}$$

■

Two-Dimensional Case in Natural Coordinates

The interpolation functions in natural coordinates for a triangle can be established in a fashion similar to the procedure for one dimension. The natural coordinates for a triangle

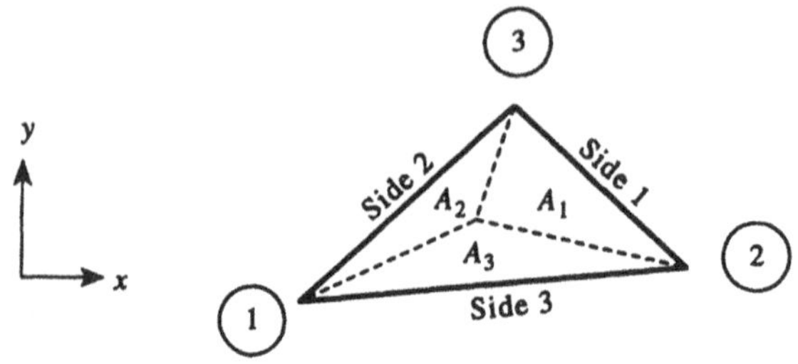

FIGURE 6.30
Triangle used to define coordinates.

are often referred to as *triangular* or *homogeneous coordinates* [Schwarz, 1980]. These are very useful in establishing shape functions referred to nodal DOF.

Consider the triangle of Fig. 6.30. We wish to locate a point within the triangle. To do so, draw lines from a given point to the three vertices, thereby dividing the triangle into three triangles of areas A_1, A_2, and A_3. The sides of the triangles are designated by the same number as the opposite vertex. The areas are identified by the number of the adjacent side. The quantities $L_i, i = 1, 2, 3,$

$$L_1 = A_1/A, \qquad L_2 = A_2/A, \qquad L_3 = A_3/A \tag{6.72}$$

where A is the area of the original triangle, are defined to be the triangular coordinates. These two-dimensional natural coordinates are similar to the one-dimensional coordinates. Note that $A_1 + A_2 + A_3 = A$ so that

$$L_1 + L_2 + L_3 = 1 \tag{6.73}$$

The relationships between the Cartesian coordinates x, y, which are the coordinates of the points in the element, and the triangular coordinates L_1, L_2, and L_3 are

$$\begin{aligned} x &= L_1 x_1 + L_2 x_2 + L_3 x_3 \\ y &= L_1 y_1 + L_2 y_2 + L_3 y_3 \end{aligned} \tag{6.74}$$

where $x_i, y_i, i = 1, 2, 3$ are the coordinates of the nodes. These can be verified at the vertices, e.g., at corner, 1, $A_2 = A_3 = 0$, $A_1 = A$ so that $L_2 = L_3 = 0$, $L_1 = 1$. Thus, $x = x_1$, as it should.

The triangular coordinates can be expressed in terms of the known locations of the vertices. From Eqs. (6.73) and (6.74),

$$\begin{bmatrix} 1 \\ x \\ y \end{bmatrix} = \begin{bmatrix} 1 & 1 & 1 \\ x_1 & x_2 & x_3 \\ y_1 & y_2 & y_3 \end{bmatrix} \begin{bmatrix} L_1 \\ L_2 \\ L_3 \end{bmatrix}$$

or, by inversion,

$$L_i = \frac{1}{2A}(\alpha_i + \beta_i x + \gamma_i y) \quad i = 1, 2, 3 \tag{6.75}$$

with

$$A = \frac{1}{2}(x_2 y_3 + x_3 y_1 + x_1 y_2 - x_2 y_1 - x_3 y_2 - x_1 y_3) = \frac{1}{2}\begin{vmatrix} 1 & 1 & 1 \\ x_1 & x_2 & x_3 \\ y_1 & y_2 & y_3 \end{vmatrix}$$

$$\alpha_1 = \begin{vmatrix} x_2 & x_3 \\ y_2 & y_3 \end{vmatrix} \qquad \beta_1 = -\begin{vmatrix} 1 & 1 \\ y_2 & y_3 \end{vmatrix} \qquad \gamma_1 = \begin{vmatrix} 1 & 1 \\ x_2 & x_3 \end{vmatrix}$$

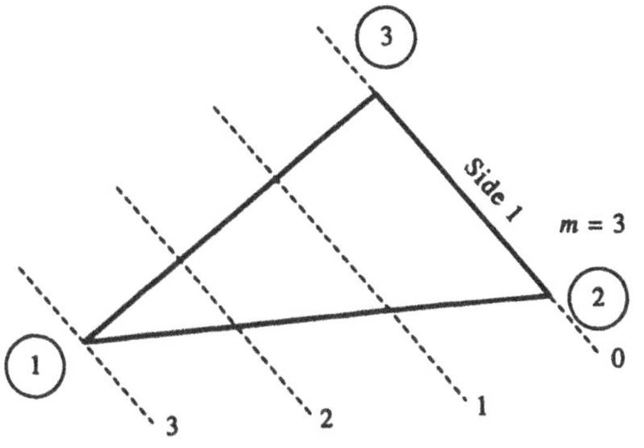

FIGURE 6.31
Subdivision of the distance between side 1 and node 1.

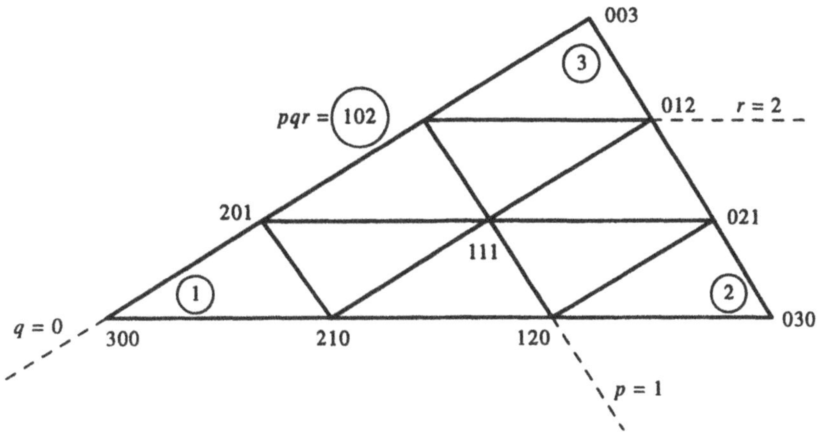

FIGURE 6.32
Labeling of a triangle, $m = 3$.

and the other coefficients $\alpha_2, \beta_2, \ldots, \gamma_3$ are defined by cyclic interchange of the subscripts in the order 1, 2, 3, e.g., $\alpha_2 = x_3 y_1 - x_1 y_3$.

The location of a point in the triangle will be identified using a special construction. As illustrated in Fig. 6.31, draw dashed lines parallel to side 1, dividing the distance between side 1 and node 1 into m equal segments. Label one of these lines as p. Draw similar sets of lines parallel to side 2, and label one of these lines as q. Similarly, label one of a set of lines parallel to side 3 as r. A point at the intersection of lines $p, q,$ and r can now be designated by the three digits $p, q,$ and r. Also, the vertices and points inside a triangle can be assigned pqr labels. A typical labeled triangle is shown in Fig. 6.32. Note that for any point, $p + q + r = m$.

Assume node point displacements are given the same pqr designation as the nodal points themselves. The triangular coordinate approach has permitted a triangle to be "subdivided" with uniquely labeled interior nodes. Trial functions of the form

$$u = \mathbf{N}\mathbf{v} = \sum^{\frac{1}{2}(m+1)(m+2)} N_{pqr} v_{pqr} \tag{6.76}$$

are to be established. For example, for $m = 2$,

$$u = N_{200}v_{200} + N_{110}v_{110} + \cdots + N_{101}v_{101}$$

The shape functions N_{pqr} must be defined such that $N_{pqr} = 1$ at point pqr and equal to zero elsewhere at other nodes. In the one-dimensional case, this was accomplished using the product of Eq. (6.71a) with Lagrangian interpolation employed in each direction. Similarly, for the two-dimensional case,

$$N_{pqr}(L_1, L_2, L_3) = N_p(L_1)\, N_q(L_2)\, N_r(L_3) \tag{6.77}$$

where $N_i(L_j)$, $j = 1, 2, 3$ and $i = p, q, r$ are given by Eq. (6.71).

There is a significant integral property of the triangular coordinates which is useful in the derivation of stiffness matrices. This is the closed form integral

$$\int_A L_1^a\, L_2^b\, L_3^c\, dA = \frac{a!b!c!}{(a+b+c+2)!}2A \tag{6.78}$$

where a is the power to which L_1 is raised, and so forth. Thus, for example, $\int_A L_j\, dA = \frac{1}{3!}2A = \frac{1}{3}A$.

These triangular coordinates are quite useful. They can be used to automatically position interior nodes. Use of these natural coordinates has the advantage that a complete polynomial of the same order as the interpolation is produced. This is quite significant and is one reason for the widespread use of triangular elements.

Three-Dimensional Case in Natural Coordinates

The natural coordinates for the three-dimensional case are volume ratios or volume coordinates. Consider the four-node tetrahedron T with the volume V shown in Fig. 6.33a. Define the side opposite to node i as S_i. Let P be a point inside the tetrahedron. Each side S_i and the point P can define an internal tetrahedron T_i with the volume V_i. Then the natural coordinate can be defined as the ratio of V_i and V, i.e.,

$$L_i = \frac{V_i}{V} \tag{6.79}$$

Note that $\sum_{i=1}^{4} L_i = 1$. The relationship between the Cartesian coordinates x, y and z and L_1, L_2, L_3, and L_4 is found to be

$$\begin{bmatrix} 1 \\ x \\ y \\ z \end{bmatrix} = \begin{bmatrix} 1 & 1 & 1 & 1 \\ x_1 & x_2 & x_3 & x_4 \\ y_1 & y_2 & y_3 & y_4 \\ z_1 & z_2 & z_3 & z_4 \end{bmatrix} \begin{bmatrix} L_1 \\ L_2 \\ L_3 \\ L_4 \end{bmatrix} \tag{6.80}$$

$$\mathbf{X} = \mathbf{N}_u \qquad \mathbf{N}^T$$

where $x_i, y_i, z_i, i = 1, 2, \ldots, 4$ are the Cartesian coordinates of the nodal points.

From Eq. (6.80), the natural coordinates L_i in terms of x, y, z are

$$\mathbf{N}^T = \mathbf{N}_u^{-1}\mathbf{X} \tag{6.81}$$

or in component form

$$L_i = \frac{1}{6V}(\alpha_i + \beta_i x + \gamma_i y + \delta_i z) \tag{6.82}$$

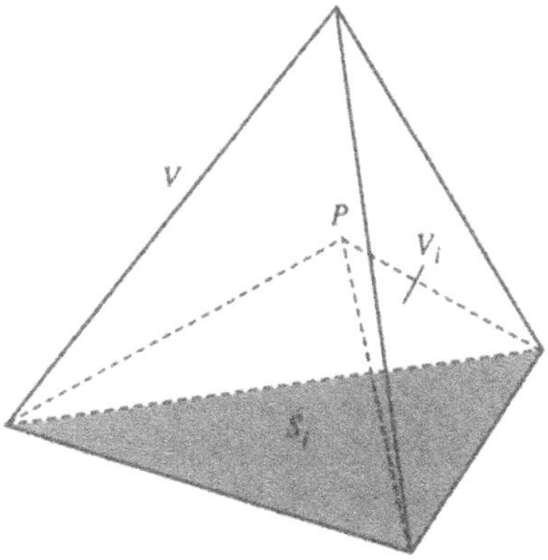

(a) Four-node tetrahedral element with internal tetrahedron defined by point P

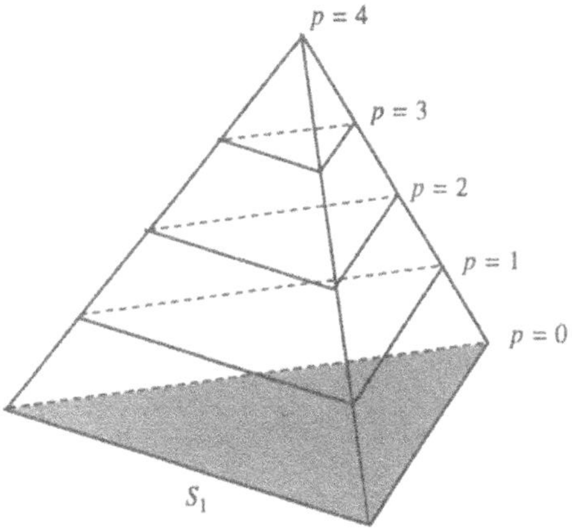

(b) Four blocks of equal thickness parallel to S_1 of a tetrahedral element

FIGURE 6.33
Tetrahedral element.

where $\alpha_i, \beta_i, \gamma_i, \delta_i, i = 1, 2, 3, 4$ are constants and $V = \frac{1}{6}\begin{bmatrix} 1 & 1 & 1 & 1 \\ x_1 & x_2 & x_3 & x_4 \\ y_1 & y_2 & y_3 & y_4 \\ z_1 & z_2 & z_3 & z_4 \end{bmatrix}$. For $i = 1$,

$$\alpha_1 = \begin{bmatrix} x_2 & x_3 & x_4 \\ y_2 & y_3 & y_4 \\ z_2 & z_3 & z_4 \end{bmatrix} \qquad \beta_1 = -\begin{bmatrix} 1 & 1 & 1 \\ y_2 & y_3 & y_4 \\ z_2 & z_3 & z_4 \end{bmatrix}$$

$$\gamma_1 = \begin{bmatrix} 1 & 1 & 1 \\ x_2 & x_3 & x_4 \\ z_2 & z_3 & z_4 \end{bmatrix} \qquad \delta_1 = -\begin{bmatrix} 1 & 1 & 1 \\ x_2 & x_3 & x_4 \\ y_2 & y_3 & y_4 \end{bmatrix}$$

and the constants for other values of i are obtained by cyclic permutation of the subscripts i. For example,

$$\alpha_2 = \begin{bmatrix} x_3 & x_4 & x_1 \\ y_3 & y_4 & y_1 \\ z_3 & z_4 & z_1 \end{bmatrix}$$

If $\mathbf{v} = [v_1 \ v_2 \ v_3 \ v_4]^T$ is the nodal displacement vector in which $v_i, i = 1, 2, \ldots, 4$, are the values of the displacement u at the nodal points, the shape function of the displacement u inside the element will be

$$u = [L_1 \quad L_2 \quad L_3 \quad L_4] \begin{bmatrix} v_1 \\ v_2 \\ v_3 \\ v_4 \end{bmatrix} \tag{6.83}$$

$$= \quad \mathbf{N} \quad \mathbf{v}$$

For the case that each side of the tetrahedron contains more than three nodes, the shape functions for the element can be obtained in a fashion similar to that explained previously for the two-dimensional case. Divide the tetrahedron into m blocks of equal thicknesses with planes parallel to S_1 as shown in Fig. 6.33b. Do the same for S_2, S_3, and S_4. Label the planes parallel to S_1 as p, and those parallel to S_2, S_3, and S_4 as q, r, and s. A point at the intersection of planes p, q, r, and s can be designated by four digits p, q, r, and s. Then the shape function of Eq. (6.83) can be written as

$$u = \mathbf{N}\,\mathbf{v} = \sum^{(m+1)(m+2)(m+3)/6} N_{pqrs}\, v_{pqrs} \tag{6.84}$$

with

$$N_{pqrs} = N_p(L_1)\, N_q(L_2)\, N_r(L_3)\, N_s(L_4)$$

The three-dimensional natural coordinates also have the property that

$$\int_V L_1^a L_2^b L_3^c L_4^d \, dV = \frac{a!b!c!d!}{(a+b+c+d+3)!} 6V \tag{6.85}$$

This property is very useful in the construction of stiffness matrices and the loading vectors.

6.5.7 Hermitian Interpolation

Elements used to model a structure undergoing bending very often involve derivatives of displacements as DOF. Thus, for a beam element the deflection and slope (first derivative of the deflection) are chosen as nodal DOF. Hermitian interpolation is useful in introducing derivative variables into the trial functions. Hermitian interpolation differs from Lagrangian interpolation (where the functions should have C^0 continuity at the interelement boundaries) in that nodal values of the interpolation functions include derivatives of the displacements, as well as the displacements themselves. The Hermitian trial functions are C^r $(r \geq 1)$ continuous at the interelement boundaries.

The shape functions for Hermitian elements are formulated in a fashion similar to that used in establishing Lagrange elements, but the Hermite polynomial replaces the Lagrange polynomial.

One-Dimensional Case in Cartesian Coordinates

To represent a function u that will satisfy conditions on u and up to its $m - 1$ derivative on the end points of an element extending from point $j = 1$ to point $j = 2$, consider the form

$$u = N_1 v_1 + N_2 v_1' + N_3 v_1'' + \cdots + N_m v_1^{(m-1)} + N_{m+1} v_2 + N_{m+2} v_2' + \cdots + N_{2m} v_2^{(m-1)}$$

$$= \sum_{j=1}^{2} \sum_{i=1}^{m} N_k(\xi)\, v_j^{(i-1)} \tag{6.86}$$

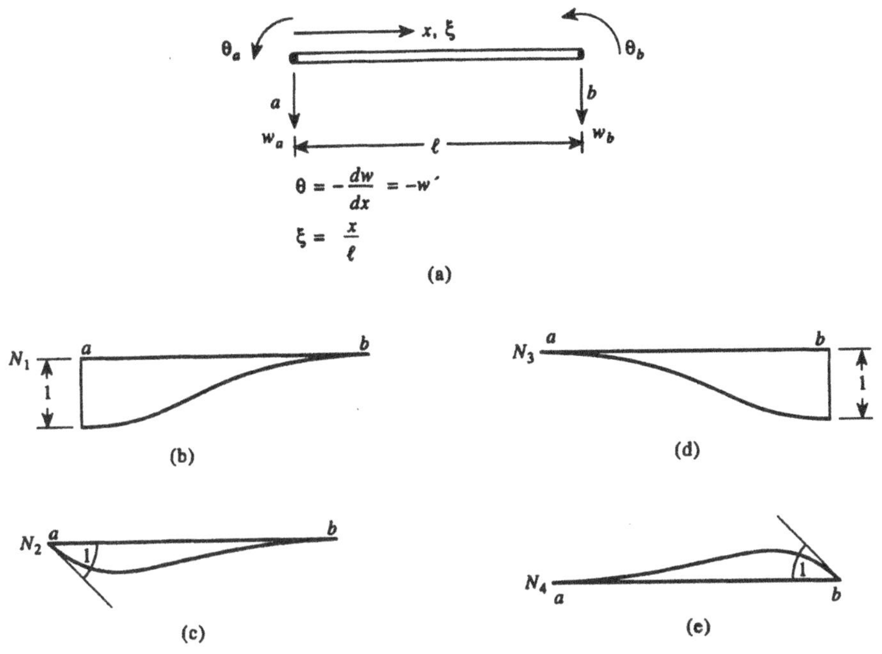

FIGURE 6.34
A beam element.

The superscripts of v denote the order of differentiation, and the subscripts refer to the end points. Also, $k = (j-1)m + i$. Similar to the Lagrangian case [Eqs. (6.62) and (6.63)], N_k must satisfy

$$\frac{d^{i-1}N_k(\xi)}{d\xi^{i-1}} = \begin{cases} 1 & \text{when } \xi = \xi_j \quad (k = (j-1)m + i) \\ 0 & \text{when } \xi \neq \xi_j \quad (k \neq (j-1)m + i) \end{cases} \tag{6.87}$$

It is assumed that $2m$ conditions are available for constructing each of the $2m$ interpolation functions N_i. Choose a polynomial series of order $2m - 1$ with $2m$ coefficients

$$N_i = \widehat{u}_1 + \widehat{u}_2\xi + \widehat{u}_3\xi^2 + \cdots + \widehat{u}_{2m}\xi^{(2m-1)} \tag{6.88}$$

A particular N_i is developed to satisfy the requirements of Eq. (6.87). This procedure leads to $2m$ equations for determining an interpolation function. A similar technique applies for the case of multiple nodal points.

For the beam element of Fig. 6.34, the interpolation functions for the displacement are derived in the same manner used to obtain Eq. (4.47b) of Chapter 4. Begin with the trial solution $w(\xi) = N_1\, w_a + N_2\, \theta_a + N_3\, w_b + N_4\, \theta_b$ and polynomial $N_i = \widehat{u}_1 + \widehat{u}_2\xi + \widehat{u}_3\xi^2 + \widehat{u}_4\xi^3$. To determine N_1, which should appear as shown in Fig. 6.34b, apply the conditions $w(0) = w_a = 1$, $\theta(0) = \theta_a = -w'(0) = 0$, $w(1) = w_b = 0$, and $\theta(1) = \theta_b = -w'(1) = 0$, giving

$$\begin{array}{lll} \xi = 0 & N_1 = 1 & w(0) = 1 = \widehat{u}_1 \\ \xi = 0 & N_1' = 0 & \theta(0) = 0 = \widehat{u}_2/\ell \\ \xi = 1 & N_1 = 0 & w(1) = 0 = \widehat{u}_1 + \widehat{u}_2 + \widehat{u}_3 + \widehat{u}_4 \\ \xi = 1 & N_1' = 0 & \theta(1) = 0 = (\widehat{u}_2 + 2\widehat{u}_3 + 3\widehat{u}_4)/\ell \end{array}$$

Note that these conditions are given in Eq. (6.87). The relations on the right-hand side can be solved to give

$$\widehat{u}_1 = 1 \qquad \widehat{u}_2 = 0 \qquad \widehat{u}_3 = -3 \qquad \widehat{u}_4 = 2$$

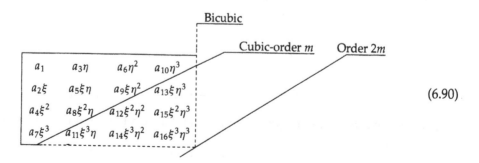

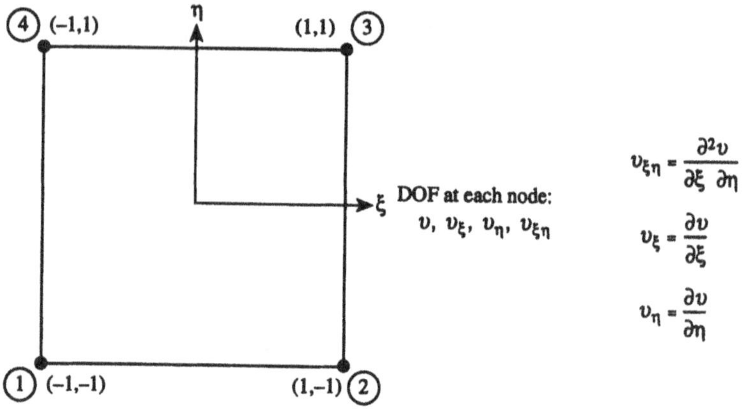

FIGURE 6.35
A rectangular element.

so that

$$N_1 = 1 - 3\xi^2 + 2\xi^3.$$

In a similar fashion, the remaining interpolation functions N_2, N_3, and N_4 of Figs. 6.34c, d, and e are found to be

$$N_2 = -\xi\ell(\xi - 1)^2, \qquad N_3 = 3\xi^2 - 2\xi^3, \qquad N_4 = -\xi^2\ell(\xi - 1)$$

These are referred to as Hermitian polynomials. The deflection trial function then becomes

$$w(\xi) = (1 - 3\xi^2 + 2\xi^3)w_a - \xi(\xi - 1)^2\ell\theta_a + (3\xi^2 - 2\xi^3)w_b - \xi^2(\xi - 1)\ell\,\theta_b$$

which corresponds to Chapter 4, Eq. (4.47b).

Two-Dimensional Case in Cartesian Coordinates

The two-dimensional Hermitian case follows closely the development of the two-dimensional Lagrangian interpolation. A trial function for the rectangular of Fig. 6.35 would be

$$u = [N_{1\xi} \quad N_{2\xi} \quad N_{3\xi} \quad N_{4\xi}] \begin{bmatrix} v_1 & v_{\eta 1} & v_4 & v_{\eta 4} \\ v_{\xi 1} & v_{\xi\eta 1} & v_{\xi 4} & v_{\xi\eta 4} \\ v_2 & v_{\eta 2} & v_3 & v_{\eta 3} \\ v_{\xi 2} & v_{\xi\eta 2} & v_{\xi 3} & v_{\xi\eta 3} \end{bmatrix} \begin{bmatrix} N_{1\eta} \\ N_{2\eta} \\ N_{3\eta} \\ N_{2\eta} \end{bmatrix} = \mathbf{N}_\xi^T\, \mathbf{R}\, \mathbf{N}_\eta \tag{6.89}$$

The corresponding polynomial contains 16 terms and is a bicubic expansion, i.e., it is formed by the multiplication of two one-dimensional cubic polynomials.

Bicubic

Cubic-order m Order $2m$

$$\begin{array}{cccc} a_1 & a_3\eta & a_6\eta^2 & a_{10}\eta^3 \\ a_2\xi & a_5\xi\eta & a_9\xi\eta^2 & a_{13}\xi\eta^3 \\ a_4\xi^2 & a_8\xi^2\eta & a_{12}\xi^2\eta^2 & a_{15}\xi^2\eta^3 \\ a_7\xi^3 & a_{11}\xi^3\eta & a_{14}\xi^3\eta^2 & a_{16}\xi^3\eta^3 \end{array} \tag{6.90}$$

EXAMPLE 6.9 Interpolation Functions for Rectangular Elements
For the normalized rectangular element, with centroidal coordinates $\xi\,\eta$, of Fig. 6.35, we will demonstrate the use of Lagrange and Hermitian polynomials in developing interpolation functions.

For four nodes, form Eq. (6.65b),

$$\begin{aligned} u(\xi,\eta) &= [N_1(\xi) \quad N_2(\xi)] \begin{bmatrix} v_1 & v_4 \\ v_2 & v_3 \end{bmatrix} \begin{bmatrix} N_1(\eta) \\ N_2(\eta) \end{bmatrix} \\ &= N_1(\xi)N_1(\eta)v_1 + N_2(\xi)N_1(\eta)v_2 + N_1(\xi)N_2(\eta)v_4 + N_2(\xi)N_2(\eta)v_3 \\ &= N_1(\xi,\eta)v_1 + N_2(\xi,\eta)v_2 + N_3(\xi,\eta)v_3 + N_4(\xi,\eta)v_4 \end{aligned} \tag{1}$$

From Eq. (6.64)

$$\begin{aligned} N_1(\xi) &= \frac{\xi-\xi_2}{\xi_1-\xi_2} = \frac{\xi-1}{-1-1} = \frac{1}{2}(1-\xi) \\ N_2(\xi) &= \frac{\xi-\xi_1}{\xi_2-\xi_1} = \frac{\xi+1}{1-(-1)} = \frac{1}{2}(1+\xi) \\ N_1(\eta) &= \frac{\eta-\eta_4}{\eta_1-\eta_4} = \frac{\eta-1}{-1-1} = \frac{1}{2}(1-\eta) \\ N_2(\eta) &= \frac{\eta-\eta_1}{\eta_4-\eta_1} = \frac{\eta+1}{1-(-1)} = \frac{1}{2}(1+\eta) \end{aligned} \tag{2}$$

so that

$$\begin{aligned} N_1(\xi,\eta) &= \frac{1}{4}(1-\xi)(1-\eta) \qquad N_2(\xi,\eta) = \frac{1}{4}(1+\xi)(1-\eta) \\ N_3(\xi,\eta) &= \frac{1}{4}(1+\xi)(1+\eta) \qquad N_4(\xi,\eta) = \frac{1}{4}(1-\xi)(1+\eta) \end{aligned} \tag{3}$$

If derivatives, such as rotations, of u are to be included at the nodes or if greater than C^0 continuity is desired at the nodes, then use Hermitian interpolation. At each node, introduce the four DOF

$$v = u, \qquad v_\xi = \frac{\partial u}{\partial \xi}, \qquad v_\eta = \frac{\partial u}{\partial \eta}, \qquad v_{\xi\eta} = \frac{\partial^2 u}{\partial \xi \partial \eta} \tag{4}$$

and assume that Eq. (6.89) applies, with $N_{i\xi} = N_i(\xi)$ and $N_{i\eta} = N_i(\eta)$. The interpolation functions are obtained by utilizing third degree Hermitian polynomials of Chapter 4, Eq. (4.47b) for $N_{i\xi}$ and $N_{i\eta}$. For example, for the term corresponding to the v_1

$$N_1(\xi,\eta) = N_{1\xi}N_{1\eta} = (1-3\xi^2+2\xi^3)(1-3\eta^2-2\eta^3) \tag{5}$$

■

6.5.8 Stiffness Matrices and Loading Vectors For Triangular and Tetrahedral Elements

This section contains the formulation of the stiffness matrices and the loading vectors for triangular and tetrahedral elements using the natural coordinates of Eqs. (6.75) and (6.82).

Triangular Element

Consider a plate under plane stress conditions. Divide the plate into three-node triangular elements of the sort shown in Fig. 6.30. Let u_x, u_y be the displacements in the global x and y directions. Traditionally, this notation is used, although in the previous chapter X and Y

would have replaced x and y. The interpolation functions for u_x and u_y are

$$\begin{aligned} u_x &= \mathbf{N}\mathbf{v}_x \\ u_y &= \mathbf{N}\mathbf{v}_y \end{aligned} \tag{6.91}$$

where

$$\mathbf{v}_x = \begin{bmatrix} u_{x1} \\ u_{x2} \\ u_{x3} \end{bmatrix} \qquad \mathbf{v}_y = \begin{bmatrix} u_{y1} \\ u_{y2} \\ u_{y3} \end{bmatrix}$$

and

$$\mathbf{N} = [L_1 \quad L_2 \quad L_3]$$

in which $u_{xi}, u_{yi}, i = 1, 2, 3$, are the values of u_x and u_y at the nodes, and L_i, $i = 1, 2, 3$ are the shape functions defined in Eq. (6.75). It is readily shown that $\mathbf{N} = [L_1 \; L_2 \; L_3]$. From Eq. (6.76) with $m = 1$,

$$u_x = N_{100}\, v_{100} + N_{010}\, v_{010} + N_{001}\, v_{001} = N_{100}\, u_{x1} + N_{010}\, u_{x2} + N_{001}\, u_{x3}$$

where N_{pqr}, with $p+q+r = 1$, is defined in Eq. (6.77). From Eq. (6.71b), $N_0(L_1) = N_0(L_2) = N_0(L_3) = 1$, $N_1(L_1) = L_1$, $N_1(L_2) = L_2$, $N_1(L_3) = L_3$. Finally, $N_1 = N_{100} = L_1$, $N_2 = N_{010} = L_2$, $N_3 = N_{001} = L_3$.

The displacement vector is

$$\mathbf{u} = \begin{bmatrix} u_x \\ u_y \end{bmatrix} = \begin{bmatrix} \mathbf{N} & \mathbf{0} \\ \mathbf{0} & \mathbf{N} \end{bmatrix} \begin{bmatrix} \mathbf{v}_x \\ \mathbf{v}_y \end{bmatrix} = \begin{bmatrix} \mathbf{N} & \mathbf{0} \\ \mathbf{0} & \mathbf{N} \end{bmatrix} \mathbf{v}^i \tag{6.92}$$

The principle of virtual work expression is [Eqs. (6.20) and (6.28)]

$$\begin{aligned} \delta W &= \int_V \delta\mathbf{u}^T \, {}_u\mathbf{D}^T \mathbf{E}\, \mathbf{D}_u\, \mathbf{u}\, dV - \int_V \delta\mathbf{u}^T\, \overline{\mathbf{p}}_V\, dV - \int_{S_p} \delta\mathbf{u}^T \overline{\mathbf{p}}\, dS \\ &= \sum_{i=1}^{M} \delta\mathbf{v}^{iT} \left\{ t \int_A \begin{bmatrix} \mathbf{N} & \mathbf{0} \\ \mathbf{0} & \mathbf{N} \end{bmatrix}^T \begin{bmatrix} {}_x\partial & 0 \\ 0 & {}_y\partial \\ {}_y\partial & {}_x\partial \end{bmatrix}^T \mathbf{E} \begin{bmatrix} \partial_x & 0 \\ 0 & \partial_y \\ \partial_y & \partial_x \end{bmatrix} \begin{bmatrix} \mathbf{N} & \mathbf{0} \\ \mathbf{0} & \mathbf{N} \end{bmatrix} dA\, \mathbf{v}^i \right. \\ &\qquad \left. - \left(t \int_V \begin{bmatrix} \mathbf{N} & \mathbf{0} \\ \mathbf{0} & \mathbf{N} \end{bmatrix}^T \overline{\mathbf{p}}_V\, dV + \int_{S_p} \begin{bmatrix} \mathbf{N} & \mathbf{0} \\ \mathbf{0} & \mathbf{N} \end{bmatrix}^T \overline{\mathbf{p}}\, dS \right) \right\} \\ &= \sum_{i=1}^{M} \delta\mathbf{v}^{iT} (\mathbf{k}^i \mathbf{v}^i - \overline{\mathbf{p}}^i) = 0 \end{aligned} \tag{6.93}$$

where $\overline{\mathbf{p}}$ is the external load on the element boundary S_p and t is the thickness of the plate. $\mathbf{E}$ is given in Eq. (1.39), and $\mathbf{D}_u$ is taken from Eq. (1.24). The stiffness matrix and loading vector for each element are given by

$$\mathbf{k}^i = t \int_A \begin{bmatrix} \mathbf{N} & \mathbf{0} \\ \mathbf{0} & \mathbf{N} \end{bmatrix}^T \begin{bmatrix} {}_x\partial & 0 \\ 0 & {}_y\partial \\ {}_y\partial & {}_x\partial \end{bmatrix}^T \mathbf{E} \begin{bmatrix} \partial_x & 0 \\ 0 & \partial_y \\ \partial_y & \partial_x \end{bmatrix} \begin{bmatrix} \mathbf{N} & \mathbf{0} \\ \mathbf{0} & \mathbf{N} \end{bmatrix} dA \tag{6.94a}$$

and

$$\overline{\mathbf{p}}^i = t\int_A \begin{bmatrix} \mathbf{N} & \mathbf{0} \\ \mathbf{0} & \mathbf{N} \end{bmatrix}^T \overline{\mathbf{p}}_V\, dA + \int_{S_p} \begin{bmatrix} \mathbf{N} & \mathbf{0} \\ \mathbf{0} & \mathbf{N} \end{bmatrix}^T \overline{\mathbf{p}}\, dS \tag{6.94b}$$

Note that in $\mathbf{k}^i$, the terms assume forms such as $\int_A (\mathbf{N}^T {}_x\partial) E_{ij} (\partial_x \mathbf{N})\, dA$, $\int_A (\mathbf{N}^T {}_y\partial) E_{ij} (\partial_x \mathbf{N})\, dA$, etc., where E_{ij} are the elements in $\mathbf{E}$. These expressions are constants and can be obtained by performing the manipulations indicated. From Eq. (6.75), $\mathbf{N}$ can be written as

$$\mathbf{N} = [L_1 \quad L_2 \quad L_3] = \frac{1}{2A}[1 \quad x \quad y] \begin{bmatrix} \alpha_1 & \alpha_2 & \alpha_3 \\ \beta_1 & \beta_2 & \beta_3 \\ \gamma_1 & \gamma_2 & \gamma_3 \end{bmatrix} \tag{6.95}$$

A typical derivative of $\mathbf{N}$ is

$$\partial_x \mathbf{N} = \frac{1}{2A}\partial_x[1 \quad x \quad y] \begin{bmatrix} \alpha_1 & \alpha_2 & \alpha_3 \\ \beta_1 & \beta_2 & \beta_3 \\ \gamma_1 & \gamma_2 & \gamma_3 \end{bmatrix} = \frac{1}{2A}[\beta_1 \quad \beta_2 \quad \beta_3] \tag{6.96}$$

Similarly

$$\partial_y \mathbf{N} = \frac{1}{2A}[\gamma_1 \quad \gamma_2 \quad \gamma_3] \tag{6.97}$$

Then the terms in the stiffness matrix can be evaluated as

$$\begin{aligned} \int_A (\mathbf{N}^T {}_x\partial) E_{ij} (\partial_x \mathbf{N})\, dA &= \int_A \begin{bmatrix} L_1 \\ L_2 \\ L_3 \end{bmatrix} {}_x\partial\, E_{ij}\, \partial_x [L_1 \quad L_2 \quad L_3]\, dA \\ &= \int_A \frac{E_{ij}}{4A^2} \begin{bmatrix} \beta_1 \\ \beta_2 \\ \beta_3 \end{bmatrix} [\beta_1 \quad \beta_2 \quad \beta_3]\, dA \\ &= \frac{E_{ij}}{4A} \begin{bmatrix} \beta_1^2 & \beta_1\beta_2 & \beta_1\beta_3 \\ \beta_2\beta_1 & \beta_2^2 & \beta_2\beta_3 \\ \beta_3\beta_1 & \beta_3\beta_2 & \beta_3^2 \end{bmatrix} \end{aligned} \tag{6.98}$$

$$\int_A (\mathbf{N}^T {}_y\partial) E_{ij} (\partial_x \mathbf{N})\, dA = \frac{E_{ij}}{4A} \begin{bmatrix} \gamma_1\beta_1 & \gamma_1\beta_2 & \gamma_1\beta_3 \\ \gamma_2\beta_1 & \gamma_2\beta_2 & \gamma_2\beta_3 \\ \gamma_3\beta_1 & \gamma_3\beta_2 & \gamma_3\beta_3 \end{bmatrix} \tag{6.99}$$

Similar expressions can be obtained for the remaining terms in $\mathbf{k}^i$ such as

$$\int_A (\mathbf{N}^T {}_x\partial) E_{ij} (\partial_y \mathbf{N})\, dA \tag{6.100}$$

Thus, the stiffness matrix $\mathbf{k}^i$ can be formed as a 6×6 matrix.

The loading vectors are calculated similarly. If the body forces $\overline{\mathbf{p}}_V = [\overline{p}_{Vx} \ \overline{p}_{Vy}]^T$, in which $\overline{p}_{Vx}$ and $\overline{p}_{Vy}$ are assumed to have a constant value $\overline{p}_V$, are to be included, the part of the loading vector due to these forces is

$$\mathbf{p}_v = t\int_A \begin{bmatrix} \mathbf{N} & \mathbf{0} \\ \mathbf{0} & \mathbf{N} \end{bmatrix}^T \overline{p}_V\, dA = \begin{bmatrix} \overline{\mathbf{p}}_{Vx} \\ \overline{\mathbf{p}}_{Vy} \end{bmatrix} \tag{6.101}$$

with

$$\overline{p}_{Vx} = \overline{p}_{Vy} = \overline{p}_V t \int_A \mathbf{N}^T\, dA = \overline{p}_V t \int_A \begin{bmatrix} L_1 \\ L_2 \\ L_3 \end{bmatrix} dA = \frac{\overline{p}_V A t}{3} \begin{bmatrix} 1 \\ 1 \\ 1 \end{bmatrix} \tag{6.102}$$

where use has been made of the integral of Eq. (6.78) with, for example, $a = 1, b = 0$, and $c = 0$.

On the element boundaries assume that the tractions $\overline{\mathbf{p}}$ have the constant value $\overline{p}$. On side 3 of Fig. 6.30, $L_3 = 0$. Then

$$\mathbf{p}_1 = \overline{p} \int_{\text{Side 3}} \begin{bmatrix} L_1 \\ L_2 \\ 0 \end{bmatrix} ds = \frac{\overline{p}\,\ell_{12}}{2} \begin{bmatrix} 1 \\ 1 \\ 0 \end{bmatrix} \tag{6.103}$$

where, as indicated in Eq. (6.70), ℓ_{12} is the length between nodes 1 and 2. Similar results are obtained for sides 1 and 2, where $L_1 = 0$ and $L_2 = 0$, respectively.

Tetrahedral Element

The development of a stiffness matrix for a tetrahedral element can follow the same procedure used for the triangular element. Rather than the displacements u_x, u_y for the triangle, use u_x, u_y, and u_z for the displacements in the global x, y, and z directions of the tetrahedron. Express the displacement vector as

$$\mathbf{u} = \begin{bmatrix} u_x \\ u_y \\ u_z \end{bmatrix} = \begin{bmatrix} \mathbf{N} & & \mathbf{0} \\ & \mathbf{N} & \\ \mathbf{0} & & \mathbf{N} \end{bmatrix} \begin{bmatrix} \mathbf{v}_x \\ \mathbf{v}_y \\ \mathbf{v}_z \end{bmatrix} = \begin{bmatrix} \mathbf{N} & & \mathbf{0} \\ & \mathbf{N} & \\ \mathbf{0} & & \mathbf{N} \end{bmatrix} \mathbf{v}^i \tag{6.104}$$

where

$$\mathbf{v}_x = \begin{bmatrix} u_{x1} \\ u_{x2} \\ u_{x3} \\ u_{x4} \end{bmatrix} \qquad \mathbf{v}_y = \begin{bmatrix} u_{y1} \\ u_{y2} \\ u_{y3} \\ u_{y4} \end{bmatrix} \qquad \mathbf{v}_z = \begin{bmatrix} u_{z1} \\ u_{z2} \\ u_{z3} \\ u_{z4} \end{bmatrix}$$

and

$$\mathbf{N} = [L_1 \quad L_2 \quad L_3 \quad L_4]$$

in which u_{xi}, u_{yi}, and u_{zi}, $i = 1, 2, 3, 4$, are the values of u_x, u_y, and u_z at the nodes, and L_i, $i = 1, 2, 3, 4$ are the shape functions defined in Eq. (6.82).

Then

$$\mathbf{N} = [L_1 \quad L_2 \quad L_3 \quad L_4] = \frac{1}{6V}[1 \quad x \quad y \quad z] \begin{bmatrix} \alpha_1 & \beta_1 & \gamma_1 & \delta_1 \\ \alpha_2 & \beta_2 & \gamma_2 & \delta_2 \\ \alpha_3 & \beta_3 & \gamma_3 & \delta_3 \\ \alpha_4 & \beta_4 & \gamma_4 & \delta_4 \end{bmatrix} \tag{6.105}$$

A typical expression in the stiffness matrix would be

$$\int_V \mathbf{N}^T {}_x\partial \, E_{ij} \, \partial_x \mathbf{N} \, dV = \int_V \begin{bmatrix} L_1 \\ L_2 \\ L_3 \\ L_4 \end{bmatrix} {}_x\partial \, E_{ij} \, \partial_x [L_1 \quad L_2 \quad L_3 \quad L_4] \, dV$$

$$= \frac{E_{ij}}{36V} \begin{bmatrix} \beta_1^2 & \beta_1\beta_2 & \beta_1\beta_3 & \beta_1\beta_4 \\ \beta_2\beta_1 & \beta_2^2 & \beta_2\beta_3 & \beta_2\beta_4 \\ \beta_3\beta_1 & \beta_3\beta_2 & \beta_3^2 & \beta_3\beta_4 \\ \beta_4\beta_1 & \beta_4\beta_2 & \beta_4\beta_3 & \beta_4^2 \end{bmatrix} \tag{6.106}$$

For the loading vector, continue to follow the scheme for the triangular element.

Triangular and tetrahedral elements are very useful in structural analysis. They can be used independently or can be combined to form quadrilateral and hexahedral elements.

In such cases, the element stiffness matrix and loading vector are the summation of the matrices of the triangular or tetrahedral elements. After the stiffness matrix and the loading vector are formed, DOF associated with the nodes not on the element boundary can be condensed out using the technique that is employed to process Eq. (6.56).

Coordinate Transformation

Although most of the derivations thus far in this chapter refer to the global xy coordinate system, some elements, e.g., the triangular element, can be more conveniently formulated in a local coordinate system. Then the stiffness matrices and loading vectors are transformed from the local system to the global system. The transformation for a triangular element will be discussed here. In Chapter 5, XYZ and xyz indicate the global and local coordinate systems, respectively. In this chapter, however, xyz are used for the global system, whereas $\tilde{x}\tilde{y}\tilde{z}$ represent the local system. Accordingly, $\mathbf{v}_j = [u_{xj}\ u_{yj}\ u_{zj}]^T$ and $\tilde{\mathbf{v}}_j = [\tilde{u}_{xj}\ \tilde{u}_{yj}\ \tilde{u}_{zj}]^T$ will be used to describe the nodal variables in the global and local coordinates at node j.

The relationship between the local $\tilde{x}\tilde{y}$ and global xy coordinate system is shown in Fig. 6.36a and that between the nodal variables in the two systems is shown in Fig. 6.36b. Note that in Fig. 6.36a one node is located at the origin of the local coordinate system and the $\tilde{x}$ axis is aligned along one side of the element. This makes the shape functions of the element much simpler to form, i.e., the expressions for α_i, β_i, and γ_i, $i = 1, 2, 3$, of Eq. (6.75) are simpler. As a consequence, it takes less effort to create the stiffness matrix using the procedure in Section 6.5.8. From Fig. 6.36b, the transformations between the displacement variables (or forces if u is replaced by p) at the jth node of element i is

$$\begin{aligned} \begin{bmatrix} \tilde{u}_{xj} \\ \tilde{u}_{yj} \end{bmatrix} &= \begin{bmatrix} \cos\alpha & \sin\alpha \\ -\sin\alpha & \cos\alpha \end{bmatrix} \begin{bmatrix} u_{xj} \\ u_{yj} \end{bmatrix} \\ \tilde{\mathbf{v}}_j^i &= \mathbf{T}_{jj}^i \qquad \mathbf{v}_j^i \end{aligned} \tag{6.107}$$

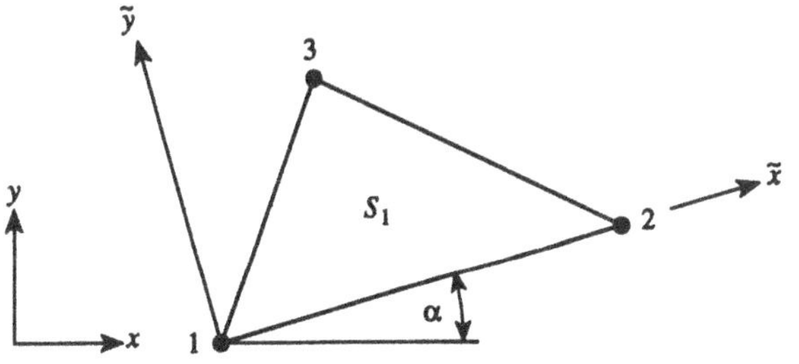

(a) Local $\tilde{x}\,\tilde{y}$ and global $x\,y$ coordinate systems for a triangular element

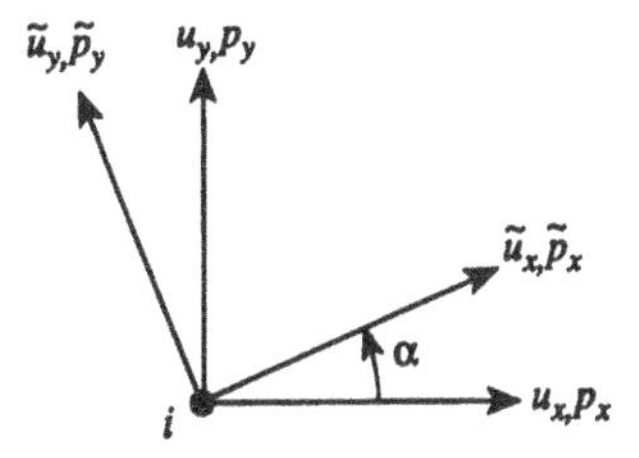

(b) Displacements and forces in the local and global coordinate systems. Positive α, the angle from global to local coordinates, is shown.

FIGURE 6.36
Coordinate transformation between the local and global systems.

Recall from Chapter 5, Eq. (5.25), that the transformation matrix $\mathbf{T}_{jj}$ has the property $\mathbf{T}_{jj}^{iT}\mathbf{T}_{jj}^{i} = \mathbf{I}$, where $\mathbf{I}$ is an identity matrix.

Arrange the entries in the nodal displacement vector $\widetilde{\mathbf{v}}^i$ as

$$\widetilde{\mathbf{v}}^i = [\widetilde{u}_{x1} \quad \widetilde{u}_{y1} \quad \widetilde{u}_{x2} \quad \widetilde{u}_{y2} \quad \widetilde{u}_{x3} \quad \widetilde{u}_{y3}]^T = [\widetilde{\mathbf{v}}_1^i \quad \widetilde{\mathbf{v}}_2^i \quad \widetilde{\mathbf{v}}_3^i]^T$$

Then from Eq. (6.107)

$$\widetilde{\mathbf{v}}^i = \begin{bmatrix} \widetilde{\mathbf{v}}_1^i \\ \widetilde{\mathbf{v}}_2^i \\ \widetilde{\mathbf{v}}_3^i \end{bmatrix} = \begin{bmatrix} \mathbf{T}_{11}^i & 0 & 0 \\ 0 & \mathbf{T}_{22}^i & 0 \\ 0 & 0 & \mathbf{T}_{33}^i \end{bmatrix} \begin{bmatrix} \mathbf{v}_1^i \\ \mathbf{v}_2^i \\ \mathbf{v}_3^i \end{bmatrix} = \mathbf{T}^i\,\mathbf{v}^i \tag{6.108}$$

or $\mathbf{v}^i = \mathbf{T}^{iT}\widetilde{\mathbf{v}}^i$, since, Eq. (5.26), $(\mathbf{T}^i)^{-1} = \mathbf{T}^{iT}$. Similarly, $\widetilde{\mathbf{p}}^i = \mathbf{T}^i\mathbf{p}^i$ and $\mathbf{p}^i = \mathbf{T}^{iT}\widetilde{\mathbf{p}}^i$. Substitution of Eq. (6.108) into the stiffness equation in the local system, $\widetilde{\mathbf{k}}^i\widetilde{\mathbf{v}}^i = \widetilde{\mathbf{p}}^i$, results in $\widetilde{\mathbf{k}}^i\widetilde{\mathbf{v}}^i = \widetilde{\mathbf{p}}^i = \widetilde{\mathbf{k}}^i\mathbf{T}^i\mathbf{v}^i - \widetilde{\mathbf{p}}^i$. Premultiply by $\mathbf{T}^{iT}$ to obtain $\mathbf{T}^{iT}\widetilde{\mathbf{k}}^i\mathbf{T}^i\mathbf{v}^i - \mathbf{T}^{iT}\widetilde{\mathbf{p}}^i = \mathbf{k}^i\mathbf{v}^i - \mathbf{p}^i$ with $\mathbf{k}^i = \mathbf{T}^{iT}\widetilde{\mathbf{k}}^i\mathbf{T}^i$ and $\mathbf{p}^i = \mathbf{T}^{iT}\widetilde{\mathbf{p}}^i$, which have the same form as Eq. (5.28).

6.6 Numerical Integration

The development of element stiffness matrices involves the evaluation of definite integrals, such as those that result from the principle of virtual work. For the elements treated in the previous section it was possible to evaluate the integrals analytically. Often, exact integration expressions are not readily obtained and numerical integration becomes essential. The numerical integration provides approximations to the integrals; however, it is contended [Zienkiewicz, 1977] that the error resulting from numerical integration may compensate for the modeling error due to the geometrical discretization of the structure. This error compensation appears to result in an improved solution.

In general, the integral can be approximated by using a simple summation of terms involving the integrand, evaluated at n specific points (integration points), and multiplied by suitable weights. Namely, for the case of one, two, or three dimensions,

$$\int_L F(\xi)\,d\xi \approx \sum_i^n W_i^{(n)} F(\xi_i) \tag{6.109}$$

$$\int_L F(\xi,\eta)\,d\xi\,d\eta \approx \sum_i^n W_i^{(n)} F(\xi_i,\eta_i) \tag{6.110}$$

$$\int_V F(\xi,\eta,\zeta)\,d\xi\,d\eta\,d\zeta \approx \sum_i^n W_i^{(n)} F(\xi_i,\eta_i,\zeta_i) \tag{6.111}$$

where $W_i^{(n)}$ is a weighting factor with superscript (n) indicating that n integration points are employed, and $F(\xi_i)$, $F(\xi_i,\eta_i)$, and $F(\xi_i,\eta_i,\zeta_i)$ are the values of the function F at the points ξ_i,η_i,ζ_i.

Numerical integration techniques tend to differ in the method of establishing values of $W_i^{(n)}$ and ξ_i (and η_i, ζ_i, as appropriate). Considerations, such as accuracy and computational efficiency, are taken into account in selecting a method. Two types of numerical integration, usually referred to as *quadrature formulas*, have been found to be particularly useful in finite element calculations. They are considered in the following sections.

6.6.1 Newton-Cotes Quadrature

Consider the one-dimensional case. With the Newton-Cotes[5] method, the n integration points ξ_i at which the function $F(\xi_i)$ is to be evaluated are established at the onset. Usually, n equally spaced integration points are chosen over the integration interval $[a, b]$. The weighting factors $W_i^{(n)}$ are determined by replacing $F(\xi)$ by a polynomial $p(\xi)$ obtained from the Lagrangian interpolation, i.e.,

$$p(\xi) = \sum_{i=1}^{n} N_i(\xi)\, F(\xi_i) \tag{6.112}$$

with $N_i(\xi)$ defined in Eq. (6.64) and note that $p(\xi_i) = F(\xi_i)$. The intention is to replace $F(\xi)$ with an approximating function that is relatively simple to integrate. Then use $\int_a^b p(\xi) d\xi$ as an approximation to $\int_a^b F(\xi) d\xi$. Integration of Eq. (6.112) over the interval $[a, b]$ gives

$$\int_a^b p(\xi)\, d\xi = \sum_{i=1}^{n} \int_a^b N_i(\xi)\, d\xi\, F(\xi_i) = \sum_{i=1}^{n} W_i^{(n)} F(\xi_i) \tag{6.113}$$

Thus,

$$W_i^{(n)} = \int_a^b N_i(\xi)\, d\xi = (b - a) C_i^{(n)} \tag{6.114}$$

where $C_i^{(n)}$ are the "weights" for the integration. Equation (6.113) is called *Newton-Cotes quadrature*.

This method permits the integral of a polynomial of order $n - 1$ to be evaluated exactly. Thus, if $F(\xi)$ is a polynomial of order $n - 1$, Eq. (6.113) gives the exact result and the error is zero. Furthermore, it can be shown that when n is odd, Eq. (6.113) permits a polynomial of order n to be integrated exactly. If $F(\xi)$ is not a polynomial, there will be an error R_n in using Eq. (6.113) to evaluate $\int_a^b F(\xi)\, d\xi$. If n is even,

$$R_n = \frac{F^{(n)}(r)}{n!} \int_a^b \Pi_n(\xi)\, d\xi \tag{6.115a}$$

where $\Pi_n(\xi) = (\xi - \xi_1)(\xi - \xi_2) \cdots (\xi - \xi_n)$, r is a point in $[a, b]$, and $F^{(n)}(r)$ is the nth derivative of $F(\xi)$. If n is odd,

$$R_n = \frac{F^{(n)}(r)}{(n+1)!} \int_a^b \xi \Pi_n(\xi)\, d\xi \tag{6.115b}$$

Usually, the higher the order of $N_i(\xi)$, the smaller the error. Table 6.6 lists the weights $C_i^{(n)}$ and errors for $n = 2$ to 6 over the integration interval $[a, b]$. Note that formulas for $n = 3$ and $n = 5$ have the same order of accuracy as the formulas for $n = 4$ and $n = 6$, respectively. For this reason, the odd formulas with $n = 3$ and $n = 5$ are used in practice.

[5]Roger Cotes (1682–1716), the son of an English minister, received his BA from Cambridge in 1702 and MA in 1706, when he was given a professorship in astronomy and natural philosophy. For almost 4 years he assisted Newton in the preparation of the second edition of Newton's *Principia*. His death of fever at 33 caused Newton to comment "Had Cotes lived we might have known something." Cotes wrote a paper on Newton's differential method in which he describes how to compute the area under a curve. The modern version of this is called the *Newton-Cotes method*. He proposed a technique similar to least squares for representing observed data. This preceded the efforts of Gauss (1795) and Legendre (1806).

TABLE 6.6

Weights for Newton-Cotes Quadrature Formulas

Newton-Cotes Integration of $F(\xi)$ over Interval $[a, b]$:

$$\int_a^b F(\xi)\, d\xi = \sum_i^n W_i^{(n)} F(\xi_i)$$

n = Number of integration points

Weighting Factor $= W_i^{(n)} = (b-a)C_i^{(n)}$

n	$C_1^{(n)}$	$C_2^{(n)}$	$C_3^{(n)}$	$C_4^{(n)}$	$C_5^{(n)}$	$C_6^{(n)}$	Upper Bound on Error R_n as a Function of the Derivative of F
2	$\frac{1}{2}$	$\frac{1}{2}$					$10^{-1}(b-a)^3F^{(2)}(r)$
3	$\frac{1}{6}$	$\frac{4}{6}$	$\frac{1}{6}$				$10^{-3}(b-a)^5F^{(4)}(r)$
4	$\frac{1}{8}$	$\frac{3}{8}$	$\frac{3}{8}$	$\frac{1}{8}$			$10^{-3}(b-a)^5F^{(4)}(r)$
5	$\frac{7}{90}$	$\frac{32}{90}$	$\frac{12}{90}$	$\frac{32}{90}$	$\frac{7}{90}$		$10^{-6}(b-a)^7F^{(6)}(r)$
6	$\frac{19}{288}$	$\frac{75}{288}$	$\frac{50}{288}$	$\frac{50}{288}$	$\frac{75}{288}$	$\frac{19}{288}$	$10^{-6}(b-a)^7F^{(6)}(r)$

* $F^{(n)}(r)$ is the nth derivative of F and r is a point in $[a, b]$.

EXAMPLE 6.10 Newton-Cotes Quadrature for $n = 2$ and $n = 3$

For $n = 2$ over $[-1, 1]$, choose equally spaced integration points $\xi_1 = -1$, $\xi_2 = 1$ to evaluate $\int_{-1}^1 F(\xi)\, d\xi$. The interpolation function is

$$p(\xi) = \sum_{i=1}^{2} N_i(\xi)\, F(\xi_i) \tag{1}$$

in which, from Eq. (6.64)

$$\begin{aligned} N_1(\xi) &= \frac{\xi-\xi_2}{\xi_1-\xi_2} = \frac{1}{2}(1-\xi) \\ N_2(\xi) &= \frac{\xi-\xi_1}{\xi_2-\xi_1} = \frac{1}{2}(1+\xi) \end{aligned} \tag{2}$$

It follows from Eq. (6.114) that the weighting factors are

$$\begin{aligned} W_1^{(2)} &= \int_{-1}^1 N_1(\xi)\, d\xi = \frac{1}{2}\int_{-1}^1 (1-\xi)\, d\xi = 1 \\ W_2^{(2)} &= \int_{-1}^1 N_2(\xi)\, d\xi = \frac{1}{2}\int_{-1}^1 (1+\xi)\, d\xi = 1 \end{aligned} \tag{3}$$

Finally, Eq. (6.113) gives the Newton-Cotes quadrature for $n = 2$ as

$$\int_{-1}^1 F(\xi)\, d\xi \approx \sum_{i=1}^{2} W_i^{(2)}\, F(\xi_i) = F(-1) + F(1) \tag{4}$$

This is the *trapezoidal rule.*

For $n = 3$ over $[-1, 1]$, choose integration points $\xi_1 = -1$, $\xi_2 = 0$, $\xi_3 = 1$. The interpolating function is

$$p(\xi) = \sum_{i=1}^{3} N_i(\xi)\, F(\xi_i) \tag{5}$$

where

$$N_1(\xi) = \frac{(\xi - \xi_2)(\xi - \xi_3)}{(\xi_1 - \xi_2)(\xi_1 - \xi_3)} = \frac{1}{2}\xi(\xi - 1)$$
$$N_2(\xi) = \frac{(\xi - \xi_1)(\xi - \xi_3)}{(\xi_2 - \xi_1)(\xi_2 - \xi_3)} = -(\xi + 1)(\xi - 1) \tag{6}$$
$$N_3(\xi) = \frac{(\xi - \xi_1)(\xi - \xi_2)}{(\xi_3 - \xi_1)(\xi_3 - \xi_2)} = \frac{1}{2}\xi(\xi + 1)$$

Then the weighting factors are

$$W_1^{(3)} = \int_{-1}^{1} N_1(\xi)\, d\xi = \frac{1}{2}\int_{-1}^{1} \xi(\xi - 1)\, d\xi = \frac{1}{3}$$
$$W_2^{(3)} = \int_{-1}^{1} N_2(\xi)\, d\xi = \int_{-1}^{1} -(\xi + 1)(\xi - 1)\, d\xi = \frac{4}{3} \tag{7}$$
$$W_3^{(3)} = \int_{-1}^{1} N_3(\xi)\, d\xi = \frac{1}{2}\int_{-1}^{1} \xi(\xi + 1)\, d\xi = \frac{1}{3}$$

Finally,

$$\int_{-1}^{1} F(\xi)\, d\xi \approx \sum_{i=1}^{3} W_i^{(3)}\, F(\xi_i) = \frac{1}{3}[F(-1) + 4F(0) + F(1)] \tag{8}$$

which is *Simpson's rule.*

These same results are listed in Table 6.6. ∎

6.6.2 Gaussian Quadrature

In Gaussian quadrature, the integration points are not fixed at the outset but are chosen to achieve the best accuracy. Since this provides better accuracy than the evenly spaced integration points of Newton-Cotes quadrature, Gaussian quadrature is the more popular method of integration. Return to the one-dimensional case. For Gaussian quadrature, it is again assumed that the integral can be approximated as a weighted sum of values of $F(\xi_i)$. If n integration points are used and the integration interval is $[-1, 1]$,

$$\int_{-1}^{1} F(\xi)\, d\xi \approx W_1^{(n)} F(\xi_1) + W_2^{(n)} F(\xi_2) + \cdots + W_n^{(n)} F(\xi_n)$$
$$= \sum_{i=1}^{n} W_i^{(n)} F(\xi_i) \tag{6.116}$$

In this formulation, both $W_i^{(n)}$ and ξ_i are unknowns to be determined. For n integration points, there will be n unknowns ξ_i and n unknowns $W_i^{(n)}$.

As in the development of the Newton-Cotes formula, use Lagrangian interpolation where $F(\xi)$ is approximated by $p(\xi)$ such that

$$p(\xi) = \sum_{i=1}^{n} N_i(\xi)\, F(\xi_i) \tag{6.117}$$

where ξ_i are still unknown. For the determination of ξ_i, define a function

$$\chi(\xi) = (\xi - \xi_1)(\xi - \xi_2)\cdots(\xi - \xi_n)$$

which is a polynomial of order n. Note that $\chi(\xi)$ is equal to zero at ξ_i. Recall that at the integration points, $p(\xi_i) = F(\xi_i)$. At intermediate points, the difference between $F(\xi)$ and $p(\xi)$ can be expressed in terms of $\chi(\xi)$. Let $F(\xi)$ be written as

$$F(\xi) = p(\xi) + \chi(\xi)(\beta_0 + \beta_1\xi + \beta_2\xi^2 + \cdots) \tag{6.118}$$

where $\beta_i, i = 1, 2, \ldots$ are appropriately chosen constants that can be used to eliminate the gaps between $F(\xi)$ and $p(\xi)$ at the intermediate points. Integrate $F(\xi)$ to obtain

$$\int_{-1}^{1} F(\xi)\,d\xi = \int_{-1}^{1} p(\xi)\,d\xi + \sum_{j=0}^{\infty} \beta_j \int_{-1}^{1} \chi(\xi)\,\xi^j\,d\xi$$

Split the final quantity into two parts

$$\sum_{j=0}^{\infty} \beta_j \int_{-1}^{1} \chi(\xi)\,\xi^j\,d\xi = \sum_{j=0}^{n-1} \beta_j \int_{-1}^{1} \chi(\xi)\xi^j\,d\xi + \sum_{j=n}^{\infty} \beta_j \int_{-1}^{1} \chi(\xi)\,\xi^j\,d\xi$$

and truncate the last part of the expansion. This gives the quadrature

$$\int_{-1}^{1} F(\xi)\,d\xi \approx \int_{-1}^{1} p(\xi)\,d\xi + \sum_{j=0}^{n-1} \beta_j \int_{-1}^{1} \chi(\xi)\,\xi^j\,d\xi \tag{6.119}$$

The first integral on the right-hand side involves a polynomial of order $n-1$, and the second integral a polynomial of order $2n-1$. Thus, the integral $\int_{-1}^{1} F(\xi)\,d\xi$ is approximated by integrating a polynomial of order $2n-1$. To improve the approximation of the integral on the left-hand side of Eq. (6.119) by the first integral on the right-hand side, the integration points are selected to make the second integral on the right-hand side of Eq. (6.119) vanish. Therefore, set

$$\int_{-1}^{1} \chi(\xi)\,\xi^j\,d\xi = 0 \quad j = 0, \ldots, n-1 \tag{6.120}$$

This gives a set of simultaneous nonlinear equations of order n for the unknown $\xi_i, i = 0, \ldots, n-1$. Return to Eq. (6.119) to obtain

$$\int_{-1}^{1} F(\xi)\,d\xi \approx \int_{-1}^{1} p(\xi)\,d\xi = \sum_{i=1}^{n} F(\xi_i) \int_{-1}^{1} N_i(\xi)\,d\xi = \sum_{i=1}^{n} W_i^{(n)} F(\xi_i) \tag{6.121}$$

where $p(\xi) = \sum_{i=1}^{n} N_i(\xi) F(\xi_i)$, $W_i^{(n)} = \int_{-1}^{1} N_i(\xi)\,d\xi$, and $N_i(\xi)$ is given in Eq. (6.64). The error of this quadrature is

$$R_n = \int_{-1}^{1} \frac{F^{(2n+1)}(r)}{(2n+1)!} \Pi_n^2(x)\,dx \tag{6.122}$$

with Π_n, r, and $F^{(2n+1)}$ having the same meaning as in Eq. (6.115).

The integration points ξ_i and weighting coefficients $W_i^{(n)}$ are given in Table 6.7 for various n. Note that they are symmetrically distributed. This will be illustrated in the following example.

The solutions (Gauss integration points) of Eq. (6.120) are equal to the roots of a Legendre[6] polynomial $P_n(\xi)$ of order n [Davis and Rabinivitz, 1975], where

$$P_0(\xi) = 1, \quad P_1(\xi) = \xi, \quad P_k(\xi) = \frac{2k-1}{k}\xi\,P_{k-1}(\xi) - \frac{k-1}{k}P_{k-2}(\xi) \quad 2 \le k \le n \tag{6.123}$$

[6]Adrian Marie Legendre (1752–1833) was a timid Frenchman whose recognition as a mathematician was suppressed by his colleague Laplace. He authored a variety of treatises on geometry, calculus, and the theory of numbers. The method of least squares appeared in 1806 in his *Nouvelles méthodes*. In 1812 Laplace provided a theoretical basis for the least squares method. Legendre was best known for his work on elliptic integrals.

TABLE 6.7

Gaussian Quadrature Formulas

Gaussian Quadrature of $F(\xi)$ over the interval $[-1, 1]$:

$$\int_{-1}^{1} F(\xi)\, d\xi = \sum_{i}^{n} W_i^{(n)} F(\xi_i)$$

n = Number of integration points

n	Configuration	Locations	ξ_i	$W_i^{(n)}$	Error R_n
1	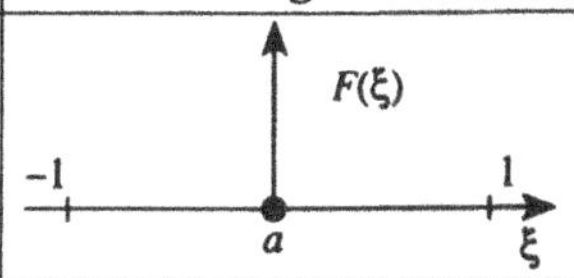	a	0	2	$\frac{1}{3}F^{(2)}(r)$*
2	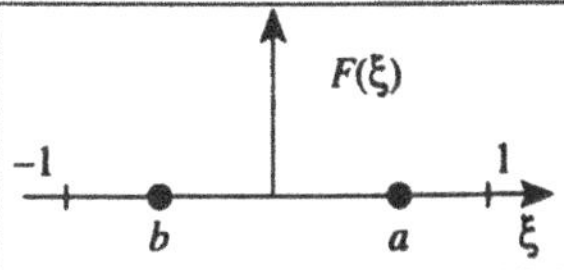	a b	$\sqrt{1/3}$ $-\sqrt{1/3}$	1 1	$\frac{1}{135}F^{(4)}(r)$
3	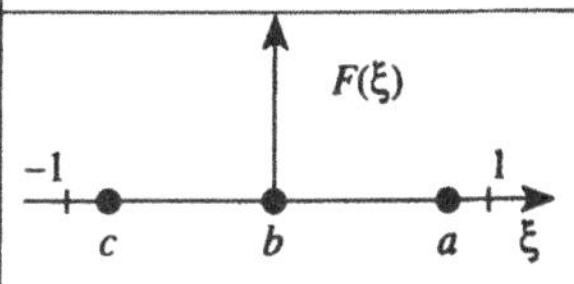	a b c	$\sqrt{3/5}$ 0 $-\sqrt{3/5}$	5/9 8/9 5/9	$\frac{1}{15750}F^{(6)}(r)$
4	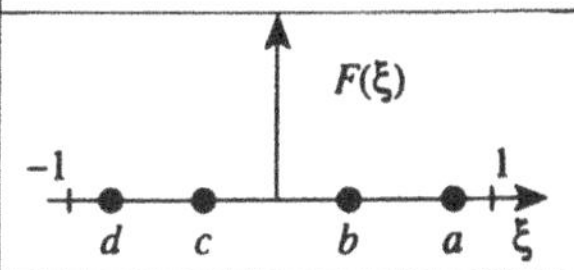	a b c d	0.86113 63116... 0.33998 10436... −0.33998 10436... −0.86113 63116...	0.34785 48451... 0.65214 51549... 0.65214 51549... 0.34785 48451...	$\frac{1}{3472875}F^{(8)}(r)$
5	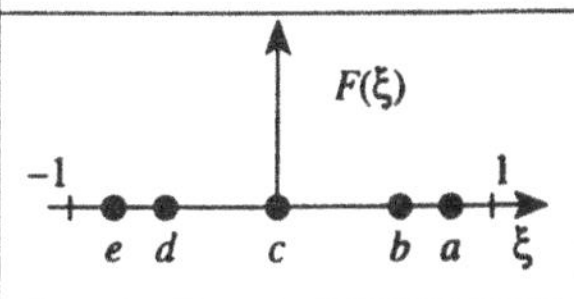	a b c d e	0.90617 98459... 0.53846 93101... 0 −0.53846 93101... −0.90617 98459 ...	0.23692 68851... 0.47862 86705... $\frac{128}{225}$ 0.47862 86705... 0.23692 68851 ...	$\frac{1}{1.2377\times10^{9}}F^{(10)}(r)$

*$F^{(n)}(r)$ is the nth derivative with respect to ξ and r is a point in $[-1, 1]$.

The n Gauss integration points of Eq. (6.123) are found by solving $P_n(\xi) = 0$ for its roots ξ_i, $i = 0, 1, \ldots, n-1$. The weighting functions are given by

$$W_i^{(n)} = \frac{2(1-\xi_i^2)}{[nP_{n-1}(\xi_i)]^2} \quad i = 1, 2, \ldots, n \tag{6.124}$$

Gaussian quadrature is the most frequently used integration procedure in finite element calculations because for the same number of integration points, the accuracy is better than that of the Newton-Cotes method.

EXAMPLE 6.11 Determination of Integration Points and Weighting Coefficients

Establish the integration points and weighting coefficients for Gaussian quadrature in the domain $[-1, 1]$ if $n = 2$.

First find the integration points. For $n = 2$, $\chi(\xi) = (\xi - \xi_1)(\xi - \xi_2)$. Use Eq.(6.123)

$$\int_{-1}^{1} (\xi - \xi_1)(\xi - \xi_2)\, d\xi = 0, \qquad \int_{-1}^{1} (\xi - \xi_1)(\xi - \xi_2)\xi\, d\xi = 0 \tag{1}$$

Upon integration, it follows that

$$\xi_1\xi_2 = -1/3, \qquad \xi_1 + \xi_2 = 0 \tag{2}$$

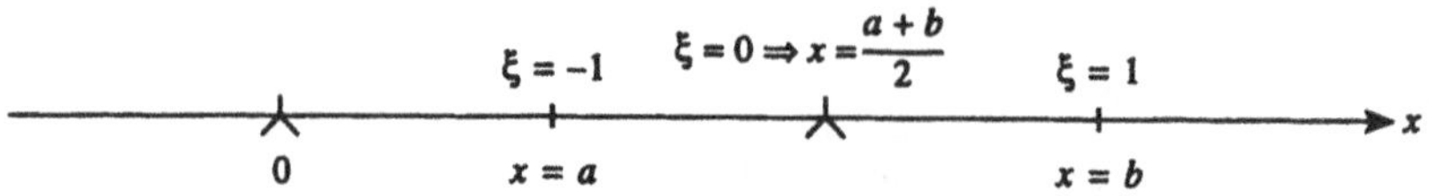

FIGURE 6.37
Transformation from $a \le x \le b$ to $-1 \le \xi \le +1$, Example 6.12.

Hence,

$$\xi_1 = -\frac{1}{\sqrt{3}}, \qquad \xi_2 = \frac{1}{\sqrt{3}} \tag{3}$$

The weighting coefficients are obtained from Eq. (6.124)

$$\begin{aligned} W_1^{(2)} &= \int_{-1}^{1} \frac{\xi - \xi_2}{\xi_1 - \xi_2} d\xi = \frac{-2\xi_2}{\xi_1 - \xi_2} = 1.0 \\ W_2^{(2)} &= \int_{-1}^{1} \frac{\xi - \xi_1}{\xi_2 - \xi_1} d\xi = \frac{-2\xi_1}{\xi_2 - \xi_1} = 1.0 \end{aligned} \tag{4}$$

Note that the integration points and weighting factors are all symmetric. Thus, the Gaussian quadrature for $n = 2$ has the form

$$\int_{-1}^{1} F(\xi)\, d\xi \approx F(\xi_1) + F(\xi_2) = F\left(-\frac{1}{\sqrt{3}}\right) + F\left(\frac{1}{\sqrt{3}}\right) \tag{5}$$

■

EXAMPLE 6.12 Application of Gaussian Quadrature
Evaluate the integral

$$\int_{1.2}^{2.8} \frac{e^x}{2} dx$$

using $n = 2$.

The integration interval $a \le x \le b$ is transformed to $-1 \le \xi \le 1$ using (Fig. 6.37)

$$x = \frac{(b-a)\xi + (b+a)}{2} = \frac{1.6\xi + 4}{2} \tag{1}$$

Then

$$e^x = e^{\frac{1.6\xi+4}{2}} \tag{2}$$

and

$$dx = \frac{(b-a)}{2} d\xi = \frac{1.6}{2} d\xi \tag{3}$$

since $a = 1.2$ and $b = 2.8$. Thus, the integral is reduced to

$$\int_{1.2}^{2.8} \frac{1}{2} e^x\, dx = \int_{-1}^{1} \frac{1.6}{4} e^{\frac{1.6\xi+4}{2}} d\xi \tag{4}$$

When $n = 2$, from Eq. (4) of the previous example, $W_1^{(2)} = 1.0$ and $W_2^{(2)} = 1.0$, and from Eq. (3), the integration points are $\xi_1 = -1/\sqrt{3} = -0.57735$ and $\xi_2 = 1/\sqrt{3} = 0.57735$. Then

$$\begin{aligned} \frac{1.6}{4} \int_{-1}^{1} e^{\frac{1.6\xi+4}{2}} d\xi &\approx \frac{1.6}{4} \left(e^{\frac{1.6\xi_1+4}{2}} + e^{\frac{1.6\xi_2+4}{2}}\right) \\ &= \frac{1.6}{4} (e^{2.4619} + e^{1.5381}) = 6.5531 \end{aligned} \tag{5}$$

The integration error is calculated using the error formula of Table 6.7 for $-1 < r < 1$. We find for $r = 1$,

$$\frac{1}{135}F^{(4)}(r) = \frac{1}{135}\frac{\partial^4}{\partial \xi^4}\left(\frac{1.6}{4}e^{\frac{1.6\xi+4}{2}}\right)\bigg|_{\xi=1} = 0.02 \tag{6}$$

and for $r = -1$,

$$\frac{1}{135}F^{(4)}(r) = \frac{1}{135}\frac{\partial^4}{\partial \xi^4}\left(\frac{1.6}{4}e^{\frac{1.6\xi+4}{2}}\right)\bigg|_{\xi=-1} = 0.004 \tag{7}$$

so that

$$0.004 < R_n < 0.02 \tag{8}$$

Thus,

$$6.549 < \int_{1.2}^{2.8} \frac{e^x}{2} dx < 6.573 \tag{9}$$

■

Rectangular and Prism Regions

Integration of a function $F(\xi, \eta)$ over a rectangular region $-1 \le \xi \le 1$, $-1 \le \eta \le 1$ can be accomplished by choosing m and h integration points in the ξ and η directions, as in the case of Fig. 6.38 for $h = m = 3$, evaluating the integral by holding η constant and integrating over ξ, and then holding ξ constant and integrating over η. This leads to

$$\int_{-1}^{+1}\int_{-1}^{+1} F(\xi, \eta)\, d\xi\, d\eta = \sum_{j=1}^{h}\sum_{i=1}^{m} W_i^{(m)} W_j^{(h)} F(\xi_i, \eta_j) \tag{6.125}$$

Table 6.7 now applies in two directions for the determination of $W_j^{(h)}$ and $W_i^{(m)}$ according to Gaussian quadrature. In this case, the total number of integration points in the domain is $h \times m = n$.

Similarly, for a right prism region, e.g., a brick configuration,

$$\int_{-1}^{+1}\int_{-1}^{+1}\int_{-1}^{+1} F(\xi, \eta, \zeta)\, d\xi\, d\eta\, d\zeta = \sum_{k=1}^{\ell}\sum_{j=1}^{m}\sum_{i=1}^{h} W_i^{(h)} W_j^{(m)} W_k^{(\ell)} F(\xi_i, \eta_j, \zeta_k) \tag{6.126}$$

where the total number of integration points is $\ell \times h \times m = n$.

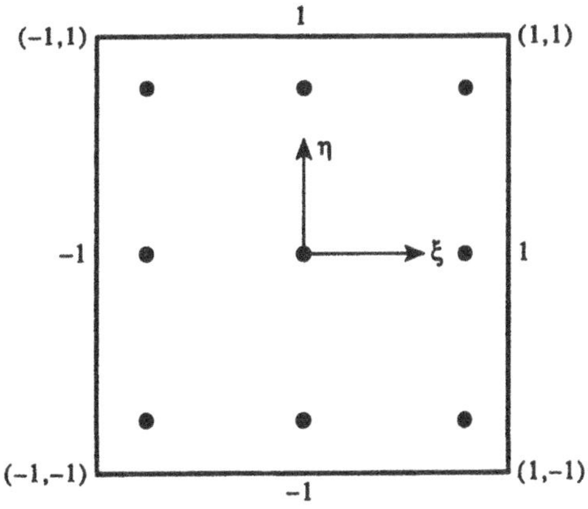

FIGURE 6.38
Integration points for a square region with $h = m = 3$.

6.6.3 Numerical Integration Over Triangular and Tetrahedral Regions

It is convenient to employ the natural coordinates defined by Eq. (6.75) and (6.82) in performing integration over triangular and tetrahedral domains. Particular closed form integrals are given by Eqs. (6.78) and (6.85). Some integrals, based on Gaussian quadrature, that are useful in finite element analyses are listed in this section. For the triangle,

$$I = \int_0^1 \int_0^{(1-L_1)} F(L_1, L_2, L_3)\, dL_1\, dL_2 = \sum_{i=1}^{n} W_i^{(n)} F(L_{1i}, L_{2i}, L_{3i}) \tag{6.127}$$

Note that the limits of integration involve the coordinate variable. Some triangular coordinates and weighting coefficients are given in Table 6.8.

TABLE 6.8

Numerical Integration Formulas for Triangles

	Location	Triangular Coordinates	Weights	Errors
Linear 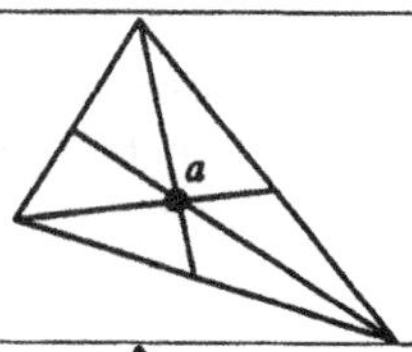	a	1/3, 1/3, 1/3	1	$R_n = O(h^2)$*
Quadratic 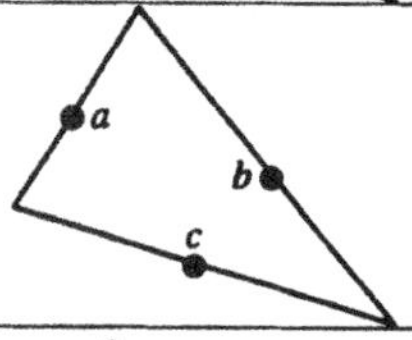	a	1/2, 1/2, 0	1/3	
	b	0, 1/2, 1/2	1/3	$R_n = O(h^3)$
	c	1/2, 0, 1/2	1/3	
Cubic 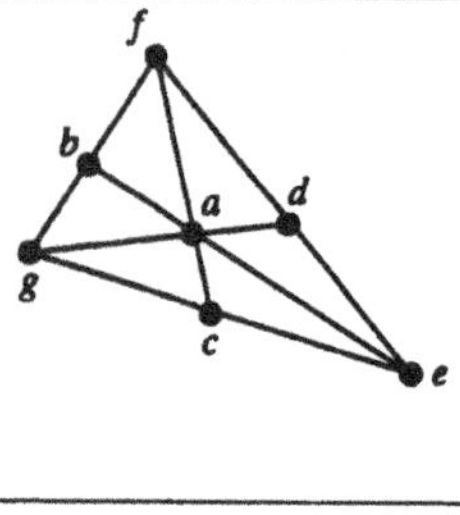	a	1/3, 1/3, 1/3	27/60	
	b	1/2, 1/2, 0		
	c	0, 1/2, 1/2	8/60	$R_n = O(h^4)$
	d	1/2, 0, 1/2		
	e	1, 0, 0		
	f	0, 1, 0	3/60	
	g	0, 0, 1		
Quintic 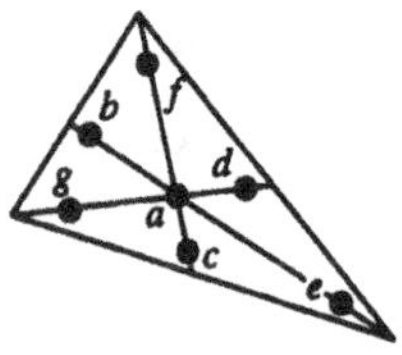	a	1/3, 1/3, 1/3,	0.225	
	b	a_1, a_2, a_2		
	c	a_2, a_1, a_2	0.13239415	
	d	a_2, a_2, a_1		$R_n = O(h^6)$
	e	a_3, a_4, a_4		
	f	a_4, a_3, a_4	0.12593918	
	g	a_4, a_4, a_3		
		with $a_1 = 0.05971587$, $a_2 = 0.47014206$, $a_3 = 0.79742699$, $a_4 = 0.10128651$		

*$O(h^n)$ means that when $h^n \to 0$, $R_n \to 0$.

TABLE 6.9

Numerical Integration Formulas for Tetrahedrons

		Location	Tetrahedral Coordinates	Weights	Errors
Linear	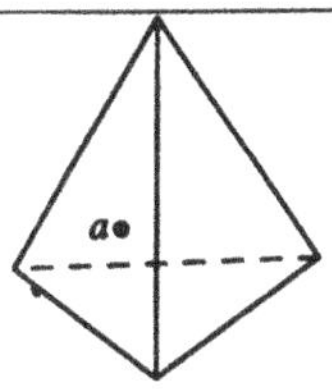	a	1/4, 1/4, 1/4, 1/4	1	$R_n = O(h^2)$
Quadratic	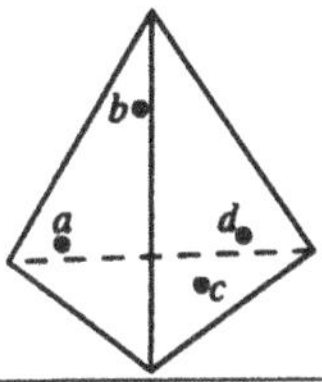	a	a_1, a_2, a_2, a_2	1/4	$R_n = O(h^3)$
		b	a_2, a_1, a_2, a_2	1/4	
		c	a_2, a_2, a_1, a_2	1/4	
		d	a_2, a_2, a_2, a_1	1/4	
			$a_1 = 0.58541020$		
			$a_2 = 0.13819660$		
Cubic	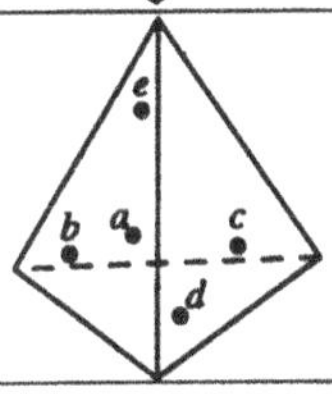	a	1/4, 1/4, 1/4, 1/4	−4/5	$R_n = O(h^4)$
		b	1/3, 1/6, 1/6, 1/6	9/20	
		c	1/6, 1/3, 1/6, 1/6	9/20	
		d	1/6, 1/6, 1/3, 1/6	9/20	
		e	1/6, 1/6, 1/6, 1/3	9/20	

For a tetrahedronal region, the integral appears as

$$I = \int_0^1 \int_0^1 \int_0^{(1-L_1-L_2)} F(L_1, L_2, L_3, L_4)\, dL_1\, dL_2\, dL_3 \\ = \sum_{i=1}^{n} W_i^{(n)} F(L_{1i}, L_{2i}, L_{3i}, L_{4i}) \tag{6.128}$$

with the values of concern given in Table 6.9.

6.7 Isoparametric Elements

The use of standard straight-sided elements to model complicated structures, especially those with curved boundaries, can be both difficult and inefficient. Many such elements with the concomitant large number of nodal displacements may be required. An appreciable reduction in the number of elements can be achieved if irregular shaped elements, such as irregular triangles, quadrilaterals, or even curved boundary elements, are used.

Although several methods for creating these kinds of elements are available, the most common approach is to establish them such that they are "parametrically" equivalent to rectilinear counterparts. That is, the irregular shaped elements are generated with a mapping of regular shaped elements.

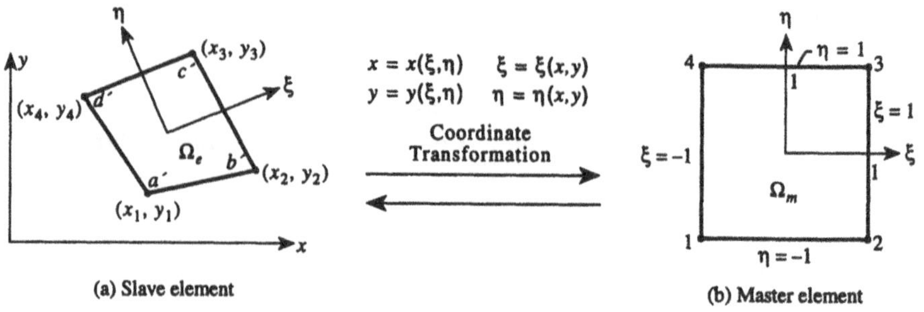

FIGURE 6.39
Isoparametric coordinate transformation.

The basic principle of *isoparametric elements* is that the interpolation functions for the displacements are also used to represent the geometry of the element. For a four-sided element, suppose the displacements u_x, u_y in the global x, y directions are expressed as

$$u_x = \sum_{i=1}^{4} N_i\, u_{xi}, \qquad u_u = \sum_{i=1}^{4} N_i\, u_{yi} \tag{6.129}$$

in which N_i are the interpolation functions, and u_{xi}, u_{yi} are the displacements at the nodes. For an isoparametric element, the geometry of the element would be represented by the same interpolation functions N_i, i.e., the global coordinate values x, y of any point in the element are

$$x = \sum_{i=1}^{4} N_i\, x_i, \qquad y = \sum_{i=1}^{4} N_i\, y_i \tag{6.130}$$

where x_i and y_i are the coordinates of the ith node in the global coordinate system.

The concept of an isoparametric element is quite useful because it can facilitate an accurate representation of irregular domains. However, the use of an isoparametric element can make it difficult to perform the integration necessary to form the element stiffness matrix and loading vector in terms of the global coordinates x and y because of the irregular shape of the element. The irregular-shaped element can be visualized as a distortion of the corresponding regular shaped element, such as the situation shown in Fig. 6.39. The integration for the element in Fig. 6.39a can be transformed to the integration in the element of Fig. 6.39b, which is much easier to implement. To do so, it is necessary to build a relationship or mapping between this distorted isoparametric element, called a *slave element*, and the corresponding regular shaped element, called a *parent* or *master element*. The finite element model is formed of the slave elements.

Consider the two-dimensional case shown in Fig. 6.39. The master element is Ω_m, and the slave element is Ω_e. The local coordinate systems (ξ, η) for these two elements have their origins at the centroids of the elements, with ξ, η varying from -1 to 1 as shown in Fig. 6.39. The coordinate transformation will map the point (ξ, η) in the master element to $x(\xi, \eta)$ and $y(\xi, \eta)$ in the slave element. From Example 6.9, Eq. (3), the interpolation functions are given by

$$\begin{aligned} N_1 &= \frac{1}{4}(1-\xi)(1-\eta), \qquad & N_2 &= \frac{1}{4}(1+\xi)(1-\eta) \\ N_3 &= \frac{1}{4}(1+\xi)(1+\eta), \qquad & N_4 &= \frac{1}{4}(1-\xi)(1+\eta) \end{aligned} \tag{6.131}$$

For the isoparametric element, the geometry, i.e., the coordinate relations, is defined by the interpolation functions for the displacement.

$$
\begin{aligned}
x &= \sum_{i=1}^{4} N_i\, x_i = \frac{1}{4}[(1-\xi)(1-\eta)x_1 + (1+\xi)(1-\eta)x_2 \\
&\quad +(1+\xi)(1+\eta)x_3 + (1-\xi)(1+\eta)x_4] \\
y &= \sum_{i=1}^{4} N_i\, y_i = \frac{1}{4}[(1-\xi)(1-\eta)y_1 + (1+\xi)(1-\eta)y_2 \\
&\quad +(1+\xi)(1+\eta)y_3 + (1-\xi)(1+\eta)y_4]
\end{aligned} \tag{6.132}
$$

For example, the line $\xi = 1$ in the master element is transformed to

$$
\begin{aligned}
x(1,\eta) &= \frac{(1-\eta)}{2}x_2 + \frac{(1+\eta)}{2}x_3 \\
&= \frac{x_2+x_3}{2} + \frac{x_3-x_2}{2}\eta \\
y(1,\eta) &= \frac{(1-\eta)}{2}y_2 + \frac{(1+\eta)}{2}y_3 \\
&= \frac{y_2+y_3}{2} + \frac{y_3-y_2}{2}\eta
\end{aligned}
$$

Thus, the line $\xi = 1$ in the master element is transformed to a line $b' - c'$ in the quadrilateral element. If the above transformation relations are invertible, line $b' - c'$ can also be transformed to $\xi = 1$ in the master element. Similarly, lines $c' - d'$, $d' - a'$, and $a' - b'$ can be tranformed to $\eta = 1$, $\xi = -1$, and $\eta = -1$ in the master element. Thus, the complete geometry can be transformed.

Using the coordinate transformation, the derivatives in the global coordinate system, where the slave element resides, can also be transformed to those in the master element. In the calculation of the element stiffness matrix and loading vector for the slave element, the derivatives $\partial/\partial x$, $\partial/\partial y$ are required. From the chain rule of differentiation,

$$
\begin{aligned}
\frac{\partial}{\partial x} &= \frac{\partial}{\partial \xi}\frac{\partial \xi}{\partial x} + \frac{\partial}{\partial \eta}\frac{\partial \eta}{\partial x} \\
\frac{\partial}{\partial y} &= \frac{\partial}{\partial \xi}\frac{\partial \xi}{\partial y} + \frac{\partial}{\partial \eta}\frac{\partial \eta}{\partial y}
\end{aligned} \tag{6.133}
$$

To evaluate $\partial/\partial x$ and $\partial/\partial y$, the derivatives $\frac{\partial \xi}{\partial x}, \frac{\partial \xi}{\partial y}, \frac{\partial \eta}{\partial x}$ and $\frac{\partial \eta}{\partial y}$ are needed. However, the explicit relationship $\xi = \xi(x, y)$ and $\eta = \eta(x, y)$ in Fig. 6.39a are difficult to obtain. This differs from the case of Fig. 6.10 where, for a rectangular element with the origin at the lower left corner, $\xi = x/a$ and $\eta = y/b$. To obtain the desired derivatives, first establish

$$
\begin{aligned}
\frac{\partial}{\partial \xi} &= \frac{\partial}{\partial x}\frac{\partial x}{\partial \xi} + \frac{\partial}{\partial y}\frac{\partial y}{\partial \xi} \\
\frac{\partial}{\partial \eta} &= \frac{\partial}{\partial x}\frac{\partial x}{\partial \eta} + \frac{\partial}{\partial y}\frac{\partial y}{\partial \eta}
\end{aligned} \tag{6.134}
$$

or, in matrix notation,

$$
\begin{bmatrix} \frac{\partial}{\partial \xi} \\ \frac{\partial}{\partial \eta} \end{bmatrix} = \begin{bmatrix} \frac{\partial x}{\partial \xi} & \frac{\partial y}{\partial \xi} \\ \frac{\partial x}{\partial \eta} & \frac{\partial y}{\partial \eta} \end{bmatrix} \begin{bmatrix} \frac{\partial}{\partial x} \\ \frac{\partial}{\partial y} \end{bmatrix} = \mathbf{J} \begin{bmatrix} \frac{\partial}{\partial x} \\ \frac{\partial}{\partial y} \end{bmatrix} \tag{6.135a}
$$

where **J** is the *Jacobian*.[7] From Eq. (6.130), the Jacobian can be expressed as

$$\mathbf{J} = \begin{bmatrix} \sum \frac{\partial N_i}{\partial \xi} x_i & \sum \frac{\partial N_i}{\partial \xi} y_i \\ \sum \frac{\partial N_i}{\partial \eta} x_i & \sum \frac{\partial N_i}{\partial \eta} y_i \end{bmatrix} = \begin{bmatrix} J_{11} & J_{12} \\ J_{21} & J_{22} \end{bmatrix} \tag{6.135b}$$

A necessary and sufficient condition for Eq. (6.135a) to be invertible is that $\det \mathbf{J} = |\mathbf{J}| \neq 0$. This same condition must be satisfied if the coordinate relations of Eq. (6.132) are invertible. The desired derivatives ∂_x and ∂_y can be expressed as

$$\begin{bmatrix} \frac{\partial}{\partial x} \\ \frac{\partial}{\partial y} \end{bmatrix} = \mathbf{J}^{-1} \begin{bmatrix} \frac{\partial}{\partial \xi} \\ \frac{\partial}{\partial \eta} \end{bmatrix} \tag{6.136a}$$

where $\mathbf{J}^{-1}$ is the inverse of the Jacobian,

$$\mathbf{J}^{-1} = \begin{bmatrix} \frac{\partial \xi}{\partial x} & \frac{\partial \eta}{\partial x} \\ \frac{\partial \xi}{\partial y} & \frac{\partial \eta}{\partial y} \end{bmatrix} = \frac{1}{|\mathbf{J}|} \begin{bmatrix} J_{22} & -J_{12} \\ -J_{21} & J_{11} \end{bmatrix} = \begin{bmatrix} J_{11}^* & J_{12}^* \\ J_{21}^* & J_{22}^* \end{bmatrix} \tag{6.136b}$$

A comparison of Eq. (6.24) and the above relation shows that $J_{11}^* = \frac{1}{a}$, $J_{22}^* = \frac{1}{b}$, and $J_{12}^* = J_{21}^* = 0$.

The determinant of the Jacobian relates the differential change in the two coordinate systems in the sense that, from the calculus, $\int_S f(x, y) dx\, dy = \int_S g(\xi, \eta) |J|\, d\xi\, d\eta$. Here, f and g are equivalent expressions for a function in the two coordinate systems. The differential area in our quadrilateral element would be

$$dA = |\mathbf{J}|\, d\xi\, d\eta \tag{6.137}$$

In order to establish the stiffness matrix and the loading vector, return to the principle of virtual work. The principle for a thin plate can be represented by the formula developed in Section 6.4.1. The element stiffness matrix and loading vector are

$$\begin{aligned} \mathbf{k}^i &= \int_V (\mathbf{D}_u \mathbf{N})^T \mathbf{E} \mathbf{D}_u \mathbf{N}\, dV = \int_V (\mathbf{N}^T{}_u \mathbf{D}^T)\, \mathbf{E}\, (\mathbf{D}_u \mathbf{N})\, dV \\ &= t \int_A (\mathbf{N}^T{}_u \mathbf{D}^T)\, \mathbf{E}\, (\mathbf{D}_u \mathbf{N})\, dA \\ \mathbf{p}^i &= \int_{S_p} \mathbf{N}^T \overline{\mathbf{p}}\, dS \end{aligned} \tag{6.138}$$

where t is the thickness of the plate, and

$$\mathbf{N} = \begin{bmatrix} N_1\ N_2\ N_3\ N_4 & \mathbf{0} \\ \mathbf{0} & N_1\ N_2\ N_3\ N_4 \end{bmatrix}$$

The material law matrix **E** is defined in Chapter 1, Eq. (1.39). In the isoparametric sense, i.e., **N** is used for both displacements and coordinates, **D** can be altered such that the integration

[7]Carl Gustav Jacob Jacobi (1804–1851). As the son of a German Jewish banker, he was raised in a wealthy, cultured atmosphere. Jacobi was forced to privately study the works of mathematicians as the leading mathematicians then were in Paris. An exception was Gauss in Göttingen. In 1826, Jacobi left Berlin for the University of Königsberg where he joined the physicists Franz Neumann and Heinrich Dove and astronomer Friedrich Bessel. There he attacked many applied problems. His mathematical accomplishments are often compared with such predecessors as Euler. His interests were varied: he once assisted Alexander von Humboldt, who was preparing his book *Kosmos*, by proving theorems from ancient Greek mathematics.

can be performed in the (ξ, η) coordinate system. From Eq. (1.24),

$$\mathbf{D}_u = \begin{bmatrix} \partial_x & 0 \\ 0 & \partial_y \\ \partial_y & \partial_x \end{bmatrix}$$

and from Eq. (6.136), $\partial_x = J_{11}^* \partial_\xi + J_{12}^* \partial_\eta$, $\partial_y = J_{21}^* \partial_\xi + J_{22}^* \partial_\eta$. Thus,

$$\mathbf{D}_u = \begin{bmatrix} J_{11}^* \partial_\xi + J_{12}^* \partial_\eta & 0 \\ 0 & J_{21}^* \partial_\xi + J_{22}^* \partial_\eta \\ J_{21}^* \partial_\xi + J_{22}^* \partial_\eta & J_{11}^* \partial_\xi + J_{12}^* \partial_\eta \end{bmatrix} \tag{6.139}$$

The stiffness matrix is obtained by substituting this $\mathbf{D}_u$ and Eq. (6.137) into Eq. (6.138). The required integration can be performed in the (ξ, η) system.

A similar procedure can be used in three-dimensional situations. From the coordinate transformation, elements of different shapes can be obtained from a master element. Master elements of different order define different transformations and generate slave elements which have the same order as their master elements and may have more complex shapes. Figures 6.40 and 6.41 show various possible transformations. Note from these figures that

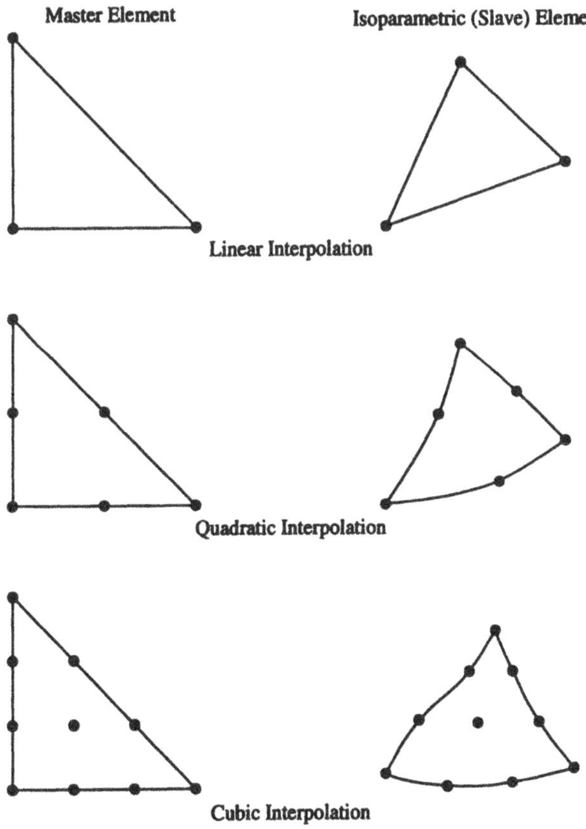

FIGURE 6.40
Triangular elements and their isoparametric forms.

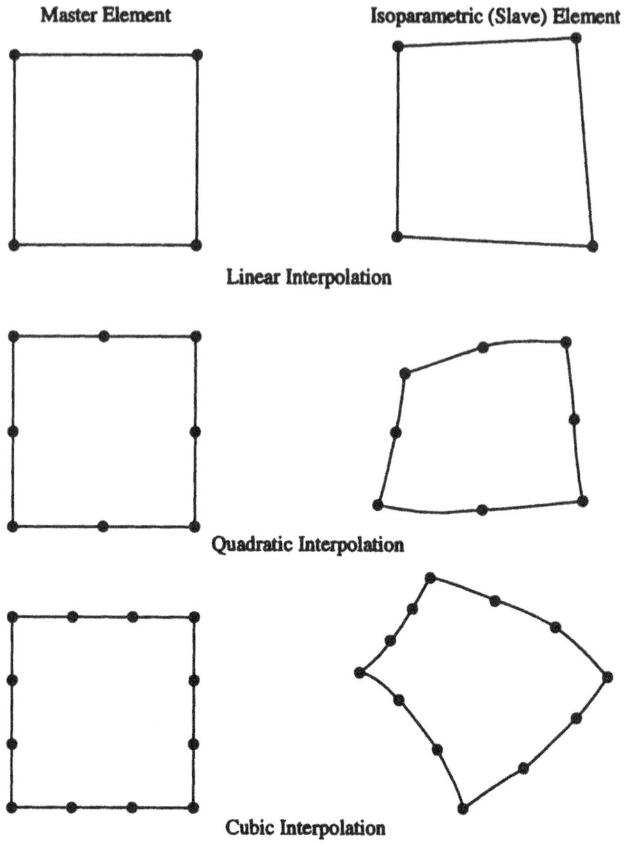

FIGURE 6.41
Quadrilateral elements and their isoparametric forms.

linear interpolation functions for triangles and rectangles lead to arbitrary triangles and quadrilaterals, respectively, and higher order interpolation functions lead to elements with curved boundaries.

It is also possible that a nonisoparametric representation for the parametric geometry can be used. If the geometric representation is of lower order than the interpolation function, the element is said to be *subparametric*. The converse defines a *superparametric* element. These parametric elements are most frequently employed in modeling three-dimensional solids. In some cases, use of these elements can contribute to modeling economy.

More isoparametric formulations for two-dimensional elements will be considered in the following sections. In particular, we will treat elements for the stretching of a plate, i.e., a flat element with in-plane loading.

6.7.1 Triangular Isoparametric Element

For a triangular master element, choose a local coordinate system with an origin at a vertex and with the remaining vertices being located by nondimensional coordinates ξ, η at (0, 1) and (1, 0) as shown in Fig. 6.42.

First derive the displacement shape function for the slave element. Suppose a linear trial function approximates the displacements in a triangular plate element undergoing

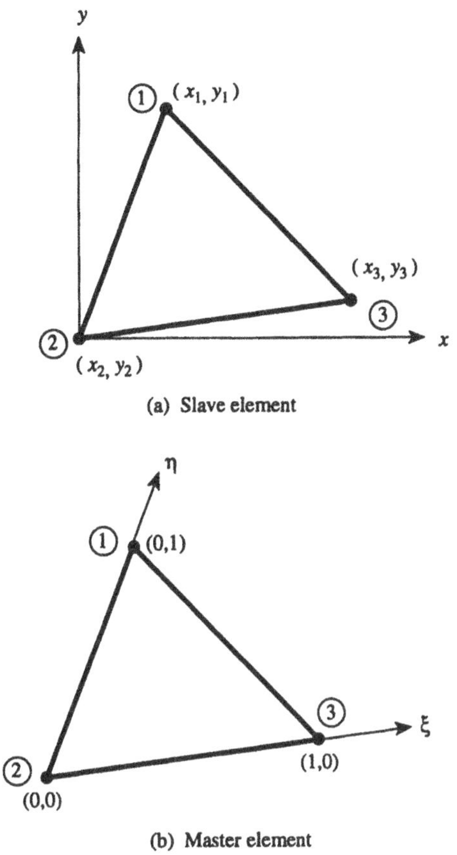

FIGURE 6.42
A triangular element.

stretching. The displacements in the x direction are approximated by

$$u_x = \mathbf{N}_u(\xi, \eta)\,\widehat{\mathbf{u}}_x \tag{6.140}$$

with the linear expression

$$\mathbf{N}_u = [1 \quad \xi \quad \eta]$$

where

$$\widehat{\mathbf{u}}_x = [\widehat{u}_{x1} \quad \widehat{u}_{x2} \quad \widehat{u}_{x3}]^T$$

contains the generalized displacements. The transformation to nodal displacement $\mathbf{v}_x$ follows the procedure described in Section 6.4.1. Define $\mathbf{v}_x = [u_{x1}\ u_{x2}\ u_{x3}]^T$. Then substitute the ξ, η coordinates of Fig. 6.42 into Eq. (6.140)

$$\begin{bmatrix} u_{x1} \\ u_{x2} \\ u_{x3} \end{bmatrix} = \begin{bmatrix} 1 & 0 & 1 \\ 1 & 0 & 0 \\ 1 & 1 & 0 \end{bmatrix} \begin{bmatrix} \widehat{u}_{x1} \\ \widehat{u}_{x2} \\ \widehat{u}_{x3} \end{bmatrix} \quad \text{or} \quad \mathbf{v}_x = \widehat{\mathbf{N}}_u\,\widehat{\mathbf{u}}_x$$

It follows that

$$\widehat{\mathbf{u}}_x = \widehat{\mathbf{N}}_u^{-1}\mathbf{v}_x = \mathbf{G}\,\mathbf{v}_x$$

with

$$\mathbf{G} = \begin{bmatrix} 0 & 1 & 0 \\ 0 & -1 & 1 \\ 1 & -1 & 0 \end{bmatrix}$$

The assumed displacement now appears as

$$u_x = \mathbf{N}_u\,\mathbf{G}\,\mathbf{v}_x = \mathbf{N}\,\mathbf{v}_x \tag{6.141}$$

Similarly, it is found that $u_y = \mathbf{N}_u\,\mathbf{G}\,\mathbf{v}_y = \mathbf{N}\,\mathbf{v}_y$. The interpolation (shape) function matrix $\mathbf{N}$ is found to be

$$\mathbf{N}_u\,\mathbf{G} = \mathbf{N} = [N_1 \quad N_2 \quad N_3] = [\eta \quad 1-\xi-\eta \quad \xi] \tag{6.142}$$

For an isoparametric element, the assumed displacements are also used to describe the geometry of the slave element. Then the coordinate transformation from the master element to the slave element is

$$\begin{aligned} x &= N_1 x_1 + N_2 x_2 + N_3 x_3 \\ y &= N_1 y_1 + N_2 y_2 + N_3 y_3 \end{aligned} \tag{6.143}$$

where $x_i, y_i, i = 1, 2, 3$ are the coordinates of the corners of the triangle forming the isoparametric element.

The element stiffness matrix is obtained by inserting the strains and stresses, derived from the assumed displacements, in the principle of virtual work. Use $\mathbf{k}^i$ of Eq. (6.138). The quantities $\mathbf{D}_u$ and $\mathbf{N}$ are obtained from the strain expressions

$$\epsilon = \begin{bmatrix} \epsilon_x \\ \epsilon_y \\ \gamma_{xy} \end{bmatrix} = \underset{\mathbf{D}_u}{\begin{bmatrix} \partial_x & 0 \\ 0 & \partial_y \\ \partial_y & \partial_x \end{bmatrix}} \underset{\mathbf{u}}{\begin{bmatrix} u_x \\ u_y \end{bmatrix}} = \underset{\mathbf{B}_u}{\mathbf{D}_u \begin{bmatrix} \mathbf{N}(\xi,\eta) & 0 \\ 0 & \mathbf{N}(\xi,\eta) \end{bmatrix}} \underset{\mathbf{v}}{\begin{bmatrix} \mathbf{v}_x \\ \mathbf{v}_y \end{bmatrix}} \tag{6.144}$$

The calculation of $\mathbf{B}_u$ involves the derivatives of $N_i(\xi, \eta)$, with respect to x and y. Since the relationship $\xi = \xi(x, y)$, $\eta = \eta(x, y)$ are complicated and difficult to use in this calculation, the procedure based on Eq. (6.136), using $\mathbf{J}^{-1}$, the inverse of the Jacobian, will be employed. For the three-node triangular element,

$$\mathbf{J} = \begin{bmatrix} \sum_i^3 \frac{\partial N_i}{\partial \xi} x_i & \sum_i^3 \frac{\partial N_i}{\partial \xi} y_i \\ \sum_i^3 \frac{\partial N_i}{\partial \eta} x_i & \sum_i^3 \frac{\partial N_i}{\partial \eta} y_i \end{bmatrix} = \begin{bmatrix} -x_2 + x_3 & -y_2 + y_3 \\ x_1 - x_2 & y_1 - y_2 \end{bmatrix}$$

The inverse of $\mathbf{J}$ can be written as

$$\mathbf{J}^{-1} = \frac{1}{2A} \begin{bmatrix} y_1 - y_2 & y_2 - y_3 \\ x_2 - x_1 & x_3 - x_2 \end{bmatrix} = \frac{1}{2A} \begin{bmatrix} \beta_3 & \beta_1 \\ \gamma_3 & \gamma_1 \end{bmatrix}$$

where $A = \frac{1}{2}|\mathbf{J}|$. Then

$$\partial_x = \frac{1}{2A}(\beta_3 \partial_\xi + \beta_1 \partial_\eta)$$

$$\partial_y = \frac{1}{2A}(-\gamma_3 \partial_\xi - \gamma_1 \partial_\eta)$$

Thus, $\mathbf{B}_u$ is evaluated as

$$\mathbf{B}_u = \mathbf{D}_u \mathbf{N} = \frac{1}{2A} \begin{bmatrix} \beta_3 \partial_\xi + \beta_1 \partial_\eta & 0 \\ 0 & -\gamma_3 \partial_\xi - \gamma_1 \partial_\eta \\ -\gamma_3 \partial_\xi - \gamma_1 \partial_\eta & \beta_3 \partial_\xi + \beta_1 \partial \eta \end{bmatrix} \begin{bmatrix} N_1 & N_2 & N_3 & 0 & 0 & 0 \\ 0 & 0 & 0 & N_1 & N_2 & N_3 \end{bmatrix}$$

$$= \frac{1}{2A} \begin{bmatrix} \beta_1 & -\beta_3 - \beta_1 & \beta_3 & 0 & 0 & 0 \\ 0 & 0 & 0 & -\gamma_1 & \gamma_3 + \gamma_1 & -\gamma_3 \\ -\gamma_1 & \gamma_3 + \gamma_1 & -\gamma_3 & \beta_1 & -\beta_3 - \beta_1 & \beta_3 \end{bmatrix} \tag{6.145}$$

Note that $\mathbf{B}_u$ does not involve ξ and η. The differential area is transformed as

$$dx\,dy = dA = |\mathbf{J}|d\xi\,d\eta = 2A\,d\xi\,d\eta$$

The stiffness matrix for the slave element is expressed as

$$\mathbf{k}^i = t\int_0^1\int_0^1 \frac{1}{4A^2}\mathbf{B}_u^T\,\mathbf{E}\,\mathbf{B}_u\,2A\,d\xi\,d\eta = \frac{t}{2}\mathbf{B}_u^T\,\mathbf{E}\,\mathbf{B}_u \tag{6.146}$$

It is seen that the formation of the stiffness matrix for the three-node triangular element does not require numerical integration. For triangular elements with more than three nodes, $\mathbf{B}_u$ involves ξ and η, and the stiffness matrix tends to be complicated to compute. As a consequence, explicit calculation of $\mathbf{k}^i$ is not feasible and numerical integration is normally employed.

Recall that a linear trial displacement was used here. Higher order trial displacements would lead to the development of elements with curved boundaries.

6.7.2 Four-Sided Isoparametric Element

Suppose the master element of Fig. 6.43a is to be mapped into the slave element of Fig. 6.43b. For the coordinate system, use an internal origin at the center for the master and slave elements with vertices defined to be at $(-1, -1)$, $(-1, 1)$, $(1, -1)$, and $(1, 1)$. The same coordinate system was employed in Fig. 6.39. Represent the shape functions, with interpolation polynomials N_i, and the element coordinates of the isoparametric element as

$$u_x = \sum_{i=1}^{g} N_i\,u_{xi} \qquad u_y = \sum_{i=1}^{g} N_i\,u_{yi}$$

$$x = \sum_{i=1}^{g} N_i\,x_i \qquad y = \sum_{i=1}^{g} N_i\,y_i$$

where g is the number of nodes. With cubic terms, the interpolation functions are given by

$$\begin{array}{l|c|c|c|c|}
 & i=5 & i=6 & i=7 & i=8 \\
N_1 = \frac{1}{4}(1-\eta)(1-\xi) & \cdots - \frac{1}{2}N_5 & \cdots & \cdots & -\frac{1}{2}N_8 \\
N_2 = \frac{1}{4}(1+\xi)(1-\eta) & \cdots - \frac{1}{2}N_5 & -\frac{1}{2}N_6 & & \\
N_3 = \frac{1}{4}(1+\eta)(1+\xi) & \cdots & -\frac{1}{2}N_6 & -\frac{1}{2}N_7 & \\
N_4 = \frac{1}{4}(1-\xi)(1+\eta) & \cdots & \cdots & -\frac{1}{2}N_7 & -\frac{1}{2}N_8 \\
\hline
N_5 = \frac{1}{2}(1-\xi^2)(1-\eta) & & & & \\
N_6 = \frac{1}{2}(1-\eta^2)(1+\xi) & & & & \\
N_7 = \frac{1}{2}(1-\xi^2)(1+\eta) & & & & \\
N_8 = \frac{1}{2}(1-\eta^2)(1-\xi) & & & &
\end{array} \tag{6.147}$$

The functions N_5, N_6, N_7, and N_8 are the interpolation functions associated with nodes 5, 6, 7, and 8 in Fig. 6.43. If these four nodes are present in the element, the element shape functions are Eq. (6.147). If fewer nodes are present, the element shape functions include

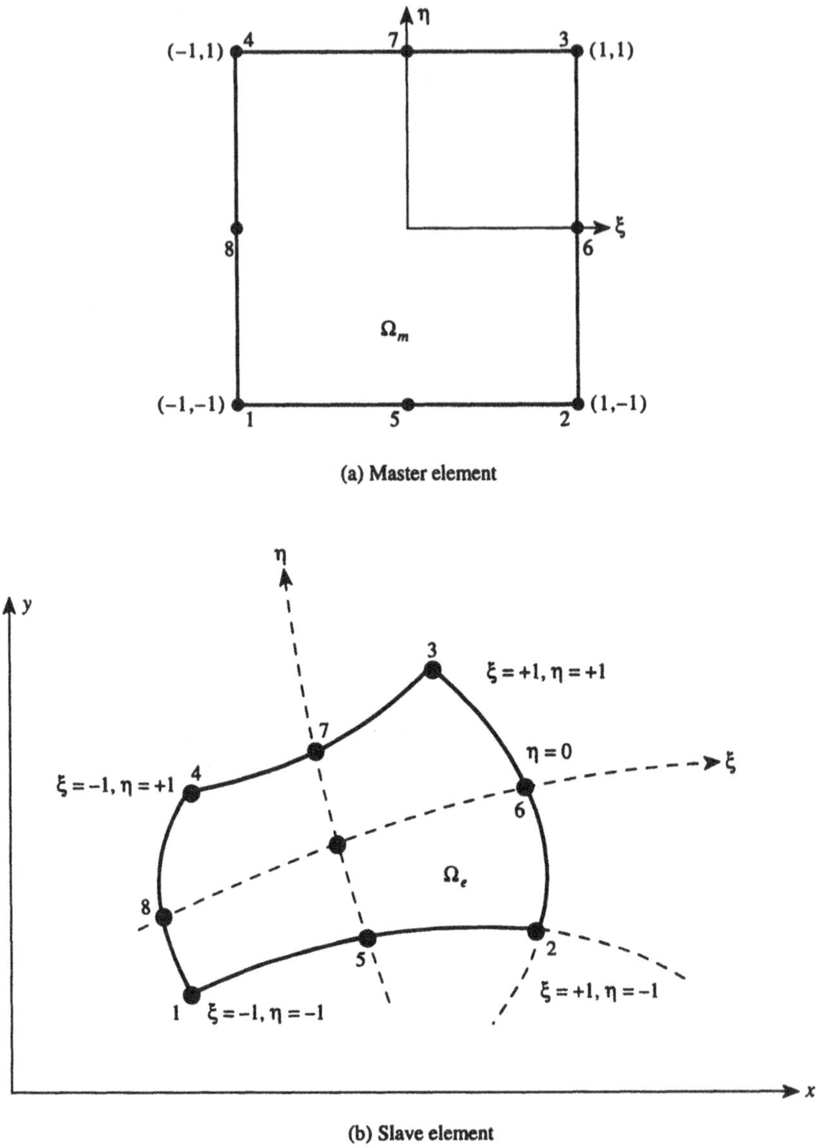

FIGURE 6.43
A four-sided element.

only the interpolation functions associated with these nodes. For example, if only node 6 is present in addition to the basic four corner nodes, the element shape functions are

$$
\begin{aligned}
N_1 &= \frac{1}{4}(1-\eta)(1-\xi), & N_2 &= \frac{1}{4}(1+\xi)(1-\eta) - \frac{1}{2}N_6 \\
N_3 &= \frac{1}{4}(1+\eta)(1+\xi) - \frac{1}{2}N_6, & N_4 &= \frac{1}{4}(1-\xi)(1+\eta), \\
N_6 &= \frac{1}{4}(1-\eta^2)(1+\xi)
\end{aligned}
\tag{6.148}
$$

These shape functions can be used to evaluate the stiffness matrices for quadrilateral elements.

EXAMPLE 6.13 Undesirable Interior Angles

It was pointed out that the coordinate transformation functions $\xi = \xi(x, y)$ and $\eta = \eta(x, y)$ must be continuous, differentiable, and invertible. In some instances, these requirements are not satisfied. This can occur if the slave elements are too distorted. This example illustrates a transformation that violates these requirements and should be avoided.

First, investigate the effect of large interior angles on the Jacobian matrix. Consider the element shown in Fig. 6.44a. Note that this element has an interior angle greater than π. We wish to map a square master element Ω_m of Fig. 6.39b into the slave element Ω_e in the global system.

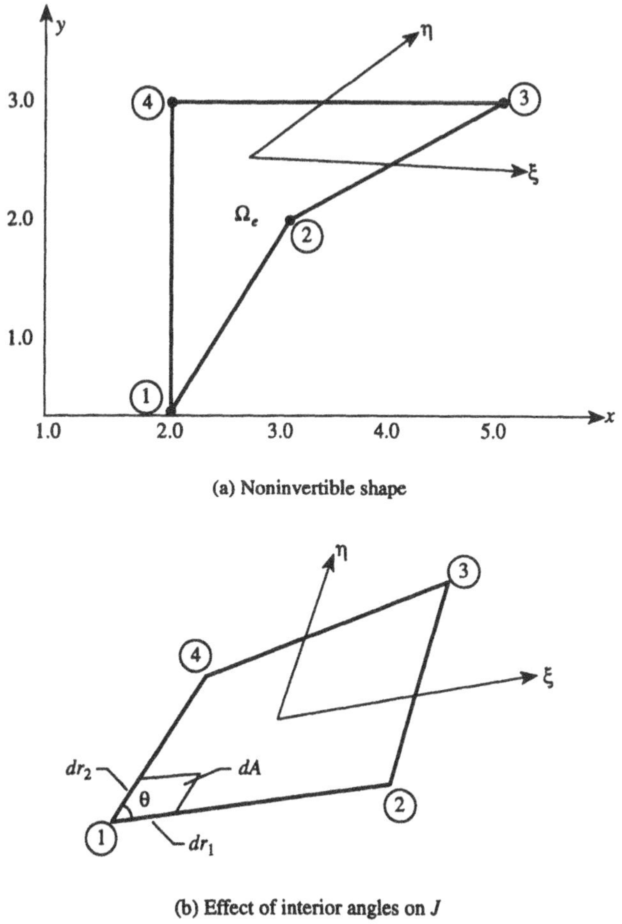

FIGURE 6.44
Inappropriate shape for the elements of Example 6.13.

Then calculate the elements of the Jacobian matrix, using the shape functions of Eq. (6.131),

$$
\begin{aligned}
\frac{\partial x}{\partial \xi} &= \sum_{i=1}^{4} \frac{\partial N_i}{\partial \xi} x_i = \frac{1}{4}[-x_1(1-\eta) + x_2(1-\eta) + x_3(1+\eta) - x_4(1+\eta)] \\
\frac{\partial x}{\partial \eta} &= \sum_{i=1}^{4} \frac{\partial N_i}{\partial \eta} x_i = \frac{1}{4}[-x_1(1-\xi) - x_2(1+\xi) + x_3(1+\xi) + x_4(1-\xi)] \\
\frac{\partial y}{\partial \xi} &= \sum_{i=1}^{4} \frac{\partial N_i}{\partial \xi} y_i = \frac{1}{4}[-y_1(1-\eta) + y_2(1-\eta) + y_3(1+\eta) - y_4(1+\eta)] \\
\frac{\partial y}{\partial \eta} &= \sum_{i=1}^{4} \frac{\partial N_i}{\partial \eta} y_i = \frac{1}{4}[-y_1(1-\xi) - y_2(1+\xi) + y_3(1+\xi) + y_4(1-\xi)]
\end{aligned} \tag{1}
$$

For element Ω_e, $x_1 = x_4 = 2$, $x_2 = 3$, $x_3 = 5$, $y_1 = 0$, $y_2 = 2$, $y_3 = y_4 = 3$

$$
\det \mathbf{J} = \begin{vmatrix} 1 + \frac{\eta}{2} & \frac{1-\eta}{2} \\ \frac{1+\xi}{2} & 1 - \frac{\xi}{2} \end{vmatrix} = \frac{3}{4}(1 + \eta - \xi) \tag{2}
$$

The determinant $\mathbf{J}$ is not non-zero everywhere. Along $\xi = 1 + \eta$, $\det \mathbf{J}$ is zero, and the Jacobian matrix is singular. This means that the coordinate transformation cannot be inverted somewhere in the element, and implies that for this element an interior angle greater than π should be avoided.

A more general investigation of the effect of certain interior angles can be conducted from the standpoint of det $\mathbf{J}$. Because the evaluation of the stiffness matrix involves dA, consider a small parallelogram area at a vertex of an element. This small area uses the two boundaries of the element as its sides (see Fig. 6.44b). Let the length of these two sides be dr_1 and dr_2. Then

$$
dA = dr_1 \cdot dr_2 \cdot \sin\theta \tag{3}
$$

Since (Eq. 6.137) $dA = \det \mathbf{J} d\xi\, d\eta$, it follows that

$$
\det \mathbf{J} = dr_1 \cdot dr_2 \cdot \sin\theta / (d\xi\, d\eta) \tag{4}
$$

It is apparent that if θ is small or close to 180°, $\det \mathbf{J}$ will be very small. Also, if θ is larger than 180°, $\det \mathbf{J}$ becomes negative. In general, an interior angle should not be too small or too large. Thus, it is often recommended that interior angles less than 30° or greater than 150° be avoided. ■

6.7.3 Instabilities

Instabilities or *spurious singular modes* in the elements occur due to deficiencies in the formulation of the elements. These instabilities are numerical phenomena, not related to the buckling treated in Chapter 11, which are discussed in numerous references on finite elements such as Cook, et al. (1989). Particular instability characteristics, which often entail *rank deficiencies*, are referred to by such names as *zero-energy modes*, *hourglass modes*, *kinematic modes*, and *mechanisms*.

To gain some insight into this problem, consider a zero-energy mode, which by definition corresponds to a displacement field that does not represent rigid body motion, yet produces zero strain energy. A stiffness matrix constructed by numerical integration is based on values obtained at the integration points of the quadrature rule. If a low order quadrature rule (few integration points) is employed and the strains happen to be zero at the integration points, a zero-energy mode occurs, leading to a stiffness matrix that is equal to zero.

6.8 Hybrid Methods

For the displacement and force methods, it is sometimes difficult to find trial solutions that satisfy the required conditions. This is particularly true for the force method. This shortcoming can be corrected with the use of extended variational theorems, in which those conditions causing difficulties are attached to the variational work expressions using Lagrange multipliers. Recall that the Lagrange multipliers can be interpreted as physical variables. These extended variational principles may be used as the basis of finite element approximations. In the literature, many of these formulations are referred to as hybrid or mixed methods [Bathe, et al., 1977; Wunderlich, 1983; Wunderlich, 1972].

The hybrid method has features of the displacement and force methods. Typically, the extended variational principle is discretized using two sets of trial solutions, one for the interior of the element and the other especially for the boundaries. There is literature of considerable size on the hybrid method (see, for example, Pian (1964) and Pian and Tong (1969)).

We choose to introduce the hybrid method by treating it as an extended force method. This is one of several available hybrid techniques. Begin with the principle of stationary complementary energy or the principle of complementary virtual work. Suppose the classical principle [Chapter 2, Eq. (2.78)]

$$-\delta W^* = \int_V \delta\boldsymbol{\sigma}^T \boldsymbol{\epsilon}\, dV - \int_{S_u} \delta\mathbf{p}^T \overline{\mathbf{u}}\, dS = 0 \tag{6.149}$$

is to be extended with the addition of the global (integral) form of the static boundary conditions ($\mathbf{p} - \overline{\mathbf{p}} = \mathbf{0}$ on S_p), i.e.,

$$\delta \int_{S_p} \mathbf{u}^T (\mathbf{p} - \overline{\mathbf{p}})\, dS = 0$$

This means that these static boundary conditions are to supplement Eq. (6.149) with the aid of the boundary displacements $\mathbf{u}$ as Lagrange multipliers. Then

$$\begin{aligned} -\delta W^* &= \int_V \delta\boldsymbol{\sigma}^T \boldsymbol{\epsilon}\, dV - \int_{S_u} \delta\mathbf{p}^T \overline{\mathbf{u}}\, dS - \delta \int_{S_p} \mathbf{u}^T (\mathbf{p} - \overline{\mathbf{p}})\, dS \\ &= \int_V \delta\boldsymbol{\sigma}^T \mathbf{E}^{-1} \boldsymbol{\sigma}\, dV - \int_{S_u} \delta\mathbf{p}^T \overline{\mathbf{u}}\, dS + \int_{S_p} \delta\mathbf{u}^T \overline{\mathbf{p}}\, dS - \int_{S_p} (\delta\mathbf{u}^T \mathbf{p} + \delta\mathbf{p}^T \mathbf{u})\, dS = 0 \end{aligned} \tag{6.150}$$

The stresses still must satisfy the conditions of equilibrium in V, i.e., $\mathbf{D}^T \boldsymbol{\sigma} + \overline{\mathbf{p}}_V = \mathbf{0}$. The extended principle of Eq. (6.150) corresponds to the hybrid functional of Chapter 2, Eq. (2.104). It should be observed that both displacements and stresses on the boundaries are unknowns, for which trial functions have to be chosen.

For a hybrid method based on an extended displacement formulation, the principle of virtual work expression [Chapter 2, Eq. (2.54)] is supplemented with an integral form of the displacement boundary condition ($\mathbf{u} - \overline{\mathbf{u}} = \mathbf{0}$ on S_u), with the boundary forces $\mathbf{p}$ as Lagrange multipliers.

The hybrid method utilizing the extended principle complementary virtual work functional of Eq. (6.150) can be implemented by selecting stress trial functions for the interior of the element and on the S_p boundary, e.g.,

$$\begin{aligned} \boldsymbol{\sigma} &= \mathbf{N}_\sigma \widehat{\boldsymbol{\sigma}} \quad \text{in} \quad V \text{ with} \\ \mathbf{p} &= \mathbf{A}^T \boldsymbol{\sigma} \quad \text{on} \quad S_p \end{aligned} \tag{6.151}$$

where $\hat{\boldsymbol{\sigma}}$ are the unknown parameters, i.e., with generalized stresses, $\mathbf{A}^T$ is the matrix of direction cosines of the normals to the boundaries [Eq. (1.57)], and $\mathbf{N}_\sigma$ contains the polynomials expressing the distribution of stresses. For the boundary displacements, choose a trial solution such as

$$\mathbf{u} = \mathbf{N}_B\,\mathbf{v} \tag{6.152}$$

where $\mathbf{v}$ are the nodal displacements, and $\mathbf{N}_B$ defines the assumed boundary displacements.

Note that when the structure is discretized into elements, the vector $\mathbf{p}$ in Eq. (6.150) contains the reaction forces on the boundaries of the elements. Hence, the second relationship of Eq. (6.151) should be applied to all elements. Also, $\mathbf{u}$ is the boundary displacement which can occur on all elements. Thus, Eq. (6.150) can be expressed as

$$\begin{aligned}\delta W^* &= \sum_i^M \left[\int_V \delta\boldsymbol{\sigma}^T \mathbf{E}^{-1} \boldsymbol{\sigma}\, dV - \int_{S_{pi}} (\delta\mathbf{u}^T\mathbf{p} + \delta\mathbf{p}^T\mathbf{u})\, dS \right] \\ &+ \sum_{j=1}^{N_1} \int_{S_{pj}} \delta\mathbf{u}^T\overline{\mathbf{p}}\, dS - \sum_{k=1}^{N_2} \int_{S_{uk}} \delta\mathbf{p}^T\overline{\mathbf{u}}\, dS = 0\end{aligned} \tag{6.153}$$

where M is the total number of elements, N_1 is the number of elements where boundary tractions are applied, N_2 is the number of elements where boundary displacements are prescribed, S_{pi} is the boundary of the ith element, and S_{pj} and S_{uk} are the boundaries of the jth and kth elements among the N_1 and N_2 elements. With the hybrid method it is common to organize the element matrix, with the help of condensation, to obtain an element stiffness matrix.

6.9 Generalized Finite Element Methods

The hybrid method of Section 6.8 is one of the most important of the generalized finite element methods. The generalized or mixed variational forms *AB*, *AD*, *CB*, and *CD* of Chapter 2, Table 2.4, can be used as the basis of further development of generalized or mixed finite element methods [Wunderlich, 1972; Wunderlich, 1983].

6.9.1 Discretization of Principles

Discretization should begin with the selection of appropriate trial functions, e.g., use

$$\mathbf{u} = \mathbf{N}_u\,\hat{\mathbf{u}} \quad \text{where} \quad \hat{\mathbf{u}} = [\hat{u}_1\ \hat{u}_2 \cdots]^T \tag{6.154}$$

for displacements and

$$\boldsymbol{\sigma} = \boldsymbol{\sigma}_p + \mathbf{N}_\sigma\hat{\boldsymbol{\sigma}} \quad (\text{or} \quad \mathbf{s} = \mathbf{s}_p + \mathbf{N}_s\hat{\mathbf{s}}) \tag{6.155}$$

for stresses (or stress resultants), where $\hat{\mathbf{u}}$ and $\hat{\boldsymbol{\sigma}}$ contain the unknown parameters, $\mathbf{N}_u$ and $\mathbf{N}_\sigma$ contain the polynomials, and $\boldsymbol{\sigma}_p$ (or $\mathbf{s}_p$) is the particular solution portion resulting from the prescribed stresses. These can be gathered together using the state vector $\mathbf{z}$, giving

$$\begin{aligned}&\mathbf{z} = \mathbf{z}_p + \mathbf{N}_z\hat{\mathbf{z}} \quad \text{with} \quad \mathbf{z} = \begin{bmatrix}\mathbf{u}\\ \boldsymbol{\sigma}\end{bmatrix} \quad \left(\text{or} \quad \mathbf{z} = \begin{bmatrix}\mathbf{u}\\ \mathbf{s}\end{bmatrix}\right) \\ &\mathbf{N}_z = \begin{bmatrix}\mathbf{N}_u & 0\\ 0 & \mathbf{N}_\sigma\end{bmatrix} \qquad \mathbf{z}_p = \begin{bmatrix}0\\ \boldsymbol{\sigma}_p\end{bmatrix}\end{aligned} \tag{6.156}$$

In order to express discrete forms of the principles of Chapter 2, Table 2.4, we will need several expressions based on Eqs. (6.154, 6.155, and 6.156) and $\mathbf{p} = \mathbf{A}^T\boldsymbol{\sigma}$ of Eq. (1.57)

$$\delta\mathbf{u}^T = \delta\hat{\mathbf{u}}^T\mathbf{N}_u^T, \qquad \delta\boldsymbol{\sigma}^T = \delta\hat{\boldsymbol{\sigma}}^T\mathbf{N}_\sigma^T, \quad \text{or} \quad \delta\mathbf{z} = \mathbf{N}_z\,\delta\hat{\mathbf{z}} \quad (\delta\mathbf{z}^T = \delta\hat{\mathbf{z}}^T\mathbf{N}_z^T),$$
$$\delta\mathbf{p}^T = \delta(\mathbf{A}^T\boldsymbol{\sigma})^T = \delta\hat{\boldsymbol{\sigma}}^T\mathbf{N}_\sigma^T\,\mathbf{A}$$

Then, the four principles of Table 2.4 can be discretized. Each of the generalized variational theorems will involve a summation over all M elements. We obtain, if $\mathbf{D}_u = \mathbf{D}_\sigma = \mathbf{D}$, for the two symmetric forms,

From CB:

$$\sum_{i=1}^{M}\left\{\delta\hat{\mathbf{z}}^T\int_{V_i}\left(\begin{bmatrix} 0 & \vdots & (\mathbf{N}_u^T\mathbf{D}^T)\mathbf{N}_\sigma \\ \cdots & \cdot & \cdots \\ \mathbf{N}_\sigma^T(\mathbf{D}\mathbf{N}_u) & \vdots & -\mathbf{N}_\sigma^T\mathbf{E}^{-1}\mathbf{N}_\sigma \end{bmatrix}\hat{\mathbf{z}} - \begin{bmatrix} -(\mathbf{N}_u^T\,\mathbf{D}^T)\boldsymbol{\sigma}_p + \mathbf{N}_u^T\,\overline{\mathbf{p}}_V \\ \cdots\cdots\cdots \\ \mathbf{N}_\sigma^T(\mathbf{E}^{-1}\boldsymbol{\sigma}_p + \boldsymbol{\epsilon}^0) \end{bmatrix}\right)dV\right.$$
$$\left. + \delta\hat{\mathbf{z}}^T\int_{S_{pi}}\begin{bmatrix} -\mathbf{N}_u^T\,\overline{\mathbf{p}} \\ \cdots \\ 0 \end{bmatrix}dS + \delta\hat{\mathbf{z}}^T\int_{S_{ui}}\left(\begin{bmatrix} 0 & \vdots & -\mathbf{N}_u^T\mathbf{P} \\ \cdots & \cdot & \cdots \\ -\mathbf{P}^T\mathbf{N}_u & \vdots & 0 \end{bmatrix}\hat{\mathbf{z}} + \begin{bmatrix} -\mathbf{N}_u^T\,\mathbf{p}_P \\ \cdots\cdots \\ \mathbf{P}^T\overline{\mathbf{u}} \end{bmatrix}\right)dS\right\}$$
$$= 0 \tag{6.157}$$

Form AD:

$$\sum_{i=1}^{M}\left\{\delta\hat{\mathbf{z}}^T\int_{V_i}\left(\begin{bmatrix} 0 & \vdots & -\mathbf{N}_u^T(\mathbf{D}^T\mathbf{N}_\sigma) \\ \cdots & \cdot & \cdots \\ -(\mathbf{N}_\sigma^T\mathbf{D})\mathbf{N}_u & \vdots & -\mathbf{N}_\sigma^T\mathbf{E}^{-1}\mathbf{N}_\sigma \end{bmatrix}\hat{\mathbf{z}} - \begin{bmatrix} \mathbf{N}_u^T(\mathbf{D}\,\boldsymbol{\sigma}_p + \overline{\mathbf{p}}_V) \\ \cdots\cdots\cdots \\ \mathbf{N}_\sigma^T(\mathbf{E}^{-1}\boldsymbol{\sigma}_p + \boldsymbol{\epsilon}^0) \end{bmatrix}\right)dV\right.$$
$$\left. + \delta\hat{\mathbf{z}}^T\int_{S_{pi}}\left(\begin{bmatrix} 0 & \vdots & \mathbf{N}_u^T\,\mathbf{P} \\ \cdots & \cdot & \cdots \\ \mathbf{P}^T\mathbf{N}_u & \vdots & 0 \end{bmatrix}\hat{\mathbf{z}} + \begin{bmatrix} \mathbf{N}_u^T(\mathbf{p}_P - \overline{\mathbf{p}}) \\ \cdots\cdots \\ 0 \end{bmatrix}\right)dS + \delta\hat{\mathbf{z}}^T\int_{S_{ui}}\begin{bmatrix} 0 \\ \cdots \\ \mathbf{P}^T\overline{\mathbf{u}} \end{bmatrix}dS\right\}$$
$$= 0 \tag{6.158}$$

where $\mathbf{P} = \mathbf{A}^T\mathbf{N}_\sigma$ and $\mathbf{p}_p = \mathbf{A}^T\boldsymbol{\sigma}_p$.

The discretized forms of generalized variational principles represent the same basic equations as the continuous forms. Depending on the variational principle, some of the fundamental equations for solids, i.e., kinematics, material law, or equilibrium, are satisfied and others are the resulting best possible approximations. The same applies to the static and geometric boundary conditions. If the trial functions satisfy one of the fundamental conditions, then the corresponding terms will fall out. For example, the boundary terms are dropped when the boundary conditions are satisfied by the trial solutions.

The system of equations for the unknowns $\hat{\mathbf{z}}$ are found using the fundamental lemma of the calculus of variations applied to Eq. (6.158), since $\delta\hat{\mathbf{z}}$ are arbitrary variations.

The forms of Eq. (6.157) lead to the element matrix for *CB*

$$\mathbf{c}^i = \begin{bmatrix} 0 & \boldsymbol{\ell}^T \\ \boldsymbol{\ell} & -\mathbf{f} \end{bmatrix}^i \quad \text{with} \quad \boldsymbol{\ell} = \int_{V_i}\mathbf{N}_\sigma^T(\mathbf{D}\mathbf{N}_u)\,dV$$
$$\mathbf{f} = \int_{V_i}\mathbf{N}_\sigma^T\mathbf{E}^{-1}\mathbf{N}_\sigma\,dV \tag{6.159}$$

For linear trial functions for all state variables, these expressions are particularly easy to integrate.

The structure of global equations depends heavily on the arrangement of the unknown variables in $\hat{\mathbf{z}}$. Often the element matrices can be assembled into a banded system of equations, with the bandwidth depending on the nodal numbering scheme. Symmetric matrices result for the forms *CB* and *AD*, but only in special cases for *AB* and *CD*. The submatrix **f** is positive definite and can be inverted.

EXAMPLE 6.14 Simply Supported Beam

As an example of the mixed method, consider the solution of a simple beam on end supports (Fig. 6.45). Let the whole beam be represented by a single element.

The state vector is

$$\mathbf{z} = [w \quad \theta \quad V \quad M]^T \tag{1}$$

The *CB* generalized variational form for a beam was developed in Chapter 2, Example 2.11, Eq. (7). If the boundary terms are assumed to be satisfied at the outset,

$$\int_0^L \delta\mathbf{z}^T \left\{ \begin{bmatrix} 0 & 0 & {}_xd & 0 \\ 0 & 0 & 1 & {}_xd \\ \cdots & & \cdots & \\ d_x & 1 & -\frac{1}{k_s GA} & 0 \\ 0 & d_x & 0 & -\frac{1}{EI} \end{bmatrix} \underbrace{\begin{bmatrix} w \\ \theta \\ \cdot\cdot \\ V \\ M \end{bmatrix}}_{\mathbf{z}} - \begin{bmatrix} \overline{p}_z \\ 0 \\ \cdot\cdot \\ 0 \\ 0 \end{bmatrix} \right\} dx = 0 \tag{2}$$

The beam element of Fig. 6.45 is of length ℓ, beginning at $x = -\ell/2$ and ending at $x = +\ell/2$. To express the axial coordinate in nondimensional form, define $\xi = 2x/\ell$. Then the element is defined in the range $-1 \le \xi \le 1$. Also, note that $d_x = (2/\ell)\, d_\xi$. In terms of the coordinate ξ, (2) becomes

$$\int_x \delta\mathbf{z}^T \left\{ \begin{bmatrix} 0 & 0 & {}_\xi d(2/\ell) & 0 \\ 0 & 0 & 1 & {}_\xi d(2/\ell) \\ (2/\ell)d_\xi & 1 & -1/k_s GA & 0 \\ 0 & (2/\ell)d_\xi & 0 & -1/EI \end{bmatrix} \begin{bmatrix} w \\ \theta \\ V \\ M \end{bmatrix} - \begin{bmatrix} \overline{p}_z \\ 0 \\ 0 \\ 0 \end{bmatrix} \right\} dx = 0 \tag{3}$$

To justify ignoring the boundary terms in (2), choose trial functions for both displacements and forces that satisfy the boundary conditions ($w = M = 0$ at $\xi = \pm 1$).

$$\underbrace{\begin{bmatrix} w \\ \theta \\ V \\ M \end{bmatrix}}_{\mathbf{z}} = \begin{bmatrix} N_w & & & \\ & N_\theta & & \\ & & N_V & \\ & & & N_M \end{bmatrix} \begin{bmatrix} \widehat{w} \\ \widehat{\theta} \\ \widehat{V} \\ \widehat{M} \end{bmatrix} = \begin{bmatrix} 1-\xi^2 & & & \\ & \xi & & \\ & & \xi & \\ & & & 1-\xi^2 \end{bmatrix} \underbrace{\begin{bmatrix} \widehat{w} \\ \widehat{\theta} \\ \widehat{V} \\ \widehat{M} \end{bmatrix}}_{\widehat{\mathbf{z}}} \tag{4}$$

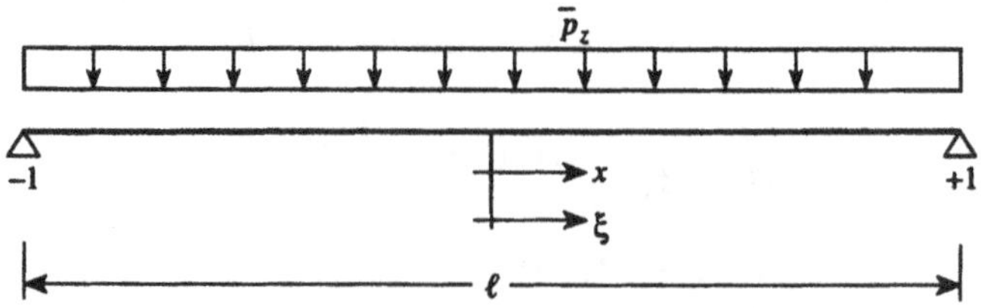

FIGURE 6.45
Notation for Example 6.14.

Insert (4) into (3), integrate, and find

$$\delta\widehat{\mathbf{z}}^T \left\{ \begin{bmatrix} 0 & 0 & (-2/\ell)\frac{4}{3} & 0 \\ 0 & 0 & \frac{2}{3} & (2/\ell)\frac{4}{3} \\ (-2/\ell)\frac{4}{3} & \frac{2}{3} & -\frac{2}{3}\frac{1}{k_s GA} & 0 \\ 0 & (2/\ell)\frac{4}{3} & 0 & -\frac{16}{15}\frac{1}{EI} \end{bmatrix} \begin{bmatrix} \widehat{w} \\ \widehat{\theta} \\ \widehat{V} \\ \widehat{M} \end{bmatrix} - \begin{bmatrix} \frac{4}{3}\overline{p}_z \\ 0 \\ 0 \\ 0 \end{bmatrix} \right\} \frac{\ell}{2} = 0 \tag{5}$$

Solve this set of equations for $\widehat{w}$, $\widehat{\theta}$, $\widehat{V}$, and $\widehat{M}$. Substitute the results into (4), giving

$$\begin{aligned} w &= \frac{\overline{p}_z \ell^2}{8}\left(\frac{\ell^2}{10EI} + \frac{1}{k_s GA}\right)(1-\xi^2) \qquad \theta = \frac{\overline{p}_z \ell^3}{20EI}\xi \\ V &= -\frac{\overline{p}_z \ell}{2}\xi \qquad M = \frac{\overline{p}_z \ell^2}{8}(1-\xi^2) \end{aligned} \tag{6}$$

These, of course, are approximate relationships. ▮

6.9.2 Flat Element with In-Plane Deformation

Consider a rectangular flat element (membrane element) undergoing in-plane deformation. An element matrix, based on a mixed variational principle, will be developed using separate trial functions for displacement and stresses.

Fundamentals

See Fig. 6.46 for the element, state variables, and notation. Suppose this element is in plane stress so that the material law is [Chapter 1, Eq. (1.40)]

$$\boldsymbol{\epsilon} = \mathbf{E}^{-1}\boldsymbol{\sigma} = \begin{bmatrix} \epsilon_x \\ \epsilon_y \\ \epsilon_{xy} \end{bmatrix} = \frac{1}{E}\begin{bmatrix} 1 & -\nu & 0 \\ -\nu & 1 & 0 \\ 0 & 0 & 2(1+\nu) \end{bmatrix}\begin{bmatrix} \sigma_x \\ \sigma_y \\ \sigma_{xy} \end{bmatrix} \tag{6.160}$$

The operator matrix $\mathbf{D}^T$ is [Chapter 1, Eq. (1.24)]

$$\mathbf{D}^T = \begin{bmatrix} \partial_x & 0 & \partial_y \\ 0 & \partial_y & \partial_x \end{bmatrix} \tag{6.161}$$

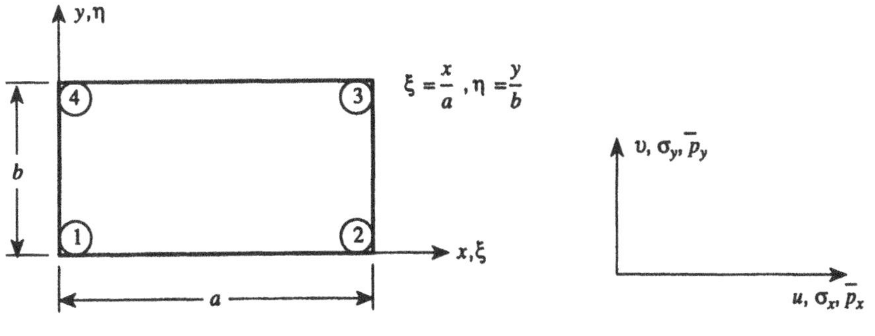

(a) Coordinates and nodal numbering (b) State variables and loading

FIGURE 6.46
A thin rectangular flat element with in-plane deformation.

and the matrix of the components of the normal vector is [Eq. (1.57)]

$$\mathbf{A}^T = \begin{bmatrix} a_x & 0 & a_y \\ 0 & a_y & a_x \end{bmatrix} \tag{6.162}$$

General Solution

As the basis of the approximation procedure, choose an extended principle of virtual work in the form of *CB* (Chapter 2). Equation *CB* is a combination of Eq. (B), the kinematics and geometric boundary conditions, and Eq. (C), the principle of virtual work. From Chapter 2, Table 2.4, form *CB* can be expressed as

$$\sum_{i=1}^{M} \left\{ \int_{V_i} \delta\mathbf{z}^T \left(\begin{bmatrix} 0 & {}_u\mathbf{D}^T \\ \mathbf{D}_u & -\mathbf{E}^{-1} \end{bmatrix} \mathbf{z} - \begin{bmatrix} \overline{\mathbf{p}}_V \\ \mathbf{0} \end{bmatrix} \right) dV + \int_{S_{pi}} \delta\mathbf{z}^T \begin{bmatrix} -\overline{\mathbf{p}} \\ \mathbf{0} \end{bmatrix} dS \right.$$
$$\left. + \int_{S_{ui}} \delta\mathbf{z}^T \left(\begin{bmatrix} \mathbf{0} & -\mathbf{A}^T \\ -\mathbf{A} & \mathbf{0} \end{bmatrix} \mathbf{z} + \begin{bmatrix} \mathbf{0} \\ \mathbf{A}\overline{\mathbf{u}} \end{bmatrix} \right) dS \right\} = 0 \tag{6.163}$$

where i is summed over all the elements (M), and the state vector $\mathbf{z}$ is given by

$$\mathbf{z} = [\mathbf{u} \quad \boldsymbol{\sigma}]^T$$

Trial Functions

Follow the development of Section 6.9.1 for three dimensions. Assume that the displacements and stresses can be represented by the trial functions.

$$\mathbf{u} = \mathbf{N}_u \hat{\mathbf{u}} \quad \text{and} \quad \boldsymbol{\sigma} = \boldsymbol{\sigma}_p + \mathbf{N}_\sigma \hat{\boldsymbol{\sigma}}$$

where $\boldsymbol{\sigma}_p$ is the particular solution portion resulting from prescribed stresses. Place these together as

$$\begin{bmatrix} \mathbf{u} \\ \boldsymbol{\sigma} \end{bmatrix} = \mathbf{z} = \mathbf{z}_p + \mathbf{N}_z \hat{\mathbf{z}} = \begin{bmatrix} \mathbf{0} \\ \boldsymbol{\sigma}_p \end{bmatrix} + \begin{bmatrix} \mathbf{N}_u & \mathbf{0} \\ \mathbf{0} & \mathbf{N}_\sigma \end{bmatrix} \begin{bmatrix} \hat{\mathbf{u}} \\ \hat{\boldsymbol{\sigma}} \end{bmatrix} \tag{6.164}$$

or

$$\mathbf{N}_z \hat{\mathbf{z}} = \begin{bmatrix} \overbrace{\mathbf{N}_{ux} \;\vdots\; \quad \;\vdots}^{\mathbf{N}_u} & & \vdots & & \vdots & \\ \cdots & \cdot & \cdots & \cdot & \cdots & \cdot & \cdots & \cdot & \cdots \\ & \vdots & \mathbf{N}_{uy} & \vdots & & \vdots & & \vdots & \\ \cdots & \cdot & \cdots & \cdot & \cdots & \cdot & \cdots & \cdot & \cdots \\ & \vdots & & \vdots & \mathbf{N}_{\sigma x} & \vdots & & \vdots & \\ \cdots & \cdot & \cdots & \cdot & \cdots & \cdot & \cdots & \cdot & \cdots \\ & \vdots & & \vdots & & \vdots & \mathbf{N}_{\sigma y} & \vdots & \\ \cdots & \cdot & \cdots & \cdot & \cdots & \cdot & \cdots & \cdot & \cdots \\ & \vdots & & \underbrace{\vdots & & \vdots & & \vdots & \mathbf{N}_{\sigma xy}}_{\mathbf{N}_\sigma} \end{bmatrix} \begin{bmatrix} \hat{\mathbf{u}}_x \\ \cdots \\ \hat{\mathbf{u}}_y \\ \cdots \\ \hat{\boldsymbol{\sigma}}_x \\ \cdots \\ \hat{\boldsymbol{\sigma}}_y \\ \cdots \\ \hat{\boldsymbol{\sigma}}_{xy} \end{bmatrix} \tag{6.165}$$

We suppose that the transformation from generalized variables to the nodal response variables has been performed, e.g., $\mathbf{N}_{ux}$ is replaced by $\mathbf{N}_{ux}\, \mathbf{G}_u$. We choose not to adjust the tilda notation, e.g., $\widehat{\mathbf{u}}_x$. For $\widehat{\mathbf{u}}_x$, let

$$\widehat{\mathbf{u}}_x = [\widehat{u}_{x1} \quad \widehat{u}_{x2} \quad \widehat{u}_{x3} \quad \widehat{u}_{x4}]^T \tag{6.166a}$$

and form the unknown state vector at nodes k as

$$\widehat{\mathbf{z}}_k = [\widehat{\mathbf{u}}_x \quad \widehat{\mathbf{u}}_y \quad \widehat{\boldsymbol{\sigma}}_x \quad \widehat{\boldsymbol{\sigma}}_y \quad \widehat{\boldsymbol{\sigma}}_{xy}]_k^T \tag{6.166b}$$

Element Matrices

Substitute Eqs. (6.165) into (6.163) to obtain the two-dimensional equivalent of Eq. (6.157). We choose to neglect the effect of the surface loads and to use trial functions that fulfill the geometric boundary conditions. Then we obtain the simplified form

$$\begin{aligned}\sum_i \Bigg\{ \delta\widehat{\mathbf{z}}^T t \int_{A_i} \begin{bmatrix} 0 & (\mathbf{N}_u^T \mathbf{D}^T)\mathbf{N}_\sigma \\ \mathbf{N}_\sigma^T(\mathbf{D}\,\mathbf{N}_u) & -\mathbf{N}_\sigma^T \mathbf{E}^{-1}\mathbf{N}_\sigma \end{bmatrix} \widehat{\mathbf{z}}\, dA \\ -\delta\widehat{\mathbf{z}}^T \int_{S_{pi}} \begin{bmatrix} \mathbf{N}_u^T \overline{\mathbf{p}} \\ 0 \end{bmatrix} dS + \delta\widehat{\mathbf{z}}^T \int_{S_{ui}} \begin{bmatrix} 0 \\ \mathbf{N}_\sigma^T \mathbf{A}\,\overline{\mathbf{u}} \end{bmatrix} dS \Bigg\} = 0\end{aligned} \tag{6.167}$$

With

$$(\mathbf{D}\mathbf{N}_u) = \mathbf{B}_u = \begin{bmatrix} \mathbf{N}_{ux,x} & | & 0 \\ - & - & - \\ 0 & | & \mathbf{N}_{uy,y} \\ - & - & - \\ \mathbf{N}_{ux,y} & | & \mathbf{N}_{uy,x} \end{bmatrix}$$

the operator matrix can be written more explicitly as

$$a\cdot b\cdot t \int_\xi \int_\eta \begin{matrix} [\;\widehat{\mathbf{u}}_x & \widehat{\mathbf{u}}_y & \widehat{\boldsymbol{\sigma}}_x & \widehat{\boldsymbol{\sigma}}_y & \widehat{\boldsymbol{\sigma}}_{xy}\;] \end{matrix} \begin{bmatrix} 0 & 0 & \mathbf{N}_{ux,x}^T\mathbf{N}_{\sigma x} & 0 & \mathbf{N}_{ux,y}^T\mathbf{N}_{\sigma xy} \\ 0 & 0 & 0 & \mathbf{N}_{uy,y}^T\mathbf{N}_{\sigma y} & \mathbf{N}_{uy,x}^T\mathbf{N}_{\sigma xy} \\ \mathbf{N}_{\sigma x}^T\mathbf{N}_{ux,x} & 0 & -\frac{1}{E}\mathbf{N}_{\sigma x}^T\mathbf{N}_{\sigma x} & \frac{\nu}{E}\mathbf{N}_{\sigma x}^T\mathbf{N}_{\sigma y} & 0 \\ 0 & \mathbf{N}_{\sigma y}^T\mathbf{N}_{uy,y} & \frac{\nu}{E}\mathbf{N}_{\sigma y}^T\mathbf{N}_{\sigma x} & -\frac{1}{E}\mathbf{N}_{\sigma y}^T\mathbf{N}_{\sigma y} & 0 \\ \mathbf{N}_{\sigma xy}^T\mathbf{N}_{ux,y} & \mathbf{N}_{\sigma xy}^T\mathbf{N}_{uy,x} & 0 & 0 & \frac{1}{G}\mathbf{N}_{\sigma xy}^T\mathbf{N}_{\sigma xy} \end{bmatrix} d\xi\, d\eta \tag{6.168}$$

Choose the same trial function for the displacements and stresses

$$\begin{aligned}\mathbf{N}_{ux} = \mathbf{N}_{uy} = \mathbf{N}_{\sigma x} = \mathbf{N}_{\sigma y} = \mathbf{N}_{\sigma xy} \\ = [\underset{①}{(1-\xi)(1-\eta)} \quad \underset{②}{\xi(1-\eta)} \quad \underset{③}{\xi\eta} \quad \underset{④}{(1-\xi)\eta}]\end{aligned} \tag{6.169}$$

The derivatives needed in Eq. (6.168) are

$$\begin{aligned}\mathbf{N}_{ux,x} = \mathbf{N}_{uy,x} &= [-(1-\eta)/a \quad (1-\eta)/a \quad \eta/a \quad -\eta/a] \\ \mathbf{N}_{ux,y} = \mathbf{N}_{uy,y} &= [-(1-\xi)/b \quad -\xi/b \quad \xi/b \quad (1-\xi)/b]\end{aligned} \tag{6.170}$$

Insertion of these trial functions into Eq. (6.168) leads to the element matrix for the CB functional, which has the same structure as Eq. (6.168).

References

Argyris, J., 1960, Energy theorems and structural analysis, *Aircraft Engineering*, 1954 and 1955, reprinted by Butterworths Scientific Publications, London.

Bathe, K.J., 1996, *Finite Element Procedures*, Prentice-Hall, Englewood Cliffs, NJ.

Bathe, K.J., 1998, What can go wrong with FEA? *Mechanical Engineering*, May, pp. 63–65.

Bathe, K.J., Oden, J.T. and Wunderlich, W. (Eds.), 1977, *Formulations and Computational Algorithms in Finite Element Analysis*, MIT Press, Boston.

Bazeley, G.P., Cheung, Y.K., Irons, B.M. and Zienkiewicz, O.C., 1965, Triangular elements in bending, conforming and non-conforming solutions, *Proc. Conf. Matrix Meth. Struct. Mechan.*, Wright-Patterson Air Force Base, OH.

Clough, R.W., 1960, The finite element method in plane stress analysis, *Proc. 2nd ASCE Conf. Electronic Computation*, Pittsburgh, PA.

Clough, R.W., 1990, Original formulation of the finite element method, *J. Finite Elem. in Anal. Des.*, Vol. 7, pp. 89–101.

Cook, R.D., Malkus, D.S. and Plesha, M.E., 1989, *Concepts and Applications of Finite Element Analysis*, 3rd ed., Wiley, New York.

Courant, R., 1943, Variational methods for the solution of problems of equilibrium and vibrations, *Bull. Am. Math. Soc.*, Vol. 49, pp. 1–43.

Davis, J.J., and Rabinivitz, P., 1975, *Methods of Numerical Integration*, Academic Press, NY.

Felippa, C.A., 1994, An appreciation of R. Courant's Variational methods for the solution of problems of equilibrium and vibrations 1943, *Int. J. for Numer. Methods in Eng.*, Vol. 37, pp. 2159–2187.

Hetenyi, M., 1946, *Beams on Elastic Foundations*, University of Michigan Press, Ann Arbor, MI.

Gupta, K.K. and Meek, J.L., 1996, A brief history of the beginnings of the finite element method, *Int. J. for Numer. Methods in Eng.*, Vol. 39, pp. 3761–3774.

MacNeal, R.H., and Harder, R.L., 1985, A proposed standard set of problems to test finite element accuracy, *J. Finite Elem. in Anal. and Des.*, Vol. 1, pp. 3–20.

Melosh, R. J., 1963, Basis for derivation of matrices for direct stiffness method, *AIAA J.*, Vol. 1, pp. 1631–1637.

Oden, J.T., 1990, The best FEM, *J. Finite Elem. in Anal. and Des.*, Vol. 7, pp. 103–114.

Pian, T.H.H., 1964, Derivation of element stiffness matrices by assumed stress distributions, *AIAA Journal*, 2, 1333–1336.

Pian, T.H.H. and Tong, P., 1969, Basis of finite element methods for solid continua, *Int. J. for Numer. Methods in Eng.*, Vol 1, pp. 3–23.

Pilkey, W.D., 1994, *Formulas for Stress, Strain, and Structural Matrices*, Wiley, New York.

Prager, W., and Synge, J.L., 1947, Approximation in elasticity based on the concept of function space, *Quart. J. Appl. Math.*, Vol. 5, pp. 214–269.

Robinson, J., 1985, *Early FEM Pioneers*, Robinson and Associates, Dorset, England.

Schwarz, H.R., 1980, *Methode der finiten Elemente*, Verlag B.G. Teubner, Stuttgart.

Stein, E., and Wunderlich, W., 1973, Finite-Element-Methoden als Variationsverfahre der Elastostatik, In *Finite Elemente in der Statik*, Verlag Wilhelm Ernst & Sohn, Berlin.

Szabo, B., and Babuska, I., 1991, *Finite Element Analysis*, Wiley, NY.

Turner, M., Clough, R.W., Martin, H.C. and Topp, L.J., 1956, Stiffness and deflection analysis of complex structures, *J. Aeronaut. Sci.*, Vol. 23, pp. 805–823.

Wunderlich, W. 1972, Discretization of element stiffness matrices by a generalized variational approach, In *Hydromechanically Loaded Shells*, University Press of Hawaii, Honolulu.

Wunderlich, W., Stein, E. and Bathe, K. (Eds.), 1981, *Nonlinear Finite Element Analysis in Structural Mechanics*, Spring-Verlag, Berlin.

Wunderlich, W. 1983, Mixed models for plates and shells: Principles—elements—examples, In *Hybrid and Mixed Finite Element Models*, Atturi, S.N., Galagher, R.H., Zienkeiwicz, O.C. (Eds.), Wiley, New York.

Wunderlich, W., and Redanz, W., 1995, Die Methode der Finiten Elemente, in *Der Ingenieurbau: Grundwissen*, W. Ernst & Sohn, Berlin.

Zienkiewicz, O.C., 1977, *The Finite Element Method in Engineering Science*, 3rd ed., McGraw-Hill, New York,

Problems

6.1 Suppose the structure of Fig. 6.14a, with $E = 30\ \mathrm{GN/m^2}$ and $\nu = 0$, is stiffened with steel rods placed along the outer edge of the structure. If the rods have circular cross-sections with 0.2 m diameters, compute the distribution of displacements and stresses throughout the structure. Use the structural properties, e.g., dimensions, of Example 6.2. For the steel rods, $E = 207\ \mathrm{GN/m^2}$, $\nu = 0.3$.

Question: Should these rods be beams or extension bars? If they are beams, how should the mismatch in DOF between the slopes at the nodes of the beams and the translations at the nodes of the planar elements be handled?

Answer: For rods treated as bars with longitudinal motion only.

Node Number	u_x (m)	u_y (m)
1	-9.163×10^{-6}	0.0
2	-6.397×10^{-6}	-2.737×10^{-5}
3	0.0	-3.522×10^{-5}
4	3.747×10^{-7}	-8.572×10^{-6}
5	-9.155×10^{-8}	-2.885×10^{-5}
6	0.0	-3.844×10^{-5}
7	8.926×10^{-6}	-1.146×10^{-5}
8	6.455×10^{-6}	-3.558×10^{-5}
9	0.0	-4.917×10^{-5}

6.2 Assume that the thin structure of Fig. 6.14a is stiffened with diagonal rods, as well as bars placed along the outer edge. All of the bars are made of steel ($E = 207\ \mathrm{GN/m^2}$, $\nu = 0.3$) with 0.2 m diameters. Continue to use the structural properties of Example 6.2. Compute the distribution of displacement and stresses throughout the structure. Note that the solution to this problem will involve local-to-global coordinate transformations.

6.3 Use the displacement method to compute the deformed profile of the structure shown in Fig. P6.3.

6.4 The structure shown in Fig. P6.4a is modeled with two plane stress elements. Obtain the shape functions for the elements and calculate the element stiffness matrices.

Hint: Use the element (local) coordinate system as shown in Fig. P6.4b and assume that

$$u_x = \hat{u}_1 + \hat{u}_2 x + \hat{u}_3 y, \qquad u_y = \hat{u}_4 + \hat{u}_5 x + \hat{u}_6 y$$

Answer:

$$\begin{bmatrix} u_x \\ u_y \end{bmatrix} = \begin{bmatrix} N_1 & 0 & N_2 & 0 & N_3 & 0 \\ 0 & N_1 & 0 & N_2 & 0 & N_3 \end{bmatrix} \begin{bmatrix} u_{x1} \\ u_{y1} \\ u_{x2} \\ u_{y2} \\ u_{x3} \\ u_{y3} \end{bmatrix}$$

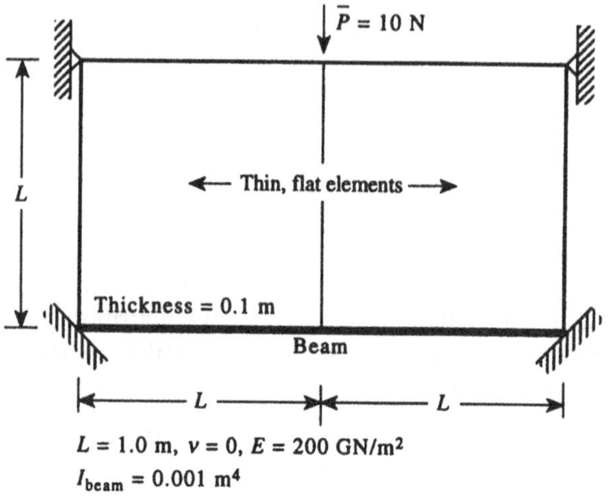

FIGURE P6.3

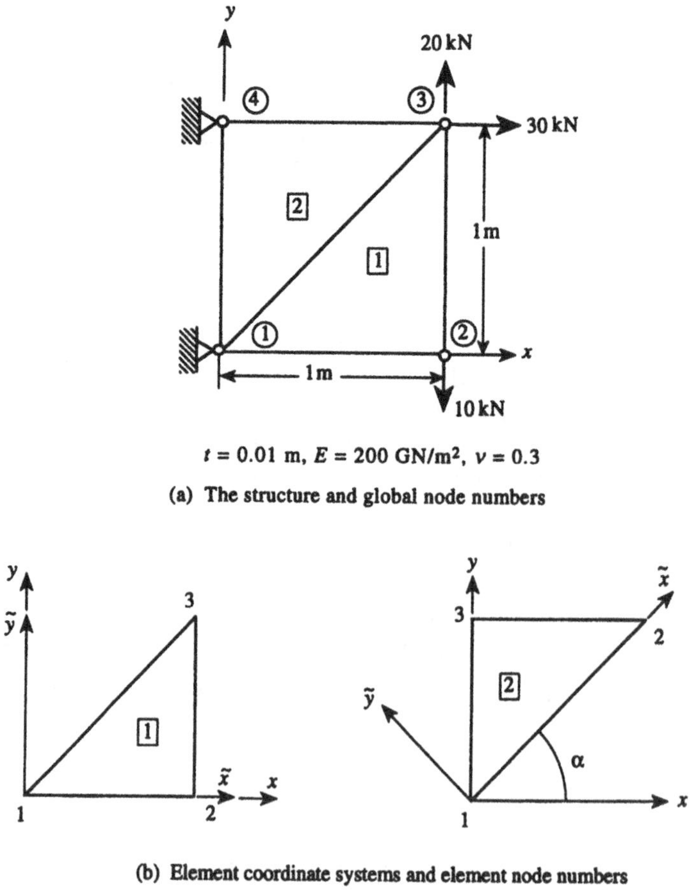

FIGURE P6.4

where

$$N_1 = \frac{y_3(x_2 - x) + y(x_3 - x_2)}{x_2 y_3}, \qquad N_2 = \frac{xy_3 - yx_3}{x_2 y_3}, \qquad N_3 = \frac{y}{y_3}$$

6.5 Use the stiffness matrices obtained in Problem 6.4 to calculate the responses of the structure of Fig. P6.4a.

Hint: The element stiffness matrices in the local coordinate systems have to be transformed into the global xy coordinate system using $\mathbf{k}^i = \mathbf{T}^{iT}\, \tilde{\mathbf{k}}^i\, \mathbf{T}^i$, where $\mathbf{k}^i$ and $\tilde{\mathbf{k}}^i$ are the element stiffness matrices in the global and local coordinate systems, respectively.

$$\mathbf{T}^i = \begin{bmatrix} \mathbf{T}^i_{11} & & \\ & \mathbf{T}^i_{22} & \\ & & \mathbf{T}^i_{33} \end{bmatrix} \qquad \mathbf{T}^i_{jj} = \begin{bmatrix} \cos\alpha & \sin\alpha \\ -\sin\alpha & \cos\alpha \end{bmatrix}$$

where α is the angle between the x (global coordinate) and the $\tilde{x}$ (local) coordinate (Fig. P6.4b) for the ith element. The global node number and the global DOF are related by

Global Node	Global DOF No.	
No.	u_x	u_y
1	1	2
2	3	4
3	5	6
4	7	8

The incidence table is

Element	Node No.		
Number	1	2	3
1	1	2	3
2	1	3	4

and the element numbers and the corresponding global DOF are related by

<table>
<tr><th rowspan="2">Element No.</th><th colspan="6">Element Node Numbers</th></tr>
<tr><th colspan="2">1</th><th colspan="2">2</th><th colspan="2">3</th></tr>
<tr><td rowspan="2">1</td><td colspan="2">1</td><td colspan="2">2</td><td colspan="2">3</td></tr>
<tr><td>1</td><td>2</td><td>3</td><td>4</td><td>5</td><td>6</td></tr>
<tr><td rowspan="2">2</td><td colspan="2">1</td><td colspan="2">3</td><td colspan="2">4</td></tr>
<tr><td>1</td><td>2</td><td>5</td><td>6</td><td>7</td><td>8</td></tr>
</table>

Answer: $u_{x2} = 0.00132$ mm $\quad u_{y2} = -0.00435$ mm
$u_{x3} = 0.02169$ mm $\quad u_{y3} = 0.00997$ mm

6.6 Investigate the completeness and compatibility of the trial functions for the elements shown in Fig. P6.6.

1. $u = \hat{u}_2\xi + \hat{u}_3\eta$
2. $u = \hat{u}_2\xi + \hat{u}_3\eta + \hat{u}_4\xi\eta$

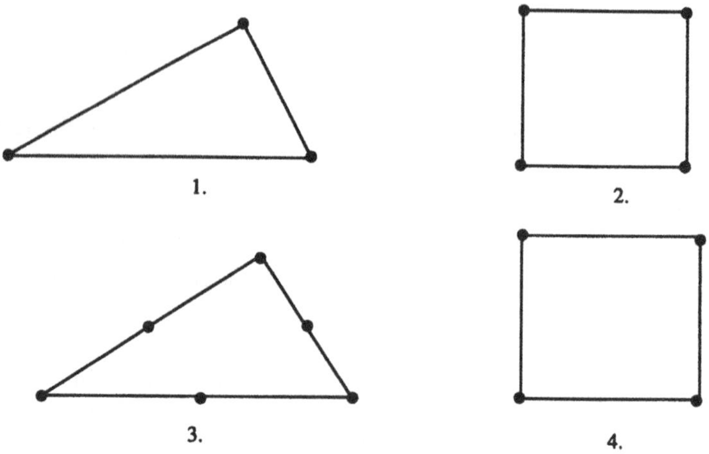

FIGURE P6.6

3. $u = \hat{u}_2\xi + \hat{u}_3\eta + \hat{u}_4\xi^2 + \hat{u}_5\eta^2$
4. $u = \hat{u}_1 + \hat{u}_2 \sin(\frac{\pi\xi}{2}) + \hat{u}_3 \sin(\frac{\pi\eta}{2}) + \hat{u}_4 \sin(\frac{\pi\xi}{2}) \sin(\frac{\pi\eta}{2})$

6.7 Find the interpolation functions $\mathbf{N} = [N_1 \quad N_2 \quad N_3]^T$ for a three node extension bar element, using the trial function $u = \hat{u}_1 + \hat{u}_2\xi + \hat{u}_3\xi^2$.

6.8 Use the formulation of Section 6.5.4 to find the expression for the stiffness matrix of Eq. (6.56) for the element of Fig. 6.10. Also, show that an incompatibility occurs between elements 1 and 2 of Fig. 6.14b.

Hint: Use the shape functions $\mathbf{N}_1$ and $\mathbf{N}_2$ of Eq. (6.55). Then

$$\mathbf{u} = \begin{bmatrix} u_x \\ u_y \end{bmatrix} = \begin{bmatrix} \mathbf{N}_1 & \mathbf{N}_2 \\ \mathbf{N}_1 & \mathbf{N}_2 \end{bmatrix} \begin{bmatrix} \mathbf{v}_x \\ \hat{\mathbf{u}}_2 \end{bmatrix} = \hat{\mathbf{N}}\hat{\mathbf{v}}$$

where $\hat{\mathbf{N}}$ is a 2×6 matrix. From Eq. (6.25), $\mathbf{B} = \mathbf{DN}$, where $\mathbf{B}$ is 3×6. Use Eq. (6.31) to find $\mathbf{k}^i$. To check compatibility, calculate the displacement of the mid-span point of the common edge of element 1 and element 2. Compare the results.

6.9 Find expressions for the coefficients $\alpha_2, \beta_2, \gamma_2, \alpha_3, \beta_3$, and γ_3 of Eq. (6.75). Also find the same quantities for Eq. (6.82).

6.10 Derive an interpolation $\mathbf{N}$ of $\mathbf{u} = \mathbf{Nv}$ for the 8 node cubic element of Fig. P6.10.

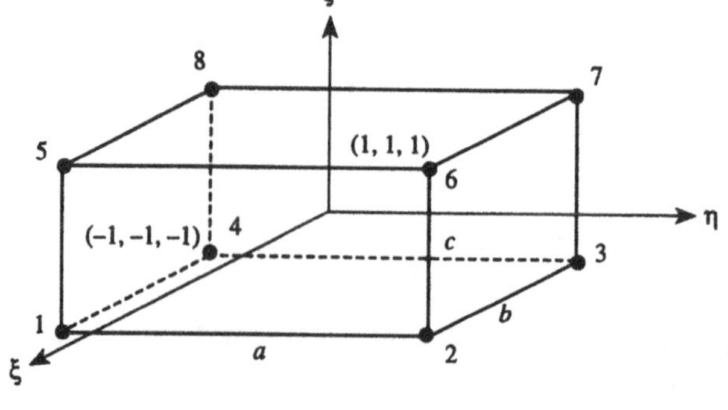

FIGURE P6-10

Hint: Assume u_x, u_y, and u_z to be of the form

$$\widehat{u}_1 + \widehat{u}_2\xi + \widehat{u}_3\eta + \widehat{u}_4\xi\eta + \widehat{u}_5\zeta + \widehat{u}_6\xi\zeta + \widehat{u}_7\eta\zeta + \widehat{u}_8\xi\eta\zeta$$

Answer: $N_i = \frac{1}{8}(1+\xi\xi_i) + (1+\eta\eta_i)(1+\zeta\zeta_i) \quad i = 1, 2, \ldots, 8$

6.11 Find a set of interpolation functions **N** of $\mathbf{u} = \mathbf{N}\mathbf{v}$ for the eight-node, 16-DOF plane stress element of Fig. P6.11. Also, apply Lagrangian interpolation to the same problem.

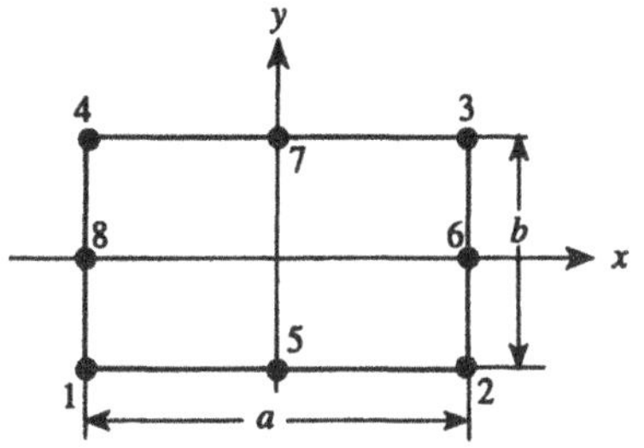

FIGURE P6.11

Answer: For a coordinate system at the center of the element and $\xi = 2x/a$, $\eta = 2y/b$, with u_x or u_y assumed to be $u_x(\xi, \eta) = \widehat{u}_1 + \widehat{u}_2\xi + \widehat{u}_3\eta + \widehat{u}_4\xi^2 + \widehat{u}_5\xi\eta + \widehat{u}_6\eta^2 + \widehat{u}_7\xi^2\eta + \widehat{u}_8\xi\eta^2$, the interpolation function u would be

$$\begin{aligned}
u(\xi, \eta) = &-\frac{1}{4}(1-\xi)(1-\eta)(1+\xi+\eta)u_{x1} \\
&-\frac{1}{4}(1+\xi)(1-\eta)(1-\xi+\eta)u_{x2} - \frac{1}{4}(1+\xi)(1+\eta)(1-\xi-\eta)u_{x3} \\
&-\frac{1}{4}(1-\xi)(1+\eta)(1+\xi-\eta)u_{x4} + \frac{1}{2}(1-\xi^2)(1-\eta)u_{x5} \\
&+\frac{1}{2}(1-\eta^2)(1+\xi)u_{x6} + \frac{1}{2}(1-\xi^2)(1+\eta)u_{x7} \\
&+\frac{1}{2}(1-\eta^2)(1-\xi)u_{x8}
\end{aligned}$$

6.12 Find an interpolation function for the one-dimensional element shown in Fig. P6.12. Use a natural coordinate system.

FIGURE P6.12

Answer: $N_1 = \frac{1}{3}L_1(4L_1-1)(2L_1-1)(4L_1-3)$, $N_2 = \frac{16}{3}L_1(4L_1-1)(2L_1-1)L_2$, $N_3 = 4L_1L_2(4L_1-1)(4L_2-1)$, $N_4 = \frac{16}{3}L_1L_2(4L_2-1)(2L_2-1)$, $N_5 = \frac{1}{3}L_2(4L_2-1)(2L_2-1)(4L_2-3)$

6.13 Find the interpolation function corresponding to nodes 1 and 2 of the triangular element shown in Fig. P6.13. Use a two-dimensional natural coordinate system.

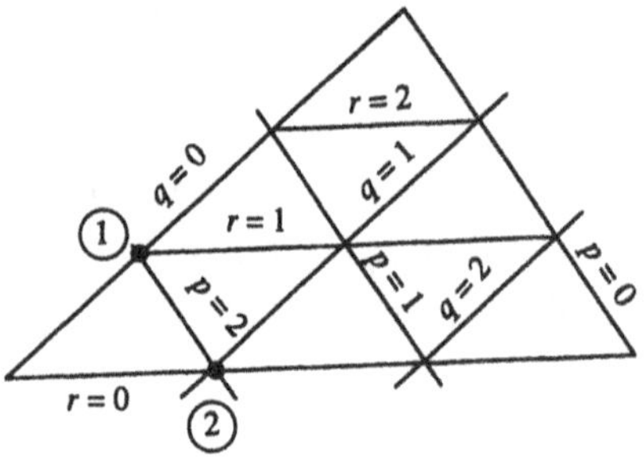

FIGURE P6.13

Hint: $N_2(L_1) = \frac{3}{2}L_1(3L_1 - 1)$, $N_1(L_3) = 3L_3$, $N_1(L_2) = 3L_2$, $N_0(L_2) = N_0(L_3) = 1$

Answer: $N_1 = \frac{9}{2}L_1L_3(3L_1 - 1)$, $N_2 = \frac{9}{2}L_2L_1(3L_1 - 1)$

6.14 Construct the interpolation function for the element shown in Fig. P6.14. The interpolation should be such that $w(\xi) = N_1w_1 + N_2w_2 + N_3\theta_2 + N_4w_3$, where $\theta_2 = -w'(\xi)|_{\xi=\xi_2}$.

Answer: $N_1 = \frac{1}{2}(\xi^2 - \xi^3)$, $N_2 = 1 + \frac{1}{2}(\xi^2 + \xi^3)$, $N_3 = \xi - \xi^3$, $N_4 = \frac{1}{2}(\xi^2 + \xi^3)$

FIGURE P6.14

6.15 Suppose we have a structure as shown in Fig. P6.15a where a beam is connected to a plane stress thin plate undergoing in-plane deformation. Beam and plane stress plate elements of thickness t are to be used to analyze this structure. At the intersection between the beam and the plate, a transition or "blending" element of the form of Fig. P6.15b can be employed. Obtain the **B** matrix for this element, where $\mathbf{k}^i = aht\int_0^1\int_0^1 \mathbf{B}^T\mathbf{E}\mathbf{B}\,d\xi\,d\eta$.

Hint: The displacements in the element can be

$$u_x = N_1u_{x1} - \frac{h}{2}\eta N_1\theta_1 + N_2u_{x2} + N_3u_{x3}, \qquad u_y = N_1u_{y1} + N_2u_{y2} + N_3u_{y3}$$

where $N_1 = \frac{1}{2}(1-\xi)$, $N_2 = \frac{1}{4}(1+\xi)(1-\eta)$, $N_3 = \frac{1}{4}(1+\xi)(1+\eta)$ and $\mathbf{D}_u$ is given in Example 6.1.

6.16 Construct the stiffness matrix for a two-node beam element with shear deformation effects taken into account. Use $\mathbf{k}^i = \ell\int_0^1 \mathbf{B}^T\mathbf{E}\mathbf{B}\,d\xi$.

Hint: Use linear interpolation shape functions for both the deflection and the rotation.

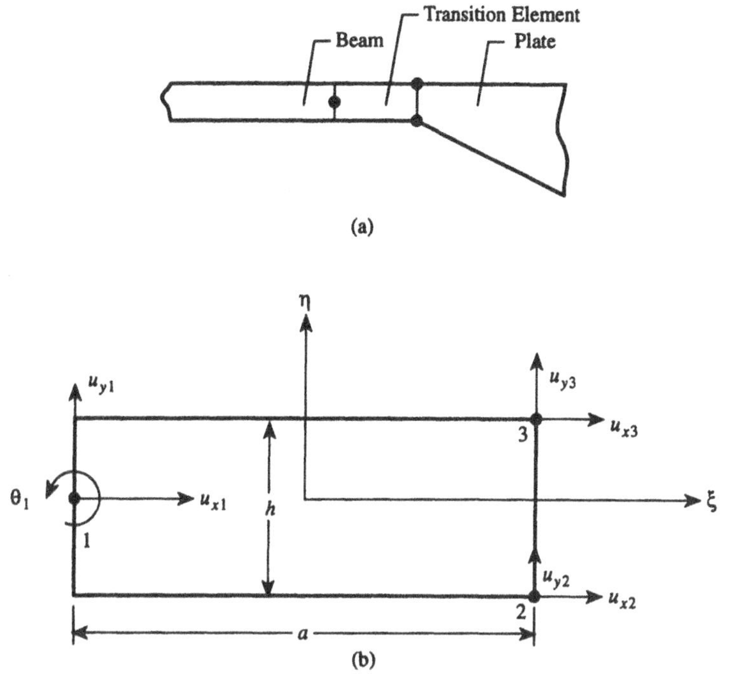

FIGURE P6.15

6.17 Show that a stiffness matrix for a homogeneous, isotropic plane strain solid can be changed to one for plane stress by replacing E by $E(1-\nu^2)$ and then ν by $\nu/(1+\nu)$.

6.18 For the tetrahedron of Section 6.5.8, show that

$$\int_V \mathbf{N}^T{}_y\partial E_{ij}\partial_x\mathbf{N}\,dV = \frac{E_{ij}}{36V}\begin{bmatrix} \gamma_1\beta_1 & \gamma_1\beta_2 & \gamma_1\beta_3 & \gamma_1\beta_4 \\ \gamma_2\beta_1 & \gamma_2\beta_2 & \gamma_2\beta_3 & \gamma_2\beta_4 \\ \gamma_3\beta_1 & \gamma_3\beta_2 & \gamma_3\beta_3 & \gamma_3\beta_4 \\ \gamma_4\beta_1 & \gamma_4\beta_2 & \gamma_4\beta_3 & \gamma_4\beta_4 \end{bmatrix}$$

6.19 Derive the interpolation functions of Eq. (6.147) beginning with the trial function

$$u_x = \widehat{u}_1 + \widehat{u}_2\xi + \widehat{u}_3\eta + \widehat{u}_4\xi\eta + \widehat{u}_5\xi^2 + \widehat{u}_6\eta^2 + \widehat{u}_7\eta\xi^2 + \widehat{u}_8\eta^2\xi$$

Let u_y be of the same form with $\widehat{u}_i$ replaced by $\widehat{v}_i$.

6.20 Derive the interpolation functions for the element shown in Fig. P6.20. This type of element has been useful in the analysis of stress concentration and crack problems.

Hint: Begin with the same trial functions used in Problem 6.19.

6.21 Find an interpolation function for a one-dimensional element with nodes at x_i and x_j, and 1 DOF per node. Begin with $u = \widehat{u}_1 + \widehat{u}_2 x$.

Answer: $\mathbf{N} = \left[\frac{x_j - x}{x_j - x_i} \quad \frac{x - x_i}{x_j - x_i}\right]$

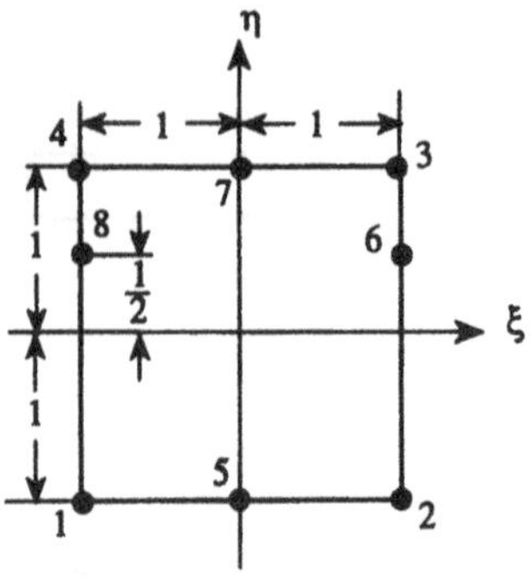

FIGURE P6.20

6.22 Show that $N_i(L_k)$ of Eq. (6.71) is obtained from the Lagrangian polynomials of Eq. (6.64) by a transformation from Cartesian to natural coordinates.

6.23 Use one-, two-, and three-point Gaussian quadrature to find values of the following integrals. Compare your results with the exact solutions.

$$1.\ \int_0^{\frac{\pi}{2}} \cos x\,dx, \quad 2.\ \int_{-1}^{1} (x^2 + \frac{1}{2}x^4)dx, \quad 3.\ \int_2^8 \frac{1}{x}dx$$

Hint: Change of variables: 1. $x = \frac{\pi}{4}(\xi + 1)$, 3. $x = 3\xi + 5$

Answer:

1. $n = 1 : 1.1107$ $n = 2 : 0.9985$, $n = 3 : 1.000008$
2. $n = 1 : 0$ $n = 2 : 0.778$, $n = 3 : 0.86667$
3. $n = 1 : 1.2$ $n = 2 : 1.3636$, $n = 3 : 1.3837$

6.24 Use the Legendre polynomial formulas to determine the Gauss integration points and weighting coefficients for Gaussian quadrature in the domain $[-1, 1]$ if $n = 2$.

Hint: From Eq. (6.123), $P_2(\xi) = \frac{3}{2}\xi^2 - \frac{1}{2} = 0$ gives $\xi_i = \pm 1/\sqrt{3}$. From Eq. (6.124) $W_i^{(2)} = 2(1 - 1/3)/(2 \cdot 1/\sqrt{3})^2 = 1$

Answer: See Example 6.11.

6.25 Use Newton-Cotes and Gaussian quadrature to integrate $\int_{-1}^{1} \frac{1}{1+\xi^2} d\xi$. The exact answer is $\pi/2 \approx 1.5708$. Try two and three integration points. Calculate the errors for each.

Answer: For $n = 2$, Newton-Cotes and Gaussian give 1 and 1.5 with errors of 36% and 4.5%, respectively. For $n = 3$, the integrals are 1.66 and 1.58 with errors of 6% and 0.8%.

6.26 Use two- and three-point Newton-Cotes and Gaussian quadrature to integrate $\int_A (x^2 + y^2)dA$ for the area shown in Fig. P6.26.

Hint: Make the coordinate transformation

$$x = \frac{1}{4}[(1 - \xi)(1 - \eta)x_1 + (1 + \xi)(1 - \eta)x_2 + 2(1 + \eta)x_3]$$
$$y = \frac{1}{4}[(1 - \xi)(1 - \eta)y_1 + (1 + \xi)(1 - \eta)y_2 + 2(1 + \eta)y_3]$$

and use $dA = dx\,dy = |J|\,d\xi\,d\eta$.

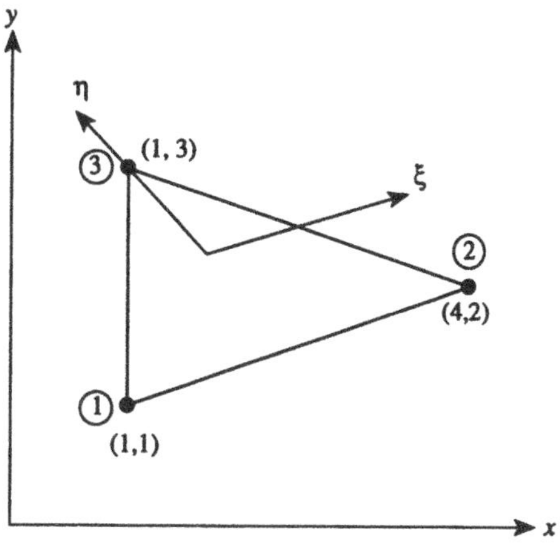

FIGURE P6.26

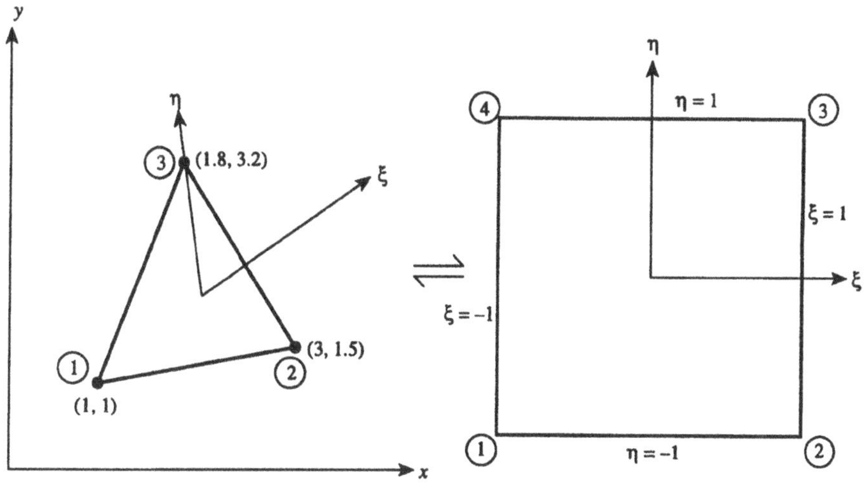

FIGURE P6.27

6.27 Derive the stiffness matrix of the triangular element shown in Fig. P6.27 by mapping it to a four-node square element.

Hint: The coordinate transformation is

$$x = \frac{1}{4}[(1-\xi)(1-\eta)x_1 + (1+\xi)(1-\eta)x_2 + 2(1+\eta)x_3]$$

$$y = \frac{1}{4}[(1-\xi)(1-\eta)y_1 + (1+\xi)(1-\eta)y_2 + 2(1+\eta)y_3]$$

For this coordinate transformation line 3–4 of the square element corresponds to point 3 of the triangle. Use $N_1 = \frac{1}{4}(1-\xi)(1-\eta)$, $N_2 = \frac{1}{4}(1+\xi)(1-\eta)$, $N_3 = \frac{1}{2}(1+\eta)$.

6.28 Construct the shape function $N_i(\xi) = \sin[\pi(a_i + b_i\xi + c_i\xi^2)]$ for the truss element shown in Fig. P6.28a. Use this shape function and two elements to calculate the displacement at $x = L/2$ of the bar shown in Fig. P6.28b. Compare the results to the exact solution which is obtained from $du/dx = \overline{p}_x/(EA)$.

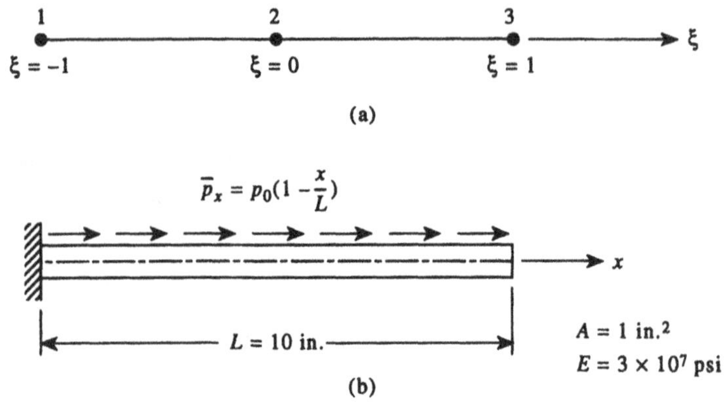

FIGURE P6.28

6.29 Form the loading vector for the in-plane deformation of the plane stress plate element shown in Fig. P6.29 with a concentrated load $\overline{P}_y = 100$ applied at the centroid of the element.

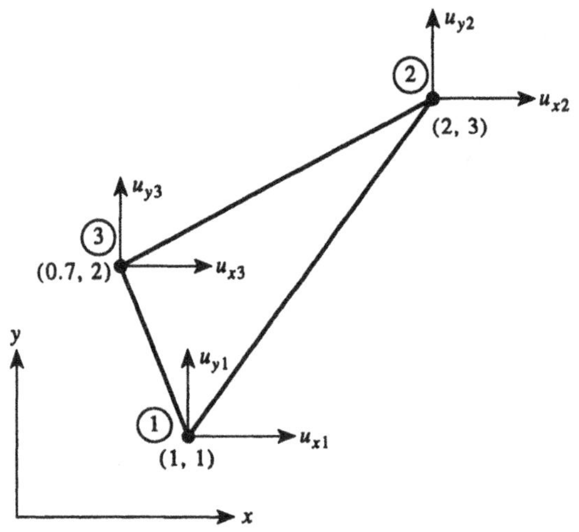

FIGURE P6.29

6.30 Construct a generalized *CB* functional for a bar subject to extension and compression. Introduce trial functions that satisfy fixed–fixed boundary conditions. Find the axial displacement and force due to a uniformly distributed axial force.

7

Direct Variational and Weighted Residual Methods: Classical Trial Function Methods

In most of this chapter, we will consider trial function methods for solving the governing equations in their differential (local) form. This contrasts with the previous chapter where a trial function technique—the finite element method—was employed to solve the integral (global) form of the governing equations. It will be shown that some of the trial function methods of this chapter can also be applied to problems formulated in global form.

The key to the successful solution of the governing equations lies in the ability to develop a reliable and robust approximate method. The trial function methods represent one class of techniques which appear to possess these characteristics for solving structural mechanics problems. This approach is of particular interest because, as seen in the previous chapter, it can be extended to large-scale systems for which the trial function is applied to each of the elements composing the system.

In the trial function methods, the unknown solution is approximated by a set of basis functions containing constants or functions. These constants or functions are chosen by a variety of criteria to provide the best approximation of the trial function family to the correct solution. Finlayson (1972) contains an in-depth study of trial function methods.

7.1 Governing Differential Equations

The fundamental differential equations for linear elastic solids are provided in Chapter 1. Differential equations are frequently expressed in operator form, e.g., $\mathbf{D}^T \mathbf{E}\, \mathbf{D}\mathbf{u} = \overline{\mathbf{p}}$. Differential operators, such as $\mathbf{D}^T \mathbf{E}\, \mathbf{D}$, can be classified as being elliptic, parabolic, and hyperbolic in form [Norrie and de Vries, 1973]. Generally, the static or stability problems of the mechanics of solids are so-called boundary value problems, which are elliptic in type, while *initial value problems* correspond to parabolic and hyperbolic operators.

We wish to employ general expressions for the governing differential equations describing the behavior of solids. Suppose u is the dependent variable or $\mathbf{u}$ the vector of dependent variables that are to be computed by solving the governing differential equations. These governing equations can be of the form

$$\mathbf{L}\mathbf{u} = \overline{\mathbf{f}} \quad \text{in} \quad V \tag{7.1}$$

subject to boundary conditions

$$\mathbf{B}\mathbf{u} = \overline{\mathbf{g}} \quad \text{on} \quad S \tag{7.2}$$

where **L** and **B** are matrix differential operators, V is the volume of the body with boundary S, and $\overline{\mathbf{f}}$ and $\overline{\mathbf{g}}$ are prescribed vectors. In general, quantities with a superscript bar are prescribed. Typical operators are the Laplacian

$$L = \nabla^2 = \frac{\partial^2}{\partial x_1^2} + \frac{\partial^2}{\partial x_2^2} + \frac{\partial^2}{\partial x_3^2} \tag{7.3}$$

or the biharmonic operator

$$L = \nabla^4 = \nabla^2\nabla^2 = \frac{\partial^4}{\partial x_1^4} + 2\frac{\partial^4}{\partial x_1^2 \partial x_2^2} + \frac{\partial^4}{\partial x_2^4} \tag{7.4}$$

In structural mechanics, frequently occurring governing equations include [Chapter 1, Eq. (1.83)]

$$\begin{aligned} \mathbf{D}^T \mathbf{E}\, \mathbf{D}\mathbf{u} &= -\overline{\mathbf{p}}_V \quad \text{in} \quad V \\ \mathbf{u} &= \overline{\mathbf{u}} \quad \text{on} \quad S_u \\ \mathbf{p} &= \overline{\mathbf{p}} \quad \text{on} \quad S_p \end{aligned} \tag{7.5a}$$

for a linear solid, and [Chapter 1, Eq. (1.127)]

$$\begin{aligned} (EIw'')'' &= \overline{p}_z \quad \text{on} \quad 0 \le x \le L \\ w &= \overline{w}, \quad \theta = \overline{\theta} \quad \text{on} \quad S_u \\ V &= \overline{V}, \quad M = \overline{M} \quad \text{on} \quad S_p \end{aligned} \tag{7.5b}$$

for a beam, where $'' = d^2/dx^2$.

The operators **L** and **B** are *linear* if **u** and its derivatives occur linearly. If a linear operator has **u** and **v** in its domain V,

$$\mathbf{L}(c_1\mathbf{u} + c_2\mathbf{v}) = \mathbf{L}(c_1\mathbf{u}) + \mathbf{L}(c_2\mathbf{v}) = c_1\mathbf{L}\mathbf{u} + c_2\mathbf{L}\mathbf{v} \tag{7.6}$$

where c_1 and c_2 are constant scalars. If the highest derivatives of **u** appear linearly in the operator, while one or more other terms are nonlinear, the operator is said to be *quasi-linear.*

An operator is *symmetric* or *self-adjoint* if for any **v** and **u** which satisfy the same homogeneous boundary conditions

$$\int_V \mathbf{v}^T \mathbf{L}\mathbf{u}\, dV = \int_V \mathbf{u}^T \mathbf{L}\mathbf{v}\, dV \tag{7.7}$$

and *positive* if

$$\int_V \mathbf{u}^T \mathbf{L}\mathbf{u}\, dV \ge 0 \tag{7.8}$$

The operator is *positive definite* if it is positive and $\int \mathbf{u}^T \mathbf{L}\mathbf{u}\, dV \equiv 0$ only when $\mathbf{u} \equiv 0$. See Problem 7.1 for an example of a self-adjoint, positive definite operator. A symmetric matrix is obtained from a self-adjoint operator, and a positive definite matrix is obtained from a positive definite operator. The key governing equations of solid mechanics have symmetric and positive definite operators for many physically meaningful boundary conditions.

7.2 Trial Functions

Suppose the governing differential equations of motion subject to prescribed boundary conditions are to be solved for the single dependent variable u or the vector $\mathbf{u}$. For example, $\mathbf{u} = [u_1\ u_2\ u_3]^T = [u\ v\ w]^T$ for three-dimensional solids. For the methods of concern here, $\mathbf{u}$ is approximated by trial solutions $\tilde{\mathbf{u}}$ of the familiar form

$$\tilde{u}(x) = \mathbf{N}_u(x)\hat{\mathbf{u}} \quad \text{or} \quad \tilde{u}(x) = \sum_{i=1}^{m} \hat{u}_i N_{ui}(x) \tag{7.9a}$$

for a scalar u and

$$\tilde{\mathbf{u}}(x) = \mathbf{N}_u(x)\hat{\mathbf{u}} \tag{7.9b}$$

for a vector $\mathbf{u}$, where N_{ui} are linearly independent chosen functions called *trial, basis,* or *shape functions*. As indicated in previous chapters, $\mathbf{N}_u(x)$ is a matrix (or row vector, if appropriate) formed of N_{ui}, and $\hat{\mathbf{u}} = [\hat{\mathbf{u}}_0\ \hat{\mathbf{u}}_1\ \hat{\mathbf{u}}_2\ \hat{\mathbf{u}}_3 \ldots \hat{\mathbf{u}}_m]^T$ are unknown or free parameters (scalar values, and in some cases functions) that are to be determined in some good "fit" sense. If $\mathbf{u}$ and $\tilde{\mathbf{u}}$ are $p \times 1$ vectors and $\hat{\mathbf{u}}$ is a $pm \times 1$ vector, then $\mathbf{N}_u(x)$ is a $p \times pm$ matrix. For each r, $1 \leq r \leq p$, the rth row of $\mathbf{N}_u(x)$ has m contiguous non-zero elements, the first of which is the $(mr - m + 1)$th entry in the row. The trial functions should be chosen such that the approximation improves as the number of terms in the solution increases. As discussed in Section 7.3.9, convergence can be defined in terms of $\tilde{u} \to u$ as $m \to \infty$. If any desired accuracy can be obtained by simply increasing the number of terms in the linear sum of Eq. (7.9a), the set of functions N_{ui} is said to be *complete*. For convergence, it is usually essential that the N_{ui} be chosen to be members of a complete set of functions. If the problem formulation involves an mth order derivative, the trial function must be $(m - 1)$ times continuously differentiable in the domain of concern.

One of the most familiar trial function approximations is the Fourier[1] series, where the trial functions are composed of sine and cosine terms. Under appropriate conditions, many functions can be conveniently and accurately represented by Fourier series in which the trial functions form a complete set.

The trial function methods are often classified into *interior* and *boundary* procedures. In the case of the interior method, the N_{ui} are chosen to satisfy the boundary conditions [Eq. (7.2)], so that $\tilde{\mathbf{u}}$ satisfies the boundary conditions for all $\hat{u}_i$. This is usually the easiest procedure for most problems. Here, the $\tilde{\mathbf{u}}$ is to be determined such that the differential equations are satisfied in some approximate sense. In the boundary method, $\tilde{\mathbf{u}}$ is chosen to satisfy the governing differential equations [Eq. (7.1)], but not the boundary conditions. The problem is reduced to that of selecting the parameters $\hat{u}_i$ such that the boundary conditions are approximated. Another possibility is the combination of these methods for which the trial solution $\tilde{\mathbf{u}}$ satisfies neither the boundary conditions nor the differential equations.

The methods of this chapter, which use trial functions of the form of Eq. (7.9), can be categorized as belonging to either residual (or weighted residual) or variational methods.

[1]Jean Baptiste Joseph Fourier (1768–1830) was a French physicist who, as the son of a tailor, was educated by the Benedictines. His lack of good birth precluded his receiving a commission in the scientific corps of the French army. He did, however, obtain a military lectureship in mathematics. He backed the revolution and accompanied Napoleon on his 1798 Eastern expedition. He was given the post of governor of Lower Egypt, where he wrote several papers on mathematics. After the 1801 French capitulation to Britain, Fourier returned to France and began his experiments on the propagation of heat. His 1822 publication *Théorie analytique de la chaleur,* which dealt with the flow of heat, contained the familiar contention that any continuous or discontinuous function of a variable can be expanded in a trigonometric series of multiples of the variable.

7.3 Residual Methods

In the residual method, the constants (functions) $\widehat{u}_i$ are chosen such that an error term or a *residual* is zero at selected points, zero in an average sense, or minimized in some fashion. For Eqs. (7.1) and (7.2), the residuals for the interior and boundary, respectively, can be expressed as

$$\begin{aligned}\mathbf{R}_V &= \mathbf{L}\widetilde{\mathbf{u}} - \overline{\mathbf{f}} \\ \mathbf{R}_S &= \mathbf{B}\widetilde{\mathbf{u}} - \overline{\mathbf{g}}\end{aligned} \quad \text{or} \quad \mathbf{R} = \mathbf{L}\widetilde{\mathbf{u}} - \overline{\mathbf{f}} \tag{7.10}$$

where the $p \times 1$ vector $\mathbf{u}$ has been replaced by the approximate $\widetilde{\mathbf{u}}$, and $\mathbf{R}_V$ and $\mathbf{R}_S$ are $p \times 1$ vectors. As shown, often the subscripts are dropped. For solids and beams, Eq. (7.10) would be

$$\mathbf{R}_V = \mathbf{D}^T \mathbf{E}\, \mathbf{D}\widetilde{\mathbf{u}} + \overline{\mathbf{p}}_V \qquad R_V = (EI\widetilde{w}'')'' - \overline{p}_z \tag{7.11}$$

$$\mathbf{R}_S = \widetilde{\mathbf{u}} - \overline{\mathbf{u}} \quad \text{as} \quad \widetilde{\mathbf{p}} - \overline{\mathbf{p}} \qquad R_S = \begin{Bmatrix} \widetilde{w} - \overline{w} & \text{as} & \widetilde{V} - \overline{V} \\ \widetilde{\theta} - \overline{\theta} & & \widetilde{M} - \overline{M} \end{Bmatrix} \tag{7.12}$$

As indicated in Section 7.2, the trial solutions are usually chosen to satisfy either the boundary conditions or the governing equations. In either case, one of the residuals would be zero. For example, with the interior method, if $\mathbf{N}_u$ is chosen to satisfy all of the boundary conditions, then $\mathbf{R}_S = 0$.

There are many techniques for selecting $\widehat{u}_i$ to minimize or make the residuals zero. Several procedures are described in the following sections. Frequently, the selection of the $\widehat{u}_i$ is based on a scalar R expressed in the form

Interior Method

$$\int_V W_j\, h_1(R_V)\, dV = 0 \quad j = 1, 2, \ldots, m \tag{7.13a}$$

Boundary Method

$$\int_S W_j\, h_2(R_S)\, dV = 0 \quad j = 1, 2, \ldots, m \tag{7.13b}$$

where h_1, h_2 are prescribed functions of R_V, R_S; the W_j are weights or independent *weighting functions*, and m is equal to the number of unknown coefficients $\widehat{u}_j$ in Eq. (7.9a).

In most of the subsequent discussion, it suffices to adopt the interior residual expressions

$$\int_V W_j\, h(R)\, dV = 0 \quad j = 1, 2, \ldots, m \tag{7.14a}$$

where the subscripts associated with h and R have been dropped.

Equation (7.14) is a set of m simultaneous equations to be solved for the coefficients $\widehat{u}_i$, $i = 1, 2, \ldots, m$ of the trial solution $\widetilde{u}(x)$.

For a general p-dimensional problem, the residual $\mathbf{R}$ is a $p \times 1$ vector. If the mapping h is chosen such that $h(\mathbf{R})$ is also a $p \times 1$ vector, then each element of $h(\mathbf{R})$ satisfies m conditions similar to Eq. (7.14a). The weighted residual method in this case may be expressed as

$$\int_V \mathbf{W}^{(k)} h(\mathbf{R})\, dV = 0 \quad k = 1, 2, \ldots, m \tag{7.14b}$$

where $\mathbf{W}^{(k)}$ is a diagonal matrix with p non-zero entries $W_1^{(k)}, W_2^{(k)}, \ldots W_p^{(k)}$. These are mp equations to be solved for mp coefficients $\widehat{\mathbf{u}}_i$, $1 \le i \le mp$.

The weighted-residual method and its variations are sometimes called *error distribution, projection,* or *assumed mode* methods.

7.3.1 Collocation

One method for selecting $\widetilde{u}_i$, termed *collocation,* is to select as many points (in domain V) as there are unknown parameters $\widehat{u}_i$ and then determine the parameters such that the residual is zero at these points. Thus, for this point fitting technique, a scalar residual R is set equal to zero at m points (in V). This leads to m simultaneous algebraic equations with $\widehat{u}_i$ as unknowns. The collocation points are selected to spread reasonably evenly over V.

In terms of Eq. (7.14), the collocation approach is equivalent to setting

$$h(R) = R, \qquad W_j = \delta(x - x_j) \quad j = 1, 2, \ldots, m$$

where δ is the Dirac[2] delta function which by definition satisfies

$$\int_{-\infty}^{\infty} \delta(x - x_j) R(x)\, dx = R(x_j) \tag{7.15}$$

The locations x_j, $j = 1, 2, \ldots, m$ are the chosen collocation points in V.

For a linear elastic solid in, say, $p = 3$ dimensional space, [Eq. (7.11)], $\mathbf{R} = \mathbf{D}^T \mathbf{E}\, \mathbf{D} \widetilde{\mathbf{u}} + \overline{\mathbf{p}}_V$. Use the trial solutions $\widetilde{\mathbf{u}} = \mathbf{N}_u \widehat{\mathbf{u}}$, and suppose the collocation points in V are x_j, $j = 1, 2, \ldots, m$. Here, $\widetilde{\mathbf{u}}$ is $p \times 1$ (3×1), $\mathbf{N}_u$ is $p \times pm$ $(3 \times 3m)$ and $\widehat{\mathbf{u}}$ is $pm \times 1$ $(3m \times 1)$. Then the collocation method leads to a system of equations

$$\mathbf{D}^T \mathbf{E}\, \mathbf{D}\, \mathbf{N}_u(x_j) \widehat{\mathbf{u}} + \overline{\mathbf{p}}_V(x_j) = \mathbf{R}(x_j) = 0 \tag{7.16}$$

or

$$\mathbf{k}_u \qquad \widehat{\mathbf{u}} - \mathbf{p}_u \qquad = 0$$

The $3m$ free parameters $\widehat{\mathbf{u}}$ are found by solving

$$\mathbf{k}_u \widehat{\mathbf{u}} = \mathbf{p}_u \tag{7.17}$$

As expressed here, $\mathbf{k}_u$ appears to correspond to a stiffness matrix. This is only a formal comparison, however, since the elements of $\widehat{\mathbf{u}}$ are not ordinarily equal to nodal displacements, as they would be for a stiffness matrix.

EXAMPLE 7.1 Beam with Linearly Varying Load

Consider the fixed-hinged beam of Fig. 7.1 which is subjected to a linearly varying load. This beam has been treated in several other chapters.

For a beam with constant EI, the governing differential equation is

$$EI \frac{d^2}{dx^2} \frac{d^2 w}{dx^2} = \overline{p}_z$$

Use $\widetilde{w} = \mathbf{N}_u(x) \widehat{\mathbf{w}}$. The collocation method leads to the formulation

$$R(x_j) = EI \frac{d^4}{dx^4} \widetilde{w}(x_j) - \overline{p}_z(x_j) = EI \frac{d^4}{dx^4} \mathbf{N}_u(x_j) \widehat{\mathbf{w}} - \overline{p}_z(x_j) = 0 \quad j = 1, 2, \ldots, m \tag{1}$$

[2]Paul Adrien Maurice Dirac (1902–1984) was born in Bristol, England and educated at Bristol and Cambridge Universities. He served at both British and American universities, including post-retirement at Florida State University. Somewhat after Born and Jordan, he formulated a general theory for quantum mechanics. In 1927 he studied at Göttingen University, where he spent considerable time with J.R. Oppenheimer, also a student. Furthermore, he had the opportunity to interact with Born there. In 1933 he, along with Schrödinger, was awarded the Nobel Prize in Physics.

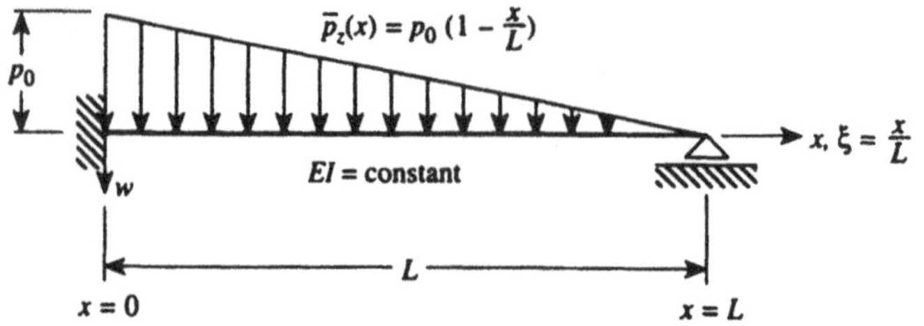

FIGURE 7.1
Beam with ramp loading.

Use a single parameter polynomial approximation to the deflection w, say

$$\widetilde{w} = \widehat{w}_1(a_1 + a_2\xi + a_3\xi^2 + a_4\xi^3) \quad \xi = x/L \tag{2}$$

where $\widehat{w}_1$ is the free parameter and $a_i, i = 1, 2, 3, 4$ are parameters that will be identified such that (2) satisfies the boundary conditions. Certainly (2) will never satisfy the left end condition $w_{\xi=0} = 0$ unless a_1 is set equal to zero. Then

$$\widetilde{w} = \widehat{w}_1(a_2\xi + a_3\xi^2 + a_4\xi^3) \tag{3}$$

Similarly, set $a_2 = 0$ in order to satisfy the boundary condition $w'_{\xi=0} = 0$. We are now left with

$$\widetilde{w} = \widehat{w}_1(a_3\xi^2 + a_4\xi^3) \tag{4}$$

which, as just shown, fulfills the boundary conditions at $\xi = 0$, and a_3, a_4 can be chosen such that $\widetilde{w}$ satisfies the boundary conditions at $\xi = 1$. However, (1) involves $\mathbf{N}_u^{iv}$, and the fourth derivative of (4) is zero. Thus, add a fourth-power term so that

$$\widetilde{w} = \widehat{w}_1(a_3\xi^2 + a_4\xi^3 + a_5\xi^4) \tag{5}$$

The left end conditions are still satisfied. From the right end conditions,

$$\begin{aligned} w_{\xi=1} &= \widehat{w}_1(a_3 + a_4 + a_5) = 0 \\ w''_{\xi=1} &= \widehat{w}_1(2a_3 + 6a_4 + 12a_5)/L^2 = 0 \end{aligned} \tag{6}$$

These demand that

$$\begin{aligned} a_3 + a_4 + a_5 &= 0 \\ 2a_3 + 6a_4 + 12a_5 &= 0 \end{aligned} \tag{7}$$

Any number of combinations of a_3, a_4, and a_5 will satisfy (7). For example, $a_3 = 3$, $a_4 = -5$, and $a_5 = 2$. Then

$$\widetilde{w} = \widehat{w}_1(3\xi^2 - 5\xi^3 + 2\xi^4) \tag{8}$$

This satisfies all of the boundary conditions. We could have chosen a trial function that does not satisfy all of the boundary conditions; however, it is better to select a trial function for the collocation method that satisfies as many of the boundary conditions as possible.

The applied loading can be expressed as

$$\overline{p}_z = p_0(1 - \xi), \qquad \xi = x/L \tag{9}$$

Choose a single collocation point at $\xi = 1/2$, i.e., $x = L/2$. Substitute (8) and (9) into (1), and note that $\widetilde{w}^{iv}(x) = 48\widehat{w}_1/L^4$. Then, from (1)

$$EI\frac{48}{L^4}\widehat{w}_1 - \frac{p_0}{2} = 0 \tag{10}$$

and

$$\widehat{w}_1 = \frac{p_0 L^4}{96\,EI} \tag{11}$$

From (8), the approximate deflection is

$$\widetilde{w} = \frac{p_0 L^4}{96\,EI}(3\xi^2 - 5\xi^3 + 2\xi^4) \tag{12}$$

which can be compared to the exact solution, which was derived in the previous chapters.

$$w = \frac{p_0 L^4}{120\,EI}(4\xi^2 - 8\xi^3 + 5\xi^4 - \xi^5) \tag{13}$$

A comparison of the deflection of (12) and (13) gives

	Deflection w		
$\xi = x/L$	**Exact ($\frac{p_0L^4}{EI}$)**	**Approximate ($\frac{p_0L^4}{EI}$)**	**Error (%)**
1/4	1.196×10^{-3}	1.22×10^{-3}	2.0
1/2	2.34375×10^{-3}	2.604×10^{-3}	11.1
3/4	1.83105×10^{-3}	2.19×10^{-3}	19.6

∎

Orthogonal Collocation

The collocation method, which only requires that the residual be evaluated at the collocation points, would appear to be the simplest of weighted-residual methods. An improvement to this procedure is to select judiciously the collocation points. A proper choice makes the computations more convenient and the results more accurate. One such method, which is discussed in several textbooks, is the orthogonal collocation technique proposed by Lanczos[3] (1939). With this method, the collocation points are chosen to be the roots of orthogonal polynomials such as the Legendre polynomials [Chapter 6, Eq. (6.123)]. Further simplicity is achieved if the constants $\widehat{u}_i$ in the trial solutions are replaced by the values of the trial solution at the collocation points. These values then become the coefficients to be determined.

[3]Cornelius Lanczos was born (Kornél Löwry) (1893–1974) and educated in Hungary. After many years of working in German universities, he moved to Purdue University in 1931. After World War II he joined the engineering staff at Boeing Airplane Co., then the National Bureau of Standards, and then returned to the academic world at UCLA. After a brief stint at North American Aviation, in 1954 he took a position in Ireland as a professor of physics at the Dublin Institute for Advanced Studies. He authored the books *The Variational Principles of Mechanics*, *Applied Analysis*, *Linear Differential Operators*, *Albert Einstein and the Cosmic World Order*, *Discourse on Fourier Series*, *Numbers Without End*, *Space Through the Ages*, and *Einstein Decade: 1905–1915*.

7.3.2 Least Squares Collocation

The estimation of parameters using least squares is usually attributed to Gauss's work of 1795. However, since this work was not published until 1809, there was some controversy because Legendre published similar concepts in 1806.

A useful application of the least squares technique is to couple it with collocation. This method minimizes the sum of the squares of the residuals at the collocation points. In this case, the number of collocation points is not necessarily equal to the number of free parameters. Suppose n collocation points are selected at x_j, $j = 1, 2, \ldots, n$, then the least squares method requires that

$$\sum^{n} R(x_j)^2 \qquad \text{be a minimum} \tag{7.18a}$$

or

$$\mathbf{C}^T\mathbf{C} \qquad \text{be a minimum} \tag{7.18b}$$

where $\mathbf{C}^T = [R(x_1)\, R(x_2)\, \ldots\, R(x_n)]$. With the trial solution of Eq. (7.9) as the approximation of u, $R(x_j) = L\widetilde{u}(x_j) - f(x_j)$. If L is linear, $R(x_j)$ is a linear equation in terms of $\widehat{u}_i$, $i = 1, 2, \ldots, m$. Let this equation be $R(x_j) = \mathbf{a}_j^T\widehat{\mathbf{u}} - b_j$, then

$$\mathbf{C} = \mathbf{A}\widehat{\mathbf{u}} - \mathbf{B}$$

where

$$\mathbf{A} = \begin{bmatrix} \mathbf{a}_1^T \\ \mathbf{a}_2^T \\ \vdots \\ \mathbf{a}_n^T \end{bmatrix} \quad \text{and} \quad \mathbf{B} = \begin{bmatrix} b_1 \\ b_2 \\ \vdots \\ b_n \end{bmatrix}$$

Equation (7.18b) becomes

$$\mathbf{C}^T\mathbf{C} = \widehat{\mathbf{u}}^T\mathbf{A}^T\,\mathbf{A}\,\widehat{\mathbf{u}} - 2\,\widehat{\mathbf{u}}^T\mathbf{A}^T\,\mathbf{B} + \mathbf{B}^T\mathbf{B}$$

The necessary condition to make $\mathbf{C}^T\mathbf{C}$ a minimum is $\frac{\partial}{\partial\widehat{\mathbf{u}}}(\mathbf{C}^T\,\mathbf{C}) = 0$, i.e.,

$$\mathbf{A}^T\mathbf{A}\,\widehat{\mathbf{u}} = \mathbf{A}^T\mathbf{B} \tag{7.19}$$

This constitutes a set of simultaneous linear equations from which $\widehat{\mathbf{u}}$ can be determined. For a vector residual $\mathbf{R}$ the minimization process is repeated p times.

EXAMPLE 7.2 Beam with Linearly Varying Load

Return to the beam of Fig. 7.1 and use the same trial function as in Example 7.1.

$$\widetilde{w} = \widehat{w}_1(3\xi^2 - 5\xi^3 + 2\xi^4) \qquad \xi = x/L \tag{1}$$

Substitute this into the beam governing equation, and use $x_1 = L/4$, $x_2 = L/2$, and $x_3 = 3L/4$ as the collocation points. The residuals at these points are

$$R(x_1) = EI\frac{48}{L^4}\widehat{w}_1 - \frac{3p_0}{4}$$
$$R(x_2) = EI\frac{48}{L^4}\widehat{w}_1 - \frac{p_0}{2} \tag{2}$$

and

$$R(x_3) = EI\frac{48}{L^4}\widehat{w}_1 - \frac{p_0}{4}$$

Then

$$\mathbf{A} = \frac{EI}{L^4}\begin{bmatrix} 48 \\ 48 \\ 48 \end{bmatrix} \quad \text{and} \quad \mathbf{B} = p_0 \begin{bmatrix} 3/4 \\ 1/2 \\ 1/4 \end{bmatrix} \tag{3}$$

And from Eq. (7.19), the linear equation for $\widehat{w}_1$ is

$$\left(\frac{48EI}{L^4}\right)^2 \cdot 3\widehat{w}_1 = \frac{48EI}{L^4} \cdot \frac{3}{2} p_0 \quad \text{or} \quad \widehat{w}_1 = \frac{p_0 L^4}{96EI} \tag{4}$$

This is the same result as that obtained in Example 7.1. ∎

7.3.3 Minimax Method

One possibility of minimizing the residual is to find the free parameters $\widehat{\mathbf{u}}$ such that the maximum residual at selected points is minimized (L_∞ norm). This approach is often referred to as the *min-max, minimax, minimum absolute error,* or *Tchebychev*[4] *fit method.*

Suppose the residual R is sampled at $x = x_j,\ j = 1, 2, \ldots, n$ locations (in V). Often, it makes sense to determine these points in the same fashion as with orthogonal collocation. Then with the minimax method the coefficients $\widehat{u}_i$ are selected such that the maximum of the absolute value of these residuals is a minimum, i.e.,

$$\max_{j=1,2,\ldots,n} |R(x_j)| \text{ is a minimum}$$

or

$$\min_{j=1,2,\ldots,n} \ \max |R(x_j)| = \min_{j=1,2,\ldots,n} \ \max |R_j| \tag{7.20}$$

As with the least squares method, n is not necessarily equal to m, the number of unknown coefficients.

This is a useful method since it can be reduced to a problem in linear programming. To convert the problem defined by Eq. (7.20) to a linear programming form, set an unknown number ϕ equal to the (unknown) maximum value of $R_j,\ j = 1, 2, \ldots, n$, that is, let

$$\phi = \max |R_j|, \quad j = 1, 2, \ldots, n \tag{7.21}$$

This is equivalent to requiring that R_j satisfy

$$|R_j| \le \phi \quad \text{or} \quad -\phi \le R_j \le \phi \tag{7.22a}$$

or

$$R_j - \phi \le 0 \quad \text{and} \quad \phi + R_j \ge 0 \tag{7.22b}$$

Now the min-max approximation problem is one of finding the unknown coefficients $\widehat{u}_i$ such that

$$\phi \qquad \text{is minimized} \tag{7.23}$$

[4]Patnutil Lvovich Tchebychev (1821–1894) was a Russian mathematician, who left an imprint in many areas of mathematics. These included the theories of integrals and numbers, quadratic forms, polynomials, motion theorems, and rectilinear motion. His collected works were published in two volumes which appeared in French in 1900 and 1907.

subject to the conditions (constraints) of Eq. (7.22b). This problem statement is readily cast into a standard linear programming format.

The minimax method, as well as the least squares method, can be extended to apply to more complicated problems taking advantage of the option to introduce constraints. For example, structures with offset supports are readily analyzed. As with the other weighted-residual methods, the minimax method can be applied to systems wherein separate particular trial solutions are established for each element forming the system [Park and Pilkey, 1982].

7.3.4 Subdomain Method

Suppose the domain or region (V) is subdivided into as many subdomains as there are free (unknown) parameters. Then choose the parameters so that the average value of the residual over each subdomain is zero. Thus, if there are mp subdomains V_j, the integral of the residual over each subdomain is set equal to zero, i.e.,

$$\int_{V_j} \mathbf{R}\, dV = 0 \quad j = 1, 2, \ldots, mp \tag{7.24a}$$

With $\mathbf{R} = \mathbf{R}_V$ of Eq. (7.10) and the trial solution of Eq. (7.9),

$$\int_{V_j} (\mathbf{L}\mathbf{N}_u\widehat{\mathbf{u}} - \overline{\mathbf{f}})\, dV = 0 \quad j = 1, 2, \ldots, mp \tag{7.24b}$$

This leads to mp simultaneous equations for the $\widehat{u}_i$, $i = 1, 2, \ldots, mp$ unknowns.

In terms of Eq. (7.14), this *subdomain method* is equivalent to setting

$$h(\mathbf{R}) = \mathbf{R}, \quad W_j = \begin{cases} 1 \text{ if } x \text{ is in the } j\text{th subdomain} \\ 0 \text{ if } x \text{ is outside the } j\text{th subdomain} \end{cases} \tag{7.25}$$

where x is the coordinate in V. This approach is sometimes referred to as the method of Biezeno[5] and Koch[6] (1923) or the *method of integral relations*. The subdomains can be chosen to be continuously adjacent, overlapping, separated, of equal size, or of different sizes.

EXAMPLE 7.3 Beam with Linearly Varying Load

In the case of a beam, the linear simultaneous equations corresponding to the subdomain method are readily obtained. For a simple beam,

$$R = \frac{d^2}{dx^2} EI \frac{d^2\widetilde{w}}{dx^2} - \overline{p}_z \tag{1}$$

Use $\widetilde{w} = \mathbf{N}_u\widehat{\mathbf{w}}$, so that Eq. (7.24) leads to linear simultaneous equations

$$\underbrace{\int_{\ell_j} \frac{d^2}{dx^2} EI\, \mathbf{N}_u''\, dx}_{\mathbf{k}_u}\ \underbrace{\widehat{\mathbf{w}}}_{\widehat{\mathbf{w}}} - \underbrace{\int_{\ell_j} \overline{p}_z(x)\, dx}_{\mathbf{p}_u} = 0 \quad j = 1, 2, \ldots, m \tag{2}$$

or $\mathbf{k}_u\widehat{\mathbf{w}} = \mathbf{p}_u$, where the beam is considered to be formed of m segments ℓ_j.

---

[5]Cornelius Benjamin Biezeno (1888–1975) was a mechanical engineer who held for many years a chair in applied mechanics at the University of Technology in Delft, Holland. He co-authored with R. Grammel the influential, monumental treatise *Technische Dynamik*, in which his disdain for energy principles is evident. It would appear as though he abhorred the notion of "virtual" work in so an exact a science as engineering. He preferred to rely on geometric concepts.

[6]J.J. Koch was a student and later colleague of Biezeno. Together they were a very effective research team in applied mechanics. Koch had a reputation for having a fine-tuned physical intuition for mechanical phenomena.

For the beam of Fig. 7.1, suppose there is a single domain, and Eq. (8) of Example 7.1 is used to approximate the displacement. Then Eq. (7.24) becomes

$$\int_0^1 \left[EI\frac{48}{L^4}\widehat{w}_1 - p_0(1-\xi) \right] d\xi = 0 \tag{3}$$

i.e.,

$$EI\frac{48}{L^4}\widehat{w}_1 - \frac{p_0}{2} = 0 \tag{4}$$

This is the same result obtained in Example 7.1. ∎

7.3.5 Orthogonality Methods

A variety of techniques can be classified as *orthogonality methods*. Define a set of linearly independent functions $\Psi_j(x)$, $r = 1, 2, \ldots, n$, in domain V. The integrals

$$\int_V \Psi_j R \, dV = 0 \quad j = 1, 2, \ldots, n \tag{7.26a}$$

or

$$\int_V \mathbf{\Psi R} \, dV = 0$$

which form a system of n equations, are referred to as the *orthogonality conditions*. For a p-dimensional problem, let $\Psi^{(k)}$ be a $p \times p$ diagonal matrix of functions for $k = 1, 2, \ldots n$. The orthogonality method requires that

$$\int_V \Psi^{(k)}\mathbf{R}\, dV = 0 \quad k = 1, 2, \ldots n \tag{7.26b}$$

with the total number of equations being np. Equations (7.26) are obtained from Eq. (7.14) by choosing h to be the identity mapping. There are several useful methods, e.g., Galerkin's,[7] least squares, and method of moments, which employ the orthogonality condition of Eq. (7.26).

7.3.6 Galerkin's Method

If Ψ_j in Eq. (7.26a) are chosen to be the m trial functions N_{ui}, then

$$\int_V N_{ui} R \, dV = \int_V N_{ui}\left[L\left(\sum \widehat{u}_i N_{ui}\right) - \overline{f} \right] dV = 0 \quad i = 1, 2, \ldots, m \tag{7.27}$$

For a p-dimensional problem, Galerkin's method requires the mp conditions $\int N_{u1i} R_1 \, dV = \cdots = \int N_{u2i} R_p \, dV = 0,\ i = 1, 2, \ldots, m$ or $\int_V \mathbf{N}_u^T \mathbf{R}\, dV = 0$, i.e.,

$$\begin{aligned} \int_V \mathbf{N}_u^T (\mathbf{L N}_u \widehat{\mathbf{u}} - \overline{\mathbf{f}})\, dV = \int_V \mathbf{N}_u^T (\mathbf{L N}_u)\, dV\, \widehat{\mathbf{u}} - \int_V \mathbf{N}_u^T \overline{\mathbf{f}}\, dV &= \mathbf{0} \\ \mathbf{k}_u \quad \widehat{\mathbf{u}} - \quad \mathbf{p}_u \quad &= \mathbf{0} \end{aligned} \tag{7.28}$$

[7] Boris Grigorievich Galerkin (1871–1945) was a Russian engineer, graduate of the Petersburg Technological Institute. He lectured at several colleges in the St. Petersburg area, including Leningrad University, where he became dean of the structural engineering department. He contributed to several challenging problems, such as the curvature of thin plates. His approximate solution of differential equations is utilized today for the solution of many applied mechanics problems.

With *Galerkin's (1915) method* or the *Bubnov*[8] *(1913)-Galerkin method*, these mp equations are used to find the mp unknown coefficients $\hat{u}_i$, $i = 1, 2, \ldots, mp$, in $\tilde{u}(x)$. Often, the N_{ui} are chosen to be members of a complete set of functions. The trial solutions are chosen to satisfy all boundary conditions.

Since Galerkin's method, the most popular of the weighted-residual techniques, will be applied extensively to solids in Section 7.4, no examples will be treated here.

7.3.7 Method of Moments

In what is called the *method of moments* [Yamada, 1950], Eq. (7.26) is used with Ψ_j, $j = 1, 2, \ldots, n$ selected as the first n members of a complete set of functions. Such series as an ordinary polynomial, trigonometric, and Tchebychev polynomial can be complete. If an ordinary polynomial is used, then $\Psi_j = x^{j-1}$, $j = 1, 2, \ldots, n$. Successively higher "moments" of the residual are required to be zero. Note that for the first approximation, i.e., $\Psi_1 = 1$, the method of moments is the same as the subdomain method with the subdomain equal to the whole domain. The method of moments is sometimes referred to as the *integral method* of von Karman[9] (1921) and Pohlhausen[10] (1921).

7.3.8 Least Squares

In Section 7.3.2, the least squares method was used to minimize the sum of the squares of the residuals at some selected points. In this section, the integral form of the least squares method is considered. Here, the integral (over domain V) of the weighted square of the residual is required to be a minimum. That is,

$$\int_V R^2 \, dV \qquad \text{is minimized} \tag{7.29}$$

Choose $W_j(x)$ to be positive, so that the integrand is positive. Substitution of Eq. (7.9) into Eq. (7.29) gives an integral containing $\hat{\mathbf{u}}$ as unknowns. For a p-dimensional residual vector $\mathbf{R}$ write Eq. (7.29) in matrix form, as

$$\begin{aligned}\int_V \mathbf{R}^T \mathbf{W} \mathbf{R} \, dV &= \int_V (\mathbf{L}\mathbf{N}_u\hat{\mathbf{u}} - \bar{\mathbf{f}})^T \mathbf{W} (\mathbf{L}\mathbf{N}_u\hat{\mathbf{u}} - \bar{\mathbf{f}}) \, dV \\ &= \int_V \left[\hat{\mathbf{u}}^T (\mathbf{L}\mathbf{N}_u)^T \mathbf{W}\,\mathbf{L}\,\mathbf{N}_u\hat{\mathbf{u}} - 2\hat{\mathbf{u}}^T (\mathbf{L}\mathbf{N}_u)^T \mathbf{W}\bar{\mathbf{f}} + \bar{\mathbf{f}}^T \mathbf{W}\bar{\mathbf{f}}\right] dV \\ &= \text{minimum}\end{aligned} \tag{7.30}$$

where $\mathbf{R}$ is a $pm \times 1$ vector and $\mathbf{W}$ is a diagonal weighting matrix with positive elements. The necessary condition that the minimum in Eq (7.30) be achieved is found by setting the

[8]I.G. Bubnov was a Russian shipbuilding engineer. In 1913 he published the fundamentals of what is known today as the Bubnov-Galerkin Method. Galerkin generalized the method in 1915.

[9]Theodore von Karman (1881–1963) was born in Hungary and received his PhD from the University of Göttingen, Germany, in 1908. He moved to the United States in 1930 and became director of the Guggenheim Aeronautical Lab and the Jet Propulsion Lab. In 1924 he co-authored the book *General Aerodynamic Theory* and in 1935 he developed the theory of supersonic drag, now called the Karman vortex trail. He contributed significantly to thermodynamics, aerodynamics, and hydrodynamics and is recognized for his pioneering efforts in the development of high speed aircraft.

[10]Karl Pohlhausen was a German scientist who studied at the University of Göttingen school of applied mechanics. He was a student of Prandtl and, after World War I, followed von Karman from Göttingen to Aachen. He moved to Wright Field in the United States after World War II and worked in fluid mechanics.

derivative of the integral with respect to $\widehat{\mathbf{u}}$ equal to zero. Thus

$$\frac{\partial}{\partial \widehat{\mathbf{u}}}\int_V [\widehat{\mathbf{u}}^T(\mathbf{LN}_u)^T\mathbf{W}\,\mathbf{LN}_u\,\widehat{\mathbf{u}} - 2\widehat{\mathbf{u}}^T(\mathbf{LN}_u)^T\mathbf{W}\overline{\mathbf{f}} + \overline{\mathbf{f}}^T\mathbf{W}\overline{\mathbf{f}}]\,dV = 0 \tag{7.31}$$

or

$$\int \frac{\partial}{\partial \widehat{\mathbf{u}}}[\widehat{\mathbf{u}}^T(\mathbf{LN}_u)^T\mathbf{W}\,\mathbf{LN}_u\widehat{\mathbf{u}} - 2\widehat{\mathbf{u}}^T(\mathbf{LN}_u)^T\mathbf{W}\overline{\mathbf{f}} + \overline{\mathbf{f}}^T\mathbf{W}\overline{\mathbf{f}}]\,dV \tag{7.32}$$

$$\begin{aligned} &= \underbrace{2\int_V (\mathbf{LN}_u)^T\mathbf{W}\,\mathbf{LN}_u\,dV}\ \widehat{\mathbf{u}} - \int_V (\mathbf{LN}_u)^T\mathbf{W}\overline{\mathbf{f}}\,dV \\ &\qquad 2(\mathbf{k}_u \qquad\qquad \widehat{\mathbf{u}} - \qquad \mathbf{p}_u) \qquad = 0 \end{aligned} \tag{7.33}$$

These are mp simultaneous linear equations which can be used to find the mp unknown $\widehat{u}_i$. Frequently, the weighting functions W_j are set equal to unity.

Equation (7.32) can be considered as a special case of the orthogonality integral of Eq. (7.26) if $\mathbf{\Psi}_i$ is set equal to $(\mathbf{LN}_u)^T\mathbf{W}$. Also, the least squares approach can be obtained from Eq. (7.14b) by using

$$h(\mathbf{R}) = \mathbf{R} \quad \text{and} \quad \mathbf{W} \text{ replaced by } \frac{\partial \mathbf{R}}{\partial \widehat{\mathbf{u}}}\mathbf{W} \tag{7.34}$$

EXAMPLE 7.4 Beam with Linearly Varying Load

For a uniform beam, $R = EIw^{iv} - \overline{p}_z$. With $\widetilde{w} = \mathbf{N}_u\widehat{\mathbf{w}}$ and W_i selected to be 1, the least square relations would be

$$\begin{aligned} &\int_0^L \mathbf{N}_u^{ivT}(EI)\mathbf{N}_u^{iv}\,dx\,\widehat{\mathbf{w}} - \int_0^L \mathbf{N}_u^{ivT}\overline{p}_z\,dx = 0 \\ &\qquad \mathbf{k}_u \qquad\qquad\qquad \widehat{\mathbf{w}} - \qquad \mathbf{p}_u \qquad = 0 \end{aligned} \tag{1}$$

or

$$\mathbf{k}_u\widehat{\mathbf{w}} = \mathbf{p}_u \tag{2}$$

For the beam with linearly varying load in Fig. 7.1, use the trial function of Eq. (8) of Example 7.1, $\widetilde{w} = \mathbf{N}_u\widehat{\mathbf{w}} = (3\xi^2 - 5\xi^3 + 2\xi^4)\widehat{w}_1$. Then $\mathbf{N}_u^{iv} = 48/L^4$ and (1) becomes

$$\int_0^L 48\frac{EI}{L^4}\left[48\frac{EI}{L^4}\widehat{w}_1 - p_0(1 - x/L)\right]dx = 48\frac{EI}{L^4}\left[\left(EI\frac{48}{L^4}\right)\widehat{w}_1 L - \left(p_0 L - \frac{p_0 L}{2}\right)\right] = 0 \tag{3}$$

Then $\widehat{w}_1 = p_0L^4/(96EI)$ and the approximate deflection is

$$\widetilde{w} = \frac{p_0 L^4}{96EI}(3\xi^2 - 5\xi^3 + 2\xi^4) \tag{4}$$

Note that this is the same as Eq. (12) of Example 7.1 ∎

7.3.9 Symmetry and Convergence of the Weighted Residual Methods

Most of the approximate methods discussed in this chapter lead to a system of linear equations in the unknown coefficients $\widehat{u}_i$, e.g.,

$$\mathbf{k}_u\widehat{\mathbf{u}} = \mathbf{p}_u \tag{7.35a}$$

Once this equation is solved for $\hat{\mathbf{u}}$, the approximate series solution

$$\tilde{u}(x) = \mathbf{N}_u(x)\hat{\mathbf{u}} \tag{7.35b}$$

is formed. We must, of course, address such important computational considerations as the ease of solving Eq. (7.35a) and the convergence characteristics of the resulting series $\tilde{u}$ [Eq. (7.35b)].

Symmetry of the Equations for the Unknown Coefficients

Matrix equations of the form of Eq. (7.35a) can be solved with fewer computations if the matrix $\mathbf{k}_u$ is symmetric. The integral expressions for the various residual methods can be identified readily in terms of the elements k_{ij} of the matrix $\mathbf{k}_u$. For example, for collocation,

$$k_{ij} = LN_{uj}(x_i) \qquad p_i = f(x_i) \tag{7.36a}$$

for the subdomain method,

$$k_{ij} = \int_{V_i} LN_{uj}\, dV \qquad p_i = \int_{V_i} f\, dV \tag{7.36b}$$

for Galerkin's method,

$$k_{ij} = \int_V N_{ui} LN_{uj}\, dV \qquad p_i = \int_V N_{ui} f\, dV \tag{7.36c}$$

and for the least squares method,

$$k_{ij} = \int_V LN_{ui} LN_{uj}\, dV \qquad p_i = \int_V LN_{ui} f\, dV \tag{7.36d}$$

As can be seen from these relations, for the collocation (and minimax) and subdomain procedures, the matrix $\mathbf{k}_u$ is not in general symmetric i.e., $k_{ij} \neq k_{ji}$. This is also the case for Galerkin's method. The matrix $\mathbf{k}_u$ is always symmetric for "quadratic" formulations such as the least squares approach.

Stability and Convergence

The term *accuracy* refers to the closeness of a solution to the true or exact solution. *Stability* refers to the growth of error as a computation proceeds. In an unstable computation, truncation, roundoff, or other errors accumulate such that the progress toward the true solution is overcome or swamped by the error. *Convergence* refers to achieving progressive closeness to a particular solution as successive solutions are computed as a parameter is changed. In such a calculation, typically, the number of terms in a trial solution is adjusted. Convergence is also used in reference to an iterative computational procedure. In such iterative techniques, the results for a particular computation become the starting point for the next computation. As this procedure is repeated, it is said to be convergent if the difference between successive results becomes smaller. Convergence studies for many trial function methods are available in several references, e.g., Finlayson (1972).

7.3.10 Weak Formulation and Boundary Conditions

In the previous sections, weighted-residual formulations such as

$$\int_V \mathbf{W}(\mathbf{L}\tilde{\mathbf{u}} - \bar{\mathbf{f}})\, dV = 0 \tag{7.37}$$

were utilized to compute the coefficients $\hat{u}_i$ of the trial functions. Alternative forms of Eq. (7.37) can be obtained by using integration by parts or the Gauss integral theorem to form integrals such as

$$\int_V (\mathbf{WL\tilde{u}} - \mathbf{W\bar{f}})\, dV = \int_V \mathbf{L}^1\mathbf{W}\mathbf{L}^2\mathbf{\tilde{u}}\, dV + \int_S \mathbf{L}^3\mathbf{W}\mathbf{L}^4\mathbf{\tilde{u}}\, dS - \int_V \mathbf{W\bar{f}}\, dV \tag{7.38}$$

in which $\mathbf{L}^1$, $\mathbf{L}^2$, $\mathbf{L}^3$, and $\mathbf{L}^4$ are differential operators, and S is the boundary. In this alternative formulation, the differentiation of $\mathbf{\tilde{u}}$ is usually of lower order than that appearing in $\mathbf{L}$, so that the requirement of the order of continuity on $\mathbf{\tilde{u}}$ can be "weakened". Hence, this is called a *weak formulation* or *weak form*. Such a formulation can be advantageous because it can make the choice of the trial functions easier. The operators $\mathbf{L}^1$ and $\mathbf{L}^3$ on the weighting function $\mathbf{W}$ involve differentiations; consequently, the continuity requirements on $\mathbf{W}$ are more severe than before. Thus, $\mathbf{W}$ must have C^{r-1} continuity where r is the highest derivative in $\mathbf{L}^1$ and $\mathbf{L}^3$. This requirement can be met by choosing appropriate $\mathbf{W}$, e.g., choosing $\mathbf{W}$ to be equal to $\mathbf{N}_u$ of Eq. (7.9).

To illustrate the fundamentals of a weak formulation, consider a linearly elastic beam with the governing equation

$$\frac{d^2}{dx^2} EI \frac{d^2 w}{dx^2} - \overline{p}_z = 0 \tag{7.39}$$

Cast this into the weighted-residual formulation

$$\int_L W_j \left(\frac{d^2}{dx^2} EI \frac{d^2 \widetilde{w}}{dx^2} - \overline{p}_z \right) dx = 0 \tag{7.40}$$

Integrate the left-hand integral by parts twice to find

$$\int_L \left(EI \frac{d^2 W_j}{dx^2} \frac{d^2 \widetilde{w}}{dx^2} - W_j \overline{p}_z \right) dx + W_j \frac{d}{dx} EI \frac{d^2 \widetilde{w}}{dx^2} \Big|_0^L + \frac{dW_j}{dx} EI \frac{d^2 \widetilde{w}}{dx^2} \Big|_0^L = 0 \tag{7.41}$$

Equation (7.41) represents the weak formulation. The order of $\widetilde{w}$ is lowered, and boundary terms have appeared. In the case of the boundary element method of Chapter 9, integration by parts is continued until all of the derivatives are switched from w to W_j, leaving only boundary terms involving the unknown $\widetilde{w}$.

In practice, in the weak formulation, W_j is often chosen to have the same physical meaning as w. When trial solutions of the form of Eq. (7.9) are used, Eq. (7.41) becomes a set of simultaneous linear equations. Because of the boundary terms, these equations tend to become complicated. Hence, it is often desirable to make W_j and the trial solution satisfy certain boundary conditions to eliminate the boundary terms from Eq. (7.41). Designate W_j and its derivative in the boundary terms, i.e., W_j and dW_j/dx, as the *forced* or *essential boundary conditions* of Eq. (7.41) and the remaining factors, $\frac{d}{dx} EI \frac{d^2 w}{dx^2} = -V$ (shear force) and $EI \frac{d^2 w}{dx^2} = -M$ (bending moment), as the *natural boundary conditions*. In a displacement formulation forced boundary conditions are the displacement conditions, whereas the natural boundary conditions are the force or static boundary conditions. These definitions of forced and natural boundary conditions are the same as in Appendix I.

7.4 Variational Methods

The trial function methods can also be approached from the viewpoint of a variational principle. Recall that in Chapter 2 the integral (global) forms of the differential equations represented the three fundamental relationships of solid mechanics—kinematics, material law, and equilibrium. It is not surprising that the integrals of the variational principles can be used directly in the same fashion as weighted-residual integrals. In fact, use of these integrals as the basis of the approximation is referred to as a *direct variational method.* The most popular direct variational method, the so-called Ritz's[11] method, employs the integral relations of the principle of stationary potential energy, or, more generally, those of the principle of virtual work. It will be shown in this section that Galerkin's method can be considered to be either a standard weighted residual method or to be based on a variational principle.

As with the previous methods, the discretization of the variational integrals leads to a system of algebraic equations. The unknowns in the trial solution are to be obtained such that a variational functional is made stationary. This procedure leads to a best approximation of certain characteristics, e.g., equilibrium or boundary conditions, of the problem.

7.4.1 Ritz's Method

One of the most frequently used approximate methods in mechanics is the *method of Ritz.* The Ritz method can be used in conjunction with the principle of virtual work which requires that $\delta W = 0$. This leads to a solution which satisfies approximately the conditions of equilibrium and the static boundary conditions. With a known material law, *the chosen approximate displacements must satisfy the kinematic conditions and displacement boundary conditions, i.e., they must be kinematically admissible.* Thus, at the outset, it is required that the assumed displacements $\tilde{\mathbf{u}}$ satisfy

$$\epsilon = \mathbf{D}\tilde{\mathbf{u}} \text{ in } V \qquad \tilde{\mathbf{u}} = \overline{\mathbf{u}} \text{ on } S_u \tag{7.42}$$

and then the free parameters $\hat{\mathbf{u}}$ are determined such that the conditions of equilibrium and the static boundary conditions are approximated as closely as possible. It should be apparent that the Ritz method is closely related to the displacement method.

From Chapter 2, we know that for a continuum the virtual work can be expressed as

$$-\delta W = -\delta W_i - \delta W_e = \int_V \delta\epsilon^T \sigma \, dV - \int_V \delta\mathbf{u}^T \overline{\mathbf{p}}_V \, dV - \int_{S_p} \delta\mathbf{u}^T \overline{\mathbf{p}} \, dS = 0 \tag{7.43}$$

For a beam, with no shear deformation, from Chapter 2, Example 2.7 or Chapter 4, Eq. (4.52),

$$-\delta W = \int_0^L \delta w'' EI \, w'' \, dx - \int_0^L \overline{p}_z \, \delta w \, dx - \underset{\text{on } S_p}{[\overline{V}\,\delta w + \overline{M}\,\delta\theta]_0^L} = 0 \tag{7.44}$$

[11] Walter Ritz (1878–1909) was born in Switzerland, son of the artist Raphael Ritz. He studied in Zurich, Switzerland, and Göttingen, Germany, where he obtained his doctorate in 1902. He then worked in Leyden, Paris, and Tübingen. In 1908, he returned to Göttingen and remained there until his untimely death.

An approximate displacement of the form $\widetilde{w}(x) = \mathbf{N}_u\widehat{\mathbf{w}}$ is chosen. Then $\delta\theta = -\delta w' = -\mathbf{N}_u'\delta\widehat{\mathbf{w}}$, $\delta\widetilde{w}''(x) = \mathbf{N}_u''(x)\,\delta\widehat{\mathbf{w}}$, and $\delta W = 0$ becomes

$$
\begin{aligned}
-\delta W &= \int_0^L (\delta\widehat{\mathbf{w}}^T \mathbf{N}_u''^T EI \mathbf{N}_u'' \widehat{\mathbf{w}})\,dx - \int_0^L \delta\widehat{\mathbf{w}}^T \mathbf{N}_u^T\, \overline{p}_z\,dx - \delta\widehat{\mathbf{w}}^T \left[\mathbf{N}_u^T\overline{V} - \mathbf{N}_u'^T\overline{M}\right]_0^L \\
&= \delta\widehat{\mathbf{w}}^T \left\{ \underbrace{EI\int_0^L \mathbf{N}_u''^T(x)\mathbf{N}_u''(x)\,dx}_{\mathbf{k}_u}\ \widehat{\mathbf{w}} - \underbrace{\int_0^L \mathbf{N}_u^T(x)\overline{p}_z(x)\,dx + \left[\mathbf{N}_u^T\overline{V} - \mathbf{N}_u'^T\overline{M}\right]_0^L}_{\overline{\mathbf{p}}_u} \right\} = 0
\end{aligned}
\tag{7.45}
$$

From this form, it is apparent that $\mathbf{k}_u$ is inherently symmetric. Finally, we obtain the system of equations

$$\mathbf{k}_u\,\widehat{\mathbf{w}} = \mathbf{p}_u \tag{7.46}$$

for the free parameters $\widehat{\mathbf{w}}$.

EXAMPLE 7.5 Beam with Linearly Varying Loading

Return to the fixed-hinged beam of Fig. 7.1. The geometrical (displacement) boundary conditions on S_u are

$$w(0) = 0, \qquad -\theta(0) = w'(0) = 0, \qquad w(L) = 0 \tag{1}$$

and the static boundary condition is

$$M(L) = 0 \tag{2}$$

on S_p. We will begin with a two-parameter approximate deflection

$$w(\xi) = N_{u1}(\xi)\widehat{w}_1 + N_{u2}(\xi)\widehat{w}_2 = [N_{u1}(\xi) \quad N_{u2}(\xi)] \begin{bmatrix} \widehat{w}_1 \\ \widehat{w}_2 \end{bmatrix} = \mathbf{N}_u\widehat{\mathbf{w}} \tag{3}$$

Since the variational principle of Eq. (7.45) contains second order derivatives, it is necessary to select an approximate deflection containing at least a third order polynomial. Choose

$$
\begin{aligned}
N_{u1}(\xi) &= a_1 + a_2\xi + a_3\xi^2 + a_4\xi^3 \\
N_{u2}(\xi) &= b_1 + b_2\xi + b_3\xi^2 + b_4\xi^3 + b_5\xi^4
\end{aligned}
\tag{4}
$$

The approximation (3) must satisfy the geometrical boundary conditions, and the constants a_i, $i = 1, 2, 3, 4$ and b_i, $i = 1, 2, 3, 4, 5$ will be chosen to accomplish this. Thus,

$$
\begin{aligned}
w(0) = 0 &\quad \text{requires that } a_1 = b_1 = 0 \\
w'(0) = 0 &\quad \text{requires that } a_2 = b_2 = 0 \\
w(L) = 0 &\quad \text{requires that } a_3 + a_4 = 0 \\
&\qquad\qquad b_3 + b_4 + b_5 = 0
\end{aligned}
\tag{5}
$$

We can now select any values for a_i, b_i that satisfy these relations, e.g.,

$$a_3 = -a_4 = 1, \qquad b_3 = b_5 = 1, \qquad b_4 = -2 \tag{6}$$

Thus,

$$\mathbf{N}_u = [(\xi^2 - \xi^3) \quad (\xi^2 - 2\xi^3 + \xi^4)] \tag{7}$$

and, since $d\xi = dx/L$,

$$\mathbf{N}_u'' = \frac{1}{L^2}[(2 - 6\xi) \quad (2 - 12\xi + 12\xi^2)] \tag{8}$$

Substitute (7) and (8) in $\delta W = 0$ of Eq. (7.45), getting

$$-\delta W = \delta \widehat{\mathbf{w}}^T \left\{ \frac{EI}{L^3} \int_0^1 \begin{bmatrix} 2-6\xi \\ 2-12\xi+12\xi^2 \end{bmatrix} [2-6\xi \quad 2-12\xi+12\xi^2]\, d\xi \begin{bmatrix} \widehat{w}_1 \\ \widehat{w}_2 \end{bmatrix} \right\}$$
$$- p_0 L \int_0^1 \begin{bmatrix} \xi^2 - \xi^3 \\ \xi^2 - 2\xi^3 + \xi^4 \end{bmatrix} (1-\xi)\, d\xi = 0$$

where $\overline{p}_z$ was taken to be equal to $p_0(1-\xi)$. The final term in Eq. (7.45), $[\mathbf{N}_u^T\, \overline{V} - \mathbf{N}_u'^T \overline{M}]_0^L$, is zero since from (7) $\mathbf{N}_u^T$ is zero at $x = 0$ and $x = L$, $\mathbf{N}_u'^T$ is zero at $x = 0$, and from (2) M is zero at $x = L$.

Then,

$$\mathbf{k}_u = \frac{EI}{L^3} \int_0^1 \begin{bmatrix} (4-24\xi+36\xi^2) & (4-36\xi+96\xi^2-72\xi^3) \\ (4-36\xi+96\xi^2-72\xi^3) & (4-48\xi+192\xi^2-288\xi^3+144\xi^4) \end{bmatrix} d\xi$$
$$= \frac{EI}{L^3} \begin{bmatrix} 4 & 0 \\ 0 & 0.8 \end{bmatrix} \tag{9}$$

$$\mathbf{p}_u = p_0 L \int_0^1 \begin{bmatrix} \xi^2 - 2\xi^3 + \xi^4 \\ \xi^2 - 3\xi^3 + 3\xi^4 - \xi^5 \end{bmatrix} d\xi = p_0 L \begin{bmatrix} 1/30 \\ 1/60 \end{bmatrix} \tag{10}$$

The unknown parameters are obtained from

$$\frac{EI}{L^3} \begin{bmatrix} 4 & 0 \\ 0 & 0.8 \end{bmatrix} \begin{bmatrix} \widehat{w}_1 \\ \widehat{w}_2 \end{bmatrix} - p_0 L \begin{bmatrix} 1/30 \\ 1/60 \end{bmatrix} = 0 \tag{11}$$

as

$$\begin{bmatrix} \widehat{w}_1 \\ \widehat{w}_2 \end{bmatrix} = \begin{bmatrix} 1/120 \\ 1/48 \end{bmatrix} \frac{p_0 L^4}{EI} \tag{12}$$

These parameters provide the deflection, moment, and shear force

$$\begin{aligned} \widetilde{w}(x) &= \frac{p_0 L^4}{EI} \left[\frac{1}{120}(\xi^2 - \xi^3) + \frac{1}{48}(\xi^2 - 2\xi^3 + \xi^4) \right] \\ &= \frac{p_0 L^4}{240EI}[7\xi^2 - 12\xi^3 + 5\xi^4] \\ M(x) &= -EI\widetilde{w}'' = -\frac{p_0 L^2}{120}(7 - 36\xi + 30\xi^2) \\ V(x) &= -EI\widetilde{w}''' = \frac{p_0 L}{30}(8 - 15\xi) \end{aligned} \tag{13}$$

Note that the boundary condition $M(L) = 0$ is not satisfied. Observe that according to this approximate solution, the linearly varying applied load leads to a quadratically varying moment and a linearly varying shear force. If more accurate results are desired, it would appear appropriate to begin with an assumed deflection containing higher order polynomials.

As a second possibility, choose a trial solution with

$$\begin{aligned} N_{u1}(\xi) &= (\xi^2 - \xi^3) \\ N_{u2}(\xi) &= (\xi^3 - \xi^4) \\ N_{u3}(\xi) &= (\xi^4 - \xi^5) \end{aligned} \tag{14}$$

Here, N_{u1} is the same as the previous case, the combination $b_1 = b_2 = b_3 = 0$, $b_4 = 1$, and $b_5 = -1$ substituted in (4) leads to N_{u2}, and N_{u3} is a higher order polynomial which satisfies the required displacement boundary conditions. Our trial deflection is

$$w(\xi) = [(\xi^2 - \xi^3) \quad (\xi^3 - \xi^4) \quad (\xi^4 - \xi^5)] \begin{bmatrix} \widehat{w}_1 \\ \widehat{w}_2 \\ \widehat{w}_3 \end{bmatrix} \tag{15}$$

For this assumption, we obtain

$$\mathbf{k}_u = \frac{EI}{L^4} \int_0^1 \begin{bmatrix} 4 - 24\xi + 36\xi^2 & 12\xi - 60\xi^2 + 72\xi^3 & 24\xi^2 - 112\xi^3 + 120\xi^4 \\ & 36\xi^2 - 144\xi^3 + 144\xi^4 & 72\xi^3 - 264\xi^4 + 240\xi^5 \\ \text{Symmetric} & & 144\xi^4 - 480\xi^5 + 400\xi^6 \end{bmatrix} dx \tag{16}$$

$$\mathbf{p}_u = \int_0^1 \overline{p}_z(x)\mathbf{N}_u^T dx = \int_0^1 p_0(1-\xi) \begin{bmatrix} \xi^2 - \xi^3 \\ \xi^3 - \xi^4 \\ \xi^4 - \xi^5 \end{bmatrix} dx = p_0 L \begin{bmatrix} 1/30 \\ 1/60 \\ 1/105 \end{bmatrix} \tag{17}$$

The equations for $\widehat{\mathbf{w}}$ are

$$\underbrace{\frac{EI}{L^3} \begin{bmatrix} 4 & 4 & 4 \\ 4 & 24/5 & 26/5 \\ 4 & 26/5 & 208/35 \end{bmatrix}}_{\mathbf{k}_u} \underbrace{\begin{bmatrix} \widehat{w}_1 \\ \widehat{w}_2 \\ \widehat{w}_3 \end{bmatrix}}_{\widehat{\mathbf{w}}} \;=\; \underbrace{p_0 L \begin{bmatrix} 1/30 \\ 1/60 \\ 1/105 \end{bmatrix}}_{\mathbf{p}_u} \tag{18}$$

which provides the solution

$$\widehat{\mathbf{w}} = \frac{p_0 L^4}{120EI} [4 \quad -4 \quad 1]^T \tag{19}$$

Let us compare both of the above results with the exact solution.

A. Exact solution:

$$w(\xi) = \frac{p_0 L^4}{120EI} (4\xi^2 - 8\xi^3 + 5\xi^4 - \xi^5)$$

B. Trial Deflection with $m = 2$:

$$w(\xi) = \frac{p_0 L^4}{120EI} (3.5\xi^2 - 6\xi^3 + 2.5\xi^4)$$

C. Trial Deflection with $m = 3$:

$$w(\xi) = \frac{p_0 L^4}{120EI} (4\xi^2 - 8\xi^3 + 5\xi^4 - \xi^5)$$

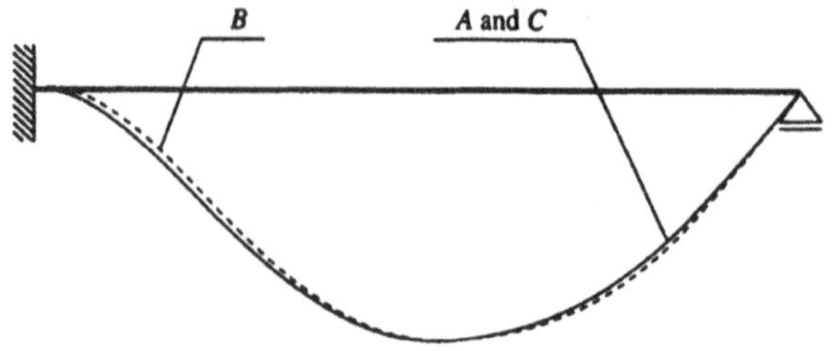

	$w_m EI/p_0 L^4$	Location ξ_0
A	0.002385	0.5528
B	0.002416	0.5683
C	0.002385	0.5528

w_m = maximum deflection
ξ_0 = location of w_m

(a) Deflection

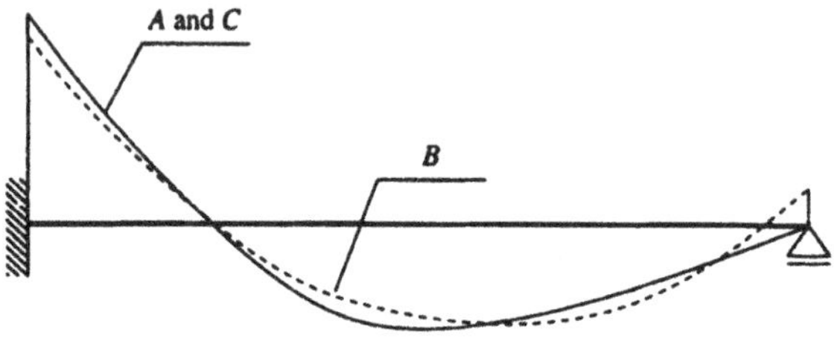

$\frac{M}{p_0 L^2}$	Location $\xi = 0$	$\xi = \xi_0$	$\xi = 1$
A	$-\frac{1}{15}$	$\frac{1}{33.54}$	0
B	$-\frac{1}{17.14}$	$\frac{1}{31.83}$	$-\frac{1}{120}$
C	$-\frac{1}{15}$	$\frac{1}{33.54}$	0

(b) Bending moment

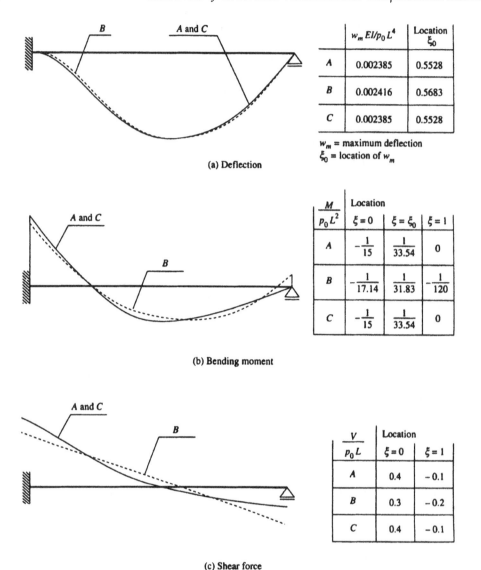

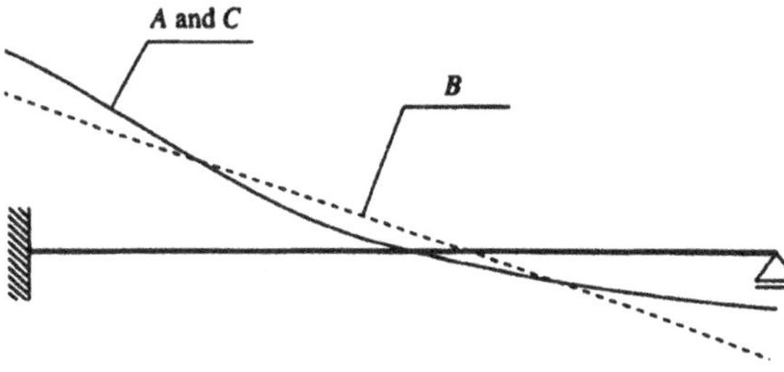

$\frac{V}{p_0 L}$	Location $\xi = 0$	$\xi = 1$
A	0.4	−0.1
B	0.3	−0.2
C	0.4	−0.1

(c) Shear force

FIGURE 7.2
Ritz method for a fixed-hinged beam.

The trial function for case C led to the exact solution. The results are summarized in Fig. 7.2. ∎

EXAMPLE 7.6 A Field Theory Problem: Torsion

Apply the Ritz Method to find the warping function, torsional constant, and stresses for an elliptical cross-section (Fig. 7.3) of a uniform bar subjected to a pure torque.

From the principle of virtual work (Chapter 2, Example 2.8), generalized to include the applied distributed torque $\overline{m}_x$,

$$-\delta W = \int_V (\tau_{xy}\, \delta\gamma_{xy} + \tau_{xz}\, \delta\gamma_{xz})\, dV + \int_L \overline{m}_x\, \delta\phi\, dx = 0 \tag{1}$$

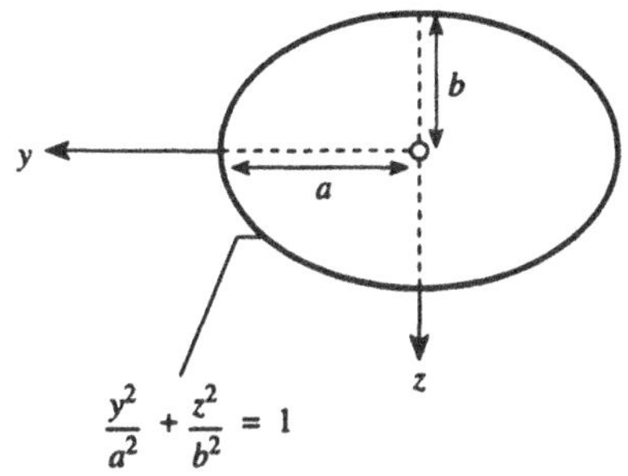

FIGURE 7.3
Bar of elliptical cross-section.

Introduction of the strains and stresses of Eqs. (1.142) and (1.143) leads to

$$-\int_V G\left[\left\{\left(\frac{\partial\omega}{\partial y}+z\right)\frac{\partial}{\partial x}\delta\phi+\phi'\frac{\partial}{\partial y}\delta\omega\right\}\phi'\left(\frac{\partial\omega}{\partial y}+z\right)\right.$$
$$\left.+\left\{\left(\frac{\partial\omega}{\partial z}-y\right)\frac{\partial}{\partial x}\delta\phi+\phi'\frac{\partial}{\partial z}\delta\omega\right\}\phi'\left(\frac{\partial\omega}{\partial z}-y\right)\right]dV+\int_L \overline{m}_x\,\delta\phi\,dx=0 \tag{2}$$

This can be separated into two terms that are equal to zero, one containing $\delta\phi$ and the other with $\delta\omega$. The $\delta\omega$ terms appear as

$$\int_V G\phi'^2\left(\frac{\partial}{\partial y}\delta\omega\frac{\partial\omega}{\partial y}+\frac{\partial}{\partial z}\delta\omega\frac{\partial\omega}{\partial z}+\frac{\partial}{\partial y}\delta\omega\, z-\frac{\partial}{\partial z}\delta\omega\, y\right)dV$$
$$=G\int_L \phi'^2 dx\int_A\left[\left(\frac{\partial}{\partial y}\delta\omega\,\frac{\partial\omega}{\partial y}+\frac{\partial}{\partial z}\delta\omega\,\frac{\partial\omega}{\partial z}\right)+\left(\frac{\partial}{\partial y}\delta\omega\, z-\frac{\partial}{\partial z}\delta\omega\, y\right)\right]dA=0 \tag{3}$$

Thus, the equation that can be solved for ω, and that will permit the cross-sectional characteristics to be computed, takes the form

$$\int_A\left[\left(\frac{\partial}{\partial y}\delta\omega\,\frac{\partial\omega}{\partial y}+\frac{\partial}{\partial z}\delta\omega\,\frac{\partial\omega}{\partial z}\right)+\left(\frac{\partial}{\partial y}\delta\omega\, z-\frac{\partial}{\partial z}\delta\omega\, y\right)\right]dA=0 \tag{4}$$

or

$$\int_A \delta\omega[({}_y\partial\,\partial_y+{}_z\partial\,\partial_z)\,\omega+({}_y\partial\, z-{}_z\partial\, y)]\,dA=0 \tag{5}$$

Suppose the trial solution is

$$\widetilde{\omega}=\widehat{\omega}yz \tag{6}$$

This volume integral should now be a (cross-sectional) surface integral. Substitution of (6) in (5) gives

$$\delta\widehat{\omega}\int_A[z^2(\widehat{\omega}+1)+y^2(\widehat{\omega}-1)]\,dA=0 \tag{7}$$

or

$$\int_A[z^2(\widehat{\omega}+1)+y^2(\widehat{\omega}-1)]\,dA=0 \tag{8}$$

For an ellipse,

$$\int_A z^2\,dA=ab^3\pi/4,\qquad \int_A y^2\,dA=a^3b\pi/4$$

and

$$\frac{ab^3\pi}{4}(\widehat{\omega}+1)+\frac{a^3b\pi}{4}(\widehat{\omega}-1)=0\quad\text{or}\quad\widehat{\omega}=\frac{a^2-b^2}{a^2+b^2} \tag{9}$$

Thus

$$\widetilde{\omega} = \frac{a^2 - b^2}{a^2 + b^2} yz \tag{10}$$

The $\delta\phi$ terms of (2) appear as

$$\int \frac{\partial}{\partial x}\delta\phi\left[\left(\frac{\partial\omega}{\partial y} + z\right)^2 + \left(\frac{\partial\omega}{\partial z} - y\right)^2\right] G\phi'\, dV - \int_L \overline{m}_x\, \delta\phi\, dx = 0 \tag{11}$$

Rewrite this as

$$\int_L \frac{\partial}{\partial x}\delta\phi \int_A \left\{\left[\left(\frac{\partial\omega}{\partial y}\right)^2 + \left(\frac{\partial\omega}{\partial z}\right)^2 + z\frac{\partial\omega}{\partial y} - y\frac{\partial\omega}{\partial z}\right]\right.$$
$$\left. + \left(z\frac{\partial\omega}{\partial y} - y\frac{\partial\omega}{\partial y} + y^2 + z^2\right)\right\} dA\, G\phi'\, dx - \int_L \overline{m}_x\, \delta\phi\, dx = 0 \tag{12}$$

The integral

$$\int_A \left[\left(\frac{\partial\omega}{\partial y}\right)^2 + \left(\frac{\partial\omega}{\partial z}\right)^2 + z\frac{\partial\omega}{\partial y} - y\frac{\partial\omega}{\partial z}\right] dA \tag{13}$$

can be shown to be zero. To do so, use the displacement relationship $\partial^2\omega/\partial y^2 + \partial^2\omega/\partial z^2 = 0$ of Eq. (1.151), Green's integral theorem of Appendix II, and the surface condition of Eq. (1.153). Introduce J of Eq. (1.154)

$$J = \int_A \left[y\left(y - \frac{\partial\omega}{\partial z}\right) + z\left(z + \frac{\partial\omega}{\partial y}\right)\right] dA \tag{14}$$

into (12), giving

$$\int_L \frac{d}{dx}\delta\phi\, GJ\, \frac{d\phi}{dx}\, dx - \int_L \overline{m}_x\, \delta\phi\, dx = 0 \tag{15}$$

Integrate by parts and use appropriate boundary conditions to find

$$\int_L \delta\phi\left(\frac{d}{dx}GJ\,\frac{d\phi}{dx}\right) dx = -\int_L \overline{m}_x\, \delta\phi\, dx \tag{16}$$

or

$$\frac{d}{dx}GJ\,\frac{d\phi}{dx} = -\overline{m}_x \tag{17}$$

This is the traditional equation of motion for torsion of a rod.

If $\widetilde{\omega}$ of (10) is inserted in (14)

$$J = \frac{a^3 b^3 \pi}{a^2 + b^2} \tag{18}$$

The stresses become

$$\begin{aligned} \tau_{xy} &= G\gamma_{xy} = -G\phi'(z + \partial\omega/\partial y) = -\frac{2zM_t}{ab^3\pi} \\ \tau_{xz} &= G\gamma_{xz} = -G\phi'(-y + \partial\omega/\partial z) = -\frac{2yM_t}{a^3 b\pi} \end{aligned} \tag{19}$$

■

These example problems illustrate several interesting characteristics of the Ritz method, especially that of its simplicity and preciseness. Also, when the trial functions span the complete domain, the resulting system of equations tends to be full in the sense that $\mathbf{k}_u$ is not banded. Even more important is that a Ritz approximation predicts the stiffness of a system to be higher than the actual stiffness. As a consequence, some static responses tend to be underestimated and such dynamic responses as natural frequencies would be upper bounds.

A Functional Form

The Ritz method, as it is often presented, involves the derivatives of a functional, frequently a weak form, with respect to the free parameters. This is equivalent to the direct use of the principle of virtual work. Suppose a potential exists for both the internal and external forces. Then the principle of virtual work can be replaced by the principle of stationary potential energy. This requires that the potential energy, utilizing kinematically admissible displacements, be stationary, i.e., $\delta\Pi = 0$. This can be expressed as

$$\delta\Pi = \frac{\partial\Pi}{\partial\widehat{u}_i}\delta\widehat{u}_i = \frac{\partial\Pi}{\partial\widehat{u}_1}\delta\widehat{u}_1 + \frac{\partial\Pi}{\partial\widehat{u}_2}\delta\widehat{u}_2 + \cdots + \frac{\partial\Pi}{\partial\widehat{u}_m}\delta\widehat{u}_m = 0 \tag{7.47}$$

Since the variations $\delta\widehat{u}_i$ are arbitrary, this relation is equivalent to

$$\frac{\partial\Pi}{\partial\widehat{u}_1} = 0, \qquad \frac{\partial\Pi}{\partial\widehat{u}_2} = 0, \ldots \qquad \frac{\partial\Pi}{\partial\widehat{u}_m} = 0 \tag{7.48}$$

i.e.,

$$\frac{\partial\Pi}{\partial\widehat{\mathbf{u}}} = 0 \tag{7.49}$$

which are a system of simultaneous equations that can be solved for $\widehat{u}_1, \ldots, \widehat{u}_m$. For linear elastic structures, $\Pi(\widehat{u}_1, \ldots, \widehat{u}_m)$ will be quadratic in $\widehat{u}_i$, so that the simultaneous equations of Eq. (7.48) will be linear algebraic equations for $\widehat{u}_i$, i.e.,

$$\mathbf{k}_u\widehat{\mathbf{u}} = \mathbf{p}_u \tag{7.50}$$

For a continuum with the kinematic conditions,

$$\boldsymbol{\epsilon} = \mathbf{D}\mathbf{u} \text{ in } V \quad \text{and} \quad \mathbf{u} = \overline{\mathbf{u}} \text{ on } S_u$$

the potential energy is [Chapter 2, Eq. (2.64)]

$$\Pi = \frac{1}{2}\int_V \boldsymbol{\epsilon}^T\mathbf{E}\boldsymbol{\epsilon}\, dV - \int_V \mathbf{u}^T\overline{\mathbf{p}}_V\, dV - \int_{S_p} \mathbf{u}^T\overline{\mathbf{p}}\, dS \tag{7.51}$$

and for a beam,

$$\Pi = \frac{1}{2}\int_0^L EI(w'')^2 dx - \int_0^L \overline{p}_z w\, dx - [\overline{V}w - \overline{M}w']_{\text{on } S_p} \tag{7.52}$$

For the beam, use a trial function

$$\widetilde{w}(x) = \mathbf{N}_u\widehat{\mathbf{w}}$$

Since $\mathbf{N}_u\widehat{\mathbf{w}} = \widehat{\mathbf{w}}^T\mathbf{N}_u^T$, the potential energy becomes

$$\Pi(\widehat{\mathbf{w}}) = \widehat{\mathbf{w}}^T\left[\frac{1}{2}EI\int_0^L \mathbf{N}_u''^T\mathbf{N}_u''\, dx\, \widehat{\mathbf{w}} - \int_0^L \mathbf{N}_u^T\overline{p}_z\, dx\right] \tag{7.53}$$

where it has been assumed that no applied forces occur on the boundaries. From $\partial\Pi/\partial\widehat{\mathbf{w}} = 0$, we obtain the set of equations

$$\begin{array}{lll} EI\int_0^L \mathbf{N}_u''^T\mathbf{N}_u''\,dx\,\widehat{\mathbf{w}} - \int_0^L \overline{p}_z\mathbf{N}_u^T dx = 0 \\ \qquad \mathbf{k}_u \qquad\quad \widehat{\mathbf{w}} - \qquad \mathbf{p}_u \qquad = 0 \end{array} \tag{7.54}$$

or

$$\mathbf{k}_u\widehat{\mathbf{w}} = \mathbf{p}_u$$

As is to be expected, we have obtained the same relations found from the principle of virtual work with $\delta W = 0$.

Solution of Differential Equations

As described above, the Ritz method uses an approximation function of the form of Eq. (7.9) to satisfy $\delta W = 0$ or to make a functional, Π, stationary. The successful application of the Ritz method depends on the construction of the expression $\delta W = 0$ or the functional Π, frequently a weak form, for which the derivatives, with respect to $\widehat{u}_i$, are taken to make it stationary. For the case that the operator $\mathbf{L}$ of Eq. (7.1) is linear, self-adjoint and positive definite and the essential boundary conditions are homogeneous, a functional of the form

$$\Pi(\mathbf{u}) = \int_L (\mathbf{u}^T\mathbf{L}\mathbf{u} - 2\mathbf{u}^T\overline{\mathbf{f}})\,dx \tag{7.55}$$

can be obtained. Although the expression of Eq. (7.55) is written for one-dimensional problems, it applies to three-dimensional problems if the longitudinal coordinate x is replaced by the volume V. It can be shown (Problem 7.18) that the stationary value of this integral, $\delta\Pi(\mathbf{u}) = 0$, is equivalent to solving $\mathbf{L}\mathbf{u} - \overline{\mathbf{f}} = \mathbf{0}$. Thus, to solve the differential equations $\mathbf{L}\mathbf{u} - \overline{\mathbf{f}} = 0$, form the functional $\Pi(\mathbf{u})$ of Eq. (7.55), employ a trial solution of Eq. (7.9), take derivatives of Π with respect to $\widehat{u}_i$ as in Eq. (7.48), and solve the resulting set of simultaneous linear algebraic equations for $\widehat{u}_i$.

7.4.2 Galerkin's Method

Galerkin's method is frequently treated from a variational viewpoint. For the mechanics of solids, Galerkin's method is widely used and is considered to be one of the most viable techniques available. It applies to problems for which the governing equations are expressed in differential equation (local) form. This includes boundary value and eigenvalue problems, such as static, stability, vibration, and even geometrically nonlinear problems. Trial solutions which satisfy all boundary conditions are normally required by the method.

Return to the weighted-residual formulation of Section 7.3.6. For a linear elastic solid, the residual is given by Eq. (7.11) as

$$\mathbf{R} = \mathbf{D}^T\mathbf{E}\mathbf{D}\widetilde{\mathbf{u}} + \overline{\mathbf{p}}_V \tag{7.56}$$

Galerkin's expression of Eq. (7.28) becomes

$$\int_V \mathbf{N}_u^T(\mathbf{D}^T\mathbf{E}\mathbf{D}\mathbf{N}_u\widehat{\mathbf{u}} + \overline{\mathbf{p}}_V)\,dV = 0 \tag{7.57a}$$

or

$$\begin{array}{lll} \int_V \mathbf{N}_u^T\mathbf{D}^T\mathbf{E}\mathbf{D}\mathbf{N}_u\,dV\,\widehat{\mathbf{u}} + \int_V \mathbf{N}_u^T\overline{\mathbf{p}}_V\,dV = 0 \\ \qquad \mathbf{k}_u \qquad\qquad \widehat{\mathbf{u}} - \qquad \mathbf{p}_u \qquad = 0 \end{array} \tag{7.57b}$$

Thus, the unknowns $\widehat{\mathbf{u}}$ are obtained by solving

$$\mathbf{k}_u \widehat{\mathbf{u}} = \mathbf{p}_u \tag{7.58}$$

In the case of a uniform beam, $R = (EI\widetilde{w}^{iv}) - \overline{p}_z$ and $\widetilde{w} = \mathbf{N}_u \widehat{\mathbf{w}}$, so that Galerkin's relations would be

$$\begin{aligned} \int_0^L \mathbf{N}_u^T (EI\,\mathbf{N}_u^{iv})\, dx\, \widehat{\mathbf{w}} - \int_0^L \mathbf{N}_u^T \overline{p}_z\, dx &= 0 \\ \mathbf{k}_u \qquad\quad \widehat{\mathbf{w}} - \qquad \mathbf{p}_u \qquad &= 0 \end{aligned} \tag{7.59a}$$

or

$$\mathbf{k}_u\, \widehat{\mathbf{w}} = \mathbf{p}_u \tag{7.59b}$$

Remember that it is necessary that the basis functions $\mathbf{N}_u$ satisfy *all* of the boundary conditions. They must also be sufficiently differentiable.

Galerkin's method can also be viewed from the standpoint of a variational technique. Recall from Chapter 1 or 2 that the solid continuum relations of Eq. (7.5a), $\mathbf{D}^T \mathbf{E}\, \mathbf{D}\mathbf{u} + \overline{\mathbf{p}}_V = 0$, is an equilibrium expression. Thus, the residual (Eq. 7.56) $\mathbf{R} = \mathbf{D}^T \mathbf{E}\, \mathbf{D}\widetilde{\mathbf{u}} + \overline{\mathbf{p}}_V$ is an "out-of-balance" force. A reasonable variational integral would seem to be

$$\int_V \delta \widetilde{\mathbf{u}}^T \mathbf{R}\, dV = 0 \tag{7.60}$$

since this is of the form of work. Thus,

$$\int_V \delta \widetilde{\mathbf{u}}^T (\mathbf{D}^T \mathbf{E} \mathbf{D}\, \widetilde{\mathbf{u}} + \overline{\mathbf{p}}_V)\, dV = 0 \tag{7.61}$$

With $\widetilde{\mathbf{u}} = \mathbf{N}_u \widehat{\mathbf{u}}$, Eq. (7.61) becomes

$$\int_V \delta \widehat{\mathbf{u}}^T \mathbf{N}_u^T (\mathbf{D}^T \mathbf{E} \mathbf{D}\, \mathbf{N}_u\, \widehat{\mathbf{u}} + \overline{\mathbf{p}}_V)\, dV = 0 \tag{7.62a}$$

or

$$\delta \widehat{\mathbf{u}}^T \Bigg\{ \underbrace{\left[\int_V \mathbf{N}_u^T (\mathbf{D}^T \mathbf{E} \mathbf{D}\, \mathbf{N}_u)\, dV \right]}_{\mathbf{k}_u} \widehat{\mathbf{u}} + \underbrace{\left[\int_V \mathbf{N}_u^T \overline{\mathbf{p}}_V\, dV \right]}_{-\mathbf{p}_u} \Bigg\} = 0 \tag{7.62b}$$

so that $\widehat{\mathbf{u}}$ can be obtained from the familiar linear relationship

$$\mathbf{k}_u\, \widehat{\mathbf{u}} = \mathbf{p}_u \tag{7.63}$$

These are the same expressions given in Eq. (7.57), which were based on Galerkin's weighted residual method.

Galerkin's relations of Eq. (7.59) for a uniform beam are readily derived from a variational integral. Begin with

$$\int_0^L (EIw^{iv} - \overline{p}_z)\, \delta w\, dx = 0 \tag{7.64}$$

which, with $\widetilde{w} = \mathbf{N}_u \widehat{\mathbf{w}}$, becomes

$$\delta \widehat{\mathbf{w}}^T \left[\left(\int_0^L \mathbf{N}_u^T EI\, \mathbf{N}_u^{iv}\, dx \right) \widehat{\mathbf{w}} - \int_0^L \mathbf{N}_u^T\, \overline{p}_z\, dx \right] = 0 \tag{7.65a}$$

Thus,

$$\mathbf{k}_u \widehat{\mathbf{w}} = \mathbf{p}_u \tag{7.65b}$$

as is given in Eq. (7.59).

EXAMPLE 7.7 Beam with Linearly Varying Loading

Consider again the fixed-hinged beam of Fig. 7.1.

Use the single parameter polynomial approximation that was derived in Example 7.1 ($m = 1$)

$$\widetilde{w} = \mathbf{N}_u \widehat{\mathbf{w}} = (3\xi^2 - 5\xi^3 + 2\xi^4)\widehat{w}_1 \tag{1}$$

As noted in Example 7.1, this satisfies all of the boundary conditions. Galerkin's relationship of Eq. (7.65) in terms of ξ appears as

$$\delta \widehat{\mathbf{w}}^T \left[\underbrace{\left(\int_0^L EI\, \mathbf{N}_u^T \mathbf{N}_u^{iv}\, dx \right)}_{\mathbf{k}_u} \widehat{\mathbf{w}} - \underbrace{p_0 \int_0^L (1-\xi) \mathbf{N}_u^T\, dx}_{\mathbf{p}_u} \right] = 0 \tag{2}$$

Introducing $\mathbf{N}_u^{iv} = 48/L^4$.

$$\int_0^1 \delta\widehat{w}_1 \frac{48}{L^4} EI(3\xi^2 - 5\xi^3 + 2\xi^4)\widehat{w}_1 L\, d\xi - \int_0^1 \delta\widehat{w}_1\, p_0(1-\xi)(3\xi^2 - 5\xi^3 + 2\xi^4)L\, d\xi = 0 \tag{3}$$

leads to

$$\frac{48EI}{L^3} \cdot \frac{3}{20}\widehat{w}_1 = \frac{1}{15} p_0 L \quad \text{or} \quad \widehat{w}_1 = \frac{p_0 L^4}{108EI}$$

Finally, the approximate deflection is

$$\widetilde{w} = \frac{p_0 L^4}{108EI}(3\xi^2 - 5\xi^3 + 2\xi^4) \tag{4}$$

Let's continue this problem by using an assumed deflection with two terms ($m = 2$). Introduce

$$\widetilde{w} = \widehat{w}_1 \underbrace{(3\xi^2 - 5\xi^3 + 2\xi^4)}_{N_{u1}} + \widehat{w}_2 \underbrace{(\xi^2 + \xi^3 - 4\xi^4 + 2\xi^5)}_{N_{u2}} \tag{5}$$

Note that all boundary conditions are still satisfied. We proceed to establish Galerkin's condition (2) with $\mathbf{N}_u = [N_{u1} \;\; N_{u2}]$, $\widehat{\mathbf{w}} = [\widehat{w}_1 \;\; \widehat{w}_2]^T$. The matrices $\mathbf{k}_u$ and $\mathbf{p}_u$ are found to be

$$\mathbf{k}_u = \frac{EI}{L^4} \int_0^1 \underbrace{\begin{bmatrix} 3\xi^2 - 5\xi^3 + 2\xi^4 \\ \xi^2 + \xi^3 - 4\xi^4 + 2\xi^5 \end{bmatrix}}_{\mathbf{N}_u^T} \underbrace{[48 \quad -96 + 240\xi]}_{\mathbf{N}_u^{iv}} L\, d\xi$$

$$= \frac{EI}{L^3} \begin{bmatrix} 7.2 & 5.6 \\ 5.6 & 5.3714 \end{bmatrix} \tag{6}$$

$$\mathbf{p}_u = p_0 L \begin{bmatrix} 0.06 \\ 0.0476 \end{bmatrix} \tag{7}$$

Finally, $\mathbf{k}_u \widehat{\mathbf{w}} = \mathbf{p}_u$ gives

$$\widehat{\mathbf{w}} = \begin{bmatrix} \widehat{w}_1 \\ \widehat{w}_2 \end{bmatrix} = \begin{bmatrix} 0.0125 \\ -0.00417 \end{bmatrix} p_0 \frac{L^4}{EI} \tag{8}$$

From $\widetilde{w} = \mathbf{N}_u \widehat{\mathbf{w}}$, the approximate deflection would be

$$\begin{aligned}\widetilde{w} &= p_0 \frac{L^4}{EI}[0.0125(3\xi^2 - 5\xi^3 + 2\xi^4) - 0.00417(\xi^2 + \xi^3 - 4\xi^4 + 2\xi^5)] \\ &= p_0 \frac{L^4}{120EI}[4\xi^2 - 8\xi^3 + 5\xi^4 - \xi^5] \end{aligned} \tag{9}$$

Use of Hermitian polynomials would also permit an approximate $\widetilde{w}$ to be established which satisfies all boundary conditions as is required by the Galerkin procedure. If fifth degree Hermitian polynomials (Fig. 7.4) are used, the approximate deflection could be

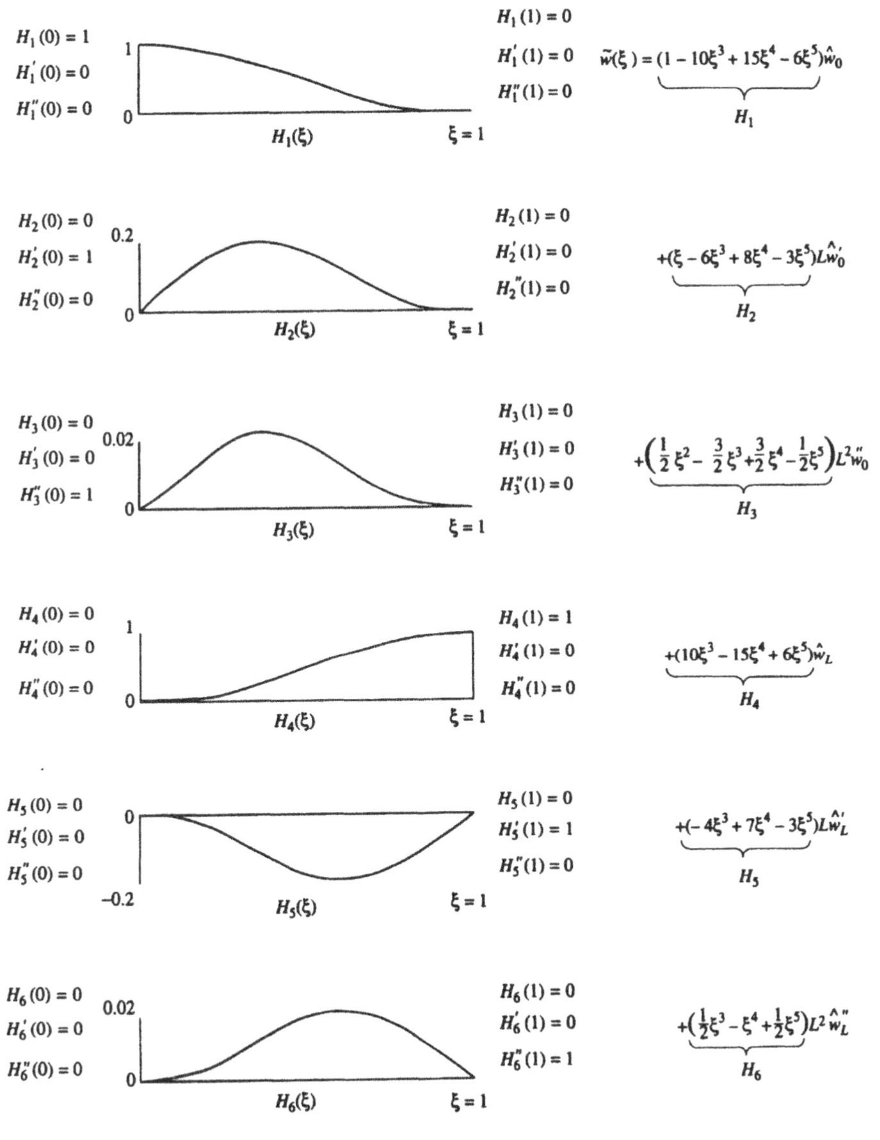

FIGURE 7.4
Fifth degree Hermitian polynomials, $\xi = x/L$.

expressed as

$$\widetilde{w}(x) = H_1\widehat{w}_0 + H_2\widetilde{w}'_0 + H_3\widehat{w}''_0 + H_4\widehat{w}_L + H_5\widehat{w}'_L + H_6\widehat{w}''_L = \mathbf{N}_u\widehat{\mathbf{w}} = \widehat{\mathbf{w}}^T\mathbf{N}_u^T \tag{10}$$

with $\widehat{w}_0 = \widetilde{w}|_{x=0}$, $\widehat{w}_L = \widetilde{w}|_{x=L}$, etc. Chapter 4, Section 4.4.2 may be helpful in understanding the use of Hermitian polynomials. Expressions for H_i^j as functions of ξ can be found in standard references or derived using the procedures of Chapter 4. Because of the boundary conditions of the beam (Fig. 7.1), $\widehat{w}_0 = \widehat{w}'_0 = \widehat{w}_L = \widehat{w}''_L = 0$. Thus,

$$\widetilde{w} = \mathbf{N}_u\widehat{\mathbf{w}} = H_3\widehat{w}''_0 + H_5\widehat{w}'_L = \widehat{\mathbf{w}}^T\mathbf{N}_u^T \tag{11}$$

with

$$\mathbf{N}_u = \left[\left(\frac{1}{2}\xi^2 - \frac{3}{2}\xi^3 + \frac{3}{2}\xi^4 - \frac{1}{2}\xi^5\right) \quad (-4\xi^3 + 7\xi^4 - 3\xi^5)\right]$$

$$\widehat{\mathbf{w}} = \begin{bmatrix} \widehat{w}''_0 \\ \widehat{w}'_L \end{bmatrix}$$

Insertion of (11) into (2) leads to the system of equations for $\widehat{\mathbf{w}}$,

$$\frac{EI}{L^3}\begin{bmatrix} 3 & 4 \\ 4 & 192 \end{bmatrix}\widehat{\mathbf{w}} = \begin{bmatrix} 1/6 \\ -8/6 \end{bmatrix} p_0 L \tag{12}$$

which has the solution

$$\widehat{\mathbf{w}} = \frac{p_0 L^4}{120EI}[8 \quad -1]^T \tag{13}$$

Substitution of this expression in (11) gives

$$\widetilde{w} = \frac{p_0 L^4}{120EI}(4\xi^2 - 8\xi^3 + 5\xi^4 - \xi^5) \tag{14}$$

The results can be compared with the exact solution, which has been derived several times in this work.

A. Exact solution:

$$\widetilde{w} = \frac{p_0 L^4}{120EI}(4\xi^2 - 8\xi^3 + 5\xi^4 - \xi^5) \tag{15}$$

B. Trial deflection with $m = 1$:

$$\widetilde{w} = \frac{p_0 L^4}{108EI}(3\xi^2 - 5\xi^3 + 2\xi^4) \tag{16}$$

C. Trial deflection with $m = 2$:

$$\widetilde{w} = \frac{p_0 L^4}{120EI}(4\xi^2 - 8\xi^3 + 5\xi^4 - \xi^5) \tag{17}$$

D. Trial deflection based on fifth degree Hermitian polynomials:

$$\widetilde{w} = \frac{p_0 L^4}{120EI}(4\xi^2 - 8\xi^3 + 5\xi^4 - \xi^5) \tag{18}$$

Clearly, this Hermitian polynomial leads to the exact deflection.

	Comparison of Galerkin's Solution			
	Deflection $\widetilde{w} = \beta \frac{p_0 L^2}{100EI}$	**Moment** $\widetilde{M} = \beta p_0 L^2$		**Shear Force** $\widetilde{V} = \beta p_0 L/10$
	β for $\xi = 0.5$	β for $\xi = 0$	β for $\xi = 0.5$	β for $\xi = 1$
Exact Soln. A	0.234375	−0.0666	0.029166	−1
Method B	0.231481	−0.05555	0.027777	−1.666
Methods C and D	0.234375	−0.0666	0.029166	−1

Note that for the $m = 1$ case, the error tends to grow in moving from the deflection to the moment to the shear force. ∎

7.4.3 Kantorovitch's Method

A variation of the Ritz method is the *Kantorovitch*[12] *method* or the *method of lines* which can be applied to reduce the partial differential equations of boundary value problems to the solution of lower dimensional (even one-dimensional) boundary value problems. In this one-dimensional case, the method involves the solution of ordinary differential equations.

The method still begins with the trial solution $\widetilde{u} = \sum_{i=1}^{m} N_{ui}\, \widehat{u}_i$, but now the $\widehat{u}_i$ are functions of one of the independent variables. In the case of a two-dimensional problem, a possible trial solution is

$$\widetilde{u}(x, y) = \sum_{i=1}^{m} N_{ui}(y)\, \widehat{u}_i(x) \tag{7.66}$$

The unknown $\widehat{u}_i(x)$ of Eq. (7.66) are determined so that a variational principle is satisfied. Instead of the system of linear algebraic equations of the Ritz method, we obtain a boundary value problem of a system of ordinary differential equations for the unknown function $\widehat{u}_i(x)$. In setting up the problem, it is not necessary to choose N_{ui} to satisfy the x boundary conditions, since they may be prescribed as boundary conditions for the functions $\widehat{u}_i$.

EXAMPLE 7.8 Kantorovitch's Method

Consider the torsion of a prismatic bar with the Prandtl stress function defined as

$$\tau_{xy} = G\phi' \frac{\partial \psi}{\partial z} \quad \text{and} \quad \tau_{xz} = -G\phi' \frac{\partial \psi}{\partial y} \tag{1}$$

Note that the definition of ψ differs from that in Chapter 1, Eq. (1.155) in that the constant $G\phi'$ is included. This definition will simplify the compatibility condition. Equation (1.159)

[12]Leonid Vitaljevich Kantorovitch (1912–1986) was born in Russia, educated in Leningrad, and worked in Siberia. He entered Leningrad University in the school of mechanics and mathematics at the age of 14 and was appointed as a full professor while still a teenager. By the age of 18, when he received a degree, he had published 11 scientific papers. In 1939 he published "Mathematical Methods of Organization and Planning of Production" which outlined the simplex method of linear programming, an important technique that until recently was thought to have been developed in the West several years later. The hot and cold war stifled the appreciation of much of his work. His achievements in functional analysis and approximate methods are widely recognized. Probably his contributions to the foundations of modern economic theory are his most important accomplishments. He was awarded a Nobel Prize in 1975.

of Chapter 1 becomes

$$\frac{\partial^2 \psi}{\partial y^2} + \frac{\partial^2 \psi}{\partial z^2} = -2 \quad \text{with } \psi = 0 \text{ on the boundary} \tag{2}$$

Use the Kantorovitch method to find the approximate solution for ψ for a rectangular cross-section of $|y| \leq 2$, $|z| \leq 1$. Choose

$$\widetilde{\psi} = (1 - z^2)\widehat{\psi}_1(y) \tag{3}$$

which satisfies the boundary condition for $z = \pm 1$. Use a functional of the form of Eq. (7.55),

$$\Pi = \int_A \left[\psi \left(\frac{\partial^2 \psi}{\partial y^2} + \frac{\partial^2 \psi}{\partial z^2} \right) + 4\psi \right] dy\, dz \tag{4}$$

From Green's theorem [Appendix II, Eq. (II.3)], which has the form $\int_A (-\partial R/\partial z + \partial T/\partial y) dA = \oint_S (-Ra_z + Ta_y) dS$, where S is the boundary of the cross-section, and a_z, a_y are the direction cosines of the boundary, the first term on the right-hand side of (4) can be written as

$$\int_A \left[\psi \left(\frac{\partial^2 \psi}{\partial y^2} + \frac{\partial^2 \psi}{\partial z^2} \right) \right] dy\, dz = \int_A \left[\frac{\partial}{\partial y} \left(\psi \frac{\partial \psi}{\partial y} \right) - \frac{\partial}{\partial z} \left(-\psi \frac{\partial \psi}{\partial z} \right) \right] dy\, dz - \int_A \left[\left(\frac{\partial \psi}{\partial y} \right)^2 + \left(\frac{\partial \psi}{\partial z} \right)^2 \right] dy\, dz \tag{5}$$

In our problem, $T = \psi \frac{\partial \psi}{\partial y}$ and $R = -\psi \frac{\partial \psi}{\partial z}$, and since $\psi = 0$ at the boundary, it follows that

$$\int_A \left[\frac{\partial}{\partial y} \left(\psi \frac{\partial \psi}{\partial y} \right) - \frac{\partial}{\partial z} \left(-\psi \frac{\partial \psi}{\partial z} \right) \right] dy\, dz = \int_S \psi \left[\frac{\partial \psi}{\partial y} a_y + \frac{\partial \psi}{\partial z} a_z \right] dy\, dz = 0 \tag{6}$$

Thus, Eq. (4) becomes

$$\Pi = -\int_A \left[\left(\frac{\partial \psi}{\partial y} \right)^2 + \left(\frac{\partial \psi}{\partial z} \right)^2 - 4\psi \right] dz\, dy \tag{7}$$

Substitution of (3) into (7) leads to

$$\Pi = -\int_{-2}^{2} \int_{-1}^{1} [(1 - z^2)^2 \, (\widehat{\psi}_1')^2 + 4z^2 \widehat{\psi}_1^2 - 4\widehat{\psi}_1 (1 - z^2)] \, dz\, dy \tag{8}$$

where the prime indicates a derivative with respect to y.

Integrate with respect to z to obtain

$$\Pi = -\int_{-2}^{2} \left[\frac{16}{15} (\widehat{\psi}_1')^2 + \frac{8}{3} \widehat{\psi}_1^2 - \frac{16}{3} \widehat{\psi}_1 \right] dy \tag{9}$$

By comparison of (9) with Appendix I, Eq. (I.1),

$$F(x, u, u') = \frac{16}{15} (\widehat{\psi}_1')^2 + \frac{8}{3} \widehat{\psi}_1^2 - \frac{16}{3} \widehat{\psi}_1 \tag{10}$$

with $u = \widehat{\psi}_1$ and $x = y$. The Euler equation can be obtained from Appendix I, Eq. (I.13) as

$$\frac{\partial F}{\partial \widehat{\psi}_1} - \frac{d}{dy} \frac{\partial F}{\partial \widehat{\psi}_1'} = 0 \tag{11}$$

i.e., the Euler equation for (9) is

$$\widehat{\psi}_1'' - \frac{5}{2}\widehat{\psi}_1 = -\frac{5}{2} \tag{12}$$

with boundary conditions $\widehat{\psi}_1 = 0$ at $y = \pm 2$. The solution to (12) is

$$\widehat{\psi}_1(y) = \frac{1}{2}\left(1 - \frac{\sinh\sqrt{10}}{\sinh\sqrt{40}}\sinh\sqrt{5/2}\,y\right) \tag{13}$$

and the final approximate solution is given by (3) with $\widehat{\psi}_1$ of (13). ∎

7.4.4 Extended Methods

Extended Ritz's Method

Since Ritz's method can be based on the principle of virtual work, it is required that the trial functions satisfy the displacement boundary conditions, i.e., $\mathbf{u} = \overline{\mathbf{u}}$ on S_u. It is possible to extend this principle and, hence, the method, to relax the conditions to be satisfied by the trial functions. Then this extended Ritz's method provides a solution that fulfills the displacement boundary conditions approximately, as well as the conditions of equilibrium and the static boundary conditions.

In order to change the formulation such that the trial solution need not satisfy the displacement boundary conditions, we will supplement the virtual work expressions with the global form of $\mathbf{u} - \overline{\mathbf{u}} = 0$ on S_u. Thus, we will add the integral

$$\Delta W = \int_{S_u} \mathbf{p}^T(\mathbf{u} - \overline{\mathbf{u}})\, dS \tag{7.67}$$

to the work expression. Then for a continuum, the virtual work of Eq. (7.43) can be extended to

$$-\delta W = \int_V \delta\epsilon^T\,\sigma\, dV - \int_V \delta\mathbf{u}^T\overline{\mathbf{p}}_V\, dV - \int_{S_p} \delta\mathbf{u}^T\overline{\mathbf{p}}\, dS - \delta\int_{S_u} \mathbf{p}^T(\mathbf{u} - \overline{\mathbf{u}})\, dS = 0 \tag{7.68a}$$

and for a beam, Eq. (7.44) would be adjusted to become

$$\begin{aligned}-\delta W = &\int_0^L EIw''\delta w''dx - \int_0^L \overline{p}_z\,\delta w\, dx - \underset{\text{on } S_p}{[\overline{V}\,\delta w + \overline{M}\,\delta\theta]_0^L}\\ &- \underset{\text{on } S_u}{\delta[V(w - \overline{w}) + M(\theta - \overline{\theta})]_0^L} = 0\end{aligned} \tag{7.68b}$$

We will convert the beam expression into a form expressed in terms of displacements only. For a uniform beam, set $\theta = -w'$, $M = -EI\,w''$, and $V = -EI\,w'''$. Also, set all applied displacements $(\overline{w}, \overline{\theta})$ and forces $(\overline{V}, \overline{M})$ on the boundaries equal to zero. This reduces the variational principle to the form

$$\begin{aligned}-\delta W = &\int_0^L EI\,w''(x)\,\delta w''(x)\,dx - \int_0^L \overline{p}_z(x)\,\delta w(x)\,dx + [\delta w'''(x)\,EI\,w(x) + w'''(x)\,EI\,\delta w(x)\\ &- \delta w''(x)\,EI\,w'(x) - w''(x)\,EI\,\delta w'(x)]_0^L = 0\end{aligned} \tag{7.69}$$

The terms in brackets are the virtual work expressions

$$[-\delta V\, w - V\,\delta w - \delta M\,\theta - M\,\delta\theta]_0^L$$

on that portion of the boundary where the displacement boundary conditions occur, i.e., on S_u. Introduce the trial solution $\widetilde{w}(x) = \mathbf{N}_u\widehat{\mathbf{w}}$, and note that

$$\delta\widetilde{w} = \mathbf{N}_u\,\delta\widehat{\mathbf{w}} = \delta\widehat{\mathbf{w}}^T\mathbf{N}_u^T$$

Then

$$-\delta W = \delta\widehat{\mathbf{w}}^T\left\{EI\left[\int_0^L \mathbf{N}_u''^T\,\mathbf{N}_u''\,dx + \left[\mathbf{N}_u'''^T\,\mathbf{N}_u + \mathbf{N}_u^T\,\mathbf{N}_u''' - \mathbf{N}_u''^T\,\mathbf{N}_u' - \mathbf{N}_u'^T\,\mathbf{N}_u''\right]_{0\ \text{on } S_u}^{L}\right]\widehat{\mathbf{w}}\right.$$
$$\left. - \int_0^L \mathbf{N}_u^T\,\overline{p}_z\,dx\right\} = 0$$

Let

$$\mathbf{R} = EI\left[\mathbf{N}_u^T\,\mathbf{N}_u''' - \mathbf{N}_u'^T\,\mathbf{N}_u''\right]_0^L \tag{7.70}$$

which corresponds to the term $[-V\delta w - M\delta\theta]_0^L$ on S_u. Then

$$\delta W = \delta\widehat{\mathbf{w}}^T\left\{\left[\underbrace{EI\int_0^L \mathbf{N}_u''^T\,\mathbf{N}_u''\,dx}_{\mathbf{k}_u} + (\mathbf{R}+\mathbf{R}^T)\right]\widehat{\mathbf{w}} - \underbrace{\int_0^L \mathbf{N}_u^T\,\overline{p}_z\,dx}_{\mathbf{p}_u}\right\} = 0 \tag{7.71a}$$

Thus, the free parameters $\widehat{\mathbf{w}}$ can be found from

$$[\mathbf{k}_u + (\mathbf{R}+\mathbf{R}^T)]\,\widehat{\mathbf{w}} - \mathbf{p}_u = 0 \tag{7.71b}$$

EXAMPLE 7.9 Beam with Linearly Varying Loading

Consider again the beam of Fig. 7.1. This time we will employ a trial function that does not necessarily satisfy the displacement boundary conditions.

We begin with a polynomial trial solution in the form

$$\widetilde{w}(x) = \mathbf{N}_u\widehat{\mathbf{w}} = [\xi \quad \xi^2 \quad \xi^3 \quad \xi^4]\begin{bmatrix}\widehat{w}_1\\ \widehat{w}_2\\ \widehat{w}_3\\ \widehat{w}_4\end{bmatrix} \tag{1}$$

Although it is not necessary that any boundary conditions be satisfied, note that the condition $w(0) = 0$ is fulfilled by (1). With $d\xi = dx/L$ the needed derivatives become

$$\begin{aligned}\mathbf{N}_u' &= \frac{1}{L}[1 \quad 2\xi \quad 3\xi^2 \quad 4\xi^3]\\ \mathbf{N}_u'' &= \frac{1}{L^2}[0 \quad 2 \quad 6\xi \quad 12\xi^2]\\ \mathbf{N}_u''' &= \frac{1}{L^3}[0 \quad 0 \quad 6 \quad 24\xi]\end{aligned} \tag{2}$$

which are obtained using

$$\mathbf{N}_u' = \frac{d\mathbf{N}_u}{dx} = \frac{d\mathbf{N}_u}{d\xi}\frac{d\xi}{dx}$$

Substitute these relationships into Eq. (7.71a)

$$\mathbf{k}_u = \frac{EI}{L^3}\int_0^1 \left\{ \begin{bmatrix} 0 \\ 2 \\ 6\xi \\ 12\xi^2 \end{bmatrix} [0 \quad 2 \quad 6\xi \quad 12\xi^2] \right\} d\xi$$

$$= \frac{EI}{L^3}\begin{bmatrix} 0 & 0 & 0 & 0 \\ 0 & 4 & 6 & 8 \\ 0 & 6 & 12 & 18 \\ 0 & 8 & 18 & \frac{144}{5} \end{bmatrix} \tag{3}$$

$$\mathbf{p}_u = p_0 L \int_0^1 \begin{bmatrix} \xi \\ \xi^2 \\ \xi^3 \\ \xi^4 \end{bmatrix} (1-\xi)\, d\xi = p_0 L \begin{bmatrix} 1/6 \\ 1/12 \\ 1/20 \\ 1/30 \end{bmatrix} \tag{4}$$

Turn now to the boundary relation **R** of Eq. (7.70). Recall that the terms in **R** are taken from $[-V\,\delta w - M\,\delta\theta]_0^L$ on S_u. The S_u boundary conditions are $w(0) = 0$, $\theta(0) = 0$, and $w(L) = 0$, and, hence, only those terms corresponding to $\delta w(0)$, $\delta\theta(0)$, and $\delta w(L)$ are retained in **R**. Thus, **R** becomes

$$\mathbf{R} = EI\left[\mathbf{N}_u^T(L)\,\mathbf{N}_u'''(L) - \mathbf{N}_u^T(0)\mathbf{N}_u'''(0) + \mathbf{N}_u'^T(0)\,\mathbf{N}_u''(0)\right]$$

$$= \frac{EI}{L^3}\left\{ \begin{bmatrix} 1 \\ 1 \\ 1 \\ 1 \end{bmatrix} [0 \quad 0 \quad 6 \quad 24] - \begin{bmatrix} 0 \\ 0 \\ 0 \\ 0 \end{bmatrix} [0 \quad 0 \quad 6 \quad 0] + \begin{bmatrix} 1 \\ 0 \\ 0 \\ 0 \end{bmatrix} [0 \quad 2 \quad 0 \quad 0] \right\}$$

$$= \frac{EI}{L^3}\begin{bmatrix} 0 & 2 & 6 & 24 \\ 0 & 0 & 6 & 24 \\ 0 & 0 & 6 & 24 \\ 0 & 0 & 6 & 24 \end{bmatrix} \tag{5}$$

The system of equations for $\widehat{\mathbf{w}}$, i.e., $[\mathbf{k}_u + (\mathbf{R} + \mathbf{R}^T)]\widehat{\mathbf{w}} - \overline{\mathbf{p}}_u = 0$, becomes

$$\underbrace{\frac{EI}{L^3}\begin{bmatrix} 0 & 2 & 6 & 24 \\ 2 & 4 & 12 & 32 \\ 6 & 12 & 24 & 48 \\ 24 & 32 & 48 & 76.8 \end{bmatrix}}_{\mathbf{k}} \underbrace{\begin{bmatrix} \widehat{w}_1 \\ \widehat{w}_2 \\ \widehat{w}_3 \\ \widehat{w}_4 \end{bmatrix}}_{\widehat{\mathbf{w}}} = \underbrace{\begin{bmatrix} 1/6 \\ 1/12 \\ 1/20 \\ 1/30 \end{bmatrix} p_0 L}_{\mathbf{p}_u} \tag{6}$$

which has the solution

$$\widehat{\mathbf{w}} = \frac{p_0 L^4}{120EI}\begin{bmatrix} -0.4167 \\ 4.0000 \\ -5.2916 \\ 1.8229 \end{bmatrix} \tag{7}$$

As a second case, use a trial function with five unknown parameters

$$\widetilde{w}(x) = [1 \quad \xi \quad \xi^2 \quad \xi^3 \quad \xi^4]\,[\widehat{w}_1 \quad \widehat{w}_2 \quad \widehat{w}_3 \quad \widehat{w}_4 \quad \widehat{w}_5]^T \tag{8}$$

In contrast to the trial function of (1), here, none of the displacement boundary conditions are satisfied. We find

$$\mathbf{N}_u' = \frac{1}{L}[0 \quad 1 \quad 2\xi \quad 3\xi^2 \quad 4\xi^3]$$
$$\mathbf{N}_u'' = \frac{1}{L^2}[0 \quad 0 \quad 2 \quad 6\xi \quad 12\xi^2] \tag{9}$$
$$\mathbf{N}_u''' = \frac{1}{L^3}[0 \quad 0 \quad 0 \quad 6 \quad 24\xi]$$

$$\mathbf{k}_u = \frac{EI}{L^3}\int_0^1 \begin{bmatrix} 0 \\ 0 \\ 2 \\ 6\xi \\ 12\xi^2 \end{bmatrix} [0 \quad 0 \quad 2 \quad 6\xi \quad 12\xi^2]\, d\xi = \frac{EI}{L^3}\begin{bmatrix} 0 & 0 & 0 & 0 & 0 \\ 0 & 0 & 0 & 0 & 0 \\ 0 & 0 & 4 & 6 & 8 \\ 0 & 0 & 6 & 12 & 18 \\ 0 & 0 & 8 & 18 & \frac{144}{5} \end{bmatrix} \tag{10}$$

Completion of the $\mathbf{R}$ and $\mathbf{p}_u$ matrices leads to the equations

$$\frac{EI}{L^3}\begin{bmatrix} 0 & 0 & 0 & 0 & 24 \\ 0 & 0 & 2 & 6 & 24 \\ 0 & 2 & 4 & 12 & 32 \\ 0 & 6 & 12 & 24 & 48 \\ 24 & 24 & 32 & 48 & 76.8 \end{bmatrix}\begin{bmatrix} \widehat{w}_1 \\ \widehat{w}_2 \\ \widehat{w}_3 \\ \widehat{w}_4 \\ \widehat{w}_5 \end{bmatrix} = \begin{bmatrix} 1/2 \\ 1/6 \\ 1/12 \\ 1/20 \\ 1/30 \end{bmatrix} p_0 L \tag{11}$$

with the solution

$$\widehat{\mathbf{w}} = \frac{p_0 L^4}{120EI}[-2.17 \quad 5.00 \quad 4.00 \quad -8.00 \quad 2.50]^T \tag{12}$$

As a third case, use of a trial function with six free parameters

$$\widetilde{w}(x) = [1 \quad \xi \quad \xi^2 \quad \xi^3 \quad \xi^4 \quad \xi^5]\,[\widehat{w}_1 \quad \widehat{w}_2 \quad \widehat{w}_3 \quad \widehat{w}_4 \quad \widehat{w}_5 \quad \widehat{w}_6]^T \tag{13}$$

leads to the generalized displacements

$$\widehat{\mathbf{w}} = \frac{p_0 L^4}{120EI}[0 \quad 0 \quad 4 \quad -8 \quad 5 \quad -1]^T \tag{14}$$

In summary, the results are

A. Exact solution is equal to D.

B. First Trial Deflection:

$$\widetilde{w}(\xi) = \frac{p_0 L^4}{120EI}(-0.4167\xi + 4\xi^2 - 5.2916\xi^3 + 1.8229\xi^4) \tag{15}$$

C. Second Trial Deflection:

$$\widetilde{w}(\xi) = \frac{p_0 L^4}{120EI}(-2.17 + 5\xi + 4\xi^2 - 8\xi^3 + 2.5\xi^4) \tag{16}$$

D. Third Trial Deflection:

$$\widetilde{w}(\xi) = \frac{p_0 L^4}{120EI}(4\xi^2 - 8\xi^3 + 5\xi^4 - \xi^5) \tag{17}$$

For each case, the values of the displacements at the boundaries are

A. The same as D.

B.
$$\widetilde{w}(0) = 0 \qquad \text{(satisfied by the trial solution)}$$
$$\widetilde{w}'(0) = -0.0035\frac{p_0 L^3}{EI} \neq 0, \tag{18}$$
$$\widetilde{w}(1) = 0.000955\frac{p_0 L^4}{EI} \neq 0,$$

C.
$$\widetilde{w}(0) = -0.01805\frac{p_0 L^4}{EI} \neq 0,$$
$$\widetilde{w}'(0) = 0.0416\frac{p_0 L^3}{EI} \neq 0, \tag{19}$$
$$\widetilde{w}(1) = 0.01\frac{p_0 L^4}{EI} \neq 0,$$

D.
$$\widetilde{w}(0) = 0, \quad \widetilde{w}'(0) = 0, \quad \widetilde{w}(1) = 0 \tag{20}$$

The displacement, moment, and shear force are compared in Fig. 7.5. Note that case B, which employed a trial function that satisfied one of the displacement boundary conditions, led to better results than case C, whose trial function satisfied no displacement boundary conditions. The higher order polynomial of case D was the best of all, since it gave the exact solution. ∎

Extended Galerkin's Method

Recall from Section 7.4.2 that Galerkin's method requires that the trial solution satisfy all boundary conditions, i.e., both displacement and force boundary conditions. These constraints can be relaxed by including these conditions in the global representation of the fundamental equations. We begin with the global form of the equations of equilibrium, along with the displacement and force boundary conditions. For a continuum

$$-\int_V \delta \mathbf{u}^T(\mathbf{D}^T\boldsymbol{\sigma} + \overline{\mathbf{p}}_V)\, dV + \int_{S_p} \delta \mathbf{u}^T(\mathbf{p} - \overline{\mathbf{p}})\, dS - \int_{S_u} \delta \mathbf{p}^T(\mathbf{u} - \overline{\mathbf{u}})\, dS = 0 \tag{7.72}$$

and for a beam

$$\int_0^L \delta w(x)[EIw^{iv}(x) - \overline{p}_z(x)]\, dx + [\delta w(V - \overline{V}) + \delta\theta(M - \overline{M})]_0^L{}_{\text{on } S_p}$$
$$- [\delta M(\theta - \overline{\theta}) + \delta V(w - \overline{w})]_0^L{}_{\text{on } S_u} = 0 \tag{7.73}$$

Assume that all applied boundary forces and displacements are zero, i.e., $\overline{V} = \overline{M} = \overline{\theta} = \overline{w} = 0$. Also, since

$$\theta = -\frac{dw}{dx} \qquad M = -EIw'' \qquad V = -EIw'''$$

Eq. (7.73) can be written as

$$\int_0^L \delta w(x)[EIw^{iv}(x)]\, dx - \int_0^L \delta w(x)\, \overline{p}_z(x)\, dx + [\delta w(-EIw'''(x)) + (-\delta w')(-EIw'')]_0^L{}_{\text{on } S_p}$$
$$- [(-w')(-EI\,\delta w'') + w(-EI\,\delta w''')]_0^L{}_{\text{on } S_u} = 0 \tag{7.74}$$

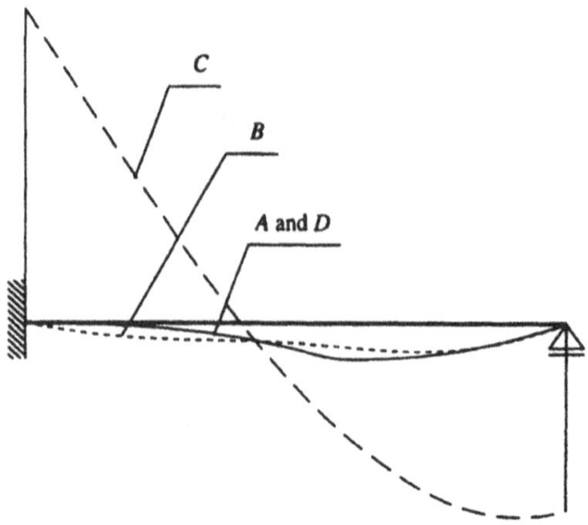

	$w_m EI/p_0 L^4$	Location ξ_0
B	0.002395	0.7257
C	0.011011	0.8000
A,D	0.002385	0.5528

w_m = maximum deflection
ξ_0 = location of w_m

(a) Deflection

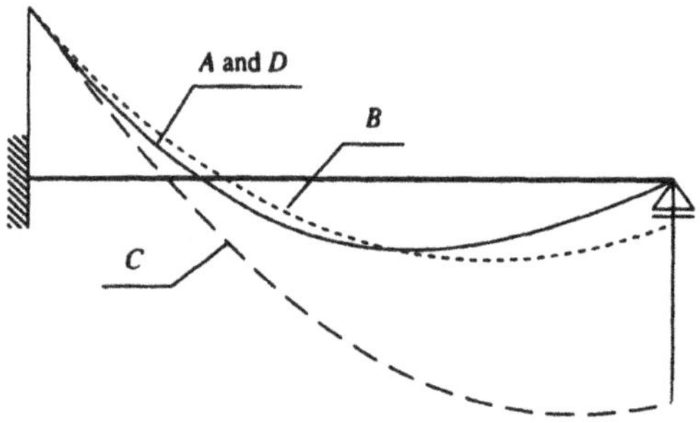

$\frac{M}{p_0 L^2}$	Location $\xi = 0$	$\xi = \xi_0$	$\xi = 1$
B	$-\frac{1}{15}$	$\frac{1}{33.08}$	$\frac{1}{64}$
C	$-\frac{1}{15}$	$\frac{1}{10.71}$	$\frac{1}{12}$
A,D	$-\frac{1}{15}$	$\frac{1}{33.54}$	0

(b) Bending moment

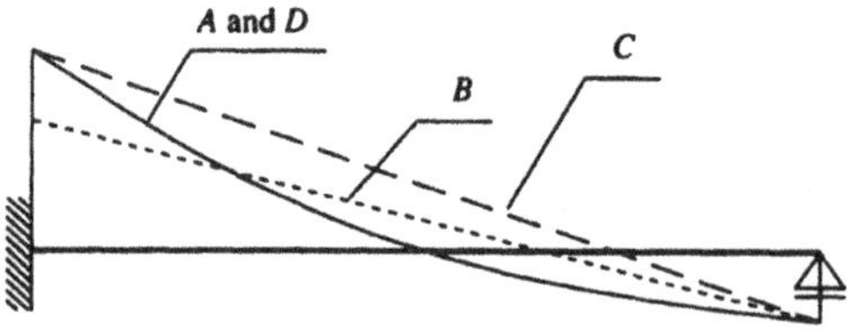

$\frac{V}{p_0 L}$	Location $\xi = 0$	$\xi = 1$
B	0.2646	- 0.1
C	0.4	- 0.1
A,D	0.4	- 0.1

(c) Shear force

FIGURE 7.5
Extended Ritz method for a fixed-hinged beam with linearly varying loading.

Introduce the trial functions $\widetilde{w} = \mathbf{N}_u \widehat{\mathbf{w}}$

$$\delta \widehat{\mathbf{w}}^T \Bigg\{ \Bigg[\underbrace{\int_0^L \mathbf{N}_u^T EI \mathbf{N}_u^{iv}\, dx}_{k_u} + \underbrace{EI \left[-\mathbf{N}_u^T \mathbf{N}_u''' + \mathbf{N}_u'^T \mathbf{N}_u'' \right]_0^L}_{\mathbf{R}_p \ \text{on } S_p}$$

$$+ \underbrace{EI \left[-\mathbf{N}_u''^T \mathbf{N}_u' + \mathbf{N}_u'''^T \mathbf{N}_u \right]_0^L}_{\mathbf{R}_u \ \text{on } S_u} \Bigg] \widehat{\mathbf{w}} - \underbrace{\int_0^L \mathbf{N}_u^T \overline{p}_z\, dx}_{\mathbf{p}_u} \Bigg\} = 0 \tag{7.75}$$

For a particular case, retain only those terms corresponding to the actual boundary conditions for the problem. We see that the free parameters $\widehat{\mathbf{w}}$ can be found from

$$[\mathbf{k}_u + (\mathbf{R}_p + \mathbf{R}_u)]\widehat{\mathbf{w}} = \mathbf{p}_u \tag{7.76}$$

EXAMPLE 7.10 Beam with Linearly Varying Loading

To illustrate the use of a trial function which does not satisfy all boundary conditions, we return again to the beam of Fig. 7.1. As the first choice of a trial function, use

$$w = \mathbf{N}_u\,\widehat{\mathbf{w}} = [\xi \quad \xi^2 \quad \xi^3 \quad \xi^4]\begin{bmatrix}\widehat{w}_1\\ \widehat{w}_2\\ \widehat{w}_3\\ \widehat{w}_4\end{bmatrix} \tag{1}$$

This approximate displacement satisfies only the boundary condition $w(0) = 0$, but not the conditions $\theta(0) = w(L) = M(L) = 0$. Substitute Eq. (2) of Example 7.9, along with $\mathbf{N}_u^{iv} = (1/L^4)\,[0\ \ 0\ \ 0\ \ 24]$, into Eq. (7.75)

$$\mathbf{k}_u = \frac{EI}{L^4}\int_0^1 L\begin{bmatrix}\xi\\ \xi^2\\ \xi^3\\ \xi^4\end{bmatrix}[0 \quad 0 \quad 0 \quad 24]\,d\xi = \frac{EI}{L^4}\begin{bmatrix}0&0&0&12\\0&0&0&8\\0&0&0&6\\0&0&0&24/5\end{bmatrix} \tag{2}$$

$$\mathbf{p}_u = p_0 L\int_0^1 \begin{bmatrix}\xi\\ \xi^2\\ \xi^3\\ \xi^4\end{bmatrix}(1-\xi)\,d\xi = p_0 L\begin{bmatrix}1/6\\1/12\\1/20\\1/30\end{bmatrix} \tag{3}$$

We still need to compute the boundary terms $\mathbf{R}_u$ and $\mathbf{R}_p$. The formation of these conditions is illustrated in Fig. 7.6. We find

$$\mathbf{R}_u = \big[\underbrace{\mathbf{N}_u'''^T(1)\,\mathbf{N}_u(1)}_{S_u \text{ at } \xi=1}\ \underbrace{+\mathbf{N}_u''^T(0)\,\mathbf{N}_u'(0) - \mathbf{N}_u'''^T(0)\,\mathbf{N}_u(0)}_{S_u \text{ at } \xi=0}\big]\,EI \tag{4}$$

$$\mathbf{R}_p = \underbrace{\big[\mathbf{N}_u'^T(1)\,\mathbf{N}_u''(1)\big]EI}_{S_p \text{ at } \xi=1} \tag{5}$$

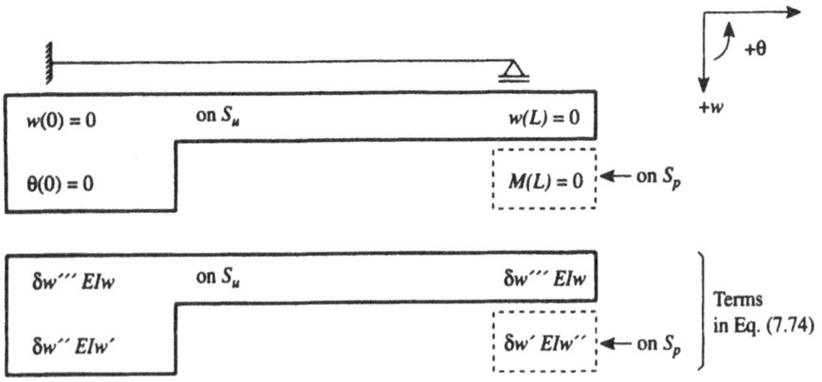

FIGURE 7.6
Boundary displacements and forces for the beam of Fig. 7.1 and corresponding boundary terms for the extended variational form.

Numerical expressions for $\mathbf{R}_u$ and $\mathbf{R}_p$ are computed as

$$\mathbf{R}_u = \frac{EI}{L^3}\left[\begin{bmatrix}0\\0\\6\\24\end{bmatrix}[1\ \ 1\ \ 1\ \ 1] + \begin{bmatrix}0\\2\\0\\0\end{bmatrix}[1\ \ 0\ \ 0\ \ 0] - \begin{bmatrix}0\\0\\6\\0\end{bmatrix}[0\ \ 0\ \ 0\ \ 0]\right]$$

$$= \frac{EI}{L^3}\left[\begin{bmatrix}0&0&0&0\\0&0&0&0\\6&6&6&6\\24&24&24&24\end{bmatrix} + \begin{bmatrix}0&0&0&0\\2&0&0&0\\0&0&0&0\\0&0&0&0\end{bmatrix} - \mathbf{0}\right] = \frac{EI}{L^3}\begin{bmatrix}0&0&0&0\\2&0&0&0\\6&6&6&6\\24&24&24&24\end{bmatrix} \tag{6}$$

$$\mathbf{R}_p = \frac{EI}{L^3}\begin{bmatrix}1\\2\\3\\4\end{bmatrix}[0\ \ 2\ \ 6\ \ 12] = \frac{EI}{L^3}\begin{bmatrix}0&2&6&12\\0&4&12&24\\0&6&18&36\\0&8&24&48\end{bmatrix}$$

Thus, the matrix on the left-hand side of Eq. (7.76) is

$$\mathbf{k} = \mathbf{k}_u + \mathbf{R}_u + \mathbf{R}_p = \frac{EI}{L^3}\begin{bmatrix}0&2&6&24\\2&4&12&32\\6&12&24&48\\24&32&48&76.8\end{bmatrix} \tag{7}$$

We find that the $\widehat{\mathbf{w}}$ are the same as obtained in Example 7.9 for the same trial function i.e., case B.

As a second case, we choose to use the trial function of case C of Example 7.9. Here,

$$\widetilde{w} = [1\ \ \xi\ \ \xi^2\ \ \xi^3\ \ \xi^4][\widehat{w}_1\ \ \widehat{w}_2\ \ \widehat{w}_3\ \ \widehat{w}_4\ \ \widehat{w}_5]^T \tag{8}$$

which satisfies none of the boundary conditions. We find

$$\begin{aligned}\mathbf{N}_u &= [1\ \ \xi\ \ \xi^2\ \ \xi^3\ \ \xi^4]\\ \mathbf{N}_u' &= \frac{1}{L}[0\ \ 1\ \ 2\xi\ \ 3\xi^2\ \ 4\xi^3]\\ \mathbf{N}_u'' &= \frac{1}{L^2}[0\ \ 0\ \ 2\ \ 6\xi\ \ 12\xi^2]\\ \mathbf{N}_u''' &= \frac{1}{L^3}[0\ \ 0\ \ 0\ \ 6\ \ 24\xi]\\ \mathbf{N}_u^{iv} &= \frac{1}{L^4}[0\ \ 0\ \ 0\ \ 0\ \ 24]\end{aligned} \tag{9}$$

Then compute

$$\mathbf{k}_u = \frac{EI}{L^3}\int_0^1 \begin{bmatrix}1\\\xi\\\xi^2\\\xi^3\\\xi^4\end{bmatrix}[0\ \ 0\ \ 0\ \ 0\ \ 24]\,d\xi = \frac{EI}{L^3}\begin{bmatrix}0&0&0&0&24\\0&0&0&0&12\\0&0&0&0&8\\0&0&0&0&6\\0&0&0&0&24/5\end{bmatrix} \tag{10}$$

$$\mathbf{p}_u = p_0 L\int_0^1 \begin{bmatrix}1\\\xi\\\xi^2\\\xi^3\\\xi^4\end{bmatrix}(1-\xi)\,d\xi = \begin{bmatrix}1/2\\1/6\\1/12\\1/20\\1/30\end{bmatrix} p_0 L \tag{11}$$

$$\mathbf{R}_u + \mathbf{R}_p = \frac{EI}{L^3}\left[\begin{bmatrix}0\\0\\0\\6\\24\end{bmatrix}[1\ 1\ 1\ 1\ 1] + \begin{bmatrix}0\\0\\2\\0\\0\end{bmatrix}[0\ 1\ 0\ 0\ 0] - \begin{bmatrix}0\\0\\0\\6\\0\end{bmatrix}[1\ 0\ 0\ 0\ 0]\right.$$

$$\left. + \begin{bmatrix}0\\1\\2\\3\\4\end{bmatrix}[0\ 0\ 2\ 6\ 12]\right] = \frac{EI}{L^3}\begin{bmatrix}0 & 0 & 0 & 0 & 0\\ 0 & 0 & 2 & 6 & 12\\ 0 & 2 & 4 & 12 & 24\\ 0 & 6 & 12 & 24 & 42\\ 24 & 24 & 32 & 48 & 72\end{bmatrix}$$

This leads to the same system of equations as obtained for case C of Example 7.9.

For the third and final trial function, consider that of case D of Example 7.9. As might be expected, this leads to the same results found for Ritz's method in Example 7.9. ∎

It is of interest that the extended Ritz and Galerkin methods gave the same results for the same trial solutions for the beam in Examples 7.9 and 7.10. This is not surprising since both of these variational methods are set up such that the same restrictions, i.e., no boundary conditions need to be satisfied, are imposed on their trial solutions and both involve the global form of the equilibrium equations. Also, for a beam, it is readily shown that

$$[\mathbf{k}_u + (\mathbf{R} + \mathbf{R}^T)]_{\text{Ritz}} = [\mathbf{k}_u + (\mathbf{R}_p + \mathbf{R}_u)]_{\text{Galerkin}} \tag{7.77}$$

To do so, integrate $\mathbf{k}_{u\ \text{Ritz}}$ twice by parts. Thus

$$\begin{aligned}\mathbf{k}_{u\ \text{Ritz}} &= EI\int_0^L \mathbf{N}_u''^T \mathbf{N}_u''\, dx\\ &= EI\int_0^L \mathbf{N}_u^T \mathbf{N}_u^{iv}\, dx + EI\left[\mathbf{N}_u'^T\, \mathbf{N}_u'' - \mathbf{N}_u^T\, \mathbf{N}_u'''\right]_0^L\end{aligned} \tag{7.78}$$

Adding this to $\mathbf{R} + \mathbf{R}^T$, with $\mathbf{R}$ from Eq. (7.70), gives $\mathbf{k}_u + \mathbf{R}_p + \mathbf{R}_u$, where $\mathbf{k}_u$, $\mathbf{R}_p$, and $\mathbf{R}_u$ are given by Eq. (7.75).

In general, the nonextended Ritz and Galerkin methods can be expected to give different results.

7.4.5 Trefftz's Method: A Boundary Method

The classical trial function methods and the finite element method, which can be treated as an extension of the classical trial function methods wherein trial solutions apply to elements into which the system has been divided, sometimes encounter difficulties when the domain (volume) is extremely large or when singularities occur in some of the variables (Fig. 7.7). Frequently, exact solutions exist for the differential equations in the volume and, sometimes, there are solutions that take the singularity into account. In such cases, it may be useful to use a boundary rather than an interior method.

Often for the boundary method, the selection of trial functions is not straightforward. In general, singular functions such as Green's functions can be employed. These lead to an approximation in the form of a set of integral equations, which are the basis of an important computational technique, the *boundary element method* (BEM) or the *boundary finite element method*, which is considered in Chapter 9.

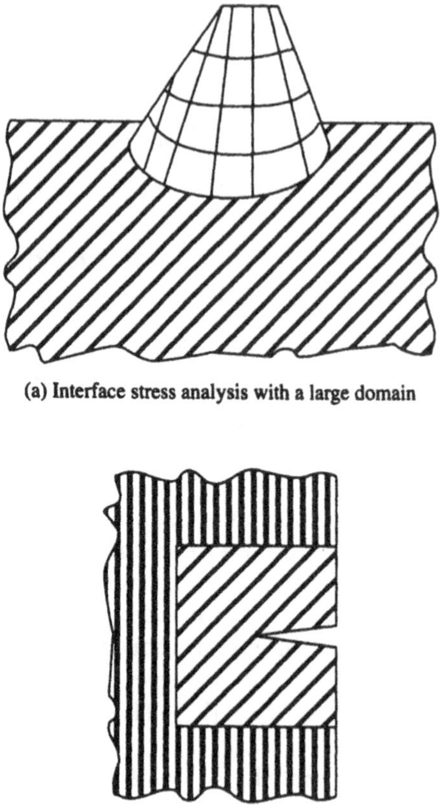

(a) Interface stress analysis with a large domain

(b) Crack problem with singular responses, e.g., stresses.

FIGURE 7.7
Configurations for which the boundary methods have been found to be useful.

In this section, a somewhat simpler and less generally applicable boundary method will be presented. This is sometimes referred to as the Trefftz[13] method [Trefftz, 1926]. This method utilizes trial functions which satisfy the governing differential equations and, through a work expression, lead to an approximation of the boundary conditions. In the Ritz method, a variational principle, using trial functions which satisfy the displacement boundary conditions and the kinematic conditions, provides an approximation to the conditions of equilibrium. However, with the Trefftz method, by employing trial functions which satisfy the differential equations of equilibrium as well as the kinematic conditions, all of the boundary conditions are approximated with the aid of a variational expression. The global form of the boundary conditions will constitute the variational expression.

For the Trefftz method, the trial solution must satisfy

$$\begin{aligned} \mathbf{D}^T\boldsymbol{\sigma} + \mathbf{p}_V &= 0 \\ \mathbf{E}^{-1}\boldsymbol{\sigma} = \boldsymbol{\epsilon} &= \mathbf{Du} \end{aligned} \quad \text{in } V \tag{7.79}$$

that is, all of the governing differential equations in V are to be satisfied. A variational expression in terms of boundary integrals of the boundary conditions $\mathbf{u} = \overline{\mathbf{u}}$ on S_u and

[13]Erich Trefftz (1888–1937), son of a Leipzig, Germany merchant, studied in Aachen, Göttingen, and Strassburg. He received his doctor's degree at Strassburg in 1913 with his research based on a suggestion by R. von Mises. In 1919, he became a professor of applied mathematics in Aachen. In 1922, he accepted a professorship in applied mechanics in Dresden, a post that he held until his death. His work on applied mechanics dealt chiefly with hydrodynamics, the theory of vibrations, and elasticity.

$\mathbf{p} = \overline{\mathbf{p}}$ on S_p is

$$\delta J = \int_{S_p} \delta \mathbf{u}^T (\mathbf{p} - \overline{\mathbf{p}})\, dS + \int_{S_u} \delta \mathbf{p}^T (\overline{\mathbf{u}} - \mathbf{u})\, dS = 0 \tag{7.80}$$

for a continuum and

$$\delta J = \underbrace{[\delta w(V - \overline{V}) + \delta\theta(M - \overline{M})]_0^L}_{\text{on } S_p} + \underbrace{[\delta V(\overline{w} - w) + \delta M(\overline{\theta} - \theta)]_0^L}_{\text{on } S_u} = 0 \tag{7.81}$$

for a beam.

Since Eqs. (7.80) and (7.81) are simply global forms of the boundary conditions, as well as constituting a form of a variational principle, the Trefftz method like Galerkin's method can be applied to problems for which a variational principle does not necessarily exist.

For Trefftz's method, use the trial solution

$$\widetilde{\mathbf{u}} = \mathbf{N}_p + \mathbf{N}_u \widehat{\mathbf{u}} \tag{7.82}$$

where $\mathbf{N}_p$ is a vector of particular solutions of the differential equations for the problem. The term $\mathbf{N}_u \widehat{\mathbf{u}}$ is a set of linearly independent functions satisfying the homogeneous differential equations. The parameters $\widehat{\mathbf{u}}$ are to be chosen to approximate the boundary conditions.

For a beam, choose a trial solution of the form

$$\widetilde{w}(x) = \mathbf{N}_p + \mathbf{N}_u \widehat{\mathbf{w}} = w_0(x) + \mathbf{N}_u(x)\, \widehat{\mathbf{w}} \tag{7.83}$$

Substitute the trial solution in the variational expression of Eq. (7.81). Note that $\delta w(x) = \mathbf{N}_u(x)\, \delta\widehat{\mathbf{w}} = \delta\widehat{\omega}^T \mathbf{N}_u^T$, and use $\theta = -w'$, $M = -EIw''$, and $V = -EIw'''$

$$\begin{aligned}\delta J &= \underbrace{\left[\delta\widehat{\mathbf{w}}^T \mathbf{N}_u^T (-EI)(w''' - \overline{w}''') - \delta\widehat{\mathbf{w}}^T \mathbf{N}_u'^T (-EI)(w'' - \overline{w}'')\right]_0^L}_{\text{on } S_p} \\ &\quad + \underbrace{\left[-EI\,\delta\widehat{\mathbf{w}}^T \mathbf{N}_u'''^T (\overline{w} - w) + EI\,\delta\widehat{\mathbf{w}}^T \mathbf{N}_u''^T (\overline{w}' - w')\right]_0^L}_{\text{on } S_u} = 0\end{aligned}$$

or

$$\delta\widehat{\mathbf{w}}^T \Big\{ \underbrace{\left[\mathbf{N}_u^T (w''' - \overline{w}''') - \mathbf{N}_u^T (w'' - \overline{w}'')\right]_0^L}_{\text{on } S_p} + \underbrace{\left[\mathbf{N}_u'''^T (\overline{w} - w) - \mathbf{N}_u''^T (\overline{w}' - w')\right]_0^L}_{\text{on } S_u} \Big\} = 0 \tag{7.84}$$

These relations can be used to determine the unknown parameters $\widehat{\mathbf{w}}$.

EXAMPLE 7.11 Beam with Linearly Varying Loading

Apply Trefftz's method to the beam of Fig. 7.1.

As can be observed from Ritz's method solutions for this beam, the particular solution can be chosen as

$$w_0(x) = \frac{p_0 L^4}{120EI}(5\xi^4 - \xi^5) \tag{1}$$

To verify the appropriateness of this $w_0(x)$, substitute it into $EIw^{iv} = \overline{p}_z$. For the $\mathbf{N}_u(x)\widehat{\mathbf{w}}$ term of the trial solution, select the simple polynomial

$$\mathbf{N}_u \widehat{\mathbf{w}} = \xi \widehat{w}_1 + \xi^2 \widehat{w}_2 = [\xi \quad \xi^2] \begin{bmatrix} \widehat{w}_1 \\ \widehat{w}_2 \end{bmatrix} \tag{2}$$

so that

$$\widetilde{w}(x) = \xi \widehat{w}_1 + \xi^2 \widehat{w}_2 + \frac{p_0 L^4}{120EI}(5\xi^4 - \xi^5) \tag{3}$$

A complete solution for a simple beam with no loading would be a third order polynomial with a term including ξ^3. Hence, it should not be surprising when the trial solution proposed here using (2) does not lead to the exact solution. However, note that the polynomial of (2) is indeed a solution of the homogeneous beam relationship $w^{iv} = 0$.

In Eq. (7.84) [or Eq. (7.81)] retain only those terms corresponding to the actual boundary conditions for the problem. Also, since there are no applied forces or displacements on the boundaries, $\overline{w} = \overline{\theta} = \overline{M} = \overline{V} = 0$. Equation (7.81) then reduces to

$$\delta\theta(L)\,M(L) - \delta V(L)\,w(L) + \delta V(0)\,w(0) + \delta M(0)\,\theta(0) = 0 \tag{4}$$

or, in the notation of Eq. (7.84),

$$\mathbf{N}_u^{\prime T}(L)\,w''(L) + \mathbf{N}_u^{\prime\prime\prime T}(L)\,w(L) - \mathbf{N}_u^{\prime\prime\prime T}(0)\,w(0) + \mathbf{N}_u^{\prime\prime T}(0)\,w'(0) = 0 \tag{5}$$

To use (5), we will need

$$\begin{aligned}
\mathbf{N}_u &= [\xi \quad \xi^2] & \widetilde{w}(x) &= [\xi \quad \xi^2]\widehat{\mathbf{w}} + \frac{p_0 L^4}{120EI}(5\xi^4 - \xi^5) \\
\mathbf{N}'_u &= \frac{1}{L}[1 \quad 2\xi] & \widetilde{w}'(x) &= \frac{1}{L}[1 \quad 2\xi]\widehat{\mathbf{w}} + \frac{p_0 L^3}{24EI}(4\xi^3 - \xi^4) \\
\mathbf{N}''_u &= \frac{1}{L^2}[0 \quad 2] & \widetilde{w}''(x) &= \frac{1}{L^2}[0 \quad 2]\widehat{\mathbf{w}} + \frac{p_0 L^2}{6EI}(3\xi^2 - \xi^3) \\
\mathbf{N}'''_u &= \frac{1}{L^3}[0 \quad 0]
\end{aligned} \tag{6}$$

Substitute these expressions in (5).

$$\begin{bmatrix} 1 \\ 2 \end{bmatrix}\left\{[0 \quad 2]\widehat{\mathbf{w}} + \frac{p_0 L^4}{3EI}\right\} + \begin{bmatrix} 0 \\ 2 \end{bmatrix}[1 \quad 0]\widehat{\mathbf{w}} = 0$$

or

$$\left\{\begin{bmatrix} 0 & 2 \\ 0 & 4 \end{bmatrix} + \begin{bmatrix} 0 & 0 \\ 2 & 0 \end{bmatrix}\right\}\widehat{\mathbf{w}} = -\begin{bmatrix} 1/3 \\ 2/3 \end{bmatrix}\frac{p_0 L^4}{EI} \tag{7}$$

Finally, we obtain a system of equations for $\widehat{\mathbf{w}}$,

$$\begin{bmatrix} 0 & 2 \\ 2 & 4 \end{bmatrix}\begin{bmatrix} \widehat{w}_1 \\ \widehat{w}_2 \end{bmatrix} = \begin{bmatrix} -1/3 \\ -2/3 \end{bmatrix}\frac{p_0 L^4}{EI} \tag{8}$$

which has the solution

$$\begin{bmatrix} \widehat{w}_1 \\ \widehat{w}_2 \end{bmatrix} = \frac{p_0 L^4}{EI}\begin{bmatrix} 0 \\ -1/6 \end{bmatrix} \tag{9}$$

The approximate deflection then becomes

$$\widetilde{w}(x) = \frac{p_0 L^4}{120EI}(-20\xi^2 + 5\xi^4 - \xi^5) \tag{10}$$

It is clear that this solution does not satisfy all boundary conditions. For example, $w(L) \neq 0$. Therefore, we will try to improve the trial solution.

For a second trial solution, use

$$\widetilde{w} = \mathbf{N}_u\widehat{\mathbf{w}} = \xi\widehat{w}_1 + \xi^2\widehat{w}_2 + \xi^3\widehat{w}_3 = [\xi \quad \xi^2 \quad \xi^3]\begin{bmatrix} \widehat{w}_1 \\ \widehat{w}_2 \\ \widehat{w}_3 \end{bmatrix} \tag{11}$$

so that the complete trial function is

$$\widetilde{w}(x) = \xi \widehat{w}_1 + \xi^2 \widehat{w}_2 + \xi^3 \widehat{w}_3 + \frac{p_0 L^4}{120EI}(5\xi^4 - \xi^5) \tag{12}$$

We will need the derivatives

$$\begin{aligned} \mathbf{N}'_u &= \frac{1}{L}[1 \quad 2\xi \quad 3\xi^2] \qquad w'(x) = \frac{1}{L}[1 \quad 2\xi \quad 3\xi^2]\widehat{\mathbf{w}} + \frac{p_0 L^3}{24EI}(4\xi^3 - \xi^4) \\ \mathbf{N}''_u &= \frac{1}{L^2}[0 \quad 2 \quad 6\xi] \qquad w''(x) = \frac{1}{L^2}[0 \quad 2 \quad 6\xi]\widehat{\mathbf{w}} + \frac{p_0 L^2}{6EI}(3\xi^2 - \xi^3) \end{aligned} \tag{13}$$

Introduction of these expressions into Eq. (7.84) in the form

$$\mathbf{N}_u'^T(1)\, w''(1) + \mathbf{N}_u'''^T(1)\, w(1) - \mathbf{N}_u'''^T(0)\, w(0) + \mathbf{N}_u''^T(0)\, w'(0) = 0 \tag{14}$$

leads to

$$\begin{bmatrix} 1 \\ 2 \\ 3 \end{bmatrix} \left\{ [0 \ 2 \ 6]\widehat{\mathbf{w}} + \frac{p_0 L^4}{3EI} \right\} + \begin{bmatrix} 0 \\ 0 \\ 6 \end{bmatrix} \left\{ [1 \quad 1 \quad 1]\widehat{\mathbf{w}} + \frac{p_0 L^4}{30EI} \right\} + \begin{bmatrix} 0 \\ 2 \\ 0 \end{bmatrix} [1 \quad 0 \quad 0]\widehat{\mathbf{w}} = 0$$

or

$$\left\{ \begin{bmatrix} 0 & 2 & 6 \\ 0 & 4 & 12 \\ 0 & 6 & 18 \end{bmatrix} + \begin{bmatrix} 0 & 0 & 0 \\ 0 & 0 & 0 \\ 6 & 6 & 6 \end{bmatrix} + \begin{bmatrix} 0 & 0 & 0 \\ 2 & 0 & 0 \\ 0 & 0 & 0 \end{bmatrix} \right\} \widehat{\mathbf{w}} = \begin{bmatrix} -1/3 \\ -2/3 \\ -1/5 - 1 \end{bmatrix} \frac{p_0 L^4}{EI}$$

This reduces to the expression

$$\begin{bmatrix} 0 & 2 & 6 \\ 2 & 4 & 12 \\ 6 & 12 & 24 \end{bmatrix} \begin{bmatrix} \widehat{w}_1 \\ \widehat{w}_2 \\ \widehat{w}_3 \end{bmatrix} = \begin{bmatrix} -1/3 \\ -2/3 \\ -6/5 \end{bmatrix} \frac{p_0 L^4}{EI} \tag{15}$$

which gives the free parameters

$$\begin{bmatrix} \widehat{w}_1 \\ \widehat{w}_2 \\ \widehat{w}_3 \end{bmatrix} = \frac{p_0 L^4}{EI} \begin{bmatrix} 0 \\ 1/30 \\ -1/15 \end{bmatrix} \tag{16}$$

This solution corresponds to the polynomial

$$\widetilde{w}(x) = \frac{p_0 L^4}{120EI}(4\xi^2 - 8\xi^3 + 5\xi^4 - \xi^5) \tag{17}$$

which, as we learned earlier, is the exact solution for the displacement.

To compare the results, we label the solutions as

A. Exact solution given by (17). Also the same as case C
B. First Trial Deflection: $\widetilde{w}(x) = p_0 L^4/(120EI)(-20\xi^2 + 5\xi^4 - \xi^5)$
C. Second Trial Deflection: $\widetilde{w}(x) = p_0 L^4/(120EI)(4\xi^2 - 8\xi^3 + 5\xi^4 - \xi^5)$

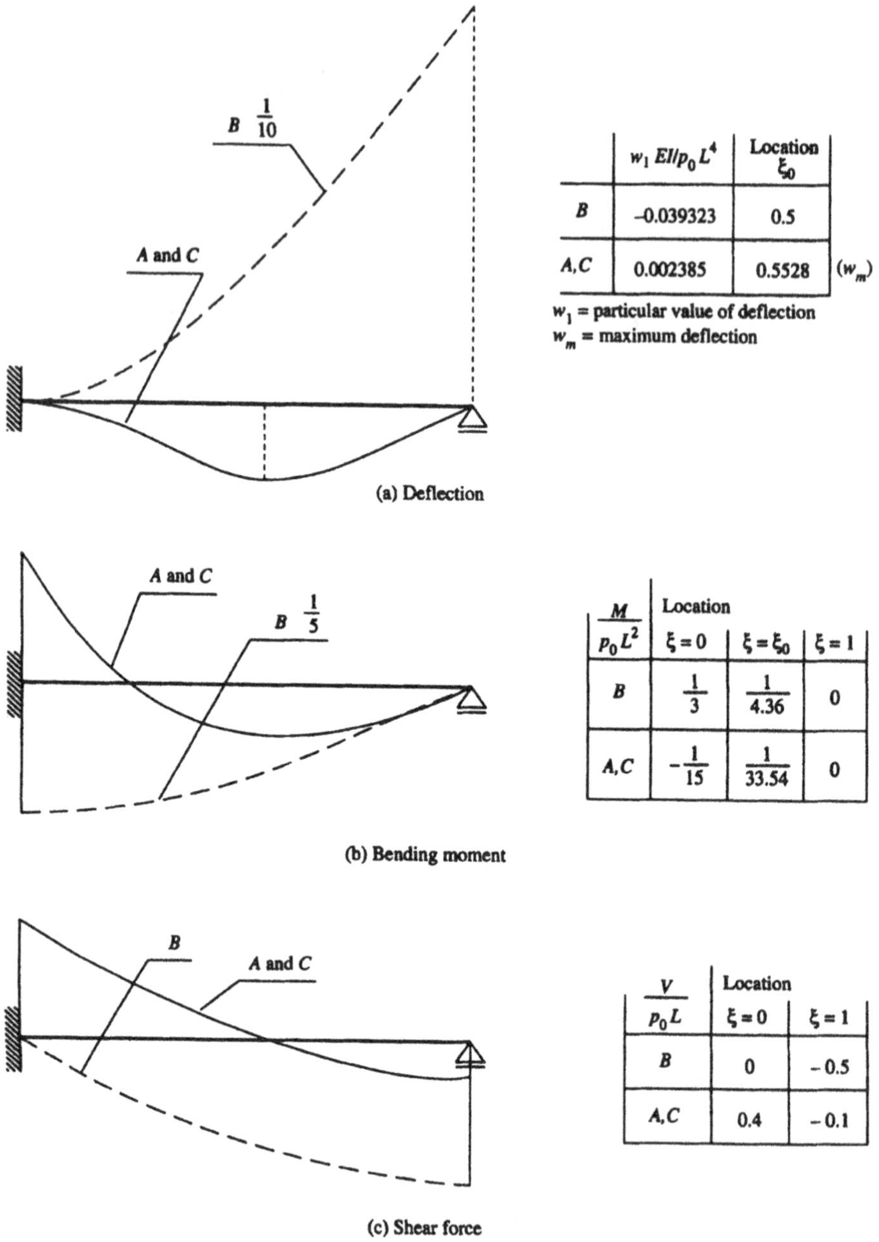

	$w_1\, EI/p_0 L^4$	Location ξ_0	
B	−0.039323	0.5	
A,C	0.002385	0.5528	(w_m)

w_1 = particular value of deflection
w_m = maximum deflection

$\frac{M}{p_0 L^2}$	Location $\xi = 0$	$\xi = \xi_0$	$\xi = 1$
B	$\frac{1}{3}$	$\frac{1}{4.36}$	0
A,C	$-\frac{1}{15}$	$\frac{1}{33.54}$	0

$\frac{V}{p_0 L}$	Location $\xi = 0$	$\xi = 1$
B	0	−0.5
A,C	0.4	−0.1

FIGURE 7.8
Trefftz's method for a fixed-hinged beam.

The boundary values are

$$\text{B. } w(0) = 0, \qquad w'(0) = 0, \qquad w(L) = -0.13\frac{p_0 L^4}{EI} \neq 0, \qquad M(L) = 0$$
$$\text{C. } w(0) = 0, \qquad w'(0) = 0, \qquad w(L) = 0, \qquad M(L) = 0$$

This displacement, moment, and shear force are compared in Fig. 7.8. ∎

Trefftz's method is not universally applicable, since solutions are not always available for the governing differential equations. Also, the method leads to algebraic equations that may be less banded than those of some other methods.

7.5 Trial Function Methods for Solids Divided into Elements: The Method of Finite Elements

Thus far in this chapter, we have utilized trial functions that apply for the whole body. The selection of appropriate trial solutions for complicated bodies, e.g., two- and three-dimensional configurations, is not simple. Furthermore, for a formulation in terms of displacements, the chosen trial solution is usually expected to approximate forces (e.g., moments or stresses), as well as the displacements. This goal is often not easy to achieve. As a further complication, the use of what would appear to be an adequate trial solution can lead to numerical instabilities. These difficulties can usually be circumvented by subdividing the body into *elements* and utilizing a separate trial solution in each element. This is, of course, the procedure that forms the basis of the *finite element method,* the powerful computational technology that was considered in-depth in the previous chapter. Basically, Ritz's method is employed to generate stiffness matrices for the elements. And in certain instances, weighted-residual methods can be used, provided that interelement conditions are properly taken into account. Then the techniques discussed in Chapters 5 and 6 for assembling a system stiffness matrix are utilized to create a set of global equilibrium equations.

7.5.1 Analog Solutions of Differential Equations

Substantial generic finite element software has been developed for the solution of various problems, especially for structural mechanics problems. Although the finite element method is fundamentally a variationally based technique, as shown in Chapter 2, differential equations can often be shown to be equivalent to variational forms. Since many different kinds of problems are described by the same kind of differential equations, it is often not necessary to create a computer program for each problem. For example, the differential equation for the one-dimensional static heat distribution and that for the extension of straight bars have the same form except that the coefficients of the derivatives of the dependent variables and the nonhomogenous terms have different physical meanings. Problems which have this characteristic are said to be *analogical.* A wide variety of problems can be solved using computer software suitable for analogical problems. The primary concern for this solution method is the interpretation and exchange of the coefficients of the derivatives and the nonhomogeneous terms between the analogical problems. Special attention to the boundary conditions is often necessary.

EXAMPLE 7.12 The Analogy Between the Heat Equation and the Extension of a Straight Bar
The governing differential equation for one-dimensional heat transfer under stationary conditions is

$$K\frac{d^2T}{dx^2} = Q \tag{1}$$

with boundary conditions, where T is the temperature, K is the thermal conductivity, and Q is the heat source. The differential equation for the extension of a straight bar is (Chapter 1,

Eq. 1.136)

$$EA\frac{d^2u}{dx^2} = -\overline{p}_x \tag{2}$$

with boundary conditions, where u is the axial displacement, E is Young's modulus, A is the area of the cross section, and $\overline{p}_x$ is the applied axial distributed traction. By comparison, the correspondence between these problems is

Heat Equation		Extension Equation
Temperature T	<– – – – – –>	Displacement u
K	<– – – – – –>	EA
Heat Source Q	<– – – – – –>	Load $-\overline{p}_x$

In addition to this correspondence between the terms in the differential equations, the relationship between the boundary conditions must be established. For the extension bar, the boundary conditions can be

1. A fixed end or an end with a prescribed displacement, $u = 0$ or $u = \overline{u}$
2. A free end, $du/dx = 0$

For the heat equation, the boundary conditions can be classified as

1. Prescribed temperature at the boundary, $T = 0$ or $T = \overline{T}$
2. Insulated, $dT/dx = 0$

If the bar is attached to a spring with the spring constant k at one end, the boundary condition at this end can be written as $EA\,du/dx = ku$. This is analogical to the convection boundary condition for the heat equation which has the form $dT/dx = \pm hT$, where h (>0) is a constant. The sign + is for the left end and − is for the right end.

With these equivalent coefficients, the heat equation can be solved by using a computer program developed for analyzing an extension bar. In doing so, the values of K and Q are input to the entries for EA and P. If quantities E and A are entered separately, A can be assigned a value first and then let $E = K/A$. In setting up the first value, care should be taken to assure that it is within the range that the computer program can accept. The boundary conditions are handled similarly. After executing the program, the computed u corresponds to the temperature distribution T, with appropriate units. ∎

7.5.2 Stiffness Matrices

As in Example 7.6 for the Ritz method, we can use the field problem of torsion of a bar to illustrate the development of stiffness matrices for a finite element solution for the warping functions across the cross section of a bar.

From Eqs. (4) and (5) of Example 7.6, the equation that can be solved for the warping function ω, and that will permit the cross-sectional characteristics of a bar in torsion to be computed, takes the form

$$\int_A \left[\left(\frac{\partial}{\partial y}\delta\omega\,\frac{\partial\omega}{\partial y} + \frac{\partial}{\partial z}\delta\omega\frac{\partial\omega}{\partial z}\right) + \left(\frac{\partial}{\partial y}\delta\omega\, z - \frac{\partial}{\partial z}\delta\omega\, y\right)\right] dA = 0 \tag{7.85}$$

or

$$\int_A \delta\omega[({}_y\partial\partial_y + {}_z\partial\partial_z)\omega + ({}_y\partial\, z - {}_z\partial\, y)]\,dA = 0 \tag{7.86}$$

The warping function $\omega(y, z)$ can be computed using a finite element analysis. For each element, approximate the warping function ω by

$$\omega(y, z) = \sum_j N_j \omega_j = \mathbf{N}\boldsymbol{\omega} = \boldsymbol{\omega}^T \mathbf{N}^T \tag{7.87}$$

where N_j are the shape functions that form the vector $\mathbf{N}$ and ω_j are the nodal values of ω which form the vector $\boldsymbol{\omega}$.

The derivations of ω are

$$\frac{\partial \omega}{\partial y} = \frac{\partial \mathbf{N}}{\partial y}\boldsymbol{\omega} \qquad \frac{\partial \omega}{\partial z} = \frac{\partial \mathbf{N}}{\partial z}\boldsymbol{\omega} \tag{7.88}$$

and

$$\delta\omega = \mathbf{N}\delta\boldsymbol{\omega} = \delta\boldsymbol{\omega}^T\mathbf{N}^T \qquad \frac{\partial}{\partial y}\delta\omega = \delta\boldsymbol{\omega}^T\frac{\partial \mathbf{N}^T}{\partial y} \qquad \frac{\partial}{\partial z}\delta\omega = \delta\boldsymbol{\omega}^T\frac{\partial \mathbf{N}^T}{\partial z} \tag{7.89}$$

Substitute these into Eq. (7.85)

$$\int_A \delta\boldsymbol{\omega}^T\left[\left(\frac{\partial \mathbf{N}^T}{\partial y}\frac{\partial \mathbf{N}}{\partial y} + \frac{\partial \mathbf{N}^T}{\partial z}\frac{\partial \mathbf{N}}{\partial z}\right)\boldsymbol{\omega} + \left(z\frac{\partial \mathbf{N}^T}{\partial y} - y\frac{\partial \mathbf{N}^T}{\partial z}\right)\right] dA = \delta\boldsymbol{\omega}^T(\mathbf{k}^i\boldsymbol{\omega} - \mathbf{p}^i) = 0 \tag{7.90}$$

where $\mathbf{k}^i$ and $\mathbf{p}^i$ are the element stiffness and "quasi-loading" vector, respectively, defined as

$$\begin{aligned}\mathbf{k}^i &= \int_A \left(\frac{\partial \mathbf{N}^T}{\partial y}\frac{\partial \mathbf{N}}{\partial y} + \frac{\partial \mathbf{N}^T}{\partial z}\frac{\partial \mathbf{N}}{\partial z}\right) dA = \int_A \left(\mathbf{N}^T{}_y\partial\ \partial_y\mathbf{N} + \mathbf{N}^T{}_z\partial\ \partial_z\mathbf{N}\right) dA \\ \mathbf{p}^i &= -\int_A \left(z\frac{\partial \mathbf{N}^T}{\partial y} - y\frac{\partial \mathbf{N}^T}{\partial z}\right) dA = -\int_A \left(z\mathbf{N}^T{}_y\partial - y\mathbf{N}^T{}_z\partial\right) dA\end{aligned} \tag{7.91}$$

These expressions can be used to define the stiffness matrix and "quasi-loading" vector for particular elements. The same relationships for $\mathbf{k}^i$ and $\mathbf{p}^i$ are derived in Pilkey (2002) using Galerkin's method.

References

Biezeno, C.B. and Koch, J.J., 1923, Overeen Nieuwe Methode ter Brerkening van Vlokke Platen met Toepassing op. Eukele voor de Technik Belangrijke Belastingsgevallen, *Ing. Grav.*, Vol. 38, pp. 25–36.

Bubnov, I.G., 1913, *Sborn. Inta Inzh. Putei Soobshch*, Vol. 81, USSR All Union Special Planning Office (SPB).

Courant, R., 1943, Variational methods for the solution of problems of equilibrium and vibration, *Bull. Am. Math. Soc.*, Vol. 49, pp. 1–23.

Finlayson, B.A., 1972, *The Method of Weighted Residuals and Variational Principles*, Academic Press, NY.

Galerkin, B.G., 1915, *Vestn. Inzh. Tech.* (in Russian), Vol. 19, p. 897 (translation 63-18924, NTIS).

Kantorovitch, L.V. and Krylov, V.I., 1956, *Approximate Methods of Higher Analysis*, Verlag der Wissenschaften, Berlin, or Interscience, NY, 1964.

Lanczos, C., 1939, Trigonometric interpolation of empirical and analytical functions, *J. Math. Phys.*, Vol. 17, pp. 123–199.

Mikhlin, S.G., 1964, *Variational Methods in Mathematical Physics*, Pergamon Press, Oxford.

Norrie, D.H. and de Vries, G., 1973, *The Finite Element Method*, Academic Press, New York.

Park, I.B. and Pilkey, W.D., 1982, The Minimax Finite Element Method, *ASCE J. Struct. Div.*, Vol. 108, pp. 998–1011.
Pohlhausen, K., 1921, The approximate integration of the differential equation for the laminar boundary layer, *Z. Angew. Math. Mech.*, Vol. 1, pp. 252–269 (translation AD 645 784, NTIS).
Rayleigh, L. (J.W. Strutt), 1873, Some general theorems relating to vibrations, *Proc. London Math. Soc.*, Vol. 4, pp. 357–368.
Ritz, W., 1909, Über eine Neue Methode zur Lösung gewisser Variationsprobleme der Mathematischen Physik, *J. Reine Angew. Math.*, Vol. 135, pp. 1–61.
Trefftz, E., 1926, Ein Gegenstück zum Ritzschen Verfahren, *Proc. 2nd Int. Congress Appl. Mech.*, Zürich.
Von Karman, T., 1921, Über Laminare und Turbulente Reibung, *Z. Angew. Math. Mech.*, Vol. 1, pp. 233–252 (translation NACA Tech. Memo 1092, 1946).
Yamada, H., 1950, A method of approximate integration of the laminar boundary layer equation, *Rep. Res. Inst. Fluid Eng.*, Vol. 6, pp. 87–98.

Problems

7.1 Show that the differential operator for an Euler-Bernoulli beam $L = \frac{d^2}{dx^2} EI \frac{d^2}{dx^2}$ is both self-adjoint and positive definite.

Hint:

$$\int_0^L vLw\, dV = \int_0^L v \frac{d^2}{dx^2} EI \frac{d^2 w}{dx^2} dx$$
$$= -vV|_0^L + \theta M|_0^L + \int_0^L \frac{d^2 w}{dx^2} EI \frac{d^2 v}{dx^2} dx$$
$$= \int_0^L w \frac{d^2}{dx^2} EI \frac{d^2 v}{dx^2} dx = \int_0^L wLv\, dV$$

Therefore, L is self-adjoint.
Since $\int_0^L vLw\ dV = \int_0^L EI \frac{d^2w}{dx^2}\frac{d^2v}{dx^2} dx$, $\int_0^L wLw\, dV = \int_0^L EI (\frac{d^2w}{dx^2})^2 dx$ so that L is positive definite.

7.2 Solve the differential equation $u'' + u = -x$ for $0 \leq x \leq 1$, with $u(0) = u(1) = 0$. Use collocation and compare your answer with the exact solution, which is $u(x) = \sin x / \sin 1 - x$.

Hint: Choose a simple trial solution such as $\tilde{u} = \hat{u}_1 x(1-x)$ which satisfies the boundary conditions, as would $x^2(1-x)$, $x^3(1-x)$, etc.

$$R = L\tilde{u} - \overline{f} = -2\hat{u}_1 + \hat{u}_1 x(1-x) + x, \quad \text{where } L = d_x^2 + 1,\ \overline{f} = -x$$

For a single collocation point at $x = 1/2$, $R_{x=1/2} = -2\hat{u}_1 + \frac{1}{4}\hat{u}_1 + 1/2$.

Results:

	Solution u	
x	Exact	Collocation
1/4	0.044	0.054
1/2	0.070	0.071
3/4	0.060	0.054

7.3 Are the approximate and exact solutions identical at the collocation points for the collocation method? Explain.

Hint: See the results of Problem 7.2.

7.4 Solve the partial differential equation $\partial_x^2 u + \partial_y^2 u = -1$ with the boundary conditions $u = 0$ on $x = \pm 1$ and $y = \pm 1$. Use a boundary residual method with collocation.

Hint: A particular solution is $-(x^2 + y^2)/4$. Two complementary functions are $1, x^4 - 6x^2y^2 + y^4$. A trial solution, which satisfies the differential equation, would be $\tilde{u} = -(x^2 + y^2)/4 + \hat{u}_1 + \hat{u}_2(x^4 - 6x^2y^2 + y^4)$. Choose $\hat{u}_1, \hat{u}_2$ to satisfy $\tilde{u} = 0$ on the square $x = \pm 1, y = \pm 1$. For collocation points $(x_1, y_1) = (1, 0)$ and $(x_2, y_2) = (1, 1)$, $\hat{u}_1 = 3/10$, $\hat{u}_2 = -1/20$. This gives $\tilde{u} = 0.3$ at $x = y = 0$, as compared to the exact solution of $u = 0.2947$.

7.5 Use the subdomain interior method to solve the problem posed in Problem 7.2.

Hint: For the trial solution, use (Problem 7.2) $\tilde{u} = \hat{u}_1 x(1 - x)$, $R = -2\hat{u}_1 + \hat{u}_1 x(1 - x) + x$. For a single subdomain $0 \le x \le 1$, $\int_0^1 R(x)dx = 0 = -11/6\hat{u}_1 + 1/2 = 0$ gives $\hat{u}_1 = 3/11$, so that $\tilde{u} = 3x(1 - x)/11$.

Results:

	Solution u	
x	**Exact**	**Subdomain**
1/4	0.044	0.051
1/2	0.070	0.068
3/4	0.060	0.051

7.6 Use Galerkin's method to solve the differential equation of Problem 7.2.

Hint: For the trial solution $\tilde{u} = \hat{u}_1 x(1 - x)$,

$$\int_0^1 x(1 - x)R(x)\,dx = \int_0^1 x(1 - x)[-2\hat{u}_1 + \hat{u}_1 x(1 - x) + x]\,dx = 0$$

gives $\hat{u}_1 = 5/18$ and $\tilde{u} = 5x(1 - x)/18$.

Results:

	Solution u	
x	**Exact**	**Galerkin**
1/4	0.044	0.052
1/2	0.070	0.069
3/4	0.060	0.052

7.7 Study the solution of the boundary value problem (with a partial differential equation)

$$\partial_x^2 u + \partial_y^2 u = 0, \qquad 0 < x < 1, \qquad 0 < y < \infty$$

with boundary conditions $u(0, y) = u(1, y) = 0$ for $y > 0$, $u(x, 0) = x(1 - x)$, and $u(x, y \to \infty) = 0$ for $0 \le x \le 1$.

(a) Use the trial solution $\tilde{u} = \hat{u}_1(y)x(1 - x)$. Find the boundary conditions on $\hat{u}_1$ and the residual function.

(b) Apply collocation along $x = 1/3$.

Answer: $\hat{u}_1 = e^{-3y}$. Note $\hat{u}_1 = \hat{u}_1(y)$.

(c) Use Galerkin's method with the trial solution

$$\widetilde{u} = \widehat{u}_1(y)x(1-x) + \widehat{u}_2(y)\, x^2(1-x)^2$$

Answer:

$$\widehat{u}_1(y) = 0.8035e^{-3.1416y} + 0.1965e^{-10.1059y}$$

$$\widehat{u}_2(y) = 0.9105(e^{-3.1416y} - e^{-10.1059y})$$

7.8 Solve the differential equation $u'' + u = x$ for $0 \leq x \leq 1$, with $u(0) = u'(1) = 0$. Use Galerkin's method and compare your results with the exact solution $u = x - \sin x/\cos 1$.

Hint: The two term approximating solution $\widetilde{u}'(x) = (1-x)(\widehat{u}_1 + \widehat{u}_2 x)$, with

$$\widetilde{u}(x) = [x - x^2/2 \quad x^2/2 - x^3/3] \begin{bmatrix} \widehat{u}_1 \\ \widehat{u}_2 \end{bmatrix} = \mathbf{N}_u \widehat{\mathbf{u}}$$

satisfies the boundary conditions. Form $R = L\widetilde{u} - \overline{f}$, which leads to the Galerkin orthogonality condition, with $L = d^2/dx^2 + 1$ and $\overline{f} = x$,

$$\int_0^1 \mathbf{N}_u^T L \mathbf{N}_u \, dx \, \widehat{\mathbf{u}} - \int_0^1 \mathbf{N}_u^T \overline{f} \, dx = 0 \qquad \text{or}$$

$$\underset{\mathbf{k}_u}{\begin{bmatrix} 72 & 17 \\ 119 & 58 \end{bmatrix}} \underset{\widehat{\mathbf{u}}}{\begin{bmatrix} \widehat{u}_1 \\ \widehat{u}_2 \end{bmatrix}} = \underset{\mathbf{p}_u}{\begin{bmatrix} -75 \\ -147 \end{bmatrix}} \quad \text{giving } \widehat{u}_1 = -0.8597 \text{ and } \widehat{u}_2 = -0.7706$$

Note that $\mathbf{k}_u$ is not symmetric. Finally, $\widetilde{u}(x) = -0.8597x + 0.0446x^2 + 0.2569x^3$.

Results:

	Solution u	
x	Exact	Galerkin
1/4	−0.2079	−0.2081
1/2	−0.3873	−0.3866
3/4	−0.5115	−0.5115
1	−0.5574	−0.5582

7.9 Use the method of moments to find a solution of the nonlinear differential equation $[(1 + au)u']' = 0$ with boundary conditions $u(0) = 0$, $u(1) = 1$. Choose a polynomial as a trial solution.

Hint: A form of a polynomial that satisfies the boundary conditions is

$$\widetilde{u}(x) = x + \sum_{i=1}^{m} \widehat{u}_i (x^{i+1} - x)$$

The residual is $R = (1 + a\widetilde{u})\widetilde{u}'' + a(\widetilde{u}')^2$.

As a first approximation, choose $\psi_1 = 1$, $\widetilde{u} = x + \widehat{u}_1(x^2 - x)$. For $a = 1$,

$$\int_0^1 R\, dx = 0 = \int_0^1 [1 + x + \widehat{u}_1(x^2 - x)] 2\widehat{u}_1 \, dx + \int_0^1 [1 + \widehat{u}_1(2x-1)]^2 \, dx$$

This gives $\widehat{u}_1 = -0.333$. As a second approximation, use $\widetilde{u} = x + \widehat{u}_1(x^2 - x) + \widehat{u}_2(x^3 - x)$. The integrals $\int_0^1 R\,dx = 0$, $\int_0^1 xR\,dx = 0$ give two nonlinear algebraic equations for $\widehat{u}_1$ and $\widehat{u}_2$. Thus, $\widetilde{u}(x) = 3x^2/2 - 3x^2/4 + x^3/4$.

Results:

	Solution u		
x	Exact	First Approx. Method of Moments	Second Approx. Method of Moments
1/4	0.323	0.313	0.332
1/2	0.581	0.583	0.594
3/4	0.803	0.813	0.809

7.10 Consider the problem posed in Problem 7.9. This system represents the steady state heat conduction across a slab with a conductivity of $1 + au$, where u is a nondimensional temperature differential.

(a) Solve the problem with the method of collocation for $a = 1$. Compare your result with the exact solution.

(b) Same as step (a), but use Galerkin's method. Be careful as this is a nonlinear system.

7.11 Use a least squares approach to solve the differential equation of Problem 7.2.

Hint: With the trial function $\widetilde{u} = \widehat{u}_1 x(1 - x)$, which satisfies the boundary conditions, $R = -2\widehat{u}_1 + \widehat{u}_1 x(1 - x) + x$, $\partial R/\partial \widehat{u}_1 = -2 + x(1 - x)$.

With

$$\int_V R\frac{\partial R}{\partial \widehat{u}_1} dV = \int_0^1 [-2\widehat{u}_1 + \widehat{u}_1 x(1 - x) + x][-2 + x(1 - x)]\,dx = 0$$

This gives $\widehat{u}_1 = 55/202$ and $\widetilde{u}(x) = 55x(1 - x)/202$

Results:

	Solution u	
x	Exact	Last Squares
1/4	0.044	0.051
1/2	0.070	0.068
3/4	0.060	0.051

7.12 Show that under certain conditions a self-adjoint operator Galerkin's method leads to symmetric equations in the undetermined coefficients $\widehat{\mathbf{u}}$.

7.13 Show that equations for the coefficients $\widehat{\mathbf{u}}$, as obtained by the least squares method, are both symmetric and positive definite.

7.14 Suppose a simply supported beam of length L is subjected to the distributed load $\overline{p}_z = \sin(\pi x/L)$. Calculate the distribution of bending moment using three weighted-residual methods, e.g., collocation, least squares, and Galerkin. Compare the results with the exact solution. Let $L = 1$.

Hint: Recall that $d^2M/dx^2 = -\overline{p}_z$.

7.15 Suppose a beam of unit length with unit distributed load ($\overline{p}_z = 1$ force/length) is fixed on both ends and rests on an elastic foundation. The governing equation would be $EId^4w/dx^4 + kw = -\overline{p}_z$, where k is the foundation modulus.

Let $k = EI = 1$. Find the distribution of deflection using collocation and Galerkin's method, and compare the results with the exact solution.

7.16 Find the torsional properties of a rectangular cross-section of width $2a$ (y direction) and height $2b$ (z direction). Use the Prandtl stress function in conjunction with the principle of complementary virtual work.

Hint: Use the Ritz method. A possible trial solution is $\tilde{\psi} = \hat{\psi}(a^2 - y^2)(b^2 - z^2)$, which satisfies $\psi = 0$ on the boundary (Chapter 1). From Eq. (4) of Chapter 2, Example 2.9, represent the principle of complementary virtual work by

$$\int_A (\nabla^2\psi + 2G\phi')\delta\psi \, dA = 0$$

This gives

$$\hat{\psi} = G\phi'(5/4)/(a^2 + b^2), \qquad M_t = 2\int_A\int \tilde{\psi}\, dy\, dz = \frac{40}{9}\frac{a^3b^3}{a^2 + b^2}G\phi'$$

Comparision for $b/a = 1$: $M_t/G\phi'|_{\text{Ritz}} = 2.22$, $M_t/G\phi'|_{\text{exact}} = 2.25$

7.17 Find the displacements and forces along the beam of Fig. P7.17 using the simple Ritz method and then the extended Ritz method. Represent the applied distributed loading as a quadratic polynomial $\overline{p}_z = \alpha_1 x^2 + \alpha_2 x + \alpha_3$, where $\alpha_i, i = 1, 2, 3$ are constants determined so that this $\overline{p}_z$ fits the loading distribution of the figure.

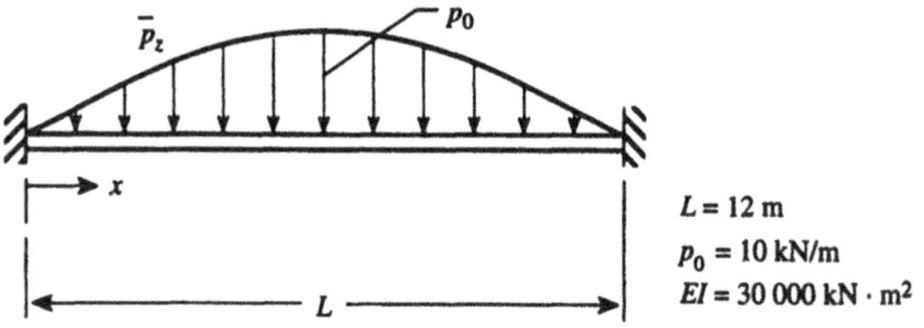

FIGURE P7.17

Answer: Loading: $\alpha_1 = -4p_0/L^2$, $\alpha_2 = 4p_0/L$, $\alpha_3 = 0$.
For the trial solution,

$$\begin{aligned}\tilde{w} &= (a_1 + a_2\xi + a_3\xi^2 + a_4\xi^3 + a_5\xi^4 + a_6\xi^5)\hat{w}_1 \\ &\quad + (b_1 + b_2\xi + b_3\xi^2 + b_4\xi^3 + b_5\xi^4 + b_6\xi^5 + b_7\xi^6)\hat{w}_2\end{aligned}$$

using $a_1 = a_2 = b_1 = b_2 = 0$, $a_3 = -a_4 = -a_5 = 1$, $b_3 = b_4 = 1$, $b_5 = -2$, and $b_6 = -b_7 = -3$ to meet the displacement boundary conditions, the simple Ritz method gives

$$\begin{aligned}\tilde{w} &= \frac{p_0 L^4}{29\,370\, EI}(919\xi^2 - 1277\xi^3 - 740\xi^4 + 1635\xi^5 - 537\xi^6) \\ &= 0.2163\xi^2 - 0.3005\xi^3 - 0.1742\xi^4 + 0.3848\xi^5 - 0.1264\xi^6\end{aligned}$$

For the trial solution,

$$\tilde{w}(x) = \xi^2\hat{w}_1 + \xi^3\hat{w}_2 + \xi^4\hat{w}_3 + \xi^5\hat{w}_4 + \xi^6\hat{w}_5$$

the extended Ritz method gives

$$\begin{aligned}\widetilde{w} &= \frac{p_0 L^4}{90EI}(3\xi^2 - 5\xi^3 + 3\xi^5 - \xi^6) \\ &= 0.2304\xi^2 - 0.384\xi^3 + 0.2304\xi^5 - 0.0768\xi^6\end{aligned}$$

which is equal to the exact solution.

7.18 Prove that the stationary value of the functional

$$\Pi(u) = \int_L (uLu - 2u\overline{f})\, dx$$

is equivalent to the solution of the differential equation $Lu - \overline{f} = 0$.

Hint: Introduce [Appendix I, Eq. (I.3)] $\widehat{u} = u + \epsilon\eta$. Then

$$\begin{aligned}\Pi(\widehat{u}) &= \int_L [(u + \epsilon\eta)L(u + \epsilon\eta) - 2(u + \epsilon\eta)\overline{f}]\, dx \\ &= \Phi(\epsilon). \quad \text{Set } \left.\frac{d\Phi(\epsilon)}{d\epsilon}\right|_{\epsilon=0} = 0\end{aligned}$$

If L is linear, $\frac{d}{d\epsilon}L(u + \epsilon\eta) = L\frac{d}{d\epsilon}(u + \epsilon\eta)$. First, carry out the differentiation and then set $\epsilon = 0$.

$$\int_L (uL\eta + \eta Lu - 2\eta\overline{f})\, dx = 0$$

If L is self-adjoint, $\int_L 2(Lu - \overline{f})\eta\, dx = 0$ and the fundamental lemma of Appendix I implies that $Lu - \overline{f} = 0$.

7.19 Prove that the stationary value of the functional

$$\Pi(\mathbf{u}) = \int_V (\mathbf{u}^T\mathbf{L}\mathbf{u} - 2\mathbf{u}^T\overline{\mathbf{f}})\, dV$$

is equivalent to the solution of the differential equation $\mathbf{L}\mathbf{u} - \overline{\mathbf{f}} = 0$.

Hint: Follow the procedure outlined in the hint of Problem 7.18.

7.20 Find the response of the beam of Fig. P7.20 using the simple Galerkin's method and the extended Galerkin's method. Compare the results with the exact solution. The applied distributed load should be represented by a second order polynomial.

Hint: See Problem 7.17 for the loading distribution.

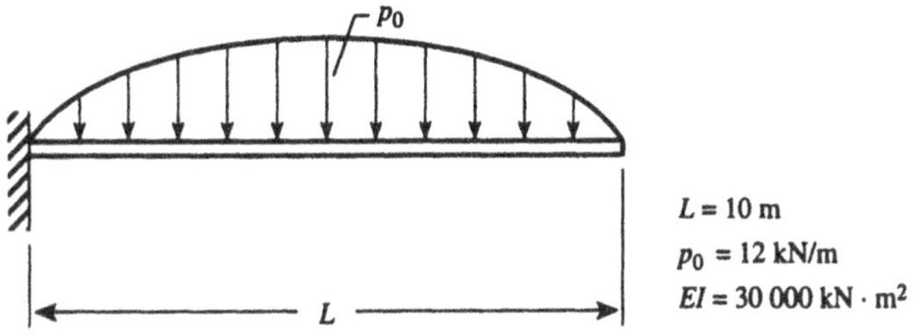

FIGURE P7.20

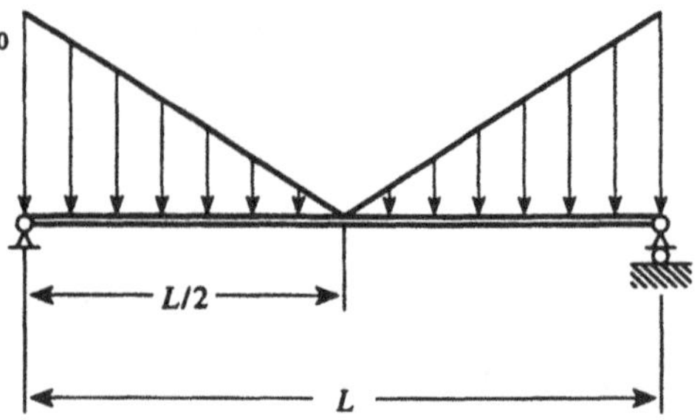

FIGURE P7.21

7.21 A simply supported beam carries the load shown in Fig. P7.21. Use a two-term trigonometric series to calculate the approximate midspan deflection. Use the Ritz and the Galerkin methods and compare to the exact solution.

7.22 Use the Trefftz method to determine the displacement and internal forces along the beam of Fig. P7.22.

Answer: For the trial solution,

$$\widetilde{w} = \mathbf{N}_p + \mathbf{N}_u \widehat{\mathbf{w}} = \frac{p_0 L^4}{360EI}\xi^6 + \xi\widehat{w}_1 + \xi^2\widehat{w}_2 + \xi^3\widehat{w}_3$$

the Trefftz method gives

$$\widetilde{w} = \frac{p_0 L^4}{360EI}(3\xi^2 - 4\xi^3 + \xi^6) = 0.0576\xi^2 - 0.0768\xi^3 + 0.0192\xi^6$$

which is also the exact beam theory solution.

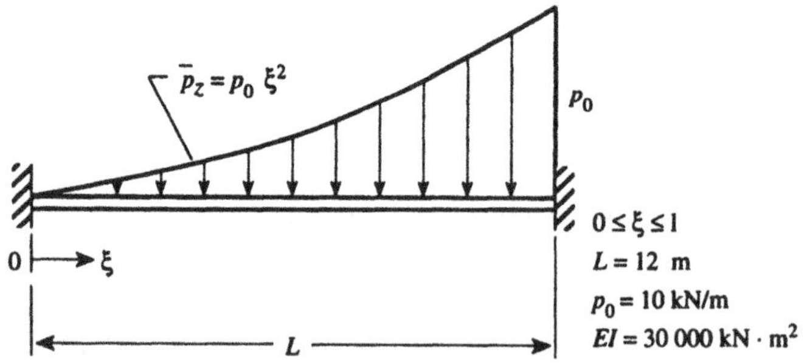

FIGURE P7.22

7.23 Compare the Trefftz method with that of Ritz. How do these approaches differ from Galerkin's method? If you wish, use the beam and results of the previous problem to assist in preparing your answer.

7.24 From Example 7.8 the force form of the torsion problem involves the governing equation

$$\frac{\partial^2\psi}{\partial z^2} + \frac{\partial^2\psi}{\partial y^2} = -2$$

Find the twisting moment M_t corresponding to the above equation being satisfied in a 6 × 4(z by y) rectangular region. Also calculate the peak shear stress. Use a trigonometric trial solution.

Hint: Recall that $\psi = 0$ on the boundary. Let the origin of the y, z coordinate system be in the center of the region. The fact that the solution should be symmetric about the z and y coordinate axes should be reflected in the choice of the trial functions. A possible choice is to use $\psi = \sum_i \psi_i N_{ui}$, with

$$N_{ui} = \cos\frac{j\pi z}{6}\cos\frac{k\pi y}{4} \quad j = 1, 3, \ldots \quad k = 1, 3, \ldots$$

Answer: $(M_t)_{\text{exact}} = 76.4,\ \tau_{\max} = 2.96$

7.25 Redo the torsion example of Problem 7.24. Use a trial solution that satisfies the $y = +2$ and -2 boundary conditions. Choose, for example, $\widetilde{\psi} = \sum_i \widehat{\psi}_i N_{ui}$, with

$$N_{u1} = (4 - y^2), \qquad N_{u2} = (4 - y^2)x^2, \qquad N_{u3} = (4 - y^2)y^2$$
$$N_{u4} = (4 - y^2)x^2y^2, \qquad N_{u5} = (4 - y^2)x^4 \ldots$$

7.26 The structural member shown in Fig. P7.26 consists of a thin linearly elastic rectangular sheet of unit thickness attached to two edge bars. It is required to support the longitudinal distributed force $\overline{p}(x)$ on the bars. Assume the height of the sheet does not change. Use the Ritz method to describe the stress distribution in the sheet. Assume a displacement pattern of the form

$$\widetilde{u} = N_{u1}\widehat{u}_1 = \widehat{u}_1 \sin\frac{\pi x}{2a}\cosh\frac{k\pi y}{2b} \qquad v = 0$$

and $\widehat{u}_1$ is the parameter to be determined. The bars have a constant cross-section A and the thickness of the sheet is t.

Hint: Let $\widetilde{u} = N_{u1}\widehat{u}_1$ and

$$\delta W = \delta\widehat{u}_1\Bigg\{\widehat{u}_1\Bigg[\underbrace{\int_{-a}^{a}[N_{u1'x}^2(x,b) + N_{u1/x}^2(x,-b)]\,EA\,dx}_{\text{Bars}} + \underbrace{\int_{-a}^{a}\int_{-b}^{b} N_{u1/x}^2(x,y)E\,dx\,dy}_{\text{Plate}}\Bigg]$$
$$-\underbrace{\int_{-a}^{a}[N_{u1}(x,b) + N_{u1}(x,-b)]\overline{p}\,dx}_{\text{Applied Loading}}\Bigg\} = 0$$

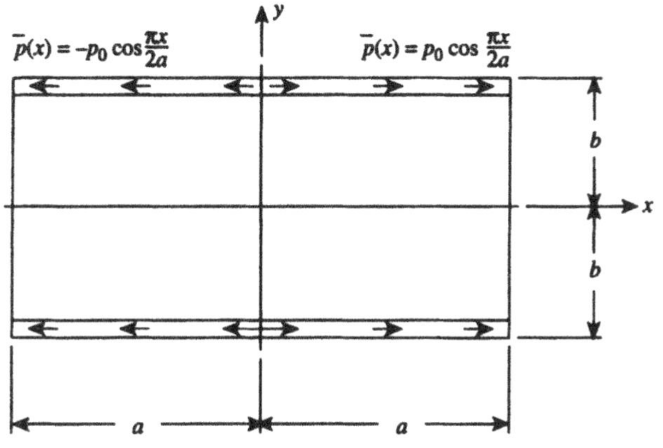

FIGURE P7.26

Answer:

$$\widehat{u}_1 = \frac{32a^2 p_0 \cosh\frac{k\pi}{2}}{E\pi^3[2\cosh^2\frac{k\pi}{2} + k\pi + 2\sinh k\pi]}$$

7.27 Consider a square thin element with in-plane loading. If the origin of the coordinates is at the center of the element, then the element covers the area $-1 \le x \le 1$, $-1 \le y \le 1$. The y edges are fixed and the x edges ($x = \pm 1$) are loaded with

$$\overline{N}_x = (1 - y^2)tE/(1 + \nu)$$

Compute the displacement and stress distribution for the element.

Hint: Assume the case of plane stress. Choose trial functions that satisfy the y edge ($y = \pm 1$) conditions, e.g., use $a(1 - y^2)$, where $a = x, x^3, xy^2$ for the x direction and y, x^2y, y^3 for the y direction.

Answer: Check your result with such evident conditions as

$$\begin{aligned} \sigma_x &= E/(1+\nu) && \text{at } x = \pm 1,\ y = 0 \\ \sigma_x &= 0 && \text{at } x = \pm 1,\ y = 1 \end{aligned}$$

7.28 Solve the boundary value problem of Problem 7.7 using the finite element method.

7.29 Suppose the domain of interest for the problem of Problem 7.2 is broken up into three elements. Now solve the problem again using finite element methodology.

7.30 Derive the element stiffness matrix and loading vector for the differential equation

$$\frac{d^2u}{dx^2} + 3u = e^{-x}$$

using the Ritz method.

7.31 Consider a simple supported beam of length L subjected to the distributed load $\overline{p}_z = \sin(\pi x/L)$. Compute the distribution of deflection using the finite element method with stiffness matrices developed using Ritz's method. Compare your results with the exact solution.

7.32 Use the finite element method to find the deflection along a beam of unit length on a elastic foundation. Suppose the loading is uniformly distributed of magnitude $|\overline{p}_z| = 1$ and the ends are fixed. Also, let $k = EI = 1$. Use Ritz's method to develop an element stiffness matrix. Compare this stiffness matrix with one developed using the transfer matrix method. Let the beam be formed of four elements.

7.33 Find the displacement and forces along the beam of Fig. P7.17 using a Ritz-based finite element method for a three-element model.

Answer: See Problem 7.17.

7.34 Find the response of the beam of Fig. P7.20 using the finite element method with a model made of three elements. Use Ritz's method to establish the stiffness matrix. Compare the results with the exact solution.

Hint: Care should be taken in forming the loading vector

$$\mathbf{p}^i = \mathbf{G}^T \int_0^1 \mathbf{N}_u^T \overline{p}_z^i \, \ell \, d\xi$$

where ℓ is the length of the element. Approximate $\overline{p}_z$ on the whole span of the beam by

$$\overline{p}_z = -\frac{4p_0}{L^2}x^2 + \frac{4p_0}{L}x$$

Let $x = \xi\ell + x_a$, where x_a is the x coordinate of the left end of the element. Then $\overline{p}_z^i = 4p_0[a_i\xi^2 + b_i\xi + c_i]$ where

$$a_i = -\left(\frac{\ell}{L}\right)^2 \qquad b_i = \frac{\ell}{L} - \frac{2\ell x_a}{L^2} \qquad c_i = \frac{x_a}{L} - \frac{x_a^2}{L^2}$$

Answer: Exact solution: $w = \dfrac{p_0}{EI}\left(-\dfrac{x^6}{90L^2} + \dfrac{x^5}{30L} - \dfrac{Lx^3}{9} + \dfrac{L^2x^2}{6}\right)$

7.35 Use a finite element procedure to solve the torsion problem of Problem 7.24.

7.36 Use a beam with distributed load $\overline{p}_z(x)$, a fixed left end, and a pinned right end to demonstrate the equivalence of the Galerkin and Ritz methods.

Hint: Begin with the potential energy

$$\Pi = \int_0^L \left[\frac{EI}{2}\left(\frac{d^2w}{dx^2}\right)^2 - \overline{p}_z w\right] dx - Mw'|_{x=L}$$

Introduce $w = N_u\widehat{w}$, and set $\partial\Pi/\partial\widehat{w}_i = 0$ for Ritz's method. Use integration by parts to find the Galerkin formulation.

Answer: For the beam problem, if the trial function satisfies all of the boundary conditions (essential and natural), Ritz's method is equivalent to Galerkin's method. In general, Galerkin's method can be employed even if the operator of the governing differential equation does not exhibit particular properties, such as being positive definite. As a result, it might seem that the Galerkin method is more generally applicable and should be favored over Ritz's method. On the other hand, in the case where the operator is linear, symmetric, and positive definite, a weak form can be formulated easily and Ritz's method is preferable. The requirement on the order of the continuity of the trial solution need not be so high for Ritz's method as for Galerkin's method. Also, the trial solution for Ritz's method only needs to satisfy the displacement (essential) boundary conditions, while for Galerkin's method, both the displacement and the static (natural) conditions must be satisfied.

8

The Finite Difference Method

It has been emphasized in this treatise that most problems in the mechanics of solids, when approached from an analytical viewpoint, can be represented locally by a system of differential equations or globally by an equivalent variational functional. Often the variational functional is formed first and used to establish the governing equations. In any event, once the mathematical (analytical) model has been established, solutions to particular problems are sought. Such solutions, which must satisfy the governing equations including the boundary conditions, can be derived from the functional or from the differential equations. As is immediately evident, contemporary mathematical methods can provide the exact solution to only the simplest forms of the governing equations. Hence, advantage is taken of the availability of the powerful digital computer by recasting problems in algebraic form. In so doing, the mathematical *continuum model*, which requires an infinite number of degrees of freedom (DOF) for its description, is replaced by a *discrete model*, which utilizes a finite number of DOF. This transformation usually involves some sort of approximation.

The trial function approximations of Chapters 6 and 7 are popular methods for implementing the necessary discretization. Another method, which is sometimes classified as a subset of the finite element approach, is the *finite difference method*. This is the subject of this chapter.

The finite difference method can be used to solve the governing differential equations directly or can be applied to the variational functional to obtain a solution. Both approaches, which are often characterized as mathematical discretizations in contrast to the more physically oriented discretization of the finite element method, will be discussed in this chapter.

8.1 Fundamentals

Consider a one-dimensional problem with governing differential equations supplemented by boundary conditions. This is usually referred to as a boundary value problem. Suppose we seek to determine a function $u(x)$, which satisfies the governing equation and boundary conditions, over the interval $0 \leq x \leq L$. With the finite difference method, a *grid* or *mesh* of points and corresponding intervals are established. Finite difference approximations to the derivatives in the mathematical model are set up at these points which are usually equally spaced. This results in a system of algebraic equations that can be solved for u at the grid points of x.

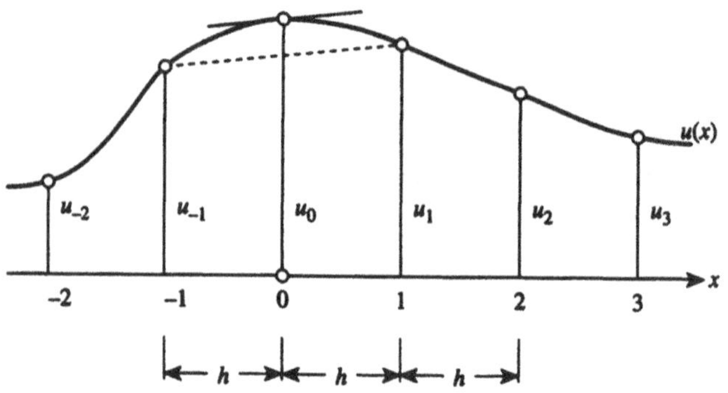

FIGURE 8.1
Notation for finite difference expressions.

Discretization of the model begins by forming discrete approximations to the derivatives. Suppose derivatives of the function $u(x)$ are to be established. Begin with the mesh of uniformly spaced points shown in Fig. 8.1 and form finite difference approximations to the derivatives.

The *central difference* quotient is defined as

$$u_0' \approx \frac{u_1 - u_{-1}}{2h} \tag{8.1}$$

where the tangent of u_0 is replaced by the chord between the values of u at -1 and 1 (see Fig. 8.1). The quantity h is the length of the interval between two successive grid points, i.e., $h = x_{k+1} - x_k$. Similarly, the difference expressions of higher order can be formed, e.g.,

$$u_0'' \approx \frac{1}{h}\left[u'\left(\frac{h}{2}\right) - u'\left(\frac{-h}{2}\right)\right] = \frac{1}{h^2}(u_{-1} - 2u_0 + u_1) \tag{8.2}$$

where $u'\left(-\frac{h}{2}\right) \approx \frac{u_0 - u_{-1}}{h}$, $u'\left(\frac{h}{2}\right) \approx \frac{u_1 - u_0}{h}$

$$u_0''' \approx \frac{1}{2h}[u''(h) - u''(-h)] = \frac{1}{2h^3}(-u_{-2} + 2u_{-1} - 2u_1 + u_2)$$
$$u_0^{iv} \approx \frac{1}{h^2}[u''(h) - 2u''(0) + u''(-h)] = \frac{1}{h^4}(u_{-2} - 4u_{-1} + 6u_0 - 4u_1 + u_2)$$

Using such difference expressions for particular derivatives, an ordinary differential equation is replaced by a system of algebraic equations. The solution of the system of equations yields values of u at particular locations, i.e., $u_k = u(x_k)$.

8.1.1 Derivation of Finite Difference Formulas

The finite difference expressions of Eqs. (8.1) and (8.2) were obtained by geometric considerations. These same expressions can be derived in several other ways. Consider again a typical point 0 and its immediate neighbors -1 and $+1$ or k, $k-1$, and $k+1$ (Fig. 8.2). To derive second order finite difference formulas, fit a second order polynomial (a parabola) to the ordinates of the three points. That is, use

$$u = a_0 + a_1 x + a_2 x^2$$

In order to fit this polynomial through the three points, the constants $a_0, a_1,$ and a_2 are

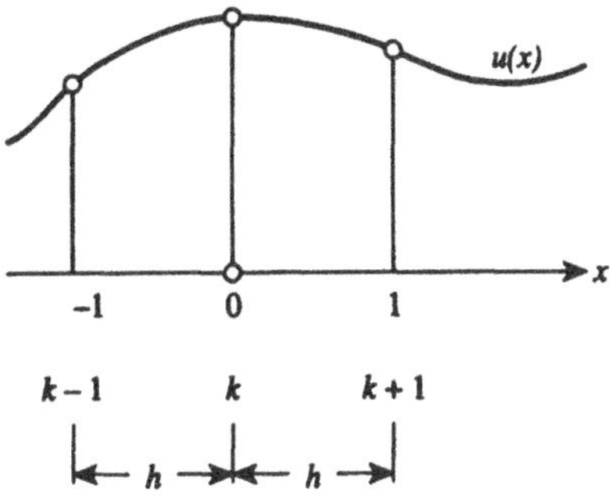

FIGURE 8.2
Notation for derivation of difference expressions.

determined such that

$$
\begin{aligned}
u &= u_{k-1} \quad \text{for } x = x_k - h \\
u &= u_k \quad \text{for } x = x_k \\
u &= u_{k+1} \quad \text{for } x = x_k + h
\end{aligned}
$$

Apply these conditions to the polynomial and obtain

$$u = u_k + \frac{(u_{k+1} - u_{k-1})}{2h}(x - x_k) + \frac{(u_{k-1} - 2u_k + u_{k+1})}{2h^2}(x - x_k)^2 \tag{8.3a}$$

with derivatives

$$u' = \frac{(u_{k+1} - u_{k-1})}{2h} + \frac{(u_{k-1} - 2u_k + u_{k+1})}{h^2}(x - x_k) \tag{8.3b}$$

$$u'' = \frac{(u_{k-1} - 2u_k + u_{k+1})}{h^2} \tag{8.3c}$$

By prescribing a value for x, Eqs. (8.3b and c) provide the first and second derivatives at any point along the polynomial.

As an example, let $x = x_k$ to obtain

$$
\begin{aligned}
u'_k &= \frac{1}{2h}(-u_{k-1} + u_{k+1}) \\
u''_k &= \frac{1}{h^2}(u_{k-1} - 2u_k + u_{k+1})
\end{aligned} \tag{8.4}
$$

These are the *central difference approximations* for u' and u''. These are referred to as second order because of the use of the second order polynomial and also because of the truncation error present. A measure of the errors associated with the finite difference technique is considered in the following section.

To establish *forward difference formulas,* set $x = x_{k-1}$ in Eqs. (8.3b and c)

$$
\begin{aligned}
u'_{k-1} &= \frac{1}{2h}(-3u_{k-1} + 4u_k - u_{k+1}) \\
u''_{k-1} &= \frac{1}{h^2}(u_{k-1} - 2u_k + u_{k+1})
\end{aligned}
$$

Or if $k - 1$ is shifted to k, these become

$$
\begin{aligned}
u'_k &= \frac{1}{2h}(-3u_k + 4u_{k+1} - u_{k+2}) \\
u''_k &= \frac{1}{h^2}(u_k - 2u_{k+1} + u_{k+2})
\end{aligned} \tag{8.5}
$$

These are the usual second order forward difference formulas.

In a similar fashion, beginning with $x = x_{k+1}$, the second order *backward difference formulas* are found to be

$$u'_k = \frac{1}{2h}(u_{k-2} - 4u_{k-1} + 3u_k)$$
$$u''_k = \frac{1}{h^2}(u_{k-2} - 2u_{k-1} + u_k) \tag{8.6}$$

Other finite difference formulas are derived in the same fashion. Miller (1979) lists central, forward, backward, and skewed formulas for up to sixth order for uniformly spaced points.

8.1.2 Use of a Taylor Series to Derive Difference Expressions and to Study Truncation Error

A Taylor[1] series can be used to develop the finite difference approximations for the derivatives of the function $u(x)$. Taylor's formula appears as

$$u(x) = u_0 + \frac{x}{1!}u'_0 + \frac{x^2}{2!}u''_0 + \cdots + \frac{x^m}{m!}u_0^{(m)} \tag{8.7}$$

For $x << 1$, the influence of higher powers of x becomes ever smaller. A Taylor series applies to functions that are continuously differentiable at least up to the level involved in the remainder.

Suppose the first derivative at point 0 is to be approximated using u_0 and the u values of neighboring points. Thus, let u'_0 be of the form

$$u'_0 = \frac{1}{h}(-a_{-1}u_{-1} - a_0u_0 - a_1u_1) + R$$

or

$$hu'_0 + a_{-1}u_{-1} + a_0u_0 + a_1u_1 = 0 + R \tag{8.8}$$

where R is the discretization (truncation) error which will be neglected in the approximate formula for u'_0. The Taylor series will yield an estimation of the error in the approximation. The coefficients a_{-1}, a_0, and a_1 are to be determined such that the error R is made as small as possible. In the first expression, the minus signs were arbitrarily introduced. In terms of the Taylor expansion,

$$u_{-1} = u_0 - \frac{h}{1!}u'_0 + \frac{h^2}{2!}u''_0 - \frac{h^3}{3!}u'''_0 + \cdots$$
$$u_1 = u_0 + \frac{h}{1!}u'_0 + \frac{h^2}{2!}u''_0 + \frac{h^3}{3!}u'''_0 + \cdots$$

Substitution of these relations into Eq. (8.8) gives

$$hu'_0 + a_{-1}u_{-1} + a_0u_0 + a_1u_1 = (a_{-1} + a_0 + a_1)u_0 + (-a_{-1} + 1 + a_1)\frac{h}{1!}u'_0$$
$$+ (a_{-1} + 0 + a_1)\frac{h^2}{2!}u''_0 + (-a_{-1} + 0 + a_1)\frac{h^3}{3!}u'''_0 + \cdots$$

[1]Brook Taylor (1685–1731) was a Cambridge-educated British mathematician who experienced ill health in his later years. As a product of a brief period of high mathematical productivity, in 1715 he published the two books *Methodus incrementorum directa et inversa* and *Linear Perspective*. He began to write on religion and philosophy, resulting in a book *Contemplatio philosophica* that appeared posthumously in 1793. In 1712 he wrote a letter to J. Machin in which he described the expansion of a function in an infinite series, an idea he had during a conversation in a coffee-house. Although there is some dispute as to who discovered "Taylor's series" first, there is little question that Taylor made his discovery independently.

Since from Eq. (8.8) this expression is equal to zero (plus R), the coefficients of the first three terms may be set equal to zero, i.e.,

$$\left.\begin{aligned} a_{-1}+a_0+a_1&=0\\ -a_{-1}+1+a_1&=0\\ a_{-1}+a_1&=0 \end{aligned}\right\} \quad \text{or} \quad a_{-1}=1/2, \quad a_0=0, \quad a_1=-1/2$$

The remaining terms provide a measure of the discretization error R. Then, Eq. (8.8) becomes

$$hu_0'+\frac{1}{2}u_{-1}+0-\frac{1}{2}u_1=-\frac{h^3}{3!}u_0'''+\cdots$$

or

$$u_0'=\frac{1}{2h}(-u_{-1}+0+u_1)-\frac{h^2}{6}u_0'''+\cdots$$

Thus, the truncation error is

$$R=-\frac{h^2}{6}u_0'''+\cdots \tag{8.9}$$

This is often expressed as $R = \text{Error} = O(h^2)$, where the symbol O means terms of a certain *order* or higher. The lowest power of h in R gives the *order* of the approximation or of the error. Thus, we say that the first order derivative is of second order accuracy, and this finite difference formula is a second order approximation. *The higher the power of h, the better the approximation and the faster the convergence of the approximation to the true value.*

8.1.3 Computational Molecules

A convenient notation for expressing difference relations is referred to as *computational molecules* or *difference stars*. For example, the first central difference derivative of u (Eq. 8.1), with respect to x, can be written as

$$u_k'=\frac{1}{2h_x}\boxed{-1 \quad \boxed{0} \quad 1}\,u+R \tag{8.10}$$

$$\begin{array}{ccc} |- & -\ \circ\ - & -| \\ k-1 & k & k+1 \end{array}$$

where $\square$ indicates the location of k, the *reference point*. Also, h has been replaced by h_x in order to make clear that the derivative with respect to x is being approximated. In the y direction,

$$u_k'=\frac{1}{2h_y}\boxed{\begin{array}{c}1\\ \\ \boxed{0}\\ \\ -1\end{array}}\,u+R \qquad \begin{array}{l} --\,k+1\\ \quad |\\ \quad \circ\ \ k\\ \quad |\\ --\,k-1\end{array} \tag{8.11}$$

Similar molecules hold for higher derivatives.

In the central difference formulas of Eqs. (8.10) and (8.11), the reference point is located at the center of the node (grid) points included in the difference expression. The central difference formulas (or central difference stars) are not symmetric for odd derivatives, e.g.,

$$u_k'=\frac{1}{2h}\boxed{-1 \quad \boxed{0} \quad 1}\,u+R \tag{8.12}$$

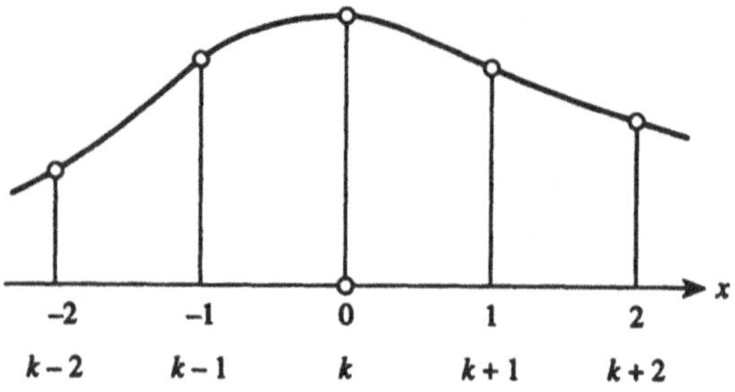

FIGURE 8.3
Notation for an improved difference scheme.

but are symmetric for even derivatives, e.g.,

$$u_k'' = \frac{1}{h^2}\boxed{1 \quad \boxed{-2} \quad 1}\, u + R \tag{8.13}$$

If the reference point is located such that more grid points are on one side than on the other, then we have either the forward or backward difference approximation, e.g.,

$$u_k' = \frac{1}{2h}\boxed{\boxed{-3} \quad 4 \quad -1}\, u + R \tag{8.14}$$

Central difference expressions yield more exact results than noncentral approximations, although in some cases, such as when boundary conditions are incorporated into a solution, we choose to work with noncentral expressions.

8.1.4 Higher Order Schemes

To improve the difference equations, additional grid or node points are included. In so doing, the order of the error (approximation) can be raised. For example, for the case of the grid points of Fig. 8.3 being used for a first order derivative, the following expression can be derived

$$u_k' = \frac{1}{12h}\boxed{1 \quad -8 \quad \boxed{0} \quad 8 \quad -1}\, u + \frac{h^4}{30} u_k^{v} + \cdots \tag{8.15}$$

This higher order difference expression along with a variety of other finite difference relations are given in Table 8.1.

8.1.5 Multiple Position Difference Method

A variation of the difference procedures is the *multiple position difference method*, wherein the derivatives at several points appear in a single expression. This technique seems to provide better approximations than the usual finite difference formulas.

As an example of this technique, suppose we wish to derive an expression containing the derivative u' and the function u at three points in the same equations. As in Section 8.1.2, we begin with

$$h(A_{-1}u_{-1}' + A_0 u_0' + A_1 u_1') + (a_{-1}u_{-1} + a_0 u_0 + a_1 u_1) = 0 + R \tag{8.16}$$

For a central difference expression, this can be simplified to

$$h(u_{-1}' + A_0 u_0' + u_1') + a_1(-u_{-1} + u_1) = 0 + R \tag{8.17}$$

TABLE 8.1

Difference Expressions

Reference Point 0 and Configuration	**Difference Relation**	**Order of the Approximation**
	$u'_k = \frac{1}{2h}\left[-1\ \boxed{0}\ +1\right]u$	$-\frac{h^2}{6}u'''_k + \cdots$
	$u'_k = \frac{1}{2h}\left[\boxed{-3}\ 4\ -1\right]u$	$+\frac{h^2}{3}u'''_k + \cdots$
	$u'_k = \frac{1}{6h}\left[-2\ \boxed{-3}\ 6\ -1\right]u$	$-\frac{h^3}{12}u^{iv}_k + \cdots$
	$u'_k = \frac{1}{12h}\left[1\ -8\ \boxed{0}\ 8\ -1\right]u$	$-\frac{h^4}{30}u^{v}_k + \cdots$
	$u'_k = \frac{1}{12h}\left[-3\ \boxed{-10}\ 18\ -6\ +1\right]u$	$-\frac{h^4}{20}u^{v}_k + \cdots$
	$u'_k = \frac{1}{12h}\left[\boxed{-25}\ 48\ -36\ 16\ -3\right]u$	$-\frac{h^4}{5}u^{v}_k + \cdots$
	$u''_k = \frac{1}{h^2}\left[1\ \boxed{-2}\ 1\right]u$	$-\frac{h^2}{12}u^{iv}_k + \cdots$
	$u''_k = \frac{1}{h^2}\left[\boxed{2}\ -5\ 4\ -1\right]u$	$+\frac{11h^2}{12}u^{vi}_k + \cdots$
	$u''_k = \frac{1}{12h^2}\left[-1\ 16\ \boxed{-30}\ 16\ -1\right]u$	$+\frac{h^4}{90}u^{vi}_k + \cdots$
	$u''_k = \frac{1}{12h^2}\left[11\ \boxed{-20}\ 6\ 4\ -1\right]u$	$-\frac{h^3}{12}u^{v}_k + \cdots$
	$u''_k = \frac{1}{12h^2}\left[\boxed{35}\ -104\ 114\ -56\ 11\right]u$	$-\frac{5h^3}{6}u^{v}_k + \cdots$
	$u'''_k = \frac{1}{2h^3}\left[-1\ 2\ \boxed{0}\ -2\ 1\right]u$	$-\frac{h^2}{4}u^{v}_k + \cdots$
	$u'''_k = \frac{1}{8h^3}\left[1\ -8\ 13\ \boxed{0}\ -13\ 8\ -1\right]u$	$+\frac{7h^4}{120}u^{vii}_k + \cdots$
	$u^{iv}_k = \frac{1}{h^4}\left[1\ -4\ \boxed{6}\ -4\ 1\right]u$	$-\frac{h^2}{6}u^{vi}_k + \cdots$
	$u^{iv}_k = \frac{1}{6h^4}\left[-1\ 12\ -39\ \boxed{56}\ -39\ 12\ -1\right]u$	$+\frac{7h^4}{240}u^{viii}_k + \cdots$

The Taylor series representations of the variables in these relations are

$$
\begin{aligned}
u'_{-1} &= u'_0 - \frac{h}{1!}u''_0 + \frac{h^2}{2!}u'''_0 - \frac{h^3}{3!}u^{iv}_0 + \frac{h^4}{4!}u^{v}_0 - \cdots \\
u'_0 &= u'_0 \\
u'_1 &= u'_0 + \frac{h}{1!}u''_0 + \frac{h^2}{2!}u'''_0 + \frac{h^3}{3!}u^{iv}_0 + \frac{h^4}{4!}u^{v}_0 + \cdots \\
u'_{-1} &= u'_0 - \frac{h}{1!}u''_0 + \frac{h^2}{2!}u'''_0 - \frac{h^3}{3!}u^{iv}_0 + \frac{h^4}{4!}u^{v}_0 - \cdots \\
u'_0 &= u'_0 \\
u'_1 &= u'_0 + \frac{h}{1!}u''_0 + \frac{h^2}{2!}u'''_0 + \frac{h^3}{3!}u^{iv}_0 + \frac{h^4}{4!}u^{v}_0 + \cdots
\end{aligned}
\tag{8.18}
$$

$$u_{-1} = u_0 - \frac{h}{1!}u_0' + \frac{h^2}{2!}u_0'' - \frac{h^3}{3!}u_0''' + \frac{h^4}{4!}u_0^{iv} - \frac{h^5}{5!}u_0^{v} + \cdots$$
$$u_0 = u_0$$
$$u_1 = u_0 + \frac{h}{1!}u_0' + \frac{h^2}{2!}u_0'' + \frac{h^3}{3!}u_0''' + \frac{h^4}{4!}u_0^{iv} + \frac{h^5}{5!}u_0^{v} + \cdots$$

Insertion of Eq. (8.18) into the left-hand side of Eq. (8.17) yields

$$\begin{aligned} h(u'_{-1} + A_0u_0' + u_1') + a_1(-u_{-1} + u_1) \\ = [0]u_0 + [1 + A_0 + 1 - (-a_1) + a_1]\frac{h}{1!}u_0' + [0]\frac{h^2}{2!}u_0'' \\ + [3 + 3 + -(-a_1) + a_1]\frac{h^3}{3!}u_0''' + [0]\frac{h^4}{4!}u_0^{iv} \\ + [5 + 5 - (-a_1) + a_1]\frac{h^5}{5!}u_0^{v} + \end{aligned}$$

The coefficients are found to be $A_0 = 4,\ a_1 = -3$ and the discretization error is

$$R = (10 - 3 - 3)\frac{h^5}{5!}u_0^{v} + \cdots = \frac{4h^5}{5!}u_0^{v} + \cdots \tag{8.19}$$

Then Eq. (8.17) appears as

$$h(u'_{-1} + 4u_0' + u_1') - 3(-u_{-1} + u_1) = 0 + \frac{4h^5}{5!}u_0^{v} + \cdots \tag{8.20}$$

or in the star notation,

$$\frac{1}{6}\boxed{1 \quad \boxed{4} \quad 1}\,u' = \frac{1}{2h}\boxed{-1 \quad \boxed{0} \quad 1}\,u + \frac{h^4}{180}u_0^{v} + \cdots \tag{8.21}$$

Expressions for higher derivatives can be found in a similar manner. Some formulas for multiple position differences are listed in Table 8.2.

8.1.6 Solution of Ordinary Differential Equations

Approximate solutions to differential equations are derived using discrete equations obtained by replacing the derivatives in the differential equations by finite difference expressions. The region under consideration is divided into a grid, and, typically, a difference expression is employed for each interior grid point. The problem of finding an unknown continuous function is replaced by a problem of solving a set of algebraic equations for discrete values. The inclusion of boundary conditions in a finite difference solution presents several problems, although few difficulties are encountered if the boundary conditions are only prescribed values of u. More difficult to satisfy are boundary conditions involving derivatives, a not uncommon occurrence.

Fictitious Nodes Beyond the Boundary

If central difference approximations are used, it is possible to define the desired function at node points outside of the boundary in order that variables such as derivatives may be determined at or near the boundary. Such fictitious points can often be avoided by using appropriate forward or backward differences at the boundaries. The incorporation

TABLE 8.2

Multiple Position Difference Expressions

Reference Point 0 and Configuration*	Difference Relation	Order of the Approximation
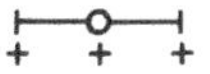	$1/6\,\begin{bmatrix}1 & \boxed{4} & 1\end{bmatrix} u' = 1/2h \begin{bmatrix}-1 & \boxed{0} & 1\end{bmatrix} u$	$-\frac{h^4}{180} u_k^{v} + \cdots$
	$1/5\,\begin{bmatrix}1 & \boxed{3} & 1\end{bmatrix} u' = 1/60h \begin{bmatrix}-1 & -28 & \boxed{0} & 28 & 1\end{bmatrix} u$	$-\frac{h^6}{2100} u_k^{vii} + \cdots$
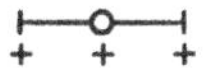	$1/12\,\begin{bmatrix}1 & \boxed{10} & 1\end{bmatrix} u'' = 1/h^2 \begin{bmatrix}1 & \boxed{-2} & 1\end{bmatrix} u$	$+\frac{h^4}{240} u_k^{vi} + \cdots$
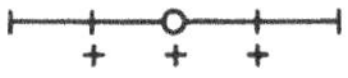	$1/15\,\begin{bmatrix}2 & \boxed{11} & 2\end{bmatrix} u'' = 1/20h^2 \begin{bmatrix}1 & 16 & \boxed{-34} & 16 & 1\end{bmatrix} u$	$-\frac{23h^6}{75600} u_k^{viii} + \cdots$
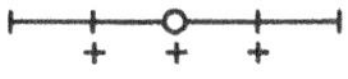	$1/4\,\begin{bmatrix}1 & \boxed{2} & 1\end{bmatrix} u''' = 1/2h^3 \begin{bmatrix}-1 & 2 & \boxed{0} & -2 & 1\end{bmatrix} u$	$-\frac{h^4}{240} u_k^{vii} + \cdots$
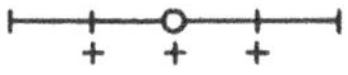	$1/6\,\begin{bmatrix}1 & \boxed{4} & 1\end{bmatrix} u^{iv} = 1/h^4 \begin{bmatrix}1 & -4 & \boxed{6} & -4 & 1\end{bmatrix} u$	$+\frac{h^4}{720} u_k^{viii} + \cdots$
	$3h^2 u_k'' + 11hu_k' + 1/6 \begin{bmatrix}\boxed{85} & -108 & 27 & -4\end{bmatrix} u$	$= 0 + R(h^5)$
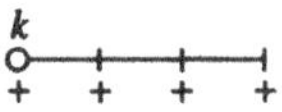	$h^2 \begin{bmatrix}\boxed{7} & -1 & -1 & 1\end{bmatrix} u'' + 30hu_k' + \begin{bmatrix}\boxed{43} & -60 & 21 & -4\end{bmatrix} u$	$= 0 + R(h^6)$
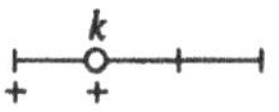	$12hu_{k-1}' + 3h^4 u_k^{iv} + 2 \begin{bmatrix}11 & \boxed{-18} & 9 & -2\end{bmatrix} u$	$= 0 + R(h^5)$
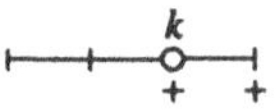	$3h^4 u_k^{iv} - 12hu_{k+1}' + 2 \begin{bmatrix}-2 & 9 & \boxed{-18} & 11\end{bmatrix} u$	$= 0 + R(h^5)$
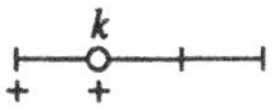	$12h^2 u_{k-1}'' - 11h^4 u_k^{iv} + 12 \begin{bmatrix}-2 & \boxed{5} & -4 & 1\end{bmatrix} u$	$= 0 + R(h^5)$
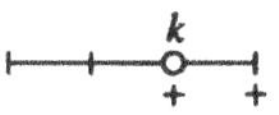	$-11h^4 u_k^{iv} + 12h^2 u_{k+1}'' + 12 \begin{bmatrix}1 & -4 & \boxed{5} & -2\end{bmatrix} u$	$= 0 + R(h^5)$

*The symbol + indicates the points whose derivatives are used in the formulas.

of derivative boundary conditions can cause the system of finite difference equations to become unsymmetric, which can be important if efficiency is of concern.

8.1.7 Convergence

In Section 8.1.6, the problem of solving differential equations is reduced to an algebraic problem in which a set of simultaneous equations is solved. Often, this will require a computer solution. The error of the finite differences approximation decreases as the mesh size decreases. More precisely, the error is proportional to a power of the mesh size. Hence, if an exact solution is not available for comparison, *convergence* of a finite difference solution is ascertained by adjusting the mesh size. The solution accuracy can also be improved by switching to a higher order difference scheme.

Consult a textbook on the finite difference method for more formal and complete discussions on convergence characteristics of finite difference approximate solutions.

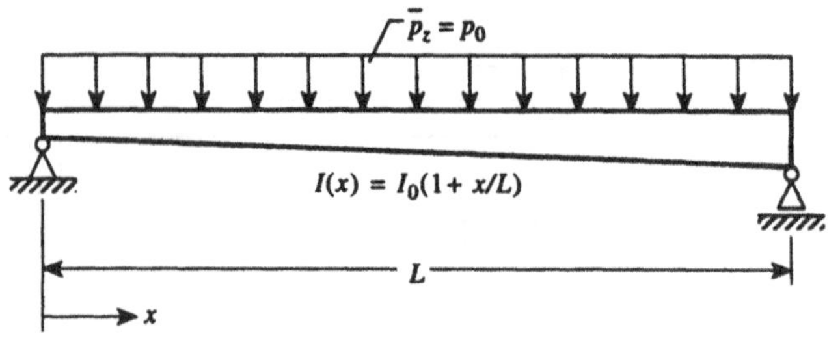

FIGURE 8.4
Beam with a variable cross-section.

EXAMPLE 8.1 Beam of Variable Cross-Section

A simply supported beam will be employed to illustrate the use of finite difference approximations in the solution of differential equations. Choose the beam of Fig. 8.4 with a cross-sectional area that varies with a moment of inertia $I(x) = I_0(1 + \frac{x}{L})$, where I_0 is the moment of inertia at the left end. We wish to compute the deflection at the quarter points along this uniformly loaded beam.

Finite difference expressions can be used to integrate the familiar fourth order beam theory equation or the simpler second order relationship

$$EI(x)\frac{d^2w}{dx^2} = -M \tag{1}$$

We choose to utilize this expression. For the statically determinate beam of Fig. 8.4, the bending moment is observed to be

$$M = \frac{p_0}{2}x(L-x) = \frac{p_0L^2}{2}\xi(1-\xi) \tag{2}$$

where $\xi = x/L$. Of course, this expression satisfies the static boundary conditions

$$M(0) = M(L) = 0 \tag{3}$$

Substitute M of (2) in (1) to obtain

$$\frac{d^2w}{dx^2} = -\frac{p_0L^2}{2EI_0}[(x/L)(1-x/L)/(1+x/L)] = -\frac{p_0L^2}{2EI_0}[\xi(1-\xi)/(1+\xi)] \tag{4}$$

The problem is now one of integrating (4) subject to the displacement boundary conditions

$$w(0) = w(L) = 0 \tag{5}$$

It is convenient to normalize w using

$$u = w/(p_0L^4/2EI_0) \tag{6}$$

so that (4) becomes

$$u'' = -\xi(1-\xi)/(1+\xi) \qquad ' = \frac{d}{d\xi} \tag{7}$$

with the boundary conditions

$$u(0) = u(1) = 0 \tag{8}$$

In establishing (7), use is made of

$$d/dx = (d/d\xi)(d\xi/dx) = (1/L)(d/d\xi) \tag{9}$$

The exact solution to (7) and (8), which can be found by simple integration, gives

$$u_{\xi=1/4} = 0.013149, \quad u_{\xi=1/2} = 0.017601, \quad u_{\xi=3/4} = 0.012026 \tag{10}$$

which will be useful for comparison.

Simple Central Difference Technique

Use the central difference form of Eq. (8.2)

$$u_i'' = \left(\frac{1}{h^2}\right)(u_{i-1} - 2u_i + u_{i+1}) \tag{11}$$

Divide the span of the beam into M segments of equal length. With $\xi = x/L$, the length of the beam is 1.

Case a (Fig. 8.5a, M = 2)

Begin with just two segments (M). Then $h = \text{length}/M = 1/2 = 0.5$. For the configuration of Fig. 8.5a.

$$u_1'' = \frac{1}{0.5^2}(u_0 - 2u_1 + u_2) = 4(0 - 2u_1 - 0) = -8u_1 \tag{12}$$

From (7), with $\xi = 0.5$ and $u_1'' = -8u_1$,

$$-8u_1 = -0.5(1 - 0.5)/(1 + 0.5) \tag{13}$$

or

$$u_1 = 0.020833 \tag{14}$$

It follows from the exact solution (10) that the error incurred with this simple configuration is

$$\frac{0.020833 - 0.017601}{0.017601} \times 100 = 18.4\% \tag{15}$$

Case b (Fig. 8.5b, M = 4)

As a second case, divide the beam into four segments. Then $M = 4$, $h = 0.25$. From (7) and (11),

$$\begin{aligned}
i = 1: \quad u_1'' &= (1/0.25^2)(u_0(=0) - 2u_1 + u_2) \\
&= -0.25(1 - 0.25)/(1 + 0.25) = -0.150 \\
i = 2: \quad u_2'' &= (1/0.25^2)(u_1 - 2u_2 + u_3) \\
&= -0.5(1 - 0.5)/(1 + 0.5) = -0.166667 \\
i = 3: \quad u_3'' &= (1/0.25^2)(u_2 - 2u_3 + u_4(=0)) \\
&= -0.75(1 - 0.75)/(1 + 0.75) = -0.107143
\end{aligned} \tag{16}$$

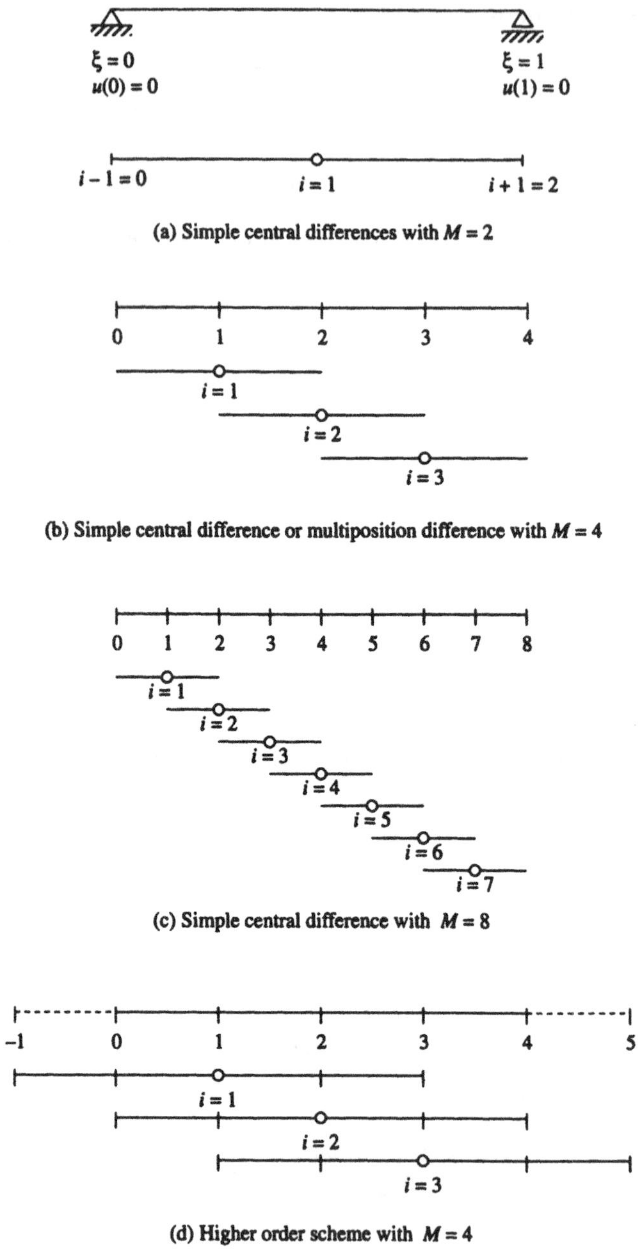

FIGURE 8.5
Various discretizations for Example 8.1.

Multiply by 0.25^2, and write the relations in matrix form

$$\underbrace{\begin{bmatrix} -2 & 1 & 0 \\ 1 & -2 & 1 \\ 0 & 1 & -2 \end{bmatrix}}_{\mathbf{K}} \underbrace{\begin{bmatrix} u_1 \\ u_2 \\ u_3 \end{bmatrix}}_{\mathbf{U}} = -10^{-3} \underbrace{\begin{bmatrix} 9.375 \\ 10.417 \\ 6.697 \end{bmatrix}}_{\mathbf{P}} \tag{17}$$

This set of equations, which would appear to be similar to stiffness equations, gives

$$u_1 = 0.013917 \quad u_2 = 0.018452 \quad u_3 = 0.012567 \tag{18}$$

At $\xi = 1/2$, the error is 4.8%.

Case c (Fig. 8.5c, M = 8)

For eight segments, $M = 8$, $h = 1/8$. Equations (7) and (11) lead to

$$\underset{\mathbf{K}}{\begin{bmatrix} -2 & 1 & & & & & \\ 1 & -2 & 1 & & & & \\ & 1 & -2 & 1 & & & \\ & & 1 & -2 & 1 & & \\ & & & 1 & -2 & 1 & \\ & & & & 1 & -2 & 1 \\ & & & & & 1 & -2 \end{bmatrix}} \underset{\mathbf{U}}{\begin{bmatrix} u_1 \\ u_2 \\ u_3 \\ u_4 \\ u_5 \\ u_6 \\ u_7 \end{bmatrix}} = -10^{-3} \underset{\mathbf{P}}{\begin{bmatrix} 14.583 \\ 18.519 \\ 22.727 \\ 22.222 \\ 19.231 \\ 14.286 \\ 7.778 \end{bmatrix}} \tag{19}$$

which gives, for example, $u_2 = 0.013333$, $u_4 = 0.017817$, and $u_6 = 0.12165$. The error at midspan ($\xi = 1/2$) is 1.2%.

Case d (Generalization)

Equation (19) is readily extended to M segments. Then $h = 1/M$, and from (11),

$$u_i'' = \frac{1}{\left(\frac{1}{M}\right)^2}(u_{i-1} - 2u_i + u_{i+1}) \tag{20}$$

Define the coordinate $\xi_i = 1/M$, $i = 1, 2, \ldots M$. From the differential equation (7),

$$u_i'' = -\xi_i(1 - \xi_i)/(1 + \xi_i) \tag{21}$$

From (20) and (21),

$$u_{i-1} - 2u_i + u_{i+1} = \frac{-1}{M^2}\frac{(i/M)(1 - i/M)}{1 + i/M} = \frac{-1}{M^3}\frac{i(M-i)}{M+i} \tag{22}$$

The boundary conditions are $u_0 = u_M = 0$. Equation (22) provides the system of stiffness-like equations

$$\underset{\mathbf{K}}{\begin{bmatrix} -2 & 1 & & & & & \\ 1 & -2 & 1 & & \mathbf{0} & & \\ & 1 & -2 & 1 & & & \\ & & \cdot & & & & \\ & & & & \cdot & & \\ & & & & & \cdot & \\ & \mathbf{0} & & & 1 & -2 & 1 \\ & & & & & 1 & -2 \end{bmatrix}} \underset{\mathbf{U}}{\begin{bmatrix} u_1 \\ u_2 \\ u_3 \\ \cdot \\ \cdot \\ \cdot \\ u_{M-2} \\ u_{M-1} \end{bmatrix}} = -(1/M^3) \underset{\mathbf{P}}{\begin{bmatrix} 1(M-1)/(M+1) \\ 2(M-2)/(M+2) \\ 3(M-3)/(M+3) \\ \cdot \\ \cdot \\ \cdot \\ (M-2)2/(M+(M-2)) \\ (M-1)1/(M+(M-1)) \end{bmatrix}} \tag{23}$$

which can be solved for u_i for a prescribed value of M.

Some important observations can be made. Matrix **K** is banded, symmetric, positive definite, and *tridiagonal* (non-zero elements occur only on the main diagonal and on one subdiagonal above and below). The solution can be efficiently computed using a solution algorithm specially designed to solve such a system of equations. Recall (Table 8.1) that the error of this finite difference approximation is $0(h^2)$. This implies that a reduction in the grid size will lead to a reduction in the error incurred and an improvement in the solution accuracy.

Accuracy Comparison for Several Values of M

ξ	Exact	$M=2$ (% error)	$M=4$ (% error)	$M=8$ (% error)	$M=16$ (% error)
0.25	0.013149		0.013917 (5.8%)	0.013333 (1.5%)	0.01320 (0.37%)
0.50	0.017601	0.020833 (18.4%)	0.018452 (4.8%)	0.017817 (1.2%)	0.01765 (0.31%)
0.75	0.012026		0.012567 (4.6%)	0.012165 (1.2%)	0.012062 (0.29%)

(24)

It is apparent that the greater the number of subdivisions, the smaller the error. An accuracy to within about 0.1% would be achieved with $M=32$. Although the error of the finite difference procedure decreases with increasing M, other errors, e.g., roundoff error, can be expected to increase.

Higher Order Schemes

If additional grid points are included in the difference equations, improved results can be expected. That is, for the same number of beam subdivisions, more accurate results will be obtained. It can also be reasoned that for the same degree of accuracy, a higher order technique requires fewer subdivisions of the structure than the simple central difference technique.

From Table 8.1, we choose a central difference expression

$$u_i'' = \frac{1}{12h^2}[-u_{i-2} + 16u_{i-1} - 30u_i + 16u_{i+1} - u_{i+2}] \tag{25}$$

which involves five grid points. As an example, use four subdivisions ($M = 4$) as shown in Fig. 8.5d. Then

$$\xi_i = i/M = i/4.$$

From (7) and (25),

$$\begin{aligned}
i = 1:\quad u_1'' &= \frac{1}{12(0.25)^2}[-u_{-1} + 16u_0 - 30u_1 + 16u_2 - u_3] \\
&= -(0.25)0.75/1.25 = -0.15 \\
i = 2:\quad u_2'' &= \frac{1}{12(0.25)^2}[-u_0 + 16u_1 - 30u_2 + 16u_3 - u_4] \\
&= -(0.5)0.5/1.5 = -0.166667 \\
i = 3:\quad u_3'' &= \frac{1}{12(0.25)^2}[-u_1 + 16u_2 - 30u_3 + 16u_4 - u_5] \\
&= -(0.75)0.25/1.75 = -0.107143
\end{aligned} \tag{26}$$

We have three equations for the seven unknowns u_{-1}, u_0, u_1, u_2, u_3, u_4, and u_5. As shown in Fig. 8.5d, fictitious nodes -1 and 5 beyond the ends of the beam have been included.

We know that the displacements at $\xi = 0$ and 1 must be zero. Thus, $u_0 = u_4 = 0$. Now, we have three equations for five unknowns. Two additional conditions will be required. Recall that the static boundary conditions are $M(\xi = 0) = M(\xi = 1) = 0$ or $u_0'' = u_4'' = 0$. The central difference expressions are

$$
\begin{aligned}
u_0'' &= (1/h^2)(u_{-1} - 2u_0 + u_1) \\
u_4'' &= (1/h^2)(u_3 - 2u_4 + u_5)
\end{aligned}
\tag{27}
$$

Since $u_0'' = u_4'' = u_0 = u_4 = 0$, (27) gives

$$u_{-1} = -u_1 \quad \text{and} \quad u_5 = -u_3 \tag{28}$$

Now (26) provides three equations for the three unknowns u_1, u_2, and u_3.

The somewhat cumbersome treatment of the fictitious node points can be avoided by using forward and backward difference expressions near the boundary. Thus, if we use (Table 8.1)

$$u_1'' = (1/12h^2)(11u_0 - 20u_1 + 6u_2 + 4u_3 - u_4) \tag{29a}$$

for $i = 1$ of (26), and

$$u_3'' = (1/12h^2)(-u_0 + 4u_1 + 6u_2 - 20u_3 + 11u_4) \tag{29b}$$

for $i = 3$ of (26), then no nodes occur beyond the ends of the beam. In this case, with $u_0 = u_4 = 0$, we again have three equations for three unknowns.

From (28) and (26), we obtain the stiffness-like equations

$$
\underset{\mathbf{K}}{\begin{bmatrix} -29 & 16 & -1 \\ 16 & -30 & 16 \\ -1 & 16 & -29 \end{bmatrix}}
\underset{\mathbf{U}}{\begin{bmatrix} u_1 \\ u_2 \\ u_3 \end{bmatrix}}
\underset{=}{=} -
\underset{\mathbf{P}}{\begin{bmatrix} 0.1125 \\ 0.1250 \\ 0.08035 \end{bmatrix}}
\tag{30}
$$

which have the solutions

$$u_1 = 0.013183, \quad u_2 = 0.017617, \quad u_3 = 0.012033 \tag{31}$$

These correspond to errors of 0.27%, 0.095%, and 0.086%, respectively. This is a significant improvement over the use of the simple central difference formula for $M = 4$.

The better results of the improved difference expression of a higher order scheme are to be expected, since the error is $0(h^4)$, while that for the simple central difference is $0(h^2)$.

For an arbitrary M, the higher order difference technique, using the central difference expression (25), provides the general expression

$$u_{i-2} + 16u_{i-1} - 30u_i + 16u_{i+1} - u_{i+2} = -12i(M - i)/M^3(M + i) \tag{32}$$

This will lead to a banded matrix of bandwidth 5.

Multiple Position Difference Method

Choose from Table 8.2 the multiple difference expression for the second derivative

$$u_{i-1}'' + 10u_i'' + u_{i+1}'' = \frac{12}{h^2}(u_{i-1} - 2u_i + u_{i+1}) \tag{33}$$

Choose four subdivisions (Fig. 8.5b) so that $M = 4$, $h = 1/4$, and $\xi_i = i/4$. From (33),

$$
\begin{aligned}
i = 1: &\quad u_0'' + 10u_1'' + u_2'' = (12)(16)(u_0 - 2u_1 + u_2) \\
i = 2: &\quad u_1'' + 10u_2'' + u_3'' = (12)(16)(u_1 - 2u_2 + u_3) \\
i = 3: &\quad u_2'' + 10u_3'' + u_4'' = (12)(16)(u_2 - 2u_3 + u_4)
\end{aligned}
\tag{34}
$$

Of course, the displacement boundary conditions are $u_0 = u_4 = 0$. The derivatives on the left-hand side are replaced by the values obtained from (7) with $M = 4$, $i = 0, 1, 2, 3$, and 4. This gives

$$u_0'' = 0, \quad u_1'' = -0.15, \quad u_2'' = -0.16667, \quad u_3'' = -0.107143, \quad u_4'' = 0$$

Substitution of these values in (34) yields

$$\underset{\mathbf{K}}{\begin{bmatrix} -1 & 2 & 0 \\ 1 & -2 & 1 \\ 0 & 1 & -2 \end{bmatrix}} \underset{\mathbf{U}}{\begin{bmatrix} u_1 \\ u_2 \\ u_3 \end{bmatrix}} \underset{=}{=} \underset{\mathbf{P}}{-10^{-3} \begin{bmatrix} 8.68 \\ 10.02 \\ 6.45 \end{bmatrix}} \tag{35}$$

with the solution $u_1 = 0.013132$, $u_2 = 0.0175843$, and $u_3 = 0.0120155$. The associated errors of −0.13%, −0.095%, and −0.082%, respectively, indicate that only a slight improvement occurs in the results relative to the higher order scheme.

The general expression for this beam, using the multiple position difference form of (33), would be

$$u_{i-1} - 2u_i + u_{i+1} = -(h^2/12)\left[\frac{\xi_i(1-\xi_{i-1})}{1+\xi_{i-1}} + 10\frac{\xi_i(1-\xi_i)}{1+\xi_i} + \frac{\xi_{i+1}(1-\xi_{i+1})}{1+\xi_{i+1}}\right] \tag{36}$$

Accuracy Comparison

Results for the three techniques for $M = 4$ and 8 are summarized below

		Simple Central Differences		Higher Order Schemes		Multiple Position Differences	
ξ	Exact	$M = 4$ (% error)	$M = 8$ (% error)	$M = 4$ (% error)	$M = 8$ (% error)	$M = 4$ (% error)	$M = 8$ (% error)
0.25	0.013149	0.013917 (5.8%)	0.013333 (1.5%)	0.013183 (0.27%)	0.013152 (0.018%)	0.013132 (−0.13%)	0.013148 (−0.0076%)
0.50	0.017601	0.018462 (4.8%)	0.017817 (1.2%)	0.017617 (0.095%)	0.017602 (0.0047%)	0.0175843 (−0.095%)	0.017600 (−0.0066%)
0.75	0.012026	0.012567 (4.6%)	0.012165 (1.2%)	0.012033 (0.086%)	0.012027 (0.0055%)	0.0120155 (−0.082%)	0.0120255 (−0.0055%)

(37)

Several conclusions can be drawn from these results. The simple central difference formula provides a result within 5% with relative efficiency (bandwidth = 3). This accuracy may be adequate for many purposes. The error decreases somewhat if the more complicated methods are employed. For $M = 4$, the errors for the higher order and multiple position methods are about 1/50 the error of the simple central difference. For $M = 8$, the change is about 1/200. ∎

EXAMPLE 8.2 Beam with Linearly Varying Loading

Apply finite differences to obtain the response of the beam of Fig. 8.6. This beam has been used as an example in several earlier chapters.

Divide the beam into five segments, so that $h = L/5$. The local governing equation is

$$EIw^{iv}(x) = \overline{p}_z(x) = p_0(1 - x/L) \tag{1}$$

with boundary conditions $w(0) = 0$, $\theta(0) = 0$, $w(L) = 0$, and $M(L) = 0$. At particular points, (1) becomes

$$EIw_i^{iv} = \overline{p}_i \tag{2}$$

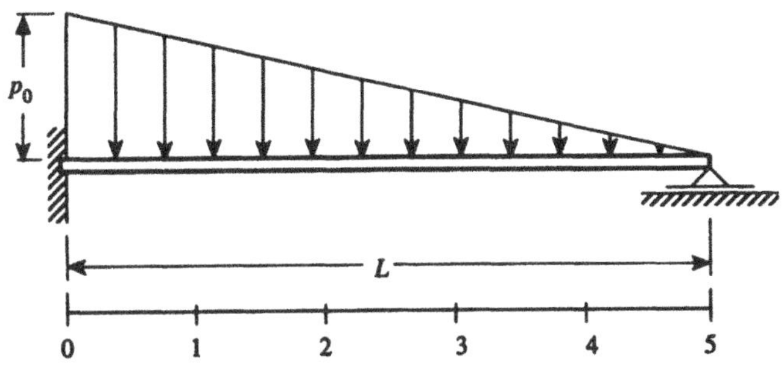

FIGURE 8.6
Fixed, simply supported beam with ramp loading.

Use a simple central difference formula from Table 8.1 (Eq. 8.2)

$$w_i^{iv} = (1/h^4)(w_{i-2} - 4w_{i-1} + 6w_i - 4w_{i+1} + w_{i+2}) \tag{3}$$

For $\overline{p}_i$ of (2), we simply use the value of the distributed load at node i. From (2) and (3), we obtain the system of equations

$$\begin{aligned}
i = 1:&\quad w_{-1} - 4w_0 + 6w_1 - 4w_2 + w_3 &&= (4/5)(p_0 h^4)/EI \\
i = 2:&\quad w_0 - 4w_1 + 6w_2 - 4w_3 + w_4 &&= (3/5)(p_0 h^4)/EI \\
i = 3:&\quad w_1 - 4w_2 + 6w_3 - 4w_4 + w_5 &&= (2/5)(p_0 h^4)/EI \\
i = 4:&\quad w_2 - 4w_3 + 6w_4 - 4w_5 + w_6 &&= (1/5)(p_0 h^4)/EI
\end{aligned} \tag{4}$$

The boundary conditions will reduce the number of unknowns in (4). Since the displacements are zero at the ends, $w_0 = w_5 = 0$. The displacement w_{-1} at the fictitious node can be identified using the slope condition $w_0' = 0(\theta_0 = 0)$. The simple central difference formula for w_0' is

$$w_0' = (1/2h)(-w_{-1} + w_1) \tag{5}$$

The condition $w_0' = 0$ gives

$$w_{-1} = w_1 \tag{6}$$

From the moment boundary condition $w_5'' = 0$ and Eq. (8.4), we find

$$w_5'' = 0 = (1/h^2)(w_4 - 2w_5 + w_6) \quad \text{or} \quad w_6 = -w_4 \tag{7}$$

Physically, conditions (6) and (7) would appear as in Fig. 8.7.

Equation (4) reduces to

$$\begin{bmatrix} 7 & -4 & 1 & 0 \\ -4 & 6 & -4 & 1 \\ 1 & -4 & 6 & -4 \\ 0 & 1 & -4 & 5 \end{bmatrix} \begin{bmatrix} w_1 \\ w_2 \\ w_3 \\ w_4 \end{bmatrix} = \begin{bmatrix} 4 \\ 3 \\ 2 \\ 1 \end{bmatrix} \frac{p_0}{5} \frac{h^4}{EI} \tag{8}$$

Although this matrix is symmetric, in general finite difference matrices are not. The solution to (8) is

$$\begin{bmatrix} w_1 \\ w_2 \\ w_3 \\ w_4 \end{bmatrix} = \begin{bmatrix} 0.001243 \\ 0.002553 \\ 0.002793 \\ 0.001788 \end{bmatrix} \frac{p_0 L^4}{EI} \tag{9}$$

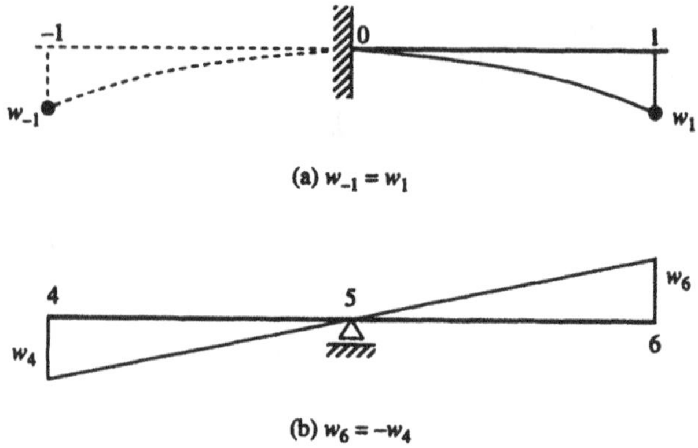

FIGURE 8.7
The fictitious boundary conditions in Example 8.2.

From Chapter 5, the exact values are

$$\begin{bmatrix} w_1 \\ w_2 \\ w_3 \\ w_4 \end{bmatrix} = \begin{bmatrix} 0.000864 \\ 0.002048 \\ 0.002352 \\ 0.001536 \end{bmatrix} \frac{p_0 L^4}{EI} \begin{matrix} (44\%) \\ (25\%) \\ (19\%) \\ (16\%) \end{matrix} \tag{10}$$

where the percent error of the approximate value is given in parentheses.

The errors in the displacement calculation are immense. Better accuracy would be obtained with a finer grid or by using an improved or multiple position difference formula.

The bending movement and shear force can be obtained from the equations

$$M = -EIw'' \quad \text{and} \quad V = -EIw''' \tag{11}$$

From Table 8.1, simple central derivatives for M and V at node i are

$$\begin{aligned} M_i &= -EI(1/h^2)(w_{i-1} - 2w_i + w_{i+1}) \\ V_i &= -EI(1/2h^2)(-w_{i-2} + 2w_{i-1} - 2w_{i+1} + w_{i+2}) \end{aligned} \tag{12}$$

These lead to the following results

	Bending Moment			Shear Force		
ξ	$\frac{M \text{ exact}}{p_0 L^2}$	$\frac{M \text{ fnt diff}}{p_0 L^2}$	% diff	$\frac{V \text{ exact}}{p_0 L}$	$\frac{V \text{ fnt diff}}{p_0 L}$	% diff
0	−0.0667	−0.06215	6.8	0.4	0.40238	0.6
0.4	0.024	0.02675	11.5	0.08	0.082	2.5
0.6	0.0293	0.3113	6.1	−0.02	−0.0179	10.3
1.0	0	0	—	−0.10	−0.0979	2.1

■

The above example dealt with a member with distributed loading. In solving the differential equation $EIw^{iv} = \overline{p}$ for a beam with a concentrated force $\overline{P}$ at point i, it is often convenient to model the concentrated force as a distributed loading. Usually, $\overline{p}_i = \overline{P}/h$ will suffice.

8.2 Partial Differential Equations

Plate, shell, and continuum problems can be governed by partial differential equations in two or more independent variables. The problem of obtaining an approximate solution using finite differences to such partial differential equations is similar in principle to the one-dimensional problems considered heretofore. Suppose a problem is to be solved over a rectangular xy region. Divide this surface into grids using equally spaced lines parallel to the x axis and similar lines parallel to the y axis as shown in Fig. 8.8. The value of the dependent variable u in a partial differential equation is to be calculated at the node points x_i, y_j of this mesh, i.e., $u_{i,j} = u(x_i, y_j)$ is to be determined. To accomplish this, we replace the partial derivatives at x_i, y_j with difference quotients and solve the resulting system of algebraic equations.

8.2.1 Difference Formulas for Partial Derivatives

Two-dimensional finite difference quotients can be built using one-dimensional expressions. From Table 8.1, the simple central difference formula for $\partial^2 u/\partial x^2$ would be

$$\partial^2 u/\partial x^2 = (1/h_x^2)[u_{i+1,j} - 2u_{i,j} + u_{i-1,j}] = (1/h_x^2)\boxed{1 \quad \boxed{-2} \quad 1}\,u \tag{8.22a}$$

while the expression for $\partial u/\partial y$ is

$$\partial u/\partial y = (1/2h_y)[u_{i,j+1} - u_{i,j-1}] = (1/2h_y)\boxed{\begin{matrix} 1 \\ \boxed{0} \\ -1 \end{matrix}}\,u \tag{8.22b}$$

where the u on the right-hand side symbolically represents appropriate nodal values of u.

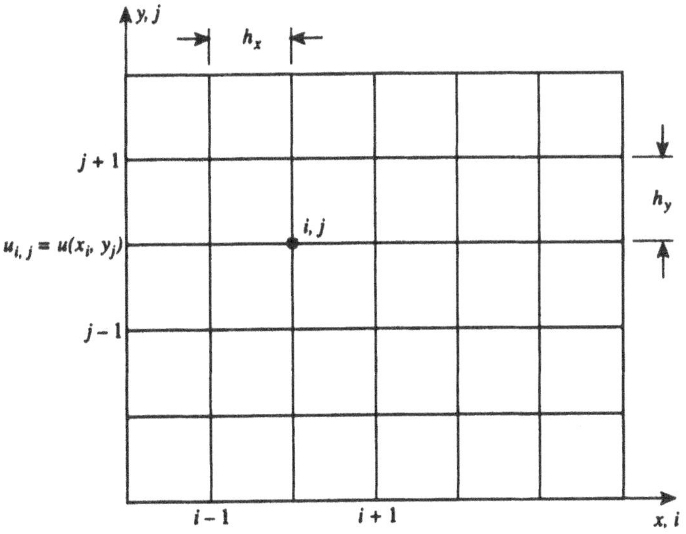

FIGURE 8.8
Notation for finite differences in two dimensions.

A mixed derivative is readily formed, e.g., for $\partial^2 u/\partial y\,\partial x(i,\,j)$ or, simply, $\partial^2 u/\partial y\,\partial x$,

$$
\begin{aligned}
\frac{\partial^2 u}{\partial y\,\partial x} &= (\partial/\partial y)(1/2h_x)(u_{i+1,j} - u_{i-1,j}) \\
&= \frac{(u_{i+1,j+1} - u_{i+1,j-1})/2h_y - (u_{i-1,j+1} - u_{i-1,j-1})/2h_y}{2h_x} \\
&= \frac{1}{4h_xh_y}(u_{i+1,j+1} - u_{i+1,j-1} - u_{i-1,j+1} + u_{i-1,j-1}) \\
&= 1/(4h_xh_y)\begin{array}{|ccc|}\hline -1 & 0 & 1 \\ 0 & \boxed{0} & 0 \\ 1 & 0 & -1 \\ \hline \end{array}\,u \quad \begin{array}{l} j+1 \\ j \\ j-1 \end{array} \\
&\qquad\qquad\quad \begin{array}{ccc} i-1 & i & i+1 \end{array}
\end{aligned}
\tag{8.23}
$$

The same result is obtained using the computational molecules of Eqs. (8.10) and (8.11) and forming the products

$$
(1/2h_x)\begin{array}{|ccc|}\hline -1 & \boxed{0} & 1 \\ \hline \end{array}\,u
$$

$$
\frac{\partial^2 u}{\partial y\,\partial x} = (1/2h_y)\begin{array}{|c|}\hline 1 \\ \boxed{0} \\ -1 \\ \hline \end{array} = 1/(4h_xh_y)\begin{array}{|ccc|}\hline -1\cdot 1 & 0\cdot 1 & 1\cdot 1 \\ -1\cdot 0 & \boxed{0\cdot 0} & 1\cdot 0 \\ -1\cdot(-1) & 0\cdot(-1) & 1\cdot(-1) \\ \hline \end{array}\,u
$$

$$
= 1/(4h_xh_y)\begin{array}{|ccc|}\hline -1 & 0 & 1 \\ 0 & \boxed{0} & 0 \\ 1 & 0 & -1 \\ \hline \end{array}\,u
$$

Similar reasoning gives

$$
(1/h_x^2)\begin{array}{|ccc|}\hline 1 & \boxed{-2} & 1 \\ \hline \end{array}\,u
$$

$$
\frac{\partial^3 u}{\partial x^2\partial y} = (1/2h_y)\begin{array}{|c|}\hline 1 \\ \boxed{0} \\ -1 \\ \hline \end{array} = 1/(2h_x^2h_y)\begin{array}{|ccc|}\hline -1 & -2 & 1 \\ 0 & \boxed{0} & 0 \\ -1 & 2 & -1 \\ \hline \end{array}\,u \quad \begin{array}{l} j+1 \\ j \\ j-1 \end{array}
\tag{8.24}
$$

$$
\begin{array}{ccc} i-1 & i & i+1 \end{array}
$$

$$
(1/h_x^2)\begin{array}{|ccc|}\hline 1 & \boxed{-2} & 1 \\ \hline \end{array}\,u
$$

$$
\frac{\partial^4 u}{\partial x^2\partial y^2} = (1/h_y^2)\begin{array}{|c|}\hline 1 \\ \boxed{-2} \\ 1 \\ \hline \end{array} = 1/(h_x^2h_y^2)\begin{array}{|ccc|}\hline 1 & -2 & 1 \\ -2 & \boxed{4} & -2 \\ 1 & -2 & 1 \\ \hline \end{array}\,u
\tag{8.25}
$$

Difference stars for equations containing summations of derivatives are also readily formed. For example, for $h_x = h_y = h$,

$$
\begin{aligned}
\frac{\partial^2 u}{\partial x^2} + \frac{\partial^2 u}{\partial y^2} &= \Delta u = (1/h^2)\begin{array}{|ccc|}\hline 1 & \boxed{-2} & 1 \\ \hline\end{array}\,u + (1/h^2)\begin{array}{|c|}\hline 1 \\ \boxed{-2} \\ 1 \\ \hline\end{array}\,u \\
&= (1/h^2)\begin{array}{|ccc|}\hline & 1 & \\ 1 & \boxed{-2\ -2} & 1 \\ & 1 & \\ \hline\end{array}\,u = 1/h^2\begin{array}{|ccc|}\hline 0 & 1 & 0 \\ 1 & \boxed{-4} & 1 \\ 0 & 1 & 0 \\ \hline\end{array}\,u
\end{aligned}
\tag{8.26}
$$

$$
\begin{aligned}
&\frac{\partial^4 u}{\partial x^4} + 2\frac{\partial^4 u}{\partial x^2 \partial y^2} + \frac{\partial^4 u}{\partial y^4} = \Delta\Delta u = \\
&(1/h^4)\begin{array}{|ccccc|}\hline 1 & -4 & \boxed{6} & -4 & 1 \\ \hline\end{array}\,u + (1/h^4)\begin{array}{|ccc|}\hline 2 & -4 & 2 \\ -4 & \boxed{8} & -4 \\ 2 & -4 & 2 \\ \hline\end{array}\,u + (1/h^4)\begin{array}{|c|}\hline 1 \\ -4 \\ \boxed{6} \\ -4 \\ 1 \\ \hline\end{array}\,u \\
&= (1/h^4)\begin{array}{|ccccc|}\hline 0 & 0 & 1 & 0 & 0 \\ 0 & 2 & -8 & 2 & 0 \\ 1 & -8 & \boxed{20} & -8 & 1 \\ 0 & 2 & -8 & 2 & 0 \\ 0 & 0 & 1 & 0 & 0 \\ \hline\end{array}\,u
\end{aligned}
\tag{8.27}
$$

The difference operation of Eq. (8.27) can also be obtained with the help of a two-dimensional Taylor series. Improved difference quotients can be established, including multiple position differences (see, for example, Zurmühl (1957)).

EXAMPLE 8.3 Torsional Stresses on the Cross-Section of a Bar

Consider the torsion of a prismatic bar with the square cross-section shown in Fig. 8.9a. The force form of the governing equation is [Chapter 1, Eq. (1.159)]

$$
\frac{\partial^2 \psi}{\partial y^2} + \frac{\partial^2 \psi}{\partial z^2} = -2 \quad \text{with } \psi = 0 \quad \text{on the boundary} \tag{1}
$$

in which ψ is the Prandtl stress function defined in Eq. (1) of Example 7.8. This is a normalized version of the stress function of Eq. (1.155).

First establish the grid system in the form shown in Fig. 8.9b. Let the grid points be equally spaced in the j and k directions with $h_y = h_z = 1/3$. Let $\psi_{j,k}$, $j, k = 1, 2, 3, 4$, be the values of ψ at the nodal points j, k. The boundary conditions for ψ are

$$
\begin{array}{lcl}
\psi_{j,1} = 0 & & \psi_{1,k} = 0 \\
\psi_{j,4} = 0 & \text{and} & \psi_{4,k} = 0 \\
j = 1, 2, 3, 4 & & k = 1, 2, 3, 4
\end{array}
\tag{2}
$$

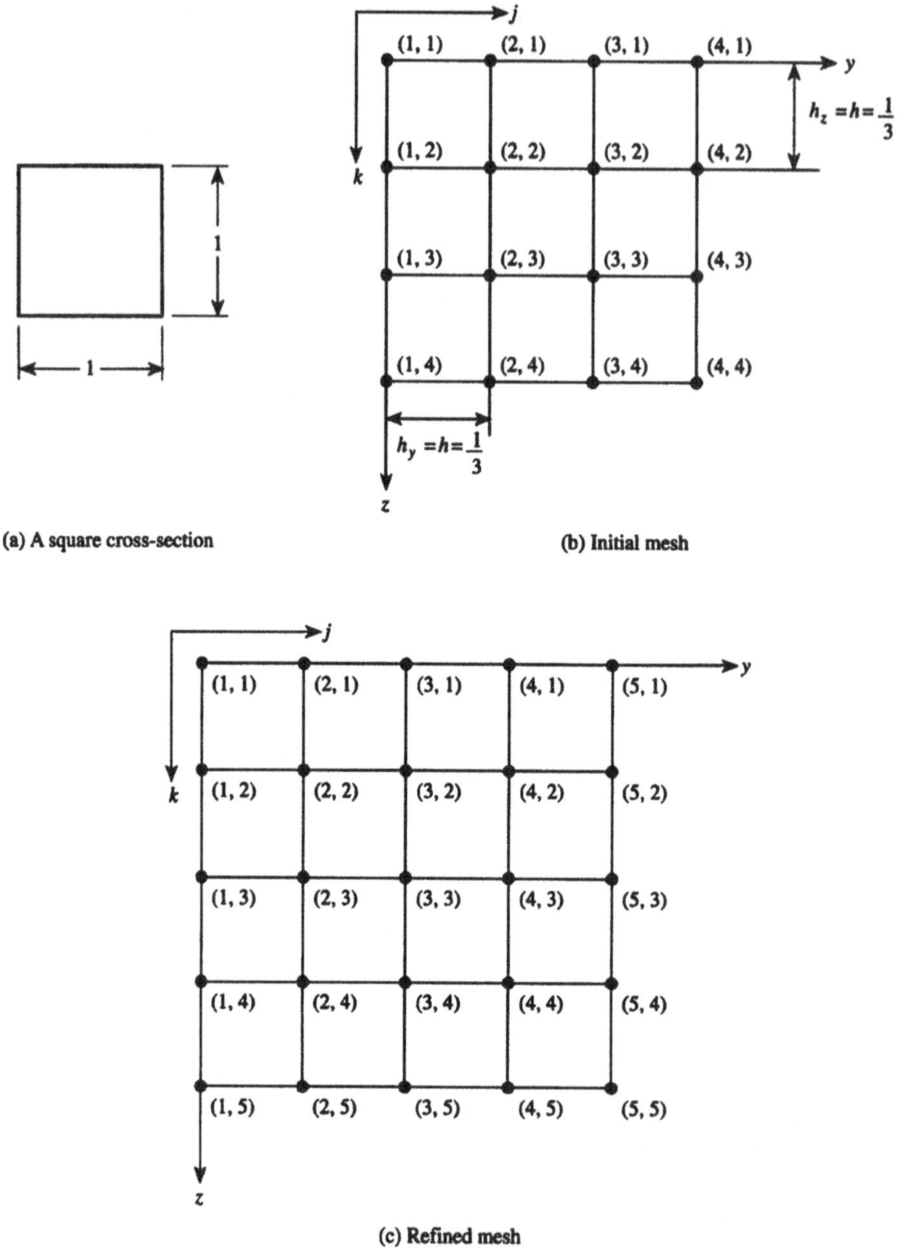

FIGURE 8.9
The square cross-section and the meshes of Example 8.3.

Use of Eq. (8.26) and these boundary conditions lead to the four linear equations

$$
\begin{aligned}
-4\psi_{2,2} + \psi_{2,3} + \psi_{3,2} &= -0.22222 \\
\psi_{2,2} - 4\psi_{2,3} + \psi_{3,3} &= -0.22222 \\
\psi_{2,2} - 4\psi_{3,2} + \psi_{3,3} &= -0.22222 \\
\psi_{2,3} + \psi_{3,2} - 4\psi_{3,3} &= -0.22222
\end{aligned}
\tag{3a}
$$

or

$$\begin{bmatrix} -4 & 1 & 1 & 0 \\ 1 & -4 & 0 & 1 \\ 1 & 0 & -4 & 1 \\ 0 & 1 & 1 & -4 \end{bmatrix} \begin{bmatrix} \psi_{2,2} \\ \psi_{2,3} \\ \psi_{3,2} \\ \psi_{3,3} \end{bmatrix} = \begin{bmatrix} -0.2222 \\ -0.2222 \\ -0.2222 \\ -0.2222 \end{bmatrix} \tag{3b}$$

$$\mathbf{K} \qquad \mathbf{U} = \mathbf{P}$$

The solution of (3) provides the values of ψ at the nodal points

$$\begin{bmatrix} \psi_{2,2} \\ \psi_{2,3} \\ \psi_{3,2} \\ \psi_{3,3} \end{bmatrix} = \begin{bmatrix} 0.11111\ (8\%) \\ 0.11111\ (8\%) \\ 0.11111\ (8\%) \\ 0.11111\ (8\%) \end{bmatrix} \tag{4}$$

The percentage in parenthesis following the values of ψ indicates the error of the finite difference solution.

Refine the grid mesh to that shown in Fig. 8.9c, and continue to use the finite difference scheme of Eq. (8.26). This leads to the system of linear equations

$$\begin{bmatrix} -4 & 1 & 0 & 1 & 0 & 0 & 0 & 0 & 0 \\ 1 & -4 & 1 & 0 & 1 & 0 & 0 & 0 & 0 \\ 0 & 1 & -4 & 0 & 0 & 1 & 0 & 0 & 0 \\ 1 & 0 & 0 & -4 & 1 & 0 & 1 & 0 & 0 \\ 0 & 1 & 0 & 1 & -4 & 1 & 0 & 1 & 0 \\ 0 & 0 & 1 & 0 & 1 & -4 & 0 & 0 & 1 \\ 0 & 0 & 0 & 1 & 0 & 0 & -4 & 1 & 0 \\ 0 & 0 & 0 & 0 & 1 & 0 & 1 & -4 & 1 \\ 0 & 0 & 0 & 0 & 0 & 1 & 0 & 1 & -4 \end{bmatrix} \begin{bmatrix} \psi_{2,2} \\ \psi_{2,3} \\ \psi_{2,4} \\ \psi_{3,2} \\ \psi_{3,3} \\ \psi_{3,4} \\ \psi_{4,2} \\ \psi_{4,3} \\ \psi_{4,4} \end{bmatrix} = \begin{bmatrix} -0.125 \\ -0.125 \\ -0.125 \\ -0.125 \\ -0.125 \\ -0.125 \\ -0.125 \\ -0.125 \\ -0.125 \end{bmatrix} \tag{5}$$

$$\mathbf{K} \qquad \mathbf{U} = \mathbf{P}$$

Note that the matrix is banded and symmetric. The solution to the linear equations is

$$\begin{bmatrix} \psi_{2,2} \\ \psi_{2,3} \\ \psi_{2,4} \\ \psi_{3,2} \\ \psi_{3,3} \\ \psi_{3,4} \\ \psi_{4,2} \\ \psi_{4,3} \\ \psi_{4,4} \end{bmatrix} = \begin{bmatrix} 0.0859 & (5.15\%) \\ 0.10937 & (4.3\%) \\ 0.0859 & (5.15\%) \\ 0.10937 & (4.3\%) \\ 0.140625 & (4.5\%) \\ 0.10937 & (4.3\%) \\ 0.0859 & (5.15\%) \\ 0.10937 & (4.3\%) \\ 0.0859 & (5.15\%) \end{bmatrix} \tag{6}$$

It can be seen that the accuracy improves as the mesh of the grid is refined.

After the stress function is calculated, the shear stresses on the cross-section can be obtained. From Chapter 1, Eq. (1.155) or, better still, from Eq. (1) of Chapter 7, Example 7.8,

$$\tau_{xy} = G\phi' \frac{\partial \psi}{\partial z} \qquad \tau_{xz} = -G\phi' \frac{\partial \psi}{\partial y} \tag{7}$$

From an appropriate finite difference scheme, the nodal stresses can be calculated. For example, with the simple central difference formulas, the shear stresses at point 3,3 of

Fig. 8.9c can be expressed as

$$\tau_{xy} = G\phi' \frac{\partial \psi}{\partial z} = \frac{G\phi'}{2h}(\psi_{3,4} - \psi_{3,2}) \qquad \tau_{xz} = -G\phi' \frac{\partial \psi}{\partial y} = -\frac{G\phi'}{2h}(\psi_{4,3} - \psi_{2,3}) \tag{8}$$

■

8.2.2 Variable Mesh Size

If the region of concern is irregular in shape so that it would be difficult to base the finite difference approximations on a uniform mesh, special accommodation can be made. Standard references on finite differences describe methods of creating nonuniform meshes. For example, a Taylor series can be used to assist in accounting for a point on a boundary curve that intersects the mesh at other than a grid point. The integral-based finite differences procedure of Section 8.3 is very convenient for approximating differential equations for irregularly shaped regions with a nonuniform mesh.

8.3 Variationally Based Finite Differences

As we have observed, the conventional finite difference equations are formed by replacing the derivatives of the variables in the governing differential equations by their difference quotients. This leads to a system of linear algebraic equations for the values of the variables at the mesh points. Such finite difference formulations are usually restricted to a regular mesh because, in most instances, an irregular mesh leads to a nonsymmetric matrix for the linear system of equations. Furthermore, considerable difficulty can occur with the conventional finite difference method in incorporating the boundary conditions. Not only are special procedures, such as the use of fictitious points, sometimes necessary, but all boundary conditions, i.e., both the force and displacement conditions, regardless of the topology on which they occur must be approximated by the differencing scheme. In recent years, it has been demonstrated that finite difference approximations can also be derived from a variational approach if the derivatives of the variables in the variational functional are replaced by corresponding difference quotients [Brush and Almroth, 1975; Bushnell, 1973; Bushnell and Almroth, 1971; Griffin and Kellogg, 1967; Griffin and Varga, 1963]. The variational principle then leads to a system of linear algebraic equations for the variables at the mesh points. Symmetric matrices can be generated because the equations are derived from a variational functional. Another advantage of the variational approach is that it may not be necessary to enforce some of the boundary conditions, since often not all of the boundary conditions are included in the variational formulation.

To be more specific about a variationally based finite differences approach, which is sometimes called the *finite difference energy method*, consider the use of the principle of virtual work. The derivatives in the integrals for internal and external work are replaced by difference expressions. Since the integrands are then piecewise constant, the integrals are easily evaluated, especially by the integration procedures sometimes employed with finite element calculations. The principle of virtual work, $\delta W = 0$, leads to a system of linear algebraic equations for the unknown displacements at the nodes. Since invoking the principle of virtual work implies that the conditions of equilibrium and the force boundary conditions will be satisfied as well as possible, the algebraic difference equations must satisfy only the displacement boundary conditions. This is in contrast to the usual finite difference method that must satisfy all boundary conditions. The principle of virtual work formulation assures that the coefficient matrix for the system of algebraic equations is symmetric and positive definite.

This coefficient matrix for the system of equations can be formed by assembling it in the same fashion as a global stiffness matrix is assembled. Indeed, the variationally based finite difference procedure is quite similar to the finite element method and, in some circumstances, can even lead to the same equations [Bushnell, 1973]. Sometimes, finite differences are treated as being a special case of finite element methods. It follows, that one could establish physical element models that correspond to the finite difference discretization.

Although the finite difference and finite element methods can be considered to be related, the finite element procedure is clearly the most dominant of the methods in use in structural mechanics today and has been successfully implemented into powerful, general purpose computer programs. Several example problems will be used to illustrate the development and application of variationally based finite differences.

EXAMPLE 8.4 Beam with Linearly Varying Loading

Consider the beam of Fig. 8.6 that was treated in Example 8.2 with conventional finite differences.

The principle of virtual work, for a beam with no concentrated loads applied on its ends, takes the form

$$\delta W = \int_0^L EI\, w''\, \delta w'' dx - \int_0^L \overline{p}_z\, \delta w\, dx = 0 \tag{1}$$

Division of the integration into six segments leads to

$$\delta W = \sum_{i=0}^{5} \int_0^{L_i} (EI\, w_i''\, \delta w_i'' - \overline{p}_z\, \delta w_i)\, dx = 0$$

Let w'' and EI be constant over each interval and let L_0 and L_5 be $L/10$ and L_1 through L_4 be $L/5$. Upon integration over the length, the expression for the principle of virtual work becomes

$$\delta W = \left(\frac{L}{10}\right)\delta W_0 + \left(\frac{L}{5}\right)(\delta W_1 + \delta W_2 + \delta W_3 + \delta W_4) + \left(\frac{L}{10}\right)\delta W_5 = 0 \tag{2}$$

with

$$\delta W_i = \delta w_i''(EI)w_i'' - \overline{p}_i\, \delta w_i \tag{3}$$

Insertion of the central difference expression

$$w_i'' = \left(\frac{1}{h^2}\right)(w_{i-1} - 2w_i + w_{i+1}) \tag{4}$$

into (3) gives

$$\delta W_i = [\delta w_{i-1} - 2\,\delta w_i + \delta w_{i+1}]\frac{EI}{\left(\frac{L}{5}\right)^4}[w_{i-1} - 2w_i + w_{i+1}] - \overline{p}_i\, \delta w_i \tag{5}$$

The quantity $\overline{p}_i$ can be assigned the value of the distributed load at point i. Thus, $\overline{p}_0 = p_0$, $\overline{p}_1 = 4p_0/5$, $\overline{p}_2 = 3p_0/5$, $\overline{p}_3 = 2p_0/5$, $\overline{p}_4 = p_0/5$, and $\overline{p}_5 = 0$. The displacement boundary conditions are $w_0 = w_0' = w_5 = 0$. The condition $w_0' = 0$ leads to $w_{-1} = w_1$. The virtual work relation (2) becomes

$$\begin{aligned}\delta W = [EI/(L/5)^3]\{&\delta w_1[(7w_1 - 4w_2 + w_3) - (4/5)p_0(L/5)]\\ &+ \delta w_2[(-4w_1 + 6w_2 - 4w_3 + w_4) - (3/5)p_0(L/5)]\\ &+ \delta w_3[(w_1 - 4w_2 + 6w_3 - 4w_4) - (2/5)p_0(L/5)]\\ &+ \delta w_4[(w_2 - 4w_3 + 6w_4 + w_6) - (1/5)p_0(L/5)]\\ &+ \delta w_6[(w_4 + w_6)]\} = 0\end{aligned} \tag{6}$$

In matrix notation, $\delta W = \delta \mathbf{V}^T(\mathbf{KV} - \overline{\mathbf{P}}) = 0$. Then $\mathbf{KV} = \overline{\mathbf{P}}$ takes the form

$$\underset{\mathbf{K}}{\begin{bmatrix} 7 & -4 & 1 & 0 & 0\\ -4 & 6 & -4 & 1 & 0\\ 1 & -4 & 6 & -4 & 0\\ 0 & 1 & -4 & 6 & 1\\ 0 & 0 & 0 & 1 & 1 \end{bmatrix}} \underset{\mathbf{V}}{\begin{bmatrix} w_1\\ w_2\\ w_3\\ w_4\\ w_5 \end{bmatrix}} = \underset{\overline{\mathbf{P}}}{\begin{bmatrix} 4/5\\ 3/5\\ 2/5\\ 1/5\\ 0 \end{bmatrix} \frac{p_0(L/5)^4}{EI}} \tag{7}$$

From the final row, $w_4 + w_6 = 0$ or $w_6 = -w_4$. Insert this relationship in the fourth row of (7). Observe that this system of symmetric equations reduces to Eq. (8) of Example 8.2. Hence, the displacements of Eq. (9) of Example 8.2 are obtained again. Furthermore, this variationally based solution gives the same moments and shear forces as the conventional solution of Example 8.2. This means that the variational approach led to a solution that satisfies exactly the force boundary condition, $M_5 = 0$, although this condition was not explicitly imposed. It is unreasonable, of course, to expect the force boundary condition always to be satisfied exactly, but a best fit is provided automatically by the variational approach. As in the case of the classical trial function solution of Chapter 8, use of a principle extended by appending boundary condition terms to the fundamental global form can ease the task of applying boundary conditions.

The above procedure is readily generalized to apply to problems with many nodes, i.e., with many DOF. To see that this is the case, it is useful to repeat the above solution using matrix notation. From (5), we can express δW_i as

$$\begin{aligned}\delta W_i &= \underset{\delta \mathbf{v}^{iT}}{[\delta w_{i-1} \quad \delta w_i \quad \delta w_{i+1}]} \begin{bmatrix} 1\\ -2\\ 1 \end{bmatrix} \frac{EI}{h^4} [1 \quad -2 \quad 1] \underset{\mathbf{v}^i}{\begin{bmatrix} w_{i-1}\\ w_i\\ w_{i+1} \end{bmatrix}}\\ &= \delta \mathbf{v}^{iT} \underset{\mathbf{k}^i}{\frac{EI}{h^4} \begin{bmatrix} 1 & -2 & 1\\ -2 & 4 & -2\\ 1 & -2 & 1 \end{bmatrix}} \mathbf{v}^i = \delta \mathbf{v}^{iT} \mathbf{k}^i \mathbf{v}^i\end{aligned} \tag{8}$$

where $\mathbf{v}^i$ is a vector of nodal displacements, and $\mathbf{k}^i$ can be considered to be a stiffness matrix. Note that $\mathbf{k}^i$ is symmetric. ∎

EXAMPLE 8.5 Torsional Stresses on the Cross-Section of a Bar

Consider the torsion of the same prismatic bar as in Example 8.3. Calculate the Prandtl stress function on the cross-section using a variationally based approach.

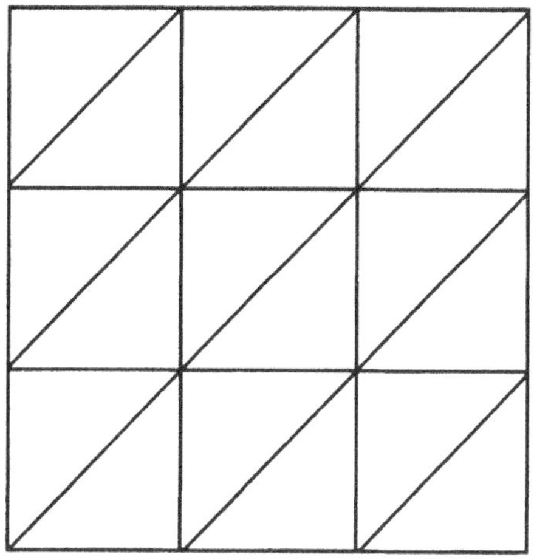

FIGURE 8.10
The triangular mesh used in Example 8.5.

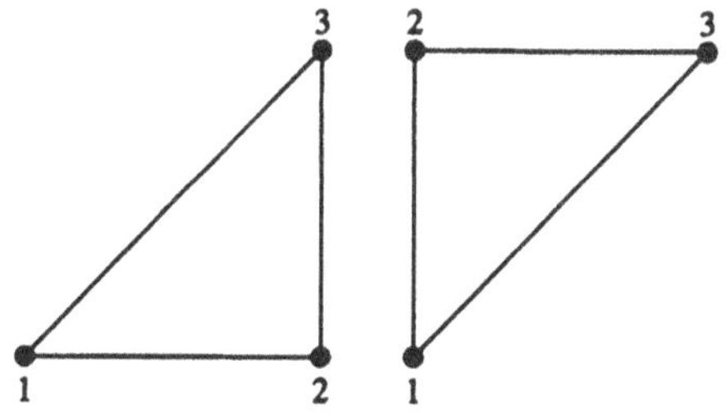

FIGURE 8.11
The numbering of the nodes of a triangle.

As in Example 8.3, the governing differential equation is given by

$$\frac{\partial^2 \psi}{\partial y^2} + \frac{\partial^2 \psi}{\partial z^2} = -2 \qquad \text{with} \quad \psi = 0 \quad \text{on the boundary} \tag{1}$$

in which ψ is the Prandtl stress function.

The corresponding global form can be based on a complementary principle such as the principle of complementary virtual work which, using the normalized version of the stress function of Chapter 1, Eq. (1.155), has the form (Eq. (3) of Chapter 2, Example 2.9)

$$\delta W^* = \int \left(\frac{\partial \psi}{\partial z} \frac{\partial}{\partial z} \delta\psi + \frac{\partial \psi}{\partial y} \frac{\partial}{\partial y} \delta\psi - 2\,\delta\psi \right) dy\, dz = 0 \tag{2}$$

Use the gridwork of Fig. 8.9b, but now divide the squares into triangles as indicated in Fig. 8.10, and number the nodes as shown in Fig. 8.11. The triangles occur in two orientations (Figs. 8.11a and b). Equation (2) involves an integration over an area, and it is desirable that the integrand be constant. From the simple central difference scheme, the derivatives of ψ in a triangle are constant, and $\delta\psi$ in the triangle can be taken to be the average of the values of $\delta\psi$ at the corner nodes. Then the integrand in (2) becomes constant, which helps facilitate the integration. Other kinds of meshes also achieve this [Pian, 1971], but are more complicated than the triangles. From the central difference scheme in Table 8.1, the derivatives $\partial\psi/\partial y$

and $\partial\psi/\partial z$, as well as $\partial_y\,\delta\psi$ and $\partial_z\,\delta\psi$, in the triangle of Fig. 8.11a can be represented as

$$\begin{aligned}\frac{\partial\psi}{\partial y} &= \frac{\psi_2-\psi_1}{h}, & \frac{\partial\psi}{\partial z} &= \frac{\psi_3-\psi_2}{h}\\ \frac{\partial}{\partial y}\delta\psi &= \frac{\delta\psi_2-\delta\psi_1}{h}, & \frac{\partial}{\partial z}\delta\psi &= \frac{\delta\psi_3-\delta\psi_2}{h}\end{aligned} \tag{3}$$

and those in Fig. 8.11b can be represented as

$$\begin{aligned}\frac{\partial\psi}{\partial y} &= \frac{\psi_3-\psi_2}{h}, & \frac{\partial\psi}{\partial z} &= \frac{\psi_2-\psi_1}{h}\\ \frac{\partial}{\partial y}\delta\psi &= \frac{\delta\psi_3-\delta\psi_2}{h}, & \frac{\partial}{\partial z}\delta\psi &= \frac{\delta\psi_2-\delta\psi_1}{h}\end{aligned} \tag{4}$$

The total complementary virtual work is the sum of the virtual work for each of the triangles k. Thus,

$$\delta W^* = \sum_k \int \left(\frac{\partial\psi}{\partial z}\frac{\partial}{\partial z}\partial\psi + \frac{\partial\psi}{\partial y}\frac{\partial}{\partial y}\delta\psi - 2\delta\psi\right) dy\,dz = 0 \tag{5}$$

Replace the derivatives in (5) by their approximations, and replace $\delta\psi$ by the average of the values of $\delta\psi$ at the three corners of the triangles. The total complementary virtual work of the kth triangle then becomes

$$\begin{aligned}\delta W^* = \sum_k A\Bigg[&\left(\frac{\psi_2^k-\psi_1^k}{h}\right)\left(\frac{\delta\psi_2^k-\delta\psi_1^k}{h}\right)\\ &+\left(\frac{\psi_3^k-\psi_2^k}{h}\right)\left(\frac{\delta\psi_3^k-\delta\psi_2^k}{h}\right)+\frac{2}{3}\left(\delta\psi_1^k+\delta\psi_2^k+\delta\psi_3^k\right)\Bigg]\end{aligned} \tag{6}$$

where A is the area of the kth triangle, and ψ_i^k, $i = 1, 2, 3$ are the values of ψ at the corners of the kth triangle. Rewrite Eq. (6) as

$$\delta W^* = \sum_k \delta\boldsymbol{\psi}^{kT}(\mathbf{k}^k\boldsymbol{\psi}^k - \mathbf{p}^k) = 0 \tag{7}$$

with

$$\boldsymbol{\psi}^k = \begin{bmatrix}\psi_1^k & \psi_2^k & \psi_3^k\end{bmatrix}^T \qquad \mathbf{k}^k = \frac{A}{h^2}\begin{bmatrix}-1 & 1 & 0\\ 1 & -2 & -1\\ 0 & -1 & -1\end{bmatrix} \qquad \mathbf{p}^k = \frac{2A}{3}\begin{bmatrix}1\\1\\1\end{bmatrix}$$

Equation (7) looks like the stiffness equations found in the finite element method, and in fact, it can be treated in the same manner as the stiffness equations. Thus, the summation process can be performed in the same way as assembling the element stiffness matrices into the global equation as described in Chapter 5, and the boundary conditions can also be imposed similarly. After the summation, Eq. (7) can be written as

$$\delta\boldsymbol{\psi}^T(\mathbf{K}\boldsymbol{\psi} - \mathbf{P}) = 0 \tag{8}$$

in which

$$\boldsymbol{\psi} = \begin{bmatrix}\psi_{2,2} & \psi_{2,3} & \psi_{3,2} & \psi_{3,3}\end{bmatrix}^T$$

and **K, P** are the same as those in Eq. (3) of Example 8.3. Also, $\psi_{i,j}$ is the value of ψ at the nodal point i, j.

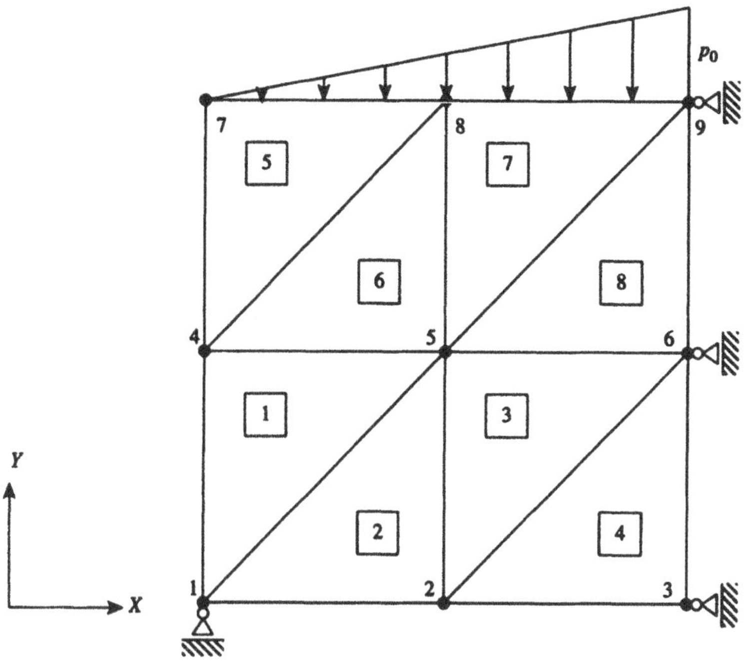

FIGURE 8.12
Finite difference mesh for the thin sheet in Example 8.6.

Observe again how similar the variational approach for finite differences is to the variational method for finite elements. ∎

EXAMPLE 8.6 A Flat, In-Plane Loaded Structure
Consider the in-plane loaded flat structure of Chapter 6, Example 6.2. Use variationally based finite differences to calculate the displacements of the nodal points.

Because of the symmetry of the loading and the supports, as well as the structure, consider only half of the structure, with the loading and supports of the form of Chapter 6, Fig. 6.14b. Discretize the model with nine nodes and eight triangles as shown in Fig. 8.12. Treat this as a two-dimensional plane stress problem, and use the principle of virtual work. For each triangle, the virtual work is given by (Chapter 2, Example 2.5)

$$\begin{aligned}\delta W &= \int_V \delta\boldsymbol{\epsilon}^T \boldsymbol{\sigma}\, dV - \int_V \delta\mathbf{u}^T \overline{\mathbf{p}}_V\, dV - \int_{S_p} \delta\mathbf{u}^T \overline{\mathbf{p}}\, ds \\ &= \int_A \delta\boldsymbol{\epsilon}^T \mathbf{s}\, dA - \int_A \delta\mathbf{u}^T \overline{\mathbf{p}}_V\, dA - \int_{S_p} \delta\mathbf{u}^T \overline{\mathbf{p}}\, ds \qquad (1)\end{aligned}$$

where $\mathbf{s}$ is defined in Eq. (2) of Chapter 2, Example 2.5. Neglect the body force $\overline{\mathbf{p}}_V$, and substitute the kinematical relations (Chapter 1, Eq. 1.24) and material law (Chapter 1, Eq. 1.39) for the two-dimensional plane stress problem in (1)

$$\begin{aligned}\delta W &= \int_A \delta\mathbf{u}^T{}_u\mathbf{D}^T\, \mathbf{E}\, \mathbf{D}_u\, \mathbf{u}\, dA - \int_{S_p} \delta\mathbf{u}^T \overline{\mathbf{p}}\, ds \\ &= \int_A \delta(\mathbf{D}\mathbf{u})^T \mathbf{E}\, \mathbf{D}\, \mathbf{u}\, dA - \int_{S_p} \delta\mathbf{u}^T \overline{\mathbf{p}}\, ds \qquad (2)\end{aligned}$$

where

$$\mathbf{D} = \begin{bmatrix} \partial_x & 0 \\ 0 & \partial_y \\ \partial_y & \partial_x \end{bmatrix} \quad \mathbf{E} = \frac{Et}{1-\nu^2}\begin{bmatrix} 1 & \nu & 0 \\ \nu & 1 & 0 \\ 0 & 0 & \frac{1-\nu}{2} \end{bmatrix} \quad \text{and} \quad \mathbf{u} = [u_x \quad u_y]^T$$

in which u_x and u_y are the displacements in the x and y directions, and t is the thickness of the plate. Expand the first term on the right-hand side of (2)

$$\int_A \delta\mathbf{u}^T{}_u\mathbf{D}^T\mathbf{E}\,\mathbf{D}_u\mathbf{u}\,dA$$
$$= \int_A [\delta u_x \quad \delta u_y]\frac{Et}{1-\nu^2}\begin{bmatrix} {}_x\partial\,\partial_x + {}_y\partial\frac{1-\nu}{2}\partial_y & {}_x\partial\nu\,\partial_y + {}_y\partial\frac{1-\nu}{2}\partial_x \\ {}_y\partial\nu\,\partial_x + {}_x\partial\frac{1-\nu}{2}\partial_y & {}_y\partial\,\partial_y + {}_x\partial\frac{1-\nu}{2}\partial_x \end{bmatrix}\begin{bmatrix} u_x \\ u_y \end{bmatrix} dA \tag{3}$$

where ${}_x\partial$ means that the derivative $\partial/\partial x$ is taken of the variable to the left and ∂_x is the usual derivative $\partial/\partial x$ of the variable to the right.

Employ the central finite difference scheme, representing the first derivatives of u_x with respect to x and y for the triangle corresponding to Fig. 8.11a as

$$\begin{aligned} \frac{\partial u_x}{\partial x} &= \frac{u_{x2}-u_{x1}}{h} = [u_{x1} \quad u_{x2} \quad u_{x3}]\begin{bmatrix} -1 \\ 1 \\ 0 \end{bmatrix}\frac{1}{h} \\ \frac{\partial u_x}{\partial y} &= \frac{u_{x3}-u_{x2}}{h} = [u_{x1} \quad u_{x2} \quad u_{x3}]\begin{bmatrix} 0 \\ -1 \\ 1 \end{bmatrix}\frac{1}{h} \end{aligned} \tag{4}$$

Similar formulas apply for $\partial u_y/\partial x$ and $\partial u_y/\partial y$. Substitute (4) and the formulas for $\partial u_y/\partial x$ and $\partial u_y/\partial y$ into (3). This leads to

$$\int_A \delta\mathbf{u}^T{}_u\mathbf{D}^T\mathbf{E}\,\mathbf{D}_u\mathbf{u}\,dA = \delta\mathbf{v}^T E^*\begin{bmatrix} \mathbf{k}_{11} & \mathbf{k}_{12} \\ \mathbf{k}_{21} & \mathbf{k}_{22} \end{bmatrix}\mathbf{v} = \delta\mathbf{v}^T\mathbf{k}\mathbf{v}$$

where A is the area of the triangle, $E^* = tEA/(1-\nu^2)h^2$, and

$$\mathbf{v} = [u_{x1} \quad u_{x2} \quad u_{x3} \quad u_{y1} \quad u_{y2} \quad u_{y3}]^T$$

$$\begin{aligned} \mathbf{k}_{11} &= \left[\begin{bmatrix} -1 \\ 1 \\ 0 \end{bmatrix}[-1 \quad 1 \quad 0] + \frac{1-\nu}{2}\begin{bmatrix} 0 \\ -1 \\ 1 \end{bmatrix}[0 \quad -1 \quad 1]\right] \\ &= \begin{bmatrix} 1 & -1 & 0 \\ -1 & \gamma & -\beta \\ 0 & -\beta & \beta \end{bmatrix} \\ \mathbf{k}_{12} &= \left[\nu\begin{bmatrix} -1 \\ 1 \\ 0 \end{bmatrix}[0 \quad -1 \quad 1] + \frac{1-\nu}{2}\begin{bmatrix} 0 \\ -1 \\ 1 \end{bmatrix}[-1 \quad 1 \quad 0]\right] \\ &= \begin{bmatrix} 0 & \nu & -\nu \\ \beta & -\beta & \nu \\ -\beta & \beta & 0 \end{bmatrix} \\ \mathbf{k}_{21} &= \left[\nu\begin{bmatrix} 0 \\ -1 \\ 1 \end{bmatrix}[-1 \quad 1 \quad 0] + \frac{1-\nu}{2}\begin{bmatrix} -1 \\ 1 \\ 0 \end{bmatrix}[0 \quad -1 \quad 1]\right] \end{aligned} \tag{5}$$

$$= \begin{bmatrix} 0 & \beta & -\beta \\ \nu & -\beta & \beta \\ -\nu & \nu & 0 \end{bmatrix}$$

$$\mathbf{k}_{22} = \left[\begin{bmatrix} 0 \\ -1 \\ 1 \end{bmatrix} [0 \quad -1 \quad 1] + \frac{1-\nu}{2} \begin{bmatrix} -1 \\ 1 \\ 0 \end{bmatrix} [-1 \quad 1 \quad 0] \right]$$

$$= \begin{bmatrix} \beta & -\beta & 0 \\ -\beta & \gamma & -1 \\ 0 & -1 & -1 \end{bmatrix}$$

with $\beta = (1-\nu)/2$ and $\gamma = (3-\nu)/2$.
Then $\mathbf{k}$ becomes

$$\mathbf{k} = E^* \begin{bmatrix} 1 & -1 & 0 & 0 & \nu & -\nu \\ -1 & \gamma & -\beta & \beta & -\beta & \nu \\ 0 & -\beta & \beta & -\beta & \beta & 0 \\ 0 & \beta & -\beta & \beta & -\beta & 0 \\ \nu & -\beta & \beta & -\beta & \gamma & -1 \\ -\nu & \nu & 0 & 0 & -1 & 0 \end{bmatrix} \tag{6}$$

Note that $\mathbf{k}$ is symmetric.

For the triangle in Fig. 8.11b, the central difference expressions for $\partial u_x/\partial x$ and $\partial u_x/\partial y$ would be

$$\begin{aligned} \frac{\partial u_x}{\partial x} &= \frac{u_{x3} - u_{x2}}{h} = [u_{x1} \quad u_{x2} \quad u_{x3}] \begin{bmatrix} 0 \\ -1 \\ 1 \end{bmatrix} \frac{1}{h} \\ \frac{\partial u_x}{\partial y} &= \frac{u_{x2} - u_{x1}}{h} = [u_{x1} \quad u_{x2} \quad u_{x3}] \begin{bmatrix} -1 \\ 1 \\ 0 \end{bmatrix} \frac{1}{h} \end{aligned} \tag{7}$$

Similar expressions can be obtained for $\partial u_y/\partial x$ and $\partial u_y/\partial y$. Substitute these expressions into (3), and obtain $\mathbf{k}$ as

$$\mathbf{k} = E^* \begin{bmatrix} \beta & \beta & 0 & 0 & \beta & -\beta \\ -\beta & \gamma & -1 & \nu & -\beta & \beta \\ 0 & -1 & 1 & -\nu & \nu & 0 \\ 0 & \nu & -\nu & 1 & -1 & 0 \\ \beta & -\beta & \nu & -1 & \gamma & -\beta \\ -\beta & \beta & 0 & 0 & -\beta & \beta \end{bmatrix} \tag{8}$$

This matrix is also symmetric.

Now calculate the loading vector. The tractions are applied on the side of the plate connecting nodes 7, 8, and 9 and can be lumped to these nodes. For triangle 5, the traction distribution on the side connecting nodes 7 and 8 is shown in Fig. 8.13a. From the conditions of equilibrium, the equivalent resultant force F_5 is $hp_0/4$ acting $2h/3$ away from node 7. The nodal reaction forces, which can be viewed as the loading on these nodes, are obtained

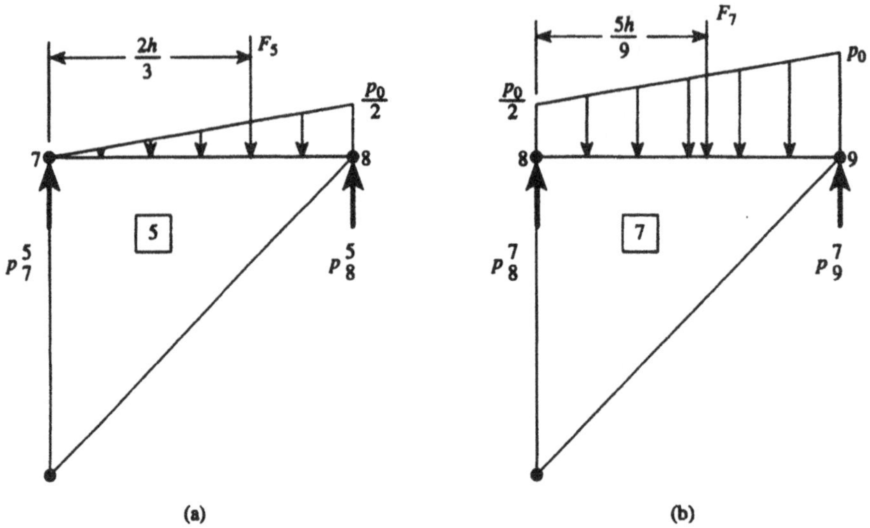

FIGURE 8.13
Loading on triangles 5 and 7.

from equilibrium requirements as

$$p_7^5 = \frac{h}{12} p_0 \qquad p_8^5 = \frac{h}{6} p_0 \tag{9}$$

For triangle 7, the traction distribution on the side connecting nodes 8 and 9 is a trapezoid. The loading on nodes 8 and 9 is found to be

$$p_8^7 = \frac{4h}{12} p_0 \qquad p_9^7 = \frac{5h}{12} p_0 \tag{10}$$

The principle of virtual work for the entire plate is formed by summing the virtual work of each triangle. An assembly process, following that described in Chapter 5, leads to the global stiffness matrix and to the global load vector. The boundary conditions

$$u_{y1} = u_{x3} = u_{x6} = u_{x9} = 0 \tag{11}$$

should be imposed, and the appropriate rows and columns deleted in the global matrix **K**. Then we obtain a system of linear equations for the solution of the nodal displacements

$$\mathbf{KV} = \mathbf{P} \tag{12}$$

where

$$\begin{aligned}
\mathbf{V} &= [u_{x1} \quad u_{x2} \quad u_{x4} \quad u_{x5} \quad u_{x7} \quad u_{x8} \quad u_{y2} \quad u_{y3} \quad u_{y4} \quad u_{y5} \quad u_{y6} \quad u_{y7} \quad u_{y8} \quad u_{y9}]^T \\
\mathbf{P} &= [0 \quad 0 \cdots P_7 \quad P_8 \quad P_9]^T \\
&= \left[0 \quad 0 \cdots p_7^5 \quad p_8^5 + p_8^7 \quad p_9^7\right]^T
\end{aligned}$$

and

$$
\mathbf{K} = E^* \begin{bmatrix}
2\gamma & & & & & & & & & & & & & \\
-1 & 2\gamma & & & & & & & & & & & & \\
-\beta & 0 & 2\gamma & & & & \text{Symmetric} & & & & & & & \\
0 & -2\beta & -2 & \gamma & & & & & & & & & & \\
0 & 0 & -\beta & 0 & \gamma & & & & & & & & & \\
0 & 0 & 0 & -2\beta & -1 & 2\gamma & & & & & & & & \\
\nu & -\beta & 0 & 0 & 0 & 0 & 2\gamma & & & & & & & \\
0 & \nu & 0 & 0 & 0 & 0 & -\beta & \gamma & & & & & & \\
\beta & 0 & -\beta & \alpha & \nu & -\alpha & 0 & 0 & 2\gamma & & & & & \\
-\alpha & \alpha & \alpha & -2\beta & 0 & \alpha & -2 & 0 & -2\beta & 4\gamma & & & & \\
0 & -\alpha & 0 & \alpha & 0 & 0 & 0 & -1 & 0 & -2\beta & 2\gamma & & & \\
0 & 0 & \beta & 0 & -\beta & \nu & 0 & 0 & -1 & 0 & 0 & \gamma & & \\
0 & 0 & -\alpha & \alpha & \beta & -\beta & 0 & 0 & 0 & -2 & 0 & -\beta & \alpha & \\
0 & 0 & 0 & -\alpha & 0 & \alpha & 0 & 0 & 0 & 0 & -1 & 0 & -\beta & \gamma
\end{bmatrix}
$$

with $E^* = tAE/(1-\nu^2)h^2$, $\alpha = (1+\nu)/2$, $\beta = (1-\nu)/2$, and $\gamma = (3-\nu)/2$.

From the dimensions of the plate in Chapter 6, Fig. 6-14 and the mesh layout in Fig. 8.12, we have $h = 1$. Assign the numerical values $t = 0.2$ m, $E = 30\ \mathrm{GN/m^2}$, $\nu = 0.0$.

Solve equation (12) to compute the displacements at the nodal points

	u_x	u_y
1	-2.5×10^{-4}	0
2	-2.02×10^{-4}	-0.44×10^{-4}
3	0	-0.55×10^{-4}
4	-0.059×10^{-4}	-0.183×10^{-4}
5	-0.0648×10^{-4}	-0.48×10^{-4}
6	0	0.062×10^{-4}
7	0.15×10^{-4}	-0.25×10^{-4}
8	0.11×10^{-4}	-0.55×10^{-4}
9	0	-0.74×10^{-4}

The displacement pattern is shown in Fig. 8.14. It is similar to that of the finite element solution in Chapter 6, but the numerical values are lower than those for the finite element method. This means that the finite difference method appears to "stiffen" the structure. For a better result, a finer mesh should be employed. ∎

It is, of course, possible to employ global (integral) formulations other than the principle of virtual work as the basis of finite difference approximations. For example, such weighted residual approaches as Galerkin's method are quite suitable. If the Hellinger-Reissner functional of Chapter 2 is utilized, both displacement and force variables occur in this mixed variational functional and, hence, in the derived finite difference equations. See Noor and Schnobrich (1973), Noor, et al. (1973), and Pian (1971) for finite difference formulations utilizing mixed methods.

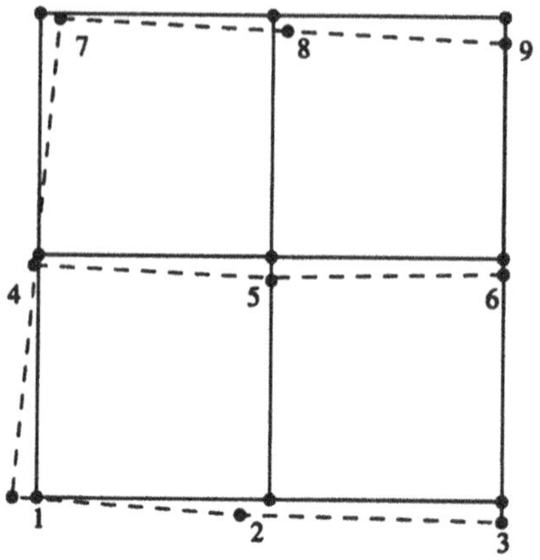

FIGURE 8.14
Displacement pattern of the plate with phase strain in Example 8.6.

References

Brush, D.O. and Almroth, B.O., 1975, *Buckling of Bars, Plates, and Shells*, McGraw-Hill, New York.

Bushnell, D., 1973, Finite difference energy models versus finite element models: Two variational approaches in one computer program, in Fenves, S.J., Perrone, N., Robinson J. and Schnobrich, W.C. (Eds.) *Numerical and Computer Methods in Structural Mechanics*, Academic Press, New York.

Bushnell, D. and Almroth, B.O., 1971, Finite difference energy method for non-linear shell structures, *J. Comput. Struct.*, Vol. 1, pp. 361–387.

Griffin, D.S. and Kellogg, R.B., 1967, A numerical solution for axially symmetrical and plane elasticity problems, *Int. J. Solids Struct.*, Vol. 3, pp. 781–794.

Griffin, D.S. and Varga, R.S., 1963, Numerical solution of plane elasticity problems, *J. Soc. Indust. Appl. Math.*, Vol. 11, pp. 1046–1060.

Miller, R.E., Jan. 1979, Finite difference formulas for two-point boundary value problems, *Dept. of Theoretical and Applied Mechanics, T & A.M., Report No. 431*, University of Illinois, Urbana.

Noor, A.K. and Schnobrich, W.C., 1973, On improved finite difference discretization procedures, *Variational Methods in Engineering*, Vol. II, Southampton University Press, Southampton.

Noor, A.K., Stephens, W.B. and Fulton, R.E., 1973, An improved numerical process for solution of solid mechanics problems, *J. Comput. Struct.*, Vol. 3, pp. 1397–1437.

Pian, T.H.H., 1971, *Variational formulations of numerical methods in solids continua*, Study No. 5, University of Waterloo, Ontario.

Zurmühl, R., 1957, Behandlung der Plattenaufgabe nach dem verbesserten Differenzenverfahren, *ZAMM*, Vol. 37, pp. 1–16.

Problems

8.1 Use finite difference approximations to solve the problem

$$\frac{d^2u}{dx^2} + u = 0, \quad u(0) = 1, \quad u(1) = 0$$

Compare your results with the exact solution.

8.2 Solve the problem

$$\frac{d^2u}{dx^2} - u = 0, \quad u(0) = 0, \; u(1) = 1$$

Answer: Exact solution gives $u_{x=1/3} = 0.29$, $u_{x=2/3} = 0.61$

8.3 Solve the problem

$$\frac{d^2u}{dx^2} - u = 0, \quad u(0) = 0, \; u'(1) = 1$$

Answer: Exact solution gives $u_{x=1/3} = 0.22$, $u_{x=2/3} = 0.46$, $u_{x=1} = 0.76$

8.4 Use a Taylor series to derive the difference expressions

$$u_0' = \frac{1}{12h}(u_{-2} - 8u_{-1} + 8u_1 - u_2)$$

and

$$u_0' = \frac{1}{12h}(-3u_{-1} - 10u_0 + 18u_1 - 6u_2 + u_3)$$

8.5 Use the multiple position difference method to solve the differential equation

$$u'' - 2u' - 3u = 3x + 1, \quad \text{subject to} \quad u(0) = 0, \; u'(0) = 1$$

Answer: You can compare your result with the exact solution $u = \frac{1}{6}e^{3x} - \frac{1}{2}e^{-x} - x + \frac{1}{3}$

8.6 The governing equation for the extension of a straight uniform bar is $u'' + \overline{p}_x/EA = 0$. The central difference quotient about point i leads to

$$(EA/h)(-u_{i-1} + 2u_i - u_{i+1}) = \overline{p}_{xi}h$$

Show that this relationship resembles that obtained using a structural analysis of a straight bar formed of two elements with nodes at $i-1$, i, and $i+1$.

Hint: For $h = \ell$, the global stiffness relation for a two-element, three-node bar would be

$$\frac{EA}{\ell}\begin{bmatrix} 1 & -1 & \\ -1 & 1+1 & -1 \\ & -1 & 1 \end{bmatrix}\begin{bmatrix} u_{i-1} \\ u_i \\ u_{i+1} \end{bmatrix} = \ell \begin{bmatrix} \overline{p}_{x,i-1} \\ \overline{p}_{x,i} \\ \overline{p}_{x,i+1} \end{bmatrix}$$

Note that the finite difference relation is the same as the central equation of the global structural equations.

8.7 Suppose a fixed-fixed beam of length L has a moment of inertia

$$I(x) = 2I_0\ [1 - 2(x/L)(1 - x/L)]$$

The applied loading of magnitude p_0 is uniformly distributed. Let $EI_0 = 30\,000\ \text{kN}\cdot\text{m}^2$, $L = 12$ m, $p_0 = 10$ kN/m. Find the deflection at $x = L/4$ using a finite difference solution. Verify your answer by using a variationally based finite difference solution.

Answer: $w(L/4) \approx 0.01$ m

8.8 Consider a beam of variable cross-section with the applied loading $\overline{p}_z = p_0(1+\xi)$, where $\xi = x/L$. The moment of inertia varies as $I(x) = I_0(9\xi^2 - 6\xi + 1)$. Find the deflection at $x = L/4$ using (a) a simple finite difference mesh and (b) an improved finite difference discretization

Answer: $w(L/4) \approx 0.9aL^2$, $a = p_0L^2/(6EI_0)$

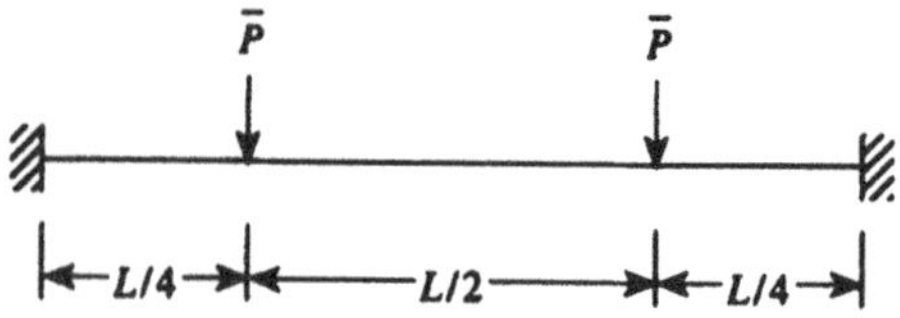

FIGURE P8.9

8.9 Find the deflection at $x = L/2$ of the beam of Fig. P8.9. Let $\overline{P} = 5\,\text{kN}$, $EI = 30\,000\,\text{kN}\cdot\text{m}^2$, and $L = 10$ m. Use a variationally based finite difference scheme.

8.10 Show that use of certain types of variational formulations of finite differences can assure symmetry and positive definiteness of the coefficient matrix of the algebraic (stiffness) equations.

8.11 Use simple central differences to find the deflections along the beam of Fig. P8.11.

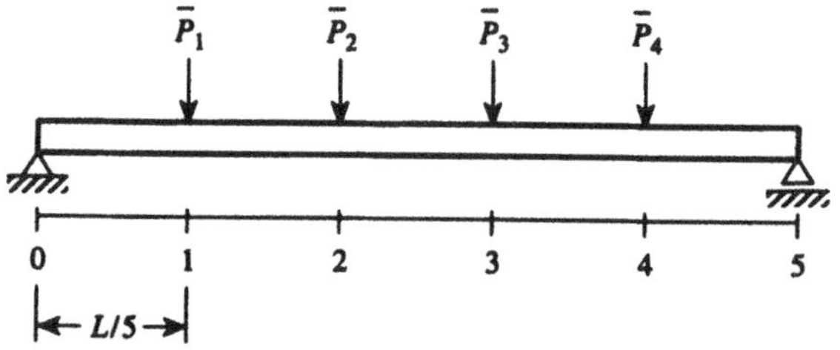

FIGURE P8.11

Answer:

$$\begin{bmatrix} 5 & -4 & 1 & 0 \\ -4 & 6 & -4 & 1 \\ 1 & -4 & 6 & -4 \\ 0 & 1 & -4 & 5 \end{bmatrix} \begin{bmatrix} w_1 \\ w_2 \\ w_3 \\ w_4 \end{bmatrix} = (L^3/125EI) \begin{bmatrix} \overline{P}_1 \\ \overline{P}_2 \\ \overline{P}_3 \\ \overline{P}_4 \end{bmatrix}$$

8.12 A simply supported beam of length L is subjected to a uniformly distributed load of magnitude p_0. Find the distribution of deflection along the beam using

(a) conventional finite differences
(b) variationally based finite differences

Answer: Use a subdivision of $h = L/6$. The conventional approach with central differences gives $w_1 = w_7 = 0$ and

$$\begin{bmatrix} 5 & -4 & 1 & 0 & 0 \\ -4 & 6 & -4 & 1 & 0 \\ 1 & -4 & 6 & -4 & 1 \\ 0 & 1 & -4 & 6 & -4 \\ 0 & 0 & 1 & -4 & 5 \end{bmatrix} \begin{bmatrix} w_1 \\ w_2 \\ w_3 \\ w_4 \\ w_5 \end{bmatrix} = (p_0h^4/EI) \begin{bmatrix} 1 \\ 1 \\ 1 \\ 1 \\ 1 \end{bmatrix}$$

The principle of stationary potential energy leads to the same result with a segment matrix

$$(EI/h^3)\begin{bmatrix} 2 & -4 & 2 \\ -4 & 8 & -4 \\ 2 & -4 & 2 \end{bmatrix}\begin{bmatrix} w_{k-1} \\ w_k \\ w_{k+1} \end{bmatrix} = \begin{bmatrix} 0 \\ 2p_0 h \\ 0 \end{bmatrix}$$

In this case, the force boundary conditions are satisfied exactly by the variational approach.

8.13 Suppose a thick elastic spherical shell contains a gas exerting a pressure p on the inner wall. The radial (r) displacement u is governed by the differential equation

$$\frac{d}{dr}\left(\frac{du}{dr} + \frac{u}{r}\right) = 0$$

subject to the boundary conditions $du/dr = -p$ at $r = r_{\text{inner}}$, $du/dr = 0$ at $r = r_{\text{outer}}$. Discretize this problem and set up a finite difference solution.

8.14 Use the principle of virtual work to implement a finite difference for the beam of Fig. P8.14. Employ a simple central difference expression on the grid shown.

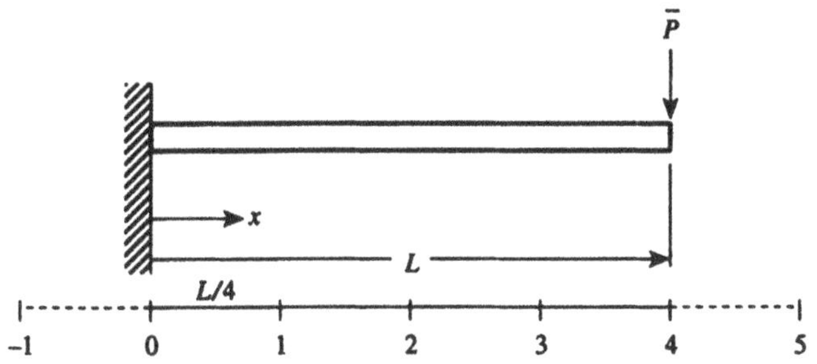

FIGURE P8.14

Answer: After application of the displacement boundary conditions, $w_0 = 0$, $w_0' = (w_1 - w_{-1})/h = 0$,

$$(64EI/L^4)\begin{bmatrix} 7 & -4 & 1 & & \\ -4 & 6 & -4 & 1 & \\ 1 & -4 & 5.5 & -3 & 0.5 \\ & 1 & -3 & 3 & 1 \\ & & 0.5 & -1 & 0.5 \end{bmatrix}\begin{bmatrix} w_1 \\ w_2 \\ w_3 \\ w_4 \\ w_5 \end{bmatrix} = \begin{bmatrix} 0 \\ 0 \\ 0 \\ \overline{P} \\ 0 \end{bmatrix}$$

This can be solved for w_1, w_2, w_3, and w_4.

9

The Boundary Element Method

The boundary element method (BEM) has emerged as an effective computational technique for the solution of a wide class of applied mechanics problems. The interior or domain type of computational methods, such as the finite element method, involve discretization of both the interior and the boundaries (surface) of a domain, whereas the BEM requires discretization along the boundaries only. Normally, this leads to a reduction of the dimensionality of the problem. BEM solutions have been found to be quite accurate, especially when the domain is infinite or semi-infinite, such as often occurs with stress concentration or crack problems. The method is particularly appropriate for linear problems. Extensions into the nonlinear range are possible, but at the expense of some of the special advantages of the method.

9.1 Beams

To introduce the boundary element method, we begin with a beam example, using the extended Galerkin method as a starting point. For a beam, the extended Galerkin method is based on the relationship [Chapter 7, Eq. (7.73)]

$$\int_0^L \delta w(EIw^{iv} - \overline{p}_z)\,dx + [\delta w(V - \overline{V}) + \delta\theta(M - \overline{M})]_{S_p} - [\delta M(\theta - \overline{\theta}) + \delta V(w - \overline{w})]_{S_u} = 0 \tag{9.1}$$

Application of integration by parts to the first term of Eq. (9.1) gives

$$\begin{aligned}\int_0^L \delta w\, EIw^{iv}\,dx &= [\delta w\, EIw''']_{S_u+S_p} - \int_0^L \delta w'\, EIw'''dx \\ &= -[\delta w\, V]_{S_u+S_p} - \int_0^L \delta w'\, EIw'''dx\end{aligned}$$

Continuation of this process applied to the integral on the right-hand side leads to

$$\int_0^L \delta w\, EIw^{iv}dx = -[\delta w\, V + \delta\theta\, M]_{S_u+S_p} + \int_0^L \delta w''\, EIw''dx$$

and eventually to

$$\int_0^L \delta w\, EI w^{iv}\, dx = -[\delta w\, V + \delta\theta\, M]_{S_u+S_p} + [\delta V\, w + \delta M \theta]_{S_u+S_p} + \int_0^L \delta w^{iv}\, EI w\, dx \tag{9.2}$$

Substitution of Eq. (9.2) into Eq. (9.1) gives

$$\begin{aligned}&\int_0^L \delta w^{iv}\, EI w\, dx - \int_0^L \delta w \overline{p}_z\, dx - [\delta w\, V + \delta\theta\, M]_{S_u} + [\delta V\, w + \delta M \theta]_{S_p} \\ &\quad - [\delta w\, \overline{V} + \delta\theta\, \overline{M}]_{S_p} + [\delta V \overline{w} + \delta M\, \overline{\theta}]_{S_u} = 0\end{aligned} \tag{9.3}$$

Combine the prescribed and reaction forces on S_p and S_u into a single bracket. Treat the displacements similarly. Then, Eq. (9.3) can be written as

$$\int_0^L \delta w^{iv}\, EI w\, dx - \int_0^L \delta w \overline{p}_z\, dx - [\delta w\, V + \delta\theta\, M]_{S_u+S_p} + [\delta V\, w + \delta M \theta]_{S_u+S_p} = 0$$

or

$$\begin{aligned}\int_0^L \delta w^{iv}\, EI w\, dx = &\int_0^L \delta w \overline{p}_z\, dx + \delta w_L V_L + \delta\theta_L M_L - \delta w_0 V_0 - \delta\theta_0 M_0 \\ &- \delta V_L w_L - \delta M_L \theta_L + \delta V_0 w_0 + \delta M_0 \theta_0\end{aligned} \tag{9.4}$$

where

$$\delta w_L = \delta w|_{x=L}, \qquad \delta V_0 = \delta V|_{x=0}, \qquad \text{etc.}$$

Choose δw such that the differential equation

$$\delta w^{iv}\, EI = \delta(\xi, x) \tag{9.5}$$

is satisfied. Here, $\delta(\xi, x)$ is the Dirac delta function, which is useful in the representation of concentrated loads, with the fundamental property

$$\int_{-\infty}^{+\infty} \delta(\xi, x)\, dx = 1$$

Also, the sifting property of the delta function for a finite domain would be

$$\int_0^L f(x)\, \delta(\xi, x)\, dx = f(\xi) \tag{9.6}$$

Note that δw is a function involving two points, one is a source point ξ on the beam and another is a point x which corresponds to the integration variable x in Eq. (9.4). Substitution of Eq. (9.5) into Eq. (9.4) will result in an expression for the displacement at point ξ, so that the left-hand side of Eq. (9.4) is a function of ξ, and so is the right-hand side. Let the solution to Eq. (9.5), i.e., the *kern function* (the *fundamental solution* or *Green's function*), be represented as [Stern, 1989]

$$\begin{aligned}\delta w &= w^*(\xi, x) = \frac{1}{12EI}[\text{sgn}(x-\xi)](x-\xi)^3 = w^* \\ \delta\theta &= -\frac{d}{dx} w^*(\xi, x) = -\frac{1}{4EI}[\text{sgn}(x-\xi)](x-\xi)^2 = \theta^* \\ \delta M &= -EI\frac{d^2}{dx^2} w^*(\xi, x) = -\frac{1}{2}[\text{sgn}(x-\xi)](x-\xi) = M^* \\ \delta V &= -EI\frac{d^3}{dx^3} w^*(\xi, x) = -\frac{1}{2}[\text{sgn}(x-\xi)] = V^*\end{aligned} \tag{9.7}$$

where w^*, θ^*, M^*, and V^* are the responses of an infinite beam at point x due to a unit force at point ξ and

$$\mathrm{sgn}(x-\xi) = \begin{cases} 1 & \text{when } x > \xi \\ -1 & \text{when } x < \xi \end{cases}$$

An arbitrary integration constant in the expression for w^* has been omitted.

Note that x and ξ can be interchanged, as Maxwell's reciprocal theorem [Chapter 3, Eq. (3.38)] asserts that for linear problems the responses at point x due to a unit force at point ξ are equal to the responses at point ξ due to a unit force at x. It should be noted that when the positions of x and ξ are interchanged, x becomes the source point and ξ becomes the integration variable. Then, the derivatives in the expressions for θ^*, M^*, and V^* should be taken with respect to ξ and dx of Eq. (9.4) should be changed to $d\xi$, $\delta w_L = \delta w|_{\xi=L}$, etc. Equation (9.4) will then become the expression for the displacement at point x.

With Eq. (9.5), the integral on the left-hand side of Eq. (9.4) becomes

$$\int_0^L \delta w^{iv}\, EI\, w\, dx = \int_0^L \delta(\xi, x)\, w(x)\, dx = w(\xi)$$

so that Eq. (9.4), with the help of Eq. (9.7), can be transformed to

$$\begin{aligned} w(\xi) = & \int_0^L w^*(\xi, x)\, \overline{p}_z(x)\, dx + V_0^* w_0 + M_0^* \theta_0 - V_L^* w_L - M_L^* \theta_L \\ & - w_0^* V_0 - \theta_0^* M_0 + w_L^* V_L + \theta_L^* M_L \end{aligned} \tag{9.8}$$

with

$$\delta V_0 = \delta V|_{x=0} = V^*(\xi, 0) = V_0^*, \qquad M_0^* = M^*(\xi, 0), \quad V_L^* = V^*(\xi, L), \quad \text{etc.}$$

The starred terms in Eq. (9.8) can be written as, for example,

$$\begin{aligned} V_0^* &= V^*(\xi, 0) = -\frac{1}{2}[\mathrm{sgn}(0-\xi)] = \frac{1}{2} \\ M_0^* &= M^*(\xi, 0) = -\frac{1}{2}[\mathrm{sgn}(0-\xi)](0-\xi) = -\frac{\xi}{2} \\ w_L^* &= w^*(\xi, L) = \frac{1}{12EI}[\mathrm{sgn}(L-\xi)](L-\xi)^3 = \frac{1}{12EI}(L-\xi)^3 \end{aligned}$$

Substitute these relationships into Eq. (9.8) to obtain

$$\begin{aligned} w(\xi) = & \int_0^L w^*(\xi, x)\, \overline{p}_z(x)\, dx + \frac{1}{2} w_0 - \frac{\xi}{2}\theta_0 + \frac{1}{2} w_L + \frac{(L-\xi)}{2}\theta_L \\ & - \frac{\xi^3}{12EI} V_0 - \frac{\xi^2}{4EI} M_0 + \frac{(L-\xi)^3}{12EI} V_L - \frac{(L-\xi)^2}{4EI} M_L \end{aligned} \tag{9.9}$$

Equation (9.9) is the expression for the deflection at point ξ along the beam axis. It can be seen that the deflection is a function of ξ only and x is not involved. The slope should be expressed as $\theta(\xi) = -dw/d\xi$. The derivative of w with respect to ξ is given by

$$\begin{aligned} w'(\xi) = & \frac{dw}{d\xi} = -\theta(\xi) \\ = & -\frac{1}{2}\theta_0 - \frac{1}{2}\theta_L - \frac{\xi^2}{4EI} V_0 - \frac{\xi}{2EI} M_0 - \frac{(L-\xi)^2}{4EI} V_L + \frac{(L-\xi)}{2EI} M_L \\ & + \int_0^L w^{*\prime}(\xi, x)\, \overline{p}_z(x)\, dx \end{aligned} \tag{9.10}$$

where $w^{*\prime}(\xi, x) = \frac{d}{d\xi} w^*(\xi, x)$. Equations (9.9) and (9.10) are the expressions for the responses along the beam in terms of the Green's function and the variables on the boundary. These are fundamental relationships on which the boundary element formulation is based. In the boundary element formulation, the unknown variables on the boundary are found first and then the responses along the beam are calculated. In Eqs. (9.9) and (9.10) the point ξ can be at any location along the beam. If the boundary variables are to be calculated, ξ has to be moved to the boundary so that Eqs. (9.9) and (9.10) involve the boundary variables only. When the point ξ is moved to the left end of the beam ($\xi = 0$), Eqs. (9.9) and (9.10), with the help of Eq. (9.7), become

$$w_0 = \frac{1}{2}w_0 + \frac{1}{2}w_L + \frac{L}{2}\theta_L + \frac{L^3}{12EI}V_L - \frac{L^2}{4EI}M_L + \frac{1}{12EI}\int_0^L x^3\overline{p}_z(x)\,dx$$

$$-\theta_0 = -\frac{1}{2}\theta_0 - \frac{1}{2}\theta_L - \frac{L^2}{4EI}V_L + \frac{L}{2EI}M_L - \frac{1}{4EI}\int_0^L x^2\overline{p}_z(x)\,dx$$

i.e.,

$$-\frac{1}{2}w_0 + \frac{1}{2}w_L + \frac{L}{2}\theta_L + \frac{L^3}{12EI}V_L - \frac{L^2}{4EI}M_L + \frac{1}{12EI}\int_0^L x^3\overline{p}_z(x)\,dx = 0$$

$$\frac{1}{2}\theta_0 - \frac{1}{2}\theta_L - \frac{L^2}{4EI}V_L + \frac{L}{2EI}M_L - \frac{1}{4EI}\int_0^L x^2\overline{p}_z(x)\,dx = 0$$

Similarly, moving ξ to the point $\xi = L$ leads to

$$\frac{1}{2}w_0 - \frac{L}{2}\theta_0 - \frac{1}{2}w_L - \frac{L^3}{12EI}V_0 - \frac{L^2}{4EI}M_0 - \frac{1}{12EI}\int_0^L (x-L)^3\,\overline{p}_z(x)\,dx = 0$$

$$-\frac{1}{2}\theta_0 + \frac{1}{2}\theta_L - \frac{L^2}{4EI}V_0 - \frac{L}{2EI}M_0 + \frac{1}{4EI}\int_0^L (x-L)^2\,\overline{p}_z(x)\,dx = 0$$

or in matrix form

$$\underset{\mathbf{H}}{\begin{bmatrix} -1 & 0 & 1 & 1 \\ 0 & -1 & 0 & 1 \\ -1 & L & 1 & 0 \\ 0 & -1 & 0 & 1 \end{bmatrix}} \underset{\mathbf{V}}{\begin{bmatrix} w_0 \\ \theta_0 \\ w_L \\ \theta_L \end{bmatrix}} \underset{=}{=} \underset{\mathbf{G}}{\begin{bmatrix} 0 & 0 & -\frac{L^3}{6EI} & \frac{L^2}{2EI} \\ 0 & 0 & -\frac{L^2}{2EI} & \frac{L}{EI} \\ -\frac{L^3}{6EI} & -\frac{L^2}{2EI} & 0 & 0 \\ \frac{L^2}{2EI} & \frac{L}{EI} & 0 & 0 \end{bmatrix}} \underset{\overline{\mathbf{P}}^*}{\begin{bmatrix} V_0 \\ M_0 \\ V_L \\ M_L \end{bmatrix}}$$

$$\underset{+}{-\frac{1}{6EI}} \underset{\mathbf{B}}{\begin{bmatrix} \int_0^L x^3\overline{p}_z\,dx \\ 3\int_0^L x^2\overline{p}_z\,dx \\ \int_0^L (x-L)^3\overline{p}_z\,dx \\ 3\int_0^L (x-L)^2\,\overline{p}_z\,dx \end{bmatrix}} \tag{9.11}$$

Equation (9.11) is a system of linear equations containing boundary variables only. This is the desired result of the boundary element formulation for the beam problem. The boundary points for the beam serve as boundary elements and the linear equations of Eq. (9.11) will be the relationships between the variables on these elements. The boundary elements consisting of boundary points contrasts with the situation for finite elements for beams where the elements are segments along the beam. This difference results in a reduction of the dimensionality of the problem when using a boundary element approach, in that the solution of a one dimensional beam problem is transformed to a solution of linear equations involving boundary points, with no consideration of in-span nodes.

The solution of Eq. (9.11) provides values of the unknown boundary variables. A well-posed beam problem consists of four known boundary variables and the same number of unknown boundary variables. Equation (9.11) can be prepared for solution by moving the known quantities in $\mathbf{V}$ and $\overline{\mathbf{P}}^*$ together with the corresponding elements in the $\mathbf{H}$ and $\mathbf{G}$ matrices to the right-hand side of the equation and the unknown quantities are moved to the left. A tractable system of linear equations $\mathbf{AX} = \mathbf{F}$ in terms of the unknown variables $\mathbf{X}$ is obtained and the unknown variables can be calculated. Once the unknown boundary variables are determined, they can be substituted into Eqs. (9.9) and (9.10) to find the responses along the beam.

It should be observed that simple manipulation of Eq. (9.11) leads to stiffness and transfer matrices. For example, multiplication of Eq. (9.11) by $\mathbf{G}^{-1}$ results in

$$\mathbf{G}^{-1}\mathbf{H}\mathbf{V} = \overline{\mathbf{P}}^* + \mathbf{G}^{-1}\mathbf{B}$$

i.e.,

$$\overline{\mathbf{P}}^* = \mathbf{K}\mathbf{V} - \overline{\mathbf{P}}^0 \tag{9.12}$$

where $\overline{\mathbf{P}}^*$ contains the shear forces and moments on both boundaries, $\mathbf{V}$ contains the boundary displacements, $\mathbf{K} = \mathbf{G}^{-1}\mathbf{H}$, and $\overline{\mathbf{P}}^0 = \mathbf{G}^{-1}\mathbf{B}$. This corresponds to Chapter 5, Eq. (5.100). Thus, Eq. (9.12) is the stiffness equation of the beam problem. It can be verified that the stiffness matrix $\mathbf{K}$ obtained in this manner is the same as that given in Chapter 4. Since only one element is involved in this beam problem, the global stiffness matrix $\mathbf{K}$ of Eq. (9.12) is equal to the element stiffness matrix $\mathbf{k}^i$.

EXAMPLE 9.1 Beam with Linearly Varying Load

Consider the familiar beam of Chapter 7, Fig. 7.1 with loading $\overline{p}_z(x) = p_0(1 - x/L)$.

The solution should begin with the determination of the unknown boundary variables. For this beam the known boundary conditions are

$$w_0 = w_L = 0, \quad \theta_0 = 0, \quad M_L = 0 \tag{1}$$

and the boundary variables to be calculated are θ_L, V_0, M_0, and V_L. Equation (9.11) is to be used to find these unknown variables. Substitute (1) and $\overline{p}_z = p_0(1 - x/L)$ into Eq. (9.11), giving

$$\underset{\mathbf{A}}{\begin{bmatrix} L & 0 & 0 & \frac{L^3}{6EI} \\ 1 & 0 & 0 & \frac{L^2}{2EI} \\ 0 & \frac{L^3}{6EI} & \frac{L^2}{2EI} & 0 \\ 1 & -\frac{L^2}{2EI} & -\frac{L}{EI} & 0 \end{bmatrix}} \underset{\mathbf{X}}{\begin{bmatrix} \theta_L \\ V_0 \\ M_0 \\ V_L \end{bmatrix}} \underset{=}{=} \underset{\mathbf{F}}{-\frac{p_0}{6EI}\begin{bmatrix} \frac{L^4}{20} \\ \frac{L^3}{4} \\ -\frac{L^4}{5} \\ \frac{3L^3}{4} \end{bmatrix}} \tag{2}$$

Solve this system of equations to find

$$V_0 = \frac{2}{5}p_0L, \quad M_0 = -\frac{1}{15}p_0L^2, \quad V_L = -\frac{p_0L}{10}, \quad \theta_L = \frac{p_0L^3}{120EI} \tag{3}$$

These results are consistent with those obtained in Chapter 3, Example 3.13. Substitution of the boundary values of (3) into Eq. (9.9) leads to an expression for the deflection along the beam. For example, at $\xi = L/2$,

$$w(L/2) = 0.00234375\frac{p_0 L^4}{EI} \tag{4}$$

which is the exact solution. ∎

9.2 Poisson's and Laplace's Equations

In the previous section, the basic relationship (Eq. 9.8) for the boundary element formulation is the expression of the displacement at any point along the beam in terms of the displacement, slope, moment, and shear force at the boundaries of the beam. In the case of two- and three-dimensional problems, the relationship for the boundary element formulation is similar to the one for beams in that the unknown variables are expressed in terms of the unknown variables and other quantities on the boundary. This leads to boundary integral equations.

In this section, an integral equation formulation and boundary element solution for field theory problems represented by the Poisson's and Laplace's equations will be derived.

9.2.1 Direct Formulation

Poisson's equation has the form

$$\nabla^2 u = b \quad \text{inside the domain } V$$

and

$$\begin{aligned} u &= \overline{u} \quad \text{on } S_u \\ q &= \overline{q} \quad \text{on } S_q \end{aligned} \tag{9.13}$$

where b is the nonhomogeneous term of the Poisson's equation, S_u and S_q are the parts of the boundary of V on which u and $q = \partial u/\partial n$, the derivative of u with respect to the outer normal of the boundary, are prescribed, respectively. Also

$$\nabla^2 = \sum_{i=1}^{3} \frac{\partial^2}{\partial x_i^2} \quad \text{for three-dimensional problems}$$

and

$$\nabla^2 = \sum_{i=1}^{2} \frac{\partial^2}{\partial x_i^2} \quad \text{for two-dimensional problems}$$

By definition, $S = S_u + S_q$. For Laplace's equation, set $b = 0$. Use the extended Galerkin method as the basis for a boundary element formulation. Write Eq. (9.13) in the equivalent integral form*

$$\int_V \delta u(\nabla^2 u - b)\, dV + \int_{S_u} \delta\frac{\partial u}{\partial n}(u - \overline{u})\, dS - \int_{S_q} \delta u(q - \overline{q})\, dS = 0 \tag{9.14}$$

*Refer to the procedures of Chapter 2, Section 2.2 for establishing global integral equations from the local, differential equations of static admissibility.

The first term of Eq. (9.14) can be written as

$$\begin{aligned}\int_V \delta u(\nabla^2 u)\, dV &= \int_V \delta u\, u_{,ii}\, dV\\ &= \int_V \left[\frac{\partial}{\partial x_1}\left(\delta u \frac{\partial u}{\partial x_1}\right) + \frac{\partial}{\partial x_2}\left(\delta u \frac{\partial u}{\partial x_2}\right) + \frac{\partial}{\partial x_3}\left(\delta u \frac{\partial u}{\partial x_3}\right)\right] dV\\ &\quad - \int_V \left(\frac{\partial u}{\partial x_1}\frac{\partial}{\partial x_1}\delta u + \frac{\partial u}{\partial x_2}\frac{\partial}{\partial x_2}\delta u + \frac{\partial u}{\partial x_3}\frac{\partial}{\partial x_3}\delta u\right) dV\\ &= \int_V [(\delta u\, u_{,i})_{,i} - (u_{,i}\, \delta u_{,i})]\, dV \end{aligned} \tag{9.15}$$

From Gauss' integral theorem [Eq. (II.8) of Appendix II], the first term in the final integral on the right-hand side of Eq. (9.15) can be written as

$$\int_V (\delta u\, u_{,i})_{,i}\, dV = \int_S (\delta u\, u_{,i}) a_i\, dS = \int_S \delta u\, (u_{,i}\, a_i)\, dS$$

where a_i is the direction cosine of the outer normal of the boundary with respect to the x_i axis. Since $u_{,i}\, a_i = \partial u/\partial n$ [e.g., Gipson, 1987]

$$\int_S \delta u\, (u_{,i}\, a_i)\, dS = \int_S \delta u \frac{\partial u}{\partial n}\, dS \tag{9.16}$$

The second term in the final integral of Eq. (9.15) can be processed as

$$\begin{aligned}\int_V u_{,i}\, \delta u_{,i}\, dV &= \int_V (u\, \delta u_{,i})_{,i}\, dV - \int_V u(\nabla^2 \delta u)\, dV\\ &= \int_V (u\, \delta u_{,i})_{,i}\, dV - \int_V u\, \delta u_{,ii}\, dV \end{aligned} \tag{9.17}$$

Again using Gauss' theorem on the first term on the right-hand side of Eq. (9.17) and following the same procedure as that for manipulating Eq. (9.15) results in

$$\int_V (u\, \delta u_{,i})_{,i}\, dV = \int_S u \frac{\partial}{\partial n} \delta u\, dS = \int_S u\, \delta \frac{\partial u}{\partial n}\, dS \tag{9.18}$$

Substitute Eqs. (9.16), (9.17), and (9.18) into Eq. (9.15) to find

$$\int_V \delta u(\nabla^2 u)\, dV = \int_S \left[\delta u \frac{\partial u}{\partial n} - u \frac{\partial}{\partial n} \delta u\right] dS + \int_V u(\nabla^2 \delta u)\, dV \tag{9.19}$$

Substitute Eq. (9.19) into Eq. (9.14) to obtain

$$\int_V u(\nabla^2 \delta u)\, dV = \int_V b\, \delta u\, dV - \int_{S_q} \overline{q}\, \delta u\, dS - \int_{S_u} q\, \delta u\, dS + \int_{S_q} u \frac{\partial}{\partial n} \delta u\, dS + \int_{S_u} \overline{u} \frac{\partial}{\partial n} \delta u\, dS \tag{9.20a}$$

or

$$\int_V u(\nabla^2 \delta u)\, dV = \int_V b\, \delta u\, dV - \int_S q\, \delta u\, dS + \int_S u \frac{\partial}{\partial n} \delta u\, dS \tag{9.20b}$$

where q and u in the second and third integrals of the right-hand side of Eq. (9.20b) represent the variables u and q on the boundary, and they satisfy the boundary conditions $u = \overline{u}$ on S_u and $q = \overline{q}$ on S_q.

Let

$$\nabla^2 \delta u = -2\alpha\pi\, \delta(\xi, x) \tag{9.21}$$

where ξ is a source point inside the domain V, x is the integration point, $\alpha = 1$ for two-dimensional problems, and $\alpha = 2$ for three-dimensional problems. As is common in the boundary element literature, the symbols ξ and x are used to indicate points, but not coordinates for a multi-dimensional problem. The weighting function δu is now a function relating two points ξ and x. This is the same as δw for the beam problem. From the properties of the Dirac delta function of Eq. (9.6), substitution of Eq. (9.21) into Eq. (9.20b) will result in the expression of u at the source point ξ. It should be noted that the positions of ξ and x in Eq. (9.21) can be interchanged, meaning that x will be the source point and ξ will be the integration point, and ξ should be involved in dV and dS of Eq. (9.20) instead of x. From partial differential equation theory [e.g., Haberman, 1987], the expressions for δu satisfying Eq. (9.21) are

$$\delta u = \delta u(\xi, x) = \frac{1}{r(\xi, x)} \tag{9.22a}$$

for three-dimensional problems and

$$\delta u = \delta u(\xi, x) = \ln \frac{1}{r(\xi, x)} \tag{9.22b}$$

for two-dimensional problems, where $r(\xi, x) = [\sum(x_j - \xi_j)^2]^{1/2} = \sqrt{r_j r_j}$ with $r_j = x_j - \xi_j$ is the distance from the point ξ to x, and x_j and ξ_j are the coordinates of x and ξ, respectively. The quantities in Eqs. (9.22) are the *fundamental solutions (Green's Functions)* for Poisson's and Laplace's problems.

From the properties of the delta function $\delta(\xi, x)$ of Eq. (9.6), and if δu is replaced by u^* and $\frac{\partial}{\partial n}\delta u$ by $\frac{\partial u^*}{\partial n} = q^*$, Eq. (9.20) can be written as

$$2\alpha\pi\ u(\xi) = -\int_V b(x)\, u^*(\xi, x)\, dV(x) + \int_S q(x)\, u^*(\xi, x)\, dS(x) - \int_S u(x)\, q^*(\xi, x)\, dS(x) \tag{9.23}$$

where

$$\begin{aligned} q^* &= \frac{\partial u^*}{\partial n} = \frac{\partial u^*}{\partial x_i} a_i = \frac{\partial}{\partial x_i}\left(\frac{1}{r}\right) a_i = \frac{\partial}{\partial x_i}\left(\frac{1}{\sqrt{\sum(x_j - \xi_j)^2}}\right) a_i \\ &= -\frac{1}{\sum(x_j - \xi_j)^2} \frac{\partial}{\partial x_i} \sqrt{\sum(x_j - \xi_j)^2}\, a_i \\ &= -\frac{1}{\sum(x_j - \xi_j)^2} \frac{(x_i - \xi_i)}{\sqrt{\sum(x_j - \xi_j)^2}} a_i \\ &= -\frac{r_i a_i}{r^3} \end{aligned}$$

for three-dimensional problems and

$$\begin{aligned} q^* &= \frac{\partial u^*}{\partial n} = \frac{\partial u^*}{\partial x_i} a_i = \frac{\partial}{\partial x_i}\left(\ln \frac{1}{r}\right) a_i = \frac{\partial}{\partial x_i}\left(\ln \frac{1}{\sqrt{\sum(x_j - \xi_j)^2}}\right) a_i \\ &= \frac{-1}{\sqrt{\sum(x_j - \xi_j)^2}} \frac{\partial}{\partial x_i} \sqrt{\sum(x_j - \xi_j)^2}\, a_i \\ &= \frac{-1}{\sqrt{\sum(x_j - \xi_j)^2}} \frac{(x_i - \xi_i)}{\sqrt{\sum(x_j - \xi_j)^2}} a_i = -\frac{r_i a_i}{r^2} \end{aligned}$$

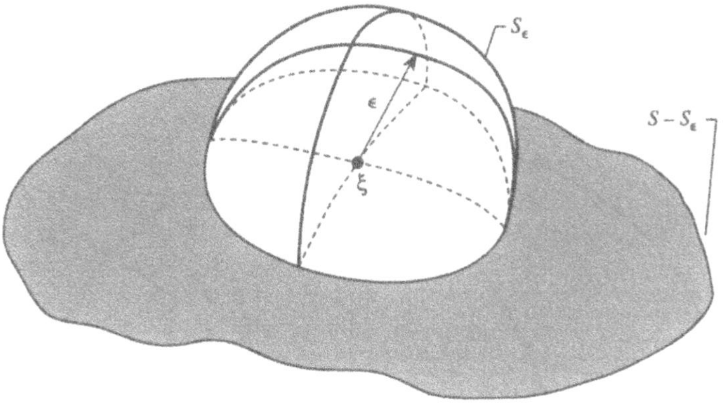

FIGURE 9.1
The boundary when ξ is moved to the boundary.

for two-dimensional problems, in which $r_i = (x_i - \xi_i)$ is the component of r in the x_i direction. Equation (9.23) has the property that the unknown variable u inside of or on the boundary of the domain is expressed in terms of u and its derivative $q = \partial u/\partial n$ on the boundary as well as in terms of the volume integral involving the known quantity b. If ξ is moved to the boundary, Eq. (9.23) will be a relationship involving unknown variables on the boundary only. When the point ξ is on the boundary, however, singularities in u^* and q^* develop and they must be given special consideration. These singularities occur because when ξ is on the boundary, r of Eqs. (9.22) becomes zero at $x = \xi$; consequently, the integrands of the boundary integrals of Eq. (9.23) involving u^* and q^* become singular.

To investigate how to overcome the singularity, assume that the boundary can be represented as shown in Fig. 9.1 where

$$S = (S - S_\epsilon) + S_\epsilon$$

in which S_ϵ is a hemispherical surface of radius ϵ. The second integral on the right-hand side of Eq. (9.23) can be written as

$$\int_S qu^*\,dS = \lim_{\epsilon\to 0}\int_{S-S_\epsilon} qu^*\,dS + \lim_{\epsilon\to 0}\int_{S_\epsilon} qu^*\,dS \tag{9.24}$$

Consider the first integral on the right-hand side of (Eq. 9.24). Since no singularity occurs on the part of the boundary $S - S_\epsilon$, the integral has no change when $\epsilon \to 0$, i.e.,

$$\lim_{\epsilon\to 0}\int_{S-S_\epsilon} qu^*\,dS = \int_S qu^*\,dS$$

For the second integral of (Eq. 9.24), substitution of the expression for $\delta u = u^* = 1/r$ of Eq. (9.22a) for three-dimensional problems into this integral leads to

$$\lim_{\epsilon\to 0}\int_{S_\epsilon} qu^*\,dS = \lim_{\epsilon\to 0} q\int_{S_\epsilon}\frac{1}{\epsilon}\,dS$$

From Fig. 9.2, $dS = \epsilon^2 \sin\varphi\,d\varphi\,d\theta$. Then

$$\lim_{\epsilon\to 0} q\int_{S_\epsilon}\frac{1}{\epsilon}\,dS = \lim_{\epsilon\to 0} q\frac{1}{\epsilon}\epsilon^2\int_0^{2\pi} d\theta\int_0^{\pi/2}\sin\varphi\,d\varphi = \lim_{\epsilon\to 0} q\,2\pi\epsilon = 0 \tag{9.25}$$

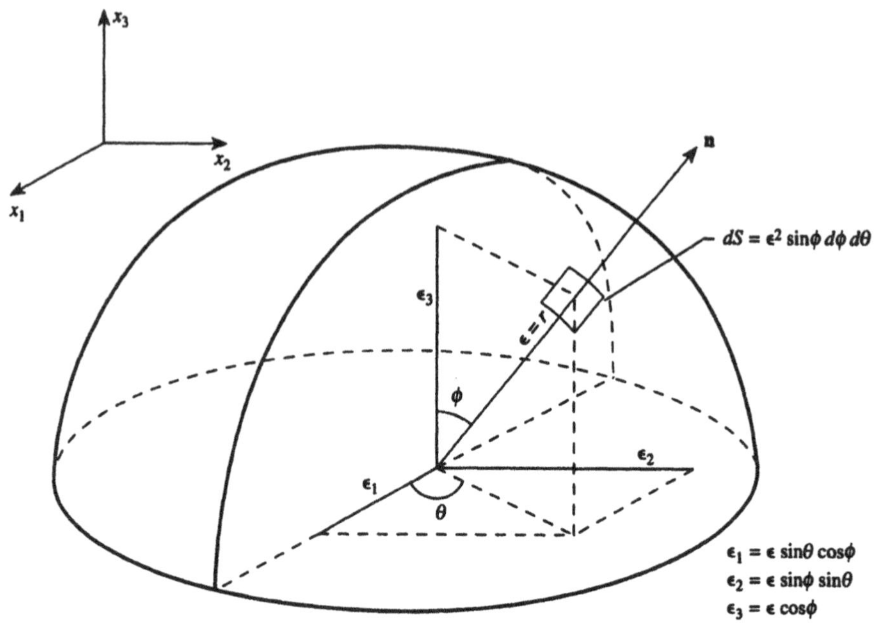

FIGURE 9.2
Hemispherical surface with radius ϵ.

Thus moving ξ to the boundary leads to no contribution from the second integral on the right-hand side of Eq. (9.23).

In the case of the third integral of Eq. (9.23),

$$\int_S u\, q^*\, dS = \lim_{\epsilon \to 0} \int_{S-S_\epsilon} u\, q^*\, dS + \lim_{\epsilon \to 0} \int_{S_\epsilon} u\, q^*\, dS \tag{9.26}$$

and

$$\lim_{\epsilon \to 0} \int_{S-S_\epsilon} u\, q^*\, dS = \int_S u\, q^*\, dS$$

For a hemispherical surface of Fig. 9.2, $a_i = \frac{r_i}{r}$. Then, with $r = \epsilon, r_i = \epsilon_i, q^*$ can be written as

$$q^* = \frac{\partial u^*}{\partial n} = \frac{\partial}{\partial n}\left(\frac{1}{\epsilon}\right) = -\frac{\epsilon_i a_i}{\epsilon^3} = -\frac{\epsilon_i}{\epsilon^3}\frac{\epsilon_i}{\epsilon} = -\frac{\epsilon_1^2 + \epsilon_2^2 + \epsilon_3^2}{\epsilon^4}$$

From the expressions for ϵ_1, ϵ_2, and ϵ_3 of Fig. 9.2, it can be seen that

$$\epsilon_1^2 + \epsilon_2^2 + \epsilon_3^2 = \epsilon^2$$

Thus, for a hemispherical surface,

$$q^* = \frac{\partial u^*}{\partial n} = -\frac{1}{\epsilon^2}$$

Then the second integral on the right-hand side of Eq. (9.26) becomes

$$\lim_{\epsilon \to 0} \int_{S_\epsilon} u\, q^*\, dS = \lim_{\epsilon \to 0}\left(-\int_{S_\epsilon} u \frac{1}{\epsilon^2}\, dS\right) = \lim_{\epsilon \to 0}\left(-u \int_{S_\epsilon} \frac{1}{\epsilon^2}\, dS\right)$$

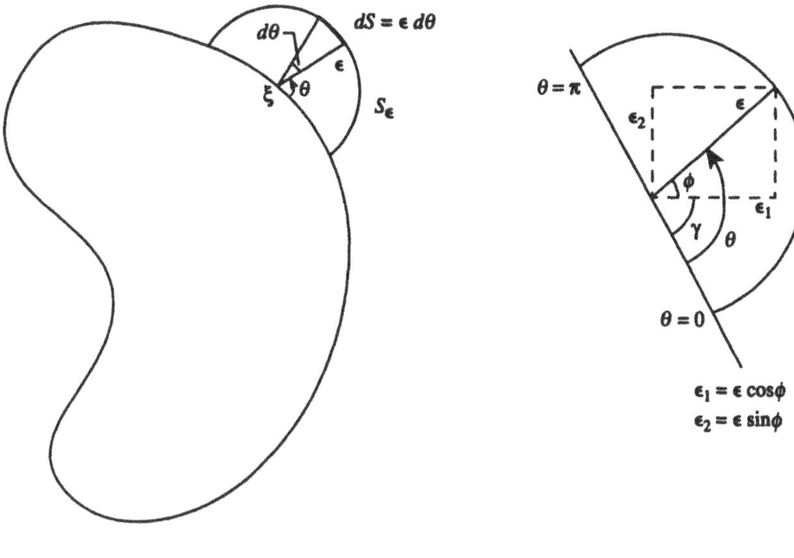

FIGURE 9.3
Two-dimensional boundary when ξ is moved to the boundary.

Substitute the expression in Fig. 9.2 for dS into the above equation to obtain

$$\lim_{\epsilon \to 0}\left(-u\int_{S_\epsilon}\frac{1}{\epsilon^2}\,dS\right) = \lim_{\epsilon \to 0}\left(-u\frac{1}{\epsilon^2}\epsilon^2\int_0^{2\pi}d\theta\int_0^{\pi/2}\sin\varphi\,d\varphi\right)$$

$$= \lim_{\epsilon \to 0}\left(-u\frac{2\pi\epsilon^2}{\epsilon^2}\right) = -2\pi u \tag{9.27}$$

For the two-dimensional case, utilize the geometry and notation of Fig. 9.3. Then the integration on the augmented segment for the second integral in Eq. (9.23) appears as

$$\lim_{\epsilon \to 0}\int_{S_\epsilon} q\ln\frac{1}{\epsilon}\,dS = -\lim_{\epsilon \to 0} q\int_0^{\pi}\epsilon\ln(\epsilon)\,d\theta$$

where the relationships $\ln\frac{1}{\epsilon} = -\ln\epsilon$ and $dS = \epsilon\,d\theta$ (Fig. 9.3) have been used. Note that $\lim_{\epsilon\to 0}\epsilon\ln(\epsilon) = 0$, so this integration has no contribution to Eq. (9.23) when ξ moves to the boundary.

Now consider the third integral on the right-hand side of Eq. (9.23). From the geometry of Fig. 9.3, it can be seen that

$$\epsilon_i a_i = \epsilon_1\cos\phi + \epsilon_2\sin\phi = \epsilon$$

where the expressions for ϵ_1 and ϵ_2 of Fig. 9.3 have been introduced. Hence

$$q^* = \frac{\partial u^*}{\partial n} = \frac{\partial}{\partial n}\left(\ln\frac{1}{\epsilon}\right) = -\frac{1}{\epsilon}\frac{\partial\epsilon}{\partial n} = -\frac{\epsilon_i a_i}{\epsilon^2} = -\frac{\epsilon}{\epsilon^2} = -\frac{1}{\epsilon}$$

$$\lim_{\epsilon \to 0}\int_{S_\epsilon} u\,q^*\,dS = \lim_{\epsilon \to 0} u\int_0^{\pi}\left(-\frac{1}{\epsilon}\right)\epsilon\,d\theta = -u\int_0^{\pi}d\theta = -\pi u$$

It is evident that the integrals contribute $-\pi u$ to Eq. (9.23) as $\epsilon \to 0$ for the two-dimensional case. Substitute the contributions for the two $(-\pi u)$ and three $(-2\pi u)$ dimensional problems into Eq. (9.23) to find the case of ξ on the boundary

$$\alpha\pi\, u(\xi) + \int_S uq^*\, dS = -\int_V bu^*\, dV + \int_S qu^*\, dS \tag{9.28}$$

Rewrite Eq. (9.28), using Eq. (9.23), to include the case of ξ inside V

$$cu(\xi) + \int_S uq^*\, dS = -\int_V bu^*\, dV + \int_S qu^*\, dS \tag{9.29}$$

with

$$c = \begin{cases} 2\alpha\pi & \text{when } \xi \text{ is inside } V \\ \alpha\pi & \text{when } \xi \text{ is on the boundary } S \end{cases}$$

and $\alpha = 1$ for two-dimensional problems and $\alpha = 2$ for three-dimensional problems. In this derivation, the boundary is assumed to be smooth at point ξ. If the boundary is not smooth, the constant c will not be equal to $\alpha\pi$. Actually, the constant c for the case of ξ on the boundary does not need to be evaluated explicitly, as it can be determined indirectly. The unknown physical quantities u and q occur only on the boundary in Eq. (9.28). Since both unknowns are calculated directly, this formulation is called the *direct formulation*. Equation (9.28) can be solved by the boundary element method, which begins with discretization of the boundary into elements. The procedure is explained in detail in Section 9.2.2.

EXAMPLE 9.2 Torsion of a Prismatic Bar

Develop the direct boundary integral formulation for the torsion problem of a prismatic bar.

The equilibrium equation for a bar under uniform torsional moment is [Chapter 1, Eq. (1.144a)]

$$\frac{\partial \tau_{xy}}{\partial y} + \frac{\partial \tau_{xz}}{\partial z} = 0 \quad \text{or} \quad \tau_{1j,j} = 0 \quad j = 2, 3 \tag{1}$$

and the boundary conditions are [Chapter 1, Eq. (1.145)]

$$\tau_{xy}\, a_y + \tau_{xz}\, a_z = 0 \quad \text{or} \quad \tau_{1j}\, a_j = 0 \quad j = 2, 3 \tag{2}$$

at the lateral surface of the bar, where a_j is the direction cosine of the normal to the boundary with respect to the x_j axis. The coordinates y and z in Eqs. (1.144) and (1.145) become x_2 and x_3 here. Write (1) and (2) in terms of the warping function ω to obtain [Chapter 1, Eqs. (1.151) and (1.153)]

$$\begin{aligned} &\nabla^2\omega = 0 && \text{inside the cross-section} \\ &\overline{q} = \frac{\partial \omega}{\partial n} = x_2 a_3 - x_3 a_2 && \text{on the boundary} \end{aligned} \tag{3}$$

where $q = \partial\omega/\partial n$ is the derivative of ω with respect to the outer normal of the boundary. The bar over q indicates a prescribed condition. The extended Galerkin's formula for this problem would take the form

$$\int_A \omega^*(\nabla^2\omega)\, dA - \int_S \omega^*(q - \overline{q})\, dS = 0 \tag{4}$$

where A is the area of the cross-section and S is the boundary. The first term of (4) can be written as

$$\int_A \omega^*(\nabla^2\omega)\,dA = \int_A \omega^*\omega_{,ii}\,dA = \int_A [(\omega^*\omega_{,i})_{,i} - \omega^*_{,i}\omega_{,i}]\,dA$$
$$= \int_A \left[\frac{\partial}{\partial x_2}\left(\omega^*\frac{\partial\omega}{\partial x_2}\right) + \frac{\partial}{\partial x_3}\left(\omega^*\frac{\partial\omega}{\partial x_3}\right)\right] dA - \int_A \left(\frac{\partial\omega^*}{\partial x_2}\frac{\partial\omega}{\partial x_2} + \frac{\partial\omega^*}{\partial x_3}\frac{\partial\omega}{\partial x_3}\right) dA \tag{5}$$

Use the Gauss' integral theorem [Appendix II, Eq. (II.8)] on the first term of the right-hand side of (5) to obtain

$$\int_A \left[\frac{\partial}{\partial x_2}\left(\omega^*\frac{\partial\omega}{\partial x_2}\right) + \frac{\partial}{\partial x_3}\left(\omega^*\frac{\partial\omega}{\partial x_3}\right)\right] dA = \int_A (\omega^*\omega_{,i})_{,i}\,dA = \int_S \omega^*\omega_{,i}\,a_i\,dS$$
$$= \int_S \omega^*\frac{\partial\omega}{\partial n}\,dS = \int_S \omega^* q\,dS \quad i = 2, 3 \tag{6}$$

Since this is a two-dimensional problem, the volume V in Appendix II, Eq. (II.8) is changed to area A and S now represents the boundary of the cross-section. The second term of the right-hand side of (5) can be transformed as

$$\int_A \left(\frac{\partial\omega^*}{\partial x_2}\frac{\partial\omega}{\partial x_2} + \frac{\partial\omega^*}{\partial x_3}\frac{\partial\omega}{\partial x_3}\right) dA = \int_A \omega^*_{,i}\omega_{,i}\,dA$$
$$= -\int_A \omega(\nabla^2\omega^*)\,dA + \int_A (\omega\,\omega^*_{,i})_{,i}\,dA$$
$$= -\int_A \omega(\nabla^2\omega^*)\,dA + \int_S \omega\,q^*\,dA \quad i = 2, 3 \tag{7}$$

where $q^* = \frac{\partial\omega^*}{\partial n}$. Gauss' theorem is used again in (7). Substitution of (6) and (7) into (5) leads to

$$\int_A \omega^*(\nabla^2\omega)\,dA = \int_A \omega(\nabla^2\omega^*)\,dA + \int_S \left(\omega^*\frac{\partial\omega}{\partial n} - \omega\frac{\partial\omega^*}{\partial n}\right) dS \tag{8}$$

Then (4) becomes

$$\int_A \omega(\nabla^2\omega^*)\,dA + \int_S \left(\overline{q}\omega^* - \omega\frac{\partial\omega^*}{\partial n}\right) dS = 0 \tag{9}$$

To find the fundamental solution ω^*, a singular function, let

$$\nabla^2\omega^* = -2\pi\ \delta(\xi, x) \tag{10}$$

where ξ and x are two points in the cross-section. The solution to (10) is (Eq. (9.22b) or Haberman, 1987)

$$\omega^* = \ln\frac{1}{r} \tag{11}$$

which has the derivative, with respect to the outer normal of the cross-section (Eq. 9.23)

$$\frac{\partial\omega^*}{\partial n} = -\frac{1}{r^2}\,r_i a_i \quad i = 2, 3 \tag{12}$$

where r_i is the component of r in the x_i direction and a_i is the direction cosine of the outer normal with respect to the x_i axis. Substitute (10), (11), and (12) into (9), along with

$\ln(1/r) = -\ln r$, to obtain

$$2\pi\,\omega(\xi) + \int_S \bar{q}\,\ln r\,dS - \int_S \omega\frac{r_i a_i}{r^2}\,dS = 0 \tag{13}$$

where ξ is a point inside the cross-section and the quantities in the integrands of the line integrals are referred to point x on the boundary.

To process the boundary integrals, move the point ξ to the boundary, and utilize the geometry and notation of Fig. 9.3. Use the procedure of Section 9.2.1 to evaluate the contributions to the integral equation of the integration along the semicircle of Fig. 9.3. Then (13) can be written as

$$c\,\omega(\xi) + \left[\int_S \bar{q}\,\ln r\,dS - \int_S \omega\frac{r_i a_i}{r^2}\,dS\right] = 0 \tag{14}$$

with

$$c = \begin{cases} 2\pi & \text{inside the cross-section} \\ \pi & \text{on the boundary, if smooth at } \xi \\ \text{Constant} & \text{on the boundary, if not smooth at } \xi \end{cases}$$

This boundary integral equation can be solved using the methodology of Section 9.2.2.

The torsional constant is found from [Chapter 1, Eq. (1.154)]

$$\begin{aligned} J &= \int_A \left(x_2^2 + x_3^2 - x_2\frac{\partial\omega}{\partial x_3} + x_3\frac{\partial\omega}{\partial x_2}\right) dA \\ &= I_p + \int_A \left[\frac{\partial}{\partial x_3}(-x_2\omega) + \frac{\partial}{\partial x_2}(x_3\omega)\right] dA \end{aligned}$$

where I_p is the polar moment of inertia. Use Gauss' theorem [Eq. (II.8) of Appendix II], where, for a two-dimensional problem, the volume V in Eq. (II.8) becomes the boundary S of the cross-section.

$$\int_A \left[\frac{\partial}{\partial x_3}(-x_2\omega) + \frac{\partial}{\partial x_2}(x_3\omega)\right] dA = \int_S (-x_2\omega\,a_3 + x_3\omega\,a_2)\,dS$$

Thus,

$$J = I_p + \int_S \omega(-x_2 a_3 + x_3 a_2)\,dS \tag{15}$$

The values of ω and its derivatives at point ξ inside the cross-section can be calculated using (13)

$$\omega(\xi) = \frac{1}{2\pi}\int_S \left(-\bar{q}\,\ln r + \omega\frac{r_i a_i}{r^2}\right) dS \tag{16}$$

$$\begin{aligned} \frac{\partial\omega(\xi)}{\partial\xi_2} &= \frac{1}{2\pi}\int_S \left[-\bar{q}\frac{\partial}{\partial\xi_2}\ln r + \omega\frac{\partial}{\partial\xi_2}\left(\frac{r_i a_i}{r^2}\right)\right] dS \\ &= \frac{1}{2\pi}\int_S \left[-\bar{q}\frac{1}{r}\frac{\partial r}{\partial\xi_2} + \omega\frac{1}{r^4}\left(r^2\frac{\partial}{\partial\xi_2}r_i a_i - r_i a_i\frac{\partial}{\partial\xi_2}r^2\right)\right] \\ &= \frac{1}{2\pi}\int_S \left(\bar{q}\frac{r_2}{r^2} + \omega\frac{r_2^2 a_2 + 2r_2 r_3 a_3 - r_3^2 a_2}{r^4}\right) dS \end{aligned} \tag{17}$$

Similarly

$$\frac{\partial \omega(\xi)}{\partial \xi_3} = \frac{1}{2\pi} \int_S \left(\overline{q} \frac{r_3}{r^2} + \omega \frac{r_3^2 a_3 + 2 r_2 r_3 a_2 - r_2^2 a_3}{r^4} \right) dS \tag{18}$$

The derivatives are taken with respect to ξ_i because the variable for $\omega(\xi)$ is ξ and the derivatives at point ξ are of interest. With $\partial\omega/\partial\xi_2$ and $\partial\omega/\partial\xi_3$ known, the stresses can be calculated from Chapter 1, Eqs. (1.142) and (1.143) at any point inside the cross-section.

Alternatively, a different approach can be employed to compute the derivatives of ω with respect to ξ_2 and ξ_3. When they are needed at a specific point inside the cross-section, first calculate the values of ω at four surrounding points. These points can be used as the vertices of a quadrilateral inside which the values of ω can be interpolated using shape functions. Then the derivatives can be calculated from these interpolation functions. For some cases the derivatives are calculated more accurately with the interpolation functions than from Eqs. (17) and (18). ■

9.2.2 Boundary Element Formulation

An analytical solution of the integral relation of Eq. (9.29) is exceedingly difficult to find. Consequently, numerical methods are employed. The basic steps involved are

1. Discretize the boundary S into elements over which approximate shape functions for u and q, as appropriate, are defined.
2. Introduce these elements into Eq. (9.29) to obtain a system of linear equations.
3. Impose the boundary conditions, and solve these equations.
4. Find the variables u and q, as desired, inside the body.

The boundary (surface) of the body can be discretized into elements (boundary elements) in the form of Figs. 9.4 and 9.5. For a two-dimensional analysis, the elements are normally straight lines (constant or linear elements) or curves (quadratic or higher order). For a three-dimensional analysis, the elements are usually quadrilaterals and triangles. Linear and quadratic elements, which are often chosen to be isoparametric elements, are shown in Fig. 9.6. The shape functions for these elements are the same as those for an element of the same order used in a finite element analysis. Thus, the assumed distributions of the variables in Eq. (9.29) can be written in terms of an element shape function $\mathbf{N}^i$ as

$$u = \mathbf{N}^i \mathbf{v}^i \tag{9.30}$$

$$q = \mathbf{N}^i \mathbf{q}^i \tag{9.31}$$

where $\mathbf{v}^i$ and $\mathbf{q}^i$ are the nodal values of u and q of the ith element, with dimensions equal to the number of nodes in the element. For the constant element, which has one node at its centroid, the unknown variable throughout the element is assumed to be equal to the variable at the nodal point. This kind of element has only a single DOF for each element and, hence, is the simplest element.

Substitute Eqs. (9.30) and (9.31) into Eq. (9.29) to obtain

$$cu(\xi_j) + \sum_{i=1}^{M} \left\{ \int_{S_i} q^* \mathbf{N}^i \, dS \right\} \mathbf{v}^i = \sum_{i=1}^{M} \left\{ \int_{S_i} u^* \mathbf{N}^i \, dS \right\} \mathbf{q}^i - \sum_{s=1}^{S} \left\{ \int_{V_s} b u^* \, dV \right\} \tag{9.32}$$

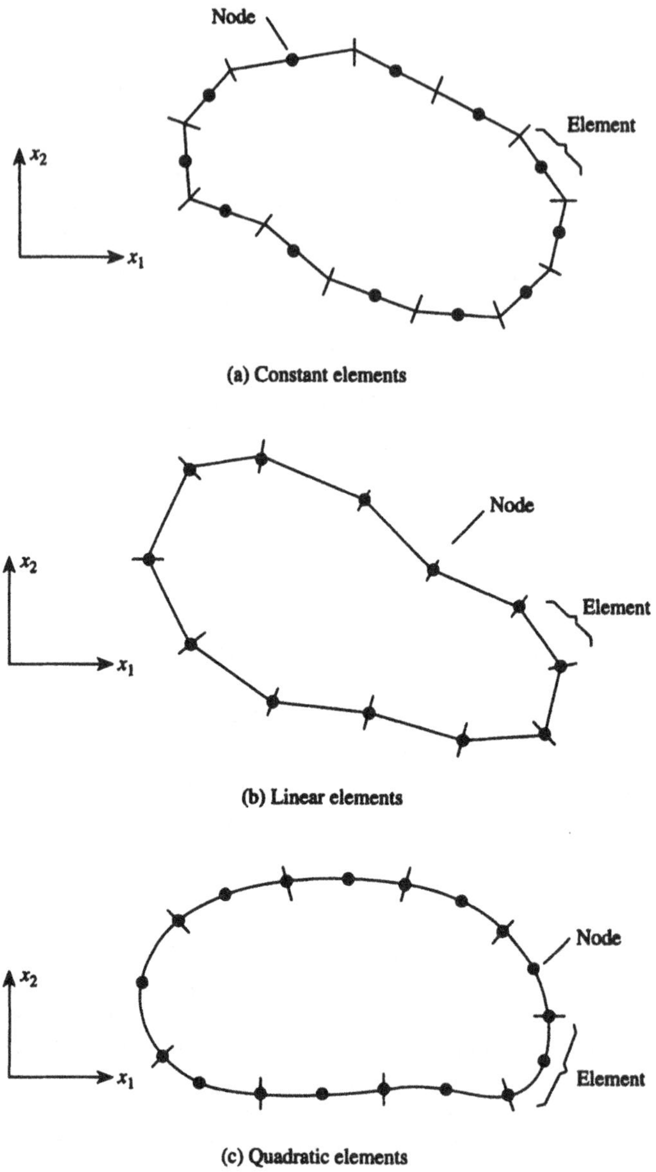

FIGURE 9.4
Boundary element discretization for two-dimensional problems.

For a two-dimensional problem with $u^* = \ln(1/r)$ and $q^* = (-r_i a_i/r^2)$

$$cu(\xi_j) + \sum_{i=1}^{M}\left\{\int_{S_i}\left(-\frac{r_k a_k}{r^2}\right)\mathbf{N}^i\,dS\right\}\mathbf{v}^i = \sum_{i=1}^{M}\left\{\int_{S_i}\ln\left(\frac{1}{r}\right)\mathbf{N}^i\,dS\right\}\mathbf{q}^i - \sum_{s=1}^{S}\left\{\int_{V_s} b\,\ln\left(\frac{1}{r}\right)dV\right\} \tag{9.33}$$

where c is given in Eq. (9.29), j is the number of a node in the global numbering scheme, ξ_j is the point on which node j is located, S_i is the length of the ith element, r is the distance from the point ξ_j to a point on S_i, r_k is a component of r in the direction of a coordinate axis, a_k is a direction cosine of the boundary, M is the number of boundary elements, and S

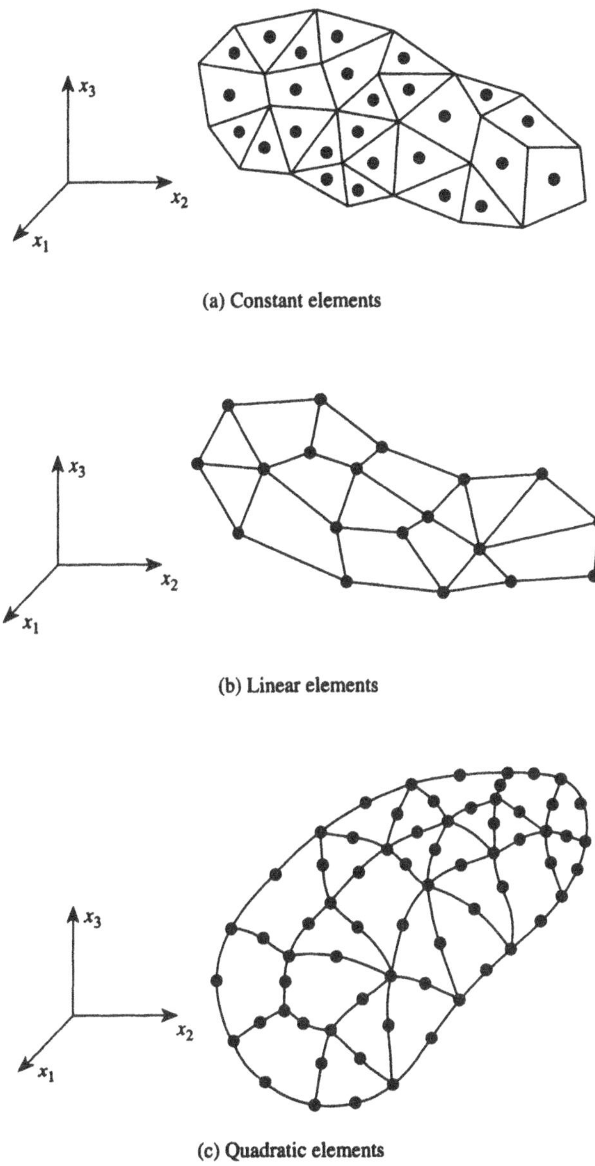

FIGURE 9.5
Boundary element discretization for three-dimensional problems.

is the number of "cells" with volume (area) V_s which are the blocks to form the total body and are needed for the volume (area) integration. Note that these cells are different from the elements in the finite element method in that they involve no unknown nodal values and are used for integration purposes only. Substitution of the expressions $u^* = 1/r$ and $q^* = -r_k a_k / r^3$ into Eq. (9.32) leads to an expression for three-dimensional problems. In this case, S_i becomes the area of the element.

Let $u(\xi_j) = v_j$, where v_j is the nodal value of u in the global numbering scheme. After the numerical integration of Eqs. (9.32), a system of linear equations is obtained

$$c v_j + \sum_{i=1}^{M} \widehat{\mathbf{H}}^{ij} \mathbf{v}^i = \sum_{i=1}^{M} \widehat{\mathbf{G}}^{ij} \mathbf{q}^i + B_j \tag{9.34}$$

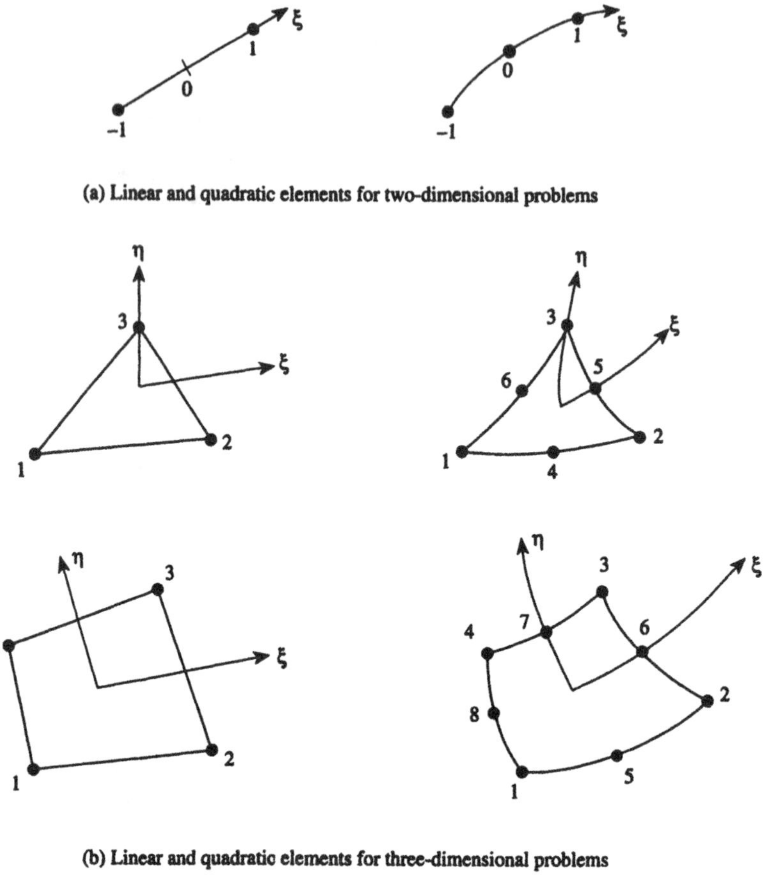
ξ
1
0
-1
ξ
1
0
-1
(a) Linear and quadratic elements for two-dimensional problems
η
3
ξ
2
1
η
3
ξ
5
6
2
1
4
η
3
ξ
4
2
1
η
3
ξ
7
6
4
8
2
1
5
(b) Linear and quadratic elements for three-dimensional problems

FIGURE 9.6
Boundary elements for two- and three-dimensional problems.

where, for a 2D problem

$$\widehat{\mathbf{H}}^{ij} = -\int_{S_i} \frac{r_k a_k}{r^2} \mathbf{N}^i \, dS \qquad \widehat{\mathbf{G}}^{ij} = \int_{S_i} \ln\left(\frac{1}{r}\right) \mathbf{N}^i \, dS$$

$$B_j = -\sum_{s=1}^{S} \left\{ \int_{V_s} b \ln\left(\frac{1}{r}\right) dV \right\}$$

and for a 3D problem

$$\widehat{\mathbf{H}}^{ij} = -\int_{S_i} \frac{r_k a_k}{r^3} \mathbf{N}^i \, dS \qquad \widehat{\mathbf{G}}^{ij} = \int_{S_i} \frac{1}{r} \mathbf{N}^i \, dS$$

$$B_j = -\sum_{s=1}^{S} \left\{ \int_{V_s} b \frac{1}{r} dV \right\}$$

Note that $\widehat{\mathbf{H}}^{ij}$, $\widehat{\mathbf{G}}^{ij}$ and B_j involve r which is the distance from the nodal point j to points on S_i. For different nodal points, r takes on different values and, hence, $\widehat{\mathbf{H}}^{ij}$, $\widehat{\mathbf{G}}^{ij}$ and B_j vary as j is changed. Also, the second term on the left-hand side of Eq. (9.34) involves all of the nodal values of u on the boundary. Furthermore, the first term on the right-hand side of Eq. (9.34) involves all of the nodal values of q on the boundary.

For each node j, the element matrices $\widehat{\mathbf{H}}^{ij}$ and $\widehat{\mathbf{G}}^{ij}$ are assembled to form the global matrices $\widehat{\mathbf{H}}_j$ and $\widehat{\mathbf{G}}_j$ such that

$$\begin{aligned}\widehat{\mathbf{H}}_j &= \sum_{i=1}^{M} \widehat{\mathbf{H}}^{ij} = [\widehat{\mathbf{H}}_{j1}\, \widehat{\mathbf{H}}_{j2} \cdots \widehat{\mathbf{H}}_{jj} \cdots \widehat{\mathbf{H}}_{jN}] \\ \widehat{\mathbf{G}}_j &= \sum_{i=1}^{M} \widehat{\mathbf{G}}^{ij} = [\widehat{\mathbf{G}}_{j1}\, \widehat{\mathbf{G}}_{j2} \cdots \widehat{\mathbf{G}}_{jj} \cdots \widehat{\mathbf{G}}_{jN}]\end{aligned} \tag{9.35}$$

The summation process is similar to that given in Chapter 5 for the stiffness matrices, i.e., the entries in $\widehat{\mathbf{H}}^{ij}$ or $\widehat{\mathbf{G}}^{ij}$ corresponding to a common node are added to form an entry in $\widehat{\mathbf{H}}_j$ or $\widehat{\mathbf{G}}_j$ for the node. Since the second term on the left-hand side and the first term on the right-hand side of Eq. (9.34) involve all the nodal values of u and q, $\widehat{\mathbf{H}}_j$ and $\widehat{\mathbf{G}}_j$ are of the order of $1 \times N$, where N is the total number of nodes. Introduce global vectors containing nodal values of u and q, $\mathbf{V} = [v_1\, v_2 \cdots v_j \cdots v_N]^T$, $\mathbf{P} = [q_1\, q_2 \cdots q_j \cdots q_N]^T$. The assembly process results in a linear equation for node j

$$cv_j + \widehat{\mathbf{H}}_j \mathbf{V} = \widehat{\mathbf{G}}_j \mathbf{P} + B_j$$

Here, c can be replaced by c_j because the value of c depends on the smoothness of the boundary at node j. Put these equations together according to the global node order for all N nodes to form the global equation

$$\mathbf{CV} + \widehat{\mathbf{H}}\mathbf{V} = \widehat{\mathbf{G}}\mathbf{P} + \mathbf{B} \tag{9.36}$$

where

$$\widehat{\mathbf{H}} = \begin{bmatrix} \widehat{\mathbf{H}}_1 \\ \widehat{\mathbf{H}}_2 \\ \vdots \\ \widehat{\mathbf{H}}_j \\ \vdots \\ \widehat{\mathbf{H}}_N \end{bmatrix}, \quad \widehat{\mathbf{G}} = \begin{bmatrix} \widehat{\mathbf{G}}_1 \\ \widehat{\mathbf{G}}_2 \\ \vdots \\ \widehat{\mathbf{G}}_j \\ \vdots \\ \widehat{\mathbf{G}}_N \end{bmatrix}, \quad \text{and} \quad \mathbf{B} = \begin{bmatrix} \widehat{B}_1 \\ \widehat{B}_2 \\ \vdots \\ \widehat{B}_j \\ \vdots \\ \widehat{B}_N \end{bmatrix}$$

Matrices $\mathbf{H}$ and $\mathbf{G}$ are of order $N \times N$ and $\mathbf{B}$ is a vector of order $N \times 1$. Matrix $\mathbf{C} = \text{diag}(c\, c \cdots c)$ is an $N \times N$ diagonal matrix. Let $\mathbf{H} = \mathbf{C} + \widehat{\mathbf{H}}$ and $\mathbf{G} = \widehat{\mathbf{G}}$ so that this equation becomes

$$\mathbf{HV} = \mathbf{GP} + \mathbf{B} \tag{9.37}$$

Note that the diagonal elements of $\mathbf{H}$ involve the constant c, which is equal to $\alpha\pi$ in Eq. (9.29) when ξ is on a smooth boundary and equal to another value when the boundary at ξ is not smooth. The constant c is often cumbersome to handle. A strategy to circumvent the treatment of this constant is to avoid the explicit evaluation of the diagonal elements of $\mathbf{H}$.

Examine Eq. (9.37) in some detail. It can be seen that matrices $\mathbf{H}$ and $\mathbf{G}$ depend on the fundamental solution, the shape functions, and the contour of the boundary. They do not depend on $\mathbf{V}$, $\mathbf{P}$, and $\mathbf{B}$. The diagonal entries H_{ii} of $\mathbf{H}$, which involve the cumbersome c, can be determined by assigning judiciously chosen values to $\mathbf{V}$, $\mathbf{P}$, and $\mathbf{B}$ in Eq. (9.36). Suppose we have $\mathbf{V} = \mathbf{I}$, which is a vector with unit entries. When the variable $\mathbf{V}$ is unity over the whole region, its derivatives are zero. Thus, $\mathbf{P}$, which contains nodal values of q, is zero. Also set $b = 0$, leading to $\mathbf{B} = \mathbf{0}$. Substitute these assumptions into Eq. (9.37) to obtain

$$\mathbf{H\,I} = \mathbf{0} \tag{9.38}$$

or

$$\begin{bmatrix} H_{11} & H_{12} & \cdots & H_{1N} \\ H_{21} & H_{22} & \cdots & \vdots \\ \vdots & \vdots & \cdots & \vdots \\ H_{N1} & H_{N2} & \cdots & H_{NN} \end{bmatrix} \begin{bmatrix} 1 \\ 1 \\ \vdots \\ 1 \end{bmatrix} = \mathbf{0}$$

This is a system of homogeneous linear equations. The ith equation in this system is

$$\sum_{j=1}^{N} H_{ij} = 0$$

Thus, the diagonal entries H_{ii} of the matrix $\mathbf{H}$ are

$$H_{ii} = -\sum_{\substack{j=1 \\ j\neq i}}^{N} H_{ij} \tag{9.39}$$

Impose the u and q boundary conditions on Eq. (9.37) and rearrange the equations such that the unknown quantities are on the left-hand side and all the known quantities are on the right. The final system of linear equations can appear as

$$\mathbf{AX} = \mathbf{F} \tag{9.40}$$

where $\mathbf{X}$ contains all of the unknown nodal values of u and q, and $\mathbf{F}$ contains the known quantities. The solution of Eq. (9.40) provides the nodal values of u and q on the boundary.

It is of interest to compare some characteristics of Eq. (9.40) with the global equations of the finite element method. The global stiffness matrix $\mathbf{K}$ of the finite element method is a banded symmetric matrix for which some special techniques can be employed to solve the equations. The boundary element method, on the other hand, leads to a full, unsymmetric matrix $\mathbf{A}$. Usually, such matrices are less efficient to solve than the banded symmetric ones. Since the dimensionality of the problem is reduced by 1 relative to that of the finite element method, the order of $\mathbf{A}$ is much smaller than that of $\mathbf{K}$ for the same problem, and hence the boundary element method is generally more efficient than the finite element method, especially for a small surface-to-volume ratio. Also, the unknowns of the displacement method of finite elements are nodal displacements only. A postprocessing routine is necessary to compute the reaction forces at the locations where the displacements are prescribed. For the boundary element method, however, $\mathbf{X}$ in Eq. (9.40) contains both u, often the nodal displacements, and q, often the tractions on the boundary, so that the reaction forces are calculated at the same time as the boundary displacements. This saves some effort in the postprocessing computations.

More details of boundary element formulations are provided in Brebbia and Dominguez (1992), Brebbia, et al. (1984), and Gipson (1987).

EXAMPLE 9.3 Calculation of the Torsional Constant

Use the direct boundary element method to calculate the torsional constant J of a 1×1 square cross section.

The formulation for the direct boundary element method for the pure torsion of a prismatic bar is given in Example 9.2, where the formula to calculate the torsional constant is provided by Eq. (15).

To find the torsional constant, the warping function ω on the boundary of the cross-section will be obtained first. Constant elements and linear elements will be used in the computation. For the constant element, the warping function over the whole element is assumed to be constant and is equal to the value of the warping function at the node which is located at the center of the element. Thus the warping function is approximated by

$$\omega = \omega_i \tag{1}$$

on each element i, where ω_i are nodal values of the warping function of the ith node (element). Substitute (1) into Eq. (14) of Example 9.2 to obtain

$$c\omega_j + \sum_{i=1}^{M} \widehat{H}^{ij}\,\omega_i = \sum_{i=1}^{M} \widehat{G}^{ij}\overline{q}_i \quad j = 1, 2, \ldots M \tag{2}$$

where

$$\widehat{H}^{ij} = -\int_{S_i} \frac{r_k a_k}{r^2}\,dS, \qquad \widehat{G}^{ij} = -\int_{S_i} \ln r\,dS$$

c is given in Eq. (14) of Example 9.2, r is the distance from node j to node i, r_k is the component of r in the x_k, $k = 2, 3$, direction, a_k is the direction cosine of the outer normal of the element with respect to the x_k axis, M is the number of elements, and $\overline{q}_i = x_2a_3 - x_3a_2$ (Eq. (3) of Example 9.2) at the nodal points. Since the shape function for the constant elements is unity, the matrix $\widehat{\mathbf{H}}^{ij}$ in Eq. (9.34) becomes a single number $\widehat{H}^{ij}$ in (2). Add c to $\widehat{H}^{ii}$ so that $\widehat{H}^{ij}$ becomes H^{ij}. Also change the notation $\widehat{G}^{ij}$ to G^{ij}. After every ω_i is spanned and the element matrices assembled, (2) becomes

$$\mathbf{HV} = \mathbf{GP} \tag{3}$$

where

$$\mathbf{V} = [\omega_1 \quad \omega_2 \quad \cdots \quad \omega_N]^T$$

$$\mathbf{P} = [\overline{q}_1 \quad \overline{q}_2 \quad \cdots \quad \overline{q}_N]^T$$

$$\mathbf{H} = \begin{bmatrix} H_{11} & H_{12} & \cdots & H_{1N} \\ \vdots & \vdots & \cdots & \vdots \\ H_{N1} & H_{N2} & \cdots & H_{NN} \end{bmatrix} \quad \text{and}$$

$$\mathbf{G} = \begin{bmatrix} G_{11} & G_{12} & \cdots & G_{1N} \\ \vdots & \vdots & \cdots & \vdots \\ G_{N1} & G_{N2} & \cdots & G_{NN} \end{bmatrix}$$

where N is the number of nodes. Note that for the torsion problem, the quantity q, the derivative of the warping function with respect to the outer normal, on the boundary is known, and $\mathbf{B} = \mathbf{0}$.

To avoid evaluating the diagonal elements of $\mathbf{H}$, the technique of Eq. (9.39) can be used, i.e.,

$$H_{ii} = -\sum_{\substack{j=1\\ j\neq i}}^{M} H_{ij} \tag{4}$$

The solution of (3) results in the warping function ω on the boundary.

For the linear element, the shape function for ω takes the form of Eq. (9.30), $u = \mathbf{N}^i\mathbf{v}^i$, and $\bar{q} = \mathbf{N}^i\mathbf{q}^i$ where

$$\mathbf{N}^i = [N_1 \quad N_2] \quad \text{with} \quad N_1 = \frac{1}{2}(1-\xi), \ N_2 = \frac{1}{2}(1+\xi)$$
$$\mathbf{v}^i = [\omega_1 \quad \omega_2]^T, \quad \bar{\mathbf{q}}^i = [\bar{q}_1 \quad \bar{q}_2]^T$$

This element is shown in Fig. 9.6a.

Follow the procedure from Eq. (9.32) to (9.34) to obtain

$$c\omega_j = \sum_{i=1}^{M} \widehat{H}^{ij} v^i = \sum_{i=1}^{M} \widehat{G}^{ij} \bar{q}^i$$

$$\text{where} \quad \widehat{H}^{ij} = -\int_{S_i} \frac{r_k a_k}{r^2} N^i \, dS$$
$$\widehat{G}^{ij} = -\int_{S_i} \ln\left(\frac{1}{r}\right) N^i \, dS.$$

After the warping function is found, the torsional constant is calculated using Eq. (15) of Example 9.2. The results of the computations are shown in the following table. The numbers in the brackets indicate the percentage error.

Torsional Constant

Number of Elements	**Constant Element**	**Linear Element**
12	0.1366 (2.8)	0.1540 (9.55)
20	0.1393 (0.9)	0.1456 (3.57)
28	0.14008 (0.35)	0.1432 (1.87)
40	0.1404 (0.126)	0.1418 (0.87)
Exact Solution	0.140577	

Although the accuracy of both elements is reasonable in this case, the constant element gives better results than the linear element. ∎

9.2.3 Indirect Formulation

The unknown variables in the boundary integral equation in the previous sections are the variables that can be used directly for the computation of the desired physical quantities. Such a formulation is referred to as being *direct*. In an alternative boundary element formulation, the *indirect formulation*, the variables sought are not directly the physical variables. Thus the unknown variables in the indirect formulation will not be u and q but some other quantities which are used to express u and q. The physical variables are calculated after these quantities are found.

The indirect formulation starts from the investigation of the solution of the equation $\nabla^2 u' = 0$ in the space outside V which is denoted by V'. Multiply $\nabla^2 u'$ by $\delta u = u^*$ of Eq. (9.22), with x in V' and ξ in V, and integrate over V' to obtain

$$\int_{V'} u^*(\nabla^2 u')\, dV' = \int_{V'} \left[\frac{\partial}{\partial x}\left(u^* \frac{\partial u'}{\partial x}\right) + \frac{\partial}{\partial y}\left(u^* \frac{\partial u'}{\partial y}\right) + \frac{\partial}{\partial z}\left(u^* \frac{\partial u'}{\partial z}\right)\right] dV'$$
$$- \int_{V'} \left(\frac{\partial u'}{\partial x}\frac{\partial u^*}{\partial x} + \frac{\partial u'}{\partial y}\frac{\partial u^*}{\partial y} + \frac{\partial u'}{\partial z}\frac{\partial u^*}{\partial z}\right) dV' \tag{9.41}$$

Follow the procedure from Eq. (9.15) to (9.19) to reduce Eq. (9.41) to

$$\int_{V'} u^*(\nabla^2 u')\, dV' - \int_{V'} u'(\nabla^2 u^*)\, dV' = \int_S (u^* q' - u' q^*)\, dS$$

where $q^* = \frac{\partial u^*}{\partial n}$, $q' = \frac{\partial u'}{\partial n'}$, the derivative of u' with respect to the outer normal of V'. Since both u' and u^* satisfy $\nabla^2 u' = 0$ and $\nabla^2 u^* = 0$ in V' [$x \neq \xi$, so $\delta(x, \xi) = 0$ in Eq. (9.21)], the above equation can be written as

$$\int_S u' q^*\, dS - \int_S q' u^*\, dS = 0 \tag{9.42}$$

At the common boundary of V and V', i.e., on S, assume that $u = u'$. Subtract Eq. (9.42) from Eq. (9.29) to obtain a different form of integral equation. For a point ξ inside V, this equation is written as

$$2\alpha\pi\ u(\xi) = -\int_V b u^*\, dV + \int_S (q - q') u^*\, dS \tag{9.43a}$$

Let $(q - q') = \sigma$, which is called the *density* of u, then Eq. (9.43a) becomes

$$2\alpha\pi\ u(\xi) = -\int_V b u^*\, dV + \int_S \sigma u^*\, dS \tag{9.43b}$$

Alternatively, if it is assumed that $q = q'$ on S, subtraction of Eq. (9.42) from Eq. (9.29) leads to

$$2\alpha\pi\ u(\xi) = -\int_S (u - u') q^* dS - \int_V b u^* dV$$

or

$$2\alpha\pi\ u(\xi) = -\int_S \mu q^*\, dS - \int_V b u^*\, dV \tag{9.44}$$

where $\mu = (u - u')$ is called a *dipole* which is a term related to potential theory [Brebbia, et al., 1984].

Equations (9.43b) and (9.44) can be used to solve Poisson's and Laplace's equations. For the Dirichlet problems, where u is prescribed on the whole boundary, the integral equations can be formed by moving ξ of Eqs. (9.43b) and (9.44) to the boundary. Using the same procedure as for the direct method, ξ is moved to the boundary, and it is found that Eq. (9.43b) is not changed, while Eq. (9.44) becomes

$$2\alpha\pi\ u(\xi) = \alpha\pi\mu(\xi) + \int_S \mu(x) \frac{\partial u^*(x, \xi)}{\partial n}\, dS - \int_V b u^*\, dV \tag{9.45}$$

With u already known (prescribed) on the boundary, the density σ and dipole μ in Eqs. (9.43b) and (9.45) can be found on the boundary, and then u inside the domain can be recovered.

For the Neumann[1] problems, where $q = \partial u/\partial n$ is prescribed on the boundary, the problems can be solved by using Eq. (9.43b). Take the derivative with respect to the outer normal at point ξ to obtain

$$2\alpha\pi\, q(\xi) = -\int_V b \frac{\partial u^*(x, \xi)}{\partial n}\, dV + \int_S \sigma(x) \frac{\partial u^*(x, \xi)}{\partial n}\, dS \tag{9.46}$$

[1] Karl Gottfried Neumann (1832–1925) was a German mathematician, spending most of his active career in Leipzig. He worked in many fields of mathematics and mechanics. He was involved in the development of potential theory and is credited with founding logarithmic potentials.

Move ξ to the boundary, and use the procedure given in Section 9.2.1 to show that the contribution of the second term of the right-hand side of the above equation is $-\alpha\pi\sigma(\xi)$. Thus Eq. (9.46) becomes

$$2\alpha\pi\, q(\xi) = -\alpha\pi\sigma(\xi) + \int_S \sigma(x)\, q^* dS - \int_V bq^*\, dV \tag{9.47}$$

Note that in Eqs. (9.45) and (9.47), the unknowns σ and μ appear both inside and outside the integrals. This kind of integral equation is called the *Fredholm*[2] *integral equation of the second kind.* In Eq. (9.43b) the unknown σ appears only inside the integral sign, so that it belongs to the class of integral equations called *Fredholm integral equations of the first kind.* In this indirect formulation, the unknown variables σ and μ in these equations are not the variables u and q that are of primary interest. The variables u and q are evaluated after σ and μ are found.

For Dirichlet problems, which are solved using Eqs. (9.43b) and (9.45), u on the left-hand sides of these equations is known because the point ξ is on the boundary. These relations can then be solved to evaluate σ and μ. For Neumann problems based on Eq. (9.47), the known boundary conditions of q at point ξ on the boundary are substituted into Eq. (9.47) and then the equation is solved for μ. After the values of σ and μ are found, they are substituted into Eqs. (9.43b) and (9.44), as appropriate, to find u inside the domain V.

The integral equations of Eqs. (9.43b), (9.45), and (9.47) can be solved using the boundary element procedure given in Section 9.2.2.

EXAMPLE 9.4 Indirect Formulation of the Torsional Bar

Study the indirect boundary element formulation for the pure torsion of a prismatic bar.

The governing differential equation and boundary condition for the warping function on the cross-section of a bar under pure torsion are given in Eq. (3) of Example 9.2. The governing equation is a two-dimensional partial differential equation with a Neumann boundary condition. For this problem, the warping function can be represented in the same fashion as Eq. (9.43b) with $b = 0$, i.e., for a point ξ inside the cross-section,

$$2\pi\, \omega(\xi) = \int_S \sigma\, u^*(\xi, x)\, dS \tag{1}$$

where ξ is a point inside the cross-section, x is a point on the boundary, and $u^* = \ln(1/r)$ is the Green's function given in Eq. (9.22b) for two-dimensional problems. Take the derivative of (1), with respect to the outer normal of the cross-section at point ξ, to obtain

$$\begin{aligned} 2\pi \frac{\partial \omega(\xi)}{\partial n} &= 2\pi \bar{q} = \int_S \sigma \frac{\partial u^*(\xi, x)}{\partial n}\, dS = \int_S \sigma \frac{\partial}{\partial n} \ln\left(\frac{1}{r}\right) dS \\ &= \int_S \sigma \left[a_2 \frac{\partial}{\partial x_2} \ln\left(\frac{1}{r}\right) + a_3 \frac{\partial}{\partial x_3} \ln\left(\frac{1}{r}\right)\right] dS \\ &= -\int_S \sigma \frac{r_i a_i}{r^2}\, dS \quad i = 2, 3 \end{aligned} \tag{2}$$

where a_i is the direction cosine of the outer normal with respect to direction x_i, $i = 2, 3$ and the bar over q indicates a prescribed condition.

[2] Erik Ivar Fredholm (1866–1927) was a Swedish professor of mathematics and physics at Stockholm University. He initiated the modern theory of integral equations, which are named after him.

The point ξ needs to be moved to the boundary to establish a boundary element formulation. Also, the right-hand side of (2) has to be given special attention, since a singularity develops when ξ is on the boundary. Treat the boundary of the cross-section in the same fashion as in Example 9.2 (Fig. 9.3) and perform the integration.

$$\int_{S_\epsilon} \sigma \frac{r_i a_i}{r^2} \, dS = \int_0^\pi \sigma \frac{\epsilon}{\epsilon^2} \epsilon \, d\theta = \int_0^\pi \sigma \, d\theta = \pi\sigma$$

Thus,

$$-\lim_{\epsilon \to 0} \left[\int_{S-S_\epsilon} \sigma \frac{r_i a_i}{r^2} \, dS + \int_{S_\epsilon} \sigma \frac{r_i a_i}{r^2} \, dS \right] = \int_S \sigma \frac{\partial u^*}{\partial n} \, dS - \pi\sigma \tag{3}$$

Then (2) becomes

$$2\pi\overline{q} = -\pi\sigma + \int_S \sigma \frac{\partial u^*}{\partial n} \, dS \tag{4}$$

After the unknown density σ on the boundary is calculated, the warping function on the boundary as well as inside the cross-section can be obtained from (1). The torsional constant can be found from Eq. (15) of Example 9.2. The derivatives of ω with respect to ξ_2 and ξ_3 can be expresed as

$$\frac{\partial \omega}{\partial \xi_2} = \frac{1}{2\pi} \int_S \sigma \frac{\partial u^*}{\partial \xi_2} \, dS \tag{5}$$

$$\frac{\partial \omega}{\partial \xi_3} = \frac{1}{2\pi} \int_S \sigma \frac{\partial u^*}{\partial \xi_3} \, dS \tag{6}$$

After these quantities are found, the stresses on the cross-section can be calculated using Chapter 1, Eqs. (1.142) and (1.143). ∎

9.3 Linear Elasticity

The boundary element formulations for two- and three-dimensional elasticity problems are given in this section. For the two-dimensional case, the formulation is for plane strain problems. For plane stress problems, the plane strain relationships with an adjustment in material constants (Chapter 1, Section 1.3.1) can be employed.

9.3.1 Basic Relations

We begin with the extended Galerkin's formula of Chapter 7, Eq. (7.72)

$$-\int_V \mathbf{u}^{*^T} (\mathbf{D}^T \boldsymbol{\sigma} + \overline{\mathbf{p}}_V) \, dV + \int_{S_p} \mathbf{u}^{*^T} (\mathbf{p} - \overline{\mathbf{p}}) \, dS - \int_{S_u} \mathbf{p}^{*^T} (\mathbf{u} - \overline{\mathbf{u}}) \, dS = 0 \tag{9.48a}$$

or

$$-\int_V (\sigma_{ij,j} + \overline{p}_{Vi}) u_i^* \, dV + \int_{S_p} (p_i - \overline{p}_i) u_i^* \, dS - \int_{S_u} (u_i - \overline{u}_i) p_i^* \, dS = 0 \tag{9.48b}$$

when $\delta\mathbf{u}$ and $\delta\mathbf{p}$ have been replaced by the weighting functions $\mathbf{u}^*$ and $\mathbf{p}^*$, as is customary in boundary element theory. This general relationship is often considered as the fundamental principle underlying the boundary element method.

It is of interest that this fundamental weighted-residual relationship can be derived in a variety of ways. Probably the most common derivation is based on Betti's reciprocal theorem of Chapter 3, Eq. (3.36), i.e.,

$$\int_V \overline{p}_{Vi} u_i^* \, dV + \int_S p_i u_i^* \, dS = \int_V \overline{p}_{Vi}^* u_i \, dV + \int_S p_i^* u_i \, dS \tag{9.49}$$

Consider the boundary S to be divided into the two parts S_p and S_u, with $\mathbf{u} = \overline{\mathbf{u}}$ on S_u and $\mathbf{p} = \overline{\mathbf{p}}$ on S_p. Then write the surface integrals of Eq. (9.49) as

$$\int_S p_i u_i^* \, dS = \int_{S_p} \overline{p}_i u_i^* \, dS + \int_{S_u} p_i u_i^* \, dS$$

$$\int_S p_i^* u_i \, dS = \int_{S_p} p_i^* u_i \, dS + \int_{S_u} p_i^* \overline{u}_i \, dS$$

Introduce these relations and the equilibrium conditions $\overline{p}_{Vi}^* = -\sigma_{ij,j}^*$ into Eq. (9.49) to obtain

$$\int_V \sigma_{ij,j}^* u_i \, dV + \int_V \overline{p}_{Vi} u_i^* \, dV = -\int_{S_p} \overline{p}_i u_i^* \, dS - \int_{S_u} p_i u_i^* \, dS + \int_{S_p} p_i^* u_i \, dS + \int_{S_u} p_i^* \overline{u}_i \, dS \tag{9.50}$$

Apply Gauss' integral theorem, in the form of Eq. (II.9) of Appendix II, twice to the first integral of Eq. (9.50) to obtain

$$\begin{aligned}\int_V \sigma_{ij,j}^* \, u_i \, dV &= \int_S p_i^* u_i \, dS - \int_V \sigma_{ij}^* \, u_{i,j} \, dV \\ &= \int_S p_i^* u_i \, dS - \int_V \sigma_{ij} \, u_{i,j}^* \, dV \\ &= \int_S p_i^* u_i \, dS - \int_S p_i u_i^* \, dS + \int_V \sigma_{ij,j} \, u_i^* \, dV \end{aligned} \tag{9.51}$$

Here, use was made of Chapter 3, Eq. (3.31) and the relationship [Eqs. (3.29) and (3.33)]

$$\sigma_{ij}^* u_{i,j} = \sigma_{ij}^* \epsilon_{ij} = \sigma_{ij} \epsilon_{ij}^* = \sigma_{ij} u_{i,j}^*$$

Remember that $S = S_p + S_u$ and insert Eq. (9.51) into Eq. (9.50). This leads directly to Eq. (9.48), the extended Galerkin's formula.

9.3.2 Fundamental Solutions

The boundary element formulation of the linear elasticity problem starts with Eq. (9.50) which is then converted into an integral equation with unknowns u_i, p_i on the boundary. To do so, the first step is to isolate u_i from the first term of Eq. (9.50). Let a concentrated force $\delta(\xi, x)$ in an arbitrary direction be applied at point ξ, so that the equilibrium equation is

$$\sigma_{ij,j}^* = -\delta(\xi, x) \, a_i \tag{9.52}$$

where ξ and x stand for two points inside the body, a_i is the direction cosine between the concentrated force and the x_i direction. In the derivations that follow, quantities with superscript * are due to the force $\delta(\xi, x) \, a_i$. Insertion of Eq. (9.52) in Eq. (9.50) and use of the sifting property of the Dirac delta function of Eq. (9.6) leads to the isolation of u_i, and Eq. (9.50) will be an expression of the displacement u_i in the domain in terms of the boundary unknowns p_i and u_i.

It can be seen that σ_{ij}^* in Eq. (9.52) are the stresses at point x due to a unit point load at point ξ in the direction of i. Hence, the displacements and tractions with superscript *, which are called the fundamental solutions of the problem, in Eq. (9.50), are the quantities at point x due to the same force. The fundamental solution u_i^* can be obtained from the displacement form of the equilibrium equations, which for a solid subjected to a body force $\mathbf{p}_V = [\overline{p}_{V1} \;\; p_{V2} \;\; \overline{p}_{V3}]^T = [\delta(\xi, x)a_1 \;\; \delta(\xi, x)a_2 \;\; \delta(\xi, x)a_3]^T$ can be expressed as [Chapter 1, Eq. (1.85)]

$$\nabla^2 u_i + \frac{1}{1-2\nu} u_{k,ki} + \frac{\overline{p}_{Vi}}{G} = 0$$

or in vector form,

$$\left(\nabla^2 + \frac{1}{1-2\nu}\nabla \operatorname{div}\right)\mathbf{u} + \frac{\overline{\mathbf{p}}_V}{G} = \mathbf{0} \tag{9.53}$$

in which ∇ and div are the operators

$$\nabla = \begin{bmatrix} \frac{\partial}{\partial x_1} & \frac{\partial}{\partial x_2} & \frac{\partial}{\partial x_3} \end{bmatrix} \qquad \operatorname{div} = \frac{\partial}{\partial x_1} + \frac{\partial}{\partial x_2} + \frac{\partial}{\partial x_3}$$

and

$$\mathbf{u} = [u_1 \quad u_2 \quad u_3]^T$$

A simple way to find the fundamental solution is to express Eq. (9.53) in terms of a vector $\mathbf{g}$, called the *Galerkin vector*. This procedure is treated in theory of elasticity texts such as Boresi and Chong (1987). In order to find an expression for $\mathbf{g}$, define another vector $\mathbf{S} = [S_1 \;\; S_2 \;\; S_3]^T$ at each point P in a volume V with piecewise smooth surfaces such that

$$\mathbf{S} = \int_V \frac{\mathbf{u}}{r}\, dV \tag{9.54a}$$

where r is the distance from point P to the differential volume dV. In component form,

$$S_i = \int_V \frac{u_i}{r}\, dV \tag{9.54b}$$

It can be verified by differentiating Eq. (9.54) that [Brebbia, et al., 1984]

$$\nabla^2 \mathbf{S} = -4\pi \mathbf{u} \tag{9.55}$$

Since div $\mathbf{S}$ ($=\nabla \cdot \mathbf{S}$, a scalar) and $\nabla \times \mathbf{S}$ (a vector) are independent quantities, i.e., a scalar and a vector are not related, they can be assigned to any scalar and vector quantities. Let

$$\nabla \cdot \mathbf{S} = -\frac{4\pi}{\lambda} \operatorname{div} \mathbf{g} \tag{9.56a}$$

$$\nabla \times \mathbf{S} = -4\pi(\nabla \times \mathbf{g}) \tag{9.56b}$$

where $\lambda = 2(1-\nu)/(1-2\nu)$, $\cdot$ and $\times$ are the dot and cross-product operators of vectors with

$$\nabla \cdot \mathbf{S} = \begin{bmatrix} \frac{\partial}{\partial x_1} & \frac{\partial}{\partial x_2} & \frac{\partial}{\partial x_3} \end{bmatrix} \cdot [S_1 \quad S_2 \quad S_3]^T$$

$$= \frac{\partial S_1}{\partial x_1} + \frac{\partial S_2}{\partial x_2} + \frac{\partial S_3}{\partial x_3}$$

$$\nabla \times \mathbf{S} = \begin{bmatrix} \frac{\partial S_3}{\partial x_2} - \frac{\partial S_2}{\partial x_3} & \frac{\partial S_1}{\partial x_3} - \frac{\partial S_3}{\partial x_1} & \frac{\partial S_2}{\partial x_1} - \frac{\partial S_1}{\partial x_2} \end{bmatrix}$$

and **g** is the Galerkin vector. It can be verified that a vector **S** has the property

$$\nabla \times (\nabla \times \mathbf{S}) = \nabla(\nabla \cdot \mathbf{S}) - \nabla^2 \mathbf{S}$$

or

$$\nabla^2 \mathbf{S} = \nabla(\nabla \cdot \mathbf{S}) - \nabla \times (\nabla \times \mathbf{S})$$

Substitute Eqs. (9.55) and (9.56) into this identity

$$4\pi \mathbf{u} = \nabla\left(\frac{4\pi}{\lambda}\text{div}\,\mathbf{g}\right) - \nabla \times [4\pi(\nabla \times \mathbf{g})]$$

or

$$\mathbf{u} = \frac{1}{\lambda}\nabla\,\text{div}\,\mathbf{g} - \nabla \times (\nabla \times \mathbf{g}) \tag{9.57}$$

Since $\nabla \times (\nabla \times \mathbf{g}) = \nabla\,\text{div}\,\mathbf{g} - \nabla^2 \mathbf{g}$, Eq. (9.57) can be written as

$$\mathbf{u} = \left(\nabla^2 - \frac{1}{2(1-\nu)}\nabla\,\text{div}\right)\mathbf{g} \tag{9.58a}$$

or

$$u_i = g_{i,kk} - \frac{1}{2(1-\nu)} g_{k,ik} \tag{9.58b}$$

This is the desired property of the Galerkin vector.

Substitution of Eq. (9.58) into Eq. (9.53) results in a biharmonic equation

$$\nabla^4 \mathbf{g} + \frac{\overline{\mathbf{p}}_V}{G} = \mathbf{0} \quad \text{or} \quad \nabla^4 g_i + p_{Vi}/G = 0 \tag{9.59}$$

This is the equilibrium equation expressed in terms of the Galerkin vector. Once the expressions for the components of the Galerkin vector are obtained, the fundamental solutions can be found. Equation (9.59) in component form with $\overline{\mathbf{p}}_V^* = [\delta(\xi, x)a_1 \;\; \delta(\xi, x)a_2 \;\; \delta(\xi, x)a_3]^T$ becomes

$$\nabla^4 g_i^* + \frac{1}{G}\delta(\xi, x)a_i = 0 \tag{9.60}$$

Here g_i^* has replaced g_i of Eq. (9.59) because $\overline{p}_{Vi}$ has been replaced by $\overline{p}_{Vi}^*$. The solution of Eq. (9.60) leads to the desired fundamental solutions.

Write Eq. (9.60) as

$$\nabla^2 F_i + \frac{1}{G}\delta(\xi, x)a_i = 0 \tag{9.61}$$

with

$$F_i = \nabla^2 g_i^*$$

The solution to Eq. (9.61) is [Haberman, 1987]

$$F_i = \nabla^2 g_i^* = \frac{1}{4\pi r G} a_i \tag{9.62}$$

for three-dimensional and

$$F_i = \nabla^2 g_i^* = \frac{1}{2\pi G} \ln\left(\frac{1}{r}\right) a_i \tag{9.63}$$

for two-dimensional problems, where $r = \sqrt{\sum (x_k - \xi_k)^2}$ is the distance between ξ and x. Here, ξ_k and x_k, $k = 1, 2, 3$ for three-dimensional problems and $k = 1, 2$ for two-dimensional problems, are the coordinates of the points ξ and x.

Equations (9.62) and (9.63) can be solved for g_i^*. Note that Eqs. (9.62) and (9.63) are functions of r only. Hence, the solutions to Eq. (9.62) and (9.63) will be functions of r only, i.e., they will be spherically symmetric for three-dimensional problems and rotationally symmetric for two-dimensional problems. For the three-dimensional problems, the operator ∇^2 is

$$\nabla^2 = \frac{1}{r^2}\frac{d}{dr}\left(r^2 \frac{d}{dr}\right)$$

so that Eq. (9.62) becomes

$$\nabla^2 g_i^* = \frac{1}{r^2}\frac{d}{dr}\left(r^2 \frac{dg_i^*}{dr}\right) = \frac{1}{4\pi r G} a_i$$

Integration leads to

$$g_i^* = \left(\frac{r}{8\pi G} + \frac{C_1}{r} + C_2\right) a_i$$

At $r = 0$, the displacement should be finite, and so should be g_i^*. Hence, C_1 must be equal to zero. Furthermore, since only the derivatives of g_i^* appear in Eq. (9.58), and differentiation of g_i^* will cancel C_2, the constant C_2 can be set to zero here. Thus,

$$g_i^* = \frac{r}{8\pi G} a_i = \widehat{g} a_i \tag{9.64}$$

with $\widehat{g} = r/(8\pi G)$. For the two-dimensional case, the operator ∇^2 is

$$\nabla^2 = \frac{1}{r}\frac{d}{dr}\left(r \frac{d}{dr}\right)$$

Then Eq. (9.63) appears as

$$\nabla^2 g_i^* = \frac{1}{r}\frac{d}{dr}\left(r \frac{dg_i^*}{dr}\right) = \frac{1}{2\pi G} \ln\left(\frac{1}{r}\right) a_i \tag{9.65}$$

Integrate Eq. (9.65), and process the integration constants in the same manner as for the three-dimensional case to obtain

$$g_i^* = \frac{1}{8\pi G} r^2 \ln\left(\frac{1}{r}\right) a_i = \widehat{g} a_i \tag{9.66}$$

with

$$\widehat{g} = \frac{1}{8\pi G} r^2 \ln\left(\frac{1}{r}\right) \tag{9.67}$$

Similar to the definitions of σ_{ij}^* of Eq. (9.52), let u_{ji}^* be the displacement in the i direction due to the force $\delta(\xi, x) a_j$. Thus the resultant displacements in the i direction due to all the forces $\delta(\xi, x)a_j$, $j = 1, 2, 3$ for three-dimensional problems and $j = 1, 2$ for two-dimensional problems, can be written as

$$u_i^* = u_{ji}^* a_j \tag{9.68}$$

In the rectangular coordinate system,

$$a_i a_j = \delta_{ij} = \begin{cases} 1 & \text{when } i = j \\ 0 & \text{when } i \neq j \end{cases}$$

Then $u_{ji}^* = u_i^* a_j$. From Eq. (9.58b),

$$u_{ji}^* = \left(g_{i,kk}^* - \frac{1}{2(1-\nu)} g_{k,ik}^* \right) a_j = (g_i^* a_j)_{,kk} - \frac{1}{2(1-\nu)} (g_k^* a_j)_{,ik}$$

Since, from Eqs. (9.64) and (9.66), $g_i^* = \widehat{g}\, a_i$ so that $g_i^* a_j = \widehat{g}\, a_i a_j = \widehat{g} \delta_{ij}$. Thus

$$u_{ji}^* = \widehat{g}_{,kk} \delta_{ij} - \frac{1}{2(1-\nu)} \widehat{g}_{,ik}\, \delta_{kj} \tag{9.69}$$

For three-dimensional problems, substitute Eq. (9.64) into Eq. (9.69) to obtain

$$u_{ji}^* = \widehat{g}_{,kk}\, \delta_{ij} - \frac{1}{2(1-\nu)} \widehat{g}_{,ik}\, \delta_{kj} = \frac{1}{16\pi G(1-\nu)} [2(1-\nu) r_{,kk}\, \delta_{ij} - r_{,ik}\, \delta_{kj}] \tag{9.70}$$

The derivatives of r are taken with reference to the coordinates of x_i, i.e.,

$$r_{,i} = \frac{\partial r}{\partial x_i} \qquad r_{,j} = \frac{\partial r}{\partial x_j} \tag{9.71a}$$

and

$$r_{,i} = \frac{\partial}{\partial x_i} \sqrt{\sum (x_k - \xi_k)^2} = \frac{(x_i - \xi_i)}{\sqrt{\sum (x_k - \xi_k)^2}} = \frac{r_i}{r} \tag{9.71b}$$

where $r_i = (x_i - \xi_i)$ is the projection of r on the x_i axis. From the relationships of Eq. (9.71),

$$\frac{\partial r^2}{\partial x_i^2} = \frac{1}{r^2} \left(r \frac{\partial r_i}{\partial x_i} - r_i \frac{\partial r}{\partial x_i} \right) = \frac{r^2 - r_i^2}{r^3}$$

Hence

$$r_{,kk} = \sum_{k=1}^{3} \frac{\partial^2 r}{\partial x_k^2} = \sum_{k=1}^{3} \frac{r^2 - r_k^2}{r^3} = \frac{2}{r}$$

Also

$$r_{,ik} = \frac{\partial^2 r}{\partial x_i \partial x_k} = \frac{\partial}{\partial x_k} \left(\frac{\partial r}{\partial x_i} \right) = \frac{\partial}{\partial x_k} \left(\frac{r_i}{r} \right) = \frac{1}{r^2} \left(r \frac{\partial r_i}{\partial x_k} - r_i \frac{\partial r}{\partial x_k} \right)$$

i.e.,

$$r_{,ik} = \begin{cases} -\frac{1}{r} r_{,i}\, r_{,k} & \text{when } i \neq k \\ \frac{1}{r}(1 - r_{,i}\, r_{,i}) & \text{when } i = k \end{cases}$$

Then Eq. (9.70) can be written as

$$u^*_{ji} = \frac{1}{16\pi G(1-\nu)}[-r_{,ij}] = \frac{1}{16\pi G(1-\nu)r} r_{,i} r_{,j} \tag{9.72a}$$

for $i \neq j$ and

$$\begin{aligned} u^*_{ji} &= \frac{1}{16\pi G(1-\nu)}[2(1-\nu)r_{,kk} - r_{,ij}] \\ &= \frac{1}{16\pi G(1-\nu)}\left[4(1-\nu)\frac{1}{r} - \frac{1}{r}(1 - r_{,i} r_{,i})\right] \\ &= \frac{1}{16\pi G(1-\nu)r}[(3-4\nu) + r_{,i} r_{,i}] \end{aligned} \tag{9.72b}$$

for $i = j$. Combine Eqs. (9.72a) and (9.72b) to obtain

$$u^*_{ji} = \frac{1}{16\pi(1-\nu)Gr}[(3-4\nu)\delta_{ij} + r_{,i} r_{,j}] \tag{9.73}$$

For two-dimensional plane strain problems, a similar procedure leads to

$$u^*_{ji} = \frac{1}{8\pi(1-\nu)G}\left[(3-4\nu)\ln\left(\frac{1}{r}\right)\delta_{ij} + r_{,i} r_{,j}\right] \tag{9.74}$$

The tractions p^*_{ji} can be obtained by substituting the displacements into the strain-displacement relations and the constitutive equations of Chapter 1, Eqs. (1.19) and (1.34), to obtain

$$\sigma^*_{ij} = S^*_{kij} a_k \tag{9.75}$$

with

$$S^*_{kij} = -\frac{1}{r^\alpha}[(1-2\nu)(\delta_{ki} r_{,j} + \delta_{kj} r_{,i} - \delta_{ij} r_{,k}) + \beta r_{,i} r_{,j} r_{,k}]$$

where $\alpha = 2, 1$, and $\beta = 3, 2$ for three-dimensional problems and two-dimensional plane strain problems, respectively. The tractions are [Chapter 1, Eq. (1.58)]

$$p^*_i = \sigma^*_{ij} a_j \quad \text{on the boundary} \tag{9.76}$$

where here (and for the following two equations) a_j is the direction cosine of the outer normal to the boundary. Then from

$$p^*_i = p^*_{ji} a_j \tag{9.77}$$

it follows that

$$p^*_{ji} = \frac{-1}{4\alpha\pi(1-\nu)r^\alpha}\left\{[(1-2\nu)\delta_{ij} + \beta r_{,i} r_{,j}]\frac{\partial r}{\partial n} - (1-2\nu)(r_{,i} a_j - r_{,j} a_i)\right\} \tag{9.78}$$

where $\partial r/\partial n$ is the derivative of r with respect to the outer normal of the boundary.

9.3.3 Integral Equation Formulation

Substitute Eqs. (9.52), (9.68), and (9.77) into Eq. (9.50) to obtain

$$-u_i a_i + \int_V \overline{p}_{Vi} u^*_{ji}\, a_j\, dV = -\int_{S_p} \overline{p}_i u^*_{ji}\, a_j\, dS - \int_{S_u} p_i u^*_{ji}\, a_j\, dS + \int_{S_p} u_i p^*_{ji}\, a_j\, dS + \int_{S_u} \overline{u}_i p^*_{ji}\, a_j\, dS$$

Exchange the positions of i, j in the above integrals to find

$$-u_i a_i + \int_V \overline{p}_{Vj} u^*_{ij}\, a_i\, dV = -\int_{S_p} \overline{p}_j u^*_{ij}\, a_i\, dS - \int_{S_u} p_j u^*_{ij}\, a_i\, dS + \int_{S_p} u_j p^*_{ij}\, a_i\, dS + \int_{S_u} \overline{u}_j p^*_{ij}\, a_i\, dS \tag{9.79}$$

The a_i's can be factored out of Eq. (9.79) with the result

$$-u_i(\xi) + \int_V \overline{p}_{Vj} u^*_{ij}\, dV = -\int_{S_p} u^*_{ij} \overline{p}_j\, dS - \int_{S_u} u^*_{ij} p_j\, dS + \int_{S_p} p^*_{ij} u_j\, dS + \int_{S_u} p^*_{ij} \overline{u}_j\, dS \tag{9.80a}$$

or

$$u_i(\xi) + \int_S p^*_{ij} u_j\, dS = \int_S u^*_{ij} p_j\, dS + \int_V \overline{p}_{Vj} u^*_{ij}\, dV \tag{9.80b}$$

where $S = S_u + S_p$, u_j and p_j are the displacements and the tractions on the whole boundary. Equation (9.80) is known as *Somigliana's*[3] *identity*. It should be observed that the integral containing $\overline{p}_{Vj}$ does not involve any unknowns.

Since the quantities in the integrals all refer to a point x inside the body or on the boundary, i.e.,

$$\overline{p}_{Vj} u^*_{ij}\, dV = \overline{p}_{Vj}(x)\, u^*_{ij}(\xi, x)\, dV(x)$$
$$u^*_{ij} p_j\, dS = u^*_{ij}(\xi, x)\, p_j(x)\, dS(x)$$
$$p^*_{ij} u_j\, dS = p^*_{ij}(\xi, x)\, u_j(x)\, dS(x)$$

Eq. (9.80) is a relationship between the displacement at a point ξ inside the boundary and the displacements $u_j(x)$ and tractions $p_j(x)$ on the boundary. Equation (9.80) is the theory of elasticity equivalence of Eq. (9.9) for beams. These are the integral equations on which the boundary element method is based. In Eq. (9.80) the unknowns appear both inside and outside the boundary integrals, a property that characterizes a Fredholm integral equation of the second kind. The aim of the boundary element method is first to calculate the unknown displacements and tractions on the boundary, then compute the displacements inside the boundary, and, finally, calculate the stresses.

[3] Carlos Somigliana (1860–1955) was an Italian physicist who specialized in geophysics. In 1892, he began a career as a professor of mathematical physics in Pavia and Turin.

9.3.4 Points on the Boundary

In order to calculate the unknown values of p_i and u_i in Eq. (9.80), it is necessary to move the point ξ to the boundary, so that all the unknowns in Eq. (9.80) are on the boundary, and a system of linear equations can be formed to solve for the unknowns. When the point ξ is on the boundary, however, singularities develop, and they must be given special consideration. Assume that the boundary can be represented as shown in Fig. 9.1, where $S = (S - S_\epsilon) + S_\epsilon$, in which S_ϵ is a hemispherical surface of radius ϵ.

The first integral on the right-hand side of Eq. (9.80b) can be written as

$$\int_S u_{ij}^* \, p_j \, dS = \lim_{\epsilon\to 0} \int_{S-S_\epsilon} u_{ij}^* \, p_j \, dS + \lim_{\epsilon\to 0} \int_{S_\epsilon} u_{ij}^* \, p_j \, dS$$

Since on the part of the boundary $S - S_\epsilon$, u_{ij}^* does not involve any singularity, the first integral will not be altered when $\epsilon \to 0$, i.e.,

$$\lim_{\epsilon\to 0} \int_{S-S_\epsilon} u_{ij}^* \, p_j \, dS = \int_S u_{ij}^* \, p_j \, dS \tag{9.81}$$

Assume that S_ϵ is very small, so that p_j can be treated as being constant on S_ϵ. Then

$$\lim_{\epsilon\to 0} \int_{S_\epsilon} u_{ij}^* \, p_j \, dS = p_j \lim_{\epsilon\to 0} \int_{S_\epsilon} u_{ij}^* \, dS \tag{9.82}$$

Although the two-dimensional case will be used to study this integral, similar arguments hold for three-dimensional cases. As shown in Fig. 9.3 and from the relationships given in Eq. (9.71),

$$dS = \epsilon \, d\theta, \quad r_{,1} = \epsilon_{,1} = \frac{\epsilon_1}{\epsilon} = \cos\varphi, \quad r_{,2} = \epsilon_{,2} = \frac{\epsilon_2}{\epsilon} = \sin\varphi \tag{9.83}$$

Substitute Eqs. (9.74) and (9.83) into Eq. (9.82) to obtain

$$\begin{aligned}
\lim_{\epsilon\to 0} \int_{S_\epsilon} u_{ij}^* \, dS &= \lim_{\epsilon\to 0} \int_{S_\epsilon} \frac{1}{8\pi(1-\nu)G} \left[(3-4\nu)\ln\frac{1}{\epsilon}\delta_{ij} + \epsilon_{,i}\,\epsilon_{,j}\right] dS \\
&= \lim_{\epsilon\to 0} \int_{S_\epsilon} \frac{1}{8\pi(1-\nu)G} \left[(3-4\nu)\ln\frac{1}{\epsilon}\delta_{ij} + \cos\varphi \, \sin\varphi\right] \epsilon \, d\theta \\
&= \frac{1}{8\pi(1-\nu)G} \left[(3-4\nu)\lim_{\epsilon\to 0} \int_{S_\epsilon} \epsilon \ln\frac{1}{\epsilon} \, d\theta \, \delta_{ij} + \lim_{\epsilon\to 0} \int_{S_\epsilon} \cos\varphi \, \sin\varphi \, \epsilon \, d\theta\right]
\end{aligned}$$

For the second integral here, $\theta = \gamma + \varphi$ (Fig. 9.3), so that $d\theta = d\varphi$, and when θ varies from 0 to π, φ varies from $-\gamma$ to $\pi - \gamma$. Then

$$\lim_{\epsilon\to 0} \int_{S_\epsilon} \cos\varphi \, \sin\varphi \, \epsilon \, d\theta = \lim_{\epsilon\to 0} \epsilon \int_{-\gamma}^{\pi-\gamma} \cos\varphi \, \sin\varphi \, d\varphi = 0$$

For the first integral,

$$\lim_{\epsilon\to 0} \int_{S_\epsilon} \epsilon \ln\frac{1}{\epsilon} \, d\theta = -\lim_{\epsilon\to 0} \frac{\ln\epsilon}{\frac{1}{\epsilon}} \int_{S_\epsilon} d\theta = -\lim_{\epsilon\to 0} \frac{\frac{d}{d\epsilon}(\ln\epsilon)}{\frac{d}{d\epsilon}\left(\frac{1}{\epsilon}\right)} \int_{S_\epsilon} d\theta = \lim_{\epsilon\to 0}(\epsilon) \int_{S_\epsilon} d\theta = 0$$

Then, in Eq. (9.82),

$$\lim_{\epsilon\to 0}\int_{S_\epsilon} u^*_{ij}\, dS = 0 \tag{9.84}$$

so that moving ξ to the boundary has no effect on this integral.

Consider the integral on the left-hand side of Eq. (9.80b),

$$\int_S p^*_{ij}\, u_j\, dS = \lim_{\epsilon\to 0}\int_{S-S_\epsilon} p^*_{ij}\, u_j\, dS + \lim_{\epsilon\to 0}\int_{S_\epsilon} p^*_{ij}\, u_j\, dS \tag{9.85}$$

In the three-dimensional space, with $r = \epsilon$, the final integral of Eq. (9.85) can be developed as

$$\lim_{\epsilon\to 0}\int_{S_\epsilon} p^*_{ij}\, u_j\, dS = \lim_{\epsilon\to 0}\int_{S_\epsilon} u_j\left\{[(1-2\nu)\delta_{ij} + 3\epsilon_{,i}\epsilon_{,j}]\frac{\partial\epsilon}{\partial n} - (1-2\nu)(\epsilon_{,i}a_j - \epsilon_{,j}a_i)\right\}$$
$$\times\frac{-1}{8\pi(1-\nu)\epsilon^2}\, dS \tag{9.86}$$

The expression for p^*_{ij} of Eq. (9.78) has been used here. Let the boundary be smooth at ξ, so that S_ϵ is a hemispherical surface of Fig. 9.1, where θ varies from 0 to 2π and φ varies from 0 to $\pi/2$. For this hemispherical surface, from the relationships of Eq. (9.71) and Fig. 9.2,

$$\frac{\partial\epsilon}{\partial n} = 1,\quad a_i = \frac{\partial\epsilon}{\partial x_i} = \frac{\epsilon_i}{\epsilon},\quad \epsilon_{,i}\, a_j - \epsilon_{,j}\, a_i = 0,\quad dS = \epsilon^2 \sin\varphi\, d\varphi\, d\theta$$

and (Problem 9.11)

$$\int_{S_\epsilon}\frac{\partial\epsilon}{\partial x_i}\frac{\partial\epsilon}{\partial x_j}\, dS = 0 \quad \text{for} \quad i \neq j$$

Then, from Eq. (9.86),

$$\lim_{\epsilon\to 0}\int_{S_\epsilon} p^*_{ij}\, u_j\, dS = \lim_{\epsilon\to 0}\int_{S_\epsilon}\frac{-1}{8\pi(1-\nu)} u_j[(1-2\nu) + 3(\epsilon_{,j})^2]\frac{1}{\epsilon^2}\epsilon^2 \sin\varphi\, d\varphi\, d\theta$$
$$= \lim_{\epsilon\to 0}\int_{S_\epsilon}\frac{-1}{8\pi(1-\nu)} u_i[(1-2\nu) + 3(\epsilon_{,i})^2]\sin\varphi\, d\varphi\, d\theta$$

It follows from the relationship

$$\int_{S_\epsilon}\sin\varphi\, d\varphi\, d\theta = \int_0^{2\pi} d\theta\int_0^{\pi/2}\sin\varphi\, d\varphi = 2\pi$$

and from Fig. 9.2, that

$$\int_{S_\epsilon}(\epsilon_{,i})^2 \sin\varphi\, d\varphi\, d\theta = \begin{cases}\int_0^{2\pi}\cos^2\theta\, d\theta\int_0^{\pi/2}\sin^3\varphi\, d\varphi & i=1\\ \int_0^{2\pi}\sin^2\theta\, d\theta\int_0^{\pi/2}\sin^3\varphi\, d\varphi & i=2\\ \int_0^{2\pi} d\theta\int_0^{\pi/2}\cos^2\varphi\,\sin\varphi\, d\varphi & i=3\end{cases}$$
$$= \frac{2\pi}{3}$$

Thus,

$$\lim_{\epsilon\to 0}\int_{S_\epsilon} p^*_{ij}\, u_j\, dS = \lim_{\epsilon\to 0}\int_{S_\epsilon} \frac{-1}{8\pi(1-\nu)} u_i[(1-2\nu)+3\epsilon_{,i}\epsilon_{,i}]\sin\varphi\, d\varphi\, d\theta$$
$$= \frac{-1}{8\pi(1-\nu)} u_i\left[(1-2\nu)2\pi + 3\frac{2\pi}{3}\right] = -\frac{1}{2}u_i \tag{9.87}$$

Note that when ξ is at a location which is not smooth, the coefficient of u_i may not be $-1/2$, but can assume other values (Problem 9.12). Actually, it will be seen later that this value does not need to be determined explicitly.

The final integral equation on the boundary is

$$c_i\, u_i(\xi) + \int_S u_j\, p^*_{ij}\, dS = \int_S p_j\, u^*_{ij}\, dS + \int_V \overline{p}_{Vj}\, u^*_{ij}\, dV \tag{9.88a}$$

or in matrix form

$$\mathbf{c}\mathbf{u}(\xi) + \int_S \mathbf{p}^*\mathbf{u}\, dS = \int_S \mathbf{u}^*\mathbf{p}\, dS + \int_V \mathbf{u}^*\,\overline{\mathbf{p}}_V\, dV \tag{9.88b}$$

where

$$\mathbf{c} = \begin{cases} \text{diag}(c_1\ \ c_2\ \ c_3) & \text{for three-dimensional problems} \\ \text{diag}(c_1\ \ c_2) & \text{for two-dimensional problems} \end{cases}$$

$$c_i = \begin{cases} 1 & \text{when } \xi \text{ is inside the boundary} \\ \frac{1}{2} & \text{when } \xi \text{ is on smooth boundary} \quad i = 1,2,3 \\ \text{Other value} & \text{when } \xi \text{ is at a corner} \end{cases}$$

$$\mathbf{u} = \begin{cases} [u_1\ \ u_2\ \ u_3]^T & \text{for three-dimensional problems} \\ [u_1\ \ u_2]^T & \text{for two-dimensional problems} \end{cases}$$

$$\mathbf{p} = \begin{cases} [p_1\ \ p_2\ \ p_3]^T & \text{for three-dimensional problems} \\ [p_1\ \ p_2]^T & \text{for two-dimensional problems} \end{cases}$$

are the displacement and traction vectors at the boundary,

$$\overline{\mathbf{p}}_V = \begin{cases} [\overline{p}_{V_1}\ \ \overline{p}_{V_2}\ \ \overline{p}_{V_3}]^T & \text{for three-dimensional problems} \\ [\overline{p}_{V_1}\ \ \overline{p}_{V_2}]^T & \text{for two-dimensional problems} \end{cases}$$

is the body force vector and

$$\mathbf{p}^* = \begin{bmatrix} p^*_{11} & p^*_{12} & p^*_{13} \\ p^*_{21} & p^*_{22} & p^*_{23} \\ p^*_{31} & p^*_{32} & p^*_{33} \end{bmatrix} \quad \mathbf{u}^* = \begin{bmatrix} u^*_{11} & u^*_{12} & u^*_{13} \\ u^*_{21} & u^*_{22} & u^*_{23} \\ u^*_{31} & u^*_{32} & u^*_{33} \end{bmatrix} \quad \text{for three-dimensional problems}$$

$$\mathbf{p}^* = \begin{bmatrix} p^*_{11} & p^*_{12} \\ p^*_{21} & p^*_{22} \end{bmatrix} \quad \mathbf{u}^* = \begin{bmatrix} u^*_{11} & u^*_{12} \\ u^*_{21} & u^*_{22} \end{bmatrix} \quad \text{for two-dimensional problems}$$

are the fundamental solution coefficient matrices.

Equation (9.88) constitutes the basic integral equation for the boundary element formulation.

9.3.5 Boundary Element Formulation

As in the case of the Poisson's and Laplace's equations, the integral equations from the linear elasticity theory have to be solved by numerical methods. The basic steps involved in this approach are similar to those for Poisson's and Laplace's equations, i.e.,

1. Discretize the boundary S into elements over which approximate displacement and traction shape functions are defined. See Figs. 9.4, 9.5, and Section 9.2.2.
2. Introduce these elements and shape functions into Eq. (9.88) to obtain a system of linear equations.
3. Impose the boundary conditions and solve these equations.
4. Find the displacements and stresses inside the body.

The assumed displacements and tractions are written in terms of an element shape function $\mathbf{N}^i$ as

$$\mathbf{u} = \mathbf{N}^i \mathbf{v}^i \tag{9.89}$$

$$\mathbf{p} = \mathbf{N}^i \mathbf{p}^i \tag{9.90}$$

where c can be replaced by c_j because it can be depend on the smoothness of the boundary at node j. $\mathbf{v}^i$ and $\mathbf{p}^i$ are the nodal displacements and tractions of the ith element, with dimensions $3 \times g$ for three dimensions and $2 \times g$ for two dimensions, in which g is the number of nodes in the element.

Substitute Eqs. (9.89) and (9.90) into Eq. (9.88) to obtain

$$\mathbf{c}\,\mathbf{v}_j + \sum_{i=1}^{M} \left\{ \int_{S_i} \mathbf{p}^* \mathbf{N}^i \, dS \right\} \mathbf{v}^i = \sum_{i=1}^{M} \left\{ \int_{S_i} \mathbf{u}^* \mathbf{N}^i \, dS \right\} \mathbf{p}^i + \sum_{s=1}^{S} \left\{ \int_{V_s} \mathbf{u}^* \overline{\mathbf{p}}_V \, dV \right\} \tag{9.91}$$

where $\mathbf{c}$ can be replaced by $\mathbf{c}_j$ because it can depend on the smoothness of the boundary at node j, $\mathbf{v}_j$ are the values of the displacements at node j in the global numbering system, S_i is the area of the ith element, M is the number of boundary elements, and S is the number of "cells" which are the blocks to form the total body and are needed only for the volume integration of the final integral which contains no unknown quantities. After the numerical integration of the integrals in Eq. (9.91), a system of linear equations is obtained.

$$\mathbf{c}\,\mathbf{v}_j + \sum_{j=1}^{M} \widehat{\mathbf{H}}^{ij} \mathbf{v}^i = \sum_{j=1}^{M} \widehat{\mathbf{G}}^{ij} \mathbf{p}^i + \mathbf{B}_j \tag{9.92}$$

where

$$\widehat{\mathbf{H}}^{ij} = \int_{S_i} \mathbf{p}^* \mathbf{N}^i \, dS \qquad \widehat{\mathbf{G}}^{ij} = \int_{S_i} \mathbf{u}^* \mathbf{N}^i \, dS$$

and

$$\mathbf{B}_j = \sum_{s=1}^{S} \int_{V_s} \mathbf{u}^* \overline{\mathbf{p}}_V \, dV$$

Use the same technique to assemble the element matrices as in Eq. (9.36) to form the global equation for node j as

$$\mathbf{c}\,\mathbf{v}_j + \widehat{\mathbf{H}}_j \mathbf{V} = \widehat{\mathbf{G}}_j \mathbf{P} + \mathbf{B}_j \tag{9.93}$$

and place these equations together to form the global equation

$$\mathbf{CV} + \widehat{\mathbf{H}}\mathbf{V} = \widehat{\mathbf{G}}\mathbf{P} + \mathbf{B}$$

where $\mathbf{C} = \text{diag}(\mathbf{c}\ \ \mathbf{c}\ \ \cdots\ \ \mathbf{c})$ is a diagonal matrix which is of the order of $3N \times 3N$ for three-dimensional problems and $2N \times 2N$ for two-dimensional problems, where N is the total number of nodes for the system. Let $\mathbf{H} = \mathbf{C} + \widehat{\mathbf{H}}$ and $\widehat{\mathbf{G}} = \mathbf{G}$, so that the above equation becomes

$$\mathbf{HV} = \mathbf{GP} + \mathbf{B} \tag{9.94}$$

for the whole system, where $\mathbf{V}$ and $\mathbf{P}$ contain all the nodal displacements and tractions,

$$\mathbf{H} = \begin{bmatrix} \mathbf{H}_{11} & \mathbf{H}_{12} & \cdots & \mathbf{H}_{1N} \\ \vdots & & & \\ \mathbf{H}_{N1} & \mathbf{H}_{N2} & \cdots & \mathbf{H}_{NN} \end{bmatrix}$$

and

$$\mathbf{G} = \begin{bmatrix} \mathbf{G}_{11} & \mathbf{G}_{12} & \cdots & \mathbf{G}_{1N} \\ \vdots & & & \\ \mathbf{G}_{N1} & \mathbf{G}_{N2} & \cdots & \mathbf{G}_{NN} \end{bmatrix}$$

Note that the matrices $\mathbf{H}$ and $\mathbf{G}$ depend on the fundamental solution, the shape functions and the contour of the boundary only and are independent of the applied forces and boundary conditions. In other words, they will not be changed if the applied forces and boundary conditions are changed. The diagonal submatrices $\mathbf{H}_{ii}$ of $\mathbf{H}$, which involves the cumbersome $\mathbf{c}$ of Eq. (9.88), can be evaluated by imposing specific displacement, force, and boundary conditions on the body. This is in the same situation as that of Eq. (9.37). Hence, the diagonal elements of $\mathbf{H}$ can be evaluated using

$$\mathbf{H}_{ii} = -\sum_{\substack{j=1 \\ j\neq i}}^{N} \mathbf{H}_{ij}$$

Impose the boundary conditions of displacements and tractions on Eq. (9.94) and move all the unknown quantities to the left-hand side and all the known quantities to the right. The final system of linear equations can appear as

$$\mathbf{AX} = \mathbf{F} \tag{9.95}$$

The solution of Eq. (9.95) provides all the nodal displacements and tractions on the boundary.

The characteristics of the global equations of the boundary element method have been compared to those of the finite element method in Section 9.2.2. The comparisons apply to linear elasticity problems as well.

See the references for more details of the boundary element formulation for linear elasticity problems.

9.3.6 Displacements and Stresses Inside the Body

With all the boundary displacements and tractions known, the displacements at a point ξ inside the body can be calculated using Eq. (9.91). Note that in this case, from Eq. (9.88), when ξ is inside the body, $c_i = 1$ and hence $\mathbf{c} = \mathbf{I}$, an identity matrix. Thus the displacements $\mathbf{u}(\xi)$ are

$$\mathbf{u}(\xi) = \sum_{j=1}^{M} \left\{ \int_{S_j} \mathbf{u}^* \mathbf{N}\, dS \right\} \mathbf{p}^j - \sum_{j=1}^{M} \left\{ \int_{S_j} \mathbf{p}^* \mathbf{N}\, dS \right\} \mathbf{v}^j + \sum_{s=1}^{S} \left\{ \int_{V_S} \mathbf{u}^* \overline{\mathbf{p}}_V\, dV \right\} \tag{9.96}$$

For the isotropic material, the stresses are expressed in terms of the displacements as [Chapter 1, Eq. (1.35)]

$$\sigma_{ij} = \frac{E\nu}{(1+\nu)(1-2\nu)}\delta_{ij}\,\epsilon_{kk} + 2G\epsilon_{ij} = \frac{2G\nu}{1-2\nu}\delta_{ij}\,u_{k,k} + G(u_{i,j} + u_{j,i}) \tag{9.97}$$

Substitution of Eq. (9.96) into (9.97) leads to

$$\sigma_{ij} = \int_S D_{kij}\,p_k\,dS - \int_S S_{kij}u_k\,dS + \int_V D_{kij}\overline{p}_{Vk}\,dV \tag{9.98}$$

where

$$D_{kij} = \frac{1}{r^\alpha}\{(1-2\nu)(\delta_{ki}r_{,j} + \delta_{kj}r_{,i} - \delta_{ij}r_{,k}) + \beta r_{,i}r_{,j}r_{,k}\}\frac{1}{4\alpha\pi(1-\nu)}$$

$$S_{kij} = \frac{2G}{r^\beta}\left\{\beta\frac{\partial r}{\partial n}[(1-2\nu)\delta_{ij}r_{,k} + \nu(\delta_{ik}r_{,j} + \delta_{jk}r_{,i}) - \gamma r_{,i}r_{,j}r_{,k}] + \beta\nu(a_i r_{,j}r_{,k} + a_j r_{,i}r_{,k})\right.$$
$$\left. + (1-2\nu)(\beta a_k r_{,i}r_{,j} + a_j\delta_{ik} + a_i\delta_{jk}) - (1-4\nu)a_k\delta_{ij}\right\}\frac{1}{4\pi\alpha(1-\nu)}$$

in which $\alpha = 2, 1$, $\beta = 3, 2$, and $\gamma = 5, 4$ for three-dimensional problems and two-dimensional plane strain problems, respectively, and r is the distance from the point where the stress is computed to the boundary. The a_j are the direction cosines of the outer normal n of boundary S.

9.4 Computational Considerations: Interpolation Functions and Element Matrices

In Sections 9.2 and 9.3, the integral equations and the methods of boundary discretization for the boundary element method are developed. In this section, some computational aspects will be considered.

The equations used for the construction of the element matrices are given in Eqs. (9.34) and (9.92). Since the shape functions for the boundary element method can be the same as those used for the finite element method, the shape functions given in Chapter 6 for one- and two-dimensional problems (except those which use nodal derivatives as nodal variables) can be used as the shape functions for the boundary elements of two- and three-dimensional problems. Among these shape functions, the isoparametric type of shape function deserves more attention here since they are widely used in the boundary element method.

The concept of isoparametric elements is that the same shape functions are used to describe both the displacement inside the element and the geometry of the element. For two-dimensional problems, the most widely used elements are linear and quadratic elements (Fig. 9.6a) with the shape functions

Linear elements:

$$N_1 = \frac{1}{2}(1-\xi) \qquad N_2 = \frac{1}{2}(1+\xi)$$

Quadratic elements:

$$N_1 = \frac{1}{2}\xi(\xi-1) \qquad N_2 = (1-\xi)(1+\xi) \qquad N_3 = \frac{1}{2}\xi(1+\xi) \tag{9.99}$$

For three-dimensional problems (Fig. 9.6b):
Linear elements [Chapter 6, Eq. (6.131)]:

$$N_1 = \frac{1}{4}(1-\xi)(1-\eta) \qquad N_2 = \frac{1}{4}(1+\xi)(1-\eta)$$
$$N_3 = \frac{1}{4}(1+\xi)(1+\eta) \qquad N_4 = \frac{1}{4}(1-\xi)(1+\eta) \tag{9.100}$$

Quadratic elements [Eq. (6.147)]:

$$N_1 = \frac{1}{4}(1-\xi)(1-\eta) - \frac{1}{2}(N_5+N_8) \qquad N_2 = \frac{1}{4}(1+\xi)(1-\eta) - \frac{1}{2}(N_5+N_6)$$
$$N_3 = \frac{1}{4}(1+\xi)(1+\eta) - \frac{1}{2}(N_6+N_7) \qquad N_4 = \frac{1}{4}(1-\xi)(1+\eta) - \frac{1}{2}(N_7+N_8)$$
$$N_5 = \frac{1}{2}(1-\xi^2)(1-\eta) \qquad N_6 = \frac{1}{2}(1-\eta^2)(1+\xi)$$
$$N_7 = \frac{1}{2}(1-\xi^2)(1+\eta) \qquad N_8 = \frac{1}{2}(1-\eta^2)(1-\xi)$$

Shape functions for higher order elements can be obtained using the methods provided in Chapter 6. The displacement is then expressed as

$$u = \sum_{i=1}^{g} N_i u_i \tag{9.101}$$

where g is the number of nodes in an element and u_i is the displacement of node i. The geometry of an isoparametric element can be expressed using the same shape functions as those used for the displacements, i.e.,

$$x_1 = \sum_{i=1}^{g} N_i x_{1i}, \qquad x_2 = \sum_{i=1}^{g} N_i x_{2i}, \qquad x_3 = \sum_{i=1}^{g} N_i x_{3i} \tag{9.102}$$

where x_{1i}, x_{2i}, and x_{3i} are the coordinates of node i. After the substitution of the quantities in Eqs. (9.101) and (9.102) into Eqs. (9.34) and (9.92), the integrand in the integrals of the expressions for $\widehat{\mathbf{H}}^{ij}$ and $\widehat{\mathbf{G}}^{ij}$ become functions of ξ and η, which vary from −1 to 1. Since all of the parts that compose the elements of the matrices $\widehat{\mathbf{H}}^{ij}$ and $\widehat{\mathbf{G}}^{ij}$ are expressed in the same ξ, η coordinate system, the differential area dS (or length for two-dimensional problems) should be expressed in terms of ξ and η also. This expression can be obtained from the relationship given in Fig. 9.7. For two-dimensional problems (Fig. 9.7a).

$$dS = \left[dx_1^2 + dx_2^2\right]^{1/2} = \left[\left(\frac{dx_1}{d\xi}\right)^2 + \left(\frac{dx_2}{d\xi}\right)^2\right]^{1/2} d\xi \tag{9.103}$$

Substitution of the relationships of Eq. (9.102) for the coordinates x_1 and x_2 into the above equation leads to an expression for dS in terms of ξ.

For three-dimensional problems, the differential area can be expressed as a function of the two vectors tangential to the ξ and η axes (Fig. 9.7b). Vector $\mathbf{r}_1$ can be written as

$$\mathbf{r}_1 = [dx_1 \quad dx_2 \quad dx_3]$$

Since the orientation of $\mathbf{r}_1$ is tangential to the ξ axis (Fig. 9.7b) and the magnitude of $\mathbf{r}_1$ is very small, dx_1, dx_2, and dx_3 are functions of ξ only. Hence, $\mathbf{r}_1$ can be written as

$$\mathbf{r}_1 = \left[\frac{\partial x_1}{\partial \xi} \quad \frac{\partial x_2}{\partial \xi} \quad \frac{\partial x_3}{\partial \xi}\right] d\xi$$

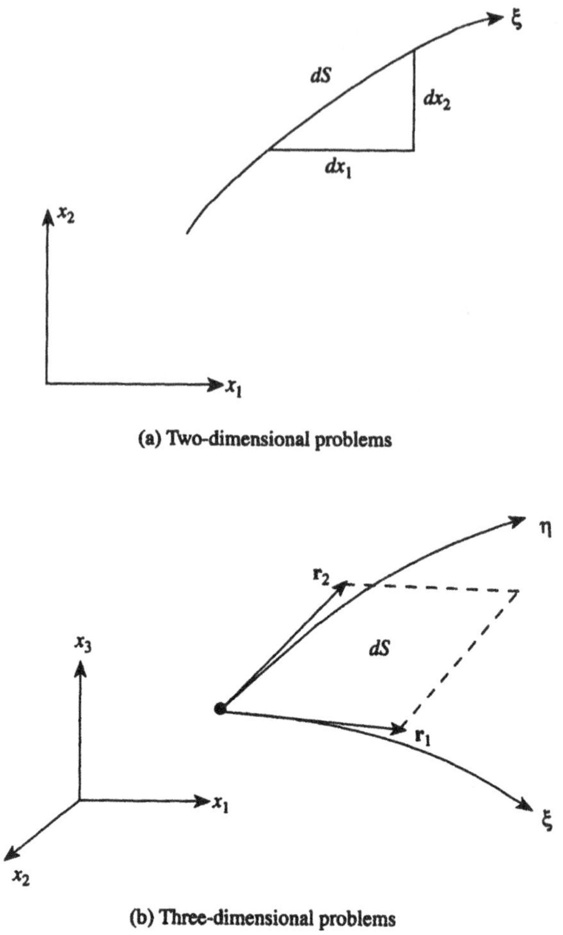

FIGURE 9.7
Differential areas for two- and three-dimensional problems.

Similarly,

$$\mathbf{r}_2 = \begin{bmatrix} \frac{\partial x_1}{\partial \eta} & \frac{\partial x_2}{\partial \eta} & \frac{\partial x_3}{\partial \eta} \end{bmatrix} d\eta$$

From vector algebra, the magnitude of the cross-product of two vectors is equal to the area of the parallelogram formed with these vectors as two sides. Thus,

$$dS = |\mathbf{r}_1 \times \mathbf{r}_2| = \left(g_1^2 + g_2^2 + g_3^2\right)^{1/2} d\xi\, d\eta \tag{9.104}$$

with

$$g_1 = \frac{\partial x_2}{\partial \xi}\frac{\partial x_3}{\partial \eta} - \frac{\partial x_3}{\partial \xi}\frac{\partial x_2}{\partial \eta}$$

$$g_2 = \frac{\partial x_3}{\partial \xi}\frac{\partial x_1}{\partial \eta} - \frac{\partial x_1}{\partial \xi}\frac{\partial x_3}{\partial \eta}$$

$$g_3 = \frac{\partial x_1}{\partial \xi}\frac{\partial x_2}{\partial \eta} - \frac{\partial x_2}{\partial \xi}\frac{\partial x_1}{\partial \eta}$$

Substitute Eqs. (9.103) and (9.104) into (9.34) and (9.92), respectively, to find the expressions for $\widehat{\mathbf{H}}^{ij}$ and $\widehat{\mathbf{G}}^{ij}$ in terms of ξ and η. The integrations can be performed using the Gauss quadrature described in Chapter 6.

For constant elements, the shape function for the displacement is a constant. Use of a constant shape function for the geometry of the element may cause inaccuracies in forming the element matrices. For these elements, the geometry of the element may be represented by the shape functions for linear or quadratic elements. This is similar to the use of superparametric elements in the finite element method.

At corners where two elements meet, the derivatives q may be discontinuous. If there is a node at the corner, the determination of q may be quite difficult. To alleviate this problem, introduce a *discontinuous element*. Split the corner node into two nodes and shift each of these two nodes a small distance into one of the two adjacent elements forming the corner. The two elements are now discontinuous. Each of these two nodes belongs to different elements and the values u and q can be determined using Eq. (9.32). When the discontinuous elements are assembled into a system with the conventional continuous elements, for each discontinuous element, add one to the total number of nodes.

There are a variety of alternatives available for treating the boundary element problem. For example, return to Eq. (9.88)

$$\mathbf{cu} + \int_S \mathbf{p}^* \mathbf{u}\, dS - \int_S \mathbf{u}^* \mathbf{p}\, dS - \int_V \mathbf{u}^* \, \overline{\mathbf{p}}_V \, dV = \mathbf{0} \tag{9.105}$$

If an exact solution can be found, it must satisfy Eq. (9.88) exactly. For the boundary element solution, the substitution of the approximate shape functions of Eqs. (9.89) and (9.90) into Eq. (9.105) does not make the left hand side of the equation zero but results in a residual. Let this residual be $\mathbf{R}_i$ for node i, then

$$\mathbf{R}_i = \mathbf{H}_i \mathbf{V} - \mathbf{G}_i \mathbf{P} - \mathbf{B}_i \tag{9.106}$$

where $\mathbf{H}_i$, $\mathbf{G}_i$ and $\mathbf{B}_i$ are obtained after assembling the element matrices $\mathbf{H}^{ij}$, $\mathbf{G}^{ij}$ and $\mathbf{B}_i$ of Eq. (9.93). Various techniques of treating this residual are given in Chapter 7. For example, the collocation method is quite suitable. The shape functions can be employed as approximate solutions to the problem, and the nodes as the collocation points. The standard collocation method introduced in Chapter 7 can be employed to find the solution at the collocation (nodal) points. Also, the least square collocation method can be used here.

Another possibility is the use of the minimax method. The objective of the minimax formulation of this problem is to make the maximum residual max $|\mathbf{R}_i|$ a minimum. In Chapter 7, Eq. (7.21) the maximum residual is denoted by ϕ, a scalar. In Eq. (9.106), however, the residual is a vector, so ϕ is replaced by the norm of a vector, i.e.,

$$\phi = \max |\mathbf{R}_i|$$

Then the minimax problem is stated as: Find the unknown elements in $\mathbf{V}$ and $\mathbf{P}$ such that the vector

$$\phi \quad \text{is minimized}$$

under the conditions $\mathbf{R}_i - \phi \leq 0$ and $\mathbf{R}_i + \phi \geq 0$. This is in the form of Chapter 7, Eq. (7.23). Minimizing ϕ means that the components of ϕ are minimized. A linear programming computer program can be used to solve this problem.

The minimax formulation is especially useful when the problem statement includes constraints on certain displacements and tractions. For example, when the prescribed boundary conditions contain the condition that at some points the displacements are required to be

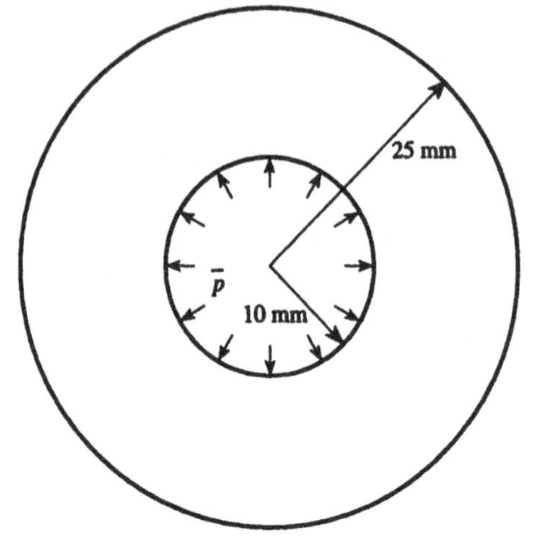

FIGURE 9.8
A hollow cylinder with internal pressure.

kept below prescribed values. These conditions can then be added to the constraint conditions of the minimax formulation.

EXAMPLE 9.5 Boundary Element Solution for a Thick Cylinder under Internal Pressure
Compute the displacements of the thick cylinder of Fig. 9.8 under an internal pressure of $\overline{p} = 0.1\ \text{GN/m}^2$. The modulus of elasticity of the material is $E = 200\ \text{GN/m}^2$, and Poisson's ratio is 0.25.

This problem can be treated as a plane strain problem (Section 1.3.1). The displacements on a point ξ inside the body or on the boundary are expressed as [Eq. (9.88)]

$$c_i u_i(\xi) + \int_S u_j p_{ij}^* \, dS = \int_S p_j u_{ij}^* \, dS + \int_A \overline{p}_{Vj} u_{ij}^* \, dA \tag{1}$$

or

$$\mathbf{c}\mathbf{u}(\xi) + \int_S \mathbf{p}^* \mathbf{u} \, dS = \int_S \mathbf{u}^* \mathbf{p} \, dS + \int_A \mathbf{u}^* \overline{\mathbf{p}}_V \, dA \tag{2}$$

where A is the area of the cross-section of the cylinder, and $i, j = 1, 2$. The terms u_{ij}^* and p_{ij}^* are given in Eq. (9.74) and (9.78). The displacement form of the governing differential equation is [Eq. (9.53)]

$$\nabla^2 u_i + \frac{1}{1 - 2\nu} u_{k,ki} + \frac{\overline{p}_{Vi}}{G} = 0 \qquad i, k = 1, 2 \tag{3}$$

Express the displacement in terms of the Galerkin vector [Eq. (9.58)]

$$u_i = g_{i,kk} - \frac{1}{2(1 - \nu)} g_{k,ik} \qquad i, k = 1, 2 \tag{4}$$

where g_i is the component of the Galerkin vector. Substitute (4) into (3) to obtain

$$\nabla^4 g_i + \frac{\overline{p}_{Vi}}{G} = 0 \tag{5}$$

Let

$$\overline{p}_{Vi} = \delta(\xi, x)\, a_i \tag{6}$$

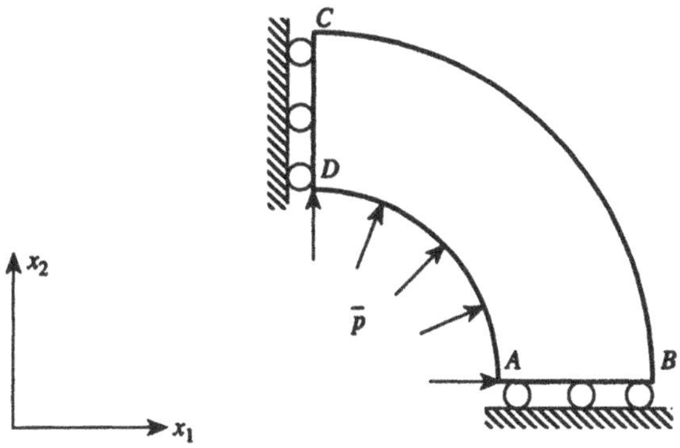

FIGURE 9.9
Modeling of the structure.

The solution of (5) for the two-dimensional plane strain problem takes the form

$$g_i = \frac{1}{8\pi G} r^2 a_i \ln \frac{1}{r} \qquad i = 1, 2 \tag{7}$$

Substitute (7) into (4) and note that when the loading takes the form given by (6), u_i becomes u_i^*, the fundamental solution. Utilize the relationship of Eq. (9.68) to find the expression for u_{ji}^* of Eq. (9.74). Follow the procedure from Eq. (9.74) to Eq. (9.78) to obtain the expression for p_{ji}^*.

Due to the symmetry of the structure and the loading, only one quarter of the cylinder needs to be modeled. This is shown in Fig. 9.9. The prescribed boundary conditions are (Fig. 9.10a):

$$\begin{aligned} &\overline{u}_2 = 0 && \text{and} \quad \overline{p}_1 = 0 && \text{on side} \quad A-B \\ &\overline{p}_i = 0 && i = 1, 2 && \text{on side} \quad B-C \\ &\overline{u}_1 = 0 && \text{and} \quad \overline{p}_2 = 0 && \text{on side} \quad C-D \\ &\overline{p}_i = \overline{p} a_i && i = 1, 2 && \text{on side} \quad D-A \end{aligned} \tag{8}$$

Also, $\overline{p}_{Vi} = 0$ for this problem. The unknown variables to be calculated are (Fig. 9.10b):

$$\begin{aligned} &u_1 && \text{and} \quad p_2 && \text{on side} \quad A-B \\ &u_i && i = 1, 2 && \text{on side} \quad B-C \\ &u_2 && \text{and} \quad p_1 && \text{on side} \quad C-D \\ &u_i && i = 1, 2 && \text{on side} \quad D-A \end{aligned} \tag{9}$$

Use quadratic elements for the boundary discretization. The geometry, displacements, and tractions are modeled with this kind of element. Substitute the quadratic shape functions of Eq. (9.99) into (1) to obtain Eq. (9.91) and follow the manipulations from Eq. (9.92) to Eq. (9.94) to form the system equations

$$\mathbf{HV} = \mathbf{GP} \tag{10}$$

Substitute the boundary conditions of (8) into (10) and move all the terms involving unknown quantities to the left-hand side and all the terms involving known quantities to the

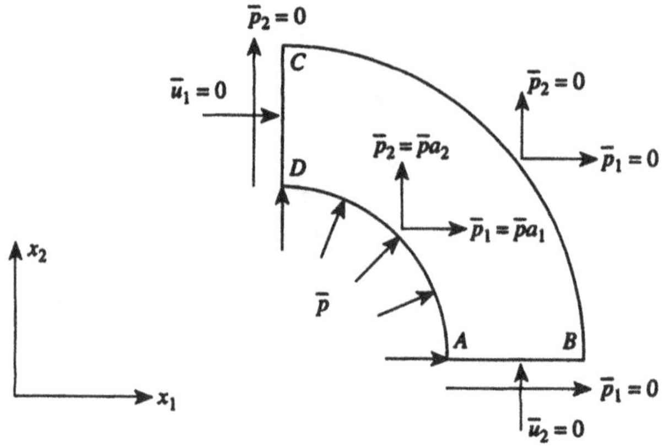

(a) Boundary conditions

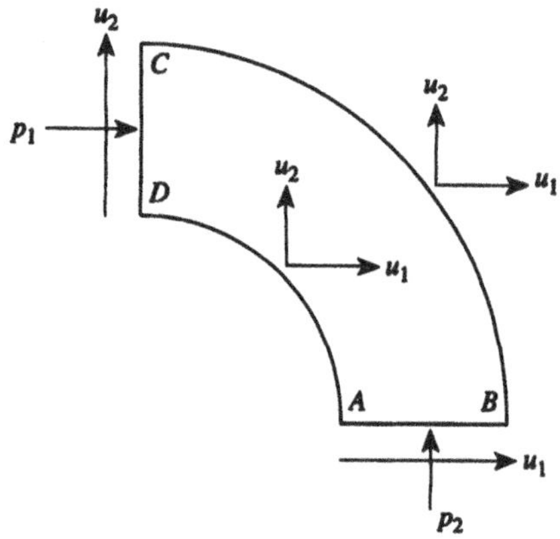

(b) Unknown variables

FIGURE 9.10
Boundary conditions and unknown variables for Example 9.5.

right-hand side of (10). This leads to a system of linear equations of the form of Eq. (9.95) which can be solved for the quantities listed in (9).

The displacements at points a, b and c in the x_1 direction, computed using the three meshes of Fig. 9.11, are given in the following table. The results are the numbers in the table multiplied by 10^{-3} mm.

Nodes	Mesh 1	Mesh 2	Mesh 3	Exact Solution
a	7.8781	8.0246	8.0350	8.0325
b	5.1668	5.2845	5.2928	5.2912
c	4.3896	4.4570	4.4631	4.4625

The boundary element solutions appear to converge to the exact solution. When a coarse mesh like mesh 1 is used, the error of the boundary element solution is within 2.5%, and when the mesh is refined to mesh 3, the error of the boundary element solution is reduced to less than 0.015%. Since only 30 nodes are used in mesh 3, it would appear that the boundary element solution is quite efficient for this accurate solution. ∎

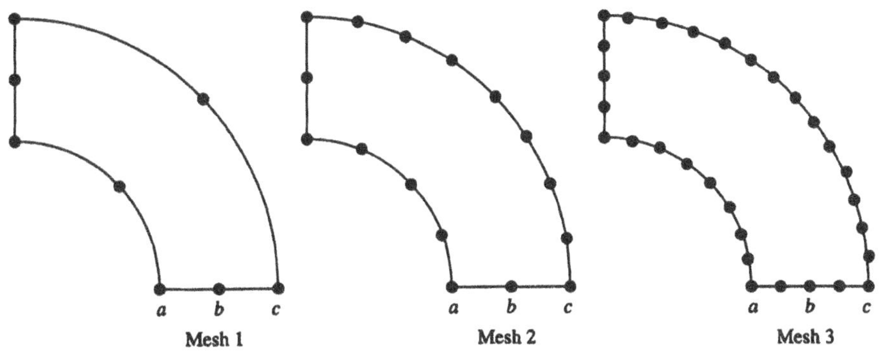

FIGURE 9.11
Boundary element meshes.

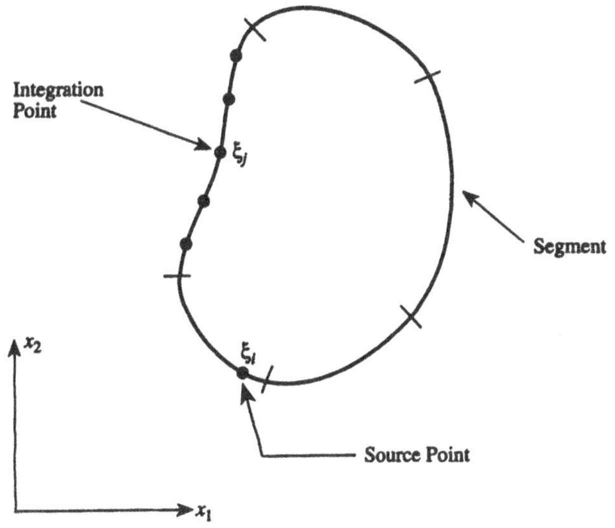

FIGURE 9.12
Boundary divided into segments.

9.5 Direct Integration of Integral Equations

A viable alternative, especially for two-dimensional problems, to the boundary element method and the finite element method is available. This is the solution of integral equations by the direct application of numerical integration schemes on the boundary of the domain. Use a two-dimensional direct formulation of the Laplace equation [Eq. (9.28) with $b = 0$] as an example. Gauss quadrature will be employed in the numerical solution.

The fundamental strategy of the direct integration method is to divide the boundary of the domain into m segments (Fig. 9.12) and use Gauss numerical integration quadrature in each segment. The total boundary may involve sharp corners which makes the application of Gauss quadrature difficult. Division of the whole boundary into relatively smooth segments can avoid this problem. The integral on the whole boundary is the sum of the integrals on the segments. Then Eq. (9.28) becomes

$$\pi u(\xi_i) + \sum_{\ell=1}^{m} \int_{S_\ell} u q^* \, dS = \sum_{\ell=1}^{m} \int_{S_\ell} q u^* \, dS \tag{9.107}$$

where S_ℓ is the length of the ℓth segment and ξ_i is a Gauss point on the boundary. The variables at the Gauss integration points are the unknowns of the problem. The first step for using the Gauss quadrature is to parameterize the integrands of the integrals in Eq. (9.107) so that the integration limits are from −1 to 1. Then Eq. (6.116) of Chapter 6 can be used for each of the integrals, i.e.,

$$\int_{S_\ell} u q^* \, dS = \int_{S_\ell} u \frac{-r_k a_k}{r^2} \, dS \approx -\sum_{j=1}^{n} u_j \frac{r_k(\xi_j) a_k(\xi_j)}{r(\xi_j)^2} W_j^{(n)} \tag{9.108a}$$

and

$$\int_{S_\ell} q u^* \, dS = \int_{S_\ell} q \ln\left(\frac{1}{r}\right) dS \approx \sum_{j=1}^{n} q_j \ln \frac{1}{r(\xi_j)} W_j^{(n)} \tag{9.108b}$$

where n is the number of Gauss points used, ξ_j represents a Gauss point on segment S_ℓ, u_j and q_j are the values of u and q at the Gauss points, $W_j^{(n)}$ is the weighting coefficient taken from Table 6.7, $r(\xi_j)$ is the distance from point ξ_i of Eq. (9.107) to the Gauss point ξ_j, and $r_k(\xi_j)$ is the component of $r(\xi_j)$ in the x_k direction. Note that when $r = 0$, the integrals in Eqs.(9.108) become singular. The singularity involved in Eq. (9.108a) will be treated later. For the integral in Eq. (9.108b) involving q, the integrand $q \ln(1/r)$ is singular when $r \to 0$, i.e., when $\xi_j \to \xi_i$ in Fig. 9.12, and numerical integration cannot be used. To overcome this problem, a special technique is employed. Further divide the kth segment containing ξ_i into n subsegments (Fig. 9.13) with the integration points at the center of these subsegments. If the unknowns are assumed to be constant on these subsegments, the integration of Eq. (9.108b) can be of the form

$$\int_{S_\ell} q u^* \, dS = \int_{S_\ell} q \ln\left(\frac{1}{r}\right) dS = \sum_{j=1}^{n} q_j \int_{S_{\ell_j}} \ln\left(\frac{1}{r}\right) dS$$

where S_{ℓ_j} is the jth subsegment of the kth segment, and q_j is the value of q on the jth

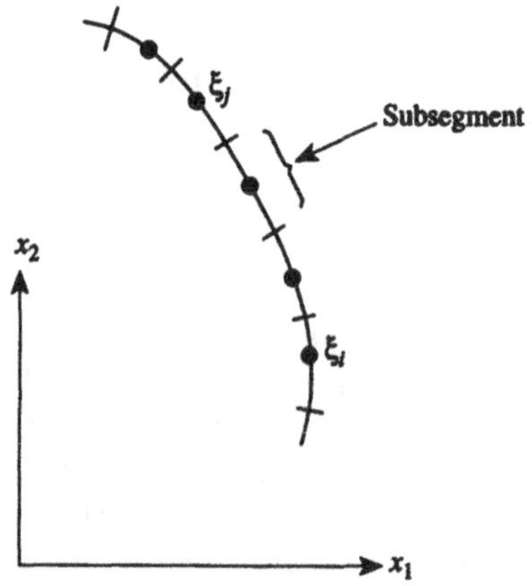

FIGURE 9.13
When ξ_j and ξ_i fall in the same segment, divide the segment into subsegments.

subsegment. Thus, the numerical integration becomes

$$\int_S q \ln\left(\frac{1}{r}\right) dS = \underbrace{\sum_{\substack{\ell=1\\ \ell\neq k}}^{m} \left[\sum_{j=1}^{n} q_j \ln\frac{1}{r(\xi_j)} W_j^{(n)}\right]}_{\text{When the segments do not contain } \xi_i} + \underbrace{\sum_{j=1}^{(n)} q_j \int_{S_{\ell_j}} \ln\left(\frac{1}{r}\right) dS.}_{\text{When the segments contain } \xi_i} \tag{9.109}$$

Substitute Eqs. (9.108) and (9.109) into Eq. (9.107) and note that the point ξ_i in parenthesis of the first term at the left-hand side of Eq. (9.107) should refer to a Gauss point. This leads to

$$\begin{aligned} u(\xi_i) &- \sum_{\ell=1}^{m}\sum_{j=1}^{n} u_j \frac{r_k(\xi_j)\, a_k(\xi_j)}{\pi r(\xi_j)^2} W_j^{(n)} \\ &= \frac{1}{\pi}\underbrace{\sum_{\substack{\ell=1\\ \ell\neq k}}^{m}\left[\sum_{j=1}^{n} q_j \ln\frac{1}{r(\xi_j)} W_j^{(n)}\right]}_{\text{When the segments do not contain } \xi_i} + \underbrace{\sum_{j=1}^{n} q_j \int_{S_{\ell_j}} \ln\left(\frac{1}{r}\right) dS}_{\text{When the segments contain } \xi_i} \end{aligned} \tag{9.110}$$

For each ξ_i, Eq. (9.110) reduces to a linear equation

$$\mathbf{H}_i\mathbf{V} = \mathbf{G}_i\mathbf{P}$$

where **V** and **P**, which are $N \times 1$ vectors, with N defined as the number of integration points on the boundary, contain all the values of u and q at the Gauss points. The elements of $\mathbf{H}_i$ and $\mathbf{G}_i$ are the coefficients of u_j and q_j in Eq. (9.110). When all the integration points are spanned, the global equation

$$\mathbf{HV} = \mathbf{GP} \tag{9.111}$$

is formed. The matrices **H** and **G** are of order $N \times N$. Note that the singularity in Eq. (9.108) occurs at the evaluation of the diagonal elements of **H**. This singularity can be avoided by using the same technique as that used to process the **H** matrix in Sections 9.2.2 and 9.3.5., i.e., the diagonal elements are not evaluated when the matrix **H** is formed. After all of the non-diagonal elements of **H** are calculated, the diagonal elements of **H** can be computed using Eq. (9.39).

Since no boundary elements and, hence, no shape functions are involved in the integration, the direct integration method appears to be more efficient than the boundary element method.

It should be noted that for the direct integration method, the segments can be quite large so that the discretization of the boundary of the domain can be very simple. The total number of integration points is $N = m \times n$, where m is the number of segments and n is the number of integration points on each segment. Since relatively few segments are needed, the number N can be small and hence the computation tends to be efficient. Also, since the Gauss integration points are not located at the ends of the segments, no integration point is located at the sharp convex and concave corners. Thus, the need to determine the normal derivative with respect to the outer normal at the sharp corners is avoided.

EXAMPLE 9.6 Torsion Problem

The two cross-sectional shapes of Fig. 9.14 will be used to illustrate the accuracy of this numerical procedure. Suppose a twisting moment of magnitude 100 occurs at the cross-sections. For these cross-sections, the boundaries are divided into segments as shown in Fig. 9.14. An equal number of Gauss integration points is employed on each segment.

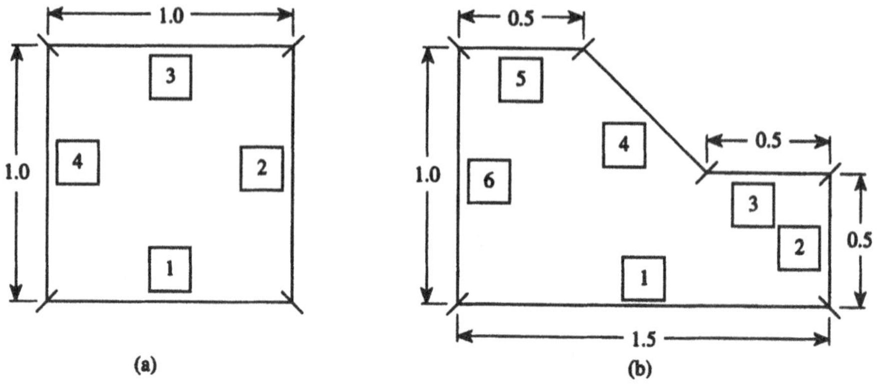

FIGURE 9.14
Cross-sections for Example 9.6.

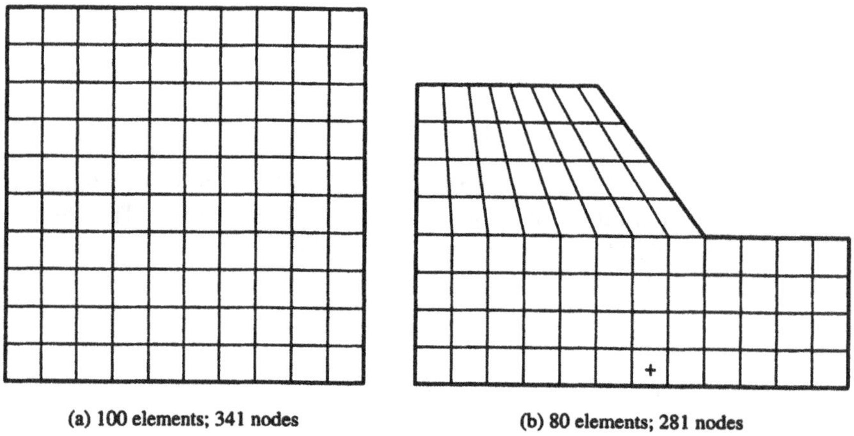

FIGURE 9.15
Finite element meshes for the cross-sections of Fig. 9.14. All elements are 8-node isoparametric elements.

The integral equation of Eq. (14) of Example 9.2 will be integrated using Gauss quadrature. The torsional constant and shear stresses are to be calculated.

The results of the computation are compared to the results using finite elements with the meshes of Fig. 9.15. The elements used are eight node isoparametric elements. The results of the direct integration and finite elements are shown in Tables 9.1 and 9.2.

It is seen that the integration of the integral equation exhibits good accuracy for the computation of the torsional constant and stresses. These results converge to the exact value or the finite element results, although the stress results show some fluctuation in the process of converging.

The results of direct integration for the square cross-section of Fig. 9.14a are also compared in Table 9.1 to a boundary element solution of the same equation. For the boundary element solution, constant and linear elements are employed. The number of elements in each computation is taken to be the same as the number of integration points on the whole boundary although the locations of the nodes of the elements are different from those of the integration points. The number of the boundary segments is not necessarily the same as that of the boundary elements but the size of the system of linear equations is kept the same.

TABLE 9.1

Results for the Cross-Section of Fig. 9.14a

Number of Integration Points/Nodes	Torsional Constant			Maximum Stress*		
	Direct Integration	BEM		Direct Integration	BEM	
		Constant	Linear		Constant	Linear
12	0.1422	0.1366	0.1540	403.1	440.53	435.84
20	0.1411	0.1393	0.1456	448.8	467.1	462.57
28	0.14077	0.14008	0.1432	462.8	465.9	464.3
40	0.14074	0.1404	0.1418	474.6	473.75	472.74
56	0.14065	0.1405	0.1418	473.8	473.36	472.74
72	0.140627	0.14055	0.14096	475.89	474.95	474.7
FEM Results		0.14058			482.42	
Exact** Solution		0.140577			482.16	

* The maximum stress is taken at the middle point of each side of the square.
**From Boresi and Chong (1987).

TABLE 9.2

Results for the Cross-Section of Fig. 9.14b

Integration Points	Torsional Constant	Maximum Stress
18	0.1333	371.2
30	0.1344	412.9
42	0.13418	431.4
60	0.13414	401.3
84	0.13405	409.6
FEM Results*	0.13400	408.97

* The maximum stress is taken at the centroid of the finite element with the sign + shown in Fig. 9.15.

The comparison shows that the results for direct integration of the torsional constant are superior to those of the boundary element method for most of the schemes. For the stress computations, the results from the direct integration are not as accurate as those from the boundary element method for the first few boundary discretization schemes. However, as the number of integration points increases, the stress results for direct integration are about the same as those obtained from the boundary element method. Since no element and hence no shape function is involved in the discretization of the boundary, the direct integration method involves less computational effort in forming the system matrices than the boundary element method and hence is a more efficient procedure for this two-dimensional problem. ∎

References

Boresi, A.P. and Chong, K.P., 1987, *Elasticity in Engineering Mechanics*, Elsevier, Amsterdam, Netherlands.

Brebbia, C.A. and Dominguez, J., 1992, *Boundary Elements, An Introductory Course*, 2nd ed., McGraw-Hill, New York.

Brebbia, C.A., Telles, J.C.F. and Wrobel, L.C., 1984, *Boundary Element Techniques*, Springer-Verlag, Berlin.
Gipson, G.S., 1987, *Boundary Element Fundamentals*, Computational Mechanics Publications, Southhampton, UK.
Haberman, R., 1987, *Elementary Applied Partial Differential Equations*, Prentice-Hall, Englewood Cliffs, New Jersey.
Stern, M., 1989, Static analysis of beams, plates and shells, in Beskos, D.E., Ed., *Boundary Element Methods in Structural Analysis*, ASCE, New York.

Problems

9.1 Obtain the stiffness and transfer matrices from Eq. (9.11). Check your results with the matrices of Chapter 4.

9.2 Use the same procedure as that for beams to derive the boundary element formulation for bars with axial loading.

Hint: The fundamental solution is

$$u^* = \frac{1}{2}|x - \xi|$$

9.3 The governing differential equation for the small transverse motion of a tight string with applied distributed load $\overline{p}_z$ is $d^2w/dx^2 = -\overline{p}_z/N$, where N is the tensile axial force. The boundary conditions are $w(0) = w(L) = 0$. Begin with the extended Galerkin's method formula. Integrate by parts to obtain the boundary integral equation. Find the fundamental solution by solving $d^2w/dx^2 = -\delta(\xi, x)/N$

Answer: The fundamental solution is $w^* = \frac{1}{2}|x - \xi|$

9.4 Begin with the equation $d^2u/dx^2 + u + x = 0$, with boundary conditions $u(0) = 1$ and $u(1) = 1$. Obtain the boundary integral equation from the extended Galerkin's formula. Find the exact boundary values $q_0 = du/dx|_{x=0}$ and $q_1 = du/dx|_{x=1}$.

Hint: The fundamental solution is $u^* = A \cos x + B \sin x$.

Answer: $q_0 = 1/\sin 1 - 1,\ q_1 = \cos 1/\sin 1 - 1$

9.5 Use the boundary element method to calculate the response at $\xi = L/2$ of a bar of length L with both ends constrained. The bar is loaded by a linearly varying axial load $\overline{p}_x = p_0(1 - x/L)$.

Answer: $u_{L/2} = -p_0L^2/(16EA)$

9.6 Verify that Eq. (9.22) satisfies Eq. (9.21).

Hint: Substitute Eq. (9.22) into the left-hand side of Eq. (9.21). Then investigate $x \neq \xi$ and $x = \xi$.

9.7 Use four constant elements for a 1×1 square cross section of a bar under torsion. Form the linear boundary element equations by hand calculations. Solve for the torsional constant.

Hint: The values of the warping function at the nodes are not unique, so the value at one of the nodes can be set to zero at the outset. The exact result for the torsional constant is 0.1406.

9.8 For a Poisson's equation $\nabla_1^2 u = -b$ with

$$\nabla_1^2 = k_1 \frac{\partial}{\partial x_1} + k_2 \frac{\partial}{\partial x_2} \qquad \text{for two-dimensional problems}$$

$$\nabla_1^2 = k_1 \frac{\partial}{\partial x_1} + k_2 \frac{\partial}{\partial x_2} + k_3 \frac{\partial}{\partial x_3} \qquad \text{for three-dimensional problems}$$

derive a boundary integral equation.

Hint: Begin with the transformation $x_i = x_i/\sqrt{k_i}$ and then use Eqs. (9.22) and (9.29).

Answer: The boundary integral equation is the same as Eq. (9.29), with the definitions

$$u^* = \frac{1}{\sqrt{k_1 k_2}} \ln \frac{1}{r} \qquad q^* = k_1 \frac{\partial u^*}{\partial x_1} + k_2 \frac{\partial u^*}{\partial x_2}$$

$$r = \sqrt{\frac{1}{k_1}(x_1 - \xi_1)^2 + \frac{1}{k_2}(x_2 - \xi_2)^2} \quad \text{for two-dimensional problems}$$

9.9 Show that when the homogeneous term b of a Poisson's equation satisfies $\nabla^2 b \equiv 0$, there exists a relationship

$$\int_V bu^* \, dV = \int_S \left(b\frac{\partial v^*}{\partial n} - v^* \frac{\partial b}{\partial n} \right) dS$$

where
$v^* = r^2/4[\ln 1/r + 1]$ for two-dimensional problems,
$v^* = r/4$ for three-dimensional problems.

Hint: Start from $\int_V (b\nabla^2 v^* - v^* \nabla^2 b)\, dV$ and recognize that $\nabla^2 v^* = u^*$.

9.10 Fill in the details of the derivation of the relations in Eqs. (9.48) to (9.51).

9.11 Verify that

$$\lim_{\epsilon \to 0} \int_{S_\epsilon} \frac{\partial \varepsilon}{\partial x_i} \frac{\partial \varepsilon}{\partial x_j} \, dS = 0$$

Hint: Use the relationship shown in Fig. 9.2 and obtain formulas for $\varepsilon_i/\varepsilon$, φ and θ. Also, note that $dS = \varepsilon^2 \sin\varphi \, d\theta \, d\varphi$.

9.12 Find the coefficients c_i of u_i of Eq. (9.88) when ξ is located at the vertex of a cone.

Hint: The range of the variation of φ is changed.

9.13 Calculate the elements H_{12}, H_{13}, H_{21}, H_{31} of the **H** matrix of Eq. (9.94) for a square plane strain region of 1×1. Discretize the boundary of the square region into 4 linear elements.

9.14 Calculate the elements G_{12}, G_{13}, G_{21}, G_{31} of the **G** matrix of Eq. (9.94) for the square plane strain region for Problem 9.13. Let the boundary element discretization be the same as in Problem 9.13.

9.15 Use two point Gauss quadrature to integrate the integral equation for the torsion problem of a 1×1 square cross-section. Form the system matrix and calculate the torsional constant.

Section C

Formulations for Dynamic and Stability Problems

10

Dynamic Responses

In the previous chapters, the responses have been "static" as the load does not vary with time. In discrete form, the governing equations are $\mathbf{KV} = \overline{\mathbf{P}}$, in which $\mathbf{K}$ is the global stiffness matrix of the structure, $\mathbf{V}$ is the displacement vector, and $\overline{\mathbf{P}}$ is the loading vector. If the force changes with time, the relation $\mathbf{KV} = \overline{\mathbf{P}}$ does not adequately describe the movement of the structure. The inclusion of inertia forces changes the governing equations to $\mathbf{M}\ddot{\mathbf{V}} + \mathbf{KV} = \overline{\mathbf{P}}$ in which $\mathbf{M}$ is the mass matrix. An analysis leads to a dynamic response.

Dynamic response problems fit into two broad classes. In one, the natural frequencies of the vibration and the corresponding mode shapes are desired. In the other, the motion of a structure subject to a prescribed load is sought.

10.1 Mass

A dynamic loading generates accelerations

$$\frac{\partial^2 \mathbf{u}}{\partial t^2} = \ddot{\mathbf{u}}$$

in a structure. This acceleration field can be considered to produce d'Alembert[1] inertia forces, e.g., $\overline{\mathbf{p}}_V = -\gamma \ddot{\mathbf{u}}$, where γ is the mass density, in the direction opposite to the acceleration. If inertia forces are included, the principle of virtual work of Chapter 6, Eq. (6.20)

$$-\delta W = \int_V \delta \boldsymbol{\epsilon}^T \boldsymbol{\sigma}\, dV - \int_V \delta \mathbf{u}^T \overline{\mathbf{p}}_V\, dV - \int_{S_p} \delta \mathbf{u}^T \overline{\mathbf{p}}\, dS = 0$$

becomes

$$-\delta W = \int_V \delta \boldsymbol{\epsilon}^T \boldsymbol{\sigma}\, dV + \int_V \delta \mathbf{u}^T \gamma \ddot{\mathbf{u}}\, dV - \int_{S_p} \delta \mathbf{u}^T \overline{\mathbf{p}}\, dS = 0 \tag{10.1}$$

[1]Jean Le Rond D'Alembert (1717–1783) was named after a church in Paris, France where he, as the illegitimate son of a society hostess, was abandoned. His step parents were of sufficient means to ensure that he received a formal education. He graduated from Mazarin College in 1735 and published in 1743 the book *Traité de dynamique*, in which the well-known *d'Alembert Principle* was proposed. He is considered to be the founder of the theory of partial differential equations.

For a linearly elastic solid, with $\boldsymbol{\sigma} = \mathbf{E}\,\mathbf{D}_u\mathbf{u}$ and $\delta\boldsymbol{\epsilon}^T = \delta(\mathbf{D}_u\mathbf{u})^T = \delta\mathbf{u}^T{}_u\mathbf{D}^T$ [Chapter 2, Eqs. (2.57) and (2.58a)], Eq. (10.1) becomes

$$-\delta W = \int_V \delta\mathbf{u}^T{}_u\mathbf{D}^T\mathbf{E}\,\mathbf{D}_u\,\mathbf{u}\,dV + \int_V \delta\mathbf{u}^T\gamma\ddot{\mathbf{u}}\,dV - \int_{S_p} \delta\mathbf{u}^T\overline{\mathbf{p}}\,dS = 0 \tag{10.2}$$

where $\mathbf{D}_u$ and ${}_u\mathbf{D}$ are operator matrices containing derivatives that operate on variables to the right and left, respectively.

Introduce an approximate $\mathbf{u}$, using the trial function approach of Chapters 4 and 6. The interpolation function representation of the displacements $\mathbf{u} = [u\ v\ w]^T$ and the second time derivatives for the ith element are

$$\begin{aligned}\mathbf{u} &= \mathbf{N}\,\mathbf{v}^i\\ \ddot{\mathbf{u}} &= \mathbf{N}\,\ddot{\mathbf{v}}^i\end{aligned} \tag{10.3}$$

where $\ddot{\mathbf{v}}^i = \partial^2\mathbf{v}^i/\partial t^2$ is the nodal acceleration and $\mathbf{N}$ contains shape functions. Then, for a structure modeled as M elements,

$$\begin{aligned}-\delta W &= \sum_{i=1}^{M}\delta\mathbf{v}^{iT}\left[\int_V \mathbf{N}^T{}_u\mathbf{D}^T\mathbf{E}\,\mathbf{D}_u\mathbf{N}\,dV\,\mathbf{v}^i + \int_V \gamma\mathbf{N}^T\mathbf{N}\,dV\,\ddot{\mathbf{v}}^i - \int_{S_p}\mathbf{N}^T\overline{\mathbf{p}}\,dS\right]\\ &= 0\end{aligned} \tag{10.4}$$

in which $\overline{\mathbf{p}}$ is the loading, including boundary tractions. This can be written as

$$\sum_{i=1}^{M}\delta\mathbf{v}^{iT}(\mathbf{k}^i\mathbf{v}^i + \mathbf{m}^i\ddot{\mathbf{v}}^i - \overline{\mathbf{p}}^i) = 0 \tag{10.5}$$

where

$$\mathbf{m}^i = \int_V \gamma\mathbf{N}^T\mathbf{N}\,dV \qquad \mathbf{k}^i = \int_V \mathbf{N}^T{}_u\mathbf{D}^T\mathbf{E}\,\mathbf{D}_u\mathbf{N}\,dV$$

and

$$\overline{\mathbf{p}}^i = \int_{S_p}\mathbf{N}^T\overline{\mathbf{p}}\,dS$$

This $\mathbf{m}^i$ is the definition of an element *mass matrix*. The matrix $\mathbf{k}^i$ and the vector $\overline{\mathbf{p}}^i$ are the element stiffness matrix and loading vector of Chapter 6, Eqs. (6.30) and (6.36).

10.1.1 Consistent Mass Matrix

One of the most widely used mass matrices is the *consistent mass matrix*. It is "consistent" in the sense that the same shape functions are used to develop the mass matrix as are employed for the stiffness matrix. Normally, this means that the polynomial shape functions developed for a static response are to be used to form the acceleration (mass) for a dynamic response.

To illustrate a consistent mass matrix, consider a uniform beam. The principle of virtual work can be expressed as (Chapter 2, Example 2.7)

$$-\delta W = \int_x \delta\mathbf{u}^T({}_u\mathbf{D}^T\mathbf{E}\,\mathbf{D}_u\mathbf{u} - \overline{\mathbf{p}})\,dx - [\delta\mathbf{u}^T\overline{\mathbf{s}}]_0^L = 0 \tag{10.6}$$

where $\overline{\mathbf{p}}$ is a vector of applied loadings. The inertia force would be $-\rho(\partial^2 w/\partial t^2) = -\rho\ddot{w}$, where ρ is the mass per unit length and w is the deflection, so that the principle of virtual work for a beam model with no shear deformation becomes

$$-\delta W = \int_x \delta w\,_x\partial^2 EI\,\partial_x^2 w\,dx + \int_x \delta w\,\rho\ddot{w}\,dx - \int_x \delta w\,\overline{p}_z\,dx - [\overline{M}\,\delta\theta + \overline{V}\,\delta w]_a^b = 0 \quad (10.7)$$

Introduce the shape functions for the static response of a beam [Chapter 4, Eq. (4.47)] $w = \mathbf{N}\,\mathbf{v}^i$, where $\mathbf{v}^i = [w_a\ \theta_a\ w_b\ \theta_b]^T$ and

$$\mathbf{N} = \begin{bmatrix} 1-3\xi^2+2\xi^3 \\ (-\xi+2\xi^2-\xi^3)\ell \\ 3\xi^2-2\xi^3 \\ (\xi^2-\xi^3)\ell \end{bmatrix}^T \quad (10.8)$$

giving the principle of virtual work of Eq. (10.7) for M elements in the form

$$-\delta W = \sum_{i=1}^{M} \delta\mathbf{v}^{iT}\left[\underbrace{\int_x \mathbf{N}_{,xx}^T EI\mathbf{N}_{,xx}\,dx}_{\mathbf{k}^i}\,\mathbf{v}^i + \underbrace{\int_x \rho\,\mathbf{N}^T\,\mathbf{N}\,dx}_{\mathbf{m}^i}\,\ddot{\mathbf{v}}^i - \underbrace{\int_x \mathbf{N}^T\overline{p}_z\,dx}_{\overline{\mathbf{p}}^i}\right] = 0 \quad (10.9)$$

where $\mathbf{N}_{,xx} = \partial_x^2\mathbf{N} = \partial^2\mathbf{N}/\partial x^2$, and the boundary terms have been ignored. Substitute $\mathbf{N}$ of Eq. (10.8) into $\mathbf{m}^i = \int_x \rho\mathbf{N}^T\mathbf{N}\,dx$ and carry out the integration. Then, the consistent mass matrix for the beam element of Fig. 10.1 becomes

$$\mathbf{m}^i = \int_0^\ell \rho\mathbf{N}^T\mathbf{N}\,dx = \frac{\rho\ell}{420}\begin{bmatrix} 156 & -22\ell & 54 & 13\ell \\ -22\ell & 4\ell^2 & -13\ell & -3\ell^2 \\ 54 & -13\ell & 156 & 22\ell \\ 13\ell & -3\ell^2 & 22\ell & 4\ell^2 \end{bmatrix} \quad (10.10a)$$

Note that this consistent mass matrix is a full symmetric matrix.

If axial motion of the beam element is considered, the nodal displacement vector is $\mathbf{v} = [u_a\ w_a\ \theta_a\ u_b\ w_b\ \theta_b]^T$ and the shape function is developed in Chapter 4, Example 4.1. The consistent mass matrix is

$$\mathbf{m}^i = \int_0^\ell \rho\mathbf{N}^T\mathbf{N}\,dx = \frac{\rho\ell}{420}\begin{bmatrix} 140 & 0 & 0 & 70 & 0 & 0 \\ 0 & 156 & -22\ell & 0 & 54 & 13\ell \\ 0 & -22\ell & 4\ell^2 & 0 & -13\ell & -3\ell^2 \\ 70 & 0 & 0 & 140 & 0 & 0 \\ 0 & 54 & -13\ell & 0 & 156 & 22\ell \\ 0 & 13\ell & -3\ell^2 & 0 & 22\ell & 4\ell^2 \end{bmatrix} \quad (10.10b)$$

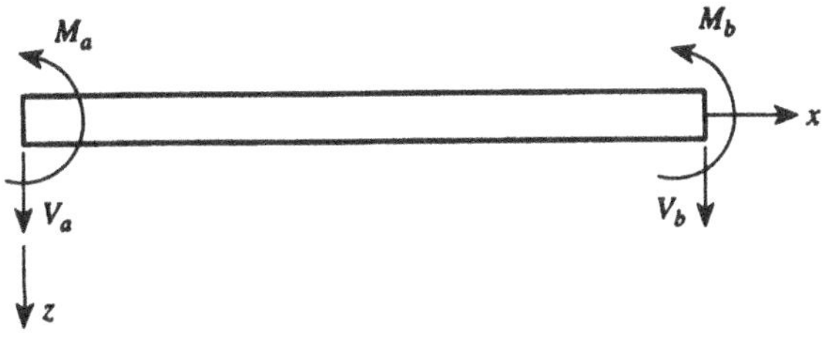

FIGURE 10.1
A uniform beam segment with in-plane bending about the y axis. Sign Convention 2.

In this development of the consistent mass matrix, only the inertia associated with the transverse deflection w of the beam centroidal axis is considered. Normally, this approximation leads to adequate precision for slender beams. For deeper beams, neglecting the inertia due to the rotation of the beam cross-section may cause some error. To take the rotary inertia of the cross-section into account, begin with the beam segment in Chapter 1, Fig. 1.15. If u_0 in Chapter 1, Eq. (1.98) is ignored, the displacement of the area dA in the x direction is $u = z\theta$, where $\theta = -\partial w/\partial x$. Then the acceleration of the area is $\ddot{u} = z\ddot{\theta}$. The integral of the moment of the force $-\gamma\ddot{u}(= -\gamma z\ddot{\theta})$ over the cross section is $-\gamma I\ddot{\theta}$, with $I = \int_A z^2\, dA$, where A is the area of the cross-section. If r_y is the radius of gyration and $\rho = \gamma A$ is the mass per unit length, the inertia moment can be written as $-\rho r_y^2\ddot{\theta}$, since $r_y^2 = I/A$. The inertia moment can be taken as a d'Alembert distributed moment applied on the beam element, so that an external virtual work term $\int_x \delta\theta\rho r_y^2\ddot{\theta}\, dx$ should be added in the expression of the principle of virtual work of Eq. (10.7). Use the trial function $\theta = -\partial w/\partial x = -\partial \mathbf{N}/\partial x\, \mathbf{v}$ to obtain the consistent matrix due to the rotary inertia of the cross-section

$$\mathbf{m}_r^i = \int_\ell \frac{\partial}{\partial x}\mathbf{N}^T\rho r_y^2\frac{\partial}{\partial x}\mathbf{N}\, dx = \frac{\rho r_y^2}{30\ell}\begin{bmatrix} 36 & -3\ell & -36 & -3\ell \\ -3\ell & 4\ell^2 & 3\ell & -\ell^2 \\ -36 & 3\ell & 36 & 3\ell \\ -3\ell & -\ell^2 & 3\ell & 4\ell^2 \end{bmatrix} \tag{10.10c}$$

Often this is expressed with $\rho r_y^2 = \gamma I$.

Since rotary inertia does not affect axial motion, $\mathbf{m}_r^i$ expanded to include axial motion would appear as

$$\mathbf{m}_r^i = \frac{\rho r_y^2}{30\ell}\begin{bmatrix} 0 & 0 & 0 & 0 & 0 & 0 \\ 0 & 36 & -3\ell & 0 & -36 & -3\ell \\ 0 & -3\ell & 4\ell^2 & 0 & 3\ell & -\ell^2 \\ 0 & 0 & 0 & 0 & 0 & 0 \\ 0 & -36 & 3\ell & 0 & 36 & 3\ell \\ 0 & -3\ell & -\ell^2 & 0 & 3\ell & 4\ell^2 \end{bmatrix} \tag{10.10d}$$

Although the consistent mass matrix, which is based on "static" shape functions, is the most frequently employed mass matrix, more accurate mass matrices can be computed by using shape functions that more closely represent the dynamic response. For example, use of the exact dynamic shape functions in $w = \mathbf{N}\mathbf{v}$ leads to an exact mass matrix. This topic is considered in some detail in Section 10.3.1.

After the consistent mass or other mass matrices for the elements are established, the global mass matrix can be assembled in the same fashion as the global stiffness matrix (Chapter 5).

EXAMPLE 10.1 Consistent Mass Matrices for a Frame

Find the element and global consistent mass matrices of the frame shown in Fig. 10.2. This same frame has been treated earlier in Chapter 5, Example 5.5. The element properties are $E = 200\ \text{GN/m}^2$, $\gamma = 7800\ \text{kg/m}^3$, and

$$\begin{aligned} \text{Elements 1 and 2:}\ I &= 2356\ \text{cm}^4 = 2.356 \times 10^{-5}\ \text{m}^4, \\ A &= 32\ \text{cm}^2 = 32 \times 10^{-4}\ \text{m}^2 \\ \text{Element 3:}\ I &= 5245\ \text{cm}^4 = 5.245 \times 10^{-5}\ \text{m}^4, \\ A &= 66\ \text{cm}^2 = 66 \times 10^{-4}\ \text{m}^2 \end{aligned} \tag{1}$$

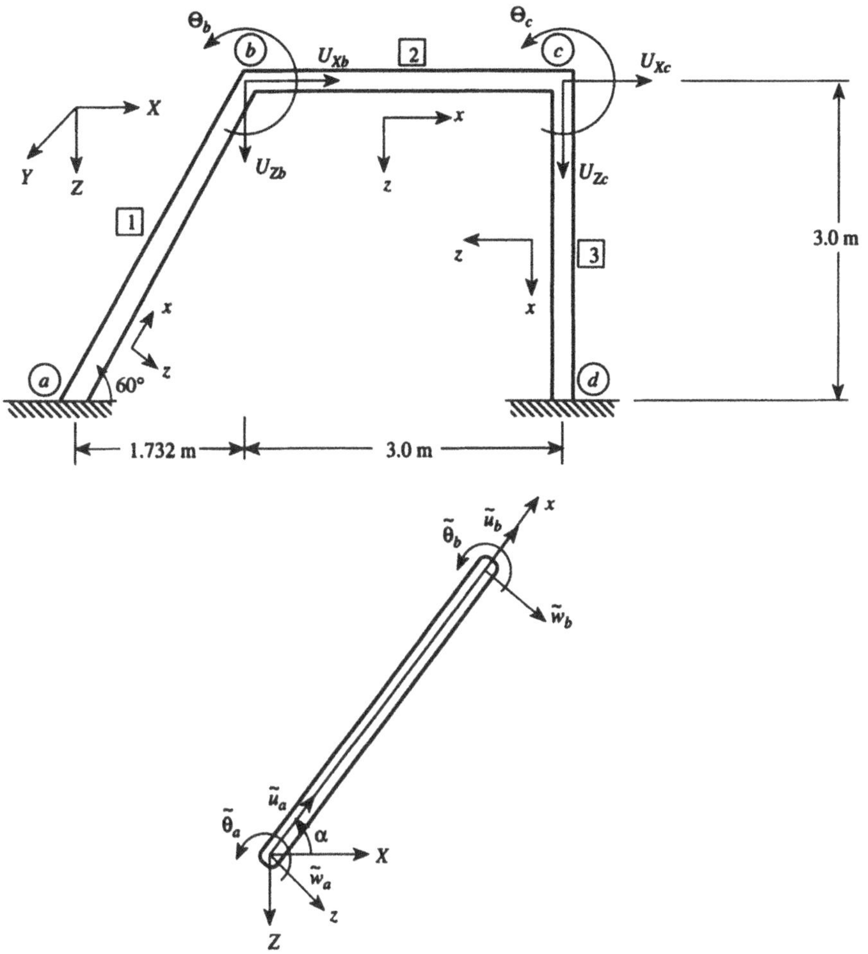

FIGURE 10.2
A plane frame.

The displacement vectors for each element are

$$\begin{aligned}\tilde{\mathbf{v}}^1 &= [\tilde{u}_a \quad \tilde{w}_a \quad \tilde{\theta}_a \quad \tilde{u}_b \quad \tilde{w}_b \quad \tilde{\theta}_b]^T \\ \tilde{\mathbf{v}}^2 &= [\tilde{u}_b \quad \tilde{w}_b \quad \tilde{\theta}_b \quad \tilde{u}_c \quad \tilde{w}_c \quad \tilde{\theta}_c]^T \\ \tilde{\mathbf{v}}^3 &= [\tilde{u}_c \quad \tilde{w}_c \quad \tilde{\theta}_c \quad \tilde{u}_d \quad \tilde{w}_d \quad \tilde{\theta}_d]^T\end{aligned} \tag{2}$$

The element stiffness matrices $\tilde{\mathbf{k}}^i$ in the local coordinate systems, the transformation matrices $\mathbf{T}^i$, and the system stiffness matrix $\mathbf{K}$ are derived in Example 5.5. The corresponding consistent mass matrices will be derived in this example.

Equations (10.10b and d) give expressions for the element mass matrices corresponding to translatory [Eq. (10.10b)] and rotary [Eq. (10.10d)] inertia. Represent the sum of these two mass matrices by $\mathbf{m}^i$, the total element mass matrix. For element 1:

$$\begin{aligned}&\ell = \sqrt{1.732^2 + 3^2} = 3.464 \text{ m}, \quad \rho = \gamma A = 7800 \times 32 \times 10^{-4} = 24.96 \text{ kg/m}, \\ &r_y = \sqrt{I/A} = \sqrt{2.356 \times 10^{-5}/32 \times 10^{-4}} = 8.581 \times 10^{-2} \text{ m}, \\ &\alpha = 60^\circ\end{aligned} \tag{3}$$

$$\tilde{\mathbf{m}}^1 = \begin{bmatrix} 28.82048 & 0.00000 & 0.00000 & 14.41024 & 0.00000 & 0.00000 \\ 0.00000 & 32.17792 & -15.70660 & 0.00000 & 11.05280 & 9.25193 \\ 0.00000 & -15.70660 & 9.96561 & 0.00000 & -9.25193 & -7.43177 \\ 14.41024 & 0.00000 & 0.00000 & 28.82048 & 0.00000 & 0.00000 \\ 0.00000 & 11.05280 & -9.25193 & 0.00000 & 32.17792 & 15.70660 \\ 0.00000 & 9.25193 & -7.43177 & 0.00000 & 15.70660 & 9.96561 \end{bmatrix} \tag{4}$$

This mass matrix referred to the global coordinate system becomes

$$\mathbf{m}^1 = \mathbf{T}^{1T}\tilde{\mathbf{m}}^1\mathbf{T}^1$$
$$= \begin{bmatrix} 31.33856 & 1.45381 & -13.60231 & 11.89216 & -1.45381 & 8.01241 \\ 1.45381 & 29.65984 & -7.85330 & -1.45381 & 13.57088 & 4.62597 \\ -13.60231 & -7.85330 & 9.96561 & -8.01241 & -4.62597 & -7.43177 \\ 11.89216 & -1.45381 & -8.01241 & 31.33856 & 1.45381 & 13.60231 \\ -1.45381 & 13.57088 & -4.62597 & 1.45381 & 29.65984 & 7.85330 \\ 8.01241 & 4.62597 & -7.43177 & 13.60231 & 7.85330 & 9.96561 \end{bmatrix} \tag{5}$$

where $\mathbf{T}^1$ is from Eq. (2) of Example 5.5.

For element 2: $\ell = 3$ m, $\rho = 24.96$ kg/m, $r_y = 8.581 \times 10^{-2}$ m, $\alpha = 0°$

$$\tilde{\mathbf{m}}^2 = \begin{bmatrix} 24.96000 & 0.00000 & 0.00000 & 12.48000 & 0.00000 & 0.00000 \\ 0.00000 & 27.88609 & -11.78524 & 0.00000 & 9.55391 & 6.93476 \\ 0.00000 & -11.78524 & 6.49180 & 0.00000 & -6.93476 & -4.83209 \\ 12.48000 & 0.00000 & 0.00000 & 24.96000 & 0.00000 & 0.00000 \\ 0.00000 & 9.55391 & -6.93476 & 0.00000 & 27.88609 & 11.78524 \\ 0.00000 & 6.93476 & -4.83209 & 0.00000 & 11.78524 & 6.49180 \end{bmatrix} \tag{6}$$

Since $\alpha = 0$, $\mathbf{m}^2 = \tilde{\mathbf{m}}^2$ for element 2.

For element 3: $\ell = 3$, $\rho = 51.48$ kg/m, $r_y = 8.915 \times 10^{-2}$ m, $\alpha = -90°$

$$\tilde{\mathbf{m}}^3 = \begin{bmatrix} 51.48000 & 0.00000 & 0.00000 & 25.74000 & 0.00000 & 0.00000 \\ 0.00000 & 57.52709 & -24.31006 & 0.00000 & 19.69291 & 14.29994 \\ 0.00000 & -24.31006 & 13.40137 & 0.00000 & -14.29994 & -9.96920 \\ 25.74000 & 0.00000 & 0.00000 & 51.48000 & 0.00000 & 0.00000 \\ 0.00000 & 19.69291 & -14.29994 & 0.00000 & 57.52709 & 24.31006 \\ 0.00000 & 14.29994 & -9.96920 & 0.00000 & 24.31006 & 13.40137 \end{bmatrix} \tag{7}$$

This element mass matrix referred to the global coordinate system becomes

$$\mathbf{m}^3 = \mathbf{T}^{3T}\tilde{\mathbf{m}}^3\mathbf{T}^3$$
$$= \begin{bmatrix} 57.52709 & 0.00000 & 24.31006 & 19.69291 & 0.00000 & -14.29994 \\ 0.00000 & 51.48000 & 0.00000 & 0.00000 & 25.74000 & 0.00000 \\ 24.31006 & 0.00000 & 13.40137 & 14.29994 & 0.00000 & -9.96920 \\ 19.69291 & 0.00000 & 14.29994 & 57.52709 & 0.00000 & -24.31006 \\ 0.00000 & 25.74000 & 0.00000 & 0.00000 & 51.48000 & 0.00000 \\ -14.29994 & 0.00000 & -9.96920 & -24.31006 & 0.00000 & 13.40137 \end{bmatrix} \tag{8}$$

Follow the procedure outlined in Example 5.5, Eqs. (6) to (9), to assemble the global mass matrix. The global displacement vector, for which the displacement boundary conditions have been imposed, is $\mathbf{V} = [U_{Xb}\ U_{Zb}\ \Theta_b\ U_{Xc}\ U_{Zc}\ \Theta_c]^T$. The global matrix, after the columns corresponding to constrained DOF have been set equal to zero and the rows corresponding to the unknown reactions are ignored, is given by

$$\mathbf{M} = \begin{bmatrix} 56.29856 & 1.45381 & 13.60231 & 12.48000 & 0.00000 & 0.00000 \\ 1.45381 & 57.54593 & -3.93193 & 0.00000 & 9.55391 & 6.93476 \\ 13.60231 & -3.93193 & 16.45741 & 0.00000 & -6.93476 & -4.83209 \\ 12.48000 & 0.00000 & 0.00000 & 82.48709 & 0.00000 & 24.31006 \\ 0.00000 & 9.55391 & -6.93476 & 0.00000 & 79.36609 & 11.78523 \\ 0.00000 & 6.93476 & -4.83209 & 24.31006 & 11.78523 & 19.89318 \end{bmatrix} \quad (9)$$

■

10.1.2 Lumped Mass Matrix

An alternative to the consistent mass matrix approach is to establish the mass matrix of a structure by forming the lumped mass matrix, i.e., to consider the mass of each element to be concentrated at its nodes. For the structural elements, such as beams, plates, and shells, the lumped mass matrix can be formed by moving the mass surrounding a node to that node. For example, for the beam element shown in Fig. 10.3, half of the mass of the beam is lumped at node a and the other half is lumped at node b. Then,

$$m_a = m_b = \rho\ell/2 \quad (10.11a)$$

Often, only the mass associated with the translational DOF is considered.

If axial motion of the beam element is included, the mass matrix, with the corresponding displacement vector $\mathbf{v} = [u_a\ w_a\ \theta_a\ u_b\ w_b\ \theta_b]^T$, is given by

$$\mathbf{m}^i = \begin{bmatrix} m_a & & & & & \\ & m_a & & & & \\ & & 0 & & & \\ & & & m_b & & \\ & & & & m_b & \\ & & & & & 0 \end{bmatrix} = \frac{\rho\ell}{2}\begin{bmatrix} 1 & & & & & \\ & 1 & & & & \\ & & 0 & & & \\ & & & 1 & & \\ & & & & 1 & \\ & & & & & 0 \end{bmatrix} \quad (10.11b)$$

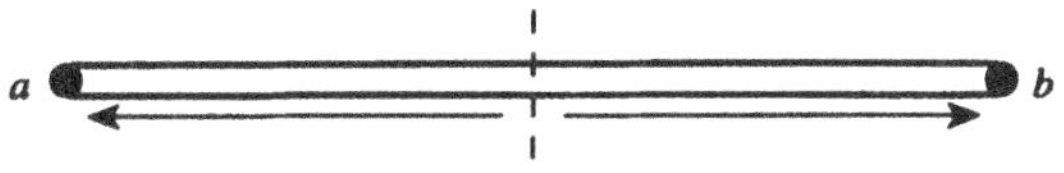

(a) Translational lumped mass

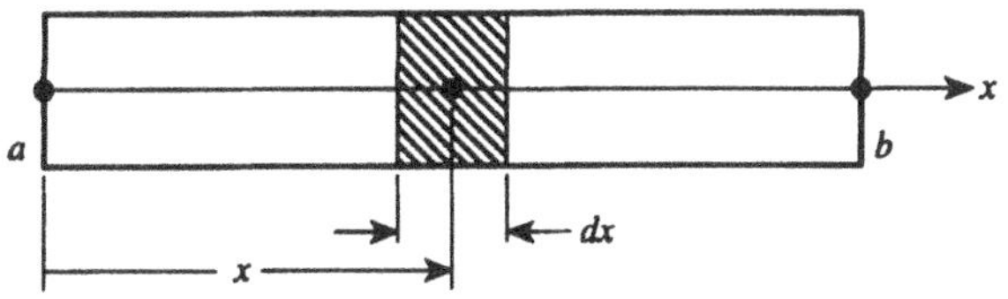

(b) Rotary lumped mass

FIGURE 10.3
Lumping the mass at the ends of a beam element.

Occasionally, for elements such as beams and plates, the mass associated with the rotational DOF is taken into account. For a beam element with two nodes, consider a "disk" of length dx, at coordinate x from node a, as shown in Fig. 10.3b. The rotary moment of inertia of this "disk" about node a is

$$dm_r = \rho(I/A + x^2)\, dx$$

where ρ is the mass density (mass per unit length) of the element, and A is the area of the beam cross-section. Because the rotary inertia of a half of an element is lumped at each node, we have

$$m_{ra} = m_{rb} = \int_0^{\ell/2} dm_r = \frac{\rho\ell}{2}\left(\frac{I}{A} + \frac{\ell^2}{12}\right) = \frac{\rho\ell}{2}\left(r_y^2 + \frac{\ell^2}{12}\right) \tag{10.11c}$$

The mass matrix to supplement Eq. (10.11b) becomes

$$\mathbf{m}_r^i = \begin{bmatrix} 0 & & & & & \\ & 0 & & & & \\ & & m_{ra} & & & \\ & & & 0 & & \\ & & & & 0 & \\ & & & & & m_{rb} \end{bmatrix} = \frac{\rho\ell}{2}\left(\frac{I}{A} + \frac{\ell^2}{12}\right)\begin{bmatrix} 0 & & & & & \\ & 0 & & & & \\ & & 1 & & & \\ & & & 0 & & \\ & & & & 0 & \\ & & & & & 1 \end{bmatrix} \tag{10.11d}$$

For the two- or three-dimensional solid elements, the formulation of the lumped mass matrix is not as straightforward. It is common in finite element computer programs to perform the integration involved in $\mathbf{m}^i$ of Eq. (10.5) by using numerical integration schemes, such as the Newton-Cotes or Gauss quadratures described in Chapter 6, Section 6.6. See, for example, Fergusson and Pilkey (1992). Usually, the numerical integration procedure, including the number and location of integration points, employed for the integration needed for the stiffness matrix [$\mathbf{k}^i$ of Eq. (10.5)] is used for the integration of $\mathbf{m}^i$ of Eq. (10.5). In all cases, the numerical integration leads to a discrete mass matrix $\mathbf{m}^i$ and in some cases numerical integration can provide the same consistent mass matrices formed by analytical integration. See Problems 10.4, 10.5, and 10.6.

An alternative to using the same numerical integration for $\mathbf{m}^i$ as employed for $\mathbf{k}^i$, is to form the mass matrix by using only the element nodes as the integration points. This is referred to as nodal quadrature. As shown in Chapter 6, the interpolation (shape) functions $\mathbf{N}$ in $\mathbf{m}^i$ of Eq. (10.5) exhibit the property

$$N_i(x_k, y_k, z_k) = \begin{cases} 0 & i \neq k \\ 1 & i = k \end{cases}.$$

Hence, if the nodes are the only integration points, a diagonal mass matrix is obtained because the value of the product of the shape functions N_i and $N_j (i \neq j)$ at any nodal point is zero. The main diagonal elements in this lumped mass matrix are

$$\begin{aligned} m_{ii} &= \sum_k^n W_k^{(n)} N_i^2(x_k, y_k, z_k)\gamma(x_k, y_k, z_k) = W_i^{(n)} N_i^2(x_i, y_i, z_i)\gamma(x_i, y_i, z_i) \\ &= W_i^{(n)}\gamma(x_i, y_i, z_i) \end{aligned} \tag{10.12a}$$

where n is the number of nodes in the element; $W_k^{(n)}$ are the weights; and x_k, y_k, and z_k are the coordinates at the nodes. For the isoparametric elements,

$$\begin{aligned} m_{ii} &= \sum_k^n W_k^{(n)} N_i^2(\xi_k, \eta_k, \zeta_k) \det \mathbf{J}(\xi_k, \eta_k, \zeta_k)\gamma(\xi_k, \eta_k, \zeta_k) \\ &= W_i^{(n)} N_i^2(\xi_i, \eta_i, \zeta_i) \det \mathbf{J}(\xi_i, \eta_i, \zeta_i)\gamma(\xi_i, \eta_i, \zeta_i) \\ &= W_i^{(n)} \det \mathbf{J}(\xi_i, \eta_i, \zeta_i)\gamma(\xi_i, \eta_i, \zeta_i) \end{aligned} \tag{10.12b}$$

where ξ_k, η_k, and ζ_k are the coordinates of nodes in the master element; and $\mathbf{J}$ is the Jacobian.

EXAMPLE 10.2 Lumped Mass Matrix of a Plane Element

For the plane stress element of Chapter 6, Fig. 6.10, find a lumped mass matrix based on the shape functions used in Section 6.4.1.

From Eq. (6.19), Chapter 6

$$\mathbf{u} = \begin{bmatrix} u_x \\ u_y \end{bmatrix} = \begin{bmatrix} N_1\ N_2\ N_3\ N_4 & 0 \\ 0 & N_1\ N_2\ N_3\ N_4 \end{bmatrix} \begin{bmatrix} u_{x1} \\ u_{x2} \\ u_{x3} \\ u_{x4} \\ u_{y1} \\ u_{y2} \\ u_{y3} \\ u_{y4} \end{bmatrix}$$

$$= \begin{bmatrix} \mathbf{N}_x & \mathbf{0} \\ \mathbf{0} & \mathbf{N}_y \end{bmatrix} \begin{bmatrix} \mathbf{v}_x \\ \mathbf{v}_y \end{bmatrix} = \mathbf{N}\,\mathbf{v} \tag{1}$$

where

$$N_1 = (1-\xi)(1-\eta) = \left(1 - \frac{x}{a}\right)\left(1 - \frac{y}{b}\right) \qquad N_3 = \xi\eta = \frac{x}{a}\frac{y}{b}$$

$$N_2 = \xi(1-\eta) = \frac{x}{a}\left(1 - \frac{y}{b}\right) \qquad N_4 = (1-\xi)\eta = \left(1 - \frac{x}{a}\right)\frac{y}{b}$$

The element mass matrix of Eq. (10.5) can be expressed for the plane stress case as

$$\mathbf{m}^i = \int_V \gamma \mathbf{N}^T \mathbf{N}\, dV = t\gamma \int_0^b \int_0^a \mathbf{N}^T \mathbf{N}\, dx\, dy \tag{2}$$

where t is the thickness of the element, and γ is the mass density of the material. Substitute the expressions for $\mathbf{N}$ of (1) into (2),

$$\begin{aligned} \mathbf{m}^i &= t\gamma \int_0^b \int_0^a \begin{bmatrix} \mathbf{N}_x^T & \mathbf{0} \\ \mathbf{0} & \mathbf{N}_y^T \end{bmatrix} \begin{bmatrix} \mathbf{N}_x & \mathbf{0} \\ \mathbf{0} & \mathbf{N}_y \end{bmatrix} dx\, dy = t\gamma \int_0^b \int_0^a \begin{bmatrix} \mathbf{N}_x^T \mathbf{N}_x & \mathbf{0} \\ \mathbf{0} & \mathbf{N}_y^T \mathbf{N}_y \end{bmatrix} dx\, dy \\ &= t\gamma ab \begin{bmatrix} \mathbf{m}^* & \mathbf{0} \\ \mathbf{0} & \mathbf{m}^* \end{bmatrix} \end{aligned} \tag{3}$$

where

$$\mathbf{N}_x^T\mathbf{N}_x = \mathbf{N}_y^T\mathbf{N}_y = \begin{bmatrix} N_1N_1 & & & \\ N_2N_1 & N_2N_2 & \text{Symmetric} & \\ N_3N_1 & N_3N_2 & N_3N_3 & \\ N_4N_1 & N_4N_2 & N_4N_3 & N_4N_4 \end{bmatrix}$$

Simple integration of (3) leads to the consistent mass matrix with

$$\mathbf{m}^* = \begin{bmatrix} 1/9 & 1/18 & 1/36 & 1/18 \\ & 1/9 & 1/18 & 1/36 \\ & & 1/9 & 1/18 \\ & \text{Symmetric} & & 1/9 \end{bmatrix}$$

To form a lumped mass matrix, choose the four nodes ($k = 1, 2, 3, 4$) as integration points. Since the shape functions exhibit the property

$$N_i(x_k, y_k) = \begin{cases} 1 & i = k \\ 0 & i \neq k \end{cases} \quad (i \text{ and } k = 1, 2, 3, 4)$$

as utilized in Eq. (10.12a), upon integration (3) becomes a diagonal matrix,

$$\mathbf{m}^i = \text{diag}(m_{11}\ m_{22}\ m_{33}\ m_{44} \mid m_{11}\ m_{22}\ m_{33}\ m_{44}) \tag{4}$$

where

$$m_{ii} = t\gamma \int_0^b \int_0^a N_i^2(x, y)\,dx\,dy = t\gamma \sum_k^4 W_k^{(4)} N_i^2(x_k, y_k) = t\gamma W_i^{(4)} \quad i = 1, 2, 3, 4$$

To calculate $m_{ii}(i = 1, 2, 3, 4)$, the integration rule of Eq. (6.125), Chapter 6, is used to find the weights $W_i^{(4)}$. Because the four nodes of the element are spaced equally in the x and y coordinate directions, two in each direction, Newton-Cotes quadrature can be used. From Eq. (6.125), the entries of the mass matrix of (4) are

$$m_{ii} = t\gamma \int_0^b \int_0^a N_i^2(x, y)\,dx\,dy = t\gamma \sum_\alpha^2 \sum_\beta^2 W_\alpha^{(2)} W_\beta^{(2)} N_i^2(x_\alpha, y_\beta) \tag{5}$$

The $W_\alpha^{(2)}$ and $W_\beta^{(2)}$ can be obtained from Table 6.6. In the x direction $W_\alpha^{(2)} = (a - 0)C_\alpha^{(2)} = \frac{1}{2}a\,(\alpha = 1, 2)$; in the y direction $W_\beta^{(2)} = (b - 0)C_\beta^{(2)} = \frac{1}{2}b(\beta = 1, 2)$. By comparison of the expressions for m_{ii} in (4) and (5),

$$W_k^{(4)} = W_\alpha^{(2)} W_\beta^{(2)} = \frac{1}{2}a\frac{1}{2}b = \frac{1}{4}ab \quad (k = 1, 2, 3, 4,\ \alpha \text{ and } \beta = 1, 2)$$

so that

$$m_{ii} = t\gamma W_i^{(4)} = \frac{1}{4}abt\gamma \quad (i = 1, 2, 3, 4) \tag{6}$$

and the element lumped mass matrix becomes

$$\mathbf{m}^i = \frac{1}{4}abt\gamma \begin{bmatrix} 1 &&&&&&& \\ & 1 &&&&&& \\ && 1 &&&&& \\ &&& 1 &&&& \\ &&&& 1 &&& \\ &&&&& 1 && \\ &&&&&& 1 & \\ &&&&&&& 1 \end{bmatrix} \tag{7}$$

It is evident that this matrix can be obtained by distributing equally the total mass $(abt\gamma)$ of the element at its four nodes for u_x and u_y. If nodes are added to the edges (between the current nodes) of the element, the lumped mass will not be this equally distributed mass matrix. ∎

After the element lumped mass matrix is formed, the global lumped mass matrix can be assembled by the process described in Chapter 5 for the stiffness matrix. See Example 10.1. The global mass matrix can also be formulated directly. The total mass associated with a global node of the structure is the summation of all the masses (or rotary inertias) lumped from the adjacent elements connected to that node. The advantage of lumped mass is that the resulting mass matrix is diagonal, and, hence, there is no dynamic coupling between the masses, i.e.,

$$\mathbf{M} = \begin{bmatrix} m_1 & 0 & & 0 & & 0 \\ 0 & m_2 & & 0 & & 0 \\ & & \ddots & & & \\ 0 & 0 & & m_i & & 0 \\ & & & & \ddots & \\ 0 & 0 & & 0 & & m_{n_d} \end{bmatrix} \tag{10.13}$$

where m_i is the mass (or rotary inertia) at DOF i, and n_d is the number of DOF. It should be noted that often some of the diagonal elements of this mass matrix may be zero, and the remaining diagonal elements are all positive. Hence, the mass matrix is positive semi-definite. Although the "lumped mass" model is only an approximation to the real mass distribution of the structure, normally, it leads to quite satisfactory results.

The lumped mass matrix is positive semi-definite when zeros occur on the diagonal, whereas the consistent mass (element and global) matrices are positive definite. The zeroes on the diagonal can complicate certain numerical algorithms. It is clear that a lumped mass matrix would require less storage space than a consistent mass matrix, which on the element level is full. It is also more economical to form and manipulate.

EXAMPLE 10.3 Lumped Mass Matrices for a Frame

Find the global lumped mass matrix of the frame of Fig. 10.2 and Example 10.1. The consistent mass matrices were derived in Example 10.1.

In general, the global lumped mass matrix can be computed using the transformations of Example 10.1 by replacing the element consistent mass matrices with the element lumped mass matrices. Alternatively, the global lumped mass matrix for the frame can be formed by lumping the mass at the ends of each element of the frame. In this alternative approach the lumped mass associated with each (translational or rotary) global DOF can be calculated by summing the mass of the adjoining beam elements at each node. If the global mass matrix is based on lumping the mass at only nodes $a, b, c,$ and d of the frame of Fig. 10.2, a rather crude approximation may be anticipated. For node a, from Eq. (10.11)

$$\begin{aligned} m_a &= \frac{1}{2}\rho\ell = \frac{1}{2} \times 24.96 \times 3.464 = 43.231 \text{ kg} \\ m_{ra} &= \frac{\rho\ell}{2}\left(r_y^2 + \frac{\ell^2}{12}\right) = \frac{1}{2} \times 24.96 \times 3.464\left[(8.581 \times 10^{-2})^2 + \frac{3.464^2}{12}\right] \\ &= 43.547 \text{ kg} \cdot \text{m}^2 \end{aligned} \tag{1}$$

For node b,

$$m_b = \left(\frac{1}{2}\rho\ell\right)_{\text{element1}} + \left(\frac{1}{2}\rho\ell\right)_{\text{element2}} = 43.231 + 37.44 = 80.67 \text{ kg}$$

$$\begin{aligned} m_{rb} &= \frac{\rho\ell}{2}\left(r_y^2 + \frac{\ell^2}{12}\right)_{\text{element1}} + \frac{\rho\ell}{2}\left(r_y^2 + \frac{\ell^2}{12}\right)_{\text{element2}} \\ &= 43.547 + 28.356 = 71.90 \text{ kg}\cdot\text{m}^2 \end{aligned} \tag{2}$$

Similarly, for node c, $m_c = 114.66$ kg, $m_{rc} = 86.88$ kg $\cdot$ m^2 and for node d, $m_d = 77.22$ kg, $m_{rd} = 58.529$ kg $\cdot$ m^2.

The global displacement vector is

$$\mathbf{V} = [U_{Xa}\ U_{Za}\ \Theta_a\ U_{Xb}\ U_{Zb}\ \Theta_b\ U_{Xc}\ U_{Zc}\ \Theta_c\ U_{Xd}\ U_{Zd}\ \Theta_d]^T \tag{3}$$

and the global mass matrix is

$$\mathbf{M} = \text{diagonal}(m_a\ m_a\ m_{ra}\ m_b\ m_b\ m_{rb}\ m_c\ m_c\ m_{rc}\ m_d\ m_d\ m_{rd}) \tag{4}$$

After the boundary conditions are imposed, the displacement vector is

$$\mathbf{V} = [U_{Xb}\ U_{Zb}\ \Theta_b\ U_{Xc}\ U_{Zc}\ \Theta_c]^T \tag{5}$$

and the corresponding global mass matrix is

$$\begin{aligned} \mathbf{M} &= \text{diagonal}(m_b\ m_b\ m_{rb}\ m_c\ m_c\ m_{rc}) \\ &= \text{diagonal}(80.67\ 80.67\ 71.90\ 114.66\ 114.66\ 86.88) \end{aligned} \tag{6}$$

■

10.1.3 Alternatives for the Formation of the Mass Matrix

Many schemes have been proposed for the formation of mass matrices. For example, a lumped mass matrix can be formed by using a *diagonal mass matrix* approach. This kind of matrix is usually constructed from the consistent mass matrix. One method for accomplishing this is [Cook, 1981]:

1. Compute only the diagonal coefficients of the consistent mass matrix.
2. Compute m, the total mass of the element.
3. Compute s as the sum of the diagonal coefficients m_{ii} associated with translation (but not rotation).
4. Scale the diagonal coefficients m_{ii} by multiplying them by the ratio m/s.

This kind of element is also particularly applicable to structures whose translational DOF are mutually parallel, such as occurs for beam and plate elements.

Another alternative form of the mass matrix is a linear combination of the consistent and the lumped mass matrices

$$\mathbf{m} = \alpha\mathbf{m}_{\text{consistent}} + \beta\mathbf{m}_{\text{lumped}} \tag{10.14}$$

This $\mathbf{m}$ is referred to as a *non-consistent* or *high-order* mass matrix. Sometimes [Hughes, 1987], a simple average is employed so that $\alpha = \beta = \frac{1}{2}$. For a two-node bar element, this leads to

$$\mathbf{m}_{\text{high-order}} = \frac{1}{2}(\mathbf{m}_{\text{consistent}} + \mathbf{m}_{\text{lumped}}) = \frac{\rho\ell}{12}\begin{bmatrix} 5 & 1 \\ 1 & 5 \end{bmatrix} \tag{10.15}$$

where, from Eq. (10.10b),

$$\mathbf{m}_{\text{consistent}} = \frac{\rho\ell}{6}\begin{bmatrix} 2 & 1 \\ 1 & 2 \end{bmatrix}$$

and, from Eq. (10.11b),

$$\mathbf{m}_{\text{lumped}} = \frac{\rho\ell}{2}\begin{bmatrix} 1 & 0 \\ 0 & 1 \end{bmatrix}$$

A study [Kim, 1993] contends that $\alpha = \frac{2}{5}$ and $\beta = \frac{3}{5}$ gives the best results for the axial vibration of a bar, while $\alpha = \frac{1}{9}$ and $\beta = \frac{8}{9}$ leads to the best results for a cantilevered beam.

The literature contains various proposals for using numerical integration schemes for generating mass matrices for two and three dimensional elements. As mentioned earlier, the most common approach is simply to use the numerical integration scheme employed for the integration of $\mathbf{k}^i$ to perform the integration to form $\mathbf{m}^i$. See the references cited above for further discussions concerning the establishment of approximate mass matrices.

10.2 Reduction of Degrees of Freedom

For a complex vibrating system, some of the DOF may have little influence on the dynamic behavior of the system. For example, when one mass in a two DOF system is much smaller than the other mass, the system can often be treated as a single-DOF system. Dynamic response problems tend to be so complicated numerically that it is accepted as a basic premise that any plausible reduction in the DOF should be implemented. Some DOF can complicate a computational procedure, as is often the case for a lumped mass matrix with zero diagonal elements. These zero elements have little effect on the dynamic response of the system. Certainly, it is reasonable to simply neglect the small and zero masses and then deal with the remainder of the mass matrix. A more accurate method of achieving this is called *kinematic condensation*. This is similar to static condensation procedures (Chapter 5), which are used in structural analysis to reduce the size of the stiffness matrix. Use the two-DOF system of Fig. 10.4 to illustrate this procedure.

The governing equation for this system is

$$\mathbf{M}\ddot{\mathbf{V}} + \mathbf{K}\mathbf{V} = \mathbf{0} \tag{10.16}$$

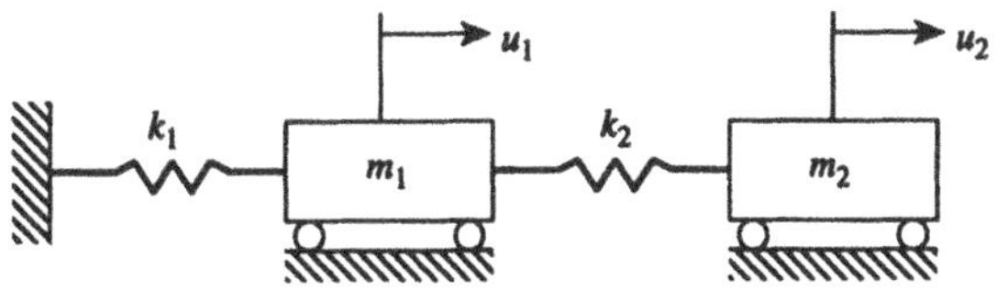

FIGURE 10.4
A two-DOF system.

where, with boundary conditions taken into account,

$$\mathbf{M} = \begin{bmatrix} m_1 & 0 \\ 0 & m_2 \end{bmatrix} = \begin{bmatrix} m_{11} & 0 \\ 0 & m_{22} \end{bmatrix}$$

$$\mathbf{K} = \begin{bmatrix} k_1 + k_2 & -k_2 \\ -k_2 & k_2 \end{bmatrix} = \begin{bmatrix} k_{11} & k_{12} \\ k_{21} & k_{22} \end{bmatrix}$$

and

$$\mathbf{V} = \begin{bmatrix} u_1 \\ u_2 \end{bmatrix}$$

Assume that m_{22} is very small compared to m_{11}, and, hence, the DOF associated with m_{22} is to be condensed out. Before treating kinematic condensation, consider briefly static condensation. The displacement equations for the static response of a general two-DOF system can be written as

$$\begin{bmatrix} k_{11} & k_{12} \\ k_{21} & k_{22} \end{bmatrix} \begin{bmatrix} u_1 \\ u_2 \end{bmatrix} = \begin{bmatrix} P_1 \\ P_2 \end{bmatrix} \tag{10.17}$$

The second equation of Eq. (10.17) can be used to express u_2 in terms of u_1

$$u_2 = \frac{P_2}{k_{22}} + T u_1 \tag{10.18}$$

where $T = -k_{21}/k_{22}$. Use this relationship in the first equation in Eq. (10.17) to eliminate u_2 (back substitution) to obtain

$$(k_{11} + k_{12} T) u_1 = P_1 - \frac{k_{12}}{k_{22}} P_2 \tag{10.19}$$

This is referred to as a statically condensed relation.

Now, return to the dynamic case where the equations of motion are given by Eq. (10.16). If the relationship between u_1 and u_2 during the dynamic response is assumed to be the same as for the static case, then the variable u_2 can be eliminated from Eq. (10.16). That is, the static relationship of Eq. (10.18) will be assumed to apply for dynamic motion. Set $P_2 = 0$, so that Eq. (10.18) provides

$$u_2 = T u_1 \quad \text{and} \quad \ddot{u}_2 = T \ddot{u}_1 \tag{10.20}$$

Substitute Eq. (10.20) into Eq. (10.16) giving

$$\begin{bmatrix} m_{11} & 0 \\ 0 & m_{22} \end{bmatrix} \begin{bmatrix} \ddot{u}_1 \\ T\ddot{u}_1 \end{bmatrix} + \begin{bmatrix} k_{11} & k_{12} \\ k_{21} & k_{22} \end{bmatrix} \begin{bmatrix} u_1 \\ T u_1 \end{bmatrix} = \begin{bmatrix} 0 \\ 0 \end{bmatrix} \tag{10.21}$$

Premultiply Eq. (10.21) by $[1 \quad T]$ to obtain

$$[1 \quad T] \begin{bmatrix} m_{11} & 0 \\ 0 & m_{22} \end{bmatrix} \begin{bmatrix} \ddot{u}_1 \\ T\ddot{u}_1 \end{bmatrix} + [1 \quad T] \begin{bmatrix} k_{11} & k_{12} \\ k_{21} & k_{22} \end{bmatrix} \begin{bmatrix} u_1 \\ T u_2 \end{bmatrix} = [1 \quad T] \begin{bmatrix} 0 \\ 0 \end{bmatrix} \tag{10.22}$$

or

$$\overline{m} \ddot{u}_1 + \overline{k} u_1 = 0 \tag{10.23}$$

where

$$\overline{m} = m_{11} + m_{22}T^2 = m_{11} + m_{22}\left(\frac{k_{21}}{k_{22}}\right)^2$$

$$\overline{k} = k_{11} + Tk_{21} + Tk_{12} + T^2k_{22} = k_{11} - \frac{k_{21}k_{12}}{k_{22}}$$

Equation (10.23) contains only one variable u_1, and, hence, the original two-DOF system is reduced to a single-DOF system.

In this procedure, Eq. (10.18) from the static case is used to reduce the DOF of the dynamic equations. This means that it is supposed that DOF 2 is related to DOF 1 through the static stiffness properties of the structure. The inertia property of DOF 2 is ignored, which implies that the structural inertia of DOF 2 does not affect the deformation shape of the system. Note that this is not the same as simply setting $m_2 = 0$. An application of the static relation of Eq. (10.18) to the dynamic case was introduced by Guyan (1965), and, hence, this kinematic condensation is often referred to as *Guyan reduction.*

Reduction of the number of DOF in a system can be considered in terms of dependent and independent DOF. In the above case, u_1 is the independent (*active* or *master*) coordinate, while u_2 is the dependent (or *slave*) coordinate. In practice, it is common to let the translation DOF be the master coordinates and the rotary DOF be the slave coordinates.

Consider now a multiple DOF system. Define $\mathbf{V}_a$ to be the active or master set of dynamic DOF and $\mathbf{V}_0$ to be the omitted set of DOF of the system, i.e., the DOF to be condensed. Rewrite the governing equations

$$\mathbf{M}\ddot{\mathbf{V}} + \mathbf{K}\mathbf{V} = \mathbf{0} \tag{10.24}$$

with

$$\mathbf{V} = \begin{bmatrix} \mathbf{V}_a \\ \mathbf{V}_0 \end{bmatrix}$$

This form is achieved by interchanging some rows and columns of the original **M** and **K** in Eq. (10.16). Partition the resulting **M** and **K** as

$$\mathbf{M} = \begin{bmatrix} \mathbf{M}_{aa} & \mathbf{M}_{a0} \\ \mathbf{M}_{0a} & \mathbf{M}_{00} \end{bmatrix} \quad \text{and} \quad \mathbf{K} = \begin{bmatrix} \mathbf{K}_{aa} & \mathbf{K}_{a0} \\ \mathbf{K}_{0a} & \mathbf{K}_{00} \end{bmatrix} \tag{10.25}$$

Then Eq. (10.24) appears as

$$\begin{bmatrix} \mathbf{M}_{aa} & \mathbf{M}_{a0} \\ \mathbf{M}_{0a} & \mathbf{M}_{00} \end{bmatrix} \begin{bmatrix} \ddot{\mathbf{V}}_a \\ \ddot{\mathbf{V}}_0 \end{bmatrix} + \begin{bmatrix} \mathbf{K}_{aa} & \mathbf{K}_{a0} \\ \mathbf{K}_{0a} & \mathbf{K}_{00} \end{bmatrix} \begin{bmatrix} \mathbf{V}_a \\ \mathbf{V}_0 \end{bmatrix} = \begin{bmatrix} \mathbf{0} \\ \mathbf{0} \end{bmatrix} \tag{10.26}$$

Note for a lumped mass model that $\mathbf{M}_{a0} = \mathbf{M}_{0a} = \mathbf{0}$. For Guyan reduction, we will neglect the dynamic effect due to the mass associated with $\mathbf{V}_0$. Then the second equation of Eq. (10.26) provides

$$\mathbf{K}_{0a}\mathbf{V}_a + \mathbf{K}_{00}\mathbf{V}_0 = \mathbf{0}$$

This relation can be solved for $\mathbf{V}_0$ in terms of $\mathbf{V}_a$ to obtain

$$\mathbf{V}_0 = -\mathbf{K}_{00}^{-1}\mathbf{K}_{0a}\mathbf{V}_a \tag{10.27}$$

Rewrite this as

$$\begin{bmatrix} \mathbf{V}_a \\ \mathbf{V}_0 \end{bmatrix} = \mathbf{T}\mathbf{V}_a, \quad \mathbf{T} = \begin{bmatrix} \mathbf{I} \\ -\mathbf{K}_{00}^{-1}\mathbf{K}_{0a} \end{bmatrix} \tag{10.28}$$

With substitution of Eq. (10.28) into Eq. (10.26), and premultiplication of the result by $\mathbf{T}^T$, the mass term in Eq. (10.26) appears as

$$\begin{aligned}
&\mathbf{T}^T \begin{bmatrix} \mathbf{M}_{aa} & \mathbf{M}_{a0} \\ \mathbf{M}_{0a} & \mathbf{M}_{00} \end{bmatrix} \mathbf{T}\ddot{\mathbf{V}}_a \\
&= \left[\mathbf{M}_{aa} - \mathbf{K}_{0a}^T \left(\mathbf{K}_{00}^{-1}\right)^T \mathbf{M}_{0a} - \mathbf{M}_{a0}\mathbf{K}_{00}^{-1}\mathbf{K}_{0a} + \mathbf{K}_{0a}^T \left(\mathbf{K}_{00}^{-1}\right)^T \mathbf{M}_{00}\mathbf{K}_{00}^{-1}\mathbf{K}_{0a}\right]\ddot{\mathbf{V}}_a \\
&= \overline{\mathbf{M}}\ddot{\mathbf{V}}_a
\end{aligned} \tag{10.29a}$$

Similarly,

$$\mathbf{T}^T\mathbf{K}\mathbf{T}\mathbf{V}_a = \left(\mathbf{K}_{aa} - \mathbf{K}_{a0}\mathbf{K}_{00}^{-1}\mathbf{K}_{0a}\right)\mathbf{V}_a = \overline{\mathbf{K}}\mathbf{V}_a \tag{10.29b}$$

Equation (10.26) becomes

$$\overline{\mathbf{M}}\ddot{\mathbf{V}}_a + \overline{\mathbf{K}}\mathbf{V}_a = \mathbf{0} \tag{10.30}$$

Equation (10.30) is the governing equation of motion with condensed dynamic DOF. In the case of a lumped mass model, the mass matrix $\overline{\mathbf{M}}$ may not be a diagonal matrix anymore because of the matrix operations in Eq. (10.29a).

10.3 Free Vibration Analysis

The governing equations for the "free" motion of a structure are

$$\mathbf{M}\ddot{\mathbf{V}} + \mathbf{K}\mathbf{V} = \mathbf{0} \tag{10.31}$$

The motion is referred to as being free, since there are no applied loadings. Often in this section, it will be assumed that $\mathbf{M}$ and $\mathbf{K}$ are the mass and stiffness matrices after the boundary conditions are imposed and, if desired, Guyan reduction is performed.

By assuming *harmonic motion*,

$$\mathbf{V} = \boldsymbol{\phi}\sin\omega t \tag{10.32}$$

the *natural frequencies* ω and the corresponding *mode shapes* $\boldsymbol{\phi}$ can be computed from the *generalized* eigenvalue problem

$$\omega^2\mathbf{M}\boldsymbol{\phi} = \mathbf{K}\boldsymbol{\phi} \tag{10.33}$$

or

$$(\mathbf{K} - \omega^2\mathbf{M})\boldsymbol{\phi} = \mathbf{0} \tag{10.34}$$

Because $\boldsymbol{\phi}$ is nontrivial,

$$|\mathbf{K} - \omega^2\mathbf{M}| = 0 \tag{10.35}$$

or, with $\omega^2 = \lambda$,

$$|\mathbf{K} - \lambda\mathbf{M}| = 0 \tag{10.36}$$

Equation (10.36) is the *characteristic equation*, and λ is called the *eigenvalue* of the equation. The solid is said to respond in the *mode* corresponding to a particular frequency.

For structures with a line-like geometry, the transfer matrix method is sometimes useful for finding the natural frequencies and mode shapes. The use of this method for eigenvalue problems is discussed in Chapter 11, Stability. Unfortunately, the transfer matrix method requires a numerical determinant search and often encounters numerical difficulties, especially in the calculation of the higher frequencies.

There are a variety of effective and efficient algorithms for attacking the eigenvalue problem of Eq. (10.34) directly without resorting to solving Eq. (10.35). In fact, reliable software for solving Eq. (10.34) is readily available. Often, eigenvalue problem computer programs require that Eq. (10.34) be converted to the so-called *standard* eigenvalue problem

$$(\mathbf{A} - \lambda\mathbf{I})\mathbf{Y} = \mathbf{0} \tag{10.37}$$

where $\mathbf{I}$ is the unit diagonal matrix, $\mathbf{A}$ is symmetric, and $\mathbf{Y}$ is the *eigenvector*. Eigenvalues λ that satisfy this relationship are said to be the eigenvalues of $\mathbf{A}$. To obtain the standard form, premultiply Eq. (10.34) by $\mathbf{M}^{-1}$.

$$(\mathbf{M}^{-1}\mathbf{K} - \lambda\mathbf{I})\boldsymbol{\phi} = \mathbf{0} \tag{10.38}$$

However, $\mathbf{M}^{-1}\mathbf{K}$ is, in general, not symmetric so further manipulations are required to achieve a standard form. In particular, Cholesky decomposition (see a linear algebra book) of $\mathbf{M}$ or $\mathbf{K}$ is performed, resulting in the standard form.

A zero eigenvalue λ_i should be obtained for each possible rigid body motion of a structure that is not completely supported. Since the mass can hold the structure together, a singular stiffness matrix $\mathbf{K}$ is more palatable for a dynamic problem than for a static solution, although some operations may not be suitable. For a real, symmetric, and nonsingular $\mathbf{K}$, the rank of $\mathbf{M}$ is equal to the number of nonzero independent eigenvalues of Eq. (10.34) and the number of frequencies is equal to the number of unrestrained nodal displacements, provided that $\mathbf{M}$ is formed of consistent mass element matrices.

The different kinds of mass matrices normally lead to slightly different eigenvalues. There does not appear to be a definitive indication in the literature that a particular mass matrix is the best for all problems. Since the consistent mass matrix is positive definite, it reduces the risk of computational difficulties. The eigenvalues computed using the consistent mass matrix are higher than the exact values. This kind of mass matrix appears to be quite accurate for flexural problems, such as for beams and shells. Lumped mass matrices are easy to form and manipulate and use less computer storage. The eigenvalues from this mass matrix usually approach the exact value from below.

Note that the mode shapes $\boldsymbol{\phi}$ are "shapes" of the equations and give the relative magnitude of the DOF, not the absolute values. This follows from $\boldsymbol{\phi}$ being the solution to a set of homogeneous equations (Eq. 10.34). The natural frequencies and mode shapes provide a fundamental description of the response of the vibrating system.

EXAMPLE 10.4 Free Vibration Analysis of a Spring Mass System

Find the natural frequencies of the spring mass system of Fig. 10.5.

The governing equation for the free vibration is

$$\begin{bmatrix} m & 0 \\ 0 & 2m \end{bmatrix}\begin{bmatrix} \ddot{u}_1 \\ \ddot{u}_2 \end{bmatrix} + \begin{bmatrix} 2k & -k \\ -k & 2k \end{bmatrix}\begin{bmatrix} u_1 \\ u_2 \end{bmatrix} = \mathbf{0} \tag{1}$$
$$\mathbf{M} \qquad \ddot{\mathbf{V}} \quad + \qquad \mathbf{K} \qquad \mathbf{V} \quad = \mathbf{0}$$

where the boundary conditions have been taken into account.

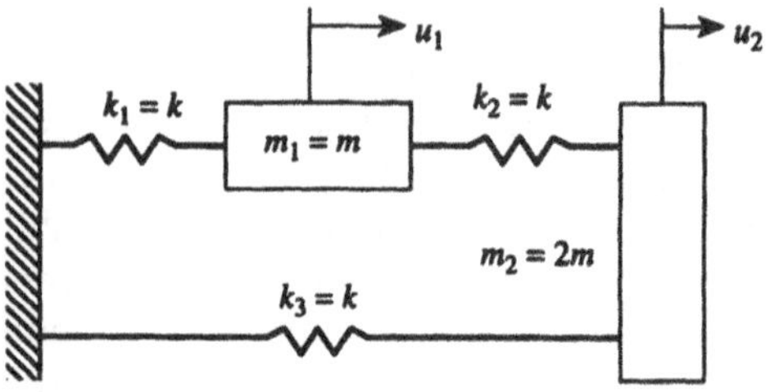

FIGURE 10.5
The two-DOF system for Examples 10.4 and 10.9.

The characteristic equation of the form of Eq. (10.36) ($|\mathbf{K} - \lambda \mathbf{M}| = 0$) is

$$\begin{bmatrix} 2h - \lambda & -h \\ -h & 2h - 2\lambda \end{bmatrix} = 0 \tag{2}$$

where $h = k/m$, and $\lambda = \omega^2$. This reduces to a polynomial

$$\lambda^2 - 3h\lambda + \frac{3}{2}h^2 = 0 \tag{3}$$

Solve (3) for λ, giving $\lambda_{1,2} = (3 \pm \sqrt{3})h/2$, or

$$\omega_1 = 0.796266\sqrt{k/m} \quad \text{and} \quad \omega_2 = 1.538188\sqrt{k/m} \tag{4}$$

Substitute ω_1 and ω_2 into Eq. (10.33) to find the mode shapes $\boldsymbol{\phi}_1$ and $\boldsymbol{\phi}_2$. Because $\boldsymbol{\phi}_1$ and $\boldsymbol{\phi}_2$ are shapes, i.e., they are relative magnitudes of the DOF obtained from the homogeneous equations $\omega^2\mathbf{M}\boldsymbol{\phi} - \mathbf{K}\boldsymbol{\phi} = \mathbf{0}$, they can be normalized (scaled) by giving a specific value to one element in each $\boldsymbol{\phi}$ and then make the other elements have the same ratio with this element as before. If the first elements in $\boldsymbol{\phi}_1$ and $\boldsymbol{\phi}_2$ are set to 1, then

$$\boldsymbol{\phi}_1 = \begin{bmatrix} 1.000 \\ 1.366025 \end{bmatrix} \quad \text{and} \quad \boldsymbol{\phi}_2 = \begin{bmatrix} 1.000 \\ -0.366025 \end{bmatrix} \tag{5}$$

If m_2 is made very small as compared to m_1, say, $m_2 = \frac{1}{10}m_1$, it would appear reasonable to ignore m_2. The system becomes a single DOF system with the governing differential equation

$$m\ddot{u} + \frac{3}{2}ku = 0 \tag{6}$$

The natural frequency is $\sqrt{1.5k/m}$, whereas the exact first natural frequency with $m_2 = \frac{1}{10}m$ can be calculated as $\omega_1 = \sqrt{1.46k/m}$. Then the error of this approximation for the eigenvalue is

$$\frac{1.5 - 1.46}{1.46} = 2.74\% \tag{7}$$

We can improve this approximation by employing Guyan reduction. From Eq. (10.18), $T = 0.5$, and from Eq. (10.23),

$$\overline{m} = m + 0.1 \times 0.25\, m = 1.025\, m, \quad \overline{k} = 2k - \frac{1}{2}k = 1.5\, k \tag{8}$$

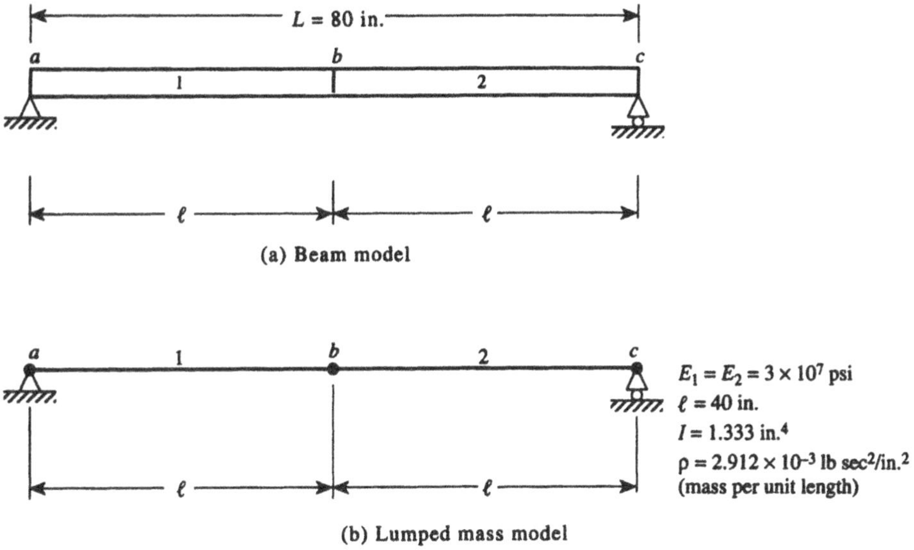

FIGURE 10.6
A uniform, simply supported beam.

Then the governing differential equation is

$$1.025\, m\ddot{u} + 1.5\, ku = 0 \tag{9}$$

and, thus, the natural frequency is $\sqrt{1.4634k/m}$ with an error for the eigenvalue of

$$\frac{1.4634 - 1.46}{1.46} = 0.23\% \tag{10}$$

■

EXAMPLE 10.5 Eigenvalues for a Beam Using a Consistent Mass Matrix Model
Use the displacement method to find the natural frequencies of the beam of Fig. 10.6. Model the beam with two elements of equal length.

Begin by assembling the global stiffness and mass matrices. The element stiffness matrix $\mathbf{k}^i$ is given by Chapter 4, Eq. (4.12), while the consistent mass matrix $\mathbf{m}^i$ is given by Eq. (10.10a).

$$\mathbf{k}^i = \frac{EI}{\ell^3}\begin{bmatrix} 12 & -6\ell & -12 & -6\ell \\ -6\ell & 4\ell^2 & 6\ell & 2\ell^2 \\ -12 & 6\ell & 12 & 6\ell \\ -6\ell & 2\ell^2 & 6\ell & 4\ell^2 \end{bmatrix} = \begin{bmatrix} \mathbf{k}_{aa}^1 & \mathbf{k}_{ab}^1 \\ \mathbf{k}_{ba}^1 & \mathbf{k}_{bb}^1 \end{bmatrix} = \begin{bmatrix} \mathbf{k}_{bb}^2 & \mathbf{k}_{bc}^2 \\ \mathbf{k}_{cb}^2 & \mathbf{k}_{cc}^2 \end{bmatrix} \quad (i = 1, 2) \tag{1}$$

$$\mathbf{m}^i = \frac{\rho\ell}{420}\begin{bmatrix} 156 & -22\ell & 54 & 13\ell \\ -22\ell & 4\ell^2 & -13\ell & -3\ell^2 \\ 54 & -13\ell & 156 & 22\ell \\ 13\ell & -3\ell^2 & 22\ell & 4\ell^2 \end{bmatrix} = \begin{bmatrix} \mathbf{m}_{aa}^1 & \mathbf{m}_{ab}^1 \\ \mathbf{m}_{ba}^1 & \mathbf{m}_{bb}^1 \end{bmatrix} = \begin{bmatrix} \mathbf{m}_{bb}^2 & \mathbf{m}_{bc}^2 \\ \mathbf{m}_{cb}^2 & \mathbf{m}_{cc}^2 \end{bmatrix} \tag{2}$$

The global stiffness matrix is assembled as

$$\mathbf{K} = \begin{bmatrix} \mathbf{k}_{aa}^1 & \mathbf{k}_{ab}^1 & \mathbf{0} \\ \mathbf{k}_{ba}^1 & \mathbf{k}_{bb}^1 + \mathbf{k}_{bb}^2 & \mathbf{k}_{bc}^2 \\ \mathbf{0} & \mathbf{k}_{cb}^2 & \mathbf{k}_{cc}^2 \end{bmatrix} = \frac{EI}{\ell^3} \begin{bmatrix} 12 & -6\ell & -12 & -6\ell & 0 & 0 \\ -6\ell & 4\ell^2 & 6\ell & 2\ell^2 & 0 & 0 \\ -12 & 6\ell & 24 & 0 & -12 & -6\ell \\ -6\ell & 2\ell^2 & 0 & 8\ell^2 & 6\ell & 2\ell^2 \\ 0 & 0 & -12 & 6\ell & 12 & 6\ell \\ 0 & 0 & -6\ell & 2\ell^2 & 6\ell & 4\ell^2 \end{bmatrix} \tag{3}$$

The corresponding global displacement vector is $\mathbf{V} = [w_a\ \theta_a\ w_b\ \theta_b\ w_c\ \theta_c]^T$ or, for harmonic motion, $\mathbf{V} = \boldsymbol{\phi}$. Similarly, the global mass matrix is

$$\mathbf{M} = \begin{bmatrix} \mathbf{m}_{aa}^1 & \mathbf{m}_{ab}^1 & \mathbf{0} \\ \mathbf{m}_{ba}^1 & \mathbf{m}_{bb}^1 + \mathbf{m}_{bb}^2 & \mathbf{m}_{bc}^2 \\ \mathbf{0} & \mathbf{m}_{cb}^2 & \mathbf{m}_{cc}^2 \end{bmatrix} = \frac{\rho\ell}{420} \begin{bmatrix} 156 & -22\ell & 54 & 13\ell & 0 & 0 \\ -22\ell & 4\ell^2 & -13\ell & -3\ell^2 & 0 & 0 \\ 54 & -13\ell & 312 & 0 & 54 & 13\ell \\ 13\ell & -3\ell^2 & 0 & 8\ell^2 & -13\ell & -3\ell^2 \\ 0 & 0 & 54 & -13\ell & 156 & 22\ell \\ 0 & 0 & 13\ell & -3\ell^2 & 22\ell & 4\ell^2 \end{bmatrix} \tag{4}$$

The frequencies can be determined by solving the generalized linear eigenvalue problem of Eq. (10.34). Equation (10.34) is obtained by applying the displacement boundary conditions to $\boldsymbol{\phi}$ and ignoring the rows in $(\mathbf{K} - \lambda\mathbf{M})\boldsymbol{\phi} = 0$ corresponding to the unknown reactions.

Thus, with $w_a = w_c = 0$, $\boldsymbol{\phi} = [w_a\ \theta_a\ w_b\ \theta_b\ w_c\ \theta_c]^T$ reduces to $\boldsymbol{\phi}_y = [\theta_a\ w_b\ \theta_b\ \theta_c]^T$ and the eigenvalue problem becomes

$$\left\{ \frac{EI}{\ell^3} \begin{bmatrix} 4\ell^2 & 6\ell & 2\ell^2 & 0 \\ 6\ell & 24 & 0 & -6\ell \\ 2\ell^2 & 0 & 8\ell^2 & 2\ell^2 \\ 0 & -6\ell & 2\ell^2 & 4\ell^2 \end{bmatrix} - \lambda \begin{bmatrix} 4\ell^2 & -13\ell & -3\ell^2 & 0 \\ -13\ell & 312 & 0 & 13\ell \\ -3\ell^2 & 0 & 8\ell^2 & -3\ell^2 \\ 0 & 13\ell & -3\ell^2 & 4\ell^2 \end{bmatrix} \frac{\rho\ell}{420} \right\} \begin{bmatrix} \theta_a \\ w_b \\ \theta_b \\ \theta_c \end{bmatrix} = \mathbf{0} \tag{5}$$

Use of a standard eigenvalue solution procedure will lead to the desired frequencies.

An alternative technique for finding the frequencies is to establish the characteristic equation from the determinant of the coefficients of $(\mathbf{K} - \lambda\mathbf{M})\boldsymbol{\phi} = \mathbf{K}_{\text{dyn}}\boldsymbol{\phi} = \mathbf{0}$, where $\mathbf{K}_{\text{dyn}}$ is referred to as the *dynamic stiffness matrix*. The boundary conditions must be taken into account in establishing the determinant. Let

$$\mathbf{K} - \omega^2\mathbf{M} = \mathbf{K}_{\text{dyn}} = [D_{ij}] \quad i, j = 1, 2, \ldots, 6, \tag{6}$$

and apply the displacement boundary conditions ($w_a = w_c = 0$). This leads to

$$\begin{bmatrix} D_{11} & D_{12} & D_{13} & D_{14} & D_{15} & D_{16} \\ D_{21} & D_{22} & D_{23} & D_{24} & D_{25} & D_{26} \\ D_{31} & D_{32} & D_{33} & D_{34} & D_{35} & D_{36} \\ D_{41} & D_{42} & D_{43} & D_{44} & D_{45} & D_{46} \\ D_{51} & D_{52} & D_{53} & D_{54} & D_{55} & D_{56} \\ D_{61} & D_{62} & D_{63} & D_{64} & D_{65} & D_{66} \end{bmatrix} \begin{bmatrix} w_a = 0 \\ \theta_a \\ w_b \\ \theta_b \\ w_c = 0 \\ \theta_c \end{bmatrix} = \mathbf{0} \tag{7}$$

The first and fifth equations correspond to unknown reactions. The remaining equations appear as

$$\begin{bmatrix} D_{22} & D_{23} & D_{24} & D_{26} \\ D_{32} & D_{33} & D_{34} & D_{36} \\ D_{42} & D_{43} & D_{44} & D_{46} \\ D_{62} & D_{63} & D_{64} & D_{66} \end{bmatrix} \begin{bmatrix} \theta_a \\ w_b \\ \theta_b \\ \theta_c \end{bmatrix} = \mathbf{0} \tag{8}$$

The characteristic equation is obtained from the determinant of the coefficients of (8), i.e.,

$$\nabla = \begin{bmatrix} D_{22} & D_{23} & D_{24} & D_{26} \\ D_{32} & D_{33} & D_{34} & D_{36} \\ D_{42} & D_{43} & D_{44} & D_{46} \\ D_{62} & D_{63} & D_{64} & D_{66} \end{bmatrix} = 0 \tag{9}$$

This relationship can be obtained directly from (5).

From (6), i.e., $D_{ij} = K_{ij} - \omega^2 M_{ij}$,

$$\begin{aligned}
D_{22} &= 3.9999 \times 10^6 - 1.7752\omega^2 \\
D_{32} &= D_{23} = 1.49996 \times 10^5 + 0.1442\omega^2 \\
D_{33} &= 1.4999 \times 10^4 - 0.0865\omega^2 \\
D_{42} &= D_{24} = 1.9999 \times 10^6 + 1.3314\omega^2 \\
D_{43} &= D_{34} = 0 \\
D_{44} &= 7.9998 \times 10^6 + 3.5505\omega^2 \\
D_{62} &= D_{26} = 0 \\
D_{36} &= D_{63} = -1.49996 \times 10^5 - 0.1442\omega^2 \\
D_{64} &= D_{46} = D_{42} \\
D_{66} &= D_{22}
\end{aligned} \tag{10}$$

Substitution of (10) into (9) and use of factorization leads to two equations:

$$\omega^4 - 41.0093 \times 10^5\omega^2 + 13.3967 \times 10^{10} = 0$$

and

$$\omega^4 - 14.1631 \times 10^6\omega^2 + 8.7036 \times 10^{12} = 0 \tag{11}$$

The roots of these equations are

$$\begin{aligned}
\omega_1 &= 181.47 \text{ rad/sec} & f_1 &= \tfrac{\omega_1}{2\pi} = 28.88 \text{ cycles/sec (Hz)} \\
\omega_2 &= 802.37 \text{ rad/sec} & f_2 &= 127.70 \text{ cycles/sec} \\
& \qquad \text{or} & & \\
\omega_3 &= 2016.93 \text{ rad/sec} & f_3 &= 321.00 \text{ cycles/sec} \\
\omega_4 &= 3676.86 \text{ rad/sec} & f_4 &= 585.19 \text{ cycles/sec}
\end{aligned} \tag{12}$$

These can be compared to the exact natural frequencies (Example 10.8) of $\omega_1 = 180.74$, $\omega_2 = 722.96$, $\omega_3 = 1626.66$, and $\omega_4 = 2891.84$.

Note that the consistent mass matrix leads to higher frequencies than the "exact" values. Also, notice that the error in the approximate frequency grows for the higher modes. In order to make the consistent mass-based frequencies more accurate, more elements should be included in the model. ∎

The above example shows that in an eigenvalue analysis using consistent mass matrices, errors may develop. The errors may fall into two categories: roundoff error and discretization error. We will examine the discretization error here. It can be shown [Hughes, 1987] that when the consistent mass and stiffness matrices are derived from the principle of virtual work, the approximate eigenvalues satisfy

$$\lambda_\ell \leq \lambda_\ell^h \leq \lambda_\ell + ch^{2(k-r)}\lambda_\ell^{(k+1)/(r+1)} \tag{10.39}$$

where λ_ℓ is the ℓth exact eigenvalue, λ_ℓ^h is the approximate value for λ_ℓ from the finite element solution, h is the mesh parameter which is the diameter of the smallest circle that contains the largest element of the mesh, c is a constant independent of h and $\ell, r+1$ is the highest order derivative appearing in the functional of the principle of virtual work of Eq. (10.4), and k is the degree of the complete polynomial appearing in the element shape functions. It can be seen from Eq. (10.39) that $\lambda_\ell^h \to \lambda_\ell$ when $h \to 0$, and the rate of convergence depends on h, k, and r. For a certain h, the accuracy of the approximation of the eigenvalue deteriorates for higher modes. This is because $0 < \lambda_1 \leq \lambda_2 \leq \lambda_3 \ldots$, and when ℓ becomes larger, $ch^{2(k-r)}\lambda_\ell^{(k+1)/(r+1)}$ also becomes larger and λ_ℓ^h has a larger interval in which to vary. Thus, in order to make the higher frequencies more accurate, more elements, i.e., a finer mesh, should be included in the model.

EXAMPLE 10.6 Eigenvalues for a Beam Using a Lumped Mass Model

Use the displacement method to find the natural frequencies of the beam of Fig. 10.6, using the lumped parameter model discussed in Section 10.1.2.

Begin by assembling the global stiffness and mass matrices. For a lumped mass model of a beam element of length ℓ, with the mass concentrated equally at the two ends, without taking into account the mass associated with the rotational DOF,

$$\mathbf{m}^i = \frac{\rho\ell}{2}\begin{bmatrix} 1 & 0 & 0 & 0 \\ 0 & 0 & 0 & 0 \\ 0 & 0 & 1 & 0 \\ 0 & 0 & 0 & 0 \end{bmatrix} \qquad i = 1, 2 \tag{1}$$

where $\mathbf{v} = [w_a\ \theta_a\ w_b\ \theta_b]^T$, and ρ is the mass per unit length. Assemble the system mass matrix by following the procedure outlined in Example 10.5. This leads to the assembled mass matrix

$$\mathbf{M} = \frac{\rho\ell}{2}\begin{bmatrix} 1 & 0 & 0 & 0 & 0 & 0 \\ 0 & 0 & 0 & 0 & 0 & 0 \\ 0 & 0 & 1+1 & 0+0 & 0 & 0 \\ 0 & 0 & 0+0 & 0+0 & 0 & 0 \\ 0 & 0 & 0 & 0 & 1 & 0 \\ 0 & 0 & 0 & 0 & 0 & 0 \end{bmatrix} \tag{2}$$

The global stiffness matrix is still given by Eq. (3) of Example 10.5.

Apply the displacement boundary conditions to $(\mathbf{K} - \lambda\mathbf{M})\boldsymbol{\phi} = \mathbf{0}$, and ignore the rows corresponding to the unknown reactions. This leads to the generalized linear eigenvalue problem

$$\left\{ \frac{EI}{\ell^3} \begin{bmatrix} 4\ell^2 & 6\ell & 2\ell^2 & 0 \\ 6\ell & 24 & 0 & -6\ell \\ 2\ell^2 & 0 & 8\ell^2 & 2\ell^2 \\ 0 & -6\ell & 2\ell^2 & 4\ell^2 \end{bmatrix} - \lambda \begin{bmatrix} 0 & 0 & 0 & 0 \\ 0 & 2 & 0 & 0 \\ 0 & 0 & 0 & 0 \\ 0 & 0 & 0 & 0 \end{bmatrix} \frac{\rho\ell}{2} \right\} \begin{bmatrix} \theta_a \\ w_b \\ \theta_b \\ \theta_c \end{bmatrix} = \mathbf{0} \tag{3}$$

where $\lambda = \omega^2$. The frequencies can be found from this relationship using a standard eigenvalue solution procedure.

We find

$$\omega_1 = 179.408 \text{ rad/sec} \quad \text{or} \quad f_1 = \frac{\omega_1}{2\pi} = 28.55 \text{ Hz} \tag{4}$$

As expected, the fundamental frequency derived using the lumped mass model is lower than the exact value of $\omega_1 = 180.74$ rad/sec. Note that only one frequency is obtained. This is expected, since for this simple model only one mass is able to move, i.e., this is a single DOF system. ∎

EXAMPLE 10.7 Natural Frequencies of a Frame

Compute the natural frequencies of the frame shown in Fig. 10.2.

Divide the frame into three beam elements. This is a crude model for a dynamic analysis, and the resulting frequencies may not be very accurate. The consistent mass model of Example 10.1, with the stiffness matrix of Eq. (11), Chapter 5, Example 5.5, leads to the natural frequencies

$$\begin{aligned} \omega_1 &= 183.8\ [159.5] \text{ rad/sec} & \text{or} \quad f_1 &= 29.3\ [25.4] \text{ Hz} \\ \omega_2 &= 711.6\ [381.2] & \text{or} \quad f_2 &= 113.26\ [60.7] \\ \omega_3 &= 1415.6\ [503.8] & \text{or} \quad f_3 &= 225.30\ [80.2] \\ \omega_4 &= 1587.3\ [1332.9] & \text{or} \quad f_4 &= 252.62\ [212.1] \\ \omega_5 &= 2505.2\ [1963.8] & \text{or} \quad f_5 &= 398.72\ [312.5] \\ \omega_6 &= 3686.0\ [2252.3] & \text{or} \quad f_6 &= 586.65\ [358.5] \end{aligned} \tag{1}$$

Shown in square brackets are the natural frequencies found using the lumped mass model of Example 10.3 with rotary inertia. It is clear that the frequencies calculated using the lumped mass model are lower than those obtained using the consistent mass model, especially for the higher modes. If more elements are included in the model, the consistent and lumped mass results are closer to each other.

If the masses associated with the rotational DOF are condensed out, the consistent [lumped] mass model gives the natural frequencies

$$\begin{aligned} \omega_1 &= 184.4\ [159.8] \text{ rad/sec} & \text{or} \quad f_1 &= 29.4\ [25.4] \text{ Hz} \\ \omega_2 &= 1555.6\ [1279.4] & \text{or} \quad f_2 &= 247.59\ [203.6] \\ \omega_3 &= 2400.0\ [1919.0] & \text{or} \quad f_3 &= 381.98\ [305.4] \\ \omega_4 &= 3244.3\ [2178.4] & \text{or} \quad f_4 &= 516.34\ [346.7] \end{aligned} \tag{2}$$

Apart from the lowest natural frequency, it appears as though condensing out the rotary DOF has a major effect on the natural frequencies. However, it should be noted that the second and third frequencies of (1) do not appear in (2), as these frequencies correspond to rotary DOF which have been condensed out. ∎

10.3.1 Dynamic Stiffness Matrices and Exact Mass Matrices

The stiffness matrix is normally obtained by substituting the shape functions $\mathbf{u} = \mathbf{N}\,\mathbf{v}$ in the internal virtual work relationship (Eq. 10.2)

$$\int_V \delta \mathbf{u}^T{}_u \mathbf{D}^T \mathbf{E}\, \mathbf{D}_u\, \mathbf{u}\, dV$$

giving

$$\mathbf{k}^i = \int_V \mathbf{N}^T{}_u \mathbf{D}^T \mathbf{E}\, \mathbf{D}_u \mathbf{N}\, dV \tag{10.40}$$

The potential accuracy of a structural analysis depends in part on the potential accuracy of the element stiffness matrix $\mathbf{k}^i$, which depends on the shape function $\mathbf{N}$. Normally, the more representative $\mathbf{N}$ is of the exact solution, the better the $\mathbf{k}^i$. For example, if $\mathbf{N}$ is the exact solution to the local (differential) governing equations for an element, the element stiffness matrix $\mathbf{k}^i$ will be exact. For a beam, the usual stiffness matrix [Chapter 4, Eq. (4.12)] is exact only for static responses since it is formed using $\mathbf{N}$ composed of static response polynomials (Eq. 10.8), which are the exact solution of the static differential equations for a beam.

Beams and bars are structural members for which exact solutions to the governing differential equations are readily available. If dynamic effects are included, the beam equations [Chapter 1, Eqs. (1.133)] become

$$\begin{aligned}
\frac{\partial w}{\partial x} &= -\theta + \frac{V}{k_s GA} \\
\frac{\partial \theta}{\partial x} &= \frac{M}{EI} + \frac{\overline{M}_T}{EI} \\
\frac{\partial V}{\partial x} &= kw + \rho \frac{\partial^2 w}{\partial t^2} - \overline{p}_z(x,t) \\
\frac{\partial M}{\partial x} &= V + (k^* - P)\theta + \rho r_y^2 \frac{\partial^2 \theta}{\partial t^2} - \overline{c}(x,t)
\end{aligned} \tag{10.41}$$

where, in addition to the inclusion of inertia, several other terms such as the effects of elastic foundations, applied moments, and thermal loading are displayed. The inertia force per unit length of beam in the x direction is $-\rho\ddot{w} = -\rho(\partial^2 w/\partial t^2)$, while the inertia moment is $-\rho r_y^2(\partial^2\theta/\partial t^2)$, which is derived in Section 10.1.1. These, together with $\overline{p}_z$, $\overline{c}$, and $\overline{M}_T$ act on the beam as the applied loading. In these equations, w, θ, V, and M are the deflection, rotation, shear force, and moment at a cross-section. Also,

k_s = Shear form factor
A = Area of the cross-section
G = Modulus of elasticity in shear
E = Young's modulus of elasticity
I = Moment of inertia taken about the neutral axis
$\overline{M}_T = \int_A E\,\alpha\,\Delta T\, z\, dA$ = Magnitude of distributed thermal moment
ΔT = Temperature change
α = Coefficient of thermal expansion
k = Elastic foundation modulus (force/length^2)
ρ = Mass density (mass per unit length)
$\overline{p}_z$ = Distributed load (force/length)
k^* = Rotary spring constant (force-length/radian)

TABLE 10.1

Generalized Dynamic Stiffness Matrix $\mathbf{k}^i_{\text{dyn}}$ for a Beam Element

$$\begin{bmatrix} V_a \\ M_a \\ V_b \\ M_b \end{bmatrix} = \mathbf{k}^i_{\text{dyn}} \begin{bmatrix} w_a \\ \theta_a \\ w_b \\ \theta_b \end{bmatrix} - \overline{\mathbf{p}}^i$$

$$\mathbf{p} \quad = \mathbf{k}^i_{\text{dyn}} \quad \mathbf{v}^i \quad - \overline{\mathbf{p}}^i$$

The expressions for $\mathbf{k}^i_{\text{dyn}}$ and $\overline{\mathbf{p}}^i$ are given as $\mathbf{k}^i$ and $\overline{\mathbf{p}}^i$ in Chapter 4, Table 4.4. The definitions of λ, η, and ζ are expanded to include dynamic-related properties.

$$\lambda = (k - \rho\omega^2)/EI$$
$$\eta = (k - \rho\omega^2)/(k_s GA)$$
$$\zeta = (N - k^* + \rho r_y^2 \omega^2)/EI$$

where

ω is the natural frequency (rad/sec)
ρ is the mass per unit length (mass/length)
r_y is the radius of gyration (length)

Other definitions are given in Chapter 4, Tables 4.3 and 4.4.

P = Axial force

r_y = Radius of gyration

$\overline{c}$ = Magnitude of the distributed applied moment (force-length/length)

These relations, which include the effects of shear deformation and rotary inertia, as well as bending, are the *Timoshenko beam equations.* The governing equations are reduced to those for a *Rayleigh*[2] *beam* (bending, rotary inertia) by setting $1/(k_s GA)$ equal to zero. For a *shear beam* (bending, shear deformation), set

$$\rho r_y^2 \frac{\partial^2 \theta}{\partial t^2} = 0$$

For free vibrations, set the applied loadings $\overline{M}_T$, $\overline{p}_z$, and $\overline{c}$ equal to zero in Eq. (10.41), and let the motion be harmonic, e.g., let $w(x, t) = w(x) \sin \omega t$. The resulting equations can be solved exactly and placed in stiffness matrix format. In Chapter 4, this was accomplished by solving the equations in transfer matrix form and then converting the results to a stiffness matrix. This procedure led to the stiffness matrix of Table 10.1. Since the stiffness matrix of Table 10.1 includes the effects of inertia, it is an element dynamic stiffness matrix, $\mathbf{k}^i_{\text{dyn}}$.

The exact stiffness matrix of Table 10.1 can also be obtained by using the exact solution of the free vibration form of Eq. (10.41) to compose $\mathbf{u} = \mathbf{N}\,\mathbf{v}$. Substitution of exact shape functions $\mathbf{N}$ in Eq. (10.40) leads to this exact stiffness matrix.

The element dynamic stiffness matrix can be assembled in the same fashion as the usual stiffness matrix, giving the global dynamic stiffness matrix $\mathbf{K}_{\text{dyn}}$. The characteristic equation, from which the natural frequencies can be computed, would take the form

$$\det \mathbf{K}_{\text{dyn}}(\omega) = 0 \tag{10.42}$$

where the boundary conditions have been employed and resulted in a reduced $\mathbf{K}_{\text{dyn}}$. A numerical determinant search can be utilized to find the frequencies.

[2]Lord Rayleigh or John William Strutt (1842–1919) was an English physicist and chemist and a Nobel Laureate (1904) for his discovery of argon. He served as a professor at Cambridge University and the Royal Institution in London. His contributions were in acoustics, hydrodynamics, and optics.

Fergusson and Pilkey (1992 and 1993) review the literature on the use of dynamic stiffness matrices.

EXAMPLE 10.8 Eigenvalues for a Beam Using a Dynamic Stiffness Matrix

Use the dynamic stiffness matrix to compute the frequencies of the beam of Examples 10.5 and 10.6. Since an exact stiffness matrix is to be employed, this can be treated as a single element beam. However, we choose to model the beam as two elements as shown in Fig. 10.6.

The stiffness relationship for the ith element ($i = 1$ or 2) is

$$\mathbf{k}^i_{\text{dyn}}\mathbf{v}^i = \mathbf{p}^i \tag{1}$$

where the dynamic stiffness matrix $\mathbf{k}^i_{\text{dyn}}$ is taken from Table 10.1 with

$$\lambda = (k_z - \rho\omega^2)/(EI) = -\rho\omega^2/(EI) < 0 \tag{2}$$

$\eta = (k_z - \rho\omega^2)/(GAk_s) = 0$ if shear deformation is not included ($1/GAk_s = 0$), $\zeta = (P - k^* + \rho r_y^2\omega^2)/(EI) = 0$ since $P = 0, k^* = 0$ and rotary effects are to be ignored.

To complete the stiffness matrix of Table 10.1, it is necessary to refer to functions provided in Chapter 4, Table 4.3. For $\lambda < 0$, Case 1 of Table 4.3 provides

$$\begin{aligned} e_1 &= (a^2A + b^2B)/g \qquad & e_2 &= (aC + bD)/g \\ e_3 &= (A - B)/g & e_4 &= (C/a - D/b)/g \end{aligned} \tag{3}$$

in which

$$\begin{aligned} a^2 &= \sqrt{(\zeta+\eta)^2/4 - \lambda} - (\zeta - \eta)/2 = \sqrt{\frac{\rho}{EI}}\omega \\ b^2 &= \sqrt{(\zeta+\eta)^2/4 - \lambda} + (\zeta - \eta)/2 = \sqrt{\frac{\rho}{EI}}\omega = a^2 \\ g &= a^2 + b^2 = 2\sqrt{\frac{\rho}{EI}}\omega \\ A &= \cosh(a\ell) \qquad B = \cos(b\ell) = \cos(a\ell) \\ C &= \sinh(a\ell) \qquad D = \sin(b\ell) = \sin(a\ell) \end{aligned} \tag{4}$$

Insertion of (4) into (3) gives

$$\begin{aligned} e_1 &= \frac{1}{2}[\cosh(a\ell) + \cos(a\ell)] \qquad e_3 = \frac{1}{2\omega}\sqrt{\frac{EI}{\rho}}[\cosh(a\ell) - \cos(a\ell)] \\ e_2 &= \frac{1}{2\sqrt{\omega}}\left(\frac{EI}{\rho}\right)^{1/4}[\sinh(a\ell) + \sin(a\ell)] \\ e_4 &= \frac{1}{2\omega\sqrt{\omega}}\left(\frac{EI}{\rho}\right)^{3/4}[\sinh(a\ell) - \sin(a\ell)] \end{aligned} \tag{5}$$

and

$$\nabla = e_3^2 - (e_2 - 0)(e_4 - 0) = e_3^2 - e_2e_4 = \frac{1}{2\omega^2}\left(\frac{EI}{\rho}\right)[1 - \cosh(a\ell)\cos(a\ell)]$$

Substitution of (5) into Table 10.1 provides

$$\begin{aligned}
k_{11}^i &= (EI/\nabla)(e_1e_2 + \lambda e_3e_4)\\
k_{12}^i &= (EI/\nabla)(e_1e_3 - e_2^2)\\
k_{13}^i &= -(EI/\nabla)e_2\\
k_{14}^i &= -(EI/\nabla)e_3\\
k_{21}^i &= k_{12}^i\\
k_{22}^i &= (EI/\nabla)(e_3e_2 - e_1e_4)\\
k_{23}^i &= -k_{14}^i\\
k_{24}^i &= (EI/\nabla)e_4\\
k_{31}^i &= k_{13}^i, \quad k_{32}^i = k_{23}^i\\
k_{33}^i &= k_{11}^i\\
k_{34}^i &= (EI/\nabla)(e_1e_3 + \lambda e_4^2)\\
k_{41}^i &= k_{14}^i, \quad k_{42}^i = k_{24}^i, \quad k_{43}^i = k_{34}^i\\
k_{44}^i &= k_{22}^i
\end{aligned} \tag{6}$$

where the subscript dyn in Table 10.1 has been dropped.

The assembled dynamic stiffness matrix appears as

$$\mathbf{K}_{\text{dyn}} = \begin{bmatrix} k_{11}^1 & k_{12}^1 & k_{13}^1 & k_{14}^1 & 0 & 0\\ k_{21}^1 & k_{22}^1 & k_{23}^1 & k_{24}^1 & 0 & 0\\ k_{31}^1 & k_{32}^1 & k_{33}^1 + k_{11}^2 & k_{34}^1 + k_{12}^2 & k_{13}^2 & k_{14}^2\\ k_{41}^1 & k_{42}^1 & k_{43}^1 + k_{21}^2 & k_{44}^1 + k_{22}^2 & k_{23}^2 & k_{24}^2\\ 0 & 0 & k_{31}^2 & k_{32}^2 & k_{33}^2 & k_{34}^2\\ 0 & 0 & k_{41}^2 & k_{42}^2 & k_{43}^2 & k_{44}^2 \end{bmatrix} \tag{7}$$

and the corresponding global displacement vector is

$$\mathbf{V} = [w_a \;\; \theta_a \;\; w_b \;\; \theta_b \;\; w_c \;\; \theta_c]^T \tag{8}$$

Introduce the displacement boundary conditions $w_a = w_c = 0$, and delete the rows for the reactions corresponding to these prescribed displacements. The characteristic equation (Eq. 10.42) then becomes

$$\begin{bmatrix} k_{22}^1 & k_{23}^1 & k_{24}^1 & 0\\ k_{32}^1 & k_{33}^1 + k_{11}^2 & k_{34}^1 + k_{12}^2 & k_{14}^2\\ k_{42}^1 & k_{43}^1 + k_{21}^2 & k_{44}^1 + k_{22}^2 & k_{24}^2\\ 0 & k_{41}^2 & k_{42}^2 & k_{44}^2 \end{bmatrix} = 0 \tag{9}$$

A frequency search applied to this relationship leads to the natural frequencies:

$$\begin{aligned}
\omega_1 &= 180.74 \text{ rad/sec}\\
\omega_2 &= 722.96 \text{ rad/sec}\\
\omega_3 &= 1626.66 \text{ rad/sec}\\
\omega_4 &= 2891.84 \text{ rad/sec}
\end{aligned}$$

This dynamic stiffness matrix has yielded the exact natural frequencies. This is because no approximate, e.g., consistent or lumped mass, modeling was employed. Use of exact mass modeling permits a coarser mesh, i.e., fewer elements, to be employed in the model. ∎

A determinant search, as would be required by Eq. (10.42) or Eq. (9) in Example 10.8, is often a numerically cumbersome, inefficient process that perhaps should be avoided. The lack of efficiency is due to the need to compute the value of the determinant for each trial value of the frequency. An in-depth review of the problems associated with the use of the dynamic stiffness matrix is given in Kim (1993). The determinant search can be avoided by establishing an eigenvalue problem in which the structural matrices are not functions of the frequency parameter ω. This was the case for the eigenvalue problem utilizing the static stiffness matrix and the consistent mass matrix, or the lumped mass matrix which was presented earlier in this section.

Further study of the accuracy of structural matrices along with the corresponding frequencies is merited. Begin with a frequency dependent mass matrix obtained in a fashion similar to the formation of the dynamic stiffness matrix $\mathbf{k}^i_{\text{dyn}}$. Place the exact (frequency dependent) shape function $\mathbf{N}$ in (Eq. 10.5).

$$\mathbf{m}^i = \int_V \gamma \mathbf{N}^T \mathbf{N}\, dV \tag{10.43}$$

or, in the case of a beam element,

$$\mathbf{m}^i = \int_a^b \rho \mathbf{N}^T \mathbf{N}\, dx \tag{10.44}$$

The same exact mass matrix is obtained by differentiating the element dynamic stiffness matrix by the frequency parameter ω [Fergusson and Pilkey, 1992 and Richards and Leung, 1977]

$$\mathbf{m}^i = -\frac{\partial \mathbf{k}^i_{\text{dyn}}}{\partial \omega^2} \tag{10.45}$$

It is possible to economize somewhat in the calculation of a frequency-dependent mass matrix by constructing a "quasi-static" mass matrix $\widetilde{\mathbf{m}}^i$ defined as

$$\widetilde{\mathbf{m}}^i = \int_V \mathbf{N}_0^T \gamma \mathbf{N}\, dV \tag{10.46}$$

where $\mathbf{N}_0$ is an element static shape function such as the $\mathbf{N}$ given in Eq. (10.8). In this notation with the static shape function denoted by $\mathbf{N}_0$, the consistent mass matrix of Section 10.1.1 is given by

$$\int_V \mathbf{N}_0^T \gamma \mathbf{N}_0\, dV \tag{10.47}$$

Define a frequency-dependent stiffness matrix $\mathbf{k}^i(\omega)$ in terms of the dynamic stiffness matrix $\mathbf{k}^i_{\text{dyn}}(\omega)$ and a frequency-dependent mass matrix $\mathbf{m}^i(\omega)$ as

$$\mathbf{k}^i_{\text{dyn}}(\omega) = \mathbf{k}^i(\omega) - \omega^2 \mathbf{m}^i(\omega) \tag{10.48}$$

Symbolic manipulation software is frequently helpful in implementing the operations required in forming the frequency-dependent structural matrices. Assemble the global matrices $\mathbf{M}(\omega)$ and $\mathbf{K}(\omega)$ using the element matrices $\mathbf{m}^i(\omega)$ and $\mathbf{k}^i(\omega)$. The eigenvalue problem then is embodied in

$$[\mathbf{K}(\omega) - \omega^2 \mathbf{M}(\omega)]\boldsymbol{\phi} = \mathbf{0} \tag{10.49}$$

A simple iterative solution of this problem appears to be effective in converging rapidly to a precise eigenvalue solution.

$$
\begin{aligned}
[\mathbf{K}(0) - \omega^2\mathbf{M}(0)]\boldsymbol{\phi} &= 0 \Rightarrow \omega = \omega_1^0, \omega_2^0, \omega_3^0, \ldots \boldsymbol{\phi} = \boldsymbol{\phi}^0 \\
\left[\mathbf{K}\left(\omega_1^0\right) - \omega^2\mathbf{M}\left(\omega_1^0\right)\right]\boldsymbol{\phi} &= 0 \Rightarrow \omega = \omega_1^1, \omega_2^1, \omega_3^1, \ldots \boldsymbol{\phi} = \boldsymbol{\phi}^1 \\
&\cdots \\
\left[\mathbf{K}\left(\omega_1^{j-1}\right) - \omega^2\mathbf{M}\left(\omega_1^{j-1}\right)\right]\boldsymbol{\phi} &= 0 \Rightarrow \omega = \omega_1^j, \omega_2^j, \omega_3^j, \ldots \boldsymbol{\phi} = \boldsymbol{\phi}^j
\end{aligned}
\tag{10.50}
$$

where the superscript j indicates the eigensolution for the jth iteration. The frequencies $\omega_1^0, \omega_2^0, \omega_3^0, \ldots$ are the same as would be found using a consistent mass matrix and the usual static stiffness matrix.

This approach, although significantly less efficient than solving the problem of Eq. (10.34), is more efficient than the determinant search required to solve Eq. (10.42). This iterative technique of Eq. (10.50) has the advantage of being able to utilize highly reliable, standard eigenvalue solvers that should result in an accurate set of frequencies and mode shapes. Moreover, the frequency-dependent mass and stiffness matrices permit a model to be employed with fewer (larger) elements than is possible with consistent or lumped mass matrices. Sometimes, more accurate higher eigenvalues are obtained from Eq. (10.50) if $\mathbf{K}$ and $\mathbf{M}$ are evaluated at ω_n^{j-1}, where $n > 1$, rather than at the lowest natural frequency ω_1^{j-1}.

The iterative scheme of Eq. (10.50) can be replaced by a higher order, e.g., quadratic, generalized eigenvalue problem, using matrices expanded in series [Fergusson and Pilkey, 1992]. Expand the mass and stiffness matrices in Taylor series to obtain

$$
\mathbf{m}^i = \sum_{n=0}^{\infty} \mathbf{m}_n\,\omega^{2n} \quad \text{or} \quad \tilde{\mathbf{m}}^i = \sum_{n=0}^{\infty} \tilde{\mathbf{m}}_n\,\omega^{2n} \quad \mathbf{k}^i = \sum_{n=0}^{\infty} \mathbf{k}_n\,\omega^{2n}
\tag{10.51}
$$

The dynamic stiffness matrix would be expanded similarly, giving

$$
\mathbf{k}^i_{\text{dyn}} = \sum_{n=0}^{\infty} (\mathbf{k}_{\text{dyn}})_n\,\omega^{2n}
\tag{10.52}
$$

It follows from Eq. (10.48) that the dynamic stiffness matrix expansion terms would be defined as

$$
(\mathbf{k}_{\text{dyn}})_n = \mathbf{k}_n - \mathbf{m}_{n-1}, \qquad n \geq 1
\tag{10.53}
$$

A simple relationship between the terms $\mathbf{m}_0, \mathbf{m}_1, \ldots$ and $\mathbf{k}_1, \mathbf{k}_2, \ldots$, etc., is useful in that only a few terms in these frequency expansions need be determined. From Fergusson and Pilkey (1992),

$$
(n+1)\mathbf{k}_{n+1} = n\mathbf{m}_n = n(n+1)\tilde{\mathbf{m}}_n = -n(n+1)(\mathbf{k}_{\text{dyn}})_{n+1}, \qquad n \geq 1
\tag{10.54}
$$

Also,

$$
\begin{aligned}
(\mathbf{k}_{\text{dyn}})_0 &= \mathbf{k}_0 \quad \text{(the traditional stiffness matrix)} \\
(\mathbf{k}_{\text{dyn}})_1 &= -\mathbf{m}_0 \quad \text{(the traditional consistent mass matrix)} \\
\mathbf{k}_1 &= 0
\end{aligned}
\tag{10.55}
$$

There is a sizeable literature on the solution of higher order eigenvalue problems.

10.3.2 Orthogonality of Mode Shapes

A very useful property of mode shapes is *orthogonality*. This is an orthogonality with respect to the mass matrix **M** and also with respect to the stiffness matrix **K**. To derive the orthogonality relations, consider two distinct solutions ω_r^2, $\boldsymbol{\phi}_r$ and ω_s^2, $\boldsymbol{\phi}_s$ of the eigenvalue problem of Eq. (10.33). These solutions satisfy

$$\mathbf{K}\boldsymbol{\phi}_r = \omega_r^2\mathbf{M}\boldsymbol{\phi}_r \tag{10.56a}$$

$$\mathbf{K}\boldsymbol{\phi}_s = \omega_s^2\mathbf{M}\boldsymbol{\phi}_s \tag{10.56b}$$

Premultiply both sides of Eq. (10.56a) by $\boldsymbol{\phi}_s^T$ and both sides of Eq. (10.56b) by $\boldsymbol{\phi}_r^T$. This gives

$$\boldsymbol{\phi}_s^T\mathbf{K}\boldsymbol{\phi}_r = \omega_r^2\boldsymbol{\phi}_s^T\mathbf{M}\boldsymbol{\phi}_r \tag{10.57a}$$

$$\boldsymbol{\phi}_r^T\mathbf{K}\boldsymbol{\phi}_s = \omega_s^2\boldsymbol{\phi}_r^T\mathbf{M}\boldsymbol{\phi}_s \tag{10.57b}$$

Next transpose Eq. (10.57b), use the property that matrices **M** and **K** are symmetric, and subtract from Eq. (10.57a). This gives

$$\left(\omega_r^2 - \omega_s^2\right)\boldsymbol{\phi}_s^T\mathbf{M}\boldsymbol{\phi}_r = 0 \tag{10.58}$$

Because, in general, the natural frequencies are distinct, $\omega_r \neq \omega_s$, Eq. (10.58) implies that

$$\boldsymbol{\phi}_s^T\mathbf{M}\boldsymbol{\phi}_r = 0 \quad r \neq s \tag{10.59}$$

which is the orthogonality condition of the mode shapes. Note that the orthogonality is with respect to the mass matrix **M**, which assumes the role of a weighting matrix. Insert Eq. (10.59) into Eq. (10.57a) and find

$$\boldsymbol{\phi}_s^T\mathbf{K}\boldsymbol{\phi}_r = 0 \tag{10.60}$$

If $r = s$, set

$$\boldsymbol{\phi}_s^T\mathbf{M}\boldsymbol{\phi}_s = M_s \tag{10.61a}$$

From Eq. (10.57)

$$\boldsymbol{\phi}_s^T\mathbf{K}\boldsymbol{\phi}_s = K_s = \omega_s^2 M_s \tag{10.61b}$$

where the quantities M_s and K_s are scalars, with M_s designated as the *generalized mass*. Division of $\boldsymbol{\phi}_s$ by $(M_s)^{1/2}$ leads to the *scaled mode shape* $\boldsymbol{\phi}_s$, and

$$\boldsymbol{\phi}_s^T\mathbf{M}\boldsymbol{\phi}_r = \delta_{sr}, \quad \boldsymbol{\phi}_s^T\mathbf{K}\boldsymbol{\phi}_r = \delta_{sr}\omega_s^2 \tag{10.62}$$

where δ_{sr} is the Kronecker delta,

$$\delta_{sr} = \begin{cases} 1 & s = r \\ 0 & s \neq r \end{cases} \tag{10.63}$$

The mode shapes are often assembled as

$$\mathbf{\Phi} = [\boldsymbol{\phi}_1 \quad \boldsymbol{\phi}_2 \quad \cdots \quad \boldsymbol{\phi}_{n_d}] \tag{10.64}$$

where n_d is the number of DOF. Then Eq. (10.61) can be expressed as

$$\mathbf{\Phi}^T\mathbf{M}\mathbf{\Phi} = \mathbf{M}_{n_d}, \quad \mathbf{\Phi}^T\mathbf{K}\mathbf{\Phi} = \mathbf{K}_{n_d} \tag{10.65a}$$

where $\mathbf{M}_{n_d}$ is a diagonal (generalized mass) matrix with non-zero elements $M_1, M_2, \ldots, M_{n_d}$

$$\mathbf{M}_{n_d} = \begin{bmatrix} M_1 & & & \mathbf{0} \\ & M_2 & & \\ & & \ddots & \\ \mathbf{0} & & & M_{n_d} \end{bmatrix} \tag{10.65b}$$

and

$$\mathbf{K}_{n_d} = \begin{bmatrix} K_1 & & & \mathbf{0} \\ & K_2 & & \\ & & \ddots & \\ \mathbf{0} & & & K_{n_d} \end{bmatrix} = \begin{bmatrix} \omega_1^2 M_1 & & & \mathbf{0} \\ & \omega_2^2 M_2 & & \\ & & \ddots & \\ \mathbf{0} & & & \omega_{n_d}^2 M_{n_d} \end{bmatrix} \tag{10.65c}$$

If the scaled mode shapes are used,

$$\mathbf{\Phi}^T\mathbf{M}\mathbf{\Phi} = \mathbf{I}, \quad \mathbf{\Phi}^T\mathbf{K}\mathbf{\Phi} = \mathbf{\Lambda} \tag{10.66a}$$

where $\mathbf{I}$ is a unit matrix with each diagonal element equal to one and with all other elements equal to zero. Also,

$$\mathbf{\Lambda} = \begin{bmatrix} \omega_1^2 & & & \\ & \omega_2^2 & & \\ & & \ddots & \\ & & & \omega_{n_d}^2 \end{bmatrix} \tag{10.66b}$$

The quantity $\mathbf{\Phi}^{-1}$ is required frequently in a vibration analysis. The orthogonality property of the mode shapes is useful in computing $\mathbf{\Phi}^{-1}$ without performing matrix inversion. To show this, premultiply Eq. (10.65a) by $\mathbf{M}_{n_d}^{-1}$

$$\mathbf{M}_{n_d}^{-1}\mathbf{\Phi}^T\mathbf{M}\mathbf{\Phi} = \mathbf{I}$$

Postmultiply both sides of this relationship by $\mathbf{\Phi}^{-1}$, and since $\mathbf{\Phi}\mathbf{\Phi}^{-1} = \mathbf{I}$, it follows that

$$\mathbf{\Phi}^{-1} = \mathbf{M}_{n_d}^{-1}\mathbf{\Phi}^T\mathbf{M} \tag{10.67a}$$

Equation (10.67a) provides a useful formula for the computation of $\mathbf{\Phi}^{-1}$. In the case of scaled mode shapes, from Eq. (10.66a),

$$\mathbf{\Phi}^{-1} = \mathbf{\Phi}^T\mathbf{M} \tag{10.67b}$$

10.4 Forced Response

The response of a structural system to prescribed time-dependent loading $\overline{\mathbf{P}}(t)$ involves the solution of

$$\mathbf{M}\ddot{\mathbf{V}} + \mathbf{K}\mathbf{V} = \overline{\mathbf{P}} \tag{10.68}$$

This is an ordinary differential equation in time that can be integrated directly. Textbooks on vibrations or structural dynamics describe a variety of time integration techniques for solving Eq. (10.68), some of which are presented in Section 10.5. For linear problems, however, the most frequently used technique in practice is the modal superposition method that employs the free vibration responses. Natural frequencies and mode shapes are often calculated at the start of a dynamic analysis of a linear structural system, so it is a relatively simple procedure to compute the transient response using modal superposition. This method will be described briefly here.

10.4.1 Modal Superposition Method

The equations of motion for the undamped systems studied thus far exhibit *coupling* for various DOF. In some cases, the coupling is through the stiffness terms, e.g., Eq. (3) of Example 10.6, while in other cases, the coupling is in the inertia terms, e.g., Eq. (5) of Example 10.5 (where neither the stiffness matrix **K** nor the mass matrix **M** is diagonal). The former is usually called *static* or *stiffness coupling* and the latter is called *dynamic coupling*. By properly choosing coordinates (DOF), a system can be rendered uncoupled. The process of expressing the equations of motion in terms of different coordinates is called a *coordinate transformation*. The set of coordinates for which the equations of motion are completely uncoupled is called the *principal, normal, modal,* or *natural coordinates*. The principal coordinates are useful in simplifying response calculations.

It is possible to find principal coordinates for any linear system. Let **V** be the coordinates in which the equations of motion are coupled, and $\mathbf{q} = [q_1 \;\; q_2 \;\; \cdots \;\; q_{n_d}]^T$ are the principal coordinates, where n_d is the number of DOF. Also, let

$$\mathbf{V} = \mathbf{R}\mathbf{q} \tag{10.69}$$

Choose

$$\mathbf{R} = [\boldsymbol{\phi}_1 \quad \boldsymbol{\phi}_2 \quad \cdots \quad \boldsymbol{\phi}_{n_d}] = \boldsymbol{\Phi} \tag{10.70}$$

where $\boldsymbol{\phi}_i$, the mode shapes, are solutions of

$$\omega_i^2 \mathbf{M}\boldsymbol{\phi}_i = \mathbf{K}\boldsymbol{\phi}_i$$

Substitute Eq. (10.69) into Eq. (10.68), and premultiply by $\mathbf{R}^T$ to obtain

$$\mathbf{R}^T\mathbf{M}\mathbf{R}\ddot{\mathbf{q}} + \mathbf{R}^T\mathbf{K}\mathbf{R}\mathbf{q} = \mathbf{R}^T\overline{\mathbf{P}} \tag{10.71}$$

From the orthogonality condition for mode shapes [Eq. (10.65a)],

$$\mathbf{R}^T\mathbf{M}\mathbf{R} = \boldsymbol{\Phi}^T\mathbf{M}\boldsymbol{\Phi} = \mathbf{M}_{n_d} \quad \text{and} \quad \mathbf{R}^T\mathbf{K}\mathbf{R} = \boldsymbol{\Phi}^T\mathbf{K}\boldsymbol{\Phi} = \mathbf{K}_{n_d}$$

in which $\mathbf{M}_{n_d}$ and $\mathbf{K}_{n_d}$ are the diagonal matrices of Eqs. (10.65b) and (10.65c). It is thus concluded that use of Eq. (10.69) with **R** defined by Eq. (10.70), uncouples Eq. (10.68), and the uncoupled differential equations appear as

$$\mathbf{M}_{n_d}\ddot{\mathbf{q}} + \mathbf{K}_{n_d}\mathbf{q} = \mathbf{P} \quad \text{or} \quad M_i\ddot{q}_i + K_i q_i = P_i \quad i = 1, 2, \ldots, n_d \tag{10.72}$$

in which $\mathbf{P} = \mathbf{\Phi}^T\overline{\mathbf{P}}$, $M_i = \boldsymbol{\phi}_i^T\mathbf{M}\boldsymbol{\phi}_i$, $K_i = \boldsymbol{\phi}_i^T\mathbf{K}\boldsymbol{\phi}_i$, and $P_i = \boldsymbol{\phi}_i^T\overline{\mathbf{P}}$.

It is apparent that the coupled governing equations for our n_d DOF system have been replaced by n_d equations [Eq. (10.72)], each of which has the same form as the governing equations for a single-DOF system. From the theory of ordinary differential equations, we know that the solution to these equations is

$$\mathbf{q} = \mathbf{A}\mathbf{q}(0) + \mathbf{B}\dot{\mathbf{q}}(0) + \mathbf{F} \tag{10.73a}$$

where

$$\mathbf{q}(0) = \begin{bmatrix} q_1(0) \\ q_2(0) \\ \vdots \\ q_{n_d}(0) \end{bmatrix} = \mathbf{R}^{-1} \begin{bmatrix} V_{10} \\ V_{20} \\ \vdots \\ V_{n_d0} \end{bmatrix} = \mathbf{\Phi}^{-1}\mathbf{V}(0) = \mathbf{\Phi}^{-1}\overline{\mathbf{V}}_0 \tag{10.73b}$$

$$\dot{\mathbf{q}}(0) = \begin{bmatrix} \dot{q}_1(0) \\ \dot{q}_2(0) \\ \vdots \\ \dot{q}_{n_d}(0) \end{bmatrix} = \mathbf{R}^{-1} \begin{bmatrix} \dot{V}_{10} \\ \dot{V}_{20} \\ \vdots \\ \dot{V}_{n_d0} \end{bmatrix} = \mathbf{\Phi}^{-1}\dot{\mathbf{V}}(0) = \mathbf{\Phi}^{-1}\dot{\overline{\mathbf{V}}}_0 \tag{10.73c}$$

$$\mathbf{A} = \begin{bmatrix} \cos\omega_1 t & & & \mathbf{0} \\ & \cos\omega_2 t & & \\ & & \ddots & \\ \mathbf{0} & & & \cos\omega_{n_d} t \end{bmatrix} \qquad \mathbf{B} = \begin{bmatrix} \frac{\sin\omega_1 t}{\omega_1} & & & \mathbf{0} \\ & \frac{\sin\omega_2 t}{\omega_2} & & \\ & & \ddots & \\ \mathbf{0} & & & \frac{\sin\omega_{n_d} t}{\omega_{n_d}} \end{bmatrix}$$

and

$$\mathbf{F} = \begin{bmatrix} F_1 & & & \mathbf{0} \\ & F_2 & & \\ & & \ddots & \\ \mathbf{0} & & & F_{n_d} \end{bmatrix} \quad \text{with} \quad F_i = \frac{1}{M_i\,\omega_i}\int_0^t P_i(\tau)\sin\omega_i(t-\tau)\,d\tau$$

and $V_{10}, V_{20}, \ldots, V_{n_d0}$, and $\dot{V}_{10}, \dot{V}_{20}, \ldots, \dot{V}_{n_d0}$ are the prescribed initial conditions (displacements and velocities) of the system which are represented by the vectors $\mathbf{V}(0) = \overline{\mathbf{V}}_0$ and $\dot{\mathbf{V}}(0) = \dot{\overline{\mathbf{V}}}_0$.

In scalar form, the solution to Eq. (10.72) is

$$q_i = q_i(0)\cos\omega_i t + \dot{q}_i(0)\frac{\sin\omega_i t}{\omega_i} + \frac{1}{M_i\,\omega_i}\int_0^t P_i(\tau)\sin\omega_i(t-\tau)\,d\tau \quad i = 1, 2, \ldots, n_d \tag{10.74a}$$

$$\begin{aligned} q_i(0) &= \boldsymbol{\phi}_i^T\frac{\mathbf{M}\overline{\mathbf{V}}_0}{M_i} = \left(\sum_j^{n_d}\sum_k^{n_d} m_{jk}\,\phi_{ij}\,V_{k0}\right)\Big/ M_i \\ \dot{q}_i(0) &= \boldsymbol{\phi}_i^T\frac{\mathbf{M}\dot{\overline{\mathbf{V}}}_0}{M_i} = \left(\sum_j^{n_d}\sum_k^{n_d} m_{jk}\,\phi_{ij}\,\dot{V}_{k0}\right)\Big/ M_i \end{aligned} \tag{10.74b}$$

in which ϕ_{ij} is the jth element in $\boldsymbol{\phi}_i$ and m_{jk} is the element in the jth row and the kth column of $\mathbf{M}$. Substitution of Eq. (10.73) or (10.74) into Eq. (10.69) provides the complete physical response $\mathbf{V}$ of an arbitrarily loaded n_d DOF system.

If the scaled mode shapes are used to form $\mathbf{R}$, then the coordinate transformation of Eq. (10.69) results in

$$\ddot{\mathbf{q}} + \boldsymbol{\Lambda}\mathbf{q} = \mathbf{P} \quad \text{or} \quad \ddot{q}_i + \omega_i^2 q_i = P_i \tag{10.75}$$

where $\boldsymbol{\Lambda}$ is the diagonal matrix of Eq. (10.66b). That is, the diagonal elements are the eigenvalues of the system. Also, $\mathbf{P} = \mathbf{R}^T\overline{\mathbf{P}}$ and $P_i = \boldsymbol{\phi}_i^T\overline{\mathbf{P}}$. The solution for each uncoupled equation is

$$q_i = q_i(0)\cos\omega_i t + \dot{q}_i(0)\frac{\sin\omega_i t}{\omega_i} + \frac{1}{\omega_i}\int_0^t P_i(\tau)\sin\omega_i(t-\tau)\,d\tau \tag{10.76}$$

where the initial conditions are obtained in the same way as in Eq. (10.73) or (10.74), with $\boldsymbol{\Phi}^{-1}$ as the inverse of the matrix containing the scaled mode shapes.

Recall that the modal superposition solution is formed from mode shapes that have unknown amplitudes. However, in spite of this indeterminate characteristic of the mode shapes, the general forced response for a dynamic system is fully determined. This can be shown by multiplying the mode shapes $\boldsymbol{\phi}_i$ by a constant and then noting from the formula for q_i of Eq. (10.73) that this constant cancels out in Eq. (10.69).

It has been demonstrated here that the modal superposition method can be used to compute the vibration responses of systems subjected to arbitrary loading. It can also be used to find the steady-state sinusoidal response, in which case the effect of damping is very important, especially if the exciting frequency is one of the natural frequencies of the system. In general, the principal coordinates for $\mathbf{M}$ and $\mathbf{K}$ are not able to decouple the damping matrix $\mathbf{C}$ in a viscous damping model of the form

$$\mathbf{M}\ddot{\mathbf{V}} + \mathbf{C}\dot{\mathbf{V}} + \mathbf{K}\mathbf{V} = \overline{\mathbf{P}} \tag{10.77}$$

because $\boldsymbol{\Phi}^T\mathbf{C}\boldsymbol{\Phi}$ is not a diagonal matrix unless $\mathbf{C}$ is in the form of *proportional damping*, i.e., $\mathbf{C} = \alpha\mathbf{M} + \beta\mathbf{K}$, where α and β are constants. Hence, the equations of motion for a damped system usually cannot be decoupled using the normal modes of the corresponding undamped system. One way of incorporating damping into the modal superposition method is the use of *modal damping*. Since the governing equation of motion in principal coordinates ($M_i\ddot{q}_i + K_i q_i = P_i$) is similar to the equations of motion for a single-DOF spring mass system, it would seem reasonable that damping can be assigned to each mode leading to equations of the form of a single DOF system with a dashpot. That is, assign damping C_i to the ith mode to form the equations

$$M_i\ddot{q}_i + C_i\dot{q}_i + K_i q_i = P_i \tag{10.78}$$

which replaces Eq. (10.72). It is more common to assign a "damping ratio" ζ_i to each mode with the resulting modal equations appearing as

$$\ddot{q}_i + 2\zeta_i\omega_i\dot{q}_i + \omega_i^2 q_i = P_i/M_i \tag{10.79}$$

where $\zeta_i = C_i/(2M_i\omega_i) = C_i/C_{cr}$ with C_{cr}, the *critical damping*, defined as $C_{cr} = 2\sqrt{M_i K_i}$. Modal damping is one form of proportional damping. The assignment of damping should be based on test data if available or else on engineering judgment and previous experience with similar systems.

Since the generalized masses M_i of Eq. (10.72) are defined in terms of the mode shapes which contain an arbitrary constant, proper selection of the constant will permit M_i to be defined (scaled) such that $M_i = 1$. Then Eq. (10.79), appears as

$$\ddot{q}_i + 2\zeta_i\omega_i\dot{q}_i + \omega_i^2 q_i = P_i \tag{10.80}$$

This equation is readily solved.

The homogeneous form of Eq. (10.80) has the solution

$$q_i = A_1 e^{\alpha_1 t} + A_2 e^{\alpha_2 t}$$

with

$$\alpha_{1,2} = \frac{1}{2}\left(-2\zeta_i\omega_i \pm \sqrt{4\zeta_i^2\omega_i^2 - 4\omega_i^2}\right) = \omega_i\left(-\zeta_i \pm \sqrt{\zeta_i^2 - 1}\right)$$

when

$\zeta_i < 1$, q_i is underdamped and oscillates
$\zeta_i > 1$, q_i is overdamped and exponentially decays
$\zeta_i = 1$, q_i is critically damped, $C_i = C_{cr}$

10.4.2 Summary of the Modal Superposition Method

The dynamic response of any linear vibration system modeled with n_d DOF can be computed using the modal superposition procedure just described. The procedure can be summarized in the following steps.

STEP 1. Formulation of Equations of Motion
For the class of problems considered here, the equations of motion can be expressed in the matrix form

$$\mathbf{M}\ddot{\mathbf{V}} + \mathbf{K}\mathbf{V} = \overline{\mathbf{P}} \tag{10.81}$$

with initial conditions $\mathbf{V}(0) = \overline{\mathbf{V}}_0$ and $\dot{\mathbf{V}}(0) = \dot{\overline{\mathbf{V}}}_0$.

STEP 2. Free Vibration Analysis
The mode shape vectors $\boldsymbol{\phi}_i$ and corresponding natural frequencies ω_i are computed by solving the generalized eigenvalue problem

$$\mathbf{K}\boldsymbol{\phi}_i = \omega_i^2\mathbf{M}\boldsymbol{\phi}_i \tag{10.82}$$

STEP 3. Compute Generalized Mass and Loading
The generalized mass M_i and generalized force P_i for the ith mode are computed using

$$\begin{aligned} M_i &= \boldsymbol{\phi}_i^T\mathbf{M}\boldsymbol{\phi}_i \quad \text{or} \quad \mathbf{M}_{n_d} = \boldsymbol{\Phi}^T\mathbf{M}\boldsymbol{\Phi} \\ P_i &= \boldsymbol{\phi}_i^T\overline{\mathbf{P}} \quad \text{or} \quad \mathbf{P} = \boldsymbol{\Phi}^T\overline{\mathbf{P}} \end{aligned} \tag{10.83}$$

for all modes of interest. If the scaled mode shapes are used, $\mathbf{M}_{n_d} = \mathbf{I}$.

STEP 4. Obtain Uncoupled Equations of Motion

The uncoupled equations for the ith modal coordinate q_i are

$$\ddot{q}_i + 2\zeta_i\omega_i\dot{q}_i + \omega_i^2 q_i = P_i(t)/M_i \tag{10.84}$$

Note that modal damping has been included here. For undamped responses, set $\zeta_i = 0$.

STEP 5. Express the Initial Conditions in Modal Coordinates

With the assistance of the orthogonality relationship for the mode shapes, the modal initial conditions $q_i(0)$ and $\dot{q}_i(0)$ can be computed in terms of the physical initial conditions using

$$\begin{aligned} q_i(0) &= \boldsymbol{\phi}_i^T \frac{\mathbf{M}\,\mathbf{V}(0)}{M_i} \\ \dot{q}_i(0) &= \boldsymbol{\phi}_i^T \frac{\mathbf{M}\,\dot{\mathbf{V}}(0)}{M_i} \end{aligned} \tag{10.85}$$

STEP 6. Compute Modal Responses to Initial Conditions

The modal coordinate due to initial conditions is, from the complementary solution to Eq. (10.84), for $\zeta_i < 1$,

$$q_{i_I}(t) = e^{-\zeta_i\omega_i t}\left[\frac{\dot{q}_i(0) + q_i(0)\zeta_i\omega_i}{\omega_i\sqrt{1-\zeta_i^2}}\sin\omega_i\sqrt{1-\zeta_i^2}\,t + q_i(0)\cos\omega_i\sqrt{1-\zeta_i^2}\,t\right] \tag{10.86}$$

If the forced responses are started from zero initial conditions, Steps 5 and 6 can be bypassed.

STEP 7. Compute Modal Responses to Applied Loading

The modal coordinate due to an applied loading, which is the particular solution of Eq. (10.84), is

$$q_{i_P}(t) = \frac{1}{M_i\omega_i\sqrt{1-\zeta_i^2}}\int_0^t P_i(\tau)e^{-\zeta_i\omega_i(t-\tau)}\sin\omega_i\sqrt{1-\zeta_i^2}\,(t-\tau)\,d\tau \tag{10.87}$$

For the case of harmonic excitation, i.e., $P_i = \overline{P}_0 \sin\omega t$, the integration in Eq. (10.87) can be carried out analytically, giving

$$q_{i_P}(t) = \frac{(\overline{P}_0/M_i)\omega_i^2}{\left[\left(1-\left(\frac{\omega}{\omega_i}\right)^2\right)^2 + 4\zeta_i^2\left(\frac{\omega}{\omega_i}\right)^2\right]^{1/2}}\sin(\omega t - \theta_i) \tag{10.88}$$

with

$$\theta_i = \tan^{-1}\left[\frac{2\zeta_i\omega_i\omega}{\omega_i^2 - \omega^2}\right]$$

The total modal response is given by

$$q_i(t) = q_{i_I}(t) + q_{i_P}(t) \tag{10.89}$$

STEP 8. Compute Physical Displacement Responses by Modal Superposition

The response in terms of **V** is found using

$$\mathbf{V} = \mathbf{\Phi q} = \sum q_i(t)\,\boldsymbol{\phi}_i \tag{10.90}$$

STEP 9. Compute Other Responses

Responses other than displacements, e.g., accelerations and elastic forces, are frequently of interest in design. Accelerations are computed by taking derivatives of the displacement responses. Thus,

$$\ddot{\mathbf{V}}(t) = \sum \boldsymbol{\phi}_i\,\ddot{q}_i(t) \tag{10.91}$$

In some cases $\ddot{q}_i(t)$ can be obtained readily from Eq. (10.84).

EXAMPLE 10.9 Forced Response of Two-DOF System

Compute the response of the system of Fig. 10.5. A suddenly applied horizontal force $\overline{P}_2 = \overline{P}_0$ is imposed on mass 2 (see Fig. 10.7).

Follow the procedure outlined in this section.

Step 1. Set up the governing equations for the free vibration

$$\begin{bmatrix} m & 0 \\ 0 & 2m \end{bmatrix} \begin{bmatrix} \ddot{u}_1 \\ \ddot{u}_2 \end{bmatrix} + \begin{bmatrix} 2k & -k \\ -k & 2k \end{bmatrix} \begin{bmatrix} u_1 \\ u_2 \end{bmatrix} = \mathbf{0}$$
$$\mathbf{M} \quad \ddot{\mathbf{V}} \quad + \quad \mathbf{K} \quad \mathbf{V} = \mathbf{0} \tag{1}$$

with the initial conditions

$$\mathbf{V}(0) = \dot{\mathbf{V}}(0) = \mathbf{0} \tag{2}$$

Step 2. From Example 10.4, it is already known that $\omega_1 = 0.796266\sqrt{k/m}$ and $\omega_2 = 1.538188\sqrt{k/m}$ and the mode shapes are found in Eq. (5) of Example 10.4.

From Eq. (10.61a)

$$M_1 = \boldsymbol{\phi}_1^T \mathbf{M} \boldsymbol{\phi}_1 = [1.000 \quad 1.366025] \begin{bmatrix} m & 0 \\ 0 & 2m \end{bmatrix} \begin{bmatrix} 1.000 \\ 1.366025 \end{bmatrix}$$
$$= 4.732049\, m, \qquad M_2 = \boldsymbol{\phi}_2^T \mathbf{M} \boldsymbol{\phi}_2 = 1.267949\, m \tag{3}$$

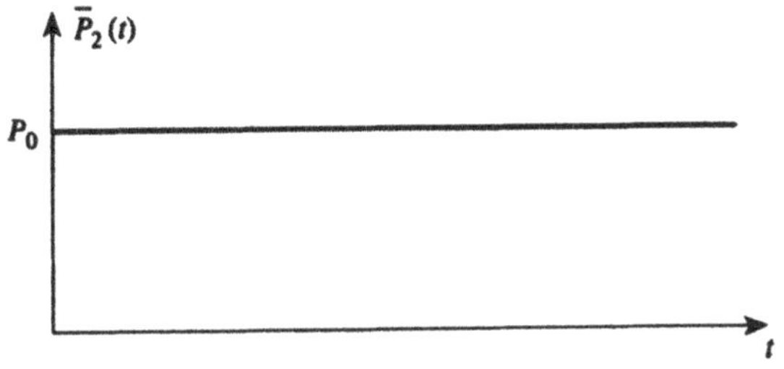

FIGURE 10.7
Loading for Example 10.9.

Division of the mode shapes by $(M_s)^{1/2}$, $s = 1, 2$, leads to scaled mode shapes

$$\boldsymbol{\phi}_1 = \frac{1}{\sqrt{m}}\begin{bmatrix} 0.459701 \\ 0.627963 \end{bmatrix} \quad \text{and} \quad \boldsymbol{\phi}_2 = \frac{1}{\sqrt{m}}\begin{bmatrix} 0.888074 \\ -0.325057 \end{bmatrix} \tag{4}$$

Then

$$\boldsymbol{\Phi} = \frac{1}{\sqrt{m}}\begin{bmatrix} 0.459701 & 0.888074 \\ 0.627963 & -0.325057 \end{bmatrix} \tag{5}$$

Step 3. Compute the generalized mass and loading matrix. For the scaled mode shapes, the generalized mass matrix is a unit matrix, and the load matrix is

$$\mathbf{P} = \boldsymbol{\Phi}^T\overline{\mathbf{P}} = \frac{P_0}{\sqrt{m}}\begin{bmatrix} 0.627963 \\ -0.325057 \end{bmatrix} \qquad t \geq 0 \tag{6}$$

Steps 4, 5, and 6. Obtain the uncoupled equations of motion. The ith ($i = 1, 2$) equation is

$$\ddot{q}_i + \omega_i^2 q_i = P_i(t) \tag{7}$$

and the initial conditions are calculated from Eq. (10.85) as $\mathbf{q}(0) = \dot{\mathbf{q}}(0) = \mathbf{0}$.

Step 7. Compute the modal response due to the applied loading. From Eq. (10.87) with $\zeta_i = 0$,

$$\begin{aligned} q_1(t) &= 0.627963\frac{P_0}{\sqrt{m}}\frac{1}{\omega_1^2}\int_0^t \sin\omega_1(t-\tau)\,d\tau \\ &= 0.627963\frac{P_0}{\omega_1^2\sqrt{m}}(1-\cos\omega_1 t) \\ q_2(t) &= -0.325057\frac{P_0}{\omega_2^2\sqrt{m}}(1-\cos\omega_2 t) \end{aligned} \tag{8}$$

Step 8. From Eq. (10.90), the physical displacement response is computed as $\mathbf{V} = \boldsymbol{\Phi}\mathbf{q}$. Thus,

$$\begin{aligned} V_1 = u_1\frac{P_0}{m}&\left[0.459701 \times 0.62727963\frac{1}{\omega_1^2}(1-\cos\omega_1 t) - 0.888074 \times 0.325057\frac{1}{\omega_2^2}(1-\cos\omega_2 t)\right] \\ &= \frac{P_0}{k}\left[0.455295\left(1-\cos 0.796266\sqrt{\frac{k}{m}}t\right) - 0.122\left(1-\cos 1.538188\sqrt{\frac{k}{m}}t\right)\right] \end{aligned} \tag{9}$$

and

$$\begin{aligned} V_2 = u_2\frac{P_0}{m}&\left[0.627963^2\frac{1}{\omega_1^2}(1-\cos\omega_1 t) + 0.325057^2\frac{1}{\omega_2^2}(1-\cos\omega_2 t)\right] \\ &= \frac{P_0}{k}\left[0.621945\left(1-\cos 0.796266\sqrt{\frac{k}{m}}t\right) + 0.44658\left(1-\cos 1.538188\sqrt{\frac{k}{m}}t\right)\right] \end{aligned} \tag{10}$$

■

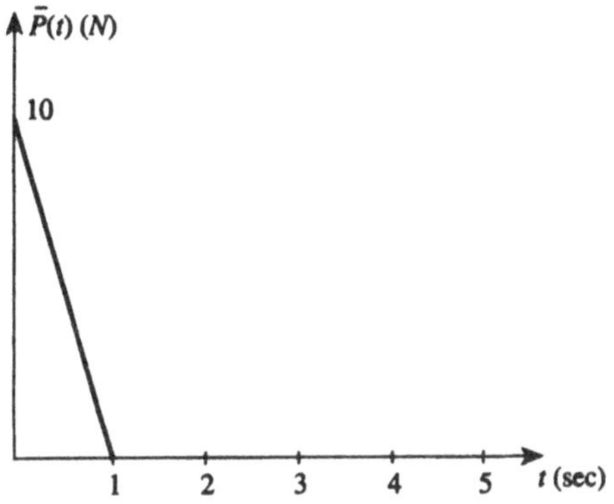

FIGURE 10.8
Loading for Examples 10.10, 10.11, and 10.12.

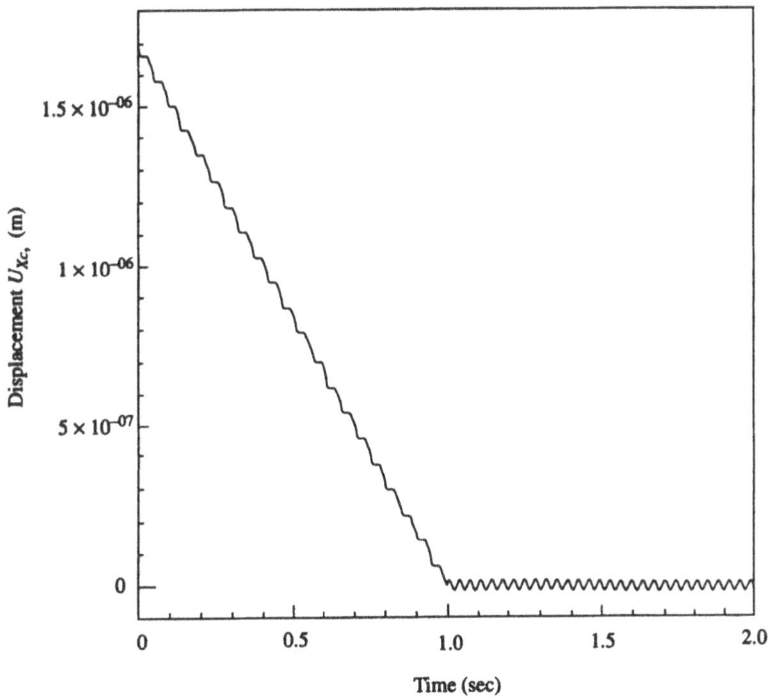

FIGURE 10.9
Displacement of the frame of Example 10.10 at node c in the X direction.

EXAMPLE 10.10 Transient Response of a Frame

Calculate the response U_{Xc} of the frame of Fig. 10.2 due to the applied force of Fig. 10.8 at node b in the X direction. Assume 2% modal damping ($\zeta_i = 0.02$) for all modes of the structure, and that the structure is initially at rest.

The solution procedure for this problem follows the steps outlined in this section. The reduced stiffness matrix and consistent mass matrix, in which the degrees associated with the rotational DOFs are condensed out, are used, and the natural frequencies are found in Example 10.7. Use the modal superposition method to find the transient response. This leads to the uncoupled equations of motion of the form of Eq. (10.84) and the associated initial conditions. After solving these equations, Eq. (10.90) is used to compute the physical response. The response $U_{Xc}(t)$ of the structure is shown in Fig. 10.9. ∎

10.4.3 Base Excitation

Suppose a prescribed displacement $\overline{U}$ is applied to the base of the frame of Fig. 10.2. This displacement is such that the base translates as $\overline{U}(t)$ in the X and Z directions. The nodal displacements $\mathbf{V}'$ would then be

$$\begin{aligned}\mathbf{V}' &= [\overline{U} + U_{Xb} + \overline{U} + U_{Zb} + \Theta_b \overline{U} + U_{Xc} + \overline{U} + U_{Zc} + \Theta_c]^T \\ &= \mathbf{V} + \mathbf{a}^T \overline{U}\end{aligned} \tag{10.92}$$

where $\mathbf{a}^T = [1\ 1\ 0\ 1\ 1\ 0]^T$ and $\mathbf{V} = [U_{Xb}\ \ U_{Zb}\ \ \Theta_b\ \ U_{Xc}\ \ U_{Zc}\ \ \Theta_c]^T$ are the displacements relative to the moving base.

Suppose there are no applied forces. Then, the governing equilibrium equations would be of the form $\mathbf{M}\ddot{\mathbf{V}}' + \mathbf{KV} = \mathbf{0}$, since Newton's second law applies to absolute accelerations of the masses. With Eq. (10.92) this leads to

$$\mathbf{M}\ddot{\mathbf{V}} + \mathbf{KV} = -\mathbf{M}\mathbf{a}^T \ddot{\overline{U}} \tag{10.93}$$

Thus, the equations of motion including base input have the same general form as Eq. (10.68) except the forcing function takes the special form

$$\overline{\mathbf{P}} = -\mathbf{M}\mathbf{a}^T \ddot{\overline{U}} \tag{10.94}$$

The physical responses are still given by Eq. (10.81). It follows that Eq. (10.84) would now appear as

$$\ddot{q}_i + 2\zeta_i \omega_i \dot{q}_i + \omega_i^2 q_i = -\boldsymbol{\phi}_i^T \frac{\mathbf{M}\mathbf{a}^T \ddot{\overline{U}}}{M_i} \tag{10.95}$$

The quantity

$$MPF_i = \boldsymbol{\phi}_i^T \frac{\mathbf{M}\mathbf{a}^T}{M_i} \tag{10.96}$$

is often referred to as the *modal participation factor* for the ith mode. To be more precise, for the case at hand, this is a translational modal participation factor. This factor depends on the mode shapes, the mass distribution, and the direction of input. As indicated in Eq. (10.95), the magnitude of the excitation force, and, hence, the modal response, is directly proportional to the modal participation factor associated with the mode. Thus, q_i can be calculated as the product of MPF_i and the solution of Eq. (10.95) due to $\ddot{\overline{U}}$ only.

10.5 Direct Integration of the Equations of Motion

The equations of motion

$$\mathbf{M}\ddot{\mathbf{V}} + \mathbf{C}\dot{\mathbf{V}} + \mathbf{KV} = \overline{\mathbf{P}} \tag{10.97}$$

can be solved using integration directly without employing modal superposition, which applies for linear responses only. Such *direct integration methods* use step-by-step numerical integration.

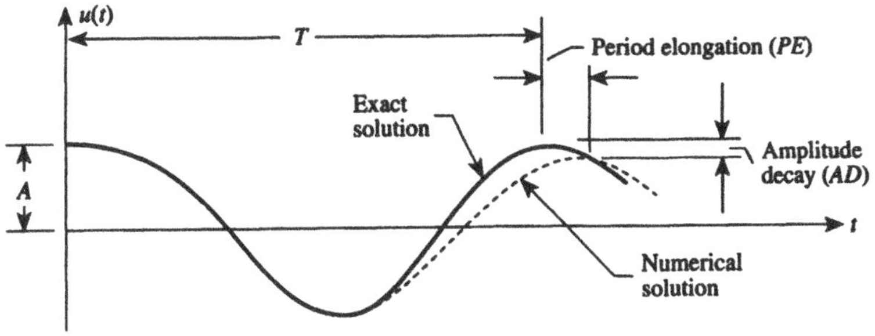

FIGURE 10.10
Period elongation and amplitude decay errors.

The direct integration methods begin with some known initial conditions, e.g., displacement $\mathbf{V}_0$, velocity $\dot{\mathbf{V}}_0$, or acceleration $\ddot{\mathbf{V}}_0$ at time $t = 0$. The integration scheme establishes an approximate solution $\mathbf{V}_1$, $\dot{\mathbf{V}}_1$, and $\ddot{\mathbf{V}}_1$ at time $t = \Delta t$, $\mathbf{V}_2$, $\dot{\mathbf{V}}_2$, $\ddot{\mathbf{V}}_2$ at $t = 2\,\Delta t$, etc., where Δt is a suitably selected time increment. In essence, the direct integration methods try to satisfy the equations of motion (10.97) at discrete time intervals Δt. Each method employs a different assumed variation of displacement, velocity, and acceleration within the time interval.

One important aspect in using a step-by-step integration method is the selection of an appropriate time step Δt to be used in the integration. On the one hand, the time step must be small enough to obtain the accurate solution. On the other hand, the time step should not be smaller than necessary in order to reduce the computation cost. As will be shown in Section 10.5.5, the value chosen affects the stability and accuracy of the solution.

The errors involved in the direct integration method may result from truncation, i.e., from the use of lower order finite difference approximations to the derivatives of continuous variables, from instability, i.e., from amplification of errors in previous time steps into later time steps, and from computational roundoff. These errors may introduce a shift in period and amplitude of responses, i.e., period elongation and amplitude decay errors. Figure 10.10 shows a sketch of these errors.

In theory, the modal superposition and direct integration methods are unrelated techniques for solving Eq. (10.97). In practice, however, the numerical stability of the direct integration methods depends on the modal characteristics (Section 10.3) of the system. *Instability* here implies *numerical instability* and means the computed solution will grow with time even after the external excitations are removed.

The commonly used direct integration methods can be, in general, classified as *explicit* or *implicit*. The explicit formulation utilizes difference expressions for velocity and acceleration which are inserted in the governing equations of motion (Eq. 10.97) at time $t = n\,\Delta t$. In contrast, for the implicit formulation, the difference expressions are placed in Eq. (10.97) at time $t = (n+1)\Delta t$. The most common difference expressions used with the explicit method are the central differences, whereas the Houbolt method, the Wilson θ method, the Newmark β method, and other newly developed methods are used in conjunction with the implicit form of Eq. (10.97).

10.5.1 Central Difference Method

In principle, any finite difference expression that approximates the acceleration and velocity in terms of displacements can be used to solve the equation of motion of Eq. (10.97).

One procedure that is often effective is the central difference method in which the velocity and acceleration are expressed as

$$\dot{\mathbf{V}}_n = \frac{1}{2\,\Delta t}[\mathbf{V}_{n+1} - \mathbf{V}_{n-1}], \qquad \ddot{\mathbf{V}}_n = \frac{1}{\Delta t^2}[\mathbf{V}_{n+1} - 2\mathbf{V}_n + \mathbf{V}_{n-1}] \tag{10.98}$$

where subscripts $n-1$, n, and $n+1$ represent the time $(n-1)\Delta t$, $n\,\Delta t$, and $(n+1)\Delta t$, respectively, with $n = 0, 1, \ldots$.

The displacement at time $(n+1)\Delta t$ is obtained by considering Eq. (10.97) at time $n\,\Delta t$, i.e.,

$$\mathbf{M}\ddot{\mathbf{V}}_n + \mathbf{C}\dot{\mathbf{V}}_n + \mathbf{K}\mathbf{V}_n = \overline{\mathbf{P}}_n \tag{10.99}$$

Substitute the relations for $\ddot{\mathbf{V}}_n$ and $\dot{\mathbf{V}}_n$ of Eq. (10.98) into Eq. (10.99), giving

$$\left(\frac{\mathbf{M}}{\Delta t^2} + \frac{\mathbf{C}}{2\,\Delta t}\right)\mathbf{V}_{n+1} = \overline{\mathbf{P}}_n - \left(\mathbf{K} - \frac{2\mathbf{M}}{\Delta t^2}\right)\mathbf{V}_n - \left(\frac{\mathbf{M}}{\Delta t^2} - \frac{\mathbf{C}}{2\,\Delta t}\right)\mathbf{V}_{n-1}$$

or

$$\widehat{\mathbf{K}}\mathbf{V}_{n+1} = \mathbf{P}_{n+1} \tag{10.100}$$

with

$$\widehat{\mathbf{K}} = \frac{\mathbf{M}}{\Delta t^2} + \frac{\mathbf{C}}{2\,\Delta t}$$

$$\mathbf{P}_{n+1} = \overline{\mathbf{P}}_n - \left(\mathbf{K} - \frac{2\mathbf{M}}{\Delta t^2}\right)\mathbf{V}_n - \left(\frac{\mathbf{M}}{\Delta t^2} - \frac{\mathbf{C}}{2\,\Delta t}\right)\mathbf{V}_{n-1}$$

The constant matrix $\widehat{\mathbf{K}}$ is called an *effective stiffness matrix*, and $\mathbf{P}_{n+1}$ is the *effective load* at time step $n+1$. Equation (10.100) is a set of linear equations and can be used to find $\mathbf{V}_{n+1}$. Since $\mathbf{M}$ and $\mathbf{C}$ are diagonal matrices, it is not necessary to solve a system of simultaneous equations. Note that $\widehat{\mathbf{K}}$ is not a function of $\mathbf{K}$.

In these relations, the equation for $\mathbf{V}_{n+1}$ involves $\mathbf{V}_n$ and $\mathbf{V}_{n-1}$. Therefore, in order to calculate the solution at the first time step Δt, a special starting procedure must be used. For example, use the second relationship of Eq. (10.98) with $n = 0$ to derive an expression for $\mathbf{V}_{-1}$. With $\dot{V}_0 = (V_1 - V_0)/\Delta t$, this leads to

$$\mathbf{V}_{-1} = \mathbf{V}_0 - \Delta t\dot{\mathbf{V}}_0 + \frac{\Delta t^2}{2}\ddot{\mathbf{V}}_0 \tag{10.101}$$

where $\ddot{\mathbf{V}}_0$ can be obtained directly from Eq. (10.97) with $\mathbf{V}_0$ and $\dot{\mathbf{V}}_0$ known. Table 10.2 summarizes the procedure for computer implementation.

Stability and Accuracy

Consider a single-DOF system so that $\mathbf{K}$, $\mathbf{C}$, $\mathbf{M}$, and $\overline{\mathbf{P}}$ reduce to k, c, m, and $\overline{p}$. Rearrange Eq. (10.100), and set $c = 0$, $\overline{p} = 0$.

$$\mathbf{X}_{n+1} = \mathbf{A}\mathbf{X}_n \tag{10.102}$$

where

$$\mathbf{X}_{n+1} = \begin{bmatrix} V_{n+1} \\ V_n \end{bmatrix} \qquad \mathbf{X}_n = \begin{bmatrix} V_n \\ V_{n-1} \end{bmatrix}$$

and

$$\mathbf{A} = \begin{bmatrix} 2 - \omega^2\,\Delta t^2 & -1 \\ 1 & 0 \end{bmatrix} \qquad \omega^2 = k/m \tag{10.103}$$

Matrix $\mathbf{A}$ is called the *amplification matrix*. Stability and accuracy of an integration algorithm depend on the eigenvalues of this amplification matrix, and in order to have a stable

TABLE 10.2
Procedure for Computer Implementation of the Central Difference Method

Initial Step

1. Select Δt
2. Calculate the constants related to Δt
   $c_1 = 1/\Delta t^2, c_2 = 1/\Delta t, c_3 = 2c_1$, and $c_4 = c_2/2$
3. Initialize $\mathbf{V}_0$, $\dot{\mathbf{V}}_0$, and $\ddot{\mathbf{V}}_0$
4. Calculate $\mathbf{V}_{-1}$ from Eq. (10.101)
5. Calculate the effective stiffness matrix of Eq. (10.100)
   $\widehat{\mathbf{K}} = c_1\mathbf{M} + c_4\mathbf{C}$
6. Decompose* (triangularize)
   $\widehat{\mathbf{K}} = \mathbf{LDL}^T$

At Each Step n:

1. Calculate the effective load from Eq. (10.100)
   $\mathbf{P}_{n+1} = \overline{\mathbf{P}}_n - (\mathbf{K} - c_3\mathbf{M})\mathbf{V}_n - (c_1\mathbf{M} - c_4\mathbf{C})\mathbf{V}_{n-1}$
2. Solve $\widehat{\mathbf{K}}\mathbf{V}_{n+1} = \mathbf{P}_{n+1}$ using $\widehat{\mathbf{K}} = \mathbf{LDL}^T$ (matrix decomposition)

* $\widehat{\mathbf{K}}$ can be decomposed into $\mathbf{LDT}^T$. $\mathbf{L}$ is a lower triangular matrix (elements only on the diagonal and below) with the diagonal elements equal to 1, and $\mathbf{D}$ is a diagonal matrix.

solution, the spectral radius, i.e., the maximum eigenvalue of matrix **A**, has to be smaller than 1. This can be readily proved. It follows from Eq. (10.102) that

$$\mathbf{X}_n = \mathbf{A}\mathbf{X}_{n-1} \tag{10.104}$$

Substitute Eq. (10.104) into Eq. (10.102)

$$\mathbf{X}_{n+1} = \mathbf{A}^2\mathbf{X}_{n-1} \tag{10.105}$$

Thus, in general,

$$\mathbf{X}_{n+1} = \mathbf{A}^{n+1}\mathbf{X}_0 \tag{10.106}$$

From the theory of matrices,

$$\mathbf{A} = \mathbf{Z}\boldsymbol{\lambda}\mathbf{Z}^{-1} \tag{10.107}$$

where $\boldsymbol{\lambda}$ are the eigenvalues of **A**, which are assumed to be distinct, and each column of **Z** is a corresponding eigenvector, whose elements are finite.

Also, from the theory of matrix analysis,

$$\mathbf{A}^{n+1} = \mathbf{Z}\boldsymbol{\lambda}^{n+1}\mathbf{Z}^{-1} \tag{10.108}$$

Since the elements of **Z** are bounded, the elements of $\mathbf{A}^{n+1}$ will be bounded provided $|\lambda_i| \leq 1.0, i = 1, 2, \ldots, n$, i.e., if $|\lambda_i| \leq 1.0$, $\mathbf{X}_{n+1}$ cannot be very large or approach infinity, and the integration is stable. The condition of $|\lambda_i| \leq 1.0$ leads to the critical time step for stability

$$\Delta t \leq \Delta t_{\text{crit}} = \frac{2}{\omega} \tag{10.109}$$

TABLE 10.3

Critical Time Step Estimates for the Central Difference Method for Various Elements

Element	Type of Mass Matrix	Critical Time Step Δt
Two-node bar	Lumped	ℓ/c
Two-node bar	Consistent	$\ell/\sqrt{3}c$
Two-node beam	Lumped	$\min\{\ell/c, (\ell/c_s)[1 + A/I(\ell/2)^2]^{-1/2}\}$
Four-node quadrilateral (Chapter 6, Section 6.7.2)	Consistent	$2/c_d g^{1/2}$

where

ℓ = Element length

$c = \sqrt{E/\gamma}$, the bar wave velocity

γ = Mass density

$c_s = G\,k_s/\gamma$, the beam shear wave velocity

I = Moment of inertia

A = Area of the beam cross-section

$c_d^2 = G(4G - E)/[(3G - E)/\gamma]$

$g = 4/A^2 \sum_{i=1}^{2} \sum_{j=1}^{4} B_{ij}\, B_{ij}$

in which B_{ij} is an entry in

$$\boldsymbol{B} = \frac{1}{2}\begin{bmatrix} (y_2 - y_4) & (y_3 - y_1) & (y_4 - y_2) & (y_1 - y_3) \\ (x_4 - x_2) & (x_1 - x_3) & (x_2 - x_4) & (x_3 - x_1) \end{bmatrix}$$

and A is the area of the quadrilateral element.

In order to have a stable solution using the central difference explicit integration algorithm, the time step Δt used for integration must be smaller than the critical time step as given by Eq. (10.109). When the frequency gets higher, the allowable time step will decrease. The conditionally stable characteristic is the main disadvantage in using an explicit integration method.

For different types of finite elements, the critical time step is different. Table 10.3 lists these time steps corresponding to different elements. These time step estimates are derived in Hughes (1987).

Equation (10.109) is often employed for multi-DOF systems in which $\Delta t_{\text{crit}} = 2/\omega_{\max}$, where $\omega_{\max}$ is the largest natural frequency of the system, i.e., the n_dth natural frequency of the n_d DOF model of the structure.

10.5.2 Houbolt Method

Houbolt (1950) used the following finite difference equations for velocity and accelerations at time $t = (n+1)\Delta t$

$$\begin{aligned} \ddot{\mathbf{V}}_{n+1} &= \frac{1}{(\Delta t)^2}[2\mathbf{V}_{n+1} - 5\mathbf{V}_n + 4\mathbf{V}_{n-1} - \mathbf{V}_{n-2}] \\ \dot{\mathbf{V}}_{n+1} &= \frac{1}{6\Delta t}[11\mathbf{V}_{n+1} - 18\mathbf{V}_n + 9\mathbf{V}_{n-1} - 2\mathbf{V}_{n-2}] \end{aligned} \tag{10.110}$$

These equations were obtained from consideration of a cubic curve that passes through four successive ordinates.

In order to obtain the solution at time $(n+1)\Delta t$, we employ the equations of motion of Eq. (10.97) at time $(n+1)\Delta t$. Substitute Eq. (10.110) into Eq. (10.97) and arrange all of the known terms on the right-hand side. This leads to

$$\left(\frac{2\mathbf{M}}{(\Delta t)^2}+\frac{11\mathbf{C}}{6\,\Delta t}+\mathbf{K}\right)\mathbf{V}_{n+1}=\overline{\mathbf{P}}_{n+1}+\left(\frac{5\mathbf{M}}{(\Delta t)^2}+\frac{3\mathbf{C}}{\Delta t}\right)\mathbf{V}_n-\left(\frac{4\mathbf{M}}{(\Delta t)^2}+\frac{3\mathbf{C}}{2\,\Delta t}\right)\mathbf{V}_{n-1}$$
$$+\left(\frac{\mathbf{M}}{(\Delta t)^2}+\frac{\mathbf{C}}{3\,\Delta t}\right)\mathbf{V}_{n-2} \qquad (10.111)$$

or

$$\widehat{\mathbf{K}}\mathbf{V}_{n+1}=\mathbf{P}_{n+1}$$

where

$$\widehat{\mathbf{K}}=\frac{2\mathbf{M}}{(\Delta t)^2}+\frac{11\mathbf{C}}{6\,\Delta t}+\mathbf{K}$$

$$\mathbf{P}_{n+1}=\overline{\mathbf{P}}_{n+1}+\left(\frac{5\mathbf{M}}{(\Delta t)^2}+\frac{3\mathbf{C}}{\Delta t}\right)\mathbf{V}_n-\left(\frac{4\mathbf{M}}{(\Delta t)^2}+\frac{3\mathbf{C}}{2\,\Delta t}\right)\mathbf{V}_{n-1}+\left(\frac{\mathbf{M}}{(\Delta t)^2}+\frac{\mathbf{C}}{3\,\Delta t}\right)\mathbf{V}_{n-2}$$

As with the central difference method, this formulation needs a special starting procedure. Houbolt used the formulas for the derivatives at the third point of the four successive points along the cubic curve. The formulas give the following values for $\mathbf{V}_{-1}$ and $\mathbf{V}_{-2}$:

$$\begin{aligned}\mathbf{V}_{-1}&=(\Delta t)^2\ddot{\mathbf{V}}_0-\mathbf{V}_1+2\mathbf{V}_0\\ \mathbf{V}_{-2}&=6\,\Delta t\dot{\mathbf{V}}_0+6(\Delta t)^2\ddot{\mathbf{V}}_0-8\mathbf{V}_1+9\mathbf{V}_0\end{aligned} \qquad (10.112)$$

The algorithm for implementation of this method is outlined in Table 10.4.

To examine the stability of the Houbolt method, study the single DOF case used for the central difference method, where $\mathbf{K}, \mathbf{C}, \mathbf{M}$, and $\overline{\mathbf{P}}$ are replaced by k, c, m, and $\overline{p}$.

TABLE 10.4
Procedure for Computer Implementation of the Houbolt Method

Initial Step

1. Select Δt
2. Calculate the constants related to Δt
   $c_1=2/\Delta t^2 \quad c_2=11/6\,\Delta t \quad c_3=5/\Delta t^2 \quad c_4=3/\Delta t$
   $c_5=-2c_1 \quad c_6=-c_4/2 \quad c_7=c_1/2 \quad c_8=c_4/9$
3. Initialize $\mathbf{V}_0$, $\dot{\mathbf{V}}_0$, and $\ddot{\mathbf{V}}_0$
4. Calculate $\mathbf{V}_{-1}$ and $\mathbf{V}_{-2}$ from Eq. (10.112)
5. Form the effective stiffness matrix $\widehat{\mathbf{K}}=c_1\mathbf{M}+c_2\mathbf{C}+\mathbf{K}$
6. Decompose (triangularize) $\widehat{\mathbf{K}}=\mathbf{LDL}^T$

At Each Step:

1. Calculate $\mathbf{P}_{n+1}=\overline{\mathbf{P}}_{n+1}+(c_3\mathbf{M}+c_4\mathbf{C})\mathbf{V}_n+(c_5\mathbf{M}+c_6\mathbf{C})\mathbf{V}_{n-1}+(c_7\mathbf{M}+c_8\mathbf{C})\mathbf{V}_{n-2}$
2. Solve $\widehat{\mathbf{K}}\mathbf{V}_{n+1}=\mathbf{P}_{n+1}$

Assume that $c = 0$ and $\overline{p} = 0$, and write Eq. (10.111) as

$$\mathbf{X}_{n+1} = \mathbf{A}\mathbf{X}_n \tag{10.113}$$

where

$$\mathbf{X}_{n+1} = \begin{bmatrix} V_{n+1} \\ V_n \\ V_{n-1} \end{bmatrix} \qquad \mathbf{X}_n = \begin{bmatrix} V_n \\ V_{n-1} \\ V_{n-2} \end{bmatrix}$$

and

$$A = \begin{bmatrix} 5/(2+\omega^2\,\Delta t^2) & -4/(2+\omega^2\,\Delta t^2) & 1/(2+\omega^2\,\Delta t^2) \\ 1 & 0 & 0 \\ 0 & 1 & 0 \end{bmatrix} \qquad \omega^2 = k/m \tag{10.114}$$

To study the stability criterion, follow the procedure outlined in Eqs. (10.104) and (10.108). The same stability condition, $|\lambda_i| \leq 1$, can be obtained. It can be shown that the three eigenvalues λ_1, λ_2, and λ_3 of $\mathbf{A}$ are always less than 1, and, hence, the stability condition $|\lambda_i| \leq 1$ with $i = 1, 2, 3$ is always satisfied. It is concluded that the Houbolt method is always stable regardless of the size of the time step Δt used for integration. An integration scheme with this property, which is a characteristic of implicit integration schemes, is said to be *unconditionally stable*. The time step Δt to be used for integration, however, will be selected in order to achieve proper accuracy.

10.5.3 Newmark Method

In the *Newmark method* or *Newmark's β method*, two parameters, γ and β, are introduced to indicate how much of the acceleration enters into the relations for velocity and displacement at the end of the time interval. The relations that are adopted are

$$\mathbf{V}_{n+1} = \mathbf{V}_n + \Delta t\,\dot{\mathbf{V}}_n + \left(\frac{1}{2} - \beta\right)(\Delta t)^2\,\ddot{\mathbf{V}}_n + \beta(\Delta t)^2\,\ddot{\mathbf{V}}_{n+1} \tag{10.115a}$$

$$\dot{\mathbf{V}}_{n+1} = \dot{\mathbf{V}}_n + (1-\gamma)\Delta t\,\ddot{\mathbf{V}}_n + \gamma\,\Delta t\,\ddot{\mathbf{V}}_{n+1} \tag{10.115b}$$

Equation (10.115a) can be used to express $\ddot{\mathbf{V}}_{n+1}$ in terms of $\mathbf{V}_{n+1}$. Substitution of this relation into Eq. (10.115b) gives $\dot{\mathbf{V}}_{n+1}$ in terms of $\mathbf{V}_{n+1}$. The recurrence relationship for $\mathbf{V}_{n+1}$ then can be obtained by substituting these two relations into the equations of motion at time $(n+1)\Delta t$. The result is

$$\widehat{\mathbf{K}}\mathbf{V}_{n+1} = \mathbf{P}_{n+1} \tag{10.116}$$

where the effective stiffness $\widehat{\mathbf{K}}$ and effective loading vector $\mathbf{P}_{n+1}$ are given by

$$\widehat{\mathbf{K}} = \mathbf{K} + \frac{1}{\beta(\Delta t)^2}\mathbf{M} + \frac{\gamma}{\beta}\frac{1}{\Delta t}\mathbf{C}$$

$$\mathbf{P}_{n+1} = \mathbf{M}\left(\frac{1}{\beta(\Delta t)^2}\mathbf{V}_n + \frac{1}{\beta}\frac{1}{\Delta t}\dot{\mathbf{V}}_n + \left(\frac{1}{2\beta} - 1\right)\ddot{\mathbf{V}}_n\right)$$

$$+\mathbf{C}\left(\frac{\gamma}{\beta}\frac{1}{\Delta t}\mathbf{V}_n + \left(\frac{\gamma}{\beta} - 1\right)\dot{\mathbf{V}}_n + \left(\frac{\gamma}{\beta} - 2\right)\frac{\Delta t}{2}\ddot{\mathbf{V}}_n\right) + \overline{\mathbf{P}}_{n+1} \tag{10.117}$$

TABLE 10.5
Procedure for Computer Implementation of the Newmark Method

Initial Step

1. Select Δt, γ, and β
2. Calculate the constants

$$c_0 = 1/(\beta(\Delta t)^2) \qquad c_1 = (\gamma/\beta)\,(1/\Delta t) \qquad c_2 = (1/\gamma)c_1$$
$$c_3 = (\tfrac{1}{2} - \beta)/\beta \qquad c_4 = \gamma/\beta - 1 \qquad c_5 = (\tfrac{1}{2} - \beta)(\tfrac{\gamma\Delta t}{\beta}) + (\gamma - 1)\Delta t$$

3. Initialize $\mathbf{V}_0$, $\dot{\mathbf{V}}_0$, and $\ddot{\mathbf{V}}_0$
4. Form the effective stiffness matrix $\widehat{\mathbf{K}} = \mathbf{K} + c_0\mathbf{M} + c_1\mathbf{C}$
5. Decompose (triangularize) $\widehat{\mathbf{K}} = \mathbf{LDL}^T$

At Each Step

1. Calculate $\mathbf{P}_{n+1} = \mathbf{M}(c_0\mathbf{V}_n + c_2\dot{\mathbf{V}}_n + c_3\ddot{\mathbf{V}}_n) + \mathbf{C}(c_1\mathbf{V}_n + c_4\dot{\mathbf{V}}_n + c_5\ddot{\mathbf{V}}_n) + \overline{\mathbf{P}}_{n+1}$
2. Solve $\widehat{\mathbf{K}}\mathbf{V}_{n+1} = \mathbf{P}_{n+1}$
   Calculate $\dot{\mathbf{V}}_{n+1}$ and $\ddot{\mathbf{V}}_{n+1}$ from Eq. (10.115)

Once $\mathbf{V}_{n+1}$ is obtained from Eq. (10.116), the corresponding velocity and acceleration vectors can be computed using

$$\dot{\mathbf{V}}_{n+1} = \dot{\mathbf{V}}_n + (1-\gamma)(\Delta t)\ddot{\mathbf{V}}_n + \frac{\gamma}{\beta}\frac{1}{\Delta t}\left[\mathbf{V}_{n+1} - \mathbf{V}_n - (\Delta t)\dot{\mathbf{V}}_n - \left(\frac{1}{2} - \beta\right)\ddot{\mathbf{V}}_n(\Delta t)^2\right] \tag{10.118a}$$

$$\ddot{\mathbf{V}}_{n+1} = \frac{1}{\beta(\Delta t)^2}\left[\mathbf{V}_{n+1} - \mathbf{V}_n + (\Delta t)\dot{\mathbf{V}}_n + \left(\frac{1}{2} - \beta\right)(\Delta t)^2\ddot{\mathbf{V}}_n\right] \tag{10.118b}$$

The procedure for computer implementation is outlined in Table 10.5.

The stability and accuracy of the Newmark method, which is controlled by two parameters β and γ, can be evaluated by examining the eigenvalues associated with the amplification matrix. This process, however, becomes tedious and lengthy because of the parameters β and γ. Newmark (1959) established the stability and accuracy of this integration scheme by examining a single-DOF system without damping, and demonstrated that the method is unconditionally stable for $\gamma = \frac{1}{2}$ and $\beta \geq \frac{1}{4}$.

10.5.4 Wilson θ Method

In the Wilson θ method [Bathe and Wilson, 1973], the acceleration is assumed to be linear from time $n\,\Delta t$ to time $(n+\theta)\Delta t$, with $\theta \geq 1.0$ as shown in Fig. 10.11. With this assumption, the acceleration at any time $\tau\,\Delta t$, where $0 \leq \tau \leq \theta$, can be obtained by linear interpolation, i.e.,

$$\ddot{\mathbf{V}}_{n+\tau} = \ddot{\mathbf{V}}_n + \frac{\tau}{\theta}(\ddot{\mathbf{V}}_{n+\theta} - \ddot{\mathbf{V}}_n)$$

Integrate with respect to $\tau\,\Delta t$ to obtain

$$\dot{\mathbf{V}}_{n+\tau} = \dot{\mathbf{V}}_n + \ddot{\mathbf{V}}_n\tau\,\Delta t + \frac{(\tau\,\Delta t)^2}{2\theta\,\Delta t}(\ddot{\mathbf{V}}_{n+\theta} - \ddot{\mathbf{V}}_n) \tag{10.119}$$

where the first term on the right-hand side is the constant of integration evaluated at $\tau = 0$. Also,

$$\mathbf{V}_{n+\tau} = \mathbf{V}_n + \dot{\mathbf{V}}_n\tau\,\Delta t + \frac{(\tau\,\Delta t)^2}{2}\ddot{\mathbf{V}}_n + \frac{(\tau\,\Delta t)^3}{6\theta\,\Delta t}(\ddot{\mathbf{V}}_{n+\theta} - \ddot{\mathbf{V}}_n) \tag{10.120}$$

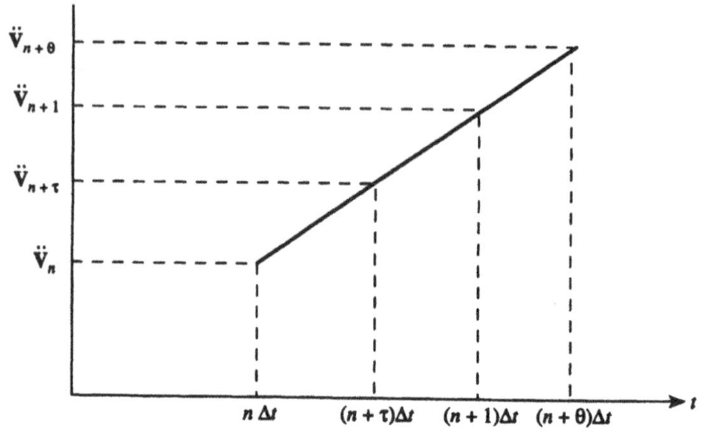

FIGURE 10.11
Linear interpolation of the acceleration for the Wilson θ method.

Equations (10.119) and (10.120) give at time $(n+\theta)\Delta t$

$$\dot{\mathbf{V}}_{n+\theta} = \dot{\mathbf{V}}_n + \frac{\theta\,\Delta t}{2}(\ddot{\mathbf{V}}_{n+\theta} + \ddot{\mathbf{V}}_n) \tag{10.121}$$

$$\mathbf{V}_{n+\theta} = \mathbf{V}_n + \theta\,\Delta t\,\dot{\mathbf{V}}_n + \frac{\theta^2\,\Delta t^2}{6}(\ddot{\mathbf{V}}_{n+\theta} + 2\ddot{\mathbf{V}}_n) \tag{10.122}$$

To obtain the solution at time $(n+1)\Delta t$, solve for $\ddot{\mathbf{V}}_{n+\theta}$ and $\dot{\mathbf{V}}_{n+\theta}$ from Eqs. (10.121) and (10.122) and substitute into the equation of motion of Eq. (10.97) at time $(n+\theta)\Delta t$, i.e.,

$$\mathbf{M}\ddot{\mathbf{V}}_{n+\theta} + \mathbf{C}\dot{\mathbf{V}}_{n+\theta} + \mathbf{K}\mathbf{V}_{n+\theta} = \overline{\mathbf{P}}_{n+\theta} \tag{10.123}$$

with

$$\overline{\mathbf{P}}_{n+\theta} = \overline{\mathbf{P}}_n + \theta(\overline{\mathbf{P}}_{n+1} - \overline{\mathbf{P}}_n).$$

A linear load vector is employed here because the accelerations are assumed to vary linearly. Equation (10.123) reduces to

$$\widehat{\mathbf{K}}\mathbf{V}_{n+\theta} = \mathbf{P}_{n+\theta} \tag{10.124}$$

where

$$\widehat{\mathbf{K}} = \mathbf{K} + \frac{6\mathbf{M}}{(\theta\,\Delta t)^2} + \frac{3}{\theta\,\Delta t}\mathbf{C}$$

and

$$\mathbf{P}_{n+\theta} = \overline{\mathbf{P}}_n + \theta(\overline{\mathbf{P}}_{n+1} - \overline{\mathbf{P}}_n) + \mathbf{M}\left(\frac{6}{(\theta\,\Delta t)^2}\mathbf{V}_n + \frac{6}{\theta\,\Delta t}\dot{\mathbf{V}}_n + 2\ddot{\mathbf{V}}_n\right)$$
$$+\mathbf{C}\left(\frac{3}{\theta\,\Delta t}\mathbf{V}_n + 2\dot{\mathbf{V}}_n + \frac{\theta\,\Delta t}{2}\ddot{\mathbf{V}}_n\right)$$

After $\mathbf{V}_{n+\theta}$ is found, the solution at time $(n+1)\Delta t$ can be obtained from Eqs. (10.121) and (10.122) by setting $\tau = 1$. The algorithm for this technique is outlined in Table 10.6.

The stability of the Wilson θ method can be investigated using the single-DOF case with $c = 0$, $\overline{p} = 0$, and the magnification matrix $\mathbf{A}$, which can be obtained from Eqs. (10.119) and (10.120) with $\tau = 1$ as

$$\mathbf{X}_{n+1} = \mathbf{A}\mathbf{X}_n \tag{10.125}$$

TABLE 10.6
Procedure for Computer Implementation of the Wilson θ Method

Initial Step

1. Select Δt, and let $\theta = 1.4$
2. Calculate the constants related to Δt
   $c_0 = 6/(\theta\Delta t)^2 \quad c_1 = 3/\theta\Delta t \quad c_2 = 2a_1$
   $c_3 = \theta\Delta t/2 \quad c_4 = c_0/\theta \quad c_5 = -c_2/\theta$
   $c_6 = 1 - 3/\theta \quad c_7 = \Delta t/2 \quad c_8 = \Delta t^2/6$
3. Initialize $\mathbf{V}_0$, $\dot{\mathbf{V}}_0$, and $\ddot{\mathbf{V}}_0$
4. Form the effective stiffness matrix
   $\widehat{\mathbf{K}} = \mathbf{K} + c_0\mathbf{M} + c_1\mathbf{C}$
5. Decompose (triangularize) $\widehat{\mathbf{K}} = \mathbf{LDL}^T$

At Each Step

1. Calculate $\mathbf{P}_{n+\theta} = \overline{\mathbf{P}}_n + \theta(\overline{\mathbf{P}}_{n+1} - \overline{\mathbf{P}}_n) + \mathbf{M}(c_0\,\mathbf{V}_n + c_2\dot{\mathbf{V}}_n + 2\ddot{\mathbf{V}}_n) +$
   $\mathbf{C}(c_1\mathbf{V}_n + 2\dot{\mathbf{V}}_n + c_3\ddot{\mathbf{V}}_n)$
2. Solve for the displacement
   $\mathbf{LDL}^T\mathbf{V}_{n+\theta} = \mathbf{P}_{n+\theta}$
3. Calculate displacements, velocities, and accelerations at $t + \Delta t$
   $\ddot{\mathbf{V}}_{n+1} = c_4(\mathbf{V}_{n+\theta} - \mathbf{V}_n) + c_5\dot{\mathbf{V}}_n + c_6\ddot{\mathbf{V}}_n$
   $\dot{\mathbf{V}}_{n+1} = \dot{\mathbf{V}}_n + c_7(\ddot{\mathbf{V}}_{n+1} + \ddot{\mathbf{V}}_n)$
   $\mathbf{V}_{n+1} = \mathbf{V}_n + \Delta t\,\dot{\mathbf{V}}_n + c_8(\ddot{\mathbf{V}}_{n+1} + 2\ddot{\mathbf{V}}_n)$

where

$$\mathbf{X}_{n+1} = \begin{bmatrix} \ddot{V}_{n+1} \\ \dot{V}_{n+1} \\ V_{n+1} \end{bmatrix} \qquad \mathbf{X}_n = \begin{bmatrix} \ddot{V}_n \\ \dot{V}_n \\ V_n \end{bmatrix}$$

$$\mathbf{A} = \begin{bmatrix} 1 - \beta\theta^2/3 - 1/\theta & -\beta\theta/\Delta t & -\beta/\Delta t^2 \\ \Delta t(1 - 1/2\theta - \beta\theta^2/6) & 1 - \beta\theta/\Delta t & -\beta/2\,\Delta t \\ \Delta t^2(1/2 - 1/6\theta - \beta\theta^2/18) & \Delta t(1 - \beta\theta/6) & 1 - \beta/6 \end{bmatrix}$$

and

$$\beta = \left(\frac{\theta}{\omega^2\,\Delta t^2} + \frac{\theta^3}{6}\right)^{-1}$$

The characteristic equation corresponding to the eigenvalue problem of matrix **A** can be obtained from Eq. (10.125). It can be seen that the Wilson θ method is unconditionally stable if $\theta \geq 1.37$. In practice, $\theta = 1.4$ is usually employed.

EXAMPLE 10.11 Transient Response of a Frame

The frame of Fig. 10.2 is chosen to illustrate the results of the use of different integration schemes. The system mass and stiffness matrices **M** and **K** are given by Eq. (9) of Example 10.1 and Eq. (11) of Chapter 5, Example 5.5. The damping matrix **C** is assumed to be proportional to the mass matrix, 0.000625 **M**. The external force is applied at node b in the X direction, with the time-dependence shown in Fig. 10.8. The central difference and Wilson θ methods are used to calculate the transient response of the beam.

Central Difference Method

Use Eq. (10.100) to calculate the response. From Eq. (10.109), the critical time step is

$$\Delta t_{\text{crit}} = \frac{2}{\omega_{\max}} = \frac{2}{3686.032} = 0.5426 \times 10^{-3}$$

where $\omega_{\max}$ is taken to be ω_6 in Eq. (1) of Example 10.7. Different time steps ($\Delta t = 0.5 \times 10^{-3}$, $\Delta t = 0.6 \times 10^{-3}$ sec) are used in the calculation. The displacement of node c in the X direction is shown in Figs. 10.12 and 10.13. It can be seen that for the time step 0.5×10^{-3}, the integration is stable up to 0.012 sec, but for the time step 0.6×10^{-3}, the integration is very unstable. The accumulated error makes the absolute value of the results extremely large.

Wilson θ Method

The calculations use Eq. (10.124). Employ the same time steps as used above for the central difference method, as well as $\Delta t = 0.1 \times 10^{-2}$. The displacement of node c in the X direction is shown in Figs. 10.12, 10.13 and 10.14. The results are stable. ∎

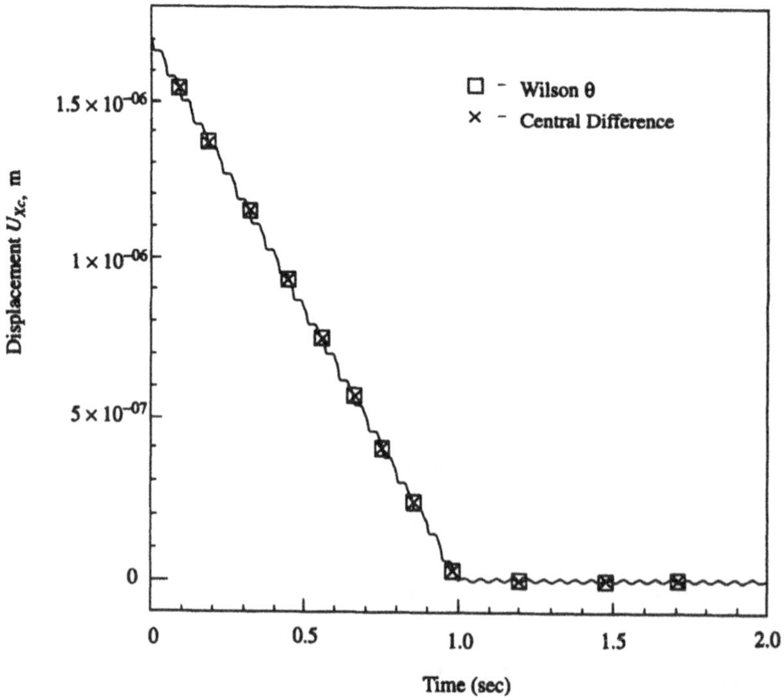

FIGURE 10.12
Comparison of the central difference method with the Wilson θ method ($\Delta t = 0.5 \times 10^{-3}$ sec).

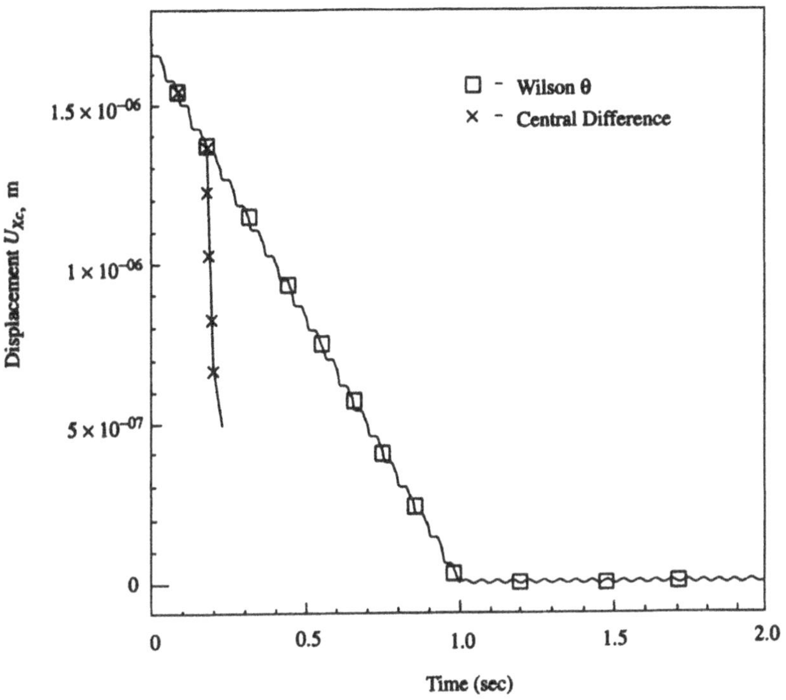

FIGURE 10.13
Comparison of the central difference method with the Wilson θ method ($\Delta t = 0.6 \times 10^{-3}$ sec).

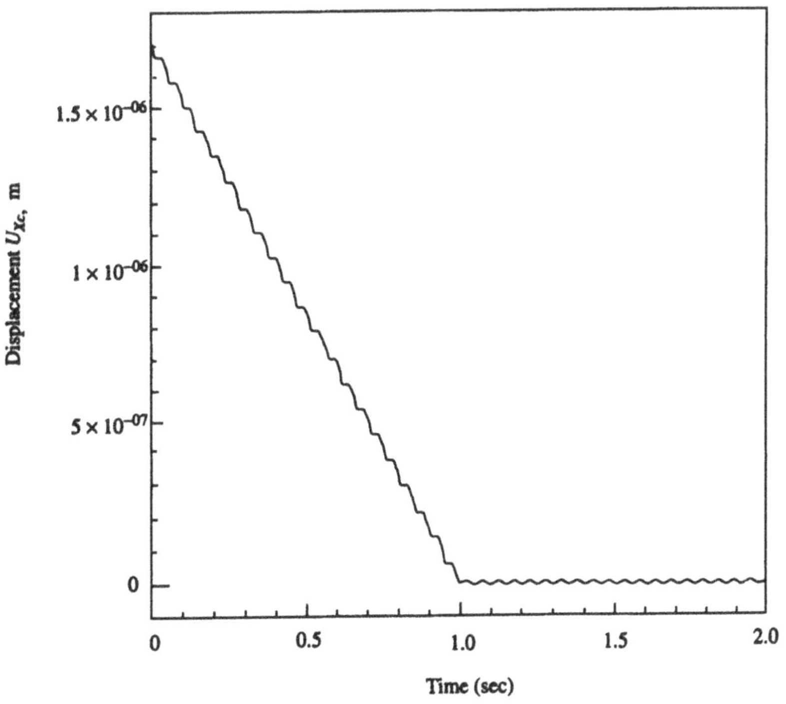

FIGURE 10.14
The Wilson θ method ($\Delta t = 0.1 \times 10^{-2}$ sec).

10.5.5 General Remarks

The most fundamental decisions involved in using the direct integration method concern the choice of (1) the appropriate integration method and (2) the appropriate integration step. These choices play an important role in the control of the cost of the analysis. The efficiency, stability, and accuracy should be taken into account for the effective use of a direct integration method.

Efficiency

The basic difference between the explicit and implicit methods is that the explicit integration method evaluates the equilibrium at $t = n\,\Delta t$, while the implicit integration method evaluates the equilibrium at $t = (n+1)\Delta t$. Both of these methods need to solve a set of simultaneous linear equations $\widehat{\mathbf{K}}\mathbf{V}_{n+1} = \mathbf{P}_{n+1}$, where $\widehat{\mathbf{K}}$ is the effective stiffness matrix. No iteration at a particular time step is needed. In the implicit method, the stiffness matrix $\mathbf{K}$ is in $\widehat{\mathbf{K}}$, while in the explicit method, $\widehat{\mathbf{K}}$ is not a function of $\mathbf{K}$. If the finite element model is very large with a large bandwidth, it can be more efficient to use the explicit method with a lumped mass matrix, so that the effective stiffness matrix does not need to be triangularized. Of course, the time step should be smaller than the critical time step.

Stability

Some remarks about the stability of the different integration methods are provided in Sections 10.5.1 through 10.5.4. In these discussions, only a single DOF system is analyzed; however, the results are assumed to apply to a multi-DOF system. The validity of this assumption follows from the relationship between the mode superposition and the direct integration methods. For a system governed by the equations

$$\mathbf{M}\ddot{\mathbf{V}} + \mathbf{C}\dot{\mathbf{V}} + \mathbf{K}\mathbf{V} = \overline{\mathbf{P}} \tag{10.126}$$

and with a proportional damping matrix, i.e., $\mathbf{C} = \alpha\mathbf{M} + \beta\mathbf{K}$, the principal coordinate transformation leads to

$$\ddot{\mathbf{q}} + \boldsymbol{\Delta}\dot{q} + \boldsymbol{\Lambda}\mathbf{q} = \mathbf{P} \tag{10.127a}$$

or

$$\ddot{q}_i + 2\zeta_i\omega_i\dot{q}_i + \omega_i^2 q_i = P_i, \quad i = 1, 2, \ldots, n_d \tag{10.127b}$$

where $\boldsymbol{\Delta}$ and $\boldsymbol{\Lambda}$ are diagonal matrices with the diagonal elements $2\zeta_i\omega_i$ and ω_i^2, respectively. If these "single-DOF-like" equations are integrated directly, the results would be the same as for the integration of Eq. (10.126) with the same time step. It is reasoned then that the stability criteria (developed in Section 10.5.1 for single DOF systems) for Eq. (10.127) apply to Eq. (10.126) as well. For the central difference method, the integration is conditionally stable. The integration time step should be smaller than the critical time step $\Delta t_{\text{crit}} = 2/\omega_{\max}$, where $\omega_{\max}^2$ is the largest diagonal element in $\boldsymbol{\Lambda}$. For the implicit integration methods, the integration is unconditionally stable.

Accuracy

The analysis of accuracy provides the rationale for selecting the time step Δt. "Accuracy" refers to the difference between the numerical solution and the exact solution when the numerical solution process is stable. For the central difference method, the time step Δt must be smaller than Δt_{crit} in order for the integration scheme to be stable. When $\Delta t < \Delta t_{\text{crit}}$, the result of the integration is usually fairly accurate, i.e., the difference between the numerical solution and exact solution is small. For the implicit integration, which is numerically

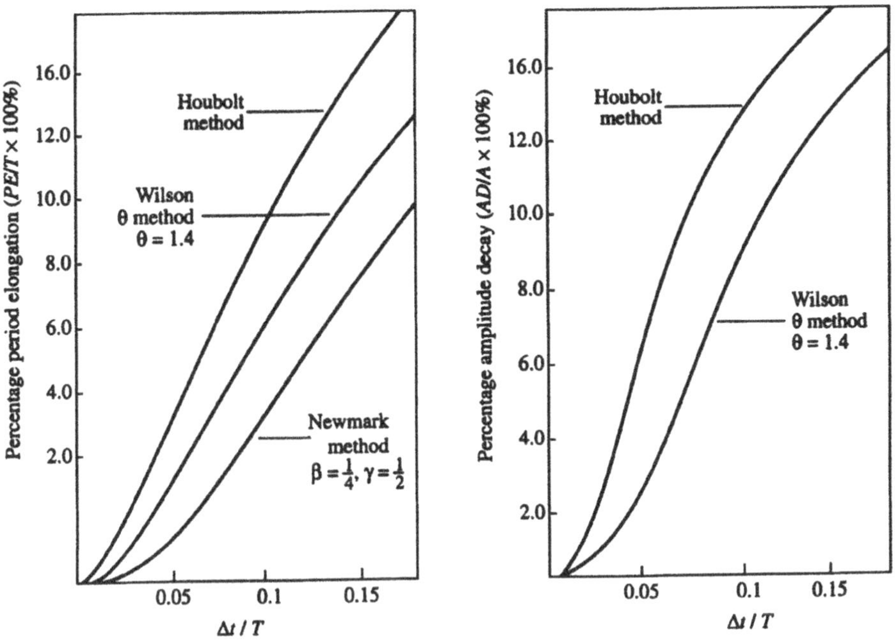

FIGURE 10.15
Percentage period elongations (PE) and amplitude decays (AD). See Fig. 10.10 for definitions.

stable, the selection of Δt should be based on the analysis of the accuracy of each particular method.

It is known that large integration steps may cause period elongation (PE) and amplitude decay (AD) (Fig. 10.10). This can be demonstrated by the integration of

$$\ddot{q}_i + \omega_i^2 q_i = 0 \tag{10.128}$$

$$q_i(0) = 1.0, \qquad \dot{q}_i(0) = 0 \tag{10.129}$$

which is one of the equations in Eq. (10.127) with $\zeta_i = 0$ and $P_i = 0$. The result is shown in Fig. 10.15. The curves in Fig. 10.15 indicate that, in general, when $\Delta t/T$ is smaller than 0.01, where T is the period corresponding to Eq. (10.128), the integration is accurate using any of the integration methods. But when $\Delta t/T$ is larger than 0.01, the different integration schemes exhibit different characteristics. For a given $\Delta t/T$, the Wilson θ method with $\theta = 1.4$ introduces less amplitude decay and period elongation than the Houbolt method, and the Newmark method only introduces period elongation, but no amplitude decay. See Bathe (1982) for further discussions of accuracy.

10.6 Dynamic Analysis Based on Ritz Vectors

If the modal superposition method of Section 10.4.1 with the usual free vibration mode shapes is applied to structures under earthquakes forces, it is found in the references by Wilson that the results are not as accurate as those obtained with basis vectors that take into account the spatial distribution of the dynamic loading. These basis vectors, called *Ritz vectors*, have been used in wave propagation and foundation analyses [Bayo and Wilson, 1985].

The numerical determination of Ritz vectors takes less computational work than needed to calculate mode shapes. This reduces the numerical effort needed in time-domain solutions of complex structural problems. In addition, the method possesses the advantages of static condensation, Guyan reduction, and higher mode truncation [Wilson, 1985]. The examples studied in Wilson, et al. (1982) have shown that a dynamic analysis based on Ritz vectors gives more accurate results, with fewer vectors, than when the usual mode shapes are employed.

Suppose the applied force $\overline{\mathbf{P}}$ in the governing equations of Eq. (10.97) can be expressed as a product of a vector of spatial distributions of loading and a function of time. That is,

$$\mathbf{M}\ddot{\mathbf{V}} + \mathbf{C}\dot{\mathbf{V}} + \mathbf{K}\mathbf{V} = \overline{\mathbf{P}} = \mathbf{F}(s)g(t) \tag{10.130}$$

The time-independent vector $\mathbf{F}(s)$ represents the spatial distribution of the external force and $g(t)$ is a function of time only. Equation (10.130) is reduced in size by using a set of global Ritz vectors $\mathbf{x}_1, \mathbf{x}_2, \ldots, \mathbf{x}_n$. Let the number of entries of $\mathbf{V}$ be n_d and let $\mathbf{X}$ be the $n_d \times n$ matrix whose ith column is $\mathbf{x}_i$. The displacement vector $\mathbf{V}$ is written in terms of the Ritz vectors as

$$\mathbf{V} = \mathbf{X}\mathbf{y} \tag{10.131}$$

where $\mathbf{y}$ is a vector with n elements. When Eq. (10.131) is substituted into Eq. (10.130) and the results premultiplied by $\mathbf{X}^T$, the following set of differential equations is obtained

$$\mathbf{M}_1\ddot{\mathbf{y}} + \mathbf{C}_1\dot{\mathbf{y}} + \mathbf{K}_1\mathbf{y} = \mathbf{F}_1 g(t) \tag{10.132}$$

The number of scalar differential equations in Eq. (10.132) is equal to n, the number of Ritz vectors chosen in the expansion of Eq. (10.131). In Eq. (10.132)

$$\mathbf{M}_1 = \mathbf{X}^T\mathbf{M}\mathbf{X} \tag{10.133}$$

$$\mathbf{K}_1 = \mathbf{X}^T\mathbf{K}\mathbf{X} \tag{10.134}$$

$$\mathbf{C}_1 = \mathbf{X}^T\mathbf{C}\mathbf{X} \tag{10.135}$$

$$\mathbf{F}_1 = \mathbf{X}^T\mathbf{F} \tag{10.136}$$

The condensed relationship of Eq. (10.132) can be solved by one of the direct step-by-step integration methods discussed in Section 10.5. To complete the derivation of Eq. (10.132), Ritz vectors will now be explicitly defined. The first Ritz vector $\mathbf{x}_1^*$ is taken as the displacement vector obtained from a static analysis with $\mathbf{F}(s)$ as the external force and is found by solving the linear algebraic equation

$$\mathbf{K}\mathbf{x}_1^* = \mathbf{F}(s) \tag{10.137}$$

for $\mathbf{x}_1^*$. The Ritz vector $\mathbf{x}_1$, which is the first column of the matrix $\mathbf{X}$, is found from $\mathbf{x}_1^*$ by normalizing $\mathbf{x}_1^*$ with respect to the mass matrix $\mathbf{M}$

$$\mathbf{x}_1 = \frac{\mathbf{x}_1^*}{\sqrt{\mathbf{x}_1^{*T}\mathbf{M}\mathbf{x}_1^*}} \tag{10.138}$$

The normalized vector $\mathbf{x}_1$ has the property

$$\mathbf{x}_1^T\mathbf{M}\mathbf{x}_1 = 1$$

The other Ritz vectors are defined by linear algebraic equations in which $\mathbf{M}$ is multiplied by the previously calculated Ritz vector and the resulting vector is used as the external force for a static analysis. Thus, for $i = 2, 3, \ldots, n$, the equation

$$\mathbf{K}\mathbf{x}_i^* = \mathbf{M}\mathbf{x}_{i-1} \tag{10.139}$$

is solved for $\mathbf{x}_i^*$. Next, the vector $\mathbf{x}_i^*$ is orthogonalized and normalized with respect to $\mathbf{M}$. These operations are carried out by calculating for $j = 1, 2, \ldots, i-1$ the scalars

$$c_j = \mathbf{x}_j^T \mathbf{M}\mathbf{x}_i^* \tag{10.140}$$

followed by

$$\mathbf{x}_i^{**} = \mathbf{x}_i^* - \sum_{j=1}^{i-1} c_j \mathbf{x}_j \tag{10.141}$$

and finally

$$\mathbf{x}_i = \frac{\mathbf{x}_i^{**}}{\sqrt{\mathbf{x}_i^{**T}\mathbf{M}\mathbf{x}_i^{**}}} \tag{10.142}$$

Since the Ritz vectors $\mathbf{x}_i$, $1 \le i \le n$, are orthogonalized and normalized with respect to the mass matrix $\mathbf{M}$, the condensed mass matrix $\mathbf{M}_1$ given by Eq. (10.133) is diagonal. The matrices $\mathbf{K}_1$ and $\mathbf{C}_1$ are, in general, full. Once the matrices $\mathbf{M}_1$ and $\mathbf{K}_1$ are calculated, it is possible to find approximate values for the lowest n natural frequencies and the corresponding mode shape vectors by solving the generalized eigenvalue problem stated in Eq. (10.82) with $\mathbf{M}$ replaced with $\mathbf{M}_1$ and $\mathbf{K}$ with $\mathbf{K}_1$.

EXAMPLE 10.12 Ritz Vector Method for a Frame

Return to the frame shown in Fig. 10.2. The system mass and stiffness matrices $\mathbf{M}$ and $\mathbf{K}$ are given by Eq. (9) of Example 10.1 and Eq. (11) of Chapter 5, Example 5.5.

As in Example 10.11 the damping matrix $\mathbf{C}$ is assumed to be proportional to the mass matrix, $\mathbf{C} = 0.000625\,\mathbf{M}$. Force $\overline{\mathbf{P}}(t)$ of Fig. 10.8 is applied at node b in the X direction. The system governing equation is given by Eq. (10.130) with $\mathbf{V} = [U_{Xb} \quad U_{Zb} \quad \Theta_b \quad U_{Xc} \quad U_{Zc} \quad \Theta_c]^T$, and, for the loading applied at node b in the X direction, $\mathbf{F} = [1 \quad 0 \quad 0 \quad 0 \quad 0 \quad 0]^T$. Also, the function $g(t)$ is expressed in Fig. 10.8.

We choose to use Ritz vectors to condense this six-DOF system to a three-DOF system. From Eq. (10.137), the first Ritz vector $\mathbf{x}_1^*$ is the solution of $\mathbf{K}\mathbf{x}_1^* = \mathbf{F}$. This gives

$$\begin{aligned}\mathbf{x}_1^* = [&1.74034 \times 10^{-7} \quad 0.97015 \times 10^{-7} \quad -0.02479 \times 10^{-7} \\ &1.71722 \times 10^{-7} \quad 0.00789 \times 10^{-7} \quad -0.43951 \times 10^{-7}]^T\end{aligned} \tag{1}$$

Normalize $\mathbf{x}_1^*$ according to Eq. (10.138) to find the first Ritz vector.

$$\begin{aligned}\mathbf{x}_1 = [&0.77242 \times 10^{-1} \quad 0.43058 \times 10^{-1} \quad -0.01100 \times 10^{-1} \\ &0.76216 \times 10^{-1} \quad 0.00350 \times 10^{-1} \quad -0.19507 \times 10^{-1}]^T\end{aligned} \tag{2}$$

Follow Eqs. (10.139) to (10.142) to find the next two Ritz vectors.

$$\begin{aligned}\mathbf{x}_2 = [&-0.90063 \times 10^{-1} \quad 0.03594 \times 10^{-1} \quad 1.79391 \times 10^{-1} \\ &-0.11120 \times 10^{-1} \quad -0.01446 \times 10^{-1} \quad 2.12710 \times 10^{-1}]^T\end{aligned} \tag{3}$$

$$\begin{aligned}\mathbf{x}_3 = [&0.51388 \times 10^{-1} \quad -0.24340 \times 10^{-1} \quad -2.07095 \times 10^{-1} \\ &-0.37934 \times 10^{-1} \quad 0.06251 \times 10^{-1} \quad 1.34613 \times 10^{-1}]^T\end{aligned} \tag{4}$$

Form the transformation matrix $\mathbf{X} = [\mathbf{x}_1 \ \ \mathbf{x}_2 \ \ \mathbf{x}_3]$. Use this matrix, along with $\mathbf{M}$ of Eq. (9) of Example 10.1 and $\mathbf{K}$ of Eq. (11) of Chapter 5, Example 5.5, to compute $\mathbf{M}_1$, $\mathbf{K}_1$, $\mathbf{C}_1$, and $\mathbf{F}_1$, using Eqs. (10.133) through (10.136)

$$\mathbf{M}_1 = \begin{bmatrix} 1.00000 & 0 & 0 \\ 0 & 1.00000 & 0 \\ 0 & 0 & 1.00000 \end{bmatrix} \tag{5}$$

$$\mathbf{K}_1 = \begin{bmatrix} 3.42824 \times 10^4 & -3.99727 \times 10^4 & 2.28076 \times 10^4 \\ -3.99727 \times 10^4 & 3.19139 \times 10^6 & -1.82093 \times 10^6 \\ 2.28075 \times 10^4 & -1.82093 \times 10^6 & 2.71161 \times 10^6 \end{bmatrix} \tag{6}$$

$$\mathbf{C}_1 = \begin{bmatrix} 0.000625 & 0 & 0 \\ 0 & 0.000625 & 0 \\ 0 & 0 & 0.000625 \end{bmatrix} \tag{7}$$

$$\mathbf{F}_1 = [0.77242 \times 10^{-1} \quad -0.90063 \times 10^{-1} \quad 0.51388 \times 10^{-1}] \tag{8}$$

As expected, this condensed mass matrix $\mathbf{M}_1$, is diagonal. The condensed system equations are given by Eq. (10.132), $\mathbf{M}_1\ddot{\mathbf{y}} + \mathbf{C}_1\dot{\mathbf{y}} + \mathbf{K}_1\mathbf{y} = \mathbf{F}_1 g(t)$. Integrate this relation and calculate the displacements $\mathbf{V}$ using Eq. (10.131).

The Wilson θ method applied to the Ritz vector equations leads to the same results as given in Example 10.11 for the Wilson θ method. ∎

References

Bathe, K.J., 1982, *Finite Element Procedures in Engineering Analysis*, Prentice-Hall, Englewood Cliffs, NJ.

Bathe, K.J. and Wilson, E.L., 1973, Stability and accuracy analysis of the direct integration method, *Earthquake Eng. Struct. Dynam.*, Vol. 1, pp. 283–291.

Bayo, E.P. and Wilson, E.L., 1984, Use of Ritz vectors in wave propogation and foundation response, *Earthquake Eng. Struct. Dynam.*, Vol. 12, pp. 499–505.

Cook, R.D., 1981, *Concepts and Applications of Finite Element Analysis*, 2nd ed., Wiley, NY.

Fergusson, N. and Pilkey, W.D., 1992, Frequency-dependent element mass matrices, *J. Appl. Mech.*, Vol. 59, pp. 136–139.

Fergusson, N. and Pilkey, W.D., 1993, Literature review of variants of the dynamic stiffness method, Part 1: The dynamic element method, *Shock and Vib. Digest*, Vol. 25(2), pp. 3–12.

Fergusson, N. and Pilkey, W.D., 1993, Literature review of variants of the dynamic stiffness method, Part 2: Frequency-dependent matrix and other corrective methods, *Shock and Vib. Digest*, Vol. 25(4), pp. 3–10.

Fried, I. and Malkus, D.S., 1975, Finite element mass matrix lumping by numerical integration with no convergence rate loss, *Int. J. Solids and Struct.*, Vol. 11, pp. 461–466.

Guyan, R.J., 1965, Reduction of stiffness and mass matrices, *AIAA J.*, Vol. 3, p. 380.

Houbolt, J.C., 1950, A recurrence matrix solution for the dynamic response of elastic aircraft, *J. Aeronaut. Sci.*, Vol. 17, pp. 540–550.

Hughes, T., 1987, *The Finite Element Method*, Prentice-Hall, Englewood Cliffs, NJ.

Kim, K., 1993, A review of mass matrices for eigenproblems, *Comp. Struct.*, Vol. 46, pp. 1041–1048.

Newmark, N.M., 1959, A method of computation for structural dynamics, *Proc. ASCE*, Vol. 85 (EM3), Part 1, pp. 67–94.

Richards, T.H. and Leung, Y.T., 1977, An accurate method in structural vibration analysis, *J. Sound Vib.*, Vol. 55, pp. 363–376.
Wilson, E.L., 1985, A new method of dynamic analysis for linear and nonlinear systems, *J. Finite Elem. in Anal. Des.*, Vol. 1, pp. 21–23.
Wilson, E.L., Yuan, M.W. and Dickens, J.M., 1982, Dynamic analysis by direct superposition of Ritz vectors, *Earthquake Eng. Struct. Dynam.*, Vol. 10, pp. 813–823.

Problems

Structural Matrices

10.1 Find the element stiffness and consistent mass matrices and the global stiffness and mass matrices for the frame of Fig. P10.1.

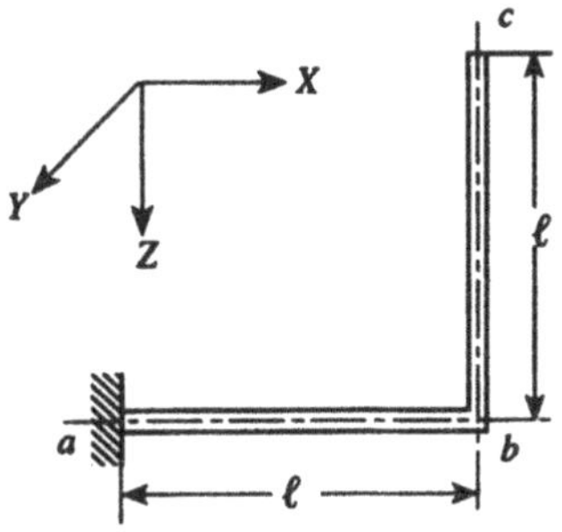

$\ell = 10$ in.
$A = 1$ in.^{2}
$I = 0.083$ in.^{4}
$E = 1 \times 10^7$ psi
$\nu = 0.33$
$\rho = 2.59 \times 10^{-4}$ lb-sec^2/in.^{2}

FIGURE P10.1
L-shaped frame.

Answer: The global stiffness and mass matrices, after invoking the boundary conditions, are

$$[U_{Xb} \quad U_{Zb} \quad \Theta_b \quad U_{Xc} \quad U_{Zc} \quad \Theta_c]$$

$$\mathbf{K} = 830\begin{bmatrix} 1216.8 & 0 & 60 & -12 & 0 & 60 \\ 0 & 1216.8 & -60 & 0 & -1204.8 & 0 \\ 60 & -60 & 800 & -60 & 0 & 200 \\ -12 & 0 & -60 & 12 & 0 & -60 \\ 0 & -1204.8 & 0 & 0 & 1204.8 & 0 \\ 60 & 0 & 200 & -60 & 0 & 800 \end{bmatrix}$$

$$\mathbf{M} = \frac{\rho\ell}{420}\begin{bmatrix} 296 & 0 & 220 & 54 & 0 & -130 \\ 0 & 296 & 220 & 0 & 70 & 0 \\ 220 & 220 & 800 & 130 & 0 & -300 \\ 54 & 0 & 130 & 156 & 0 & -220 \\ 0 & 70 & 0 & 0 & 140 & 0 \\ -130 & 0 & -300 & -220 & 0 & 400 \end{bmatrix}$$

10.2 Find the lumped mass matrices (element and global) of the frame in Problem 10.1 with and without rotary inertia.

Answer: Global matrix, based on physical lumping of mass at the ends of two elements, without rotary inertia $\mathbf{M} \approx 1.259 \times 10^{-3}$ diag (2 2 0 1 1 0) where $\mathbf{V} = [U_{Xb}\ U_{Zb}\ \Theta_b\ U_{Xc}\ U_{Zc}\ \Theta_c]^T$ with rotary inertia $\mathbf{M} \approx 1.259 \times 10^{-3}$ diag (2 2 17 1 1 8.5)

10.3 Use Guyan reduction to condense out the DOF related to the rotation for the frame in Problem 10.1. Find the stiffness, consistent, and lumped mass matrices.

Answer:

$$[U_{Xb} \quad U_{Zb} \quad U_{Xc} \quad U_{Zc}]$$

$$\mathbf{K} = 8.3 \times 10^5 \begin{bmatrix} 1.2065 & -0.0026 & -0.0017 & 0 \\ -0.0026 & 1.2117 & 0.0026 & -1.2048 \\ -0.0017 & 0.0026 & 0.0017 & 0 \\ 0 & -1.2048 & 0 & 1.2048 \end{bmatrix}$$

$$\mathbf{M} = \frac{\rho\ell}{420} \begin{bmatrix} 315.3469 & -35.8776 & 64.6531 & 0 \\ -35.8776 & 267.1020 & -9.1224 & 70.0000 \\ 64.6531 & -9.1224 & 115.3469 & 0 \\ 0 & 70.0000 & 0 & 140.0000 \end{bmatrix}$$

Lumped mass matrix without rotary inertia:

$$\mathbf{M} = 1.259 \times 10^{-3} \begin{bmatrix} 2.0572 & 0 & 0 & 0 \\ 0 & 2.0572 & 0 & 0 \\ 0 & 0 & 1.0286 & 0 \\ 0 & 0 & 0 & 1.0286 \end{bmatrix}$$

Lumped mass matrix with rotary inertia:

$$\mathbf{M} = 1.259 \times 10^{-3} \begin{bmatrix} 2.2321 & 0.0159 & -0.1749 & 0 \\ 0.0159 & 2.2003 & -0.0159 & 0 \\ -0.1749 & -0.0159 & 1.2035 & 0 \\ 0 & 0 & 0 & 1.0286 \end{bmatrix}$$

10.4 Use three point Gauss quadrature to find the mass matrix for beams using the shape functions of Eq. (10.8)

Answer:

$$\mathbf{m}^i = \frac{\rho\ell}{1200} \begin{bmatrix} 444 & -62\ell & 156 & 38\ell \\ -62\ell & 11\ell^2 & -38\ell & -9\ell^2 \\ 156 & -38\ell & 444 & 62\ell \\ 38\ell & -9\ell^2 & 62\ell & 11\ell^2 \end{bmatrix}$$

10.5 Approximate mass matrices can be obtained in many ways. Show that the higher order matrix of Eq. (10.15) can be produced by employing a special numerical integration scheme with the integration points at $\pm\sqrt{2/3}$ and the weights equal to one in the integral for the element mass matrix of Eq. (10.5).

Hint: Use the shape functions of Chapter 4, Example 4.1.

10.6 Use four point Gauss quadrature to find the mass matrix for beams using the shape functions of Eq. (10.8)

Answer: Same as Eq. (10.10a), i.e., the usual consistent mass matrix for beams is obtained.

10.7 Find a mass matrix for the two-dimensional eight node element of Fig. P10.7 with thickness t and mass density γ.

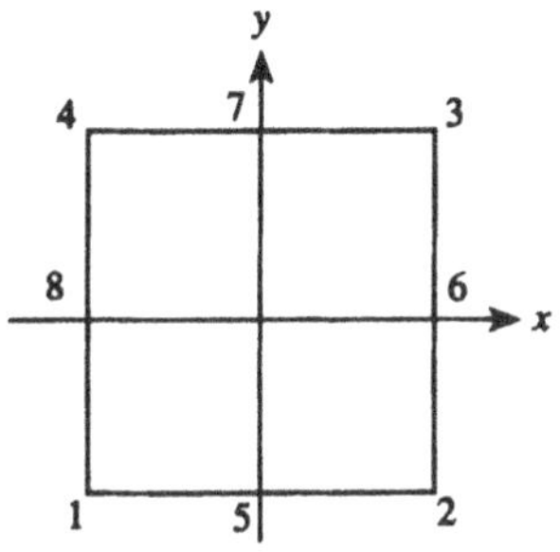

FIGURE P10.7

Natural Frequencies

10.8 Find expressions for the first two natural frequencies of the spring-mass system of Fig. P10.8.

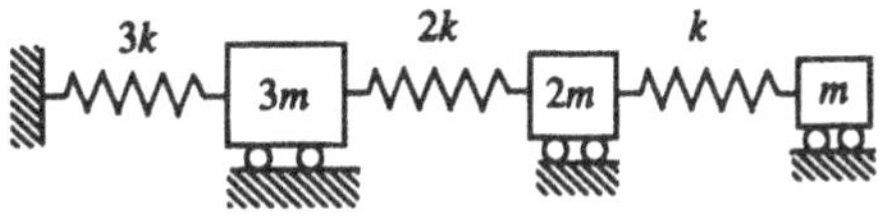

FIGURE P10.8

10.9 For $k = 100\ \text{lb/in.}$ and $m = 1\ \text{lb} \cdot \text{sec}^2/\text{in.}$, find the two lowest frequencies of the three DOF system of Fig. P10.8.

Answer: 0.87 Hz, 1.81 Hz.

10.10 Find three natural frequencies and mode shapes for a uniform extension bar fixed at the left end and free at the right end.

Answer: Exact result for first mode: $\omega_1 = \pi/2\sqrt{EA/(\rho L^2)}$, $\phi_1 = \sin \pi x/(2L)$

10.11 Find three natural frequencies for the longitudinal motion of the rod of Fig. P10.11.

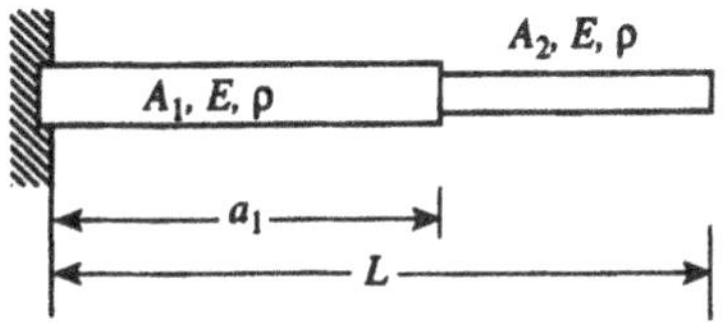

FIGURE P10.11

10.12 Find the first three natural frequencies of a cantilevered beam.

Answer: Exact results: $\omega_1 = 1.875^2\sqrt{EI/(\rho L^4)}$, $\omega_2 = 4.694^2\sqrt{EI/(\rho L^4)}$, $\omega_3 = 7.855^2\sqrt{EI/(\rho L^4)}$.

10.13 Suppose a cantilevered extension bar of mass/length ρ carries a mass m at the free end. Find four natural frequencies and mode shapes.

Answer: Exact result for first mode with $\rho L/m = 1$; $\omega_1 = 0.860\sqrt{EA/(\rho L^2)}$, $\phi_1 = \sin 0.860\frac{x}{L}$.

10.14 Find the natural frequencies of a clamped free beam using a determinant search and the dynamic stiffness matrix of Table 10.1. Suppose the cross-section is circular of diameter d. Let $L = 1$ m, $d = 0.1$ m, $E = 207$ GPa, $\nu = 0.3$ and $\gamma = 7800$ kg/m^3.

Answer: $\omega_i = \lambda_i^2\sqrt{(EI/\gamma AL^4)}$ with $\lambda_1 = 1.8699$, $\lambda_2 = 4.6065$, $\lambda_3 = 7.5313$.

10.15 Find the fundamental natural frequency of the system shown in Fig. P10.15. Let $L = 20$ in., $m = 10$ lb $\cdot$ sec^2/in., $E = 10^7$ psi, $I = 2$ in.^{4}, and the weight density of the beam material equals 1.5 lb/in.

(a) Model the beam with distributed mass.
(b) Neglect the weight of the beam.
(c) Lump all the weight of the beam at midspan and treat the system as a two-DOF system.
(d) Use Guyan reduction to reduce the system to a single-DOF model.

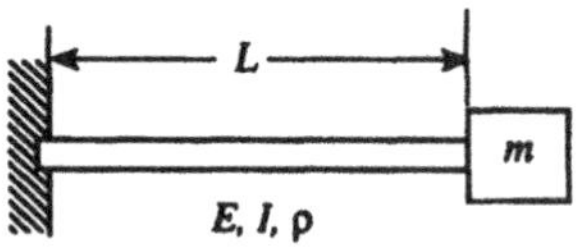

FIGURE P10.15

Answer: (b) 27.39 rad/sec (c) 27.46 rad/sec (d) 27.37 rad/sec

10.16 Find the first ten natural frequencies and mode shapes for the bending of the stepped beam of Fig. P10.16. Use consistent mass matrices. Also, condense out the rotational DOF. Find the first five natural frequencies and mode shapes.

Answer:

Mode No.	Consistent Mass	Mode No.	Condensed Case
1	6.251946×10	1	62.520
2	1.058243×10^2	2	105.840
3	2.222744×10^2	3	222.369
4	3.616329×10^2	4	361.800
5	5.487418×10^2	5	548.834
6	6.676266×10^3		
7	7.858440×10^3		
8	8.973045×10^3		
9	1.012506×10^4		
10	1.097755×10^4		

10.17 Find the natural frequencies of the frame of Fig. P10.1. Use the mass and stiffness matrices obtained in Problem 10.1.

Answer: $\omega_1 = 663$, $\omega_2 = 1807$, $\omega_3 = 9819$, $\omega_4 = 25\,385$, $\omega_5 = 28\,777$, $\omega_6 = 59\,439$.

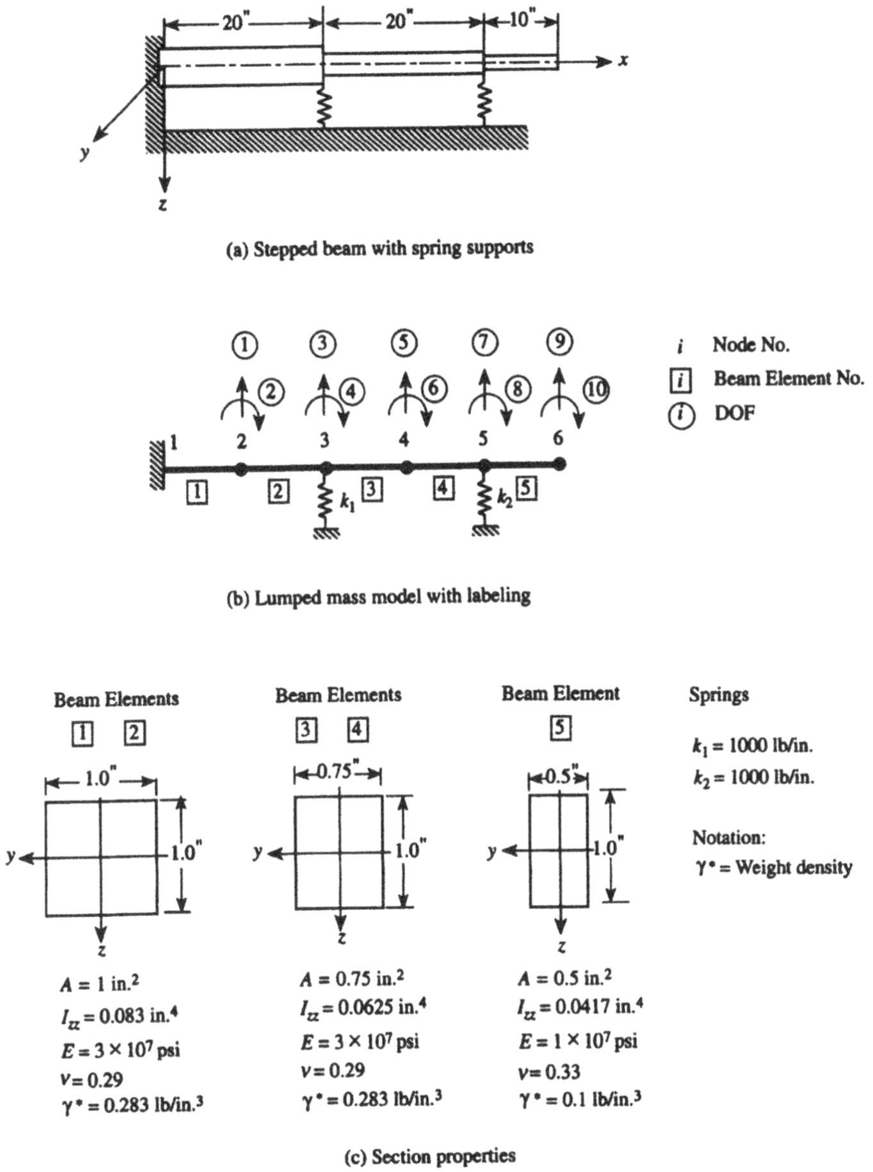

FIGURE P10.16

10.18 Use the lumped mass matrix of Problem 10.2 to find the natural frequencies of the frame of Fig. P10.1.

Answer: $\omega_1 = 546, \omega_2 = 1479, \omega_3 = 5801, \omega_4 = 8606, \omega_5 = 20\,047, \omega_6 = 34\,536.$

10.19 Use the condensed stiffness and mass matrices of Problem 10.3 to find the natural frequencies of the frame of Fig. P10.1.

Answer:

Consistent mass matrix		Lumped mass matrix	
$\omega_1 = 663,$	$\omega_2 = 1825,$	$\omega_1 = 559,$	$\omega_2 = 1499$
$\omega_3 = 24\,150,$	$\omega_4 = 52\,574,$	$\omega_3 = 19\,663,$	$\omega_4 = 34\,045$

10.20 Find the first three natural frequencies of a fixed-pinned beam of length L. Use both consistent and lumped mass matrices. Compare your results with exact frequencies. Discuss means of convergence to the exact results.

Answer: Exact Frequencies:

$$\omega_1 = 15.4182\sqrt{EI/\rho L^2}\ \text{rad/s},$$
$$\omega_2 = 49.9651\sqrt{EI/\rho L^2}$$
$$\omega_3 = 104.2461\sqrt{EI/\rho L^2}.$$

10.21 Consider the axial motion of a uniform rod fixed at the left end and free at the right end. The cross-sectional area is A and the length L. Use the stiffness matrix for extensions of Chapter 4 and a consistent mass matrix. Compare the frequencies obtained using one element and a two element model with the exact solution.

Answer:

$\omega_i = C\sqrt{E/\gamma L^2}$

$\gamma =$ mass/volume

	C		
Mode	Exact	1 Element	2 Elements
1	1.571	1.732	1.610
2	4.712		5.628

10.22 For the transverse motion of a clamped-free uniform beam, compare the first two natural frequencies obtained using one, two, and three element models with the exact solution. Use the stiffness matrix for beam bending from Chapter 4 and a consistent mass matrix.

Answer: $\omega_i = C^2\sqrt{EI/\rho L^4}$

		C		
Mode	Exact	1 Element	2 Elements	3 Elements
1	1.8751	1.880	1.8754	1.8751
2	4.6941	5.900	4.7130	4.7041

Transient Responses

10.23 Use the modal superposition method to compute the longitudinal response of a bar fixed at the left end and free at the right end, subject to the sudden application of a uniformly distributed longitudinal force of intensity $\overline{p}_x$ (force/length).

Answer:

$$u(x,t) = \frac{16L^2\overline{p}_x}{\pi^3 c^2 \rho} \sum_{i=1,3,5}^{\infty} \frac{1}{i^3} \sin\frac{i\pi x}{L}\left(1 - \cos\frac{i\pi ct}{2L}\right)$$
$$c = \sqrt{EA/\rho}$$

10.24 Find the mode shapes of the frame in Problem 10.1. Verify their orthogonality conditions and normalize the mode shapes.

10.25 Calculate the X direction displacement of point c of the frame in Fig. P10.1 using the modal superposition method. The applied force is shown in Fig. 10.8, and acts at point c in the X direction. Assume 2% modal damping for all modes of the structure and that the structure is initially at rest.

Answer: Max X direction displacement at node $c = 0.0188$ in.

10.26 Use the central difference method to solve Problem 10.25. Find the critical time step.

10.27 Use the Houbolt method to solve Problem 10.25.

10.28 Use the Newmark method to solve Problem 10.25.

10.29 Use the Wilson-θ method to solve Problem 10.25.

Ritz Vectors

10.30 Use the Ritz vector method to solve Problem 10.25. Assume the damping matrix is proportional to the mass matrix, and is equal to 0.000625 **M**. Condense the system equations from 6 DOF to 1, 2, 3, 4, 5, and 6 DOF, respectively. Compare the displacement at c in the X direction with the results of Problems 10.27, 10.28, or 10.29.

10.31 The amplitude of the time-varying load acting on the fixed-pinned beam of Chapter 7, Fig. 7.1 varies linearly over the span. Let the time variation be sinusoidal so that the external load may be written as $F(s)\sin\Omega t$, where $F(s)$ represents the spatial distribution. The beam is 1600 mm long and has a square 60 mm × 60 mm cross-section. The density of the beam is 7860 kg/m^3 and its modulus of elasticity is 200 GPa. Also, $p_0 = 2$ kN/m.

(a) Assemble the global structural matrices and apply the boundary conditions to obtain 7 × 7 global stiffness and consistent mass matrices for the beam with 4 elements of equal length.

(b) Find the natural frequencies and the mode shapes of the beam using the matrices of part (a).

(c) Let the loading frequency Ω be equal to one-third the lowest natural frequency of the beam. Calculate the displacement of the midpoint of the beam using the modal properties found in part (b).

(d) Determine 3 × 3 stiffness and mass matrices for the beam using 3 Ritz vectors.

(e) Calculate the 3 natural frequencies corresponding to the matrices found in part (d) and compare with the results of part (b).

(f) Calculate the displacement of the midpoint of the beam using the modal properties of the matrices of part (d) and compare with the results of part (c).

11

Stability Analysis

Stability is not a clearly defined concept for all disciplines, but some simple analogies can be used to illustrate an intuitive notion of stability. Often we think in terms of the dynamics of a rigid body as shown in Fig. 11.1. There, slight disturbances to the equilibrium positions can be demonstrated in terms of stability. If the ball tends to return to the bottom of the trough, as the result of a slight disturbance, the equilibrium position is said to be *stable.* However, as in the case of Fig. 11.1b, if a small disturbance leads to a finite motion of the ball, the critical condition is called *unstable.* If the ball remains at the same vertical level, the equilibrium configuration is referred to as being *neutral.*

Many phenomena exhibit instabilities. Although most instability studies in structural mechanics deal with the elastic buckling of structural members and systems, there are several other areas of considerable interest such as plastic stability, creep instability, which is time dependent, and thermal stability, which is temperature dependent. Methods for the study of stability include the equilibrium method and the energy method.

Traditionally, the problem of buckling is to ascertain the conditions for which a structure in equilibrium is no longer stable. There is usually a parameter P, such as an applied load, for which the structure remains stable if P is small enough and becomes unstable for sufficiently large values of P. In the stable state, there is a unique configuration for each value of P. At a particular value of P, denoted P_{cr} for a *critical value* or *buckling load,* the structure ceases to be stable. In stability analyses of structures, we wish to find the equilibrium configurations under specified levels of applied loadings and to determine which of these are stable.

In linear structural mechanics, displacements are proportional to applied loads. Buckling, however, is characterized by an instability in which an inordinate increase in displacement can result from a small increase in applied load. As a consequence, buckling is a topic that belongs to nonlinear, rather than linear, mechanics.

EXAMPLE 11.1 Introduction to Some Stability Concepts

Apply the conditions of equilibrium to the rigid rod system of Fig. 11.2. Sum moments about point A of the displaced configuration to obtain

$$\sum M_A = 0 : \overline{P}L \sin\phi + \overline{H}L \cos\phi = k\phi \tag{1}$$

or

$$\overline{P} = \frac{k\phi - \overline{H}L \cos\phi}{L \sin\phi} \tag{2}$$

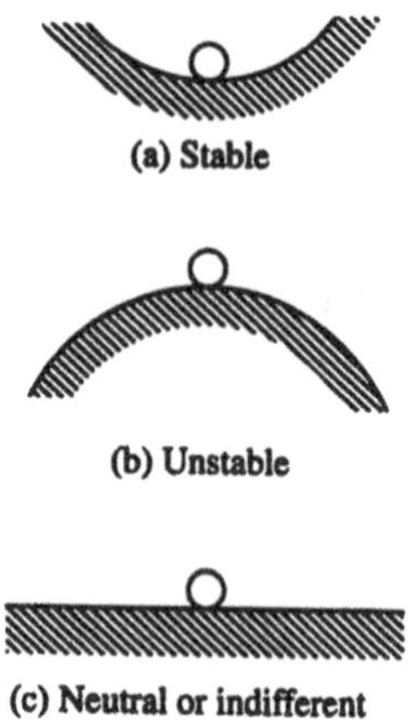

FIGURE 11.1
Stability of equilibrium positions.

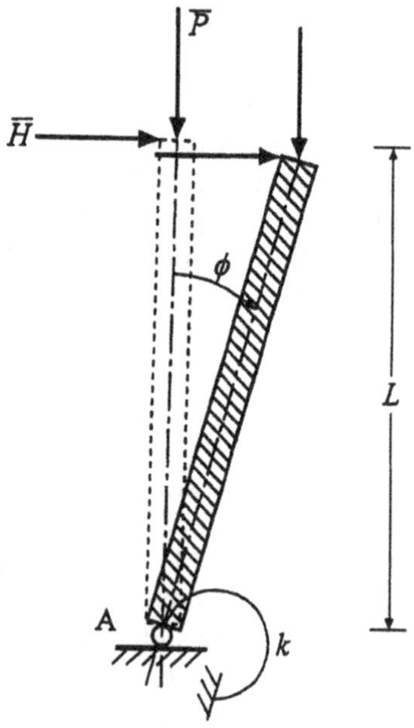

FIGURE 11.2
A rigid rod with spring.

Note that $\overline{P}$ is a nonlinear function of ϕ. To observe the response, plot $\overline{P}$ vs. ϕ. As shown for the case of non-zero $\overline{H}$ in Fig. 11.3a, typically, the displacement ϕ varies smoothly as $\overline{P}$ increases, i.e., there is a unique and stable state of equilibrium for each $\overline{P}$.

An interesting case occurs for $\overline{H} = 0$. Then

$$\overline{P} = \frac{k\phi}{L \sin \phi} \tag{3}$$

which is always greater than k/L. On the $\phi = 0$ axis, it would appear that the $\overline{P}$ cannot be evaluated. However, note that in the limit as $\phi \to 0$, $\overline{P}$ is equal to k/L. Thus, with the rod in a vertical position (at $\phi = 0$), $\overline{P}$ can be applied and increased until reaching the value $\overline{P} = k/L$, whereupon the bar could follow the path defined by (3). Then a small change in $\overline{P}$ can cause a major change in ϕ. The phenomenon where there is more than one equilibrium path is called *bifurcation*. The value $\overline{P} = k/L$ is the bifurcation point (see Fig. 11.3b).

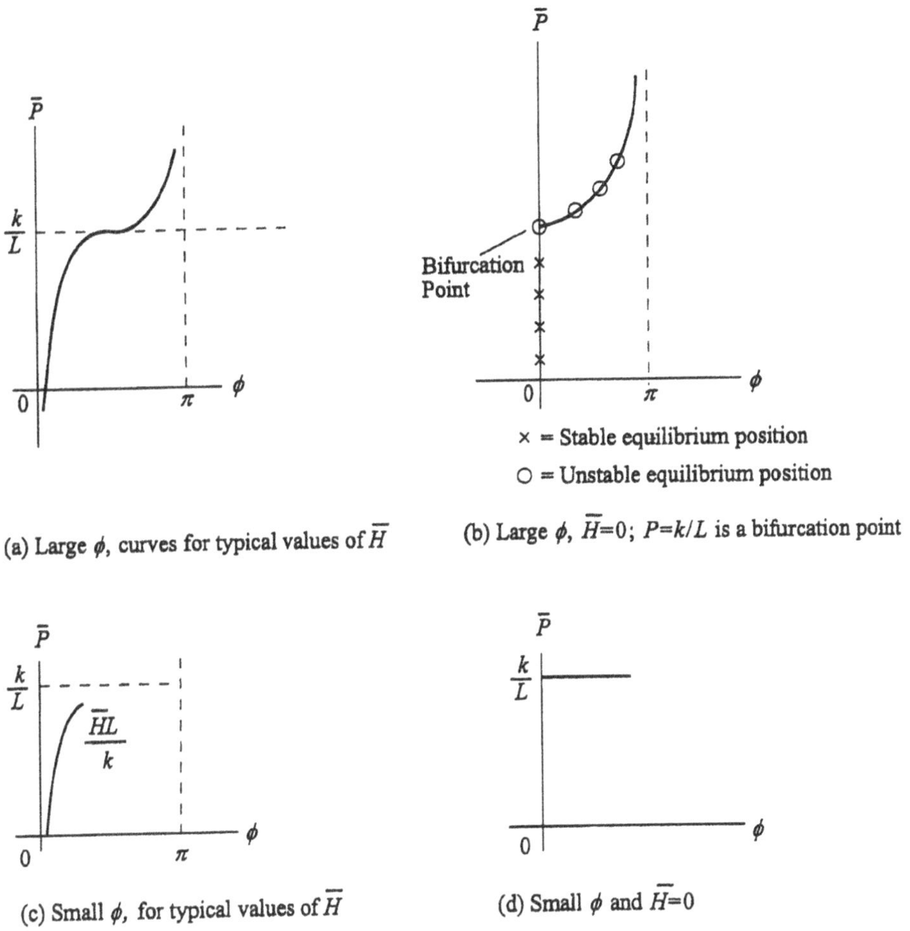

FIGURE 11.3
$\overline{P}$ vs. ϕ curves for the bar of Fig. 11.2.

Similar results are found if only small values of ϕ are to be considered. Then $\cos\phi \approx 1$, $\sin\phi \approx \phi$, so that $\overline{P} = (k\phi - \overline{H}L)/L\phi$. This can be written as

$$(k - \overline{P}L)\phi = \overline{H}L \tag{4}$$

which appears to be in a "stiffness-like" form. In (4), k corresponds to the conventional stiffness, and $\overline{P}L$ is the so-called differential (or geometric or stress) stiffness, that occurs in stability theory. If $\overline{P}$ in (4) approaches k/L, then small values at applied load $\overline{H}$ will make ϕ approach infinity. The plots of Figs. 11.3c and d show that the small ϕ theory indicates the same trends as the more accurate nonlinear theory of Figs. 11.3a and b. The bifurcation point of $\overline{P} = k/L$ is evident in Fig. 11.3d.

This example contains the characteristics of more difficult stability problems. The case of Fig. 11.3a, where $\overline{P}$ is a nonlinear function of ϕ and a unique state of equilibrium exists for each $\overline{P}$, would be analyzed from the standpoint of stresses if this were an elastic body. This is sometimes said to be a *third order theory*. It is characterized by the nonlinear function of ϕ, and that equilibrium is taken at the deformed state. When $\overline{H} = 0$, as in Fig. 11.3b, the bifurcation phenomenon can be identified.

For small ϕ, another stress analysis could be utilized in the case exhibited in Fig. 11.3c. If the influence of the nonlinearities are only approximately taken into account, this is often referred to as *second order theory*. Most of the stability solutions in this chapter are of second order.

First order theory is studied by basing equilibrium on an undeformed geometry, which is the case for this example when $\phi = 0$. ∎

11.1 Criteria for Stability

11.1.1 Energy Criterion

Consider an energy-based definition of the concept of stability. Begin by investigating the potential energy Π of a body as a perturbation moves it from a fundamental or basic configuration to an adjacent configuration.

$$\underset{\substack{\text{Basic}\\ \text{Configuration}}}{\mathring{\Pi}} \quad + \quad \underset{\substack{\text{Perturbation}\\ \text{or Incremental}\\ \text{Change}}}{\Delta\Pi} \quad = \quad \underset{\substack{\text{Neighboring}\\ \text{(Adjacent)}\\ \text{Configuration}}}{\widetilde{\Pi}} \tag{11.1}$$

The energy of the adjacent configuration can be written as a function of the energy of the basic configuration

$$\widetilde{\Pi} = \mathring{\Pi} + \Delta\Pi = \mathring{\Pi} + \delta\mathring{\Pi} + \frac{1}{2}\delta^2\mathring{\Pi} + \cdots \tag{11.2}$$

where $\delta\mathring{\Pi}, \delta^2\mathring{\Pi}, \ldots$ are the first, second, ... variations of $\mathring{\Pi}$. From the principle of stationary potential energy [Chapter 2, Eq. (2.65)], $\delta\mathring{\Pi} = 0$ for a body in equilibrium. Then the incremental change can be expressed as

$$\Delta\Pi = \widetilde{\Pi} - \mathring{\Pi} = \frac{1}{2}\delta^2\mathring{\Pi} + \cdots \tag{11.3}$$

It is evident that the state of the potential due to the perturbation is described by the second variation of the energy of the basic configuration. Bazant and Cedolin (1991) and Pflüger (1964) provide a thorough development of the second variation of potential energy as a stability criterion.

The various states of stability are illustrated in Fig. 11.4. For the transition from the fundamental to the adjacent configuration to remain stable, i.e., for the body to be in a stable state of equilibrium, any arbitrary perturbation should lead to an increase in potential energy, i.e., $\delta^2\mathring{\Pi} > 0$. If a particular perturbation can be found that leads to a decrease in potential energy, kinetic energy can be released, and the body is in an unstable state of equilibrium. Here, $\widehat{\delta}^2\mathring{\Pi} < 0$, where $\widehat{\delta}$ denotes the particular or special variation that leads to instability. For the border case, wherein no change in potential energy takes place, $\delta^2\mathring{\Pi} = 0$ or $\widehat{\delta}^2\mathring{\Pi} = 0$ is called the neutral or indifferent configuration.

In the classical stability theory, the state of the neutral configuration serves as the criterion for a critical load. This neutral configuration corresponds to a nonunique state of equilibrium that occurs for at least a particular perturbation.

In the case of a neutral or indifferent configuration, $\widetilde{\Pi} = \mathring{\Pi}$, so that

$$\delta(\widetilde{\Pi} - \mathring{\Pi}) = \delta(\Delta\Pi) = 0 \tag{11.4}$$

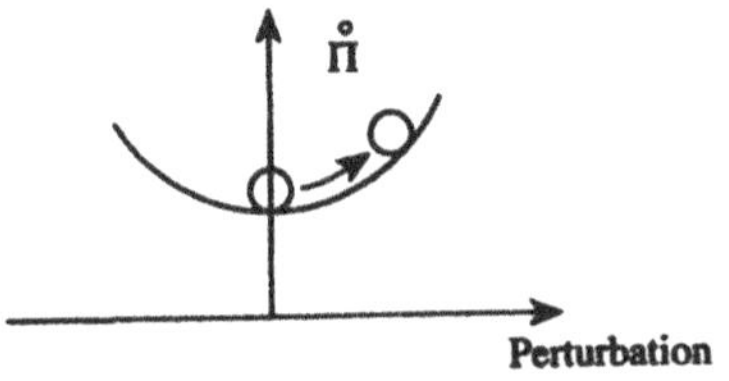

(a) Stability: $\delta^2 \mathring{\Pi} > 0$ for an arbitrary variation δ.

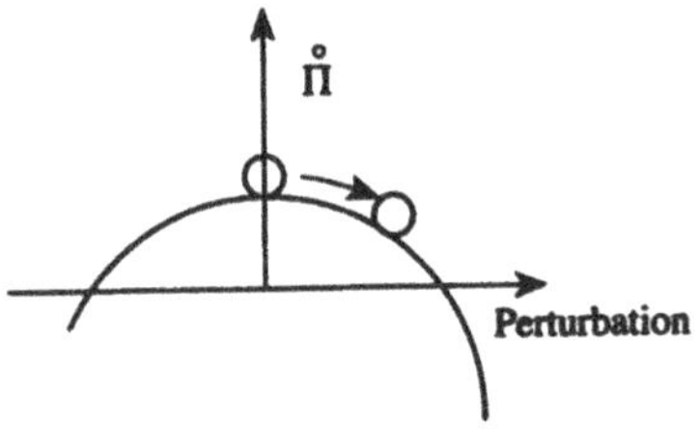

(b) Instability: $\hat{\delta}^2 \mathring{\Pi} < 0$ for a particular variation $\hat{\delta}$.

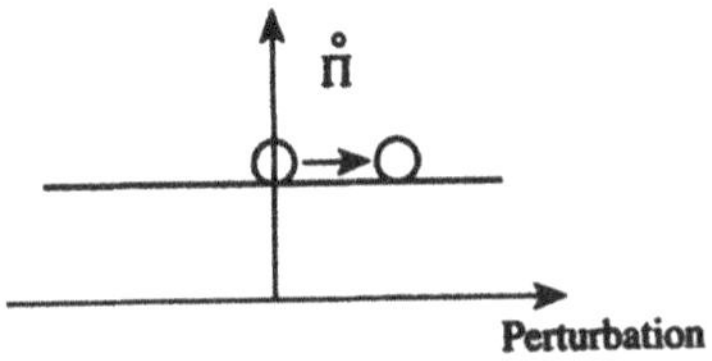

(c) Neutral configuration or indifferent case: $\delta^2 \mathring{\Pi} = 0$ or $\hat{\delta}^2 \mathring{\Pi} = 0$.

FIGURE 11.4
Stability and instability from the standpoint of energy.

Suppose the particular variation that corresponds to the critical load is $\widehat{\delta}$. Then

$$\delta(\Delta\Pi) = \delta\left(\widehat{\delta}\mathring{\Pi} + \frac{1}{2}\widehat{\delta}^2\mathring{\Pi} + \cdots\right) = \delta\left(\frac{1}{2}\widehat{\delta}^2\mathring{\Pi}\right) = 0 \tag{11.5}$$

or

$$\delta(\Delta\Pi) = \delta\left(\frac{1}{2}\widehat{\delta}^2\mathring{\Pi}\right) = 0 \tag{11.6}$$

Equation (11.6) can be used as the criterion for computing the critical load. It holds only for conservative loading.

EXAMPLE 11.2 A Simple Example
Return to the weightless, rigid rod with the elastic base attachment of Fig. 11.2. The potential energy for $\overline{H} = 0$ is given by

$$\begin{aligned}\Pi &= \text{(strain energy of the spring)} - \text{(potential energy of the applied loading)} \\ &= \frac{1}{2}k\phi^2 - \overline{P}L(1 - \cos\phi)\end{aligned} \tag{1}$$

For equilibrium $\delta\mathring{\Pi} = 0$ or

$$k\phi - \overline{P}L\sin\phi = 0 \tag{2}$$

which corresponds to the value of $\overline{P}$ of Eq. (3) of Example 11.1.

To study the stability of the system, note that for $\phi = 0$

$$\delta^2 \mathring{\Pi} = k - \overline{P}L \tag{3}$$

Thus,

$$k - \overline{P}L \begin{cases} > 0 & \text{stable} \\ = 0 & \text{critical} \\ < 0 & \text{unstable} \end{cases} \tag{4}$$

We conclude that $\overline{P} = P_{cr} = k/L$ is the critical level of the applied loading. For example, the rod in its vertical position is stable for $\overline{P} < P_{cr}$. ∎

11.1.2 General Criterion

As already demonstrated by the examples given in the last section, stability problems in structural mechanics can be viewed from several standpoints and placed in distinct categories.

In addition to the energy criterion just described, the classical equilibrium method is based on the description of the structural response with differential equations and leads, for example, to the Euler load of columns. In these differential equations, some nonlinear deformation terms are included that are linearized by additional assumptions. The associated eigenvalue problem is solved, with a load parameter introduced. The bifurcation of equilibrium is defined as the stability limit and is calculated as the lowest value of the load factor for which the eigenvalue determinant is zero. The energy criterion and the classical equilibrium categories for stability are related (Pflüger, 1964), both being based on descriptions of the structure without imperfections.

From a practical standpoint, it is important to consider structures with imperfections in geometry. Of particular interest is the identification of load characteristics leading to deformations that exceed prescribed limits. Regions are sought corresponding to relative extrema for the load-deformation diagrams and their particular effects and conditions for a change in strategies. These are the *limit point* or *snap-through* problems. Insight into this with respect to the stability of a structure can be obtained from a mechanical or a mathematical viewpoint.

To use a static formulation, the applied loading must be conservative. For nonconservative systems, such as systems wherein the change in direction of the loading is taken into account, a dynamic formulation must be employed. A kinetic stability criterion needs to be introduced, utilizing the dynamic equation of motion (Ziegler, 1977). To obtain a tractable solution with this approach normally involves a major effort.

With proper definitions, the various methods for formulating a stability analysis can be based on a common fundamental foundation. This formulation is of general character and has to be specialized for particular physical problems. The steps for implementing this generalized criterion are summarized below. The terms in parentheses in the different steps are referring to the application of the general stability criterion in structural mechanics.

Summary of the General Criterion for Stability

- Establish the unperturbed state (in equilibrium)—the Fundamental state.
- Disturbance (of equilibrium) yields perturbed state—the Neighboring state.
- Establish a characteristic (change of displacement) that is a critical feature of the state.

- Stability behavior: Behavior of the critical characteristic (change of displacement) due to the disturbance. The level of the characteristic is selected such that it should not be exceeded.
- Undisturbed state is stable: The critical characteristic (change of displacement) does not exceed the criterion level fixed, for example, by design code. Otherwise the system is unstable, (the limit being the critical load).

In the first step, a fundamental state of the system is defined. Imposing a disturbance on this fundamental state leads to a neighboring disturbed state. Then a characteristic property is established that appropriately describes the stability problem. For this property, a measure is defined that should not be exceeded. In observing the transition to the neighboring state, the stability or instability of the system can be assessed.

With the selection of the critical characteristic, the general stability criterion is specified and can be applied to a structure. It is, of course, necessary to utilize a structural analysis to determine the behavior of stability characteristics of a structure.

It is to be noted that this criterion applies for the investigation of stability for perfect as well as for imperfect structures. The stability concept thus is not restricted to the bifurcation of equilibrium cases (Leipholz, 1968).

For a complete treatment of the stability behavior of structures, a fully nonlinear investigation is necessary. This is not covered in this book. Here, the presentation is restricted primarily to structures composed of one-dimensional members with in-plane loading and to the so-called theory of second order. Approximate methods can be employed to include the most influential nonlinear effects due to imperfections. As a special case, the homogeneous solution, i.e., the determinant set equal to zero, leads to the eigenvalues and provides the bifurcation level.

11.2 Local Equations for a Beam Column

11.2.1 Definition of Transverse Forces and Longitudinal Forces

Stability analyses of structures are, in general, based on nonlinear characterizations of the problem. Therefore, in the formulation of the basic equations, the choice of the reference configuration plays an important role. The variables need to be referred to the deformed or to the undeformed configuration.

In the case of a beam problem (Fig. 11.5), stresses and stress resultants are usually defined with respect to the cross-section of the deformed structure, whereas the displacements normally are measured with respect to the undeformed configuration. Therefore, in addition

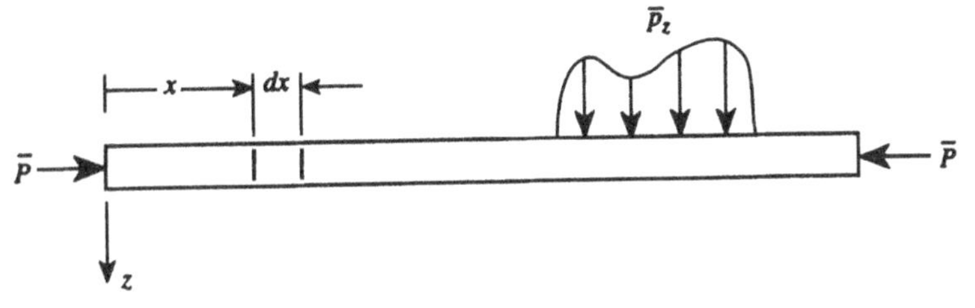

FIGURE 11.5
Beam column.

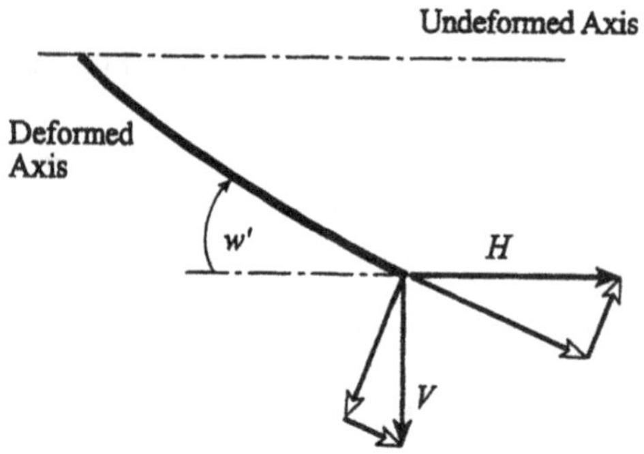

FIGURE 11.6
Transformation of forces aligned along the undeformed axes with components along axes for the deformed bar.

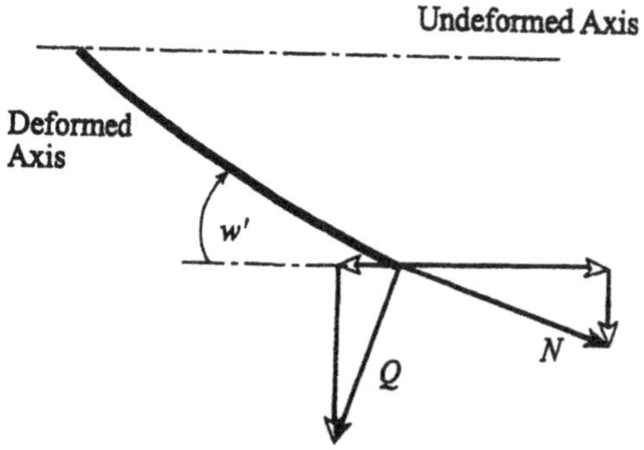

FIGURE 11.7
Transformation of forces aligned along the deformed axes with components along axes for the undeformed bar.

to the usual stress resultants, fictitious forces are introduced that are conjugate to the directions of the displacements in the sense that they perform work together. Assume small rotations between the two axes so that $\sin w' \approx \tan w' \approx w'$ and $\cos w' \approx 1$. In the framework of the theory of second order, the linear transformation of the stress resultants H and V into components N and Q with respect to the deformed axes is given by (Fig. 11.6).

$$\begin{aligned} N &= H\cos w' + V\sin w' \quad \text{or} \quad N \approx H + Vw' \\ Q &= -H\sin w' + V\cos w' \quad \text{or} \quad Q \approx -Hw' + V \end{aligned} \tag{11.7}$$

In matrix notation, with the addition of the moment,

$$\begin{bmatrix} N \\ Q \\ M \end{bmatrix} = \begin{bmatrix} \cos w' & \sin w' & 0 \\ -\sin w' & \cos w' & 0 \\ 0 & 0 & 1 \end{bmatrix} \begin{bmatrix} H \\ V \\ M \end{bmatrix} \tag{11.8}$$

The transformation of the stress resultants N and Q of Fig. 11.7 into components H and V along the axes corresponding to the undeformed bar is

$$\begin{aligned} H &= N\cos w' - Q\sin w' \quad \text{or} \quad H \approx N - Qw' \\ V &= N\sin w' + Q\cos w' \quad \text{or} \quad V \approx Nw' + Q \end{aligned} \tag{11.9}$$

In matrix notation

$$\begin{bmatrix} H \\ V \\ M \end{bmatrix} = \begin{bmatrix} \cos w' & -\sin w' & 0 \\ \sin w' & \cos w' & 0 \\ 0 & 0 & 1 \end{bmatrix} \begin{bmatrix} N \\ Q \\ M \end{bmatrix} \tag{11.10}$$

Further assume that

$$Vw' \ll H \quad \text{and} \quad Qw' \ll N \tag{11.11}$$

Then the relationships between the two sets of stress resultants N and Q and V and H become

$$\begin{aligned} H &\approx N \\ V &\approx Q + w'N \\ Q &\approx V - w'H \approx V - w'N \end{aligned} \tag{11.12}$$

If initial imperfections w_0' are to be taken into account, replace w' by $w' + w_0'$.

11.2.2 Fundamental Equations (Differential Form)

In a first approximation to a fully nonlinear analysis, the most important terms need to be retained. For beam problems, an engineering-type approach is the so-called theory of second order.

Conditions of Equilibrium

The equilibrium conditions take into account the change from undeformed to the deformed geometry by introducing the transformation of Eq. (11.10) with the assumptions of Eq. (11.12). The primary difference occurs in the summation of the moments (Fig. 11.8 and

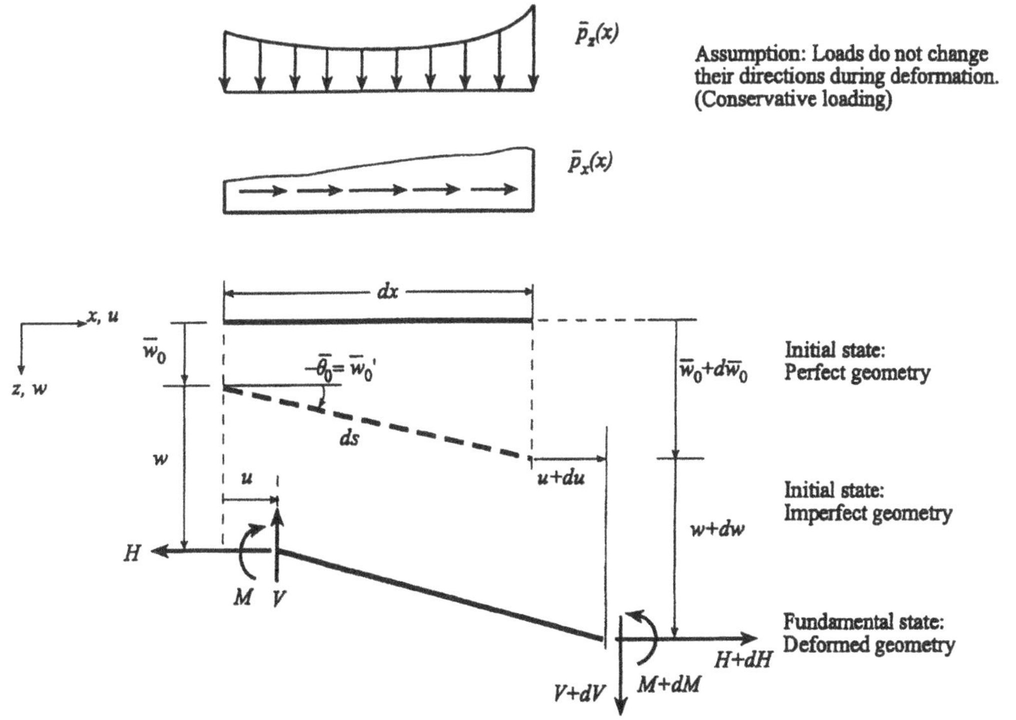

FIGURE 11.8
Undeformed and deformed states of a bar.

the following equation)

$$\sum M = 0: -M + (M + dM) - V\,dx + H(dw + d\overline{w}_0) = 0 \tag{11.13a}$$

$$dM - V\,dx + H(dw + d\overline{w}_0) = 0 \tag{11.13b}$$

$$\frac{dM}{dx}dx - V\,dx + H\frac{dw + d\overline{w}_0}{dx}dx = 0 \tag{11.13c}$$

$$\frac{dM}{dx} - \left[V - H\left(\frac{dw}{dx} + \frac{d\overline{w}_0}{dx}\right)\right] = 0 \tag{11.13d}$$

where, from Eq. (11.12)

$$Q = \left[V - H\left(\frac{dw}{dx} + \frac{d\overline{w}_0}{dx}\right)\right] \tag{11.14}$$

The equilibrium of the forces in the horizontal and vertical directions

$$\sum F_{\text{hor}} = 0: \frac{dH}{dx} - \overline{p}_x = 0 \tag{11.15}$$

$$\sum F_{\text{vert}} = 0: \frac{dV}{dx} + \overline{p}_z = 0 \tag{11.16}$$

The term $H\frac{dw}{dx}$ in Eq. (11.13d) introduces a nonlinearity with respect to the longitudinal force, which for compressive loads may lead to stability failure. The other difference as compared with the usual equations for a beam (for instance those given in Chapter 4) is the inclusion of the initial imperfections $\overline{w}_0$ (given displacements) and $\overline{\theta}_0$ (given rotations) that describe the initial deviations from perfect geometry. Because these initial imperfections are difficult to avoid in practice, they need to be considered in describing the realistic behavior of most structures. The kinematical equations and the material law are assumed to be linear. Additional nonlinearities may be introduced in the formulation, but are omitted here.

Kinematics

From Chapter 1 the kinematical equations for a beam are

$$\begin{aligned} \kappa &= \frac{d\theta}{dx} \\ \varepsilon &= \frac{du}{dx} \\ \gamma &= \frac{dw}{dx} + \theta \end{aligned} \tag{11.17}$$

For an Euler-Bernoulli beam, the shear deformation effects are ignored so that $\gamma = 0$, and the kinematical relations are reduced to

$$\theta = -\frac{dw}{dx} \quad \text{and} \quad \kappa = -\frac{d^2w}{dx^2} \tag{11.18}$$

Material Law

In the theory of second order it is presumed that the unknown longitudinal strains contained in N are known quantities, obtained from the linear analysis as the approximations

$$N = EA\varepsilon \approx N^0 \tag{11.19}$$

If $\overline{\kappa}$ and $\overline{\varepsilon}$ are the imposed curvature and strain respectively, due, for example, to a temperature change, the material laws appear as

$$\begin{aligned} M &= EI(\kappa - \overline{\kappa}) = EI\left(-\frac{d^2w}{dx^2} - \overline{\kappa}\right) \\ N &= EA(\varepsilon - \overline{\varepsilon}) = EA\left(\frac{du}{dx} - \overline{\varepsilon}\right) \\ Q &= k_s GA\gamma \end{aligned} \tag{11.20}$$

Of course, the final expression is zero for Euler-Bernoulli beams. The shear force V follows from Eqs. (11.12), (11.19), and (11.20).

$$V = Q - H\theta \approx Q - N\theta = k_s GA\gamma - (EA\varepsilon)\theta \approx -N^0\theta \tag{11.21}$$

Governing Equations

These fundamental equations can be combined to form a set of first order or higher order differential equations. The first order equations are:

Extension:

$$\frac{d}{dx}\begin{bmatrix} u \\ H \end{bmatrix} = \begin{bmatrix} 0 & \frac{1}{EA} \\ 0 & 0 \end{bmatrix}\begin{bmatrix} u \\ H \end{bmatrix} + \begin{bmatrix} \overline{\varepsilon} \\ -\overline{p}_x \end{bmatrix} \tag{11.22}$$

Bending:

$$\begin{aligned} \frac{d}{dx}\begin{bmatrix} w \\ \theta \\ V \\ M \end{bmatrix} &= \begin{bmatrix} 0 & -1 & 0 & 0 \\ 0 & 0 & 0 & \frac{1}{EI} \\ 0 & 0 & 0 & 0 \\ 0 & H & 1 & 0 \end{bmatrix}\begin{bmatrix} w \\ \theta \\ V \\ M \end{bmatrix} + \begin{bmatrix} 0 \\ \overline{\kappa} \\ -\overline{p}_z \\ H\overline{\theta}_0 \end{bmatrix} \\ \frac{d\mathbf{z}}{dx} &= \qquad\qquad \mathbf{A} \qquad\qquad\quad \mathbf{z} \quad + \quad \overline{\mathbf{P}} \end{aligned} \tag{11.23}$$

In higher order form:

Extension:

$$\begin{aligned} \frac{d}{dx}\left(EA\frac{du}{dx}\right) &= -\overline{p}_x + EA\frac{d\overline{\varepsilon}}{dx} \\ H &= EA\frac{du}{dx} - EA\overline{\varepsilon} \end{aligned} \tag{11.24}$$

Bending:

$$\begin{aligned} \frac{d^2}{dx^2}\left(EI\frac{d^2w}{dx^2}\right) - H\frac{d^2w}{dx^2} &= \overline{p}_z - EI\frac{d^2\overline{\kappa}}{dx^2} - H\frac{d\overline{\theta}_0}{dx} \\ V &= -\frac{d}{dx}EI\frac{d^2w}{dx^2} + H\frac{dw}{dx} - EI\frac{d\overline{\kappa}}{dx} - H\overline{\theta}_0 \\ M &= -EI\frac{d^2w}{dx^2} - EI\overline{\kappa} \\ \theta &= -\frac{dw}{dx} \end{aligned} \tag{11.25}$$

These equations are uncoupled in the linear theory, but are coupled here by

$$H\frac{dw}{dx} \approx N\frac{dw}{dx} = EA\frac{du}{dx}\frac{dw}{dx} \tag{11.26}$$

This is a nonlinear term that reveals the nonlinear character of the theory of stability. In the framework of the theory of second order it is assumed that the normal force is known, for it can be estimated or provided by a linear analysis of the structure

$$H\frac{dw}{dx} \approx N^0\frac{dw}{dx} \tag{11.27}$$

Upon solution of the equations, which have been linearized by this assumption, the bounds on the effect of this assumption on the solution accuracy can be checked. If needed, perform a second analysis with an updated value of the normal force. This becomes an iterative process, which is typical for nonlinear solutions.

11.2.3 Solution of the Governing Equations

For the case when the coefficients are constant, it is possible to obtain an exact analytical solution. The system of the first order differential equations of Eq. (11.23) has the solution (Eq. 4.90)

$$\mathbf{z}(x) = e^{\mathbf{A}x}\mathbf{z}(0) + e^{\mathbf{A}x}\int_0^x e^{-\mathbf{A}t}\,\overline{\mathbf{P}}(t)\,dt = \mathbf{U}(x)\,\mathbf{z}(0) + \overline{\mathbf{z}} \tag{11.28}$$

$$\mathbf{U}(x) = e^{\mathbf{A}x} = \mathbf{I} + \frac{1}{1!}\mathbf{A}x + \frac{1}{2!}\mathbf{A}^2x^2 + \frac{1}{3!}\mathbf{A}^3x^3 + \frac{1}{4!}\mathbf{A}^4x^4 + \cdots \tag{11.29}$$

Similarly, the loading terms can be expanded as

$$\overline{\mathbf{z}}(x) = x\left(\mathbf{I} + \frac{1}{1!}\mathbf{A}x + \frac{1}{2!}\mathbf{A}^2x^2 + \frac{1}{3!}\mathbf{A}^3x^3 + \frac{1}{4!}\mathbf{A}^4x^4 + \cdots\right)\overline{\mathbf{P}} \tag{11.30}$$

This solution as a matrix series has the advantage that, in general, it can be applied also in the case of non-constant coefficients of the differential equations. Then, the terms of the series expansion can be obtained in a straightforward manner, as, for example, in the case of the stability of curved beams (Wunderlich and Beverungen, 1977).

Alternatively, the solution of the homogeneous part of the higher order relationships of Eq. (11.25)

$$EI\frac{d^4w}{dx^4} - N^0\frac{d^2w}{dx^2} = 0 \tag{11.31}$$

becomes, with $\varepsilon^2 = \ell^2\frac{|N^0|}{EI}$,

Compression,

$$\begin{aligned} N^0 &= -\overline{P} \\ w(x) &= C_1 + C_2\frac{x}{\ell} + C_3\cos\varepsilon\frac{x}{\ell} + C_4\sin\varepsilon\frac{x}{\ell} \end{aligned} \tag{11.32}$$

Tension,

$$\begin{aligned} N^0 &= \overline{P} \\ w(x) &= C_1 + C_2\frac{x}{\ell} + C_3\cosh\varepsilon\frac{x}{\ell} + C_4\sinh\varepsilon\frac{x}{\ell} \end{aligned} \tag{11.33}$$

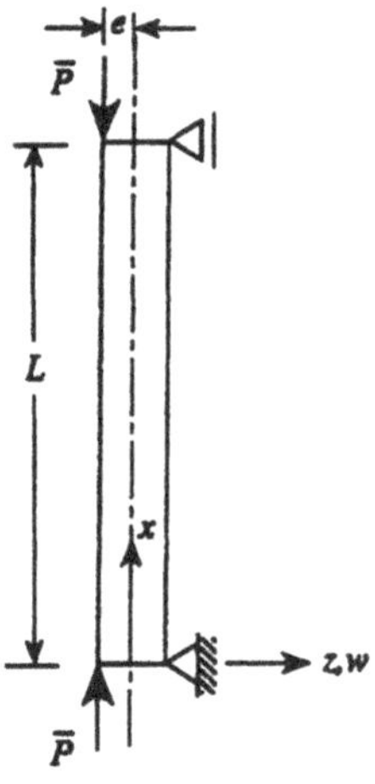

FIGURE 11.9
Simply supported column with eccentrically applied axial force.

These are closed form solutions composed of functions that are defined analytically. It should be kept in mind, however, that these functions are also defined through a power series expansion and are computed as such in practical calculations on the computer. The approach using the matrix series appears to be more direct and more generally applicable.

EXAMPLE 11.3 Stability of a Column with Eccentrically Applied Axial Load
To study the stability of a beam, consider the simply supported column of Fig. 11.9 with an eccentrically applied axial force.

The boundary conditions are

$$w(0) = 0, \quad M(0) = \overline{P}e, \quad w(L) = 0, \quad M(L) = \overline{P}e \tag{1}$$

Apply these boundary conditions to the solution of Eq. (11.32), with $N^0 = -\overline{P}$, so that $\varepsilon^2 = L^2\overline{P}/EI$.

$$\begin{matrix} w(0) = 0 \\ w(L) = 0 \\ M(0) = \overline{P}e \\ M(L) = \overline{P}e \end{matrix} \quad \begin{bmatrix} 1 & 0 & 1 & 0 \\ 1 & 1 & c(L) & s(L) \\ 0 & 0 & EI\varepsilon^2/L^2 & 0 \\ 0 & 0 & EI\varepsilon^2 c(L)/L^2 & EI\varepsilon^2 s(L)/L^2 \end{bmatrix} \begin{bmatrix} C_1 \\ C_2 \\ C_3 \\ C_4 \end{bmatrix} = \begin{bmatrix} 0 \\ 0 \\ e\varepsilon^2 EI/L^2 \\ e\varepsilon^2 EI/L^2 \end{bmatrix} \quad \begin{matrix} s(x) = \sin \varepsilon \frac{x}{L} \\ c(x) = \cos \varepsilon \frac{x}{L} \end{matrix} \tag{2}$$

The vector on the right-hand side of this relationship contains the homogenous terms that would be zero for a pure column (with no eccentricity). The constants C_1, C_2, C_3, and C_4, found by solving (2) are substituted into Eq. (11.32) giving

$$w(x) = e\left(\frac{1 - \cos \varepsilon}{\sin \varepsilon} \sin \varepsilon \frac{x}{L} + \cos \varepsilon \frac{x}{L} - 1\right) \tag{3}$$

The maximum value of the deflection

$$w(L/2) = e\left(\frac{1}{\cos \varepsilon/2} - 1\right) \tag{4}$$

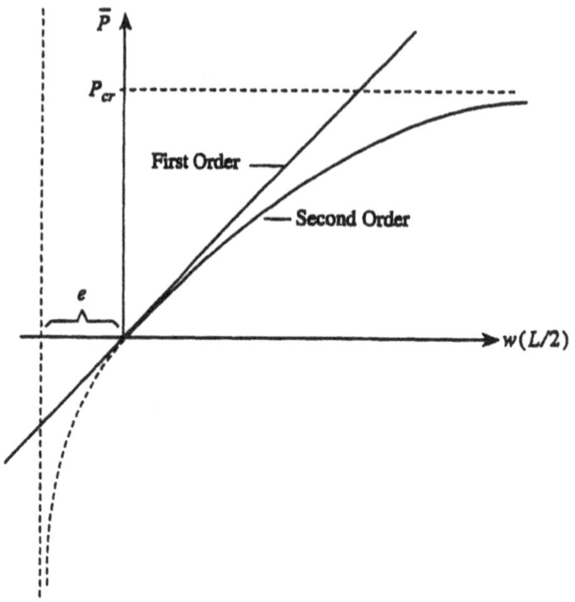

FIGURE 11.10
Midbeam deflections as a function of $\overline{P}$.

is plotted in Fig. 11.10 as the second order solution. Also illustrated is a first order solution, which is the deflection due to moments $\overline{P}e$ applied to the ends of the beam, with no consideration given to the change in influence of the axial load $\overline{P}$ on the response as the deflection changes, i.e., the terms involving H are ignored in Eq. (11.23) or (11.25). Note that the first order solution is linear, while that for the second order case is nonlinear.

If the critical or unstable situation is associated with inordinately large deformations, then (3) and (4) can be used to define the axial load leading to bifurcation. Note from (3) that if $\sin\varepsilon = 0$ or from (4) if $\cos\varepsilon/2 = 0$, the deflection grows without limit. These equalities occur if $\varepsilon = \pi$. Then

$$\overline{P} = P_{cr} = \frac{\pi^2 EI}{L^2} \tag{5}$$

This critical value of $\overline{P}$ defines the horizontal line of Fig. 11.10. Other values of ε can also lead to critical responses. In the case of $\sin\varepsilon = 0$, ε can assume the values $n\pi$, where $n = 0, 1, 2, 3, \ldots$. However, for $n = 0$, the axial force P must also be zero and the mechanical problem has been altered (a "trivial" solution). Under normal circumstances, for $n = 2, 3, \ldots$, the physical problem is of little interest, since the structure has already been subjected to a critical deformation (at $n = 1$). ∎

11.2.4 Element Matrices Using Analytical Solutions

The closed form solutions with analytical functions of Eqs. (11.32) and (11.33) are for a bar with constant bending stiffness EI, for which second order effects are taken into account.

From Eq. (11.23) the other state variables can be expressed in closed form, too.

$$\begin{bmatrix} w(x) \\ \theta(x) \\ V(x) \\ M(x) \end{bmatrix} = \begin{bmatrix} w \\ -w' \\ -EIw''' + N^0 w' \\ -EIw'' \end{bmatrix}$$

$$= \begin{bmatrix} 1 & x/\ell & \vdots & c(x) & s(x) \\ 0 & -1/\ell & \vdots & (\varepsilon/\ell)s(x) & -(\varepsilon/\ell)c(x) \\ \cdots & \cdots & & \cdots\cdots & \cdots\cdots \\ 0 & N^0/\ell & \vdots & 0 & 0 \\ 0 & 0 & \vdots & EI(\varepsilon/\ell)^2 c(x) & EI(\varepsilon/\ell)^2 s(x) \end{bmatrix} \begin{bmatrix} C_1 \\ C_2 \\ C_3 \\ C_4 \end{bmatrix}$$

$$= \begin{bmatrix} \mathbf{N}_u \\ \cdots \\ \mathbf{N}_s \end{bmatrix} \mathbf{c} \tag{11.34}$$

where for compression

$$c(x) = \cos \varepsilon \frac{x}{\ell} \qquad s(x) = \sin \varepsilon \frac{x}{\ell} \tag{11.35}$$

and for tension

$$c(x) = \cosh \varepsilon \frac{x}{\ell} \qquad s(x) = \sinh \varepsilon \frac{x}{\ell}$$

The four free parameters C_1 to C_4 can now be transformed such that they express quantities with direct physical meaning, either through the four state variables at one end of the bar $\mathbf{z}(0) = \mathbf{z}_a = [w_a \ \theta_a \ V_a \ M_a]^T$, or through the four displacement variables $\mathbf{v} = [w_a \ \theta_a \ w_b \ \theta_b]^T$ in the manner of the definitions of the transfer matrix or the stiffness matrix, respectively, see Figure 11.11. Thus, this mathematical solution of Eq. (11.34)

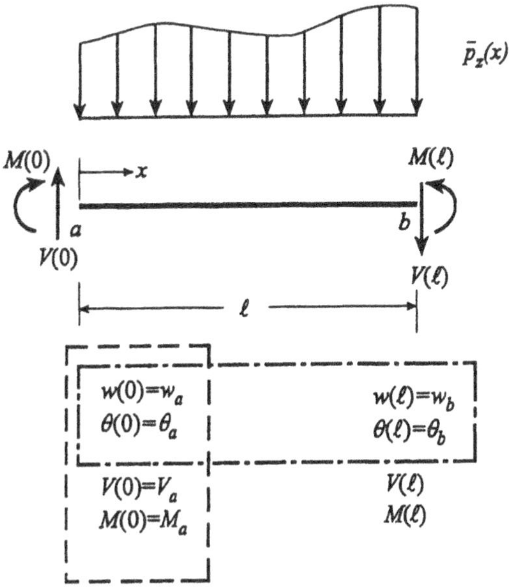

FIGURE 11.11
Combination of state variables into $\mathbf{z}_a$ for transfer matrices or $\mathbf{v}$ for stiffness matrices.

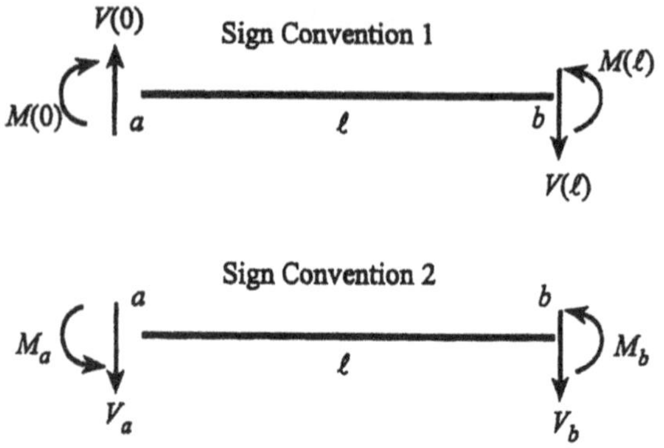

FIGURE 11.12
Sign conventions for beams.

can be rearranged leading either to the stiffness matrix $\mathbf{k}$ or to the transfer matrix $\mathbf{U}$ of the bar including second order effects. For the transformation, the solutions contained in $\mathbf{N}_u$ and $\mathbf{N}_s$ have to be evaluated at the ends of the bar. These discrete values are combined into the matrices $\widehat{\mathbf{N}}_u$ and $\widehat{\mathbf{N}}_s$ (Sign Convention 1 for the transfer matrix) or into $\widehat{\mathbf{N}}_v$ and $\widehat{\mathbf{N}}_p$ (Sign Convention 2 for the stiffness matrix) (Fig. 11.12). For the stiffness matrix

$$\mathbf{v} = \begin{bmatrix} \mathbf{v}_a \\ \mathbf{v}_b \end{bmatrix} = \begin{bmatrix} w(0) \\ \theta(0) \\ \cdots \\ w(\ell) \\ \theta(\ell) \end{bmatrix} = \begin{bmatrix} \widehat{\mathbf{N}}_u(0) \\ \cdots \\ \widehat{\mathbf{N}}_u(\ell) \end{bmatrix} \mathbf{c} = \widehat{\mathbf{N}}_v \mathbf{c} \tag{11.36}$$

$$\mathbf{p} = \begin{bmatrix} \mathbf{p}_a \\ \mathbf{p}_b \end{bmatrix} = \begin{bmatrix} -V(0) \\ -M(0) \\ \cdots \\ V(\ell) \\ M(\ell) \end{bmatrix} = \begin{bmatrix} -\widehat{\mathbf{N}}_s(0) \\ \cdots \\ \widehat{\mathbf{N}}_s(\ell) \end{bmatrix} \mathbf{c} = \widehat{\mathbf{N}}_p \mathbf{c} \tag{11.37}$$

The stiffness matrix is defined as

$$\mathbf{p} = \mathbf{k}\,\mathbf{v} \tag{11.38}$$

Substitute the value of the analytical solution at both edges of the bar, Eqs. (11.36) and (11.37), into Eq. (11.38)

$$\widehat{\mathbf{N}}_p \mathbf{c} = \mathbf{k}\,\widehat{\mathbf{N}}_v \mathbf{c} \tag{11.39}$$

where the conversion from forces in Sign Convention 1 to Sign Convention 2 has been taken into account. It follows that

$$\mathbf{k} = \widehat{\mathbf{N}}_p\,\widehat{\mathbf{N}}_v^{-1} \tag{11.40}$$

The transfer matrix is defined by

$$\mathbf{z}_b = \mathbf{U}\mathbf{z}_a \tag{11.41}$$

and is expressed in terms of the values of the analytical solution on both ends of the element using Eqs. (11.34), (11.36) and (11.37)

$$\begin{bmatrix} \mathbf{v}_b \\ \cdots \\ \mathbf{s}_b \end{bmatrix} = \underbrace{\begin{bmatrix} \widehat{\mathbf{N}}_u(\ell) \\ \cdots \\ \widehat{\mathbf{N}}_s(\ell) \end{bmatrix}}_{\widehat{\mathbf{N}}_z(\ell)} \mathbf{c} = \mathbf{U} \underbrace{\begin{bmatrix} \widehat{\mathbf{N}}_u(0) \\ \cdots \\ \widehat{\mathbf{N}}_s(0) \end{bmatrix}}_{\widehat{\mathbf{N}}_z(0)} \mathbf{c} = \mathbf{U}\mathbf{z}_a \tag{11.42}$$

Therefore, the transition between the mathematical unknowns of the solution and the mechanical degrees of freedom at the ends is given by

$$\mathbf{U} = \widehat{\mathbf{N}}_z(\ell)\,\widehat{\mathbf{N}}_z^{-1}(0) \tag{11.43}$$

The transfer matrix, as well as the stiffness matrix, are given explicitly in the following two sections, accompanied by the respective load vectors for some load cases. In the response expressions of this section, the nonlinear second order effect is characterized by the specific parameter

$$\varepsilon = \ell\sqrt{\frac{|N^0|}{EI}} \tag{11.44}$$

Basically, it contains the *slenderness ratio* ℓ/r and the longitudinal strains N^0/EA of the beam column with its radius of gyration $r = \sqrt{I/A}$.

11.2.5 Transfer Matrix for a Beam (Theory of Second Order)

The basic relationship for transfer matrices of the ith element, as described in Chapter 4, is given by

$$\mathbf{z}^i(x) = \mathbf{U}^i(x)\mathbf{z}^i(0) + \overline{\mathbf{z}}^i(x) \tag{11.45}$$

The state variables with Sign Convention 1 are shown in Fig. 11.13a. The state variable vector $\mathbf{z}^i$ and loading vector $\overline{\mathbf{z}}^i$ are given by

$$\mathbf{z}^i(x) = \begin{bmatrix} u(x) \\ w(x) \\ \theta(x) \\ N(x) \\ V(x) \\ M(x) \end{bmatrix}^i \qquad \overline{\mathbf{z}}^i(x) = \begin{bmatrix} \overline{u}(x) \\ \overline{w}(x) \\ \overline{\theta}(x) \\ \overline{N}(x) \\ \overline{V}(x) \\ \overline{M}(x) \end{bmatrix}^i \tag{11.46}$$

Use of Eq. (11.43) leads to the following transfer matrices, which include second order effects.

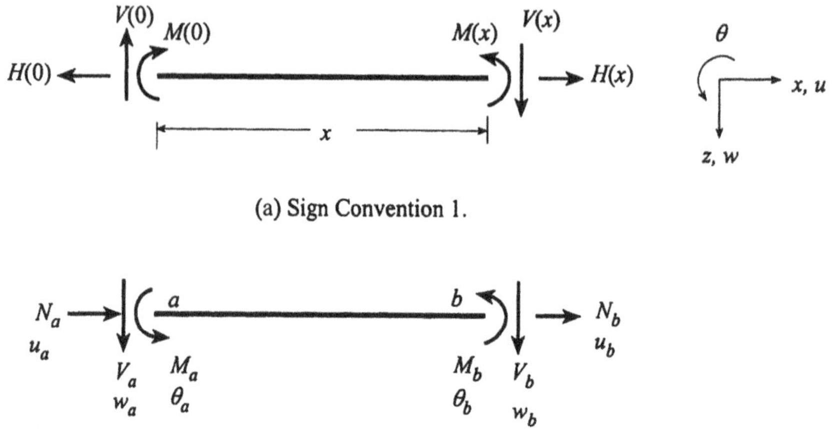

(a) Sign Convention 1.

(b) Sign Convention 2.

FIGURE 11.13
Element of a bar with state variables.

Bar in Compression ($N^0 < 0$)

$$\varepsilon = \ell \cdot \sqrt{\frac{|N^0|}{EI}}$$

$$\mathbf{U}_{x=\ell} = \begin{bmatrix} 1 & 0 & 0 & \vdots & \frac{\ell}{EA} & 0 & 0 \\ 0 & 1 & -\frac{\ell \sin\varepsilon}{\varepsilon} & \vdots & 0 & -\frac{\ell^3}{EI}\frac{(\varepsilon - \sin\varepsilon)}{\varepsilon^3} & -\frac{\ell^2}{EI}\frac{(1-\cos\varepsilon)}{\varepsilon^2} \\ 0 & 0 & \cos\varepsilon & \vdots & 0 & \frac{\ell^2}{EI}\frac{(1-\cos\varepsilon)}{\varepsilon^2} & \frac{\ell}{EI}\frac{\sin\varepsilon}{\varepsilon} \\ \cdots & \cdots & \cdots & \cdots & \cdots & \cdots & \cdots \\ 0 & 0 & 0 & \vdots & 1 & 0 & 0 \\ 0 & 0 & 0 & \vdots & 0 & 1 & 0 \\ 0 & 0 & -\frac{EI}{\ell}\varepsilon\sin\varepsilon & \vdots & 0 & \frac{\ell\sin\varepsilon}{\varepsilon} & \cos\varepsilon \end{bmatrix} \tag{11.47}$$

Bar in Tension ($N^0 > 0$)

$$\varepsilon = \ell \cdot \sqrt{\frac{N^0}{EI}}$$

$$\mathbf{U}_{x=\ell} = \begin{bmatrix} 1 & 0 & 0 & \vdots & \frac{\ell}{EA} & 0 & 0 \\ 0 & 1 & -\frac{\ell \sinh\varepsilon}{\varepsilon} & \vdots & 0 & \frac{\ell^3}{EI}\frac{(\varepsilon - \sinh\varepsilon)}{\varepsilon^3} & \frac{\ell^2}{EI}\frac{(1-\cosh\varepsilon)}{\varepsilon^2} \\ 0 & 0 & \cosh\varepsilon & \vdots & 0 & -\frac{\ell^2}{EI}\frac{(1-\cosh\varepsilon)}{\varepsilon^2} & \\ \cdots & \cdots & \cdots & \cdots & \cdots & \cdots & \cdots \\ 0 & 0 & 0 & \vdots & 1 & 0 & 0 \\ 0 & 0 & 0 & \vdots & 0 & 1 & 0 \\ 0 & 0 & -\frac{EI}{\ell}\varepsilon\sinh\varepsilon & \vdots & 0 & \frac{\ell\sinh\varepsilon}{\varepsilon} & \cosh\varepsilon \end{bmatrix} \tag{11.48}$$

These transfer matrices were also derived in Chapter 4 in a different notation and are tabulated in a rather general form in Table 4.3. For the corresponding load vectors, see also Table 4.3.

11.2.6 Stiffness Matrix for a Beam (Theory of Second Order)

The basic relationship for stiffness matrices of the ith element, as described in Chapter 4, is given by

$$\mathbf{p}^i = \mathbf{k}^i \mathbf{v}^i - \overline{\mathbf{p}}^{i0}$$

The definition of the state variables using Sign Convention 2 is shown in Fig. 11.13b referred to the undeformed state. They form the vector of the force variables $\mathbf{p}^i$ and the displacement variables $\mathbf{v}^i$ for element i.

$$\mathbf{p}^i = \begin{bmatrix} N_a \\ V_a \\ M_a \\ N_b \\ V_b \\ M_b \end{bmatrix}^i \qquad \mathbf{v}^i = \begin{bmatrix} u_a \\ w_a \\ \theta_a \\ u_b \\ w_b \\ \theta_b \end{bmatrix}^i \tag{11.49}$$

Any number of methods can be employed to obtain the stiffness matrix $\mathbf{k}^i$ for a beam element subjected to an axial force. For example, Chapter 4, Eq. (4.11) can be used to convert the transfer matrices of Eqs. (11.47) and (11.48) into stiffness matrices. Also the development of Eq. (11.40) can be utilized to obtain the following stiffness matrices which include second order effects:

Bar in Compression $(N^0 < 0)$

$$\varepsilon = \ell\sqrt{\frac{N^0}{EI}}$$

$$\begin{bmatrix} N_a \\ V_a \\ M_a \\ N_b \\ V_b \\ M_b \end{bmatrix} = \left[\begin{array}{ccc|ccc} EA/\ell & 0 & 0 & -EA/\ell & 0 & 0 \\ & [2(A'+B')-D']EI/\ell^3 & -(A'+B')EI/\ell^2 & 0 & -[2(A'+B')-D']EI/\ell^3 & -(A'+B')EI/\ell^2 \\ & & A'EI/\ell & 0 & (A'+B')EI/\ell^2 & B'EI/\ell \\ \hline & & & EA/\ell & 0 & 0 \\ & \text{Symmetric} & & & [2(A'+B')-D']EI/\ell^3 & (A'+B')EI/\ell^2 \\ & & & & & A'EI/\ell \end{array}\right] \begin{bmatrix} u_a \\ w_a \\ \theta_a \\ u_b \\ w_b \\ \theta_b \end{bmatrix}$$

$$\mathbf{p}^i \quad = \quad \mathbf{k}^i \quad \mathbf{v}^i \tag{11.50}$$

where the parameters A', B', D' are (Chwalla, 1959)

$$A' = \frac{\varepsilon(\sin\varepsilon - \varepsilon\cos\varepsilon)}{2(1-\cos\varepsilon) - \varepsilon\sin\varepsilon} \qquad A' = 4 - \frac{2}{15}\varepsilon^2 - \frac{11}{6300}\varepsilon^4 - \cdots$$

$$B' = \frac{\varepsilon(\varepsilon - \sin\varepsilon)}{2(1-\cos\varepsilon) - \varepsilon\sin\varepsilon} \qquad B' = 2 + \frac{1}{30}\varepsilon^2 + \frac{13}{12\,600}\varepsilon^4 - \cdots$$

$$A' + B' = \frac{\varepsilon^2(1-\cos\varepsilon)}{2(1-\cos\varepsilon) - \varepsilon\sin\varepsilon} \qquad A' + B' = 6 - \frac{1}{10}\varepsilon^2 - \frac{1}{1400}\varepsilon^4 - \cdots$$

$$D' = \varepsilon^2 = \ell^2\frac{|N|}{EI}$$

$$2(A'+B') - D' = \frac{\varepsilon^3\sin\varepsilon}{2(1-\cos\varepsilon) - \varepsilon\sin\varepsilon} \qquad 2(A'+B') - D' = 12 - \frac{6}{5}\varepsilon^2 - \frac{1}{700}\varepsilon^4 - \cdots \tag{11.51}$$

It is clear that this matrix reduces to the stiffness matrix for an Euler-Bernoulli matrix as $\varepsilon = \ell\sqrt{|N^0|/EI} \to 0$. Then $A' = 4$, $B' = 2$, $D' = 0$. The expansions in ε of Eq. (11.51) permit the formation of stiffness matrices expressed in terms of polynomials in ε, rather than in terms of transcendental functions.

The applied distributed loading $\overline{\mathbf{p}}^{i0}$ of the basic relationship $\mathbf{p}^i = \mathbf{k}^i\mathbf{v}^i - \mathbf{p}^{i0}$ can be obtained by integration of Eqs. (11.23) or from a power series expansion. Loading functions for several distributed loadings are listed in Tables 4.4 and 11.1. See Example 11.5 for $\overline{\mathbf{p}}^{i0}$ for some prescribed imperfections.

Bar in Tension ($N^0 > 0$)

$$\begin{bmatrix} N_a \\ V_a \\ M_a \\ N_b \\ V_b \\ M_b \end{bmatrix} = \left[\begin{array}{ccc:ccc} EA/\ell & 0 & 0 & -EA/\ell & 0 & 0 \\ & [2(A'+B')+D']EI/\ell^3 & -(A'+B')EI/\ell^2 & 0 & -[2(A'+B')+D']EI/\ell^3 & -(A'+B')EI/\ell^2 \\ & & A'EI/\ell & 0 & (A'+B')EI/\ell^2 & B'EI/\ell \\ \hdashline & & & EA/\ell & 0 & 0 \\ & \text{Symmetric} & & & [2(A'+B')+D']EI/\ell^3 & (A'+B')EI/\ell^2 \\ & & & & & A'EI/\ell \end{array}\right] \begin{bmatrix} u_a \\ w_a \\ \theta_a \\ u_b \\ w_b \\ \theta_b \end{bmatrix}$$

$$\mathbf{p}^i = \mathbf{k}^i \mathbf{v}^i \tag{11.52}$$

where

$$A' = \frac{\varepsilon(\sinh\varepsilon - \varepsilon\cosh\varepsilon)}{2(\cosh\varepsilon - 1) - \varepsilon\sinh\varepsilon} \qquad A' = 4 + \frac{2}{15}\varepsilon^2 - \frac{11}{6300}\varepsilon^4 + \cdots$$

$$B' = \frac{\varepsilon(\varepsilon - \sinh\varepsilon)}{2(\cosh\varepsilon - 1) - \varepsilon\sinh\varepsilon} \qquad B' = 2 - \frac{1}{30}\varepsilon^2 + \frac{13}{12\,600}\varepsilon^4 - \cdots$$

$$A' + B' = \frac{\varepsilon^2(1 - \cosh\varepsilon)}{2(\cosh\varepsilon - 1) - \varepsilon\sinh\varepsilon} \qquad A' + B' = 6 + \frac{1}{10}\varepsilon^2 - \frac{1}{1400}\varepsilon^4 + \cdots$$

$$D' = \varepsilon^2 = \ell^2\frac{N}{EI}$$

$$2(A' + B') + D' = \frac{-\varepsilon^3\sinh\varepsilon}{2(\cosh\varepsilon - 1) - \varepsilon\sinh\varepsilon} \qquad 2(A' + B') + D' = 12 + \frac{6}{5}\varepsilon^2 - \frac{1}{700}\varepsilon^4 + \cdots \tag{11.53}$$

Note that the axial force N occurs in both the matrix $\mathbf{k}^i$ (through $\varepsilon = \ell\sqrt{|N^0|/EI}$) and the force vector $\mathbf{p}^i$. Use of this matrix in a stability study of frames is discussed later in this chapter.

The corresponding load vectors are readily calculated, or can be taken from Tables 11.1 or 4.3.

The parameters A', B', $A' + B'$, D', $2(A' + B') \pm D'$, are plotted in Fig. 11.14 as functions of ε (Chwalla, 1959).

TABLE 11.1

Functions $\mathbf{p}^{i0}$ at Beam Boundaries for applied Distributed Transverse Loading

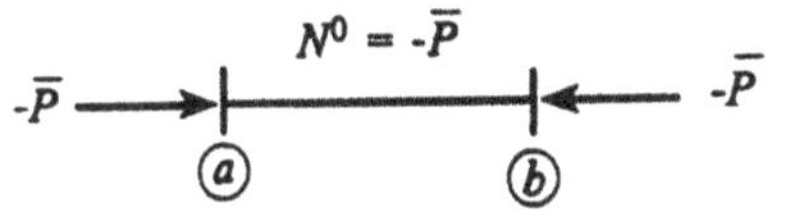

Moments	$\bar{M}_a^0$	$\bar{M}_b^0$	$\bar{M}_a^0$
p_0 (uniform)	$-\dfrac{p_0\ell^2}{2(A'+B')}$	$\dfrac{p_0\ell^2}{2(A'+B')}$	$-\dfrac{p_0\ell^2}{2A'}$
p_a (triangular)	$-\dfrac{p_a\ell^2}{6\varepsilon^2}(6+B'-2A')$	$\dfrac{p_a\ell^2}{6\varepsilon^2}(2B'-A')$	$-\dfrac{p_a\ell^2}{3\varepsilon^2}(3-C')$
p_b (triangular)	$-\dfrac{p_b\ell^2}{6\varepsilon^2}(2B'-A')$	$\dfrac{p_b\ell^2}{6\varepsilon^2}(6+B'-2A')$	$-\dfrac{p_b\ell^2}{6\varepsilon^2}\left(\dfrac{3\varepsilon^2}{A'}-6+2C'\right)$
Vertical Forces	$\bar{V}_a^0$	$\bar{V}_b^0$	$\bar{V}_a^0$
p_0 (uniform)	$\dfrac{p_0\ell}{2}$	$\dfrac{p_0\ell}{2}$	$\dfrac{p_0\ell}{2}+\dfrac{p_0\ell}{2A'}$
p_a (triangular)	$\dfrac{p_a\ell}{3}+\dfrac{p_a\ell}{6\varepsilon^2}(A'+B'-6)$	$\dfrac{p_a\ell}{6}-\dfrac{p_a\ell}{6\varepsilon^2}(A'+B'-6)$	$\dfrac{p_a\ell}{3}+\dfrac{p_a\ell}{3\varepsilon^2}(3-C')$
p_b (triangular)	$-\dfrac{p_b\ell}{6}-\dfrac{p_b\ell}{6\varepsilon^2}(A'+B'-6)$	$\dfrac{p_b\ell}{3}+\dfrac{p_b\ell}{6\varepsilon^2}(A'+B'-6)$	$\dfrac{p_b\ell}{6}+\dfrac{p_b\ell}{6}\left(\dfrac{3}{A'}-\dfrac{6}{\varepsilon^2}+\dfrac{2C'}{\varepsilon^2}\right)$

These reactions contain second order effects and should be applied for large compressive force (high ε)

$$C=\frac{\varepsilon^2\sin\varepsilon}{\sin\varepsilon-\varepsilon\cos\varepsilon}=3-\frac{1}{5}\varepsilon^2-\frac{1}{175}\varepsilon^4-\ldots$$

11.3 Variationally Based Stability Analysis (Theory of Second Order)

11.3.1 Principle of Virtual Work

The energy criterion for stability of Section 11.1.1 was given by $\delta(\frac{1}{2}\hat{\delta}^2\hat{\Pi})=0$, which is equivalent to $\delta(\Delta\Pi)=0$ and states that the first variation of the incremental change in energy is zero. It is convenient to develop the theory in terms of the virtual work W rather than the potential energy change $\Delta\Pi$. Thus, the stability criterion will be

$$-\delta(\Delta W)=0 \tag{11.54}$$

where $\delta(\Delta W)$ is the virtual work corresponding to the incremental change.

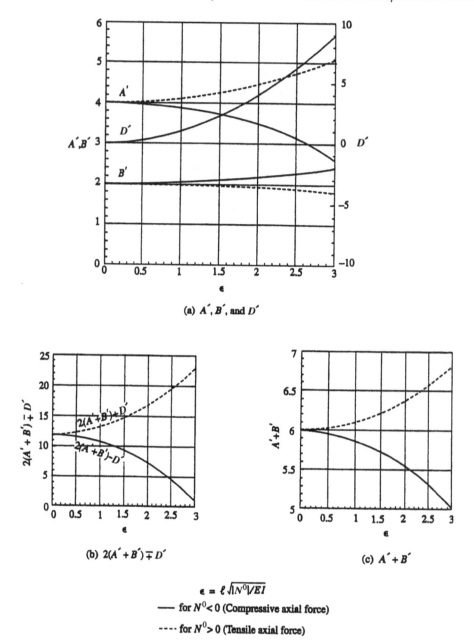

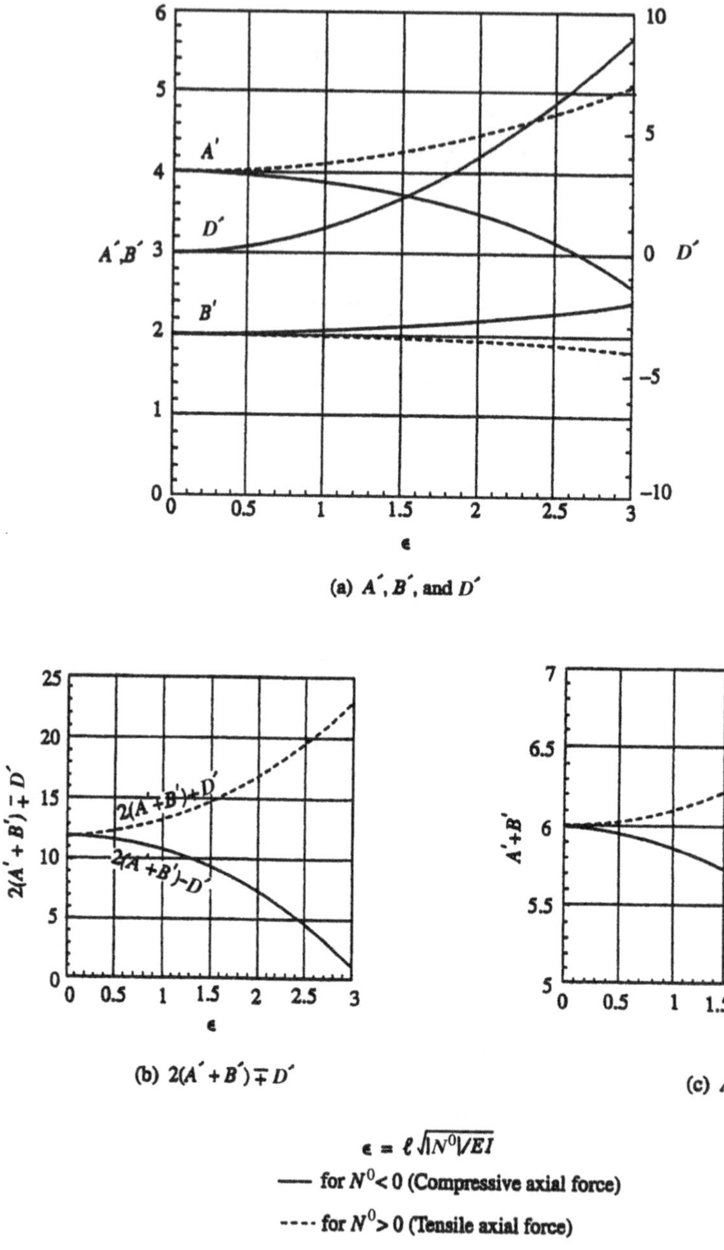

FIGURE 11.14
Parameters for stiffness matrices.

If virtual work expressions are employed for stability problems and other nonlinear theories, the analysis can be based on the methods already given in previous chapters, especially on the displacement method of Chapters 4 to 6. In this chapter, the displacement approach is demonstrated for structures with one-dimensional members, especially beam columns and frames including second order effects.

A variety of approximations for these systems can be formulated beginning with the first order theory, a linear theory. An engineering approximation in which the fundamental

TABLE 11.2

Overview of Stress and Stability Analyses of a Beam

Conditions	Theory of First Order	Theory of Second Order	Theory of Third Order
Equilibrium based on:	Undeformed system	Deformed system, with only those terms with the greatest influence considered	Deformed system
Strains	Linear $\varepsilon_x = u'$ $\kappa = 1/\rho = w''$	Nonlinear (in part) $\varepsilon_x = u' + \frac{1}{2}(w')^2$ $1/\rho = w''$	Nonlinear $\varepsilon_x = u' + \frac{1}{2}(w')^2$ $1/\rho = w''/(1+(w')^2)^{3/2}$
Force-displacement diagram	Linear	Nonlinear	Nonlinear
Superposition	Valid	Not valid	Not valid
Governing differential equations	Linear	Linearized	Nonlinear
Failure due to:	Stress or displacement level	Critical load	Critical load
Bifurcation of equilibrium	Not possible	Homogeneous problem yields eigenvalue	Also can study post-critical behavior

state is taken from the linear theory, such as that given in Section 11.2.2 and then using linearized equations, is the second order theory. The homogeneous equations render the bifurcations of equilibrium and the critical loads. This approach is emphasized in this chapter. A full-fledged nonlinear analysis is the third order theory and is based on an incremental formulation. Some properties of these methods are summarized in Table 11.2. The nonlinear strains in this table are obtained from Chapter 1, Eq. (1.15) in which the square of $\partial u/\partial x$ is neglected for the second order theory.

In the second order theory, the nonlinear effects are treated in a single increment, with the full loading in which the forces are known from a previous state. The values of these forces are then compared with the corresponding results of the analysis. If a desired relative accuracy is not reached, the analysis has to be repeated with the new forces as initial values. Thus, for each iteration, the analysis is based on the whole virtual work ($\Delta W \approx W$).

The desired relations arise from the global form corresponding to the differential equations of Section 11.2.2 using the forces V and H referred to the undeformed configuration. The principle of the virtual work then appears as

$$\begin{aligned} -\delta W &= \int (H\,\delta u' + V\,\delta w' + M\,\delta\theta')\,dx \\ &\quad - \int (\overline{p}_x\,\delta u + \overline{p}_z\,\delta w + H\overline{w}_0'\,\delta w + EI\overline{\kappa}\,\delta\theta')\,dx - [\overline{V}\,\delta w + \overline{M}\,\delta\theta]_b^a \\ &= 0 \end{aligned} \tag{11.55}$$

Introduce the approximations of Eq. (11.12), leading to $H \approx N = EAu'$, $V = Q + Nw' \approx Nw' = EAu'w'$ for a beam in which the shear terms in the material law for Q are neglected.

Based on the assumption that N is taken from the first order analysis $N \approx N^0$, the familiar second order theory is obtained. The internal virtual work then becomes (Wunderlich, 1976)

$$
\begin{aligned}
-\delta W_i &= \int (\delta u' EAu' + \delta w' N^0 w' + \delta w'' EI w'')\, dx \\
&= \int_x \delta \mathbf{u}^T \underbrace{\begin{bmatrix} {}_xd\, EA\, d_x & \vdots & 0 \\ \cdots\cdots & \vdots & \cdots\cdots \\ 0 & \vdots & {}_xd\, N^0 d_x + {}_xd^2\, EI\, d_x^2 \end{bmatrix}}_{\mathbf{k}^D} \underbrace{\begin{bmatrix} u \\ \cdots \\ w \end{bmatrix}}_{\mathbf{u}} dx
\end{aligned}
\tag{11.56}
$$

To include the effect of imperfections, replace w' with $w' + \overline{w}_0'$. The external virtual work is given by

$$
\delta W_e = \int (\underbrace{\overline{p}_x\, \delta u}_{\substack{\text{Horizontal}\\ \text{Distributed}\\ \text{Load}}} + \underbrace{\overline{p}_z\, \delta w}_{\substack{\text{Vertical}\\ \text{Distributed}\\ \text{Load}}} - \underbrace{N^0 \overline{w}_0'\, \delta w'}_{\text{Imperfections}} + \underbrace{EI\overline{\kappa}\, \delta\theta'}_{\substack{\text{Imposed}\\ \text{Curvature}}})\, dx \; \underbrace{- [\overline{V}\, \delta w + \overline{M}\, \delta\theta]_b^a}_{\substack{\text{Concentrated Loads}\\ \text{at the Boundaries}}} \tag{11.57}
$$

For the conditions of equilibrium to hold,

$$
-\delta W = -\delta W_i - \delta W_e = 0 \tag{11.58}
$$

with the internal virtual work of Eq. (11.56) and the external virtual work from Eq. (11.57). If shear deformation is taken into account, the internal virtual work relationship would appear as (Chapter 2, Example 2.7).

$$
-\delta W = \int_x \delta [u \quad w \quad \theta] \begin{bmatrix} {}_xd\, EA\, d_x & \vdots & & \vdots & \\ \cdots\cdots & \cdots & \cdots\cdots & \cdots & \cdots \\ & \vdots & \begin{matrix} {}_xd\, N^0 d_x \\ + {}_xd\, k_s GA\, d_x \end{matrix} & \vdots & {}_xd\, k_s GA \\ \cdots\cdots & \cdots & \cdots\cdots & \cdots & \cdots \\ & \vdots & k_s GA\, d_x & \vdots & \begin{matrix} {}_xd\, EI\, d_x \\ + k_s GA \end{matrix} \end{bmatrix} \begin{bmatrix} u \\ w \\ \theta \end{bmatrix} dx = 0 \tag{11.59}
$$

where $N^0 > 0$ for tensile axial forces. This expression is applied in Example 11.12, where an appropriate bifurcation load is found with and without consideration of shear deformation.

11.3.2 Beam Element Matrices (Theory of Second Order)

The principle of virtual work can also be used to generate the element stiffness matrices for stability analyses. For a beam the principle of virtual work takes the form of Eq. (11.58) with the extension term deleted. With the introduction of the shape function $w = \mathbf{N}_u \mathbf{G} \mathbf{v} = \mathbf{N} \mathbf{v}$, the element contribution for an element of length ℓ with compressive axial force N^0 is

$$-\delta W = \delta \mathbf{v}^T \left\{ \int_\ell \mathbf{N}''^T EI \mathbf{N}'' \, dx - N^0 \int_\ell \mathbf{N}'^T \mathbf{N}' \, dx \right\} \mathbf{v}$$

$$= \delta \mathbf{v}^T \left\{ \mathbf{G}^T \left[\int_\ell \mathbf{N}_u''^T EI \mathbf{N}_u'' \, dx \right] \mathbf{G} - N^0 \mathbf{G}^T \left[\int_\ell \mathbf{N}_u'^T \mathbf{N}_u' \, dx \right] \mathbf{G} \right\} \mathbf{v} \tag{11.60}$$

Define for element i

$$\begin{aligned} \mathbf{k}_{\text{lin}}^i &= \int_\ell \mathbf{N}''^T EI \mathbf{N}'' \, dx = \mathbf{G}^T \int_\ell \mathbf{N}_u''^T EI \mathbf{N}_u'' \, dx \, \mathbf{G} \\ \mathbf{k}_{\text{geo}}^i &= -N^0 \int_\ell \mathbf{N}'^T \mathbf{N}' \, dx = -N^0 \mathbf{G}^T \int_\ell \mathbf{N}_u'^T \mathbf{N}_u' \, dx \, \mathbf{G} \end{aligned} \tag{11.61}$$

Matrix $\mathbf{k}_{\text{lin}}^i$ is the same element stiffness matrix that was discussed in Chapters 4 and 5. This stiffness matrix is used in the static analysis of structures. If the same (static) shape function $\mathbf{N}$ is used in forming $\mathbf{k}_{\text{geo}}^i$ as in forming $\mathbf{k}_{\text{lin}}^i$, then $\mathbf{k}_{\text{geo}}^i$, which is called the *geometric stiffness matrix*, is said to be *consistent*. The matrix $\mathbf{k}_{\text{geo}}^i$ is also known as the *initial stress stiffness matrix*, *stability coefficient matrix*, or simply the *stress stiffness matrix*.

In this section, we will study the development of the element matrices $\mathbf{k}_{\text{lin}}^i$ and $\mathbf{k}_{\text{geo}}^i$ and applications to simple beam models. The assembly into system matrices of these element matrices will be discussed in Section 11.4, along with stability analyses of complex structures.

One of the most common assumed displacements is a trial function formed of Hermitian polynomials. The Hermitian polynomial shape functions of Chapter 4, Eq. (4.47b) $w = w_a H_1 + \ell\theta_a H_2 + w_b H_3 + \ell\theta_b H_4$ led to the exact element stiffness matrix for the static response in Chapter 4. As mentioned above, the use of the same (static response) polynomial to form the stability matrix $\mathbf{k}_{\text{geo}}^i$ provides a "consistent" matrix. Introduce (Chapter 4, Eq. (4.47a))

$$w = \mathbf{N}\mathbf{v} = \underbrace{[1 \quad \xi \quad \xi^2 \quad \xi^3]}_{\mathbf{N}_u} \underbrace{\begin{bmatrix} 1 & 0 & 0 & 0 \\ 0 & -1 & 0 & 0 \\ -3 & 2 & 3 & 1 \\ 2 & -1 & -2 & -1 \end{bmatrix}}_{\mathbf{G}} \underbrace{\begin{bmatrix} w_a \\ \ell\theta_a \\ w_b \\ \ell\theta_b \end{bmatrix}}_{\mathbf{v}} \tag{11.62}$$

with

$$\mathbf{N}_u' = [0 \quad 1 \quad 2\xi \quad 3\xi^2]\frac{1}{\ell}$$

$$\mathbf{N}_u'' = [0 \quad 0 \quad 2 \quad 6\xi]\frac{1}{\ell^2}$$

into Eq. (11.61). This gives the stiffness matrix $\mathbf{k}_{\text{lin}}^i$

$$\mathbf{k}_{\text{lin}}^i = \begin{array}{c} [\, w_a \quad \ell\theta_a \quad w_b \quad \ell\theta_b \,] \\ \begin{bmatrix} 12 & -6 & -12 & -6 \\ -6 & 4 & 6 & 2 \\ -12 & 6 & 12 & 6 \\ -6 & 2 & 6 & 4 \end{bmatrix} \end{array} \frac{EI}{\ell^3} \tag{11.63}$$

which is the same matrix derived in Chapter 4. The consistent geometric stiffness matrix becomes

$$\mathbf{k}^i_{\text{geo}} = -N^0 \begin{array}{c} [\ w_a \quad \ell\theta_a \quad w_b \quad \ell\theta_b\ ] \\ \begin{bmatrix} 6/5 & -1/10 & -6/5 & -1/10 \\ -1/10 & 2/15 & 1/10 & -1/30 \\ -6/5 & 1/10 & 6/5 & 1/10 \\ -1/10 & -1/30 & 1/10 & 2/15 \end{bmatrix} \end{array} \frac{1}{\ell} \tag{11.64}$$

which is independent of the material properties of the element.

This geometric stiffness matrix can be obtained from a series expansion of the stiffness matrix of Eq. (11.50). If axial extension is ignored and the displacement vector $\mathbf{v}^i = [w_a \ \ \ell\theta_a \ \ w_b \ \ \ell\theta_b]^T$ is introduced, the stiffness matrix of Eq. (11.50) appears as

$$\underset{\mathbf{p}^i}{\begin{bmatrix} V_a \\ M_a/\ell \\ V_b \\ M_b/\ell \end{bmatrix}} = \underset{\mathbf{k}^i_{\text{total}}}{\frac{EI}{\ell^3}\begin{bmatrix} 2(A'+B')-D' & -(A'+B') & -[2(A'+B')-D'] & -(A'+B') \\ & A' & A'+B' & B' \\ \text{Symmetric} & & 2(A'+B')-D' & A'+B' \\ & & & A' \end{bmatrix}} \underset{\mathbf{v}^i}{\begin{bmatrix} w_a \\ \ell\theta_a \\ w_b \\ \ell\theta_b \end{bmatrix}}$$

Expand this matrix using the series expressions for A', B', and D' of Eq. (11.51). The first two terms appear as

$$\mathbf{k}^i_{\text{total}} = \underset{\mathbf{k}^i_{\text{lin}}}{\begin{bmatrix} 12 & -6 & -12 & -6 \\ & 4 & 6 & 2 \\ \text{Symmetric} & & 12 & 6 \\ & & & 4 \end{bmatrix} \frac{EI}{\ell^3}} \underset{\mathbf{k}^i_{\text{geo}}}{- N^0 \begin{bmatrix} 6/5 & -1/10 & -6/5 & -1/10 \\ & 2/15 & 1/10 & -1/30 \\ \text{Symmetric} & & 6/5 & 1/10 \\ & & & 2/15 \end{bmatrix} \frac{1}{\ell}} + \cdots \tag{11.65}$$

or

$$\mathbf{k}^i_{\text{total}} = \mathbf{k}^i_{\text{lin}} + \mathbf{k}^i_{\text{geo}} + \cdots$$

Thus, the first term of the series expansion representation of Eq. (11.50) is the usual static stiffness matrix and the second term is the consistent geometric stiffness matrix.

The expressions for linear and geometric stiffness matrices for beam elements with second order effects are given in Table 11.3, in which boundary conditions other than the standard clamped ones are included. For the corresponding load vector, the loading terms consistent with linear theory, as in Tables 4.2 and 4.4, can be employed.

The accuracy of the approximate series solution relative to the analytical solution of Section 11.2.6 is of interest. The series representation of the analytical solution may be truncated at various powers of the parameter ε. In Fig. 11.15, the solutions for different levels of truncation are shown together with the solution for a beam (with fixed right end) according to linear first order theory. For constant EI, the exact solution for deflection w is a cubic polynomial. Although only small differences occur for the deflection w,

TABLE 11.3

Linear and Geometric Stiffness Matrices for a Beam with Various Boundary Conditions (In these matrices N^0 is in tension. For compression, replace N^0 by $-N^0$)

Fixed-Fixed

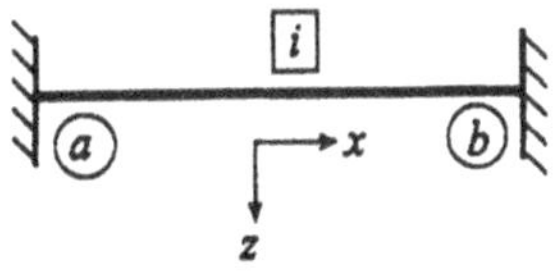

$$\begin{bmatrix} V_a \\ M_a \\ \cdots \\ V_b \\ M_b \end{bmatrix} = \left\{ \frac{EI}{\ell^3} \begin{bmatrix} 12 & -6\ell & \vdots & -12 & -6\ell \\ -6\ell & 4\ell^2 & \vdots & 6\ell & 2\ell^2 \\ \cdots & \cdots & \vdots & \cdots & \cdots \\ -12 & 6\ell & \vdots & 12 & 6\ell \\ -6\ell & 2\ell^2 & \vdots & 6\ell & 4\ell^2 \end{bmatrix} + \frac{N^0}{\ell} \begin{bmatrix} \frac{6}{5} & -\frac{1}{10}\ell & \vdots & -\frac{6}{5} & -\frac{1}{10}\ell \\ -\frac{1}{10}\ell & \frac{2}{15}\ell^2 & \vdots & \frac{1}{10}\ell & -\frac{1}{30}\ell^2 \\ \cdots & \cdots & \vdots & \cdots & \cdots \\ -\frac{6}{5} & -\frac{1}{10}\ell & \vdots & \frac{6}{5} & \frac{1}{10}\ell \\ -\frac{1}{10}\ell & -\frac{1}{30}\ell^2 & \vdots & \frac{1}{10}\ell & \frac{2}{15}\ell^2 \end{bmatrix} \right\} \begin{bmatrix} w_a \\ \theta_a \\ \cdots \\ w_b \\ \theta_b \end{bmatrix} - \begin{bmatrix} \overline{V}_a^0 \\ \overline{M}_a^0 \\ \cdots \\ \overline{V}_b^0 \\ \overline{M}_b^0 \end{bmatrix}$$

Fixed-Hinged

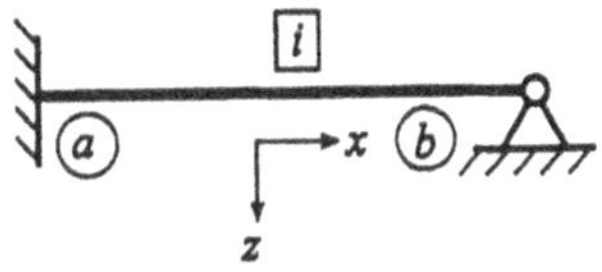

$$\begin{bmatrix} V_a \\ M_a \\ \cdots \\ V_b \end{bmatrix} = \left\{ \frac{EI}{\ell^3} \begin{bmatrix} 3 & -3\ell & \vdots & -3 \\ -3\ell & 3\ell^2 & \vdots & 3\ell \\ \cdots & \cdots & \cdots & \cdots \\ -3 & 3\ell & \vdots & 3 \end{bmatrix} + \frac{N^0}{\ell} \begin{bmatrix} \frac{6}{5} & -\frac{1}{5}\ell & \vdots & -\frac{6}{5} \\ -\frac{1}{5}\ell & \frac{1}{5}\ell^2 & \vdots & \frac{1}{5}\ell \\ \cdots & \cdots & \vdots & \cdots \\ -\frac{6}{5} & -\frac{1}{5}\ell & \vdots & \frac{6}{5} \end{bmatrix} \right\} \begin{bmatrix} w_a \\ \theta_a \\ \cdots \\ w_b \end{bmatrix} - \begin{bmatrix} \overline{V}_a^0 \\ \overline{M}_a^0 \\ \cdots \\ \overline{V}_b^0 \end{bmatrix}$$

Hinged-Fixed

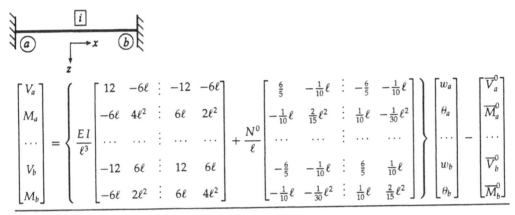

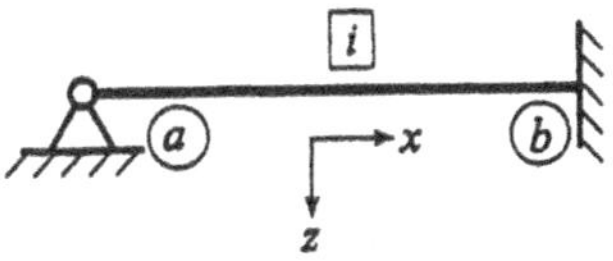

$$\begin{bmatrix} V_a \\ \cdots \\ V_b \\ M_b \end{bmatrix} = \left\{ \frac{EI}{\ell^3} \begin{bmatrix} 3 & \vdots & -3 & -3\ell \\ \cdots & \vdots & \cdots & \cdots \\ -3 & \vdots & 3 & 3\ell \\ -3\ell & \vdots & 3\ell & 3\ell^2 \end{bmatrix} + \frac{N^0}{\ell} \begin{bmatrix} \frac{6}{5} & \vdots & -\frac{6}{5} & -\frac{1}{5}\ell \\ \cdots & \cdots & \cdots & \cdots \\ -\frac{6}{5} & \vdots & \frac{6}{5} & \frac{1}{5}\ell \\ -\frac{1}{5}\ell & \vdots & \frac{1}{5}\ell & \frac{1}{5}\ell^2 \end{bmatrix} \right\} \begin{bmatrix} w_a \\ \cdots \\ w_b \\ \theta_b \end{bmatrix} - \begin{bmatrix} \overline{V}_a^0 \\ \cdots \\ \overline{V}_a^0 \\ \overline{M}_b^0 \end{bmatrix}$$

the deviations are larger for the rotation, which is the first derivative of w. On the other hand, the comparison also shows that the approximation for the geometric matrix up to ε^2 is relatively good, as it covers with one term about half of the possible difference between a linear first order theory and the analytical solution, including second order effects.

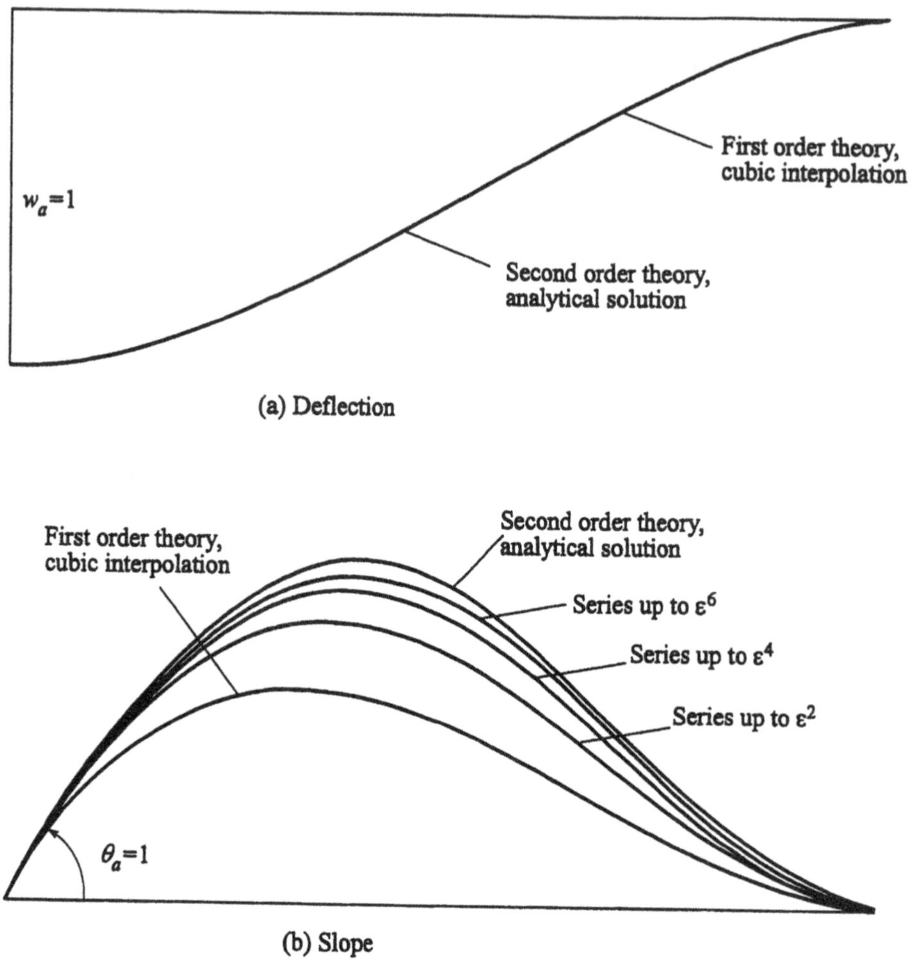

FIGURE 11.15
Comparison of different interpolations.

Several methods are available for evaluating the element matrix for a beam element in the framework of second order theory. An overview is given in Table 11.4. The principle of virtual work provides an excellent basis for the exact or approximate evaluation of the stiffness matrix for the beam element. The results are given in Section 11.2.6, Eqs. (11.50) and (11.52) for the exact, and in Section 11.3.2, Eq. (11.65) for the approximate stiffness matrix using trial functions. On the other hand, numerically exact transfer matrices can be obtained by integrating the corresponding system of first order differential equations of Eq. (11.28). They can be transformed into the corresponding numerically exact stiffness matrix using Eq. (4.11). Alternatively, the transformation given in Eq. (4.79) can be used to calculate a transfer matrix equivalent to the approximate linear and geometric stiffness matrices. The transfer and stiffness matrices given in Tables 4.3 and 4.4 of Chapter 4 also include second order effects and were obtained using exact analytical solutions, and, for this reason, give the same information as the transfer and stiffness matrices of Eqs. (11.50) and (11.52), although in a different notation.

TABLE 11.4

Evaluation Methods for Element Matrices

Definitions:	
Transfer Matrix $\mathbf{U}$ $\mathbf{z}_b = \mathbf{U}\,\mathbf{z}_a + \bar{\mathbf{z}}$ $\begin{bmatrix} \mathbf{v}_a \\ \mathbf{s}_a \end{bmatrix}$	Stiffness Matrix $\mathbf{k}$ $\mathbf{p} = \mathbf{k}\,\mathbf{v} - \bar{\mathbf{p}}^0$ $\begin{bmatrix} \mathbf{v}_a \\ \mathbf{v}_b \end{bmatrix}$
Basic Equations:	
Differential Equations	Principle of Virtual Work
Solution by:	

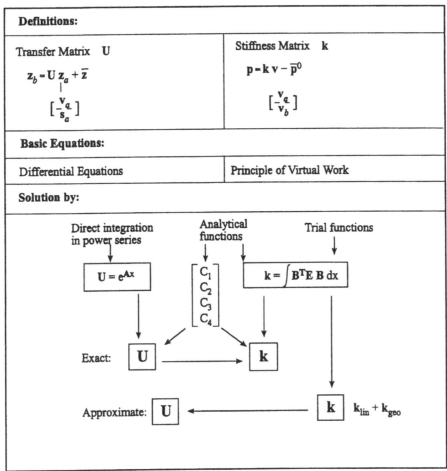

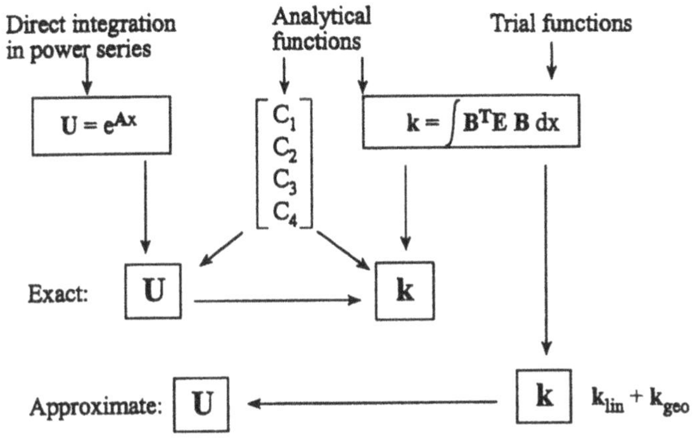

11.4 System Analysis (Theory of Second Order)

To evaluate the stability of structural systems, the displacement method of Chapters 4, 5 and 6 can be applied, although heretofore the displacement method was utilized for linear systems. For the system analysis, the element stiffness matrices $\mathbf{k}^i_{\text{lin}}$ (such as $\mathbf{k}^i_{\text{lin}}$ of Eq. (11.63) for beams) are assembled to form the system stiffness matrix $\mathbf{K}_{\text{lin}}$. In a similar fashion, the geometric stiffness matrices $\mathbf{k}^i_{\text{geo}}$ of the elements (such as the beam $\mathbf{k}^i_{\text{geo}}$ of Eq. (11.64)) are assembled into the system geometric stiffness matrix $\mathbf{K}_{\text{geo}}$.

It is important to remember, however, that, due to the nonlinear character of the problem, the superposition of different load cases is not possible. The analysis has to be performed by a realistic combination of the applied loading multiplied by a scaling multiplier called the *load factor*, which has to be kept above a certain safety limit (e.g., 1.75 or 1.5). In addition, initially a linear analysis needs to be performed to obtain the variables of the fundamental

TABLE 11.5

Calculation Procedure

Define the system, cross-sectional properties, length.

Theory of First order

- Basic load Cases
- Linear calculations (Fundamental state)
  Stress resultants of the fundamental state, e.g. N^0

Theory of Second Order

- Choose load combinations with imperfections such as initial deformations and initial curvatures, and factors of safety
- Calculation steps
  1. N^0 for each element
  2. Stiffness matrix and load vector for each element
  3. Assemble the system stiffness matrix and the system load vector
  4. Solve the system equations
  5. Compute the member stress resultants
  6. Compare with the values of Step 1.
  7. If necessary, proceed through Steps 1 to 6 using the improved values of N^0. Thus, implement an interation procedure.
  8. Post processing calculations:
     Max M, corresponding Q and N.
     Max N, corresponding M, Q, etc.

---

state, which have to be known in the sense of Eq. (11.19) to linearize the basic equations or the principle of virtual work. The steps of the procedure are summarized in Table 11.5.

EXAMPLE 11.4 Influence of Geometric Imperfections on a Framework

In this example, the influence of imperfections on the behavior of the framework of Fig. 11.16a is investigated. The structure consists of three beam column elements that are rigid with respect to extension but flexible for bending.

In addition to the applied loading, it is assumed that element 1 is initially inclined and has an initial parabolic curvature, both imperfections caused by the manufacturing process. Element 3 experiences the same initial inclination, see Fig. 11.16b.

As shown in Fig. 11.17a, these imperfections can be replaced by equivalent loads or reactions at the ends of the elements. For the case of the parabolic shape of the imperfection, the equivalent end loads are obtained by setting the midspan moment $M = Nw_0$ of the imperfection equal to the midspan moment $M = \overline{q}\ell^2/8$ for an element with end reaction forces and a uniformly distributed load $\overline{q}$. This gives the equivalent load magnitude $\overline{q} = 8Nw_0/\ell^2$ and end reactions $4Nw_0/\ell$ shown in Fig. 11.17a. For an initial inclination, the resulting moment $N\theta_0\ell$ leads to the reaction $N\theta_0$. The corresponding loading terms for the forces and moments at the element ends are given in Fig. 11.17b. These loading terms are employed during the process of assembling the nodal forces in the system analysis.

The results of this analysis are provided in Table 11.6, in which various values of the state variables are given for three cases: linear analysis and second-order analyses with and without imperfections. The distribution of the stress resultants for the three cases are shown in Fig. 11.18.

Comparison of the results shows that the linear theory underestimates the deflections and stress resultants and could lead to an unsafe design. In addition, there are substantial differences between the results with and without the prescribed imperfections. From Fig. 11.18,

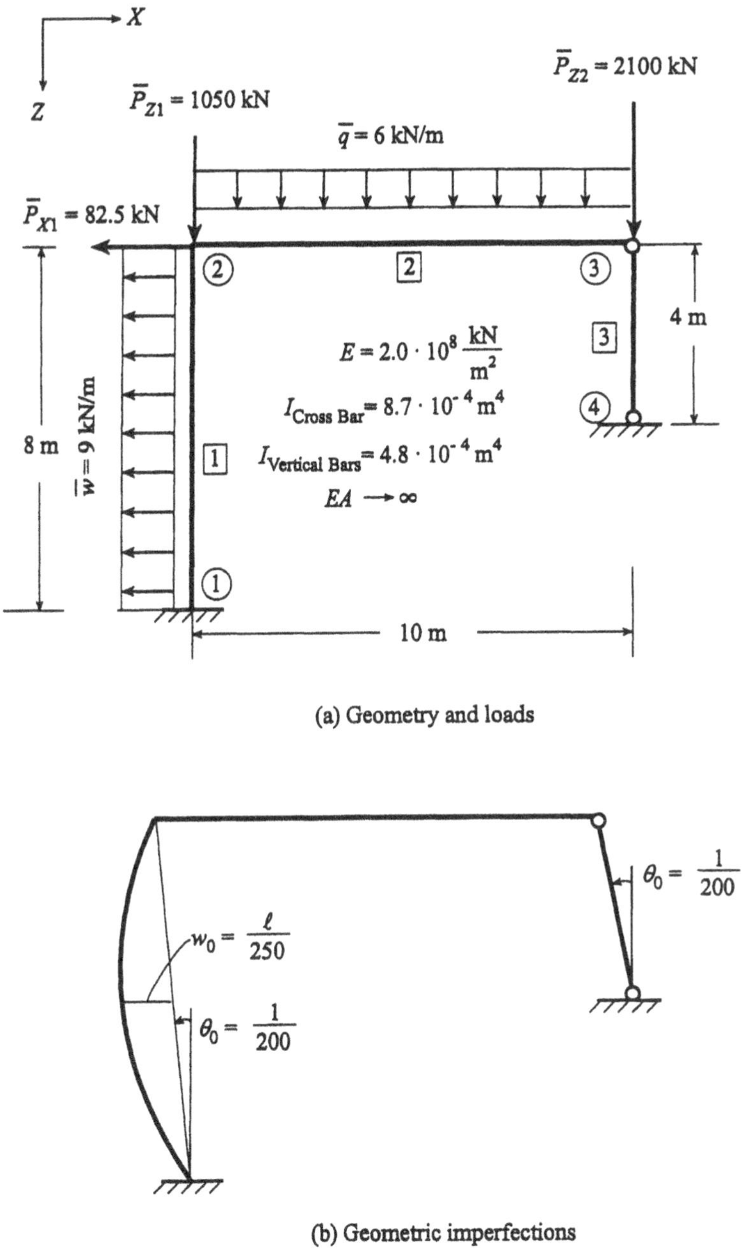

FIGURE 11.16
Influence of imperfections on a framework.

it is apparent that the moments and shearing forces are affected significantly by the second order effects and the imperfections. It is concluded that, not only does the influence of second order effects have to be taken into account to obtain realistic results and safe designs, but also initial imperfections play an important role and have to be considered. For this reason, in some design codes, limits on imperfections are prescribed to assure a safe design. ∎

(a) Imperfections and equivalent load

Imperfection	$\bar{M}_a^0$	$\bar{M}_b^0$	$\bar{V}_b^0$	$\bar{V}_b^0$	Element Forces
Parabolic w_0	$-\dfrac{4}{A'+B'}Nw_0$	$\dfrac{4}{A'+B'}Nw_0$	0	0	
Inclination θ_0	0	0	$-N\theta_0$	$N\theta_0$	

(b) Contribution to element load vector

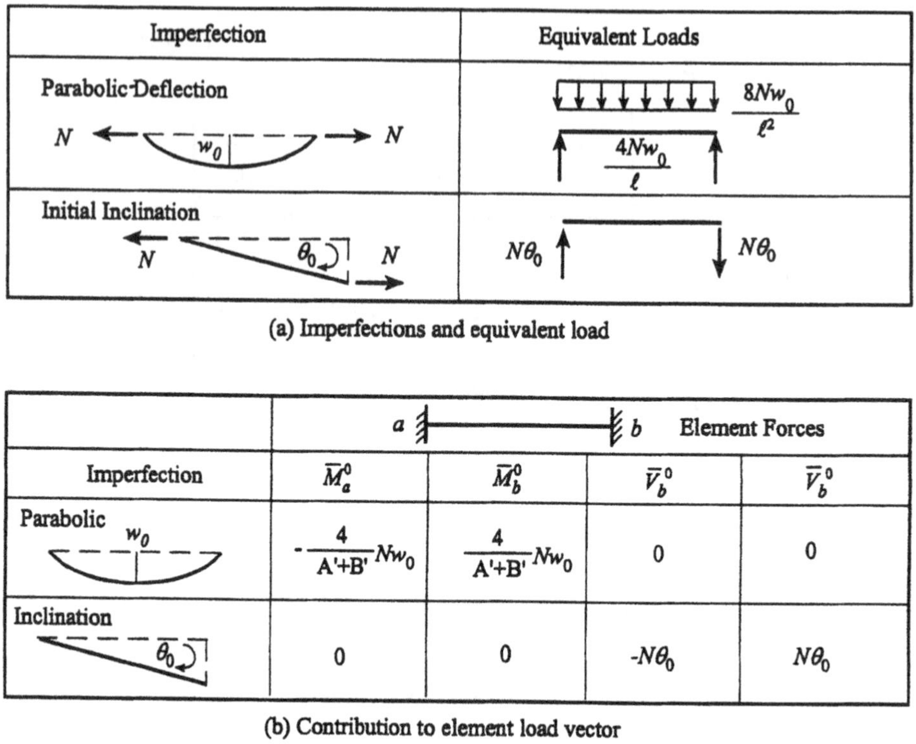

FIGURE 11.17
Introduction of imperfections.

TABLE 11.6

Comparison of the Results

	First Order (Linear)	Second Order	
		Without Imperfections	With Imperfections
U_2	−0.07831	−0.15417	−0.17387
Θ_2	0.00553	0.01210	0.013556
M_1^1	584.2	1055.5	1211.2
M_2^1	−363.8	−704.8	−780.0
V_1^1	−154.5	−233.9	−253.9
V_2^1	−82.5	−161.9	−181.9
V_2^2	66.4	100.5	108.0
V_3^2	6.4	40.5	48.0
V_3^3	0.0	79.4	99.4
N^1	−1116.4	−1150.5	−1158.0
N^2	0.0	−79.4	−99.4
N^3	−2093.6	−2059.5	−2052.0

Subscripts and superscripts refer to nodes and elements, respectively. For example, U_2 is the horizontal displacement of node 2.

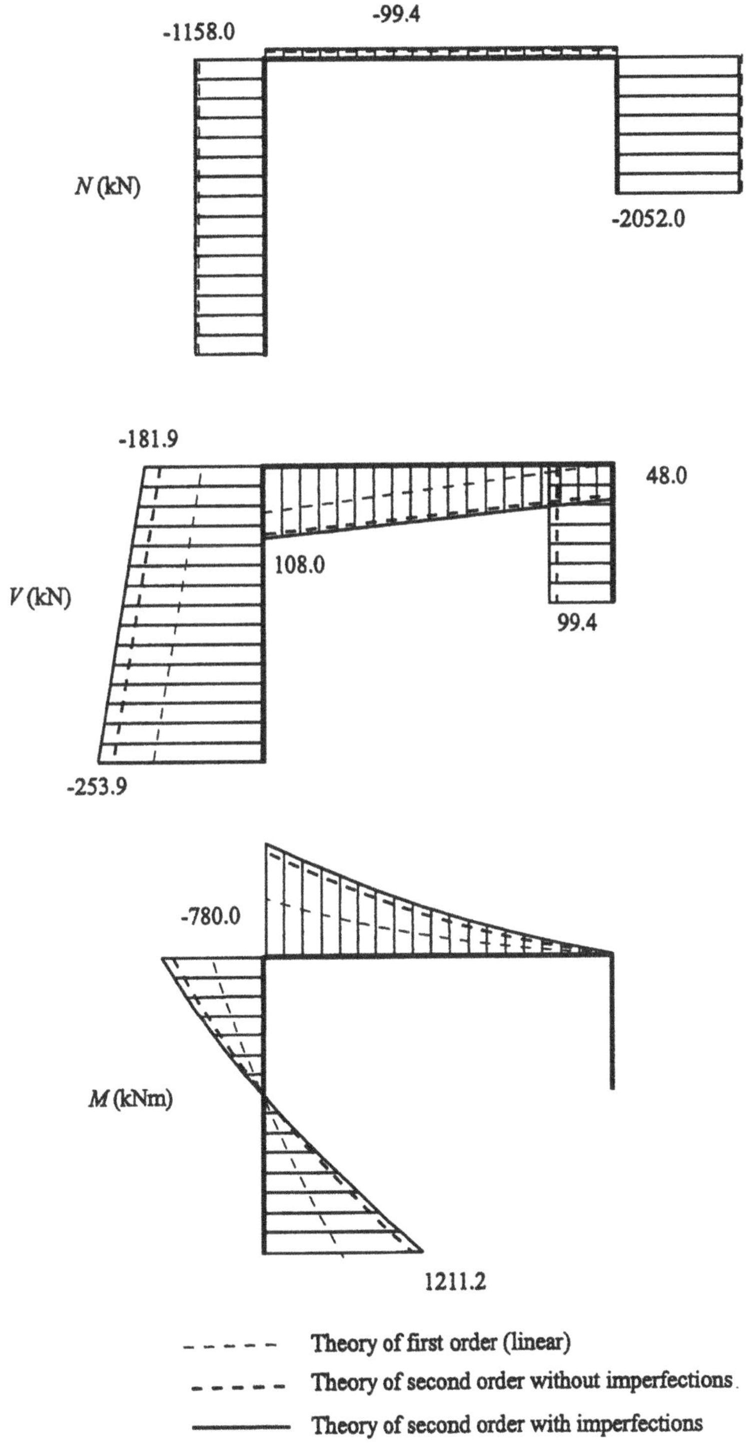

FIGURE 11.18
Distribution of the forces and moments.

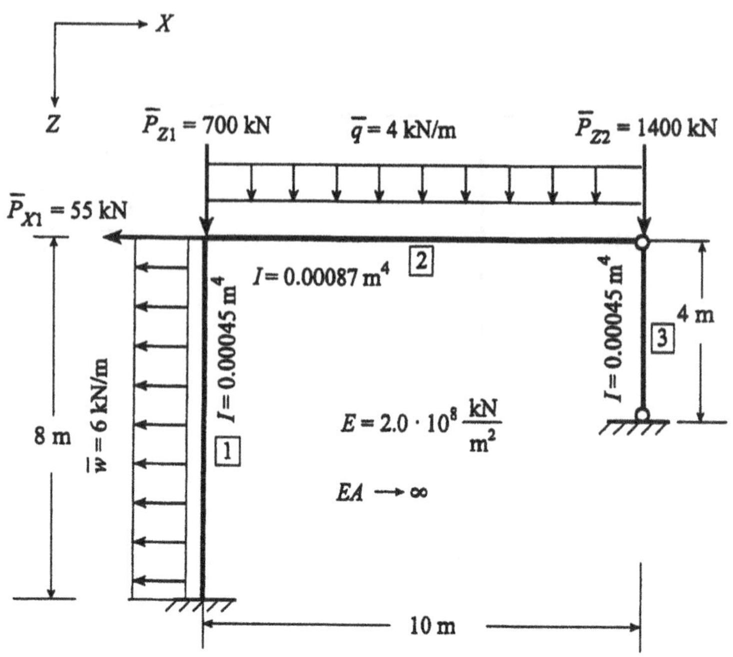

FIGURE 11.19
Geometry and loading of a framework.

EXAMPLE 11.5 System Analysis of a Framework with Second Order Effects
In the previous example, the results of first order and second order calculations are compared to emphasize the importance of the imperfections on stability behavior. In this example, a second order analysis of a framework is demonstrated in full detail. Choose the framework of Fig. 11.19, where it is assumed that the beam elements are rigid with respect to extension.

The calculation procedure of an approximate geometrically nonlinear analysis, including second order effects, is given in Table 11.5. In the initial integration step, which for practical problems very often is sufficient, the stress resultants of the linear analysis are taken as the fundamental state of the stability analysis. For this example, a linear analysis of this framework gives $N^0_{\text{element1}} = -1116.4$ kN, $N^0_{\text{element2}} = 0$, $N^0_{\text{element3}} = -2093.6$ kN.

Load Combinations and Unknowns of the Displacement Method
In addition to the geometric description of the framework, the basic load cases and possible imperfections have to be selected. Because the superposition of the results of different load cases does not apply for a nonlinear problem, a specific load combination has to be fixed. We choose to introduce a factor of safety of 1.5 (see Fig. 11.20). For both the linear and the second order calculations, the displacement method can be employed as described in Chapter 5 with the global (X, Y, Z) and the local (x, y, z) coordinates. Follow the usual procedure of the displacement method and choose the nodal unknowns as shown in Fig. 11.21, considering the specific boundary conditions of the problem. In general, three local degrees of freedom are assigned to each node of this planar system: a horizontal displacement (U), a vertical displacement (W), and a rotation (Θ). These are either unknown or prescribed by the boundary conditions, for instance, for node 1 with clamped end conditions all three degrees of freedom have to vanish, as shown in Fig. 11.21. Because the extensional stiffness EA is assumed to be infinite in all members, the values of the responses of nodes 2 and 3 are not independent. Hence, U_2 and U_3 are equal, and the vertical displacements of nodes 2 and 3 are zero. Due to these assumptions, only two nodal values remain as unknowns: U_2 and Θ_2.

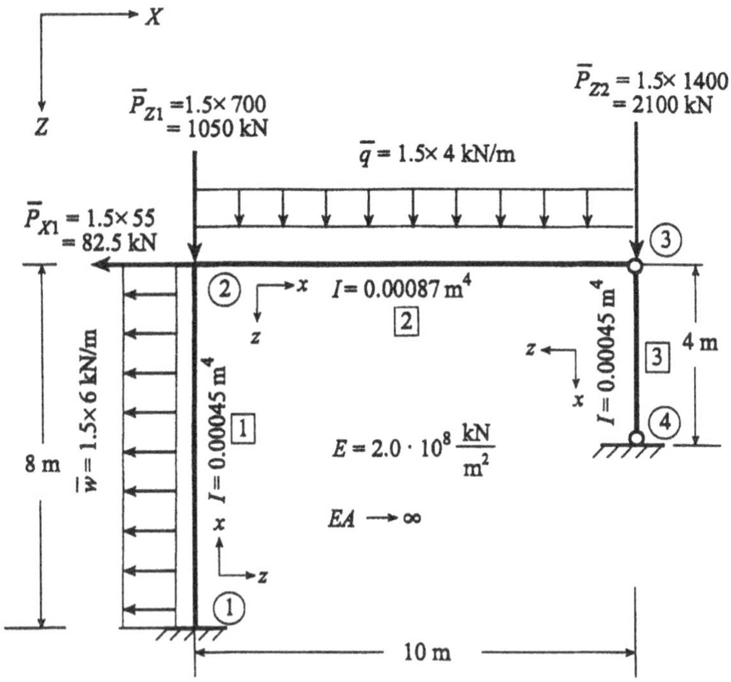

FIGURE 11.20
Introduction of a factor of safety of 1.5.

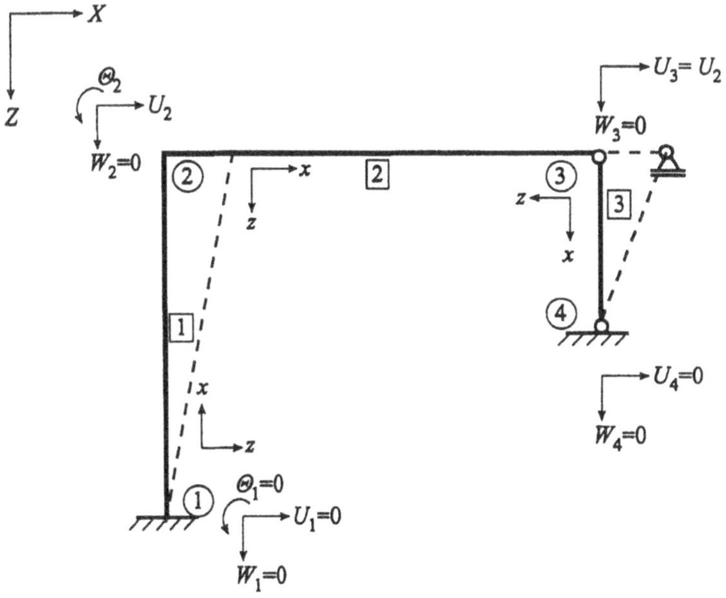

FIGURE 11.21
Unknowns and prescribed degrees of freedom.

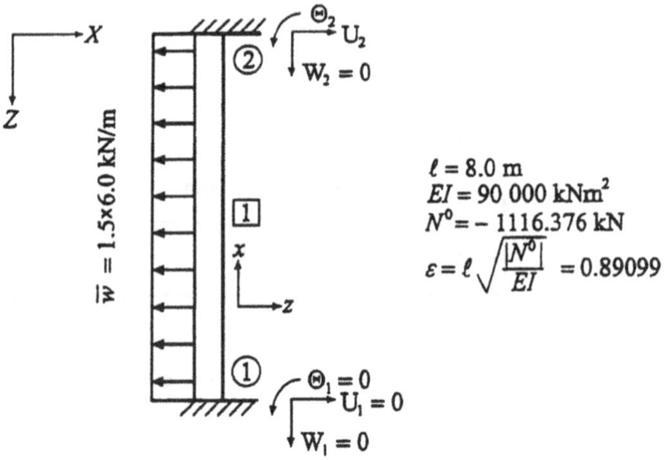

FIGURE 11.22
Element 1 with local and global coordinates.

From the calculation procedure of Table 11.5, the next step is the formation of the stiffness matrices and load vectors for each element. First, they are calculated with respect to the local coordinates of each element. The transformation to the global coordinate systems follows. This leads to the element matrices and load vectors which can be assembled into the system stiffness matrix and system load vector.

As mentioned, the element matrices of the linear system (first order) have to be supplemented with the geometric parts of the stiffness matrices to include second order effects.

We now proceed to form the linear and geometric parts of the stiffness matrices and vectors for the three elements of the system. The initial step of a second order anlaysis requires the calculation of the normal forces N^0 in the elements. These are usually obtained from a linear analysis.

Stiffness Matrices and Load Vector of Element 1

For element 1, a beam column, substitude the values of Fig. 11.22 into Eq. (11.63). For the displacement vector arranged as $\mathbf{v}^i = [w_a\ \theta_a\ w_b\ \theta_b]^T$ we obtain (Table 11.3)

$$\mathbf{k}^1_{\text{lin}} = \frac{EI}{\ell^3}\begin{bmatrix} 12 & -6\ell & 12 & -6\ell \\ -6\ell & 4\ell^2 & 6\ell & 2\ell^2 \\ -12 & 6\ell & 12 & 6\ell \\ -6\ell & 2\ell^2 & 6\ell & 4\ell^2 \end{bmatrix} = \begin{bmatrix} 2109.38 & -8437.5 & -2109.38 & -8437.50 \\ & 45\,000.0 & 8437.5 & 22\,500.0 \\ & & 2109.38 & 8437.50 \\ \text{Symmetric} & & & 45\,000.0 \end{bmatrix} \tag{1}$$

From Eq. (11.64)

$$\mathbf{k}^1_{\text{geo}} = \frac{N^0}{\ell}\begin{bmatrix} \frac{6}{5} & -\frac{\ell}{10} & -\frac{6}{5} & -\frac{\ell}{10} \\ & \frac{2\ell^2}{15} & \frac{\ell}{10} & -\frac{\ell^2}{30} \\ \text{Symmetric} & & \frac{6}{5} & \frac{\ell}{10} \\ & & & \frac{2\ell^2}{15} \end{bmatrix}$$

$$= \begin{bmatrix} -167.46 & 111.64 & 167.46 & 111.64 \\ & -1190.80 & -111.64 & 297.70 \\ & & -167.46 & -111.64 \\ \text{Symmetric} & & & -1190.80 \end{bmatrix} \tag{2}$$

Then, sum these parts of the local stiffness matrix of element 1.

$$\mathbf{k}_{\text{total}}^{1} = \mathbf{k}^{1} = \mathbf{k}_{\text{lin}}^{1} + \mathbf{k}_{\text{geo}}^{1}$$

$$= \begin{matrix} (u_{X1}) & (\theta_{Y1}) & (u_{X2}) & (\theta_{Y2}) & & \\ \left[\begin{matrix} 1941.92 \\ \\ \text{Symmetric} \\ \\ \end{matrix}\right. & \begin{matrix} -8325.86 \\ 43\,809.20 \\ \\ \\ \end{matrix} & \begin{matrix} -1941.92 \\ 8325.86 \\ 1941.92 \\ \\ \end{matrix} & \left.\begin{matrix} -8325.86 \\ 22\,797.70 \\ 8325.86 \\ 43\,809.20 \end{matrix}\right] & \begin{matrix} (V_1^1) \\ (M_1^1) \\ (V_2^1) \\ (M_2^1) \end{matrix} & \begin{matrix} (F_{X1}^1) \\ (M_{Y1}^1) \\ (F_{X2}^1) \\ (M_{Y2}^1) \end{matrix} \\ U_1 = 0 & \Theta_1 = 0 & U_2 & \Theta_2 & & \end{matrix} \tag{3}$$

The transformation of the local stiffness matrices of element 1 to global coordinates is accomplished by setting $w_2 = U_2$. The displacements and forces in parentheses above and to the right of the matrix of (3) indicate the local and global variables to assist in identifying where the rows and columns fit when assembling the nodal equilibrium equations in the summation process to obtain the system stiffness matrix.

In contrast to the expressions arising from the analytical solution, the loading vector for the approximation with the geometric stiffness does not contain geometrically nonlinear terms. From Eq. (4.73)

$$-\overline{\mathbf{p}}^{10} = \overline{w}\begin{bmatrix} -\ell/2 \\ \ell^2/12 \\ -\ell/2 \\ -\ell^2/12 \end{bmatrix} = \begin{bmatrix} 36.00 \\ -48.00 \\ 36.00 \\ 48.00 \end{bmatrix} \begin{matrix} (V_1^{10}) \\ (M_1^{10}) \\ (V_2^{10}) \\ (M_2^{10}) \end{matrix} \quad \begin{matrix} (F_{X1}^{10}) \\ (M_{Y1}^{10}) \\ (F_{X2}^{10}) \\ (M_{Y2}^{10}) \end{matrix} \tag{4}$$

Stiffness Matrix and Load Vector of Element 2

For element 2, a beam, the longitudinal force of the linear fundamental state is zero. This value is obtained from the linear analysis of the framework as the initial step in the analysis. Thus, only the elastic part of the stiffness matrix is of importance and is evaluated for the system fixed at one end with a moment release at the other. This stiffness matrix can be taken from the fixed-hinged case of Table 11.3. Use the element properties from Fig. 11.23

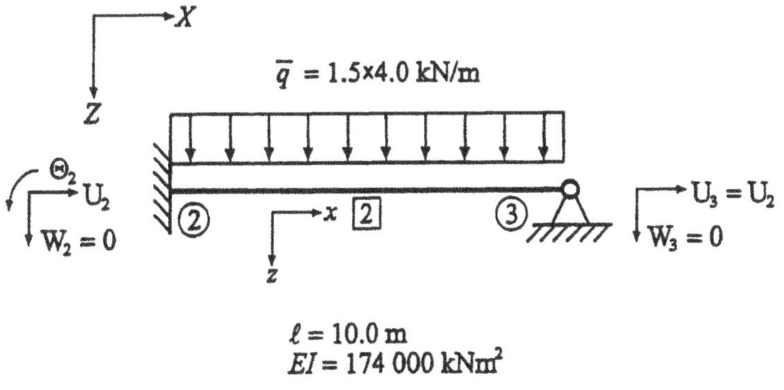

FIGURE 11.23
Element 2 with local global coordinates.

to obtain

$$\mathbf{k}^2 = \begin{bmatrix} 3\frac{EI}{\ell^3} & -3\frac{EI}{\ell^2} & -3\frac{EI}{\ell^3} \\ -3\frac{EI}{\ell^2} & 3\frac{EI}{\ell} & 3\frac{EI}{\ell^2} \\ -3\frac{EI}{\ell^3} & 3\frac{EI}{\ell^2} & 3\frac{EI}{\ell^3} \end{bmatrix} = \begin{array}{c} \begin{array}{ccc} (w_{Z2}) & (\theta_{Z2}) & (w_{Z3}) \end{array} \\ \begin{bmatrix} 522 & -5220 & -520 \\ & 52\,200 & 5220 \\ \text{Symmetric} & & 522 \end{bmatrix} \\ \begin{array}{ccc} W_2 = 0 & \Theta_2 & W_3 = 0 \end{array} \end{array} \begin{array}{cc} (V_2^2) & (F_{Z2}^2) \\ (M_2^2) & (M_{Y2}^2) \\ (V_3^2) & (F_{Z3}^2) \end{array} \tag{5}$$

For this element, the global and local coordinate system coincide. The loading vector is given by

$$-\overline{\mathbf{p}}^{20} = \overline{q} \begin{bmatrix} -5\ell/8 \\ \ell^2/8 \\ -3\ell/8 \end{bmatrix} = \begin{bmatrix} -37.5 \\ 75.0 \\ -22.5 \end{bmatrix} \begin{array}{cc} (V_2^{20}) & (F_{Z2}^{20}) \\ (M_2^{20}) & (M_{Y2}^{20}) \\ (V_3^{20}) & (F_{Z3}^{20}) \end{array} \tag{6}$$

Stiffness Matrix of Element 3

Element 3 has only hinges on both ends and no transverse loads and, hence, no bending. This bar acts like a truss member, influencing the bending of the system by second order effects only (Fig. 11.24). However, these second order effects contained in the geometric matrix contribute to the change in the moments in other members. The horizontal inclination of the top of the truss in the direction of the unknown displacement U_2 (Fig. 11.21) corresponds to a horizontal reaction force for element 3. With the value N^0/ℓ it contributes through the geometric stiffness to the equilibrium of the horizontal forces for the determination of the unknown U_2.

$$\mathbf{k}_{\text{geo}}^2 = -2093.63 \begin{bmatrix} 1/4 & -1/4 \\ -1/4 & 1/4 \end{bmatrix} = \begin{array}{c} \begin{array}{cc} (u_{X3}) & (u_{X4}) \end{array} \\ \begin{bmatrix} -523.406 & 523.406 \\ 523.406 & -523.406 \end{bmatrix} \\ \begin{array}{cc} U_3 = U_2 & U_4 = 0 \end{array} \end{array} \begin{array}{cc} (V_3^3) & (F_{X3}^3) \\ (V_4^3) & (F_{X4}^3) \end{array} \tag{7}$$

The transformation of the stiffness matrix of element 3 to global coordinates is accomplished by setting $X = -z$.

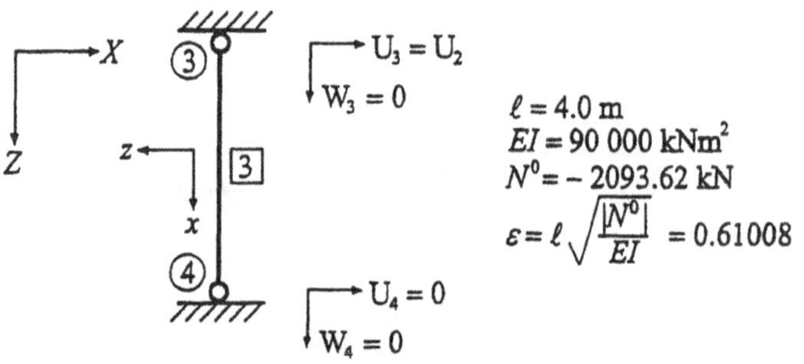

FIGURE 11.24
Element 3 with local and global coordinates.

Formation of the System Matrix

To form the global stiffness matrix, assemble the stiffness matrices of the elements with the usual process of the displacement method as described in Chapter 5. The geometric matrices are handled in the same manner as the linear matrices. The superposition is shown below, starting with the topological description of the frame in the incidence table, followed by an expression of the assembly as equilibrium at the nodes. For details, the reader is referred to Chapter 5, Example 5.5, where the same procedure is described.

Incidence Table

Element	(u_{Xa})	(w_{Za})	(θ_{Ya})	(u_{Xb})	(w_{Zb})	(θ_{Yb})
1	0	0	0	1	0	2
2	0	0	2	0	0	0
3	1	0	0	0	0	0

(8)

From the incidence table we obtain information as to which degrees of freedom of a particular element contribute to the formation of equilibrium for the unknown nodal displacements. In this example, the unknown horizontal displacement U_2 and the unknown rotation Θ_2 at node 2 are determined by the equilibrium conditions for the conjugate horizontal forces F_{X2} and moments M_{Y2} at this node.

		1	2	
		U_2	Θ_2	
		1941.92	8325.86	Element 1
1	F_{X2}			Element 2
		−523.41		Element 3
		8325.86	43 809.20	Element 1
2	M_{Y2}		52 200.00	Element 2
				Element 3

(9)

In this 2 × 2 matrix the contributions of the three elements can be clearly distinguished when compared with the element matrices of (3), (5), and (7). The first row gives the equilibrium of the forces in the horizontal direction at node 2, while the second describes the equilibrium of the moments at the same node. Summation of the element stiffnesses renders the stiffness matrix of the system:

$$\mathbf{K}_{\text{total}} = \begin{bmatrix} 1418.51 & 8325.86 \\ 8325.86 & 96\,009.20 \end{bmatrix} \tag{10}$$

From Eq. (5.100), the addition of the loading vectors takes the form

$$\overline{\mathbf{P}} = \overline{\mathbf{P}}^{*} + \overline{\mathbf{P}}^{0} \tag{11}$$

$$\mathbf{P}_{\text{total}} = \begin{bmatrix} \overset{\overline{\mathbf{P}}^{*}}{-82.5} & \overset{\overline{\mathbf{P}}^{0}}{-36.00 + 0.0} \\ 0.0 & -48.00 - 75.0 \end{bmatrix} = \begin{bmatrix} -118.50 \\ -123.00 \end{bmatrix} \tag{12}$$

where $\overline{\mathbf{P}}^{*}$ contains the loads at the nodes (−82.5, 0) applied directly at node 2 and the vector $\overline{\mathbf{P}}^{0}$ contains the horizontal force and the sum of the two end moments (−36.00 + 0.0) and (−48.00 − 75.0) arising from the distributed loads on the two bars adjacent to node 2.

Solution of the system of equations $\mathbf{KV} = \overline{\mathbf{P}}$ gives

$$\mathbf{V} = \begin{bmatrix} -0.15482228 \\ 0.01214497 \end{bmatrix} \begin{matrix} (U_2) \\ (\Theta_2) \end{matrix} \tag{13}$$

Determination of the Member Stress Resultants

From these two nodal values, the displacements and rotations at the element edges can be determined.

The forces and moments for the elements can be obtained using

$$\mathbf{p}^i = \mathbf{k}^i\mathbf{v}^i - \overline{\mathbf{p}}^{i0} \tag{14}$$

For element 1, from (13)

$$\mathbf{v}^1 = \begin{bmatrix} 0.0 \\ 0.0 \\ -0.1542228 \\ 0.01214497 \end{bmatrix} \begin{matrix} (u_{X1}) \\ (\theta_{Y1}) \\ (u_{X2}) \\ (\theta_{Y2}) \end{matrix} \tag{15}$$

From (3) and (15)

$$\mathbf{k}^1\mathbf{v}^1 = \begin{bmatrix} 199.535 \\ -1012.15 \\ -199.535 \\ -756.968 \end{bmatrix} \begin{matrix} (V_1^1) & (F_{X1}^1) \\ (M_1^1) & (M_{Y1}^1) \\ (V_2^1) & (F_{X2}^1) \\ (M_2^1) & (M_{Y2}^1) \end{matrix} \tag{16}$$

Introduce (4)

$$\begin{matrix} \begin{bmatrix} 199.535 \\ -1012.15 \\ -199.535 \\ -756.968 \end{bmatrix} & + & \begin{bmatrix} 36.00 \\ -48.00 \\ 36.00 \\ 48.00 \end{bmatrix} & = & \begin{bmatrix} 235.535 \\ -1060.152 \\ -163.535 \\ -708.968 \end{bmatrix} \\ \mathbf{k}^1\mathbf{v}^1 & - & \overline{\mathbf{p}}^{10} & = & \mathbf{p}^1 \end{matrix} \begin{matrix} (V_1^1) & (F_{X1}^1) \\ (M_1^1) & (M_{Y1}^1) \\ (V_2^1) & (F_{X2}^1) \\ (M_2^1) & (M_{Y2}^1) \end{matrix} \tag{17}$$

For element 2

$$\mathbf{v}^2 = \begin{bmatrix} 0.0 \\ 0.01214497 \\ 0.0 \end{bmatrix} \begin{matrix} (w_2) & (w_{Z2}) \\ (\theta_2) & (\theta_{Y2}) \\ (w_3) & (w_{Z3}) \end{matrix} \tag{18}$$

From (5) and (18)

$$\mathbf{k}^2\mathbf{v}^2 = \begin{bmatrix} -63.3968 \\ 633.9675 \\ 63.3968 \end{bmatrix} \begin{matrix} (V_2^2) & (F_{Z2}^2) \\ (M_2^2) & (M_{Y2}^2) \\ (V_3^2) & (F_{Z3}^2) \end{matrix} \tag{19}$$

$$\begin{matrix} \begin{bmatrix} -63.3968 \\ 633.9675 \\ 63.3968 \end{bmatrix} & + & \begin{bmatrix} -37.5 \\ 75.0 \\ -22.5 \end{bmatrix} & = & \begin{bmatrix} -100.897 \\ 708.968 \\ 40.897 \end{bmatrix} \\ \mathbf{k}^2\mathbf{v}^2 & - & \overline{\mathbf{p}}^{20} & = & \mathbf{p}^2 \end{matrix} \begin{matrix} (V_2^2) & (F_{Z2}^2) \\ (M_2^2) & (M_{Y2}^2) \\ (V_3^2) & (F_{Z3}^2) \end{matrix} \tag{20}$$

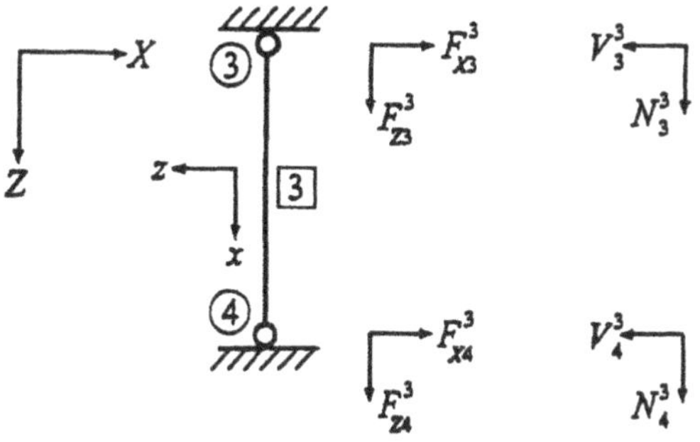

FIGURE 11.25
Element 3 forces.

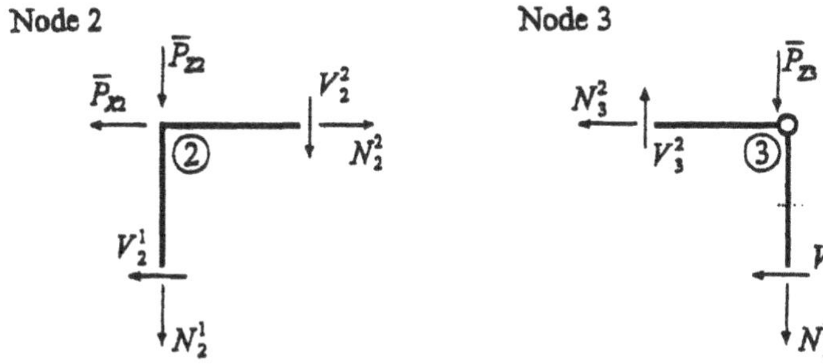

FIGURE 11.26
Determination of longitudinal forces.

For element 3 (Fig. 11.25)

$$\mathbf{v}^3 = \begin{bmatrix} -0.15482228 \\ 0.0 \end{bmatrix} \begin{matrix} (u_{X3}) & (-w_3) \\ (u_{X4}) & (-w_4) \end{matrix} \tag{21}$$

From (7) and (21)

$$\mathbf{k}^3\mathbf{v}^3 = \begin{bmatrix} 81.035 \\ -81.035 \end{bmatrix} \begin{matrix} (F^3_{X3}) & (-V^3_3) \\ (F^3_{X4}) & (-V^3_4) \end{matrix} = -\mathbf{p}^3 \tag{22}$$

Whereas the transverse forces V and the moments M can be obtained using the stiffness relations of the element, the longitudinal forces have to be determined by direct nodal equilibrium. See Fig. 11.26. Sign Convention 1 will be utilized.

$$\begin{aligned}
&\textstyle\sum F_X = 0 : N^2_2 - \overline{P}_{X2} - V^1_2 = 0 && \textstyle\sum F_X = 0 : N^2_3 + V^3_3 = 0 \\
&\quad N^2_2 = 82.5 - 163.535 = -81.035 && \quad N^2_3 = -81.035 \\
&\textstyle\sum F_Z = 0 : N^1_2 + V^2_2 + \overline{P}_{Z2} = 0 && \textstyle\sum F_Z = 0 : N^3_3 - V^2_3 + \overline{P}_{Z3} = 0 \\
&\quad N^1_2 = -1050 - 100.897 = -1151.897 && \quad N^3_3 = 40.887 - 2100 = -2059.113
\end{aligned} \tag{23}$$

Iteration with Improved Values of N

In the next step of the calculation procedure of Table 11.5 the results obtained by the initial analysis with second order effects have to be compared with the results of the linear (first order) analysis of the frame. The forces N^i are given in the summary of results of Table 11.7. For element 1 we obtain from the linear analysis the value of $N = -1116.4$ kN resulting

TABLE 11.7

Comparison of the Results

	First Order	1. Iteration		2. Iteration	
	(Linear)	Analytical	Approximate	Analytical	Approximate
U_2	−0.07831	−0.15477	−0.15482	−0.15417	−0.15423
Θ_2	0.00553	0.01213	0.01214	0.01210	0.01211
M_1^1	584.2	1060.4	1060.2	1055.5	1055.3
M_2^1	−363.8	−708.4	−709.0	−704.8	−705.4
V_1^1	−154.5	−235.5	−235.5	−233.9	−233.9
V_2^2	−82.5	−163.5	−163.5	−161.9	−161.9
V_2^2	66.4	100.8	100.9	100.5	100.5
V_3^2	6.4	40.8	40.9	40.5	40.5
V_3^3	0.0	81.0	81.0	79.4	79.4
N^1	−1116.4	1150.8	−1150.9	−1150.5	−1150.5
N^2	0.0	−81.0	−81.0	−79.4	−79.4
N^3	−2093.6	−2059.2	−2059.1	−2059.5	−2059.5

in a parameter $\varepsilon = 0.90466$. This compares with the new result $N = -1150.9$ kN and $\varepsilon = 0.90466$. The differences appear to be small, depending on the desired accuracy and the stiffness of the system. For small differences, a second iteration is not necessary. This may not be the case for systems in which larger deflections are caused by low stiffness, for example, for a guyed mast. Table 11.7 also shows the approximate solution resulting from a second iteration with updated normal forces. These results confirm that, for this example, only small differences are obtained.

In design, usually the moments are of primary importance, and it can be seen from the results that for M, second order effects may have a rather large influence. It follows that, for systems with highly compressed members, a stability analysis is essential from the standpoint of safety.

In Table 11.7, the columns labeled "Approximate" correspond to the use of geometric matrices as explained in this example. The solutions based on the exact matrices, such as those of Table 4.4, are provided in the columns labeled "Analytical." The differences are quite small, which is typical for the case when the parameter ε is not larger than 1. This criterion can always be satisfied by reducing the length of the elements in the sense of the finite element method. ∎

11.5 Bifurcation of Equilibrium

11.5.1 Conditions for Bifurcation

The critical loads associated with a bifurcation of the equilibrium can be obtained from a special case of the general nonlinear approach. The homogeneous part of the corresponding system of algebraic equations $\mathbf{KV} = \mathbf{0}$ representing equilibrium of the system can be solved only for specific solutions, the eigenvalues of the system. The eigenvalues can be determined from the roots of

$$\det \mathbf{K} = \mathbf{0} \tag{11.66}$$

They represent the critical values of the solution path. For a structure, they give the conditions of buckling of either the whole system or of its members. The critical buckling load follows from the lowest eigenvalue. The corresponding eigenvector or modeshape can be visualized as the buckling shape or buckling pattern. In addition to the lowest eigenvalue and corresponding eigenvector, the higher ones may be of interest. This is the case, for instance, if symmetric and antisymmetric buckling exists.

Another method of determining the buckling load is to assemble separately the linear and the geometric stiffness matrices and to associate a factor λ (a multiplier for the fundamental state) with the geometric matrix to form the homogeneous system

$$\mathbf{KV} = (\mathbf{K}_{\text{lin}} + \lambda \mathbf{K}_{\text{geo}})\mathbf{V} = \mathbf{0} \tag{11.67}$$

This possibility is applicable only if the stiffness can be separated into linear and geometric parts, which is not the case for analytical solutions, because they usually lead to transcendental functions or higher power series.

11.5.2 Determination of the Buckling Loads Using Analytical Solutions

For the analytical solutions of Section 11.2.4, the condition det $\mathbf{K} = \mathbf{0}$ leads to equations with transcendental functions with, in general, an infinite number of roots. The determination of the corresponding eigenvalues and eigenvectors is rather involved [Zurmühl, 1963], but gives the full spectrum of information. This procedure is applied mostly for beams and small structures. The following two examples will demonstrate the procedure to set up the problem and find the roots. In addition to the resulting two eigenvalues, which represent the first and second bifurcation (or buckling) loads, the corresponding eigenvectors are given. These buckling shapes portray the buckling behavior of the structures.

EXAMPLE 11.6 Buckling of a Hinged-Hinged Beam

Use the condition det $\mathbf{K} = 0$ to find the first two bifurcation loads for a hinged-hinged column. See Fig. 11.27a. The lowest bifurcation load is referred to as the *Euler load* of this beam column. More specifically, these are four cases of Euler loads corresponding to four different boundary conditions. The hinged-hinged beam is usually labeled as Euler case 4.

The kinematic boundary conditions $w_a = 0$ and $w_b = 0$ applied to the stiffness equations of Eq. (11.50), in which the longitudinal motion terms are ignored, gives

$$\begin{bmatrix} V_a \\ M_a \\ \cdots \\ V_b \\ M_b \end{bmatrix} = \frac{EI}{L^3} \left[\begin{array}{c:c:c:c} [2(A'+B')-D'] & -(A'+B')L & -[2(A'+B')-D'] & -(A'+B')L \\ -(A'+B')L & A'L^2 & (A'+B')L & B'L^2 \\ \cdots & \cdots\ \cdots & \cdots & \cdots\ \cdots \\ -[2(A'+B')-D'] & (A'+B')L & [2(A'+B')-D'] & (A'+B')L \\ -(A'+B')L & B'L^2 & (A'+B')L & A'L^2 \end{array}\right] \begin{bmatrix} w_a = 0 \\ \theta_a \\ \\ w_b = 0 \\ \theta_b \end{bmatrix} \tag{1}$$

The first and third rows and columns vanish. Equation (11.66) corresponds to the determinant of the four remaining terms being zero. Then

$$\det \mathbf{K} = 0 = EIL(A'^2 - B'^2) = 0 \tag{2}$$

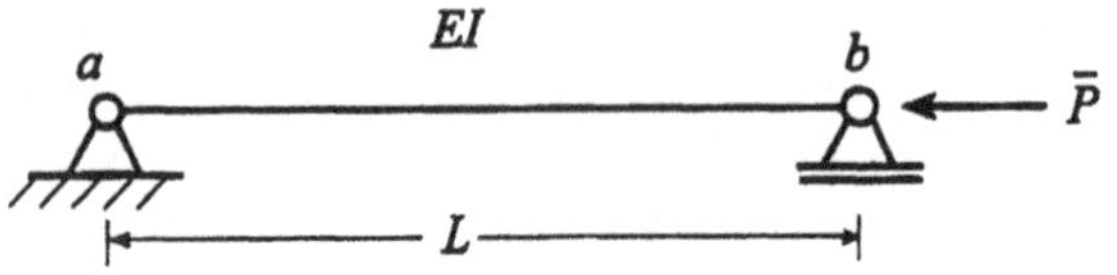

(a)

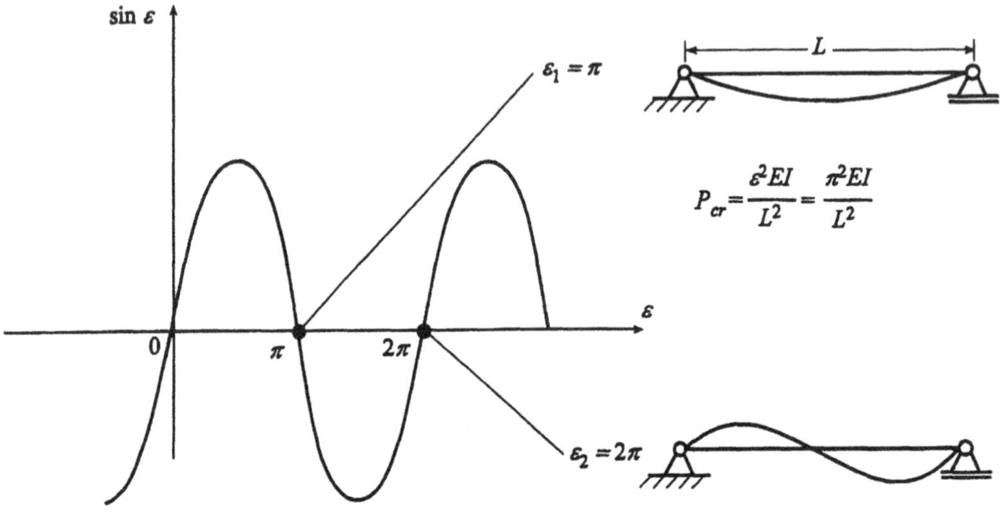

(b) The first two eigenvalues and modeshapes

FIGURE 11.27
Column with hinged-hinged boundaries, Euler Case 4.

Equation (2), with the definitions of A' and B' of Eq. (11.51), becomes

$$\varepsilon \sin \varepsilon = 0 \quad \text{or} \quad \sin \varepsilon = 0 \tag{3}$$

The roots of this transcendental equation are displayed in Fig. 11.27b. With $\varepsilon_1 = \pi$ and $\varepsilon_2 = 2\pi$ as the first two eigenvalues. The lowest bifurcation load (first eigenvalue), which corresponds to instability (buckling) of the beam, is from $\varepsilon_1 = \pi$,

$$P_{\text{cr}} = \frac{\varepsilon^2 EI}{L^2} = \frac{\pi^2 EI}{L^2} \tag{4}$$

The corresponding eigenshapes (mode shapes or buckling shapes) are depicted in Fig. 11.27b. ∎

EXAMPLE 11.7 Buckling of Fixed-Hinged Beam

Find the first two bifurcation loads (the lowest eigenvalues) of the fixed-hinged column of Fig. 11.28a. This corresponds to Euler case 2.

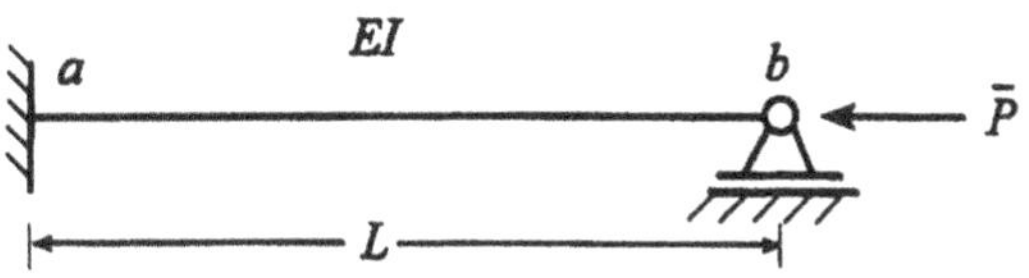

(a)

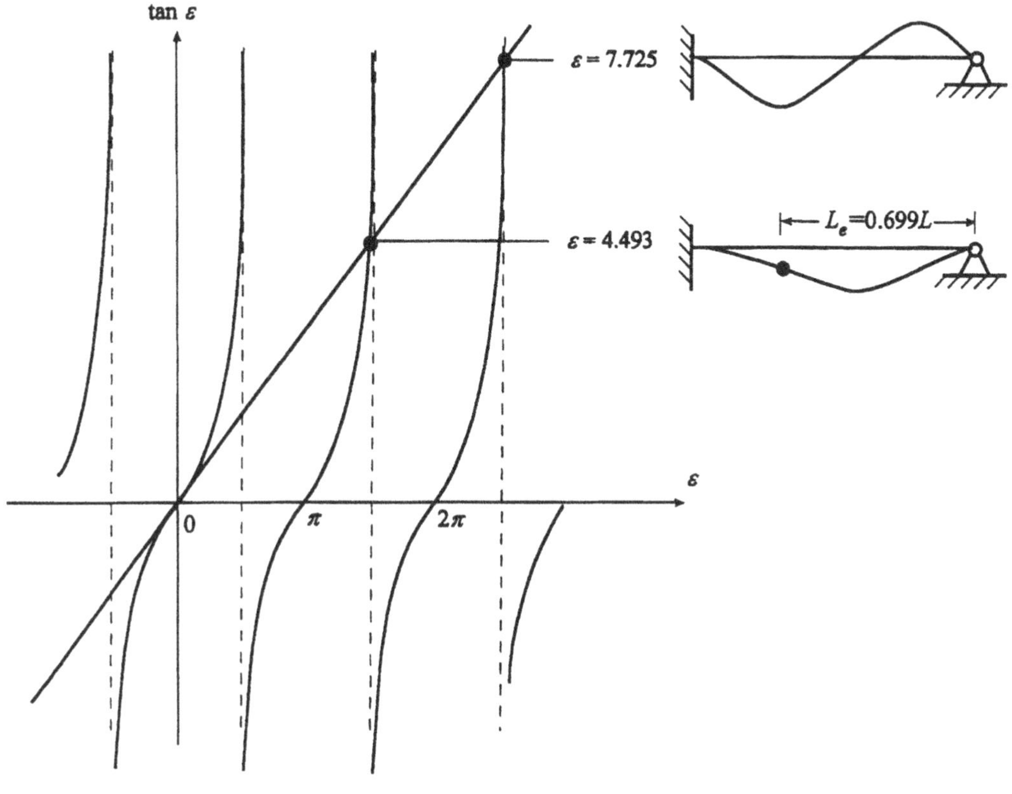

(b) The first two eigenvalues and modeshapes

FIGURE 11.28
Buckling of a fixed-hinged beam, Euler Case 2.

The displacement boundary conditions $w_a = 0$, $\theta_a = 0$, and $w_b = 0$ applied to the bending equations of the stiffness relations of Eq. (11.50) provide

$$\begin{bmatrix} V_a \\ M_a \\ \cdots \\ V_b \\ M_b \end{bmatrix} = \frac{EI}{L^3} \begin{bmatrix} [2(A'+B')-D'] & \vdots & -(A'+B')L & \vdots & -[2(A'+B')-D'] & \vdots & -(A'+B')L \\ -(A'+B')L & \vdots & A'L^2 & \vdots & (A'+B')L & \vdots & B'L^2 \\ \cdots & \cdots & \cdots & \cdots & \cdots & \cdots & \cdots \\ -[2(A'+B')-D'] & \vdots & (A'+B')L & \vdots & [2(A'+B')-D'] & \vdots & (A'+B')L \\ -(A'+B')L & \vdots & B'L^2 & \vdots & (A'+B')L & \vdots & A'L^2 \end{bmatrix} \begin{bmatrix} w_a = 0 \\ \theta_a = 0 \\ \\ w_b = 0 \\ \theta_b \end{bmatrix} \tag{1}$$

The first three columns and rows vanish and the condition det $\mathbf{K} = 0$ becomes $EIA'/L = 0$. Since EI/L is not zero, the quantity A' can be zero for particular values (eigenvalues) of ε.

$$A' = \frac{\varepsilon(\sin\varepsilon - \varepsilon\cos\varepsilon)}{2(1-\cos\varepsilon) - \varepsilon\sin\varepsilon} = 0 \tag{2}$$

Thus,

$$\varepsilon\sin\varepsilon - \varepsilon^2\cos\varepsilon = 0 \quad \text{or} \quad \tan\varepsilon = \varepsilon \tag{3}$$

Equation (3) is plotted in Fig. 11.28b and the first two eigenvalues identified. Normally, the structure will buckle for the first eigenvalue $\varepsilon_1 = 4.493$. This gives

$$P_{cr} = \frac{\varepsilon^2 EI}{L^2} = \frac{\pi^2 EI}{(0.699L)^2} = 2.05\frac{\pi^2 EI}{L^2} \tag{4}$$

where π is introduced so that P_{cr} appears in the same form as the Euler load of the previous example. The quantity $0.699L$, which is referred to as the *equivalent length* L_e, is shown in Fig. 11.28b.

The equivalent length is often used in the analysis of frameworks when an approximate calculation involving the buckling load of an equivalent column replaces a more rigorous, complex analysis. ∎

EXAMPLE 11.8 Column with Moment Release

Find the critical axial force of the column with a moment release shown in Fig. 11.29.

A simple approach to solving this problem involves establishing as one of the boundary conditions the relationship between the deflection and shear force at the moment release. Consider the free-body diagram of the link between the hinges of Fig. 11.30. Summation of moments gives $\overline{P}w(L) = VL_1$ or

$$V = \frac{\overline{P}w(L)}{L_1} \tag{1}$$

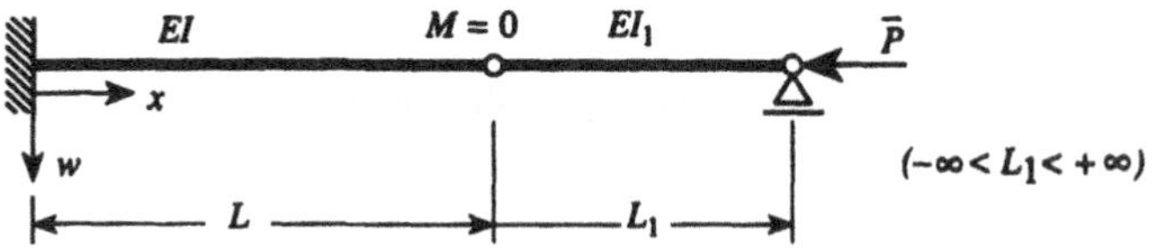

FIGURE 11.29
Column with moment release.

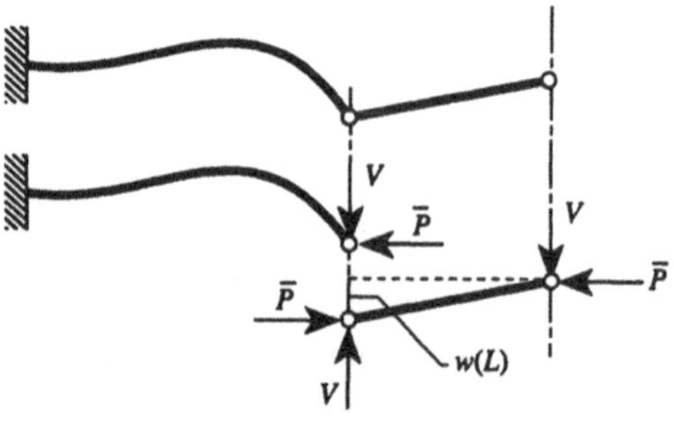

FIGURE 11.30
Free-body diagram.

From Eq. (11.23), with $N = \overline{P}$

$$V = -EI\frac{d^3w}{dx^3} - \overline{P}\frac{dw}{dx} \tag{2}$$

where N is compressive. From Eqs. (1) and (2) at $x = L$,

$$\left[EI\frac{d^3w}{dx^3} + \overline{P}\frac{dw}{dx} + \overline{P}\frac{w}{L_1}\right]_{x=L} = 0 \tag{3}$$

This and the relations $w(0) = w'(0) = 0$, $M(L) = 0$ are the boundary conditions for the column.

From Eq. (11.32), for $0 \le x \le L$,

$$\begin{aligned}
w &= C_1 + C_2\frac{x}{L} + C_3\cos\varepsilon\frac{x}{L} + C_4\sin\varepsilon\frac{x}{L} \\
w' &= C_2 - C_3\frac{\varepsilon}{L}\sin\varepsilon\frac{x}{L} + C_4\frac{\varepsilon}{L}\cos\varepsilon\frac{x}{L} \\
w'' &= -C_3\left(\frac{\varepsilon}{L}\right)^2\cos\varepsilon\frac{x}{L} - C_4\left(\frac{\varepsilon}{L}\right)^2\sin\varepsilon\frac{x}{L} \qquad \varepsilon^2 = L^2\frac{\overline{P}}{EI} \\
w''' &= C_3\left(\frac{\varepsilon}{L}\right)^2\sin\varepsilon\frac{x}{L} - C_4\left(\frac{\varepsilon}{L}\right)^3\cos\varepsilon\frac{x}{L}
\end{aligned} \tag{4}$$

The boundary conditions $w(0) = 0$, $w'(0) = 0$, and $w''(L) = 0$ and (3) imposed on (4) become, with the abbreviations $s = \sin\varepsilon$, $c = \cos\varepsilon$,

$$\begin{bmatrix} 1 & 0 & 1 & 0 \\ 0 & 1 & 0 & \varepsilon \\ 0 & 0 & -c & -s \\ 1 & 1 + L_1/L & c & s \end{bmatrix}\begin{bmatrix} C_1 \\ C_2 \\ C_3 \\ C_4 \end{bmatrix} = 0 \tag{5}$$

The critical axial force is obtained by setting the determinant of these relations equal to zero. Thus,

$$\det\begin{bmatrix} 1 & 0 & 1 & 0 \\ 0 & 1 & 0 & \varepsilon \\ 0 & 0 & -c & -s \\ 1 & 1 + L_1/L & c & s \end{bmatrix} = 0 = -\frac{\varepsilon}{L}(1 + L_1/L)c + s \tag{6}$$

or

$$\tan\varepsilon = \varepsilon(1 + L_1/L) \tag{7}$$

This characteristic equation is plotted in Fig. 11.31, where the straight lines projecting radially from 0 represent the right-hand side of (7). The curved lines in Fig. 11.31 are plots of $\tan\varepsilon$. The cases for which $L_1 < 0$ correspond to segments of length L_1 lying to the left of the moment release. For particular ranges of values L_1, the buckling loads can be determined.

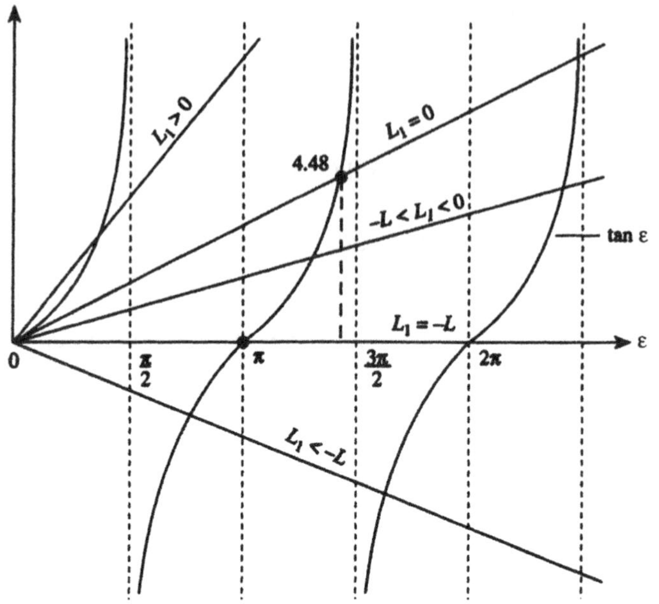

FIGURE 11.31
The characteristic equation.

We find

$$\begin{aligned}
&\text{for } L_1 > 0: \quad \varepsilon < \frac{\pi}{2} \quad \text{or} \quad P_{cr} < \frac{\pi^2 EI}{4L^2} \\
&\text{for } L_1 = 0: \quad \varepsilon\pi = 1.4303\pi \quad \text{or} \quad P_{cr} = \frac{2.05\pi^2 EI}{L^2} \\
&\text{for } -L < L_1 < 0: \quad \pi < \varepsilon < 1.4303\pi \quad \text{or} \quad \frac{\pi^2 EI}{L^2} < P_{cr} < \frac{2.05\pi^2 EI}{L^2} \\
&\text{for } L_1 = -L: \quad \varepsilon = \pi \quad \text{or} \quad P_{cr} = \frac{\pi^2 EI}{L^2} \\
&\text{for } L_1 < -L: \quad \frac{\pi}{2} < \varepsilon < \pi \quad \text{or} \quad \frac{\pi^2 EI}{4L^2} < P_{cr} < \frac{\pi^2 EI}{L^2} \\
&\text{for } L_1 = \pm\infty: \quad \varepsilon = \frac{\pi}{2} \quad \text{or} \quad P_{cr} = \frac{\pi^2 EI}{4L^2}
\end{aligned} \tag{8}$$

The buckling load is plotted in Fig. 11.32a as a function of L_1.

The buckled configurations are displayed in Fig. 11.32b. The buckled configurations of columns with hinged ends, which have the same buckling forces as those of the original structure of Fig. 11.30, are shown. ∎

EXAMPLE 11.9 Column of Variable Cross-Section

Find the buckling load for the stepped column of Fig. 11.33.

Since the transfer matrix for a column element has been established (Eq. 11.47), columns formed of elements of different geometries are readily analyzed for the conditions of instability. The transfer matrix method of analysis developed in Chapter 5 can be applied to such columns as that of Fig. 11.33.

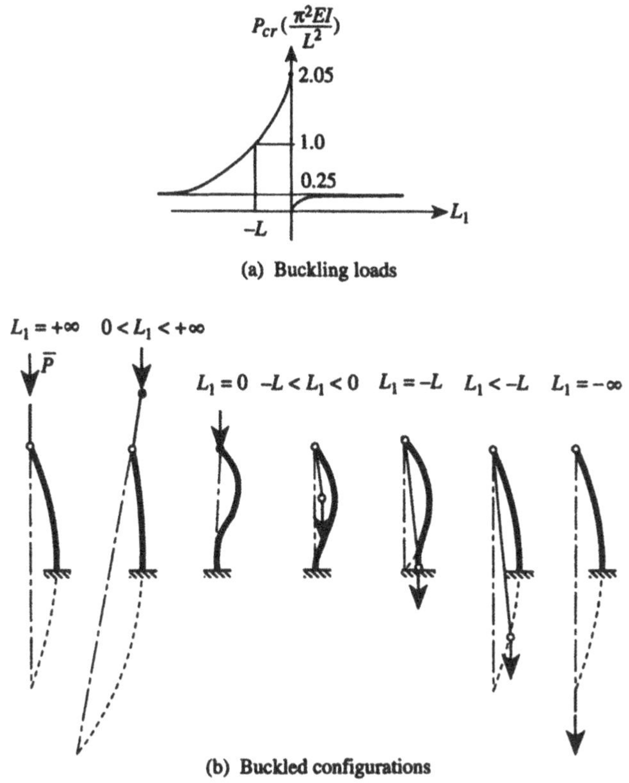

FIGURE 11.32
Buckling loads and buckled configurations. The equivalent columns with hinged ends are shown with dashed lines.

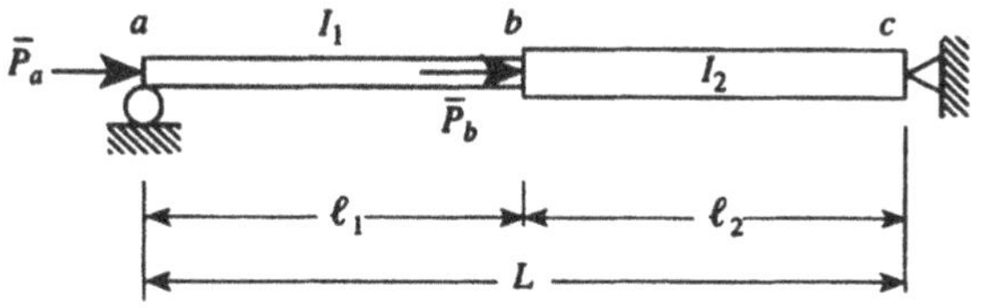

FIGURE 11.33
A stepped column.

The global transfer matrix is defined by

$$\mathbf{z}_L = \mathbf{U}\mathbf{z}_a \tag{1}$$

with

$$\mathbf{U} = \mathbf{U}^2\mathbf{U}^1 \tag{2}$$

where $\mathbf{U}^1$ is given by Eq. (11.47) with ℓ, EI, and N replaced by ℓ_1, EI_1, and $\overline{P}_a$, respectively. Also, $\mathbf{U}^2$ is given by Eq. (11.47) using ℓ_2, EI_2, and $\overline{P}_a + \overline{P}_b$. Note that the axial force N in element 2 is $\overline{P}_a + \overline{P}_b$, not just $\overline{P}_b$.

The column is simply supported on both ends, so that the boundary conditions are

$$w_{x=a} = M_{x=a} = w_{x=L} = M_{x=L} = 0 \tag{3}$$

These conditions applied to (1) appear as

$$\begin{bmatrix} w=0 \\ \theta \\ V \\ M=0 \end{bmatrix}_L = \begin{bmatrix} & \vdots & U_{w\theta} & U_{wV} & \vdots & \\ \cdots & \vdots & \cdots & \cdots & \vdots & \cdots \\ \cdots & \vdots & \cdots & \cdots & \vdots & \cdots \\ & \vdots & U_{M\theta} & U_{MV} & \vdots & \end{bmatrix} \begin{bmatrix} w=0 \\ \theta \\ V \\ M=0 \end{bmatrix}_a \tag{4}$$

$$\mathbf{z}_b = \mathbf{U}^i \qquad \mathbf{z}_a$$

or

$$\begin{aligned} 0 &= \theta_a U_{w\theta} + V_a U_{wV} \\ 0 &= \theta_a U_{M\theta} + V_a U_{MV} \end{aligned} \tag{5}$$

The determinant (set equal to zero) of these homogeneous equations constitutes the characteristic equation. That is,

$$\nabla = U_{w\theta} U_{MV} - U_{M\theta} U_{wV} = 0 \tag{6}$$

leads to the buckling load. The relation of (6) becomes

$$\frac{\ell_1^2}{\varepsilon_1^2} - \frac{\varepsilon_1^2 LEI_1/\overline{P}_b\ell_1 + \ell_1^2}{\varepsilon_1 \tan\varepsilon_1} - \ell_2^2\frac{I_1/I_2}{\varepsilon_2^2} - \frac{\varepsilon_2^2 LEI_2/\overline{P}_b\ell_2 - \ell_2^2 I_1/I_2}{\varepsilon_2 \tan\varepsilon_2} = 0 \tag{7}$$

where

$$\varepsilon_1^2 = \ell_1^2 \frac{\overline{P}_a}{EI_1} \qquad \varepsilon_2^2 = \ell_2^2 \frac{\overline{P}_a + \overline{P}_b}{EI_2}$$

Combinations of $\overline{P}_a$ and $\overline{P}_b$ that satisfy $\nabla = 0$ define the conditions of instability. Normally, $\overline{P}_a$ and $\overline{P}_b$ are not independent. Typically, $\overline{P}_b$ is known to be proportional to $\overline{P}_a$, i.e., $\overline{P}_b = \beta\overline{P}_a$ where β is a known constant. Then (6) is the characteristic equation for a single unknown, the lowest value of which is the buckling load.

The buckling loads for complicated stepped columns must be found numerically. In such cases, a numerical search for the lowest root of $\nabla = 0$ is performed. Typically, this process begins by evaluating ∇ for an estimated buckling load that is believed to be below the actual buckling load. Increase the estimate, and repeat the evaluation of ∇. Continue the process until ∇ changes sign. The desired buckling load, which lies between the two previous estimates, is then found to a prescribed accuracy by utilizing a nonlinear equation solver such as Newton-Raphson.[1] See the references in Chapter 10 for a review of various techniques for conducting numerical determinant searches. ∎

11.5.3 Determination of the Buckling Loads Using Geometric Stiffness Matrices

In those cases where a closed form solution of the condition det $\mathbf{K} = \mathbf{0}$ is not available or is too complex, the linear and geometric stiffness matrices can be used to obtain the first root of equations developed using the displacement method. The assembled stiffness matrices of the system can then be written as Eq. (11.67), $\mathbf{KV} = (\mathbf{K}_{\text{lin}} + \lambda\mathbf{K}_{\text{geo}})\mathbf{V} = \mathbf{0}$ in which λ is

[1] Joseph D. Raphson (1648–1715) was a British disciple of Newton. Indeed, it is said that his devotion to Newton clouded his assessment of the notation used for the calculus in that he favored Newton's notation over Leibniz's. This viewpoint hampered the adoption of the Leibniz calculus notation in England for over a century. In 1690 he published a method for approximating the real roots of a numerical equation. It supplanted a technique proposed by Newton and is now referred to as the Newton-Raphson method.

a factor for the fundamental state corresponding to a certain combination of the loading. Sometimes λ is referred to as the *load factor*. It can, for example, represent the influence of the fundamental state in the form $\lambda = \varepsilon^2 = \ell^2 N^0/EI$.

The critical load λ_{cr} (lowest eigenvalue of the system) can be obtained relatively simply by the so-called inverse vector-iteration method (See, for example, Zumühl, 1963)

$$\mathbf{K}_{\text{lin}} \mathbf{V}_i = -\lambda_{i-1} \, \mathbf{K}_{\text{geo}} \, \mathbf{V}_{i-1} \tag{11.68}$$

which always converges toward the lowest eigenvalue $\lambda_{\min}$. The iteration is started by an arbitrary vector $\mathbf{V}_0$ and λ_0 and should be scaled with respect to a constant length for each step, e.g., $|\mathbf{V}_i| = 1$:

$$\mathbf{V}_i = \mathbf{V}_i/a \quad \text{with } a = \sqrt{\mathbf{V}_i^T \mathbf{V}_i} \tag{11.69}$$

The columns of Examples 11.6 and 11.7 will be considered again to demonstrate the procedure. It can be observed that the buckling loads obtained will be higher than the exact value. This is due to the approximate nature of the geometric matrix. The results may be improved by subdividing the model of the structures into more elements.

EXAMPLE 11.10 Column Buckling

For the hinged-hinged column of Example 11.6 (Fig. 11.27) the buckling load was found as the lowest root of

$$A'^2 - B'^2 = 0 \tag{1}$$

This led to

$$P_{\text{cr}} = \frac{\varepsilon^2 EI}{L^2} \tag{2}$$

with $\varepsilon^2 = \pi^2 = 9.870$. If A' and B' are expanded in terms of ε^2, then (Eq. 11.51)

$$A'^2 - B'^2 = \left(4 - \frac{2}{15}\varepsilon^2\right)^2 - \left(2 + \frac{1}{30}\varepsilon^2\right)^2 = 0 \tag{3}$$

where only terms up to ε^2 are retained. Equation (3) becomes

$$\frac{1}{60}\varepsilon^4 - \frac{6}{5}\varepsilon^2 + 12 = 0 \tag{4}$$

giving $\varepsilon^2 = 12$ and $\varepsilon_1 = 3.464$. The corresponding critical load is, as expected, higher than the exact value of (2). For design, note that this approximation is on the unsafe side.

For the fixed-hinged column of Example 11.7 (Fig. 11.28), the critical load is obtained from

$$A' = 0 \tag{5}$$

leading to (2) with $\varepsilon^2 = 2.05\pi^2$ or $\varepsilon_1 = 4.493$. If the expansions of Eq. (11.51) are introduced into (5), we find

$$A' = 4 - \frac{2}{15}\varepsilon^2 = 0 \tag{6}$$

when terms up to ε^2 are retained. This leads to $\varepsilon^2 = 30$ or the lowest root of $\varepsilon_1 = 5.477$. If terms up to ε^4 are retained

$$A' = 4 - \frac{2}{15}\varepsilon^2 - \frac{11}{6300}\varepsilon^4 = 0 \tag{7}$$

TABLE 11.8

Buckling Load Convergence as the Number of Elements Increases

$$\varepsilon^2 = \frac{|\bar{P}|}{EI}\ell^2$$

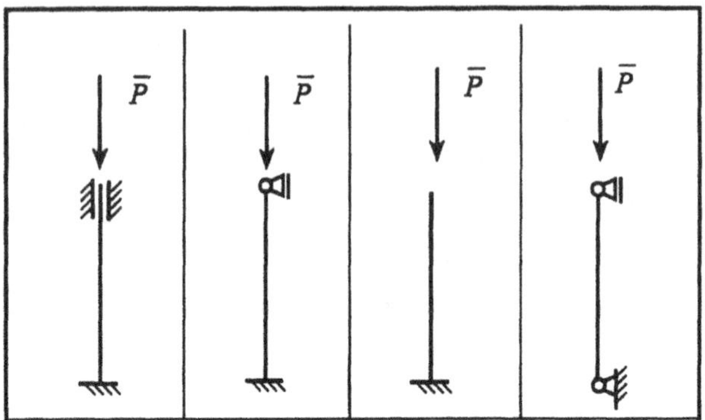

Number of Elements				
	Exact Solution for ε^2			
	39.478	20.191	2.467	9.870
	Approximate Solution for ε^2			
1	—	30	2.486	12
2	40	20.686	2.480	9.945
4	39.775	20.323	2.467	9.874

with a lowest root of $\varepsilon_1 = 4.80$, which differs from the exact value of ε_1 by 7%. Note that, as more terms are retained, the solution improves, always approaching the exact value from above.

Improved solutions are obtained by increasing the number of elements used to model the beam. Table 11.8 illustrates this type of convergence, where for each element terms up to and including ε^2 are retained. The four so-called *Euler cases* of boundary conditions are shown in order in the table. For cases 2 and 4, the exact values were given in Examples 11.7 and 11.6, respectively. Note that in all cases the solutions approach the exact buckling load from above. ∎

EXAMPLE 11.11 Critical Loads and Buckling Shapes of a Framework

Return to the framework of the Example 11.5 to evaluate the critical loads for which bifurcations of the equilibrium state occurs. The fundamental state is taken either from the second order analysis (for instance, after the first iteration) or approximately from the linear analysis. It is also of interest to learn whether the latter possibility is applicable for this type of structure.

The geometry and load combinations are shown in Fig. 11.34, in which the loads are multiplied by the common factor λ, the value of which is to be determined by the conditions of Eq. (11.66), det $\mathbf{K} = 0$.

In Fig. 11.35, the load level for bifurcation of equilibrium is shown in the general load displacement diagram. The relation between the load factor λ and the deflection, chosen

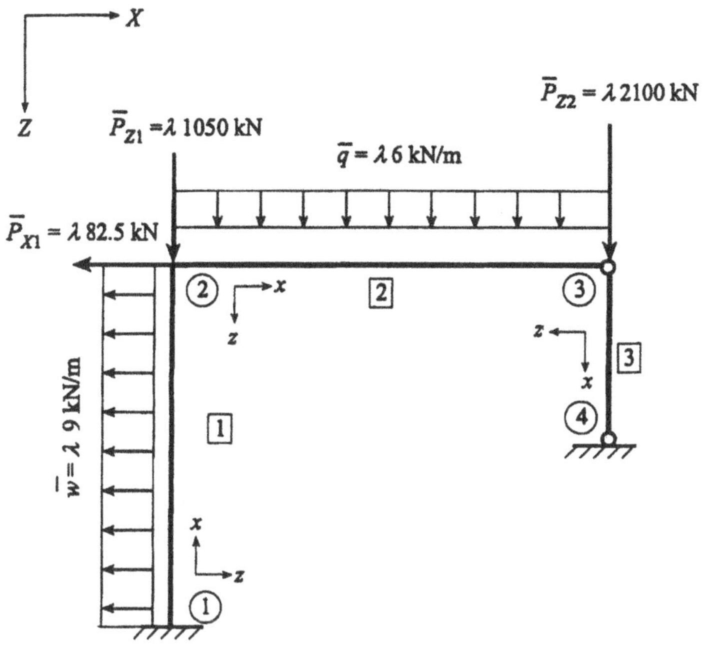

FIGURE 11.34
Geometry and load combinations.

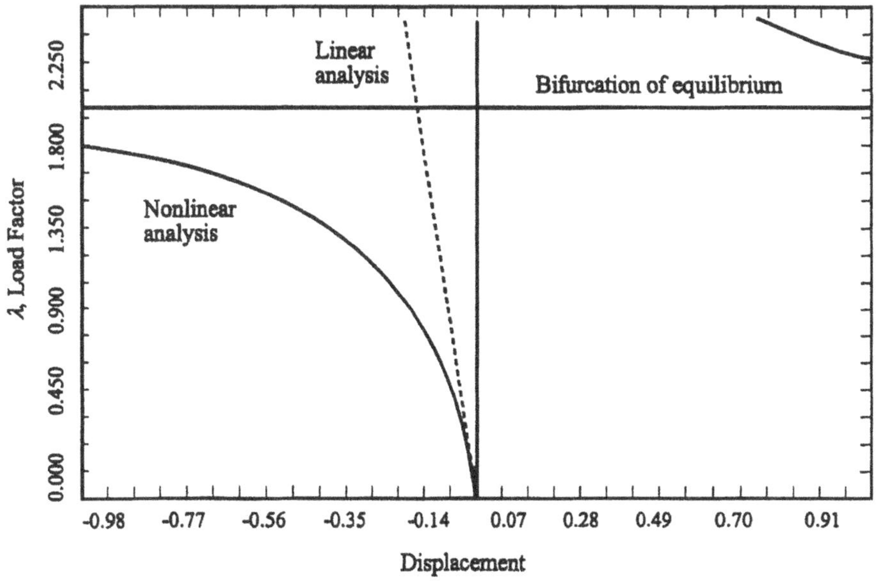

FIGURE 11.35
Load factor λ versus displacement U_2.

to be the critical characteristic in the sense of the criterion of stability of Section 11.1.2, is also given for the linear and the geometrically nonlinear analyses. The bifurcation level in Fig. 11.35 is related to the value of 2.03, which is the first root of the condition det $\mathbf{K} = 0$.

Higher roots can also be evaluated, and it is advisable to use the approach employing analytical solutions or higher power series to obtain reliable results for the higher roots.

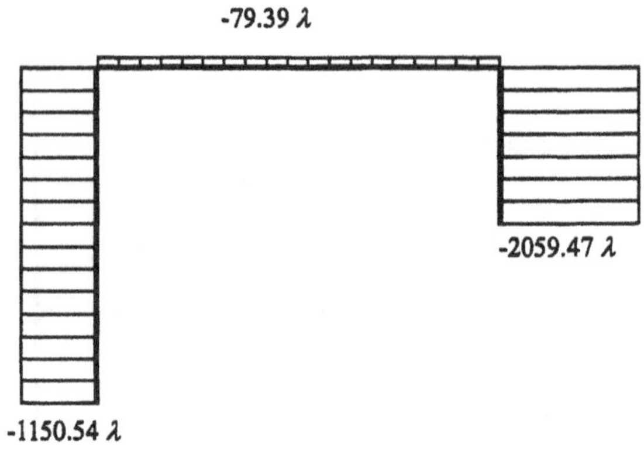

FIGURE 11.36
Longitudinal forces N, with second order effects after the 2nd iteration.

For the evaluation of the system matrix, the approach demonstrated in Example 11.5 can be employed. For the fundamental state, we use either the results of the linear analysis or the results of the second order analysis. Figure 11.36 contains the distribution of longitudinal forces with second order effects as given in the last column in Table 11.7.

The element matrices are formed in the same manner as in Example 11.5: the linear part of the stiffness is the same, but the geometric matrices are slightly different due to the choice of the fundamental state. In this case, the second order analysis solution after the second iteration was chosen.

To form the system stiffnesses, the element matrices have to be assembled separately for the linear and the second order parts. As noted in Example 11.5, Eq. (9), the first row in the assembled matrices gives the equilibrium of the forces in the horizontal direction at node 2, while the second describes the equilibrium of the moments at the same node.

Assembly of the elastic stiffness matrix:

	U_2	Θ_2	
F_{X2}	2109.38	8437.50	Element 1
			Element 2
			Element 3
M_{Y2}	8437.50	45 000.00	Element 1
		52 200.00	Element 2
			Element 3

(1)

$$\mathbf{K}_{\text{lin}} = \begin{bmatrix} 2109.38 & 8437.50 \\ 8437.50 & 97\,200.00 \end{bmatrix}$$

Assembly of the geometric stiffness matrix:

	U_2	Θ_2	
F_{X2}	−172.58	−115.05	Element 1
			Element 2
	−514.87		Element 3
M_{Y2}	−115.05	−1227.24	Element 1
		−158.79	Element 2
			Element 3

(2)

$$\mathbf{K}_{\text{geo}} = \begin{bmatrix} -687.45 & -115.05 \\ -115.05 & -1386.03 \end{bmatrix}$$

The homogenous system of equations

$$(\mathbf{K}_{\text{lin}} + \lambda\mathbf{K}_{\text{geo}})\mathbf{V} = 0 \tag{3}$$

has a nontrivial solution only for det $\mathbf{K} = 0$

$$\det\left[\begin{bmatrix} 2109.38 & 8437.50 \\ 8437.50 & 97\,200.00 \end{bmatrix} + \lambda\begin{bmatrix} -687.45 & -115.05 \\ -115.05 & -1386.03 \end{bmatrix}\right] = 0 \tag{4}$$

or

$$\det\begin{bmatrix} 2109.38 - \lambda 687.45 & 8437.50 \quad - \lambda 115.05 \\ 8437.50 - \lambda 115.05 & 97\,200.00 - \lambda 1386.03 \end{bmatrix} = 0 \tag{5}$$

Evaluation of the determinant of the system matrix renders the algebraic equation for λ with the first two roots corresponding to the number of unknowns of the system.

$$939\,582\lambda^2 - 67\,801\,922\lambda + 133\,839\,844 = 0 \tag{6}$$

or

$$\lambda^2 - 72.162\lambda + 142.446 = 0 \tag{7}$$

See Fig. 11.37. This provides the roots

$$\begin{aligned} \lambda_1 &= 2.031 \\ \lambda_2 &= 70.13 \end{aligned} \tag{8}$$

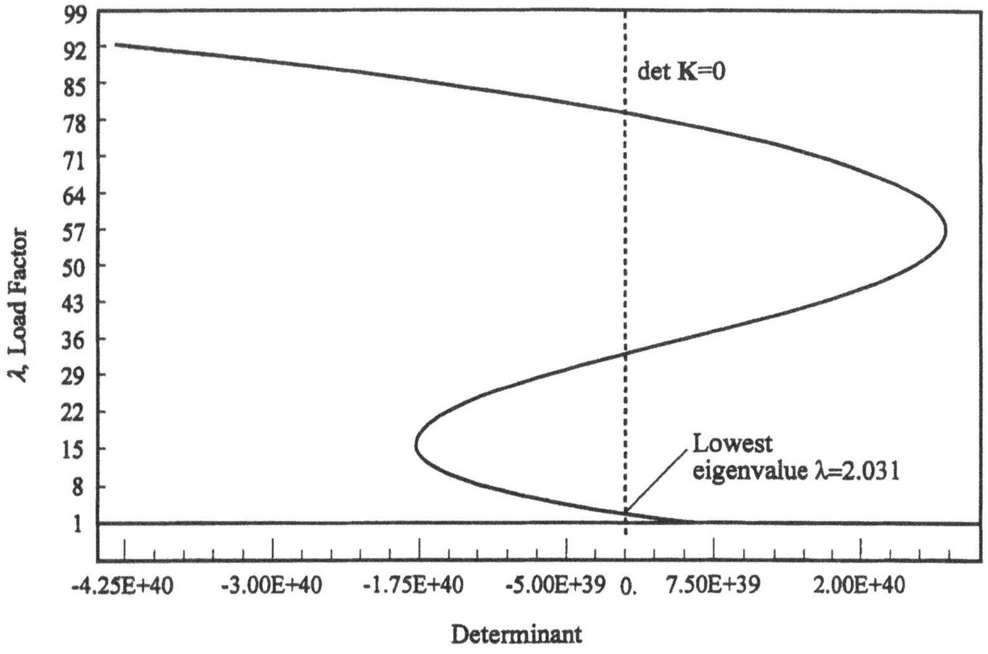

FIGURE 11.37
The value of det **K** using the geometric stiffness matrix.

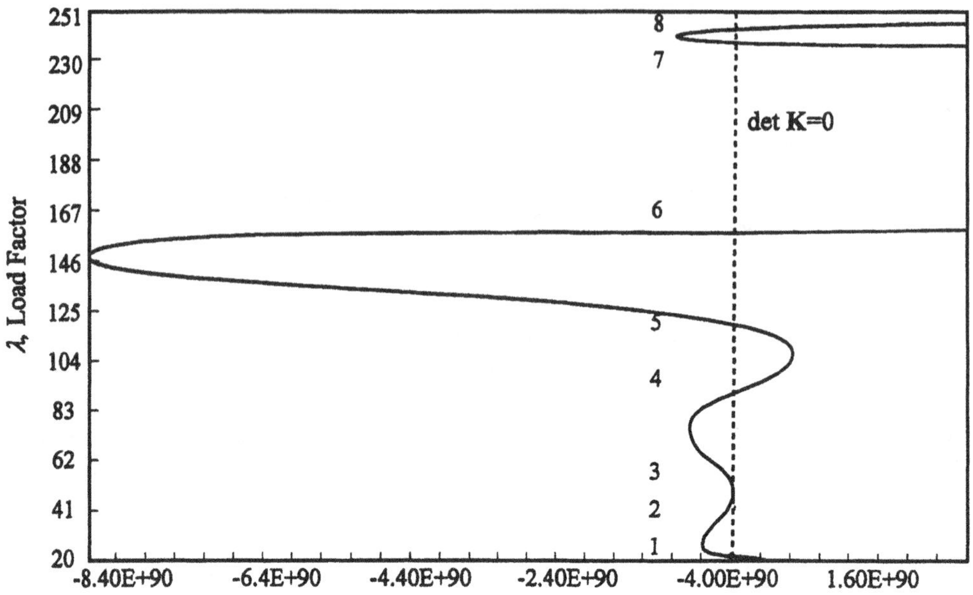

FIGURE 11.38
The value of det **K** using the analytical stiffness matrix.

Thus, the lowest eigenvalue, which is the critical load factor, is 2.031. The lowest eigenvalue identifies the critical load, in the sense that the bifurcation of equilibrium is reached when the load combination is increased by a factor of 2.031.

If the same procedure is performed with the linear solution as the fundamental state, a load factor of 2.024 is obtained. From this, we can conclude that for frameworks it is sufficient to investigate the lowest level of bifurcations of equilibrium using a linear fundamental state. This conclusion cannot necessarily be drawn for higher eigenvalues, which for a global investigation of the system may also be of practical interest. For the determination of higher eigenvalues, we choose here the second order solution computed using analytical functions or higher power series. In this case, the root search of the determinant has to be performed with more sophisticated tools [Zurmühl, 1963].

For this framework, the first eight eigenvalues were obtained as the roots of the analytical system matrix the values of which are given in Fig. 11.39. Comparison of the second root obtained by using the geometric matrix only (Fig. 11.37) and the one resulting from the solution with analytical functions as provided in Fig. 11.38, shows that the approximate approach is only appropriate for the first eigenvalue, i.e., for the lowest critical load of the system. Both of these figures diagram the determinant det **K** versus λ.

Whereas the eigenvalues are associated with the critical loads, the corresponding eigenvector describes the buckling mode associated with it. The amplitudes of a buckling shape are determined relative to an appropriately chosen value. These shapes (Fig. 11.39) can provide insight into the buckling behavior of the framework. We observe that the first buckling shape is connected with a horizontal displacement of the whole frame, while the second eigenvalue is associated with the local buckling of the bar-column element 3, and the next eigenvalue with element 3 interacting with element 1, etc. It is noteworthy that the investigation of bar 3 as a single member (local buckling) yields the same value.

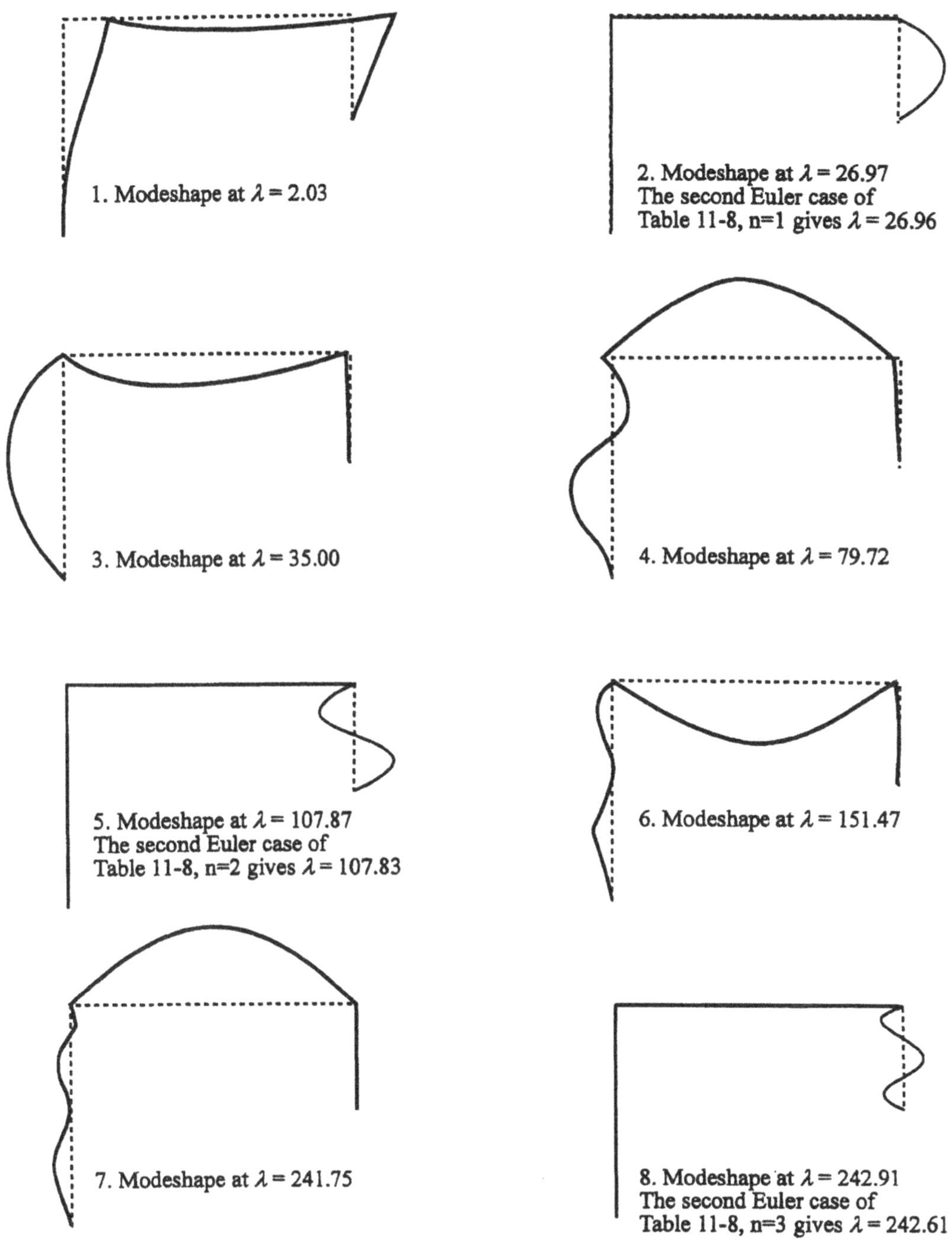

FIGURE 11.39
Eigenforms (buckling modeshapes) corresponding to the eigenvalues.

From this example, it is evident that the knowledge or estimation of the buckling shapes is very useful and gives additional insight:

- Provided that the effective length of its members can be estimated, the analysis of the system can be replaced by approximate calculations of the individual members.

- As demonstrated in Example 11.4, imperfections may reduce the critical loads substantially. Therefore, a safe design involves the imposition of possible imperfections that lead to the lowest critical loads. However, the shapes and the amplitudes corresponding to the imperfections are usually not known in advance and have to be assumed. The buckling mode corresponding to the lowest eigenvalue of the system is considered to be the worst shape, where the amplitudes can be taken from appropriate design codes.

∎

EXAMPLE 11.12 Simply Supported Beam with Shear Deformation

Find the buckling load of the beam of Fig. 11.40a. Assume displacements of the form

$$\begin{aligned} u &= C_1(x + L/2) && \rightarrow u' = C_1 \\ w &= C_2(L^2/4 - x^2) && \rightarrow w' = -2C_2 x \\ \theta &= \quad C_3 x && \rightarrow \theta' = C_3 \end{aligned} \tag{1}$$

which are sketched in Fig. 11.40b. These displacements satisfy all displacement boundary conditions ($u(-L/2) = 0,\ w(-L/2) = w(L/2) = 0$). Substitution of these trial functions into the principle of virtual work expression of Eq. (11.59) gives

$$\int_{-L/2}^{L/2} \begin{bmatrix} EA & 0 & 0 \\ 0 & (-2x)(-\overline{P})(-2x) & (-2x)k_s GAx \\ & +(-2x)k_s GA(-2x) & \\ 0 & x\,k_s GA(-2x) & EI + xk_s GAx \end{bmatrix} dx \begin{bmatrix} C_1 \\ C_2 \\ C_3 \end{bmatrix} = \begin{bmatrix} 0 \\ 0 \\ 0 \end{bmatrix} \tag{2}$$

where N^0 has been replaced by the compressive axial force $-\overline{P}$. Integration of this

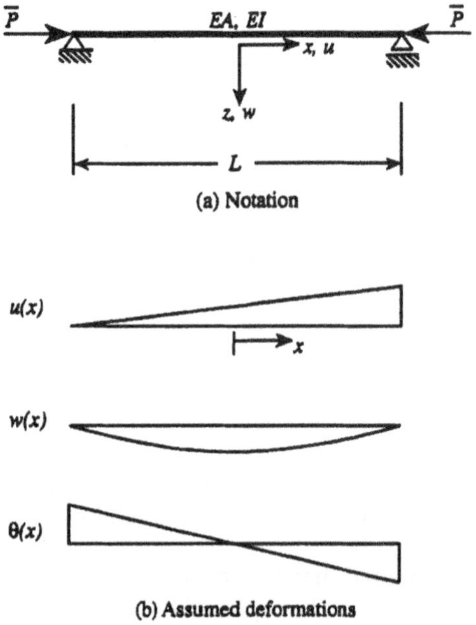

FIGURE 11.40
A simply supported beam with shear deformation.

expression leads to

$$\begin{bmatrix} EAL & 0 & 0 \\ 0 & (L^3/3)(k_sGA - \overline{P}) & -(L^3/6)k_sGA \\ 0 & -(L^3/6)k_sGA & EIL + (L^3/12)k_sGA \end{bmatrix} \begin{bmatrix} C_1 \\ C_2 \\ C_3 \end{bmatrix} = \begin{bmatrix} 0 \\ 0 \\ 0 \end{bmatrix} \tag{3}$$

From det $\mathbf{K} = 0$, the critical load is found to be

$$P_{cr} = \frac{12EI/L^2}{1 + \frac{1}{k_sGA}12EI/L^2} \tag{4}$$

This value is lower than when shear deformation is not considered. In fact,

$$P_{cr}|_{\text{Shear}} = \frac{P_{cr}|_{\text{No Shear}}}{1 + P_{cr}|_{\text{No Shear}}/k_sGA} \tag{5}$$

■

EXAMPLE 11.13 Variable Axial Force and Variable Cross-Section

Consider a beam element with moment of inertia and axial force that vary linearly along the beam (see Fig. 11.41).

The moment of inertia is assumed to vary as

$$EI = EI_0 - EI_0\frac{x}{L}\widehat{\beta} = EI_0(1 - \xi\widehat{\beta}) \tag{1}$$

with $\xi = x/L$, and similarly, the axial force is taken to be

$$\overline{P}(\xi) = \overline{P}_0(1 - \xi\beta) \tag{2}$$

where $\widehat{\beta}$ and β are prescribed. The virtual work relationship can be written as

$$\begin{aligned} -\delta W = \delta\mathbf{v}^T\Big\{&\mathbf{G}^T\int[\mathbf{N}_u''^T\mathbf{N}_u''EI_0\,dx]\mathbf{G} - \widehat{\beta}\mathbf{G}^T\left[\int\mathbf{N}_u''^T\mathbf{N}_u''\frac{x}{L}EI_0\,dx\right]\mathbf{G} \\ &-\mathbf{G}^T\left[\int\mathbf{N}_u'^T\mathbf{N}_u'(\overline{P}_0)\,dx\right]\mathbf{G} + \beta\mathbf{G}^T\left[\int\mathbf{N}_u'^T\mathbf{N}_u'\frac{x}{L}(\overline{P}_0)dx\right]\mathbf{G}\Big\}\mathbf{v} \\ = 0& \end{aligned} \tag{3}$$

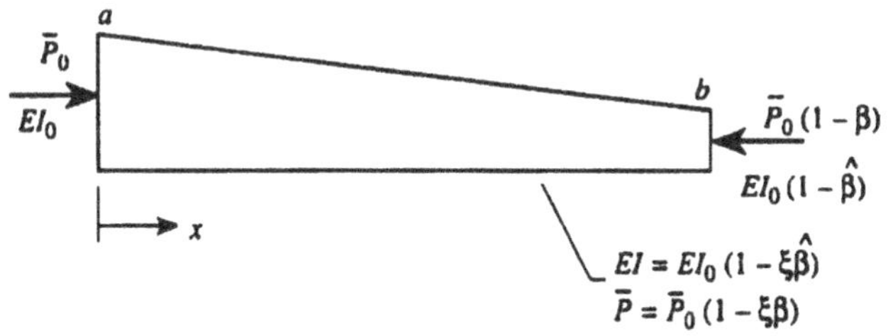

FIGURE 11.41
Beam element of variable cross-section with a variable axial force.

Use the Hermitian polynomials as assumed displacements and find the stiffness matrix for this single element system to be

$$
\mathbf{K} = \mathbf{k}^1_{\text{lin}} + \mathbf{k}^1_{\text{geo}}
$$

$$
= \begin{bmatrix}
\begin{array}{c} 12-6\widehat{\beta} \\ -1.2\lambda+0.6\beta\lambda \end{array} & \vdots & \begin{array}{c} -6+2\widehat{\beta} \\ +0.1\lambda-0.1\beta\lambda \end{array} & \vdots & \begin{array}{c} -12+6\widehat{\beta} \\ +1.2\lambda-0.6\beta\lambda \end{array} & \vdots & \begin{array}{c} -6+4\widehat{\beta} \\ +0.1\lambda \end{array} \\
\cdots & & \cdots & & \cdots & & \cdots \\
\begin{array}{c} -6+2\widehat{\beta} \\ +0.1\lambda-0.1\beta\lambda \end{array} & \vdots & \begin{array}{c} 4-\widehat{\beta} \\ -2\lambda/15+\beta\lambda/30 \end{array} & \vdots & \begin{array}{c} 6-2\widehat{\beta} \\ -0.1\lambda+0.1\beta\lambda \end{array} & \vdots & \begin{array}{c} 2-\widehat{\beta} \\ +\lambda/30-\beta\lambda/60 \end{array} \\
\cdots & & \cdots & & \cdots & & \cdots \\
\begin{array}{c} -12+6\widehat{\beta} \\ +1.2\lambda-0.6\beta\lambda \end{array} & \vdots & \begin{array}{c} 6-2\widehat{\beta} \\ -0.1\lambda+0.1\beta\lambda \end{array} & \vdots & \begin{array}{c} 12-6\widehat{\beta} \\ -1.2\lambda+0.6\beta\lambda \end{array} & \vdots & \begin{array}{c} 6-4\widehat{\beta} \\ -0.1\lambda \end{array} \\
\cdots & & \cdots & & \cdots & & \cdots \\
\begin{array}{c} -6+4\widehat{\beta} \\ +0.1\lambda \end{array} & \vdots & \begin{array}{c} 2-\widehat{\beta} \\ +\lambda/30-\beta\lambda/60 \end{array} & \vdots & \begin{array}{c} 6-4\widehat{\beta} \\ -0.1\lambda \end{array} & \vdots & \begin{array}{c} 4-3\widehat{\beta} \\ -2\lambda/15+0.1\beta\lambda \end{array}
\end{bmatrix} \frac{EI_0}{L^3} \tag{4}
$$

where $\lambda = \varepsilon^2 = \overline{P}_0 L^2/EI_0$. To find the buckling load, apply the boundary conditions to form a reduced **K** and solve det **K** = 0. This, if desired, can be represented for this single element model in two terms in the form

$$
\det\left(\mathbf{k}'_{\text{lin}} + \mathbf{k}'_{\text{geo}}\right) = 0 \tag{5}
$$

where $\mathbf{k}'_{\text{lin}}$ and $\mathbf{k}'_{\text{geo}}$ are matrices reduced by application of the boundary conditions. ∎

EXAMPLE 11.14 A Stepped Column

The stepped column of Fig. 11.42a will be used to illustrate several of the techniques for computing buckling loads for a structural system. The boundary and in-span conditions are $w_a = w_c = w_d = 0$, $M_a = M_d = 0$.

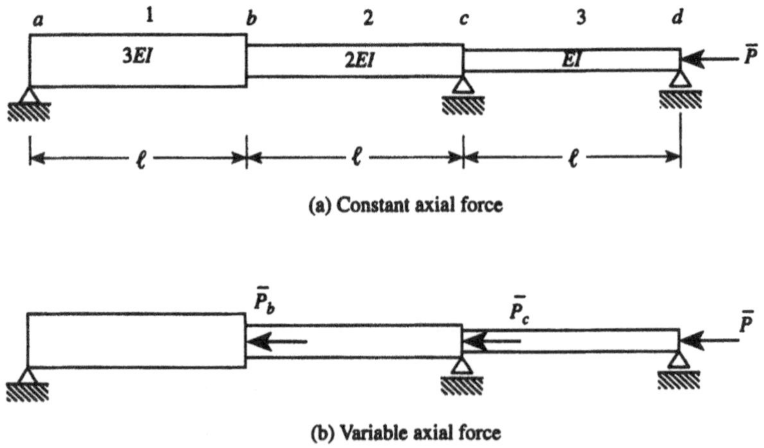

FIGURE 11.42
A stepped column.

Transfer Matrix Method

Use the transfer matrix of Eq. (11.47), along with the methods of Chapter 5 for incorporating the in-span support, to develop a global transfer matrix. The boundary conditions applied to the global transfer matrix equations lead to the characteristic equation and the critical axial load of

$$P_{cr} = 7.064 \frac{EI}{\ell^2} \tag{1}$$

This is the exact critical load as no approximation, other than approximations involved in engineering beam theory, is made.

Displacement Method Related Techniques

The goal here is to set up the eigenvalue problem of Eq. (11.67). First, establish the linear stiffness matrix $\mathbf{K}_{\text{lin}}$ by assembling the element stiffness matrices $\mathbf{k}^i_{\text{lin}} i = 1, 2, 3$ of Eq. (11.63) to obtain

$$\begin{bmatrix} V_a \\ M_a/\ell \\ V_b \\ M_b/\ell \\ V_c \\ M_c/\ell \\ V_d \\ M_d/\ell \end{bmatrix} = \frac{EI}{\ell^3} \begin{bmatrix} 36 & -18 & -36 & -18 & 0 & 0 & 0 & 0 \\ -18 & 12 & 18 & 6 & 0 & 0 & 0 & 0 \\ -36 & 18 & 60 & 6 & -24 & -12 & 0 & 0 \\ -18 & 6 & 6 & 20 & 12 & 4 & 0 & 0 \\ 0 & 0 & -24 & 12 & 36 & 6 & -12 & -6 \\ 0 & 0 & -12 & 4 & 6 & 12 & 6 & 2 \\ 0 & 0 & 0 & 0 & -12 & 6 & 12 & 6 \\ 0 & 0 & 0 & 0 & -6 & 2 & 6 & 4 \end{bmatrix} \begin{bmatrix} w_a \\ \ell\theta_a \\ w_b \\ \ell\theta_b \\ w_c \\ \ell\theta_c \\ w_d \\ \ell\theta_d \end{bmatrix} \tag{2}$$

Next, the global geometric stiffness matrix $\mathbf{K}_{\text{geo}}$ should be assembled using the element geometric stiffness matrices $\mathbf{k}^i_{\text{geo}}$.

Use the consistent geometric stiffness matrix of Eq. (11.64). Assemble the global matrix for the beam. Apply the displacement boundary conditions. Then the eigenvalue problem appears as

$$\left\{ \underset{\mathbf{K}_{\text{lin}}}{\begin{bmatrix} 12 & 18 & 6 & 0 & 0 \\ 18 & 60 & 6 & -12 & 0 \\ 6 & 6 & 20 & 4 & 0 \\ 0 & -12 & 4 & 12 & 2 \\ 0 & 0 & 0 & 2 & 4 \end{bmatrix}} \underset{+}{-\lambda} \underset{\mathbf{K}_{\text{geo}}}{\begin{bmatrix} 2/15 & 1/10 & -1/30 & 0 & 0 \\ 1/10 & 12/5 & 0 & -1/10 & 0 \\ -1/30 & 0 & 4/15 & -1/30 & 0 \\ 0 & -1/10 & -1/30 & 4/15 & -1/30 \\ 0 & 0 & 0 & -1/30 & 2/15 \end{bmatrix}} \right\} \underset{\mathbf{V} = \mathbf{0}}{\mathbf{V} = \mathbf{0}} \tag{3}$$

where $\lambda = \varepsilon^2 = \overline{P}\ell^2/EI$. This provides a critical load of magnitude

$$P_{cr} = 7.298 \frac{EI}{\ell^2} \tag{4}$$

which is above the "exact" value of (1).

Varying Axial Force

Suppose this stepped column is subject to axial forces applied at b and c, as shown in Fig. 11.42b, in addition to $\overline{P}$ applied at the right end. The solution procedure remains the same, except now the axial force in element 1 is $\overline{P} + \overline{P}_c + \overline{P}_b$, in element 2 is $\overline{P} + \overline{P}_c$, and in element 3 is $\overline{P}$. In order to find the buckling load, normally it is assumed that the axial forces remain in fixed proportion to each other, e.g., suppose $\overline{P}_b = \alpha\overline{P}$ and $\overline{P}_c = \beta\overline{P}$. Then, the buckling load $\overline{P}$ is calculated. ∎

11.5.4 Determination of the Buckling Loads Using Other Approximate Methods

There are several approximate methods of interest for finding a critical load. These include a simple iteration procedure, finite differences, and a weighted-residual approach.

An Iteration Technique (Stodola's,[2] Picard's or Stodola-Vianello's Method)

The fourth order differential equation for a beam with a compressive axial force $\overline{P}$, but no transverse loading $\overline{p}_z$, is given by [Eq. (11.31)]

$$EIw^{iv} + \overline{P}w'' = 0$$

For constant EI and $\overline{P}$, integration gives

$$EIw'' + \overline{P}w = C_1 + C_2 x$$

With the introduction of redefined constants $\overline{C}_1$ and $\overline{C}_2$ this expression can be rewritten as

$$w'' = -\frac{\overline{P}}{EI}(w + \overline{C}_1 + \overline{C}_2 x)$$

Integration gives

$$w = -\frac{\overline{P}}{EI}\left[\int_0^x \int_0^x (w + \overline{C}_1 + \overline{C}_2 x)\, dx\, dx + \overline{C}_3 x + \overline{C}_4\right] \tag{11.70}$$

An iteration scheme can be designed, which converges after a number of iterations to a "good" approximation, beginning with an assumed w on the right-hand side. From Eq. (11.70) for the $(n-1)$th iteration

$$\widetilde{w}^{n-1} = -\frac{P_{cr}}{EI}\left[\int_0^x \int_0^x (\widetilde{w}^{n-1} + \overline{C}_1 + \overline{C}_2 x)\, dx\, dx + \overline{C}_3 x + \overline{C}_4\right] \tag{11.71}$$

Define

$$\widetilde{w}^{n} = -\left[\int_0^x \int_0^x (\widetilde{w}^{n-1} + \overline{C}_1 + \overline{C}_2 x)\, dx\, dx + \overline{C}_3 x + \overline{C}_4\right] \tag{11.72}$$

From Eqs. (11.71) and (11.72)

$$\widetilde{w}^{n-1} = \frac{P_{cr}}{EI}\widetilde{w}^{n} \tag{11.73}$$

which provides an approximation for the eigenvalue P_{cr}, i.e.,

$$P_{cr} = EI\frac{\widetilde{w}^{n-1}}{\widetilde{w}^{n}} \tag{11.74}$$

the constants $\overline{C}_i$, $i = 1, 2, 3, 4$ are determined at each iteration step using the boundary conditions.

[2]Aurel Stodola (1859–1942) was a Hungarian-born engineer, specializing in the development of steam and gas turbines. He served on the faculty of the University of Zurich, Switzerland, for more than 35 years.

EXAMPLE 11.15 A Fixed-Hinged Column

Use the iteration procedure to compute the critical load for the fixed-hinged column of Fig. 11.28.

Begin by determining the constants of integration using the boundary conditions

$$\begin{aligned} w &= w' = 0 \quad \text{at } x = 0 \\ w &= w'' = 0 \quad \text{at } x = L \end{aligned} \tag{1}$$

We find from Eq. (11.70) that

$$\begin{aligned} w_{x=0} &= 0 \quad \text{gives } \overline{C}_4 = 0 \\ w'_{x=0} &= 0 \quad \text{gives } \overline{C}_3 = 0 \\ w_{x=L} &= 0 \quad \text{gives } \int_0^L \int_0^x w\,dx\,dx + \frac{1}{6}\overline{C}_2 L^3 + \frac{1}{2}\overline{C}_1 L^2 = 0 \\ w''_{x=L} &= 0 \quad \text{gives } \overline{C}_2 L = -\overline{C}_1 \end{aligned} \tag{2}$$

Then

$$\overline{C}_2 = \frac{3}{L^3}\int_0^L \int_0^x w\,dx\,dx \qquad \overline{C}_1 = -\frac{3}{L^2}\int_0^L \int_0^x w\,dx\,dx \tag{3}$$

and

$$w = -\frac{\overline{P}}{EI}\left[\int_0^x \int_0^x w\,dx\,dx + \frac{x^3}{2L^3}\int_0^L \int_0^x w\,dx\,dx - \frac{3x^2}{2L^2}\int_0^L \int_0^x w\,dx\,dx\right] \tag{4}$$

As an initial assumed displacement, use a polynomial that satisfies the boundary conditions. For example, use the function of Chapter 7, Eq. (12), Example 7.1:

$$\widetilde{w}^0 = C(3L^2x^2 - 5Lx^3 + 2x^4) \tag{5}$$

From Eq. (11.72)

$$\begin{aligned} \widetilde{w}^1 &= -C\left[\int_0^x \int_0^x (3L^2x^2 - 5Lx^3 + 2x^4)\,dx\,dx + \frac{x^3}{2L^3}\int_0^L \int_0^x (3L^2x^2 - 5Lx^3 + 2x^4)\,dx\,dx \right. \\ &\quad \left. - \frac{3x^2}{2L^2}\int_0^L \int_0^x (3L^2x^2 - 5Lx^3 + 2x^4)\,dx\,dx\right] \\ &= -\frac{C}{60}[-6x^2L^4 + 2x^3L^3 + 15x^4L^2 - 15x^5L + 4x^6] \end{aligned} \tag{6}$$

From Eq. (11.74),

$$\begin{aligned} P_{\text{cr}} = EI\frac{\widetilde{w}^{n-1}}{\widetilde{w}^n} = \frac{EI\widetilde{w}^0}{\widetilde{w}^1} &= -\frac{60(3L^2x^2 - 5Lx^3 + 2x^4)}{-6x^2L^4 + 2x^3L^3 + 15x^4L^2 - 15x^5L + 4x^6}EI \\ &= -\frac{60(3L^2 - 5Lx + 2x^2)}{-6L^4 + 2xL^3 + 15x^2L - 15x^3L + 4x^4}EI \\ &= \frac{60(3L - 2x)EI}{6L^3 + 4L^2x - 11Lx^2 + 4x^3} \end{aligned} \tag{7}$$

The critical load can be determined by utilizing particular values of x. The results can be compared to the exact value

$$P_{\text{cr}} = 20.1907\frac{EI}{L^2} \tag{8}$$

From (7), we find

x	0	$L/4$	$L/2$	$3L/4$	L
$\frac{P_{cr}L^2}{EI}$	30.00	23.53	20.87	20.00	20.00
% Error	48.58	16.53	3.18	−0.94	−0.94

(9)

The mean value would be

$$P_{cr} = \frac{1}{5}(30 + 23.53 + 20.87 + 20.00 + 20.00)\frac{EI}{L^2} = 22.88\frac{EI}{L^2} \tag{10}$$

which differs from the correct value by 13.32%.

One method of avoiding the problem of Eq. (11.74) leading to a buckling load that depends on x, is to take the integral of $\widetilde{w}^{n-1}$ and $\widetilde{w}^n$. Then

$$P_{cr} = EI\frac{\int_0^L \widetilde{w}^{n-1}\,dx}{\int_0^L \widetilde{w}^n\,dx} \tag{11}$$

In the case of this example,

$$\int_0^L \widetilde{w}^0\,dx = C\int_0^L (3L^2x^2 - 5Lx^3 + 2x^4)\,dx = C\left(1 - \frac{5}{4} + \frac{2}{5}\right)L^5 = \frac{3}{20}CL^5$$

$$\int_0^L \widetilde{w}^1\,dx = \frac{-C}{60}\int_0^L (-6x^2L^4 + 2x^3L^3 + 15x^4L^2 - 15x^5L + 4x^6)\,dx \tag{12}$$

$$= \frac{-C}{60}\left(-2 + \frac{1}{2} + 3 - \frac{5}{2} + \frac{4}{7}\right)L^7 = \frac{C}{140}L^7$$

so that from (11),

$$P_{cr} = EI\frac{\int_0^L \widetilde{w}^0\,dx}{\int_0^L \widetilde{w}^1\,dx} = \frac{3\cdot 140}{20}\frac{EI}{L^2} = 21.00\frac{EI}{L^2} \tag{13}$$

which is 4.01% in error.

Another possibility is provided by the *Rayleigh quotient*

$$P_{cr} = \frac{\int_0^L EI w''^2\,dx}{\int_0^L w'^2\,dx} \tag{14}$$

which is derived for eigenvalue problems in elementary vibration textbooks. Insert $\widetilde{w}^1$ of (6) in (14) to obtain

$$P_{cr} = \frac{\int_0^L EI w''^2\,dx}{\int_0^L w'^2\,dx} = \frac{12EI\int_0^L(-L^4 + L^3x + 15L^2x^2 - 25Lx^3 + 10x^4)^2\,dx}{\int_0^L(-12L^4x + 6L^3x^2 + 60L^2x^3 - 75Lx^4 + 24x^5)^2\,dx}$$

$$= 20.243\frac{EI}{L^2} \tag{15}$$

The error associated with this approximation is 0.11%.

A Second Iteration

To observe the effect of further iterations, return to (4) and determine $\widetilde{w}^2$,

$$\widetilde{w}^2 = \frac{C}{60}\left[\int_0^x\int_0^x (4x^6 - 15Lx^5 + 15L^2x^4 + 2L^3x^3 - 6L^4x^2)\,dx\,dx + \frac{x^3}{2L^3}\int_0^L\int_0^x (4x^6 - 15Lx^5 + 15L^2x^4 + 2L^3x^3 - 6L^4x^2)\,dx\,dx - \frac{3x^2}{2L^2}\int_0^L\int_0^x (4x^6 - 15Lx^5 + 15L^2x^4 + 2L^3x^3 - 6L^4x^2)\,dx\,dx\right]$$

$$= \frac{C}{8400}(10x^8 - 50Lx^7 + 70L^2x^6 + 14L^3x^5 - 70L^4x^4 - 13L^5x^3 + 39L^6x^2) \tag{16}$$

From Eq. (11.74),

$$P_{cr} = -\frac{140(4x^4 - 15Lx^3 + 15L^2x^2 + 2L^3x - 6L^4)}{(10x^6 - 50Lx^5 + 70L^2x^4 + 14L^3x^3 - 70L^4x^2 - 13L^5x + 39L^6)}\frac{EI}{L^2} \tag{17}$$

This can be evaluated, giving

x	0	$L/4$	$L/2$	$3L/4$	L
$\frac{P_{cr}L^2}{EI}$	21.54	21.03	20.40	20.08	20.00
% Error	6.67	4.17	1.06	−0.55	−0.94

(18)

The mean value would be

$$\frac{P_{cr}L^2}{EI} = \frac{1}{5}(21.54 + 21.03 + 20.40 + 20.08 + 20.00) = 20.61 \quad (2.09\%\ \text{Error}) \tag{19}$$

From (11)

$$P_{cr} = 20.38\frac{EI}{L^2} \quad (0.92\%\ \text{Error}) \tag{20}$$

and from the Rayleigh quotient of (14),

$$P_{cr} = 20.196\frac{EI}{L^2} \quad (0.007\%\ \text{Error}) \tag{21}$$

■

Finite Differences

As is to be expected, the finite difference expressions of Chapter 8 can be employed to compute the critical loading. Simply use the finite difference relations to discretize the governing equations [Eq. (11.31)]. Apply the boundary conditions and then find the critical load from a characteristic equation formed of the determinant of the finite difference equations.

EXAMPLE 11.16 A Fixed-Hinged Column

Use finite difference equations to compute the critical load for the fixed-hinged column of Fig. 11.28.

The governing differential equation for a beam with axial force and with $\overline{p}_z = 0$ is [Eq. (11.31)]

$$w^{iv} + \alpha^2 w'' = 0, \qquad \alpha^2 = \overline{P}/EI \tag{1}$$

From Chapter 8, Table 8.1, the central difference expressions for w^{iv} and w'' are

$$w_k'' = \left(\frac{1}{h^2}\right)(w_{k-1} - 2w_k + w_{k+1}) \tag{2a}$$

$$w_k^{iv} = \left(\frac{1}{h^4}\right)(w_{k-2} - 4w_{k-1} + 6w_k - 4w_{k+1} + w_{k+2}) \tag{2b}$$

where h is length of interval.

The boundary conditions of $w_0' = 0$ and $w_{x=L}'' = 0$ applied to the central difference expressions for w' and $w''[w_0' = \frac{1}{2h}(w_1 - w_{-1})$ and (2a)] give $w_1 = w_{-1}$ and $w_{k+1} = -w_{k-1}$, respectively (see Fig. 11.43).

Two Segments ($M = 2$, Fig. 11.44)

In this case, $h = \text{length}/M = L/2$, and at node 1, the substitution of (2) into (1) gives

$$\frac{1}{h^4}(w_{-1} - 4w_0 + 6w_1 - 4w_2 + w_3) + \frac{\alpha^2}{h^2}(w_0 - 2w_1 + w_2) = 0 \tag{3}$$

With the relations $w_{-1} = w_1$, $w_3 = -w_1$, and the boundary conditions $w_0 = w_2 = 0$

$$\frac{1}{h^4}6w_1 - \frac{\alpha^2}{h^2}2w_1 = 0 \quad \text{or} \quad \left(\frac{3}{h^2} - \alpha^2\right)w_1 = 0. \tag{4}$$

This gives $\alpha^2 = 3/h^2 = P/EI$. Finally, we conclude that

$$P_{\text{cr}} = \frac{3EI}{h^2} = \frac{12EI}{L^2} \tag{5}$$

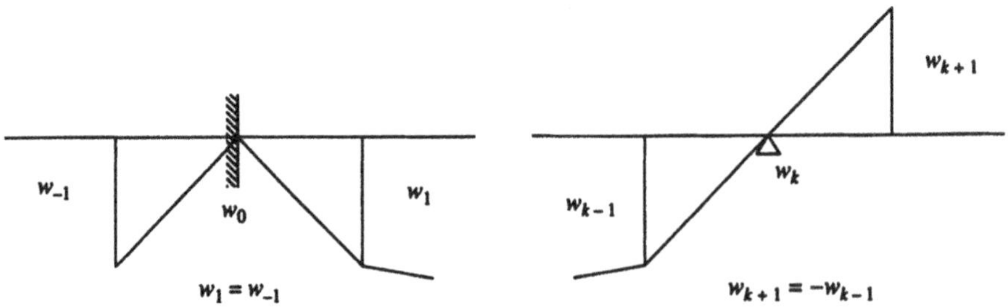

FIGURE 11.43
Finite difference boundary conditions for the beam of Fig. 11.28.

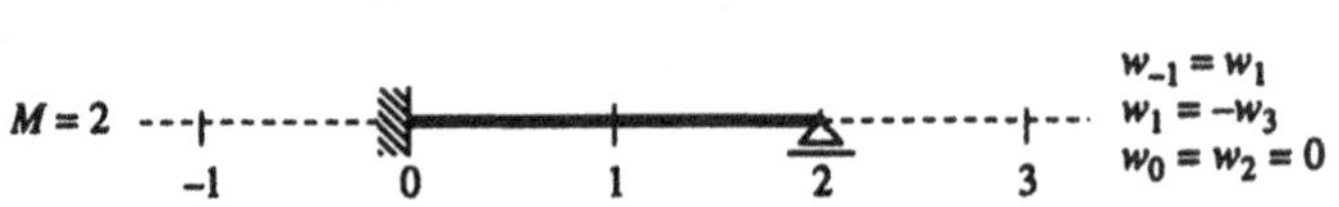

FIGURE 11.44
Central differences with two segments.

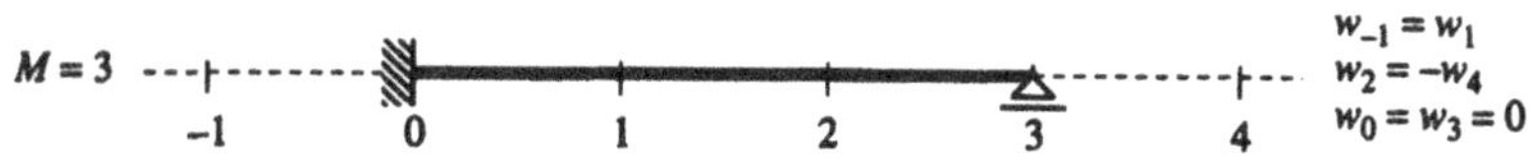

FIGURE 11.45
Central differences with three segments.

Three Segments ($M = 3$, Fig. 11.45)

For three segments, h = length/M = $L/3$. Then, with (1) discretized using (2) at nodes 1 and 2,

$$\begin{aligned}\frac{1}{h^4}(w_{-1} - 4w_0 + 6w_1 - 4w_2 + w_3) + \frac{\alpha^2}{h^2}(w_0 - 2w_1 + w_2) = 0 \\ \frac{1}{h^4}(w_0 - 4w_1 + 6w_2 - 4w_3 + w_4) + \frac{\alpha^2}{h^2}(w_1 - 2w_2 + w_3) = 0\end{aligned} \tag{6}$$

These reduce to

$$(7w_1 - 4w_2) + \alpha^2 h^2(-2w_1 + w_2) = 0, \qquad (-4w_1 + 5w_2) + \alpha^2 h^2(w_1 - 2w_2) = 0 \tag{7}$$

With $\lambda = \alpha^2 h^2$,

$$\underset{\mathbf{K}}{\begin{bmatrix} 7 - 2\lambda & -4 + \lambda \\ -4 + \lambda & 5 - 2\lambda \end{bmatrix}} \underset{\mathbf{U}}{\begin{bmatrix} w_1 \\ w_2 \end{bmatrix}} = \underset{\mathbf{0}}{\begin{bmatrix} 0 \\ 0 \end{bmatrix}} \tag{8}$$

The determinant of these equations provides the characteristic equation from which the critical load can be calculated. Thus,

$$\begin{vmatrix} 7 - 2\lambda & -4 + \lambda \\ -4 + \lambda & 5 - 2\lambda \end{vmatrix} = 0 \quad \text{or} \quad 19 - 16\lambda + 3\lambda^2 = 0 \tag{9}$$

Then $\lambda_{1,2} = 16/6 \pm \sqrt{28/36}$ or $\lambda_1 = 3.549$, $\lambda_2 = 1.785$. The lower value of λ leads to the critical load.

$$P_{cr} = 1.785 \cdot 9\frac{EI}{L^2} = 16.063\frac{EI}{L^2} \tag{10}$$

Collatz (1960) suggests that an improved eigenvalue can be obtained from two ($n = 2$ and $n = 3$) finite difference calculations using

$$P_{cr} = P_1 + \frac{h_1^2}{h_2^2 - h_1^2}(P_1 - P_2) \tag{11}$$

For the beam of this example,

$$P_{cr} = \left(12 + \frac{9}{5}4.063\right)\frac{EI}{L^2} = 19.31\frac{EI}{L^2} \tag{12}$$

This is 4.4% in error relative to the exact solution of $P_{cr} = 20.19EI/L^2$.

Four Segments ($M = 4$, Fig. 11.46)

For four segments, $h = L/4$, and at nodes 1, 2, and 3,

$$\begin{aligned}\frac{1}{h^4}(w_{-1} - 4w_0 + 6w_1 - 4w_2 + w_3) + \frac{\alpha^2}{h^2}(w_o - 2w_1 + w_2) = 0 \\ \frac{1}{h^4}(w_0 - 4w_1 + 6w_2 - 4w_3 + w_4) + \frac{\alpha^2}{h^2}(w_1 - 2w_2 + w_3) = 0 \\ \frac{1}{h^4}(w_1 - 4w_2 + 6w_3 - 4w_4 + w_5) + \frac{\alpha^2}{h^2}(w_2 - 2w_3 + w_4) = 0\end{aligned} \tag{13}$$

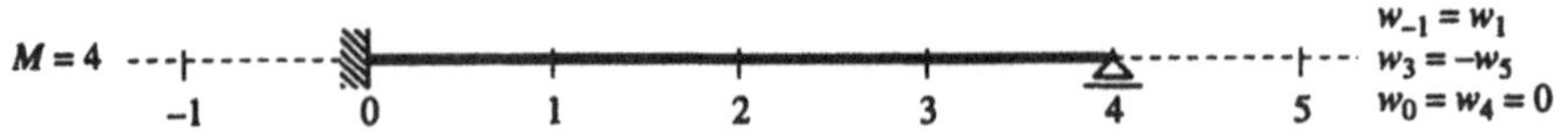

FIGURE 11.46
Central differences with four segments.

so that

$$\begin{aligned}(7w_1 - 4w_2 + w_3) + \lambda(-2w_1 + w_2) &= 0\\(-4w_1 + 6w_2 - 4w_3) + \lambda(w_1 - 2w_2 + w_3) &= 0\\(w_1 - 4w_2 + 5w_3) + \lambda(w_2 - 2w_3) &= 0\end{aligned} \tag{14}$$

and

$$\underbrace{\begin{bmatrix}(7-2\lambda) & (-4+\lambda) & (1+0)\\(-4+\lambda) & (6-2\lambda) & (-4+\lambda)\\(1+0) & (-4+\lambda) & (5-2\lambda)\end{bmatrix}}_{\mathbf{K}}\underbrace{\begin{bmatrix}w_1\\w_2\\w_3\end{bmatrix}}_{\mathbf{U}} = \underbrace{\begin{bmatrix}0\\0\\0\end{bmatrix}}_{\mathbf{0}} \tag{15}$$

Then $44 - 68\lambda + 30\lambda^2 - 4\lambda^3 = 0$, which gives $\lambda_1 = 1.111$. The critical load would then be

$$P_{cr} = 1.111\frac{16EI}{L^2} = 17.78\frac{EI}{L^2} \tag{16}$$

With Collatz's improvement,

$$P_{cr} = (16.06 + 3.93)\frac{EI}{L^2} = 19.99\frac{EI}{L^2} \tag{17}$$

This differs from the exact solution by 1.0%. ∎

Galerkin's Method

Any of the weighted-residual procedures of Chapter 7 can also be employed to compute buckling loads. To illustrate this, consider Galerkin's method which begins with the global form of the governing differential equation (Eq. (11.31)$(EIw'')'' - N^0w'' = 0$) for a beam element in the form

$$\int_a^b \delta w(EIw'')''\,dx - \int_a^b \delta w\, N^0 w''\,dx = 0 \tag{11.75}$$

and utilizes a trial function that satisfies all boundary conditions. With $w = \mathbf{N}_u\mathbf{G}\mathbf{v}$,

$$\delta\mathbf{v}^T\left\{\mathbf{G}^T\left[\int \mathbf{N}_u^T EI\mathbf{N}_u^{iv}\,dx\right]\mathbf{G} - \mathbf{G}^T\left[\int \mathbf{N}_u^T N^0\mathbf{N}_u''\,dx\right]\mathbf{G}\right\}\mathbf{v} = 0 \tag{11.76}$$

EXAMPLE 11.17 A Fixed-Hinged Column

For the fixed-hinged column of Fig. 11.28, choose the assumed displacement $w = \widehat{w}_1(\xi^4 - 2.5\xi^3 + 1.5\xi^2)$, $\xi = x/L$ and note that this satisfies the force boundary condition $M_b = 0$, as well as all displacement boundary conditions ($w_a = \theta_a = w_b = 0$). For this trial solution,

$$\begin{aligned}\mathbf{v} &= w_1\\\mathbf{N}_u &= [\xi^2 \quad \xi^3 \quad \xi^4]\\\mathbf{N}_u'' &= [2 \quad 6\xi \quad 12\xi^2]/L^2\\\mathbf{N}_u^{iv} &= [0 \quad 0 \quad 24]/L^4\\\mathbf{G} &= [1.5 \quad -2.5 \quad 1]^T\end{aligned} \tag{1}$$

Substitution of these expressions into Eq. (11.76), with $N^0 = -\overline{P}$, leads to

$$-\delta W = \delta \mathbf{v}^T [EI\mathbf{G}^T \mathbf{k}_{EI} \mathbf{G} + \overline{P}\mathbf{G}^T \mathbf{k}_P \mathbf{G}]\mathbf{v} = 0 \tag{2}$$

with

$$\mathbf{k}_P = \int_0^1 \mathbf{N}_u^T \mathbf{N}_u'' L\, d\xi = \frac{1}{L^2}\int_0^1 \begin{bmatrix} 2\xi^2 & 6\xi^3 & 12\xi^4 \\ 2\xi^3 & 6\xi^4 & 12\xi^5 \\ 2\xi^4 & 6\xi^5 & 12\xi^6 \end{bmatrix} L\, d\xi = \frac{1}{L}\begin{bmatrix} \frac{2}{3} & \frac{3}{2} & \frac{12}{5} \\ \frac{1}{2} & \frac{6}{5} & 2 \\ \frac{2}{5} & 1 & \frac{12}{7} \end{bmatrix} \tag{3}$$

$$\mathbf{k}_{EI} = \int_0^1 \mathbf{N}_u^T \mathbf{N}_u^{iv} L\, d\xi = \frac{1}{L^4}\int_0^1 \begin{bmatrix} 0 & 0 & 24\xi^2 \\ 0 & 0 & 24\xi^3 \\ 0 & 0 & 24\xi^4 \end{bmatrix} L\, d\xi = \frac{1}{L^3}\begin{bmatrix} 0 & 0 & 8 \\ 0 & 0 & 6 \\ 0 & 0 & 24/5 \end{bmatrix} \tag{4}$$

This gives $1.8EIL - \overline{P}\, 0.0857\, L^3 = 0$ or

$$P_{\text{cr}} = \frac{1.8}{0.0857}\frac{EI}{L^2} = 21\frac{EI}{L^2} \tag{5}$$

Consider a two-term assumed displacement that satisfies all the boundary conditions.

$$w = (3\xi^2/2 - 5\xi^3/2 + \xi^4)\widehat{w}_1 + (4\xi^3/3 - 7\xi^4/3 + \xi^5)\widehat{w}_2 \tag{6}$$

For this assumed displacement,

$$\mathbf{v} = \begin{bmatrix} \widehat{w}_1 \\ \widehat{w}_2 \end{bmatrix} \qquad G = \begin{bmatrix} 3/2 & 0 \\ -5/2 & 4/3 \\ 1 & -7/3 \\ 0 & 1 \end{bmatrix} \tag{7}$$

$$\begin{aligned} \mathbf{N}_u &= [\xi^2 \quad \xi^3 \quad \xi^4 \quad \xi^5] \\ \mathbf{N}_u' &= [2\xi \quad 3\xi^2 \quad 4\xi^3 \quad 5\xi^4]/L \\ \mathbf{N}_u'' &= [2 \quad 6\xi \quad 12\xi^2 \quad 20\xi^3]/L^2 \\ \mathbf{N}_u''' &= [0 \quad 6 \quad 24\xi \quad 60\xi^2]/L^3 \\ \mathbf{N}_u^{iv} &= [0 \quad 0 \quad 24 \quad 120\xi]/L^4 \end{aligned} \tag{8}$$

With these expressions and $\lambda = \varepsilon^2 = \overline{P}L^2/EI$, Eq. (11.76) leads to

$$\begin{bmatrix} 1.80 & 0.80 \\ 0.80 & 0.6095 \end{bmatrix} - \lambda \begin{bmatrix} 0.0857 & 0.0429 \\ 0.0429 & 0.0254 \end{bmatrix} = \begin{bmatrix} 0 \\ 0 \end{bmatrix} \tag{9}$$

or $\lambda^2 - 8714\lambda + 1359.2 = 0$. Then

$$P_{\text{cr}} = 20.35\frac{EI}{L^2} \tag{10}$$

as compared to the exact solution of $P_{\text{cr}} = 20.19EI/L^2$. Note that Galerkin's method appears to provide an upper bound to the exact solution. ∎

References

Bazant, Z.P. and Cedolin, L., 1991, *Stability of Structures*, Oxford University Press, Oxford.
Chwalla, E., 1959, Hilfstafeln zur Berechnung von Spannungsproblemen der Theorie II. Ordnung und von Knickproblemen, Stahlbau Verlag, Köln.
Collatz, L., 1960, *The Numerical Treatment of Differential Equations*, 3rd ed., Springer-Verlag, Berlin.
Leipholz, H., 1968, *Stabilitätstheorie*, Teubner Verlag, Stuttgart.
Pflüger, A., 1964, *Stabilitätsprobleme der Elastostatik*, 2nd ed., Springer-Verlag, Berlin.
Pilkey, W.D., 1994, *Formulas for Stress, Strain and Structural Matrices*, Wiley, New York.
Wunderlich, W., 1976., Zur computerorientierten Formulierung von Stabilitätsproblemen, (Computer oriented Forumulation of Problems in Stability), in *Festschrift W. Zerna, Institut KIB*, Werner-Verlag, Düsseldorf, pp. 111–119.
Wunderlich, W. and Beverungen, G., 1977, Geometrisch nichtlineare Theorie und Berechnung eben gekrümmter Stäbe (Geometrically nonlinear theory and analysis of plane, curved rods), *Bauingenieur*, Vol. 52, pp. 225–237.
Ziegler, H., 1977, *Principles of Structural Stability*, 2nd ed., Birkhäuser Verlag Basel and Stuttgart.
Zurmühl, R., 1963, *Praktische Mathematik*, 4th ed., Springer-Verlag, Berlin.

Problems

11.1 Start with the governing differential equation $EId^4w/dx^4 + Nd^2w/dx^2 = \overline{p}_z$ and derive the stiffness matrix for a beam subject to an axial load N.

Hint: First derive a transfer matrix [Eq. (11.47)] and then convert it to a stiffness matrix.

Answer: Eq. (11.50)

11.2 Consider a column free on one end and fixed on the other. Distinguish between first and second order theory analyses.

11.3 Find the axial force P_{cr} that will buckle a column that is fixed at one end and hinged at the other.

Answer: $P_{cr} = 20.19EI/L^2$

11.4 A column is fixed at one end and guided at the other end. Find the critical axial force. Indicate how the critical load changes if a hinged support replaces the fixed end.

Answer: Fixed end $P_{cr} = \pi^2 EI/L^2$, Hinged end $P_{cr} = \pi^2 EI/(4L^2)$

11.5 A horizontal beam of length $2L$ is hinged at both ends and rests on a rigid support at $x = 1.5L$ from the left end. Find the critical axial force.

Answer: $P_{cr} = 5.89EI/L^2$

11.6 Suppose the moment of inertia of a beam of length L varies with the axial coordinate x as $I = I_0(1 + \beta \sin \pi x/L)$, where β is a known constant and I_0 is a nominal moment of inertia. Use Galerkin's method to find the critical axial load. Begin with the trial solution

$$\widetilde{w} = \sum_{i=1}^{m} \sin \frac{i\pi x}{L}$$

Hint: Use identities of the sort

$$\frac{1}{L}\int_0^L \sin\frac{i\pi x}{L}\sin\frac{k\pi x}{L}\,dx = \begin{cases} 0 & \text{if } i \neq k \\ 1/2 & \text{if } i = k \end{cases}$$

If $\beta = 0$, the cross-section is uniform, and for $m = 1$ you should obtain $P_{cr} = \pi^2 E I_0/L^2$. If $\beta \neq 0$ and $m = 1$, i.e., only one term is retained,

$$P_{cr} = \frac{\pi^2}{L^2} E I_0\left(1 + \frac{8\beta}{3\pi}\right)$$

This is less than 1.5% too high for $\beta = 1$.

11.7 Find the critical axial load $\overline{P}$ in the stepped column of Fig. P11.7.

Answer: $P_{cr} \approx 4.3 E I/L^2$

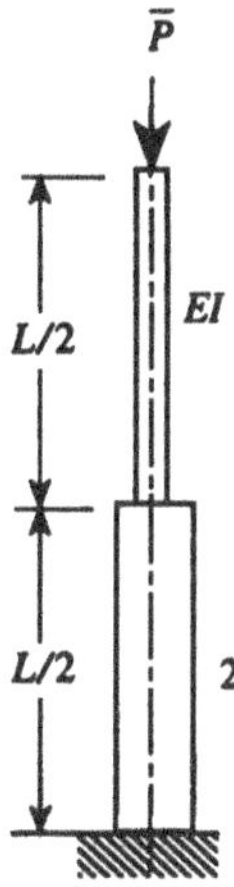

FIGURE P11.7

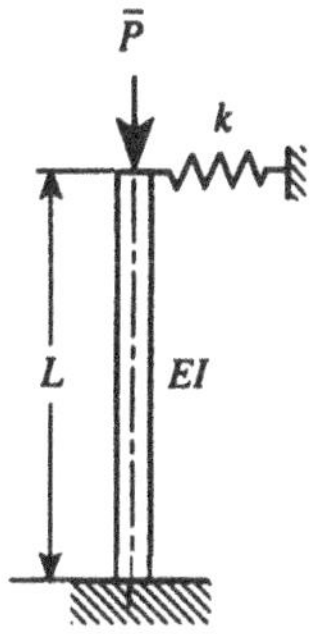

FIGURE P11.8

11.8 Find the buckling load of the column of Fig. P11.8, with a spring support at one end.

Answer: $\overline{P}_{cr} = \varepsilon_0 E I/L^2$ where ε_0 is the lowest non-zero solution of $\tan \varepsilon_0 = \varepsilon_0(1 - \frac{\varepsilon_0}{k}\frac{EI}{L^2})$

11.9 Find the critical load for the spring supported column with constant EI shown in Fig. P11.9.

11.10 For a fixed-fixed column of length L, calculate the buckling load using a four-element model.

Answer: $P_{cr} = 39.47 E I/L^2$

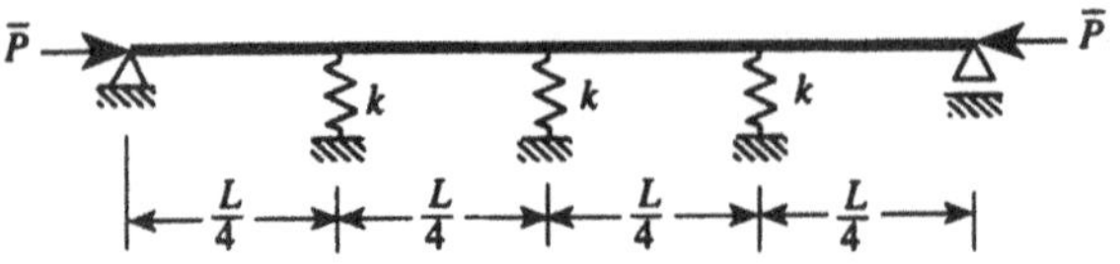

FIGURE P11.9

11.11 Determine the critical load $\overline{P}$ for the stepped column of Fig. P11.11, with a varying axial force. Assume $\overline{P}_b = m\overline{P}$, where m is a prescribed constant of proportionality.

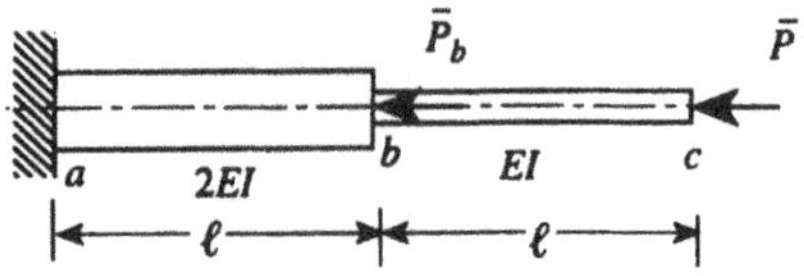

FIGURE P11.11

11.12 A beam column with axial force $\overline{P}$ is subject to a concentrated transverse force $\overline{P}_1$ at midspan $(L/2)$. If both ends are fixed, find the deflection at midspan.

Answer: $w(L/2) = \dfrac{\overline{P}_1 L^3}{192\, EI}\left[\dfrac{12(2 - 2\cos\beta - \beta\sin\beta)}{\beta^3 \sin\beta}\right] \quad \beta = \dfrac{L}{2}\left(\dfrac{\overline{P}}{EI}\right)^{1/2}$

11.13 Find the critical value of $\overline{P}$ of the frame of Fig. P11.13. Use consistent geometric stiffness matrices.

Hint: Only the vertical bar is subject to an axial force.

Answer: $P_{cr} \approx 12\,700$ kN

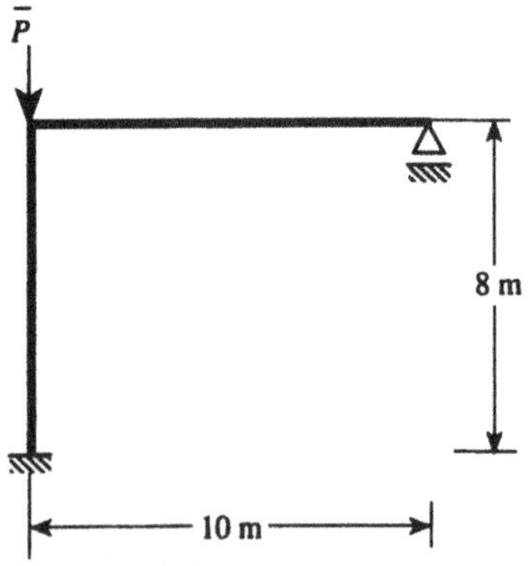

FIGURE P11.13

11.14 For the frame of Fig. P11.14, use det **K** = 0 to find the value of $\overline{M}^*$ that buckles the frame

Answer: $\overline{M}^*_{cr} = 33.535$ MNm.

11.15 Solve problem 11.14 using $(\mathbf{K}_{lin} + \lambda\mathbf{K}_{geo})\mathbf{V} = 0$.

Answer: $\overline{M}_{cr} = 41.587$ MNm.

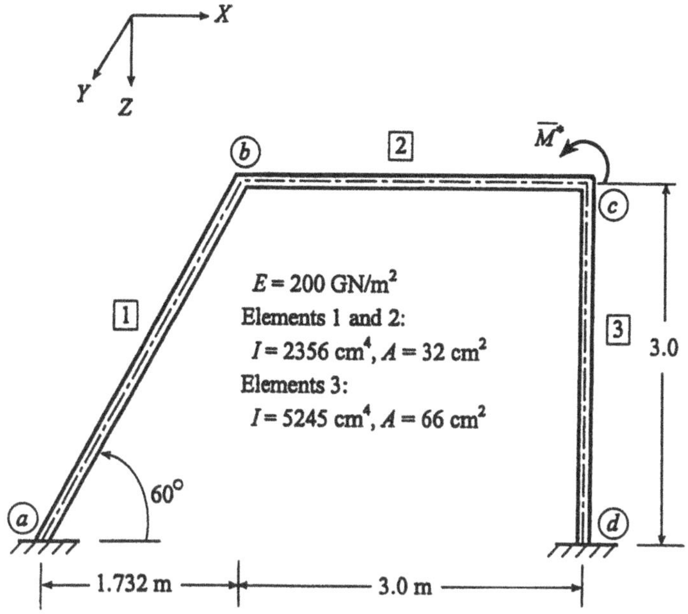

FIGURE P11.14

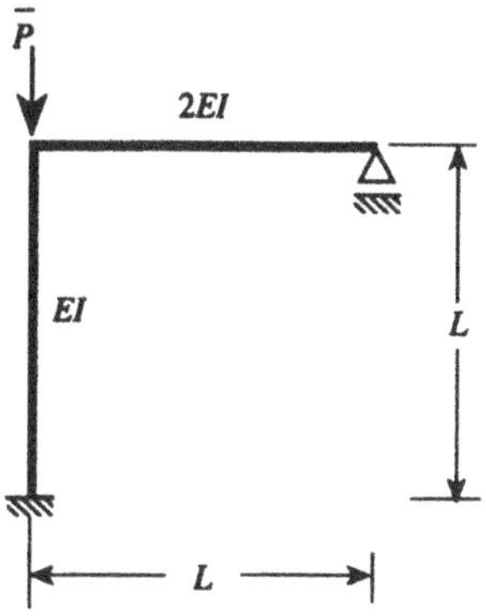

FIGURE P11.16

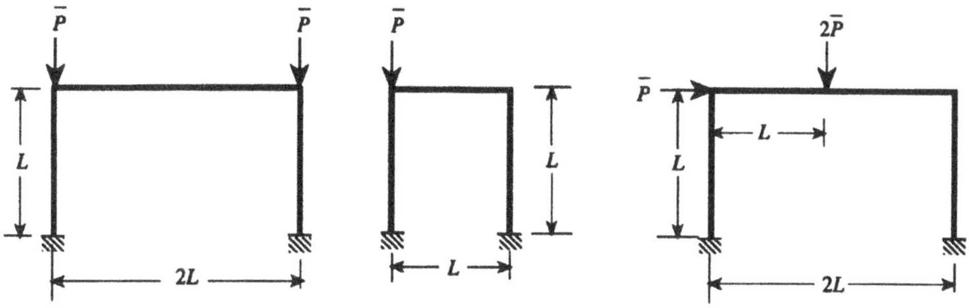

FIGURE P11.17

11.16 Find the force $\overline{P}$ corresponding to buckling of the frame of Fig. P11.16.

Answer: $P_{cr} = 30.7\ EI/L^2$.

11.17 Find the critical force $\overline{P}$ for the frames shown in Fig. P11.17. All members have the same EI.

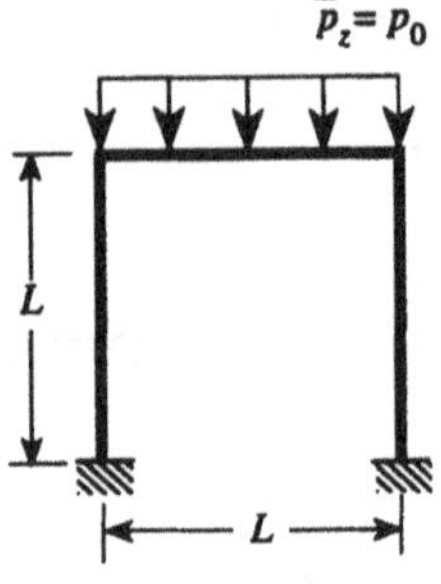

FIGURE P11.18

11.18 Determine the critical value of p_0 for the frame of Fig. P11.18. All members have the same EI.

Section D

Bars and Plates

12

Beams

The straight beam represented by engineering beam theory has been used as an example throughout this work. This theory takes into account the effects of extension, bending, and shear deformation. The local and global forms of the straight beam equations were derived in Chapters 1 and 2, respectively. It is the intention here to present rather complete equations for straight beams. The derivations are based on the reduction of three-dimensional elasticity equations to the appropriate beam theory. The governing equations can be integrated and solved, using the displacement method, to find the displacements, including rotations, and the corresponding forces (stress resultants) along the member. Furthermore, analyses are presented for the cross-sectional properties needed for the study of beams. Also, analytical expressions for the distribution of normal and shear stresses on the cross-section are discussed. Computational methods for calculating cross-sectional properties and stress distributions for arbitrary cross-sectional shapes are presented.

12.1 Displacements and Forces in Straight Bars

We begin with the derivation of a linear theory in which the three-dimensional continuum is reduced to a combination of a two-dimensional cross-sectional problem and a one-dimensional longitudinal analysis, see Wunderlich (1977). Both the cross-sectional and the longitudinal analyses are substantially simpler to carry out than the original three-dimensional problem.

The coordinate system remains the same as for simple beams (Fig. 12.1a). Positive forces and moments are shown in Figs. 12.1a and b, respectively. Note that these definitions are an extension of Sign Convention 1. The displacements corresponding to these forces are shown in Fig. 12.2.

12.1.1 Virtual Work

By definition, the dimensions of a bar normal to the axial (longitudinal) coordinate are very small relative to the length of the bar. The influence of the strains associated with these directions will be neglected. This means that the principle of virtual work relation $\delta W = 0$

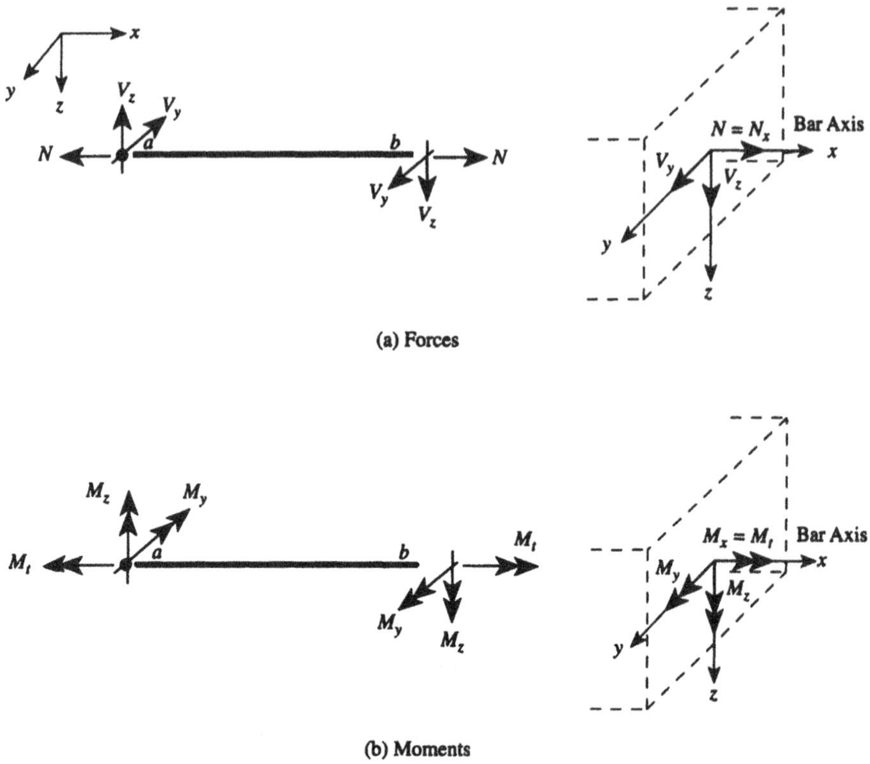

FIGURE 12.1
Positive forces and moments are shown. This corresponds to Sign Convention 1 for simple beams.

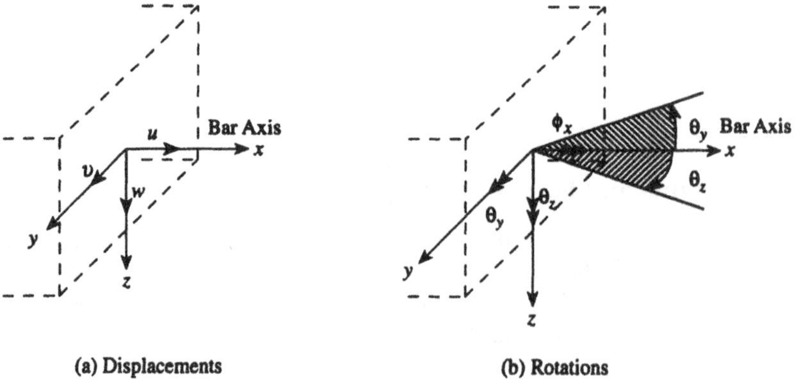

FIGURE 12.2
Positive translations and rotations are shown.

takes the form [Chapter 2, Eq. (2.54)]

$$-\delta W = \int_V (\sigma_x \,\delta\epsilon_x + \tau_{xy}\,\delta\gamma_{xy} + \tau_{xz}\,\delta\gamma_{xz} + \underbrace{\sigma_y\,\delta\epsilon_y + \sigma_z\,\delta\epsilon_z + \tau_{yz}\,\delta\gamma_{yz}}_{\text{Neglected}})\,dV$$
$$- \int_{S_p} (\delta u_x \overline{p}_x + \delta u_y \overline{p}_y + \delta u_z \overline{p}_z)\,dS = 0 \tag{12.1}$$

The body force terms of Eq. (2.54) have been ignored.

12.1.2 Kinematical Relationships

For a continuum (Chapter 1), three displacements $\mathbf{u} = [u_x \ u_y \ u_z]^T$ are necessary to describe the motion of an arbitrary point. More quantities are required to adequately describe the spatial motion of a point on the axis of a bar. These are three translations (u, v, w), three rotations $(\phi_x, \theta_y, \theta_z)$, and one warping parameter (ψ). Thus, the fundamental kinematic variables are

$$\text{Continuum} \qquad\qquad \text{Bar Axis}$$

$$\begin{bmatrix} u_x \\ u_y \\ u_z \end{bmatrix} \Longrightarrow \begin{bmatrix} u \\ v \\ w \\ \phi_x \\ \theta_y \\ \theta_z \\ \psi \end{bmatrix} \begin{array}{l} \left.\begin{array}{l} \\ \\ \\ \end{array}\right\} \text{Displacements} \\ \left.\begin{array}{l} \\ \\ \\ \end{array}\right\} \text{Rotations} \\ \left.\right\} \text{Warping parameter} \end{array} \tag{12.2}$$

In Chapter 1, Eq. (1.98), the axial displacement of a point on a cross-section due to bending deformation in the xz plane was found on the basis of plane cross-sections remaining plane to be

$$u_x(x, z) = u(x) + z\theta_y \tag{12.3}$$

where u was the axial displacement due to stretching at the centroid, z was measured from the centroid, and θ_y was the rotation about the y axis. It will be assumed that a bar in space with bending, extension, and twisting responds in a similar fashion.

Assume that the displacement u_x, u_y, and u_z at some point $P(y, z)$ on the cross-section are related to the bar axis variables $(u, v, w, \phi_x, \theta_y, \theta_z,$ and $\psi)$ at another point Q on the cross-section by

$$\begin{aligned} u_x &= u - y\theta_z + z\theta_y + \omega(y, z)\psi \\ u_y &= v - z\phi_x \\ u_z &= w + y\phi_x \end{aligned} \tag{12.4}$$

where x, y, z are the coordinates of point $P(y, z)$ relative to Q (Fig. 12.3). Observe that $u = u(x)$, $v = v(x)$, and $w = w(x)$ do not vary throughout a given cross-section. The terms $y\theta_z$, $z\theta_y$, $z\phi_x$, and $y\phi_x$ vary linearly. The function $\omega(y, z)$ is referred to as the *warping function.*

The relationships of Eq. (12.4) can be considered to be natural extensions of Eq. (12.3). The geometrical explanations for the selection of these relationships follow.

Displacement u_x

Axial Force

The contribution of the axial force N to the displacement u_x is simply

$$u_x(N) = u \tag{12.5a}$$

where $u = u(x)$, the displacement of point Q, does not vary over the cross-section normal to the x axis.

Bending

From Figs. 12.3a and b, the axial displacement for a point P at y, z in an element of length dx is

$$du_x = -y\,d\theta_z + z\,d\theta_y$$

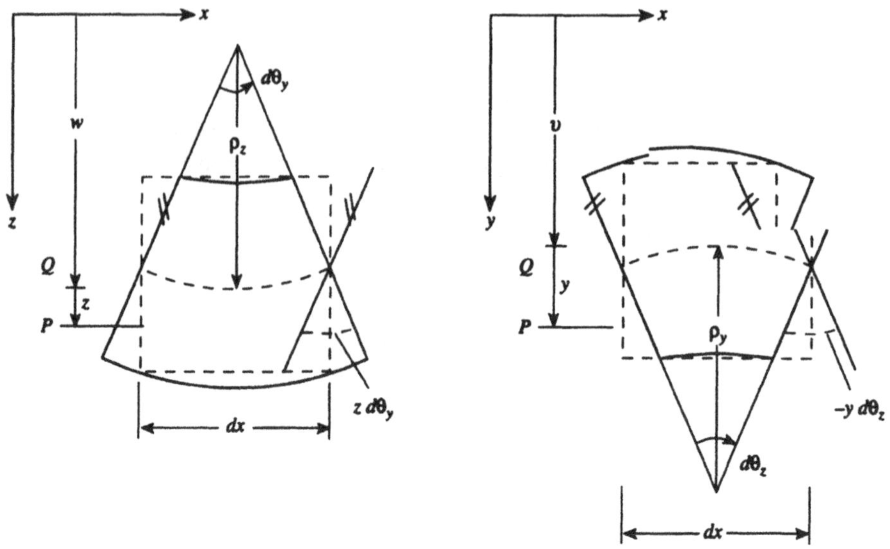

(a) Bending of an element of length dx in the xz plane

(b) Bending of an element of length dx in the xy plane

FIGURE 12.3
Bending in xz and xy planes.

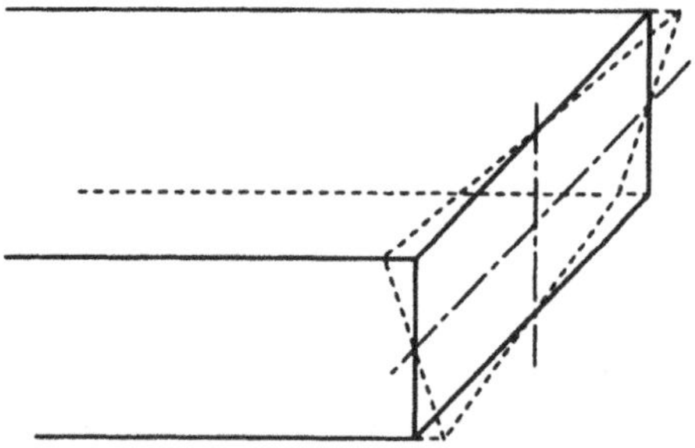

FIGURE 12.4
Warping.

where $d\theta_z$ and $d\theta_y$ are rotations about the pole point Q, and y, z are the coordinates of point P relative to Q. Upon integration,

$$u_x(M) = -y\theta_z + z\theta_y \tag{12.5b}$$

is the displacement of P generated by bending. The rotations θ_y, θ_z, like u, v, and w, are functions of x only and are independent of the cross-sectional coordinates y and z.

Warping

Warping is the out-of-plane displacement or distortion illustrated in Fig. 12.4 that can occur as a bar is twisted. To characterize the displacement shown in Fig. 12.4, it is at least necessary to employ a function that is bilinear over the cross-section, say $\omega(y, z)$.

Thus, warping contributes

$$u_x(M_\omega) = \omega(y, z)\psi \tag{12.5c}$$

to the axial displacements, where u_x is indicated to be due to a warping moment M_ω which will be considered later. This relationship is similar to that of Chapter 1, Eq. (1.141) for simple torsion. The kinematic parameter for warping, ψ, is a function of the x (and not y, z) coordinate, i.e., $\psi = \psi(x)$.

Total Displacement

The sum of the contributions of Eqs. (12.5a, b, and c) leads to the total displacement u_x as given in Eq. (12.4), i.e.,

$$u_x = u - y\theta_z + z\theta_y + \omega(y, z)\psi$$

Displacements u_y, u_z

Suppose a particular cross-section is not distorted in its own plane, i.e., the shape of a cross-section does not change while the bar is being bent. Furthermore, suppose twisting occurs about point Q. If the displacements of Q along the y, z axes are v, w, then point $P(y, z)$ displaces (Fig. 12.5)

$$\begin{aligned} u_y &= v - z\phi_x \\ u_z &= w + y\phi_x \end{aligned} \tag{12.6}$$

where, again, y and z are the coordinates of point P relative to point Q, and $\phi_x = \phi$ is the angle of twist.

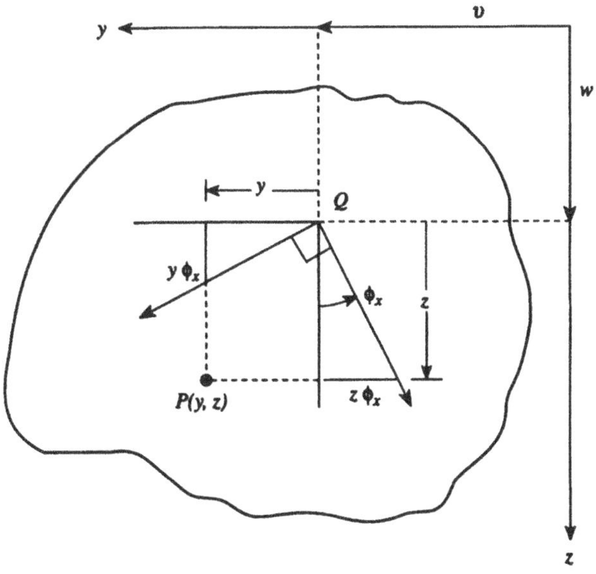

FIGURE 12.5
Twisting.

12.1.3 Strain-Displacement Relations

Substitution of the displacements of Eq. (12.4) into the linear strain-displacement relations for the continuum [Chapter 1, Eq. (1.18)] leads to

$$\begin{array}{llll} \text{Continuum Representation of a Bar} & & \text{Bar variables} & \\ \epsilon_x = \dfrac{\partial u_x}{\partial x} & = & u' & -y\theta_z' + z\theta_y' + \omega\psi' \\ & = & \epsilon & -y\kappa_z + z\kappa_y + \omega\kappa_\omega. \\ & & \text{Extension} & \text{Curvatures} \end{array} \tag{12.7a}$$

$$\begin{aligned} \gamma_{xy} = \frac{\partial u_y}{\partial x} + \frac{\partial u_x}{\partial y} &= v' - \theta_z - z\phi_x' + \frac{\partial\omega}{\partial y}\psi \\ &= v' - \theta_z - \left(z + \frac{\partial\omega}{\partial y}\right)\phi_x' + \frac{\partial\omega}{\partial y}\gamma_{tS} \\ &= \underset{\substack{\text{Shear}\\\text{Force}}}{\gamma_y} - \underset{\substack{\text{Shear Strain from}\\\text{Primary}\\\text{Torsion}}}{\left(z + \frac{\partial\omega}{\partial y}\right)\gamma_{tP}} + \underset{\substack{\text{Warping}\\\text{Torsion}}}{\frac{\partial\omega}{\partial y}\gamma_{tS}} \end{aligned} \tag{12.7b}$$

$$\begin{aligned} \gamma_{xz} = \frac{\partial u_z}{\partial x} + \frac{\partial u_x}{\partial z} &= w' + \theta_y + y\phi_x' + \frac{\partial\omega}{\partial z}\psi. \\ &= \underset{\substack{\text{Shear}\\\text{Force}}}{\gamma_z} - \underset{\substack{\text{Shear Strain from}\\\text{Primary}\\\text{Torsion}}}{\left(-y + \frac{\partial\omega}{\partial z}\right)\gamma_{tP}} + \underset{\substack{\text{Warping}\\\text{Torsion}}}{\frac{\partial\omega}{\partial z}\gamma_{tS}} \end{aligned} \tag{12.7c}$$

where the strains and curvatures of the bar axis are defined as $\epsilon = u'$, $\kappa_z = \theta_z'$, $\kappa_y = \theta_y'$, $\kappa_\omega = \psi'$, $\psi = \gamma_{tS} - \phi_x'$, $\gamma_y = v' - \theta_z$, $\gamma_{tP} = \phi_x'$, and $\gamma_z = w' + \theta_y$. The prime indicates the partial derivative with respect to x.

12.1.4 Internal Virtual Work

From Eq. (12.1), the internal virtual work is given by

$$-\delta W_i = \int_V (\sigma_x\,\delta\epsilon_x + \tau_{xy}\,\delta\gamma_{xy} + \tau_{xz}\,\delta\gamma_{xz})\,dV. \tag{12.8}$$

Substitution of the strain expressions of Eq. (12.7) into Eq. (12.8) leads to

$$\begin{aligned} -\delta W_i = \int_x\int_A \Big\{ &\sigma_x\,\delta(u' - y\theta_z' + z\theta_y' + \omega\psi') + \tau_{xy}\left[\delta(v' - \theta_z) + \delta\left(\frac{\partial\omega}{\partial y}\psi - z\phi_x'\right)\right] \\ &+ \tau_{xz}\left[\delta(w' + \theta_y) + \delta\left(\frac{\partial\omega}{\partial z}\psi + y\phi_x'\right)\right]\Big\}\,dA\,dx \end{aligned}$$

$$
\begin{aligned}
&= \int_x \Bigg\{ \underbrace{\left[\int_A \sigma_x \, dA \right]}_{N} \delta u' && \text{Axial Force} \\
&+ \underbrace{\left[-\int_A \sigma_x y \, dA \right]}_{M_z} \delta\theta_z' + \underbrace{\left[\int_A \sigma_x z \, dA \right]}_{M_y} \delta\theta_y' + \underbrace{\left[\int_A \sigma_x \omega \, dA \right]}_{M_\omega \text{ (Bimoment)}} \delta\psi' && \text{Moments} \\
&+ \underbrace{\left[\int_A \tau_{xy} \, dA \right]}_{V_y} \delta(v' - \theta_z) + \underbrace{\left[\int_A \tau_{xz} \, dA \right]}_{V_z} \delta(w' + \theta_y) && \text{Shear Forces} \\
&+ \underbrace{\left\{ \int_A \left[\tau_{xz}\left(y - \frac{\partial \omega}{\partial z} \right) - \tau_{xy}\left(z + \frac{\partial \omega}{\partial y} \right) \right] dA \right\}}_{M_{tP}} \delta\phi_x' && \text{Primary Torsional Moment} \\
&+ \underbrace{\left[\int_A \left(\tau_{xy} \frac{\partial \omega}{\partial y} + \tau_{xz} \frac{\partial \omega}{\partial z} \right) dA \right]}_{M_{tS}} \delta(\phi_x' + \psi) \Bigg\} \, dx && \text{Secondary Torsional Moment (from warping shear)}
\end{aligned}
\tag{12.9}
$$

The definitions of the stress resultants (internal forces and moments) have been identified in Eq. (12.9) as

$$
N = \int_A \sigma_x \, dA, \qquad M_z = -\int_A \sigma_x y \, dA, \qquad M_y = \int_A \sigma_x z \, dA
$$

$$
M_\omega = \int_A \sigma_x \omega \, dA, \qquad V_y = \int_A \tau_{xy} \, dA, \qquad V_z = \int_A \tau_{xz} \, dA
$$

$$
M_{tP} = \int_A \left[\tau_{xz}\left(y - \frac{\partial \omega}{\partial z} \right) - \tau_{xy}\left(z + \frac{\partial \omega}{\partial y} \right) \right] dA
$$

$$
M_{tS} = \int_A \left[\tau_{xy} \frac{\partial \omega}{\partial y} + \tau_{xz} \frac{\partial \omega}{\partial z} \right] dA
$$

Thus, Eq. (12.9) leads to the internal virtual work expression for the bar

$$
-\delta W_i = \int_x \{ N\delta\epsilon + M_z \, \delta\kappa_z + M_y \, \delta\kappa_y + M_\omega \, \delta\kappa_\omega + \underline{V_y \, \delta\gamma_y} + \underline{V_z \, \delta\gamma_z} + M_{tP} \, \delta\gamma_{tP} + \underline{M_{tS} \, \delta\gamma_{tS}} \} \, dx \tag{12.10}
$$

The stress resultants and corresponding strains are summarized in the following table:

	Stress Resultant	Strain
Longitudinal	N	$\epsilon = u'$
Bending	M_z, M_y	$\kappa_z = \theta_z'$, $\kappa_y = \theta_y'$
Torsion	M_{tP}, M_{tS}, M_ω	$\gamma_{tP} = \phi_x'$, $\gamma_{tS} = \phi_x' + \psi$, $\kappa_\omega = \psi'$
Direct Shear	V_y, V_z	$\gamma_y = v' - \theta_z$, $\gamma_z = w' + \theta_y$

Simplifying Kinematic Assumptions

Bending: Bernoulli's Hypothesis

Neglecting the transverse shear strains, i.e.,

$$
\begin{aligned}
\gamma_y &= v' - \theta_z = 0 \\
\gamma_z &= w' + \theta_y = 0
\end{aligned}
\tag{12.11}
$$

corresponds to Bernoulli's hypothesis of plane cross-sections remaining plane and orthogonal to the axis during bending. For such cases, i.e., when shear deformation is not taken into account,

$$\theta_y = -w' \qquad \theta_z = v' \tag{12.12}$$

Torsion: Wagner's Hypothesis

Wagner's hypothesis is the assumption that a cross-sectional area retains its shape during torsion, i.e., it remains undistorted. This hypothesis is equivalent to specifying that warping torsion generated shear strains can be neglected, i.e.,

$$\gamma_{tS} = \phi_x' + \psi = 0 \quad \text{or} \quad \psi = -\phi_x' \tag{12.13}$$

Internal Virtual Work

The underlined terms of Eq. (12.10) contain the strains (γ_y, γ_z, and γ_{tS}) that are zero for the kinematic assumptions just discussed. If these terms are left out and the definitions $\epsilon = u'$, $\kappa_z = \theta_z'$, $\kappa_y = \theta_y'$, $\kappa_\omega = \psi'$, and $\gamma_{tP} = \phi_x'$ are introduced in Eq. (12.10), the internal virtual work will contain five terms rather than eight. Thus,

$$-\delta W_i = \int_x (N\,\delta u' + M_z\,\delta v'' - M_y\,\delta w'' - M_\omega\,\delta\phi_x'' + M_{tP}\,\delta\phi_x')\,dx \tag{12.14}$$

Introduction of the Material Law

Return to the internal virtual work of Eq. (12.8), and introduce the material law in order to obtain an expression in terms of strains. Thus, substitution of Hooke's law into Eq. (12.8) gives

$$-\delta W_i = \int_V (E\epsilon_x\,\delta\epsilon_x + G\gamma_{xy}\,\delta\gamma_{xy} + G\gamma_{xz}\,\delta\gamma_{xz})\,dV \tag{12.15}$$

Neglect the shear terms related to γ_y, γ_z, and γ_{tS}, and introduce the strain displacement relations of Eq. (12.7) into Eq. (12.15). This leads to

$$\begin{aligned} -\delta W_i = \int_x \int_A \Big\{ & E(\epsilon - y\kappa_z + z\kappa_y + \omega\kappa_\omega)\delta(\epsilon - y\kappa_z + z\kappa_y + \omega\kappa_\omega) \\ & + G\left(z + \frac{\partial\omega}{\partial y}\right)\gamma_{tP}\left(z + \frac{\partial\omega}{\partial y}\right)\delta\gamma_{tP} + G\left(-y + \frac{\partial\omega}{\partial z}\right)\gamma_{tP}\left(-y + \frac{\partial\omega}{\partial z}\right)\delta\gamma_{tP}\Big\}\,dA\,dx \end{aligned}$$

For constant E and G over the cross-section, this expression can be written as

$$\begin{aligned} -\delta W_i = \int_x \Bigg\{ & E\Bigg\{\underbrace{\left[\int_A dA\right]}_{A}\epsilon - \underbrace{\left[\int_A y\,dA\right]}_{I_y}\kappa_z + \underbrace{\left[\int_A z\,dA\right]}_{I_z}\kappa_y + \underbrace{\left[\int_A \omega\,dA\right]}_{I_\omega}\kappa_\omega\Bigg\}\delta\epsilon \\ & - E\Bigg\{\underbrace{\left[\int_A y\,dA\right]}_{I_y}\epsilon - \underbrace{\left[\int_A yy\,dA\right]}_{I_{yy}}\kappa_z + \underbrace{\left[\int_A yz\,dA\right]}_{I_{yz}}\kappa_y + \underbrace{\left[\int_A y\omega\,dA\right]}_{I_{y\omega}}\kappa_\omega\Bigg\}\delta\kappa_z \end{aligned}$$

$$+E\left\{\underbrace{\left[\int_A z\,dA\right]}_{I_z}\epsilon-\underbrace{\left[\int_A zy\,dA\right]}_{I_{zy}}\kappa_z+\underbrace{\left[\int_A zz\,dA\right]}_{I_{zz}=I}\kappa_y+\underbrace{\left[\int_A z\omega\,dA\right]}_{I_{z\omega}}\kappa_\omega\right\}\delta\kappa_y$$

$$+E\left\{\underbrace{\left[\int_A \omega\,dA\right]}_{I_\omega}\epsilon-\underbrace{\left[\int_A \omega y\,dA\right]}_{I_{\omega y}}\kappa_z+\underbrace{\left[\int_A \omega z\,dA\right]}_{I_{\omega z}}\kappa_y+\underbrace{\left[\int_A \omega\omega\,dA\right]}_{I_{\omega\omega}}\kappa_\omega\right\}\delta\kappa_\omega$$

$$\left.+G\underbrace{\int_A\left[\left(z+\frac{\partial\omega}{\partial y}\right)^2+\left(-y+\frac{\partial\omega}{\partial z}\right)^2\right]dA}_{J_t}\,\gamma_{tP}\,\delta\gamma_{tP}\right\}dx. \tag{12.16}$$

This provides definitions of the cross-sectional characteristics for arbitrary coordinate axes y and z. Thus

$$\begin{aligned}
A&=\int_A dA\ [L^2] & I_y&=\int_A y\,dA\ [L^3] & I_z&=\int_A z\,dA\ [L^3]\\
I_\omega&=\int_A \omega\,dA\ [L^4] & I_{yz}&=\int_A yz\,dA\ [L^4] & I_{y\omega}&=\int_A y\omega\,dA\ [L^5]\\
I_{z\omega}&=\int_A z\omega\,dA\ [L^5] & I_{yy}&=\int_A y^2\,dA\ [L^4] & I_{zz}&=\int_A z^2\,dA\ [L^4]\\
I_{\omega\omega}&=\int_A \omega^2\,dA\ [L^6] & J_t&=\int_A\left[\left(z+\frac{\partial\omega}{\partial y}\right)^2+\left(-y+\frac{\partial\omega}{\partial z}\right)^2\right]dA\ [L^4]
\end{aligned} \tag{12.17}$$

where the dimensions are in brackets [ ], with L indicating length. These cross-sectional properties are displayed in Fig. 12.6a [Bornscheuer, 1952].

Special cross-sectional axes are of interest. If the cross-sectional coordinates y, z are the centroidal axes y_C, z_C, the first moments of the areas are zero, i.e.,

$$\begin{aligned}
I_{\overline{y}}&=0 & \overline{y}&=y-y_C & y_C&=I_y/A\\
I_{\overline{z}}&=0 & \overline{z}&=z-z_C & z_C&=I_z/A.
\end{aligned} \tag{12.18a}$$

The corresponding ω coordinate is ω_C, and Eq. (12.18a) is supplemented with the condition

$$I_{\overline{\omega}}=0 \qquad \overline{\omega}=\omega-\omega_C \qquad \omega_C=I_\omega/A. \tag{12.18b}$$

These relations substituted into Eq. (12.17) lead to

$$\begin{aligned}
I_{\overline{y}\,\overline{z}}&=I_{yz}-\frac{I_yI_z}{A}\ [L^4] & I_{\overline{y}\,\overline{\omega}}&=I_{y\omega}-\frac{I_yI_\omega}{A}\ [L^5]\\
I_{\overline{z}\,\overline{\omega}}&=I_{z\omega}-\frac{I_zI_\omega}{A}\ [L^5] & I_{\overline{y}\,\overline{y}}&=I_{yy}-\frac{I_yI_y}{A}\ [L^4]\\
I_{\overline{z}\,\overline{z}}&=I_{zz}-\frac{I_zI_z}{A}\ [L^4] & I_{\overline{\omega}\,\overline{\omega}}&=I_{\omega\omega}-\frac{I_\omega I_\omega}{A}\ [L^6]
\end{aligned} \tag{12.18c}$$

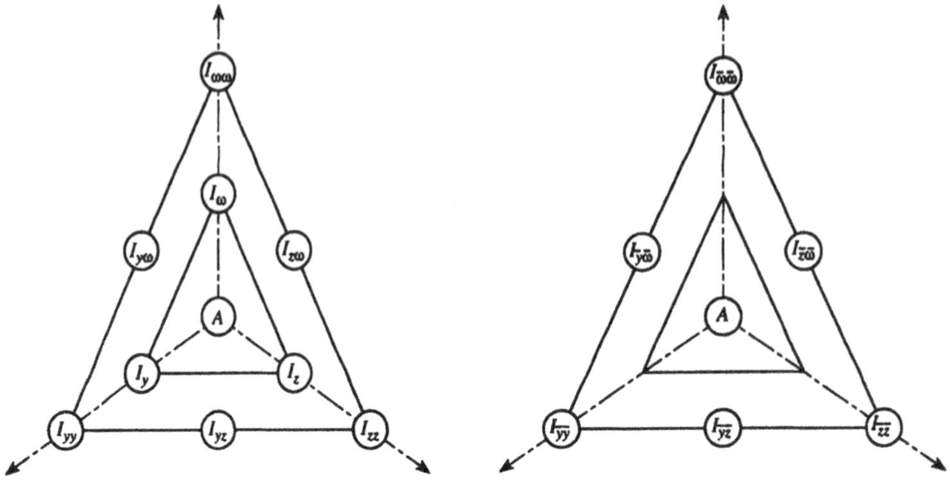

(a) Arbitrary coordinate axes y and z. (b) Centroidal coordinates y_C and z_C.

FIGURE 12.6
Representation of the cross-sectional characteristics for particular axes.

These properties are displayed in Fig. 12.6b. To understand the manipulations needed to move from Eqs. (12.18a) and (12.18b) to Eq. (12.18c), observe that

$$\begin{aligned} I_{\overline{y}\overline{z}} &= \int_A (y - y_C)(z - z_C)\, dA \\ &= \int_A yz\, dA - y_C \int_A z\, dA - z_C \int_A y\, dA + \int_A y_C z_C\, dA \\ &= I_{yz} - \frac{I_y}{A} I_z - \frac{I_z}{A} I_y + \frac{I_z}{A}\frac{I_y}{A} A \\ &= I_{yz} - \frac{I_y I_z}{A} \end{aligned}$$

If the $\overline{y}, \overline{z}$ axes are rotated through an angle α until reaching $\widetilde{y}, \widetilde{z}$ such that

$$I_{\widetilde{y}\widetilde{z}} = 0 \tag{12.19a}$$

then these are referred to as *centroidal principal axes.* The condition $I_{\widetilde{y}\widetilde{z}} = 0$ leads to the familiar formula

$$\tan 2\alpha = \frac{2I_{\overline{y}\overline{z}}}{I_{\overline{y}\overline{y}} - I_{\overline{z}\overline{z}}} \qquad \begin{aligned} \widetilde{y} &= \overline{y}\cos\alpha + \overline{z}\sin\alpha \\ \widetilde{z} &= \overline{z}\cos\alpha - \overline{y}\sin\alpha \end{aligned} \tag{12.19b}$$

and the *principal moments of inertia*

$$\left.\begin{matrix} I_{\widetilde{y}\widetilde{y}} \\ I_{\widetilde{z}\widetilde{z}} \end{matrix}\right\} = \frac{1}{2}(I_{\overline{y}\overline{y}} + I_{\overline{z}\overline{z}}) \pm \frac{1}{2}\sqrt{(I_{\overline{y}\overline{y}} - I_{\overline{z}\overline{z}})^2 + 4I_{\overline{y}\overline{z}}^2} \tag{12.19c}$$

Suppose the origin of the $\overline{y}, \overline{z}$ axes is shifted from the centroid to the origin of the $\overline{y}^*, \overline{z}^*$ axes for which

$$I_{\overline{y}^*\overline{\omega}^*} = I_{\overline{z}^*\overline{\omega}^*} = 0 \tag{12.20a}$$

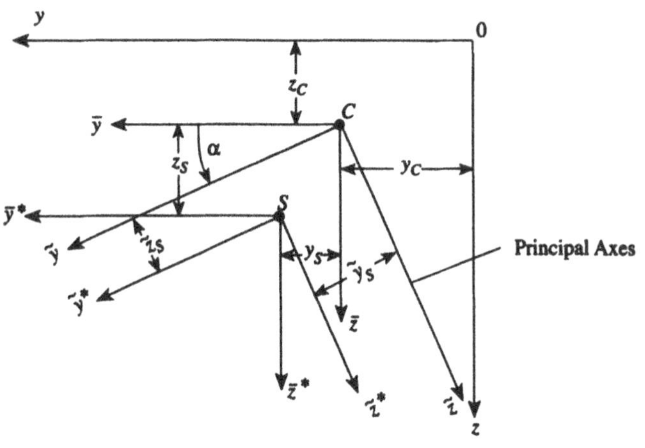

FIGURE 12.7
Coordinate systems.

This point is the *center of twist* or the *shear center*. The shear center is the point on the cross-section through which the plane of the resultant of the applied loads passes such that no twisting moments are developed. From $I_{\bar{y}^*\bar{\omega}^*} = 0$ and $I_{\bar{z}^*\bar{\omega}^*} = 0$, the distance of the shear center from the $\bar{y}$, $\bar{z}$ principal axes is

$$y_S = \frac{I_{\bar{z}\bar{\omega}}I_{\bar{y}\bar{y}} - I_{\bar{y}\bar{\omega}}I_{\bar{y}\bar{z}}}{I_{\bar{y}\bar{y}}I_{\bar{z}\bar{z}} - I_{\bar{y}\bar{z}}^2} \quad \text{in the } \bar{y} \text{ direction}$$

$$z_S = \frac{-I_{\bar{y}\bar{\omega}}I_{\bar{z}\bar{z}} + I_{\bar{z}\bar{\omega}}I_{\bar{y}\bar{z}}}{I_{\bar{y}\bar{y}}I_{\bar{z}\bar{z}} - I_{\bar{y}\bar{z}}^2} \quad \text{in the } \bar{z} \text{ direction} \tag{12.20b}$$

If the $\bar{y}^*$, $\bar{z}^*$ axes are rotated through an angle α (Fig. 12.7), which corresponds to the principal axes for the centroid, the distance of the shear center from the $\tilde{y}$, $\tilde{z}$ axes is

$$\tilde{y}_S = \frac{I_{\tilde{z}\tilde{\omega}}}{I_{\tilde{z}\tilde{z}}} \qquad \tilde{z}_S = -\frac{I_{\tilde{y}\tilde{\omega}}}{I_{\tilde{y}\tilde{y}}}. \tag{12.21}$$

Shear centers will be treated later in this chapter.

If the ω coordinate, corresponding to the principal axes $\tilde{y}^*$, $\tilde{z}^*$ with the origin at the shear center, is taken as $\tilde{\omega}^* = \tilde{\omega} + \tilde{z}_S\tilde{y} - \tilde{y}_S\tilde{z}$, then

$$I_{\tilde{\omega}^*\tilde{\omega}^*} = \int_A \tilde{\omega}^{*2}\, dA = \int_A (\tilde{\omega} + \tilde{z}_S\tilde{y} - \tilde{y}_S\tilde{z})^2\, dA$$

Substitution of Eqs. (12.20a) and (12.21) into this expression leads to

$$I_{\tilde{\omega}^*\tilde{\omega}^*} = I_{\tilde{\omega}\tilde{\omega}} - \frac{I_{\tilde{y}\tilde{\omega}}^2}{I_{\tilde{y}\tilde{y}}} - \frac{I_{\tilde{z}\tilde{\omega}}^2}{I_{\tilde{z}\tilde{z}}} \tag{12.22a}$$

Since $I_{\tilde{y}\tilde{y}} > 0$ and $I_{\tilde{z}\tilde{z}} > 0$, it can be seen from Eq. (12.22a) that $I_{\tilde{\omega}^*\tilde{\omega}^*}$ is a minimum for all choices of the $\tilde{\omega}$ coordinate. Rewrite Eq. (12.22a) as

$$I_{\tilde{\omega}^*\tilde{\omega}^*} = I_{\tilde{\omega}\tilde{\omega}} + \tilde{z}_S I_{\tilde{y}\tilde{\omega}} - \tilde{y}_S I_{\tilde{z}\tilde{\omega}} \tag{12.22b}$$

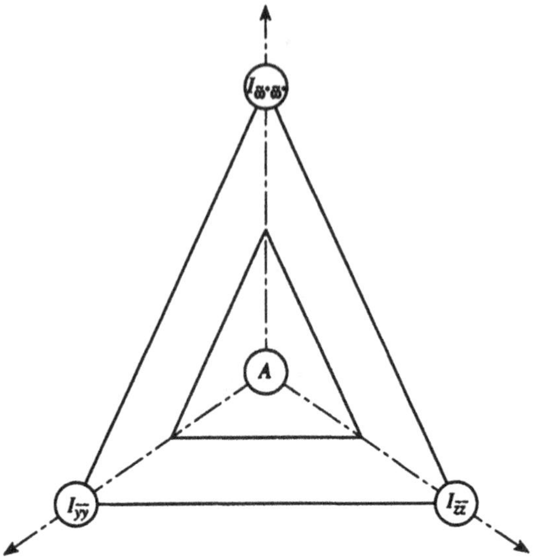

FIGURE 12.8
Representation of the cross-sectional characteristics for principal axes.

The principal moments of inertia and $I_{\bar{\omega}^*\bar{\omega}^*}$ are shown in Fig. 12.8.

The quantity J_t in Eq. (12.17) can be expressed as

$$J_t = \int_A \left(y^2 + z^2 + z\frac{\partial \omega}{\partial y} - y\frac{\partial \omega}{\partial z} \right) dA + \int_A \left[\left(\frac{\partial \omega}{\partial y} \right)^2 + \left(\frac{\partial \omega}{\partial z} \right)^2 \right] dA + \int_A \left(z\frac{\partial \omega}{\partial y} - y\frac{\partial \omega}{\partial z} \right) dA \tag{12.23}$$

The first integral in Eq. (12.23) is the torsional constant J given in Chapter 1, Eq. (1.154). The second term can be processed as

$$\int_A \left[\left(\frac{\partial \omega}{\partial y} \right)^2 + \left(\frac{\partial \omega}{\partial z} \right)^2 \right] dA = \int_A \left[\frac{\partial}{\partial y}\left(\omega\frac{\partial \omega}{\partial y} \right) + \frac{\partial}{\partial z}\left(\omega\frac{\partial \omega}{\partial z} \right) \right] dA - \int_A \omega \left(\frac{\partial^2 \omega}{\partial y^2} + \frac{\partial^2 \omega}{\partial z^2} \right) dA$$

From Gauss's integral theorem in Appendix II, Eq. (II.8),

$$\int_A \left[\frac{\partial}{\partial y}\left(\omega\frac{\partial \omega}{\partial y} \right) + \frac{\partial}{\partial z}\left(\omega\frac{\partial \omega}{\partial z} \right) \right] dA = \int_S \omega \left(\frac{\partial \omega}{\partial y} a_y + \frac{\partial \omega}{\partial z} a_z \right) dS$$

For a two-dimensional problem, the volume V in Eq. (II.8) becomes the area A. The third integral of Eq. (12.23) can be written as

$$\int_A \left(z\frac{\partial \omega}{\partial y} - y\frac{\partial \omega}{\partial z} \right) dA = \int_A \left[\frac{\partial}{\partial y}(z\,\omega) + \frac{\partial}{\partial z}(-y\,\omega) \right] dA = \int_S \omega(z\,a_y - y\,a_z)\, dS$$

Here Gauss's integral theorem is used again. Thus, the expression of J_t appears as

$$J_t = J - \int_A \omega \left(\frac{\partial^2 \omega}{\partial y^2} + \frac{\partial^2 \omega}{\partial z^2} \right) dA + \int_S \omega \left(\frac{\partial \omega}{\partial y} a_y + \frac{\partial \omega}{\partial z} a_z + z\,a_y - y\,a_z \right) dS$$

Note that ω satisfies Eqs. (1.151) and (1.153), so that the integrals on the right-hand side of the above equation vanish. This leads to

$$J_t = J = \int_A \left(y^2 + z^2 + z\frac{\partial \omega}{\partial y} - y\frac{\partial \omega}{\partial z} \right) dA \qquad (12.24)$$

The forces and displacements for the beam will be referred to these special axes. Use of these assumptions will simplify the relationships derived thus far. The internal virtual work of Eq. (12.16) now reduces to

$$-\delta W_i = \int_x (\underbrace{EA\,\epsilon\,\delta\epsilon}_{\text{Axial Extension}} + \underbrace{EI_{yy}\,\kappa_z\,\delta\kappa_z}_{\text{Bending about the z axis}} + \underbrace{EI_{zz}\,\kappa_y\,\delta\kappa_y}_{\text{Bending about the y axis}} + \underbrace{EI_{\omega\omega}\,\kappa_\omega\,\delta\kappa_\omega}_{\text{Warping Torsion}} + \underbrace{GJ_t\,\gamma_{tP}\,\delta\gamma_{tP}}_{\text{Primary Torsion}})\,dx \qquad (12.25)$$

where, for convenience, the superscripts have been dropped from the coordinates. If displacements are introduced using $\varepsilon = u'$, $\kappa_z = \theta_z' = v''$, $\kappa_y = \theta_y' = -w''$, $\kappa_\omega = \psi' = -\phi_x''$, and $\gamma_{tP} = \phi_x'$, and u, v, and w are referred to the shear center, then

$$\begin{aligned}
-\delta W_i &= \int_x (EAu_S'\,\delta\epsilon + EI_{yy}v_S''\,\delta\kappa_z - EI_{zz}w_S''\,\delta\kappa_y - EI_{\omega\omega}\phi_S''\,\delta\kappa_\omega + GJ_t\phi_S'\,\delta\gamma_{tP})\,dx \\
&= \int_x (EAu_S'\,\delta u_S' + EI_{yy}v_S''\,\delta v_S'' + EI_{zz}w_S''\,\delta w_S'' + EI_{\omega\omega}\phi_S''\,\delta\phi_S'' + GJ_t\phi_S'\,\delta\phi_S')\,dx \qquad (12.26)
\end{aligned}$$

where $\phi_S = (\phi_x)_S$, and the subscripts indicate that the quantities are with respect to the shear center. Remember that Bernoulli's and Wagner's hypotheses still apply. Comparison of Eqs. (12.14) and (12.26) shows that the net internal forces and displacements are related by

$$\begin{aligned}
EA\,u' &= N \\
EI_{yy}\,v'' &= M_z \\
-EI_{zz}\,w'' &= M_y \\
-EI_{\omega\omega}\,\phi'' &= M_\omega \\
GJ_t\,\phi' &= M_t
\end{aligned} \qquad (12.27)$$

where N, M_z, and M_y are referred to the centroidal principal axes, and $M_{tP} = M_t$ is referred to the principal center of twist (shear center). The S subscripts have been dropped. The cross-sectional property $I_{\omega\omega}$ is often represented by Γ and is called the *warping constant*.

12.1.5 External Virtual Work

Consider a cross-section with loads applied at point P as shown in Fig. 12.9. An expression for the external virtual work is given as the final integral of Eq. (12.1). For the beam, we will form the external virtual work as the load times the virtual displacement of the point of application of the load in the direction of the load. This displacement will be measured relative to the shear center.

The displacements v_P, w_P of the point (P) of application of the loading can be expressed as

$$\begin{aligned}
v_P &= v_S - \overline{z}_S\,\phi \\
w_P &= w_S + \overline{y}_S\,\phi
\end{aligned} \qquad (12.28)$$

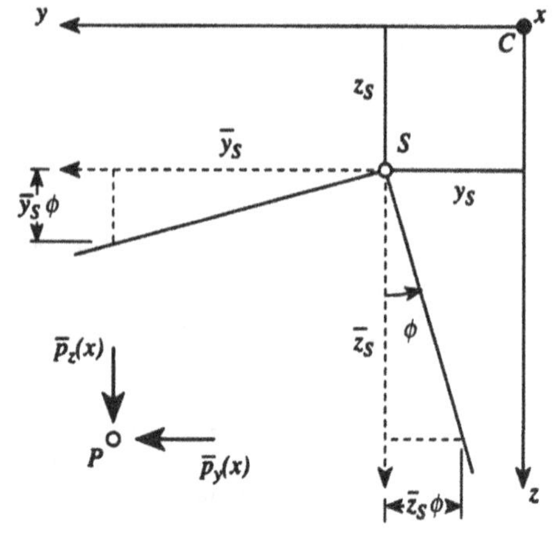

FIGURE 12.9
Applied loads.

where v_S, w_S are the displacements of the shear center, and $\overline{y}_S$, $\overline{z}_S$ are the coordinates of point P relative to the shear center. The external virtual work is

$$\delta W_e = \int_x (\overline{p}_x \delta u + \overline{p}_y \delta v_P + \overline{p}_z \delta w_P + \overline{m}_t \delta\phi)\, dx + [\,\overline{M}_z \delta v_S' - \overline{M}_y \delta w_S' - \overline{M}_\omega \delta\phi' + \overline{M}_t \delta\phi]\big|_a^b$$
$$+ [\overline{N}\,\delta u + \overline{V}_y\,\delta v_P + \overline{V}_z\,\delta w_P]\big|_a^b. \tag{12.29}$$

This expression contains the virtual work due to the distributed forces $\overline{p}_x$, $\overline{p}_y$, and $\overline{p}_z$, as well as the distributed moment $\overline{m}_t$. If concentrated applied forces are present, the appropriate integrals would be replaced by summations. Terms for concentrated applied forces on the boundaries at a and b are also given in Eq. (12.29). In terms of virtual displacements of the shear center, Eq. (12.29) becomes

$$\begin{aligned}\delta W_e = &\int_x [\overline{p}_x\, \delta u + \overline{p}_y\, \delta(v_S - \overline{z}_S\phi) + \overline{p}_z\, \delta(w_S + \overline{y}_S\phi) + \overline{m}_t\, \delta\phi]\, dx \\ &+ [\overline{M}_z\, \delta v_S' - \overline{M}_y\, \delta w_S' - \overline{M}_\omega\, \delta\phi' + \overline{N}\,\delta u + \overline{V}_y\,\delta v_S + \overline{V}_z\,\delta w_S \\ &+ (\overline{M}_t - \overline{V}_y\overline{z}_S + \overline{V}_z\overline{y}_S)\,\delta\phi]\big|_a^b.\end{aligned} \tag{12.30}$$

12.1.6 The Complete Virtual Work

From Eqs. (12.26) and (12.30), with subscript S ignored, the total virtual work appears as

$$\begin{aligned}-\delta W &= -\delta(W_i + W_e) = 0 \\ &= \int_x [(E A u')\, \delta u' + (E I_{zz} w'')\, \delta w'' + (E I_{yy} v'')\, \delta v'' + (E I_{\omega\omega}\phi'')\, \delta\phi'' + (G J_t \phi')\, \delta\phi']\, dx \\ &\quad - \int_x [\overline{p}_x\, \delta u + \overline{p}_y\, \delta v + \overline{p}_z\, \delta w + \underbrace{(\overline{m}_t - \overline{p}_y\overline{z}_S + \overline{p}_z\overline{y}_S)}_{\overline{m}_\phi}\, \delta\phi]\, dx \\ &\quad - [\overline{M}_z\, \delta v' - \overline{M}_y\, \delta w' - \overline{M}_\omega\, \delta\phi' + \overline{N}\,\delta u + \overline{V}_y\,\delta v + \overline{V}_z\,\delta w \\ &\quad + \underbrace{(\overline{M}_t - \overline{V}_y\overline{z}_S + \overline{V}_z\overline{y}_S)}_{\overline{M}_\phi}\,\delta\phi]\big|_a^b = 0.\end{aligned} \tag{12.31}$$

In matrix notation, this expression takes the form

$$\int_x \delta \mathbf{u}^T \left\{ \underbrace{\begin{bmatrix} {}_xd EA\, d_x & \vdots & \vdots & \vdots \\ \cdots & \cdots & \cdots & \cdots \\ & \vdots\ {}_xd^2 EI_{yy}\, d_x^2 & \vdots & \vdots \\ \cdots & \cdots & \cdots & \cdots \\ & \vdots & \vdots\ {}_xd^2 EI_{zz}\, d_x^2 & \vdots \\ \cdots & \cdots & \cdots & \cdots \\ & \vdots & \vdots & \vdots\ {}_xd^2 EI_{\omega\omega}\, d_x^2 \\ & \vdots & \vdots & \vdots\ + {}_xd GJ_t\, d_x \end{bmatrix}}_{\mathbf{k}^D} \underbrace{\begin{bmatrix} u \\ \cdots \\ v \\ \cdots \\ w \\ \cdots \\ \phi \end{bmatrix}}_{\mathbf{u}} - \underbrace{\begin{bmatrix} \overline{p}_x \\ \cdots \\ \overline{p}_y \\ \cdots \\ \overline{p}_z \\ \cdots \\ \overline{m}_\phi \end{bmatrix}}_{\overline{\mathbf{p}}} \right\} dx = 0 \tag{12.32}$$

where the boundary terms have not been included. The symbol ${}_xd$ used in the matrix $\mathbf{k}^D$ indicates the application of d/dx to the preceding variable $\delta \mathbf{u}^T$. As explained in Chapter 4, the differential operator matrix $\mathbf{k}^D$ forms the basis of the element stiffness matrix for the bar.

12.1.7 Governing Local Equations

The differential equation form of the governing equations can be obtained from Eq. (12.32) by utilizing integration by parts. Recall that integration by parts is the one-dimensional equivalent of the divergence theorem. The equations provided by the principle of virtual work are the equilibrium relations. We find

$$\begin{aligned} (EA\,u')' + \overline{p}_x &= 0 \\ (EI_{yy}\,v'')'' - \overline{p}_y &= 0 \\ (EI_{zz}\,w'')'' - \overline{p}_z &= 0 \\ (EI_{\omega\omega}\,\phi'')'' - (GJ_t\,\phi')' - \overline{m}_\phi &= 0 \end{aligned} \tag{12.33}$$

with the boundary conditions

$$\left.\begin{aligned} EI_{yy}v'' - \overline{M}_z &= 0 \\ EI_{zz}w'' + \overline{M}_y &= 0 \\ EI_{\omega\omega}\phi'' + \overline{M}_\omega &= 0 \end{aligned}\right\} \quad \text{at} \quad x = a \text{ and } x = b$$

and

$$\left.\begin{aligned} EA\,u' - \overline{N} &= 0 \\ (EI_{yy}\,v'')' + \overline{V}_y &= 0 \\ (EI_{zz}\,w'')' + \overline{V}_z &= 0 \\ (EI_{\omega\omega}\,\phi'')' - GJ_t\,\phi' + \overline{M}_\phi &= 0 \end{aligned}\right\} \quad \text{at} \quad x = a \text{ and } x = b$$

The first order local governing equations can be obtained from the relationships developed in this section following the procedures outlined in Chapter 1 or 2.

12.1.8 Element and System Calculations

Stiffness matrices can be derived using the procedures of Chapters 4 and 5. If the (local) governing differential equations are to be used, they are integrated, perhaps in transfer matrix form, and then transformed into element stiffness matrices. The element stiffness matrix can also be obtained from the principle of virtual work. For a system, local to global coordinate transformations are introduced, global stiffness matrices assembled, and a solution found with the displacement method.

12.2 Cross-Sectional Properties and Stress Distributions

12.2.1 Analytical Expressions

Section 12.1 contains the equations for finding the displacements and forces along bars. In the theory of bars, we can uncouple the determination of the response (displacements and forces) along the x axis from the calculation of stresses on the y, z cross-section. In this section, we will treat the problem of finding these cross-sectional stresses. A more comprehensive treatment, especially for computational methods for arbitrarily shaped cross sections, is provided in Pilkey (2002).

Stresses in Straight Bars of Solid Cross-Section

Normal Stress

In calculating the normal stress in a bar with a cross-section that is solid, i.e., not formed with thin elements, the effects of warping are usually not taken into consideration because they are small. If warping effects are ignored in Eq. (12.4), the axial displacement of an arbitrary point (x, y, z) is given by

$$u_x = u(x) - y\,\theta_z(x) + z\,\theta_y(x) \tag{12.34}$$

where $u(x)$ is the displacement of the origin of the coordinate system, and $\theta_y(x)$ and $\theta_z(x)$ are, respectively, the rotations of the cross-section about the y and z axes. The corresponding strain [Eq. (12.7a)] and stress are then

$$\begin{aligned} \epsilon_x &= \frac{du_x}{dx} = \frac{du}{dx} - y\frac{d\theta_z}{dx} + z\frac{d\theta_y}{dx} \\ \sigma_x &= E(\epsilon_x - \alpha\Delta T) = E\left(\frac{du}{dx} - y\frac{d\theta_z}{dx} + z\frac{d\theta_y}{dx} - \alpha\Delta T\right) \end{aligned} \tag{12.35}$$

where the Poisson's constant ν terms in the stress-strain relationships are ignored and thermal effects are included. This stress distribution gives rise to the following stress resultants

$$\begin{aligned} N &= \int_A \sigma_x\, dA = \int_A E\left(\frac{du}{dx} - y\frac{d\theta_z}{dx} + z\frac{d\theta_y}{dx} - \alpha\Delta T\right) dA \\ M_y &= \int_A \sigma_x z\, dA = \int_A E\left(\frac{du}{dx} - y\frac{d\theta_z}{dx} + z\frac{d\theta_y}{dx} - \alpha\Delta T\right) z\, dA \\ M_z &= -\int_A \sigma_x y\, dA = -\int_A E\left(\frac{du}{dx} - y\frac{d\theta_z}{dx} + z\frac{d\theta_y}{dx} - \alpha\Delta T\right) y\, dA \end{aligned} \tag{12.36}$$

With the definitions

$$\widehat{N} = N + N_T, \qquad N_T = \int_A E\alpha\Delta T\, dA$$
$$\widehat{M}_y = M_y + M_{Ty}, \qquad M_{Ty} = \int_A E\alpha\Delta T z\, dA \tag{12.37}$$
$$\widehat{M}_z = M_z + M_{Tz}, \qquad M_{Tz} = -\int_A E\alpha\Delta T y\, dA$$

and y, z centroidal axes so that the expressions $\int_A z\,dA = \int_A y\,dA = 0$ hold, Eq. (12.36) becomes

$$\frac{\widehat{N}}{E} = A\frac{du}{dx}$$
$$\frac{\widehat{M}_y}{E} = -I_{yz}\frac{d\theta_z}{dx} + I_{zz}\frac{d\theta_y}{dx} \tag{12.38}$$
$$\frac{\widehat{M}_z}{E} = I_{yy}\frac{d\theta_z}{dx} - I_{yz}\frac{d\theta_y}{dx}$$

The normal stress is found by solving Eqs. (12.38) for $du/dx, d\theta_z/dx, d\theta_y/dx$ and placing the resulting expressions into Eq. (12.35). This leads to

$$\sigma_x = \frac{\widehat{N}}{A} + \frac{\widehat{M}_y I_{yy} + \widehat{M}_z I_{yz}}{I_{yy}I_{zz} - I_{yz}^2} z - \frac{\widehat{M}_y I_{yz} + \widehat{M}_z I_{zz}}{I_{yy}I_{zz} - I_{yz}^2} y - E\alpha\Delta T \tag{12.39}$$

Shear Stress Due to Shear Forces

To derive the shear stress corresponding to the transverse shear force, consider the equilibrium of a slice (cross-sectional area A_0) of an element of a bar as shown in Fig. 12.10. Although in Fig. 12.10, the slice is taken to be below the location at which τ is desired, the slice could just as well be taken as being above this location. If $\tau = \tau_{xz} = \tau_{zx}$ is assumed to

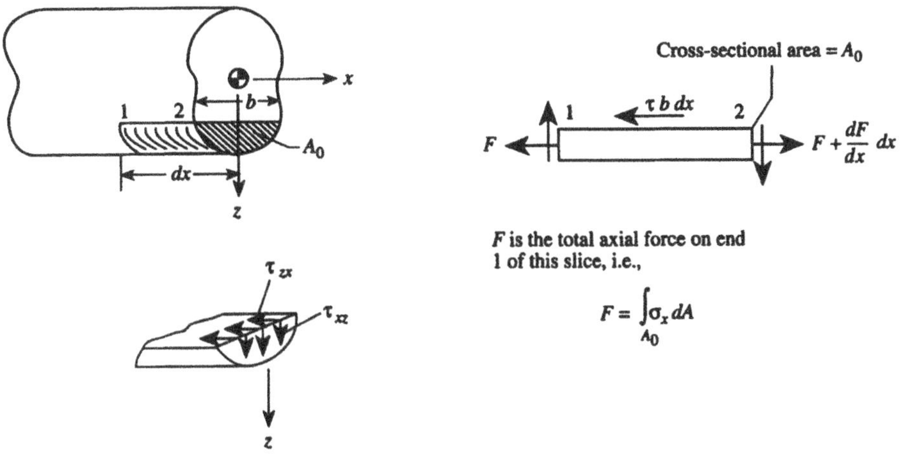

FIGURE 12.10
Calculation of shear stress.

be constant over the width b, the axial equilibrium requirements are $\tau\, b\, dx = dF$ or

$$\tau = \frac{1}{b}\frac{dF}{dx} = \frac{1}{b}\int_{A_0} \frac{d\sigma_x}{dx}\, dA \tag{12.40}$$

where $F = \int_{A_0} \sigma_x\, dA$. Substitution of σ_x from Eq. (12.39) into Eq. (12.40) yields

$$\tau = \frac{1}{b}\left[\frac{A_0}{A}\widehat{N}' + \frac{\widehat{M}'_y I_{yy} + \widehat{M}'_z I_{yz}}{I_{yy}I_{zz} - I_{yz}^2} I_z^* - \frac{\widehat{M}'_y I_{yz} + \widehat{M}'_z I_{zz}}{I_{yy}I_{zz} - I_{yz}^2} I_y^* - \int_{A_0} E(\alpha\Delta T)'\, dA\right] \tag{12.41}$$

where $()' = d/dx$, $A_0 = \int_{A_0} dA$, $I_z^* = \int_{A_0} z\, dA$, $I_y^* = \int_{A_0} y\, dA$.

The conditions of equilibrium relating the bending moments to the shear forces can be expressed as [Chapter 1, Eq. (1.113)]

$$\begin{aligned} \widehat{M}'_y &= V_z + V_{Tz} = \widehat{V}_z, \quad \text{with} \quad V_{Tz} = \frac{dM_{Ty}}{dx} \\ \widehat{M}'_z &= -V_y - V_{Ty} = -\widehat{V}_y, \quad \text{with} \quad V_{Ty} = \frac{dM_{Tz}}{dx} \end{aligned} \tag{12.42}$$

Equation (12.41) then can be written as

$$\tau = \frac{1}{b}\left[\frac{A_0}{A}\widehat{N}' + \frac{\widehat{V}_z I_{yy} - \widehat{V}_y I_{yz}}{I_{yy}I_{zz} - I_{yz}^2} I_z^* - \frac{\widehat{V}_z I_{yz} - \widehat{V}_y I_{zz}}{I_{yy}I_{zz} - I_{yz}^2} I_y^* - \int_{A_0} E(\alpha\Delta T)'\, dA\right] \tag{12.43}$$

or

$$\tau = \frac{1}{b}\left[\frac{A_0}{A}\widehat{N}' + \frac{I_y^* I_{zz} - I_z^* I_{yz}}{I_{yy}I_{zz} - I_{yz}^2}\widehat{V}_y + \frac{I_z^* I_{yy} - I_y^* I_{yz}}{I_{yy}I_{zz} - I_{yz}^2}\widehat{V}_z - \int_{A_0} E(\alpha\Delta T)'\, dA\right]$$

These expressions should be regarded as reasonable approximations, since the assumption that τ is constant over the width is often not valid. More accurate shear stresses can be calculated using a finite element implementation of theory of elasticity, as discussed later.

Shear Stress Due to Torsional Moment

For solid cross-sections, the shear stresses due to torsion are those resulting from M_{tP} (Eq. 12.9). These shear stresses are given in Chapter 1, Eq. (1.143). In calculating these stresses, the warping of the cross-section must be considered. The longitudinal displacement in the x direction due to warping is [Eq. (12.5c)]

$$u_x = \omega(y, z)\,\psi \tag{12.44a}$$

From Wagner's hypothesis, $\psi = -\phi'$. Also, if the displacements due to bending are not considered [Eq. (12.6)],

$$u_y = -\phi\, z, \qquad u_z = \phi\, y \tag{12.44b}$$

Then

$$\tau_{xy} = G\gamma_{xy} = G\left(\frac{\partial u_x}{\partial y} + \frac{\partial u_y}{\partial x}\right) = -G\phi'\left(\frac{\partial\omega}{\partial y} + z\right) \tag{12.45a}$$

$$\tau_{xz} = G\gamma_{xz} = G\left(\frac{\partial u_x}{\partial z} + \frac{\partial u_z}{\partial x}\right) = -G\phi'\left(\frac{\partial\omega}{\partial z} - y\right) \tag{12.45b}$$

which are in agreement with Chapter 1, Eqs. (1.142) and (1.143).

Although formulas for the warping function ω are available [Pilkey, 1994] for a few cross-sectional shapes, ω can be computed for arbitrary shapes as discussed in Section 12.2.2.

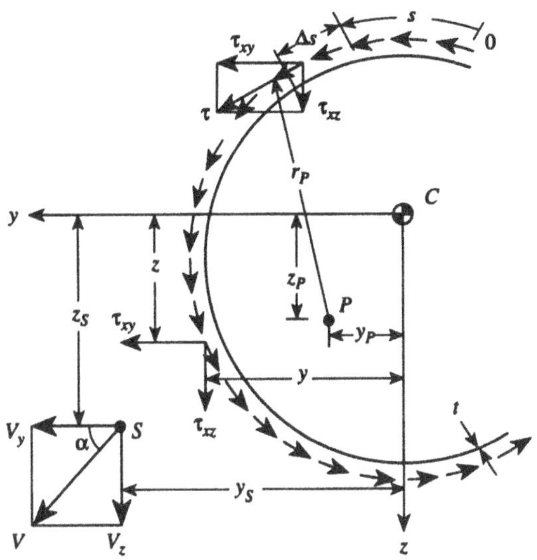

FIGURE 12.11
Thin-walled cross-section. The centroid and shear center are designated by C and S, respectively.

For circular cross-sections, where no warping occurs, the torsional constant J of Eq. (12.24) reduces to $J = \int_A (y^2 + z^2)\, dA = \int_A r^2\, dA$ and the tangential shear stress is given by $\tau = M_{tP}\, r/J$.

Stresses in Straight Thin-Walled Open Section Beams

Normal Stress

The formula for the normal stress in a thin-walled beam is derived in a manner similar to that employed for a solid bar except that the out-of-plane deformation due to warping torsion should be considered. The axial displacement of a point on the middle line of the wall profile (Fig. 12.11) is given by Eq. (12.4)

$$u_x = u - z\frac{dw}{dx} - y\frac{dv}{dx} - \omega\frac{d\phi}{dx}$$

where Bernoulli's (Eq. 12.12) and Wagner's (Eq. 12.13) hypotheses are invoked, and the normal stress is then

$$\sigma_x = E\left[\frac{du}{dx} - z\frac{d^2w}{dx^2} - y\frac{d^2v}{dx^2} - \omega\frac{d^2\phi}{dx^2} - \alpha\Delta T\right] \tag{12.46}$$

where u now is the axial displacement of the origin of a coordinate s which begins at one end (a free edge) of the middle line of this open cross-section (Fig. 12.11). The stress resultants of Eq. (12.38) become

$$\begin{aligned}\frac{\widehat{N}}{E} &= Au' - I_\omega\,\phi'' \\ \frac{\widehat{M}_y}{E} &= -I_{zz}\,w'' - I_{yz}\,v'' - I_{\omega z}\,\phi'' \\ \frac{\widehat{M}_z}{E} &= I_{yz}\,w'' + I_{yy}\,v'' - I_{\omega y}\,\phi''\end{aligned} \tag{12.47}$$

Equation (12.47) can be solved for u', v'', and w'' to obtain

$$\begin{aligned} u' &= \frac{\widehat{N}}{EA} + \frac{I_\omega}{A}\phi'' \\ v'' &= \frac{\widehat{M}_y I_{yz} + \widehat{M}_z I_{zz}}{E\left(I_{yy}I_{zz} - I_{yz}^2\right)} - \frac{I_{yz}I_{\omega z} - I_{zz}I_{\omega y}}{\left(I_{yy}I_{zz} - I_{yz}^2\right)}\phi'' \\ w'' &= -\frac{\widehat{M}_y I_{yy} + \widehat{M}_z I_{yz}}{E\left(I_{yy}I_{zz} - I_{yz}^2\right)} + \frac{I_{yy}I_{\omega z} - I_{yz}I_{\omega y}}{I_{yy}I_{zz} - I_{yz}^2}\phi'' \end{aligned} \tag{12.48}$$

However, as will be shown in the following section, the coefficients of ϕ'' in the second and third equations of Eq. (12.48) are identically zero. This is the result of $I_{\omega y} = \int_A \omega y\, dA$ and $I_{\omega z} = \int_A \omega z\, dA$ being equal to zero, which occurs if the shear center, which is defined on the next page, and the origin of the coordinate system coincide. Thus, substitution of Eq. (12.48) into Eq. (12.46) gives

$$\sigma_x = \left[\frac{\widehat{N}}{A} - \frac{\widehat{M}_y I_{yz} + \widehat{M}_z I_{zz}}{I_{yy}I_{zz} - I_{yz}^2}y + \frac{\widehat{M}_y I_{yy} + \widehat{M}_z I_{yz}}{I_{yy}I_{zz} - I_{yz}^2}z + E\left(\frac{1}{A}I_\omega - \omega\right)\phi'' - E\alpha\Delta T\right] \tag{12.49}$$

In Eq. (12.9), a bimoment is defined by $M_\omega = \int_A \sigma_x \omega\, dA$. With σ_x of Eq. (12.49)

$$\begin{aligned} M_\omega &= \int_A \left[\frac{\widehat{N}}{A} - \frac{\widehat{M}_y I_{yz} + \widehat{M}_z I_{zz}}{I_{yy}I_{zz} - I_{yz}^2}y + \frac{\widehat{M}_y I_{yy} + \widehat{M}_z I_{yz}}{I_{yy}I_{zz} - I_{yz}^2}z + E\left(\frac{1}{A}I_\omega - \omega\right)\phi'' - E\alpha\Delta T\right]\omega\, dA \\ &= \frac{\widehat{N}}{A}I_\omega + E\left(\frac{1}{A}I_\omega^2 - I_{\omega\omega}\right)\phi'' - E\alpha\, M_{\omega T} \end{aligned} \tag{12.50}$$

where $M_{\omega T} = \int_A \Delta T \omega\, dA$ and $I_{\omega\omega} = \Gamma$ is the warping constant. Note that $I_\omega = \int_A \omega\, dA$ involves the integration of ω along the coordinate s, which follows along the middle line of the wall profile as shown in Fig. 12.11. The magnitude of this integral depends on the location of the origin of s [Goodier, 1938]. There exist certain points on the contour of the wall for which, when they are used for the origin of s, I_ω vanishes. Let the origin of s be chosen such that $I_\omega = 0$ (corresponding to Eq. 12.18b). Also, let $\widehat{M}_\omega = M_\omega + E\alpha M_{\omega T}$. Then Eq. (12.50) reduces to

$$\widehat{M}_\omega = -E\, I_{\omega\omega}\, \phi'' \tag{12.51}$$

Finally, an expression for the normal stress in a thin-walled beam is obtained by replacing ϕ'' in Eq. (12.49) by Eq. (12.51)

$$\sigma_x = \left[\frac{\widehat{N}}{A} - \frac{\widehat{M}_y I_{yz} + \widehat{M}_z I_{zz}}{I_{yy}I_{zz} - I_{yz}^2}y + \frac{\widehat{M}_y I_{yy} + \widehat{M}_z I_{yz}}{I_{yy}I_{zz} - I_{yz}^2}z + \frac{\widehat{M}_\omega}{I_{\omega\omega}}\omega - E\alpha\Delta T\right] \tag{12.52}$$

Note that the part of the normal stress due to the warping of the cross-section is

$$\sigma_\omega = \frac{\widehat{M}_\omega}{I_{\omega\omega}}\omega = \frac{\widehat{M}_\omega}{\Gamma}\omega \tag{12.53}$$

Shear Stress

The expression for the shear stress is derived by substituting Eq. (12.52) into Eq. (12.40) in the form

$$\tau = \frac{1}{t}\int_{A_0} \frac{d\sigma_x}{dx}\, dA = \frac{1}{t}\int_0^s \frac{d\sigma_x}{dx} t\, ds$$
$$= \frac{A_0 \widehat{N}'}{At} + \frac{I_{yy}I_z^* - I_{yz}I_y^*}{t\left(I_{yy}I_{zz} - I_{yz}^2\right)}\widehat{M}_y' - \frac{I_{zz}I_y^* - I_{yz}I_z^*}{t\left(I_{yy}I_{zz} - I_{yz}^2\right)}\widehat{M}_z' + \frac{I_\omega^*}{tI_{\omega\omega}} M_{tS} - \frac{1}{t}\int_0^s E(\alpha\Delta T)' t\, ds \quad (12.54)$$

where t is the wall thickness, A_0 is the area between the s coordinate origin 0, a free edge, and s, the point of interest,

$$I_y^* = \int_{A_0} y\, dA, \qquad I_z^* = \int_{A_0} z\, dA, \qquad I_\omega^* = \int_{A_0} \omega\, dA = \int_0^s \omega t\, ds$$

and $M_{tS} = d\widehat{M}_\omega/dx$ is called the warping torque, or the secondary torsional moment (Eq. 12.9). In Eq. (12.54), the shear stress due to warping is

$$\tau_\omega = \frac{I_\omega^*}{tI_{\omega\omega}} M_{tS} \quad \text{or} \quad q = \tau_\omega t = \frac{I_\omega^*}{I_{\omega\omega}} M_{tS} \quad (12.55)$$

where q is called the *shear flow*.

An alternative form is provided by the substitution (Eq. 12.42)

$$\widehat{M}_y' = \widehat{V}_z, \qquad \widehat{M}_z' = -\widehat{V}_y$$

The shear stress in a thin-walled open section contains two distinct modes. The first mode of shear stress, which is due to the bending, non-uniform axial deformation (restrained warping), and the thermal gradient, is given in Eq. (12.54). The second mode is due to a pure twisting of the beam during which the cross-sections are free to warp. These stresses, which can also be expressed as shear flows, are given in Eqs. (12.45). Summation of the stresses of Eqs. (12.54) and (12.45) gives the total shear stresses on the cross-section.

Shear Center

The shear center S of a cross-section is the y, z location through which the plane of the resultant of the applied loadings must pass so that no twisting moment is developed. Hence, by definition, the twisting moment due to the shear force and the shear stress about an arbitrary point P must vanish, i.e., (Fig. 12.11)

$$\int_0^{s_L} \tau\, t\, r_p\, ds + V_z(y_S - y_P) - V_y(z_S - z_P) = 0 \quad (12.56)$$

where τ is the shear stress due to the shear forces V_y and V_z acting at the shear center S, and s_L is the total length of the wall profile.

Now insert τ from Eq. (12.54) (with the axial, thermal, and constrained warping effects neglected) into Eq. (12.56) and integrate. This leads to

$$\left(y_S - y_P + \frac{I_{yy}I_{\omega z} - I_{yz}I_{\omega y}}{I_{yy}I_{zz} - I_{yz}^2}\right)V_z - \left(z_S - z_P + \frac{I_{zz}I_{\omega y} - I_{yz}I_{\omega z}}{I_{yy}I_{zz} - I_{yz}^2}\right)V_y = 0$$

in which [Goodier, 1938] $I_{\omega z} = \int_A I_z^* \, d\omega = \int_A \omega z \, dA$, $I_{\omega y} = \int_A I_y^* \, d\omega = \int_A \omega \, y \, dA$, and $d\omega = r_p \, ds$. The quantities $I_{\omega y}$ and $I_{\omega z}$ have the same meaning here as they do in Eq. (12.16). The coordinates y_S and z_S are obtained by equating to zero the coefficients of V_y and V_z. Thus,

$$y_S = y_P - \frac{I_{yy} I_{\omega z} - I_{yz} I_{\omega y}}{I_{yy} I_{zz} - I_{yz}^2}, \qquad z_S = z_P - \frac{I_{zz} I_{\omega y} - I_{yz} I_{\omega z}}{I_{yy} I_{zz} - I_{yz}^2} \tag{12.57}$$

These give the coordinates of the shear center relative to an arbitrary point P, which often is the location of the origin of the coordinate system. Also, it follows that when P and S coincide

$$\frac{I_{yy} I_{\omega z} - I_{yz} I_{\omega y}}{I_{yy} I_{zz} - I_{yz}^2} = \frac{I_{zz} I_{\omega z} - I_{yz} I_{\omega y}}{I_{yy} I_{zz} - I_{yz}^2} = 0 \tag{12.58}$$

or $I_{\omega y} = I_{\omega z} = 0$. This conforms with Eq. (12.20a).

12.2.2 Finite Element Analysis for Cross Sections of Arbitrary Shape

Some of the equations for the stresses in Section 12.2.1 can be solved analytically only for cross-sections of regular shape such as a circle or a rectangle. The calculation of the normal stress in Eq. (12.39) is straightforward for many shapes. However, the shear stresses for bars of arbitrary cross-sectional shapes are exceedingly difficult to calculate accurately. For example, the shear stress of Eq. (12.43) is based on the assumption that the shear stress does not vary along the width b. More accurate analyses show that this is a questionable assumption and that theory of elasticity equations need to be employed. For the accurate calculations of the shear stress on cross-sections of arbitrary shape, computational techniques can be employed. The commonly used numerical procedures in the cross-sectional analyses are the finite element method and the boundary solutions, including the boundary element and the direct boundary integration methods. The boundary solution methods are introduced in Chapter 9, with example problems concerning beam cross-sectional analyses. Here, finite element analyses of the beam cross-sectional problems will be discussed briefly.

User-friendly postprocessors for calculating the distribution of cross-sectional stresses are now available for general purpose structural analysis software programs. For all but the simplest calculations, e.g., normal stresses for solid cross-sections, these postprocessors give more accurate distributions than the traditional analytical formulas. In particular, shear and thermal stresses should be calculated using these postprocessors. Furthermore these same postprocessors are useful as preprocessors to calculate cross-sectional constants that are needed as input to general purpose structural analysis programs. A typical finite element formulation will be introduced here. Detailed formulations and software are given in Pilkey (2002). See the website http://www.mae.virginia.edu/faculty/software/pilkey.php for the software.

Computation of the Warping Function ω and the Related Stresses

Recall the formulations for torsion of Chapter 1, Section 1.9, where the shear stresses τ_{xy} and τ_{xz} are expressed using both displacement and force formulations. The displacement formulation involves the warping function ω which satisfies [Chapter 1, Eq. (1.151)]

$$\frac{\partial^2 \omega}{\partial y^2} + \frac{\partial^2 \omega}{\partial z^2} = 0 \quad \text{or} \quad \nabla^2 \omega = 0 \tag{12.59}$$

with the boundary condition

$$\left(\frac{\partial \omega}{\partial z} - y \right) a_z + \left(\frac{\partial \omega}{\partial y} + z \right) a_y = 0 \tag{12.60}$$

where a_y and a_z are the direction cosines of the outer normal of the boundary. For the force formulation, the Prandtl stress function ψ (Eq. 1.155) is introduced which satisfies

$$\frac{\partial^2 \psi}{\partial y^2} + \frac{\partial^2 \psi}{\partial z^2} = -2G\phi' \quad \text{or} \quad \nabla^2 \psi = -2G\phi' \tag{12.61}$$

inside the cross-section and $\psi = C$, where C is a constant that is equal to 0 for a simply connected region on the boundary. Equation (12.59) is Laplace's equation, while Eq. (12.61) is a Poisson's equation. Since analytical solutions for these two problems can be obtained only for a few cross-sections of simple shapes, numerical methods should be used. We choose to apply the finite element method, based on the principle of virtual work, for the computation of ω and the stresses of interest.

It is shown in Chapter 7, Section 7.5.2 that the displacement formulation of Eqs. (12.59) and (12.60) in global form (the principle of virtual work) can be expressed as [Eq. (7.86)]

$$\int_A \delta\omega[({}_y\partial\partial_y + {}_z\partial\partial_z)\omega + ({}_y\partial\, z - {}_z\partial\, y)]\, dA = 0 \tag{12.62}$$

In terms of shape functions this appears as [Eq. (7.90)]

$$\int_A \delta\omega^T \left[\left(\frac{\partial \mathbf{N}^T}{\partial y}\frac{\partial \mathbf{N}}{\partial y} + \frac{\partial \mathbf{N}^T}{\partial z}\frac{\partial \mathbf{N}}{\partial z} \right) \omega + \left(z\frac{\partial \mathbf{N}^T}{\partial y} - y\frac{\partial \mathbf{N}^T}{\partial z} \right) \right] dA \tag{12.63}$$

so that the equations

$$\mathbf{k}^i \omega^i = \mathbf{p}^i \tag{12.64}$$

with [Eq. (7.91)]

$$\begin{aligned} \mathbf{k}^i &= \int_A \left(\frac{\partial \mathbf{N}^T}{\partial y}\frac{\partial \mathbf{N}}{\partial y} + \frac{\partial \mathbf{N}^T}{\partial z}\frac{\partial \mathbf{N}}{\partial z} \right) dA = \int_A (\mathbf{N}^T{}_y\partial\, \partial_y \mathbf{N} + \mathbf{N}^T{}_z\partial\, \partial_z \mathbf{N})\, dA \\ \mathbf{p}^i &= -\int_A \left(z\frac{\partial \mathbf{N}^T}{\partial y} - y\frac{\partial \mathbf{N}^T}{\partial z} \right) dA = -\int_A (z\,\mathbf{N}^T{}_y\partial - y\,\mathbf{N}^T{}_z\partial)\, dA \end{aligned} \tag{12.65}$$

need to be assembled for the cross section and solved for the distribution of ω over the cross section.

The assembled equations can be expressed as

$$\mathbf{K}\omega = \mathbf{P} \tag{12.66}$$

A three-node triangular element and a nine-node isoparametric element which we developed on the basis of Eq. (12.62) are discussed in the first edition of this book. Another presentation of this problem and its solution for ω is provided in Pilkey (2002). Given $\omega\,(y, z)$ the torsional constant J of Eq. (12.24) can be computed using

$$J = I_{yy} + I_{zz} - \boldsymbol{\omega}^T \mathbf{P} = I_{yy} + I_{zz} - \boldsymbol{\omega}^T \mathbf{K}\boldsymbol{\omega} \tag{12.67}$$

For a given torque M_t, the torsional stresses can be obtained from Eq. (12.45)

$$\begin{aligned} \tau_{xy} &= -\frac{M_t}{J}\left(\frac{\partial \omega}{\partial y} + z \right) \\ \tau_{xz} &= -\frac{M_t}{J}\left(\frac{\partial \omega}{\partial z} - y \right) \end{aligned} \tag{12.68}$$

In a computational solution, typically the stresses are calculated initially at the Gaussian integration points. These stresses can be multiplied by a smoothing matrix to find the element nodal stresses. The stresses from adjacent elements can be averaged at the element nodes.

If the cross-section is thin, it may be convenient to model it using line elements. Line elements can be obtained by collapsing the isoparametric elements into very thin (line) elements [Surana, 1979]. For example, in the limit as the thickness is reduced, the nine-node isoparametric element becomes a three-node line element.

Transverse Shear Loads Related Properties and Stresses

This topic is covered extensively in Pilkey (2002), where finite element computer programs are provided for all of the cross-sectional properties and stresses discussed here.

In Section 12.2, the expressions for the shear stresses due to transverse shear forces are obtained from the assumption that the plane of the cross-section remains plane after deformation. In reality, when shear stresses are present, the cross-section cannot remain plane and, as a consequence, some error is incurred in evaluating the shear stresses. Better solutions for the shear stresses can be obtained from the theory of elasticity. Exact solutions for the shear stresses are available only for a few beams with particular boundary conditions and loading, e.g., for a beam with the left end clamped and a transverse tip-load applied at the right end. It is also known that the distribution of the transverse shear stresses for a beam with uniformly distributed transverse applied loading is the same as for a tip-loaded beam [Mason and Herrmann, 1968]. Thus, it is reasonable to assume that the functional relationship between the internal shear force and shear strains for the tip-loaded beam applies to other cases and can form the basis for the analysis of the shear stresses on a cross-section of arbitrary shape. This leads to a warping function that can be computed using a finite element analysis. This distribution over the cross section of the warping function, which is not related to the warping function ω for torsion, can be computed and then used to calculate several shear-related cross-sectional properties, as well as the shear stresses due to the transverse shear loads. Brief discussions of two shear related properties follow.

Shear Center

As mentioned previously, the shear center is the point on the cross section through which the resultant shear force should pass if there is to be no torsion. For a cross section with two axes of symmetry, the shear center is at the centroid of the cross section. If there is one axis of symmetry, the shear center falls on this axis. If the cross section consists of two intersecting flanges, the shear center is at the intersection point.

The most common shear center equations, which are based on the theory of elasticity, are functions of a material constant, usually Poisson's ratio. Thus, in this case the shear center is not a purely geometric property of the cross section.

An alternative definition of a shear center by Trefftz (1936) does not involve the dependence on Poisson's ratio. These shear center equations are particularly suitable for thin-walled beams.

Shear Deformation Coefficients

Shear deformation coefficients have been employed for many years to improve beam deflections. In this book, the shear stiffness factor k_s was introduced in Eq. (1.109). A brief history of the various definitions of the shear coefficients is provided by Hutchinson (2001).

The approach in Pilkey (2002) is to define the coefficients by equating the strain energy for a beam based on the theory of elasticity to the strain energy for a beam represented by engineering beam theory, including shear deformation coefficients. These shear coefficients, $\alpha_{ij}, i, j = y, z$, where, for example, $\alpha_{z,z} = 1/k_s$ form a symmetric tensor. In general, the principal axes of this tensor differ from the principal axes of the tensor for the moments of inertia that appear in the beam bending equations. That is, the principal shear axes differ from the principal bending axes, with symmetric cross-sections being an exception. These shear deformation coefficients differ from the more traditional ones in that they vary with the thickness of a cross section. Whereas a serious computational problem called *shear locking* (see Chapter 13) can occur when the traditional shear deformation coefficients are employed, the use of thickness-dependent coefficients tends to counter the problem.

EXAMPLE 12.1 Finite Element Solution for Cross-Sectional Properties of a Symmetric Channel Section

A symmetric open-channel section is shown in Fig. 12.12a. Let $h = 18$ in., $b = 8$ in., $t = 1$ in., and $\nu = 1/3$. The reference coordinate system is at the midpoint of the web. The cross-sectional properties and stresses can be calculated using a finite element program that can treat a cross section of any shape (Pilkey, 2002).

The mesh chosen for calculations is shown in Fig. 12.12b. Some of the cross-sectional properties computed using the finite element program are shown in Fig. 12.13.

Many references provide formulas to approximate several of the cross-sectional properties. Typically the torsional constant J for open thin-walled cross sections is usually given by the approximate formula

$$J = \frac{St^3}{3} \tag{1}$$

where t is constant and S is the length of the median line of the cross section. For this example, $S = 34$ in. Then, $J = 11.33$ in^4. The finite element calculation gives $J = 11.29$ in^4.

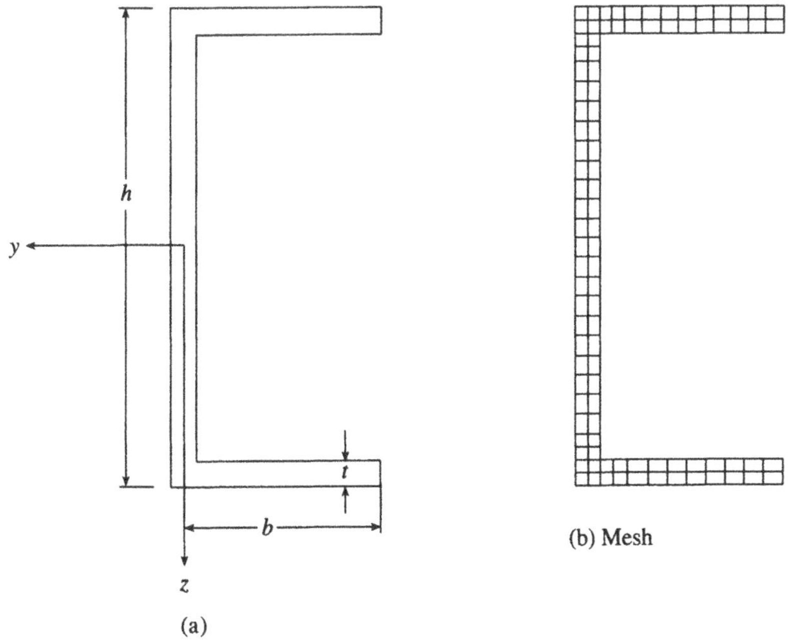

FIGURE 12.12
Channel section.

Y Centroid	1.87500
Z Centroid	0.00000
Y Shear Center	−2.86769
Z Shear Center	0.00000
Y Shear Center wrt Centroid	−4.74269
Z Shear Center wrt Centroid	0.00000
Y Shear Center wrt Centroid (Trefftz)	−4.74259
Z Shear Center wrt Centroid (Trefftz)	0.00000
Moment of Inertia I_{zz}	1781.83333
Moment of Inertia I_{yy}	342.83333
Product of Inertia I_{yz}	0.00000
Moment of Inertia I_{zzC}	1787.83333
Moment of Inertia I_{yyC}	223.30208
Product of Inertia I_{yzC}	0.00000
Polar Moment of Inertia	2011.13542
Y Radius of Gyration	7.25144
Z Radius of Gyration	2.56275
Principal Bending Angle (rad)	0.00000
Principal Bending Angle (deg)	0.00000
Principal Moment of Inertia (max)	1787.83333
Principal Moment of Inertia (min)	223.30208
Y Coordinate Extent	8.50000
Z Coordinate Extent	19.00000
Y Shear Coefficient	3.40789
Z Shear Coefficient	2.18337
YZ Shear Coefficient	0.00000
Torsional Constant	11.28862
Warping Constant wrt Shear Center	12763.15184
Warping Constant wrt Centroid	283214.57041

FIGURE 12.13
Some cross-sectional properties of a channel section.

The moment of inertia about the coordinate system shown in Fig. 12.12a is

$$I_{zz} = \frac{(6b + h)\, h^2 t}{12} \tag{2}$$

which gives $I_{zz} = 1782$ in^4. The finite element program computes 1788 in^4. The shear center is normally defined as

$$y_S = \frac{3b^2}{6b + h} \tag{3}$$

resulting in $y_S = 2.91$ in. The program finds $y_S = 2.87$ in. The warping constant of a channel section is usually listed in formula tables as

$$I_{\omega\omega} = \Gamma = \frac{b^3 h^2 t\,(3b + 2h)}{12\,(6b + h)} \tag{4}$$

leading to $\Gamma = 12{,}567$ in^6. The finite element program calculates what should be the more accurate result of $I_{\omega\omega} = \Gamma = 12{,}763$ in^6.

Normally the transverse shear calculations depend on material properties. Both the shear center and the shear deformation coefficients usually depend on Poisson's ratio. The computer results show that the Trefftz shear center, which does not depend on Poisson's ratio, is virtually the same as the shear center obtained from the transverse shear boundary value problem. ∎

References

Bornscheuer, F.W., 1952, "Systematische Darstellung des Biege-und Verdrehvorganges unter besonderer Berücksichtigung der Wölbkrafttorsion" *Der Stahlbau*, Vol. 21, p. 1.

Chang, P.Y., Thasanatorn, C. and Pilkey, W.D., 1975, Restrained warping stresses in thin-walled open sections, *J. Struct. Div.*, ASCE, Vol. 101, pp. 2467–2472.

Copper, C., 1993, "Thermoelastic Solutions for Beams," Ph.D. Thesis, University of Virginia.

Cowper, G.R., 1966, The shear coefficient in Timoshenko's beam theory, *J. Appl. Mech.*, Vol. 33, p. 2.

Goodier, J.N., 1938, On the problems of the beam and the plate in the theory of elasticity, *Trans. Royal Society of Canada*, Vol. 32, pp. 65–88.

Herrmann, L.R., 1965, Elastic torsional analysis of irregular shapes, *J. Eng. Mech. Div.*, ASCE, Vol. 91, pp. 11–19.

Hinton, E., Scott, F.C. and Ricketts, R.E., 1975, Local least squares smoothing for parabolic isoparametric elements, *Int. J. Numer. Methods Eng.*, Vol. 9, pp. 235–256.

Hutchinson, J.R., 2001, Shear coefficients for Timoshenko beam theory, *J. Appl. Mech.*, Vol. 33, pp. 335–340.

Liu, Y., Pilkey, W.D., Antes, H. and Rubenchik, V., 1993, Direct integration of the integral equations of the beam torsion problem, *Comp. & Struct.*, Vol. 48, pp. 647–652.

Mason, W.F. and Herrmann, L.R., 1968, Elastic shear analysis of general prismatic beams, *J. Eng. Mech. Div.*, ASCE, Vol. 94, pp. 965–983.

Pilkey, W.D., 1994, *Formulas for Stress, Strain, and Structural Matrices*, Wiley, NY.

Pilkey, W.D., 2002, *Analysis and Design of Elastic Beams, Computational Methods*, Wiley, NY.

Pilkey, W.D. and Liu, Y., 1993, Field theory: A two-dimensional case for not using finite or boundary elements, *Fin. Elem. in Anal. & Design*, Vol. 13, pp. 127–136.

Schramm, U., Kitis, L., Kang. W. and Pilkey, W.D., 1994, On the shear deformation coefficient in beam theory, *Fin. Elem. in Anal. & Design*, Vol. 16, pp. 141–162.

Surana, K.S., 1979, Isoparametric elements for cross-sectional properties and stress analysis of beams, *Int. J. Numer. Methods Eng.*, Vol. 14, pp. 475–497.

Timoshenko, S., and Goodier, J.N., 1951, *Theory of Elasticity*, McGraw-Hill, NY.

Trefftz, E., 1936, Uber den Schubmittelpunkt in einem durch eine Einzellast gebogenen Balken, *Z. Angew. Math. Mech.*, Vol. 15, pp. 220–225.

Wunderlich, W., 1977, Incremental formulation for geometrically nonlinear problems, in *Formulations and Computional Algorithms in Finite Element Analysis*, Bathe, K.J., Oden, J.T. and Wunderlich, W. (Eds.), MIT Press, Cambridge, MA.

Problems

Displacements and Forces

12.1 Prove that a "plane cross-section remains plane" when a straight beam is subjected only to bending moments at the end points. Assume that the shape of the cross-section remains unchanged during deformation.

12.2 Derive the governing differential equations of the beam shown in Fig. P12.2.

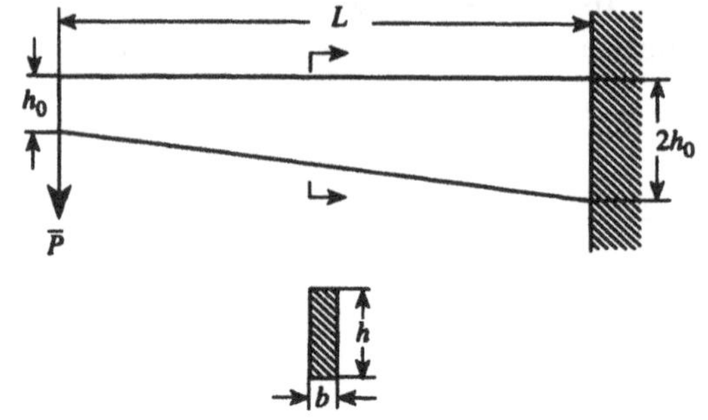

FIGURE P12.2

12.3 Compute the normal stress and shear stress at the fixed end of the beam in Fig. P12.2, assuming that $h_0 = 10$ in., $L = 120$ in., $b = 3$ in., and $\overline{P} = 20{,}000$ lb.

12.4 Show that the principal moments of inertia are defined by Eqs. (12.19b, c.)

12.5 Verify that the shear center coordinates are given by Eq. (12.20b).

12.6 Show that Eq. (12.21) represents the distance of the shear center from the $\widetilde{y}, \widetilde{z}$ axes.

12.7 Provide a detailed derivation of Eq. (12.22a).

12.8 Use the Hellinger-Reissner variational principle to obtain the governing differential equations of a straight beam.

Hint: See Chapter 2, Example 2.11.

12.9 Find a system of first-order governing differential equations for the extension, torsion, and bending of a bar.

Hint: Convert Eq. (12.33) to first order form.

12.10 Calculate the displacements of a simply-supported thin-walled beam of Fig. P12.10 at the middle point along the span. The cross-sections of the beam are assumed to be free to warp.

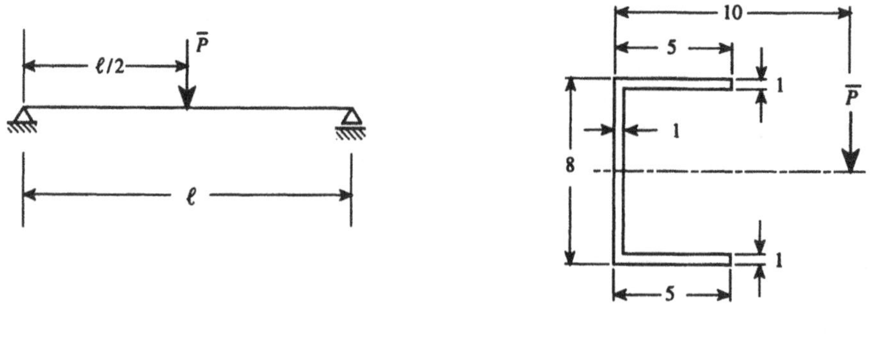

FIGURE P12.10

Cross-Sectional Properties and Stress Distributions

12.11 Calculate the normal and shear stress distributions for the beam shown in Fig. P12.11. Let $\overline{P}$ pass through the shear center. State why shear stresses found using a finite element solution are expected to be more accurate than those calculated using an analytical solution.

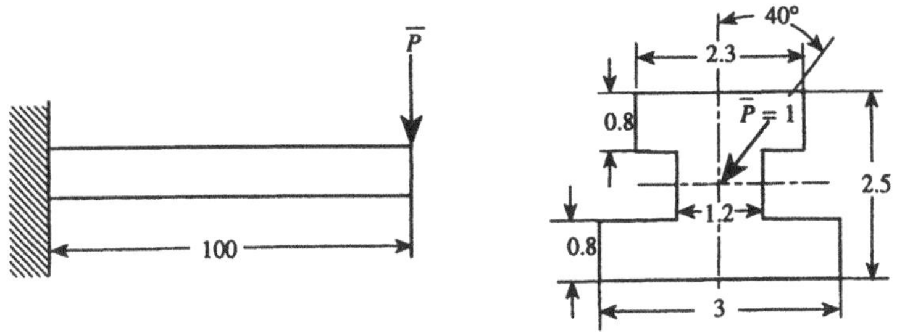

FIGURE P12.11

12.12 Figure P12.12 shows the values of the torsional constant for several configurations that are frequently given in handbooks. Use a computational solution to see how close you can come to these values.

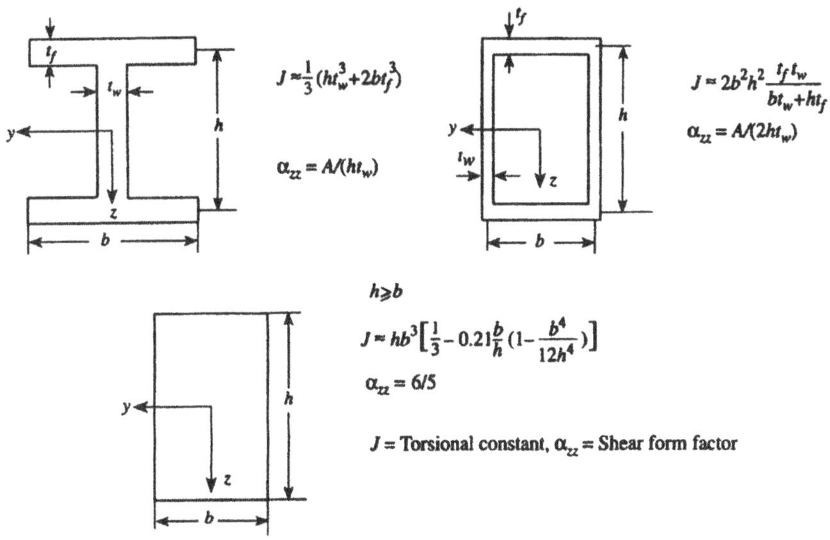

FIGURE P12.12

12.13 Verify the shear form factors of Fig. P12.12. Use a finite element solution.

12.14 Textbooks usually list the shear center for the configuration of Fig. P12.14 to be located at $y_S = 3b^2t_f/[ht_\omega + 3bt_f/2]$. Is this an accurate expression?

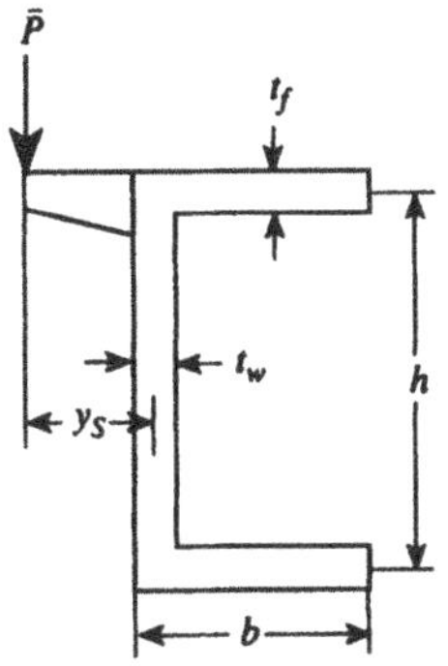

FIGURE P12.14

12.15 Calculate the locations of the shear centers of the cross-sections shown in Fig. P12.15.

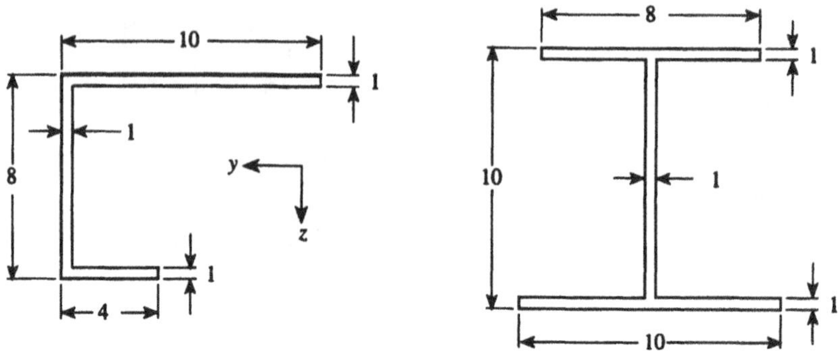

FIGURE P12.15

12.16 Find the peak normal and shear stresses for two 100 unit long cantilevered beams made of uniform beams of the cross-sections shown in Fig. P12.15. The beams are loaded with moments $\overline{M}_y = 200$ at the free ends.

13

Plates

Plates are flat structural elements with thicknesses much smaller than the other dimensions. Familiar examples of plates include flat roofs, doors, table tops, and manhole covers.

In analyzing a plate (Fig. 13.1), it is common to consider the plate to be divided into equal halves by a plane, the *midplane* or *middle surface*, parallel to the flat faces. The plate thickness t is measured normal to the middle surface. The fundamental equations of plate theory relate the displacements and forces of the middle surface to the applied loading. Plate equations will appear similar to beam equations, since beam theory can be regarded as a special case of plate theory.

Plates can be classified as being *thin* or *thick*. Most of the material presented here applies to thin plates. Roughly, for a plate to be considered as being thin, the ratio of the thickness to the shortest span length should be less than about 1/10. The plates treated here are made of materials that are homogeneous and isotropic. If material properties are the same at all locations, the material is said to be *homogeneous*, while material properties identical in all directions as viewed from a particular point are *isotropic*.

13.1 In-Plane Deformation (Stretching)

Although the primary subject of this chapter will be the bending of plates, we choose to begin this chapter with a summary of the equations developed in Chapters 1 and 2 for the in-plane deformation of a flat element (plate). This is the two-dimensional equivalent of the extension of a bar, a one-dimensional problem. If the applied loads are tensile, the plate is sometimes referred to as a membrane.

13.1.1 Cartesian Coordinate System

Kinematical Relationships

The two displacements for a thin element lying in the xy plane

$$\mathbf{u} = [u_x \quad u_y]^T \tag{13.1a}$$

and three strains

$$\epsilon = [\epsilon_x \quad \epsilon_y \quad \gamma_{xy}]^T \tag{13.1b}$$

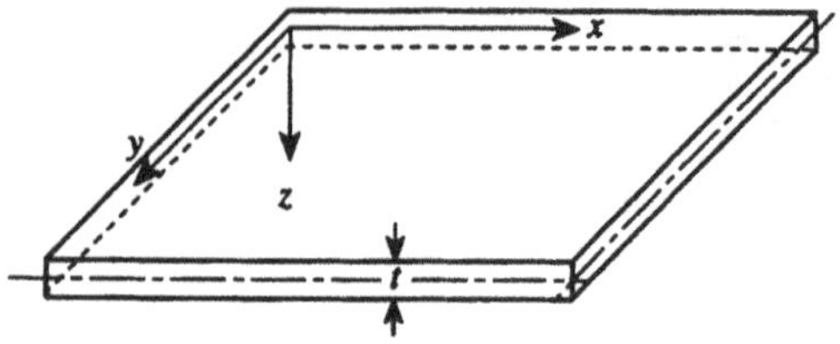

FIGURE 13.1
Coordinate system and middle surface of a plate.

are related by

$$\epsilon_x = \frac{\partial u_x}{\partial x} \qquad \epsilon_y = \frac{\partial u_y}{\partial y} \qquad \gamma_{xy} = \frac{\partial u_x}{\partial y} + \frac{\partial u_y}{\partial x}$$

or

$$\begin{bmatrix} \epsilon_x \\ \epsilon_y \\ \gamma_{xy} \end{bmatrix} = \begin{bmatrix} \partial_x & 0 \\ 0 & \partial_y \\ \partial_y & \partial_x \end{bmatrix} \begin{bmatrix} u_x \\ u_y \end{bmatrix} \tag{13.1c}$$

$$\boldsymbol{\epsilon} \quad = \mathbf{D} \text{ (or } \mathbf{D}_u) \quad \mathbf{u}$$

Material Law

For a thin structure, it is convenient to replace the stresses (σ_x, σ_y, and τ_{xy}) with *stress resultants* in force per unit length. For a thin element of thickness t, these are defined as

$$n_x = \int_{-t/2}^{t/2} \sigma_x \, dz \qquad n_y = \int_{-t/2}^{t/2} \sigma_y \, dz \qquad n_{xy} = \int_{-t/2}^{t/2} \tau_{xy} \, dz \tag{13.2}$$

These can be combined to form the vector

$$\mathbf{s} = [n_x \quad n_y \quad n_{xy}]^T$$

As explained in Chapter 1, Section 1.3.1, these stress resultants can be related to the strains $\epsilon = [\epsilon_x \ \epsilon_y \ \gamma_{xy}]^T$ based on plane stress or plane strain assumptions. In the case of plane stress, with the assumptions $\sigma_z = \tau_{zx} = \tau_{zy} = 0$, it is found that

$$\begin{bmatrix} n_x \\ n_y \\ n_{xy} \end{bmatrix} = D \begin{bmatrix} 1 & \nu & 0 \\ \nu & 1 & 0 \\ 0 & 0 & \frac{1-\nu}{2} \end{bmatrix} \begin{bmatrix} \epsilon_x \\ \epsilon_y \\ \gamma_{xy} \end{bmatrix}$$

$$\mathbf{s} \quad = \quad \mathbf{E} \quad \boldsymbol{\epsilon} \tag{13.3a}$$

where $D = Et/(1 - \nu^2)$, or

$$\begin{bmatrix} \epsilon_x \\ \epsilon_y \\ \gamma_{xy} \end{bmatrix} = \frac{1}{Et} \begin{bmatrix} 1 & -\nu & 0 \\ -\nu & 1 & 0 \\ 0 & 0 & 2(1+\nu) \end{bmatrix} \begin{bmatrix} n_x \\ n_y \\ n_{xy} \end{bmatrix} \tag{13.3b}$$

$$\boldsymbol{\epsilon} \quad = \quad \mathbf{E}^{-1} \quad \mathbf{s}$$

Conditions of Equilibrium

The equilibrium equations, which provide a relationship between the stress resultants and the body forces (force/area) $\overline{\mathbf{p}}_V = [\overline{p}_{V_x} \ \overline{p}_{V_y}]^T$ are

$$\begin{aligned} \frac{\partial n_x}{\partial x} + \frac{\partial n_{yx}}{\partial y} + \overline{p}_{Vx} &= 0 \\ \frac{\partial n_y}{\partial y} + \frac{\partial n_{xy}}{\partial x} + \overline{p}_{Vy} &= 0 \end{aligned} \qquad n_{xy} = n_{yx} \tag{13.4a}$$

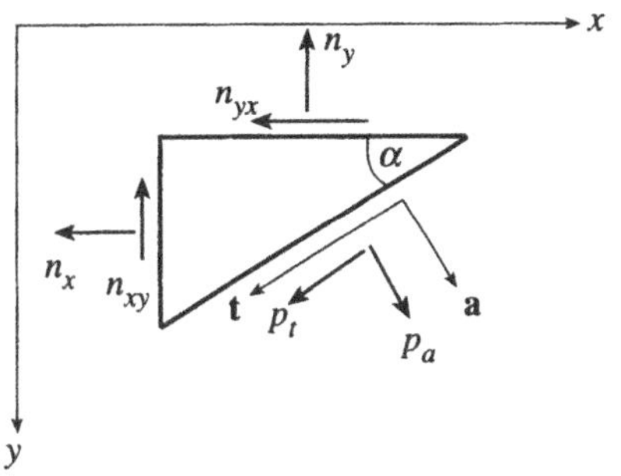

FIGURE 13.2
Surface forces.

or in matrix form,

$$\begin{bmatrix} \partial_x & 0 & \partial_y \\ 0 & \partial_y & \partial_x \end{bmatrix} \begin{bmatrix} n_x \\ n_y \\ n_{xy} \end{bmatrix} + \begin{bmatrix} \overline{p}_{Vx} \\ \overline{p}_{Vy} \end{bmatrix} = \mathbf{0}$$
$$\mathbf{D}^T \quad \mathbf{s} + \overline{\mathbf{p}}_V = \mathbf{0} \tag{13.4b}$$

Surface Forces and Boundary Conditions

The displacement boundary conditions on S_u are

$$\begin{aligned} u_x &= \overline{u}_x \\ u_y &= \overline{u}_y \end{aligned} \tag{13.5a}$$

where letters with bars over them represent prescribed or applied quantities.

The force boundary conditions occur on $S_p = S - S_u$. For a surface with normal **a** and tangent **t** (Fig. 13.2), with the surface forces p_a (normal) and p_t (tangent), the force boundary conditions are

$$\begin{aligned} p_a &= \overline{p}_a \\ p_t &= \overline{p}_t \end{aligned} \tag{13.5b}$$

The surface forces and stress resultants are related by

$$\begin{bmatrix} p_a \\ p_t \end{bmatrix} = \begin{bmatrix} \sin^2\alpha & \cos^2\alpha & 2\sin\alpha\cos\alpha \\ -\sin\alpha\cos\alpha & \sin\alpha\cos\alpha & -\cos^2\alpha + \sin^2\alpha \end{bmatrix} \begin{bmatrix} n_x \\ n_y \\ n_{xy} \end{bmatrix} \tag{13.6}$$

13.1.2 Circular Plates

A circular plate with in-plane loading is traditionally assumed to be in a state of plane stress. The displacements and strains, as well as stresses (stress resultants), should be expressed with polar coordinates. The relationships between the Cartesian and polar coordinate systems are (Fig. 13.3a)

$$\begin{aligned} x &= r\cos\phi \\ y &= r\sin\phi \end{aligned} \tag{13.7a}$$

It follows that

$$\begin{aligned} r^2 &= x^2 + y^2 \\ \phi &= \tan^{-1}\frac{y}{x} \end{aligned} \tag{13.7b}$$

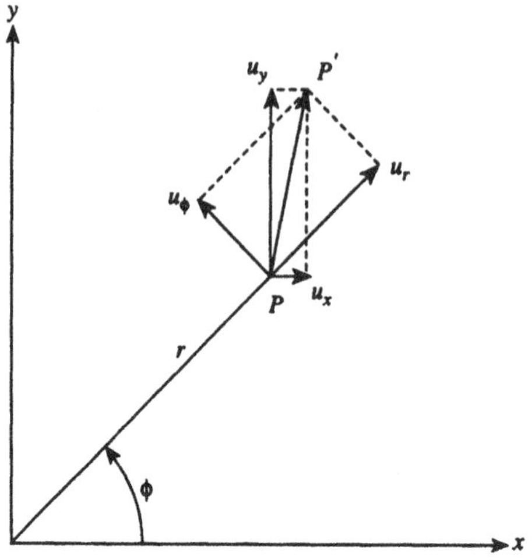

(a) Displacements in the Cartesian and polar coordinate systems

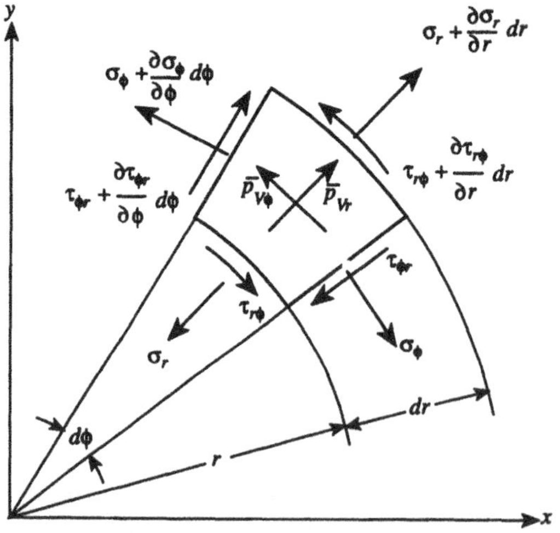

(b) Equilibrium of an element

FIGURE 13.3
Polar coordinate systems.

and

$$\frac{\partial r}{\partial x} = \cos\phi = \frac{x}{r} \qquad \frac{\partial r}{\partial y} = \sin\phi = \frac{y}{r}$$
$$\frac{\partial \phi}{\partial x} = -\frac{1}{r}\sin\phi = -\frac{y}{r^2} \qquad \frac{\partial \phi}{\partial y} = \frac{1}{r}\cos\phi = \frac{x}{r^2} \tag{13.7c}$$

These derivatives can be obtained by using the Jacobian (Problem 13.1) or directly from Eq. (13.7b). As an example of the latter case, observe from $r^2 = x^2 + y^2$ that

$$2r\frac{\partial r}{\partial x} = 2x \quad \text{so that} \quad \frac{\partial r}{\partial x} = \frac{x}{r} = \cos\phi$$

Kinematical Relationships

The displacements in a circular plate with in-plane deformation

$$\mathbf{u} = [u_r \quad u_\phi]^T \tag{13.8a}$$

and the three strains

$$\boldsymbol{\epsilon} = [\epsilon_r \quad \epsilon_\phi \quad \gamma_{r\phi}]^T \tag{13.8b}$$

are related by

$$\epsilon_r = \frac{\partial u_r}{\partial r} \qquad \epsilon_\phi = \frac{u_r}{r} + \frac{1}{r}\frac{\partial u_\phi}{\partial \phi} \qquad \gamma_{r\phi} = \frac{1}{r}\frac{\partial u_r}{\partial \phi} + \frac{\partial u_\phi}{\partial r} - \frac{1}{r}u_\phi$$

or

$$\begin{bmatrix} \epsilon_r \\ \epsilon_\phi \\ \gamma_{r\phi} \end{bmatrix} = \begin{bmatrix} \partial_r & 0 \\ 1/r & (1/r)\partial_\phi \\ (1/r)\partial_\phi & \partial_r - 1/r \end{bmatrix} \begin{bmatrix} u_r \\ u_\phi \end{bmatrix} \tag{13.8c}$$
$$\boldsymbol{\epsilon} \quad = \quad \mathbf{D}_u \quad \mathbf{u}$$

These relations can be obtained by transforming the displacements u_x, u_y of the previous section to polar coordinates. With respect to the Cartesian coordinates, the displacement components from P to P' of Fig. 13.3a are u_x and u_y, and with respect to the polar coordinates, the displacement components are u_r and u_ϕ. It can be observed in Fig. 13.3a that

$$\begin{aligned} u_x &= u_r \cos\phi - u_\phi \sin\phi \\ u_y &= u_r \sin\phi + u_\phi \cos\phi \end{aligned} \tag{13.9}$$

Substitution of the first of these relations into Eq. (13.1c) leads to

$$\begin{aligned} \epsilon_x &= \frac{\partial u_x}{\partial x} = \frac{\partial u_x}{\partial \phi}\frac{\partial \phi}{\partial x} + \frac{\partial u_x}{\partial r}\frac{\partial r}{\partial x} \\ &= \left(-\frac{\partial u_r}{\partial \phi}\cos\phi + u_r \sin\phi + \frac{\partial u_\phi}{\partial \phi}\sin\phi + u_\phi \cos\phi\right)\frac{1}{r}\sin\phi \\ &\quad + \left(\frac{\partial u_r}{\partial r}\cos\phi - \frac{\partial u_\phi}{\partial r}\sin\phi\right)\cos\phi \end{aligned}$$

When $\phi \to 0$, $\epsilon_x \to \epsilon_r$, so that

$$\epsilon_r = \lim_{\phi \to 0} \epsilon_x = \frac{\partial u_r}{\partial r}$$

The strains ϵ_ϕ and $\gamma_{r\phi}$ can be obtained in a similar fashion.

Material Law

For the circular plate of thickness t, the stress resultants $\mathbf{s} = [n_r \quad n_\phi \quad n_{r\phi}]^T$ are defined as

$$n_r = \int_{-t/2}^{t/2} \sigma_r \, dz \qquad n_\phi = \int_{-t/2}^{t/2} \sigma_\phi \, dz \qquad n_{r\phi} = \int_{-t/2}^{t/2} \tau_{r\phi} \, dz$$

The material relations of Eq. (13.3) remain valid. Thus, the relationships between these stress resultants and the strains $\boldsymbol{\epsilon} = [\epsilon_r \quad \epsilon_\phi \quad \gamma_{r\phi}]^T$ are

$$\begin{bmatrix} n_r \\ n_\phi \\ n_{r\phi} \end{bmatrix} = D \begin{bmatrix} 1 & \nu & 0 \\ \nu & 1 & 0 \\ 0 & 0 & \frac{1-\nu}{2} \end{bmatrix} \begin{bmatrix} \epsilon_r \\ \epsilon_\phi \\ \gamma_{r\phi} \end{bmatrix} \qquad D = \frac{Et}{1-\nu^2} \tag{13.10a}$$
$$\mathbf{s} \quad = \quad \mathbf{E} \quad \boldsymbol{\epsilon}$$

or

$$\begin{bmatrix} \epsilon_r \\ \epsilon_\phi \\ \gamma_{r\phi} \end{bmatrix} = \frac{1}{Et}\begin{bmatrix} 1 & -\nu & 0 \\ -\nu & 1 & 0 \\ 0 & 0 & 2(1+\nu) \end{bmatrix}\begin{bmatrix} n_r \\ n_\phi \\ n_{r\phi} \end{bmatrix} \tag{13.10b}$$

$$\boldsymbol{\epsilon} = \mathbf{E}^{-1} \quad \mathbf{s}$$

If thermal loading effects are to be included, Eqs. (13.10a) and (13.10b), with $\epsilon_0 = [\alpha\,\Delta T \;\; \alpha\,\Delta T \;\; 0]^T$, can be written as [Chapter 1, Eqs. (1.43) and (1.44)]

$$\begin{bmatrix} n_r \\ n_\phi \\ n_{r\phi} \end{bmatrix} = D\begin{bmatrix} 1 & \nu & 0 \\ \nu & 1 & 0 \\ 0 & 0 & \frac{1-\nu}{2} \end{bmatrix}\begin{bmatrix} \epsilon_r \\ \epsilon_\phi \\ \gamma_{r\phi} \end{bmatrix} - \begin{bmatrix} \alpha E \Delta T t/(1-\nu) \\ \alpha E \Delta T t/(1-\nu) \\ 0 \end{bmatrix} \tag{13.10c}$$

$$\mathbf{s} = \mathbf{E} \quad \boldsymbol{\epsilon} - \mathbf{E}\boldsymbol{\epsilon}^0$$

and

$$\begin{bmatrix} \epsilon_r \\ \epsilon_\phi \\ \gamma_{r\phi} \end{bmatrix} = \frac{1}{Et}\begin{bmatrix} 1 & -\nu & 0 \\ -\nu & 1 & 0 \\ 0 & 0 & 2(1+\nu) \end{bmatrix}\begin{bmatrix} n_r \\ n_\phi \\ n_{r\phi} \end{bmatrix} + \begin{bmatrix} \alpha\,\Delta T \\ \alpha\,\Delta T \\ 0 \end{bmatrix} \tag{13.10d}$$

$$\boldsymbol{\epsilon} = \mathbf{E}^{-1} \quad \mathbf{s} + \boldsymbol{\epsilon}^0$$

Conditions of Equilibrium

The equilibrium equations, which provide relationships between the stress resultants and the body forces (force/area $\overline{\mathbf{p}}_V = [\overline{p}_{Vr} \;\; \overline{p}_{V\phi}]^T$, are (Fig. 13.3b)

$$\begin{aligned} &\frac{\partial n_r}{\partial r} + \frac{1}{r}\frac{\partial n_{r\phi}}{\partial \phi} + \frac{(n_r - n_\phi)}{r} + \overline{p}_{Vr} = 0 \\ &\frac{1}{r}\frac{\partial n_\phi}{\partial \phi} + \frac{\partial n_{r\phi}}{\partial r} + \frac{2n_{r\phi}}{r} + \overline{p}_{V\phi} = 0 \end{aligned} \qquad n_{r\phi} = n_{\phi r} \tag{13.11a}$$

or in matrix form,

$$\begin{bmatrix} \partial_r + 1/r & -1/r & (1/r)\partial_\phi \\ 0 & (1/r)\partial_\phi & \partial_r + 2/r \end{bmatrix}\begin{bmatrix} n_r \\ n_\phi \\ n_{r\phi} \end{bmatrix} + \begin{bmatrix} \overline{p}_{Vr} \\ \overline{p}_{V\phi} \end{bmatrix} = 0 \tag{13.11b}$$

$$\mathbf{D}_s^T \quad \mathbf{s} + \overline{\mathbf{p}}_V = 0$$

Boundary Conditions

The displacement boundary conditions are

$$\begin{aligned} u_r &= \overline{u}_r \\ u_\phi &= \overline{u}_\phi \end{aligned} \qquad \text{on } S_u \tag{13.12}$$

and the force boundary conditions are

$$\begin{aligned} p_r &= \overline{p}_r \\ p_\phi &= \overline{p}_\phi \end{aligned} \qquad \text{on } S_p \tag{13.13}$$

where $\overline{p}_r$ and $\overline{p}_\phi$ are the surface forces (force/length) in the r and ϕ directions around the circumference of the disk.

First Order Governing Differential Equations

The first order governing differential equations for the in-plane loaded circular disk with axially symmetrical deformation can be obtained from the first relations of Eqs. (13.10b) and (13.11a). The variation of the displacements and internal forces are independent of ϕ for axially symmetric loading. Let $\overline{p}_r(r) = \overline{p}_{Vr}$ denote the applied radial forces (force/length^2) in the plane of the disk, including radial pressure, centrifugal, and thermal forces, all of which are assumed to be independent of ϕ. Because of the symmetry, $\gamma_{r\phi} = n_{r\phi} = 0$. Also, $\partial_\phi(\,) = 0$. From Eq. (13.8c), $\epsilon_\phi = u_r/r$, and from Eq. (13.10b), $\epsilon_\phi = (-\nu n_r + n_\phi)/Et$ or

$$n_\phi = \frac{Et}{r}u_r + \nu n_r$$

From Eqs. (13.8c) and (13.10b), $\epsilon_r = \partial_r u_r$ and $\epsilon_r = (n_r - \nu n_\phi)/Et$, or $\partial_r u_r = (n_r - \nu n_\phi)/Et = [n_r - \nu(\frac{Et}{r}u_r + \nu n_r)]/Et$. From this relation and the first condition of equilibrium of Eq. (13.11a), the first order governing equations are

$$\frac{du_r}{dr} = -\nu\frac{u_r}{r} + \frac{1-\nu^2}{Et}n_r \tag{13.14a}$$

$$\frac{dn_r}{dr} = \frac{Et}{r^2}u_r + \frac{\nu-1}{r}n_r - \overline{p}_r(r)$$

or

$$\begin{aligned}\frac{d}{dr}\begin{bmatrix} u_r \\ n_r \end{bmatrix} &= \begin{bmatrix} -\nu/r & (1-\nu^2)/Et \\ Et/r^2 & (\nu-1)/r \end{bmatrix}\begin{bmatrix} u_r \\ n_r \end{bmatrix} + \begin{bmatrix} 0 \\ -\overline{p}_r \end{bmatrix} \\ dz/dr &= \qquad\mathbf{A}\qquad\qquad \mathbf{z} \quad + \quad \overline{\mathbf{P}}\end{aligned} \tag{13.14b}$$

Note that $\mathbf{A}$ is not constant as $\mathbf{A} = \mathbf{A}(r)$. If inertia forces are included, change the ordinary derivatives $\frac{d}{dr}$ to partial derivatives $\frac{\partial}{\partial r}$ and replace $\overline{p}_r$ by $\overline{p}_r = -\rho\partial^2 u_r/\partial t^2$, where ρ is the mass per unit area.

In higher order form, for a disk with constant thickness and material properties,

$$\frac{d^2u_r}{dr^2} + \frac{1}{r}\frac{du_r}{dr} - \frac{1}{r^2}u_r = \frac{d}{dr}\left[\frac{1}{r}\frac{d}{dr}(ru_r)\right] = -\frac{1-\nu^2}{Et}\overline{p}_r \tag{13.14c}$$

Integration of this relationship gives u_r, which can then be placed in the first of Eqs. (13.14a) to find n_r.

The transfer and stiffness matrices from the solution of Eq. (13.14) are given in Table 13.1.

EXAMPLE 13.1 Rotating Disk

Find the displacement u_r and the internal force n_r of a rotating circular disk with inner and outer radii of a and b. The angular speed of the disk is Ω.

Assume that the material of the disk is homogeneous. The disk is subjected to a symmetric loading of $\overline{p}_r = \rho r\Omega^2$, where ρ is the mass density (mass/area). The displacement u_r and internal force n_r are independent of ϕ. The governing equations for the disk are given by Eq. (13.14), which are readily integrated. We find

$$\begin{aligned}\begin{bmatrix} u_r \\ n_r \end{bmatrix}_r &= \begin{bmatrix} U_{uu} & U_{un} \\ U_{nu} & U_{nn} \end{bmatrix}\begin{bmatrix} u_r \\ n_r \end{bmatrix}_{r=a} + \begin{bmatrix} u_r^0 \\ n_r^0 \end{bmatrix} \\ \mathbf{z} \quad &= \qquad \mathbf{U}^i \qquad\quad \mathbf{z}_a \quad + \quad \overline{\mathbf{z}}^i\end{aligned} \tag{1}$$

TABLE 13.1

Transfer and Stiffness Matrices of an Annular Plate with In-Plane Loading

a—Radial coordinate of the inner surface of the disc element. That is, the element begins at a and continues to r for the transfer matrix and to b for the stiffness matrix.
α—Coefficient of thermal expansion.
ρ—Mass density (mass/area).
Ω—Angular velocity of rotation of disk that leads to centrifugal loading force (radians/time).
$\overline{p}_r(r)$—Arbitrary loading intensity in r direction (force/length^2).
$\Delta T(r)$—Arbitrary temperature change.
ΔT_1—Constant temperature change.
Set $\Delta T(\xi) = 0$ if only a constant temperature change is present.
u_r and n_r—Displacement and radial in-plane force per unit length.
E—Modulus of elasticity.
ν—Poisson's ratio.
t—Thickness of disk.

$$u^0 = \frac{(r^2-a^2)}{2r}(1+\nu)\alpha\,\Delta T_1 + \frac{(1+\nu)}{r}\alpha\int_a^r \xi\,\Delta T(\xi)\,d\xi - \frac{(r^2-a^2)^2}{r}\frac{(1-\nu^2)}{8}\frac{\rho\Omega^2}{E} - \frac{(1-\nu^2)}{Et}\int_a^r\left[\eta\int_a^\eta \overline{p}_r(\xi)\,d\xi\right]d\eta$$

$$n_r^0 = -\frac{t(r^2-a^2)}{2r^2}E\alpha\,\Delta T_1 - \frac{tE\alpha}{r^2}\int_a^r \xi\,\Delta T(\xi)\,d\xi - (r^2-a^2)\frac{\rho\Omega^2 t}{4}\left[(1+\nu)+\frac{(1-\nu)}{2}\frac{(r^2+a^2)}{r^2}\right] + \frac{(1-\nu)}{r^2}\int_a^r\eta\left[\int_a^\eta \overline{p}_r(\xi)\,d\xi\right]d\eta - \int_a^r \overline{p}_r(\xi)\,d\xi$$

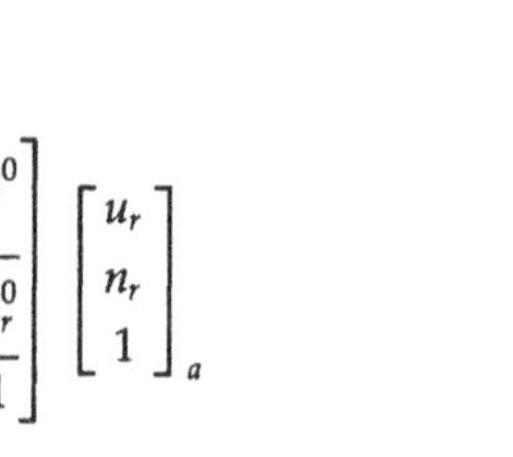

Transfer Matrix (Sign Convention 1)

$$\mathbf{U}^i\mathbf{z}_a = \left[\begin{array}{cc|c} \frac{r}{a}\left[1-\frac{1+\nu}{2}\frac{(r^2-a^2)}{r^2}\right] & \frac{1}{Et}\frac{1-\nu^2}{2}\frac{(r^2-a^2)}{r} & u^0 \\ \hline \frac{Et}{2ar^2}(r^2-a^2) & 1-\frac{1-\nu}{2}\frac{(r^2-a^2)}{r^2} & n_r^0 \\ \hline 0 & 0 & 1 \end{array}\right]\begin{bmatrix} u_r \\ n_r \\ 1 \end{bmatrix}_a$$

Stiffness Matrix (Sign Convention 2)

$$\begin{bmatrix} P_a \\ P_b \end{bmatrix} = \begin{bmatrix} k_{aa} & k_{ab} \\ k_{ba} & k_{bb} \end{bmatrix}\begin{bmatrix} u_a \\ u_b \end{bmatrix} + \begin{bmatrix} P_a^0 \\ P_b^0 \end{bmatrix}$$

$$\mathbf{p}^i \quad = \quad \mathbf{k}^i \quad \mathbf{v}^i \quad - \quad \overline{\mathbf{p}}^i$$

where

$$k_{aa} = 2\pi Et\left[\beta_0^2(1-\nu)+(1+\nu)\right]/H \qquad H = (1-\nu^2)(\beta_0^2-1)$$

$$k_{ab} = k_{ba} = -4\pi Eh\beta_0/H \qquad \beta_0 = b/a$$

$$k_{bb} = 2\pi\left[\beta_0^2(1+\nu)+(1-\nu)\right]/H \qquad P_a^0 = -k_{ab}u^0 \quad P_b^0 = 2\pi bn_r^0 - k_{bb}u^0$$

with

$$U_{uu} = \frac{r}{a}\left[1 - \frac{(1+\nu)}{2}\frac{(r^2-a^2)}{r^2}\right] \qquad U_{un} = \frac{1}{Et}\frac{(1-\nu^2)}{2}\frac{(r^2-a^2)}{r}$$

$$U_{nu} = \frac{Et}{2ar^2}(r^2-a^2) \qquad U_{nn} = 1 - \frac{(1-\nu)}{2}\frac{(r^2-a^2)}{r^2}$$

$$u_r^0 = -\frac{r^2-a^2}{r}\frac{1-\nu^2}{8}\frac{\rho\Omega^2}{E} \qquad n_r^0 = -(r^2-a^2)\frac{\rho\Omega^2 t}{4}\left[(1+\nu) + \frac{(1-\nu)}{2}\frac{(r^2+a^2)}{r^2}\right]$$

The initial parameters u_r and n_r at $r = a$ can be determined from the relation $\mathbf{z}_b = \mathbf{U}^i\mathbf{z}_a + \overline{\mathbf{z}}^i$. Note that $n_r|_{r=b} = n_r|_{r=a} = 0$. Thus, (1) can be written as

$$\begin{aligned} \begin{bmatrix} u_r \\ 0 \end{bmatrix}_{r=b} &= \begin{bmatrix} U_{uu} & U_{un} \\ U_{nu} & U_{nn} \end{bmatrix} \begin{bmatrix} u_r \\ 0 \end{bmatrix}_{r=a} + \begin{bmatrix} u_r^0 \\ n_r^0 \end{bmatrix}_{r=a} \\ \mathbf{z}_b &= \mathbf{U}^i(b) \qquad \mathbf{z}_a + \overline{\mathbf{z}}^i \end{aligned} \tag{2}$$

Then

$$u_r|_{r=a} = \left.\frac{-n_r^0}{U_{nu}}\right|_{r=b} = \frac{ab^2\rho\Omega^2}{2E}\left[(1+\nu) + \frac{(1-\nu)}{2}\frac{b^2+a^2}{b^2}\right] \tag{3}$$

With the initial parameters known, the response of the disk can be determined from (1). The corresponding loading terms and stiffness matrices are given in Table 13.1. ∎

13.1.3 Variational (Global) Relationships

The principle of virtual work for these in-plane deformation problems becomes (Chapter 2, Example 2.5)

$$\delta W = -\int_A \delta\mathbf{u}^T({}_u\mathbf{D}^T\,\mathbf{s} - \overline{\mathbf{p}}_V)\,dA + \int_{S_p} \delta\mathbf{u}^T\,\overline{\mathbf{p}}\,ds = 0 \tag{13.15}$$

where ds is an infinitesimal length along a perimeter boundary of the flat element. Upon substitution of $\mathbf{u}$, $\mathbf{s}$, $\mathbf{D}$, and $\overline{\mathbf{p}}_V$ for a rectangular element,

$$\delta W = -\int_A \delta[u_x \quad u_y]\left\{\begin{bmatrix} {}_x\partial & 0 & {}_y\partial \\ 0 & {}_y\partial & {}_x\partial \end{bmatrix}\begin{bmatrix} n_x \\ n_y \\ n_{xy} \end{bmatrix} - \begin{bmatrix} \overline{p}_{Vx} \\ \overline{p}_{Vy} \end{bmatrix}\right\} dA + \int_{S_p} \delta[u_x \quad u_y]\begin{bmatrix} \overline{p}_x \\ \overline{p}_y \end{bmatrix} ds = 0 \tag{13.16}$$

where $\overline{\mathbf{p}} = [\overline{p}_x\ \overline{p}_y]^T$ are the boundary loads. Remember that ${}_x\partial$ and ${}_y\partial$ mean that the derivatives ∂_x and ∂_y are taken on the variables to the left of ${}_x\partial$ and ${}_y\partial$. For circular disks,

$$\begin{aligned} \delta W = &-\int_A \delta[u_r \quad u_\phi]\left\{\begin{bmatrix} {}_r\partial & 1/r & (1/r)_\phi\partial \\ 0 & (1/r)_\phi\partial & {}_r\partial - 1/r \end{bmatrix}\begin{bmatrix} n_r \\ n_\phi \\ n_{r\phi} \end{bmatrix} - \begin{bmatrix} \overline{p}_{Vr} \\ \overline{p}_{V\phi} \end{bmatrix}\right\} dA \\ &+ \int_{S_p} \delta[u_r \quad u_p]\begin{bmatrix} \overline{p}_r \\ \overline{p}_\phi \end{bmatrix} ds = 0 \end{aligned} \tag{13.17}$$

where $\overline{p}_r$ and $\overline{p}_\phi$ are the applied boundary loads.

In terms of displacements, the principle of virtual work appears as

$$\delta W = -\int_A \delta\mathbf{u}^T(\mathbf{k}^D\mathbf{u} - \overline{\mathbf{p}}_V)\,dA + \int_{S_p} \delta\mathbf{u}^T\overline{\mathbf{p}}\,ds = 0 \tag{13.18}$$

with $\mathbf{k}^D = {}_u\mathbf{D}^T \mathbf{E}\, \mathbf{D}_u$. In the case of rectangular flat elements (Chapter 2, Example 2.5) with constant D,

$$\mathbf{k}^D = D\begin{bmatrix} {}_x\partial\partial_x + {}_y\partial\frac{(1-\nu)}{2}\partial_y & \vdots & {}_x\partial\nu\partial_y + {}_y\partial\frac{(1-\nu)}{2}\partial_x \\ \cdots\cdots & & \cdots\cdots \\ {}_y\partial\nu\partial_x + {}_x\partial\frac{(1-\nu)}{2}\partial_y & \vdots & {}_y\partial\partial_y + {}_x\partial\frac{(1-\nu)}{2}\partial_x \end{bmatrix} \tag{13.19a}$$

For circular plates with in-plane loading and constant D,

$$\mathbf{k}^D = D\begin{bmatrix} {}_r\partial\partial_r + {}_r\partial\frac{\nu}{r} + \frac{\nu}{r}\partial_r + {}_\phi\partial\frac{(1-\nu)}{2r^2}\partial_\phi + \frac{1}{r^2} & \vdots & {}_r\partial\frac{\nu}{r}\partial_\phi + \frac{1-\nu}{2r}\left({}_\phi\partial\partial_r - \frac{1}{r}{}_\phi\partial\right) + \frac{1}{r^2}\partial_\phi \\ \cdots\cdots & & \cdots\cdots \\ {}_\phi\partial\frac{1}{r^2} + \frac{1-\nu}{2r}\left({}_r\partial\partial_\phi - \frac{1}{r}\partial_\phi\right) + {}_\phi\partial\frac{\nu}{r}\partial_r & \vdots & {}_\phi\partial\frac{1}{r^2}\partial_\phi + \frac{1-\nu}{2}\left({}_r\partial\partial_r - {}_r\partial\frac{1}{r} - \frac{1}{r}\partial_r + \frac{1}{r^2}\right) \end{bmatrix} \tag{13.19b}$$

These relationships form the basis for the development of stiffness matrices. See Chapter 6, Section 6.4.1 for the use of $\mathbf{k}^D$ in the formulation of a solution.

13.2 Transverse Deformation of a Plate

In plate theory, the equations for a three-dimensional continuum are to be referred to the middle surface of a plate. We will begin by establishing the kinematic, material law, and equilibrium relations for the transverse deformation of a plate. These relationships will then be specialized for a plate without shear deformation effects, the so-called *Kirchhoff*[1] *plate*. Finally, a plate with shear deformation effects will be studied.

13.2.1 Kinematical Relationships

The displacements u_x, u_y, and u_z of the plate are to be expressed by the deflection w and rotations θ_x and θ_y of the plate middle surface. See Fig. 13.4 for positive displacements. We choose to use definitions of displacements and forces that are traditional with plate theory, rather than the definitions consistent with the coordinate directions used elsewhere in this work. For example, note how the definition of θ_x differs from that of Chapter 1, where $\theta = \theta_y$ corresponds to θ_x of this chapter. The middle surface is assumed to remain unstrained. Similar to beam theory, it is assumed that plane sections normal to the middle surface before bending remain plane after bending. When shear deformation is taken into account, the plane will not necessarily remain normal to the middle surface. With the plane section remaining plane assumption, the displacements appear as

$$\begin{bmatrix} u_x \\ u_y \\ u_z \end{bmatrix} = \begin{bmatrix} z & 0 & 0 \\ 0 & z & 0 \\ 0 & 0 & 1 \end{bmatrix} \begin{bmatrix} \theta_x \\ \theta_y \\ w \end{bmatrix} \tag{13.20}$$

[1]Gustav Robert Kirchhoff (1824–1887) was a German physicist who followed Bunsen, of Bunsen burner fame, to a lengthy stay as a professor at the University of Heidelberg. Kirchhoff was one of the German scientists credited with applying quality scientific methodology for overcoming the headstart in the industrial revolution of the English and French. Kirchhoff's law was fundamental to the thermodynamics of radiation, and, as interpreted by Planck, fundamental to quantum physics. His work on chemical elements with Bunsen led to the method of spectral analysis. A midcareer accident forced him to use crutches or a wheelchair and to discontinue experimental research.

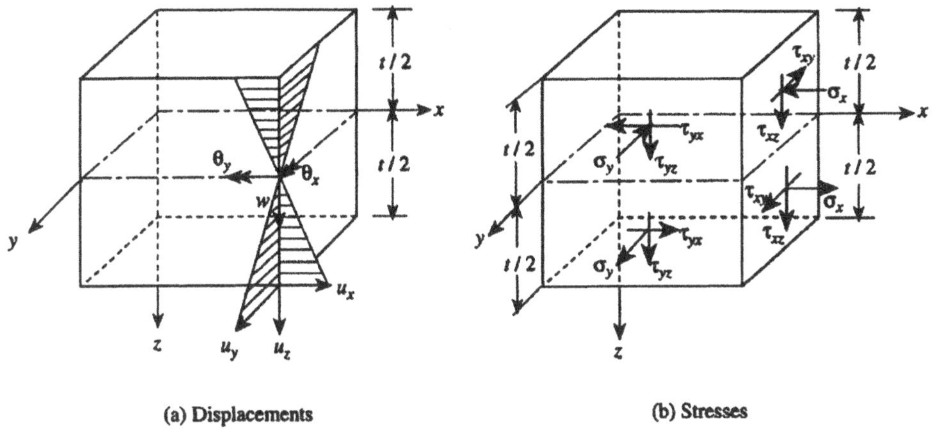

FIGURE 13.4
Displacements and stresses.

From Chapter 1, Eq. (1.20) the strains for an elastic solid are given by

$$\begin{bmatrix} \epsilon_x \\ \epsilon_y \\ \epsilon_z \\ \gamma_{xy} \\ \gamma_{xz} \\ \gamma_{yz} \end{bmatrix} = \underset{\mathbf{D}}{\begin{bmatrix} \partial_x & 0 & 0 \\ 0 & \partial_y & 0 \\ 0 & 0 & \partial_z \\ \partial_y & \partial_x & 0 \\ \partial_z & 0 & \partial_x \\ 0 & \partial_z & \partial_y \end{bmatrix}} \begin{bmatrix} u_x \\ u_y \\ u_z \end{bmatrix} \tag{13.21}$$

Note that $u_z = w$ is the deflection of the middle surface and does not vary with z. Hence, $\epsilon_z = \partial_z u_z = 0$. Also, it is assumed that the middle surface transverse displacement, the deflection w, is small compared to the thickness of the plate. Hence, the rotation or slope of the deformed surface is small. The square of the slope is then negligible in comparison to unity, so that, as in Chapter 1, Eq. (1.97), the curvature is equal to the rate of change of the rotation. Substitute the displacements of Eq. (13.20) into Eq. (13.21) to find the strains

$$\begin{bmatrix} \epsilon_x \\ \epsilon_y \\ \epsilon_z \\ \gamma_{xy} \\ \gamma_{xz} \\ \gamma_{yz} \end{bmatrix} = \begin{bmatrix} z\,\theta_{x,x} \\ z\,\theta_{y,y} \\ 0 \\ z(\theta_{x,y} + \theta_{y,x}) \\ (\theta_x + w_{,x}) \\ (\theta_y + w_{,y}) \end{bmatrix} = \begin{bmatrix} z\,\kappa_x \\ z\,\kappa_y \\ 0 \\ z\,2\kappa_{xy} \\ \gamma_{xz} \\ \gamma_{yz} \end{bmatrix} \tag{13.22}$$

From Eq. (13.22), the kinematical relations for the transverse deformation of a plate are

$$\underset{\boldsymbol{\epsilon}}{\begin{bmatrix} \kappa_x \\ \kappa_y \\ 2\kappa_{xy} \\ \gamma_{xz} \\ \gamma_{yz} \end{bmatrix}} = \underset{\mathbf{D}_u}{\begin{bmatrix} \partial_x & 0 & 0 \\ 0 & \partial_y & 0 \\ \partial_y & \partial_x & 0 \\ 1 & 0 & \partial_x \\ 0 & 1 & \partial_y \end{bmatrix}} \underset{\mathbf{u}}{\begin{bmatrix} \theta_x \\ \theta_y \\ w \end{bmatrix}} \tag{13.23}$$

13.2.2 Material Law

The material law of Chapter 1, Eq. (1.32) for a three-dimensional continuum appears as

$$
\begin{bmatrix} \epsilon_x \\ \epsilon_y \\ \epsilon_z \\ \cdots \\ \gamma_{xy} \\ \gamma_{xz} \\ \gamma_{yz} \end{bmatrix} = \frac{1}{E} \left[\begin{array}{ccc:ccc} 1 & -\nu & -\nu & & & \\ -\nu & 1 & -\nu & & \mathbf{0} & \\ -\nu & -\nu & 1 & & & \\ \hdashline & & & 2(1+\nu) & & \\ & \mathbf{0} & & & 2(1+\nu) & \\ & & & & & 2(1+\nu) \end{array}\right] \begin{bmatrix} \sigma_x \\ \sigma_y \\ \sigma_z \\ \cdots \\ \tau_{xy} \\ \tau_{xz} \\ \tau_{yz} \end{bmatrix}
$$
$$
\boldsymbol{\epsilon} = \mathbf{E}^{-1} \boldsymbol{\sigma} \tag{13.24}
$$

Recall from Eq. (13.22) that $\epsilon_z = 0$. The stress σ_z is smaller than other stress components and can be neglected. This approximation may be questionable in the neighborhood of concentrated transverse loading. With $\sigma_z = 0$, inversion of Eq. (13.24) gives

$$
\begin{bmatrix} \sigma_x \\ \sigma_y \\ \tau_{xy} \\ \cdots \\ \tau_{xz} \\ \tau_{yz} \end{bmatrix} = \frac{E}{1-\nu^2} \left[\begin{array}{ccc:cc} 1 & \nu & 0 & & \\ \nu & 1 & 0 & \mathbf{0} & \\ 0 & 0 & \frac{1-\nu}{2} & & \\ \hdashline & \mathbf{0} & & \frac{1-\nu}{2} & \\ & & & & \frac{1-\nu}{2} \end{array}\right] \begin{bmatrix} \epsilon_x \\ \epsilon_y \\ \gamma_{xy} \\ \cdots \\ \gamma_{xz} \\ \gamma_{yz} \end{bmatrix}
$$
$$
\boldsymbol{\sigma} = \mathbf{E} \boldsymbol{\epsilon} \tag{13.25}
$$

Frequently, the two expressions for τ_{xz} and τ_{yz} are written as

$$
\tau_{xz} = G\gamma_{xz}, \qquad \tau_{yz} = G\gamma_{yz} \tag{13.26}
$$

with

$$
G = \frac{E}{2(1+\nu)}
$$

Figure 13.4 shows the stress components defined in Eq. (13.25). The stresses σ_x, σ_y, τ_{xy}, τ_{xz}, and τ_{yz} are defined similarly to those in the engineering beam theory. There, it was convenient to replace cross-sectional stresses by their resultant forces. For plates, we choose to utilize forces and moments per unit length. These are the *stress resultants* (Fig. 13.5).

$$
\begin{bmatrix} m_x \\ m_y \\ m_{xy} = m_{yx} \end{bmatrix} = \int_{-\frac{t}{2}}^{\frac{t}{2}} \begin{bmatrix} \sigma_x \\ \sigma_y \\ \tau_{xy} = \tau_{yx} \end{bmatrix} z\,dz \quad \text{Bending and Twisting Moments}
$$
$$
\begin{bmatrix} q_x \\ q_y \end{bmatrix} = \int_{-\frac{t}{2}}^{\frac{t}{2}} \begin{bmatrix} \tau_{xz} \\ \tau_{yz} \end{bmatrix} dz \quad \text{Shear Forces} \tag{13.27}
$$

The signs (directions) of the moments and forces in Fig. 13.5 correspond to those of the stress components in Fig. 13.4.

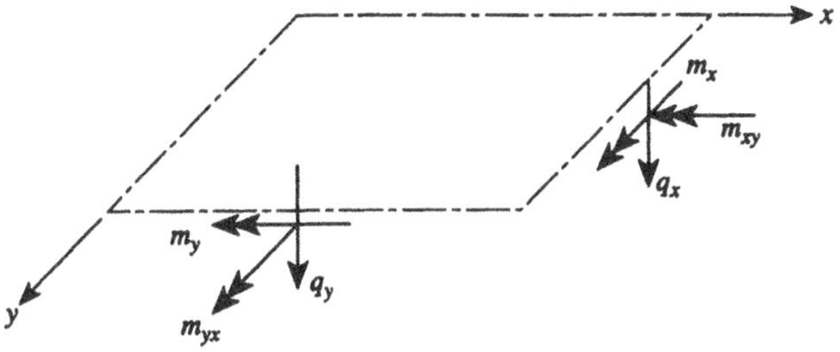

FIGURE 13.5
Positive stress resultants.

Substitute the kinematic relations of Eq. (13.22) and the material law of Eq. (13.25) into the stress resultants of Eq. (13.27). We find, for example,

$$
\begin{aligned}
m_x &= \int_{-\frac{t}{2}}^{\frac{t}{2}} \sigma_x\, z\, dz = \int_{-\frac{t}{2}}^{\frac{t}{2}} \frac{E}{1-\nu^2}(\epsilon_x + \nu\epsilon_y) z\, dz = \int_{-\frac{t}{2}}^{\frac{t}{2}} \frac{E}{1-\nu^2}(z\kappa_x + \nu z\kappa_y) z\, dz \\
&= \frac{E}{1-\nu^2}(\kappa_x + \nu\kappa_y)\int_{-\frac{t}{2}}^{\frac{t}{2}} z^2\, dz = \frac{Et^3}{12(1-\nu^2)}(\kappa_x + \nu\kappa_y)
\end{aligned}
\tag{13.28}
$$

In matrix notation, the relations for the moments are

$$
\begin{bmatrix} m_x \\ m_y \\ m_{xy} \end{bmatrix} = \frac{Et^3}{12(1-\nu^2)} \begin{bmatrix} 1 & \nu & 0 \\ \nu & 1 & 0 \\ 0 & 0 & \frac{1-\nu}{2} \end{bmatrix} \begin{bmatrix} \kappa_x \\ \kappa_y \\ 2\kappa_{xy} \end{bmatrix}
\tag{13.29}
$$

13.2.3 Conditions of Equilibrium

To derive the differential equations of equilibrium, consider an element $t\, dx\, dy$ subject to an applied load $\overline{p}_z$ (Fig. 13.6). The condition that the sum of the vertical forces is equal to zero gives

$$
\frac{\partial q_x}{\partial x} dx\, dy + \frac{\partial q_y}{\partial y} dx\, dy + \overline{p}_z\, dx\, dy = 0
\tag{13.30a}
$$

or

$$
\frac{\partial q_x}{\partial x} + \frac{\partial q_y}{\partial y} + \overline{p}_z = 0
$$

The summation of moments about the x axis set to zero yields

$$
\frac{\partial m_{xy}}{\partial x} dx\, dy + \frac{\partial m_y}{\partial y} dx\, dy - q_y\, dx\, dy = 0
\tag{13.30b}
$$

or

$$
\frac{\partial m_{xy}}{\partial x} + \frac{\partial m_y}{\partial y} - q_y = 0
$$

Similarly, the equilibrium of moments about the y axis provides the relationship

$$
\frac{\partial m_{yx}}{\partial y} + \frac{\partial m_x}{\partial x} - q_x = 0
\tag{13.30c}
$$

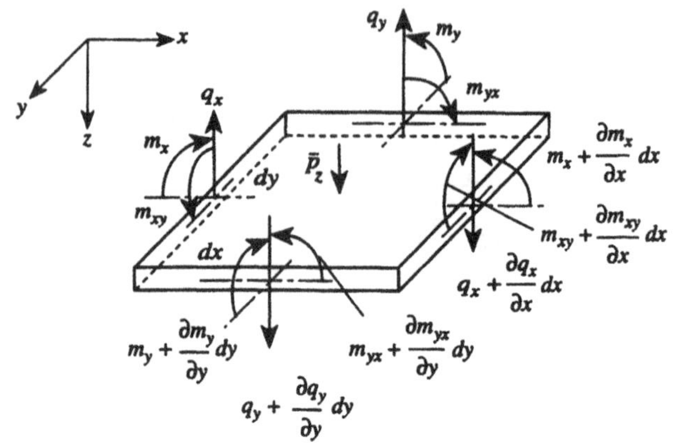

FIGURE 13.6
Equilibrium element.

Finally, it can be seen that $m_{xy} = m_{yx}$. In summary,

$$\begin{aligned} \partial_x q_x + \partial_y q_y + \overline{p}_z &= 0 \\ \partial_x m_x + \partial_y m_{yx} - q_x &= 0 \\ \partial_y m_y + \partial_x m_{xy} - q_y &= 0 \\ m_{xy} &= m_{yx} \end{aligned} \tag{13.31a}$$

or in matrix notation,

$$\underbrace{\begin{bmatrix} \partial_x & 0 & \partial_y & -1 & 0 \\ 0 & \partial_y & \partial_x & 0 & -1 \\ 0 & 0 & 0 & \partial_x & \partial_y \end{bmatrix}}_{\mathbf{D}_s^T} \underbrace{\begin{bmatrix} m_x \\ m_y \\ m_{xy} \\ q_x \\ q_y \end{bmatrix}}_{\mathbf{s}} + \underbrace{\begin{bmatrix} 0 \\ 0 \\ \overline{p}_z \end{bmatrix}}_{\overline{\mathbf{p}}} = \underbrace{\begin{bmatrix} 0 \\ 0 \\ 0 \end{bmatrix}}_{\mathbf{0}} \tag{13.31b}$$

These equations for kinematics, material law, and equilibrium can now be utilized to develop the plate theories in common use today.

13.3 Classical Plate Theory

Classical plate theory is based on the so-called Kirchhoff assumptions. In this theory, it is assumed that straight lines, initially normal to the middle surface, remain straight and undeformed and remain normal to the middle surface as the plate deforms. This implies that deformations due to transverse shear are neglected. This theory results in tractable governing equations that are comparable in their usefulness to the engineering beam theory equations.

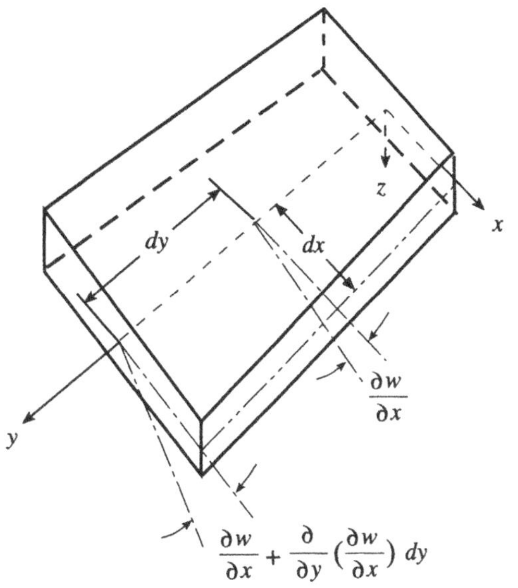

FIGURE 13.7
Twisting.

13.3.1 Rectangular Plates

Kinematical Relationships

Since shear deformation is not taken into account, $\gamma_{xz} = \gamma_{yz} = 0$, and the final two relations of Eq. (13.22) lead to

$$\begin{aligned} \theta_x &= -w_{,x} \\ \theta_y &= -w_{,y} \end{aligned} \tag{13.32}$$

Then the remaining strains of Eq. (13.22) become

$$\epsilon_x = -z\frac{\partial^2 w}{\partial x^2}, \qquad \epsilon_y = -z\frac{\partial^2 w}{\partial y^2}, \qquad \gamma_{xy} = -2z\frac{\partial^2 w}{\partial x \partial y} \tag{13.33}$$

or in terms of curvatures,

$$\begin{bmatrix} \kappa_x \\ \kappa_y \\ 2\kappa_{xy} \end{bmatrix} = \begin{bmatrix} -\partial_x^2 \\ -\partial_y^2 \\ -2\partial_x\partial_y \end{bmatrix} [w] \tag{13.34}$$
$$\boldsymbol{\epsilon} \quad = \quad \mathbf{D}_u \quad \mathbf{u}$$

The final expression of Eq. (13.34) is called the *twist* with respect to the x and y axes of the middle surface (Fig. 13.7).

Material Law

The material relationship of Eq. (13.29) applies, i.e.,

$$\begin{bmatrix} m_x \\ m_y \\ m_{xy} \end{bmatrix} = K \begin{bmatrix} 1 & \nu & 0 \\ \nu & 1 & 0 \\ 0 & 0 & \frac{1-\nu}{2} \end{bmatrix} \begin{bmatrix} \kappa_x \\ \kappa_y \\ 2\kappa_{xy} \end{bmatrix} \tag{13.35}$$
$$\mathbf{s} \quad = \quad \mathbf{E} \quad \boldsymbol{\epsilon}$$

where

$$K = \frac{Et^3}{12(1-\nu^2)}$$

which is known as the plate *stiffness*. The inverted relationship gives the curvatures in terms of the stress resultants

$$\begin{bmatrix} \kappa_x \\ \kappa_y \\ 2\kappa_{xy} \end{bmatrix} = \frac{12}{Et^3}\begin{bmatrix} 1 & -\nu & 0 \\ -\nu & 1 & 0 \\ 0 & 0 & 2(1+\nu) \end{bmatrix}\begin{bmatrix} m_x \\ m_y \\ m_{xy} \end{bmatrix} \tag{13.36}$$

$$\boldsymbol{\epsilon} \quad = \quad \mathbf{E}^{-1} \quad \mathbf{s}$$

Note that from Eq. (13.34), the relations of Eq. (13.35) in terms of displacements appear as

$$\begin{aligned} m_x &= -K\left(\partial_x^2 w + \nu \partial_y^2 w\right) \\ m_y &= -K\left(\partial_y^2 w + \nu \partial_x^2 w\right) \\ m_{xy} &= -K(1-\nu)\partial_x \partial_y w \end{aligned} \tag{13.37}$$

From the material law of Eq. (13.25), i.e.,

$$\begin{aligned} \sigma_x &= \frac{E}{1-\nu^2}(\epsilon_x + \nu\epsilon_y) \\ \sigma_y &= \frac{E}{1-\nu^2}(\epsilon_y + \nu\epsilon_x) \\ \tau_{xy} &= G\gamma_{xy} \end{aligned}$$

and Eqs. (13.33) and (13.37), the stresses can be expressed in terms of the stress resultants by

$$\begin{aligned} \sigma_x &= \frac{m_x z}{t^3/12} \\ \sigma_y &= \frac{m_y z}{t^3/12} \\ \tau_{xy} &= \frac{m_{xy} z}{t^3/12} = \tau_{yx} \end{aligned} \tag{13.38}$$

Since z is measured from the middle surface, the maximum normal stresses will occur at $z = \pm t/2$ on the top and bottom surfaces (see Fig. 13.8).

Conditions of Equilibrium

Consider the equilibrium conditions for the plate element in Fig. 13.6 in light of the assumptions of Kirchhoff plate theory. Recall that the Kirchhoff plate theory omits the effect of shear strains $\gamma_{xz} = \tau_{xz}/G$ and $\gamma_{yz} = \tau_{yz}/G$ on the bending of the plate. Vertical forces q_x and q_y are not negligible. The external load $\overline{p}_z$ is carried by these shear forces together with the moments m_x, m_y, and m_{xy}. The equilibrium conditions (Eq. 13.31) of the plate still apply.

Substitute the second and third relations of Eq. (13.31a) into the first relation of Eq. (13.31a). This gives the higher order equilibrium relation

$$\frac{\partial^2 m_x}{\partial x^2} + 2\frac{\partial^2 m_{xy}}{\partial x\,\partial y} + \frac{\partial^2 m_y}{\partial y^2} + \overline{p}_z = 0 \tag{13.39a}$$

or

$$[\partial_x^2 \quad \partial_y^2 \quad 2\partial_x\,\partial_y]\begin{bmatrix} m_x \\ m_y \\ m_{xy} \end{bmatrix} + [\overline{p}_z] = 0 \tag{13.39b}$$

$$\mathbf{D}_s^T \quad \mathbf{s} \quad + \quad \overline{\mathbf{p}} \quad = 0$$

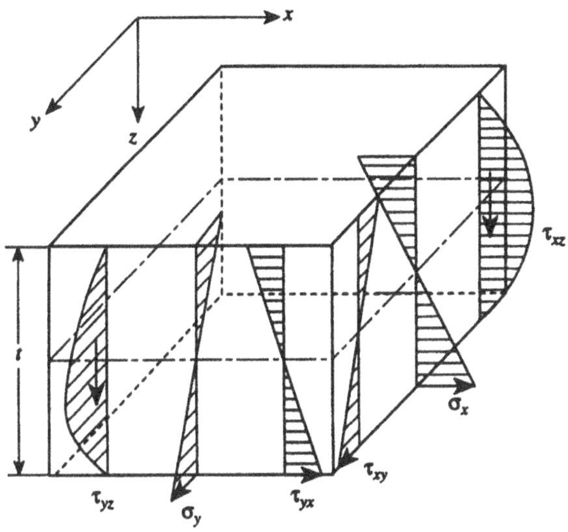

FIGURE 13.8
Stress distributions.

Other useful expressions are obtained from the second and third relations of the equilibrium conditions of Eq. (13.31a). Substitute the expressions for m_x, m_y, and m_{xy} of Eq. (13.37) into these equilibrium relations to find q_x and q_y in terms of the displacement w,

$$q_x = -K\frac{\partial}{\partial x}\left(\frac{\partial^2 w}{\partial x^2} + \frac{\partial^2 w}{\partial y^2}\right) \tag{13.40a}$$

$$q_y = -K\frac{\partial}{\partial y}\left(\frac{\partial^2 w}{\partial x^2} + \frac{\partial^2 w}{\partial y^2}\right) \tag{13.40b}$$

Local Form of the Governing Equations

The governing fourth order differential equation, expressed in terms of the displacement w, is obtained by substituting Eqs. (13.37) into Eq. (13.39)

$$\frac{\partial^4 w}{\partial x^4} + 2\frac{\partial^4 w}{\partial x^2 \partial y^2} + \frac{\partial^4 w}{\partial y^4} = \frac{\overline{p}_z}{K} \tag{13.41}$$

or

$$\nabla^4 w = \frac{\overline{p}_z}{K}$$

with

$$\nabla^4 = \nabla^2\nabla^2 = (\nabla^2)^2 \qquad \nabla^2 = \frac{\partial^2}{\partial x^2} + \frac{\partial^2}{\partial y^2} = \partial_x^2 + \partial_y^2$$

An alternative second order form of the governing equations is sometimes useful. To obtain this form define M, the *moment function* or *moment sum*, as

$$M = \frac{m_x + m_y}{(1+\nu)} = -K\left(\partial_x^2 w + \partial_y^2 w\right) = -K\nabla^2 w \tag{13.42}$$

Then the shear forces are

$$q_x = \frac{\partial M}{\partial x} \qquad q_y = \frac{\partial M}{\partial y} \tag{13.43}$$

Finally, the fourth order governing relationship of Eq. (13.41) is replaced by two equations of similar structure

$$\begin{aligned} \nabla^2 M &= \partial_x^2 M + \partial_y^2 M = -\overline{p}_z \\ \nabla^2 w &= \partial_x^2 w + \partial_y^2 w = -M/K \end{aligned} \tag{13.44}$$

Stresses

The stress components σ_x, σ_y, and τ_{xy} of Eq. (13.38) were found using the material law. The stress components σ_z, τ_{xz}, and τ_{yz} cannot be obtained from the material law, since they cannot be related to the corresponding strains, ϵ_z, γ_{xz}, and γ_{yz}, which are neglected as a result of the basic assumptions. However, the equations of equilibrium will permit these stresses to be determined. The three-dimensional equilibrium equation $\mathbf{D}^T\boldsymbol{\sigma} + \overline{\mathbf{p}}_V = \mathbf{0}$ without body forces is (Chapter 1, Eq. 1.51)

$$\begin{aligned} \partial_x \sigma_x + \partial_y \tau_{yx} + \partial_z \tau_{zx} &= 0 \\ \partial_x \tau_{xy} + \partial_y \sigma_y + \partial_z \tau_{zy} &= 0 \\ \partial_x \tau_{xz} + \partial_y \tau_{yz} + \partial_z \sigma_z &= 0 \end{aligned} \tag{13.45}$$

Upon integration, the first equation gives

$$\tau_{xz} = -\int (\partial_x \sigma_x + \partial_y \tau_{xy})\, dz + C$$

where C is an arbitrary integration constant. Equivalently,

$$\tau_{xz} = -\int_a^z (\partial_x \sigma_x + \partial_y \tau_{xy})\, dz$$

where a is an arbitrary point along z and z is the coordinate of a point along the thickness of the plate. Let $a = t/2$, then using Eqs. (13.38) and (13.31a)

$$\begin{aligned} \tau_{xz} &= \int_z^{t/2} (\partial_x \sigma_x + \partial_y \tau_{xy})\, dz = \int_z^{t/2} \frac{12z}{t^3}(\partial_x m_x + \partial_y m_{xy})\, dz = \frac{12}{t^3}\int_z^{t/2} z q_x\, dz \\ &= \frac{3q_x}{2t}[1 - (2z/t)^2] \end{aligned} \tag{13.46a}$$

Similarly,

$$\tau_{yz} = \int_z^{t/2} (\partial_y \sigma_y + \partial_x \tau_{xy})\, dz = \frac{3q_y}{2t}[1 - (2z/t)^2] \tag{13.46b}$$

Integration of the final relation of Eq. (13.45) leads to the distribution of the transverse normal stress. From Eqs. (13.45), (13.46a), (13.46b), and (13.31a)

$$\partial_z \sigma_z = -\partial_x \tau_{zx} - \partial_y \tau_{zy} = -\frac{3}{2t}\left(1 - \left(\frac{2z}{t}\right)^2\right)(\partial_x q_x + \partial_y q_y) = \frac{3}{2t}\left(1 - \left(\frac{2z}{t}\right)^2\right)\overline{p}_z$$

Then

$$\sigma_z = \frac{3\overline{p}_z}{2t}\int_z^{t/2}\left(1-\left(\frac{2z}{t}\right)^2\right)dz$$
$$= -\frac{3\overline{p}_z}{4}\left[-\frac{2}{3}+\frac{2z}{t}-\frac{1}{3}\left(\frac{2z}{t}\right)^3\right] \tag{13.46c}$$

Observe that τ_{xz} and τ_{yz} vary parabolically over the plate thickness (Fig. 13.8), while σ_z varies cubically. Furthermore, the z-directed stresses τ_{xz} and τ_{yz} tend to be very small relative to the τ_{xy} stress of Eq. (13.38).

Similar to beam shear stresses, peak plate shear stresses occur at the middle surface where $z = 0$. There

$$\tau_{xz}|_{\max} = \frac{3}{2}\frac{q_x}{t} \qquad \tau_{yz}|_{\max} = \frac{3}{2}\frac{q_y}{t} \tag{13.47}$$

Boundary Conditions

Now that local (differential) governing equations have been established, boundary conditions which the displacements and forces must satisfy will be considered. As in the case of a beam, the solution to the fourth order plate equation (Eq. 13.41) requires that two boundary conditions be satisfied at each edge of the plate. These conditions can be a combination of deflection, slope, shear force, and moment. However, unlike the beam, with the plate there appear to be two moments—a bending moment and a twisting moment. This apparent surplus of moments can be corrected by replacing the twisting moment by equivalent shear forces. In accordance with St. Venant's principle[δ] this replacement affects the stress distribution only in the neighborhood of the boundary. As will be seen later in this chapter, the problem of an apparent redundant boundary condition disappears when considering plates from the standpoint of a theory including shear deformation.

The development of an equivalent shear force was explained by Kirchhoff using an energy relationship and then mechanically by Thomson[2] and Tait[3] in 1883. Statically, the twisting moment in Fig. 13.5 can be represented by a pair of horizontal forces or equivalently by a pair of vertical forces (Fig. 13.9a). Consider the two successive elements of length dy on the $x = a$ boundary as shown in Fig. 13.9b. On one element, the twisting moment $m_{xy}\,dy$ (Fig. 13.9b) is replaced by a statically equivalent couple of equal and opposite forces m_{xy} separated by dy (Fig. 13.9c), and on the next element, the couple is formed by the forces $(m_{xy} + \partial_y m_{xy}\,dy)$. The adjoining forces $m_{xy} + \partial_y m_{xy}\,dy$ and m_{xy} have opposite signs so that their sum is $\partial_y m_{xy}\,dy$. Add this force to the shear force q_x to obtain the equivalent transverse

---

[δ]In 1855 Barre de Saint-Venant enunciated a useful principle that now bears his name. In essence, this principle can be stated as the redistribution of loading, resulting from a set of forces acting on a small region of the surface of an elastic body being replaced by a statically equivalent set of forces, causing significant changes in the stress distribution only in the neighborhood of the loading, while stresses remain essentially the same in those portions of the body located at large distances from the applied loading. By "large distances" are meant distances great in comparison with the dimensions of the surface on which the loading is applied. "Statically equivalent" sets of forces mean that the two distributions of loadings have the same resultant force and moment.

[2]William Thomson (1824–1907), the son of an Irish (and later Scottish) professor of engineering, was educated at home. He was given the title of Baron Kelvin of Largs. He was a prolific scientist and, with Helmholtz of Germany, is credited with establishing physics as a science at the beginning of the 20th century.

[3]Peter Guthrie Tait (1831–1901) was a Scottish physicist and mathematician. In dynamics, Tait backed the use of quaternions, having promised a dying Hamilton to write an elementary text on the subject. This led to a dispute over the vector method supported by J.W. Gibbs and Oliver Heaviside. In 1868, he caused further controversy by writing a pro-British book *Sketch of the History of Thermodynamics*.

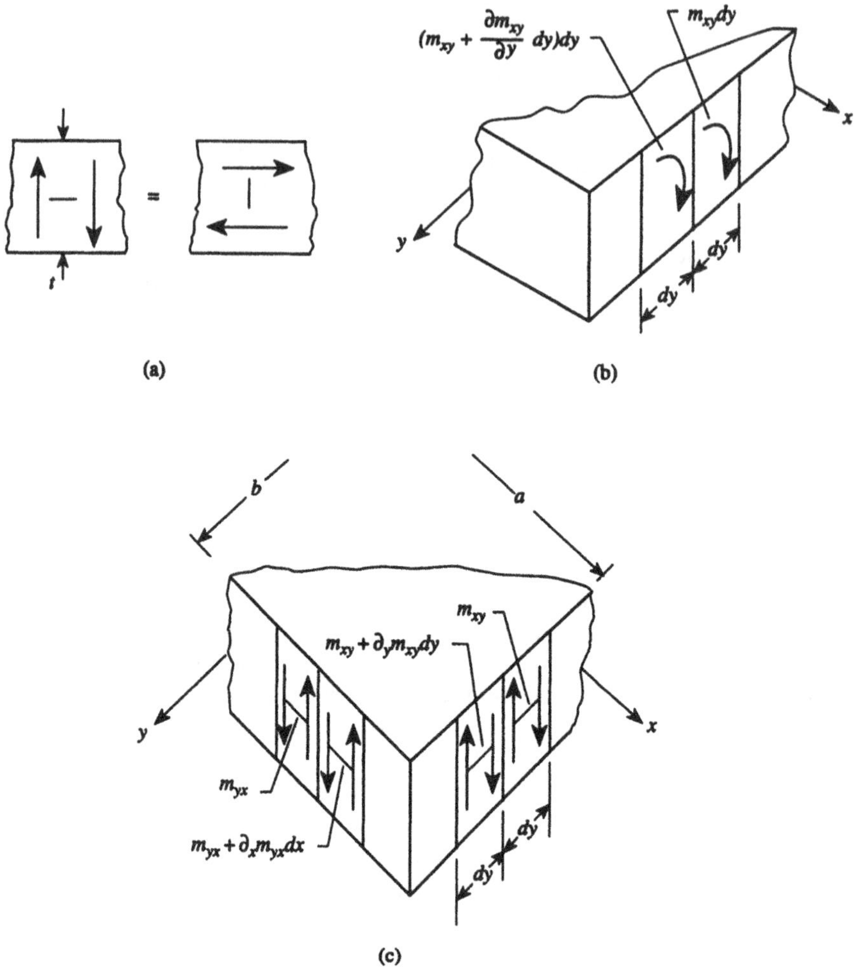

FIGURE 13.9
Edge conditions.

shear force V_x^*

$$V_x^* \, dy = q_x \, dy + \frac{\partial m_{xy}}{\partial y} dy$$

or

$$V_x^* = q_x + \frac{\partial m_{xy}}{\partial y} = -K\big[\partial_x^3 w + (2-\nu)\partial_x \partial_y^2 w\big] \tag{13.48}$$

This is the equivalent shear force per unit length along a particular value of x, which is parallel to the y axis.

Along a prescribed y value, which is parallel to the x axis, the equivalent shear force would be

$$V_y^* = q_y + \frac{\partial m_{xy}}{\partial x} = -K\big[\partial_y^3 w + (2-\nu)\partial_x^2 \partial_y w\big] \tag{13.49}$$

The boundary conditions for the fundamental edge conditions are now evident. For a boundary at $x = a$:

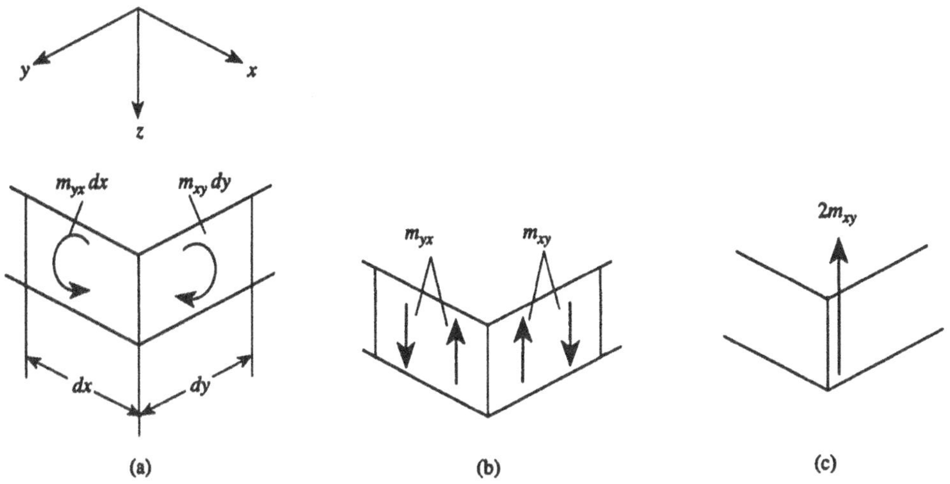

FIGURE 13.10
Corner force.

Simply Supported (Hinged or Pinned) Edge:

$$w = 0, \qquad m_x = -K\left(\partial_x^2 w + \nu \partial_y^2 w\right) = 0 \tag{13.50}$$

The condition $w = 0$ along $x = a$ implies $\partial_y w = 0$ and $\partial_y^2 w = 0$ on $x = a$. Thus, the condition $m_x = 0$ can be expressed as $\partial_x^2 w = 0$.

Fixed (Clamped or Built-In) Edge:

$$w = 0, \quad \partial_x w = 0 \tag{13.51}$$

Free Edge:

$$m_x = -K\left(\partial_x^2 w + \nu \partial_y^2 w\right) = 0, \qquad V_x^* = -K\left[\partial_x^3 w + (2 - \nu)\partial_x\, \partial_y^2 w\right] = 0 \tag{13.52}$$

Guided (Sliding) Edge:

$$\partial_x w = 0, \qquad V_x^* = -K\left[\partial_x^3 w + (2 - \nu)\partial_x\, \partial_y^2 w\right] = 0 \tag{13.53}$$

An interesting phenomenon occurs at the corners of the plate where concentrated forces may be generated. The shear force equivalent of the twisting moments m_{xy} on the edge of the plate (Fig. 13.9) leads not only to the distributed shear forces V_x^*, V_y^*, but also to concentrated forces at the corners. To understand this, consider the twisting moments adjacent to the corner as shown in Fig. 13.10a. These moments can be represented as the equivalent forces m_{xy} and m_{yx} of Fig. 13.10b. The summation of the forces at the corner leads to the corner force

$$R = 2m_{xy} \tag{13.54}$$

of Fig. 13.10c. This means that when a rectangular plate is supported in some manner, e.g., simply supported, the reaction forces include not only the equivalent shear forces V_x^* and V_y^* along the edges but also the corner forces R. If there is no support at a corner, the corner tends to rise or move downward depending on the sign of m_{xy}.

In more general nomenclature, the displacement boundary conditions on S_u are

$$w = \overline{w}, \qquad \theta_a = \overline{\theta}_a \tag{13.55}$$

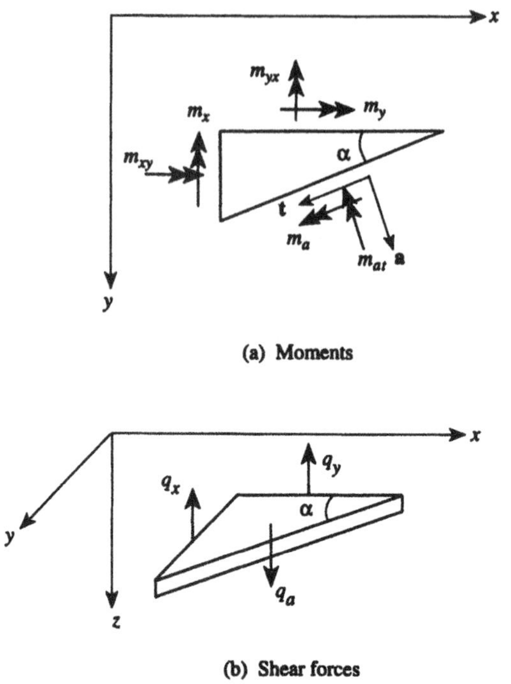

FIGURE 13.11
Surface conditions.

while the force (static) boundary conditions on $S_p = S - S_u$ are

$$m_a = \overline{m}_a, \qquad V_a^* = q_a + \partial_t m_{at} = \overline{V}_a^* \tag{13.56}$$

where (Fig. 13.11) the surface force conditions are

$$\begin{bmatrix} m_a \\ m_{at} \end{bmatrix} = \begin{bmatrix} \sin^2\alpha & \cos^2\alpha & 2\sin\alpha\,\cos\alpha \\ -\sin\alpha\cos\alpha & \sin\alpha\cos\alpha & -\cos^2\alpha + \sin^2\alpha \end{bmatrix} \begin{bmatrix} m_x \\ m_y \\ m_{xy} \end{bmatrix}$$
$$q_a = q_x \sin\alpha + q_y \cos\alpha \tag{13.57}$$

First Order Form of the Local (Differential) Governing Equations

The kinematical relations, material law, and equilibrium conditions of this section, as well as the relationships of Eqs. (13.48) and (13.37) are readily rearranged in first order form with $\theta = \theta_x$, $V = V_x^*$.

$$\begin{aligned}
\frac{\partial w}{\partial x} &= -\theta \\
\frac{\partial \theta}{\partial x} &= \nu \frac{\partial^2 w}{\partial y^2} + \frac{m_x}{K} \\
\frac{\partial V}{\partial x} &= K(1-\nu^2)\frac{\partial^4 w}{\partial y^4} - \nu \frac{\partial^2 m_x}{\partial y^2} - \overline{p}_z(x, y) \\
\frac{\partial m_x}{\partial x} &= -2K(1-\nu)\frac{\partial^2 \theta}{\partial y^2} + V
\end{aligned} \tag{13.58a}$$

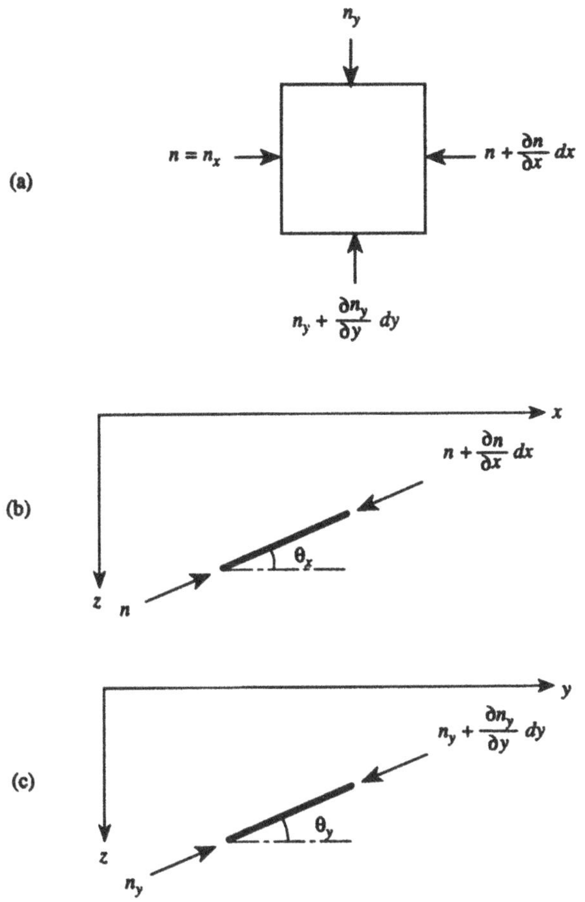

FIGURE 13.12
Plate element subjected to compression forces (force/length).

or

$$\frac{\partial}{\partial x}\begin{bmatrix} w \\ \theta \\ V \\ m_x \end{bmatrix} = \begin{bmatrix} 0 & -1 & 0 & 0 \\ \nu\partial_y^2 & 0 & 0 & 1/K \\ K(1-\nu^2)\partial_y^4 & 0 & 0 & -\nu\partial_y^2 \\ 0 & -2K(1-\nu)\partial_y^2 & 1 & 0 \end{bmatrix}\begin{bmatrix} w \\ \theta \\ V \\ m_x \end{bmatrix} + \begin{bmatrix} 0 \\ 0 \\ -\overline{p}_z \\ 0 \end{bmatrix} \tag{13.58b}$$

$$\frac{\partial \mathbf{z}}{\partial x} = \mathbf{A} \qquad \mathbf{z} + \overline{\mathbf{P}}$$

The nonlinear effect of compressive in-plane forces on bending (buckling) can be introduced in an approximate manner. Treat these forces in the sense of the theory of second order of Chapter 11. If the in-plane compression forces (force per unit length along an edge), $n_x = n$ and n_y, are applied in the middle plane of the plate as in the case of Fig. 13.12, the equilibrium conditions in the z direction would be

$$-n\sin\theta_x\, dy + \left(n + \frac{\partial n}{\partial x}\, dx\right)\left(\sin\theta_x + \frac{\partial}{\partial x}\sin\theta_x\, dx\right) dy$$
$$-n_y\sin\theta_y\, dx + \left(n_y + \frac{\partial n_y}{\partial y} dy\right)\left(\sin\theta_y + \frac{\partial}{\partial y}\sin\theta_y\, dy\right) dx = 0$$

or

$$-n\frac{\partial^2 w}{\partial x^2} - n_y\frac{\partial^2 w}{\partial y^2} - \frac{\partial n}{\partial x}\frac{\partial w}{\partial x} - \frac{\partial n_y}{\partial y}\frac{\partial w}{\partial y} = 0$$

If $\frac{\partial n}{\partial x}$ and $\frac{\partial n_y}{\partial y}$ are very small, then

$$-n\frac{\partial^2 w}{\partial x^2} - n_y\frac{\partial^2 w}{\partial y^2} = 0$$

Together with the condition of equilibrium of Eq. (13.30a), the equilibrium of forces in the z direction becomes

$$\frac{\partial q_x}{\partial x} + \frac{\partial q_y}{\partial y} - n\frac{\partial^2 w}{\partial x^2} - n_y\frac{\partial^2 w}{\partial y^2} = -\overline{p}_z \tag{13.59}$$

Furthermore, if the plate is on an elastic foundation with the elastic constant k, and the dynamic response is of interest, two terms, kw and $\rho\partial^2 w/\partial t^2$, should be added to the right side of Eq. (13.59). The quantity ρ in the inertia term is the mass per unit area and k is the modulus of the elastic foundation in force/length^3. In the dynamic case, the applied load should be written as $\overline{p}_z = \overline{p}_z(x, y, t)$. Then the governing equations of motion become

$$\begin{aligned}
\frac{\partial w}{\partial x} &= -\theta \\
\frac{\partial \theta}{\partial x} &= \nu\frac{\partial^2 w}{\partial y^2} + \frac{m_x}{K} \\
\frac{\partial V}{\partial x} &= K(1-\nu^2)\frac{\partial^4 w}{\partial y^4} + (n_y - \nu n)\frac{\partial^2 w}{\partial y^2} - \frac{1}{K}\left(n + \nu K\frac{\partial^2}{\partial y^2}\right)m_x \\
&\quad + kw + \rho\frac{\partial^2 w}{\partial t^2} - \overline{p}_z(x, y, t) \\
\frac{\partial m_x}{\partial x} &= -2(1-\nu)K\frac{\partial^2 \theta}{\partial y^2} + V
\end{aligned} \tag{13.60}$$

These governing equations can be solved to establish the transfer matrix as well as the stiffness matrix, of an element, or for special cases, these relationships will provide the complete solution, as illustrated in the following section.

Responses

Consider a rectangular plate with dimension L_x along the x direction, and L_y along the other edge. The y derivatives of the governing relationships of Eq. (13.58) can be eliminated by expanding the state variables $\mathbf{z}$ in a sine series.

$$\mathbf{z} = \begin{bmatrix} w(x, y) \\ \theta(x, y) \\ V(x, y) \\ m_x(x, y) \end{bmatrix} = \sum_{m=1}^{\infty} \begin{bmatrix} w_m(x) \\ \theta_m(x) \\ V_m(x) \\ m_m(x) \end{bmatrix} \sin\frac{m\pi y}{L_y} \tag{13.61}$$

These expansions correspond to boundary conditions of $w = 0$ and $m_y = 0$ along $y = 0$ and $y = L_y$. Hence, the plate being considered here is simply supported along $y = 0$ and $y = L_y$. Normally, use of a few terms in the series expansion will suffice to achieve acceptable accuracy.

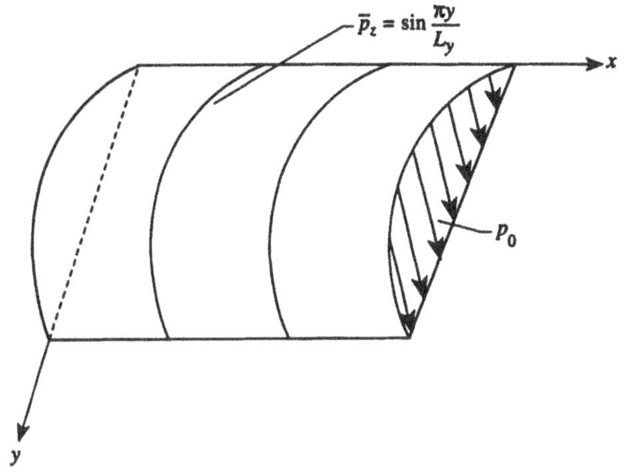

FIGURE 13.13
Loading with sinusoidal y distribution.

The loading $\overline{p}_z$ can also be expanded in a similar sine series

$$\overline{p}_z(x, y) = \sum_{m=1}^{\infty} p_m(x) \sin \frac{m\pi y}{L_y} \tag{13.62}$$

An expression for calculating $p_m(x)$ is found by multiplying both sides of Eq. (13.62) by $\sin(n\pi y/L_y)$ and integrating from $y = 0$ to $y = L_y$. Use the orthogonality conditions for trigonometric functions such as

$$\int_0^{L_y} \sin \frac{m\pi y}{L_y} \sin \frac{n\pi y}{L_y} dy = \begin{cases} 0 & m \neq n \\ \frac{L_y}{2} & m = n \end{cases}$$

to liberate $p_m(x)$.

$$p_m(x) = \frac{2}{L_y} \int_0^{L_y} \overline{p}_z(x, y) \sin \frac{m\pi y}{L_y} dy \tag{13.63}$$

Since $\overline{p}_z(x, y)$ is given, $p_m(x)$, the transformed loading function coefficient, can always be determined from this equation.

EXAMPLE 13.2 Transformed Loading Functions
Find the transformed loading function coefficients p_m if the applied loading is sinusoidal in the y direction as shown in Fig. 13.13.

The loading can be expressed as

$$\overline{p}_z(x, y) = p_0(x) \sin \frac{\pi y}{L_y} \tag{1}$$

From Eq. (13.63),

$$p_m(x) = \frac{2}{L_y} \int_0^{L_y} p_0 \sin \frac{\pi y}{L_y} \sin \frac{m\pi y}{L_y} dy = \begin{cases} p_0 & \text{if } m = 1 \\ 0 & \text{if } m > 1 \end{cases} \tag{2}$$

■

Insert the series expansions of Eqs. (13.61) and (13.62) in the governing equations of Eq. (13.58) to obtain

$$\begin{aligned}\frac{dw_m}{dx} &= -\theta_m \\ \frac{d\theta_m}{dx} &= -\nu\left(\frac{m\pi}{L_y}\right)^2 w_m + \frac{m_m}{K} \\ \frac{dV_m}{dx} &= K(1-\nu^2)\left(\frac{m\pi}{L_y}\right)^4 w_m + \nu\left(\frac{m\pi}{L_y}\right)^2 m_m - p_m(x) \\ \frac{dm_m}{dx} &= 2K(1-\nu)\left(\frac{m\pi}{L_y}\right)^2 \theta_m + V_m\end{aligned} \tag{13.64a}$$

or

$$\begin{aligned}\frac{d}{dx}\begin{bmatrix} w_m \\ \theta_m \\ V_m \\ m_m \end{bmatrix} &= \begin{bmatrix} 0 & -1 & 0 & 0 \\ -\nu\left(\frac{m\pi}{L_y}\right)^2 & 0 & 0 & 1/K \\ K(1-\nu^2)\left(\frac{m\pi}{L_y}\right)^4 & 0 & 0 & \nu\left(\frac{m\pi}{L_y}\right)^2 \\ 0 & 2K(1-\nu)\left(\frac{m\pi}{L_y}\right)^2 & 1 & 0 \end{bmatrix}\begin{bmatrix} w_m \\ \theta_m \\ V_m \\ m_m \end{bmatrix} + \begin{bmatrix} 0 \\ 0 \\ -p_m \\ 0 \end{bmatrix} \\ \frac{d\mathbf{z}}{dx} &= \qquad\qquad\qquad\qquad \mathbf{A} \qquad\qquad\qquad\qquad\quad \mathbf{z} \quad + \quad \overline{\mathbf{P}}\end{aligned} \tag{13.64b}$$

This system of ordinary differential equations can now be solved, using the techniques of Chapter 4, to obtain the transfer and the stiffness matrices.

For the case of other boundary conditions at $y = 0$ and $y = L_y$, a series expansion of the state variables and loadings can be assumed as

$$\mathbf{z} = \begin{bmatrix} w(x, y) \\ \theta(x, y) \\ V(x, y) \\ m_x(x, y) \end{bmatrix} = \sum_{m=1}^{\infty} \begin{bmatrix} w_m(x) \\ \theta_m(x) \\ V_m(x) \\ m_m(x) \end{bmatrix} \phi_m(y) \tag{13.65}$$

where $\phi_m(y)$ is selected to satisfy the boundary conditions at $y = 0$ and $y = L_y$. A possible choice for ϕ_m is to use mode shapes for a beam with the appropriate boundary conditions. These are

Case	Boundary Conditions		ϕ_m	β_m
	$y = 0$	$y = L_y$		
1	Simply Supported	Simply Supported	$\sin \beta_m y$	$\frac{m\pi}{L_y}$
2	Fixed	Simply Supported	$\cosh \beta_m y - \cos \beta_m y + E_m(\sinh \beta_m y - \sin \beta_m y)$	$\frac{(4m+1)\pi}{4L_y}$
3	Fixed	Fixed	$\cosh \beta_m y - \cos \beta_m y - E_m(\sinh \beta_m y - \sin \beta_m y)$	$\frac{(2m+1)\pi}{2L_y}$

where $E_m = (\cosh \eta_m - \cos \eta_m)/(\sinh \eta_m - \sin \eta_m)$, and $\eta_m = \beta_m L_y$. The loadings should also be expanded in the same series. Substitution of these relations into Eqs. (13.58) or (13.60) leads to governing differential equations in terms of w_m, θ_m, V_m, and m_m. For example, substitute these into Eq. (13.60) for harmonic motion (of frequency ω) to obtain

$$\frac{d\mathbf{z}}{dx} = \mathbf{A}\mathbf{z} + \overline{\mathbf{P}} \tag{13.66}$$

with

$$\mathbf{A} = \begin{bmatrix} 0 & -1 & 0 & 0 \\ \nu\beta_m^2\psi_m & 0 & 0 & 1/K \\ (n_y - \nu n)\beta_m^2\psi_m + k & 0 & 0 & -\frac{1}{K}(n + \nu K\beta_m^2\psi_m) \\ \quad +K(1-\nu^2)\beta_m^4 - \rho\omega^2 & & & \\ 0 & 2K(1-\nu)\beta_m^2\psi_m & 1 & 0 \end{bmatrix}$$

$$\overline{\mathbf{P}} = \begin{bmatrix} 0 \\ 0 \\ -p_m \\ 0 \end{bmatrix}, \quad \psi_m = \frac{\alpha}{L_y\beta_m^2}\int_0^{L_y} \phi_m''\,\phi_m\,dy, \quad p_m = \frac{\alpha}{L_y}\int_0^{L_y} \overline{p}_z(x, y, t)\phi_m\,dy$$

$$\alpha = \begin{cases} 2 \text{ Simply supported-simply supported} \\ 1 \text{ Fixed-fixed or fixed-simply supported} \end{cases}$$

Case	Boundary Conditions		ψ_m $m = 1, 2, \ldots$			
	$y = 0$	$y = L_y$	ψ_1	ψ_2	ψ_3	ψ_m m large
1	Simply Supported	Simply Supported	1	1	1	1
2	Fixed	Simply Supported	$\frac{2.9317}{\eta_1}$	$\frac{6.0686}{\eta_2}$	$\frac{9.2095}{\eta_3}$	$\frac{\eta_m - 1}{\eta_m}$
3	Fixed	Fixed	$\frac{2.6009}{\eta_1}$	$\frac{5.8634}{\eta_2}$	$\frac{8.9984}{\eta_3}$	$\frac{\eta_m - 2}{\eta_m}$

The solution of Eq. (13.66) in transfer matrix and stiffness matrix form is given in Pilkey (1994). This reference contains matrices for very general plates, including plates with arbitrary loading. Also tabulated are mass matrices for rectangular plates.

EXAMPLE 13.3 Simply Supported Rectangular Plate under Uniform Loading

Consider a rectangular plate simply supported on all edges. The dimensions of the plate are $L_x = L$ in the x direction and L_y in the y direction. The loading on the plate is uniform with the intensity $\overline{p}_z = p_0$. Since the edges at $y = 0$ and $y = L_y$ are simply supported, the state variables can be expressed as in Eq. (13.61), and the loading can also be expanded into a sine series of Eq. (13.62), with Eq. (13.63) giving

$$p_m = \begin{cases} \frac{4p_0}{m\pi} & m = 1, 3, 5, \ldots \\ 0 & m = 2, 4, 6, \ldots \end{cases} \tag{1}$$

These lead to the first order differential relation of Eq. (13.64). Follow the procedure in Chapter 4 to obtain the transfer matrix solution

$$\mathbf{z}_x = \mathbf{U}^i\mathbf{z}_0 + \overline{\mathbf{z}}^i \tag{2}$$

with

$$\mathbf{z} = [w_m \quad \theta_m \quad V_m \quad m_m]^T$$

$$\mathbf{U}^i = \begin{bmatrix} U_{ww} & U_{w\theta} & U_{wV} & U_{wm} \\ U_{\theta w} & U_{\theta\theta} & U_{\theta V} & U_{\theta m} \\ U_{Vw} & U_{V\theta} & U_{VV} & U_{Vm} \\ U_{mw} & U_{m\theta} & U_{mV} & U_{mm} \end{bmatrix}$$

and

$$\overline{\mathbf{z}}^i = \begin{bmatrix} w_m^0 & \theta_m^0 & V_m^0 & m_m^0 \end{bmatrix}^T$$

or

$$\begin{bmatrix} w_m(x) \\ \theta_m(x) \\ V_m(x) \\ m_m(x) \end{bmatrix} = \begin{bmatrix} U_{ww} & U_{w\theta} & U_{wV} & U_{wm} \\ U_{\theta w} & U_{\theta\theta} & U_{\theta V} & U_{\theta m} \\ U_{Vw} & U_{V\theta} & U_{VV} & U_{Vm} \\ U_{mw} & U_{m\theta} & U_{mV} & U_{mm} \end{bmatrix} \begin{bmatrix} w_m(0) \\ \theta_m(0) \\ V_m(0) \\ m_m(0) \end{bmatrix} + \begin{bmatrix} w_m^0(x) \\ \theta_m^0(x) \\ V_m^0(x) \\ m_m^0(x) \end{bmatrix} \tag{3}$$

in which, with $\beta_m = \frac{m\pi}{L_y}$,

$$U_{ww} = -\frac{\beta_m(1-\nu)}{2} x \sinh \beta_m x + \cosh \beta_m x$$

$$U_{w\theta} = -\frac{1}{2}\left[(1+\nu)\frac{\sinh \beta_m x}{\beta_m} + (1-\nu)x \cosh \beta_m x\right]$$

$$U_{wV} = \frac{1}{2K\beta_m^2}\left(\frac{\sinh \beta_m x}{\beta_m} - x \cosh \beta_m x\right)$$

$$U_{wm} = -\frac{x \sinh \beta_m x}{2K\beta_m}$$

$$U_{\theta w} = -\frac{\beta_m^2}{2}\left[(1+\nu)\frac{\sinh \beta_m x}{\beta_m} - (1-\nu)x \cosh \beta_m x\right]$$

$$U_{\theta\theta} = \frac{\beta_m(1-\nu)}{2} x \sinh \beta_m x + \cosh \beta_m x$$

$$U_{\theta V} = x \frac{\sinh \beta_m x}{2K\beta_m}$$

$$U_{\theta m} = \frac{1}{2K}\left[\frac{\sinh \beta_m x}{\beta_m} + x \cosh \beta_m x\right]$$

$$U_{Vw} = \frac{K\beta_m^4}{2}\left[(3-2\nu-\nu^2)\frac{\sinh \beta_m x}{\beta_m} - (1-\nu)^2 x \cosh \beta_m x\right]$$

$$U_{V\theta} = -\frac{K\beta_m^3(1-\nu)^2}{2}[x \sinh \beta_m x]$$

$$U_{VV} = -\frac{\beta_m(1-\nu)}{2} x \sinh \beta_m x + \cosh \beta_m x$$

$$U_{Vm} = \frac{\beta_m^2}{2}\left[(1+\nu)\frac{\sinh \beta_m x}{\beta_m} - (1-\nu)x \cosh \beta_m x\right]$$

$$
\begin{aligned}
U_{mw} &= \frac{K\beta_m^3(1-\nu)^2}{2} x \sinh \beta_m x \\
U_{m\theta} &= \frac{K\beta_m^2}{2}\left[(3-2\nu-\nu^2)\frac{\sinh \beta_m x}{\beta_m} + (1-\nu)^2 x \cosh \beta_m x\right] \\
U_{mV} &= \frac{1}{2}\left[(1+\nu)\frac{\sinh \beta_m x}{\beta_m} + (1-\nu) x \cosh \beta_m x\right] \\
U_{mm} &= \frac{\beta_m(1-\nu)}{2} x \sinh \beta_m x + \cosh \beta_m x
\end{aligned}
$$

and

$$
\begin{aligned}
w_m^0 &= \frac{1}{2K\beta_m^4} p_m(\beta_m x \sinh \beta_m x - 2\cosh \beta_m x + 2) \\
\theta_m^0 &= \frac{1}{2K\beta_m^4} p_m \beta_m^2 \left(\frac{\sinh \beta_m x}{\beta_m} - x \cosh \beta_m x\right) \\
V_m^0 &= p_m \left(\frac{1-\nu}{2} x \cosh \beta_m x - \frac{3-\nu}{2\beta_m} \sinh \beta_m x\right) \\
m_m^0 &= -\frac{1}{\beta_m^2} p_m \left(\frac{1-\nu}{2} \beta_m x \sinh \beta_m x + \nu \cosh \beta_m x - \nu\right)
\end{aligned}
$$

The boundary conditions for the state variables are

$$
w_m(L) = m_m(L) = w_m(0) = m_m(0) = 0. \tag{4}
$$

Substitution of these conditions into (3) leads to

$$
\begin{aligned}
[U_{w\theta}\,\theta_m(0) + U_{wV}\,V_m(0)]_{x=L} &= -\left[w_m^0\right]_{x=L} \\
[U_{m\theta}\,\theta_m(0) + U_{mV}\,V_m(0)]_{x=L} &= -\left[m_m^0\right]_{x=L}
\end{aligned} \tag{5}
$$

Thus, the initial parameters $\theta_m(0)$ and $V_m(0)$ can be determined as

$$
\begin{aligned}
\theta_m(0) &= \frac{1}{\Delta}\left(-w_m^0 U_{mV} + m_m^0 U_{wV}\right)_{x=L} \\
V_m(0) &= \frac{1}{\Delta}\left(-m_m^0 U_{w\theta} + w_m^0 U_{m\theta}\right)_{x=L}
\end{aligned} \tag{6}
$$

where

$$
\Delta = [U_{w\theta} U_{mV} - U_{m\theta} U_{wV}]_{x=L} \tag{7}
$$

Once the state variable parameters at $x = 0$ are known, the values of the state variables, as well as those of the stresses, at any location of the plate can be determined. These manipulations are rather tedious for hand calculations, but very convenient for computers.

This solution leads to the deflection

$$
\begin{aligned}
w = \sum_{m=1}^{\infty} w_m(x) \sin \frac{m\pi y}{L_y} &= \frac{4p_0 L_y^4}{\pi^5 K} \sum_m \frac{1}{m^5}\left[1 + \left(\frac{\beta_m x}{2} + \tanh \frac{\beta_m L}{2}\right.\right. \\
&\left.\left. - \frac{\beta_m L}{4\cosh^2(\beta_m L/2)}\right) \sinh \beta_m x - \left(1 + \frac{\beta_m x}{2} \tanh \frac{\beta_m L}{2}\right) \cosh \beta_m x\right] \sin \beta_m y \\
& \qquad m = 1, 3, 5, \ldots
\end{aligned} \tag{8}
$$

The series converges so rapidly that often the first term provides sufficient accuracy. ∎

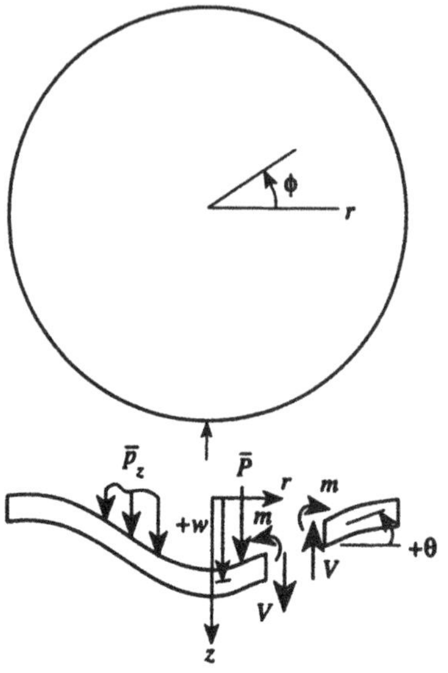

FIGURE 13.14
Polar coordinates. Positive displacement w, slope θ, moment m, shear force V, and applied loading $\bar{p}_z$. Sign Convention 1.

13.3.2 Circular Plates

For circular plates, it is convenient to switch from a rectangular to a polar coordinate system (Fig. 13.14). Equation (13.41) can simply be employed in polar coordinates with

$$\nabla^2 = \frac{\partial^2}{\partial r^2} + \frac{1}{r}\frac{\partial}{\partial r} + \frac{1}{r^2}\frac{\partial^2}{\partial \phi^2} \tag{13.67}$$

To derive this expression and other similar useful formulas in polar coordinates, begin with the relationships in Eq. (13.7). Derivatives are found by the chain rule of differentiation. For example,

$$\begin{aligned}\frac{\partial w}{\partial x} &= \frac{\partial w}{\partial r}\frac{\partial r}{\partial x} + \frac{\partial w}{\partial \phi}\frac{\partial \phi}{\partial x} = \frac{\partial w}{\partial r}\cos\phi - \frac{1}{r}\frac{\partial w}{\partial \phi}\sin\phi \\ \frac{\partial^2 w}{\partial x^2} &= \frac{\partial^2 w}{\partial r^2}\cos^2\phi - 2\frac{\partial^2 w}{\partial\phi\partial r}\frac{\sin\phi\cos\phi}{r} + \frac{\partial w}{\partial r}\frac{\sin^2\phi}{r} + 2\frac{\partial w}{\partial \phi}\frac{\sin\phi\cos\phi}{r} + \frac{\partial^2 w}{\partial \phi^2}\frac{\sin^2\phi}{r^2}\end{aligned} \tag{13.68}$$

Substitution of the derivatives of Eq. (13.68) and similar expressions into the governing equations of the previous subsections leads to the local form of the governing differential equation

$$\nabla^4 w = \frac{\bar{p}_z}{K} \tag{13.69a}$$

with $\nabla^4 = \nabla^2\nabla^2$, where ∇^2 is defined in Eq. (13.67) and $K = Et^3/[12(1-\nu^2)]$. Equation (13.69a) can be written as

$$\left(\frac{\partial^2}{\partial r^2} + \frac{1}{r}\frac{\partial}{\partial r} + \frac{1}{r^2}\frac{\partial^2}{\partial \phi^2}\right)\left(\frac{\partial^2}{\partial r^2} + \frac{1}{r}\frac{\partial}{\partial r} + \frac{1}{r^2}\frac{\partial^2}{\partial \phi^2}\right)w = \frac{\bar{p}_z}{K} \tag{13.69b}$$

The other fundamental quantities become

$$\begin{aligned}
m_r &= -K\left[\partial_r^2 w + \nu\left(\frac{1}{r}\partial_r w + \frac{1}{r^2}\partial_\phi^2 w\right)\right] \\
m_\phi &= -K\left(\frac{1}{r}\partial_r w + \frac{1}{r^2}\partial_\phi^2 w + \nu\partial_r^2 w\right) \\
m_{r\phi} &= -(1-\nu)K\left(\frac{1}{r}\partial_r\partial_\phi w - \frac{1}{r^2}\partial_\phi w\right) \\
q_r &= -K\,\partial_r\nabla^2 w \\
q_\phi &= -K\frac{1}{r}\partial_\phi\nabla^2 w
\end{aligned} \tag{13.70}$$

with the equivalent shear forces

$$\begin{aligned}
V_r^* &= V = q_r + \frac{1}{r}\partial_\phi m_{r\phi} = -K\left[\partial_r\nabla^2 w + \frac{1-\nu}{r}\partial_\phi\left(\frac{1}{r}\partial_r\partial_\phi w - \frac{1}{r^2}\partial_\phi w\right)\right] \\
V_\phi^* &= q_\phi + \partial_r m_{r\phi} = -K\left[\frac{1}{r}\partial_\phi\nabla^2 w + (1-\nu)\partial_r\left(\frac{1}{r}\partial_r\partial_\phi w - \frac{1}{r^2}\partial_\phi w\right)\right]
\end{aligned} \tag{13.71}$$

The stresses σ_r, σ_ϕ, $\tau_{r\phi}$ are given by

$$\sigma_r = \frac{12 m_r z}{t^3} \qquad \sigma_\phi = \frac{12 m_\phi z}{t^3} \qquad \tau_{r\phi} = \frac{12 m_{r\phi} z}{t^3} \tag{13.72}$$

Note that the peak stress values occur on the surfaces at $z = \pm t/2$.

The first-order governing differential equations can be obtained by rearranging Eqs. (13.70) and (13.71) and introducing the kinematic equation $\partial w/\partial r = -\theta$. Alternatively, convert Eq. (13.58) to a polar coordinate system, with the r axis coinciding with the x axis. These procedures lead to the equations

$$\frac{\partial}{\partial r}\begin{bmatrix} w \\ \theta \\ V \\ m_r \end{bmatrix} = \begin{bmatrix} 0 & -1 & 0 & 0 \\ \nu(1/r^2)\dfrac{\partial^2}{\partial\phi^2} & -\nu/r & 0 & 1/K \\ K\left[\dfrac{1}{r^4}(1-\nu^2)\dfrac{\partial^4}{\partial\phi^4} - \dfrac{2(1-\nu)}{r^4}\dfrac{\partial^2}{\partial\phi^2}\right] & -\dfrac{K(3-2\nu-\nu^2)}{r^3}\dfrac{\partial^2}{\partial\phi^2} & -1/r & -(\nu/r^2)\dfrac{\partial^2}{\partial\phi^2} \\ -\dfrac{K(3-2\nu-\nu^2)}{r^3}\dfrac{\partial^2}{\partial\phi^2} & \dfrac{K}{r^2}\left[(1-\nu^2) - 2(1-\nu)\dfrac{\partial^2}{\partial\phi^2}\right] & 1 & -(1-\nu)/r \end{bmatrix}\begin{bmatrix} w \\ \theta \\ V \\ m_r \end{bmatrix} + \begin{bmatrix} 0 \\ 0 \\ -\overline{p}_z \\ 0 \end{bmatrix}$$

$$\frac{\partial \mathbf{z}}{\partial r} = \mathbf{A}\qquad\qquad \mathbf{z} + \overline{\mathbf{P}} \tag{13.73}$$

Including the effects of an elastic foundation of modulus k, compressive in-plane forces per unit length n_r (radial) and n_ϕ (circumferential), and the dynamic response, the governing equations in first order form appear as

$$\frac{\partial \mathbf{z}}{\partial r} = \mathbf{A}\mathbf{z} + \overline{\mathbf{P}}$$

with $\mathbf{z}$ and $\overline{\mathbf{P}}$ the same as in Eq. (13.73) and

$$\mathbf{A} = \begin{bmatrix} 0 & \vdots & -1 & \vdots & 0 & \vdots & 0 \\ \cdots & & \cdots & & \cdots & & \cdots \\ (\nu/r^2)\frac{\partial}{\partial\phi^2} & \vdots & -\nu/r & \vdots & 0 & \vdots & 1/K \\ \cdots & & \cdots & & \cdots & & \cdots \\ K\left[\frac{(1-\nu^2)}{r^4}\frac{\partial^4}{\partial\phi^4} - \frac{2(1-\nu)}{r^4}\frac{\partial^2}{\partial\phi^2}\right] + \frac{1}{r^2}\frac{\partial}{\partial\phi}\left(n_\phi\frac{\partial}{\partial\phi}\right) + k + \rho\frac{\partial^2}{\partial t^2} & \vdots & -K\frac{(3-2\nu-\nu^2)}{r^3}\frac{\partial^2}{\partial\phi^2} & \vdots & -1/r & \vdots & -(\nu/r)\frac{\partial^2}{\partial\phi^2} \\ \cdots & & \cdots & & \cdots & & \cdots \\ -K\frac{(3-2\nu-\nu^2)}{r^3}\frac{\partial^2}{\partial\phi^2} & \vdots & -n_r + K\frac{(1-\nu^2)}{r^2} - 2K\frac{(1-\nu)}{r^2}\frac{\partial^2}{\partial\phi^2} & \vdots & 1 & \vdots & -(1-\nu)/r \end{bmatrix} \tag{13.74}$$

Reduction to Ordinary Differential Equations

The ϕ derivatives can be eliminated in Eqs. (13.73) and (13.74) by expanding the variables in a Fourier series

$$\mathbf{z} = \begin{bmatrix} w(r,\phi) \\ \theta(r,\phi) \\ V(r,\phi) \\ m_r(r,\phi) \end{bmatrix} = \sum_{m=0}^{\infty} \left\{ \begin{bmatrix} w_m^c(r) \\ \theta_m^c(r) \\ V_m^c(r) \\ m_m^c(r) \end{bmatrix} \cos m\phi + \begin{bmatrix} w_m^s(r) \\ \theta_m^s(r) \\ V_m^s(r) \\ m_m^s(r) \end{bmatrix} \sin m\phi \right\} \tag{13.75}$$

The loading $\overline{p}_z(r,\phi)$ should also be expanded in a Fourier series

$$\overline{p}_z(r,\phi) = \sum_{m=0}^{\infty} \left[p_m^c(r)\cos m\phi + p_m^s(r)\sin m\phi\right] \tag{13.76}$$

Multiply both sides of Eq. (13.76) by $\cos n\phi$ and integrate from $\phi = 0$ to $\phi = 2\pi$, i.e.,

$$\int_0^{2\pi} \overline{p}_z(r,\phi)\cos n\phi\, d\phi = \sum_{m=0}^{\infty}\left[\int_0^{2\pi} p_m^c(r)\cos m\phi\, \cos n\phi\, d\phi + \int_0^{2\pi} p_m^s(r)\sin m\phi\, \cos n\phi\, d\phi\right]$$

Since

$$\int_0^{2\pi} \cos m\phi\, \cos n\phi\, d\phi = \begin{cases} 2\pi & m = n = 0 \\ \pi & m = n \\ 0 & m \neq n \end{cases} \quad \text{and} \quad \int_0^{2\pi} \sin m\phi\, \cos n\phi\, d\phi = 0$$

it follows that

$$p_0^c(r) = \frac{1}{2\pi}\int_0^{2\pi} \overline{p}_z(r,\phi)\, d\phi \tag{13.77a}$$

$$p_m^c(r) = \frac{1}{\pi}\int_0^{2\pi} \overline{p}_z(r,\phi)\cos m\phi\, d\phi \qquad m > 0 \tag{13.77b}$$

Similarly, multiply both sides of Eq. (13.76) by $\sin n\phi$ and integrate and note the relationship

$$\int_0^{2\pi} \sin m\phi \ \sin n\phi \ d\phi = \begin{cases} \pi & m = n, m > 0 \\ 0 & m \neq n \end{cases}$$

Then

$$p_m^s(r) = \frac{1}{\pi}\int_0^{2\pi} \overline{p}_z(r, \phi) \sin m\phi \ d\phi \quad m > 0 \tag{13.77c}$$

Several properties of these series are of interest. For symmetrical response of a plate, m is equal to zero, and the series expansions reduce to

$$\begin{aligned} w(r,\phi) &= w(r) = w_0^c(r), & \theta(r,\phi) &= \theta(r) = \theta_0^c(r), \\ m_r(r,\phi) &= m_r(r) = m_0^c(r), & V(r,\phi) &= V(r) = V_0^c(r) \end{aligned}$$
$$\overline{p}_z(r,\phi) = \overline{p}_z(r) = p_m^c(r)$$

If $\overline{p}_z(r, \phi)$ is an odd function of ϕ, i.e., if $\overline{p}_z(r, \phi) = -\overline{p}_z(r, -\phi)$, then

$$p_m^c(r) = 0 \quad \text{for } m = 0, 1, 2, \ldots,$$

and Eq. (13.75) reduces to a sine series. For even functions of ϕ, i.e., $\overline{p}_z(r, \phi) = \overline{p}_z(r, -\phi)$.

$$p_m^s(r) = 0 \quad \text{for } m = 0, 1, 2, \ldots,$$

and Eq. (13.75) contains only cosine terms.

The Fourier series expansions for w and $\overline{p}_z$ placed in Eq. (13.69b) lead to

$$\left(\frac{d^2}{dr^2} + \frac{1}{r}\frac{d}{dr} - \frac{m^2}{r^2}\right)\left(\frac{d^2 w_m}{dr^2} + \frac{1}{r}\frac{dw_m}{dr} - \frac{m^2}{r^2}w_m\right) = p_m^j \tag{13.78a}$$
$$j = c \text{ or } s, \quad m = 0, 1, 2, \ldots$$

which is readily shown to have the complementary functions

$$\begin{aligned} m = 0:&\quad w_m = C_1 + C_2 \ln r + C_3 r^2 + C_4 r^2 \ln r \\ m = 1:&\quad w_m = C_1 r + C_2/r + C_3 r^3 + C_4 r \ln r \\ m \geq 2:&\quad w_m = C_1 r^m + C_2 r^{-m} + C_3 r^{2+m} + C_4 r^{2-m} \end{aligned} \tag{13.78b}$$

where the arbitrary constants are different for each m. With the addition of particular solutions, these functions can be used to find the response of a circular plate with arbitrary loading.

Insertion of the series expansions for $\mathbf{z}$ (Eq. 13.75) and $\overline{p}_z$ (Eq. 13.76) in $\partial_r \mathbf{z} = \mathbf{A}\mathbf{z} + \overline{\mathbf{P}}$ of Eq. (13.74) leads to ordinary governing differential equations

$$\frac{d\mathbf{z}}{dr} = \mathbf{A}\mathbf{z} + \overline{\mathbf{P}} \tag{13.79a}$$

or

$$\begin{aligned} \frac{dw_m^j}{dr} &= -\theta_m^j \qquad j = c \quad \text{or} \quad s, \quad m = 0, 1, 2, 3, \ldots \\ \frac{d\theta_m^j}{dr} &= \frac{m_m^j}{K} - \frac{\nu}{r}\theta_m^j - \frac{\nu m^2}{r^2}w_m^j \end{aligned}$$

$$\begin{aligned}\frac{dV_m^j}{dr} &= -\frac{V_m^j}{r} + \frac{\nu m^2}{r^2} m_m^j + \left\{\frac{Km^2(1-\nu)}{r^4}[2+m^2(1+\nu)] - \frac{n_\phi m^2}{r^2} - \rho\omega^2\right\} w_m^j \\ &\quad + (3+\nu)\frac{Km^2(1-\nu)}{r^3}\theta_m^j + kw_m^j - p_m^j(r) \\ \frac{dm_m^j}{dr} &= -(1-\nu)\frac{m_m^j}{r} + V_m^j + K(1-\nu)(3+\nu)\frac{m^2}{r^3}w_m^j + \left[-n + \frac{K(1-\nu)(1+\nu+2m^2)}{r^2}\right]\theta_m^j\end{aligned} \tag{13.79b}$$

where the elastic foundation term kw_m^j and the harmonic inertia term $\rho\omega^2 w_m^j$ are included. These are the governing equations for a circular plate that is symmetric geometrically about $r = 0$. Note that there is no need for the applied loading to be symmetric.

In the case of a symmetrically loaded plate, the local (differential) governing equations are no longer functions of ϕ. They are obtained from Eq. (13.79) with $m = 0$. Let $w_0^j = w, \theta_0^j = \theta, V_0^j = V$, and $m_0^j = m_r$.

$$\begin{aligned}\frac{d}{dr}\begin{bmatrix} w \\ \theta \\ V \\ m_r \end{bmatrix} &= \begin{bmatrix} 0 & -1 & 0 & 0 \\ 0 & -\nu/r & 0 & 1/K \\ k-\rho\omega^2 & 0 & -1/r & 0 \\ 0 & -n_r + K(1-\nu^2)/r^2 & 1 & -(1-\nu)/r \end{bmatrix}\begin{bmatrix} w \\ \theta \\ V \\ m_r \end{bmatrix} + \begin{bmatrix} 0 \\ 0 \\ -\overline{p}_z(r) \\ 0 \end{bmatrix} \\ \frac{d\mathbf{z}}{dr} &= \mathbf{A}(r) \qquad \mathbf{z} + \overline{\mathbf{P}}\end{aligned} \tag{13.80}$$

It is convenient to solve Eqs. (13.79) and (13.80) in transfer matrix form. In solving these differential equations remember that $\mathbf{A}$ is a function of r. These transfer matrices can be transformed into stiffness matrices using the transformation of Eq. (4.16) of Chapter 4. Pilkey (1994) contains a comprehensive collection of transfer and stiffness matrices for circular plates including statically responding plates with asymmetric distributed and concentrated loading. This reference also lists lumped and consistent mass matrices for symmetric and asymmetric motion.

In higher order form, the equations for a symmetrically loaded circular plate appear as

$$\begin{aligned}&\nabla^4\omega = \left(\frac{d^2}{dr^2} + \frac{1}{r}\frac{d}{dr}\right)\left(\frac{d^2w}{dr^2} + \frac{1}{r}\frac{dw}{dr}\right) = \frac{\overline{p}_z}{K} \\ &m_r = -K\left(\frac{d^2w}{dr^2} + \frac{\nu}{r}\frac{dw}{dr}\right) \qquad m_\phi = -K\left(\nu\frac{d^2w}{dr^2} + \frac{1}{r}\frac{dw}{dr}\right) \\ &m_{r\phi} = 0, \qquad V_\phi = 0 \\ &V_r^* = V = -K\left(\frac{d^3w}{dr^3} + \frac{1}{r}\frac{d^2w}{dr^2} - \frac{1}{r^2}\frac{dw}{dr}\right)\end{aligned} \tag{13.81}$$

EXAMPLE 13.4 Plate Element Without a Center Hole, Under Symmetrical Uniform Loading

Construct the transfer matrix for the plate element shown in Fig. 13.14. The plate is loaded by a uniform pressure $\overline{p}_z$.

Since the load is symmetric, the deformations and internal forces are symmetric. The complementary function can be taken from Eq. (13.78b) for $m = 0$. The particular solution is Cr^4, where C is a constant that can be evaluated using Eq. (13.78a). Alternatively, use the relationships of Eq. (13.81), beginning with the governing equation

$$\nabla^4 w = \left(\frac{d^2}{dr^2} + \frac{1}{r}\frac{d}{dr}\right)\left(\frac{d^2w}{dr^2} + \frac{1}{r}\frac{dw}{dr}\right) = \frac{1}{r}\frac{d}{dr}\left\{r\frac{d}{dr}\left[\frac{1}{r}\frac{d}{dr}\left(r\frac{dw}{dr}\right)\right]\right\} = \frac{\overline{p}_z}{K} \tag{1}$$

It is readily observed that integration of this equation for constant $\overline{p}_z$ gives

$$w = C_1 + C_2 \ln r + C_3 r^2 + C_4 r^2 \ln r + \frac{\overline{p}_z}{64K} r^4 \tag{2}$$

Then

$$\begin{aligned}
\theta &= -\frac{dw}{dr} = -\frac{C_2}{r} - 2C_3 r - C_4 r(2\ln r + 1) - \frac{\overline{p}_z r^3}{16K} \\
V &= -K\left(\frac{d^3 w}{dr^3} + \frac{1}{r}\frac{d^2 w}{dr^2} - \frac{1}{r^2}\frac{dw}{dr}\right) = -K\left[C_4 \frac{4}{r} + \frac{\overline{p}_z r}{2K}\right] \\
m &= -K\left(\frac{d^2 w}{dr^2} + \frac{\nu}{r}\frac{dw}{dr}\right) = -K\left\{\frac{C_2}{r^2}(\nu - 1) + C_4[(1+\nu)(2\ln r + 1) + 2]\right. \\
&\qquad \left. + 2C_3(1+\nu) + \frac{\overline{p}_z r^2}{16K}(3+\nu)\right\}
\end{aligned} \tag{3}$$

or in matrix form,

$$\begin{bmatrix} w \\ \theta \\ V \\ m \end{bmatrix} = \begin{bmatrix} 1 & \ln r & r^2 & r^2 \ln r \\ 0 & -1/r & -2r & -r(2\ln r + 1) \\ 0 & 0 & 0 & -4K/r \\ 0 & -K(\nu - 1)/r^2 & -2K(1+\nu) & -K[(1+\nu)(2\ln r + 1) + 2] \end{bmatrix} \begin{bmatrix} C_1 \\ C_2 \\ C_3 \\ C_4 \end{bmatrix} + \begin{bmatrix} \overline{p}_z r^4/(64K) \\ -\overline{p}_z r^3/(16K) \\ -\overline{p}_z r/2 \\ -\overline{p}_z r^2(3+\nu)/(16) \end{bmatrix} \tag{4}$$

In order to convert this to transfer matrix form it is necessary to express the integration constants $C_i, i = 1, 2, \ldots, 4$, in terms of the state variables w, θ, V, and m at $r = 0$. Note that all of these variables must have finite magnitudes at $r = 0$. From this condition, it can be reasoned that $C_4 = C_2 = 0$. It is of interest to note that since the deflection is symmetric, $\theta = 0$ at $r = 0$. In general, the total shear force along the perimeter of a circle of radius r is

$$2\pi r V = \int_0^{2\pi} \int_0^r \overline{p}_z(r)\, r\, dr\, d\theta = 2\pi \int_0^r \overline{p}_z(r)\, r\, dr \tag{5}$$

so that

$$V = \frac{1}{r} \int_0^r \overline{p}_z(r)\, r\, dr \tag{6}$$

It is apparent that as $r \to 0$, $V \to 0$, i.e., $V_0 = 0$. Imposition of the conditions that θ and V are zero at $r = 0$, leads again to the conclusion that $C_4 = C_2 = 0$.

With $C_4 = C_2 = 0$, (4) at $r = 0$ becomes

$$\begin{bmatrix} w(0) \\ m(0) \end{bmatrix} = \begin{bmatrix} 1 & 0 \\ 0 & -2K(1+\nu) \end{bmatrix} \begin{bmatrix} C_1 \\ C_3 \end{bmatrix} \tag{7}$$

or

$$C_1 = w(0) = w_0, \qquad C_3 = -m(0)/[2K(1+\nu)] = -m_0/[2K(1+\nu)]$$

Thus, the state variables in terms of their initial values have the form

$$\begin{aligned} w &= w_0 - m_0 \frac{r^2}{2K(1+\nu)} + \frac{\overline{p}_z r^4}{64K} \\ \theta &= m_0 \frac{r}{K(1+\nu)} - \frac{\overline{p}_z r^3}{16K} \\ V &= -\frac{\overline{p}_z r}{2} \\ m &= m_0 - \frac{\overline{p}_z r^2}{16}(3+\nu) \end{aligned} \tag{8}$$

or in matrix notation,

$$\begin{bmatrix} w \\ \theta \\ V \\ m \end{bmatrix} = \begin{bmatrix} 1 & 0 & 0 & -r^2/[2K(1+\nu)] \\ 0 & 0 & 0 & r/[K(1+\nu)] \\ 0 & 0 & 0 & 0 \\ 0 & 0 & 0 & 1 \end{bmatrix} \begin{bmatrix} w_0 \\ \theta_0 \\ V_0 \\ m_0 \end{bmatrix} + \begin{bmatrix} \overline{p}_z r^4/(64K) \\ -\overline{p}_z r^3/(16K) \\ -\overline{p}_z r/2 \\ -\overline{p}_z r^2(3+\nu)/16 \end{bmatrix} \tag{9}$$

$$\mathbf{z} \quad = \quad \mathbf{U}^i \quad \mathbf{z} \quad + \quad \overline{\mathbf{z}}^i$$

where $\mathbf{U}^i$ is the transfer matrix. ∎

EXAMPLE 13.5 Responses of a Circular Plate Without a Center Hole

Calculate the responses for a circular plate with no center hole. Suppose the symmetrical loading is uniformly distributed. The plate is clamped on the boundary $r = b$.

The responses can be calculated using Eq. (9) of Example 13.4. The initial parameters w_0, θ_0, V_0, and M_0 have to be determined first. It is already known from Example 13.4 that

$$\theta_0 = V_0 = 0 \tag{1}$$

At the fixed outer boundary where $r = b$,

$$w|_{r=b} = 0, \qquad \theta|_{r=b} = 0 \tag{2}$$

Substitution of (1) and (2) into Eq. (9) of Example 13.4 with $r = b$ leads to

$$\begin{aligned} w_0 - m_0 b^2/[2K(1+\nu)] + \overline{p}_z b^4/(64K) &= 0 \\ m_0 b/[K(1+\nu)] - \overline{p}_z b^3/(16K) &= 0 \end{aligned} \tag{3}$$

Solve these relationships to find

$$w_0 = \frac{\overline{p}_z b^4}{64K} \qquad m_0 = \frac{\overline{p}_z b^2}{16}(1+\nu) \tag{4}$$

Thus, the responses are

$$\begin{aligned} w &= w_0 - \frac{r^2}{2K(1+\nu)} m_0 + \frac{\overline{p}_z r^4}{64K} = \frac{\overline{p}_z}{64K}(r^2 - b^2)^2 \\ \theta &= \frac{r}{K(1+\nu)} m_0 - \frac{\overline{p}_z r^3}{16K} = \frac{\overline{p}_z r}{16K}(b^2 - r^2) \end{aligned}$$

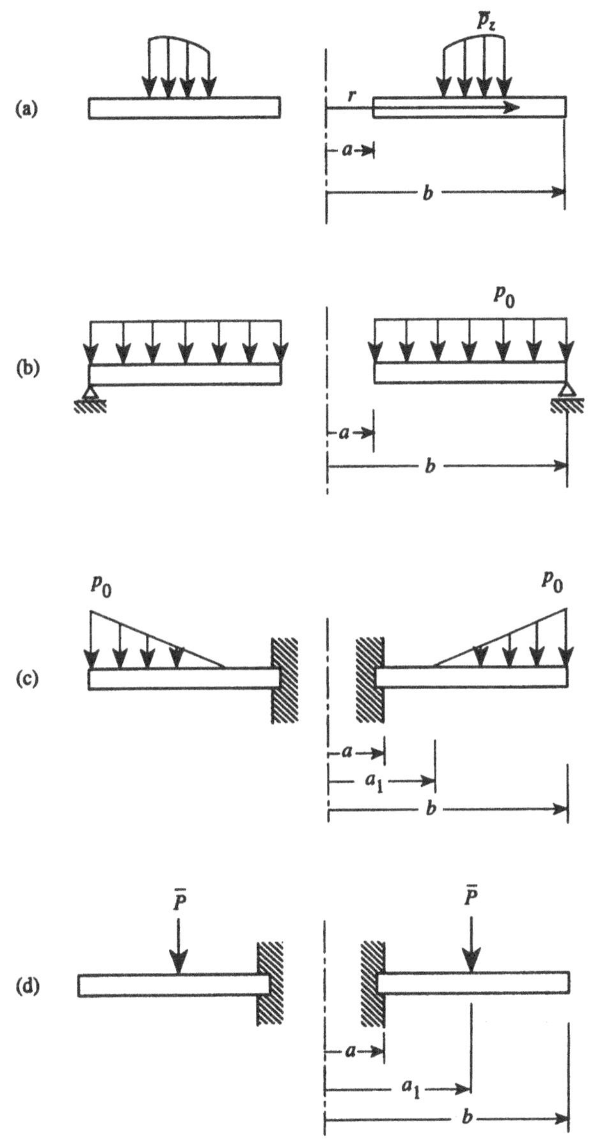

FIGURE 13.15
A symmetrically loaded plate element with a center hole.

$$V = -\overline{p}_z r/2 \tag{5}$$

$$m = m_0 - \frac{\overline{p}_z r^2(3+\nu)}{16} = \frac{\overline{p}_z}{16}[b^2(1+\nu) - r^2(3+\nu)]$$

∎

EXAMPLE 13.6 Plate Element with Center Hole

Consider a symmetrically loaded plate element with a center hole as shown in Fig. 13.15a.

Begin with Eqs. (2) and (3) of Example 13.4. For the moment assume that $\overline{p}_z$ is not constant in the radial direction and make the appropriate adjustments in Eqs. (2) and (3). Evaluate w, θ, V and m at $r = a$ and solve for the integration constants $C_i, i = 1, \ldots, 4$. Use

$w_{r=a} = w_0, \theta_{r=a} = \theta_0, V_{r=a} = V_0$, and $m_{r=a} = m_0$. Substitute these expressions for C_i into Eqs. (2) and (3) of Example 13.4 to find

$$
\begin{aligned}
w &= w_0 + \theta_0 \left[-\frac{1}{2}(1+\nu)a \ln\frac{r}{a} - (1-\nu)\frac{(r^2-a^2)}{4a} \right] + V_0 \frac{a}{4K}\left[-(a^2+r^2)\ln\frac{r}{a} + (r^2-a^2) \right] \\
&\quad + m_0 \left(\frac{a^2}{2K}\ln\frac{r}{a} - \frac{r^2-a^2}{4K} \right) + \int^r \frac{1}{r}\int r \int \frac{1}{r}\int \frac{\overline{p}_z r}{K} dr \\
&= w_0\, U_{ww} + \theta_0\, U_{w\theta} + V_0\, U_{wV} + m_0\, U_{wm} + w^0 \\
\theta &= \theta_0 \left[(1+\nu)\frac{a}{2r} + (1-\nu)\frac{r}{2a} \right] + V_0 \left[\frac{ar}{2K}\ln\frac{r}{a} - \frac{a}{4Kr}(r^2-a^2) \right] \\
&\quad + m_0 \frac{1}{2Kr}(r^2-a^2) - \frac{1}{r}\int^r r \int \frac{1}{r}\int \frac{\overline{p}_z r}{K} dr \\
&= \theta_0\, U_{\theta\theta} + V_0\, U_{\theta V} + m_0\, U_{\theta m} + \theta^0 \\
V &= V_0 \frac{a}{r} - \frac{1}{r}\int^r \overline{p}_z r\, dr = V_0\, U_{VV} + V^0 \\
m &= \theta_0 (1-\nu^2)\frac{Ka}{2}\left(\frac{1}{a^2} - \frac{1}{r^2} \right) + V_0 \frac{a}{2}\left[(1+\nu)\ln\frac{r}{a} + (1+\nu)\frac{r^2-a^2}{2r^2} \right] \\
&\quad + m_0 \left(\frac{1-\nu}{2}\frac{a^2}{r^2} + \frac{1+\nu}{2} \right) - \int^r \frac{1}{r}\int \overline{p}_z r\, dr + \frac{1-\nu}{r^2}\int^r r \int \frac{1}{r}\int \overline{p}_z r\, dr \\
&= \theta_0\, U_{m\theta} + V_0\, U_{mV} + m_0\, U_{mm} + m^0
\end{aligned} \tag{1}
$$

These relationships define the transfer matrix for a symmetrically loaded circular plate.

As a particular example, consider the plate with a uniformly distributed symmetrical load $\overline{p}_z = p_0$, beginning at the inner edge as shown in Fig. 13.15b. The loading terms of (1) become

$$
\begin{aligned}
w^0 &= \frac{p_0}{K}\int_a^r \frac{1}{r}\int r \int \frac{1}{r}\int r\, dr = \frac{p_0}{8K}\left[\frac{r^4}{8} - \frac{5a^4}{8} + \frac{a^2r^2}{2} - a^2\left(r^2 + \frac{a^2}{2} \right)\ln\frac{r}{a} \right] \\
\theta^0 &= -\frac{p_0}{K}\frac{1}{r}\int_a^r r \int \frac{1}{r}\int r\, dr = -\frac{p_0}{4K}\left(\frac{r^3}{4} - \frac{a^4}{4r} + a^2 r \ln\frac{r}{a} \right) \\
V^0 &= -\frac{p_0}{r}\int_a^r r\, dr = -\frac{p_0}{2r}(r^2-a^2) \\
m^0 &= p_0 \left(-\int_a^r \frac{1}{r}\int r\, dr + \frac{1-\nu}{r^2}\int_a^r r \int \frac{1}{r}\int r\, dr \right) \\
&= -\frac{p_0}{4}\left[\frac{3+\nu}{4}r^2 - a^2 + \frac{1}{4r^2}(1-\nu)a^4 - (1+\nu)a^2 \ln\frac{r}{a} \right]
\end{aligned} \tag{2}
$$

If the outer edge is simply supported and the inner edge is free, the boundary conditions are $m_0 = V_0 = 0$, $w|_{r=b} = 0$, and $m|_{r=b} = 0$. Then the first and the fourth equations of (1) are

$$
\begin{aligned}
w|_{r=b} &= w_0 + \theta_0 \left[-\frac{1}{2}(1+\nu)a \ln\frac{b}{a} - (1-\nu)\frac{b^2-a^2}{4a} \right] + w^0|_{r=b} = 0 \\
m|_{r=b} &= \theta_0 (1-\nu^2)\frac{Ka}{2}\left(\frac{1}{a^2} - \frac{1}{b^2} \right) + m^0|_{r=b} = 0
\end{aligned} \tag{3}
$$

The initial parameters w_0 and θ_0 can be determined by solving (3)

$$\theta_0 = \frac{p_0}{4}\left[\frac{3+\nu}{4}b^2 - a^2 + \frac{1}{4b^2}(1-\nu)a^4 - (1+\nu)a^2\ln\frac{b}{a}\right] \Big/ \left[(1-\nu)^2\frac{Ka}{2}\left(\frac{1}{a^2}-\frac{1}{b^2}\right)\right]$$

$$\begin{aligned} w_0 = &-\frac{p_0}{8K}\left[\frac{b^4}{8} - \frac{5a^4}{8} + \frac{a^2b^2}{2} - a^2\left(b^2+\frac{a^2}{2}\right)\ln\frac{b}{a}\right] \\ &-\frac{p_0}{4}\left[\frac{3+\nu}{4}b^2 - a^2 + \frac{1}{4b^2}(1-\nu)a^4 - (1+\nu)a^2\ln\frac{b}{a}\right]\left[-\frac{1}{2}(1+\nu)a\ \ln\frac{b}{a}\right. \\ &\left.-(1-\nu)\frac{(b^2-a^2)}{4a}\right] \Big/ \left[(1-\nu^2)\frac{Ka}{2}\left(\frac{1}{a^2}-\frac{1}{b^2}\right)\right] \end{aligned} \tag{4}$$

Substitution of these values for w_0 and θ_0, along with $m_0 = V_0 = 0$ in (1) provides the desired responses.

As a second example, consider a plate with the linearly increasing distributed symmetrical load shown in Fig. 13.15c. This load, which begins at $r = a_1$, is expressed as

$$\overline{p}_z(r) = \begin{cases} 0 & \text{if } r < a_1 \\ p_0\frac{r-a_1}{b-a_1} & \text{if } r \geq a_1 \end{cases} = \langle r-a_1\rangle^0 p_0\frac{r-a_1}{b-a_1} \tag{5}$$

where $\langle r-a_1\rangle^0$ is the unit step function defined as

$$\langle r-a_1\rangle^0 = \begin{cases} 0 & \text{if } r < a_1 \\ 1 & \text{if } r \geq a_1 \end{cases}$$

The loading terms of (1) become

$$\begin{aligned} w^0(r) &= \int^r \frac{1}{r}\int r\int\frac{1}{r}\int\frac{\overline{p}_z}{K}\,dr \\ &= \begin{cases} 0 & r < a_1 \\ \int_{a_1}^r \frac{1}{r}\int r\int\frac{1}{r}\int\frac{p_0 r(r-a_1)}{K(b-a_1)}\,dr & r \geq a_1 \end{cases} \\ &= \begin{cases} 0 & r < a_1 \\ \frac{p_0}{K(b-a_1)}\left[\frac{r^5}{225} - \frac{a_1r^4}{64} - \frac{a_1^3r^2}{144} + \frac{29a_1^5}{1600} + \frac{a_1^3}{8}\left(\frac{r^2}{3}+\frac{a_1^2}{10}\right)\ln\frac{r}{a_1}\right] & r \geq a_1 \end{cases} \\ &= \langle r-a_1\rangle^0\frac{p_0}{K(b-a_1)}\left[\frac{r^5}{225} - \frac{a_1r^4}{64} - \frac{a_1^3r^2}{144} + \frac{29a_1^5}{1600} + \frac{a_1^3}{8}\left(\frac{r^2}{3}+\frac{a_1^2}{10}\right)\ln\frac{r}{a_1}\right] \end{aligned} \tag{6}$$

$$\begin{aligned} \theta^0(r) &= \begin{cases} 0 & r < a_1 \\ -\frac{1}{r}\int_{a_1}^r r\int\frac{1}{r}\int\frac{p_0 r(r-a_1)}{K(b-a_1)}\,dr & r \geq a_1 \end{cases} \\ &= \begin{cases} 0 & r < a_1 \\ -\frac{p_0}{K(b-a_1)}\left[\frac{r^4}{45} - \frac{a_1r^3}{16} + \frac{a_1^3r}{36} + \frac{a_1^5}{80r} + \frac{a_1^3}{12}r\ \ln\frac{r}{a_1}\right] & r \geq a_1 \end{cases} \end{aligned} \tag{7}$$

$$\begin{aligned} V^0(r) &= \begin{cases} 0 & r < a_1 \\ -\frac{1}{r}\int_{a_1}^r \frac{p_0 r(r-a_1)}{(b-a_1)}\,dr & r \geq a_1 \end{cases} \\ &= \begin{cases} 0 & r < a_1 \\ -\frac{p_0}{(b-a_1)}\left(\frac{r^2}{3} - \frac{a_1r}{2} + \frac{a_1^3}{6r}\right) & r \geq a_1 \end{cases} \end{aligned} \tag{8}$$

$$m^0(r) = \begin{cases} 0 & r < a_1 \\ -\int_{a_1}^{r} \frac{1}{r} \int \frac{p_0 r(r-a_1)}{(b-a_1)}\,dr + \frac{1-\nu}{r^2}\int_{a_1}^{r} r \int \frac{1}{r} \int \frac{p_0 r(r-a_1)}{(b-a_1)}\,dr & r \geq a_1 \end{cases}$$

$$= -\langle r-a_1\rangle^0 \frac{p_0}{(b-a_1)}\left(\frac{4+\nu}{45}r^3 - \frac{3+\nu}{16}a_1 r^2 + \frac{4+\nu}{36}a_1^3 - \frac{1-\nu}{80}\frac{a_1^5}{r^2} + \frac{1+\nu}{12}a_1^3 \ln\frac{r}{a_1}\right) \tag{9}$$

Since the inner edge is fixed and the outer edge is free (cantilevered), the boundary conditions are $w_{r=a} = w_0 = 0$, $\theta_{r=a} = \theta_0 = 0$, $m_{r=b} = 0$, and $V_{r=b} = 0$. From (1), these conditions lead to the two equations.

$$\begin{aligned} V_{r=b} &= V_0\, U_{VV}(b) + V^0(b) = 0 \\ m_{r=b} &= V_0\, U_{mV}(b) + m_0\, U_{mm}(b) + m^0(b) = 0 \end{aligned} \tag{10}$$

which can be solved for V_0 and m_0. Equation (1) is now a fully defined response for this plate.

For a third example of a plate with a center hole, place a concentrated line load $\overline{P}$ (force/length) at $r = a_1$ as shown in Fig. 13.15d. One approach is to represent the concentrated line load $\overline{P}$ in terms of the distributed load $\overline{p}_z$ using a delta function δ in the form $\overline{p}_z(r) = \overline{P}\delta(r, a_1)$, where the delta function is characterized by the sifting property

$$\int_a^r f(r)\,\delta(r, a_1)\,dr = \begin{cases} 0 & \text{if } r < a_1 \\ f(a_1) & \text{if } r \geq a_1 \end{cases} = \langle r-a_1\rangle^0 f(a_1) \tag{11}$$

where $\langle r-a_1\rangle^0$ is the unit step function of (5). The solution is given by (1), with the loading terms (particular solutions)

$$\begin{aligned} w^0(r) &= \int_a^r \frac{1}{r}\int r \int \frac{1}{r}\int \frac{\overline{p}_z r}{K}\,dr = \int_a^r \frac{1}{r}\int r\int\frac{1}{r}\int \frac{\overline{P}r}{K}\delta(r,a_1)\,dr \\ &= \begin{cases} 0 & r < a_1 \\ \frac{\overline{P}}{K}\int_{a_1}^r \frac{1}{r}\int_{a_1}^r r \int_{a_1}^r \frac{1}{r} a_1\,dr & r \geq a_1 \end{cases} \\ &= \langle r-a_1\rangle^0 \frac{\overline{P}}{K}\int_{a_1}^r \frac{1}{r}\int_{a_1}^r r\int_{a_1}^r \frac{1}{r}a_1\,dr \\ &= \begin{cases} 0 & r < a_1 \\ \frac{\overline{P}}{K}\int_{a_1}^r \frac{1}{r}\int_{a_1}^r r\, a_1 \ln\frac{r}{a_1}\,dr & r \geq a_1 \end{cases} \\ &= \begin{cases} 0 & r < a_1 \\ \frac{\overline{P}a_1}{4K}\int_{a_1}^r \left(2r\ln\frac{r}{a_1} - r + \frac{a_1^2}{r}\right)dr & r \geq a_1 \end{cases} \\ &= \begin{cases} 0 & r < a_1 \\ \frac{\overline{P}a_1}{4K}\left[(r^2+a_1^2)\ln\frac{r}{a_1} - (r^2-a_1^2)\right] & r \geq a_1 \end{cases} \\ &= \langle r-a_1\rangle^0 \frac{\overline{P}a_1}{4K}\left[(r^2+a_1^2)\ln\frac{r}{a_1} - (r^2-a_1^2)\right] \end{aligned} \tag{12}$$

$$
\begin{aligned}
\theta^0(r) &= -\frac{1}{r}\int_a^r r\int\frac{1}{r}\int\frac{\overline{p}_z r}{K}\,dr \\
&= \begin{cases} 0 & r < a_1 \\ -\frac{\overline{P}a_1}{2K}\left[r\ln\frac{r}{a_1} - \frac{1}{2r}(r^2 - a_1^2)\right] & r \geq a_1 \end{cases}
\end{aligned} \tag{13}
$$

$$
V^0(r) = -\frac{1}{r}\int_a^r \overline{p}_z r\,dr = \begin{cases} 0 & r < a_1 \\ -\overline{P}\frac{a_1}{r} & r \geq a_1 \end{cases}
$$

$$
m^0(r) = -\int_a^r \frac{1}{r}\int \overline{p}_z r\,dr + \frac{1-\nu}{r^2}\int_a^r r\int\frac{1}{r}\int\overline{p}_z r\,dr \tag{14}
$$

$$
= \begin{cases} 0 & r < a_1 \\ -\frac{\overline{P}a_1}{2}\left[(1+\nu)\ln\frac{r}{a_1} - \frac{1-\nu}{2}\left(1 - \frac{a_1^2}{r^2}\right)\right] & r \geq a_1 \end{cases} \tag{15}
$$

The inner and outer rims are fixed and the boundary conditions $w_{r=a} = w_0 = 0, \theta_{r=a} = \theta_0 = 0$, $w_{r=b} = \theta_{r=b} = 0$ imposed on (1) lead to

$$
\begin{aligned}
w_{r=b} &= V_0\,U_{wV}(b) + m_0\,U_{wm}(b) + w^0(b) = 0 \\
\theta_{r=b} &= V_0\,U_{\theta V}(b) + m_0\,U_{\theta m}(b) + \theta^0(b) = 0
\end{aligned} \tag{16}
$$

Substitution into (1) of V_0 and m_0, which are obtained from (16), provides a fully defined response.

An alternative approach for handling concentrated applied line loads (applied force or moment) is to take advantage of the similarity between an in-span line load and the initial stress resultants V_0 and m_0 of (1). For example, the applied line load $\overline{P}$ has the same effect on the response as the initial shear force V_0. Of course $\overline{P}$ of Fig. 13.15d is applied at $r = a_1$, while V_0 occurs at $r = a$. Also, whereas $\overline{P}$ acts downwards, V_0 is (Fig. 13.14) directed upwards. Thus, from (1), the influence of V_0 on the deflection w is

$$
+V_0\frac{a}{4K}\left[-(a^2+r^2)\ln\frac{r}{a} + (r^2 - a^2)\right] \tag{17}
$$

while the contribution of the applied load $\overline{P}$ to the deflection would be

$$
-\overline{P}\frac{a_1}{4K}\left[-(a_1^2+r^2)\ln\frac{r}{a_1} + (r^2 - a_1^2)\right] = w^0 \tag{18}
$$

It is evident from (12) that this is the same result found by integration. Similar reasoning leads to the expressions for θ^0, V^0, and m^0. ▮

13.3.3 Variational (Integral) Relationships

The internal virtual work is given by

$$
\delta W_i = -\int_A \delta\boldsymbol{\epsilon}^T\mathbf{s}\,dA
$$

with $\mathbf{s} = \mathbf{E}\boldsymbol{\epsilon}$ taken from Eq. (13.35) and $\boldsymbol{\epsilon} = \mathbf{D}_u\mathbf{u}$ taken from Eq. (13.34). Supplement this with the external virtual work to find the displacement form (Eq. 2.58)

$$
\delta W = \delta W_i + \delta W_e = -\int_A \delta\mathbf{u}^T(\mathbf{k}^D\mathbf{u} - \overline{\mathbf{p}}_V)\,dA + \int_{S_p}\delta\mathbf{u}^T\overline{\mathbf{p}}\,ds = 0 \tag{13.82}
$$

with the differential stiffness operator $\mathbf{k}^D = {}_u\mathbf{D}^T\mathbf{E}\,\mathbf{D}_u$. The quantities $\overline{\mathbf{p}}_V$ and $\overline{\mathbf{p}}$ contain $\overline{p}_z$ and $\overline{p}_s$, the applied transverse loading and the loading on the edges of the plate, respectively. Since $\mathbf{u} = [w]$ and

$$\mathbf{D}_u = \begin{bmatrix} -\partial_x^2 \\ -\partial_y^2 \\ -2\partial_x\,\partial_y \end{bmatrix} \tag{13.83}$$

the virtual work becomes

$$\delta W = -\int_A \delta w \left\{ K \begin{bmatrix} {}_{xx}\partial(\partial_{xx} + \nu\,\partial_{yy}) + \\ {}_{yx}\partial\, 2(1-\nu)\partial_{xy} + \\ {}_{yy}\partial(\partial_{yy} + \nu\partial_{xx}) \end{bmatrix} w - \overline{p}_z \right\} dA + \int_{S_p} \delta w\, \overline{p}_s\, ds = 0 \tag{13.84a}$$

so that

$$\mathbf{k}^D = K[{}_{xx}\partial(\partial_{xx} + \nu\partial_{yy}) + {}_{yx}\partial\, 2(1-\nu)\partial_{xy} + {}_{yy}\partial(\partial_{yy} + \nu\partial_{xx})] \tag{13.84b}$$

13.4 Shear Deformation Effects

The history of the attempts to develop analytical models to represent the bending of a plate can be traced to the early 1800s and the work of Sophie Germain[4], Lagrange, and Poisson [Todhunter and Pearson, 1886]. These efforts resulted in what we now refer to as Kirchhoff or classical plate theory, with transverse shear strains γ_{xz} and γ_{yz} neglected. Poisson (1829) addressed the question of boundary conditions and contended that three boundary conditions should be prescribed on each edge. As mentioned in Section 13.3.1, Kirchhoff (1850) reasoned that two edge conditions are more suitable than three and defined a special shear force to reduce the number of forces on the boundary from three to two. As mentioned previously, Thomas and Tait (1883) supplemented an energy-related explanation of the equivalent shear force with a physical clarification.

The fundamentals of the plate theory became clearer when Eric Reissner (1945) introduced the effect of transverse shear strains. Mindlin[5] (1951) formulated the plate bending problem with the influence of rotary inertia and shear deformation, which is consistent with Reissner's formulation. Plate theory, including transverse shear deformation effects, which is sometimes called the *Reissner-Mindlin plate theory*, will be considered briefly here.

[4]Sophie Germain (1776–1831) was France's most significant female mathematician during her time. She was self-educated, using the library of her well-to-do father. She decided to become a mathematician, over the objections of her parents, when, at thirteen, she read of Archimedes being killed by a Roman soldier. As a teenager she obtained some lecture notes from the École Polytechnique, including some of those of Lagrangė. Using the pseudonym Le Blanc, she submitted a major homework project to Lagrange. He was so impressed that he discovered her real name and began to assist her. She initiated correspondence with scholars such as Legendre. She returned to the use of the name Le Blanc in corresponding with Gauss. He discovered her real name after she inquired of a French general as to the safety of Gauss when French troops were occupying parts of Germany.

[5]Raymond David Mindlin (1906–1987), an American civil engineer, was born in New York City and attended Columbia University for all of his degrees. He remained as a faculty member at Columbia, where he had a distinguished career including significant developments in plate theory. His 1936 PhD dissertation presented the elasticity solution of the "Mindlin Problem," the effect of a concentrated force in the interior of a semi-infinite solid.

This will lead to a system of six first order equations. Hence, three boundary conditions on each edge may be prescribed.

13.4.1 Fundamental Local Relationships

The kinematical relations of Eq. (13.23) and the equilibrium conditions of Eq. (13.31) still hold for a plate with shear deformation. The material law, however, requires special attention.

The material law of Eq. (13.29) needs to be supplemented with relationships between shear resultants q_x, q_y and the strains γ_{xz}, γ_{yz} referred to the middle surface. To find these relationships, set the internal work due to the distributed stresses and strains equal to the internal work of the stress resultants and the strains of the middle surface. For example, for q_x, the work

$$-W_i = \int_A \int_z \tau_{xz}\, \gamma_{xz}\, dz\, dA$$

should be equal to the work

$$\int_A q_x\, \gamma_{xz}\, dA$$

which is expressed in terms of middle surface variables. From the former expression, with $\gamma_{xz} = \tau_{xz}/G$,

$$\begin{aligned}\int_A \int_z \tau_{xz}\, \gamma_{xz}\, dz\, dA &= \int_A \int_z \frac{\tau_{xz}^2}{G}\, dz\, dA = \int_A \int_{-\frac{t}{2}}^{\frac{t}{2}} \frac{1}{G}\left\{\frac{3q_x}{2t}\left[1 - \left(2\frac{z}{t}\right)^2\right]\right\}^2 dz\, dA \\ &= \int_A \frac{8}{15}\left(\frac{3}{2}\frac{q_x}{t}\right)^2 \frac{t}{G} dA = \int_A \frac{6}{5}\frac{q_x^2}{Gt}\, dA = \int_A q_x \frac{6}{5}\frac{q_x}{Gt}\, dA \end{aligned} \tag{13.85}$$

where the shear stress distribution of Eq. (13.46a) was employed, i.e., $\tau_{xz} = (3q_x/2t)[1 - (2z/t)^2]$. Since, in terms of the middle surface, the work is $\int_A q_x\, \gamma_{xz}\, dA$, we conclude that by comparison with the final integral of Eq. (13.85) $(\gamma_{xz})_{\text{middle surface}} = \frac{6}{5}\frac{q_x}{Gt}$. Thus, the material law for shear becomes

$$\begin{bmatrix} q_x \\ q_y \end{bmatrix} = Gt \begin{bmatrix} 5/6 & \\ & 5/6 \end{bmatrix} \begin{bmatrix} \gamma_{xz} \\ \gamma_{yz} \end{bmatrix}. \tag{13.86}$$

The complete material law then appears as

$$\begin{bmatrix} m_x \\ m_y \\ m_{xy} \\ q_x \\ q_y \end{bmatrix} = \begin{bmatrix} K \begin{bmatrix} 1 & \nu & 0 \\ \nu & 1 & 0 \\ 0 & 0 & \frac{1-\nu}{2} \end{bmatrix} & \mathbf{0} \\ \mathbf{0} & Gt \begin{bmatrix} 5/6 & 0 \\ 0 & 5/6 \end{bmatrix} \end{bmatrix} \begin{bmatrix} \kappa_x \\ \kappa_y \\ 2\kappa_{xy} \\ \gamma_{xz} \\ \gamma_{yz} \end{bmatrix} = \begin{bmatrix} \mathbf{E}_B & \mathbf{0} \\ \mathbf{0} & \mathbf{E}_V \end{bmatrix} \epsilon \tag{13.87}$$

$$\mathbf{s} \quad = \quad \mathbf{E} \quad \epsilon$$

or

$$\begin{bmatrix} \kappa_x \\ \kappa_y \\ 2\kappa_{xy} \\ \gamma_{xz} \\ \gamma_{yz} \end{bmatrix} = \begin{bmatrix} \frac{12}{Et^3}\begin{bmatrix} 1 & -\nu & 0 \\ -\nu & 1 & 0 \\ 0 & 0 & 2(1+\nu) \end{bmatrix} & \mathbf{0} \\ \mathbf{0} & \frac{1}{Gt}\begin{bmatrix} 6/5 & 0 \\ 0 & 6/5 \end{bmatrix} \end{bmatrix} \begin{bmatrix} m_x \\ m_y \\ m_{xy} \\ q_x \\ q_y \end{bmatrix} \tag{13.88}$$

$$\epsilon \quad = \quad \mathbf{E}^{-1} \quad \mathbf{s}$$

13.4.2 Boundary Conditions

The surface stress resultants of Eq. (13.57) still apply. The displacement boundary conditions will be

$$w = \overline{w}, \qquad \theta_a = \overline{\theta}_a, \qquad \theta_t = \overline{\theta}_t \tag{13.89a}$$

and the force boundary conditions

$$m_a = \overline{m}_a, \qquad m_t = \overline{m}_t, \qquad q_a = \overline{q}_a. \tag{13.89b}$$

Thus, for this plate with shear deformation, the specification of boundary conditions is straightforward. This is in contrast to the previous model without shear deformation, where it was necessary to define an equivalent shear force.

13.4.3 Local Form of the Governing Equations

The relationships of this section can be rearranged in a variety of ways. One approach of particular interest involves defining two invariant combinations

$$\Phi \equiv \partial_x\theta_x + \partial_y\theta_y, \qquad \Psi \equiv \partial_x\theta_y - \partial_y\theta_x \tag{13.90}$$

Substitution of the kinematic relation of Eq. (13.23) and Hooke's law of Eq. (13.29) into the second and third of the equilibrium expressions of Eq. (13.31a) leads to

$$q_x = K\left(\partial_x\Phi - \frac{1-\nu}{2}\partial_y\Psi\right)$$

$$q_y = K\left(\partial_y\Phi + \frac{1-\nu}{2}\partial_x\Psi\right)$$

Substitute these relationships into the first equation of the equilibrium conditions of Eq. (13.31a) to find

$$K\nabla^2\Phi = -\overline{p}_z \qquad \nabla^2 = \partial_x^2 + \partial_y^2$$

The final two equations of the kinematic relations of Eq. (13.22), together with the expressions for γ_{xy} and γ_{xz} of Eq. (13.86), yield

$$\partial_x w = -\theta_x + \frac{K}{Gt}\frac{6}{5}\left(\partial_x\Phi - \frac{1-\nu}{2}\partial_y\Psi\right)$$

$$\partial_y w = -\theta_y + \frac{K}{Gt}\frac{6}{5}\left(\partial_y\Phi + \frac{1-\nu}{2}\partial_x\Psi\right)$$

These equations lead to a viable form of the governing equations

$$\begin{aligned} \nabla^2 w &= -\Phi + \frac{K}{Gt}\frac{6}{5}\nabla^2\Phi = -\Phi - \frac{6}{5}\frac{\overline{p}_z}{G} \\ \Psi - \frac{K}{G}\frac{6}{5}\frac{1-\nu}{2}\nabla^2\Psi &= \Psi - \frac{t^3}{10}\nabla^2\Psi = 0 \end{aligned} \tag{13.91}$$

The governing equations can also be arranged in first order form similar to Eqs. (13.58). For this plate theory, which includes shear deformation effects, there are six variables having derivatives with respect to x. The first order equations using these variables as the state variables are

$$
\begin{aligned}
\frac{\partial w}{\partial x} &= -\theta_x + \frac{6}{5}\frac{q_x}{Gt} \\
\frac{\partial \theta_x}{\partial x} &= \frac{1}{(1+\nu-\nu^2)K}\left(m_x - K\frac{\partial \theta_y}{\partial y}\right) \\
\frac{\partial \theta_y}{\partial x} &= -\frac{\partial \theta_x}{\partial y} + \frac{2}{K(1-\nu)}m_{xy} \\
\frac{\partial q_x}{\partial x} &= -Gt\frac{5}{6}\left(\frac{\partial^2 w}{\partial y^2} + \frac{\partial \theta_y}{\partial y}\right) - \overline{p}_z \\
\frac{\partial m_x}{\partial x} &= q_x - \frac{\partial m_{yx}}{\partial y} \\
\frac{\partial m_{xy}}{\partial x} &= \frac{\partial w}{\partial y} + \theta_y + \frac{\nu}{1+\nu-\nu^2}\frac{\partial m_x}{\partial y} - K\nu\frac{1-\nu}{1+\nu-\nu^2}\frac{\partial^2 \theta_y}{\partial y^2}
\end{aligned}
\tag{13.92}
$$

The first three of these equations come from the kinematic relations of Eq. (13.22) and the material law of Eq. (13.29), the final three from the equilibrium conditions of Eq. (13.31).

Note that when the shear deformation effects are included, the plane normal to the middle surface does not remain normal after deformation. This can be seen from the first relation of Eq. (13.92). Moreover, this relation implies that the rotations are independent variables and hence there are six state variables, including three displacements (w, θ_x, θ_y) and three forces (q_x, m_x, m_{xy}). There is no need to define an artificial shear force on the boundary as we did for the Kirchhoff plate theory, where it was necessary to reduce the number of forces on the boundary from three to two. Also, compare the first equation of Eqs. (13.92) with Chapter 1, Eq. (1.133a) and note that the factor 6/5 is analogous to $1/k_s$ of beam theory. If $k_s = 5/6$, the shear correction factor for plates, the first relation of Eq. (13.92) can be written as $\frac{\partial w}{\partial x} = -\theta_x + \frac{q_x}{k_s G t}$, which is similar to beam theory. It is shown in Pilkey (2002) that it may be wise to choose a different shear correction factor for very thin plates.

13.4.4 Variational Relationships

It is convenient to express the internal virtual work in two terms, one for bending and one representing shear effects

$$
-\delta W_i = \int_A \delta\boldsymbol{\epsilon}^T \mathbf{s}\, dA = \int_A \delta\boldsymbol{\kappa}^T \mathbf{m}\, dA + \int_A \delta\boldsymbol{\gamma}^T \mathbf{q}\, dA \tag{13.93}
$$

where the stress resultants are

$$
\mathbf{s} = [m_x \quad m_y \quad m_{xy} \quad q_x \quad q_y]^T = [\mathbf{m} \quad \mathbf{q}]^T
$$

and the strains are given by

$$
\boldsymbol{\epsilon} = [\kappa_x \quad \kappa_y \quad 2\kappa_{xy} \quad \gamma_{xz} \quad \gamma_{yz}]^T = [\boldsymbol{\kappa} \quad \boldsymbol{\gamma}]^T
$$

Using this notation, the material law of Eq. (13.87) can be expressed as

$$
\begin{bmatrix} \mathbf{m} \\ \mathbf{q} \end{bmatrix} = \begin{bmatrix} \mathbf{E}_B & \mathbf{0} \\ \mathbf{0} & \mathbf{E}_V \end{bmatrix} \begin{bmatrix} \boldsymbol{\kappa} \\ \boldsymbol{\gamma} \end{bmatrix} \tag{13.94}
$$

$$
\mathbf{s} \quad = \quad \mathbf{E} \quad \boldsymbol{\epsilon}
$$

where the definitions of $\mathbf{E}_B$ and $\mathbf{E}_V$ are evident in Eq. (13.87). Also, the strain-displacement relation of Eq. (13.23) can be divided into two expressions

$$\begin{aligned} \begin{bmatrix} \kappa_x \\ \kappa_y \\ 2\kappa_{xy} \end{bmatrix} &= \begin{bmatrix} \partial_x & 0 \\ 0 & \partial_y \\ \partial_y & \partial_x \end{bmatrix} \begin{bmatrix} \theta_x \\ \theta_y \end{bmatrix} \\ \boldsymbol{\kappa} &= \quad \mathbf{D}_\kappa \qquad \boldsymbol{\theta} \\ \begin{bmatrix} \gamma_{xz} \\ \gamma_{yz} \end{bmatrix} &= \begin{bmatrix} \partial_x & 1 & 0 \\ \partial_y & 0 & 1 \end{bmatrix} \begin{bmatrix} w \\ \theta_x \\ \theta_y \end{bmatrix} \\ \boldsymbol{\gamma} &= \quad \mathbf{D}_\gamma \qquad \mathbf{u} \end{aligned} \tag{13.95}$$

Finally, the internal virtual work can be expressed as

$$\begin{aligned} -\delta W_i &= \int_A \delta(\mathbf{Du})^T \mathbf{E}\, \mathbf{Du}\, dA \\ &= \int_A \delta(\mathbf{D}_\kappa \boldsymbol{\theta})^T \mathbf{E}_B\, \mathbf{D}_\kappa\, \boldsymbol{\theta}\, dA + \int_A \delta(\mathbf{D}_\gamma \mathbf{u})^T \mathbf{E}_V\, \mathbf{D}_\gamma\, \mathbf{u}\, dA \\ &= \int_A \delta\boldsymbol{\theta}^T\, {}_\kappa\mathbf{D}^T\, \mathbf{E}_B\, \mathbf{D}_\kappa\, \boldsymbol{\theta}\, dA + \int_A \delta\mathbf{u}^T\, {}_\gamma\mathbf{D}^T\, \mathbf{E}_V\, \mathbf{D}_\gamma\, \mathbf{u}\, dA \end{aligned} \tag{13.96}$$

which leads to the desired principle of virtual work expression

$$\begin{aligned} \delta W = &-K \int_A \underbrace{\delta[\theta_x \;\; \theta_y]}_{\delta\boldsymbol{\theta}^T} \underbrace{\begin{bmatrix} {}_x\partial\partial_x + {}_y\partial\frac{1-\nu}{2}\partial_y & {}_x\partial\nu\partial_y + {}_y\partial\frac{1-\nu}{2}\partial_x \\ {}_y\partial\nu\partial_x + {}_x\partial\frac{1-\nu}{2}\partial_y & {}_y\partial\partial_y + {}_x\partial\frac{1-\nu}{2}\partial_x \end{bmatrix}}_{\mathbf{k}_B^D} \underbrace{\begin{bmatrix} \theta_x \\ \theta_y \end{bmatrix}}_{\boldsymbol{\theta}} dA \\ &- \frac{5Gt}{6} \int_A \underbrace{\delta[w \;\; \theta_x \;\; \theta_y]}_{\delta\mathbf{u}^T} \underbrace{\begin{bmatrix} {}_x\partial\partial_x + {}_y\partial\partial_y & {}_y\partial & {}_x\partial \\ \partial_y & 1 & 0 \\ \partial_x & 0 & 1 \end{bmatrix}}_{\mathbf{k}_V^D} \underbrace{\begin{bmatrix} w \\ \theta_x \\ \theta_y \end{bmatrix}}_{\mathbf{u}} dA \\ &+ \int_A \underbrace{\delta[w \;\; \theta_x \;\; \theta_y]}_{\delta\mathbf{u}^T} \underbrace{\begin{bmatrix} \overline{p}_z \\ 0 \\ 0 \end{bmatrix}}_{\overline{\mathbf{p}}_V} dA + \int_{S_p} \underbrace{\delta[w \;\; \theta_x \;\; \theta_y]}_{\delta\mathbf{u}^T} \underbrace{\begin{bmatrix} \overline{p}_s \\ 0 \\ 0 \end{bmatrix}}_{\overline{\mathbf{p}}} ds \\ = &-K\left[\int_A \delta\boldsymbol{\theta}^T \mathbf{k}_B^D\, \boldsymbol{\theta}\, dA + \frac{6k_s(1-\nu)}{t^2} \int_A \delta\mathbf{u}^T \mathbf{k}_V^D\, \mathbf{u}\, dA\right] \\ &+ \int_A \delta\mathbf{u}^T\, \overline{\mathbf{p}}_V\, dA + \int_{S_p} \delta\mathbf{u}^T \overline{\mathbf{p}}\, ds = 0 \end{aligned} \tag{13.97}$$

where $k_s = 5/6$ is the shear correction factor, or

$$\begin{aligned} \delta W = &- \int_A \delta[w \;\; \theta_x \;\; \theta_y] \\ &\times \begin{bmatrix} {}_x\partial\frac{5Gt}{6}\partial_x + {}_y\partial\frac{5Gt}{6}\partial_y & {}_y\partial\frac{5Gt}{6} & {}_x\partial\frac{5Gt}{6} \\ \frac{5Gt}{6}\partial_y & {}_x\partial K\partial_x + {}_y\partial\frac{K(1-\nu)}{2}\partial_y + \frac{5Gt}{6} & {}_x\partial K\nu\partial_y + {}_y\partial\frac{K(1-\nu)}{2}\partial_x \\ \frac{5Gt}{6}\partial_x & {}_y\partial\nu K\partial_x + {}_x\partial\frac{K(1-\nu)}{2}\partial_y & {}_y\partial K\partial_y + {}_x\partial\frac{K(1-\nu)}{2}\partial_x + \frac{5Gt}{6} \end{bmatrix} \begin{bmatrix} w \\ \theta_x \\ \theta_y \end{bmatrix} dA \\ &+ \int_A \delta[w \;\; \theta_x \;\; \theta_y] \begin{bmatrix} \overline{p}_z \\ 0 \\ 0 \end{bmatrix} dA + \int_{S_p} \delta[w \;\; \theta_x \;\; \theta_y] \begin{bmatrix} \overline{p}_s \\ 0 \\ 0 \end{bmatrix} ds = 0 \end{aligned} \tag{13.98}$$

13.5 Finite Element Solutions

Thus far, the governing equations of plates in local (differential) as well as in global (integral) forms have been developed. In the previous sections, some analytical stiffness matrices are derived. These are useful in solving a certain range of plate problems, such as circular plate problems. However, if plates of arbitrary geometry are to be treated, it is necessary to use approximate (finite element) stiffness matrices. In this section, some approximate stiffness matrices for plate elements will be developed. The derivations of these elements provide a basic background in the principles underlying element development. However, there are many plate elements and only a few are addressed here. See a reference such as Zienkiewicz and Taylor (2000) for a more thorough treatment of plate elements.

Plate elements are often important in the analysis of shells. Shell elements are usually developed by

(1) Discretizing a shell theory,
(2) Discretizing three-dimensional continuum equations (degenerated shell elements),
(3) Superimposing the stiffness matrices of thin, flat elements for membrane (in-plane) effects and for plate bending.

Because of the third approach, plate elements, such as those developed in this section, are useful in the computational analysis of shells.

A simple plate element shape is the rectangle. This will be treated first, followed by triangular plate elements.

13.5.1 Rectangular Plate Element

A stiffness matrix, including shear deformation effects, for the rectangular element in Fig. 13.16 will be developed. Choose as nodal DOF the deflection w and the two slopes θ_x and θ_y. Thus,

$$\begin{aligned} \mathbf{v}^i &= [w_1\ w_2\ w_3\ w_4 \quad \theta_{x1}\ \theta_{x2}\ \theta_{x3}\ \theta_{x4} \quad \theta_{y1}\ \theta_{y2}\ \theta_{y3}\ \theta_{y4}]^T \\ &= [\quad \mathbf{w} \qquad\qquad \boldsymbol{\theta} \quad]^T \end{aligned} \tag{13.99}$$

Since shear deformation effects are to be taken into account, the shape functions for θ_x and θ_y can be chosen separately. One approach for selecting trial functions is to use those

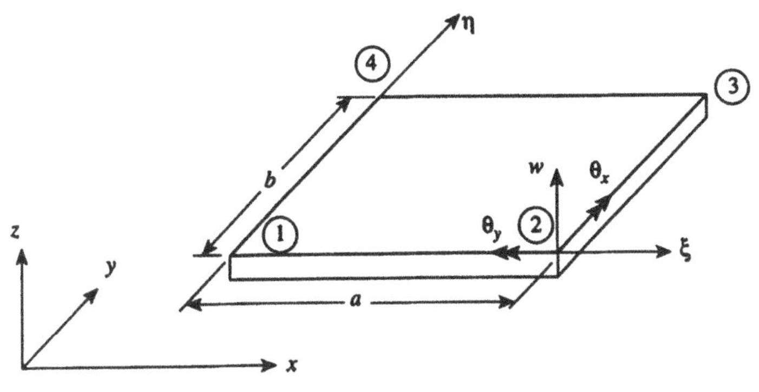

FIGURE 13.16
Rectangular element.

developed in Chapter 6, Section 6.4 for a rectangular element with in-plane deformation. Thus, introduce the bilinear shape functions

$$\mathbf{u} = \begin{bmatrix} w \\ \cdots \\ \theta_x \\ \cdots \\ \theta_y \end{bmatrix} = \begin{bmatrix} N_1\ N_2\ N_3\ N_4 & \vdots & 0 & \vdots & 0 \\ \cdots & & \cdots & & \cdots \\ 0 & \vdots & N_1\ N_2\ N_3\ N_4 & \vdots & 0 \\ \cdots & & \cdots & & \cdots \\ 0 & \vdots & 0 & \vdots & N_1\ N_2\ N_3\ N_4 \end{bmatrix} \mathbf{v}^i$$

$$= \mathbf{N}\mathbf{v}^i = \begin{bmatrix} \mathbf{N}_w & \mathbf{0} \\ \mathbf{0} & \mathbf{N}_\theta \end{bmatrix} \begin{bmatrix} \mathbf{w} \\ \boldsymbol{\theta} \end{bmatrix} \tag{13.100}$$

where

$$\mathbf{N}_w = [N_1\ N_2\ N_3\ N_4]$$

$$\mathbf{N}_\theta = \begin{bmatrix} N_1 & N_2 & N_3 & N_4 & & & 0 & \\ & & 0 & & N_1 & N_2 & N_3 & N_4 \end{bmatrix}$$

and

$$N_1 = (1-\xi)(1-\eta) \qquad N_2 = \xi(1-\eta)$$
$$N_3 = \xi\eta \qquad N_4 = \eta(1-\xi)$$

These shape functions will lead to a first order (C^0) element, for which the displacement w and the rotations θ_x and θ_y will be continuous along the element boundaries.

The stiffness matrix will be established using the principle of virtual work, with δW_i given by (Eq. 13.97)

$$-\delta W_i = K\left[\underbrace{\int_A \delta\boldsymbol{\theta}^T\, \mathbf{k}_B^D\, \boldsymbol{\theta}\, dA}_{\text{I}} + \frac{6k_s(1-\nu)}{t^2}\underbrace{\int_A \delta\mathbf{u}^T\, \mathbf{k}_V^D\, \mathbf{u}\, dA}_{\text{II}}\right] \tag{13.101}$$

where $\boldsymbol{\theta} = [\theta_x\ \theta_y]^T$, $\mathbf{k}_B^D = {}_\kappa\mathbf{D}^T\mathbf{E}_B\mathbf{D}_\kappa$, and $\mathbf{k}_V^D = {}_\gamma\mathbf{D}^T\mathbf{E}_V\mathbf{D}_\gamma$ are given in Eq. (13.97). Integral I will lead to the stiffness matrix $\mathbf{k}_B$ for bending, and integral II will provide the stiffness matrix $\mathbf{k}_V$ corresponding to shear deformation effects.

In the case of in-plane deformation, the internal virtual work (Eq. 13.18) takes a form similar to Eq. (13.101)

$$-\delta W_i = \int_A \delta\mathbf{u}^T\mathbf{k}^D\mathbf{u}\, dA \tag{13.102}$$

with $\mathbf{u} = [u_x\ u_y]^T$ and $\mathbf{k}^D = {}_u\mathbf{D}^T\mathbf{E}\,\mathbf{D}_u$. This leads to the stiffness matrix $\mathbf{k}$ for in-plane deformation.

Since $\mathbf{D}_\kappa$ (Eq. 13.95) = $\mathbf{D}_u$ (Eq. 13.1c) and $\mathbf{E}_B$ (Eq. 13.87) = $\frac{K}{D}\mathbf{E}$ (Eq. 13.10a), we conclude that $\mathbf{k}_B = \frac{K}{D}\mathbf{k} = \frac{t^2}{12}\mathbf{k}$. It is apparent that, with the factor $t^2/12$ the stiffness matrix for bending can be formed directly from the stiffness matrix (Eq. (6.34), Chapter 6), for in-plane deformation. Thus, if the displacement vector $\boldsymbol{\theta}$ of $\mathbf{k}_B$ is expanded to include $\mathbf{w}$,

$$
\mathbf{k}_B = \frac{Et}{24(1-\nu^2)}\frac{t^2}{12}
\begin{array}{c}
\begin{matrix} [w_1 & w_2 & w_3 & w_4 & \theta_{x1} & \theta_{x2} & \theta_{x3} & \theta_{x4} & \theta_{y1} & \theta_{y2} & \theta_{y3} & \theta_{y4}\ ] \end{matrix} \\
\left[\begin{array}{cccccccccccc}
0 & & & & & & & & & & & \\
 & 0 & & & & & & & & & & \\
 & & 0 & & & & & 0 & & & & \\
 & & & 0 & & & & & & & & \\
 & & & & A_{\alpha\beta} & C_{\alpha\beta} & -A_{\alpha\beta}/2 & B_{\alpha\beta} & \nu_2 & -\nu_3 & -\nu_2 & \nu_3 \\
 & & & & & A_{\alpha\beta} & B_{\alpha\beta} & -A_{\alpha\beta}/2 & \nu_3 & -\nu_2 & -\nu_3 & \nu_2 \\
 & & & & & & A_{\alpha\beta} & C_{\alpha\beta} & -\nu_2 & \nu_3 & \nu_2 & -\nu_3 \\
 & & & & & & & A_{\alpha\beta} & -\nu_3 & \nu_2 & \nu_3 & -\nu_2 \\
 & & & & & \text{Symmetric} & & & A_{\beta\alpha} & B_{\beta\alpha} & -A_{\beta\alpha}/2 & C_{\beta\alpha} \\
 & & & & & & & & & A_{\beta\alpha} & C_{\beta\alpha} & -A_{\beta\alpha}/2 \\
 & & & & & & & & & & A_{\beta\alpha} & B_{\beta\alpha} \\
 & & & & & & & & & & & A_{\beta\alpha}
\end{array}\right]
\end{array}
\tag{13.103}
$$

$$
\alpha = b/a, \qquad \beta = a/b, \qquad \nu_1 = 1-\nu, \qquad \nu_2 = 3(1+\nu), \qquad \nu_3 = 3(1-3\nu)
$$

$$
A_{\alpha\beta} = 8\alpha + 4\beta\nu_1, \qquad B_{\alpha\beta} = -4\alpha - 2\beta\nu_1, \qquad C_{\alpha\beta} = -8\alpha + 2\beta\nu_1
$$

$A_{\beta\alpha}$, $B_{\beta\alpha}$, and $C_{\beta\alpha}$ are obtained by interchanging α and β in the expressions for $A_{\alpha\beta}$, $B_{\alpha\beta}$ and $C_{\alpha\beta}$, respectively.

Turn to integral II of Eq. (13.101), which accounts for shear deformation effects. From Eqs. (13.96) and (13.101), with $\zeta = 5Gt/(6K) = 6k_s(1-\nu)/t^2$

$$
\begin{aligned}
K\zeta \int_A \delta\mathbf{u}^T \mathbf{k}_V^D\, \mathbf{u}\, dA &= K\zeta \int_A \delta(\mathbf{D}_\gamma \mathbf{u})^T \mathbf{E}_V \mathbf{D}_\gamma \mathbf{u}\, dA \\
&= K\zeta \int_A \delta(\mathbf{D}_\gamma \mathbf{N}\,\mathbf{v})^T \mathbf{E}_V\, \mathbf{D}_\gamma \mathbf{N}\,\mathbf{v}\, dA = K\zeta\, \delta\mathbf{v}^{iT} \int_A \mathbf{B}_\gamma^T \mathbf{B}_\gamma\, dA\, \mathbf{v}^i
\end{aligned}
\tag{13.104}
$$

or

$$
\mathbf{k}_V = \int_A \mathbf{B}_\gamma^T \mathbf{B}_\gamma\, dA
$$

with $\mathbf{B}_\gamma = \mathbf{D}_\gamma \mathbf{N}$. Note that with $\frac{5Gt}{6}$ factored out, $\mathbf{E}_V$ of Eq. (13.87) becomes an identity matrix. We find

$$
\mathbf{B}_\gamma = \begin{bmatrix} \mathbf{B}_{11} & \mathbf{B}_{12} & 0 \\ \mathbf{B}_{21} & 0 & \mathbf{B}_{23} \end{bmatrix} \tag{13.105}
$$

where

$$
\begin{aligned}
\mathbf{B}_{11} &= [-(1-\eta)/a \quad (1-\eta)/a \quad \eta/a \quad -\eta/a] \\
\mathbf{B}_{12} &= [(1-\eta)(1-\xi) \quad \xi(1-\eta) \quad \xi\eta \quad 1-\eta] \\
\mathbf{B}_{21} &= [-(1-\xi)/b \quad -\xi/b \quad \xi/b \quad (1-\xi)/b] \\
\mathbf{B}_{23} &= [(1-\xi)(1-\eta) \quad \xi(1-\eta) \quad \xi\eta \quad 1(1-\eta)\eta]
\end{aligned}
$$

Note that the first row ($\mathbf{B}_{11}$ and $\mathbf{B}_{12}$) of $\mathbf{B}_\gamma$ corresponds to γ_{xz} and the second row ($\mathbf{B}_{21}$ and $\mathbf{B}_{23}$) corresponds to γ_{yz}. The scheme of numerical integration of Fig. 13.17 will be employed to perform the required integration. The Gaussian integration point is chosen at the center of the element, which corresponds to $n = 1$ of Table 6.7, i.e., at $\xi = \eta = 0.5$. At this point, $\mathbf{B}_\gamma$ becomes

$$
\widehat{\mathbf{B}}_\gamma = \begin{bmatrix} -\frac{1}{2a} \ \frac{1}{2a} \ \frac{1}{2a} - \frac{1}{2a} & \frac{1}{4}\ \frac{1}{4}\ \frac{1}{4}\ \frac{1}{4} & \mathbf{0} \\ -\frac{1}{2b} - \frac{1}{2b}\ \frac{1}{2b}\ \frac{1}{2b} & \mathbf{0} & \frac{1}{4}\ \frac{1}{4}\ \frac{1}{4}\ \frac{1}{4} \end{bmatrix} \tag{13.106}
$$

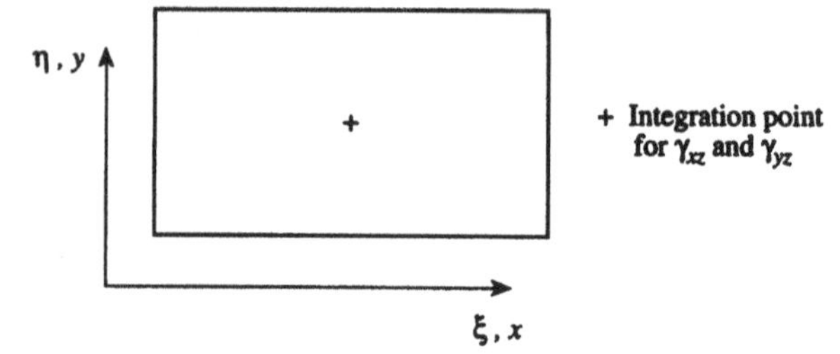

FIGURE 13.17
Numerical integration scheme.

Then $\mathbf{k}_V$ can be evaluated as

$$\mathbf{k}_V = \int_A \widehat{\mathbf{B}}_\gamma^T \, \widehat{\mathbf{B}}_\gamma \, dA \tag{13.107}$$

This gives $\mathbf{k}_V$ equal to

$$\frac{1}{24}\begin{bmatrix}
6\alpha+6\beta & -6\alpha+6\beta & -6\alpha-6\beta & 6\alpha-6\beta & -3b & -3b & -3b & -3b & -3a & -3a & -3a & -3a \\
 & 6\alpha+6\beta & 6\alpha-6\beta & -6\alpha-6\beta & 3b & 3b & 3b & 3b & -3a & -3a & -3a & -3a \\
 & & 6\alpha+6\beta & -6\alpha+6\beta & 3b & 3b & 3b & 3b & 3a & 3a & 3a & 3a \\
 & & & 6\alpha+6\beta & -3b & -3b & -3b & -3b & 3a & 3a & 3a & 3a \\
 & & & & 1.5ab & 1.5ab & 1.5ab & 1.5ab & & & & \\
 & & & & & 1.5ab & 1.5ab & 1.5ab & & \mathbf{0} & & \\
 & & & & & & 1.5ab & 1.5ab & & & & \\
 & & & & & & & 1.5ab & & & & \\
 & \text{Symmetric} & & & & & & & 1.5ab & 1.5ab & 1.5ab & 1.5ab \\
 & & & & & & & & & 1.5ab & 1.5ab & 1.5ab \\
 & & & & & & & & & & 1.5ab & 1.5ab \\
 & & & & & & & & & & & 1.5ab
\end{bmatrix} \tag{13.108}$$

$$\alpha = \frac{b}{a}, \qquad \beta = \frac{a}{b}$$

The final element stiffness matrix is the sum of $\mathbf{k}_B$ and $\mathbf{k}_V$ in the form

$$\mathbf{k}^i = K[\mathbf{k}_B + \zeta \mathbf{k}_V] \tag{13.109}$$

When the element matrices are assembled and boundary conditions imposed, the global stiffness equation appears as

$$(\mathbf{K}_B + \zeta \mathbf{K}_V)\mathbf{V} = \overline{\mathbf{P}}/K \tag{13.110}$$

From Eq. (13.104), ζ can be defined as $\zeta = 6k_s(1-\nu)/t^2$. If k_s is constant, it can be observed that for very thin plates ζ can assume a very large value. Also, as $t \to 0, \zeta \to \infty$. Consequently, the unrealistic result of $\mathbf{V} = \mathbf{0}$ can be obtained, regardless of the magnitude of $\overline{\mathbf{P}}$. This phenomenon is called *locking*, which is intended to imply that the displacement is "locked". Locking also occurs for thin beam elements when traditional shear deformation effects are taken into account and the displacement and slope are represented by independent shape functions. As discussed in Section 12.2.2, Chapter 12, for beams and plates it may be help to employ a k_s that varies with the thickness.

A classical method of alleviating the locking problem is to make the $\mathbf{K}_V$ matrix singular so that $\zeta \mathbf{K}_V$ can be finite. One way to accomplish this is to reduce the rank of the matrix $\mathbf{k}_V$ and use a low order numerical integration scheme to integrate $\mathbf{k}_V$. This is called *reduced integration.* It is known that if the number of strains at the integration points is less than the degrees of freedom available, then the singularity exists for the global stiffness matrix [Zienkiewicz, 1977]. Let h be the total number of integration points, k the number of strains used in the formation of the stiffness matrix $\mathbf{k}_V$, and j the total number of degrees of freedom in $\mathbf{V}$ (with suitable restraints against rigid body motion). Then, if $j - hk > 0$, the stiffness matrix $\mathbf{k}_V$ will be singular. Usually, a single point Gauss integration scheme is used for the formation of $\mathbf{k}_V$. After the element stiffness matrices $\mathbf{k}_V$ are assembled to form the global stiffness matrix $\mathbf{K}_V$ and the boundary conditions are applied, the matrix $\mathbf{K}_V$ is still singular and, thus, locking will be avoided. Another technique to eliminate the locking problem is to impose the Kirchhoff assumption at discrete points. This leads to an element called the *discrete Kirchhoff theory* (*DKT*) element, which will be treated later.

Variations of this 12 DOF rectangular element of this section can improve the performance of the element. For example, the addition of the twist $\partial^2 w/\partial x \partial y$ as a degree of freedom tends to improve the accuracy of an element based on Kirchhoff plate theory, but decreases the efficiency as it would now be a 16 DOF element. Rectangular elements limit the modeling options, especially near irregular boundaries. Triangular elements which are discussed in the following section, provide much greater flexibility.

13.5.2 Triangular Plate Elements

Plates with irregular boundaries require the use of nonrectangular elements, e.g., triangular elements. For example, the meshes for stress concentration regions are often modeled with triangular elements. Such elements are also used to form quadrilateral elements, as rectangular elements are not easily generalized into quadrilateral shapes. The transformation of coordinates of the type described in Chapter 6, Section 6.7 may be performed, but unfortunately the constant strain criterion is then violated [Zienhiewicz, 1977]. Typically, the quadrilateral element is treated as the composition of four three-node triangular elements. Usually, triangular elements are defined by the diagonals of the quadrilateral elements.

As in the case of rectangular elements, stretching and compression effects, obtained, for example, from a plane stress analysis, can be superimposed on the plate bending triangular

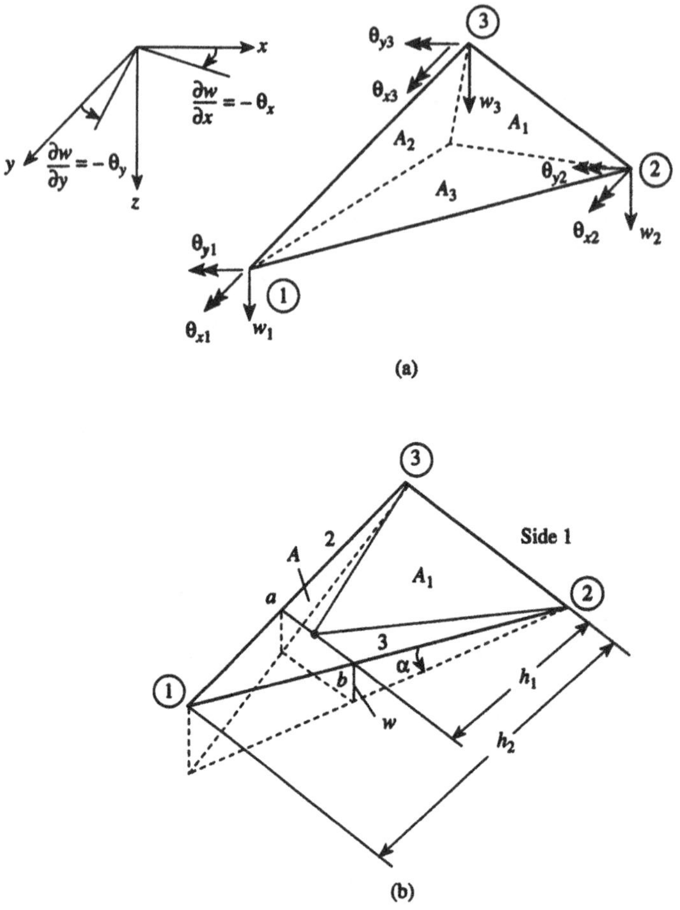

FIGURE 13.18
Triangular element and the area coordinate system.

stiffness matrix. These plate bending elements with in-plane deformation effects can be used in buckling analyses and in the modeling of curved surfaces such as occur with shells.

Derivations of three representative kinds of triangular elements are provided in this section.

Three Node, Nonconforming, Nine DOF Element

Element Variables and Trial Functions

Consider the triangular element in Fig. 13.18a and use Kirchhoff plate theory in the development of a stiffness matrix, so that no shear deformation effects are included. This element has a node at each of its three vertices, with three DOF at each node, i.e., the deflection w, the slopes $\theta_x = -\frac{\partial w}{\partial x}$, and $\theta_y = -\frac{\partial w}{\partial y}$.

As the triangular element has nine DOF, include only nine terms in a polynomial shape function. Here, an immediate difficulty arises as the complete cubic polynomial expansion contains ten terms (Chapter 6, Section 6.5.5), and any omission has to be made rather arbitrarily to retain symmetry. All ten terms could be retained and two coefficients made equal to limit the number of unknowns to nine. This leads to a serious problem, in that the matrix, corresponding to $\mathbf{N}_u$ in the form of $w = \mathbf{N}_u\widehat{\mathbf{w}}$, becomes singular for certain orientations of the triangular sides. This happens, for instance, when two sides of the triangle are parallel to the x and y axes.

An alternative shape function is to use the two-dimensional natural coordinate system described in Chapter 6. This is usually considered to be the logical choice for triangles. This approach will be followed here.

The element displacement field (shape function) can be constructed as the sum of two parts

$$w = w_1 + w_2 \tag{13.111}$$

where w_1 is the rigid body displacement, and w_2 is the deflection of the element when it is regarded as being simply supported at the nodes. For plate bending, the rigid body motions that can occur are

1. A rigid body translation in the z direction of the form $w_1 = constant$
2. A rigid body rotation about side 1 (Fig. 13.18) of the triangle. It can be observed from Fig. 13.18b that the rigid body translation in the z direction of line ab, which is parallel to side 1, is $w = h_1\alpha$, where α is the angle of rotation. Let A_1 be any triangle with a vertex on line ab. It is apparent that for points on ab, $L_1 = A_1/A = h_1/h$. Then, $w_1 = h_1\,\alpha = h\alpha L_1$, or $w_1 = (constant)L_1$.
3. A rigid body rotation about side 2 (Fig. 13.18) of the triangle of the form $w_1 = (constant)L_2$

It is apparent that the shape function for rigid body motions may be written as

$$w_1 = \widehat{w}_1 L_1 + \widehat{w}_2 L_2 + \widehat{w}_3 L_3 \tag{13.112}$$

Assign appropriately the values of $\widehat{w}_1$, $\widehat{w}_2$, and $\widehat{w}_3$, and use the condition $L_1+L_2+L_3 = 1$, to show that Eq. (13.112) satisfies the three rigid body motions. For example, if $\widehat{w}_1 = \widehat{w}_2 = \widehat{w}_3$, condition 1 is satisfied, and if $\widehat{w}_2 = \widehat{w}_3 = 0$ and $\widehat{w}_1 = \widehat{w}_3 = 0$, conditions 2 and 3 are satisfied, respectively. These three rigid body motions form a rigid body motion plane. It can be seen from Chapter 6, Section 6.5.6 that a linear combination of the natural coordinates $L_i, i = 1, 2, 3$, forms a linear function of x and y which defines a plane.

Since the element has nine DOF, a cubic polynomial can be used as the shape function, i.e., the shape function can be the linear combination of the cubic terms

$$L_1^2 L_2, \qquad L_2^2 L_3, \qquad L_3^2 L_1, \qquad L_2^2 L_1, \qquad L_3^2 L_2, \qquad L_1^2 L_3, \qquad L_1 L_2 L_3$$

Substitution of the expressions for L_1, L_2, and L_3 of Chapter 6, Eq. (6.75) into the above relationships will result in cubic polynomials in terms of x and y. Because the rigid body movement is already specified in Eq. (13.112), the cubic terms can be used to express the relative deflection expression w_2 of Eq. (13.111). It is customary [Zienkiewicz, 1977] to form these terms using such combinations as $L_2^2 L_3 + \frac{1}{2} L_1 L_2 L_3$. The first term of this combination has zero values at the nodes and zero slope along a side and the second term has zero values and slopes at all three corners. Six of these combinations constitute the relative deflection part of the displacement shape function.

$$\begin{aligned} w_2 = {} & \widehat{w}_4\left(L_3^2 L_1 + \frac{1}{2}L_1L_2L_3\right) + \widehat{w}_5\left(L_3^2 L_2 + \frac{1}{2}L_1L_2L_3\right) \\ & + \widehat{w}_6\left(L_1^2 L_3 + \frac{1}{2}L_1L_2L_3\right) + \widehat{w}_7\left(L_1^2 L_2 + \frac{1}{2}L_1L_2L_3\right) \\ & + \widehat{w}_8\left(L_2^2 L_3 + \frac{1}{2}L_1L_2L_3\right) + \widehat{w}_9\left(L_2^2 L_1 + \frac{1}{2}L_1L_2L_3\right) \end{aligned} \tag{13.113}$$

Then the shape function of the elements can be written as

$$w = w_1 + w_2 = \mathbf{N}_u \widehat{\mathbf{w}} \tag{13.114}$$

with $\mathbf{N}_u$ and $\widehat{\mathbf{w}}$ defined by Eqs. (13.112), (13.113), and (13.114). The unknowns $\widehat{w}_j$, $j = 1, 2, \ldots 9$, can be expressed in terms of the unknown nodal displacements by substituting the nodal values of w, $\theta_x = -\partial w/\partial x$ and $\theta_y = -\partial w/\partial y$ into Eq. (13.114) to form for the ith element

$$\mathbf{v}^i = \widehat{\mathbf{N}}_u \widehat{\mathbf{w}}$$

with

$$\mathbf{v}^i = [w_1 \quad \theta_{x1} \quad \theta_{y1} \cdots\cdots w_3 \quad \theta_{x3} \quad \theta_{y3}]^T$$

Then $\widehat{\mathbf{w}}$ is found to be

$$\widehat{\mathbf{w}} = \widehat{\mathbf{N}}_u^{-1} \mathbf{v}^i = \mathbf{G}\mathbf{v}^i$$

Thus, for the ith element,

$$w = \mathbf{N}_u \mathbf{G} \mathbf{v}^i = \mathbf{N}\mathbf{v}^i \tag{13.115}$$

or $w = \mathbf{N}_1\mathbf{v}_1 + \mathbf{N}_2\mathbf{v}_2 + \mathbf{N}_3\mathbf{v}_3$ where

$$\mathbf{v}_j = [w_j \quad \theta_{xj} \quad \theta_{yj}]^T \quad j = 1, 2, 3$$

and

$$\mathbf{N}_1 = \begin{bmatrix} L_1 + L_1^2 L_2 + L_1^2 L_3 - L_1 L_2^2 - L_1 L_3^2 \\ -b_3(L_1^2 L_2 + \frac{1}{2} L_1 L_2 L_3) + b_2(L_3 L_1^2 + \frac{1}{2} L_1 L_2 L_3) \\ c_3(L_1^2 L_2 + \frac{1}{2} L_1 L_2 L_3) - c_2(L_3 L_1^2 + \frac{1}{2} L_1 L_2 L_3) \end{bmatrix}^T$$

The other two functions $\mathbf{N}_2$ and $\mathbf{N}_3$ for nodes 2 and 3 are obtained by a cyclic permutation of suffixes $1 - 2 - 3$. The constants b_j and c_j, $j = 1, 2, 3$, involved in $\mathbf{N}_j$ are

$$\begin{aligned} b_1 &= x_3 - x_2 & b_2 &= x_1 - x_3 & b_3 &= x_2 - x_1 \\ c_1 &= y_2 - y_3 & c_2 &= y_3 - y_1 & c_3 &= y_1 - y_2 \end{aligned}$$

Stiffness Matrix

The principle of virtual work is used to derive the stiffness matrix and loading vector. For plate elements without shear deformation effects, this principle can be expressed as [Eq. (13.84a)]

$$-\delta W = -\delta W_i - \delta W_e = \int_A \delta w(\mathbf{k}^D w - \overline{p}_z)\, dA - \int_{S_p} \delta w\, \overline{p}_s\, ds \tag{13.116}$$

where $\mathbf{k}^D = {}_u\mathbf{D}^T\, \mathbf{E}\, \mathbf{D}_u$, and $\overline{p}_z$ is equal to the applied pressure. The expressions for $\mathbf{D}_u$ and $\mathbf{k}^D$ are given in Eqs. (13.83) and (13.84b). Introduce the shape function $\mathbf{N}$ so that the element stiffness matrix and loading vector appear as

$$\mathbf{k}^i = \int_A \mathbf{N}^T\, {}_u\mathbf{D}^T \mathbf{E} \mathbf{D}_u\, \mathbf{N}\, dA \tag{13.117}$$

$$\overline{\mathbf{p}}^i = \int_A \mathbf{N}^T \overline{p}_z\, dA \tag{13.118}$$

These matrices can be evaluated readily using the relations of Chapter 6, Eq. (6.78).

This element is based on Kirchhoff plate theory which does not take the shear deformation into account. The chief assumption is that straight lines which are initially normal to the middle surface of the plate remain straight and normal to the middle surface in the loaded

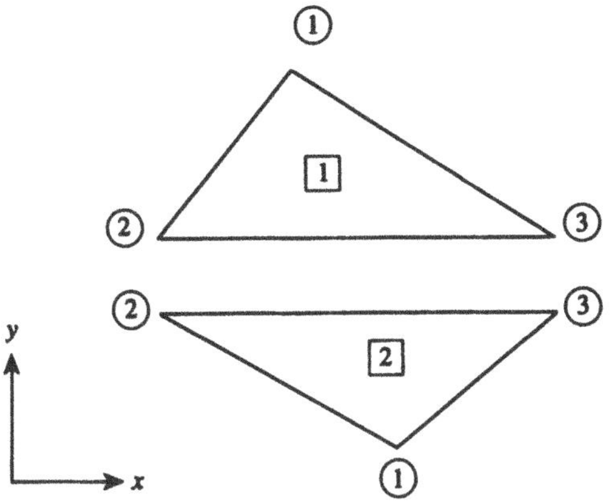

FIGURE 13.19
Edges of triangular elements.

configuration. The deformation throughout the plate is expressed solely in terms of the middle surface deflection w. The fact that the plate behavior is characterized by the single variable w has considerable advantages, but it also means that the nature of classical plate theory differs significantly from that of two- or three-dimensional elasticity theory. As far as the finite element displacement approach is concerned, C^1 continuity (Chapter 6, Section 6.5), i.e., continuity of w and its first derivatives, on interelement boundaries is required for compatibility. The requirement for C^1 continuity does complicate element development considerably. This requirement cannot, in general, be satisfied.

Use the triangular elements in Fig. 13.19 to illustrate the problem of C^1 continuity not being satisfied. The displacement w of Eq. (13.114) along side 2-3 of element 1, where $L_1 = 0$, of Fig. 13.19 can be written as a function of x, i.e.,

$$w^1 = \alpha_0 + \alpha_1 x + \alpha_2 x^2 + \alpha_3 x^3$$

This expression can be obtained by substituting $L_1 = 0$ and the relationships for L_2 and L_3 of Chapter 6, Eq. (6.75) into Eq. (13.114). Note that along this boundary, the y coordinate is constant. Similarly, the displacement along side 2–3 of element 2 is

$$w^2 = \beta_0 + \beta_1 x + \beta_2 x^2 + \beta_3 x^3$$

The parameters $\alpha_0, \alpha_1, \alpha_2,$ and α_3 can be uniquely expressed in terms of the four DOF ($w_2, w_3, \theta_{x2},$ and θ_{x3}) at nodes 2 and 3, and since these DOF are shared by the adjacent elements, we have $w^1 = w^2$. This means that the displacement and the rotation $\theta_x = -\partial w/\partial x$ are continuous along side 2–3 for elements 1 and 2. For the slope $\partial w/\partial y$ in the y direction, however, the situation is quite different. Let $L_1 = 0$ and take the derivative of w of Eq. (13.114) with respect to y. A polynomial

$$\frac{\partial w}{\partial y} = \gamma_0 + \gamma_1 x + \gamma_2 x^2 \tag{13.119}$$

is obtained for side 2–3. Since there are only two DOF (θ_y at nodes 2 and 3) related to $\partial w/\partial y$ along this side, the parameters γ_i, $i = 0, \ldots 3$, cannot be uniquely expressed in terms of these two DOF. On the other hand, if the expression of Eq. (13.119) is to be obtained from Eq. (13.115), the parameters γ_i must involve the nodal variables at node 1. Thus, $\theta_y = -\partial w/\partial y$ may be different for elements 1 and 2 of Fig. 13.19 along side 2–3. Therefore, discontinuities of normal slope, or kinks, will generally occur and the requirement of continuity of the first derivatives of the shape function is violated. This kind of element with the discontinuity of shapes at the boundary is referred to as a *nonconforming element*. The present element is one of a number of nonconforming but viable plate bending elements. The element displacement field properly represents the rigid body motion and constant strain states. Furthermore, the element passes the patch test [Bazeley, et al., 1965].

The plate elements based on the Kirchhoff plate theory, which use w, $\theta_x = -\partial w/\partial x$, and $\theta_y = -\partial w/\partial y$ as nodal DOF, are usually nonconforming. An exception is a rectangular element which has w, $w_{,x}$, $w_{,y}$ and $w_{,xy}$ as nodal unknowns. If desired, there are means for avoiding nonconforming elements. The first approach is to use the plate theory that takes shear deformation effects into consideration. Such an element was considered in the previous section, and another will be treated in the next section. For these elements, independent shape functions are used for w and θ so the shape functions for these variables are continuous at the element boundary. The second approach is to use more DOF at the nodes to make the slope continuous. For example, if two more DOF related to the slope of the deflection are added to the nodes of the triangular element of this section, the slope can be completely defined and the discontinuity can disappear. This type of triangular element is discussed later.

Discrete Kirchhoff Theory (DKT)

It is apparent from the formulation of the triangular plate bending element in the previous section that it is difficult to formulate a compatible triangular element with nine DOF using a single polynomial approximation for w. A viable approach [Batoz, et al., 1980] for formulating a compatible triangular Kirchhoff plate element is to start as though Reissner-Mindlin plate theory applies, so that the nodal variables for the deflection and the rotations are independent of each other. Then the shape functions for these quantities in a triangular element can be made continuous at the inter-element boundaries, forming a compatible C^0 element. Since the plate is very thin, the terms in the principle of virtual work corresponding to shear deformation effects are assumed to be negligible. The Kirchhoff plate theory assumptions are introduced at discrete points along the boundary of the element. The shape functions are designed to maintain compatibility, so the element is still conforming. The resulting element is called the *discrete Kirchhoff theory element*, or simply the DKT element.

The DKT element uses the same nodal variables as those in Fig. 13.18. The formulation of the stiffness matrix and the loading vector of this element starts from the principle of virtual work of Eq. (13.96). The first term of Eq. (13.96) corresponds to the bending of the plate and the second term to the effects of shear deformation, which is to be neglected. For the bending response, only the rotations appear in the expression for the principle of virtual work of Eq. (13.96), so that only the shape functions for rotations are needed in developing an element. For an element with six nodes (Fig. 13.20) the rotations are approximated by

$$\theta_x = \sum_{i=1}^{6} N_i \theta_{xi} \qquad \theta_y = \sum_{i=1}^{6} N_i \theta_{yi} \tag{13.120}$$

in which θ_{xi}, θ_{yi} are the rotations at the nodes, and nodes 4, 5, 6 are midside nodes between the corner nodes 1, 2, 3. The shape functions N_i, formed with natural coordinates as quadratic polynomials, are expressed as

$$\begin{aligned} &N_1 = L_1(2L_1 - 1) = (1 - L_2 - L_3)(1 - 2L_2 - 2L_3), \qquad N_2 = L_2(2L_2 - 1) \\ &N_3 = L_3(2L_3 - 1), \quad N_4 = 4L_2L_3, \\ &N_5 = 4L_1L_3 = 4(1 - L_2 - L_3)L_3, \qquad N_6 = 4L_1L_2 = 4(1 - L_2 - L_3)L_2 \end{aligned} \tag{13.121}$$

where $L_i, i = 1, 2, 3$, are the area coordinates defined in Chapter 6. The shape functions can be obtained from Eqs. (6.76) and (6.77). For a discrete Kirchhoff theory plate element

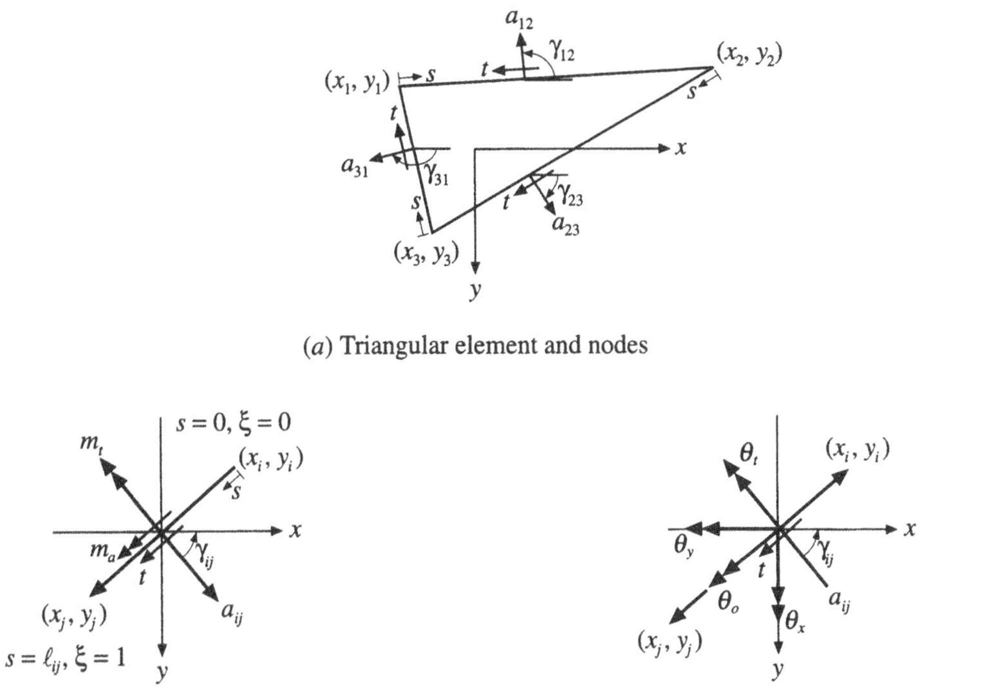

(*a*) Triangular element and nodes

(*b*) Quantities on a particular boundary

(*c*) Relation between θ_x, θ_y, and θ_a, θ_t

FIGURE 13.20
The geometry of a triangular element.

the Kirchhoff assumptions are to be imposed at particular points. For the element shown in Fig. 13.20, the assumptions imposed are

$$\gamma = \begin{bmatrix} \theta_x + w_{,x} \\ \theta_y + w_{,y} \end{bmatrix} = 0 \tag{13.122}$$

at the corner nodes and

$$\theta_{tk} = -w_{,sk} \tag{13.123}$$

at the middle nodes where $k = 4, 5, 6$. Here θ_t denotes the rotation about the normal to the boundary of the element at node k and $w_{,sk}$ is the derivative of w with respect to the s direction (Fig. 13.20), which coincides with the boundary. It is assumed that the rotation about the s direction of Fig. 13.20 at the middle nodes ($k = 4, 5, 6$) is the average value of the rotations of the end nodes, i.e.,

$$\theta_{ak} = \frac{1}{2}(\theta_{ai} + \theta_{aj}) \tag{13.124}$$

where the corner nodes are $ij = 23, 31,$ and 12. As will be seen later, Eqs. (13.123) and (13.124) will lead to relationships between the variables at the in-span nodes and the corner nodes so that the stiffness matrix will involve corner variables only.

Let the variation of w along the element boundaries be cubic. Thus, along a boundary of length ℓ_{ij} with nodes i and j at the ends (Fig. 13.20), the displacement w takes the form

$$w = a_0 + a_1 s + a_2 s^2 + a_3 s^3 \tag{13.125}$$

The derivative of w with respect to s is

$$w' = a_1 + 2a_2 s + 3a_3 s^2$$

At $s = 0$ (node i), $w_i = a_0$, $w_i' = a_1$ and at $s = \ell_{ij}$ (node j), where ℓ_{ij} is the length of side ij, $w_j = a_0 + a_1\ell_{ij} + a_2\ell_{ij}^2 + a_3\ell_{ij}^3$, $w_{,j} = a_1 + 2a_2\ell_{ij} + 3a_3\ell_{ij}^2$. The coefficients a_2 and a_3 can be obtained from the solution of these four equations. Since the coefficients $a_i, i = 0, \ldots 3$, are uniquely expressed in terms of the nodal variables which are shared by the adjacent elements, the displacement is continuous between the elements. Substitute the expressions for $a_i, i = 0, 1, 2, 3$ into Eq. (13.125) and let s assume the values at the nodal points. Form the derivative of w with respect to s at node k

$$w_{,sk} = -\frac{3}{2\ell_{ij}} w_i - \frac{1}{4} w_{,si} + \frac{3}{2\ell_{ij}} w_j - \frac{1}{4} w_{,sj} \tag{13.126}$$

Because the first term of the principle of virtual work of Eq. (13.96) does not involve w, there is no need to define an interpolation function for w on the element. However, a cubic variation for w on the boundary has been assumed for the sake of obtaining the expression for $w_{,s}$. Since w varies cubically along the element boundaries, $w_{,s}$ varies quadratically. Since $w_{,s}$ matches θ_t at the three points along each side, i.e., let $\theta_t = -w_{,s}$, the form of the quadratic function for $w_{,s}$ is determined uniquely.

To obtain the relationships between the variables θ_x, θ_y in Eq. (13.120) and the nodal variables θ_a, θ_t, use Fig. 13.20c.

$$\begin{bmatrix} \theta_x \\ \theta_y \end{bmatrix} = \begin{bmatrix} \cos\gamma_{ij} & -\sin\gamma_{ij} \\ \sin\gamma_{ij} & \cos\gamma_{ij} \end{bmatrix} \begin{bmatrix} \theta_a \\ \theta_t \end{bmatrix} = \begin{bmatrix} -\cos\gamma_{ij} & \sin\gamma_{ij} \\ -\sin\gamma_{ij} & \cos\gamma_{ij} \end{bmatrix} \begin{bmatrix} w_{,a} \\ w_{,s} \end{bmatrix} \tag{13.127}$$

and

$$\begin{bmatrix} w_{,a} \\ w_{,s} \end{bmatrix} = \begin{bmatrix} -\cos\gamma_{ij} & -\sin\gamma_{ij} \\ \sin\gamma_{ij} & -\cos\gamma_{ij} \end{bmatrix} \begin{bmatrix} \theta_x \\ \theta_y \end{bmatrix} \tag{13.128}$$

where γ_{ij} is defined in Fig. 13.20. Substitute Eqs. (13.126), (13.127), and (13.128) at each node into Eq. (13.120) to obtain the shape functions for θ_x and θ_y in terms of the nodal variables. That is, use Eqs. (13.127) and (13.128) at each node to obtain expressions for θ_{xi} and θ_{yi} and then substitute these quantities into Eq. (13.120) to find the shape functions for θ_x and θ_y. This results in

$$\theta_x = \mathbf{N}_x \mathbf{v}^i, \qquad \theta_y = \mathbf{N}_y \mathbf{v}^i \tag{13.129}$$

or $\boldsymbol{\theta} = \mathbf{N}\mathbf{v}^i$, where $\boldsymbol{\theta} = [\theta_x \ \theta_y]^T$, $\mathbf{N} = [\mathbf{N}_x \ \mathbf{N}_y]^T$, and $\mathbf{v}^i = [w_1 \, \theta_{x1} \, \theta_{y1} \, w_2 \, \theta_{x2} \, \theta_{y2} \, w_3 \, \theta_{x3} \, \theta_{y3}]^T$. Also, $\mathbf{N}_x$ and $\mathbf{N}_y$ are nine component vectors of the new shape functions for θ_x and θ_y. These shape functions are expressed as

$$\begin{aligned} N_{x1} &= 1.5(a_6 N_6 - a_5 N_5), \qquad & N_{x2} &= N_1 - b_5 N_5 - b_6 N_6 \\ N_{x3} &= -c_5 N_5 - c_6 N_6, & N_{y1} &= 1.5(d_6 N_6 - d_5 N_5) \\ N_{y2} &= N_{x3}, & N_{y3} &= N_1 - e_5 N_5 - e_6 N_6 \end{aligned} \tag{13.130}$$

where $N_i, i = 1, 2 \ldots, 6$, are the shape functions of Eq. (13.121). The functions N_{x4}, N_{x5}, and N_{x6} are obtained from the above expression by replacing N_1 by N_2 and indices 6 and 5 by 4 and 6, respectively. Similarly, the functions N_{x7}, N_{x8}, and N_{x9} are found by replacing N_1 by N_3 and indices 6 and 5 by 5 and 4, respectively. The same procedure leads to N_{y5}, N_{y6}, etc. Also

$$\begin{aligned} a_k &= -x_{ij}/\ell_{ij}^2 & d_k &= -y_{ij}/\ell_{ij}^2 \\ b_k &= \left(\frac{1}{4}x_{ij}^2 - \frac{1}{2}y_{ij}^2\right)/\ell_{ij}^2 \qquad & e_k &= \left(\frac{1}{4}y_{ij}^2 - \frac{1}{2}x_{ij}^2\right)/\ell_{ij}^2 \\ c_k &= \frac{3}{4}x_{ij}y_{ij}/\ell_{ij}^2 \end{aligned} \tag{13.131}$$

where $k = 4, 5$, and 6 correspond to the sides $ij = 23, 31$, and 12, respectively. The quantities x_{ij} and y_{ij} are defined in Fig. 13.20.

The stiffness matrix of the DKT element is formulated from the principle of virtual work of Eq. (13.96). Since the shear deformation effect is neglected, the shape functions for θ_x and θ_y are substituted into the first part of Eq. (13.96), giving

$$\begin{aligned} \delta W_i &= \int_A \delta\boldsymbol{\theta}^T{}_\kappa \mathbf{D}^T\ \mathbf{E}_B\ \mathbf{D}_\kappa \boldsymbol{\theta}\, dA = \delta\mathbf{v}^{iT} \int_A \mathbf{N}^T\ {}_\kappa\mathbf{D}^T \mathbf{E}_B\ \mathbf{D}_\kappa \mathbf{N}\, dA\ \mathbf{v}^i \\ &= \delta\mathbf{v}^{iT} \int_A \mathbf{B}^T \mathbf{E}_B \mathbf{B}\, dA\ \mathbf{v}^i = \delta\mathbf{v}^{iT} \mathbf{k}\, \mathbf{v}^i \end{aligned} \tag{13.132}$$

with

$$\mathbf{k} = \int_A \mathbf{B}^T \mathbf{E}_B \mathbf{B}\, dA \tag{13.133}$$

and

$$\mathbf{B} = \frac{1}{2A} \begin{bmatrix} & y_{31}\mathbf{N}^T_{x,L_2} + y_{12}\mathbf{N}^T_{x,L_3} & \\ & -x_{31}\mathbf{N}^T_{y,L_2} - x_{12}\mathbf{N}^T_{y,L_3} & \\ -x_{31}\mathbf{N}^T_{x,L_2} & -x_{12}\mathbf{N}^T_{x,L_3} + y_{31}\mathbf{N}^T_{y,L_2} & +y_{12}\mathbf{N}^T_{y,L_3} \end{bmatrix} \tag{13.134}$$

where A is the area of the element. The matrices $\mathbf{N}_x$, $\mathbf{N}_{x,L_2}$, $\mathbf{N}_{y,L_2}$, $\mathbf{N}_{x,L_3}$, and $\mathbf{N}_{y,L_3}$ are given by

$$\mathbf{N}_{x,L_2} = \begin{bmatrix} P_6(1-2L_2) + (P_5 - P_6)L_3 \\ -4 + 6(L_2 + L_3) + r_6(1-2L_2) - L_3(r_5 + r_6) \\ -q_6(1-2L_2) + (q_5 + q_6)L_3 \\ -P_6(1-2L_2) + L_3(P_4 + P_6) \\ -2 + 6L_2 + r_6(1-2L_2) + L_3(r_4 - r_6) \\ -q_6(1-2L_1) + L_3(q_6 - q_4) \\ -L_3(P_5 + P_4) \\ -L_3(r_5 - r_4) \\ -L_3(q_4 - q_5) \end{bmatrix} \tag{13.135a}$$

$$\mathbf{N}_{y,L_2} = \begin{bmatrix} t_6(1-2L_2) + (t_5 - t_6)L_3 \\ -q_6(1-2L_2) + L_3(q_5 + q_6) \\ -1 - r_6(1-2L_2) + (r_5 + r_6)L_3 \\ -t_6(1-2L_2) + L_3(t_4 + t_6) \\ -q_6(1-2L_2) - L_3(q_4 - q_6) \\ 1 - r_6(1-2L_2) - L_3(r_4 - r_6) \\ -L_3(t_5 + t_4) \\ -L_3(q_4 - q_5) \\ L_3(r_5 - r_4) \end{bmatrix} \tag{13.135b}$$

$$\mathbf{N}_{x,L_3} = \begin{bmatrix} -P_5(1-2L_3)-(P_6-P_5)L_2 \\ -4+6(L_2+L_3)+r_5(1-2L_3)-L_2(r_5+r_6) \\ -q_5(1-2L_3)+(q_5+q_6)L_2 \\ L_2(P_4+P_6) \\ -L_2(r_6-r_4) \\ -L_2(q_4-q_6) \\ P_5(1-2L_3)-L_2(P_5+P_4) \\ -2+6L_3+r_5(1-2L_3)+L_2(r_4-r_5) \\ -q_5(1-2L_3)-L_2(q_4-q_5) \end{bmatrix} \tag{13.135c}$$

$$\mathbf{N}_{y,L_3} = \begin{bmatrix} -t_5(1-2L_3)-(t_6-t_5)L_2 \\ -q_5(1-2L_3)+L_2(q_5+q_6) \\ -1-t_5(1-2L_3)+(r_5+r_6)L_2 \\ L_2(t_4+t_6) \\ -L_2(q_4-q_6) \\ -L_2(r_4-r_6) \\ t_5(1-2L_3)-L_2(t_5+t_4) \\ -q_5(1-2L_3)-L_2(q_4-q_5) \\ 1-r_5(1-2L_3)-L_2(r_4-r_5) \end{bmatrix} \tag{13.135d}$$

where

$$P_k = -6x_{ij}/\ell_{ij}^2 = 6a_k, \qquad t_k = -6y_{ij}/\ell_{ij}^2 = 6d_k$$

$$q_k = 3x_{ij}y_{ij}/\ell_{ij}^2 = 4c_k, \qquad r_k = 3y_{ij}^2/\ell_{ij}^2 = -4(2b_k + e_k)$$

$$k = 4, 5, 6 \quad \text{for } ij = 23, 31, 12, \text{ respectively}$$

Thus, the stiffness matrix can be evaluated.

For the loading vector, a uniform loading can be represented approximately by lumped concentrated loads applied on the nodes. Then the element loading vector corresponding to a uniform pressure load p_0 per unit middle surface area can be written as

$$\overline{\mathbf{p}} = \frac{A}{3}[p_0 \quad 0 \quad 0 \quad p_0 \quad 0 \quad 0 \quad p_0 \quad 0 \quad 0]^T \tag{13.136}$$

where A is the area of the element.

Hybrid Stress Model Element (HSM)

In addition to the displacement type of element for the plate bending problem, hybrid elements have been shown to be viable [Batoz, et al., 1980]. The concept of the hybrid element is introduced in Chapter 6, Section 6.9. A hybrid element can be based on the extended principle of complementary virtual work of Eq. (6.150) or Chapter 2, Eq. (2.104). As is shown in Chapter 6, a typical hybrid element employs two kinds of shape functions. For example, compatible displacement functions can be assumed along the interelement boundaries, in addition to an assumed stress field which satisfies the equilibrium conditions inside the element. As a result, a compatible element is formed. This kind of element is called the hybrid stress model (HSM). Another type of hybrid element is based on an assumed displacement field inside the element, with stresses assumed along interelement boundaries. See Pian and Tong (1969) for more details.

In the formulation of the HSM element, the classical Kirchhoff thin plate theory is employed. The shear stresses γ_{xz} and γ_{yz} are assumed to vanish everywhere in the plate, i.e., $\theta_x = -w_{,x}$ and $\theta_y = -w_{,y}$. The HSM element uses the same nodal DOF as the DKT element.

The extended complementary hybrid functional of Chapter 2, Eq. (2.104) is

$$\Pi_H^* = -\int_V U_0^*(\sigma)dV + \int_{S_u} \mathbf{p}^T\overline{\mathbf{u}}\, dS + \int_{S_p} \mathbf{u}^T(\mathbf{p}-\overline{\mathbf{p}})\, dS \tag{13.137}$$

where $\int_V U_0^*(\sigma)\, dV$ is the complementary energy, which can be expressed as

$$-\int_V U_0^*(\sigma)\, dV = -\int_V \frac{1}{2}\sigma^T \mathbf{E}_B^{-1}\, \sigma\, dV = -\int_{-t/2}^{t/2}\int_A \frac{1}{2}\sigma^T\, \mathbf{E}_B^{-1}\, \sigma\, dA\, dz \tag{13.138}$$

Substitute the expressions for σ_x, σ_y, and τ_{xy} of Eq. (13.27) into Eq. (13.138) to obtain

$$-\int_V U_0^*(\sigma)\, dV = -\frac{1}{2}\int_A \mathbf{s}^T\, \mathbf{E}_B^{-1}\, \mathbf{s}\, dA \tag{13.139}$$

where $\mathbf{s} = [m_x\ m_y\ m_{xy}]^T$. For a homogeneous isotropic plate, $\mathbf{E}_B$ from Eq. (13.35) leads to

$$\frac{1}{2}\mathbf{s}^T\, \mathbf{E}_B^{-1}\, \mathbf{s} = \frac{12}{Et^3}\left[(m_x+m_y)^2 + 2(1+\nu)(m_{xy}^2 - m_x m_y)\right]$$

The second and the third terms on the right hand side of Eq. (13.137) can be rearranged as

$$\int_{S_u} \mathbf{p}^T\overline{\mathbf{u}}\, dS + \int_{S_p} \mathbf{u}^T(\mathbf{p}-\overline{\mathbf{p}})\, dS = \int_S \mathbf{p}^T\, \mathbf{u}\, dS - \int_{S_p} \overline{\mathbf{p}}^T\, \mathbf{u}\, dS$$

where $S = S_u + S_p$. Then Eq. (13.137) becomes

$$\begin{aligned}\Pi_H^* &= -\int_A \frac{1}{2}\mathbf{s}^T\, \mathbf{E}_B^{-1}\, \mathbf{s}\, dA + \int_S (w\, q_a - w_{,a}\, m_a - w_{,s}\, m_t)\, dS - \int_{S_p} \overline{\mathbf{p}}^T\, \mathbf{u}\, dS \\ &= \qquad U_1 \qquad + \qquad U_2 \qquad - \int_{S_p} \overline{\mathbf{p}}^T\, \mathbf{u}\, dS\end{aligned} \tag{13.140}$$

where $U_1 = -\int U_0^*(\sigma)\, dV = -\int_A \frac{1}{2}\mathbf{s}^T\mathbf{E}_B^{-1}\mathbf{s}\, dA$ and $U_2 = \int_S (w\, q_a - w_{,a}\, m_a - w_{,s} m_t)\, dS$. The stress resultants m_a and m_t are shown in Fig. 13.20 and q_a is the shear force on the boundary. These quantities are expressed as

$$\begin{aligned} q_a &= c(m_{x,x} + m_{xy,y}) + s(m_{y,y} + m_{xy,x}) \\ m_a &= cc m_x + 2cs m_{xy} + ss m_y \\ m_t &= -sc m_x + (cc - ss)m_{xy} + sc m_y \end{aligned} \tag{13.141}$$

where c and s abbreviate $\cos\gamma$ and $\sin\gamma$, respectively, in which γ denotes the angle between the outer normal of the boundary and the x axis. The shape functions for $\mathbf{s}$, w, $w_{,a}$, and $w_{,s}$ will be given later.

The last term of Eq. (13.140) represents the loading vector. It is assumed that uniform loading can be represented by lumped loads in the form of Eq. (13.136).

The independent quantities subject to variation in the hybrid stress functional are the moment components m_x, m_y, and m_{xy} inside the element and the displacements w and $w_{,a}$ along the element boundary S with the subsidiary conditions

$$m_{x,xx} + 2m_{xy,xy} + m_{y,yy} = 0 \quad \text{in } A \tag{13.142}$$

and

$$w = \overline{w}, \quad w_{,a} = \overline{w}_{,a} \quad \text{on } S_u \tag{13.143}$$

These are the equilibrium conditions of Eq. (13.39) (without $\overline{p}_z$) and the displacement boundary conditions of Eq. (13.55). These conditions must be satisfied when using the complementary hybrid model of Eq. (13.137) to establish a finite element formulation.

There are two independent approximate shape functions used for the HSM element. One is an assumed stress (bending moments) field and the other is the displacement along the interelement boundaries. Assume that the bending moments vary linearly in the interior of the element and the displacement w varies cubically along the sides, with a linear variation of $w_{,a}$. For the bending moment, assume that

$$\mathbf{s} = \mathbf{N}_\sigma\, \hat{\boldsymbol{\sigma}} \tag{13.144}$$

where

$$\mathbf{N}_\sigma = \begin{bmatrix} \mathbf{N}_\sigma^* & 0 & 0 \\ 0 & \mathbf{N}_\sigma^* & 0 \\ 0 & 0 & \mathbf{N}_\sigma^* \end{bmatrix}$$

$$\mathbf{N}_\sigma^* = [1 \quad x \quad y]$$

and $\hat{\boldsymbol{\sigma}}$ is the 9×1 vector of generalized parameters. Note that Eq. (13.144) satisfies the equilibrium condition of Eq. (13.142).

Substitute Eq. (13.144) into Eq. (13.139) to obtain

$$U_1 = -\int_V U_0^*(\sigma)\, dV = -\frac{1}{2}\int_A \mathbf{s}^T \mathbf{E}_B^{-1} \mathbf{s}\, dA = -\frac{1}{2}\hat{\boldsymbol{\sigma}}^T \mathbf{B}\, \hat{\boldsymbol{\sigma}} \tag{13.145}$$

where

$$\mathbf{B} = \int_A \mathbf{N}_\sigma^T\, \mathbf{E}_B^{-1}\, \mathbf{N}_\sigma\, dA$$

The 9×9 matrix $\mathbf{B}$ can be written as

$$\mathbf{B} = \begin{bmatrix} c_{11}\phi & c_{12}\phi & c_{13}\phi \\ c_{21}\phi & c_{22}\phi & c_{23}\phi \\ c_{31}\phi & c_{32}\phi & c_{33}\phi \end{bmatrix} \tag{13.146}$$

where the coefficients c_{ij} are the elements in $\mathbf{E}_B^{-1}$ and

$$\phi = \int_A \mathbf{N}_\sigma^{*T}\, \mathbf{N}_\sigma^*\, dA = \int_A \begin{bmatrix} 1 & x & y \\ x & x^2 & xy \\ y & xy & y^2 \end{bmatrix} dA \tag{13.147}$$

Since $\mathbf{E}_B^{-1}$ is symmetric, $\mathbf{B}$ is also symmetric. If the centroid of the coordinate system is located at the centroid of the element, ϕ can be written as

$$\phi = \int_A \begin{bmatrix} A & 0 & 0 \\ 0 & I_{xx} & I_{xy} \\ 0 & I_{xy} & I_{yy} \end{bmatrix} dA \tag{13.148}$$

where A is the area of the element and I_{xx}, I_{yy}, and I_{yz} are the moments of inertia and product of inertia of the element with respect to the centroidal axes, i.e.,

$$
\begin{aligned}
I_{xx} &= \int_A x^2 dA = \frac{A}{12}(x_1^2 + x_2^2 + x_3^2) \\
I_{yy} &= \int_A y^2 dA = \frac{A}{12}(y_1^2 + y_2^2 + y_3^2) \\
I_{xy} &= \int_A xy\,dA = \frac{A}{12}(x_1 y_1 + x_2 y_2 + x_3 y_3)
\end{aligned}
\tag{13.149}
$$

in which $x_i, y_i, i = 1, 2, 3$, are the coordinates of the nodes.

The next important step in the formulation of the stiffness matrix is the evaluation of the second integral on the right hand side of Eq. (13.140). The integration on the whole boundary $S = 12 + 23 + 31$ can be broken into the sum of the integrals on the sides ij ($ij = 12$ or 23 or 31), i.e.,

$$
U_2 = \int_S (w\, q_a - w_{,a}\, m_a - w_{,s}\, m_t)\, dS = \sum_{i=1}^{3} U_{Sij}
$$

where

$$
U_{Sij} = \int_{ij} (w q_a - w_{,a} m_a - w_{,s}\, m_t)\, dS = \int_{ij} \begin{bmatrix} q_a & -m_a & -m_t \end{bmatrix} \begin{bmatrix} w \\ w_{,a} \\ w_{,s} \end{bmatrix} dS \tag{13.150}
$$

is the integration on side ij. The components q_a, m_a, and m_t are given in Eq. (13.141) in which

$$
c = \cos \gamma_{ij} = -y_{ij}/\ell_{ij} \qquad s = \sin \gamma_{ij} = x_{ij}/\ell_{ij}
$$

where x_{ij}, y_{ij}, and ℓ_{ij} are shown in Fig. 13.20. By means of Eqs. (13.141) and (13.144), the boundary forces q_a, m_a and m_t can be written as

$$
\begin{bmatrix} q_a \\ -m_a \\ -m_t \end{bmatrix}_{ij} = \mathbf{R}_{ij} \hat{\sigma} \tag{13.151}
$$

where

$$
\mathbf{R}_{ij} = \begin{bmatrix}
0 & c & 0 & 0 & 0 & s & 0 & s & c \\
-cc & -ccx & -ccy & -ss & -ssx & -ssy & -2cs & -2csx & -2csy \\
cs & csx & csy & -cs & -csx & -csy & -(cc - ss) & -(cc - ss)x & -(cc - ss)y
\end{bmatrix}
$$

with

$$
x = x_i - \xi x_{ij} \qquad y = y_i - \xi y_{ij} \quad \text{and} \quad \xi = s/\ell_{ij}
$$

in which s is shown in Fig. 13.20. It is seen that $\mathbf{R}_{ij}$ is a linear expression in ξ.

Since Eq. (13.150) involves the displacements w and its derivatives on the boundary, assumptions on the displacements have to be made. Use the shape function

$$
w = N_{B1}(\xi) w_i + N_{B2}(\xi) w_j + N_{B3}(\xi) w_{,si} + N_{B4}(\xi) w_{,sj} \tag{13.152}
$$

where

$$
\begin{aligned}
N_{B1} &= 1 - 3\xi^2 + 2\xi^3 & N_{B2} &= 3\xi^2 - 2\xi^3 \\
N_{B3} &= \ell_{ij}(\xi - 2\xi^2 + \xi^3) & N_{B4} &= \ell_{ij}(-\xi^2 + \xi^3)
\end{aligned}
$$

The normal slope $w_{,a}$ is assumed to vary linearly

$$w_{,a} = (1-\xi)w_{,ai} + \xi w_{,aj} \tag{13.153}$$

Use the relationships of Eqs. (13.128), (13.129), (13.152), and (13.153) to find

$$\begin{bmatrix} w \\ w_{,a} \\ w_{,s} \end{bmatrix} = \mathbf{L}_{ij}\mathbf{v}_{ij} \tag{13.154}$$

where

$$\mathbf{v}_{ij} = [w_i \quad \theta_{xi} \quad \theta_{yi} \quad w_j \quad \theta_{xj} \quad \theta_{yj}]^T$$

and

$$\mathbf{L}_{ij} = \begin{bmatrix} N_{B1} & sN_{B3} & -cN_{B3} & N_{B2} & sN_{B4} & -cN_{B4} \\ 0 & -c(1-\xi) & -s(1-\xi) & 0 & -c\xi & -s\xi \\ (N_{B1,\xi})/\ell_{ij} & (sN_{B3,\xi})/\ell_{ij} & (-cN_{B3,\xi})/\ell_{ij} & (N_{B2,\xi})/\ell_{ij} & (sN_{B4,\xi})/\ell_{ij} & (-cN_{B4,\xi})/\ell_{ij} \end{bmatrix}$$

Substitute Eqs. (13.151) and (13.154) into Eq. (13.150) to obtain

$$U_{Sij} = \widehat{\boldsymbol{\sigma}}^T \mathbf{C}_{ij}\mathbf{v}_{ij} \tag{13.155}$$

where

$$\mathbf{C}_{ij} = \int_0^{\ell_{ij}} \mathbf{R}_{ij}^T \mathbf{L}_{ij}\, dS = \ell_{ij} \int_0^1 \mathbf{R}_{ij}^T(\xi)\, \mathbf{L}_{ij}(\xi)\, d\xi$$

The matrix $\mathbf{C}_{ij}$ is of order of 9×6 and the integration can be performed in closed form. Summation of the integrals U_{Sij} for the three sides completes the integral of the second term on the right hand side of Eq. (13.140)

$$U_2 = \int_S (wq_a - w_{,a}m_a - w_{,s}m_t)\, dS = U_{S12} + U_{S23} + U_{S31} = \widehat{\boldsymbol{\sigma}}^T \mathbf{C}\,\mathbf{v} \tag{13.156}$$

with $\mathbf{C}\,\mathbf{v} = \mathbf{C}_{12}\,\mathbf{v}_{12} + \mathbf{C}_{23}\,\mathbf{v}_{23} + \mathbf{C}_{31}\,\mathbf{v}_{31}$. Note that $\mathbf{C}$ is a 9×9 matrix and $\mathbf{v}$, which contains all of the nodal variables, is a 9×1 vector. Substitution of Eqs. (13.145) and (13.156) into Eq. (13.140) (without the last term) leads to

$$\Pi_H^* = -\frac{1}{2}\widehat{\boldsymbol{\sigma}}^T \mathbf{B}\,\widehat{\boldsymbol{\sigma}} + \widehat{\boldsymbol{\sigma}}^T \mathbf{C}\,\mathbf{v} \tag{13.157}$$

Since, ultimately, the displacements are the desired variables, the expression of the stiffness equation must be in terms of $\mathbf{v}$, i.e., $\widehat{\boldsymbol{\sigma}}$ should be expressed in terms of $\mathbf{v}$. Note that the stress parameters $\widehat{\boldsymbol{\sigma}}$ are independent of $\mathbf{v}$. In order for $\boldsymbol{\Pi}_H^*$ to be stationary, set the first variation to be equal to zero [Chapter 2, Eq. (2.82)]

$$\delta\boldsymbol{\Pi}_H^* = \frac{\partial \boldsymbol{\Pi}_H^*}{\partial \widehat{\boldsymbol{\sigma}}}\delta\widehat{\boldsymbol{\sigma}} + \frac{\partial \boldsymbol{\Pi}_H^*}{\partial \mathbf{v}}\delta\widehat{\mathbf{v}} = 0 \quad \text{or} \quad \frac{\partial \boldsymbol{\Pi}_H^*}{\partial \widehat{\boldsymbol{\sigma}}} = 0 \quad \text{and} \quad \frac{\partial \boldsymbol{\Pi}_H^*}{\partial \mathbf{v}} = 0$$

Then

$$\frac{\partial \boldsymbol{\Pi}_H^*}{\partial \widehat{\boldsymbol{\sigma}}} = -\mathbf{B}\,\widehat{\boldsymbol{\sigma}} + \mathbf{C}\,\mathbf{v} = 0$$

This leads to

$$\widehat{\boldsymbol{\sigma}} = \mathbf{B}^{-1}\mathbf{C}\,\mathbf{v} \tag{13.158}$$

Substitute Eq. (13.158) into Eq. (13.157) to find

$$\Pi_H^* = -\frac{1}{2}\mathbf{v}^T\mathbf{C}^T(\mathbf{B}^{-1})^T\mathbf{B}\mathbf{B}^{-1}\mathbf{C}\mathbf{v} + \mathbf{v}^T\mathbf{C}^T(\mathbf{B}^{-1})^T\mathbf{C}\mathbf{v} = \frac{1}{2}\mathbf{v}^T\mathbf{C}^T\mathbf{B}^{-1}\mathbf{C}\mathbf{v}$$

where, because $\mathbf{B}$ is symmetric, $(\mathbf{B}^{-1})^T = (\mathbf{B}^T)^{-1} = \mathbf{B}^{-1}$. The second derivative of this Π_H^* with respect to $\mathbf{v}$ shows that (Chapter 3, Example 3.3) the stiffness matrix is given by

$$\mathbf{k}_{HSM} = \mathbf{C}^T\mathbf{B}^{-1}\mathbf{C} \tag{13.159}$$

In evaluating the stiffness matrix $\mathbf{k}_{HSM}$, the matrix $\mathbf{B}^{-1}$ has to be calculated. This matrix can be expressed as

$$\mathbf{B}^{-1} = \begin{bmatrix} e_{11}\mathbf{\Lambda} & e_{12}\mathbf{\Lambda} & e_{13}\mathbf{\Lambda} \\ e_{21}\mathbf{\Lambda} & e_{22}\mathbf{\Lambda} & e_{23}\mathbf{\Lambda} \\ e_{31}\mathbf{\Lambda} & e_{32}\mathbf{\Lambda} & e_{33}\mathbf{\Lambda} \end{bmatrix} \tag{13.160}$$

where e_{ij} are the elements in $\mathbf{E}_B$ and

$$\mathbf{\Lambda} = \boldsymbol{\phi}^{-1} = \frac{1}{A\left(I_{xx}I_{yy} - I_{xy}^2\right)} \begin{bmatrix} I_{xx}I_{yy} - I_{xy}^2 & 0 & 0 \\ 0 & AI_{yy} & -AI_{xy} \\ 0 & -AI_{xy} & AI_{xx} \end{bmatrix} \tag{13.161}$$

After the nodal displacements are found, the stress parameters can be obtained from Eq. (13.158) and the bending moments in the plate can be calculated using Eq. (13.144).

EXAMPLE 13.7 Square Plate under Concentrated Load with Clamped and Simply Supported Edges

Examine the accuracy of the DKT and HSM elements through the analysis of a square plate. Consider the square plate of sides $2a$, with either simply supported or clamped edges, shown in Fig. 13.21. A concentrated load P is applied at the center of the plate. Owing to the symmetry of the plate, only one-quarter of the plate is modeled. Two different mesh orientations (1 and 2 of Fig. 13.21) are used in the analysis. Four different sizes of meshes are considered: 2, 8, 32, and 128 triangular elements are used to form the meshes corresponding to $N = 1, 2, 4$, and 8, respectively, where N is the number of rows and columns of elements in the mesh. In all cases, simply supported as well as clamped boundary conditions are considered.

The results of the computations are shown in Fig. 13.22 to Fig. 13.25. Figures 13.22 and 13.23 show that the DKT and HSM elements are quite effective. However, mesh 2 does not appear to be very effective in modeling the clamped plate problem, since all DOF of the corner element vanish. The influence of mesh orientation on the displacement is more severe for the DKT element than for the HSM element, as it is seen that the error curves of the DKT are further apart than for the HSM element. It is apparent that the mesh orientation has a significant effect on the accuracy.

The stress resultant calculations are shown in Figs. 13.24 and 13.25. The results for the moment reaction at the center of the side of the plate are given in Fig. 13.24, where it can be seen that all of the results converge. This means that the boundary condition $m_n = 0$ can be satisfied. The moment reactions of the corner points are shown in Fig. 13.25 and very good convergence is observed. ∎

In addition to the three-node triangular elements discussed in the previous sections, higher order elements are also available. One example is a 21-DOF element shown in

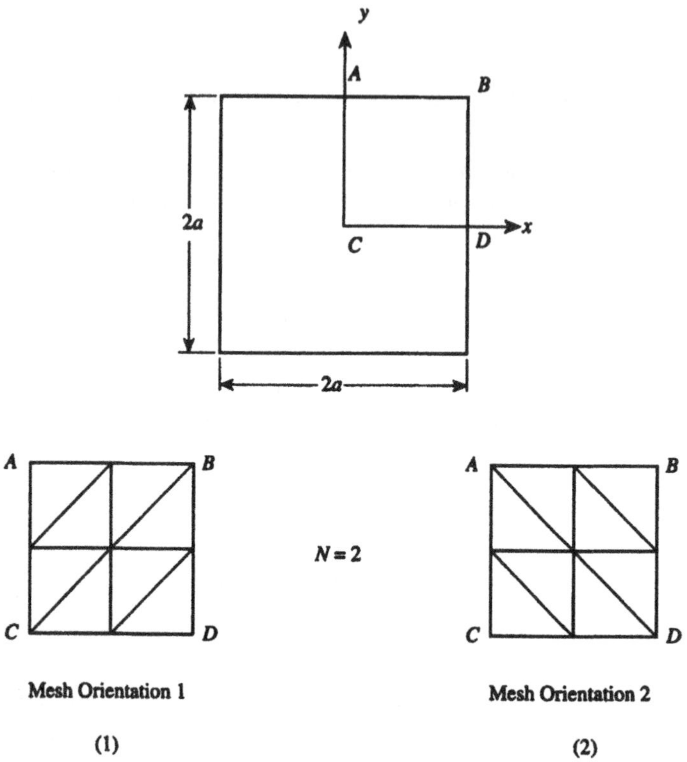

FIGURE 13.21
Square plate and mesh orientations.

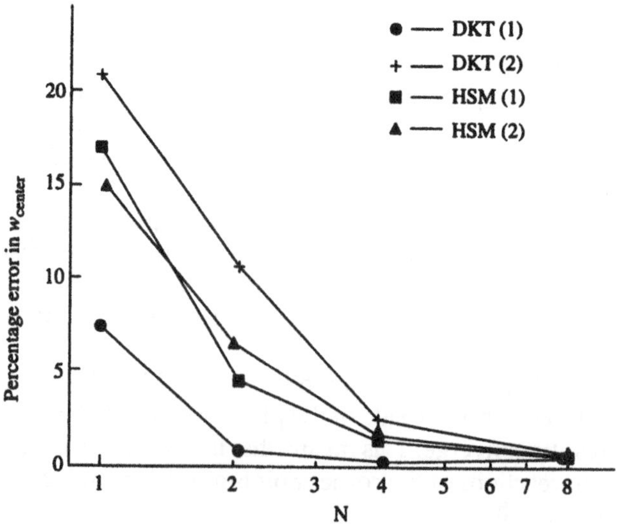

FIGURE 13.22
Simply supported plate with concentrated load. Error in deflection w_{center} at the center of the plate.

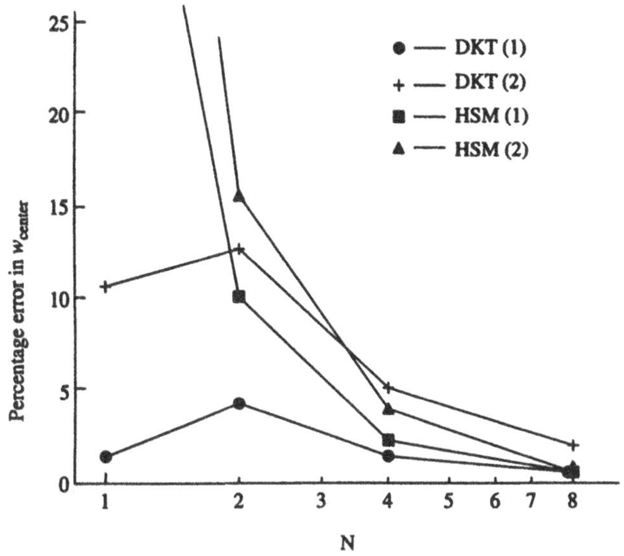

FIGURE 13.23
Clamped plate with concentrated load. Error in deflection w_{center} at the center of the plate.

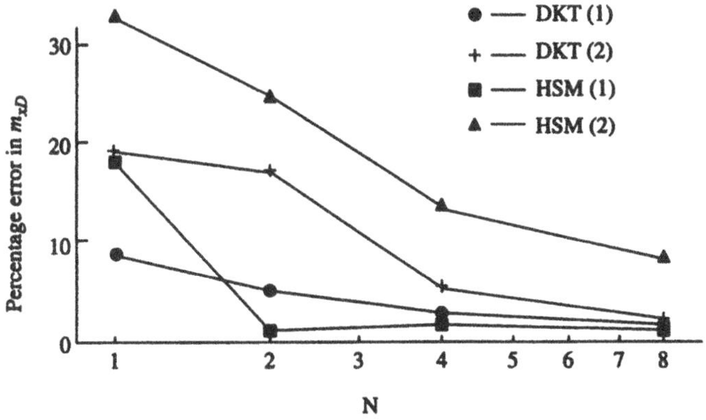

FIGURE 13.24
Clamped plate with concentrated load. Error in bending moment at the center of the side.

Fig. 13.26. The displacement w and its derivatives with respect to x and y are used as the DOF at the corner nodes and the derivative of the displacement with respect to the outer normal of the element boundary is used as the DOF at the center of each side. A complete 5th order polynomial containing 21 unknowns can serve as the trial solution. The principle of virtual work for the Kirchhoff theory can be used to derive the stiffness matrix. As a result, a compatible element can be formulated. Other kinds of triangular elements are introduced in the literature. See, for example, Bazeley, et al. (1965).

The three node DKT and HSM elements can be transformed to four node quadrilateral elements. One approach is to compose the quadrilateral element using four three node triangular elements. Another approach is to adjust the shape functions of the triangular element to form the four node quadrilateral elements. For the DKT element, change the shape functions N_i in Eq. (13.120) to the shape functions for a quadrilateral element. Equation (13.126)

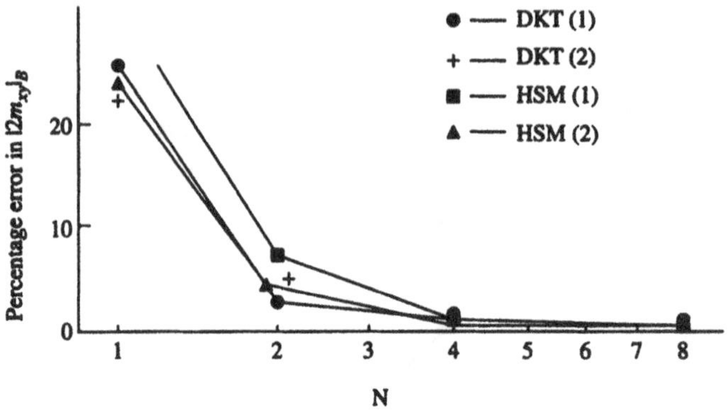

FIGURE 13.25
Simply supported plate with concentrated load. Error in a corner reaction.

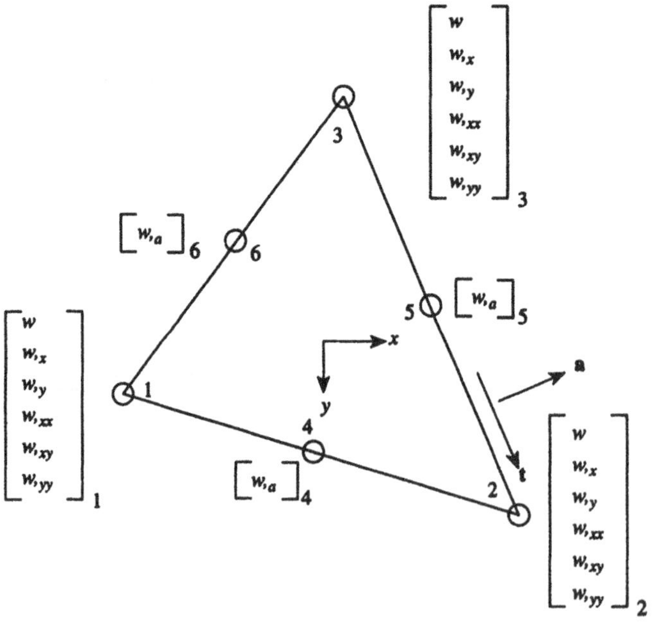

FIGURE 13.26
Nodal degrees of freedom for a 21-DOF triangular element.

is still valid for the boundary displacement of a quadrilateral element. Follow the same procedure as the triangular element to formulate the stiffness matrix for the quadrilateral element. For the HSM element, the trial functions of Eq. (13.144) can be employed for the moment distribution in the element. For the quadrilateral element, one more boundary integral U_{Sij} of Eq. (13.155) is formed and added to Eq. (13.156), leading to a 12 × 12 **C** matrix. All of the other procedures remain the same.

References

Batoz, J., Bathe, J., and Ho, L., 1980, A study of three-node triangular plate bending elements, *Int. J. for Numerical Methods in Engineering*, Vol. 15, pp. 1771–1812.

Bazeley, G.P., Cheung, Y.K., Irons, B.M., and Zienkiewicz, O.C., 1965, Triangular elements in bending-conforming and non-conforming solutions, *Proc. Conf. Matrix Methods Struct.*, Air Force Institute of Technology, Wright-Patterson Air Force Base, Dayton, OH.

Kirchhoff, G., 1850, Über das Gleichgewicht und die Bewegung einer elastischen Scheibe, *Crelles J.*, Vol. 40, pp. 51–88.

Mindlin, R.D., 1951, Influence of rotary inertia and shear on flexural motions of isotropic elastic plates, *J. Appl. Mech.*, Vol. 18, pp. 31–38.

Pian, T.H.H., and Tong. P., 1969, Basis of finite element methods for solid continua, *Int. J. for Numerical Methods in Engineering*, No. 1, pp. 3–28.

Pilkey, W.D., 1994, *Formulas for Stress, Strain, and Structural Matrices*, Wiley, New York.

Pilkey, W.D., 2002, *Analysis and Design of Elastic Beams, Computational Methods*, Wiley, New York.

Poisson, S.D., 1829, Mémoire sur l'équilibre et le mouvement des corps élastiques, *Memoirs of the Academy*, Paris, Vol. 8, pp. 357–570 and 623–627.

Reissner, E., 1945, The effect of transverse shear deformation on the bending of elastic plates, *J. Appl. Mech.*, Vol. 12, pp. A69–77.

Thomson, W. (Lord Kelvin), and Tait, P.G., 1883, *Treatise on Natural Philosophy*, Cambridge University Press, London.

Todhunter, I. and Pearson, K., 1886, *A History of the Theory of Elasticity*, Cambridge University Press, London.

Zienkiewicz, O.C., 1977, *The Finite Element Method*, 3rd ed., McGraw-Hill, New York.

Zienkiewicz, O.C. and Taylor, R.L., 2000, *The Finite Element Method, Vol. 2: Solid Mechanics*, 5th ed., Butterworth, Heinemann, London.

Problems

In-Plane Deformation

13.1 Given the relationships $x = r\cos\phi$, $y = r\sin\phi$, use the Jacobian to show that

$$\frac{\partial r}{\partial x} = \cos\phi, \quad \frac{\partial r}{\partial y} = \sin\phi, \quad \frac{\partial \phi}{\partial x} = -\frac{1}{r}\sin\phi, \quad \frac{\partial \phi}{\partial y} = \frac{1}{r}\cos\phi$$

Hint: Form

$$\begin{bmatrix} \frac{\partial}{\partial r} \\ \frac{\partial}{\partial \phi} \end{bmatrix} = \begin{bmatrix} \frac{\partial x}{\partial r} & \frac{\partial y}{\partial r} \\ \frac{\partial x}{\partial \phi} & \frac{\partial y}{\partial \phi} \end{bmatrix} \begin{bmatrix} \frac{\partial}{\partial x} \\ \frac{\partial}{\partial y} \end{bmatrix} = \mathbf{J} \begin{bmatrix} \frac{\partial}{\partial x} \\ \frac{\partial}{\partial y} \end{bmatrix} = \begin{bmatrix} \cos\phi & \sin\phi \\ -r\sin\phi & r\cos\phi \end{bmatrix} \begin{bmatrix} \frac{\partial}{\partial x} \\ \frac{\partial}{\partial y} \end{bmatrix}$$

Then

$$\begin{bmatrix} \frac{\partial}{\partial x} \\ \frac{\partial}{\partial y} \end{bmatrix} = \mathbf{J}^{-1} \begin{bmatrix} \frac{\partial}{\partial r} \\ \frac{\partial}{\partial \phi} \end{bmatrix} = \begin{bmatrix} \cos\phi & -\frac{1}{r}\sin\phi \\ \sin\phi & \frac{1}{r}\cos\phi \end{bmatrix} \begin{bmatrix} \frac{\partial}{\partial r} \\ \frac{\partial}{\partial \phi} \end{bmatrix}$$

13.2 Show that Eq. (13.11a) corresponds to the conditions of equilibrium for the element of Fig. 13.3.

13.3 Find the radial displacements and internal forces in a circular disk subject to a radial pressure of magnitude p_0(force/length) on the inner periphery at $r = a$. The outer rim at $r = b$ is free of loading.

Answer: The radial displacement is $u_r = \frac{p_0}{tE}\frac{a^2}{b^2-a^2}\left[r(1-\nu) + \frac{b^2}{r}(1+\nu)\right]$

13.4 Find the radial displacement and internal forces in a circular disk with no center hole. The disk rotates at an angular velocity Ω and is subject to a radially directed external pressure of magnitude p_0 on the periphery at radius b.

Answer: $u_r = -p_0\frac{r(1-\nu)}{Et} + \frac{p\Omega^2 r(1-\nu)}{8E}[(3+\nu)b^2 - (1+\nu)r^2]$

13.5 Thick cylinders are usually modeled as being in a state of plane strain in the axial direction, i.e., the axial strain in the cylinder is zero. The in-plane element equations of this chapter apply to the thick cylinder if the plane stress relations of this chapter are transformed to plane strain, using the plane-stress to plane-strain conversion factors of Chapter 1. Find the radial stress and displacement due to thermal loading $\Delta T(r)$ in a thick, long cylinder of radius $r = a_L$ with no center hole.

Answer: $\sigma_r = \frac{E}{1-\nu}\frac{a}{a_L^2} r \int_0^{a_L} \xi \Delta T(\xi)\, d\xi - \frac{E}{1-\nu}\frac{\alpha}{r^2}\int_0^r \xi\Delta T(\xi)\, d\xi$

Rectangular Plates

13.6 Consider a rectangular plate with dimensions L_x and L_y, which is simply supported on all sides. Expand the loading and response variables in double sine series, e.g.

$$\overline{p}_z(x, y) = \sum_{m=1}^{\infty}\sum_{n=1}^{\infty} p_{mn} \sin\frac{n\pi x}{L_x}\sin\frac{m\pi y}{L_y}$$

$$w(x, y) = \sum_{m=1}^{\infty}\sum_{n=1}^{\infty} w_{mn} \sin\frac{n\pi x}{L_x}\sin\frac{m\pi y}{L_y}$$

This is referred to as a Navier solution. Show that, in general,

$$w_{mn} = p_{mn}\Bigg/\left[K\pi^4\left(\frac{n^2}{L_x^2} + \frac{m^2}{L_y^2}\right)^2\right]$$

$$\text{where } p_{mn} = \frac{4}{L_x L_y}\int_0^{L_x}\int_0^{L_y} \overline{p}_z(x, y)\sin\frac{n\pi x}{L_x}\sin\frac{m\pi y}{L_y}\, dx\, dy$$

Hint: Use orthogonality conditions of the sort employed in deriving Eq. (13.63) from Eq. (13.62).

13.7 For the Navier solution of Problem 13.6, find the deflection, slope, moment, and shear force distributions for a uniform loading of magnitude p_0.

Answer: $p_{mn} = 16\, p_0/(\pi^2 mn)$ for m, n odd integers, $p_{mn} = 0$ if m or n or both are even.

$$w_{mn} = \frac{16\, p_0/(\pi^6 K)}{nm\left(\frac{n^2}{L_x^2} + \frac{m^2}{L_y^2}\right)^2} \qquad \text{for } m \text{ and } n \text{ odd}$$

13.8 Find the deflection in a rectangular plate simply supported on all boundaries due to a transverse concentrated force $\overline{P}$ applied at $x = c$, $y = d$.

Answer: Use the expression for w_{mn} of Problem 13.6, with

$$p_{mn} = \frac{4\overline{P}}{L_x L_y}\sin\frac{n\pi c}{L_x}\sin\frac{m\pi d}{L_y}$$

13.9 Use the Navier solutions of Problems 13.6 and 13.7 to find the response of a simply supported rectangular plate subjected to a uniformly distributed load p_0. The plate is of widths L and $2L$ in the x and y directions, respectively. Find the center deflection, maximum moments and edge reactions.

Answer: $w_{\text{max}} \simeq 0.0101\ p_0 L^4/K$. Two terms provide sufficient accuracy. Maximum moments occur at center. Four terms needed for reasonable accuracy.

$$V_{x=0} = \frac{16 p_0 L}{\pi^3} \sum_{n=1}^{\infty} \sum_{m=1}^{\infty} \frac{n^2 + (2-\nu)(m^2/4)}{m(n^2 + m^2/4)^2} \sin \frac{m\pi y}{2L} \qquad m, n = 1, 3, 5, \ldots$$

13.10 For a square plate of side L subject to the applied loading $\overline{p}_z(x, y) = p_0 \sin \frac{\pi x}{L} \sin \frac{\pi y}{L}$, find the maximum deflection, the peak bending moment per unit length, the reactions along the edges, and the downward forces necessary to prevent the corners from rising. All edges are simply supported. Plot the corner and edge forces. Verify that the total downward force equals the total upward force.

Answer:

$$w = \frac{p_0 L^4}{4K\pi^4} \sin \frac{\pi x}{L} \sin \frac{\pi y}{L}, \qquad w_{\text{max}} = p_0\, L^4/(4K\pi^4)$$

$$m_{\text{max}} = m_x|_{\substack{x=L/2\\ y=L/2}} = m_y|_{\substack{x=L/2\\ y=L/2}} = p_0 L^2 (1+\nu)/(4\pi^2)$$

$$V_{\text{along } x=0} = \frac{p_0 L}{4\pi}(3-\nu) \sin \frac{\pi y}{L}, \qquad \text{Corner forces } = 2 p_0 L^2 (1-\nu)/(4\pi^2)$$

Total upward force $= 2 p_0 L^2 (3-\nu)/\pi^2$

13.11 A simply supported rectangular plate is twice as long ($2L$) as it is wide (L). For a uniformly distributed load p_0 per unit area, find the maximum deflection.

Answer:

$$w_{\text{max}} = w_{\text{center}} = \frac{p_0\, 16 L^4 \sin \frac{m\pi}{2} \sin \frac{n\pi}{2}}{K\pi^6 mn(n^2 + m^2/4)^2} = \frac{p_0\, 16 L^4 (-1)^{(n+m-2)/2}}{K\pi^6 mn(n^2 + m^2/4)^2}$$

13.12 For a single sine series solution (a Lévy or Lévy-Nadai solution) with $w(x, y) = \sum_{m=1}^{\infty} w_m(x) \sin \frac{m\pi y}{L_y}$ and $\overline{p}_z(x, y) = \sum_{m=1}^{\infty} p_m(x) \sin \frac{m\pi y}{L_y}$, show that the governing fourth order equation for the deflection is

$$w_m^{iv} - 2\left(\frac{m\pi}{L_y}\right)^2 w_m'' + \left(\frac{m\pi}{L_y}\right)^4 w_m = \frac{p_m}{K}$$

where $' = \frac{d}{dx}$, which has the complementary solution

$$w_m = C_1 \cosh \frac{m\pi x}{L_y} + C_2 \sinh \frac{m\pi x}{L_y} + C_3 \frac{m\pi x}{L_y} \cosh \frac{m\pi x}{L_y} + C_4 \frac{m\pi x}{L_y} \sinh \frac{m\pi x}{L_y}$$

13.13 Determine the deflection of a square plate with the sinusoidal loading $\overline{p}_z = p_0 \sin \frac{\pi y}{L_y}$. The edges at $y = 0$ and $y = L_y$ are simply supported, while the other two edges are fixed. Use a single sine series solution.

Answer:

$$p_m = \begin{cases} p_0 & m = 1 \\ 0 & m > 1 \end{cases} \qquad \beta^2 = (\pi/L)^2$$

$$w = \frac{p_0}{K\beta^4}\left[1 + \frac{\beta x \sinh\pi + \cosh\pi - 1}{\pi + \sinh\pi}\sinh\beta x \right.$$
$$\left. + \frac{\beta x(\cosh\pi - 1) - \pi - \sinh\pi}{\pi + \sinh\pi}\cosh\beta x\right]\sin\frac{\pi y}{L_y}$$

13.14 For a single sine series solution, determine the transformed loading function p_m for a rectangular plate with a constant loading in the y direction. Suppose the load begins at $y = b_1$ and ends at $y = b_2$.

Answer:

$$p_m = \frac{2p_0(x)}{m\pi}\left(\cos\frac{m\pi b_1}{L_y} - \cos\frac{m\pi b_2}{L_y}\right)$$

$$\text{If } b_1 = 0, \quad b_2 = L_y: \qquad p_m = \begin{cases} \frac{4p_0(x)}{m\pi} & \text{if } m = 1, 3, 5, 7, \ldots \\ 0 & \text{if } m = 2, 4, 6, 8, \ldots \end{cases}$$

13.15 Derive the first-order relations of Eq. (13.58a).

13.16 Use an analytical solution to determine the deflection of the rectangular plate shown in Fig. P13.16 if the edges at $y = 0$ and $y = L_y$ are:

(a) simply supported

(b) fixed

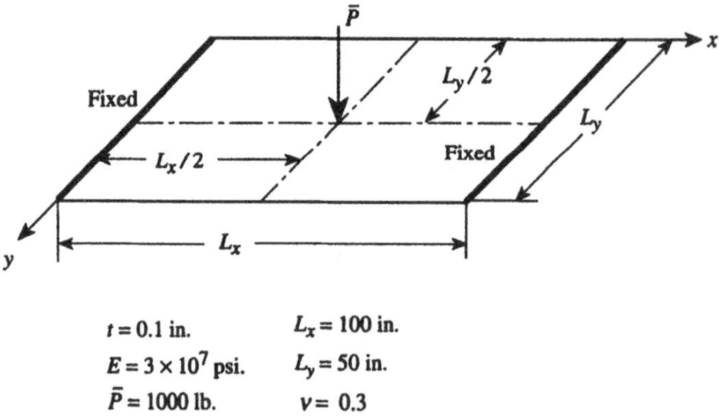

FIGURE P13.16

13.17 Find the natural frequencies ω_{mn}, $m, n = 1, 2, 3, \ldots$, for a simply supported rectangular plate.

Answer:

$$\omega_{mn} = \frac{C_{mn}}{L_y^2}\left(\frac{K}{\rho}\right)^{1/2}, \quad C_{mn} = \pi^2(m^2 + \alpha^2 n^2), \quad \alpha = L_y/L$$

13.18 Find the critical (buckling) in-plane force $(n_y)_{cr}$ in the y direction of a rectangular plate that is simply supported on all sides.

Answer:

$$(n_y)_{cr} = C\frac{\pi^2 K}{L^2}, \quad C = \frac{\left(\frac{m}{\alpha} + \frac{\alpha n^2}{m}\right)^2}{1 + \frac{n_x}{n_y}\left(\frac{\alpha n}{m}\right)^2}, \quad \alpha = L_y/L$$

n, m = number of half waves in the x and y directions

13.19 Determine the critical in-plane load $(n_y)_{cr}$ in a simply supported plate for which L_y is much less than L. Also, $n_x = 0$.

Hint: In the answer of Problem 13.18, set $n_x = 0$ and $\alpha \ll 1$. For $n = 1$, the minimum value of C [and $(n_y)_{cr}$] occurs for $m = 1$

Answer:

$$(n_y)_{cr} = \frac{K\pi^2}{L^2}\left(\frac{1}{\alpha} + \alpha\right)^2$$

Circular Plates

13.20 Find the deflection of a circular plate with no center hole and loaded with a concentrated ring load $\overline{P}$. This is a line load (force/length) extending symmetrically around the plate at radius $r = a_1$. The outer rim at $r = b$ is simply supported.

Answer: The initial parameters are

$$w_0 = -\frac{\overline{P}a_1}{4K}\left\{a_1^2 \ln\frac{b}{a_1} - \left[\frac{1-\nu}{2(1+\nu)} + 1\right](b^2 - a_1^2)\right\}$$

$$M_0 = \frac{\overline{P}a_1}{2}\left[(1+\nu)\ln\frac{b}{a_1} + \frac{1+\nu}{2}\left(1 - \frac{a_1^2}{b^2}\right)\right]$$

$$\theta_0 = V_0 = 0$$

13.21 Determine an expression for the deflection of the circular plate shown in Fig. 13.21.

Hint: Reduce the solution of Problem 13.20 to the case of a plate with a total center load $\overline{P}_{\text{total}}$. Set $\overline{P} = \overline{P}_{\text{total}}/(2\pi a_1)$ and take the limit of w as $a_1 \to 0$.

Answer:

$$w = \frac{\overline{P}_{\text{total}}}{16\pi K}\left[(b^2 - r^2)\frac{3+\nu}{1+\nu} - 2r^2 \ln\frac{b}{r}\right]$$

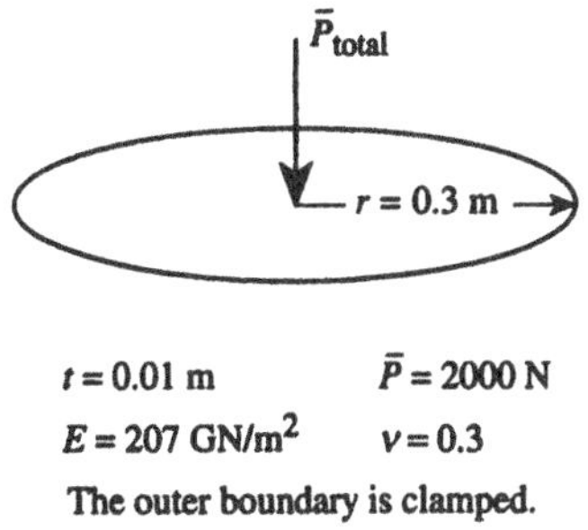

FIGURE P13.21

13.22 Derive an equation for the deflection of the circular plate with the symmetrical loading shown in Fig. P13.22. Also, find the deflection if the outer rim is simply supported.

Hint: $\overline{p}_z = p_0 r/b$

Answer: $p_0^c = p_0 r/b, \quad p_m^c = p_m^s = 0 \quad \text{for } m > 0$
For simply supported edge:

$$w = \frac{p_0}{bK}\left[\frac{1}{90}\frac{4+\nu}{1+\nu}(b^2 - r^2)r^3 - \frac{1}{225}(b^5 - r^5)\right]$$

For fixed edge:

$$w = \frac{p_0}{450K}\left(\frac{2r^5}{b} - 5b^2r^2 + 3b^4\right)$$

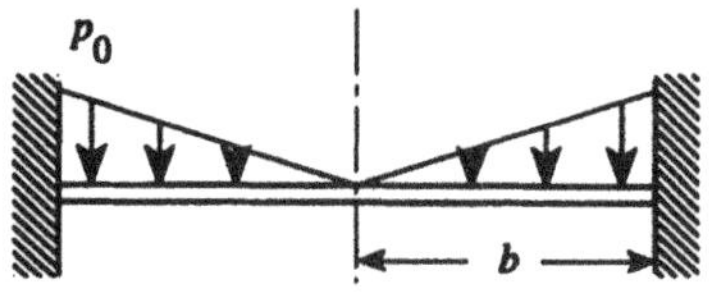

FIGURE P13.22

13.23 Find the deflection of the plate of Fig. P13.23 with a symmetrical line moment $\overline{M}_1$ applied to the inner rim.

Answer:

$$w = \frac{\overline{M}_1}{K}\left[\frac{b^2a^2}{(1-\nu)(b^2-a^2)}\ln\frac{r}{b} + \frac{a^2(r^2-b^2)}{2(1+\nu)(b^2-a^2)}\right]$$

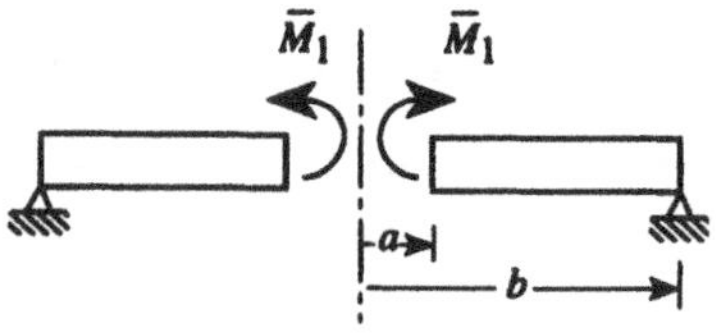

FIGURE P13.23

13.24 Derive expressions for the deflection, slope, shear force, and moment for the plate with the symmetrical loading of Fig. P13.24.

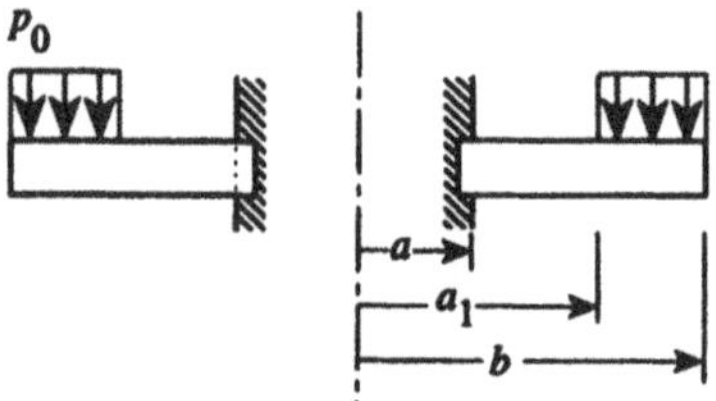

FIGURE P13.24

13.25 Determine the distribution of deflection and moment in the plate with the symmetrical line loads of Fig. 13.25.

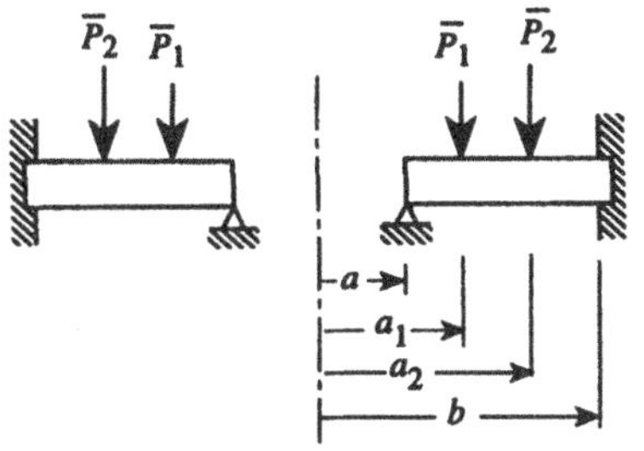

FIGURE P13.25

13.26 Find the natural frequencies ω_{mn}, $m, n = 1, 2, 3, \ldots$ for a circular plate with no central hole if the outer rim at $r = b$ is simply supported. Set $\nu = 0.3$.

Answer: $\omega_{mn} = \frac{C_{mn}}{b^2}\left(\frac{K}{\rho}\right)^{1/2}$

C_{mn}			
n \ m	0	1	2
0	4.977	13.94	25.65
1	29.76	48.51	70.14
2	74.20	102.80	134.33

13.27 Determine the natural frequencies ω_{mn} for a circular plate with no central hole if the outer edge at $r = b$ is clamped. Set $\nu = 0.3$.

Answer: $\omega_{mn} = \frac{C_{mn}}{b^2}\left(\frac{K}{\rho}\right)^{1/2}$

C_{mn}			
n \ m	0	1	2
0	10.216	21.26	34.88
1	39.77	60.82	84.58
2	89.10	120.08	153.81

13.28 Calculate the critical (buckling) in-plane radial force per unit length $(n_r)_{cr}$ applied at the outer edge $(r = b)$ of a circular plate with no center hole for (a) a simply supported outer edge and (b) a fixed outer edge. There are neither nodal circles nor nodal diameters present in the buckling mode shapes.

Answer: $n_{cr} = C\frac{K}{b^2}$ (a) $C = 4.191$, (b) $C = 14.68$

13.29 Find the critical in-plane radial force per unit length $(n_r)_{cr}$ applied at the outer edge $(r = b)$ of a circular plate with a center hole from $r = 0$ to $r = a$. (a) The outer edge is fixed and the inner edge is free. (b) The outer edge is simply supported and the inner edge is free. There are neither nodal circles nor nodal diameters present in the buckling mode shapes.

Answer: $n_{cr} = C\frac{K}{b^2}$

	C		
a/b case	0.0	0.1	0.2
(a)	14.68	14.00	13.26
(b)	4.19	4.00	3.70

13.30 Suppose a circular plate with no center hole is subject to a nonsymmetrical load $\overline{p}_z = p_0 \frac{r}{b} \cos\phi$. The outer rim at $r = b$ is hinged.

Hint: Use Eq. (13.76) with $p_m^s(r) = 0$ and $m = 1$.

Answer:

$$w = \frac{p_0 b^4}{192(3+\nu)K}\left(\frac{r}{b}\right)\left[1-\left(\frac{r}{b}\right)^2\right]\left[(7+\nu)-(3+\nu)\left(\frac{r}{b}\right)^2\right]\cos\phi$$

$$m_r = \frac{p_0 b^2}{48}(5+\nu)\left(\frac{r}{b}\right)\left[1-\left(\frac{r}{b}\right)^2\right]\cos\phi$$

Stiffness (or Transfer) Matrix Methods

13.31 Find the radial displacements and internal forces in a circular disk subject to a radial pressure of p_0(force/length) on the inner periphery at $r = a$. The disk is of thickness t_1 for $a \le r \le a_1$ and thickness t_2 for $a_1 < r \le b$. The outer rim at $r = b$ is free of loading.

13.32 A simply supported rectangular plate is twice as long ($L_x = 2L$) as it is wide ($L_y = L$). The plate rests on a midspan ($x = L$) rigid knife edge support that begins at $y = 0$ and ends at $y = L_y$.

13.33 Find the deflection, slope, shear force, and moment distributions in the symmetrically loaded circular plate of Fig. P13.33.

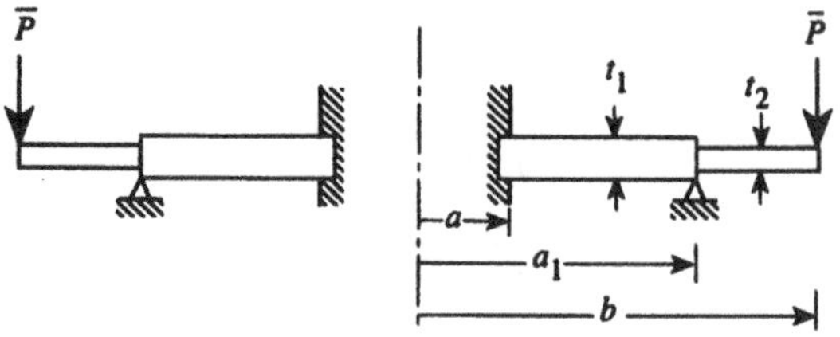

FIGURE P13.33

13.34 Find the first three natural frequencies of the plate of Fig. P13.33 if $\nu = 0.3$, $E = 207$ GN/m^2, $a = 0.5$ m, $a_1 = 1.0$ m, $b = 1.5$ m, $t_1 = 0.01$, and $t_2 = 0.005$.

Finite Elements

13.35 Consider a rectangular element with eight nodes, four of which are at the corners and four at the midpoints of the sides. Use Kirchhoff plate theory to develop a stiffness matrix for this element. Assume there are eight DOF, with four deflection DOF at the corner nodes and four rotational (slope) DOF at the midpoints of the sides. Use the displacement function

$$w = \mathbf{N}_u \widehat{\mathbf{w}} = \widehat{w}_1 + \widehat{w}_2 x + \widehat{w}_3 y + \widehat{w}_4 x^2 + \widehat{w}_5 xy + \widehat{w}_6 y^2 + \widehat{w}_7 x^3 + \widehat{w}_8 y^3$$

13.36 Set up a triangular element analogous to the rectangular element of Problem 13.35. Use six DOF.

13.37 Formulate the stiffness matrix of a four node quadrilateral element with w, θ_x, and θ_y as independent variables at the nodes. Use linear interpolation polynomials for all of these variables. Use the Reissner-Mindlin plate theory.

Hint: Follow the development of the rectangular element in Section 13.5.1.

13.38 Derive the stiffness matrix for a 12 DOF rectangular element based on Kirchhoff theory. The DOF at each node are w, $\theta_x = -\partial w/\partial x$, $\theta_y = -\partial w/\partial y$.

13.39 Calculate the deflection of the plate shown in Fig. P13.16 using the finite element method and compare your results with those of Problem 13.16.

13.40 Use a finite element solution to see how close you can approximate the circular plate results of Problem 13.21 (Fig. P13.21).

13.41 Based on rectangular elements, find the maximum deflection in a simply supported rectangular plate that is twice as long ($2L$) as it is wide (L). The load is uniformly distributed of magnitude p_0(force/area). Use numerical values to compare your solution to the answer given for Problem 13.11.

13.42 Use the finite element method to find the natural frequencies for a simply supported rectangular plate. For particular numerical values, compare your frequencies with those of Problem 13.17.

13.43 Find the natural frequencies of a circular plate with no center hole if the outer rim is simply supported. For particular numerical values, compare with the frequencies given in Problem 13.26.

13.44 Derive the expressions for the shape functions in Eq. (13.130).

13.45 Obtain the formulas for the rotations of Eq. (13.129) for a four-node quadrilateral DKT element.

13.46 Derive $\mathbf{L}_{ij}$ of Eq. (13.154) by using the relationships in Eqs. (13.128), (13.129), (13.152), and (13.153).

13.47 Give the explicit form of $\mathbf{B}$ of Eq. (13.146) and $\mathbf{B}^{-1}$ of Eq. (13.158) for a homogeneous isotropic plate.

13.48 Calculate $\mathbf{C}_{ij}$ of Eq. (13.155).

13.49 Give the explicit form of the matrix $\mathbf{C}$ of Eq. (13.157).

Section E

Appendices

Appendix I

Some Fundamentals of Variational Calculus

The calculus of variations deals with the minima (or maxima) of functionals, e.g., integrals, whereas much of the ordinary calculus is concerned with functions of single variables. The origin of the calculus of variations is usually traced back to the celebrated brachistochrone problem, which, as was the custom of the times, was publicly posed as a challenge to fellow mathematicians by Jakob Bernoulli[1] in 1696. This isoperimetric problem was to find the path lying in a vertical plane, along which a frictionless particle must slide from one point to a lower point in minimum time. The proper curve turns out to be a cycloid. Not only did Bernoulli develop a solution, but so did his brother Johann[2] and the famous scientists Leibniz[3], Newton[4], and L'Hospital[5]. Sir Isaac Newton considered the problem years after

[1] Jakob Bernoulli (1654–1705) was a member of the Bernoulli family of mathematicians and physicists. The Bernoulli family was of Dutch origin and was driven to Switzerland by the Spanish persecution. He was the older brother and mathematics instructor of Johann (John) Bernoulli. Like his brother, he was obstinate and vindictive, with concerns about inferiority. He was better known for his resourceful, carefully crafted application of the infinitesimal calculus than he was for fundamental developments in mathematics. From his writings it is clear that he comprehended well the principles of this new analysis. He appears to have been the first to develop an integral calculus and was the first to use the term integral (1690).

[2] Johann Bernoulli (1667–1748) was the tenth child in his family. He studied mathematics with Jakob in Basel and received a degree in medicine. The efforts of Johann's son Daniel (1700–1782) and his brother Jakob laid the foundation on which Euler developed the differential equation of the elastic curve. Johann was one of the more fascinating members of this productive family. His life was tainted by private and professional quarrels. For example, he tried to replace an incorrect solution of his own for a problem of isoperimetrical curves with a solution developed by his brother Jakob. Also, he once expelled his son Daniel from his house for being awarded a prize by the French Academy which he had hoped to receive himself. His 1732 developments in hydraulics were usually accepted as plagiarism from the work of his son Daniel.

[3] Gottfried Wilhelm Leibniz (1646–1716) was a German mathematician and philosopher who, along with the elder Bernoullis, is considered as the founder of modern analysis. Although Leibniz studied at the University of Leipzig, where his father was a professor of moral philosophy, he was mostly self-taught in mathematics. He earned a doctorate in law and worked for the government for most of his career. Notwithstanding the question as to whether they drew the basic concepts from Newton or developed them themselves, this continental trio is usually regarded as having functioned independently of the English mathematicians. Leibniz was as recognized a philosopher as he was a mathematician. He designed a calculating machine, and meddled in diplomacy and even proposed a Napoleon-like plan of German-French cooperation which was to include the conquest of certain North African and Asian countries. Late in life, he became embroiled in an embarrassing controversy as to whether he or Newton had discovered the differential calculus. However, there appears to be no doubt that Leibniz introduced the differential notation.

[4] Isaac Newton (1642–1727) was a British mathematician whose influence on scientific thought is probably unparalleled. He was born on Christmas day, the same day in 1642 that Galileo died. Newton's father, a farmer, died before Newton was born. Fortunately, an uncle encouraged his college education. His accomplishments in some

having refrained from serious scientific work and, it is said, formulated a solution in a single evening. In addition to initiating variational calculus, one of the solution techniques advanced geometrical optics as well.

The fundamental problem of the calculus of variations is to find a function $u(x)$ such that

$$\Pi = \int_a^b F(x, u, u')\, dx \quad \text{where } u' = \frac{du}{dx} \tag{I.1}$$

is rendered *stationary*, i.e., Π assumes an extreme value. The integral Π is said to be stationary when its first variation vanishes:

$$\delta\Pi = 0 \tag{I.2}$$

in which the operator symbol δ is introduced to indicate a "variation," a concept that will be explained shortly. Determining the stationary value of an integral like Π is similar to the problem in the calculus where a stationary value (minimum, maximum, or point of inflexion) of a function is sought. There, a function assumes a stationary value at a point at which the first derivative of the function is zero.

Necessary conditions that the integrand F must satisfy to make Π stationary will now be developed, beginning with a definition of the quantities involved. Define $u(x)$ to be a function of x, for x in the interval (a, b). Let F be a known function such as an energy density. The value of Π depends on the value of F, which, in turn, is a function of x, as well as of u and u'. The dependence on x, the independent coordinate, and on u and its first derivative is used here as an example. The functional may involve other coordinates and derivatives as well. A quantity such as Π, whose value depends on a function, is called a *functional*. It can be considered as a function that depends upon the entire distribution of one or more functions, rather than just on a number of discrete variables. The *domain* of a functional is a collection of admissible functions belonging to a class of functions in function space rather than a region in a coordinate space; thus, the particular function $u(x)$ that extremizes Π is to be found. Formally, it is necessary to require for the special functional of Eq. (I.1) that $u(x)$ be twice differentiable in x and that F be twice differentiable with respect to the variables x, u, and u'.

Let $\widehat{u}(x)$ be a family of neighboring paths of the extremizing path $u(x)$, and assume that at the end points $x = a, b$ their values coincide. Represent the $\widehat{u}$ family as

$$\begin{aligned} \widehat{u}(x, \epsilon) &= u(x) + \epsilon\eta(x) = u(x) + \delta u(x) \\ &= \text{extremizing path} + \text{variation} \end{aligned} \tag{I.3}$$

where ϵ is a small parameter, and $\delta u(x)$ is called the *variation* of $u(x)$

$$\delta u = \widehat{u}(x, \epsilon) - u(x) = \epsilon\eta(x) \tag{I.4}$$

areas were so complete that he left a barrier which had to be crossed before progress could continue. Of course, there are Newton's laws of motion, which, in part, seem to have been due to several others, including Galileo. Then there is his development of infinitesimal calculus. The bulk of his work was done between 1665 and 1686, but much of it was not available in print until several years later.

[5]Guillaume François L'Hospital (1661–1704) was a French mathematician who studied under Johann Bernoulli. It is thought that at the age of 15 he solved a cycloid problem proposed by Pascal. In 1691, Bernoulli spent several months in L'Hospital's house teaching him calculus. This permitted L'Hospital to join the exclusive club of Newton, Leibniz, and the two elder Bernoullis as those who understood the infinitesimal calculus. After L'Hospital's death, Johann Bernoulli was publicly upset about not being given enough credit for his contributions to L'Hospital's work. L'Hospital wrote the first calculus book, *Analyse des Infiniment Petits pour L'intelligence des Lignes Courbes*, which was published in 1696. Somewhat rare among mathematicians of the time, L'Hospital was characterized as being both modest and generous.

or simply

$$\delta u = \hat{u} - u$$

Here, $\eta(x)$ is a twice differentiable function of undefined amplitude with $\eta(a) = \eta(b) = 0$. Note that $\hat{u}$ coincides with u if $\epsilon = 0$.

A useful characteristic is the commutative property of the variation and derivative of u. In order to show that $\frac{d}{dx}(\delta u) = \delta\left(\frac{du}{dx}\right)$, first note from Eq. (I.4) that

$$\frac{d}{dx}(\delta u) = \hat{u}'(x, \epsilon) - u'(x) \tag{I.5}$$

Since the variation of u' is defined as

$$\delta(u') = \hat{u}'(x, \epsilon) - u'(x)$$

it can be concluded that

$$\frac{d}{dx}(\delta u) = \delta\left(\frac{du}{dx}\right) \tag{I.6}$$

It should be observed that this holds only for continuous functions possessing derivatives of the requisite order.

Although the δ (delta) operator and the differential calculus d operator are used formally in a similar fashion, they should be clearly distinguished from each other. For the function $u(x)$, the differential calculus quantity du designates the vertical distance between points on a given curve at locations of infinitesimal distance dx apart (Fig. I.1). However, δu is not associated with neighboring points on a given curve, but rather represents a small but arbitrary change in the ordinate u for a particular value of x. In Fig. I.1, at the specified location $x = c$, δu is the vertical difference between any of the $\hat{u}$ curves (B or C) and the u curve (A). Note that no δx is associated with δu.

For u specified on the boundary, the variation δu must be zero because the specified value of u does not vary at this particular value of x. As a consequence, the variation δu is zero where u is specified, and it is arbitrary elsewhere. The variation δu is said to undergo a *virtual* change.

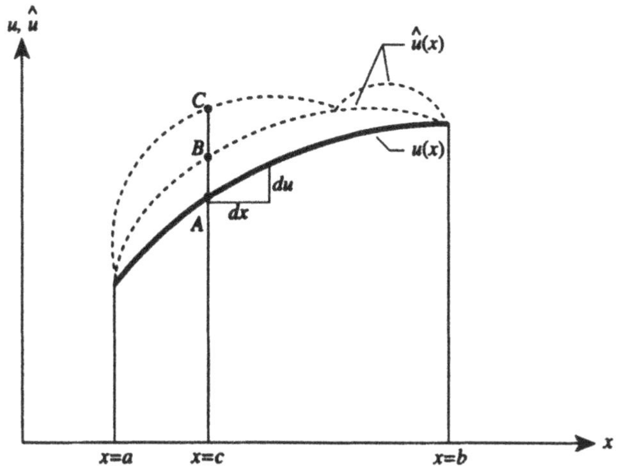

FIGURE I.1
The delta operator.

As mentioned previously, the operator δ can be used formally just as one uses the operator d. For example,

$$\delta(u')^2 = 2u'\delta u' \tag{I.7a}$$
$$\delta(u+v) = \delta u + \delta v \tag{I.7b}$$
$$\delta u = 0 \qquad \text{if } u \text{ is specified (constant)} \tag{I.7c}$$

Also,

$$\delta\left(\int u\,dx\right) = \int (\delta u)\,dx \tag{I.7d}$$

To solve the variational problem of extremizing Π, we seek the extreme value of the integral Π by considering

$$\Pi(u+\epsilon\eta) = \Phi(\epsilon) = \int_a^b F(x, u+\epsilon\eta, u'+\epsilon\eta')\,dx \tag{I.8}$$

in the limit as $\epsilon \to 0$. By using $u+\epsilon\eta$ as "admissible functions" in the sense that $\eta(a) = \eta(b) = 0$, we can reduce the problem of extremizing Π to the ordinary calculus problem of finding the extreme value of Φ, a function of the parameter ϵ. That is, since $\widehat{u} \to u$ for $\epsilon = 0$, the necessary condition for Π to be an extremum will be

$$\left.\frac{d\Phi(\epsilon)}{d\epsilon}\right|_{\epsilon=0} = 0 \tag{I.9}$$

Recall that $\widehat{u} = u + \epsilon\eta$. The derivative of Φ, with respect to ϵ, can be expressed as

$$\frac{d\Phi(\epsilon)}{d\epsilon} = \int_a^b \left(\frac{\partial F}{\partial \widehat{u}}\frac{d\widehat{u}}{d\epsilon} + \frac{\partial F}{\partial \widehat{u}'}\frac{d\widehat{u}'}{d\epsilon}\right)dx = \int_a^b \left(\eta\frac{\partial F}{\partial \widehat{u}} + \eta'\frac{\partial F}{\partial \widehat{u}'}\right)dx$$

or

$$\left.\frac{d\Phi(\epsilon)}{d\epsilon}\right|_{\epsilon=0} = \int_a^b \left(\eta\frac{\partial F}{\partial u} + \eta'\frac{\partial F}{\partial u'}\right)dx = 0 \tag{I.10}$$

Integration by parts of the second term in the integral of Eq. (I.10) and use of the end conditions $\eta(a) = \eta(b) = 0$ leads to

$$\int_a^b \eta(x)\left(\frac{\partial F}{\partial u} - \frac{d}{dx}\frac{\partial F}{\partial u'}\right)dx = 0 \tag{I.11}$$

A basic lemma states that if $\Psi(x)$ is a continuous function in $a \le x \le b$, then the relation

$$\int_a^b \eta(x)\,\Psi(x)\,dx = 0 \tag{I.12}$$

holds for arbitrary continuous functions $\eta(x)$ with continuous first derivatives if, and only if, $\Psi(x) \equiv 0$. This is valid for functions $\eta(x)$ which vanish at the ends. This result is often referred to as the *fundamental lemma of the calculus of variations*. Since $\eta(x)$ is assumed to satisfy the conditions mentioned above, it follows immediately from Eq. (I.11) that

$$\frac{\partial F}{\partial u} - \frac{d}{dx}\frac{\partial F}{\partial u'} = 0 \tag{I.13}$$

This differential equation is called the *Euler equation* associated with Π. It is a necessary condition for a function $u(x)$ to extremize the functional Π.

EXAMPLE I.1 Extension of a Bar

The total potential energy Π in a simple extension bar (length L, Young's modulus E, cross-sectional area A, axial displacement u, and axial loading $\overline{p}_x$) is

$$\Pi = \int_0^L \left[\frac{1}{2} E A (u')^2 - \overline{p}_x u\right] dx \tag{1}$$

According to the principle of stationary potential energy, a fundamental energy theorem, the kinematically admissible deformations which also satisfy equilibrium must correspond to the assumption of a stationary value of the total potential energy. Thus, for such a bar, $\delta\Pi$ vanishes when Π is given by (1).

By comparison of (1) with Eq. (I.1),

$$F(x, u, u') = \frac{1}{2} E A (u')^2 - \overline{p}_x u \tag{2}$$

To derive Euler's equation, we calculate

$$\frac{\partial F}{\partial u} = -\overline{p}_x \qquad \frac{\partial F}{\partial u'} = EA\, u'$$

Thus, from Eq. (I.13),

$$\frac{\partial F}{\partial u} - \frac{d}{dx}\frac{\partial F}{\partial u'} = -\overline{p}_x - \frac{d}{dx}(EA\, u') = 0 \tag{3}$$

or

$$\frac{d}{dx}\left(EA\,\frac{du}{dx}\right) = -\overline{p}_x \tag{4}$$

which is the governing equation for the extension of a simple bar.

Equations (1) and (4) can be viewed as being different but equivalent analytical representations of the same problem. The differential equation is sometimes referred to as the classical or *local* model and its solution u must be twice differentiable. The integral equation (1) is called the variational or *global* equation for the problem, and the solution of $\delta\Pi = 0$ is sometimes referred to as being *weak*, since this u need only be differentiable once. Moreover, although not evident here, some boundary conditions usually associated with (4) are included in the variational integral (1). ∎

It remains to define the conditions under which Π is rendered stationary when the first variation of Π vanishes. In other words, it remains to be shown that this stationary value is equivalent to the extreme value represented by Eq. (I.13). Note that $F(x, u + \epsilon\eta, u' + \epsilon\eta')$ can also be written using the operator δ as $F(x, u + \delta u, u' + \delta u')$. For a fixed x, expand F in a Taylor series about u and u' to obtain

$$F(x, u + \delta u, u' + \delta u') = F(x, u, u') + \left[\frac{\partial F}{\partial u}\delta u + \frac{\partial F}{\partial u'}\delta u'\right] \tag{I.14}$$

plus higher order terms containing $(\delta u)^2$, $(\delta u')^2$, $(\delta u)^3$, etc. Rewrite Eq. (I.14) as

$$F(x, u + \delta u, u' + \delta u') - F(x, u, u') = \left[\frac{\partial F}{\partial u}\delta u + \frac{\partial F}{\partial u'}\delta u'\right] + \text{higher order terms}$$

The left-hand side of this expression is the change in F due to the variation δu for a fixed x, i.e., it is equal to δF. The term in square brackets on the right-hand side of Eq. (I.14) is referred to as the first variation of F, the higher order terms comprise higher order variations. This first variation of a functional expression,

$$\delta F = \frac{\partial F}{\partial u}\delta u + \frac{\partial F}{\partial u'}\delta u' \tag{I.15}$$

will be used frequently. If the higher order terms are neglected, we can write the first variation of the functional Π as

$$\delta \Pi = \int_a^b \delta F \, dx = \int_a^b \left(\frac{\partial F}{\partial u}\delta u + \frac{\partial F}{\partial u'}\delta u' \right) dx \tag{I.16}$$

Integration by parts of the second term of the integrand permits δu to be factored out, and use of the conditions $\delta u = 0$ at $x = a, b$ leads to

$$\delta \Pi = \int_a^b \delta u \left(\frac{\partial F}{\partial u} - \frac{d}{dx}\frac{\partial F}{\partial u'} \right) dx \tag{I.17}$$

For $\delta \Pi = 0$ with a properly behaving δu, the Euler equation (I.13) follows from the fundamental lemma of the calculus of variations. Also, it can be reasoned that since the variations δu are arbitrary, setting the integral of Eq. (I.17) to zero and using the fundamental lemma of the calculus of variations leads to Euler's equation. Thus, finding the stationary value of Π by setting the first variation of Π equal to zero is equivalent to finding the extremal value of Π by setting $d\Pi/d\epsilon|_{\epsilon=0}$ equal to zero.

It can be shown that as with second derivatives in ordinary calculus, the second variations $\delta^2\Pi$ can be used to characterize the extremum as either a minimum or a maximum, i.e., $\delta^2\Pi > 0 \rightarrow$ minimum of Π, $\delta^2\Pi < 0 \rightarrow$ maximum of Π. Some similarities between the differential calculus and the variational calculus are summarized in Table I.1.

Finally, some remarks concerning the boundary conditions should be made. The integration by parts of the second term in the integrand of Eq. (I.10) gives

$$\int_a^b \eta' \frac{\partial F}{\partial u'} dx = \eta \frac{\partial F}{\partial u'}\bigg|_a^b - \int_a^b \eta \frac{d}{dx}\frac{\partial F}{\partial u'} dx \tag{I.18}$$

To proceed to Eq. (I.11), it is necessary to use the condition $\eta = 0$ at $x = a, b$. These are referred to as *forced boundary conditions*. If η had been left arbitrary at the boundaries, it would have been necessary to use the condition $\partial F/\partial u' = 0$ at $x = a, b$. These conditions are called the *natural boundary conditions*.

TABLE I.1

Comparison of the Ordinary and Variational Calculus

	Differential Calculus	Variational Calculus
Problem formulation	Involves a function	Involves a function of a function = functional
Necessary condition for an extreme value	First derivative = 0	First variation = 0
Result	A single value	A function
The type of extremum follows from	The second derivative	The second variation

EXAMPLE I.2 Bending of a Beam

The total potential energy of an ordinary beam is

$$\Pi = \int_0^L \left[\frac{1}{2}M\kappa - \overline{p}_z w\right] dx = \int_0^L \left[\frac{1}{2}(EIw'')w'' - \overline{p}_z w\right] dx \tag{1}$$

To apply the principle of stationary potential energy, we set $\delta\Pi = 0$. By comparison of (1) with Eq. (I.1), function F is chosen such that

$$F(x, w, w'') = \frac{1}{2}(EIw'')w'' - \overline{p}_z w \tag{2}$$

Euler's equation as expressed by Eq. (I.13) does not apply directly because here $F = F(x, w, w'')$ contains second but no first derivatives of w. It can be shown, that for such functionals, Euler's equation takes the form

$$\frac{\partial F}{\partial w} + \frac{d^2}{dx^2}\frac{\partial F}{\partial w''} = 0 \tag{3}$$

However, rather than using Euler's equation directly, the desired governing equation will be derived in the same fashion as Euler's equation was:

$$\begin{aligned}
\left.\frac{d\Phi(\epsilon)}{d\epsilon}\right|_{\epsilon=0} &= \int_0^L \left(\eta''\frac{\partial F}{\partial w''} + \eta\frac{\partial F}{\partial w}\right) dx = 0 \\
&= \int_0^L [\eta''(EIw'') - \eta\overline{p}_z]\, dx = (EIw'')\eta'\big|_0^L - \int_0^L (EIw'')'\eta'\, dx - \int_0^L \eta\overline{p}_z\, dx \\
&= (EIw'')\eta'\big|_0^L - (EIw'')'\eta\big|_0^L + \int_0^L [(EIw'')'' - \overline{p}_z]\eta\, dx \\
&= (-M\eta' + V\eta)\big|_0^L + \int_0^L [(EIw'')'' - \overline{p}_z]\eta\, dx = 0
\end{aligned} \tag{4}$$

where Chapter 1, Eq. (1.127) was used. If $\eta, \eta' = 0$ for $x = 0, L$ (the forced boundary conditions) or $M, V = 0$ for $x = 0, L$ (the natural boundary conditions), then

$$(EIw'')'' = \overline{p}_z \tag{5}$$

is Euler's equation, which is recognized to be the governing differential equation for beam theory.

Usually, governing equations such as (5) are derived using variational notation as employed in Eqs. (I.16) and (I.17). For the case of our beam with Π given by (1),

$$\begin{aligned}
\delta\Pi &= \int_0^L \delta F\, dx = \int_0^L \left(\frac{\partial F}{\partial w''}\delta w'' + \frac{\partial F}{\partial w}\delta w\right) dx = \int_0^L (EIw''\delta w'' - \overline{p}_z\,\delta w)\, dx \\
&= (EIw''\,\delta w')\big|_0^L - \int_0^L [(EIw'')'\,\delta w' - \overline{p}_z\,\delta w]\, dx \\
&= (EIw''\,\delta w')\big|_0^L - [(EIw'')'\,\delta w]\big|_0^L + \int_0^L [(EIw'')'' + \overline{p}_z]\,\delta w\, dx = 0
\end{aligned} \tag{6}$$

Thus,

$$(EIw'')'' = \overline{p}_z \quad \text{for all } x \tag{7}$$

and

$$\left.\begin{array}{lll}\delta w' = 0 & \text{or} & EIw'' = -M = 0 \\ \delta w = 0 & \text{or} & (EIw'')' = -V = 0\end{array}\right\} \quad \text{at } x = 0, L \tag{8}$$

■

Many variational problems involve subsidiary conditions. For example, find the minimum of

$$\Pi = \int_a^b F(x, u, u')\, dx \tag{I.19a}$$

subject to the restriction

$$J = \int_a^b G(x, u, u')\, dx = 0 \tag{I.19b}$$

where G is a known function. This is referred to as an *isoperimetric problem*.

One technique for treating this problem forms the basis of the computational optimization techniques called *penalty function methods.* Begin by multiplying J by a factor *(Lagrange multiplier)* λ and adding the product to the original functional Π to give a new functional H,

$$H = \Pi + \lambda J = \int_a^b F^*(x, u, u')\, dx \tag{I.20}$$

where $F^* = F + \lambda G$. Recall that the goal was to extremize Π subject to $J = 0$. It is apparent from Eq. (I.20) that extremizing Π is the same as extremizing H as long as $J = 0$. Thus, in a sense, the non-zero J's in Eq. (I.20) tend to "penalize" the process of selecting an extreme value of H. The $u(x)$ that extremizes Π and satisfies $J = 0$ must also satisfy $\delta H = 0$. The necessary condition for u to correspond to the extreme value of H is the Euler equation.

$$\frac{\partial F^*}{\partial u} - \frac{d}{dx}\frac{\partial F^*}{\partial u'} = 0 \tag{I.21}$$

Refer to standard optimization or calculus of variations texts for more thorough and rigorous treatments of constrained problems. In some variational formulations of structural mechanics, λ must have the units of force, and, as a consequence, it is then referred to as a *generalized force*.

Problems

I.1 What is the curve that joins two points in a plane such that the distance along the arc is a minimum?

Hint: Use $F = [1 + (u')^2]^{1/2}$.

Answer: $u = C_1 x + C_2$, a straight line

I.2 Derive Euler equations for the following cases.

a. $F = F(x, u')$

b. $F = F(u, u')$

Hint: For b, integrate the first term in the integral of Eq. (I.10) by parts, so that η' can be factored out of both terms. Note that $\frac{\partial F}{\partial u} = \frac{\partial F}{\partial x}\frac{\partial x}{\partial u}$.

Answer: a. $\frac{d}{dx}\left(\frac{\partial F}{\partial u'}\right) = 0$ b. $u'\frac{\partial F}{\partial u'} - F = 0$

I.3 Consider the transverse deformation of a string of length L in tension T. The transverse load $\overline{p}_z$ causes a transverse deformation w from the original line. Show that for small deformations a line element of string of length dx before loading becomes $\sqrt{1 + (w')^2}\, dx \approx [1 + \frac{1}{2}(w')^2]\, dx$ after loading. Show that the strain energy change per unit original length of string is $\frac{1}{2}T(w')^2$ and that the total potential energy of the string is

$$\Pi = \int_0^L \left[\frac{1}{2}T(w')^2\, dx - \overline{p}_z w\right] dx$$

Derive Euler's equation in the form

$$Tw'' = -\overline{p}_z$$

Appendix II

Integral Theorems

Several useful integral relationships are presented here without proofs. See a calculus textbook for details on these integral theorems.

II.1 Integration by Parts

The integration by parts formula is expressed as

$$\int_a^b u(x)\, v'(x)\, dx = u(x)\, v(x)|_a^b - \int_a^b v(x)\, u'(x)\, dx \tag{II.1}$$

or

$$\int_a^b u\, dv = (u\, v)|_a^b - \int_a^b v\, du \tag{II.2}$$

where $v' = dv/dx$ and $u' = du/dx$. For integration by parts to apply, $u(x)$ and $v(x)$ must be differentiable for $a \leq x \leq b$.

Integration by parts is used in our studies primarily to shift the derivative from one variable to another; this is what occurs in Eq. (II.1), where the derivative is shifted from v to u. Also, terms at the boundaries are produced.

II.2 Green's Theorem

Green's[1] *theorem* is usually written as

$$\int_S \left(-\frac{\partial R}{\partial z} + \frac{\partial T}{\partial y}\right) dS = \oint (-R\, a_z + T\, a_y)\, ds$$
$$= \oint \left(R\frac{dy}{ds} + T\frac{dz}{ds}\right) ds = \oint (R\, dy + T\, dz) \tag{II.3}$$

[1]George Green (1793–1841) was an English mathematician who did most of his scientific work before receiving a formal education. His knowledge of mathematics was gained by self-study. He entered Cambridge at age 40 and

where $R = R(y, z)$, $T = T(y, z)$, and $dS = dy\,dz$, a surface element. The quantity s is a coordinate along the contour of the cross-section, and the integral $\oint$ indicates that the integration is to be taken completely around the contour. Also, a_z and a_y are the direction cosines of the outward normal along the contour. For example, a_y is the cosine between the outward normal and the y axis. If $R = 0$,

$$\int_S \frac{\partial T}{\partial y}\, dS = \oint T a_y\, ds \tag{II.4}$$

Integrals that occur frequently in solid mechanics are of the form

$$\int_S Q \frac{\partial T}{\partial y}\, dS$$

In order to convert Eqs. (II.3) or (II.4) to this form, consider the identity

$$\int_S Q \frac{\partial T}{\partial y}\, dS = \int_S \frac{\partial}{\partial y}(TQ)\, dS - \int_S T \frac{\partial Q}{\partial y}\, dS \tag{II.5}$$

Substitution of Eq. (II.4), rewritten as

$$\int_S \frac{\partial}{\partial y}(TQ)\, dS = \oint TQ\, a_y\, ds$$

into the first integral on the right-hand side of Eq. (II.5) gives

$$\int_S Q \frac{\partial T}{\partial y}\, dS = \oint TQ\, a_y\, ds - \int_S T \frac{\partial Q}{\partial y}\, dS \tag{II.6}$$

This is often referred to as an integration by parts formula. The reason for this reference is clear because Green's theorem shifts the derivative from one function to another and also produces a line integral. Thus, Green's theorem does for a surface integral what integration by parts does to a one-dimensional integral. Furthermore, integration by parts can be used to derive Eq. (II.6). In the plane, Green's theorem is identical to Stokes'[2] theorem.

II.3 Gauss' Integral Theorem

In dealing with variational principles, it is often essential to transform certain integrals over a volume V into integrals over the surface S bounding the volume and vice versa. This can be accomplished using *Gauss' integral theorem.*[3]

$$\int_S \mathbf{v} \cdot \mathbf{a}\, dS = \int_V \operatorname{div} \mathbf{v}\, dV \tag{II.7}$$

graduated in 1837 only a few years before his death. In 1828, he published a study of the concept of a "potential" and its application to various branches of physics. William Thomson and Maxwell extended this concept, leading to fundamental mathematical theories of elasticity. Green introduced the integral theorem of this appendix as well as *Green's function,* which is employed in the solution of partial differential equations and is the fundamental solution utilized in the boundary element method.

[2]George Gabriel Stokes (1819–1903) was an English mathematician and a prominent member of the Cambridge school that was influential in the development of mathematics. Others included Lord Kelvin, J. Clerk Maxwell, Lord Rayleigh, J.J. Thomson, Sir Joseph Larmor, and Sir Earnest Rutherford. Stokes introduced the idea of non-uniform convergence and a theorem on critical values of the sums of periodic series.

[3]Carl Frederick Gauss (1777–1855) was a German mathematician, physicist, and astronomer. Born in Braunschweig, he was the prodigy son of a laborer and is considered as one of the greatest of mathematicians. Most of his life he was a professor at Göttingen University, where he was the director of the observatory. He made

where **v** is an arbitrary vector, **a** is the outward normal vector at a point on S, $\mathbf{v} \cdot \mathbf{a}$ represents the dot product, and div $\mathbf{v} = v_{i,i} = v_{1,1} + v_{2,2} + v_{3,3}$ is the divergence of **v**. This particular form of Gauss' theorem is often referred to as the *divergence theorem*. In solid mechanics, like Green's theorem, Gauss' theorem is sometimes referred to as a generalized form of the integration by parts. Derivations of Gauss' theorem can be found in many textbooks, including basic calculus books. In index notation, Eq. (II.7) appears as

$$\int_S v_i \, a_i \, dS = \int_V v_{i,i} \, dV \tag{II.8}$$

where a_i is a direction cosine of **a**.

Typically, we will apply Gauss' theorem to an expression such as the work relationship $p_j \, u_j = \mathbf{p}^T \mathbf{u}$, where p_j is related to the stresses by (Eq. 1.58) $p_j = \sigma_{ij} \, a_i$. Thus,

$$\begin{aligned} \int_S p_j \, u_j \, dS &= \int_S \sigma_{ij} \, a_i \, u_j \, dS = \int_S (\sigma_{ij} \, u_j) \, a_i \, dS \\ &= \int_V (\sigma_{ij} \, u_j)_{,i} \, dV = \int_V (\sigma_{ij,i} \, u_j + \sigma_{ij} \, u_{j,i}) \, dV \end{aligned} \tag{II.9}$$

substantive contributions in several areas of mathematics, as well as in astronomy. It is said that his potential as a mathematician was first recognized in primary school when his class was assigned the task of finding the sum of all the integers from 1 to 100. All but Gauss busily added the numbers and were unable to complete the necessary additions without error. Gauss, after a few moments of thought, simply wrote down the correct answer. His reasoning was: 100 plus 1 is 101, 99 plus 2 is 101, 98 plus 3 is 101, ..., 51 plus 50 is 101. Thus, there are fifty 101s, which gives 5050.

Appendix III

Summary of the Differential and Integral Forms of the Governing Equations

In this book, the basic equations for the response of structures have been given in the local (differential) form, as well as in global (integral) form. It was shown that both forms can be transformed into each other by applying integration by parts for one-dimensional problems and the Gauss integral theorem for multidimensional problems.

Chapter 2, Table 2.6 illustrates the general structure of the basic equations of Chapters 1 and 2, and as explained in Chapter 2, Section 2.5, highlights the dual character of the principle of virtual work and the principle of complementary virtual work.

The use of operator matrices makes it possible to employ the same general form of equations for different types of structures, e.g., beams and plates, in which only the entries of the matrices need to be adjusted. The matrix form of the basic equations is particularly enlightening as it exhibits more clearly the structure of the equations. For example, such characteristics as symmetry, or lack thereof, pertinent to the basic equations can be recognized easier than in long-hand or index notation [Wunderlich 1977]. Furthermore, the matrix equations provide a direct basis for numerical calculations and discretization processes like the finite element method.

This appendix provides a summary of the fundamental equations for a continuum and for several structural members.

It is intended here to illustrate and emphasize the similarity in the structure of the governing equations for a continuum and for structural members. Also, the operator matrices are summarized for a continuum and certain beam and plate theories. Most of the governing equations, such as for beams, are derived in detail elsewhere in this book:

Structure	Notation and Derivation Provided in Chapter
Continuum	1
Beams	1
Bars	12
Plate Stretching	1, 13
Plate Bending	13

III.1　Local Fundamental Equations: Differential Governing Equations

The differential equation form of the governing equations are arranged in terms of the three fundamentals: kinematic requirements, material law, and the conditions of equilibrium. The coordinate system, notation, and sign convention of Chapter 1 are employed here.

III.1.1　Kinematics

Continuum

Displacements: $\mathbf{u} = [u_x \quad u_y \quad u_z]^T$

Strains: $\epsilon = [\epsilon_x \quad \epsilon_y \quad \epsilon_z \quad \gamma_{xy} \quad \gamma_{xz} \quad \gamma_{yz}]^T$

Kinematical Equations:

$$\epsilon_x = \partial_x u_x \qquad\qquad \gamma_{xy} = \partial_y u_x + \partial_x u_y$$

$$\epsilon_y = \partial_y u_y \qquad\qquad \gamma_{xz} = \partial_z u_x + \partial_x u_z$$

$$\epsilon_z = \partial_z u_z \qquad\qquad \gamma_{yz} = \partial_z u_y + \partial_y u_z$$

or in matrix form,

$$
\begin{bmatrix} \epsilon_x \\ \epsilon_y \\ \epsilon_z \\ \gamma_{xy} \\ \gamma_{xz} \\ \gamma_{yz} \end{bmatrix}
=
\begin{bmatrix}
\partial_x & 0 & 0 \\
0 & \partial_y & 0 \\
0 & 0 & \partial_z \\
\partial_y & \partial_x & 0 \\
\partial_z & 0 & \partial_x \\
0 & \partial_z & \partial_y
\end{bmatrix}
\begin{bmatrix} u_x \\ u_y \\ u_z \end{bmatrix}
$$

$$\epsilon \quad = \quad\quad \mathbf{D} \quad\quad \mathbf{u}$$

Beams Including Axial Extension

with	without

Shear Deformation Effects

Displacements:　$\mathbf{u} = [u_o \quad w \quad \theta]^T$ $\qquad$ $\mathbf{u} = [u_o \quad w]^T$

Strains:　$\qquad\epsilon = [\epsilon_{ox} \quad \gamma \quad \kappa]^T$ $\qquad$ $\epsilon = [\epsilon_{ox} \quad \kappa]^T$

Kinematical Equations:

$$\epsilon_{ox} = d_x u_o \quad = u_o'$$

$$\gamma = d_x w + \theta = w' + \theta \qquad\qquad \epsilon_{ox} = d_x u_o = u_o'$$

$$\kappa = d_x \theta \quad = \theta' \qquad\qquad\qquad \kappa = -d_x^2 w = -w''$$

or in matrix form,

$$
\begin{bmatrix} \epsilon_{ox} \\ \gamma \\ \kappa \end{bmatrix}
=
\begin{bmatrix}
d_x & 0 & 0 \\
0 & d_x & 1 \\
0 & 0 & d_x
\end{bmatrix}
\begin{bmatrix} u_o \\ w \\ \theta \end{bmatrix}
\qquad
\begin{bmatrix} \epsilon_{ox} \\ \kappa \end{bmatrix}
=
\begin{bmatrix}
d_x & 0 \\
0 & -d_x^2
\end{bmatrix}
\begin{bmatrix} u_o \\ w \end{bmatrix}
$$

$$\epsilon \quad = \quad\quad \mathbf{D}_u \quad\quad \mathbf{u} \qquad\qquad \epsilon \quad = \quad \mathbf{D}_u \quad \mathbf{u}$$

Torsion of Bars

Circular Cross Section

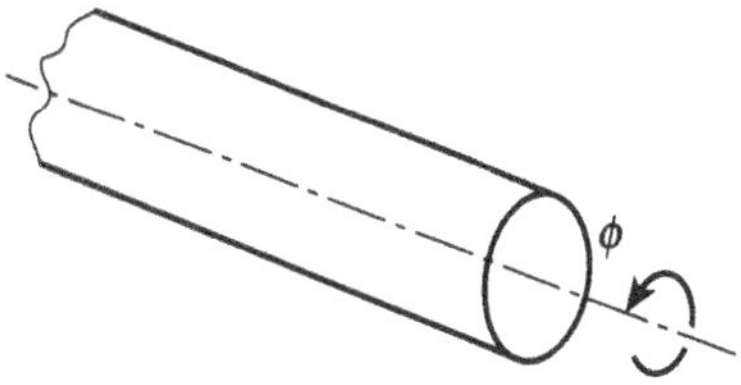

Displacements: $\mathbf{u} = \phi$

Strains: $\epsilon = \kappa_\phi$

Kinematical Equations:

$$\kappa_\phi = -d_x\phi$$

or in matrix form,

$$\begin{array}{cccc} [\kappa_\phi] & = & [-d_x] & [\phi] \\ \epsilon & = & \mathbf{D}_u & \mathbf{u} \end{array}$$

Thin-Walled Beams

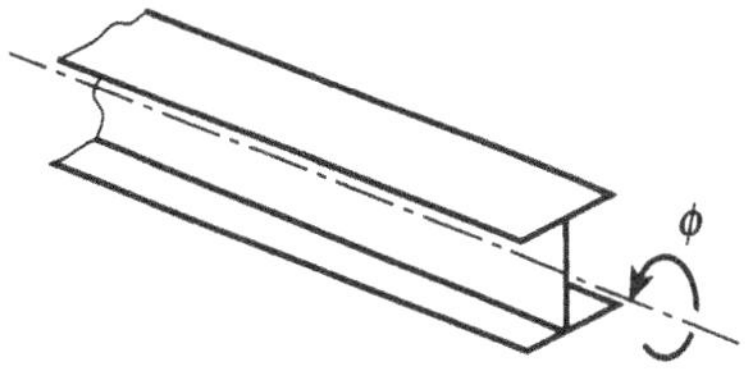

$\mathbf{u} = [\phi \quad \psi]^{\mathrm{T}}$

$\epsilon = [\kappa_\phi \quad \kappa_\omega]^{\mathrm{T}}$

$$\kappa_\phi = -d_x\phi$$

$$\kappa_\omega = d_x\psi$$

$$\begin{array}{cccc} \begin{bmatrix} \kappa_\phi \\ \kappa_\omega \end{bmatrix} & = & \begin{bmatrix} -d_x & 0 \\ 0 & d_x \end{bmatrix} & \begin{bmatrix} \phi \\ \psi \end{bmatrix} \\ \epsilon & = & \mathbf{D}_u & \mathbf{u} \end{array}$$

In-Plane Deformation (Stretching) of a Plate

Displacements: $\mathbf{u} = [u_x \quad u_y]^{\mathrm{T}}$

Strains: $\epsilon = [\epsilon_x \quad \epsilon_y \quad \gamma_{xy}]^{\mathrm{T}}$

Kinematical Equations:

$$\epsilon_x = \partial_x u_x$$

$$\epsilon_y = \partial_y u_y$$

$$\gamma_{xy} = \partial_y u_x + \partial_x u_y$$

or in matrix form,

$$\begin{array}{cccc} \begin{bmatrix} \epsilon_x \\ \epsilon_y \\ \gamma_{xy} \end{bmatrix} & = & \begin{bmatrix} \partial_x & 0 \\ 0 & \partial_y \\ \partial_y & \partial_x \end{bmatrix} & \begin{bmatrix} u_x \\ u_y \end{bmatrix} \\ \epsilon & = & \mathbf{D} & \mathbf{u} \end{array}$$

Transverse Deformation of a Plate
Rectilinear Coordinates

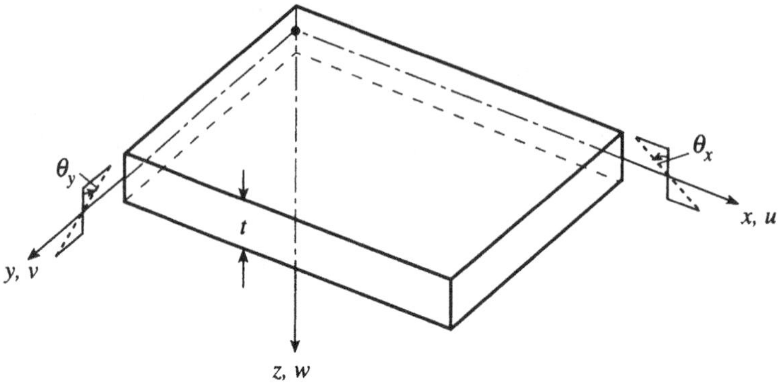

with | without

Shear Deformation Effects

With Shear Deformation Effects

Displacements: $\mathbf{u} = [w \quad \theta_x \quad \theta_y]^{\mathrm{T}}$

Strains:

$$\epsilon = [\kappa_x \quad \kappa_y \quad \kappa_{xy} \quad \gamma_{xz} \quad \gamma_{yz}]$$

Kinematical Equations:

$$\begin{aligned} \kappa_x &= \partial_x \theta_x \\ \kappa_y &= \partial_y \theta_y \\ 2\kappa_{xy} &= \partial_y \theta_x + \partial_x \theta_y \\ \gamma_{xz} &= \partial_x w + \theta_x \\ \gamma_{yz} &= \partial_y w + \theta_y \end{aligned}$$

or in matrix form,

$$\begin{bmatrix} \kappa_x \\ \kappa_y \\ 2\kappa_{xy} \\ \gamma_{xz} \\ \gamma_{yz} \end{bmatrix} = \begin{bmatrix} \partial_x & 0 & 0 \\ 0 & \partial_y & 0 \\ \partial_y & \partial_x & 0 \\ 1 & 0 & \partial_x \\ 0 & 1 & \partial_y \end{bmatrix} \begin{bmatrix} \theta_x \\ \theta_y \\ w \end{bmatrix}$$

$$\epsilon \quad = \quad \mathbf{D}_u \quad \mathbf{u}$$

Without Shear Deformation Effects

$\mathbf{u} = w$

$$\epsilon = [\kappa_x \quad \kappa_y \quad \kappa_{xy}]^{\mathrm{T}}$$

$$\begin{aligned} \kappa_x &= -\partial_x^2 w \\ \kappa_y &= -\partial_y^2 w \\ 2\kappa_{xy} &= -2\partial_x \partial_y w \\ \gamma_{xz} &= 0 \\ \gamma_{yz} &= 0 \end{aligned}$$

$$\begin{bmatrix} \kappa_x \\ \kappa_y \\ 2\kappa_{xy} \end{bmatrix} = \begin{bmatrix} -\partial_x^2 \\ -\partial_y^2 \\ -2\partial_x \partial_y \end{bmatrix} [w]$$

$$\epsilon \quad = \quad \mathbf{D}_u \quad \mathbf{u}$$

Circular Plates

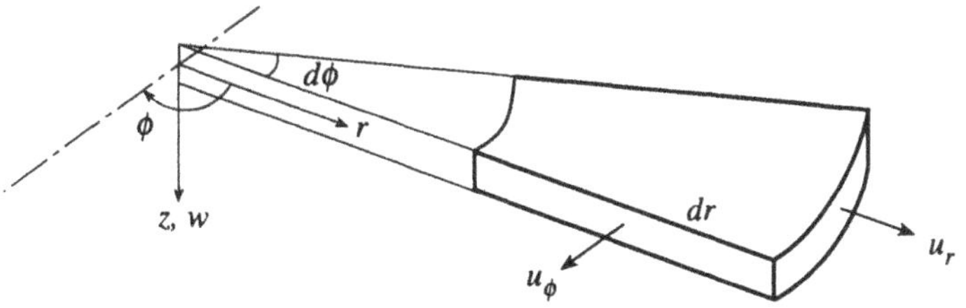

In-Plane Deformation (Stretching):

Displacements: $\mathbf{u} = [u_r \quad u_\phi]^T$

Strains: $\boldsymbol{\epsilon} = [\epsilon_r \quad \epsilon_\phi \quad \gamma_{r\phi}]^T$

Kinematical Relations:

$$\epsilon_r = \partial_r u_r$$

$$\epsilon_\phi = \frac{u_r}{r} + \frac{1}{r}\partial_\phi u_\phi$$

$$\gamma_{r\phi} = \frac{1}{r}\partial_\phi u_r + \partial_r u_\phi - \frac{1}{r}u_\phi$$

or in matrix form,

$$\begin{bmatrix} \epsilon_r \\ \epsilon_\phi \\ \gamma_{r\phi} \end{bmatrix} = \begin{bmatrix} \partial_r & 0 \\ \frac{1}{r} & \frac{1}{r}\partial_\phi \\ \frac{1}{r}\partial_\phi & \partial_r - \frac{1}{r} \end{bmatrix} \begin{bmatrix} u_r \\ u_\phi \end{bmatrix}$$

$$\boldsymbol{\epsilon} \quad = \quad \mathbf{D}_u \quad \mathbf{u}$$

Transverse Deformation:

$\mathbf{u} = w$

$\boldsymbol{\epsilon} = [\kappa_r \quad \kappa_\phi \quad \kappa_{r\phi}]^T$

$$\kappa_r = -\partial_r^2 w$$

$$\kappa_\phi = -\frac{1}{r}\partial_r w - \frac{1}{r^2}\partial_\phi^2 w$$

$$2\kappa_{r\phi} = -2\left(-\frac{1}{r}\partial_{r\phi}^2 w + \frac{1}{r^2}\partial_\phi w\right)$$

$$\begin{bmatrix} \kappa_r \\ \kappa_\phi \\ 2\kappa_{r\phi} \end{bmatrix} = \begin{bmatrix} -\partial_r^2 \\ -\frac{1}{r}\partial_r - \frac{1}{r^2}\partial_\phi^2 \\ \frac{2}{r}\partial_{r\phi}^2 - \frac{2}{r^2}\partial_\phi \end{bmatrix} [w]$$

$$\boldsymbol{\epsilon} \quad = \quad \mathbf{D}_u \quad \mathbf{u}$$

III.1.2 Material Law

Continuum

Stresses: $\boldsymbol{\sigma} = [\sigma_x \quad \sigma_y \quad \sigma_z \quad \tau_{xy} \quad \tau_{xz} \quad \tau_{yz}]^T$

Strains: $\boldsymbol{\epsilon} = [\epsilon_x \quad \epsilon_y \quad \epsilon_z \quad \gamma_{xy} \quad \gamma_{xz} \quad \gamma_{yz}]^T$

Material Law:

$$\begin{bmatrix} \sigma_x \\ \sigma_y \\ \sigma_z \\ \tau_{xy} \\ \tau_{xz} \\ \tau_{yz} \end{bmatrix} = \frac{E}{(1+\nu)(1-2\nu)} \left[\begin{array}{ccc:ccc} 1-\nu & \nu & \nu & & & \\ \nu & 1-\nu & \nu & & \mathbf{0} & \\ \nu & \nu & 1-\nu & & & \\ \hdashline & & & \frac{1-2\nu}{2} & 0 & 0 \\ & \mathbf{0} & & 0 & \frac{1-2\nu}{2} & 0 \\ & & & 0 & 0 & \frac{1-2\nu}{2} \end{array}\right] \begin{bmatrix} \epsilon_x \\ \epsilon_y \\ \epsilon_z \\ \gamma_{xy} \\ \gamma_{xz} \\ \gamma_{yz} \end{bmatrix}$$

$$\boldsymbol{\sigma} \quad = \quad \mathbf{E} \quad \boldsymbol{\epsilon}$$

or

$$\epsilon = \frac{1}{E}\underbrace{\left[\begin{array}{ccc|ccc} 1 & -\nu & -\nu & & & \\ -\nu & 1 & -\nu & & \mathbf{0} & \\ -\nu & -\nu & 1 & & & \\ \hline & & & 2(1+\nu) & 0 & 0 \\ & \mathbf{0} & & 0 & 2(1+\nu) & 0 \\ & & & 0 & 0 & 2(1+\nu) \end{array}\right]}_{\mathbf{E}^{-1}}\boldsymbol{\sigma}$$

Beams Including Axial Extension

with | without
Shear Deformation Effects

with Shear Deformation Effects	without Shear Deformation Effects
Stress Resultants:	
$\mathbf{s} = [N \quad V \quad M]^{\mathrm{T}}$	$\mathbf{s} = [N \quad M]^{\mathrm{T}}$
$\epsilon = [\epsilon_{ox} \quad \gamma \quad \kappa]$	$\epsilon = [\epsilon_{ox} \quad \kappa]^{\mathrm{T}}$
$\underbrace{\begin{bmatrix} N \\ V \\ M \end{bmatrix}}_{\mathbf{s}} = \underbrace{\begin{bmatrix} EA & & \\ & k_s GA & \\ & & EI \end{bmatrix}}_{\mathbf{E}} \underbrace{\begin{bmatrix} \epsilon_{ox} \\ \gamma \\ \kappa \end{bmatrix}}_{\epsilon}$	$\underbrace{\begin{bmatrix} N \\ M \end{bmatrix}}_{\mathbf{s}} = \underbrace{\begin{bmatrix} EA & 0 \\ 0 & EI \end{bmatrix}}_{\mathbf{E}} \underbrace{\begin{bmatrix} \epsilon_{ox} \\ \kappa \end{bmatrix}}_{\epsilon}$
$\epsilon = \underbrace{\begin{bmatrix} 1/EA & 0 & 0 \\ 0 & 1/k_s GA & 0 \\ 0 & 0 & 1/EI \end{bmatrix}}_{\mathbf{E}^{-1}} \mathbf{s}$	$\epsilon = \underbrace{\begin{bmatrix} 1/EA & 0 \\ 0 & 1/EI \end{bmatrix}}_{\mathbf{E}^{-1}} \mathbf{s}$

Torsion of Bars

Circular Cross Section

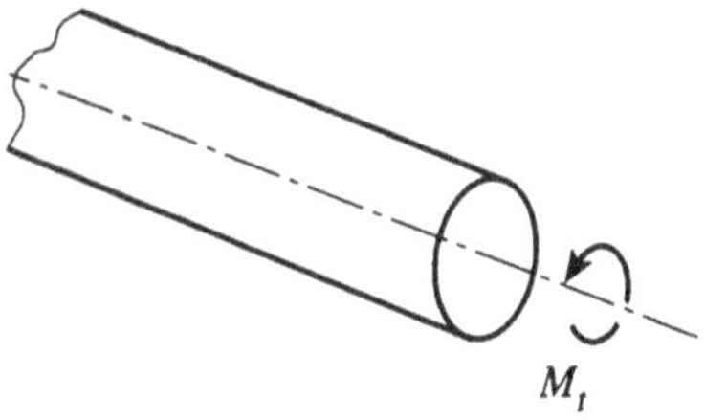

Net Forces: $\mathbf{s} = M_t$

Strains: $\boldsymbol{\epsilon} = \kappa_\phi$

Material Law: $M_t = GJ\,\kappa_\phi$

or in matrix form,

$$\begin{array}{c}[M_t] = [GJ][\kappa_\phi] \\ \mathbf{s} \ = \ \mathbf{E} \ \ \boldsymbol{\epsilon}\end{array}$$

Thin-Walled Beams

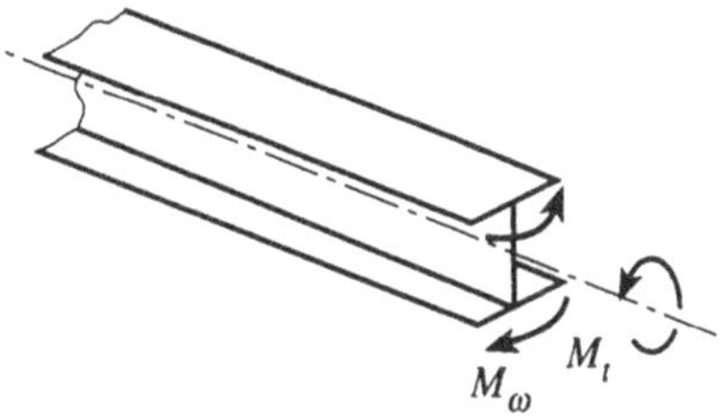

$\mathbf{s} = [M_t \quad M_\omega]^{\mathrm{T}}$

$\boldsymbol{\epsilon} = [\kappa_\phi \quad \kappa_\omega]^{\mathrm{T}}$

$M_t = GJ\,\kappa_\phi$

$M_\omega = EI_{\omega\omega}\,\kappa_\omega$

$$\underbrace{\begin{bmatrix} M_t \\ M_\omega \end{bmatrix}}_{\mathbf{s}} = \underbrace{\begin{bmatrix} GJ & 0 \\ 0 & EI_{\omega\omega} \end{bmatrix}}_{\mathbf{E}} \underbrace{\begin{bmatrix} \kappa_\phi \\ \kappa_\omega \end{bmatrix}}_{\boldsymbol{\epsilon}}$$

In-Plane Deformation of a Plate

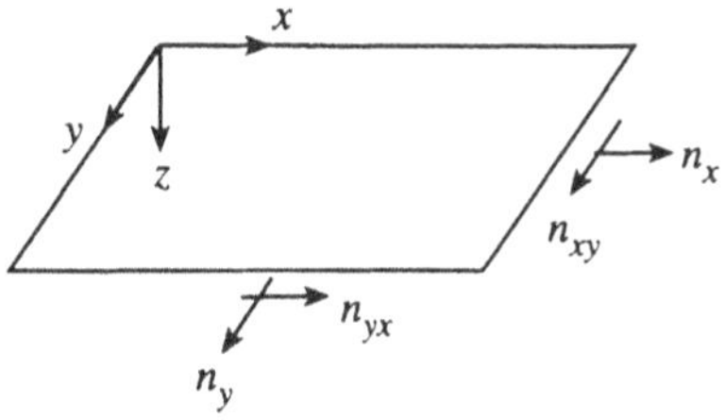

Stress Resultants:

$$\mathbf{s} = [n_x \quad n_y \quad n_{xy}]^{\mathrm{T}}$$

$$n_x = \int_{-t/2}^{t/2} \sigma_x\, dz \quad n_y = \int_{-t/2}^{t/2} \sigma_y\, dz \quad n_{xy} = \int_{-t/2}^{t/2} \tau_{xy}\, dz$$

where t is the plate thickness and $n_{xy} = n_{yx}$.

Strains:

$$\boldsymbol{\epsilon} = [\epsilon_x \quad \epsilon_y \quad \gamma_{xy}]^{\mathrm{T}}$$

where $\gamma_{xy} = \gamma_{yx}$.

Material Law

Plane Stress Condition:

$$\sigma_z = \tau_{zx} = \tau_{zy} = 0 \qquad D = \frac{Et}{1-\nu^2}$$

$$\begin{bmatrix} n_x \\ n_y \\ n_{xy} \end{bmatrix} = D \begin{bmatrix} 1 & \nu & 0 \\ \nu & 1 & 0 \\ 0 & 0 & \frac{1-\nu}{2} \end{bmatrix} \begin{bmatrix} \epsilon_x \\ \epsilon_y \\ \gamma_{xy} \end{bmatrix}$$

$$\mathbf{s} = \mathbf{E}\,\boldsymbol{\epsilon}$$

$$\begin{bmatrix} \epsilon_x \\ \epsilon_y \\ \gamma_{xy} \end{bmatrix} = \frac{1}{Et} \begin{bmatrix} 1 & -\nu & 0 \\ -\nu & 1 & 0 \\ 0 & 0 & 2(1+\nu) \end{bmatrix} \begin{bmatrix} n_x \\ n_y \\ n_{xy} \end{bmatrix}$$

$$\boldsymbol{\epsilon} = \mathbf{E}^{-1}\,\mathbf{s}$$

Plane Strain Condition:

$$\epsilon_z = \gamma_{xz} = \gamma_{yz} = 0$$

$$\begin{bmatrix} n_x \\ n_y \\ n_{xy} \end{bmatrix} = \frac{Et}{(1+\nu)(1-2\nu)} \begin{bmatrix} 1-\nu & \nu & 0 \\ \nu & 1-\nu & 0 \\ 0 & 0 & \frac{1-2\nu}{2} \end{bmatrix} \begin{bmatrix} \epsilon_x \\ \epsilon_y \\ \gamma_{xy} \end{bmatrix}$$

$$\mathbf{s} = \mathbf{E}\,\boldsymbol{\epsilon}$$

$$\begin{bmatrix} \epsilon_x \\ \epsilon_y \\ \gamma_{xy} \end{bmatrix} = \frac{1+\nu}{Et} \begin{bmatrix} 1-\nu & -\nu & 0 \\ -\nu & 1-\nu & 0 \\ 0 & 0 & 2 \end{bmatrix} \begin{bmatrix} n_x \\ n_y \\ n_{xy} \end{bmatrix}$$

$$\boldsymbol{\epsilon} = \mathbf{E}^{-1}\,\mathbf{s}$$

Transverse Deformation of a Plate

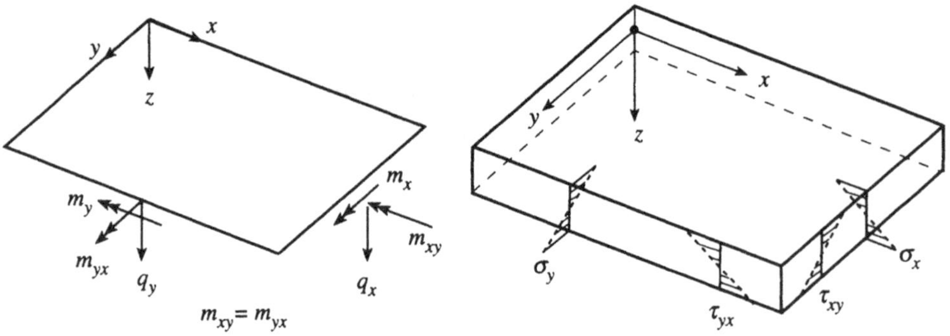

with | without
Shear Deformation Effects

With shear deformation effects

Stress Resultants:

$$\mathbf{s} = [m_x \quad m_y \quad m_{xy} \quad q_x \quad q_y]^{\mathrm{T}}$$

$$\begin{bmatrix} q_x \\ q_y \end{bmatrix} = \int_{-t/2}^{t/2} \begin{bmatrix} \tau_{xz} \\ \tau_{yz} \end{bmatrix} dz$$

Strains:

$$\boldsymbol{\epsilon} = [\kappa_x \quad \kappa_y \quad \kappa_{xy} \quad \gamma_{xz} \quad \gamma_{yz}]^{\mathrm{T}}$$

Without shear deformation effects

$$\mathbf{s} = [m_x \quad m_y \quad m_{xy}]^{\mathrm{T}}$$

$$\begin{bmatrix} m_x \\ m_y \\ m_{xy} \end{bmatrix} = \int_{-t/2}^{t/2} \begin{bmatrix} \sigma_x \\ \sigma_y \\ \tau_{xy} \end{bmatrix} z\,dx$$

$$\boldsymbol{\epsilon} = [\kappa_x \quad \kappa_y \quad \kappa_{xy}]^{\mathrm{T}}$$

Material Law

With Shear Deformation Effects

$$\begin{bmatrix} m_x \\ m_y \\ m_{xy} \\ q_x \\ q_y \end{bmatrix} = \begin{bmatrix} K\begin{bmatrix} 1 & \nu & 0 \\ \nu & 1 & 0 \\ 0 & 0 & \frac{1-\nu}{2} \end{bmatrix} & \mathbf{0} \\ \mathbf{0} & Gt\begin{bmatrix} 5/6 & 0 \\ 0 & 5/6 \end{bmatrix} \end{bmatrix} \begin{bmatrix} \kappa_x \\ \kappa_y \\ 2\kappa_{xy} \\ \gamma_{xz} \\ \gamma_{yz} \end{bmatrix}$$

$$\mathbf{s} \quad = \quad \mathbf{E} \quad \boldsymbol{\epsilon}$$

$$\begin{bmatrix} \kappa_x \\ \kappa_y \\ 2\kappa_{xy} \\ \gamma_{xz} \\ \gamma_{yz} \end{bmatrix} = \begin{bmatrix} \frac{12}{Et^3}\begin{bmatrix} 1 & -\nu & 0 \\ -\nu & 1 & 0 \\ 0 & 0 & 2(1+\nu) \end{bmatrix} & \mathbf{0} \\ \mathbf{0} & \frac{1}{Gt}\begin{bmatrix} 6/5 & 0 \\ 0 & 6/5 \end{bmatrix} \end{bmatrix} \begin{bmatrix} m_x \\ m_y \\ m_{xy} \\ q_x \\ q_y \end{bmatrix}$$

$$\boldsymbol{\epsilon} \quad = \quad \mathbf{E}^{-1} \quad \mathbf{s}$$

Plate Rigidity: $K = \frac{Et^3}{12(1-\nu^2)}$

Without Shear Deformation Effects

$$\begin{bmatrix} m_x \\ m_y \\ m_{xy} \end{bmatrix} = K\begin{bmatrix} 1 & \nu & 0 \\ \nu & 1 & 0 \\ 0 & 0 & \frac{1-\nu}{2} \end{bmatrix} \begin{bmatrix} \kappa_x \\ \kappa_y \\ 2\kappa_{xy} \end{bmatrix}$$

$$\mathbf{s} \quad = \quad \mathbf{E} \quad \boldsymbol{\epsilon}$$

$$\begin{bmatrix} \kappa_x \\ \kappa_y \\ 2\kappa_{xy} \end{bmatrix} = \frac{12}{Et^3}\begin{bmatrix} 1 & -\nu & 0 \\ -\nu & 1 & 0 \\ 0 & 0 & 2(1+\nu) \end{bmatrix} \begin{bmatrix} m_x \\ m_y \\ m_{xy} \end{bmatrix}$$

$$\boldsymbol{\epsilon} \quad = \quad \mathbf{E}^{-1} \quad \mathbf{s}$$

In-Plane Deformation of a Circular Plate

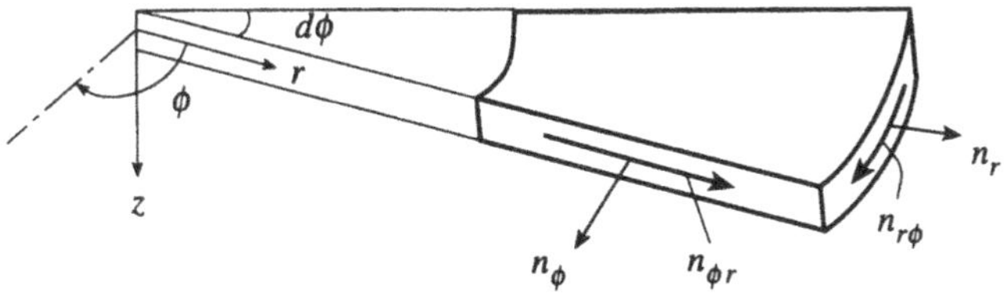

Stress Resultants: $\mathbf{s} = [n_r \quad n_\phi \quad n_{r\phi}]^{\mathrm{T}}$ $\qquad D = Et/(1-\nu^2)$

Strains: $\boldsymbol{\epsilon} = [\epsilon_r \quad \epsilon_\phi \quad \gamma_{r\phi}]^{\mathrm{T}}$

Material Law:

$$\begin{bmatrix} n_r \\ n_\phi \\ n_{r\phi} \end{bmatrix} = D \begin{bmatrix} 1 & \nu & 0 \\ \nu & 1 & 0 \\ 0 & 0 & \frac{1-\nu}{2} \end{bmatrix} \begin{bmatrix} \epsilon_r \\ \epsilon_\phi \\ \gamma_{r\phi} \end{bmatrix}$$

$$\mathbf{s} = \mathbf{E} \quad \boldsymbol{\epsilon}$$

or

$$\begin{bmatrix} \epsilon_r \\ \epsilon_\phi \\ \gamma_{r\phi} \end{bmatrix} = \frac{1}{Et} \begin{bmatrix} 1 & -\nu & 0 \\ -\nu & 1 & 0 \\ 0 & 0 & 2(1+\nu) \end{bmatrix} \begin{bmatrix} n_r \\ n_\phi \\ n_{r\phi} \end{bmatrix}$$

$$\boldsymbol{\epsilon} = \mathbf{E}^{-1} \quad \mathbf{s}$$

Transverse Deformation of a Circular Plate

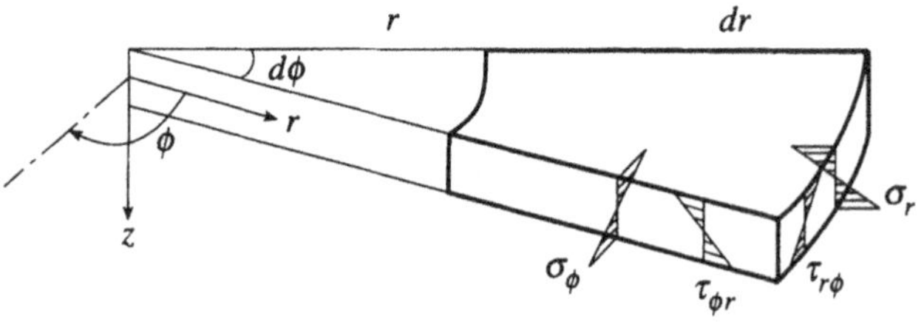

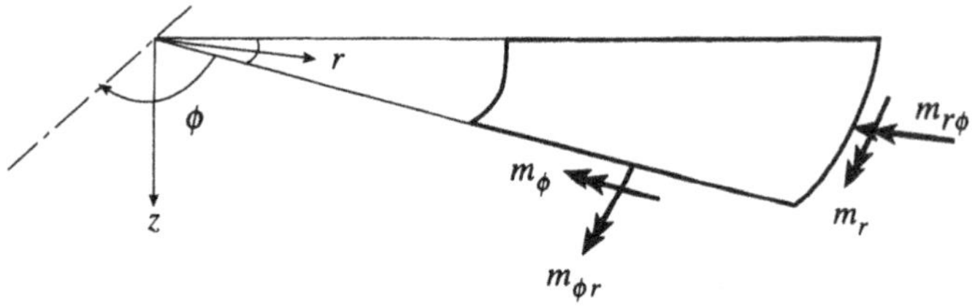

Stress Resultants: $\mathbf{s} = [m_r \quad m_\phi \quad m_{r\phi}]^T \qquad K = Et^3/(12(1-\nu^2))$

Strains: $\boldsymbol{\epsilon} = [\kappa_r \quad \kappa_\phi \quad \kappa_{r\phi}]^T$

Material Law:

$$\begin{bmatrix} m_r \\ m_\phi \\ m_{r\phi} \end{bmatrix} = K \begin{bmatrix} 1 & \nu & 0 \\ \nu & 1 & 0 \\ 0 & 0 & \frac{1-\nu}{2} \end{bmatrix} \begin{bmatrix} \kappa_r \\ \kappa_\phi \\ 2\kappa_{r\phi} \end{bmatrix}$$

$$\mathbf{s} = \mathbf{E} \quad \boldsymbol{\epsilon}$$

$$\begin{bmatrix} \kappa_r \\ \kappa_\phi \\ \kappa_{r\phi} \end{bmatrix} = \frac{12}{Et^3} \begin{bmatrix} 1 & -\nu & 0 \\ -\nu & 1 & 0 \\ 0 & 0 & 2(1+\nu) \end{bmatrix} \begin{bmatrix} m_r \\ m_\phi \\ m_{r\phi} \end{bmatrix}$$

$$\boldsymbol{\epsilon} = \mathbf{E}^{-1} \quad \mathbf{s}$$

III.1.3 Equilibrium

Continuum

Stresses: $\boldsymbol{\sigma} = [\sigma_x \quad \sigma_y \quad \sigma_z \quad \tau_{xy} \quad \tau_{xz} \quad \tau_{yz}]^{\mathrm{T}}$

Body Forces (e.g., weight): $\overline{\mathbf{p}}_V = [\overline{p}_{Vx} \quad \overline{p}_{Vy} \quad \overline{p}_{Vz}]^{\mathrm{T}}$

Conditions of Equilibrium: $\sigma_{ij,j} + \overline{p}_{Vi} = 0$

$$\partial_x\sigma_x + \partial_y\tau_{xy} + \partial_z\tau_{xz} + \overline{p}_{Vx} = 0; \qquad \tau_{xy} = \tau_{yx}$$

$$\partial_x\tau_{yx} + \partial_y\sigma_y + \partial_z\tau_{yz} + \overline{p}_{Vy} = 0; \qquad \tau_{xz} = \tau_{zx}$$

$$\partial_x\tau_{zx} + \partial_y\tau_{zy} + \partial_z\sigma_z + \overline{p}_{Vz} = 0; \qquad \tau_{yz} = \tau_{zy}$$

or in matrix form,

$$\begin{bmatrix} \partial_x & 0 & 0 & \partial_y & \partial_z & 0 \\ 0 & \partial_y & 0 & \partial_x & 0 & \partial_z \\ 0 & 0 & \partial_z & 0 & \partial_x & \partial_y \end{bmatrix} [\boldsymbol{\sigma}] + \begin{bmatrix} \overline{p}_{Vx} \\ \overline{p}_{Vy} \\ \overline{p}_{Vz} \end{bmatrix} = \begin{bmatrix} 0 \\ 0 \\ 0 \end{bmatrix}$$

$$\mathbf{D}^{\mathrm{T}} \qquad \boldsymbol{\sigma} + \overline{\mathbf{p}}_V = \mathbf{0}$$

Beams Including Axial Extension

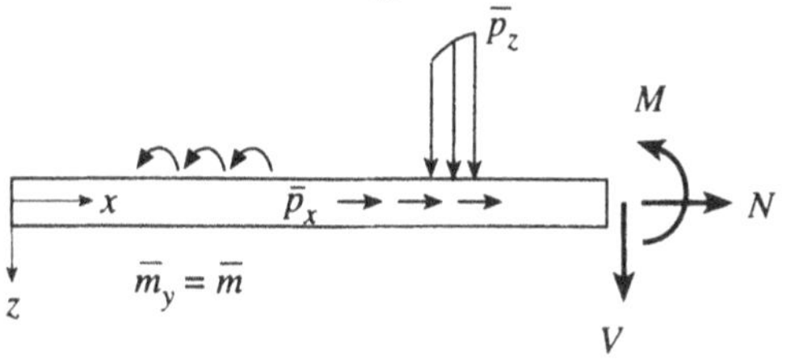

with | without

Shear Deformation Effects

With shear deformation effects	Without shear deformation effects
$\mathbf{s} = [N \quad V \quad M]^{\mathrm{T}}$	$\mathbf{s} = [N \quad M]^{\mathrm{T}}$
$\overline{\mathbf{p}} = [\overline{p}_x \quad \overline{p}_z \quad \overline{m}]^{\mathrm{T}}$	$\overline{\mathbf{p}} = [\overline{p}_x \quad \overline{p}_z]^{\mathrm{T}}$
$d_xN + \overline{p}_z = 0$	$d_xN + \overline{p}_x = 0$
$d_xV + \overline{p}_z = 0$	$d_x^2M + \overline{p}_z = 0$
$d_xM - V + \overline{m} = 0$	
$\begin{bmatrix} d_x & 0 & 0 \\ 0 & d_x & 0 \\ 0 & -1 & d_x \end{bmatrix} \mathbf{s} + \begin{bmatrix} \overline{p}_x \\ \overline{p}_z \\ \overline{m} \end{bmatrix} = \mathbf{0}$	$\begin{bmatrix} d_x & 0 \\ 0 & d_x^2 \end{bmatrix} \mathbf{s} + \begin{bmatrix} \overline{p}_x \\ \overline{p}_z \end{bmatrix} = \mathbf{0}$
$\mathbf{D}_s^{\mathrm{T}} \quad \mathbf{s} + \overline{\mathbf{p}} = \mathbf{0}$	$\mathbf{D}_s^{\mathrm{T}} \quad \mathbf{s} + \overline{\mathbf{p}} = \mathbf{0}$

Torsion of Bars

Circular Cross Section

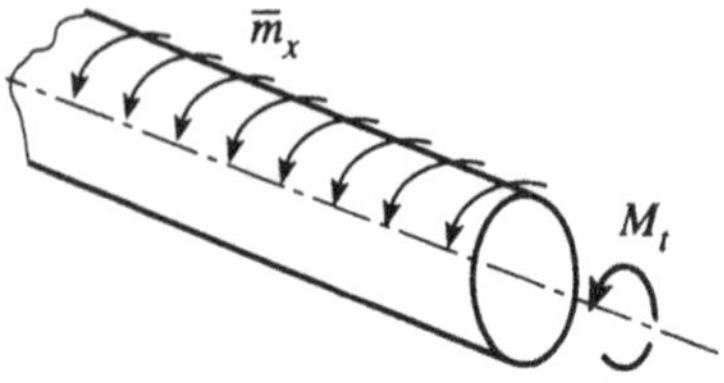

Stress Resultants: $\mathbf{s} = M_t$

Loads: $\overline{\mathbf{p}} = \overline{m}_x$

Conditions of Equilibrium:

$$\frac{dM_t}{dx} + \overline{m}_x = 0$$

or in matrix form,

$$\begin{matrix} [d_x] & [M_t] = [-\overline{m}_x] \\ \mathbf{D}_s^{\mathrm{T}} & \mathbf{s} \;=\; \overline{\mathbf{p}} \end{matrix}$$

Thin-Walled Beams

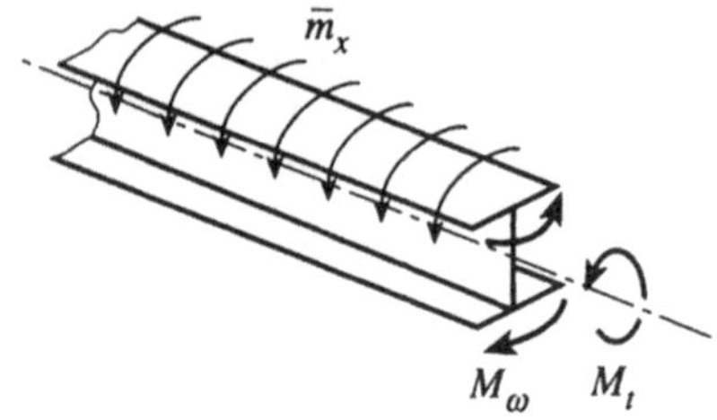

$\mathbf{s} = [M_t \quad M_\omega]^{\mathrm{T}}$

$\overline{\mathbf{p}} = \overline{m}_x$

$$\frac{dM_t}{dx} + \frac{d^2M_\omega}{dx^2} - \overline{m}_x = 0$$

$$\begin{matrix} [d_x \quad d_x^2] & \begin{bmatrix} M_t \\ M_\omega \end{bmatrix} = [\overline{m}_x] \\ \mathbf{D}_s^{\mathrm{T}} & \mathbf{s} \;=\; \overline{\mathbf{p}} \end{matrix}$$

In-Plane Deformation of a Plate

Stress Resultants: $\mathbf{s} = [n_x \quad n_y \quad n_{xy}]^{\mathrm{T}}$

Loads: $\overline{\mathbf{p}}_V = [\overline{p}_{Vx} \quad \overline{p}_{Vy}]^{\mathrm{T}}$

Conditions of Equilibrium:

$$\begin{aligned} \partial_x n_x + \partial_y n_{yx} + \overline{p}_{Vx} &= 0 \\ \partial_y n_y + \partial_x n_{xy} + \overline{p}_{Vy} &= 0 \\ n_{xy} &= n_{yx} \end{aligned}$$

or in matrix form,

$$\begin{bmatrix} \partial_x & 0 & \partial_y \\ 0 & \partial_y & \partial_x \end{bmatrix} \begin{bmatrix} n_x \\ n_y \\ n_{xy} \end{bmatrix} + \begin{bmatrix} \overline{p}_{Vx} \\ \overline{p}_{Vy} \end{bmatrix} = \mathbf{0}$$

$$\mathbf{D}^{\mathrm{T}} \quad \mathbf{s} \; + \; \overline{\mathbf{p}}_V \; = \mathbf{0}$$

Transverse Deformation of a Plate
Rectilinear Coordinates

with | without
Shear Deformation Effects

With Shear Deformation Effects

Stress Resultants:

$$\mathbf{s} = [m_x \quad m_y \quad m_{xy} \quad q_x \quad q_y]^{\mathrm{T}}$$

Surface Loading:

$$\overline{\mathbf{p}} = \overline{p}_z$$

Conditions of Equilibrium:

$$\begin{aligned} &\partial_x q_x + \partial_y q_y + \overline{p}_z = 0 \\ &\partial_x m_x + \partial_y m_{yx} - q_x = 0 \\ &\partial_y m_y + \partial_x m_{xy} - q_y = 0 \\ &m_{xy} = m_{yx} \end{aligned}$$

or in matrix form

$$\begin{bmatrix} \partial_x & 0 & \partial_y & -1 & 0 \\ 0 & \partial_y & \partial_x & 0 & -1 \\ 0 & 0 & 0 & \partial_x & \partial_y \end{bmatrix} \begin{bmatrix} m_x \\ m_y \\ m_{xy} \\ q_x \\ q_y \end{bmatrix} + \begin{bmatrix} 0 \\ 0 \\ \overline{p}_z \end{bmatrix} = \mathbf{0}$$

$$\mathbf{D}_s^{\mathrm{T}} \qquad \mathbf{s} \quad + \quad \overline{\mathbf{p}} \quad = \mathbf{0}$$

Without Shear Deformation Effects

$$\mathbf{s} = [m_x \quad m_y \quad m_{xy}]^{\mathrm{T}}$$

$$\overline{\mathbf{p}} = \overline{p}_z$$

$$\partial_x^2 m_x + 2\partial_x \partial_y m_{xy} + \partial_y^2 m_y + \overline{p}_z = 0$$

$$\begin{bmatrix} \partial_x^2 & \partial_y^2 & 2\partial_x \partial_y \end{bmatrix} \begin{bmatrix} m_x \\ m_y \\ m_{xy} \end{bmatrix} + [\overline{p}_z] = \mathbf{0}$$

$$\mathbf{D}_s^{\mathrm{T}} \qquad \mathbf{s} \quad + \quad \overline{\mathbf{p}} \quad = \mathbf{0}$$

Circular Plate

In-Plane Deformation (Stretching):

Stress Resultants:

$$\mathbf{s} = [n_r \quad n_\phi \quad n_{r\phi}]^{\mathrm{T}}$$

Forces:

$$\overline{\mathbf{p}} = [\overline{p}_{Vx} \quad \overline{p}_{Vy}]^{\mathrm{T}}$$

Conditions of Equilibrium:

$$\begin{bmatrix} \partial_r + 1/r & -1/r & (1/r)\partial_\phi \\ 0 & (1/r)\partial_\phi & \partial_r + 1/r \end{bmatrix} \begin{bmatrix} n_r \\ n_\phi \\ n_{r\phi} \end{bmatrix} + \begin{bmatrix} \overline{p}_{Vr} \\ \overline{p}_{V\phi} \end{bmatrix} = \mathbf{0}$$

$$\mathbf{D}_s^{\mathrm{T}} \qquad \mathbf{s} \quad + \quad \overline{\mathbf{p}} \quad = \mathbf{0}$$

Transverse Deformation:

$$\mathbf{s} = [m_r \quad m_\phi \quad m_{r\phi}]^{\mathrm{T}}$$

$$\overline{\mathbf{p}} = \overline{p}_z$$

$$\begin{bmatrix} \partial_r^2 & \frac{1}{r^2}\partial_\phi^2 & 2\left(\frac{1}{r^2}\partial_\phi - \frac{1}{r}\partial_{r\phi}^2\right) \end{bmatrix} \begin{bmatrix} m_r \\ m_\phi \\ m_{r\phi} \end{bmatrix} + [\overline{p}_z] = \mathbf{0}$$

$$\mathbf{D}_s^{\mathrm{T}} \qquad \mathbf{s} \quad + \quad \overline{\mathbf{p}} \quad = \mathbf{0}$$

III.1.4 Surface Forces and Boundary Conditions

On the surface, stresses can be expressed in terms of surface forces. The orientation of the surface is defined by its normal vector with direction cosines a_x, a_y, a_z. The vector of surface forces is expressed in terms of the stresses and direction cosines as

$$\begin{bmatrix} p_x \\ p_y \\ p_z \end{bmatrix} = \begin{bmatrix} a_x & 0 & 0 & a_y & a_z & 0 \\ 0 & a_y & 0 & a_x & 0 & a_z \\ 0 & 0 & a_z & 0 & a_x & a_y \end{bmatrix} \boldsymbol{\sigma}$$

$$\mathbf{p} \quad = \qquad \mathbf{A}^{\mathrm{T}} \qquad \boldsymbol{\sigma}$$

The matrix $\mathbf{A}^{\mathrm{T}}$ has the same structure as $\mathbf{D}^{\mathrm{T}}$. The derivatives in $\mathbf{D}^{\mathrm{T}}$ correspond to the projection directions in $\mathbf{A}^{\mathrm{T}}$.

Boundary Conditions

Continuum

Displacement (kinematic) Boundary Conditions on S_u, a portion of the total surface S:

$$\mathbf{u} = \overline{\mathbf{u}} \quad \text{on } S_u$$

Static Boundary Conditions on the portion $S_p = S - S_u$:

$$\mathbf{p} = \overline{\mathbf{p}} \quad \text{on } S_p$$

Beams Including Axial Extension

$$\left.\begin{aligned} u_o &= \overline{u}_o \\ w &= \overline{w} \\ \theta &= \overline{\theta} \end{aligned}\right\} \quad \text{on } S_u$$

$$\left.\begin{aligned} N &= \overline{N} \\ V &= \overline{V} \\ M &= \overline{M} \end{aligned}\right\} \quad \text{on } S_p$$

Torsion of Bars

$$\begin{aligned} \phi &= \overline{\phi} \quad && \text{on } S_u \\ M_t &= \overline{M}_t && \text{on } S_p \end{aligned}$$

Torsion of Thin Walled Beams

$$\left.\begin{aligned} \phi &= \overline{\phi} \\ \psi &= \overline{\psi} \end{aligned}\right\} \quad \text{on } S_u$$

$$\left.\begin{aligned} M_t &= \overline{M}_t \\ M_\omega &= \overline{M}_\omega \end{aligned}\right\} \quad \text{on } S_p$$

or

$$M_{to} = \overline{M}_{to} \quad \text{on } S_p \qquad M_{to} = \text{total twisting moment}$$

$$M_{to} = M_t + \frac{dM_\omega}{dx}$$

In-Plane Deformation of a Plate

$$\left.\begin{aligned} u_x &= \overline{u}_x \\ u_y &= \overline{u}_y \end{aligned}\right\} \quad \text{on } S_u$$

Static Boundary Conditions on $S_p = S - S_u$:

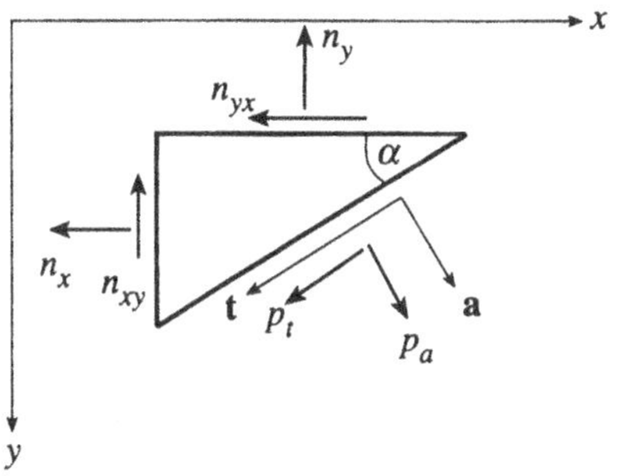

Surface Forces:

$$\begin{bmatrix} p_a \\ p_t \end{bmatrix} = \begin{bmatrix} \sin^2\alpha & \cos^2\alpha & 2\sin\alpha\cos\alpha \\ -\sin\alpha\cos\alpha & \sin\alpha\cos\alpha & -\cos^2\alpha + \sin^2\alpha \end{bmatrix} \begin{bmatrix} n_x \\ n_y \\ n_{xy} \end{bmatrix}$$

$$\mathbf{p} = \mathbf{T} \cdot \mathbf{A}^{\mathrm{T}} \mathbf{n}$$

T: Transformation Matrix

$$\begin{aligned} p_a &= \overline{p}_a \\ p_t &= \overline{p}_t \end{aligned} \quad \text{on } S_p$$

Transverse Deformation of a Plate

with	without
Shear Deformation Effects	
$w = \overline{w}$, $\theta_a = \overline{\theta}_a$, $\theta_t = \overline{\theta}_t$ on S_u	$w = \overline{w}$, $\theta_a = \overline{\theta}_a$ on S_u

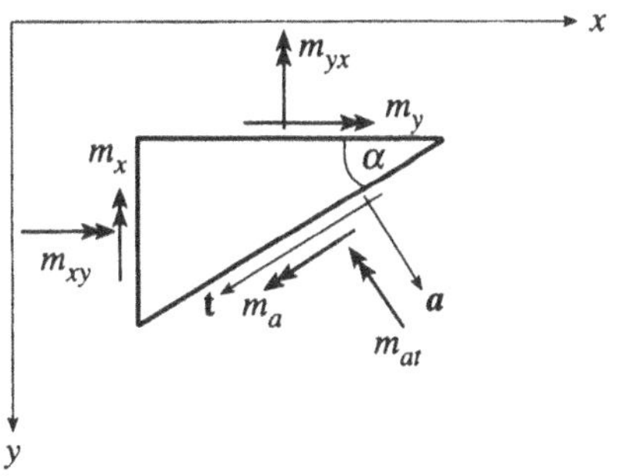

Surface Stress Resultants:

$$\begin{bmatrix} m_a \\ m_{at} \end{bmatrix} = \begin{bmatrix} \sin^2\alpha & \cos^2\alpha & 2\sin\alpha\cos\alpha \\ -\sin\alpha\cos\alpha & \sin\alpha\cos\alpha & -\cos^2\alpha + \sin^2\alpha \end{bmatrix} \begin{bmatrix} m_x \\ m_y \\ m_{xy} \end{bmatrix}$$

$$q_a = q_x \sin\alpha + q_y \cos\alpha$$

with shear deformation	without shear deformation
$m_a = \overline{m}_a$	$m_a = \overline{m}_a$
$m_{at} = \overline{m}_{at}$ on S_p	$V_a^* = \overline{q}_a + \partial_t m_{at} = \overline{V}_a^*$ on S_p
$q_a = \overline{q}_a$	Kirchhoff equivalent force

Circular Plate

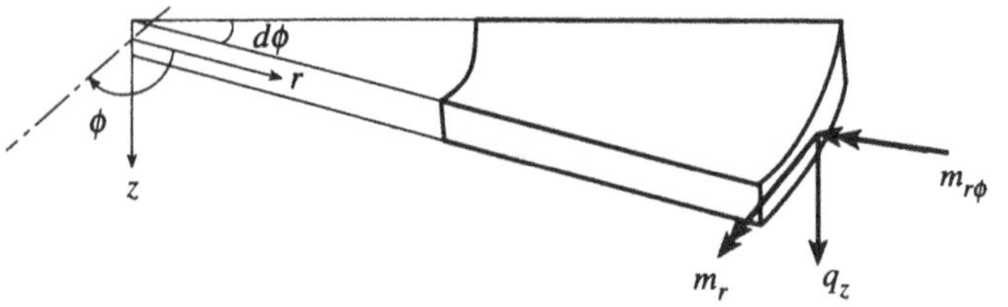

In-Plane Deformation (Stretching):	Transverse Deformation:
$u_r = \overline{u}_r$, $u_\phi = \overline{u}_\phi$ on S_u	$w = \overline{w}$, $\theta_r = \overline{\theta}_r$ on S_u
$p_r = \overline{p}_r$, $p_\phi = \overline{p}_\phi$ on S_p	$m_r = \overline{m}_r$, $V_r^* = \overline{q}_r + \frac{1}{r}\partial_\phi m_{r\phi} = \overline{V}_r^*$ on S_p
	V_r^* = Kirchhoff equivalent force

III.2 Displacement Formulation of Differential Governing Equations

The displacement formulation of the differential governing equations is obtained by substituting the kinematical relations and material law into the equilibrium equations.

Continuum

$$G\nabla^2 u_i + (\lambda + G)u_{k,ki} + \overline{p}_{Vi} = 0$$

where

$$\lambda = \frac{E\nu}{(1+\nu)(1-2\nu)}$$

$$G = \frac{E}{2(1+\nu)}$$

$$\nabla^2 = \partial_x^2 + \partial_y^2$$

Beams Including Axial Extension

with | without
Shear Deformation Effects

With shear deformation effects:

$$d_x EA d_x u_o = -\overline{p}_x$$

$$d_x k_s GA d_x w + d_x k_s GA\theta = -\overline{p}_z$$

$$d_x EI\, d_x\theta - k_s GA(d_x w + \theta) = 0$$

or in matrix notation,

$$\begin{bmatrix} d_x EA d_x & 0 & 0 \\ 0 & d_x k_s GA d_x & d_x k_s GA \\ 0 & -k_s GA d_x & d_x EI\, d_x - k_s GA \end{bmatrix} \begin{bmatrix} u_o \\ w \\ \theta \end{bmatrix} = \begin{bmatrix} -\overline{p}_x \\ -\overline{p}_z \\ 0 \end{bmatrix}$$

Without shear deformation effects:

$$d_x EA d_x u_o = -\overline{p}_x$$

$$d_x^2 EI\, d_x^2 w = \overline{p}_z$$

$$\begin{bmatrix} d_x EA d_x & 0 \\ 0 & d_x^2 EI\, d_x^2 \end{bmatrix} \begin{bmatrix} u_o \\ w \end{bmatrix} = \begin{bmatrix} -\overline{p}_x \\ \overline{p}_z \end{bmatrix}$$

Torsion of Bars

Circular Cross Section	Thin Walled Beams
$d_x GJ\, d_x \phi = -\overline{m}_x$	$E\Gamma I_{\omega\omega}\phi^{iv} - GJ\,\phi'' = \overline{m}_x$

In-Plane Deformation of a Plate

$$\partial_x D\partial_x u_x + \partial_y D_1 \partial_y u_x + \partial_x D\nu\partial_y u_y + \partial_y \frac{D}{2} D_1 \partial_x u_y + \overline{p}_{Vx} = 0$$
$$\partial_y D\nu\partial_x u_x + \partial_x \frac{D}{2} D_1 \partial_y u_x + \partial_y D\partial_y u_y + \partial_x D_1 \partial_x u_y + \overline{p}_{Vy} = 0$$

or in matrix notation, $D_1 = \frac{D}{2}(1-\nu) \quad D = Et/(1-v^2)$

$$\begin{bmatrix} \partial_x D\partial_x + \partial_y D_1 \partial_y & \partial_x D\nu\partial_y + \partial_y \frac{D}{2} D_1 \partial_x \\ \partial_y D\nu\partial_x + \partial_x \frac{D}{2} D_1 \partial_y & \partial_y D\partial_y + \partial_x D_1 \partial_x \end{bmatrix} \begin{bmatrix} u_x \\ u_y \end{bmatrix} + \begin{bmatrix} \overline{p}_{Vx} \\ \overline{p}_{Vy} \end{bmatrix} = \mathbf{0}$$

Transverse Deformation of a Plate
Rectilinear Coordinates

with Shear Deformation Effects, $K = Et^3/(12(1-v^2))$

$$\frac{5}{6}(\partial_x Gt\partial_x w + \partial_y Gt\partial_y w + \partial_x Gt\theta_x + \partial_y Gt\theta_y) + \overline{p}_z = 0$$
$$\frac{5}{6}(Gt\partial_x w + \frac{6}{5}K\partial_x^2\theta_x + \frac{3}{5}K(1-\nu)\partial_y^2\theta_x + Gt\theta_x + \frac{3}{5}K(1+\nu)\partial_x\partial_y\theta_y) = 0$$
$$\frac{5}{6}(Gt\partial_y w + \frac{3}{5}K(1+\nu)\partial_x\partial_y\theta_x + \frac{6}{5}K\partial_y^2\theta_y + \frac{3}{5}K(1-\nu)\partial_x^2\theta_y + Gt\theta_y) = 0$$

or in matrix form,

$$\frac{5}{6}\begin{bmatrix} \partial_x Gt\partial_x + \partial_y Gt\partial_y & \partial_x Gt & \partial_y Gt \\ Gt\partial_x & \frac{6}{5}K\partial_x^2 + \frac{3}{5}K(1-\nu)\partial_y^2 + Gt & \frac{3}{5}K(1+\nu)\partial_x\partial_y \\ Gt\partial_y & \frac{3}{5}K(1+\nu)\partial_x\partial_y & \frac{6}{5}K\partial_y^2 + \frac{3}{5}K(1-\nu)\partial_x^2 + Gt \end{bmatrix} \begin{bmatrix} w \\ \theta_x \\ \theta_y \end{bmatrix}$$
$$+ \begin{bmatrix} \overline{p}_z \\ 0 \\ 0 \end{bmatrix} = \mathbf{0}$$

without Shear Deformation Effects

$$K\left[\frac{\partial^4 w}{\partial x^4} + 2\frac{\partial^4 w}{\partial x^2 \partial y^2} + \frac{\partial^4 w}{dy^4}\right] = \overline{p}_z$$

Circular Plate

In-Plane Deformation (Stretching)

$$\begin{bmatrix} \partial_r^2 + 2\frac{\nu\partial_r}{r} + \partial_\phi \frac{1-\nu}{2r^2}\partial_\phi + \frac{1}{r^2} & \partial_r \frac{\nu}{r}\partial_\phi + \frac{1-\nu}{2r}\left(\partial_\phi\partial_r - \frac{1}{r}\partial_\phi\right) + \frac{1}{r^2}\partial_\phi \\ \frac{1}{r}\partial_\phi \nu \partial_r + \frac{\partial_\phi}{r^2} + \frac{1-\nu}{2r}\left(\partial_r\partial_\phi - \frac{1}{r}\partial_\phi\right) & \partial_\phi \frac{1}{r^2}\partial_\phi + \frac{1-\nu}{2}\left(\partial_r^2 - 2\partial_r \frac{1}{r} + \frac{1}{r^2}\right) \end{bmatrix} \begin{bmatrix} u_r \\ u_\phi \end{bmatrix} + \begin{bmatrix} \overline{p}_{Vr} \\ \overline{p}_{V\phi} \end{bmatrix} = \mathbf{0}$$

Transverse Deformation

$$\nabla^4 w = \frac{\overline{p}_z}{K}$$

where

$$\nabla^4 = \nabla^2\nabla^2 = \left(\partial_r^2 + \frac{1}{r}\partial_r + \frac{1}{r^2}\partial_\phi^2\right)\left(\partial_r^2 + \frac{1}{r}\partial_r + \frac{1}{r^2}\partial_\phi^2\right)$$

$$K \begin{bmatrix} \partial_r^4 + \nu\left(\frac{1}{r}\partial_r^3 + \frac{1}{r^2}\partial_r^2\partial_\phi^2\right) \\ +\left(\partial_{r\phi}^2 \frac{1}{r} - \partial_\phi \frac{1}{r^2}\right)2(1-\nu)\left(\frac{1}{r}\partial_{r\phi}^2 - \frac{1}{r^2}\partial_\phi\right) \\ +\left(\partial_r \frac{1}{r} + \partial_\phi^2 \frac{1}{r^2}\right)\left(\frac{1}{r}\partial_r + \frac{1}{r^2}\partial_\phi^2 + \nu\partial_r^2\right) \end{bmatrix} w + \overline{p}_z = 0$$

III.3 Mixed (State Variable) Form of the Differential Governing Equations

The mixed or state variable form of the differential governing equations has both forces and displacements as variables. The mixed form is a set of first order equations in the state variables.

Beams Including Axial Extension

with Shear Deformation Effects

$$d_x u = \frac{N}{EA}$$

$$d_x w = -\theta + \frac{V}{k_s GA}$$

$$d_x \theta = \frac{M}{EI}$$

$$d_x N = -\overline{p}_x$$

$$d_x V = -\overline{p}_z$$

$$d_x M = V$$

or in matrix form,

$$\frac{d}{dx}\begin{bmatrix} u \\ w \\ \theta \\ N \\ V \\ M \end{bmatrix} = \begin{bmatrix} 0 & 0 & 0 & \frac{1}{EA} & 0 & 0 \\ 0 & 0 & -1 & 0 & \frac{1}{k_s GA} & 0 \\ 0 & 0 & 0 & 0 & 0 & \frac{1}{EI} \\ 0 & 0 & 0 & 0 & 0 & 0 \\ 0 & 0 & 0 & 0 & 0 & 0 \\ 0 & 0 & 0 & 0 & 1 & 0 \end{bmatrix} \begin{bmatrix} u \\ w \\ \theta \\ N \\ V \\ M \end{bmatrix} + \begin{bmatrix} 0 \\ 0 \\ 0 \\ -\overline{p}_x \\ -\overline{p}_z \\ 0 \end{bmatrix}$$

$$\frac{d\mathbf{z}}{dx} = \mathbf{A}\,\mathbf{z} + \overline{\mathbf{P}}$$

without Shear Deformation Effects

$$d_x u = \frac{N}{EA}$$

$$d_x w = -\theta$$

$$d_x \theta = \frac{M}{EI}$$

$$d_x N = -\overline{p}_x$$

$$d_x V = -\overline{p}_z$$

$$d_x M = V$$

or in matrix form,

$$\frac{d}{dx}\begin{bmatrix} u \\ w \\ \theta \\ N \\ V \\ M \end{bmatrix} = \begin{bmatrix} 0 & 0 & 0 & \frac{1}{EA} & 0 & 0 \\ 0 & 0 & -1 & 0 & 0 & 0 \\ 0 & 0 & 0 & 0 & 0 & \frac{1}{EI} \\ 0 & 0 & 0 & 0 & 0 & 0 \\ 0 & 0 & 0 & 0 & 0 & 0 \\ 0 & 0 & 0 & 0 & 1 & 0 \end{bmatrix} \begin{bmatrix} u \\ w \\ \theta \\ N \\ V \\ M \end{bmatrix} + \begin{bmatrix} 0 \\ 0 \\ 0 \\ -\overline{p}_x \\ -\overline{p}_z \\ 0 \end{bmatrix}$$

$$\frac{d\mathbf{z}}{dx} = \mathbf{A} \quad \mathbf{z} + \overline{\mathbf{P}}$$

Torsion of Bars

Circular Cross Section

$$\frac{d\phi}{dx} = \frac{M_t}{GJ}$$

$$\frac{dM_t}{dx} = -\overline{m}_x$$

or in matrix form,

$$\frac{d}{dx}\begin{bmatrix} \phi \\ M_t \end{bmatrix} = \begin{bmatrix} 0 & \frac{1}{GJ} \\ 0 & 0 \end{bmatrix} \begin{bmatrix} \phi \\ M_t \end{bmatrix} + \begin{bmatrix} 0 \\ -\overline{m}_x \end{bmatrix}$$

$$\frac{d}{dx}\mathbf{z} = \mathbf{A} \quad \mathbf{z} + \overline{\mathbf{P}}$$

Thin-Walled Beams

$$\frac{d\phi}{dx} = -\psi$$

$$\frac{d\psi}{dx} = \frac{M_\omega}{EI_{\omega\omega}}$$

$$\frac{dM_\omega}{dx} = M_{to} + GJ\,\psi$$

$$\frac{dM_{to}}{dx} = -\overline{m}_x$$

or in matrix form,

$$\frac{d}{dx}\begin{bmatrix} \phi \\ \psi \\ M_\omega \\ M_{to} \end{bmatrix} = \begin{bmatrix} 0 & -1 & 0 & 0 \\ 0 & 0 & \frac{1}{EI_{\omega\omega}} & 0 \\ 0 & GJ & 0 & 1 \\ 0 & 0 & 0 & 0 \end{bmatrix} \begin{bmatrix} \phi \\ \psi \\ M_\omega \\ M_{to} \end{bmatrix} + \begin{bmatrix} 0 \\ 0 \\ 0 \\ -\overline{m}_x \end{bmatrix}$$

$$\frac{d}{dx}\mathbf{z} = \mathbf{A} \quad \mathbf{z} + \overline{\mathbf{p}}$$

Transverse Deformation of Plate
Rectilinear Coordinates

Without Shear Deformation

$$\frac{\partial w}{\partial x} = -\theta$$

$$\frac{\partial \theta}{\partial x} = \nu \frac{\partial^2 w}{\partial y^2} + \frac{m_x}{K}$$

$$\frac{\partial V}{\partial x} = -\overline{p} + K(1-\nu^2)\frac{\partial^4 w}{\partial y^4} - \nu \frac{\partial^2 m_x}{\partial y^2}$$

$$\frac{\partial m_x}{\partial x} = V^* - K(1-\nu)\frac{\partial^2 \theta}{\partial y^2}$$

$(V^* = q_x + \frac{\partial m_{xy}}{\partial y}$, the Kirchhoff effective shear force)

or in matrix form,

$$\frac{\partial}{\partial x}\begin{bmatrix} w \\ \theta \\ V^* \\ m_x \end{bmatrix} = \begin{bmatrix} 0 & -1 & 0 & 0 \\ \nu\partial_y^2 & 0 & 0 & \frac{1}{K} \\ K(1-\nu^2)\partial_y^4 & 0 & 0 & -\nu\partial_y^2 \\ 0 & -K(1-\nu)\partial_y^2 & 1 & 0 \end{bmatrix}\begin{bmatrix} w \\ \theta \\ V^* \\ m_x \end{bmatrix} + \begin{bmatrix} 0 \\ 0 \\ -\overline{p} \\ 0 \end{bmatrix}$$

$$\frac{\partial}{\partial x}\mathbf{z} = \mathbf{A}\,\mathbf{z} + \overline{\mathbf{P}}$$

Circular Plate

$$\frac{\partial w}{\partial r} = -\theta$$

$$\frac{\partial \theta}{\partial r} = \frac{m_r}{K} + \nu\left(\frac{1}{r^2}\frac{\partial^2 w}{\partial \phi^2} - \frac{\theta}{r}\right)$$

$$\frac{\partial V^*}{\partial r} = -\frac{V^*}{r} - \frac{\nu}{r^2}\frac{\partial^2 m_r}{\partial \phi^2} + K(1-\nu^2)\frac{1}{r^4}\frac{\partial^4 w}{\partial \phi^4} - 2K(1-\nu)\frac{1}{r^4}\frac{\partial^2 w}{\partial \phi^2}$$
$$-K(3-2\nu-\nu^2)\frac{1}{r^3}\frac{\partial^2 \theta}{\partial \phi^2} - \overline{p}_z$$

$$\frac{\partial m_r}{\partial r} = -(1-\nu)\frac{m_r}{r} + V^* - K(3-2\nu-\nu^2)\frac{1}{r^3}\frac{\partial^2 w}{\partial \phi^2} + K(1-\nu^2)\frac{\theta}{r^2}$$
$$-2K(1-\nu)\frac{1}{r^2}\frac{\partial^2 \theta}{\partial \phi^2}$$

or in matrix form,

$$\frac{\partial}{\partial r}\begin{bmatrix} w \\ \theta \\ V^* \\ m_r \end{bmatrix} = \begin{bmatrix} 0 & -1 & 0 & 0 \\ \frac{\nu}{r^2}\partial_\phi^2 & -\frac{\nu}{r} & 0 & \frac{1}{K} \\ \begin{matrix} K(1-\nu^2)\frac{1}{r^4}\partial_\phi^4 - \\ 2K(1-\nu)\frac{1}{r^4}\partial_\phi^2 \end{matrix} & -K(3-2\nu-\nu^2)\frac{1}{r^3}\partial_\phi^2 & -\frac{1}{r} & -\frac{\nu}{r^2}\partial_\phi^2 \\ -K(3-2\nu-\nu^2)\frac{1}{r^3}\partial_\phi^2 & \begin{matrix} K(1-\nu^2)\frac{1}{r^2} - \\ 2K(1-\nu)\frac{1}{r^2}\partial_\phi^2 \end{matrix} & 1 & \frac{-(1-\nu)}{r} \end{bmatrix}\begin{bmatrix} w \\ \theta \\ V^* \\ m_r \end{bmatrix} + \begin{bmatrix} 0 \\ 0 \\ -\overline{p}_z \\ 0 \end{bmatrix}$$

$$\frac{\partial}{\partial r}\mathbf{z} = \mathbf{A}\,\mathbf{z} + \overline{\mathbf{P}}$$

V^* = Kirchhoff effective shear force (per unit length) along a circular arc.

III.4 Global Fundamental Equations: Integral Form of the Governing Equations

The differential form of the governing equations are equivalent in the sense of Euler's equations of the calculus of the variations to integral relationships. These integral equations can be transformed into variational principles with the assistance of the Gauss' integral theorem, i.e., the divergence theorem.

In Chapter 2, several forms of the global governing equations are presented. Here, we choose to summarize only the principle of virtual work related forms. Recall that in integral form the conditions of equilibrium and the static boundary conditions are equivalent to the principle of virtual work.

Continuum

Local Formulation

Equilibrium:

$$\mathbf{D}^{\mathrm{T}}\boldsymbol{\sigma} + \overline{\mathbf{p}}_V = \mathbf{0} \quad \text{in } V$$

Static Boundary Conditions:

$$\mathbf{p} = \overline{\mathbf{p}} \quad \text{on } S_p$$

Global Formulation (Virtual Displacement $\delta\mathbf{u}$)

$$\int_{S_p} \delta\mathbf{u}^{\mathrm{T}}(\mathbf{p} - \overline{\mathbf{p}})\, dS = \int_V \delta\mathbf{u}(\mathbf{D}^{\mathrm{T}}\boldsymbol{\sigma} + \overline{\mathbf{p}}_V)\, dV$$

Index Notation:

$$\int_{S_p} \delta u_i (p_i - \overline{p}_i)\, dS = \int_V \delta u_i (\sigma_{ij,j} + \overline{p}_{Vi})\, dV \tag{III.1}$$

Gauss' integral theorem applied to the first term gives

$$\int_{S_p} \delta u_i\, p_i\, dS = \int_V (\sigma_{ij,j}\delta u_i + \sigma_{ij}\underbrace{\delta u_{i,j}}_{\substack{=\delta\epsilon_{ij} \\ \text{when } \epsilon = \mathbf{Du}}})\, dV - \int_{S_u} \underbrace{\delta u_i\, p_i\, dS}_{\substack{=0 \\ \text{for } \mathbf{u} = \overline{\mathbf{u}}}}$$

Substitution of this relation into Eq. (III.1) yields the Principle of Virtual Work:

$$\int_V \delta u_{i,j}\, \sigma_{ij}\, dV - \int_V \delta u_i\, \overline{p}_{Vi}\, dV - \int_{S_p} \delta u_i\, \overline{p}_i\, dS = 0$$

If the kinematic conditions $\epsilon = \mathbf{Du}$ in V and the displacement boundary conditions $\mathbf{u} = \overline{\mathbf{u}}$ on S_u are satisfied at the outset, then the principle of virtual work applies. This principle is equivalent to the conditions of equillibrium and the static boundary conditions.

In matrix notation:

$$-\delta W = \int_V \delta\mathbf{u}^{\mathrm{T}}{}_u\mathbf{D}^{\mathrm{T}}\boldsymbol{\sigma}\, dV - \int_V \delta\mathbf{u}^{\mathrm{T}}\overline{\mathbf{p}}_V\, dV - \int_{S_p} \delta\mathbf{u}^{\mathrm{T}}\overline{\mathbf{p}}\, dS = 0$$

$$\text{with} \quad \begin{aligned} &\epsilon = \mathbf{D}_u\mathbf{u} \text{ in } V \\ &\mathbf{u} = \overline{\mathbf{u}} \quad \text{on } S_u \end{aligned}$$

The subscript u to the left of $\mathbf{D}^{\mathrm{T}}$, i.e., ${}_u\mathbf{D}^{\mathrm{T}}$, signifies the application of the operator to the preceding quantity, i.e., $\delta\mathbf{u}^{\mathrm{T}}$

$$-\delta W = \int_V \delta[u_x \quad u_y \quad u_z] \left\{ \underbrace{\begin{bmatrix} {}_x\partial & 0 & 0 & {}_y\partial & {}_z\partial & 0 \\ 0 & {}_y\partial & 0 & {}_x\partial & 0 & {}_z\partial \\ 0 & 0 & {}_z\partial & 0 & {}_x\partial & {}_y\partial \end{bmatrix}}_{{}_u\mathbf{D}^{\mathrm{T}}} \boldsymbol{\sigma} - \begin{bmatrix} \overline{p}_{Vx} \\ \overline{p}_{Vy} \\ \overline{p}_{Vz} \end{bmatrix} \right\} dV$$

$$+ \int_{S_p} \delta[u_x \quad u_y \quad u_z] \begin{bmatrix} \overline{p}_x \\ \overline{p}_y \\ \overline{p}_z \end{bmatrix} dS = 0$$

For a formulation in terms of displacements, the material law is introduced:

$$\boldsymbol{\sigma} = \mathbf{E}\boldsymbol{\epsilon} = \mathbf{E}\mathbf{D}_u\mathbf{u}$$

$$-\delta W = \int_V \delta\mathbf{u} \underbrace{{}_u\mathbf{D}^{\mathrm{T}}\mathbf{E}\mathbf{D}_u}_{\underline{\mathbf{k}}^D = \text{symmetric operator matrix}} \mathbf{u}\, dV - \int_V \delta\mathbf{u}^{\mathrm{T}}\overline{\mathbf{p}}_V\, dV - \int_{S_p} \delta\mathbf{u}^{\mathrm{T}}\overline{\mathbf{p}}\, dS = 0$$

or with $G = E/(2(1+\nu))$

$-\delta W =$

$$\int_V \delta[u_x \quad u_y \quad u_z] \left\{ \underbrace{\begin{bmatrix} {}_x\partial \frac{2G(1-\nu)}{1-2\nu}\partial_x + {}_y\partial G\partial_y + {}_z\partial G\partial_z & {}_x\partial \frac{2G\nu}{1-2\nu}\partial_y + {}_y\partial G\partial_x & {}_x\partial \frac{2G\nu}{1-2\nu}\partial_z + {}_z\partial G\partial_x \\ {}_y\partial \frac{2G\nu}{1-2\nu}\partial_x + {}_x\partial G\partial_y & {}_y\partial \frac{2G(1-\nu)}{1-2\nu}\partial_y + {}_x\partial G\partial_x + {}_z\partial G\partial_z & {}_y\partial \frac{2G\nu}{1-2\nu}\partial_z + {}_z\partial G\partial_y \\ {}_z\partial \frac{2G\nu}{1-2\nu}\partial_x + {}_x\partial G\partial_z & {}_z\partial \frac{2G\nu}{1-2\nu}\partial_y + {}_y\partial G\partial_z & {}_z\partial \frac{2G(1-\nu)}{1-2\nu}\partial_z + {}_x\partial G\partial_x + {}_y\partial G\partial_y \end{bmatrix}}_{\underline{\mathbf{k}}^D} \begin{bmatrix} u_x \\ u_y \\ u_z \end{bmatrix} - \begin{bmatrix} \overline{p}_{Vx} \\ \overline{p}_{Vy} \\ \overline{p}_{Vz} \end{bmatrix} \right\} dV$$

$$- \int_{S_p} \delta[u_x \quad u_y \quad u_z] \begin{bmatrix} \overline{p}_x \\ \overline{p}_y \\ \overline{p}_z \end{bmatrix} dS = 0$$

These equations form the basis of the displacement method of analysis.

Beams Including Axial Extension
with Shear Deformation Effects

Local Formulation
Equilibrium:

$$\mathbf{D}_s^{\mathrm{T}}\mathbf{s} + \overline{\mathbf{p}} = \mathbf{0}$$

Static Boundary Conditions:

$$\mathbf{s} = \overline{\mathbf{s}} \qquad \begin{matrix} N = \overline{N} \\ V = \overline{V} \\ M = \overline{M} \end{matrix} \quad \text{on the ends}$$

Global Formulation (Virtual Displacement $\delta\mathbf{u}$)

$$\left[\delta\mathbf{u}^{\mathrm{T}}(\mathbf{s}-\overline{\mathbf{s}})\right]_0^L = \int_x \delta\mathbf{u}^{\mathrm{T}}\left(\mathbf{D}_s^{\mathrm{T}}\mathbf{s}+\overline{\mathbf{p}}\right)dx$$

In Component Form:

$$\left[\delta\begin{bmatrix}u_o & w & \theta\end{bmatrix}\begin{bmatrix}N-\overline{N}\\ V-\overline{V}\\ M-\overline{M}\end{bmatrix}\right]_0^L = \int \delta\begin{bmatrix}u_o & w & \theta\end{bmatrix}\left\{\begin{bmatrix}d_x & 0 & 0\\ 0 & d_x & 0\\ 0 & -1 & d_x\end{bmatrix}\begin{bmatrix}N\\ V\\ M\end{bmatrix}+\begin{bmatrix}\overline{p}_x\\ \overline{p}_z\\ 0\end{bmatrix}\right\}dx \qquad \text{(III.2)}$$

Integration by parts applied to the first term of Eq. (III.1) gives

$$\underbrace{\left[\delta\mathbf{u}^{\mathrm{T}}\mathbf{s}\right]_0^L}_{\text{on } S_p} = \int_x (\delta\mathbf{u}\,\mathbf{s}_{,x} + \underbrace{\delta\mathbf{u}_{,x}\,\mathbf{s}}_{=\delta\epsilon_x})\,dx - \underbrace{\left[\delta\mathbf{u}^{\mathrm{T}}\mathbf{s}\right]_0^L}_{\text{on } S_u}$$

when $\epsilon = \mathbf{Du}$ $\qquad = 0$ for $\mathbf{u} = \overline{\mathbf{u}}$ on the ends

Substitution of this relation into Eq. (III.2) yields the Principle of Virtual Work:

$$\int_x \delta\begin{bmatrix}u_o & w & \theta\end{bmatrix}\begin{bmatrix}{}_xd & 0 & 0\\ 0 & {}_xd & 0\\ 0 & 1 & {}_xd\end{bmatrix}\begin{bmatrix}N\\ V\\ M\end{bmatrix}dx - \int_x \delta\begin{bmatrix}u_o & w & \theta\end{bmatrix}\begin{bmatrix}\overline{p}_x\\ \overline{p}_z\\ 0\end{bmatrix}dx$$

$$-\left[\delta\begin{bmatrix}u_o & w & \theta\end{bmatrix}\begin{bmatrix}\overline{N}\\ \overline{V}\\ \overline{M}\end{bmatrix}\right]_0^L = 0$$

In matrix notation:

$$-\delta W = \int_x \delta\mathbf{u}^{\mathrm{T}}\,{}_u\mathbf{D}^{\mathrm{T}}\mathbf{s}\,dx - \int_x \delta\mathbf{u}^{\mathrm{T}}\overline{\mathbf{p}}\,dx - \left[\delta\mathbf{u}^{\mathrm{T}}\mathbf{s}\right]_0^L$$

with $\qquad \epsilon = \mathbf{D}_u\mathbf{u}$ throughout the beam

$\mathbf{u} = \overline{\mathbf{u}}$ on the ends

$$-\delta W = \int_x \delta\begin{bmatrix}u_o & w & \theta\end{bmatrix}\left\{\underbrace{\begin{bmatrix}{}_xd & 0 & 0\\ 0 & {}_xd & 0\\ 0 & 1 & {}_xd\end{bmatrix}}_{{}_u\mathbf{D}^T}\begin{bmatrix}N\\ V\\ M\end{bmatrix}-\begin{bmatrix}\overline{p}_x\\ \overline{p}_z\\ 0\end{bmatrix}\right\}dx - \left[\delta\begin{bmatrix}u_o & w & \theta\end{bmatrix}\begin{bmatrix}\overline{N}\\ \overline{V}\\ \overline{M}\end{bmatrix}\right]_0^L = 0$$

$$\mathbf{s} = \mathbf{E}\epsilon = \mathbf{E}\mathbf{D}_u\mathbf{u}$$

$$-\delta W = \int_x \delta\mathbf{u}^{\mathrm{T}}\,\underbrace{{}_u\mathbf{D}^{\mathrm{T}}\mathbf{E}\mathbf{D}_u}_{\mathbf{k}^D}\,\mathbf{u}\,dx - \int_x \delta\mathbf{u}^{\mathrm{T}}\overline{\mathbf{p}}\,dx - \left[\delta\mathbf{u}^{\mathrm{T}}\overline{\mathbf{s}}\right]_0^L = 0$$

or

$$-\delta W = \delta[u_0 \quad w \quad \theta] \left\{ \underbrace{\begin{bmatrix} {}_xdEAd_x & 0 & 0 \\ 0 & {}_xdk_sGAd_x & {}_xdk_sGA \\ & & {}_xdEI\,d_x \\ 0 & k_sGAd_x & +k_sGA \end{bmatrix}}_{\mathbf{k}^D} \begin{bmatrix} u_0 \\ w \\ \theta \end{bmatrix} - \begin{bmatrix} \overline{p}_x \\ \overline{p}_z \\ 0 \end{bmatrix} \right\} dx$$

$$-\left[\delta[u_0 \quad w \quad \theta] \begin{bmatrix} \overline{N} \\ \overline{V} \\ \overline{M} \end{bmatrix} \right]_0^L = 0$$

Torsion of Bars
Circular Cross Section

$$-\delta W = \int \delta\mathbf{u}^{\mathrm{T}}({}_u\mathbf{D}^{\mathrm{T}}\mathbf{s} - \overline{\mathbf{p}})\,dx - \left[\delta\mathbf{u}^{\mathrm{T}}(\mathbf{s} - \overline{\mathbf{s}})\right]_0^L = 0$$

$$-\delta W = \int \delta[\phi]\{[{}_xd][M_t] - [\overline{m}_x]\}\,dx - \left[\delta[\phi][M_t]\right]_0^L = 0$$

Transformation to a Displacement Formulation

$$-\delta W = \int_x \delta[\phi]\{[{}_xdGJ\,d_x][\phi] - [\overline{m}_x]\}\,dx - \left[\delta[\phi][\overline{M}_t]\right]_0^L$$

$$\boldsymbol{\delta}\mathbf{u}^{\mathrm{T}}\{ \quad \mathbf{k}^D \quad \mathbf{u} - \mathbf{p} \}$$

Torsion of Thin-Walled Beams

$$-\delta W = \int_x \delta\mathbf{u}^{\mathrm{T}}({}_u\mathbf{D}^{\mathrm{T}}\mathbf{s} - \overline{\mathbf{p}})\,dx - \left[\delta\mathbf{u}^{\mathrm{T}}(\mathbf{s} - \overline{\mathbf{s}})\right]_0^L = 0$$

$$-\delta W = \int_x \delta[\phi \quad \psi] \left\{ \begin{bmatrix} {}_xd & \\ & {}_xd \end{bmatrix} \begin{bmatrix} M_t \\ M_\omega \end{bmatrix} - \begin{bmatrix} \overline{m}_x \\ 0 \end{bmatrix} \right\} dx - \left[\delta[\phi \quad \psi] \begin{bmatrix} \overline{M}_t \\ 0 \end{bmatrix} \right]_0^L = 0$$

$$-\delta W = \int_x \delta[\phi \quad \psi] \left\{ \begin{bmatrix} {}_xdGJ\,d_x & 0 \\ 0 & {}_xdEI_{\omega\omega}d_x \end{bmatrix} \begin{bmatrix} \phi \\ \psi \end{bmatrix} - \begin{bmatrix} \overline{M}_t \\ 0 \end{bmatrix} \right\} dx - \left[\delta[\phi \quad \psi] \begin{bmatrix} \overline{M}_{to} \\ \overline{M}_\omega \end{bmatrix} \right]_0^L = 0$$

$$\boldsymbol{\delta}\mathbf{u}^{\mathrm{T}} \quad \{ \quad \mathbf{k}^D \quad \mathbf{u} \quad - \quad \mathbf{p} \quad \}$$

In-Plane Deformation of a Plate

$$-\delta W = \int_A \delta\mathbf{u}^{\mathrm{T}}({}_u\mathbf{D}^{\mathrm{T}}\mathbf{s} - \overline{\mathbf{p}}_V)\,dA - \int_S \delta\mathbf{u}^{\mathrm{T}}\overline{\mathbf{p}}\,ds = 0$$

$$-\delta W = \int_A \delta[u_x \quad u_y] \left\{ \begin{bmatrix} {}_x\partial & 0 & {}_y\partial \\ 0 & {}_y\partial & {}_x\partial \end{bmatrix} \begin{bmatrix} n_x \\ n_y \\ n_{xy} \end{bmatrix} - \begin{bmatrix} \overline{p}_{V_x} \\ \overline{p}_{V_y} \end{bmatrix} \right\} dA - \int_S \delta[u_x \quad u_y] \begin{bmatrix} \overline{p}_x \\ \overline{p}_y \end{bmatrix} ds = 0$$

Transformation to a Displacement Formulation:

$$-\delta W = \int_A \delta\mathbf{u}^{\mathrm{T}}(\mathbf{k}^D\mathbf{u} - \overline{\mathbf{p}}_V)\,dA - \int_{S_p} \delta\mathbf{u}^{\mathrm{T}}\overline{\mathbf{p}}\,ds = 0$$

$$\mathbf{k}^D = {}_u\mathbf{D}^{\mathrm{T}}\mathbf{E}\mathbf{D}_u$$

which for plane stress expands to:

$$-\delta W = \int_A \delta[u_x \quad u_y]\left\{\begin{bmatrix} {}_x\partial D\partial_x + {}_y\partial\frac{D(1-\nu)}{2}\partial_y & {}_x\partial\nu D\partial_y + {}_y\partial\frac{D(1-\nu)}{2}\partial_x \\ {}_y\partial\nu D\partial_x + {}_x\partial\frac{D(1-\nu)}{2}\partial_y & {}_y\partial D\partial_y + {}_x\partial\frac{D(1-\nu)}{2}\partial_x \end{bmatrix}\begin{bmatrix} u_x \\ u_y \end{bmatrix} - \begin{bmatrix} \overline{p}_{V_x} \\ \overline{p}_{V_y} \end{bmatrix}\right\}dA$$

$$- \int_{S_p} \delta[u_x \quad u_y]\begin{bmatrix} \overline{p}_x \\ \overline{p}_y \end{bmatrix} ds = 0$$

Transverse Deformation of a Plate Rectilinear Coordinates

with Shear Deformation Effects

$$-\delta W = \int_A \delta\mathbf{u}^{\mathrm{T}}(\mathbf{k}^D\mathbf{u} - \overline{\mathbf{p}}_V)\,dA - \int_{S_p} \delta\mathbf{u}^{\mathrm{T}}\overline{\mathbf{p}}\,ds = 0$$

$$-\delta W = \int_A \delta[w \quad \theta_x \quad \theta_y]\left\{\begin{bmatrix} {}_x\partial & 0 & {}_y\partial & 1 & 0 \\ 0 & {}_y\partial & {}_x\partial & 0 & 1 \\ 0 & 0 & 0 & {}_x\partial & {}_y\partial \end{bmatrix}\begin{bmatrix} m_x \\ m_y \\ m_{xy} \\ q_x \\ q_y \end{bmatrix} - \begin{bmatrix} \overline{p}_z \\ 0 \\ 0 \end{bmatrix}\right\}dA$$

$$- \int_S \delta[w \quad \theta_x \quad \theta_y]\begin{bmatrix} \overline{p}_s \\ 0 \\ 0 \end{bmatrix} ds = 0$$

$$-\delta W = \int_A \delta[w \quad \theta_x \quad \theta_y]\left\{\begin{bmatrix} {}_x\partial\frac{5}{6}Gt\partial_x + {}_y\partial\frac{5}{6}Gt\partial_y & {}_x\partial\frac{5}{6}Gt & {}_y\partial\frac{5}{6}Gt \\ \frac{5}{6}Gt\partial_x & {}_x\partial K\partial_x + {}_y\partial\frac{1-\nu}{2}K\partial_y + \frac{5}{6}Gt & {}_x\partial\nu K\partial_y + {}_y\partial\frac{1-\nu}{2}K\partial_x \\ \frac{5}{6}Gt\partial_y & {}_y\partial\nu K\partial_x + {}_x\partial\frac{1-\nu}{2}K\partial_y & {}_y\partial K\partial_y + {}_x\partial\frac{1-\nu}{2}K\partial_x + \frac{5}{6}Gt \end{bmatrix}\begin{bmatrix} w \\ \theta_x \\ \theta_y \end{bmatrix} - \begin{bmatrix} \overline{p}_z \\ 0 \\ 0 \end{bmatrix}\right\}dA$$

$$- \int_S \delta[w \quad \theta_x \quad \theta_y]\begin{bmatrix} \overline{p}_s \\ 0 \\ 0 \end{bmatrix} ds = 0$$

without Shear Deformation Effects

$$-\delta W = \int_A \delta w \left\{ \begin{bmatrix} {}_x\partial^2 & {}_y\partial^2 & 2\,{}_x\partial\,{}_y\partial \end{bmatrix} \begin{bmatrix} m_x \\ m_y \\ m_{xy} \end{bmatrix} - [\overline{p}_z] \right\} dA - \int_S \delta[w \quad \theta_n] \begin{bmatrix} \overline{p}_s \\ 0 \end{bmatrix} ds = 0$$

$\overline{p}_z$: weight and surface loads

$\overline{p}_s$: boundary-line loads in z-direction

$$-\delta W = \int_A \delta w \left\{ K \begin{bmatrix} {}_{xx}\partial(\partial_{xx} + \nu\partial_{yy}) + \\ {}_{yx}\partial 2(1-\nu)\partial_{xy} + \\ {}_{yy}\partial(\partial_{yy} + \nu\partial_{xx}) \end{bmatrix} w - \overline{p}_z \right\} dA - \int_S \delta w\, \overline{p}_s\, ds = 0$$

Circular Plate

In-Plane Deformation (Stretching):

$$-\delta W = \int_A \delta \mathbf{u}^{\mathrm{T}} \left({}_u\mathbf{D}^{\mathrm{T}}\mathbf{s} - \overline{\mathbf{p}}_V \right) dA - \int_{S_p} \delta \mathbf{u}^{\mathrm{T}} \overline{\mathbf{p}}\, ds$$

$$-\delta W = \int_A \delta[u_r \quad u_\phi] \left\{ \begin{bmatrix} {}_r\partial & \frac{1}{r} & \frac{{}_\phi\partial}{r} \\ 0 & \frac{{}_\phi\partial}{r} & {}_r\partial - \frac{1}{r} \end{bmatrix} \begin{bmatrix} n_r \\ n_\phi \\ n_{r\phi} \end{bmatrix} - \begin{bmatrix} \overline{p}_{V_r} \\ \overline{p}_{V_\phi} \end{bmatrix} \right\} dA - \int_{S_p} \delta[u_r \quad u_\phi] \begin{bmatrix} \overline{p}_r \\ \overline{p}_\phi \end{bmatrix} ds$$

Transformation to a Displacement Formulation:

$$-\delta W = \int_A \delta \mathbf{u}^{\mathrm{T}} (\mathbf{k}^D \mathbf{u} - \overline{\mathbf{p}}_V)\, dA - \int_{S_p} \delta \mathbf{u}^{\mathrm{T}} \overline{\mathbf{p}}\, ds$$

$$-\delta W = \int_A [u_r \quad u_\phi] \left\{ \left[\begin{array}{c:c} \begin{matrix} {}_r\partial\partial_r + \\ {}_r\partial\frac{\nu}{r} + \frac{\nu}{r}\partial_r \\ +{}_\phi\partial\frac{(1-\nu)}{2r^2}\partial_\phi + \frac{1}{r^2} \end{matrix} & \begin{matrix} {}_r\partial\frac{\nu}{r}\partial_\phi + \\ \frac{(1-\nu)}{2r}\left({}_\phi\partial\partial_r - \frac{1}{r}{}_\phi\partial\right) \\ +\frac{1}{r^2}\partial_\phi \end{matrix} \\ \hdashline \begin{matrix} {}_\phi\partial\frac{1}{r^2} + \\ \frac{(1-\nu)}{2r}\left({}_r\partial\partial_\phi - \frac{1}{r}\partial_\phi\right) \\ +{}_\phi\partial\frac{\nu}{r}\partial_r \end{matrix} & \begin{matrix} {}_\phi\partial\frac{1}{r^2}\partial_\phi + \\ \frac{(1-\nu)}{2}\left({}_r\partial\partial_r - {}_r\partial\frac{1}{r} - \right. \\ \left. \frac{1}{r}\partial_r + \frac{1}{r^2}\right) \end{matrix} \end{array} \right] \begin{bmatrix} u_r \\ \\ u_\phi \end{bmatrix} - \begin{bmatrix} \overline{p}_{Vr} \\ \\ \overline{p}_{V\phi} \end{bmatrix} \right\} dA$$

$$- \int_{S_p} \delta[u_r \quad u_\phi] \begin{bmatrix} \overline{p}_r \\ \overline{p}_\phi \end{bmatrix} ds$$

Transverse Deformation Without Shear Deformation Effects

$$-\delta W = \int_A \delta \mathbf{u}^{\mathrm{T}} \left({}_u\mathbf{D}^{\mathrm{T}}\mathbf{s} - \overline{\mathbf{p}} \right) dA - \int_S \delta \mathbf{u}^{\mathrm{T}} \overline{\mathbf{p}}_s\, ds = 0$$

$$-\delta W = \int_A \delta w \left\{ \begin{bmatrix} {}_r\partial^2 & {}_\phi\partial^2\frac{1}{r^2} & 2\left({}_\phi\partial\frac{1}{r^2} - {}_{r\phi}\partial\frac{1}{r}\right) \end{bmatrix} \begin{bmatrix} m_r \\ m_\phi \\ m_{r\phi} \end{bmatrix} - [\overline{p}_z] \right\} dA - \int_s \delta w\, \overline{p}_s\, ds = 0$$

$\overline{p}_s$ = boundary line loads in z-direction

Transformation to a Displacement Formulation:

$$-\delta W = \int_A \delta \mathbf{u}^{\mathrm{T}} (\mathbf{k}^D \mathbf{u} - \overline{\mathbf{p}})\, dA - \int_S \delta \mathbf{u}^{\mathrm{T}} \overline{\mathbf{p}}_s \, ds = 0$$

$$-\delta W = \int_A \delta w \left\{ K \begin{bmatrix} {}_r\partial^2 \left[\partial_r^2 + \nu\left(\frac{1}{r}\partial_r + \frac{1}{r^2}\partial_\phi^2\right)\right] \\ +\left({}_{r\phi}\partial^2 \frac{1}{r} - {}_\phi\partial \frac{1}{r^2}\right) 2(1-\nu)\left(\frac{1}{r}\partial_{r\phi}^2 - \frac{1}{r^2}\partial_\phi\right) \\ +\left({}_r\partial \frac{1}{r} + {}_\phi\partial^2 \frac{1}{r^2}\right)\left(\frac{1}{r}\partial_r + \frac{1}{r^2}\partial_\phi^2 + \nu\partial_r^2\right) \end{bmatrix} w - \overline{p}_z \right\} dA - \int_s \delta w \, \overline{p}_s \, ds = 0$$

III.5 Summary

To illustrate the common structure of the governing equations for a solid continuum and structural members, the local and global forms of the fundamental equations have been summarized in this chapter. Note the similarity in structure of the equations for a continuum and some structural members.

Reference

Wunderlich, W., 1977, Incremental formulation for geometrically nonlinear problems, in *Formulations and Computational Algorithms in Finite Element Analysis*, Bathe, K.J., Oden, J.T. and Wunderlich, W., Eds., MIT Press, Cambridge, MA.

Subject Index

C

D

E

F

I

J

K

L

M

Q

R

S

T

Bibliographic Index

L

M

N

O

P

R

S

T

V

W

Y

Printed and bound by CPI Group (UK) Ltd, Croydon, CR0 4YY
23/10/2024
01778248-0013